Consumer price index, 1010
Contract buyout, 149
Credit cards, 387, 399
Customer satisfaction, 521
Defective products, 388, 426
Discounted loan, 106, 110
Elasticity of demand, 763, 767, 931
Electrical consumption, 814
Emergency medical care, 491
Employee retention, 416
Fair market price of a bond, 109
Farm management, 261
Fuel prices, 211
Furniture production, 212
Gini index, 860
Gross domestic product, 752, 778–779
Highway construction, 313
Homeowners who move, 504
Industrial production, 236
Insurance, 63, 419, 420
International marketing, 363
Internet demand, 233
  usage, 487
Jewelry production, 219, 253
Labor negotiations, 375
Learning curve in airplane production, 26, 30
Least cost rule, 1057
Liquor sales, 762
Lot size, 696–697, 698, 715
Lumber production, 350, 351
Mass transit, 197, 225
Maximizing harvest yield, 681
Memory chip demand, 766
Money-back guarantees, 416
Movie tickets, 223
Municipal management, 212
National debt, 593, 607, 761
Net savings from new machinery, 830
Newspaper prices, 762
Nursery management, 242
Office supplies, 313
Oil demand, 766
  imports and fuel efficiency, 63
  prices, 687
Oil wells, 196, 899
Overseas manufacturing, 211
Pareto's law, 815
Pasteurization temperature, 560
Portfolio management, 492
Price discrimination, 998, 999, 1056
Process control, 461

Production accidents, 245
Productivity, 765
Pulpwood forest value, 678
Quality control, 76, 381, 398, 438
Real estate, 180, 237, 473
Reliability, 401
Renters who move, 504
Repetitive tasks, 815, 816
Research expenditures, 63
Rule of .6, 29
Sales growth, 945
  prediction, 48, 1000
  taxes, 211
Small businesses, 287
Statistical process control, 457
Super Bowl advertising, 1012
Tax revenue, 688
Taxi repairs, 520
Teenage population, 575
Term insurance, 516
Timber forest value, 667, 676
Transportation, 306, 314
Truck rentals, 495
Unemployment, 1057
Value of a building, 933
  an MBA, 986
Warranties, 900
Year-end bonuses, 225

Personal Finance and Management
Accumulation of wealth, 932
Apartment rents, 142
Art appreciation, 784, 936
Becoming a millionaire, 121
Bridge loan, 110
Car buying, 142, 143, 149
Certificates of deposit, 147
College costs, 75, 120, 142, 143
Comparing interest rates, 75
Computer financing, 143
Contest prizes, 141, 149
Cost of maintaining a home, 795
Credit cards, 142, 143, 149
Depreciating a car, 67, 747
Disposable income, 438
Furniture buying, 147
Home appreciation, 948
  buying, 121, 131, 148, 149
  financing, 144, 166
  improvements, 110
  mortgages, 136, 137
Household income, 450
Income taxes, 52, 62, 151, 161, 166, 196, 244

Lottery winnings, 121
Millionaires, 75
Musical instruments, 101
Personal wealth, 927, 956
Retirement savings, 130, 148
Solar water heaters, 796, 858
Stock market return rate, 131
Tokyo mortgages, 142
Treasury bonds, 147
Value of coin collection, 721
Zero coupon bond, 109, 118

Social and Behavioral Sciences
Absenteeism, 998
Age at first marriage, 17
  of "Iceman," 90
  of "Nutcracker Man," 97
Airline hijacking, 407
Automobile accidents, 440
Campaign expenses, 688
Cell phone usage, 733
Crime, 1010
Dating older women, 90
Demand for oil, 762
Divorces, 795
Early human ancestors, 90
Ebbinghaus model, 747
Education and income, 733
Elasticity and taxation, 719
Election costs, 77
Employment seekers, 886
Energy dependence, 245
Equal pay for equal work, 17
ESP, 420
False testimony, 363
Farm size distribution, 511
Fund raising, 875
Gap between rich and poor, 94
Gini index, 838–841
Household income, 429
How to ask an embarrassing question, 408
Income and education, 454
IQ distribution, 908–910, 913
Liquor and beer: elasticity and taxation, 762
Long term trends in higher education, 63
Longevity, 48
Lorenz curve, 841

Marriages, 815
Mass transit, 503
Mazes, 900
Medical insurance, 440
Memory, 731
Most populous country, 76
Multiple-choice tests, 420, 421
National debt, 147
Old age, 473
Per capita personal income, 16
Political advertising, 181, 294, 326, 341
   contributions, 387, 399
   fund raising, 363
Polygraph tests, 407
Population dynamics, 502
   growth, 601, 767
Practice and rest, 998
Prison terms, 900
Raising IQs, 449
Retirement income, 211
Senate committees, 375, 388
Smoking and education, 18, 576
Spread of information, 729
   of rumors, 945, 949, 957
Status, 619, 663, 986
Stevens' Law, 635
Stimulus and response, 663
Tax revenue, 714
Third-world economics, 517
Traffic accidents, 709
Unemployment, 407, 408, 449
Voting patterns, 404, 406, 502
Welfare, 709
World population growth, 76

Topics of General Interest
5-Card hands, 372
Africa population, 829
Age of cave paintings, 90
   Dead Sea Scrolls, 87
   Shroud of Turin, 90
Aging of America, 1056
Airline safety, 421
   flight path, 648, 663
Approximation of π, 914
Art appreciation, 147, 730
   gallery shows, 503
Bad deal on an apartment, 148
Baseball cards, 148, 730
Boiling point and altitude, 47
Capacity of a computer disk, 678
Carbon 14 dating, 90
China population, 860
Cigarette smoking, 769–770, 816
Class attendance, 400

College graduation rates, 517, 520
   tuition, 577
Cooling coffee, 77, 747
Cost of college education, 1013
Dam construction, 816
   sediment, 949
Decoding by majority, 420
Dinosaurs, 29
Disaster relief, 273
Dog training, 346
   years, 62, 63
Drug interception, 648
Drunk driving, 1012
Duration of telephone calls, 899
Elevator stops, 388
Emergency services, 449
Estate division, 166, 180, 197
Eternal recognition, 863
Expected tries until success, 421
Fair toss from an unfair coin, 400
Freezing of ice, 795
Friendships, 932
Fuel economy, 676
   efficiency, 461, 713
Fund raising, 948
Gambling, 353
Gompertz growth curve, 950
Grades, 16, 407
Hailstones, 576, 709
Highway safety, 986, 1038
Homework grading, 375
Horse ancestor, 97
   racing, 374, 389
   trading, 121
Ice cream prices, 815
Impact time of a projectile, 47
   velocity, 47, 606
Largest postal package, 687, 1026
Life expectancy, 18
Lives saved by seat belts, 831
Lost luggage, 477
Lotteries, 387, 425
Manhattan Island purchase, 75
Maximum height of a bullet, 606
Measurement errors, 1038
Melting ice, 716
Monty Hall problem, 388
Moore's law of computer memory, 765
Most populous country, 731
Multiple-choice tests, 400, 417
Nuclear meltdown, 76
Odd-man out, 421
Oldest dinosaur, 90
Oseola McCarty Scholarship Fund, 130

Paper stacking, 31
Permanent endowments, 894, 898–901, 954
Population, 560
   and individual birthrate, 950
   growth, 765, 766, 950
   of a city, 619, 631, 886
   of states, 72, 77
   of U.S., 722, 818–819, 828
   of world, 731, 747, 859, 1005
Postage stamp prices, 957, 1013
Raindrops, 950
Random walks and gambler's ruin, 516
Rate of growth of a circle, 630
   of a sphere, 630
Relativity, 544
Richter scale, 30
Rocket tracking, 710
Rumor accuracy, 504
Safe cars, unsafe streets, 959, 1013
SAT scores, 438, 473
Scuba diving, 971, 1038
Searching for lost plane, 426
Shared birthdays, 401
Smoking, 1010, 1011
   and longevity, 1011
   mortality rates, 1003
Snowballs, 709
Speeding, 710
Stamp appreciation, 948
St. Louis Gateway Arch, 77
Stopping distance, 47
Superconductivity, 527, 545
Survival rate, 633
Suspension bridge, 914
Temperature conversion, 16
Thermos bottle temperature, 766
Time of a murder, 932
   saved by speeding, 585
Traffic flow, 491
Tsunamis, 47
Water pressure, 47
Water wheel, 676
Waterfalls, 30
Weight of a child, 814, 858
Why four-of-a-kind beats a full-house, 372
Wind energy, 765
   speed, 93
Windchill index, 607, 963, 972, 986, 1038
Wine storage, 688
Working mothers, 492
World's largest city, 765

Finite Mathematics
AND Applied Calculus

SECOND EDITION

Geoffrey C. Berresford
Long Island University

Andrew M. Rockett
Long Island University

Houghton Mifflin Company
Boston   New York

Publisher: Jack Shira
Sponsoring Editor: Lauren Schultz
Associate Editor: Jennifer King
Editorial Associate: Kasey McGarrigle
Senior Project Editor: Tracy Patruno
Manufacturing Manager: Florence Cadran
Senior Marketing Manager: Danielle Potvin
Marketing Associate: Nicole Mollica

Cover image: Rock Formations by Lake, © DAVID BROOKOVER/Photonica

Photo credits:
Chapter 1: AP Wide World Photos; Chapter 2: Chuck Savage/Corbis; Chapter 3: Michael Keller/Corbis; Chapter 4: Schwarz Shaul/Corbis Sygma; page 247, Robert Brenner/PhotoEdit; Chapter 5: Superstock; Chapter 6: Mark Segal/Index Stock; Chapter 7: David Norton/Masterfile; Chapter 8: Firefly Productions/Corbis; Chapter 9: AP Wide World Photos; Chapter 10: Tony Page/Stone/Getty Images; Chapter 11: Jeremy Walker/Getty Images; Chapter 12: Alan Schein Photography/Corbis; Chapter 13: AP Photo/Insurance Institute for Highway Safety.

Copyright © 2005 by Houghton Mifflin Company. All rights reserved.

TI-83 is a registered trademark of Texas Instruments Incorporated, Excel, Microsoft, and Windows are either registered trademarks or trademarks of Microsoft Corporation in the United States and/or other countries.

No part of this work may be reproduced or transmitted in any form or by any means, electronic or mechanical, including photocopying and recording, or by any information storage or retrieval system without the prior written permission of Houghton Mifflin Company unless such copying is expressly permitted by federal copyright law. Address inquiries to College Permissions, Houghton Mifflin Company, 222 Berkeley Street, Boston, MA 02116-3764.

Printed in the U.S.A.

Library of Congress Control Number: 2003109911

ISBN: 0-618-37213-X

123456789-DOC-08 07 06 05 04

Contents

Preface ix
A User's Guide to Features xvii
Integrating Excel xx
Graphing Calculator Terminology xxii

Preliminaries 1

1 Functions 1

Application Preview: World Record Mile Runs 1

1.1 Real Numbers, Inequalities, and Lines 2
1.2 Exponents 18
1.3 Functions 31
1.4 Functions, Continued 40
1.5 Exponential Functions 64
1.6 Logarithmic Functions 78

Chapter Summary with Hints and Suggestions 91
Review Exercises for Chapter 1 93

Part I Finite Mathematics 99

2 Mathematics of Finance 100

Application Preview: Musical Instruments as Investments 101

2.1 Simple Interest 102
2.2 Compound Interest 111
2.3 Annuities 122
2.4 Amortization 132

iii

Chapter Summary with Hints and Suggestions   144
Review Exercises for Chapter 2   146

3 Systems of Equations and Matrices   150

Application Preview: A "Taxing" Problem   151

3.1 Systems of Two Linear Equations in Two Variables   152
3.2 Matrices and Linear Equations in Two Variables   168
3.3 Systems of Linear Equations and the Gauss–Jordan Method   182
3.4 Matrix Arithmetic   198
3.5 Inverse Matrices and Systems of Linear Equations   213
3.6 Introduction to Modeling: Leontief Models and Least Squares   226

Chapter Summary with Hints and Suggestions   238
Review Exercises for Chapter 3   241

4 Linear Programming   246

Application Preview: Managing an Investment Portfolio   247

4.1 Linear Inequalities   248
4.2 Two-Variable Linear Programming Problems   261
4.3 The Simplex Method for Standard Maximum Problems   274
4.4 Standard Minimum Problems and Duality   297
4.5* Nonstandard Problems: the Dual Pivot Element and the Two-Stage Method   315
4.6* Nonstandard Problems: Artificial Variables and the Big-M Method   329

Chapter Summary with Hints and Suggestions   344
Review Exercises for Chapter 4   346

*These two sections provide *alternate* methods for solving the *same* types of problems. The choice of which one to cover is left to the instructor.

5 Probability   352

Application Preview: The Chevalier de Meré   353

5.1 Sets, Counting, and Venn Diagrams   354

- **5.2** Permutations and Combinations   364
- **5.3** Probability Spaces   377
- **5.4** Conditional Probability and Independence   389
- **5.5** Bayes' Formula   402
- **5.6** Random Variables and Distributions   408

  Chapter Summary with Hints and Suggestions   422
  Review Exercises for Chapter 5   424

6 Statistics   428

Application Preview: Household Income   429

- **6.1** Random Samples and Data Organization   430
- **6.2** Measures of Central Tendency   441
- **6.3** Measures of Variation   451
- **6.4** Normal Distributions and Binomial Approximation   462

  Chapter Summary with Hints and Suggestions   474
  Review Exercises for Chapter 6   476

7 Markov Chains   478

Application Preview: Weather in Sri Lanka   479

- **7.1** States and Transitions   480
- **7.2** Regular Markov Chains   493
- **7.3** Absorbing Markov Chains   504

  Chapter Summary with Hints and Suggestions   518
  Review Exercises for Chapter 7   520

Part II Calculus   525

8 Derivatives and Their Uses   526

Application Preview: Temperature, Superconductivity, and Limits   527

- **8.1** Limits and Continuity   528
- **8.2** Rates of Change, Slopes, and Derivatives   545

8.3 Some Differentiation Formulas   561
8.4 The Product and Quotient Rules   577
8.5 Higher-Order Derivatives   594
8.6 The Chain Rule and the Generalized Power Rule   608
8.7 Nondifferentiable Functions   621

Chapter Summary with Hints and Suggestions   627
Review Exercises for Chapter 8   629

9 Further Applications of Derivatives   634

Application Preview: Stevens' Law of Psychophysics   635

9.1 Graphing Using the First Derivative   636
9.2 Graphing Using the First and Second Derivatives   649
9.3 Optimization   664
9.4 Further Applications of Optimization   679
9.5 Optimizing Lot Size and Harvest Size   689
9.6 Implicit Differentiation and Related Rates   698

Chapter Summary with Hints and Suggestions   711
Review Exercises for Chapter 9   712

Cumulative Review for Chapters 1, 8–9   716

10 Exponential and Logarithmic Functions   718

Application Preview: Elasticity and Social Policy   719

10.1 Review of Exponential and Logarithmic Functions   720
10.2 Differentiation of Logarithmic and Exponential Functions   733
10.3 Two Applications to Economics: Related Rates and Elasticity of Demand   751

Chapter Summary with Hints and Suggestions   763
Review Exercises for Chapter 10   765

11 Integration and Its Applications   768

Application Preview: Cigarette Smoking   769

11.1 Antiderivatives and Indefinite Integrals   770

11.2 Integration Using Logarithmic and Exponential Functions 784
11.3 Definite Integrals and Areas 796
11.4 Further Applications of Definite Integrals: Average Value and Area Between Curves 817
11.5 Two Applications to Economics: Consumers' Surplus and Income Distribution 832
11.6 Integration by Substitution 841

Chapter Summary with Hints and Suggestions 855
Review Exercises for Chapter 11 857

12 Integration Techniques and Differential Equations 862

Application Preview: Improper Integrals and Eternal Recognition 863

12.1 Integration by Parts 864
12.2 Integration Using Tables 877
12.3 Improper Integrals 887
12.4 Numerical Integration 901
12.5 Differential Equations 915
12.6 Further Applications of Differential Equations: Three Models of Growth 934

Chapter Summary with Hints and Suggestions 951
Review Exercises for Chapter 12 953

13 Calculus of Several Variables 958

Application Preview: Safe Cars, Unsafe Streets 959

13.1 Functions of Several Variables 960
13.2 Partial Derivatives 972
13.3 Optimizing Functions of Several Variables 987
13.4 Least Squares 999
13.5 Lagrange Multipliers and Constrained Optimization 1013
13.6 Total Differentials and Approximate Changes 1027
13.7 Multiple Integrals 1039

Chapter Summary with Hints and Suggestions 1053
Review Exercises for Chapter 13 1055

Cumulative Review for Chapters 1, 8–13 1059

Appendix: Normal Probabilities Using Tables   A-1
Answers to Selected Exercises   B-1
Index   I-1

Available on the Web Go to math.college.hmco.com and follow the links to this text.

G Game Theory

Application Preview: Sherlock Holmes and James Moriarty
G.1 Two-Person Games and Saddle Points
G.2 Mixed Strategies
G.3 Games and Linear Programming

Chapter Summary with Hints and Suggestions
Review Exercises for Chapter G

L Logic

L.1 Statements and Connectives
Application Preview: Grammar, Logic, and Rhetoric

L.2 Truth Tables
Application Preview: Circuits

L.3 Implications
Application Preview: "If…, then…" as a Computer Instruction

L.4 Valid Arguments
Application Preview: Fuzzy Logic and Artificial Intelligence

L.5 Quantifiers and Euler Diagrams
Application Preview: The Dog Walking Ordinance

Chapter Summary with Hints and Suggestions
Review Exercises for Chapter L
Chapter Projects and Essays

Preface

A scientific study of yawning found that more yawns occurred in calculus class than anywhere else.* This book hopes to remedy that situation. Rather than being another dry recitation of standard results, our presentation exhibits some of the many fascinating and useful applications of mathematics in business, the sciences, and everyday life. Even beyond its utility, however, mathematics has beauty, elegance, and simplicity, some of which we hope to convey.

This book is an introduction to finite mathematics and calculus and their applications to the management, social, behavioral, and biomedical sciences, and other fields. *Preliminaries* (Chapter 1) consists of a brief review of functions; *Part I—Finite Mathematics* (Chapters 2–7) covers the mathematics of finance, matrices, linear programming, finite probability, statistics, and Markov chains; *Part II—Calculus* (Chapters 8–13) covers derivatives and integrals, and concludes with material on differential equations and calculus of several variables.

CHANGES IN THE SECOND EDITION

First, what has *not* changed is the basic character of the book: simple, clear, and mathematically correct explanations of mathematics, alternating with relevant and engaging examples.

We have shortened and tightened the exposition throughout the book without omitting any topics.

We have simplified notations (for example, using an "arrow" notation for row-reducing matrices), added new illustrations (especially in the finance and linear programming chapters), expanded material comparing concepts (such as mean, median, and mode), improved the treatment of some topics (such as Bayes' formula using probability trees), rewritten some sections (such as *Rates of Change, Slopes, and Derivatives* to begin with the rates of change, as seems appropriate for a book featuring so many applications), updated almost every problem that uses real-world data, added new exercises to every chapter and in every category, and expanded many of the answers.

We have added a chapter on Markov chains. A chapter on Logic is available on a CD and the Internet at **math.college.hmco.com**, and

---

*Ronald Baenninger, "Some Comparative Aspects of Yawning in Betta splendens, Homo sapiens, Panthera leo, and Papoi spinx," *Journal of Comparative Psychology* **101**(4):349–354.

the chapter on Game Theory from our *Finite Mathematics* text is also available on the Website and the CD.

We have moved the **Projects and Essays** to the CD and the Internet at **math.college.hmco.com** along with most "Application Previews" from the first edition.

For courses using Microsoft Excel or similar software we have created (optional) **Spreadsheet Explorations.**

We have removed the scientific calculator symbol (but kept the graphing calculator symbol) since calculator use is now universal among today's college students.

We have added spaces around every equation and many mathematical expressions to make them more visible and the text more readable.

FEATURES

Realistic Applications Most courses using this book will be very "applied," and therefore this book contains an unusually large number of applications, many appearing in no other textbook. In Part I we calculate the real cost of a car loan, the fair price of a lottery ticket, and the probability that a jury will return the correct verdict; in Part II we use calculus to maximize longevity, estimate the dangers of cigarette smoking, study global warming, and evaluate strategies for controlling heroin, marijuana, and liquor sales. In this way we demonstrate that mathematics is not just manipulation of abstract symbols but is deeply connected to everyday life.

Graphing Calculators (Optional) Reading this book does *not* require a graphing calculator, but having one will simplify the calculations in many problems, and at the same time deepen understanding of mathematics by permitting you to concentrate on *concepts*. Throughout the book are **Graphing Calculator Explorations** and **Exercises,** which explore new topics, carry out "messy" calculations, or show the limitations of technology. A discussion of the essentials of graphing calculators follows this preface. For those not using a graphing calculator, the Graphing Calculator Explorations are boxed so that they can be read for enrichment. You will, however, need a calculator with keys like y^x and LN for powers and natural logarithms.

Graphing Calculator Programs (Optional) For certain topics, the computational effort may be reduced and the understanding enhanced by using the optional graphing calculator programs we created for use with this book. These programs are free and easy to obtain (see "How to Obtain Graphing Calculator Programs" later in this Preface). The topics covered are: amortization tables, matrix row operations, pivot operations, normal approximations of binomial distributions, steady state distributions for regular Markov chains, Riemann sums, Simpson's and trapezoidal approximations, and slope fields.

Spreadsheets (Optional) Each chapter now includes an Excel **Spreadsheet Exploration.** Ancillary materials for Microsoft Excel are also available (see "Supplements for the Student" later in this Preface).

Application Previews Each chapter begins with an Application Preview that presents an interesting application of the mathematics being developed. They are self-contained (although some exercises are based on them) and serve to motivate the coming material. Topics include world records in the mile run, investment portfolios, Stevens' Law of Psychophysics, cigarette smoking, and predicting personal wealth.

Practice Problems Learning requires active participation—"mathematics is not a spectator sport." Throughout the readings are short pencil-and-paper **Practice Problems** designed to consolidate understanding of one topic before another is introduced. Complete solutions to all Practice Problems are given at the end of each section.

Annotations Notes to the right of many mathematical formulas and manipulations state the results in words, emphasizing the important skill of "reading mathematics." They also provide explanation and justification for the steps in calculations, and interpretation of the results.

Enhanced Readability An elegant four-color design has been used to make the book more visually appealing and to clarify the organization. For the sake of continuity, references to earlier material have been minimized by restating results whenever they are used. Where references are necessary, explicit page numbers are provided. Space has been added around equations and other mathematical expressions within sentences to make them easier to read.

Extensive Exercises Anyone who has learned mathematics did so by solving many problems, and the exercises are the most essential part of the learning process. Exercises are graded from routine drill to significant applications. Most **Applied Exercises** have both general and specific titles, such as "Environmental Science: Pollution Control." The **Explorations and Excursions** are exercises of a more advanced nature that carry the development of certain topics beyond the level of the text. At the end of the book are answers to the odd-numbered exercises, and answers to *all* Chapter Review Exercises and Cumulative Review Exercises.

Constant Reinforcement Summaries and reviews occur frequently. **Section Summaries** briefly state essential formulas and key concepts. **Chapter Summaries** review the major developments (keyed to particular review exercises) and offer **Hints and Suggestions** that unify the chapter, give helpful reminders, and list a selection of **Review Exercises for a Practice Test** for the chapter. **Cumulative Reviews** contain exercises from groups of related chapters.

Accuracy and Proofs All of the answers and other mathematics have been checked carefully by several mathematicians. The statements of definitions and theorems are mathematically accurate. Because the treatment is applied rather than theoretical, intuitive and geometric justifications have often been preferred to formal proofs. Such a justification or proof accompanies every important mathematical idea; we never resort to phrases like "it can be shown that . . .". When proofs are given, they are correct and "honest."

Philosophy We wrote this book with several principles in mind. One is that to learn something, it is best to begin doing it as soon as possible. Therefore, the preliminary material is brief, so that students begin doing and applying real mathematics without delay. An early start allows more time during the course for interesting applications and necessary review. Another principle is that the mathematics should be done together with the applications. Consequently every section contains applications (there are no "pure math" sections).

HOW TO OBTAIN GRAPHING CALCULATOR PROGRAMS

The optional graphing calculator programs used in the text have been written for a variety of Texas Instruments Graphing Calculators (including the *TI-82, TI-83, TI-83 Plus, TI-85, TI-86, TI-89* and *TI-92*), and may be obtained for free in any of the following ways:

- If you know someone who already has the programs on a Texas Instruments graphing calculator like yours, you can easily transfer the programs from their calculator to yours using the black cable that came with the calculator and the LINK button.

- You may download the programs and instructions from the Houghton Mifflin website at **math.college.hmco.com** onto a computer. Then use TI-GRAPH LINK™ or TI-CONNECT™ (available for purchase in stores) to transfer the programs to your calculator.

- You may send a $3\frac{1}{2}$ inch or zip disk to the authors at the following address, specifying the title of your textbook, the type of your computer (IBM-compatible or Macintosh), and your type of Texas Instruments calculator. We will return a disk containing the appropriate programs and descriptive information. You then use TI-GRAPH LINK™ or TI-CONNECT™ to transfer the programs from your computer to your calculator.

- You may send your calculator (*TI-82, TI-83, TI-83 Plus, TI-85, TI-86, TI-89,* or *TI-92*) to the authors at the following address, specifying the title of your textbook. We will return it loaded with the appropriate programs and a packet of descriptive information.

Authors' Address: Dr. G. C. Berresford and Dr. A. M. Rockett
Department of Mathematics
C. W. Post Campus of Long Island University
720 Northern Boulevard
Brookville, New York 11548–1300

e-mail: **gberres@liu.edu** *or* **rockett@liu.edu**

HOW TO OBTAIN EXCEL SPREADSHEET EXPLORATIONS

The MicroSoft Excel spreadsheets used in the Spreadsheet Explorations may be obtained for free in either of the following ways:

- You may download the spreadsheet files from the Houghton Mifflin website at **math.college.hmco.com** onto a computer.
- You may send a $3\frac{1}{2}$ inch or zip disk to the authors at the above address, specifying both the title of your textbook and the type of your computer (IBM-compatible or Macintosh), and we will return it to you with the appropriate Excel files.

SUPPLEMENTS FOR THE INSTRUCTOR

Instructor's Resource Manual (with Solutions and Printed Test Bank) contains complete solutions to all problems in the text and a printed Test Bank available in computerized form on the CD. Two Chapter Tests for each chapter are also included.

HM ClassPrep with HM Testing CD-ROM combines all the resources of ClassPrep plus the algorithmically generated tests and gradebook features of HM Testing. This CD-ROM includes a variety of supplements: an electronic Instructor's Solutions Manual, Digital Figures and Tables from the book, pre-made chapter tests, the resources available on the student CD, and end of chapter Projects and Essays. Two extra chapters, one on Game Theory and one on Logic are also on the CD.

- Create and print quizzes and tests quickly and easily
- Deliver tests and record results via a LAN or the Internet
- Record and tabulate student test grades in an electronic gradebook

Eduspace® is a text-specific online learning environment that combines algorithmic tutorials with homework capabilities and classroom management functions.

- Electronically grade and record student results
- Manage a lecture-based or distance learning course online

Companion Website for Instructors offers additional teaching resources such as Digital Lesson Slides. Visit **math.college.hmco.com/instructors** and choose this textbook from the list provided.

SUPPLEMENTS FOR THE STUDENT

HM mathSpace® Tutorial CD-ROM is a new tutorial CD that allows students to practice skills and review concepts. The CD contains algorithmically generated exercises and step-by-step solutions for student practice, along with a brief introduction to Excel, Graphing Calculator Guide, and end of chapter Projects and Essays. Also included is an algebra tutorial review that provides a skill refresher.

SMARTHINKING™ live online tutoring allows students to communicate in real-time with e-structors who will help them understand difficult concepts and guide them through the problem solving process.

Eduspace® is a text-specific online learning environment that combines algorithmic tutorials with homework capabilities.

- Practice skills with additional exercises and quizzes
- Get tutorial help outside class
- Complete homework assignments online if assigned by the instructor

Excel Guide for Finite Mathematics and Applied Calculus by Revathi Narasimhan, Kean University, provides Excel information, including step-by-step examples and sample exercises.

Companion Website for Students offers additional learning resources such as the Graphing Calculator Programs that accompany this text. Visit the text-specific website at **math.college.hmco.com/students** and choose this textbook from the list provided.

Houghton Mifflin Mathematics Instructional Videos and DVDs provide a lecture series featuring Dana Mosely in which he provides careful explanations of key concepts, examples, exercises, and applications in a lecture-based format.

Student Solutions Manual contains complete solutions to all odd-numbered problems in the text and solutions to *all* Chapter Review and Cumulative Test questions.

ACKNOWLEDGMENTS

We are indebted to many people for their useful suggestions, conversations, and correspondence during the writing and revising of this book. We thank Chris and Lee Berresford, Anne Burns, Richard Cavaliere, Ruth Enoch, Theodore Faticoni, Jeff Goodman, Susah Halter, Brita and Ed Immergut, Ethel Matin, Gary Patric, Shelly Rothman, Charlene Russert, Stuart Saal, Bob Sickles, Michael Simon, John Stevenson, and all of our "Math 5–6" students at C. W. Post over past years for serving as proofreaders and critics.

We had the good fortune to have had supportive and expert editors at Houghton Mifflin: Lauren Schultz (Sponsoring Editor), Jennifer King (Associate Editor), Kasey McGarrigle (Editorial Associate) and Tracy Patruno (Senior Project Editor). They made the difficult tasks seem easy, and helped beyond words. Helen Medley (accuracy reviewer) and Deana Richmond (accuracy reviewer and solutions manual author) saved us from many mistakes and confusions. We also express our gratitude to the many others at Houghton Mifflin who made important contributions too numerous to mention. Thanks also to Janet Nuciforo and others at Nesbitt Graphics for their attention to the finer details of production.

The following reviewers have contributed greatly to the development of this text:

Faiz Al-Rubaee, *University of North Florida*; Yossef Balas, *University of Southern Mississippi*; James J. Ball, *Indiana State University*; Rajanikant Bhatt, *Texas A&M University*; Wesley T. Black, *Illinois Valley Community College*; John A. Blake, *Oakwood College*, Alabama; Marilyn Blockus, *San Jose State University*, California; William L. Blubaugh, *University of Northern Colorado*; Dr. Larry Bouldin, *Roane State Community College*, Tennessee; Bob Bradshaw, *Ohlone College*, Fremont, California; Dave Bregenzer, *Utah State University*; Prof. Stephen Brick, *University of South Alabama*; Dean Brown, *Youngstown State University*, Ohio; Jeremy Case, *Taylor University*, Indiana; Donald O. Clayton, *Madisonville Community College*, Kentucky; Charles C. Clever, *South Dakota State University*; Dr. Edith Cook, *Suffolk University*, Massachusetts; Dale L. Craft, *South Florida Community College*; Kent Craghead, *Colby Community College*, Kansas; Michael Eurgubian, *Santa Rosa Junior College*, California; Rob Farinelli, *Community College of Allegheny County*, Pennsylvania; Tom Greenwood, *Bakersfield College*, California; John Haverhals, *Bradley University*, Illinois; Randall Helmstutler, *University of Virginia*; Irvin Roy Hentzel, *Iowa State University*; David Hutchison, *Indiana State University*; Alan S. Jian, *Solano Community College*, California; Dr. Hilbert Johs, *Wayne State College*, Nebraska; Carmen M. Latterell, *University of Minnesota-Duluth*; Michael Longfritz,

Rensselear Polytechnic Institute, New York; Dr. Hank Martel, *Broward Community College,* Florida; Stephen J. Merrill, *Marquette University,* Wisconsin; Donna Mills, *Frederick Community College,* Maryland; Pat Moreland, *Cowley College,* Kansas; Sue Neal, *Wichita State University,* Kansas; Armando I. Perez, *Laredo Community College,* Texas; Thomas Riedel, *University of Louisville,* Kentucky; Catherine A. Roberts, *University of Rhode Island;* Stephen B. Rodi, *Austin Community College,* Texas; Ben Rushing, Jr., *Northwestern State University of Louisiana;* George W. Schultz, *St. Petersburg College,* Florida; Kimberly A. Smith, *University of Tennessee;* Jaak Vilms, *Colorado State University;* Elizabeth White, *Trident Technical College,* Charleston, South Carolina; Kenneth J. Word, *Central Texas College*

DEDICATION

We dedicate this book to our wives, Barbara and Kathryn, and our children, Lee, Chris, Justin, and Joshua, for their understanding and patience, without which this book would not exist.

COMMENTS WELCOMED

With the knowledge that any book can always be improved, we welcome corrections, constructive criticisms, and suggestions from every reader.

A User's Guide to Features

▼ Applications Preview

Found on every chapter opening page, Application Previews serve to motivate the chapter. They offer unique "mathematics in your world" applications or interesting historical notes. Pages with further information on the topics, and often related exercises, are referenced.

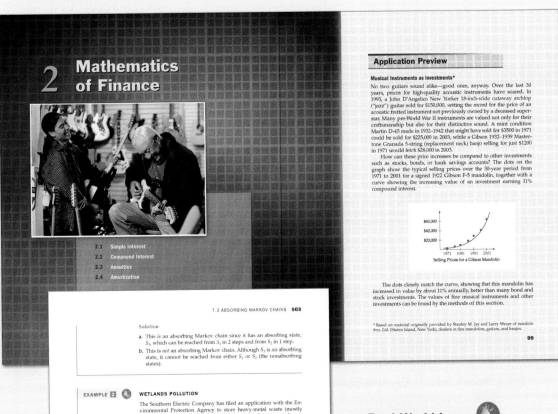

◀ Real World Icon

This globe icon marks examples in which mathematics is connected to every-day life.

xvii

▼ Practice Problems

Students can check their understanding of a topic as they read the text or do homework by working out a Practice Problem. *Complete solutions* are found at the end of each section, just after the Section Summary.

▼ Graphing Calculator Explorations

To allow for optional use of the graphing calculator, the Explorations are boxed. Exercises and examples that are designed to be done with a graphing calculator are marked with an icon.

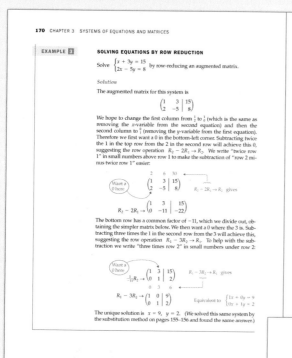

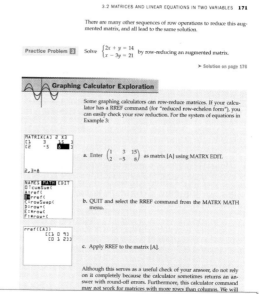

Spreadsheet Explorations ▶

Boxed for optional use, these spreadsheets enhance students' understanding of the material using Excel, an alternative for those who prefer spreadsheet technology.

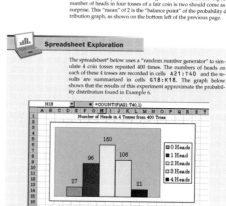

▼ Section Summary

Found at the end of every section, these summaries briefly state the main ideas of the section, providing a study tool or reminder for students. They are followed by complete solutions for the Practice Problems within that section.

▼ Applied Exercises

The applications in the exercise sets are labeled with a general title as well as a more specific title. The instructor can easily assign homework that is relevant to the students in that class.

End of Chapter Material ➤

To help students study, each chapter ends with a **Chapter Summary with Hints and Suggestions** and **Review Exercises**. The last bullet of the Hints and Suggestions lists the Review Exercises that a student could use to self-test. Both even and odd answers are supplied in the back of the book.

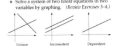

Integrating Excel

If you would like to use Excel or another spreadsheet software when working the exercises in this text, refer to the chart below. It lists exercises from many sections that you might find instructive to do with spreadsheet technology. Please note that none of these exercises are dependent on Excel. If you would like help using Excel, please consider the *Excel Guide for Finite Mathematics and Applied Calculus*, 2nd Edition, which is available from Houghton Mifflin. Additionally, the *Getting Started with Excel* chapter of the guide is available on the CD and website.

Section	Suggested Exercises
1.1	75–80
1.2	95–98
1.3	77–85
1.4	81–88
1.5	27–44
1.6	29–36
2.1	1–41
2.2	1–44
2.3	1–30
2.4	1–38
3.4	21–30 and 41–50
3.5	1–8 and 37–50
3.6	9–30
4.3	31–40
4.4	9–34
4.5	9–34
4.6	9–34
5.2	1–44
5.3	31–40
5.4	26 and 33
5.5	17
5.6	21–50
6.1	15–25
6.2	1–26
6.3	1–31
6.4	1–40
7.1	13–16 and 23–25
7.2	9–30
7.3	5–8, 13–40, 42, 44, 46, and 47

Section	Suggested Exercises
8.1	1–12 and 79–81
8.2	9–16
8.3	45–46 and 59–60
8.4	61
8.5	47–48
8.6	59
8.7	11–12
9.1	29–56, 62, and 64
9.2	27–46 and 59–63
9.3	21–37 and 43–50
9.4	1–21, 24, and 25
9.5	1–18
10.1	15–36
10.2	63–66 and 70–78
10.3	40–41
11.2	31–34 and 47–48
11.3	7–18
11.4	23–32, 55, 56, and 59
11.5	31–32
12.4	9–18 and 27–36
13.1	27–28 and 36–38
13.3	21–32
13.4	9–18 and 27–32
13.5	25–40

Graphing Calculator Terminology

The graphing calculator applications have been kept as generic as possible for use with any of the popular graphing calculators. Certain standard calculator terms are capitalized in this book and are described below. Your calculator may use slightly different terminology. The viewing or graphing **WINDOW** is the part of the Cartesian plane shown in the display screen of your graphing calculator. **XMIN** and **XMAX** are the smallest and largest x-values shown, and **YMIN** and **YMAX** are the smallest and largest y-values shown. These values can be set by using the **WINDOW** or **RANGE** command and are changed automatically by using any of the **ZOOM** operations. **XSCALE** and **YSCALE** define the distance between tick marks on the x- and y-axes.

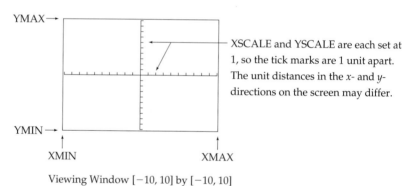

XSCALE and YSCALE are each set at 1, so the tick marks are 1 unit apart. The unit distances in the x- and y-directions on the screen may differ.

Viewing Window $[-10, 10]$ by $[-10, 10]$

The viewing window is always [XMIN, XMAX] by [YMIN, YMAX]. We will set XSCALE and YSCALE so that there are a reasonable number of tick marks (generally 2 to 20) on each axis. The x- and y-axes will not be visible if the viewing window does not include the origin.

Pixel, an abbreviation for *picture element*, refers to a tiny rectangle on the screen that can be darkened to represent a dot on a graph. Pixels are arranged in a rectangular array on the screen. In the above window, the axes and tick marks are formed by darkened pixels. The size of the screen and number of pixels vary with different calculators.

TRACE allows you to move a flashing pixel, or *cursor,* along a curve in the viewing window with the x- and y-coordinates shown at the bottom of the screen.

Useful Hint: To make the x-values in TRACE take simple values like .1, .2, and .3, choose XMIN and XMAX to be multiples of one less than the number of pixels across the screen. For example, on the *TI-83* and *TI-83 Plus*, which have 95 pixels across the screen, using an x-window like $[-9.4, 9.4]$ or $[-4.7, 4.7]$ or $[-940, 940]$ will TRACE with simpler x-values than the standard windows stated in this book.

ZOOM IN allows you to magnify any part of the viewing window to see finer detail around a chosen point. **ZOOM OUT** does the opposite, like stepping back to see a larger portion of the plane but with less detail. These and other **ZOOM** commands change the viewing window.

VALUE or **EVALUATE** finds the value of a previously entered expression at a specified x-value.

SOLVE or **ROOT** finds the x-value that solves $f(x) = 0$, equivalently, the x-intercepts of a curve. When applied to a difference $f(x) - g(x)$, it finds the x-value where the two curves meet (also done by the **INTERSECT** command).

MAX and **MIN** find the maximum and minimum values of a previously entered curve between specified x-values.

NDERIV or **DERIV** or **dy/dx** approximates the *derivative* of a function at a point. **FnInt** or **$\int f(x)dx$** approximates the definite integral of a function on an interval.

In **CONNECTED MODE** your calculator will darken pixels to connect calculated points on a graph to show it as a continuous or "unbroken" curve. However, this may lead to "false lines" in a graph that should have breaks or "jumps." False lines can be eliminated by using **DOT MODE**.

The **TABLE** command lists in table form the values of a function, just as you have probably done when graphing a curve. The x-values may be chosen by you or by the calculator.

The **Order of Operations** used by most calculators evaluates operations in the following order: first powers and roots, then operations like **LN** and **LOG,** then multiplication and division, then addition and subtraction—left to right within each level. For example, $5\wedge 2x$ means $(5\wedge 2)x$, *not* $5\wedge(2x)$. Also, $1/x + 1$ means $(1/x) + 1$, *not* $1/(x + 1)$. See your calculator's instruction manual for further information. *Be careful:* Some calculators evaluate $1/2x$ as $(1/2)x$ and some as $1/(2x)$. When in doubt, use parentheses to clarify the expression.

Much more information can be found in the manual for your graphing calculator. Other features will be discussed later as needed.

1 Functions

Moroccan runner Hicham El Guerrouj, current world record holder for the mile run, bested the record set 6 years earlier by 1.26 seconds.

- 1.1 **Real Numbers, Inequalities, and Lines**
- 1.2 **Exponents**
- 1.3 **Functions**
- 1.4 **Functions, Continued**
- 1.5 **Exponential Functions**
- 1.6 **Logarithmic Functions**

Application Preview

World Record Mile Runs

The dots on the graph below show the world record times for the mile run from 1865 to the 1999 world record of 3 minutes 43.13 seconds, set by the Moroccan runner Hicham El Guerrouj. These points fall roughly along a line, called the **regression line**. The regression line is easily found using a graphing calculator, based on a method called **least squares**, which is explained in Section 3.6 of Chapter 3 and Section 13.4 of Chapter 13. Notice that the times do not level off as you might expect, but continue to decrease.

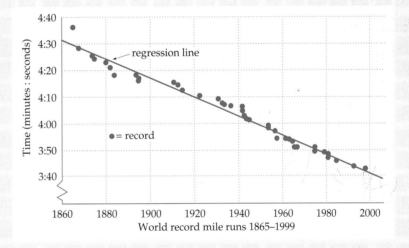

History of the Record for the Mile Run

Time	Year	Athlete	Time	Year	Athlete
4:36.5	1865	Richard Webster	4:12.6	1915	Norman Taber
4:29.0	1868	William Chinnery	4:10.4	1923	Paavo Nurmi
4:28.8	1868	Walter Gibbs	4:09.2	1931	Jules Ladoumegue
4:26.0	1874	Walter Slade	4:07.6	1933	Jack Lovelock
4:24.5	1875	Walter Slade	4:06.8	1934	Glenn Cunningham
4:23.2	1880	Walter George	4:06.4	1937	Sydney Wooderson
4:21.4	1882	Walter George	4:06.2	1942	Gunder Hägg
4:18.4	1884	Walter George	4:06.2	1942	Arne Andersson
4:18.2	1894	Fred Bacon	4:04.6	1942	Gunder Hägg
4:17.0	1895	Fred Bacon	4:02.6	1943	Arne Andersson
4:15.6	1895	Thomas Conneff	4:01.6	1944	Arne Andersson
4:15.4	1911	John Paul Jones	4:01.4	1945	Gunder Hägg
4:14.4	1913	John Paul Jones	3:59.4	1954	Roger Bannister

World Record Mile Runs (continued)

Time	Year	Athlete	Time	Year	Athlete
3:58.0	1954	John Landy	3:49.4	1975	John Walker
3:57.2	1957	Derek Ibbotson	3:49.0	1979	Sebastian Coe
3:54.5	1958	Herb Elliott	3:48.8	1980	Steve Ovett
3:54.4	1962	Peter Snell	3:48.53	1981	Sebastian Coe
3:54.1	1964	Peter Snell	3:48.40	1981	Steve Ovett
3:53.6	1965	Michel Jazy	3:47.33	1981	Sebastian Coe
3:51.3	1966	Jim Ryun	3:46.31	1985	Steve Cram
3:51.1	1967	Jim Ryun	3:44.39	1993	Noureddine Morceli
3:51.0	1975	Filbert Bayi	3:43.13	1999	Hicham El Guerrouj

Source: USA Track & Field

The equation of the regression line shown in the graph is $y = -0.356x + 257.44$, where x represents years after 1900 and y is in seconds. The regression line can be used to predict the world mile record in future years. Notice that the most recent world record would have been predicted quite accurately by this line, since the rightmost dot falls almost exactly on the line. Linear trends, however, must not be extended too far. The downward slope of this line means that it will eventually "predict" mile runs in a fraction of a second, or even in *negative* time (see page 16). *Moral:* In the real world, linear trends do not continue indefinitely. This and other topics in "linear" mathematics will be developed in Section 1.1 and thereafter.

1.1 REAL NUMBERS, INEQUALITIES, AND LINES

Introduction

In this section we will study *linear* relationships between two variable quantities—that is, relationships that can be represented by *lines*. In later sections we will study *nonlinear* relationships, which can be represented by *curves*.

When reading this book, it will be helpful (but not necessary) to have a graphing calculator. The **Graphing Calculator Explorations** show how to use a graphing calculator to explore concepts more deeply or to analyze applications in more detail. The parts of the book that require graphing calculators are marked by the symbol . If you do not have a *graphing* calculator, you will need a *scientific* or *business* calculator with keys like y^x and $\ln x$ for powers and logarithms. For those with access to a computer, the **Spreadsheet Explorations** show how spreadsheets can be used for some of the operations of a graphing calculator.

Real Numbers and Inequalities

In this book the word "number" means *real* number, a number that can be represented by a point on the number line (also called the *real line*).

The *order* of the real numbers is expressed by inequalities, with $a < b$ meaning "*a* is to the *left* of *b*," and $a > b$ meaning "*a* is to the *right* of *b*."

Inequalities

Inequality	In Words
$a < b$	*a* is less than (smaller than) *b*
$a \leq b$	*a* is less than or equal to *b*
$a > b$	*a* is greater than (larger than) *b*
$a \geq b$	*a* is greater than or equal to *b*

The inequalities $a < b$ and $a > b$ are called "strict" inequalities, and $a \leq b$ and $a \geq b$ are called "nonstrict" inequalities.

EXAMPLE 1

INEQUALITIES BETWEEN NUMBERS

a. $3 \leq 5$ b. $6 > -2$ c. $-10 < -5$

-10 is less than (smaller than) -5

Throughout this book are many **Practice Problems**—short questions designed to check your understanding of a topic before moving on to new material. Full solutions are given at the end of the section. Solve the following Practice Problem and then check your answer.

Practice Problem 1

Which number is smaller: $\dfrac{1}{100}$ or $-1{,}000{,}000$? ▶ **Solution on page 13**

Multiplying or dividing both sides of an inequality by a negative number reverses the direction of the inequality:

$-3 < 2$ but $3 > -2$ Multiplying by -1

A *double* inequality, such as $a < x < b$, means that *both* the inequalities $a < x$ and $x < b$ hold. The inequality $a < x < b$ can be interpreted graphically as "x is between a and b."

Sets and Intervals

Braces { } are read "the set of all" and a vertical bar | is read "such that."

EXAMPLE 2

INTERPRETING SETS

a. $\{x \mid x > 3\}$ means "the set of all x such that x is greater than 3."

(The set of all ↓, Such that ↑)

b. $\{x \mid -2 < x < 5\}$ means "the set of all x such that x is between -2 and 5."

Practice Problem 2

a. Write in set notation "the set of all x such that x is greater than or equal to -7."

b. Express in words: $\{x \mid x < -1\}$. ➤ Solutions on page 13

The set $\{x \mid 2 \leq x \leq 5\}$ can be expressed in *interval* notation by enclosing the endpoints 2 and 5 in square brackets: [2, 5]. The *square* brackets indicate that the endpoints are *included*. The set $\{x \mid 2 < x < 5\}$ can be written (2, 5). The *parentheses* indicate that the endpoints 2 and 5 are *excluded*. An interval is *closed* if it includes both endpoints, and *open* if it includes neither endpoint. The four types of intervals are shown below: a *solid* dot • on the graph indicates that the point is *included* in the interval; a *hollow* dot ○ indicates that the point is *excluded*.

Finite Intervals

Interval Notation	Set Notation	Graph	Type
[a, b]	$\{x \mid a \leq x \leq b\}$	•———• a b	Closed
(a, b)	$\{x \mid a < x < b\}$	○———○ a b	Open

Finite Intervals (continued)

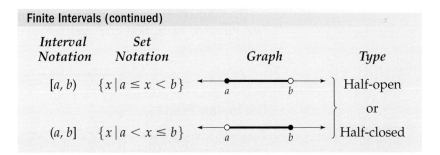

An interval may extend infinitely far to the right (indicated by the symbol ∞ for "infinity") or infinitely far to the left (indicated by $-\infty$ for "negative infinity"). Note that ∞ and $-\infty$ are not numbers, but are merely symbols to indicate that the interval extends endlessly in one direction or the other. The infinite intervals in the next box are said to be *closed* or *open* depending on whether they *include* or *exclude* their single endpoint.

Infinite Intervals

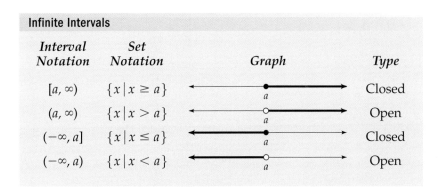

EXAMPLE 3

GRAPHING SETS AND INTERVALS

Interval Notation	Set Notation	Graph
$(-2, 5]$	$\{x \mid -2 < x \leq 5\}$	○————● $-2 \quad\quad 5$
$[3, \infty)$	$\{x \mid x \geq 3\}$	●———→ 3

We use *parentheses* rather than square brackets with ∞ and $-\infty$ since they are not actual numbers.

The interval $(-\infty, \infty)$ extends infinitely far in *both* directions (meaning the entire real line) and is also denoted by $\mathbb{R}$ (the set of all real numbers).

$$\mathbb{R} = (-\infty, \infty)$$

Cartesian Plane

Two real lines or *axes,* one horizontal and one vertical, intersecting at their zero points, define the *Cartesian plane.** The axes divide the plane into four *quadrants,* I through IV, as shown below.

Any point in the Cartesian plane can be specified uniquely by an ordered pair of numbers (x, y); x, called the *abscissa* or *x-coordinate,* is the number on the horizontal axis corresponding to the point; y, called the *ordinate* or *y-coordinate,* is the number on the vertical axis corresponding to the point.

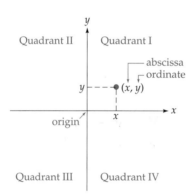

The Cartesian plane

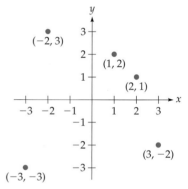

The Cartesian plane with several points. Order matters: (1, 2) is not the same as (2, 1)

Lines and Slopes

The symbol Δ (read "delta," the Greek letter D) means "the change in." For any two points (x_1, y_1) and (x_2, y_2) we define

$\Delta x = x_2 - x_1$ The change in x is the difference in the x-coordinates
$\Delta y = y_2 - y_1$ The change in y is the difference in the y-coordinates

Any two distinct points determine a line. A nonvertical line has a *slope* that measures the steepness of the line, defined as *the change in y divided by the change in x* for any two points on the line.

---

* So named because it was originated by the French philosopher and mathematician René Descartes (1596–1650). Following the custom of the day, Descartes signed his scholarly papers with his Latin name Cartesius, hence "Cartesian" plane.

Slope of Line Through (x_1, y_1) and (x_2, y_2)

$$m = \frac{\Delta y}{\Delta x} = \frac{y_2 - y_1}{x_2 - x_1}$$

Slope is the change in y over the change in x $(x_2 \neq x_1)$

The changes Δy and Δx are often called, respectively, the "rise" and the "run," with the understanding that a negative "rise" means a "fall." Slope is then "rise over run."

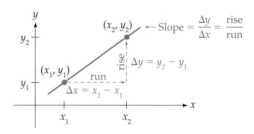

EXAMPLE 4

FINDING SLOPES AND GRAPHING LINES

Find the slope of the line through each pair of points, and graph the line.

a. $(1, 3), (2, 5)$ b. $(2, 4), (3, 1)$
c. $(-1, 3), (2, 3)$ d. $(2, -1), (2, 3)$

Solution

We use the slope formula 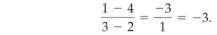 for each pair $(x_1, y_1), (x_2, y_2)$.

a. For $(1, 3)$ and $(2, 5)$ the slope is

$$\frac{5 - 3}{2 - 1} = \frac{2}{1} = 2.$$

b. For $(2, 4)$ and $(3, 1)$ the slope is

$$\frac{1 - 4}{3 - 2} = \frac{-3}{1} = -3.$$

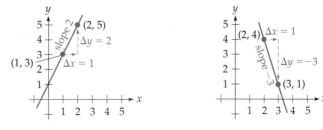

c. For $(-1, 3)$ and $(2, 3)$ the slope is $\dfrac{3-3}{2-(-1)} = \dfrac{0}{3} = 0$.

d. For $(2, -1)$ and $(2, 3)$ the slope is *undefined*: $\dfrac{3-(-1)}{2-2} = \dfrac{4}{0}$.

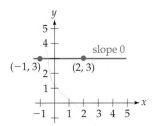

If $\Delta x = 1$, as in Examples 4a and 4b, then the slope is just the "rise," giving

$$\text{Slope} = \begin{pmatrix} \text{Amount that the line rises} \\ \text{when } x \text{ increases by } 1 \end{pmatrix}$$

Practice Problem 3 A company president is considering four different business strategies, called S_1, S_2, S_3, and S_4, each with different projected future profits. The graph on the right shows the annual projected profit for the first few years for each of the strategies. Which strategy will yield:

a. The highest projected profit in year 1?

b. The highest projected profit in the long run?   ➤ Solutions on page 13

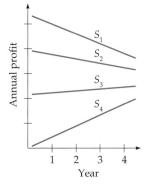

Equations of Lines

The point where a nonvertical line crosses the *y*-axis is called the *y-intercept* of the line. The *y*-intercept can be given either as the *y*-coordinate b or as the point $(0, b)$. Such a line can be expressed very simply in terms of its slope and *y*-intercept, representing the points by variable coordinates (or "variables") x and y.

Slope-Intercept Form of a Line

$y = mx + b$ $\begin{array}{l} m = \text{slope} \\ b = y\text{-intercept} \end{array}$

EXAMPLE 5

USING THE SLOPE-INTERCEPT FORM

Find an equation of the line with slope -2 and y-intercept 4, and graph it.

Solution

$$y = -2x + 4 \qquad \begin{array}{l} y = mx + b \text{ with} \\ m = -2 \text{ and } b = 4 \end{array}$$

We graph the line by first plotting the y-intercept $(0, 4)$. Using the slope $m = -2$, we plot another point 1 unit to the right and 2 units *down* from the y-intercept. We then draw the line through these two points, as shown on the left.

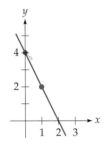

Graphing Calculator Exploration

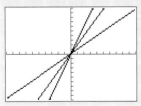

$y_1 = x$, $y_2 = 2x$, and $y_3 = 3x$ on $[-10, 10]$ by $[-10, 10]$

a. Use a graphing calculator to graph the lines $y_1 = x$, $y_2 = 2x$, and $y_3 = 3x$ simultaneously on the window $[-10, 10]$ by $[-10, 10]$. How do the graphs change as the coefficient of x increases from 1 to 2 to 3?

b. Predict what the graph of $y = 0.5x$ would look like. What about $y = -2x$? Check your predictions by graphing them.

c. Describe the graph of the line $y = mx$ for any number m.

10 CHAPTER 1 FUNCTIONS

> **Point-Slope Form of a Line**
>
> $$y - y_1 = m(x - x_1)$$
>
> (x_1, y_1) = point on the line
> m = slope

This form comes directly from the slope formula $m = \dfrac{y_2 - y_1}{x_2 - x_1}$ by replacing x_2 and y_2 by x and y, and then multiplying each side by $(x - x_1)$. It is most useful when you know the slope of the line and a point on it.

EXAMPLE 6

USING THE POINT-SLOPE FORM

Find an equation of the line through $(6, -2)$ with slope $-\frac{1}{2}$.

Solution

$$y - (-2) = -\frac{1}{2}(x - 6) \qquad \begin{array}{l} y - y_1 = m(x - x_1) \text{ with} \\ y_1 = -2,\ m = -\frac{1}{2},\ \text{and}\ x_1 = 6 \end{array}$$

$$y + 2 = -\frac{1}{2}x + 3 \qquad \text{Eliminating parentheses}$$

$$y = -\frac{1}{2}x + 1 \qquad \text{Subtracting 2 from each side}$$

Alternatively, we could have found this equation using $y = mx + b$, replacing m by the given slope $-\frac{1}{2}$, and then substituting the given $x = 6$ and $y = -2$ to evaluate b.

EXAMPLE 7

FINDING AN EQUATION FOR A LINE THROUGH TWO POINTS

Find an equation for the line through the points $(4, 1)$ and $(7, -2)$.

Solution

The slope is not given, so we calculate it from the two points.

$$m = \frac{-2 - 1}{7 - 4} = \frac{-3}{3} = -1 \qquad m = \frac{y_2 - y_1}{x_2 - x_1} \text{ with } (4, 1) \text{ and } (7, -2)$$

Then we use the point-slope formula with this slope and either of the two points.

$$y - 1 = -1(x - 4) \qquad \text{$y - y_1 = m(x - x_1)$ with slope -1 and point $(4, 1)$}$$

$$y - 1 = -x + 4 \qquad \text{Eliminating parentheses}$$

$$y = -x + 5 \qquad \text{Adding 1 to each side}$$

Practice Problem 4   Find the slope-intercept form of the line through the points (2, 1) and (4, 7).
▶ Solution on page 13

Vertical and horizontal lines have particularly simple equations: a variable equaling a constant.

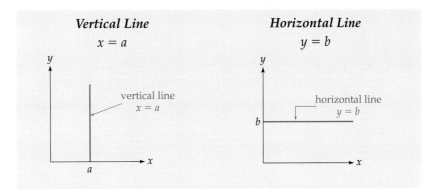

EXAMPLE 8

GRAPHING VERTICAL AND HORIZONTAL LINES

Graph the lines $x = 2$ and $y = 6$.

Solution

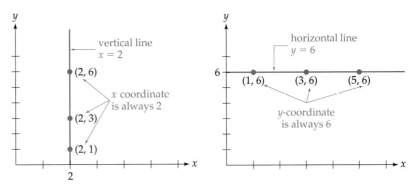

EXAMPLE 9 **FINDING EQUATIONS OF VERTICAL AND HORIZONTAL LINES**

a. Find an equation for the *vertical* line through the point (3, 2).
b. Find an equation for the *horizontal* line through the point (3, 2).

Solution

a. Vertical line $x = 3$ $x = a$, with a being the x-coordinate from (3, 2)

b. Horizontal line $y = 2$ $y = b$, with b being the y-coordinate from (3, 2)

Practice Problem 5 Find an equation for the vertical line through the point $(-2, 10)$.

▶ **Solution on next page**

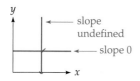

Distinguish carefully between slopes of vertical and horizontal lines:

Vertical line: slope is *undefined*.

Horizontal line: slope *is* defined and is *zero*.

There is one form that covers *all* lines, vertical and nonvertical.

General Linear Equation

$$ax + by = c$$

For constants a, b, c, with a and b not both zero

Any equation that can be written in this form is called a *linear equation*, and the variables are said to *depend linearly* on each other.

EXAMPLE 10 **FINDING THE SLOPE AND THE *y*-INTERCEPT FROM A LINEAR EQUATION**

Find the slope and y-intercept of the line $2x + 3y = 12$.

Solution   We write the line in slope-intercept form. Solving for y:

$$3y = -2x + 12$$ Subtracting $2x$ from both sides of $2x + 3y = 12$

$$y = -\frac{2}{3}x + 4$$ Dividing each side by 3 gives the slope-intercept form $y = mx + b$

Therefore, the slope is $-\frac{2}{3}$ and the y-intercept is (0, 4).

Practice Problem 6 Find the slope and y-intercept of the line $x - \dfrac{y}{3} = 2$.

➤ Solution on next page

1.1 Section Summary

An *interval* is a set of real numbers corresponding to a section of the real line. The interval is *closed* if it contains all of its endpoints, and *open* if it contains none of its endpoints.

The nonvertical line through two points (x_1, y_1) and (x_2, y_2) has slope

$$m = \frac{\Delta y}{\Delta x} = \frac{y_2 - y_1}{x_2 - x_1} \qquad x_1 \neq x_2$$

There are five equations or "forms" for lines:

$y = mx + b$	Slope-intercept form $m =$ slope, $b = y$-intercept
$y - y_1 = m(x - x_1)$	Point-slope form $(x_1, y_1) =$ point, $m =$ slope
$x = a$	Vertical line (slope undefined) $a = x$-intercept
$y = b$	Horizontal line (slope zero) $b = y$-intercept
$ax + by = c$	General linear equation

➤ **Solutions to Practice Problems**

1. $-1{,}000{,}000$ [the negative sign makes it less than (to the left of) the positive number $\frac{1}{100}$]

2. a. $\{x \mid x \geq -7\}$
   b. The set of all x such that x is less than -1

3. a. S_1
   b. S_4

4. $m = \dfrac{7 - 1}{4 - 2} = \dfrac{6}{2} = 3$ From points $(2, 1)$ and $(4, 7)$

   $y - 1 = 3(x - 2)$ Using the point-slope form with $(x_1, y_1) = (2, 1)$
   $y - 1 = 3x - 6$
   $y = 3x - 5$

5. $x = -2$

6. $x - \dfrac{y}{3} = 2$

$-\dfrac{y}{3} = -x + 2$ Subtracting x from each side

$y = 3x - 6$ Multiplying each side by -3

Slope is $m = 3$ and y-intercept is $(0, -6)$.

1.1 Exercises

Write each interval in set notation and graph it on the real line.

1. $[0, 6)$ **2.** $(-3, 5]$ **3.** $(-\infty, 2]$ **4.** $[7, \infty)$

5. Given the equation $y = 5x - 12$, how will y change if x:
  a. Increases by 3 units?
  b. Decreases by 2 units?

6. Given the equation $y = -2x + 7$, how will y change if x:
  a. Increases by 5 units?
  b. Decreases by 4 units?

Find the slope (if it is defined) of the line determined by each pair of points.

7. $(2, 3)$ and $(4, -1)$ **8.** $(3, -1)$ and $(5, 7)$
9. $(-4, 0)$ and $(2, 2)$ **10.** $(-1, 4)$ and $(5, 1)$
11. $(0, -1)$ and $(4, -1)$ **12.** $\left(-2, \tfrac{1}{2}\right)$ and $\left(5, \tfrac{1}{2}\right)$
13. $(2, -1)$ and $(2, 5)$ **14.** $(6, -4)$ and $(6, -3)$

For each equation, find the slope m and y-intercept $(0, b)$ (when they exist) and draw the graph.

15. $y = 3x - 4$ **16.** $y = 2x$
17. $y = -\tfrac{1}{2}x$ **18.** $y = -\tfrac{1}{3}x + 2$
19. $y = 4$ **20.** $y = -3$
21. $x = 4$ **22.** $x = -3$
23. $2x - 3y = 12$ **24.** $3x + 2y = 18$
25. $x + y = 0$ **26.** $x = 2y + 4$
27. $x - y = 0$ **28.** $y = \tfrac{2}{3}(x - 3)$

29. $y = \dfrac{x + 2}{3}$ **30.** $\dfrac{x}{2} + \dfrac{y}{3} = 1$

31. $\dfrac{2x}{3} - y = 1$ **32.** $\dfrac{x + 1}{2} + \dfrac{y + 1}{2} = 1$

Use a graphing calculator to graph each line. [*Note:* Your graph will depend on the viewing window you choose. Begin with a "standard" window such as $[-10, 10]$ by $[-10, 10]$, and choose a larger window (or "zoom out") if the line does not appear.]

33. $y = 2x - 8$ **34.** $y = 3x - 6$
35. $y = 7 - 3x$ **36.** $y = 5 - 2x$
37. $y = 50 - x$ **38.** $y = x - 40$

Write an equation of the line satisfying the following conditions. If possible, write your answer in the form $y = mx + b$.

39. Slope -2.25 and y-intercept 3
40. Slope $\tfrac{2}{3}$ and y-intercept -8
41. Slope 5 and passing through the point $(-1, -2)$
42. Slope -1 and passing through the point $(4, 3)$
43. Horizontal and passing through the point $(1.5, -4)$
44. Horizontal and passing through the point $\left(\tfrac{1}{2}, \tfrac{3}{4}\right)$
45. Vertical and passing through the point $(1.5, -4)$
46. Vertical and passing through the point $\left(\tfrac{1}{2}, \tfrac{3}{4}\right)$
47. Passing through the points $(5, 3)$ and $(7, -1)$
48. Passing through the points $(3, -1)$ and $(6, 0)$
49. Passing through the points $(1, -1)$ and $(5, -1)$

50. Passing through the points (2, 0) and (2, −4)

Write an equation of the form $y = mx + b$ for each line in the following graphs. [Hint: Either find the slope and y-intercept or use any two points on the line.]

51.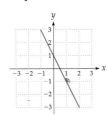

52.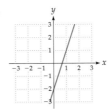

53.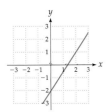

54.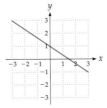

Write equations for the lines determining the four sides of each figure.

55.

56.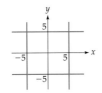

57. Show that $y - y_1 = m(x - x_1)$ simplifies to $y = mx + b$ if the point (x_1, y_1) is the y-intercept $(0, b)$.

58. Show that the linear equation $\dfrac{x}{a} + \dfrac{y}{b} = 1$ has x-intercept $(a, 0)$ and y-intercept $(0, b)$. (The x-intercept is the point where the line crosses the x-axis.)

59. Find the x-intercept $(a, 0)$ where the line $y = mx + b$ crosses the x-axis. Under what condition on m will a single x-intercept exist?

60. i. Show that the general linear equation $ax + by = c$ with $b \neq 0$ can be written as $y = -\dfrac{a}{b}x + \dfrac{c}{b}$, which is the equation of a line in slope-intercept form.

ii. Show that the general linear equation $ax + by = c$ with $b = 0$ but $a \neq 0$ can be written as $x = \dfrac{c}{a}$, which is the equation of a vertical line.

[Note: Since these steps are reversible, parts (i) and (ii) together show that the general linear equation $ax + by = c$ (for a and b not both zero) includes vertical and nonvertical lines.]

61. a. Graph the lines $y_1 = -x$, $y_2 = -2x$, and $y_3 = -3x$ on the window $[-5, 5]$ by $[-5, 5]$ (using the *negation* key $(-)$, not the *subtraction* key $-$). Observe how the coefficient of x changes the slope of the line.

b. Predict what the line $y = -9x$ would look like, and then check your prediction by graphing it.

62. a. Graph the lines $y_1 = x + 2$, $y_2 = x + 1$, $y_3 = x$, $y_4 = x - 1$, and $y_5 = x - 2$ on the window $[-5, 5]$ by $[-5, 5]$. Observe how the constant changes the position of the line.

b. Predict what the lines $y = x + 4$ and $y = x - 4$ would look like, and then check your prediction by graphing them.

APPLIED EXERCISES

63. BUSINESS: Energy Usage A utility considers demand for electricity "low" if it is below 8 mkW (million kilowatts), "average" if it is at least 8 mkW but below 20 mkW, "high" if it is at least 20 mkW but below 40 mkW, and "critical" if it is 40 mkW or more. Express these demand levels in interval notation. [Hint: The interval for "low" is [0, 8).]

64. **GENERAL: Grades** If a grade of 90 through 100 is an A, at least 80 but less than 90 is a B, at least 70 but less than 80 a C, at least 60 but less than 70 a D, and below 60 an F, write these grade levels in interval form (ignoring rounding). [*Hint:* F would be [0, 60).]

65. **ATHLETICS: Mile Run** Read the Application Preview on pages 1–2.
    a. Use the regression line $y = -0.356x + 257.44$ to predict the world record in the year 2010. [*Hint:* If x represents years after 1900, what value of x corresponds to the year 2010? The resulting y will be in seconds, and should be converted to minutes and seconds.]
    b. According to this formula, when will the record be 3 minutes 30 seconds? [*Hint:* Set the formula equal to 210 seconds and solve. What year corresponds to this x-value?]

66. **ATHLETICS: Mile Run** Read the Application Preview on pages 1–2. Evaluate the regression line $y = -0.356x + 257.44$ at $x = 720$ and at $x = 724$ (corresponding to the years 2620 and 2624). Does the formula give reasonable times for the mile record in these years? [*Moral:* Linear trends may not continue indefinitely.]

67. **BUSINESS: Corporate Profit** A company's profit increased linearly from $6 million at the end of year 1 to $14 million at the end of year 3.
    a. Use the two (year, profit) data points (1, 6) and (3, 14) to find the linear relationship $y = mx + b$ between $x =$ year and $y =$ profit.
    b. Find the company's profit at the end of 2 years.
    c. Predict the company's profit at the end of 5 years.

68. **ECONOMICS: Per Capita Personal Income** In the short run, per capita personal income (PCPI) in the United States grows approximately linearly. In 1990 PCPI was 19.1, and in 2000 it had grown to 29.7 (both in thousands of dollars). (*Source:* Bureau of Economic Analysis)
    a. Use the two given (year, PCPI) data points (0, 19.1) and (10, 29.7) to find the linear relationship $y = mx + b$ between $x =$ years since 1990 and $y =$ PCPI.
    b. Use your linear relationship to predict PCPI in 2010.

69. **GENERAL: Temperature** On the Fahrenheit temperature scale, water freezes at 32° and boils at 212°. On the Celsius (centigrade) scale, water freezes at 0° and boils at 100°.
    a. Use the two (Celsius, Fahrenheit) data points (0, 32) and (100, 212) to find the linear relationship $y = mx + b$ between $x =$ Celsius temperature and $y =$ Fahrenheit temperature.
    b. Find the Fahrenheit temperature that corresponds to 20° Celsius.

70. **ECOLOGY: Waste Disposal** The amount of municipal solid waste generated per person per day in the United States has increased approximately linearly, from 2.7 pounds in 1960 to 4.6 pounds in 2000.

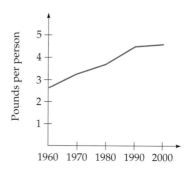

Municipal Solid Waste Generated Per Capita Per Day
(*Source:* U.S. Environmental Protection Agency)

    a. Use the two (year, pound) data points (0, 2.7) and (40, 4.6) to find the linear relationship $y = mx + b$ between $x =$ years since 1960 and $y =$ per capita waste.
    b. Use your formula to predict the amount in the year 2010.

71–72. **BUSINESS: Straight-Line Depreciation** Straight-line depreciation is a method for estimating the value of an asset (such as a piece of machinery) as it loses value ("depreciates") through use. Given the original *price* of an asset, its *useful lifetime*, and its *scrap value* (its value at the end of its useful lifetime), the value of the asset after t years is given by the formula

$$\text{Value} = (\text{price}) - \left(\frac{(\text{price}) - (\text{scrap value})}{(\text{useful lifetime})}\right) \cdot t$$

for $0 \leq t \leq (\text{useful lifetime})$

71. a. A farmer buys a harvester for $50,000 and estimates its useful life to be 20 years, after which its scrap value will be $6000. Use the formula above to find a formula for the value V of the harvester after t years, for $0 \leq t \leq 20$.
  b. Use your formula to find the value of the harvester after 5 years.
  c. Graph the function found in part (a) on a graphing calculator on the window [0, 20] by [0, 50,000]. [*Hint:* Use x instead of t.]

72. a. A newspaper buys a printing press for $800,000 and estimates its useful life to be 20 years, after which its scrap value will be $60,000. Use the formula above Exercise 71 to find a formula for the value V of the press after t years, for $0 \leq t \leq 20$.
  b. Use your formula to find the value of the press after 10 years.
  c. Graph the function found in part (a) on a graphing calculator on the window [0, 20] by [0, 800,000]. [*Hint:* Use x instead of t.]

73–74. ENVIRONMENTAL SCIENCE: Beverton-Holt Recruitment Curve Some organisms exhibit a density-dependent mortality from one generation to the next. Let $R > 1$ be the net reproductive rate (that is, the number of surviving offspring per parent), let $x > 0$ be the density of parents and y be the density of surviving offspring. The *Beverton-Holt recruitment curve* is

$$y = \frac{Rx}{1 + \left(\frac{R-1}{K}\right)x}$$

where $K > 0$ is the *carrying capacity* of the organism's environment. Notice that if $x = K$, then $y = K$.

73. Show that if $x < K$, then $x < y < K$. Explain what this means about the population size over successive generations if the initial population is smaller than the carrying capacity of the environment.

74. Show that if $x > K$, then $K < y < x$. Explain what this means about the population size over successive generations if the initial population is larger than the carrying capacity of the environment.

75. SOCIAL SCIENCE: Age at First Marriage Americans are marrying later and later. Based on data from the U.S. Bureau of the Census for the years 1980 to 2000, the median age at first marriage for men is $y_1 = 24.8 + 0.13x$, and for women it is $y_2 = 22.1 + 0.17x$, where x is the number of years since 1980.
  a. Graph these lines on the window [0, 30] by [20, 30].
  b. Use these lines to predict the median marriage ages for men and women in the year 2010. [*Hint:* Which x-value corresponds to 2010? Then use TRACE, EVALUATE, or TABLE.]
  c. Predict the median marriage ages for men and women in the year 2020.

76. SOCIAL SCIENCE: Equal Pay for Equal Work Women's pay has often lagged behind men's, although Title VII of the Civil Rights Act requires equal pay for equal work. Based on data from 1980–2000, women's annual earnings as a percent of men's can be approximated by the formula $y = 0.97x + 59.5$, where x is the number of years since 1980. (For example, $x = 10$ gives $y = 69.2$, so in 1990 women's wages were about 69.2% of men's wages.)
  a. Graph this line on the window [0, 30] by [0, 100].
  b. Use this line to predict the percentage in the year 2005. [*Hint:* Which x-value corresponds to 2005? Then use TRACE, EVALUATE, or TABLE.]
  c. Predict the percentage in the year 2010. (*Source:* U.S. Department of Labor—Women's Bureau)

77. GENERAL: Smoking and Education According to a recent study,* the probability that a smoker will quit smoking increases with the smoker's educational level. The probability (expressed as a percent) that a smoker with x years of education will quit is approximately $y = 0.831x^2 - 18.1x + 137.3$ (for $10 \le x \le 16$).

a. Graph this curve on the window [10, 16] by [0, 100].
b. Find the probability that a high school graduate smoker $(x = 12)$ will quit.
c. Find the probability that a college graduate smoker $(x = 16)$ will quit.

78. ENVIRONMENTAL SCIENCE: Wind Energy The use of wind power is growing rapidly after a slow start, especially in Europe, where it is seen as an efficient and renewable source of energy. Global wind power generating capacity for the years 1980 to 2000 is given approximately by $y = 30.5x^2 - 112x + 250$ megawatts, where x is the number of years after 1980. (One megawatt would supply the electrical needs of approximately 100 homes).

a. Graph this curve on the window [0, 30] by [0, 2500].
b. Use this curve to predict the global wind power generating capacity in the year 2005. [*Hint:* Which x-value corresponds to 2005? Then use TRACE, EVALUATE, or TABLE.]
c. Predict the global wind power generating capacity in the year 2010.

(*Source:* Worldwatch Institute)

* William Sander, "Schooling and Quitting Smoking," *The Review of Economics and Statistics* LXXVII(1):191–199, February 1995.

79–80. GENERAL: Life Expectancy The following tables give the life expectancy (years of life expected) for a newborn child born in the indicated year (Exercise 79 is for males, Exercise 80 for females). For each exercise:

a. Enter the data into a graphing calculator and make a plot of the resulting points, with Years Since 1960 on the x-axis and Life Expectancy on the y-axis.
b. Use the graphing calculator to find the linear regression line for these points. Enter the resulting function as y_1, which then estimates life expectancy based on the year of birth. Graph the points together with the regression line.
c. Use your line y_1 to estimate the life expectancy of a child born in the year 2025. (This might be your child or grandchild.) [*Hint:* What x-value corresponds to 2025?]

79.

	Birth Year (years since 1960)	Life Expectancy (male)
1960	0	66.6
1970	10	67.1
1980	20	70.0
1990	30	71.8
2000	40	74.4

80.

	Birth Year (years since 1960)	Life Expectancy (female)
1960	0	73.1
1970	10	74.7
1980	20	77.5
1990	30	78.8
2000	40	79.6

1.2 EXPONENTS

Introduction

Not all variables are related linearly. In this section we will discuss exponents, which will enable us to express many nonlinear relationships.

Positive Integer Exponents

Numbers may be expressed with exponents, as in $2^3 = 2 \cdot 2 \cdot 2 = 8$. More generally, for any positive integer n, x^n means the product of n x's.

$$x^n = \overbrace{x \cdot x \cdots x}^{n}$$

The number being raised to the power is called the *base* and the power is the *exponent*:

$$x^n \quad \begin{array}{l} \leftarrow \text{Exponent or power} \\ \leftarrow \text{Base} \end{array}$$

There are several *properties of exponents* for simplifying expressions. The first three are known, respectively, as the addition, subtraction, and multiplication properties of exponents.

Properties of Exponents

$x^m \cdot x^n = x^{m+n}$	To *multiply* powers of the same base, *add* the exponents
$\dfrac{x^m}{x^n} = x^{m-n}$	To *divide* powers of the same base, *subtract* the exponents (top exponent minus bottom exponent)
$(x^m)^n = x^{m \cdot n}$	To raise a power to a power, *multiply* the powers
$(xy)^n = x^n \cdot y^n$	To raise a product to a power, raise *each factor* to the power
$\left(\dfrac{x}{y}\right)^n = \dfrac{x^n}{y^n}$	To raise a fraction to a power, raise the numerator *and* denominator to the power

EXAMPLE 1

SIMPLIFYING EXPONENTS

a. $x^2 \cdot x^3 = x^{\overset{2+3}{5}}$ Since $x^2 \cdot x^3 = \underbrace{\overbrace{x \cdot x}^{2} \cdot \overbrace{x \cdot x \cdot x}^{3}}_{5}$

b. $\dfrac{x^5}{x^3} = x^{\overset{5-3}{2}}$ Since $\dfrac{x^5}{x^3} = \dfrac{\overbrace{x \cdot x \cdot \cancel{x} \cdot \cancel{x} \cdot \cancel{x}}^{2}}{\cancel{x} \cdot \cancel{x} \cdot \cancel{x}}$

c. $(x^3)^2 = x^6$ ← 2·3   Since $(x^3)^2 = x^3 \cdot x^3$

d. $(2w)^3 = 2^3 w^3 = 8w^3$ Since $(2w)^3 = 2w \cdot 2w \cdot 2w = 2^3 w^3$

e. $\left(\dfrac{x}{5}\right)^3 = \dfrac{x^3}{5^3} = \dfrac{x^3}{125}$ Since $\left(\dfrac{x}{5}\right)^3 = \dfrac{x}{5} \cdot \dfrac{x}{5} \cdot \dfrac{x}{5}$

f. $\dfrac{[(x^2)^3]^4}{x^5 \cdot x^7 \cdot x} = \dfrac{x^{24}}{x^{13}} = x^{11}$ Combining all the rules

2·3·4 ↑ 24   24 − 13
5 + 7 + 1

Practice Problem 1 Simplify: **a.** $\dfrac{x^5 \cdot x}{x^2}$ **b.** $[(x^3)^2]^2$ ▶ Solutions on page 28

Remember: For exponents in the form $x^2 \cdot x^3 = x^5$, *add* exponents.
For exponents in the form $(x^2)^3 = x^6$, *multiply* exponents.

Graphing Calculator Exploration

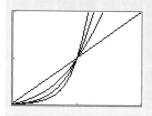

a. Use a graphing calculator to graph $y_1 = x$, $y_2 = x^2$, $y_3 = x^3$, and $y_4 = x^4$ on the viewing window [0, 2] by [0, 2]. Use TRACE to identify which curve goes with which power.

b. Which curve is highest for values of x between 0 and 1? Which is lowest?

c. Which curve is highest for values of x greater than 1? Which is lowest?

d. Predict what the curve $y = x^5$ would look like. Check your prediction by graphing it.

e. Predict which of these curves will be positive when x is negative. Check your prediction by changing the viewing window to [−2, 2] by [−2, 2].

Zero and Negative Exponents

For any number x other than zero, we define

$x^0 = 1$	x to the power 0 is 1
$x^{-1} = \dfrac{1}{x}$	x to the power -1 is 1 over x
$x^{-2} = \dfrac{1}{x^2}$	x to the power -2 is 1 over x squared
$x^{-n} = \dfrac{1}{x^n}$	x to a negative power is 1 over x to the positive power

The definitions of x^0 and x^{-n} are motivated by the following calculations.

$$1 = \frac{x^2}{x^2} = x^{2-2} = x^0 \qquad \text{The subtraction property of exponents leads to } x^0 = 1$$

$$\frac{1}{x^n} = \frac{x^0}{x^n} = x^{0-n} = x^{-n} \qquad x^0 = 1 \text{ and the subtraction property of exponents lead to } x^{-n} = \frac{1}{x^n}$$

EXAMPLE 2

SIMPLIFYING ZERO AND NEGATIVE EXPONENTS

a. $5^0 = 1$

b. $7^{-1} = \dfrac{1}{7}$

c. $3^{-2} = \dfrac{1}{3^2} = \dfrac{1}{9}$

d. $(-2)^{-3} = \dfrac{1}{(-2)^3} = \dfrac{1}{-8} = -\dfrac{1}{8}$

e. 0^0 and 0^{-3} are undefined.

Practice Problem 2 Evaluate: a. 2^0 b. 2^{-4} ▶ Solutions on page 28

A fraction to a negative power means *division* by the fraction, so we "invert and multiply."

$$\left(\frac{x}{y}\right)^{-1} = \frac{1}{\frac{x}{y}} = 1 \cdot \frac{y}{x} = \frac{y}{x}$$

Reciprocal of the original fraction

Therefore, for $x \neq 0$ and $y \neq 0$,

$$\left(\frac{x}{y}\right)^{-1} = \frac{y}{x}$$ A fraction to the power -1 is the reciprocal of the fraction

$$\left(\frac{x}{y}\right)^{-n} = \left(\frac{y}{x}\right)^{n}$$ A fraction to the negative power is the reciprocal of the fraction to the positive power

EXAMPLE 3

SIMPLIFYING FRACTIONS TO NEGATIVE EXPONENTS

a. $\left(\dfrac{3}{2}\right)^{-1} = \dfrac{2}{3}$ ↑ Reciprocal of $\dfrac{3}{2}$

b. $\left(\dfrac{1}{2}\right)^{-3} = \left(\dfrac{2}{1}\right)^{3} = \dfrac{2^3}{1^3} = 8$

Practice Problem 3

Simplify: $\left(\dfrac{2}{3}\right)^{-2}$ ➤ Solution on page 28

Roots and Fractional Exponents

We may take the square root of any *nonnegative* number, and the cube root of *any* number.

EXAMPLE 4

EVALUATING ROOTS

a. $\sqrt{9} = 3$

b. $\sqrt{-9}$ is undefined.   Square roots of negative numbers are not defined

c. $\sqrt[3]{8} = 2$

d. $\sqrt[3]{-8} = -2$ Cube roots of negative numbers *are* defined

e. $\sqrt[3]{\dfrac{27}{8}} = \dfrac{\sqrt[3]{27}}{\sqrt[3]{8}} = \dfrac{3}{2}$

There are *two* square roots of 9, namely 3 and -3, but the radical sign $\sqrt{}$ means just the *positive* one (the "principal" square root).

$\sqrt[n]{a}$ means the principal nth root of a.   Principal means the positive root if there are two

In general, we may take *odd* roots of *any* number, but *even* roots only if the number is positive or zero.

EXAMPLE 5

EVALUATING ROOTS OF POSITIVE AND NEGATIVE NUMBERS

Odd roots of negative numbers *are* defined ✓

a. $\sqrt[4]{81} = 3$ b. $\sqrt[5]{-32} = -2$ Since $(-2)^5 = -32$

Graphing Calculator Exploration

a. Use a graphing calculator to graph $y_1 = x$, $y_2 = \sqrt{x}$, $y_3 = \sqrt[3]{x}$, and $y_4 = \sqrt[4]{x}$ simultaneously on the viewing window [0, 3] by [0, 2]. Use TRACE to identify which curve goes with which root.

b. Which curve is highest for values of x between 0 and 1? Which is lowest?

c. Which curve is highest for values of x greater than 1? Which is lowest?

d. Predict what the curve $y = \sqrt[5]{x}$ would look like. Check your prediction by graphing it.

e. Which of these roots are defined for *negative* values of x? Check your answer by changing the window to [−3, 3] by [−2, 2] and using TRACE where x is negative.

Fractional Exponents

Fractional exponents are defined as follows:

$x^{\frac{1}{2}} = \sqrt{x}$ Power $\frac{1}{2}$ means the principal square root

$x^{\frac{1}{3}} = \sqrt[3]{x}$ Power $\frac{1}{3}$ means the cube root

$x^{\frac{1}{n}} = \sqrt[n]{x}$ Power $\frac{1}{n}$ means the principal nth root (for a positive integer n)

24 CHAPTER 1 FUNCTIONS

The definition of $x^{\frac{1}{2}}$ is motivated by the multiplication property of exponents:

$$\left(x^{\frac{1}{2}}\right)^2 = x^{\frac{1}{2}\cdot 2} = x^1 = x$$

Taking square roots of each side of $\left(x^{\frac{1}{2}}\right)^2 = x$ gives

$$x^{\frac{1}{2}} = \sqrt{x} \qquad \text{x to the half power means the square root of x}$$

EXAMPLE 6   **EVALUATING FRACTIONAL EXPONENTS**

a. $9^{\frac{1}{2}} = \sqrt{9} = 3$ b. $125^{\frac{1}{3}} = \sqrt[3]{125} = 5$

c. $81^{\frac{1}{4}} = \sqrt[4]{81} = 3$ d. $(-32)^{\frac{1}{5}} = \sqrt[5]{-32} = -2$

e. $\left(-\frac{27}{8}\right)^{\frac{1}{3}} = \sqrt[3]{-\frac{27}{8}} = -\frac{\sqrt[3]{27}}{\sqrt[3]{8}} = -\frac{3}{2}$

Practice Problem 4   Evaluate: a. $(-27)^{\frac{1}{3}}$ b. $\left(\frac{16}{81}\right)^{\frac{1}{4}}$ ➤ **Solutions on page 28**

To define $x^{\frac{m}{n}}$ for positive integers m and n, the exponent $\frac{m}{n}$ must be fully reduced (for example, $\frac{4}{6}$ must be reduced to $\frac{2}{3}$). Then

$$x^{\frac{m}{n}} = \left(x^{\frac{1}{n}}\right)^m = (x^m)^{\frac{1}{n}} \qquad \text{Since in both cases the exponents multiply to $\frac{m}{n}$}$$

Therefore we define:

Fractional Exponents

$$x^{\frac{m}{n}} = \left(\sqrt[n]{x}\right)^m = \sqrt[n]{x^m} \qquad \begin{array}{l}\text{$x^{m/n}$ means the mth power of}\\\text{the nth root, or equivalently,}\\\text{the nth root of the mth power}\end{array}$$

Both expressions, $\left(\sqrt[n]{x}\right)^m$ and $\sqrt[n]{x^m}$, will give the same answer. In either case the numerator determines the power and the denominator determines the root.

Power over root

EXAMPLE 7 EVALUATING FRACTIONAL EXPONENTS

a. $8^{2/3} = \sqrt[3]{8^2} = \sqrt[3]{64} = 4$ — First the power, then the root

b. $8^{2/3} = (\sqrt[3]{8})^2 = (2)^2 = 4$ — First the root, then the power

$\Big\}$ Same

c. $25^{3/2} = (\sqrt{25})^3 = (5)^3 = 125$

d. $\left(\dfrac{-27}{8}\right)^{2/3} = \left(\sqrt[3]{\dfrac{-27}{8}}\right)^2 = \left(\dfrac{-3}{2}\right)^2 = \dfrac{9}{4}$

Practice Problem 5 Evaluate: a. $16^{3/2}$ b. $(-8)^{2/3}$ ▶ Solutions on page 28

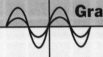

Graphing Calculator Exploration

a. Use a graphing calculator to evaluate $25^{3/2}$. [On some calculators, press 25^(3 ÷ 2).] Your answer should agree with Example 7c above.

b. Evaluate $(-8)^{2/3}$. Use the (−) key for negation, and parentheses around the exponent. Your answer should be 4. If you get an "error," try evaluating the expression as $((-8)^{1/3})^2$ or $((-8)^2)^{1/3}$. Whichever way works, remember it for evaluating negative numbers to fractional powers in the future.

EXAMPLE 8 EVALUATING NEGATIVE FRACTIONAL EXPONENTS

a. $8^{-2/3} = \dfrac{1}{8^{2/3}} = \dfrac{1}{(\sqrt[3]{8})^2} = \dfrac{1}{2^2} = \dfrac{1}{4}$ A negative exponent means the reciprocal of the number to the positive exponent, which is then evaluated as before

b. $\left(\dfrac{9}{4}\right)^{-3/2} = \left(\dfrac{4}{9}\right)^{3/2} = \left(\sqrt{\dfrac{4}{9}}\right)^3 = \left(\dfrac{2}{3}\right)^3 = \dfrac{8}{27}$

— Interpreting the power 3/2
— Reciprocal to the positive exponent
— Negative exponent

Practice Problem 6   Evaluate: **a.** $25^{-3/2}$ **b.** $\left(\dfrac{1}{4}\right)^{-1/2}$ **c.** $5^{1.3}$ [*Hint:* Use a calculator.]

➤ Solutions on page 28

Avoiding Pitfalls in Simplifying

The square root of a product is equal to the product of the square roots:
$$\sqrt{a \cdot b} = \sqrt{a} \cdot \sqrt{b}$$
However, the corresponding statement for *sums* is *not* true:
$$\sqrt{a+b} \quad \text{is } not \text{ equal to} \quad \sqrt{a} + \sqrt{b}$$
For example,

$$\underbrace{\sqrt{9+16}}_{\sqrt{25}} \neq \underbrace{\sqrt{9}}_{3} + \underbrace{\sqrt{16}}_{4} \qquad \text{The two sides are not equal: one is 5 and the other is 7}$$

Therefore, do not "simplify" $\sqrt{x^2 + 9}$ into $x + 3$. The expression $\sqrt{x^2 + 9}$ *cannot be simplified.* Similarly,
$$(x + y)^2 \quad \text{is } not \text{ equal to} \quad x^2 + y^2$$
The expression $(x + y)^2$ means $(x + y)$ times itself:
$$(x+y)^2 = (x+y)(x+y) = x^2 + xy + yx + y^2 = x^2 + 2xy + y^2$$

This result is worth remembering, since we will use it frequently in Chapter 8.

$$(x+y)^2 = x^2 + 2xy + y^2 \qquad \begin{array}{l}(x+y)^2 \text{ is the first number squared} \\ \text{plus twice the product of the numbers} \\ \text{plus the second number squared}\end{array}$$

Learning Curves in Airplane Production

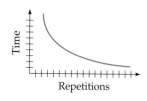
Repetitions

It is a truism that the more you practice a task, the faster you can do it. Successive repetitions generally take less time, following a "learning curve" like that on the left. Learning curves are used in industrial production. For example, it took 150,000 work-hours to build the first Boeing 707 airliner, while later planes $(n = 2, 3, \ldots, 300)$ took less time.*

$$\begin{pmatrix}\text{Time to build} \\ \text{plane number } n\end{pmatrix} = 150 n^{-0.322} \quad \text{thousand work-hours}$$

---

* A work-hour is the amount of work that a person can do in 1 hour. For further information on learning curves in industrial production, see J. M. Dutton et al., "The History of Progress Functions as a Managerial Technology," *Business History Review* **58**:204–233.

The time for the 10th Boeing 707 is found by substituting $n = 10$:

$$\begin{pmatrix} \text{Time to build} \\ \text{plane 10} \end{pmatrix} = 150(10)^{-0.322} \qquad \text{$150n^{-0.322}$ with $n = 10$}$$

$$\approx 71.46 \text{ thousand work-hours} \qquad \text{Using a calculator}$$

This shows that building the 10th Boeing 707 took about 71,460 work-hours, which is less than half of the 150,000 work-hours needed for the first. For the 100th 707:

$$\begin{pmatrix} \text{Time to build} \\ \text{plane 100} \end{pmatrix} = 150(100)^{-0.322} \qquad \text{$150n^{-0.322}$ with $n = 100$}$$

$$\approx 34.05 \text{ thousand work-hours}$$

or about 34,050 work-hours, which is less than half the time needed to build the 10th. Such learning curves are used for determining the cost of a contract to build several planes.

Notice that the learning curve graphed on the previous page decreases less steeply as the number of repetitions increases. This means that while construction time continues to decrease, it does so more slowly for later planes. This behavior, called "diminishing returns," is typical of learning curves.

1.2 Section Summary

We defined zero, negative, and fractional exponents as follows:

$$x^0 = 1 \qquad \text{for } x \neq 0$$

$$x^{-n} = \frac{1}{x^n} \qquad \text{for } x \neq 0$$

$$x^{\frac{m}{n}} = \left(\sqrt[n]{x}\right)^m = \sqrt[n]{x^m} \qquad m > 0, \ n > 0, \ \frac{m}{n} \text{ fully reduced}$$

With these definitions, the following properties of exponents hold for *all* exponents, whether integral or fractional, positive or negative.

$$x^m \cdot x^n = x^{m+n} \qquad (x^m)^n = x^{m \cdot n} \qquad \left(\frac{x}{y}\right)^n = \frac{x^n}{y^n}$$

$$\frac{x^m}{x^n} = x^{m-n} \qquad (xy)^n = x^n \cdot y^n$$

Solutions to Practice Problems

1. a. $\dfrac{x^5 \cdot x}{x^2} = \dfrac{x^6}{x^2} = x^4$
   b. $[(x^3)^2]^2 = x^{3 \cdot 2 \cdot 2} = x^{12}$

2. a. $2^0 = 1$
   b. $2^{-4} = \dfrac{1}{2^4} = \dfrac{1}{16}$

3. $\left(\dfrac{2}{3}\right)^{-2} = \left(\dfrac{3}{2}\right)^2 = \dfrac{9}{4}$

4. a. $(-27)^{1/3} = \sqrt[3]{-27} = -3$
   b. $\left(\dfrac{16}{81}\right)^{1/4} = \sqrt[4]{\dfrac{16}{81}} = \dfrac{\sqrt[4]{16}}{\sqrt[4]{81}} = \dfrac{2}{3}$

5. a. $16^{3/2} = \left(\sqrt{16}\right)^3 = 4^3 = 64$
   b. $(-8)^{2/3} = \left(\sqrt[3]{-8}\right)^2 = (-2)^2 = 4$

6. a. $25^{-3/2} = \dfrac{1}{25^{3/2}} = \dfrac{1}{(\sqrt{25})^3} = \dfrac{1}{5^3} = \dfrac{1}{125}$
   b. $\left(\dfrac{1}{4}\right)^{-1/2} = \left(\dfrac{4}{1}\right)^{1/2} = \sqrt{4} = 2$
   c. $5^{1.3} \approx 8.103$

1.2 Exercises

Evaluate each expression *without* using a calculator.

1. $(2^2 \cdot 2)^2$
2. $(5^2 \cdot 4)^2$
3. 2^{-4}
4. 3^{-3}
5. $\left(\dfrac{1}{2}\right)^{-3}$
6. $\left(\dfrac{1}{3}\right)^{-2}$
7. $\left(\dfrac{5}{8}\right)^{-1}$
8. $\left(\dfrac{3}{4}\right)^{-1}$
9. $4^{-2} \cdot 2^{-1}$
10. $3^{-2} \cdot 9^{-1}$
11. $\left(\dfrac{3}{2}\right)^{-3}$
12. $\left(\dfrac{2}{3}\right)^{-3}$
13. $\left(\dfrac{1}{3}\right)^{-2} - \left(\dfrac{1}{2}\right)^{-3}$
14. $\left(\dfrac{1}{3}\right)^{-2} - \left(\dfrac{1}{2}\right)^{-2}$
15. $\left[\left(\dfrac{2}{3}\right)^{-2}\right]^{-1}$
16. $\left[\left(\dfrac{2}{5}\right)^{-2}\right]^{-1}$
17. $25^{1/2}$
18. $36^{1/2}$
19. $25^{3/2}$
20. $16^{3/2}$
21. $16^{3/4}$
22. $27^{2/3}$
23. $(-8)^{2/3}$
24. $(-27)^{2/3}$
25. $(-8)^{5/3}$
26. $(-27)^{5/3}$
27. $\left(\dfrac{25}{36}\right)^{3/2}$
28. $\left(\dfrac{16}{25}\right)^{3/2}$
29. $\left(\dfrac{27}{125}\right)^{2/3}$
30. $\left(\dfrac{125}{8}\right)^{2/3}$
31. $\left(\dfrac{1}{32}\right)^{2/5}$
32. $\left(\dfrac{1}{32}\right)^{3/5}$
33. $4^{-1/2}$
34. $9^{-1/2}$
35. $4^{-3/2}$
36. $9^{-3/2}$
37. $8^{-2/3}$
38. $16^{-3/4}$
39. $(-8)^{-1/3}$
40. $(-27)^{-1/3}$
41. $(-8)^{-2/3}$
42. $(-27)^{-2/3}$
43. $\left(\dfrac{25}{16}\right)^{-1/2}$
44. $\left(\dfrac{16}{9}\right)^{-1/2}$
45. $\left(\dfrac{25}{16}\right)^{-3/2}$
46. $\left(\dfrac{16}{9}\right)^{-3/2}$
47. $\left(-\dfrac{1}{27}\right)^{-5/3}$
48. $\left(-\dfrac{1}{8}\right)^{-5/3}$

Use a calculator to evaluate each expression. Round answers to two decimal places.

49. $7^{0.39}$
50. $5^{0.47}$
51. $8^{2.7}$
52. $5^{3.9}$

Use a graphing calculator to evaluate each expression.

53. $(-8)^{7/3}$
54. $(-8)^{5/3}$
55. $\left[\left(\dfrac{5}{2}\right)^{-1}\right]^{-2}$
56. $\left[\left(\dfrac{3}{2}\right)^{-2}\right]^{-1}$
57. $[(4)^{-1}]^{0.5}$
58. $[(0.25)^{-1}]^{0.5}$
59. $(0.4^{-7})^{-1/7}$
60. $[(0.5^{-1})^{-2}]^{-3}$
61. $[(0.1)^{0.1}]^{0.1}$
62. $\left(1 + \dfrac{1}{1000}\right)^{1000}$
63. $\left(1 - \dfrac{1}{1000}\right)^{-1000}$
64. $(1 + 10^{-6})^{10^6}$

Simplify.

65. $(x^3 \cdot x^2)^2$
66. $(x^4 \cdot x^3)^2$
67. $[z^2(z \cdot z^2)^2 z]^3$

68. $[z(z^3 \cdot z)^2 z^2]^2$
69. $[(x^2)^2]^2$
70. $[(x^3)^3]^3$
71. $\dfrac{(ww^2)^3}{w^3 w}$
72. $\dfrac{(ww^3)^2}{w^3 w^2}$
73. $\dfrac{(5xy^4)^2}{25x^3 y^3}$
74. $\dfrac{(4x^3 y)^2}{8x^2 y^3}$
75. $\dfrac{(9xy^3 z)^2}{3(xyz)^2}$
76. $\dfrac{(5x^2 y^3 z)^2}{5(xyz)^2}$
77. $\dfrac{(2u^2 vw^3)^2}{4(uw^2)^2}$
78. $\dfrac{(u^3 vw^2)^2}{9(u^2 w)^2}$

APPLIED EXERCISES

79–80. ALLOMETRY: Dinosaurs The study of size and shape is called "allometry," and many allometric relationships involve exponents that are fractions or decimals. For example, the body measurements of most four-legged animals, from mice to elephants, obey (approximately) the following power law:

$$\left(\begin{array}{c}\text{Average body}\\ \text{thickness}\end{array}\right) = 0.4 \text{ (hip-to-shoulder length)}^{3/2}$$

where body thickness is measured vertically and all measurements are in feet. Assuming that this same relationship held for dinosaurs, find the average body thickness of the following dinosaurs, whose hip-to-shoulder length can be measured from their skeletons:

79. Diplodocus, whose hip-to-shoulder length was 16 feet.

80. Triceratops, whose hip-to-shoulder length was 14 feet.

81–82. BUSINESS: The Rule of .6 Many chemical and refining companies use "the rule of point six" to estimate the cost of new equipment. According to this rule, if a piece of equipment (such as a storage tank) originally cost C dollars, then the cost of similar equipment that is x times as large will be approximately $x^{0.6} C$ dollars. For example, if the original equipment cost C dollars, then new equipment with twice the capacity of the old equipment $(x = 2)$ would cost $2^{0.6} C = 1.516 C$ dollars — that is, about 1.5 times as much. Therefore, to increase capacity by 100% costs only about 50% more.*

* Although the rule of .6 is only a rough "rule of thumb," it can be somewhat justified on the basis that the equipment of such industries consists mainly of containers, and the cost of a container depends on its surface area (square units), which increases more slowly than its capacity (cubic units).

81. Use the rule of .6 to find how costs change if a company wants to quadruple $(x = 4)$ its capacity.

82. Use the rule of .6 to find how costs change if a company wants to triple $(x = 3)$ its capacity.

83–84. BUSINESS: The Rule of .6 (*continuation*) Use a graphing calculator to graph $y = x^{0.6}$, expressing y, the cost multiple for larger equipment, in terms of x, the size multiple. Use the viewing window [0, 5] by [0, 3].

83. By how much can a company multiply its capacity for twice the money? That is, find the value of x that satisfies $x^{0.6} = 2$. [Hint: Either use TRACE or find where $y_1 = x^{0.6}$ INTERSECTs $y_2 = 2$.]

84. Does the curve rise more steeply or less steeply as x increases? What does this mean about how rapidly cost increases as equipment size increases?

85–86. ALLOMETRY: Heart Rate It is well known that the hearts of smaller animals beat faster than the hearts of larger animals. The actual relationship is approximately

$$(\text{Heart rate}) = 250(\text{weight})^{-1/4}$$

where the heart rate is in beats per minute and the weight is in pounds. Use this relationship to estimate the heart rate of:

85. A 16-pound dog.

86. A 625-pound grizzly bear.

87–88. ALLOMETRY: Heart Rate (*continuation*) Use a graphing calculator to graph $y = 250 x^{-0.25}$, which expresses y, heartbeats per minute, in terms of x, the animal's weight. Use the viewing window [0, 200] by [0, 150].

87. Notice that the curve decreases less steeply for larger values of x. Explain what this means about how rapidly heart rate decreases as body weight increases.

88. Evaluate this formula at your own weight, x, to find your predicted heart rate. Then take your pulse and see if the numbers (roughly) agree.

89–90. BUSINESS AND PSYCHOLOGY: Learning Curves in Airplane Production Recall (pages 26–27) that the learning curve for the production of Boeing 707 airplanes is $150n^{-0.322}$ (thousand work-hours). Find how many work-hours it took to build:

89. The 50th Boeing 707.

90. The 250th Boeing 707.

91. GENERAL: Richter Scale The Richter scale (developed by Charles Richter in 1935) is widely used to measure the strength of earthquakes. Every increase of 1 on the Richter scale corresponds to a 10-fold increase in ground motion. Therefore, an increase on the Richter scale from A to B means that ground motion increases by a factor of 10^{B-A} (for $B > A$). Find the increase in ground motion between the following earthquakes:

a. The 1994 Northridge, California, earthquake, measuring 6.8 on the Richter scale, and the 1906 San Francisco earthquake, measuring 8.3. (The San Francisco earthquake resulted in 500 deaths and a 3-day fire that destroyed 4 square miles of San Francisco.)

b. The 1995 Kobe (Japan) earthquake, measuring 7.2 on the Richter scale, and the 1933 Miyagi earthquake, measuring 8.1. (The Miyagi earthquake caused a 90-foot-high tsunami, or "tidal wave," that killed 3064 people. The death toll from the Kobe earthquake was more than 5000.)

92. GENERAL: Richter Scale (*continuation*) Every increase of 1 on the Richter scale corresponds to an approximately *30-fold* increase in *energy released*. Therefore, an increase on the Richter scale from A to B means that the energy released increases by a factor of 30^{B-A} (for $B > A$).

a. Find the increase in *energy released* between the earthquakes in Exercise 91a.

b. Find the increase in *energy released* between the earthquakes in Exercise 91b.

93–94. GENERAL: Waterfalls Water falling from a waterfall that is x feet high will hit the ground with speed $\frac{60}{11}x^{0.5}$ miles per hour (neglecting air resistance).

93. Find the speed of the water at the bottom of the highest waterfall in the world, Angel Falls in Venezuela (3281 feet high).

94. Find the speed of the water at the bottom of the highest waterfall in the United States, Ribbon Falls in Yosemite, California (1650 feet high).

95–96. ENVIRONMENTAL SCIENCE: Biodiversity It is well known that larger land areas can support larger numbers of species. According to one study,* multiplying the land area by a factor of x multiplies the number of species by a factor of $x^{0.239}$. Use a graphing calculator to graph $y = x^{0.239}$. Use the viewing window [0, 100] by [0, 4].

95. Find the multiple x for the land area that leads to *double* the number of species. That is, find the value of x such that $x^{0.239} = 2$. [*Hint:* Either use TRACE or find where $y_1 = x^{0.239}$ INTERSECTs $y_2 = 2$.]

96. Find the multiple x for the land area that leads to triple the number of species. That is, find the value of x such that $x^{0.239} = 3$. [*Hint:* Either use TRACE or find where $y_1 = x^{0.239}$ INTERSECTs $y_2 = 3$.]

97. BUSINESS: Learning Curves A manufacturer of supercomputers finds that the number of work-hours required to build the 1st, the 10th, the 20th, and the 30th supercomputers are as follows:

Supercomputer Number	Work-Hours Required
1	3200
10	1900
20	1400
30	1300

* Rober H. MacArthur and Edward O. Wilson, *The Theory of Island Biogeography* (Princeton University Press, 1967).

a. Enter these numbers into a graphing calculator and make a plot of the resulting points (Supercomputer Number on the x-axis and Work-Hours Required on the y-axis).
b. Have the calculator find the power regression formula for these data, fitting a curve of the form $y = ax^b$ to the points. Enter the results in y_1. Plot the points together with the regression curve. Observe that the curve fits the points rather well.
c. Evaluate y_1 at $x = 50$ to predict the number of work-hours required to build the 50th supercomputer.

98. **GENERAL: Paper Stacking** Suppose that you take an ordinary piece of paper (about $\frac{1}{250}$ of an inch thick), cut it in half, stack the two halves, cut them in two and stack the pieces, and repeat this cutting and stacking operation many times. Each operation doubles the height of the stack, so that after a total of 25 such operations, the stack would be $2^{25} \cdot \frac{1}{250} \cdot \frac{1}{12} \cdot \frac{1}{5280}$ miles high.

a. Evaluate this height.
b. Use a graphing calculator to make a TABLE showing the height of the stack when the number of operations is 15 or more. [*Hint:* Use $y_1 = 2^x/(250 \cdot 12 \cdot 5280)$ for values of x beginning at 15.] For what number of operations is the stack over 1 mile high? over 10 miles high? over 100 miles high?

1.3 FUNCTIONS

Introduction

In the previous section we saw that the time required to build a Boeing 707 airliner will vary, depending on the number that have already been built. Mathematical relationships such as this, in which one number depends on another, are called *functions*, and are central to the study of mathematics. In this section we define and give some applications of functions.

Functions

A *function* is a rule or procedure for finding, from a given number, a new number.* If the function is denoted by f and the given number by x, then the resulting number is written $f(x)$ (read "f of x") and is called *the value of the function f at x*. The set of numbers x for which a function f is defined is called the *domain* of f, and the set of all resulting function values $f(x)$ is called the *range* of f. For any x in the domain, $f(x)$ must be a *single* number.

Function

A *function* f is a rule that assigns to each number x in a set a number $f(x)$. The set of all allowable values of x is called the *domain*, and the set of all values $f(x)$ for x in the domain is called the *range*.

---

* In this chapter the word "function" will mean *function of one variable*. In Chapters 4 and 13 we will discuss functions of more than one variable.

For example, recording the temperature at a given location throughout a particular day would define a *temperature* function:

$$f(x) = \begin{pmatrix} \text{Temperature at} \\ \text{time } x \text{ hours} \end{pmatrix} \qquad \text{Domain would be [0, 24]}$$

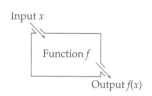

Input x

Function f

Output $f(x)$

A function f may be thought of as a numerical procedure or "machine" that takes an "input" number x and produces an "output" number $f(x)$, as shown on the left. The permissible input numbers form the *domain*, and the resulting output numbers form the *range*.

We will be mostly concerned with functions that are defined by *formulas* for calculating $f(x)$ from x. If the domain of such a function is not stated, then it is always taken to be the *largest* set of numbers for which the function is defined, called the *natural domain* of the function. To *graph* a function f, we plot all points (x, y) such that x is in the domain and $y = f(x)$. We call x the *independent variable* and y the *dependent variable*, since y *depends on* (is calculated from) x. The domain and range can be illustrated graphically.

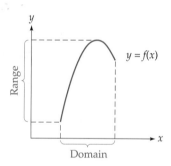

The domain of a function $y = f(x)$ is the set of all allowable x-values, and the range is the set of all corresponding y-values.

Practice Problem 1 Find the domain and range of the function graphed below.

▶ Solution on page 44

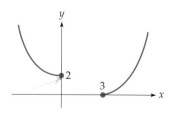

EXAMPLE 1 — FINDING THE DOMAIN AND RANGE

For the function $f(x) = \dfrac{1}{x-1}$, find:

a. $f(5)$ **b.** the domain   **c.** the range

Solution

a. $f(5) = \dfrac{1}{5-1} = \dfrac{1}{4}$ $\quad f(x) = \dfrac{1}{x-1}$ with $x=5$

$\quad f(x) = \dfrac{1}{x-1}$ is defined for all x except $x=1$.

b. Domain $= \{x \mid x \neq 1\}$

c. The graph of the function (from a graphing calculator) is shown on the right. From it, and realizing that the curve continues upward and downward (as may be verified by zooming out), it is clear that *every* y value is taken except for $y=0$ (since the curve does not touch the x-axis). Therefore:

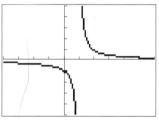

$f(x) = \dfrac{1}{x-1}$ on $[-4, 6]$ by $[-5, 5]$

Range $= \{y \mid y \neq 0\}$

May also be written $\{z \mid z \neq 0\}$ or with any other letter

The range could also be found by solving $y = \dfrac{1}{x-1}$ for x, giving $x = \dfrac{1}{y} + 1$, which again shows that y can take any value except 0.

EXAMPLE 2 — FINDING THE DOMAIN AND RANGE

For $f(x) = 2x^2 + 4x - 5$, determine:

a. $f(-3)$ **b.** the domain   **c.** the range

Solution

a. $f(-3) = 2(-3)^2 + 4(-3) - 5$ $\quad f(x) = 2x^2 + 4x - 5$ with each x replaced by -3
$\qquad\quad = 18 - 12 - 5 = 1$

b. Domain $= \mathbb{R}$ $\quad 2x^2 + 4x - 5$ is defined for *all* real numbers

34 CHAPTER 1 FUNCTIONS

c. From the graph of $f(x) = 2x^2 + 4x - 5$ on the right, the lowest y value is -7 (as can be found from TRACE or MINIMUM), and all higher y values are taken (since the curve is a parabola opening upward). Therefore:

$$\text{Range} = \{y \mid y \geq -7\}$$

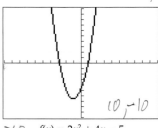

$f(x) = 2x^2 + 4x - 5$ on $[-10, 10]$ by $[-10, 10]$

Any letters may be used for defining a function or describing the domain and the range. For example, since the range $\{y \mid y \geq -7\}$ is a set of numbers, it could also be written $\{w \mid w \geq -7\}$ or, in interval notation without *any* variables, as $[-7, \infty)$.

Practice Problem 2

For $g(z) = \sqrt{z - 2}$, determine:

a. $g(27)$ b. the domain   c. the range   ➤ Solutions on page 44

For each x in the domain of a function there must be a *single* number $y = f(x)$, so the graph of a function cannot have two points (x, y) with the same x value but different y values. This leads to the following *graphical* test for functions.

Vertical Line Test for Functions

A curve in the Cartesian plane is the graph of a *function* if and only if no vertical line intersects the curve at more than one point.

EXAMPLE 3

USING THE VERTICAL LINE TEST

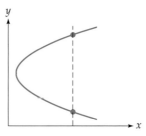

This is *not* the graph of a function of x because there is a vertical line (shown dashed) that intersects the curve twice.

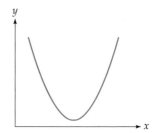

This *is* the graph of a function of x because no vertical line intersects the curve more than once.

A graph that has two or more points (x, y) with the same x-value but different y-values, such as the one on the left in Example 3, defines a *relation* rather than a function. We will be concerned exclusively with *functions,* and so we will use the terms "function," "graph," and "curve" interchangeably.

Functions can be classified into several types.

Linear Function

A *linear function* is a function that can be expressed in the form

$$f(x) = mx + b$$

with constants m and b. Its graph is a line with slope m and y-intercept b.

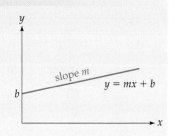

EXAMPLE 4   FINDING A COMPANY'S COST FUNCTION

An electronics company manufactures pocket calculators at a cost of $9 each, and the company's fixed costs (such as rent) amount to $400 per day. Find a function $C(x)$ that gives the total cost of producing x pocket calculators in a day.

Solution

Each calculator costs $9 to produce, so x calculators will cost $9x$ dollars, to which we must add the fixed costs of $400.

$$\underbrace{C(x)}_{\text{Total cost}} = \underbrace{9}_{\substack{\text{Unit}\\\text{cost}}}\underbrace{x}_{\substack{\text{Number}\\\text{of units}}} + \underbrace{400}_{\substack{\text{Fixed}\\\text{cost}}}$$

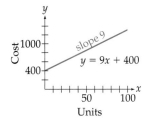

The graph of $C(x) = 9x + 400$ is a line with slope 9 and y-intercept 400, as shown on the left. Notice that the *slope* is the same as the *rate of change* of the cost (increasing at the rate of $9 per additional calculator), which is also the company's *marginal cost* (the cost of producing one more calculator is $9). For a linear cost function, the *slope,* the *rate of change,* and the *marginal cost* are always the same.

Practice Problem 3

A trucking company will deliver furniture for a charge of $25 plus 5% of the purchase price of the furniture. Find a function $D(x)$ that gives the delivery charge for a piece of furniture that costs x dollars.

handwritten: $.05x + 25$

➤ Solution on page 44

A mathematical description of a real-world situation is called a *mathematical model*. For example, the cost function $C(x) = 9x + 400$ from the previous example is a mathematical model for the cost of manufacturing calculators. In this model, x, the number of calculators, should take only whole-number values $(0, 1, 2, 3, \ldots)$, and the graph should consist of discrete dots rather than a continuous curve. Instead, we will find it easier to let x take *continuous* values, and round up or down as necessary at the end.

> **Quadratic Function**
>
> A *quadratic function* is a function that can be expressed in the form
>
> $$f(x) = ax^2 + bx + c$$
>
> with constants ("coefficients") $a \neq 0$, b, and c. Its graph is called a *parabola*.

The condition $a \neq 0$ keeps the function from becoming $f(x) = bx + c$, which would be linear. Many familiar curves are parabolas.

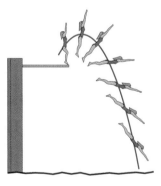

The center of gravity of a diver describes a parabola.

A stream of water from a hose takes the shape of a parabola.

The parabola $f(x) = ax^2 + bx + c$ opens *upward* if the constant a is *positive* and opens *downward* if the constant a is *negative*. The *vertex* of a parabola is its "central" point. The vertex is the *lowest* point on the parabola if it opens *up*, and the *highest* point if it opens *down*.

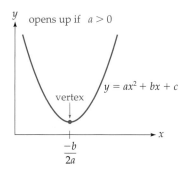
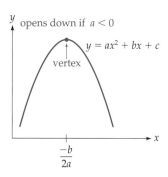

Graphing Calculator Exploration

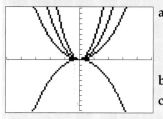

a. Graph the parabolas $y_1 = x^2$, $y_2 = 2x^2$, and $y_3 = 4x^2$ on the graphing window $[-5, 5]$ by $[-10, 10]$. Use TRACE to identify which curve goes with which formula. How does the shape of the parabola change when the coefficient of x^2 increases?
b. Graph $y_4 = -x^2$. What did the negative sign do to the parabola?
c. Predict the shape of the parabolas $y_5 = -2x^2$ and $y_6 = \frac{1}{3}x^2$. Then check your predictions by graphing the functions.

The x-coordinate of the vertex of a parabola may be found by the following formula, which will be derived in Exercise 57 on page 647.

Vertex Formula for a Parabola

The vertex of the parabola $f(x) = ax^2 + bx + c$ has x-coordinate

$$x = \frac{-b}{2a}$$

For example, the vertex of the parabola $y = 2x^2 + 4x - 5$ has x-coordinate

$$x = \frac{-4}{2 \cdot 2} = \frac{-4}{4} = -1 \qquad x = \frac{-b}{2a} \text{ with } a = 2 \text{ and } b = 4$$

as we can see from the graph on page 34. The y-coordinate of the vertex comes from evaluating $y = 2x^2 + 4x - 5$ at this x-coordinate.

EXAMPLE 5 GRAPHING A QUADRATIC FUNCTION

Graph the quadratic function $f(x) = 2x^2 - 40x + 104$.

Solution

Graphing using a graphing calculator is largely a matter of finding an appropriate viewing window, as the following three unsatisfactory windows show.

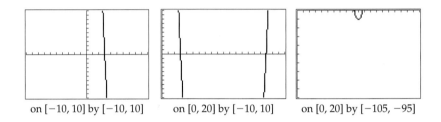

on [−10, 10] by [−10, 10]   on [0, 20] by [−10, 10]   on [0, 20] by [−105, −95]

To find an appropriate viewing window, we use the vertex formula:

$$x = \frac{-b}{2a} = \frac{-(-40)}{2 \cdot 2} = \frac{40}{4} = 10 \qquad \begin{array}{l} \text{x-coordinate of the vertex, from} \\ x = \frac{-b}{2a} \text{ with } a = 2 \text{ and } b = -40 \end{array}$$

We move a few units, say 5, to either side of $x = 10$, making the x-window [5, 15]. Using the calculator to EVALUATE the given function at $x = 10$ (or evaluating by hand) gives $y(10) = -96$. Since the parabola opens upward (the coefficient of x^2 is positive), the curve rises up from its vertex, so we select a y-interval from -96 upward, say [−96, −70]. Graphing the function on the window [5, 15] by [−96, −70] gives the following result. (Some other graphing windows are just as good.)

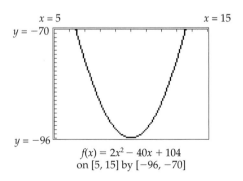

$f(x) = 2x^2 - 40x + 104$
on [5, 15] by [−96, −70]

Solving Quadratic Equations

A value of x that solves an equation $f(x) = 0$ is called a *root* of the equation, or a *zero* of the function, or an *x-intercept* of the graph of $y = f(x)$. The roots of a quadratic equation can often be found by factoring.

EXAMPLE 6 SOLVING A QUADRATIC EQUATION BY FACTORING

Solve $2x^2 - 4x = 6$.

Solution

$$2x^2 - 4x - 6 = 0 \quad \text{Subtracting 6 from each side to get zero on the right}$$

$$2(x^2 - 2x - 3) = 0 \quad \text{Factoring out a 2}$$

$$2(x - 3) \cdot (x + 1) = 0 \quad \text{Factoring } x^2 - 2x - 3$$

Equals 0 at $x = 3$ Equals 0 at $x = -1$ Finding x-values that make each factor zero

$$x = 3, \quad x = -1 \quad \text{Solutions}$$

Graphing Calculator Exploration

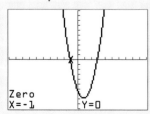

Find the solutions to the equation in Example 6 by graphing the function $f(x) = 2x^2 - 4x - 6$ and using ZERO or TRACE to find where the curve crosses the x-axis. Your answers should agree with those found in Example 6.

Practice Problem 4 Solve by factoring or graphing: $9x - 3x^2 = -30$

➤ Solution on page 44

Quadratic equations can often be solved by the "Quadratic Formula." A derivation of this formula is given on pages 43–44.

> **Quadratic Formula**
>
> The solutions to $ax^2 + bx + c = 0$ are
>
> $$x = \frac{-b \pm \sqrt{b^2 - 4ac}}{2a}$$
>
> The "plus or minus" $\pm$ sign means calculate *both* ways, first using the $+$ sign and then using the $-$ sign

In a business, it is often important to find a company's *break-even points*, the numbers of units of production where the company's costs are equal to its revenue.

EXAMPLE 7

FINDING BREAK-EVEN POINTS

A company that installs automobile compact disc (CD) players finds that if it installs x CD players per day, then its costs will be $C(x) = 120x + 4800$ and its revenue will be $R(x) = -2x^2 + 400x$ (both in dollars). Find the company's break-even points.

Solution

We set the cost and revenue functions equal to each other and solve.

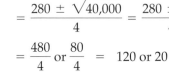

$120x + 4800 = -2x^2 + 400x$	Setting $C(x) = R(x)$
$2x^2 - 280x + 4800 = 0$	Combining all terms on one side
$x = \dfrac{280 \pm \sqrt{(-280)^2 - 4 \cdot 2 \cdot 4800}}{2 \cdot 2}$	Quadratic Formula with $a = 2$, $b = -280$, and $c = 4800$
$= \dfrac{280 \pm \sqrt{40{,}000}}{4} = \dfrac{280 \pm 200}{4}$	Working out the formula on a calculator
$= \dfrac{480}{4}$ or $\dfrac{80}{4} = 120$ or 20	

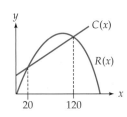

The company will break even when it installs either 20 or 120 CD players (as shown in the graph on the left).

Although it is important for a company to know where its break-even points are, most companies want to do better than break even—they want to maximize their profit. Profit is defined as *revenue minus cost* (since profit is what is left over after subtracting expenses from income).

Profit

> Profit = Revenue − Cost

EXAMPLE 8

MAXIMIZING PROFIT

For the CD installer whose daily revenue and cost functions were given in Example 7, find the number of units that maximizes profit, and the maximum profit.

Solution

The profit function is the revenue function minus the cost function.

$$P(x) = \underbrace{-2x^2 + 400x}_{R(x)} - \underbrace{(120x + 4800)}_{C(x)} \qquad \begin{array}{l} P(x) = R(x) - C(x) \text{ with} \\ R(x) = -2x^2 + 400x \text{ and} \\ C(x) = 120x + 4800 \end{array}$$

$$= -2x^2 + 280x - 4800 \qquad \text{Simplifying}$$

Since this function represents a parabola opening downward (because of the −2), it is maximized at its vertex, which is found using the vertex formula.

$$x = \frac{-280}{2(-2)} = \frac{-280}{-4} = 70 \qquad \begin{array}{l} x = \dfrac{-b}{2a} \text{ with} \\ a = -2 \text{ and } b = 280 \end{array}$$

Thus, profit is maximized when 70 CD players are installed. For the maximum profit, we substitute $x = 70$ into the profit function:

$$P(70) = -2(70)^2 + 280 \cdot 70 - 4800 \qquad \begin{array}{l} P(x) = -2x^2 + 280x - 4800 \\ \text{with } x = 70 \end{array}$$

$$= 5000 \qquad \text{Multiplying and combining}$$

Therefore, the company will maximize its profit when it installs 70 CD players per day. Its maximum profit will be $5000 per day.

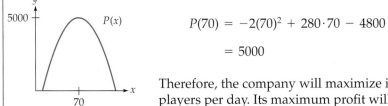

Why doesn't a company make more profit the more it sells? Because to increase its sales it must lower its prices, which eventually leads to lower profits. The relationship among the cost, revenue, and profit functions can be seen graphically as follows.

Spreadsheet Exploration

The following spreadsheet* shows the graphs of the functions $R(x)$, $C(x)$, and $P(x)$ from Examples 7 and 8. The values for x were made by entering 0 in A3 and then copying the formula =A3+10 for A4 into cells A5 through A18. The values for $R(x)$ were found by entering the formula =-2*A3*A3+400*A3 in B3 and then copying it into cells B4 through B18. Then $C(x)$ was similarly found by starting with C3 being =120*A3+4800. $P(x)$, the difference $R(x) - C(x)$, was found with D3 being =B3-C3. The chart was made by plotting the values for the three columns corresponding to the revenue, cost, and profit.

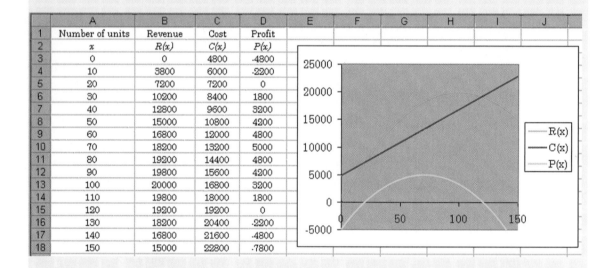

Notice that the break-even points from Example 7 ($x = 20$ and $x = 120$) correspond to a profit of zero, and that the maximum profit (at $x = 70$) occurs halfway between the two break-even points.

Not all quadratic equations have (real) solutions.

---

* See the Preface for how to obtain this and other Excel spreadsheets.

EXAMPLE 9

USING THE QUADRATIC FORMULA

Solve $\frac{1}{2}x^2 - 3x + 5 = 0$.

Solution

The Quadratic Formula with $a = \frac{1}{2}$, $b = -3$, and $c = 5$ gives

$$x = \frac{3 \pm \sqrt{9 - 4(\frac{1}{2})(5)}}{2(\frac{1}{2})} = \frac{3 \pm \sqrt{9 - 10}}{1} = 3 \pm \sqrt{-1} \quad \text{Undefined}$$

Therefore, the equation $\frac{1}{2}x^2 - 3x + 5 = 0$ *has no real solutions* (because of the undefined $\sqrt{-1}$). The geometrical reason that there are no solutions can be seen in the graph on the left: The curve never reaches the *x*-axis, so the function never equals zero.

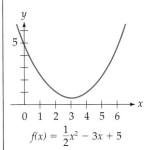

$f(x) = \frac{1}{2}x^2 - 3x + 5$

The quantity $b^2 - 4ac$, whose square root appears in the Quadratic Formula, is called the *discriminant*. If the discriminant is *positive* (as in Example 7), the equation $ax^2 + bx + c = 0$ has *two* solutions (since the square root is added and subtracted). If the discriminant is *zero*, there is only *one* root (since adding and subtracting zero gives the same answer). If the discriminant is *negative* (as in Example 9), then the equation has *no* real roots. Therefore, the discriminant being positive, zero, or negative corresponds to the parabola meeting the *x*-axis at 2, 1, or 0 points, as shown below.

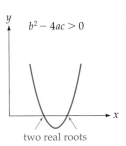
two real roots

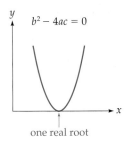
one real root

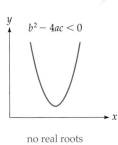
no real roots

Derivation of the Quadratic Formula

$$ax^2 + bx + c = 0 \qquad \text{The quadratic set equal to zero}$$
$$ax^2 + bx = -c \qquad \text{Subtracting } c$$
$$4a^2x^2 + 4abx = -4ac \qquad \text{Multiplying by } 4a$$

44 CHAPTER 1 FUNCTIONS

$$4a^2x^2 + 4abx + b^2 = b^2 - 4ac \quad \text{Adding } b^2$$
$$(2ax + b)^2 = b^2 - 4ac \quad \text{Since } 4a^2x^2 + 4abx + b^2 = (2ax + b)^2$$
$$2ax + b = \pm\sqrt{b^2 - 4ac} \quad \text{Taking square roots}$$
$$2ax = -b \pm \sqrt{b^2 - 4ac} \quad \text{Subtracting } b$$
$$x = \frac{-b \pm \sqrt{b^2 - 4ac}}{2a} \quad \text{Dividing by } 2a \text{ gives the Quadratic Formula}$$

1.3 Section Summary

In this section we defined and gave examples of *functions*, and saw how to find their domains and ranges. The most important characteristic of a function f is that for any given "input" number x in the domain, there is exactly one "output" number $f(x)$. This requirement is stated geometrically in the *vertical line test*, that no vertical line can intersect the graph of a function at more than one point. We then defined *linear functions* (whose graphs are lines) and *quadratic functions* (whose graphs are parabolas). We solved quadratic equations by factoring, graphing, and using the Quadratic Formula. We maximized and minimized quadratic functions using the vertex formula.

▶ **Solutions to Practice Problems**

1. Domain: $\{x \mid x \leq 0 \text{ or } x \geq 3\}$; Range: $\{y \mid y \geq 0\}$

2. a. $g(27) = \sqrt{27 - 2} = \sqrt{25} = 5$
b. Domain: $\{z \mid z \geq 2\}$
c. Range: $\{y \mid y \geq 0\}$

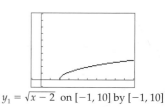

$y_1 = \sqrt{x - 2}$ on $[-1, 10]$ by $[-1, 10]$

3. $D(x) = 25 + 0.05x$

4. $9x - 3x^2 = -30$
$-3x^2 + 9x + 30 = 0$
$-3(x^2 - 3x - 10) = 0$
$-3(x - 5)(x + 2) = 0$
$x = 5, x = -2 \quad$ or from:

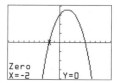

1.3 Exercises

Determine whether each graph defines a function of x.

1.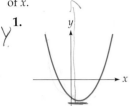
2.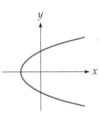
3.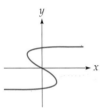
4.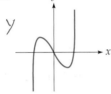
5.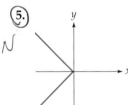
6.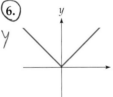
7.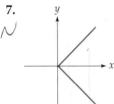
8.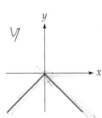

Find the domain and range of each function from its graph.

9.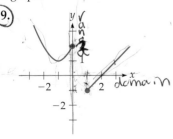
10.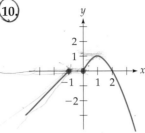

11–22. For each function:
a. Evaluate the given expression.
b. Find the domain of the function.
c. Find the range.
[*Hint:* Use a graphing calculator.]

11. $f(x) = \sqrt{x-1}$; find $f(10)$
12. $f(x) = \sqrt{x-4}$; find $f(40)$
13. $h(z) = \dfrac{1}{z+4}$; find $h(-5)$
14. $h(z) = \dfrac{1}{z+7}$; find $h(-8)$
15. $h(x) = x^{1/4}$; find $h(81)$
16. $h(x) = x^{1/6}$; find $h(64)$
17. $h(x) = x^{2/3}$; find $f(-8)$

[*Hint for Exercises 17 and 18:* You may need to enter $x^{m/n}$ as $(x^m)^{1/n}$ or as $(x^{1/n})^m$, as discussed on page 25.]

18. $f(x) = x^{4/5}$; find $f(-32)$
19. $f(x) = \sqrt{4-x^2}$; find $f(0)$
20. $f(x) = \dfrac{1}{\sqrt{x}}$; find $f(4)$
21. $f(x) = \sqrt{-x}$; find $f(-25)$
22. $f(x) = -\sqrt{-x}$; find $f(-100)$

23–30. Graph each function "by hand." [*Note:* Even if you have a graphing calculator, it is important to be able to sketch simple curves by finding a few important points.]

23. $f(x) = 3x - 2$
24. $f(x) = 2x - 3$
25. $f(x) = -x + 1$
26. $f(x) = -3x + 5$

27. $f(x) = 2x^2 + 4x - 16$
28. $f(x) = 3x^2 - 6x - 9$
29. $f(x) = -3x^2 + 6x + 9$
30. $f(x) = -2x^2 + 4x + 16$

31–34. For each quadratic function:
a. Find the vertex using the vertex formula.
b. Graph the function on an appropriate viewing window. (Answers may differ.)

31. $f(x) = x^2 - 40x + 500$
32. $f(x) = x^2 + 40x + 500$
33. $f(x) = -x^2 - 80x - 1800$
34. $f(x) = -x^2 + 80x - 1800$

35–52. Solve each equation by factoring or the Quadratic Formula, as appropriate.

35. $x^2 - 6x - 7 = 0$
36. $x^2 - x - 20 = 0$
37. $x^2 + 2x = 15$
38. $x^2 - 3x = 54$
39. $2x^2 + 40 = 18x$
40. $3x^2 + 18 = 15x$
41. $5x^2 - 50x = 0$
42. $3x^2 - 36x = 0$
43. $2x^2 - 50 = 0$
44. $3x^2 - 27 = 0$
45. $4x^2 + 24x + 40 = 4$
46. $3x^2 - 6x + 9 = 6$
47. $-4x^2 + 12x = 8$
48. $-3x^2 + 6x = -24$
49. $2x^2 - 12x + 20 = 0$
50. $2x^2 - 8x + 10 = 0$
51. $3x^2 + 12 = 0$
52. $5x^2 + 20 = 0$

53–62. Solve each equation using a graphing calculator. [*Hint:* Begin with the viewing window $[-10, 10]$ by $[-10, 10]$ or another of your choice (see Useful Hint in Graphing Calculator Terminology following the Preface) and use ZERO, SOLVE, or TRACE and ZOOM IN.] (In Exercises 61 and 62, round answers to two decimal places.)

53. $x^2 - x - 20 = 0$
54. $x^2 + 2x - 15 = 0$
55. $2x^2 + 40 = 18x$
56. $3x^2 + 18 = 15x$
57. $4x^2 + 24x + 45 = 9$
58. $3x^2 - 6x + 5 = 2$
59. $3x^2 + 7x + 12 = 0$
60. $5x^2 + 14x + 20 = 0$
61. $2x^2 + 3x - 6 = 0$
62. $3x^2 + 5x - 7 = 0$

63. Use your graphing calculator to graph the following four equations simultaneously on the viewing window $[-10, 10]$ by $[-10, 10]$:

$y_1 = 2x + 6$
$y_2 = 2x + 2$
$y_3 = 2x - 2$
$y_4 = 2x - 6$

a. What do the lines have in common and how do they differ?
b. Write the equation of another line with the same slope that lies 2 units below the lowest line. Then check your answer by graphing it with the others.

64. Use your graphing calculator to graph the following four equations simultaneously on the viewing window $[-10, 10]$ by $[-10, 10]$:

$y_1 = 3x + 4$
$y_2 = 1x + 4$
$y_3 = -1x + 4$ (Use $(-)$ to get $-1x$.)
$y_4 = -3x + 4$

a. What do the lines have in common and how do they differ?
b. Write the equation of a line through this y-intercept with slope $\frac{1}{2}$. Then check your answer by graphing it with the others.

APPLIED EXERCISES

65. **BUSINESS: Cost Functions** A lumberyard will deliver wood for $4 per board foot plus a delivery charge of $20. Find a function $C(x)$ for the cost of having x board feet of lumber delivered.

66. **BUSINESS: Cost Functions** A company manufactures bicycles at a cost of $55 each. If the company's fixed costs are $900, express the company's costs as a linear function of x, the number of bicycles produced.

67. BUSINESS: Salary An employee's weekly salary is $500 plus $15 per hour of overtime. Find a function $P(x)$ giving his pay for a week in which he worked x hours of overtime.

68. BUSINESS: Salary A sales clerk's weekly salary is $300 plus 2% of her total week's sales. Find a function $P(x)$ for her pay for a week in which she sold x dollars of merchandise.

69. GENERAL: Water Pressure At a depth of d feet underwater, the water pressure is $p(d) = 0.45d + 15$ pounds per square inch. Find the pressure at:
  a. The bottom of a 6-foot-deep swimming pool.
  b. The maximum ocean depth of 35,000 feet.

70. GENERAL: Boiling Point At higher altitudes, water boils at lower temperatures. This is why at high altitudes foods must be boiled for longer times—the lower boiling point imparts less heat to the food. At an altitude of h thousand feet above sea level, water boils at a temperature of $B(h) = -1.8h + 212$ degrees Fahrenheit. Find the altitude at which water boils at 98.6 degrees Fahrenheit. (Your answer will show that at a high enough altitude, water boils at normal body temperature. This is why airplane cabins must be pressurized—at high enough altitudes one's blood would boil.)

71–72. GENERAL: Stopping Distance According to data from the National Transportation Safety Board, a car traveling at speed v miles per hour should be able to come to a full stop in a distance of
$$D(v) = 0.055v^2 + 1.1v \quad \text{feet}$$
Find the stopping distance required for a car traveling at:

71. 40 mph.   **72.** 60 mph.

73. BIOMEDICAL: Cell Growth The number of cells in a culture after t days is given by $N(t) = 200 + 50t^2$. Find the size of the culture after:
  a. 2 days.   b. 10 days.

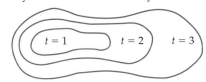

74. GENERAL: Juggling If you toss a ball h feet straight up, it will return to your hand after $T(h) = 0.5\sqrt{h}$ seconds. This leads to the *juggler's dilemma*: Juggling more balls means tossing them higher. However, the square root in the above formula means that tossing them twice as high does not gain twice as much time, but only $\sqrt{2} \approx 1.4$ times as much time. Because of this, there is a limit to the number of balls that a person can juggle, which seems to be about ten. Use this formula to find:
  a. How long will a ball spend in the air if it is tossed to a height of 4 feet? 8 feet?
  b. How high must it be tossed to spend 2 seconds in the air? 3 seconds in the air?

75. GENERAL: Impact Velocity If a marble is dropped from a height of x feet, it will hit the ground with velocity $v(x) = \frac{60}{11}\sqrt{x}$ miles per hour (neglecting air resistance). Use this formula to find the velocity with which a marble will strike the ground if it is dropped from the top of the tallest building in the United States, the 1454-foot Sears Tower in Chicago.

76. GENERAL: Tsunamis The speed of a tsunami (popularly known as a tidal wave, although it has nothing whatever to do with tides) depends on the depth of the water through which it is traveling. At a depth of d feet, the speed of a tsunami will be $s(d) = 3.86\sqrt{d}$ miles per hour. Find the speed of a tsunami in the Pacific basin where the average depth is 15,000 feet.

77–78. GENERAL: Impact Time of a Projectile If an object is thrown upward so that its height (in feet) above the ground t seconds after it is thrown is given by the function $h(t)$, find when the object hits the ground. That is, find the positive value of t such that $h(t) = 0$. Give the answer correct to two decimal places. [*Hint:* Enter the function in terms of x rather than t. Use the ZERO operation, or TRACE and ZOOM IN, or similar operations.]

77. $h(t) = -16t^2 + 45t + 5$

78. $h(t) = -16t^2 + 40t + 4$

79. BUSINESS: Break-Even Points and Maximum Profit A company that produces tracking devices for computer disk drives finds that if it produces x devices per week, its costs will

be $C(x) = 180x + 16{,}000$ and its revenue will be $R(x) = -2x^2 + 660x$ (both in dollars).
a. Find the company's break-even points.
b. Find the number of devices that will maximize profit, and the maximum profit.

80. **BUSINESS: Break-Even Points and Maximum Profit** City and Country Cycles finds that if it sells x racing bicycles per month, its costs will be $C(x) = 420x + 72{,}000$ and its revenue will be $R(x) = -3x^2 + 1800x$ (both in dollars).
a. Find the store's break-even points.
b. Find the number of bicycles that will maximize profit, and the maximum profit.

81. **BUSINESS: Break-Even Points and Maximum Profit** A sporting goods store finds that if it sells x exercise machines per day, its costs will be $C(x) = 100x + 3200$ and its revenue will be $R(x) = -2x^2 + 300x$ (both in dollars).
a. Find the store's break-even points.
b. Find the number of sales that will maximize profit, and the maximum profit.

82. **BUSINESS: Break-Even Points and Maximum Profit** A company that installs car alarm systems finds that if it installs x systems per week, its costs will be $C(x) = 210x + 72{,}000$ and its revenue will be $R(x) = -3x^2 + 1230x$ (both in dollars).
a. Find the company's break-even points.
b. Find the number of installations that will maximize profit, and the maximum profit.

83. **BIOMEDICAL: Muscle Contraction** The fundamental equation of muscle contraction is of the form $(w + a)(v + b) = c$, where w is the weight placed on the muscle, v is the velocity of contraction of the muscle, and a, b, and c are constants that depend upon the muscle and the units of measurement. Solve this equation for v as a function of w, a, b, and c.

84. **GENERAL: Longevity** According to insurance data, when a person reaches age 65, the probability of living for another x decades is approximated by the function $f(x) = -0.077x^2 - 0.057x + 1$ (for $0 \le x \le 3$). Find the probability that such a person will live for another:
a. one decade
b. two decades
c. three decades

85. **BUSINESS: Sales** The following table gives a company's annual sales (in millions of units) at the ends of its first through fourth years.

Year	Sales] (millions)
1	3.8
2	3.6
3	3.7
4	4.0

a. Enter the numbers from the table into a graphing calculator and make a plot of the resulting points (Year on the x-axis and Sales on the y-axis).
b. Have your calculator find the quadratic (parabolic) regression formula for these data. Then enter the result in y_1, which gives a formula for sales each year. Plot the points together with the regression line.
c. Predict the sales at the end of year 5 by evaluating $y_1(5)$.

1.4 FUNCTIONS, CONTINUED

Introduction

In this section we will define other useful types of functions and an important operation, the *composition* of two functions.

Polynomial Functions

A *polynomial function* (or simply a *polynomial*) is a function that can be written in the form

$$f(x) = a_n x^n + a_{n-1} x^{n-1} \pm \cdots + a_2 x^2 + a_1 x + a_0$$

where n is a nonnegative integer and $a_0, a_1, \ldots, a_n$ are (real) numbers, called *coefficients*. The *domain* of a polynomial is $\mathbb{R}$, the set of all (real) numbers. The *degree* of a polynomial is the highest power of the variable. The following are polynomials.

$f(x) = 2x^8 - 3x^7 + 4x^5 - 5$	A polynomial of degree 8 (since the highest power of x is 8)
$f(x) = -4x^2 - \frac{1}{3}x + 19$	A polynomial of degree 2 (a quadratic function)
$f(x) = x - 1$	A polynomial of degree 1 (a linear function)
$f(x) = 6$	A polynomial of degree 0 (a constant function)

Polynomials are used to model many situations in which change occurs at different rates. For example, the polynomial in the graph on the left might represent the total cost of manufacturing x units of a product. At first, costs rise quite steeply as a result of high start-up expenses, then they rise more slowly as the economies of mass production come into play, and finally they rise more steeply as new production facilities need to be built.

Polynomial equations can often be solved by factoring (just as with quadratic equations).

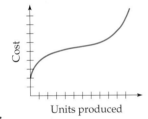

A cost function may increase at different rates at different production levels.

EXAMPLE 1

SOLVING A POLYNOMIAL EQUATION BY FACTORING

Solve $3x^4 - 6x^3 = 24x^2$

Solution

$3x^4 - 6x^3 - 24x^2 = 0$	Rewritten with all the terms on the left side
$3x^2(x^2 - 2x - 8) = 0$	Factoring out $3x^2$
$3x^2 \; (x - 4) \; (x + 2) = 0$	Factoring further
Equals zero at $x = 0$ — Equals zero at $x = 4$ — Equals zero at $x = -2$	Finding the zeros of each factor
$x = 0, \quad x = 4, \quad x = -2$	Solutions

As in this example, if a positive power of x can be factored out of a polynomial, then $x = 0$ is one of the roots.

Practice Problem 1 Solve $2x^3 - 4x^2 = 48x$ ➤ Solution on page 60

Rational Functions

The word "ratio" means fraction or quotient, and the quotient of two polynomials is called a *rational function*. The following are rational functions.

$$f(x) = \frac{4x^3 + 3x^2}{x^2 - 2x + 1} \qquad g(x) = \frac{1}{x^2 + 1} \qquad \text{A rational function is a polynomial over a polynomial}$$

The domain of a rational function is the set of all numbers for which the denominator is not zero. For example, the domain of the function on the left above is $\{x \mid x \neq 1\}$ (since $x = 1$ makes the denominator zero), and the domain of the function on the right is $\mathbb{R}$ (since $x^2 + 1$ is never zero).

Practice Problem 2 What is the domain of the rational function $f(x) = \dfrac{18}{(x + 2)(x - 4)}$?

➤ Solution on page 60

Simplifying a rational function by canceling a common factor from the numerator and the denominator can change the domain of the function, so that the "simplified" and "original" versions may not be equal (since they have different domains). For example, the rational function on the left below is not defined at $x = 1$, but the simplified version on the right *is* defined at $x = 1$, so that the two functions are technically not equal.

$$\underbrace{\frac{x^2 - 1}{x - 1}}_{\substack{\text{Not defined at } x = 1, \\ \text{so the domain is } \{x \mid x \neq 1\}}} = \frac{(x + 1)(x - 1)}{x - 1} \neq \underbrace{x + 1}_{\substack{\text{Is defined at } x = 1, \\ \text{so the domain is } \mathbb{R}}}$$

However, the functions *are* equal at every x-value *except* $x = 1$, and the graphs (shown on the following page) are the same except that the rational function omits the point at $x = 1$. We will return to this technical issue when we discuss limits in Section 8.1.

The rational function has a "missing point" at $x = 1$.

Graph of $y = \dfrac{x^2 - 1}{x - 1}$

Graph of $y = x + 1$

Piecewise Linear Functions

The rule for calculating the values of a function may be given in several parts. If each part is linear, the function is called a *piecewise linear function*, and its graph consists of "pieces" of straight lines.

EXAMPLE 2

GRAPHING A PIECEWISE LINEAR FUNCTION

Graph $f(x) = \begin{cases} 5 - 2x & \text{if } x \geq 2 \\ x + 3 & \text{if } x < 2 \end{cases}$

This notation means: Use the top formula for $x \geq 2$ and the bottom formula for $x < 2$

Solution

We graph one "piece" at a time.

Step 1: To graph the first part, $f(x) = 5 - 2x$ if $x \geq 2$, we use the "endpoint" $x = 2$ and also $x = 4$ (or any other x-value satisfying $x \geq 2$). The points are (2, 1) and (4, −3), with the y-coordinates calculated from $f(x) = 5 - 2x$. Draw the line through these two points, but only for $x \geq 2$ (from $x = 2$ to the *right*), as shown on the left.

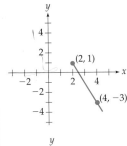

Step 2: For the second part, $f(x) = x + 3$ if $x < 2$, the restriction $x < 2$ means that the line ends just *before* $x = 2$. We mark this "missing point" (2, 5) by an "open circle" ∘ to indicate that it is *not* included in the graph (the y-coordinate comes from $f(x) = x + 3$). For a second point, choose $x = 0$ (or any other $x < 2$), giving (0, 3). Draw the line through these two points, but only for $x < 2$ (to the *left* of $x = 2$), completing the graph of the function.

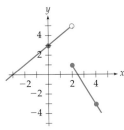

An important piecewise linear function is the *absolute value* function.

EXAMPLE 3

THE ABSOLUTE VALUE FUNCTION

The absolute value function is $f(x) = |x|$ defined as

$$f(x) = \begin{cases} x & \text{if } x \geq 0 \\ -x & \text{if } x < 0 \end{cases}$$

The second line, for *negative x*, attaches a *second* negative sign to make the result *positive*

For example, when applied to either 3 or -3, the function gives *positive* 3:

$$f(3) = 3 \qquad \text{Using the top formula (since } 3 \geq 0\text{)}$$

$$f(-3) = -(-3) = 3 \qquad \text{Using the bottom formula (since } -3 < 0\text{)}$$

To graph the absolute value function, we may proceed as in Example 2, or simply observe that for $x \geq 0$, the function gives $y = x$ (a half-line from the origin with slope 1), and for $x < 0$, it gives $y = -x$ (a half-line on the other side of the origin with slope -1), as shown in the following graph.

Absolute Value Function

$$f(x) = |x| = \begin{cases} x & \text{if } x \geq 0 \\ -x & \text{if } x < 0 \end{cases}$$

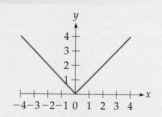

The absolute value function $f(x) = |x|$ has a "corner" at the origin.

Examples 2 and 3 show that the "pieces" of a piecewise linear function may or may not be connected.

EXAMPLE 4

GRAPHING AN INCOME TAX FUNCTION

Federal income taxes are "progressive," meaning that they take a higher percentage of higher incomes. For example, the 2001 federal income tax for a single taxpayer whose taxable income was no more than $136,750 was determined by a three-part rule: 15% of income up to $27,050, plus 27.5% of any amount over $27,050 up to $65,550, plus 30.5% of any amount over $65,550 up to $136,750. For an income of x dollars, the tax $f(x)$ may be expressed as follows:

$$f(x) = \begin{cases} 0.15x & \text{if } 0 \le x \le 27{,}050 \\ 4057.50 + 0.275(x - 27{,}050) & \text{if } 27{,}050 < x \le 65{,}550 \\ 14{,}645 + 0.305(x - 65{,}550) & \text{if } 65{,}550 < x \le 136{,}750 \end{cases}$$

Graphing this by the same technique as before leads to the following graph. The slopes 0.15, 0.275, and 0.305 are called the *marginal tax rates*.

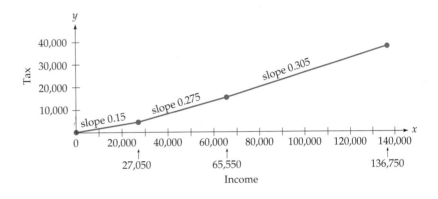

Composite Functions

Just as we substitute a *number* into a function, we may substitute a *function* into a function. For two functions f and g, evaluating f at $g(x)$ gives $f(g(x))$, called the *composition of f with g evaluated at x*.*

> **Composite Functions**
>
> The *composition* of f with g evaluated at x is $f(g(x))$.

The *domain* of $f(g(x))$ is the set of all numbers x in the domain of g such that $g(x)$ is in the domain of f. If we think of the functions f and g as "numerical machines," then the composition $f(g(x))$ may

---

* The composition $f(g(x))$ may also be written $(f \circ g)(x)$, although we will not use this notation.

be thought of as a *combined* machine in which the output of g is connected to the input of f.

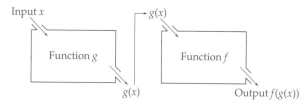

A "machine" for generating the composition of f with g. A number x is fed into the function g, and the output g(x) is then fed into the function f, resulting in f(g(x)).

EXAMPLE 5

FINDING A COMPOSITE FUNCTION

If $f(x) = x^7$ and $g(x) = x^3 - 2x$, find the composition $f(g(x))$.

Solution

$$f(g(x)) \;=\; \underbrace{[g(x)]^7}_{\substack{f(x) = x^7 \text{ with } x \\ \text{replaced by } g(x)}} \;=\; \underbrace{(x^3 - 2x)^7}_{\substack{\text{Using} \\ g(x) = x^3 - 2x}}$$

EXAMPLE 6

FINDING BOTH COMPOSITE FUNCTIONS

If $f(x) = \dfrac{x+8}{x-1}$ and $g(x) = \sqrt{x}$, find $f(g(x))$ and $g(f(x))$.

Solution

$$f(g(x)) = \frac{g(x) + 8}{g(x) - 1} = \frac{\sqrt{x} + 8}{\sqrt{x} - 1} \qquad \begin{array}{l} f(x) = \dfrac{x+8}{x-1} \text{ with } x \\ \text{replaced by } g(x) = \sqrt{x} \end{array}$$

$$g(f(x)) = \sqrt{f(x)} = \sqrt{\frac{x+8}{x-1}} \qquad \begin{array}{l} g(x) = \sqrt{x} \text{ with } x \\ \text{replaced by } f(x) = \dfrac{x+8}{x-1} \end{array}$$

The order of composition is important: $f(g(x))$ is *not* the same as $g(f(x))$. To show this, we evaluate the above $f(g(x))$ and $g(f(x))$ at $x = 4$:

$$f(g(4)) = \frac{\sqrt{4} + 8}{\sqrt{4} - 1} = \frac{2 + 8}{2 - 1} = \frac{10}{1} = 10 \qquad f(g(x)) = \frac{\sqrt{x} + 8}{\sqrt{x} - 1} \text{ at } x = 4$$

Different answers

$$g(f(4)) = \sqrt{\frac{4 + 8}{4 - 1}} = \sqrt{\frac{12}{3}} = \sqrt{4} = 2 \qquad g(f(x)) = \sqrt{\frac{x + 8}{x - 1}} \text{ at } x = 4$$

Practice Problem 3 If $f(x) = x^2 + 1$ and $g(x) = \sqrt[3]{x}$, find: **a.** $f(g(x))$, **b.** $g(f(x))$.

➤ Solutions on page 60

EXAMPLE 7

PREDICTING WATER USAGE

A planning commission estimates that if a city's population is p thousand people, its daily water usage will be $W(p) = 30p^{1.2}$ thousand gallons. The commission further predicts that the population in t years will be $p(t) = 60 + 2t$ thousand people. Express the water usage W as a function of t, the number of years from now, and find the water usage 10 years from now.

Solution

Water usage W as a function of t is the *composition* of $W(p)$ with $p(t)$:

$$W(p(t)) = 30[p(t)]^{1.2} = 30(60 + 2t)^{1.2} \qquad \begin{array}{l} W = 30p^{1.2} \text{ with } p \\ \text{replaced by } p(t) = 60 + 2t \end{array}$$

To find water usage in 10 years, we evaluate $W(p(t))$ at $t = 10$:

$$\begin{aligned} W(p(10)) &= 30(60 + 2 \cdot 10)^{1.2} & & 30(60 + 2t)^{1.2} \text{ with } t = 10 \\ &= 30(80)^{1.2} \approx 5765 & & \text{Using a calculator} \end{aligned}$$

Thousand gallons

Therefore, in 10 years the city will need about 5,765,000 gallons of water per day.

Shifts of Graphs

Sometimes the composition of two functions is just a horizontal or vertical shift of an original graph. This occurs when one of the functions is simply the addition or subtraction of a constant. The following diagrams show the graph of $y = x^2$ together with various shifts and the functions that generate them.

Vertical shifts

$y = x^2 + 5$
$y = x^2$
$y = x^2 - 5$

Horizontal shifts

$y = (x + 5)^2$ $y = x^2$ $y = (x - 5)^2$

In general, adding to or subtracting from the *x-value* means a *horizontal* shift, while adding to or subtracting from the *function* means a *vertical* shift. These same ideas hold for *any* function: given the graph of $y = f(x)$, adding or subtracting a positive number a to the function $f(x)$ or to the variable x shifts the graph as follows:

Shifts of Graphs

Function	Shift	
$y = f(x) + a$	shifted *up* by a units	} Vertical shifts
$y = f(x) - a$	shifted *down* by a units	
$y = f(x + a)$	shifted *left* by a units	} Horizontal shifts
$y = f(x - a)$	shifted *right* by a units	

Of course, a graph can be shifted both horizontally *and* vertically:

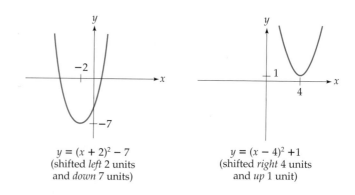

$y = (x + 2)^2 - 7$
(shifted *left* 2 units
and *down* 7 units)

$y = (x - 4)^2 + 1$
(shifted *right* 4 units
and *up* 1 unit)

Such double shifts can be applied to *any* function $y = f(x)$: the graph of $y = f(x + a) + b$ is shifted *left a* units and *up b* units (with the understanding that a *negative a* or *b* means that the direction is reversed).

Be careful: Remember that adding a *positive* number to x means a *left* shift.

Graphing Calculator Exploration

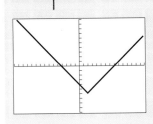

The absolute value function $y = |x|$ may be graphed on some graphing calculator as $y_1 = \text{ABS}(x)$.

a. Graph $y_1 = \text{ABS}(x - 2) - 6$ and observe that the absolute value function is shifted *right* 2 units and *down* 6 units. (The graph shown is drawn using ZOOM ZSquare.)

b. Predict the shift of $y_1 = \text{ABS}(x + 4) + 2$ and then verify your prediction by graphing the function on your calculator.

Given a function $f(x)$, to find an algebraic expression for the "shifted" function $f(x + h)$ we simply replace each occurrence of x by $x + h$.

EXAMPLE 8

FINDING $f(x + h)$ FROM $f(x)$

If $f(x) = x^2 - 5x$, find $f(x + h)$.

Solution

$f(x + h) = (x + h)^2 - 5(x + h)$ $\quad f(x) = x^2 - 5x$ with each x replaced by $x + h$

$= x^2 + 2xh + h^2 - 5x - 5h$ Expanding

Difference Quotients

The quantity $\dfrac{f(x + h) - f(x)}{h}$ will be very important in Chapter 8 when we begin studying calculus. It is called the *difference quotient*, since it is a quotient whose numerator is a difference. It gives the slope (rise over run) between the points in the curve $y = f(x)$ at x and at $x + h$.

EXAMPLE 9　FINDING A DIFFERENCE QUOTIENT

If $f(x) = x^2 - 4x + 1$, find and simplify $\dfrac{f(x+h) - f(x)}{h}$ 　　$(h \neq 0)$

Solution

$$\dfrac{f(x+h) - f(x)}{h} = \dfrac{\overbrace{(x+h)^2 - 4(x+h) + 1}^{f(x+h)} - \overbrace{(x^2 - 4x + 1)}^{f(x)}}{h}$$

$$= \dfrac{x^2 + 2xh + h^2 - 4x - 4h + 1 - x^2 + 4x - 1}{h} \quad \text{Expanding}$$

$$= \dfrac{x^2 + 2xh + h^2 - 4x - 4h + \cancel{1} - x^2 + 4x - \cancel{1}}{h} \quad \text{Canceling}$$

$$= \dfrac{2xh + h^2 - 4h}{h} = \dfrac{h(2x + h - 4)}{h} \quad \text{Factoring an } h \text{ from the top}$$

$$= \dfrac{h(2x + h - 4)}{h} = 2x + h - 4 \quad \text{Canceling } h \text{ from top and bottom (since } h \neq 0\text{)}$$

Practice Problem 4

If $f(x) = 3x^2 - 2x + 1$, find and simplify $\dfrac{f(x+h) - f(x)}{h}$.

▶ Solution on page 60

EXAMPLE 10　FINDING A DIFFERENCE QUOTIENT

If $f(x) = \dfrac{1}{x}$, find and simplify $\dfrac{f(x+h) - f(x)}{h}$ 　　$(h \neq 0)$

Solution

$$\dfrac{f(x+h) - f(x)}{h} = \dfrac{\overbrace{\dfrac{1}{x+h}}^{f(x+h)} - \overbrace{\dfrac{1}{x}}^{f(x)}}{h}$$

$$= \dfrac{1}{h}\left(\dfrac{1}{x+h} - \dfrac{1}{x}\right) \quad \text{Multiplying by } 1/h \text{ instead of dividing by } h$$

$$= \dfrac{1}{h}\left(\dfrac{x}{(x+h)x} - \dfrac{x+h}{(x+h)x}\right) \quad \text{Using the common denominator } (x+h)x$$

$$= \frac{1}{h} \cdot \frac{\overbrace{x - (x+h)}^{-h}}{(x+h)x} = \frac{1}{h} \cdot \frac{-h}{(x+h)x} \quad \text{Subtracting fractions, and simplifying}$$

$$= \frac{1}{\cancel{h}} \cdot \frac{\overset{-1}{\cancel{-h}}}{(x+h)x} = \frac{-1}{(x+h)x} \quad \text{Canceling } h \; (h \neq 0)$$

1.4 Section Summary

We have introduced a variety of functions: polynomials (which include linear and quadratic functions), rational functions, and piecewise linear functions. Examples of these are shown below. You should be able to identify these basic types of functions from their algebraic forms. We also added constants to perform horizontal and vertical *shifts* of graphs of functions, and combined functions by using the "output" of one as the "input" of the other, resulting in *composite* functions.

A Gallery of Functions

POLYNOMIALS

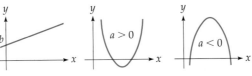

Linear function
$f(x) = mx + b$

Quadratic functions
$f(x) = ax^2 + bx + c$

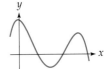

$f(x) = ax^4 + bx^3 + cx^2 + dx + e$

RATIONAL FUNCTIONS

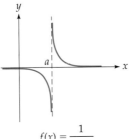

$f(x) = \dfrac{1}{x - a}$

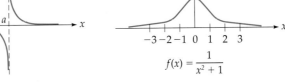

$f(x) = \dfrac{1}{x^2 + 1}$

PIECEWISE LINEAR FUNCTIONS

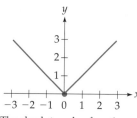

The absolute value function
$f(x) = |x|$

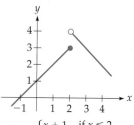

$f(x) = \begin{cases} x+1 & \text{if } x \leq 2 \\ 6-x & \text{if } x > 2 \end{cases}$

▶ Solutions to Practice Problems

1. $2x^3 - 4x^2 - 48x = 0$
   $2x(x^2 - 2x - 24) = 0$
   $2x(x+4)(x-6) = 0$
   $x = 0,\ x = -4,\ x = 6$

2. $\{x \mid x \neq -2, x \neq 4\}$

3. a. $f(g(x)) = [g(x)]^2 + 1 = \left(\sqrt[3]{x}\right)^2 + 1$ or $x^{2/3} + 1$
   b. $g(f(x)) = \sqrt[3]{f(x)} = \sqrt[3]{x^2 + 1}$ or $(x^2 + 1)^{1/3}$

4. $\dfrac{f(x+h) - f(x)}{h} = \dfrac{3(x+h)^2 - 2(x+h) + 1 - (3x^2 - 2x + 1)}{h}$

   $= \dfrac{3x^2 + 6xh + 3h^2 - 2x - 2h + 1 - 3x^2 + 2x - 1}{h}$

   $= \dfrac{h(6x + 3h - 2)}{h} = 6x + 3h - 2$

1.4 Exercises

Find the domain and range of each function graphed below.

1.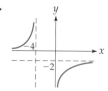

2.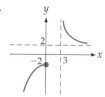

For each function in Exercises 3–10:

a. Evaluate the given expression.
b. Find the domain of the function.
c. Find the range. [*Hint:* Use a graphing calculator. You may have to ignore some false lines on the graph. Graphing in "dot mode" will also eliminate false lines.]

3. $f(x) = \dfrac{1}{x+4}$; find $f(-3)$

4. $f(x) = \dfrac{1}{(x-1)^2}$; find $f(-1)$

5. $f(x) = \dfrac{x^2}{x-1}$; find $f(-1)$

6. $f(x) = \dfrac{x^2}{x+2}$; find $f(2)$

7. $f(x) = \dfrac{12}{x(x+4)}$; find $f(2)$

8. $f(x) = \dfrac{16}{x(x-4)}$; find $f(-4)$

9. $g(x) = |x+2|$; find $g(-5)$

10. $g(x) = |x|+2$; find $g(-5)$

Solve each equation by factoring.

11. $x^5 + 2x^4 - 3x^3 = 0$
12. $x^6 - x^5 - 6x^4 = 0$
13. $5x^3 - 20x = 0$
14. $2x^5 - 50x^3 = 0$
15. $2x^3 + 18x = 12x^2$
16. $3x^4 + 12x^2 = 12x^3$
17. $6x^5 = 30x^4$
18. $5x^4 = 20x^3$
19. $3x^{5/2} - 6x^{3/2} = 9x^{1/2}$
20. $2x^{7/2} + 8x^{5/2} = 24x^{3/2}$

[Hint: First factor out a fractional power.]

Solve each equation using a graphing calculator. (For Exercises 31 and 32, round answers to two decimal places.)

21. $x^3 - 2x^2 - 8x = 0$
22. $x^3 + 2x^2 - 8x = 0$
23. $x^5 - 2x^4 - 3x^3 = 0$
24. $x^6 + x^5 - 6x^4 = 0$
25. $2x^3 = 12x^2 - 18x$
26. $3x^4 + 12x^3 + 12x^2 = 0$
27. $6x^5 + 30x^4 = 0$
28. $5x^4 + 20x^3 = 0$
29. $2x^{5/2} + 4x^{3/2} = 6x^{1/2}$
30. $3x^{7/2} - 12x^{5/2} = 36x^{3/2}$
31. $x^5 - x^4 - 5x^3 = 0$
32. $x^6 + 2x^5 - 5x^4 = 0$

Graph each function.

33. $f(x) = |x-3| - 3$
34. $f(x) = |x+2| - 2$

[Hint: Think of shifted graphs.]

35. $f(x) = \begin{cases} 2x - 7 & \text{if } x \geq 4 \\ 2 - x & \text{if } x < 4 \end{cases}$

36. $f(x) = \begin{cases} 2 - x & \text{if } x \geq 3 \\ 2x - 4 & \text{if } x < 3 \end{cases}$

37. $f(x) = \begin{cases} 8 - 2x & \text{if } x \geq 2 \\ x + 2 & \text{if } x < 2 \end{cases}$

38. $f(x) = \begin{cases} 2x - 4 & \text{if } x > 3 \\ 5 - x & \text{if } x \leq 3 \end{cases}$

Identify each function as a polynomial, a rational function, a piecewise linear function, or none of these. (Do not graph the functions; just identify their types.)

39. $f(x) = x^5$
40. $f(x) = 3|x|$
41. $f(x) = 1 - |x|$
42. $f(x) = x^4$
43. $f(x) = x + 2$

44. $f(x) = \begin{cases} 3x - 1 & \text{if } x \geq 2 \\ 1 - x & \text{if } x < 2 \end{cases}$

45. $f(x) = \dfrac{1}{x+2}$
46. $f(x) = x^2 + 9$

47. $f(x) = \begin{cases} x - 2 & \text{if } x < 3 \\ 7 - 4x & \text{if } x \geq 3 \end{cases}$

48. $f(x) = \dfrac{x}{x^2 + 9}$
49. $f(x) = 3x^2 - 2x$
50. $f(x) = x^3 - x^{2/3}$
51. $f(x) = x^2 + x^{1/2}$
52. $f(x) = 5$

53. For the functions $y_1 = 100x$, $y_2 = 10x^2$, $y_3 = x^3$:
   a. Predict which curve will be the highest for large values of x.
   b. Predict which curve will be the lowest for large values of x.
   c. Check your predictions by graphing the functions on the window [0, 12] by [0, 2000].
   d. From your graph, where do all these curves meet?

54. Graph the parabola $y_1 = 1 - x^2$ and the semicircle $y_2 = \sqrt{1 - x^2}$ on the graphing window $[-1, 1]$ by $[0, 1]$. Use ZSQUARE to make the semicircle look more like a semicircle. Use TRACE to determine which is the "inside" curve (the parabola or the semicircle) and which is the "outside" curve. These graphs show that when you graph a parabola, you should draw the curve near the vertex to be slightly more "pointed" than a circular curve.

55. For any x, the function $INT(x)$ is defined as the greatest integer less than or equal to x. For example, $INT(3.7) = 3$ and $INT(-4.2) = -5$.
   a. Use a graphing calculator to graph the function $y_1 = INT(x)$. (You may need to graph it in DOT mode to eliminate false connecting lines.)
   b. From your graph, what are the domain and range of this function?

62 CHAPTER 1 FUNCTIONS

56. a. Use a graphing calculator to graph the function $y_1 = 2\,\text{INT}(x)$. (See the previous exercise for a definition of $\text{INT}(x)$.)
   b. From your graph, what are the domain and range of this function?

For each pair $f(x)$ and $g(x)$, find **a.** $f(g(x))$, **b.** $g(f(x))$.

57. $f(x) = x^5$; $g(x) = 7x - 1$
58. $f(x) = x^8$; $g(x) = 2x + 5$
59. $f(x) = \dfrac{1}{x}$; $g(x) = x^2 + 1$
60. $f(x) = \sqrt{x}$; $g(x) = x^3 - 1$
61. $f(x) = x^3 - x^2$; $g(x) = \sqrt{x} - 1$
62. $f(x) = x - \sqrt{x}$; $g(x) = x^2 + 1$
63. $f(x) = \dfrac{x^3 - 1}{x^3 + 1}$; $g(x) = x^2 - x$
64. $f(x) = \dfrac{x^4 + 1}{x^4 - 1}$; $g(x) = x^3 + x$

65. a. Find the composition $f(g(x))$ of the two linear functions $f(x) = ax + b$ and $g(x) = cx + d$ (for constants a, b, c, and d).
   b. Is the composition of two linear functions always a linear function?

66. a. Is the composition of two quadratic functions always a quadratic function? [*Hint:* Find the composition of $f(x) = x^2$ and $g(x) = x^2$.]
   b. Is the composition of two polynomials always a polynomial?

For each function in Exercises 67–78, find and simplify $\dfrac{f(x+h) - f(x)}{h}$. (Assume $h \neq 0$.)

67. $f(x) = 5x^2$
68. $f(x) = 3x^2$
69. $f(x) = 2x^2 - 5x + 1$
70. $f(x) = 3x^2 - 5x + 2$
71. $f(x) = 7x^2 - 3x + 2$
72. $f(x) = 4x^2 - 5x + 3$
73. $f(x) = x^3$
   [*Hint:* Use $(x+h)^3 = x^3 + 3x^2h + 3xh^2 + h^3$.]
74. $f(x) = x^4$
   [*Hint:* Use $(x+h)^4 = x^4 + 4x^3h + 6x^2h^2 + 4xh^3 + h^4$.]
75. $f(x) = \dfrac{2}{x}$
76. $f(x) = \dfrac{3}{x}$
77. $f(x) = \dfrac{1}{x^2}$
78. $f(x) = \sqrt{x}$

[*Hint for Exercise 78:* Multiply top and bottom of the fraction by $\sqrt{x+h} + \sqrt{x}$.]

79. How will the graph of $y = (x+3)^3 + 6$ differ from the graph of $y = x^3$? Check by graphing both functions together.

80. How will the graph of $y = -(x-4)^2 + 8$ differ from the graph of $y = -x^2$? Check by graphing both functions together.

APPLIED EXERCISES

81. **ECONOMICS: Income Tax** The following function expresses an income tax that is 10% for incomes up to $5000, and otherwise is $500 plus 30% of income in excess of $5000.

$$f(x) = \begin{cases} 0.10x & \text{if } 0 \leq x \leq 5000 \\ 500 + 0.30(x - 5000) & \text{if } x > 5000 \end{cases}$$

   a. Calculate the tax on an income of $3000.
   b. Calculate the tax on an income of $5000.
   c. Calculate the tax on an income of $10,000.
   d. Graph the function.

82. **ECONOMICS: Income Tax** The following function expresses an income tax that is 15% for incomes up to $6000, and otherwise is $900 plus 40% of income in excess of $6000.

$$f(x) = \begin{cases} 0.15x & \text{if } 0 \leq x \leq 6000 \\ 900 + 0.40(x - 6000) & \text{if } x > 6000 \end{cases}$$

   a. Calculate the tax on an income of $3000.
   b. Calculate the tax on an income of $6000.
   c. Calculate the tax on an income of $10,000.
   d. Graph the function.

83–84. **GENERAL: Dog Years** The usual estimate that each human-year corresponds to 7 dog-years is not very accurate for young dogs, since they quickly reach adulthood. Exercises 83 and 84 give more accurate formulas for converting human-years x into dog-years. For each conversion formula:

a. Find the number of dog-years corresponding to the following amounts of human time: 8 months, 1 year and 4 months, 4 years, 10 years.
b. Graph the function.

83. $f(x) = \begin{cases} 10.5x & \text{if } 0 \le x \le 2 \\ 21 + 4(x - 2) & \text{if } x > 2 \end{cases}$

(This function expresses dog-years as $10\frac{1}{2}$ dog-years per human-year for the first 2 years and then 4 dog-years per human-year for each year thereafter.)

84. $f(x) = \begin{cases} 15x & \text{if } 0 \le x \le 1 \\ 15 + 9(x - 1) & \text{if } 1 < x \le 2 \\ 24 + 4(x - 2) & \text{if } x > 2 \end{cases}$

(This function expresses dog-years as 15 dog-years per human-year for the first year, 9 dog-years per human-year for the second year, and then 4 dog-years per human-year for each year thereafter.)

85. **BUSINESS: Insurance Reserves** An insurance company keeps reserves (money to pay claims) of $R(v) = 2v^{0.3}$, where v is the value of all of its policies, and the value of its policies is predicted to be $v(t) = 60 + 3t$, where t is the number of years from now. (Both R and v are in millions of dollars.) Express the reserves R as a function of t, and evaluate the function at $t = 10$.

86. **BUSINESS: Research Expenditures** An electronics company's research budget is $R(p) = 3p^{0.25}$, where p is the company's profit, and the profit is predicted to be $p(t) = 55 + 4t$, where t is the number of years from now. (Both R and p are in millions of dollars.) Express the research expenditure R as a function of t, and evaluate the function at $t = 5$.

87. **GENERAL: Oil Imports and Fuel Efficiency** The average fuel efficiency of all cars in America, called the *fleet mpg*, is 21.6 mpg. The amount of crude oil that would be saved if the fleet mpg were increased to a value of x is

$$S(x) = 3208 - \frac{69{,}300}{x} \quad \text{million barrels annually}$$

a. Graph the function $S(x)$ on the viewing window [21.6, 40] by [0, 2000].
b. For what fleet mpg x will the savings reach 720 million barrels, which is the amount annually imported from the Middle East OPEC nations? [*Hint:* Either ZOOM IN around the point at $y = 720$ or, better, use INTERSECT with $y_2 = 720$.] [*Note:* New cars average 29 mpg, but fleet mpg always lags. Ten manufacturers have already built and tested prototype cars getting from 78 to 138 mpg.]
(*Sources:* Federal Highway Administration and the Rocky Mountain Institute.)

88. **GENERAL: Long-Term Trends in Higher Education** The following table gives the percentage of adults in the United States who were college graduates, for the years 1940 to 2000.

Years	Years Since 1940	Percentage of College Graduates
1940	0	4
1960	20	9
1980	40	18
2000	60	26

a. Enter these data into a graphing calculator and make a plot of the resulting points (Years since 1940 on the x-axis and Percentage on the y-axis).
b. Have your calculator find the cubic (third-order polynomial) regression formula for these data. Then enter the result as y_1, which gives a formula for the percentage for each year. Plot the points together with the regression curve.
c. Use the regression formula to predict the percentage of college graduates in the year 2010 ($x = 70$).
d. Use the regression formula to predict the percentage of college graduates in the year 2060 ($x = 120$). The answer shows that polynomial predictions, if extended too far, give nonsensical results.

1.5 EXPONENTIAL FUNCTIONS

Introduction

We now introduce exponential functions, showing how they are used to model the processes of growth and decay.

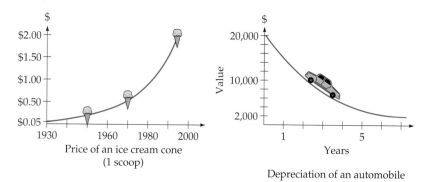

Price of an ice cream cone (1 scoop)

Depreciation of an automobile

We will also define the very important mathematical constant e.

Exponential Functions

A function that has a variable in an exponent, such as $f(x) = 2^x$, is called an *exponential function*.

$$f(x) = 2^x$$

with x as the Exponent and 2 as the Base.

The table below shows some values of the exponential function $f(x) = 2^x$, and its graph (based on these points) is shown on the right.

x	$y = 2^x$
-3	$2^{-3} = \frac{1}{8}$
-2	$2^{-2} = \frac{1}{4}$
-1	$2^{-1} = \frac{1}{2}$
0	$2^0 = 1$
1	$2^1 = 2$
2	$2^2 = 4$
3	$2^3 = 8$

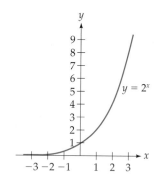

Domain of $y = 2^x$ is $\mathbb{R} = (-\infty, \infty)$ and range is $(0, \infty)$.

Clearly, the graph of the exponential function 2^x is quite different from the parabola x^2.

The exponential function $f(x) = \left(\frac{1}{2}\right)^x$ has base $\frac{1}{2}$. The following table shows some of its values, and its graph is shown to the right of the table. Notice that it is the mirror image of the curve $y = 2^x$.

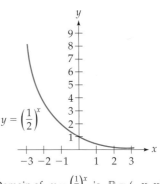

x	$y = \left(\frac{1}{2}\right)^x$
-3	$\left(\frac{1}{2}\right)^{-3} = 8$
-2	$\left(\frac{1}{2}\right)^{-2} = 4$
-1	$\left(\frac{1}{2}\right)^{-1} = 2$
0	$\left(\frac{1}{2}\right)^{0} = 1$
1	$\left(\frac{1}{2}\right)^{1} = \frac{1}{2}$
2	$\left(\frac{1}{2}\right)^{2} = \frac{1}{4}$
3	$\left(\frac{1}{2}\right)^{3} = \frac{1}{8}$

Domain of $y = \left(\frac{1}{2}\right)^x$ is $\mathbb{R} = (-\infty, \infty)$ and range is $(0, \infty)$.

We can define an exponential function $f(x) = a^x$ for any positive base a. We always take the base to be positive, so for the rest of this section *the letter a will stand for a positive constant.*

Exponential functions with bases $a > 1$ are used to model *growth*, as in populations or savings accounts, and exponential functions with bases $a < 1$ are used to model *decay*, as in depreciation. (For base $a = 1$, the graph is a horizontal line, since $1^x = 1$ for all x.)

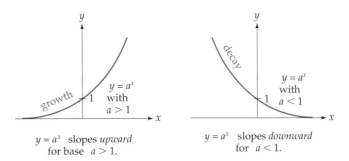

$y = a^x$ slopes *upward* for base $a > 1$.

$y = a^x$ slopes *downward* for $a < 1$.

Compound Interest

Money invested at compound interest grows exponentially. (The word "compound" means that the interest is added to the account, earning more interest.) Banks always state *annual* interest rates, but the compounding may be done more frequently. For example, if a bank offers 8% compounded quarterly, then each quarter you get 2% (one quarter of the annual 8%), so that 2% of your money is added to the account each quarter. If you begin with P dollars (the "principal"), at the end of the first quarter you would have P dollars plus 2% of P dollars:

$$\begin{pmatrix} \text{Value after} \\ \text{1 quarter} \end{pmatrix} = P + 0.02P = P \cdot (1 + 0.02)$$

Notice that increasing a quantity by 2% is the same as multiplying it by $(1 + 0.02)$. Since a year has 4 quarters, t years will have $4t$ quarters. Therefore, to find the value of your account after t years, we simply multiply the principal by $(1 + 0.02)$ a total of $4t$ times, obtaining:

$$\begin{pmatrix} \text{Value after} \\ t \text{ years} \end{pmatrix} = P \cdot \overbrace{(1 + 0.02) \cdot (1 + 0.02) \cdots (1 + 0.02)}^{4t \text{ times}}$$

$$= P \cdot (1 + 0.02)^{4t}$$

The 8%, which gave the $\frac{0.08}{4} = 0.02$ quarterly rate, can be replaced by any interest rate r (written in decimal form), and the 4 can be replaced by any number m of compounding periods per year, leading to the following general formula.

Compound Interest

For P dollars invested at interest rate r compounded m times a year,

$$\begin{pmatrix} \text{Value after} \\ t \text{ years} \end{pmatrix} = P \cdot \left(1 + \frac{r}{m}\right)^{mt}$$

r = annual rate
m = periods per year
t = number of years

For example, for monthly compounding we would use $m = 12$ and for daily compounding $m = 365$ (the number of days in the year).

EXAMPLE 1 **FINDING A VALUE UNDER COMPOUND INTEREST**

Find the value of $4000 invested for 2 years at 12% compounded quarterly.

Solution

$$4000 \cdot \left(1 + \underbrace{\frac{0.12}{4}}_{0.03}\right)^{4 \cdot 2} = 4000(1 + 0.03)^8$$

$P \cdot \left(1 + \frac{r}{m}\right)^{mt}$
with $P = 4000$,
$r = 0.12$, $m = 4$,
and $t = 2$

$$= 4000 \cdot 1.03^8 \approx 5067.08$$

Using a calculator

The value after two years will be $5067.08.

We may interpret the formula $P(1 + 0.03)^8$ intuitively as follows: multiplying the principal by $(1 + 0.03)$ means that you keep the original amount (the "1") plus some interest (the 0.03), and the exponent 8 means that this is done a total of 8 times.

Practice Problem 1

Find the value of $2000 invested for 3 years at 24% compounded monthly.
➤ Solution on page 74

Much more will be said about compound interest in Section 2.2 of Chapter 2.

Graphing Calculator Exploration

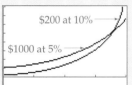

In the long run, which is more important—principal or interest rate? The graph shows the value of $1000 at 5% interest, together with the value of a mere $200 at the higher rate of 10% (both compounded annually). The fact that the (initially) lower curve eventually surpasses the higher one illustrates a general fact: the higher interest rate will eventually prevail, regardless of the size of the initial investment.

Find how soon the $200 at 10% will surpass the $1000 at 5% by graphing $y_1 = 1000(1 + 0.05)^x$ and $y_2 = 200(1 + 0.10)^x$ on the window [0, 40] by [−2000, 8000] and using INTERSECT. What if the higher rate is only 9%? (You will have to extend your window.)

Depreciation by a Fixed Percentage

Depreciation by a fixed percentage means that a piece of equipment loses a fixed percentage (say 30%) of its value each year. Losing a percentage of value is like compound interest but with a *negative* interest rate. Therefore, we use the compound interest formula with $m = 1$ (since depreciation is annual) and with r being *negative*.

EXAMPLE 2

DEPRECIATING AN ASSET

A car worth $15,000 depreciates in value by 40% each year. How much is it worth after 3 years?

Solution

The car loses 40% of its value each year, which is equivalent to an annual interest rate of *negative* 40%. The compound interest formula gives

$$15{,}000(1 - 0.40)^3 = 15{,}000(0.60)^3 = \$3240 \qquad \begin{array}{l} P(1+r/m)^{mt} \text{ with} \\ P = 15{,}000, \ r = -0.40, \\ m = 1, \text{ and } t = 3 \end{array}$$

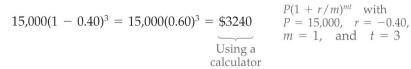

The exponential function $f(x) = 15{,}000(0.60)^x$, giving the value of the car after x years of depreciation, is graphed on the left.

Practice Problem 2 A printing press, originally worth $50,000, loses 20% of its value each year. What is its value after 4 years? ➤ **Solution on page 74**

The above graph shows that depreciation by a fixed percentage is quite different from "straight-line" depreciation. Under straight-line depreciation the same *dollar* value is lost each year, while under fixed-percentage depreciation the same *percentage* of value is lost each year, resulting in larger dollar losses in the early years and smaller dollar losses in later years. Depreciation by a fixed percentage (also called the "declining balance" method) is one type of "accelerated" depreciation. The method of depreciation that one uses depends on how one chooses to estimate value, and in practice is often determined by the tax laws.

The Number e

Imagine that a bank offers 100% interest, and that you deposit $1 for 1 year. Let us see how the value changes under different types of compounding.

For *annual* compounding, your $1 would in a year grow to $2 (the original dollar plus a dollar interest).

For *quarterly* compounding, we use the compound interest formula with $P = 1$, $r = 1$ (for 100%), $m = 4$, and $t = 1$:

$$1\left(1 + \frac{1}{4}\right)^{1 \cdot 4} = (1 + 0.25)^4 = (1.25)^4 \approx 2.44 \qquad P(1+r/m)^{mt}$$

or 2 dollars and 44 cents, an improvement of 44 cents over annual compounding.

1.5 EXPONENTIAL FUNCTIONS

For *daily* compounding, the value after a year would be

$$\left(1 + \frac{1}{365}\right)^{365} \approx 2.71 \qquad \begin{array}{l} m = 365 \text{ periods} \\ \dfrac{r}{m} = \dfrac{100\%}{365} = \dfrac{1}{365} \end{array}$$

an increase of 27 cents over quarterly compounding. Clearly, if the interest rate, the principal, and the amount of time stay the same, the value increases as the compounding is done more frequently.

In general, if the compounding is done m times a year, the value of the dollar after a year will be

$$\left(\begin{array}{c}\text{Value of \$1 after 1 year at 100\%} \\ \text{interest compounded } m \text{ times a year}\end{array}\right) = \left(1 + \frac{1}{m}\right)^m$$

The following table shows the value of $\left(1 + \dfrac{1}{m}\right)^m$ for various values of m.

Value of $1 at 100% Interest Compounded m Times a Year for 1 Year

m	$\left(1 + \dfrac{1}{m}\right)^m$	Answer (rounded)	
1	$\left(1 + \dfrac{1}{1}\right)^1$	$= 2.00000$	Annual compounding
4	$\left(1 + \dfrac{1}{4}\right)^4$	≈ 2.44141	Quarterly compounding
365	$\left(1 + \dfrac{1}{365}\right)^{365}$	≈ 2.71457	Daily compounding
10,000	$\left(1 + \dfrac{1}{10,000}\right)^{10,000}$	≈ 2.71815	
100,000	$\left(1 + \dfrac{1}{100,000}\right)^{100,000}$	≈ 2.71827	
1,000,000	$\left(1 + \dfrac{1}{1,000,000}\right)^{1,000,000}$	≈ 2.71828	Answers agree to five decimal places
10,000,000	$\left(1 + \dfrac{1}{10,000,000}\right)^{10,000,000}$	≈ 2.71828	

Notice that as m increases, the values in the right-hand column seem to be settling down to a definite value, approximately 2.71828. That is, as m approaches infinity, the limit of $\left(1 + \dfrac{1}{m}\right)^m$ is approximately 2.71828.

This particular number is very important in mathematics, and is given the name e (just as 3.14159... is given the name π). In the following definition, we use the letter n to state the definition in its traditional form.

The Constant e

$$e = \lim_{n \to \infty} \left(1 + \frac{1}{n}\right)^n = 2.71828\ldots$$

$n \to \infty$ is read "n approaches infinity." The dots mean that the decimal expansion goes on forever

The same e appears in probability and statistics in the formula for the "bell-shaped" or "normal" curve (see page 463). Its value has been calculated to many thousands of decimal places, and its value to 15 decimal places is $e \approx 2.718281828459045$.

Continuous Compounding of Interest

This kind of compound interest, the limit as the compounding frequency approaches infinity, is called *continuous* compounding. We have shown that $1 at 100% interest compounded continuously for 1 year would be worth precisely e dollars (about $2.72). The formula for continuous compound interest at other rates is as follows (a justification for it is given at the end of this section).

Continuous Compounding

For P dollars invested at interest rate r compounded continuously,

$$\begin{pmatrix}\text{Value after} \\ t \text{ years}\end{pmatrix} = Pe^{rt}$$

EXAMPLE 3 **FINDING VALUE WITH CONTINUOUS COMPOUNDING**

Find the value of $1000 at 8% interest compounded continuously for 20 years.

Solution

We use the formula Pe^{rt} with $P = 1000$, $r = 0.08$, and $t = 20$.

$$Pe^{rt} = 1000 \cdot e^{(0.08)(20)} = 1000 \cdot e^{1.6} \approx \$4953.03$$

$\underbrace{}_{P}$ $\underset{r}{\uparrow}$ $\underset{t}{\uparrow}$ $\underbrace{\phantom{e^{1.6}}}_{4.95303}$

$e^{1.6}$ is usually found using the [2nd] and [LN] keys

Intuitive Meaning of Continuous Compounding

Under quarterly compounding, your money is, in a sense, earning interest throughout the quarter, but the interest is not added to your account until the end of the quarter. Under continuous compounding, the interest is added to your account *as it is earned*, with no delay. The extra earnings in continuous compounding come from this "instant crediting" of interest, since then your interest starts earning more interest immediately.

The Function $y = e^x$

The number e gives us a new exponential function $f(x) = e^x$. This function is used extensively in business, economics, and all areas of science. The table below shows the values of e^x for various values of x. These values lead to the graph of $f(x) = e^x$ shown on the right.

x	$y = e^x$
-3	$e^{-3} \approx 0.05$
-2	$e^{-2} \approx 0.14$
-1	$e^{-1} \approx 0.37$
0	$e^0 = 1$
1	$e^1 \approx 2.72$
2	$e^2 \approx 7.39$
3	$e^3 \approx 20.09$

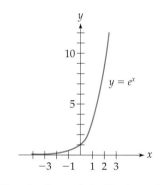

Domain of $y = e^x$ is $\mathbb{R} = (-\infty, \infty)$ and range is $(0, \infty)$.

Notice that e^x is never zero, and is positive for all values of x, even when x is negative. We restate this important observation as follows:

e to any power is positive.

The following graph shows the function $f(x) = e^{kx}$ for various values of the constant k. For positive values of k the curve rises, and for negative values of k the curve falls (as you move to the right). For higher values of k the curve rises more steeply. Each curve crosses the y-axis at 1.

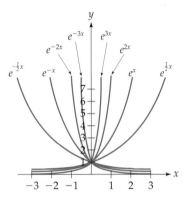

$f(x) = e^{kx}$ for various values of k.

Graphing Calculator Exploration

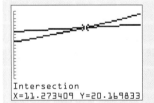

The most populous states are California and Texas, with New York third and Florida fourth but gaining. According to data from the Census Bureau, x years after 2000 the population of New York will be $19e^{0.0053x}$ and the population of Florida will be $15.9e^{0.0211x}$ (both in millions).

a. Graph these two functions on the window [0, 20] by [0, 25]. [Use the 2nd and LN keys for entering e to powers.]

b. Use INTERSECT to find the x-value where the curves intersect.

c. From your answer to part (b), in which year is Florida projected to overtake New York as the third largest state? [*Hint: x* is years after 2000.]

Exponential Growth

All exponential growth, whether continuous or discrete, has one common characteristic: The amount of growth is proportional to the size. For example, the interest that a bank account earns is proportional to the size of the account, and the growth of a population is proportional to the size of the population. This is in contrast, for example, to a person's height, which does not increase exponentially. That is, exponential growth occurs in those situations where a quantity grows *in proportion to its size*.

Justification of the Formula for Continuous Compounding

The compound interest formula Pe^{rt} is derived as follows: P dollars invested for t years at interest rate r compounded m times a year yields

$$P\left(1 + \frac{r}{m}\right)^{mt}$$ From the formula on page 66

Define $n = m/r$, so that $m = rn$. Replacing m by rn and letting m (and therefore n) approach ∞, this becomes

$$P\left(1 + \frac{r}{rn}\right)^{rnt} = P\left[\underbrace{\left(1 + \frac{1}{n}\right)^{n}}_{\text{Approaches } e \text{ as } n \to \infty}\right]^{rt} \to Pe^{rt} \qquad \text{Letting } n \to \infty$$

The limit shown on the right is the continuous compounding formula Pe^{rt}.

1.5 Section Summary

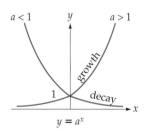

Exponential functions have exponents that involve variables. The exponential functions $f(x) = a^x$ slope *upward* or *downward* depending upon whether the base a (which must be positive) satisfies $a > 1$ or $a < 1$.

The formula $P(1 + r/m)^{mt}$ gives the value after t years of an investment of P dollars that increases at annual rate r compounded m times a year. This same formula, with a *negative* growth rate r and $m = 1$, governs depreciation by a fixed percentage.

By considering compound interest that is compounded more and more frequently, we defined a new constant e:

$$e = \lim_{n \to \infty}\left(1 + \frac{1}{n}\right)^n \approx 2.71828$$

Exponential functions with base e are used extensively in modeling many types of growth, such as the growth of populations and interest that is compounded continuously. For interest compounded *continuously*, the formula is Pe^{rt}.

> **Solutions to Practice Problems**

1. $2000\left(1 + \dfrac{0.24}{12}\right)^{12 \cdot 3} = 2000(1 + 0.02)^{36} = 2000(1.02)^{36} \approx \$4079.77$

2. $50{,}000(1 - 0.20)^4 = 50{,}000(0.8)^4 = 50{,}000(0.4096) = \$20{,}480$

1.5 Exercises

Use a calculator to evaluate, rounding to three decimal places.

1. a. e^2 b. e^{-2} c. $e^{1/2}$
2. a. e^3 b. e^{-3} c. $e^{1/3}$

Express as a power of e.

3. a. $e^5 e^{-2}$ b. $\dfrac{e^5}{e^3}$ c. $\dfrac{e^5 e^{-1}}{e^{-2} e}$

4. a. $e^{-3} e^5$ b. $\dfrac{e^4}{e}$ c. $\dfrac{e^2 e}{e^{-2} e^{-1}}$

Graph each function. If you are using a graphing calculator, make a hand-drawn sketch from the screen.

5. $y = 3^x$
6. $y = 5^x$
7. $y = \left(\dfrac{1}{3}\right)^x$
8. $y = \left(\dfrac{1}{5}\right)^x$

Calculate each value of e^x using a calculator.

9. $e^{1.74}$
10. $e^{-0.09}$

11. **e^x Versus x^n** Which curve is eventually higher, x to a power or e^x?
    a. Graph x^2 and e^x on the window [0, 5] by [0, 20]. Which curve is higher?
    b. Graph x^3 and e^x on the window [0, 6] by [0, 200]. Which curve is higher for large values of x?
    c. Graph x^4 and e^x on the window [0, 10] by [0, 10,000]. Which curve is higher for large values of x?
    d. Graph x^5 and e^x on the window [0, 15] by [0, 1,000,000]. Which curve is higher for large values of x?
    e. Do you think that e^x will exceed x^6 for large values of x? Based on these observations, can you make a conjecture about e^x and *any* power of x?

12. **Linear Versus Exponential Growth**
    a. Graph $y_1 = x$ and $y_2 = e^{0.01x}$ in the window [0, 10] by [0, 10]. Which curve is higher for x near 10?
    b. Then graph the same curves on the window [0, 1000] by [0, 1000]. Which curve is higher for x near 1000?
    A function such as y_1 represents *linear* growth, and y_2 represents *exponential* growth, and the result here is true in general: exponential growth always beats linear growth (eventually, no matter what the constants).*

---

*The realization that populations grow exponentially while food supplies grow only linearly caused the great nineteenth-century essayist Thomas Carlyle to dub economics the "dismal science." He was commenting not on how interesting economics is, but on the grim conclusions that follow from populations outstripping their food supplies.

APPLIED EXERCISES

13. **BUSINESS: Interest** Find the value of $1000 deposited in a bank at 10% interest for 8 years compounded:
    a. annually.   b. quarterly.   c. continuously.

14. **BUSINESS: Interest** Find the value of $1000 deposited in a bank at 12% interest for 8 years compounded:
    a. annually.   b. quarterly.   c. continuously.

15. **GENERAL: Interest** A loan shark lends you $100 at 2% compound interest per week (this is a *weekly*, not an annual rate).
    a. How much will you owe after 3 years?
    b. In "street" language, the profit on such a loan is known as the "vigorish" or the "vig." Find the shark's vig.

16. **GENERAL: Compound Interest** In 1626, Peter Minuit purchased Manhattan Island from the native Americans for $24 worth of trinkets and beads. Find what the $24 would be worth in the year 2000 if it had been deposited in a bank paying 5% interest compounded quarterly.

17. **PERSONAL FINANCE: Millionaires** Your rich uncle, wanting to make you a millionaire, deposited $5810 at the time of your birth in a trust fund paying 8% compounded quarterly. Will he succeed by the time you retire at age 65?

18. **PERSONAL FINANCE: College Funding** To pay for a college education, you deposit $12,000 in a tax-free education trust fund when your child is born. If the interest rate is 8% compounded daily, when your child is 18 will the fund have grown to the estimated $50,000 cost of a college education?

19. **PERSONAL FINANCE: Comparing Interest Rates** Which is better, 10% compounded quarterly or 9.9% compounded continuously? [*Hint:* Which will yield more for a deposit of $1 for one year?]

20. **PERSONAL FINANCE: Comparing Interest Rates** Which is better, 8% compounded quarterly or 7.9% compounded continuously? [*Hint:* Which will yield more for a deposit of $1 for one year?]

21. **BUSINESS: Zero-Coupon Bonds** A bond trader prices a zero-coupon bond* at $560 and the amount grows at interest rate 5.8% compounded continuously. Will it reach its "par" value of $1000 in ten years? (Continuous compounding is frequently used in bond trading.)

22. **BUSINESS: Zero-Coupon Bonds** A bond trader prices a zero-coupon bond* at $625 and the amount grows at interest rate 4.7% compounded continuously. Will it reach its "par" value of $1000 in ten years? (Continuous compounding is frequently used in bond trading.)

23. **BUSINESS: Options Trading** The Black–Scholes** formula for pricing options involves continuous compounding. If an option is now worth $2000, and its value grows at interest rate 5.5% compounded continuously, what will be its value in 8 years?

24. **BUSINESS: Options Trading** The Black–Scholes** formula for pricing options involves continuous compounding. If an option is now worth $10,000, and its value grows at interest rate 6.1% compounded continuously, what will be its value in 5 years?

25. **GENERAL: Automobile Depreciation** A $20,000 automobile depreciates by 35% each year. Find its value after:
    a. 4 years.   b. 6 months.

26. **GENERAL: Vehicle Depreciation** A $25,000 pickup truck depreciates in value by 20% each year. Find its value after:
    a. 3 years.   b. 6 months.

---

* A *zero-coupon bond* is a bond that makes no payments (coupons) until it matures, at which time it pays its "face value" of $1000. (You buy it for much less than $1000.)

** An "option" is an offer to buy or sell an asset at some time in the future. The Black–Scholes formula provides an accurate way to price options and led to the award of the Nobel Memorial Prize in 1997 to Robert Merton and Myron Scholes. (Fischer Black was deceased and therefore ineligible for the Nobel Prize.)

27. GENERAL: Population According to the United Nations Fund for Population Activities, the population of the world x years after the year 2000 will be $5.89e^{0.0175x}$ billion people (for $0 \leq x \leq 20$). Use this formula to predict the world population in the year 2010.

28. GENERAL: Population The most populous country is China, with a (2000) population of 1.26 billion, expected to grow to $1.26 \cdot 1.008^x$ billion x years later. The 2000 population of India was 1.02 billion, expected to grow to $1.02 \cdot 1.017^x$ billion x years later. According to these predictions, which population will be larger in the year 2025?

29. GENERAL: Nuclear Meltdown According to the Nuclear Regulatory Commission, the probability of a "severe core meltdown accident" at a nuclear reactor in the United States within the next n years is $1 - (0.9997)^{100n}$. Find the probability of a meltdown:

a. within 25 years.
b. within 40 years.

(The 1986 core meltdown in the Chernobyl reactor in the Soviet Union spread radiation over much of Eastern Europe, leading to an undetermined number of fatalities.)

30. GENERAL: Mosquitoes Female mosquitoes (*Culex pipiens*) feed on blood (only the females drink blood) and then lay several hundred eggs. In this way each mosquito can, on the average, breed another 300 mosquitoes in about 9 days. Find the number of great-grandchildren mosquitoes that will be descended from one female mosquito, assuming that all eggs hatch and mature.

31. ENVIRONMENTAL SCIENCE: Light According to the Bouguer–Lambert Law, the proportion of light that penetrates ordinary seawater to a depth of x feet is $e^{-0.44x}$. Find the proportion of light that penetrates to a depth of:

a. 3 feet.
b. 10 feet.

32–33. BIOMEDICAL: Drug Dosage If a dosage d of a drug is administered to a patient, the amount of the drug remaining in the tissues t hours later will be $f(t) = de^{-kt}$, where k (the "absorption constant") depends on the drug.*

32. For the immunosuppressant cyclosporine, the absorption constant is $k = 0.012$. For a dose of $d = 400$ milligrams, use the preceding formula to find the amount of cyclosporine remaining in the tissues after:

a. 24 hours.
b. 48 hours.

33. For the cardioregulator digoxin, the absorption constant is $k = 0.018$. For a dose of $d = 2$ milligrams, use the preceding formula to find the amount remaining in the tissues after:

a. 24 hours.
b. 48 hours.

34. BIOMEDICAL: Bacterial Growth A colony of bacteria in a petri dish doubles in size every hour. At noon the petri dish is just covered with bacteria. At what time was the petri dish:

a. 50% covered. [*Hint:* No calculation needed.]
b. 25% covered?

35. BUSINESS: Advertising A company finds that x days after the conclusion of an advertising campaign the daily sales of a new product are $S(x) = 100 + 800e^{-0.2x}$. Find the daily sales 10 days after the end of the advertising campaign.

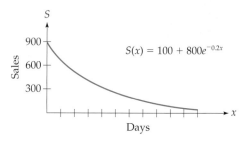

36. BUSINESS: Quality Control A company finds that the proportion of its light bulbs that will burn continuously for longer than t weeks is $e^{-0.01t}$. Find the proportion of bulbs that burn for longer than 10 weeks.

* For further details, see T. R. Harrison, ed., *Principles of Internal Medicine*, 15th ed. (New York: McGraw-Hill, 2001).

37. GENERAL: Temperature A covered mug of coffee originally at 200 degrees Fahrenheit, if left for t hours in a room whose temperature is 70 degrees, will cool to a temperature of $70 + 130e^{-1.8t}$ degrees. Find the temperature of the coffee after:
a. 15 minutes. b. half an hour.

38. BEHAVIORAL SCIENCE: Learning In certain experiments the percentage of items that are remembered after t time units is
$$p(t) = 100 \frac{1 + e}{1 + e^{t+1}}$$
Such curves are called "forgetting" curves.
Find the percentage remembered after:
a. 0 time units. b. 2 time units.

39. BIOMEDICAL: Epidemics The Reed–Frost model for the spread of an epidemic predicts that the number I of newly infected people is $I = S(1 - e^{-rx})$, where S is the number of susceptible people, r is the effective contact rate, and x is the number of infectious people. Suppose that a school reports an outbreak of measles with $x = 10$ cases, and that the effective contact rate is $r = 0.01$. If the number of susceptibles is $S = 400$, use the Reed–Frost model to estimate how many students will be newly infected during this stage of the epidemic.

40. SOCIAL SCIENCE: Election Cost The cost of winning a seat in the House of Representatives in recent years has been approximately $805e^{0.0625x}$ thousand dollars, where x is the number of years since 2000. Estimate the cost of winning a House seat in the year 2010. (*Source:* Center for Responsive Politics)

41. ATHLETICS: Olympic Games When the Olympic Games were held near Mexico City in the summer of 1968, many athletes were concerned that the high elevation would affect their performance. Air pressure decreases exponentially by 0.4% for each 100 feet of altitude, so changing the altitude by x feet means multiplying the air pressure by $(1 - 0.004)^{x/100}$. By what percentage did the air pressure decrease in moving from Tokyo (the site of the 1964 Summer Olympics, at altitude 30 feet) to Mexico City (altitude 7347 feet)?

42. BIOMEDICAL: Radioactive Contamination The core meltdown and explosions at the nuclear reactor in Chernobyl in 1986 released large amounts of strontium 90, which decays exponentially at the rate of 2.5% per year so that after x years the proportion remaining will be $(1 - 0.025)^x$. Areas downwind of the reactor will be uninhabitable for 100 years. What percent of the original strontium 90 contamination will still be present after:
a. 50 years? b. 100 years?

43. GENERAL: Population As stated earlier, the most populous state is California, with Texas second but gaining. According to the Census Bureau, x years after 2000 the population of California will be $34e^{0.013x}$ and the population of Texas will be $21e^{0.021x}$ (all in millions).
a. Graph these two functions on a calculator on the window [0, 100] by [0, 100].
b. In which year is Texas projected to overtake California as the most populous state? [*Hint:* Use INTERSECT.]

44. GENERAL: St. Louis Arch The Gateway Arch in St. Louis is built around a mathematical curve called a "catenary." The height of this catenary above the ground at a point x feet from the center line is
$$y = 688 - 31.5(e^{0.01033x} + e^{-0.01033x})$$
a. Graph this curve on a calculator on the window [−400, 400] by [0, 700].
b. Find the height of the Gateway Arch at its highest point, using the fact that the top of the arch is 5 feet higher than the top of the central catenary.

1.6 LOGARITHMIC FUNCTIONS

Introduction

In this section we will introduce *logarithmic* functions, concentrating on *common* (base 10) and *natural* (base e) logarithms. We will then use common and natural logarithms to solve problems about depreciation and carbon-14 dating. Common logarithms will be particularly useful in Chapter 2.

Common Logarithms

The word "logarithm" (abbreviated "log") means *power* or *exponent*. The number being raised to the power is called the *base* and is written as a subscript. For example, the expression

$$\log_{10} 1000 \qquad \text{Read: log (base 10) of 1000}$$

(base)

means the *exponent* to which we have to raise 10 to get 1000. Since $10^3 = 1000$, the exponent is 3, so the *logarithm* is 3.

$$\log_{10} 1000 = 3 \qquad \text{Since } 10^3 = 1000$$

Logarithms with base 10 are called *common logarithms*. For common logarithms we often omit the subscript, with base 10 understood.

Common Logarithms

$\log x = y \qquad \text{is equivalent to} \qquad 10^y = x \qquad \log x \text{ means } \log_{10} x$ (base 10)

Since $10^y = x$ is positive for every value of y, $\log x$ is defined only for *positive* values of x. The common logarithm of a number can often be found by expressing the number as a power of 10 and then taking the exponent.

EXAMPLE 1

FINDING A COMMON LOGARITHM

Evaluate $\log 100$.

Solution

$$\log 100 = y \quad \text{is equivalent to} \quad 10^y = 100 \quad \begin{array}{l} y = 2 \text{ works, so} \\ 2 \text{ is the logarithm} \end{array}$$

The logarithm y is the exponent that solves

Therefore,

$$\log 100 = 2 \qquad \text{Since} \quad 10^2 = 100$$

EXAMPLE 2

FINDING A COMMON LOGARITHM

Evaluate $\log \dfrac{1}{10}$.

Solution

$$\log \dfrac{1}{10} = y \quad \text{is equivalent to} \quad 10^y = \dfrac{1}{10} \quad \begin{array}{l} y = -1 \text{ works, so} \\ -1 \text{ is the logarithm} \end{array}$$

The logarithm y is the exponent that solves

Therefore,

$$\log \dfrac{1}{10} = -1 \qquad \text{Since} \quad 10^{-1} = \dfrac{1}{10}$$

Practice Problem 1 Evaluate $\log 10{,}000$. ▶ Solution on page 89

Graphing Calculator Exploration

```
log(100)
            2
log(1/10)
           -1
log(10000)
            4
```

On a graphing calculator, common logarithms are found using the LOG key. Verify that the last three common logarithms we considered can be found in this way.

Properties of Common Logarithms

Because logarithms are exponents, the properties of exponents can be restated as properties of logarithms. For positive numbers M and N and any number P:

Properties of Common Logarithms

1. $\log 1 = 0$ — The log of 1 is 0 (since $10^0 = 1$)
2. $\log 10 = 1$ — The log of 10 is 1 (since $10^1 = 10$)
3. $\log 10^x = x$ — The log of 10 to a power is just the power (since $10^x = 10^x$)
4. $10^{\log x} = x$ — 10 raised to the log of a number is just the number $(x > 0)$
5. $\log (M \cdot N) = \log M + \log N$ — The log of a product is the sum of the logs
6. $\log \left(\dfrac{1}{N}\right) = -\log N$ — The log of 1 over a number is minus the log of the number
7. $\log \left(\dfrac{M}{N}\right) = \log M - \log N$ — The log of a quotient is the difference of the logs
8. $\log (M^P) = P \cdot \log M$ — The log of a number to a power is the power times the log

The first two properties are simply special cases of the third (with $x = 0$ and $x = 1$). Because logs are exponents, the third property simply says that the exponent of 10 that gives 10^x is x, which is obvious when you think about it. Because $\log x$ is the power of 10 that gives x, raising 10 to that power must give x, which is the fourth property. Justifications for properties 5–8 are given on pages 87–88. Property 8 will be particularly useful in applications, and can be summarized: *Logarithms bring down exponents.*

EXAMPLE 3

USING THE PROPERTIES OF COMMON LOGARITHMS

a. $\log 10^7 = 7$ — Property 3
b. $10^{\log 13} = 13$ — Property 4
c. $\log (3 \cdot 4) = \log 3 + \log 4$ — Property 5
d. $\log \left(\dfrac{1}{4}\right) = -\log 4$ — Property 6

e. $\log\left(\frac{3}{4}\right) = \log 3 - \log 4$ Property 7

f. $\log(5^3) = 3 \log 5$ Property 8: $\log(5^3) = 3 \log 5$

EXAMPLE 4

FINDING WHEN A CAR DEPRECIATES BY HALF

A car depreciates by 20% per year. When will it be worth only half its original value?

Solution

We know from page 67 that depreciation follows the compound interest formula but with a *negative* interest rate (since it *loses* value). Since we are not told the price of the car, we let P stand for the price. The compound interest formula with negative interest rate $r = -0.20$, set equal to $0.5P$ (representing *half* the original price) gives:

$P(1 - 0.20)^t = 0.5P$	$P(1 + r/m)^{mt}$ with $r = -0.20$ and $m = 1$
$(0.80)^t = 0.5$	Canceling Ps and simplifying
$\log(0.80)^t = \log 0.5$	Taking log of each side
$t \cdot \log 0.80 = \log 0.5$	Bringing down the power (property 8)
$t = \dfrac{\log 0.5}{\log 0.80}$	Dividing by $\log 0.80$
$t \approx 3.1$	Using a calculator

The car will be worth half its value in about 3.1 years.

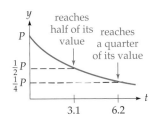

Incidentally, in another 3.1 years the car will again fall to half its value, thereby reaching *one quarter* of its original value in 6.2 years. This halving of value will continue every 3.1 years.

Graphs of Logarithmic and Exponential Functions

If a point (x, y) lies on the graph of $y = \log x$ or, equivalently, $x = 10^y$, then, reversing x and y, the point (y, x) lies on the graph of $y = 10^x$. That is, the curves $y = \log x$ and $y = 10^x$ are related by having their x- and y-coordinates *reversed*, so the curves are *mirror images* of each other across the line $y = x$.

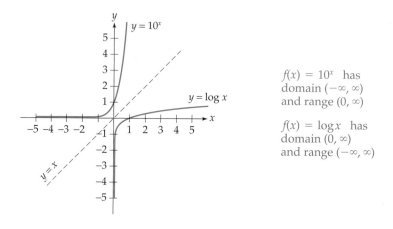

$f(x) = 10^x$ has domain $(-\infty, \infty)$ and range $(0, \infty)$

$f(x) = \log x$ has domain $(0, \infty)$ and range $(-\infty, \infty)$

This graphical relationship is equivalent to the fact that the functions $\log x$ and 10^x "undo" or "reverse" each other, as is shown by properties 3 and 4 on page 80. Such functions are called *inverse functions*:

$$\log x \quad \text{and} \quad 10^x \quad \text{are inverse functions}$$

Logarithms to Other Bases

We may calculate logarithms to bases other than 10. In fact, *any* positive number other than 1 may be used as a base for logs, using the following definition.

Base a Logarithms

$$\log_a x = y \quad \text{is equivalent to} \quad a^y = x \quad (x > 0)$$

For example,

$$\log_2 8 = 3 \quad\quad \text{Since } 2^3 = 8$$

and

$$\log_9 3 = \tfrac{1}{2} \quad\quad \text{Since } 9^{1/2} = \sqrt{9} = 3$$

Natural Logarithms

We will use only one other base, the number e (approximately 2.718) that we defined on page 70. Logarithms to the base e are called *natural* or *Napierian* logarithms.* The natural logarithm of a positive number x is written $\ln x$ ("n" for "natural") and may be found using the $\boxed{\text{LN}}$ key on a calculator.

Natural Logarithms

$\ln x = y$ is equivalent to $e^y = x$ $\ln x$ means $\log_e x$ (base e)

Practice Problem 2   Use a calculator to find $\ln 8.34$.   Solution on page 89

The properties of natural logarithms are similar to those of common logarithms but with ln instead of log and e instead of 10. For positive numbers M and N and any number P:

Properties of Natural Logarithms

1. $\ln 1 = 0$ — The natural log of 1 is 0 (since $e^0 = 1$)
2. $\ln e = 1$ — The natural log of e is 1 (since $e^1 = e$)
3. $\ln e^x = x$ — The natural log of e to a power is just the power (since $e^x = e^x$)
4. $e^{\ln x} = x$ — e raised to the natural log of a number is just the number $(x > 0)$
5. $\ln(M \cdot N) = \ln M + \ln N$ — The natural log of a product is the sum of the logs
6. $\ln\left(\dfrac{1}{N}\right) = -\ln N$ — The natural log of 1 over a number is minus the log of the number
7. $\ln\left(\dfrac{M}{N}\right) = \ln M - \ln N$ — The natural log of a quotient is the difference of the logs
8. $\ln(M^P) = P \cdot \ln M$ — The natural log of a number to a power is the power times the log

As with common logs, the first two properties are special cases of the third (with $x = 0$ and $x = 1$). The third and fourth properties have interpretations analogous to their "common" counterparts (for example, the third says that the exponent of e that gives e^x is x). As before, property 8 can be summarized: *Logarithms bring down exponents.*

---

*After John Napier (1550–1617), a Scottish mathematician who, incidentally, invented the decimal point.

EXAMPLE 5 — USING THE PROPERTIES OF NATURAL LOGARITHMS

a. $\ln e^7 = 7$ — Property 3

b. $e^{\ln 13} = 13$ — Property 4

c. $\ln (3 \cdot 4) = \ln 3 + \ln 4$ — Property 5

d. $\ln \left(\dfrac{1}{4}\right) = -\ln 4$ — Property 6

e. $\ln \left(\dfrac{3}{4}\right) = \ln 3 - \ln 4$ — Property 7

f. $\ln (5^3) = 3 \ln 5$ — Property 8: $\ln(5^3) = 3 \ln 5$

Graphing Calculator Exploration

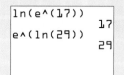

a. Evaluate $\ln e^{17}$ to verify that the answer is 17. Change 17 to other numbers (positive, negative, or zero) in order to verify that $\ln e^x = x$ for any x.

b. Evaluate $e^{\ln 29}$ to verify that the answer is 29. Change the 29 to another positive number to verify that $e^{\ln x} = x$. What about negative numbers, or zero?

Properties 3 and 4 show that $y = \ln x$ and $y = e^x$ are *inverse functions*, so their graphs are reflections of each other in the diagonal line $y = x$.

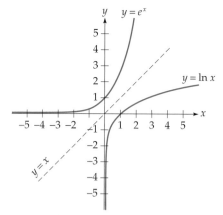

$f(x) = e^x$ has domain $(-\infty, \infty)$ and range $(0, \infty)$

$f(x) = \ln x$ has domain $(0, \infty)$ and range $(-\infty, \infty)$

e^x and $\ln x$ are inverse functions

The properties of natural logarithms are helpful for simplifying functions.

Graphing Calculator Exploration

a. Graph the function $y = \ln x^2$ on the window $[-5, 5]$ by $[-5, 5]$.
b. Change the function to $y = 2 \ln x$ and explain why the two graphs are different. (Doesn't property 8 say that the two functions should be the same?)

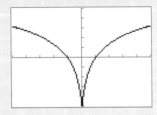 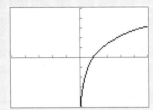

EXAMPLE 6   SIMPLIFYING A FUNCTION

$$f(x) = \ln(2x) - \ln 2$$
$$= \ln 2 + \ln x - \ln 2 \qquad \text{Since } \ln(2x) = \ln 2 + \ln x \text{ by property 5}$$
$$= \ln x \qquad \text{Canceling}$$

EXAMPLE 7   SIMPLIFYING A FUNCTION

$$f(x) = \ln\left(\frac{x}{e}\right) + 1$$
$$= \ln x - \ln e + 1 \qquad \text{Since } \ln(x/e) = \ln x - \ln e \text{ by property 7}$$
$$= \ln x - 1 + 1 \qquad \text{Since } \ln e = 1 \text{ by property 2}$$
$$= \ln x \qquad \text{Canceling}$$

EXAMPLE 8 SIMPLIFYING A FUNCTION

$$f(x) = \ln(x^5) - \ln(x^3) = 5\ln x - 3\ln x \quad \text{Bringing down exponents by property 8}$$

$$= 2\ln x \quad \text{Combining}$$

Graphing Calculator Exploration

Some advanced graphing calculators have computer algebra systems that can simplify algebraic expressions. For example, the Texas Instruments TI-89 graphing calculator simplifies logarithmic expressions, but only if the condition $x > 0$ is included so that the logarithms are defined.

Entered →

- expand(ln(x^5) − ln(x))
  - ln(x^5) − ln(x) ← Not simplified;
- expand(ln(x^5) − ln(x)) | x > 0 ← but if we require $x > 0$,
  - 4·ln(x) ← it is simplified!

Carbon-14 Dating

All living things absorb small amounts of radioactive carbon-14 from the atmosphere. When they die, the carbon-14 stops being absorbed and decays exponentially into ordinary carbon. Therefore, the proportion of carbon-14 still present in a fossil or other ancient remain can be used to estimate how old it is. The proportion of the original carbon-14 that will be present after t years is

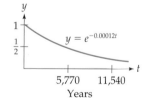

$y = e^{-0.00012t}$

$$\left(\begin{array}{c}\text{Proportion of carbon-14}\\\text{remaining after } t \text{ years}\end{array}\right) = e^{-0.00012t}$$

EXAMPLE 9

DATING BY CARBON-14

The Dead Sea Scrolls, discovered in a cave near the Dead Sea in what was then Jordan, are among the earliest documents of Western civilization. Estimate the age of the Dead Sea Scrolls if the animal skins on which some were written contain 78% of their original carbon-14.

Solution

The proportion of carbon-14 remaining after t years is $e^{-0.00012t}$. We equate this formula to the actual proportion (expressed as a decimal).

$$e^{-0.00012t} = 0.78 \qquad \text{Equating the proportions}$$

$$\ln e^{-0.00012t} = \ln 0.78 \qquad \text{Taking natural logs}$$

$$-0.00012t = \ln 0.78 \qquad \begin{array}{l}\ln e^{-0.00012t} = -0.00012t \\ \text{by property 3}\end{array}$$

$$t = \frac{\ln 0.78}{-0.00012} \approx \frac{-0.24846}{-0.00012} \approx 2071 \qquad \begin{array}{l}\text{Solving for } t \text{ and using a} \\ \text{calculator}\end{array}$$

Therefore, the Dead Sea Scrolls are approximately 2070 years old.

Graphing Calculator Exploration

Solve Example 9 by graphing $y_1 = e^{-0.00012x}$ and $y_2 = 0.78$ on the window [0, 10,000] by [0, 1] and using INTERSECT to find where they meet.

Both here and in Example 4 on page 81 we solved for the variable in the exponent by taking logarithms, but here we used *natural* logarithms and there we used *common* logarithms. Is there a difference? Not really—you can use logarithms to *any* base to solve for exponents, and the final answer will be the same (see Exercise 37). However, when e is involved, it is usually easier to use *natural* logs.

Justification of Properties 5–8 of Logarithms

Properties 1–4 of logarithms (both common and natural) were justified earlier. Properties 5–8 of common logarithms are justified as follows, with the "natural" justifications obtained simply by replacing 10 by e.

The addition law of exponents, $10^x \cdot 10^y = 10^{x+y}$, can be stated in words: "The exponent of a product is the sum of the exponents." Because logs are exponents, this can be restated as "The log of a product is the sum of the logs," which is just property 5.

The subtraction law of exponents, $10^x / 10^y = 10^{x-y}$ can be stated, "The exponent of a quotient is the difference of the exponents." This translates into "The log of a quotient is the difference of the logs," which is just property 7. Property 6 is simply a special case of property 7 with $M = 1$ (and using property 1).

The law of exponents $(10^x)^y = 10^{x \cdot y}$ says that the exponent y can be "brought down" and multiplied by the x. Because logs are exponents, this says that in $\log(M^P)$, the exponent P can be brought down and multiplied by the logarithm $\log M$, giving $P \cdot \log M$, which is just property 8.

1.6 Section Summary

Logarithms are exponents: Common logs are exponents of 10, and natural logs are exponents of e. That is,

$y = \log x$ is equivalent to $x = 10^y$ $\log x$ means $\log_{10} x$ (base 10)

$y = \ln x$ is equivalent to $x = e^y$ $\ln x$ means $\log_e x$ (base e)

Each property of logarithms is equivalent to a property of exponents.

Logarithmic Property	*Exponential Property*
$\log 1 = 0$	$10^0 = 1$
$\log(M \cdot N) = \log M + \log N$	$10^x \cdot 10^y = 10^{x+y}$
$\log\left(\dfrac{1}{N}\right) = -\log N$	$\dfrac{1}{10^y} = 10^{-y}$
$\log\left(\dfrac{M}{N}\right) = \log M - \log N$	$\dfrac{10^x}{10^y} = 10^{x-y}$
$\log(M^P) = P \cdot \log M$	$(10^x)^y = 10^{x \cdot y}$

In practice, logarithms are found using the LOG and LN keys on a calculator. The properties of exponents lead to properties of logarithms, which are listed on pages 80 and 83. Property 8, that *logs bring down exponents*, is particularly useful in applications that require solving for a variable in the exponent.

> **Solutions to Practice Problems**

1. $\log 10{,}000 = 4$ (Since $10^4 = 10{,}000$)
2. $\ln 8.34 \approx 2.121$ (Using a calculator)

1.6 Exercises

Find each logarithm *without* using a calculator.

1. a. $\log 100{,}000$ b. $\log \frac{1}{100}$ c. $\log \sqrt{10}$
2. a. $\log 1000$ b. $\log \frac{1}{1000}$ c. $\log \sqrt[3]{10}$
3. a. $\ln e^5$ b. $\ln \frac{1}{e}$ c. $\ln \sqrt[3]{e}$
4. a. $\ln e^3$ b. $\ln \frac{1}{e^2}$ c. $\ln \sqrt{e}$
5. a. $\ln 1$ b. $\ln (\ln e^e)$ c. $\ln \sqrt[3]{e^2}$
6. a. $\ln e$ b. $\ln (\ln e)$ c. $\ln \sqrt{e^3}$
7. a. $\log_4 16$ b. $\log_4 \frac{1}{4}$ c. $\log_4 2$
8. a. $\log_9 81$ b. $\log_9 3$ c. $\log_9 \frac{1}{9}$

For each logarithm:
i. Find the logarithm using a calculator, rounding your answer to three decimal places.
ii. Raise the appropriate base (either 10 or e) to the power you found in part (i) and check that the result agrees with the number in the original problem. (*Note:* For the power, include *all* of the digits that your calculator showed for part (i) to minimize the error.)

9. a. $\log 22.3$ b. $\ln 22.3$
10. a. $\log 44.9$ b. $\ln 44.9$

Use the properties of natural logarithms to simplify each function.

11. $f(x) = \ln (9x) - \ln 9$ 12. $f(x) = \ln \left(\frac{x}{2}\right) + \ln 2$
13. $f(x) = \ln (x^3) - \ln x$ 14. $f(x) = \ln (4x) - \ln 4$
15. $f(x) = \ln \left(\frac{x}{4}\right) + \ln 4$
16. $f(x) = \ln (x^5) - 3 \ln x$
17. $f(x) = \ln (e^{5x}) - 2x - \ln 1$
18. $f(x) = \ln (e^{-2x}) + 3x + \ln 1$
19. $f(x) = 8x - e^{\ln x}$
20. $f(x) = e^{\ln x} + \ln (e^{-x})$

21. Without using a calculator, sketch the graph of $f(x) = \log (x + 1)$.
22. Without using a calculator, sketch the graph of $f(x) = \ln (x + e)$.
23. Find the domain and range and graph the function $f(x) = \ln (x^2 - 1)$.
24. Find the domain and range and graph the function $f(x) = \ln (1 - x^2)$.

APPLIED EXERCISES

25. **GENERAL: Depreciation** A car depreciates by 30% per year. When will it be worth only:
    a. half its original value?
    b. one quarter its original value?

26. **GENERAL: Depreciation** A truck depreciates by 25% per year. When will it be worth only:
    a. half its original value?
    b. one quarter its original value?

27. **BUSINESS: Depreciation** An industrial printing press depreciates by 15% per year. When will it be worth two thirds of its original value?

28. BUSINESS: Depreciation
A supercomputer depreciates by 40% per year. When will it be worth only one tenth of its original value?

29–30. GENERAL: Carbon-14 Dating
The proportion of carbon-14 still present in a sample after t years is $e^{-0.00012t}$.

29. Use the preceding formula to estimate the age of the cave paintings discovered in 1994 in the Ardèche region of France if the carbon with which they were drawn contains only 2.3% of its original carbon-14. They are the oldest known paintings in the world.

30. Use the preceding formula to estimate the age of the Shroud of Turin, believed by many to be the burial cloth of Christ, from the fact that its linen fibers contained only 92.3% of their original carbon-14.

31–32. GENERAL: Potassium-40 Dating
The radioactive isotope potassium-40 is used to date very old remains. The proportion of potassium-40 that remains after t million years is $e^{-0.00054t}$. Use this function to estimate the age of the following fossils.

31. The most nearly complete skeleton of an early human ancestor ever found was discovered in Kenya in 1984. Use the above formula to estimate the age of the remains if they contained 99.91% of their original potassium-40.

32. Dating Older Women Use the above formula to estimate the age of the partial skeleton of *Australopithecus afarensis*, known as "Lucy," which was found in Ethiopia in 1977, if it had 99.82% of its original potassium-40.

33–34. BUSINESS: Depreciation
Solve the following exercises on a graphing calculator by graphing a function for the depreciated value (using x for ease of entry) together with a constant function and using INTERSECT to find where they meet. You will have to choose an appropriate window.

33. A company jet that originally cost $5 million depreciates by 25% per year. When will it be worth only $500,000, at which time it can be written off as a tax loss?

34. A $400,000 fire engine depreciates by 15% per year. When will it be worth only $40,000, at which time it will be sold for scrap?

35. GENERAL: Potassium-40 Dating Estimate the age of the oldest known dinosaur, a dog-sized creature called *Herrerasaurus* found in Argentina in 1988, if volcanic material found with it contained 88.4% of its original potassium-40. (Use the potassium-40 decay function given in the directions for Exercises 31–32.)

36. GENERAL: Carbon-14 Dating In 1991 two hikers in the Italian Alps found the frozen but well-preserved body of the most ancient human ever found, dubbed "Iceman." Estimate the age of Iceman if his grass cape contained 53% of its original carbon-14. (Use the carbon-14 decay function stated in the directions for Exercises 29–30.)

Explorations and Excursions

The following problem extends and augments the material presented in the text.

Change-of-Base Formula for Logarithms

37. Let a and b be any two bases, and let x be any positive number.

a. Give a justification for each numbered equals sign.
$$\log_a x \stackrel{1}{=} \log_a b^{\log_b x} \stackrel{2}{=} (\log_b x) \cdot (\log_a b)$$

b. Show that the result can be written as
$$\log_a x = (\log_a b) \cdot (\log_b x)$$
This is the "change-of-base" formula for logarithms, enabling one to express $\log_b x$ in terms of $\log_a x$.

c. Use the change-of-base formula in the numerator and denominator of the following fraction to justify each numbered equals sign.
$$\frac{\log_a x}{\log_a y} \stackrel{3}{=} \frac{(\log_a b) \cdot (\log_b x)}{(\log_a b) \cdot (\log_b y)} \stackrel{4}{=} \frac{\log_b x}{\log_b y}$$

This equation shows that when you are finding a *ratio* of logarithms, using one base gives the same result as using any other base. Because solving for a variable in the exponent involves calculating *ratios* of logarithms, you may do so using logarithms to *any* base.

Chapter Summary with Hints and Suggestions

Reading the text and doing the exercises in this chapter have helped you to master the following skills, which are listed by section (in case you need to review them) and are keyed to particular Review Exercises. Answers for all Review Exercises are given at the back of the book, and full solutions can be found in the Student Solutions Manual.

1.1 Real Numbers, Inequalities, and Lines

- Translate an interval into set notation and graph it on the real line. *(Review Exercises 1–4.)*

$$[a, b] \quad (a, b) \quad [a, b) \quad (a, b]$$
$$(-\infty, b] \quad (-\infty, b) \quad [a, \infty) \quad (a, \infty) \quad (-\infty, \infty)$$

- Express given information in interval form. *(Review Exercises 5–6.)*

- Find an equation for a line that satisfies certain conditions. *(Review Exercises 7–12.)*

$$m = \frac{y_2 - y_1}{x_2 - x_1} \quad y = mx + b$$
$$y - y_1 = m(x - x_1) \quad x = a \quad y = b$$
$$ax + by = c$$

- Find an equation of a line from its graph. *(Review Exercises 13–14.)*

- Use straight-line depreciation to find the value of an asset. *(Review Exercises 15–16.)*

- Use real-world data to find a regression line and make a prediction. *(Review Exercises 17–18.)*

1.2 Exponents

- Evaluate negative and fractional exponents without a calculator. *(Review Exercises 19–26.)*

$$x^0 = 1 \quad x^{-n} = \frac{1}{x^n} \quad x^{m/n} = \sqrt[n]{x^m} = \left(\sqrt[n]{x}\right)^m$$

- Evaluate an exponential expression using a calculator. *(Review Exercises 27–28.)*

1.3 Functions

- Evaluate and find the domain and range of a function. *(Review Exercises 29–32.)*

  A function f is a rule that assigns to each number x in a set (the domain) a (single) number $f(x)$. The range is the set of all resulting values $f(x)$.

- Use the vertical line test to see if a graph defines a function. *(Review Exercises 33–34.)*

- Graph a linear function. *(Review Exercises 35–36.)*

$$f(x) = mx + b$$

- Graph a quadratic function. *(Review Exercises 37–38.)*

$$f(x) = ax^2 + bx + c$$

- Solve a quadratic equation by factoring and by the Quadratic Formula. *(Review Exercises 39–42.)*

  Vertex   **x-intercepts**

  $$x = \frac{-b}{2a} \qquad x = \frac{-b \pm \sqrt{b^2 - 4ac}}{2a}$$

- Use a graphing calculator to graph a quadratic function. *(Review Exercises 43–44.)*

- Construct a linear function from a word problem or from real-life data, and then use the function in an application. *(Review Exercises 45–48.)*

- For given cost and revenue functions, calculate the break-even points and maximum profit. *(Review Exercises 49–50.)*

1.4 Functions, Continued

- Evaluate and find the domain and range of a more complicated function.
  (*Review Exercises 51–54.*)

- Solve a polynomial equation by factoring.
  (*Review Exercises 55–58.*)

- Graph a "shifted" function.
  (*Review Exercises 59–60.*)

- Graph a piecewise linear function.
  (*Review Exercises 61–62.*)

- Given two functions, find their composition.
  (*Review Exercises 63–66.*)

$$f(g(x)) \quad g(f(x))$$

- For a given function $f(x)$, find and simplify the difference quotient $\dfrac{f(x+h) - f(x)}{h}$.
  (*Review Exercises 67–68.*)

- Solve an applied problem involving the composition of functions. (*Review Exercise 69.*)

- Solve a polynomial equation.
  (*Review Exercises 70–71.*)

- Fit a curve to real-life data and make a prediction. (*Review Exercise 72.*)

1.5 Exponential Functions

- Sketch the graph of an exponential function.
  (*Review Exercises 73–74.*)

- Find the value of money invested at compound interest. (*Review Exercises 75–76.*)

$$P\left(1 + \frac{r}{m}\right)^{mt}$$

- Depreciate an asset by a fixed percentage per year. (*Review Exercises 77–78.*)

$$\left(\text{Above formula with } \begin{array}{c} m = 1 \\ \text{and a } negative \text{ value for } r \end{array}\right)$$

- Find the value of a deposit invested with continuous compounding.
  (*Review Exercises 79–80.*)

$$Pe^{rt}$$

1.6 Logarithmic Functions

- Evaluate common and natural logarithms *without* using a calculator.
  (*Review Exercises 81–82.*)

$$\log 1 = 0 \quad \log 10 = 1 \quad \log 10^x = x \quad 10^{\log x} = x$$

$$\log(M \cdot N) = \log M + \log N$$

$$\log\left(\frac{M}{N}\right) = \log M - \log N$$

$$\log\left(\frac{1}{N}\right) = -\log N \quad \log(M^P) = P \cdot \log M$$

$$\ln 1 = 0 \quad \ln e = 1 \quad \ln e^x = x \quad e^{\ln x} = x$$

$$\ln(M \cdot N) = \ln M + \ln N$$

$$\ln\left(\frac{M}{N}\right) = \ln M - \ln N$$

$$\ln\left(\frac{1}{N}\right) = -\ln N \quad \ln(M^P) = P \cdot \ln M$$

- Use the properties of natural logarithms to simplify a function. (*Review Exercises 83–84.*)

- Determine when an asset depreciates to a fraction of its value. (*Review Exercises 85–86.*)

- Estimate the age of a fossil.
  (*Review Exercises 87–88.*)

Hints and Suggestions

- (*Overview*) In reviewing this chapter, notice the difference between *geometric* objects (points, curves, etc.) and *analytic* objects (numbers, functions, etc.), and the connections between them. Descartes first made this connection: by drawing axes, he saw that points could be specified by numerical coordinates, and so *curves* could be specified by *equations* governing their coordinates. This idea connected geometry to algebra, previously distinct subjects. You should be able to express geometric objects analytically, and vice versa. For example, given a *graph* of a line, you should be able to find an *equation* for it, and given a quadratic *function*, you should be able to *graph* it.

- A graphing calculator or a computer with appropriate software can help you to *explore* a concept more fully (for example, seeing how a curve changes as a coefficient or exponent changes) and also to *solve* a problem (for example, eliminating the point-plotting aspect of graphing, or finding a regression line).

- If you don't have a graphing calculator, you should have a scientific or business calculator to carry out calculations, especially in later chapters.
- The Practice Problems help you to check your mastery of the skills presented. Complete solutions are given at the end of each section.
- The Student Solutions Manual, available separately from your bookstore, provides fully worked-out solutions to selected exercises.
- Interest rates are always *annual* (unless clearly stated otherwise) but the *compounding* may be done more frequently (in the formula, m times a year).
- When do you use the formula $P(1 + r/m)^{mt}$ and when do you use Pe^{rt}? Use Pe^{rt} if the word "continuous" occurs and use $P(1 + r/m)^{mt}$ if it does not.
- The formula Pe^{rt} has no m because there is no period to wait for the interest to be compounded—interest is added to the account *continuously*.
- For depreciation, use $P(1 + r/m)^{mt}$ with a *negative* value for r. For example, depreciating by 15% annually would mean $m = 1$ and $r = -0.15$.
- **Practice for Test:** Review Exercises, 1, 9, 11, 13, 15, 17, 19, 31, 33, 35, 37, 43, 47, 49, 55, 61, 65, 67, 71, 75, 77, 79, 81, 83, 85, and 87.

Review Exercises for Chapter 1

Practice test exercise numbers are in **green**.

1.1 Real Numbers, Inequalities, and Lines

Write each interval in set notation and graph it on the real line.

1. $(2, 5]$
2. $[-2, 0)$
3. $[100, \infty)$
4. $(-\infty, 6]$

5. **GENERAL: Wind Speed** The United States Coast Guard defines a "hurricane" as winds of at least 74 mph, a "storm" as winds of at least 55 mph but less than 74 mph, a "gale" as winds of at least 38 mph but less than 55 mph, and a "small craft warning" as winds of at least 21 mph but less than 38 mph. Express each of these wind conditions in interval form. [*Hint:* A small craft warning is [21, 38).]

6. State in interval form:
   a. The set of all positive numbers.
   b. The set of all negative numbers.
   c. The set of all nonnegative numbers.
   d. The set of all nonpositive numbers.

Write an equation of the line satisfying each of the following conditions. If possible, write your answer in the form $y = mx + b$.

7. Slope 2 and passing through the point $(1, -3)$
8. Slope -3 and passing through the point $(-1, 6)$
9. Vertical and passing through the point $(2, 3)$
10. Horizontal and passing through the point $(2, 3)$
11. Passing through the points $(-1, 3)$ and $(2, -3)$
12. Passing through the points $(1, -2)$ and $(3, 4)$

Write an equation of the form $y = mx + b$ for each line graphed below.

13.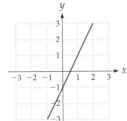

14.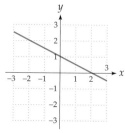

15. **BUSINESS: Straight-Line Depreciation** A contractor buys a backhoe for $25,000 and estimates its useful life to be 8 years, after which its scrap value will be $1000.
    a. Use straight-line depreciation to find a formula for the value V of the backhoe after t years, for $0 \leq t \leq 8$.
    b. Use your formula to find the value of the backhoe after 4 years.

16. **BUSINESS: Straight-Line Depreciation** A trucking company buys a satellite communication system for $78,000 and estimates its useful life to be 15 years, after which its scrap value will be $3000.
    a. Use straight-line depreciation to find a formula for the value V of the system after t years, for $0 \leq t \leq 15$.
    b. Use your formula to find the value of the system after 8 years.

17. **ECOLOGY: Sulfur Dioxide Pollution** Sulfur dioxide pollution has decreased significantly in the United States during the last 25 years, mostly because of antipollution devices on automobiles and on coal- and oil-fired power plants. The following table shows sulfur dioxide emissions (in millions of tons) in the United States from 1975 to 2000. To avoid large numbers, years are listed in the table as years since 1975.

Years Since 1975	Sulfur Dioxide Emissions
1975	0	26.0
1980	5	23.5
1985	10	21.6
1990	15	19.3
1995	20	18.2
2000	25	18.1

Sources: Worldwatch Institute and U.S. Environmental Protection Agency

a. Enter the table numbers into a graphing calculator and make a plot of the resulting points (Years Since 1975 on the x-axis and Sulfur Dioxide Emissions on the y-axis).
b. Have your calculator find the linear regression formula for these data. Then enter the result in y_1, which gives a formula for sulfur dioxide emissions in each year. Plot the points together with the regression line. How well does the line fit the data?
c. Use your formula to predict the sulfur dioxide pollution in the year 2010 (assuming that the past trend continues).

18. **SOCIAL SCIENCE: Gap Between Rich and Poor** During the last few decades, the richest 20% of the world's people have been growing richer, while the poorest 20% have been growing poorer. The probable consequences of this growing gap are not only social instability but also environmental decline, since the richest consume more wastefully while the poorest must cut down rainforests and overgraze land just to survive. The following table shows the ratio of income of the richest 20% to the poorest 20% from 1960 to 2000. To avoid large numbers, years are listed in the table as years since 1960.

	Years Since 1960	Ratio of Richest to Poorest
1960	0	30 to 1
1970	10	32 to 1
1980	20	45 to 1
1990	30	60 to 1
2000	40	72 to 1

Source: United Nations Development Programme

a. Enter the table numbers into a graphing calculator and make a plot of the resulting points [Years Since 1960 on the x-axis and the larger number in the Ratio column (the 30, 32, etc.) on the y-axis].
b. Have your calculator find the linear regression formula for these data. Then enter the result in y_1, which gives a formula for the ratio of richest to poorest in each year. Plot the points together with the regression line. How well does the line fit the data?

c. Use your formula to predict the ratio in the year 2010 (assuming that the past trend continues).

1.2 Exponents

Evaluate each expression without using a calculator.

19. $\left(\frac{1}{6}\right)^{-2}$
20. $\left(\frac{4}{3}\right)^{-1}$
21. $64^{1/2}$
22. $1000^{1/3}$
23. $81^{-3/4}$
24. $100^{-3/2}$
25. $\left(-\frac{8}{27}\right)^{-2/3}$
26. $\left(\frac{9}{16}\right)^{-3/2}$

Use a calculator to evaluate each expression. Round answers to two decimal places.

27. $3^{2.4}$
28. $12^{1.9}$

1.3 Functions

For each function in Exercises 29–32:
a. Evaluate the given expression.
b. Find the domain.

c. Find the range.

29. $f(x) = \sqrt{x - 7}$; find $f(11)$
30. $g(t) = \dfrac{1}{t + 3}$; find $g(-1)$
31. $h(w) = w^{-3/4}$; find $h(16)$
32. $w(z) = z^{-4/3}$; find $w(8)$

Determine whether each graph defines a function of x.

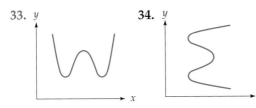

Graph each function.
35. $f(x) = 4x - 8$
36. $f(x) = 6 - 2x$
37. $f(x) = -2x^2 - 4x + 6$
38. $f(x) = 3x^2 - 6x$

Solve each equation by
a. factoring b. the Quadratic Formula.

39. $3x^2 + 9x = 0$
40. $2x^2 - 8x - 10 = 0$
41. $3x^2 + 3x + 5 = 11$
42. $4x^2 - 2 = 2$

For each quadratic function in Exercises 43–44:
a. Find the vertex using the vertex formula.
b. Graph the function on an appropriate viewing window. (Answers may vary.)

43. $f(x) = x^2 - 10x - 25$
44. $f(x) = x^2 + 14x - 15$

45. **BUSINESS: Car Rentals** A rental company rents cars for $45 per day and $0.12 per mile. Find a function $C(x)$ for the cost of a rented car driven for x miles in a day.

46. **BUSINESS: Simple Interest** If money is borrowed for a short period of time, generally less than a year, the interest is often calculated as *simple* interest, according to the formula Interest $= P \cdot r \cdot t$, where P is the principal, r is the rate (expressed as a decimal), and t is the time (in years). Find a function $I(t)$ for the interest charged on a loan of $10,000 at an interest rate of 8% for t years. Simplify your answer.

47. **GENERAL: Air Temperature** The air temperature decreases by about 1 degree Fahrenheit for each 300 feet of altitude. Find a function $T(x)$ for the temperature at an altitude of x feet if the sea level temperature is 70°.

48. **ECOLOGY: Carbon Dioxide Pollution** The burning of fossil fuels (such as oil and coal) added 27 billion tons of carbon dioxide to the atmosphere during 2000, and this annual amount is growing by 0.58 billion tons per year. Find a function $C(t)$ for the amount of carbon dioxide added during the year t years after 2000, and use the formula to find how soon this annual amount will reach 30 billion tons. [*Note:* Carbon dioxide traps solar heat, increasing the earth's temperature, and may lead to flooding of lowland areas by melting the polar ice.]
Source: U.S. Environmental Protection Agency.

49. **BUSINESS: Break-Even Points and Maximum Profit** A store that installs satellite TV receivers finds that if it installs x receivers per week, then its costs will be $C(x) = 80x + 1950$

and its revenue will be $R(x) = -2x^2 + 240x$ (both in dollars).

a. Find the store's break-even points.
b. Find the number of receivers the store should install to maximize profit, and the maximum profit.

50. BUSINESS: Break-Even Points and Maximum Profit An air conditioner outlet finds that if it sells x air conditioners per month, its costs will be $C(x) = 220x + 202{,}500$ and its revenue will be $R(x) = -3x^2 + 2020x$ (both in dollars).

a. Find the outlet's break-even points.
b. Find the number of air conditioners the outlet should sell to maximize profit, and the maximum profit.

1.4 Functions, Continued

For each function in Exercises 51–54:
a. Evaluate the given expression.
b. Find the domain.
c. Find the range.

51. $f(x) = \dfrac{3}{x(x-2)}$; find $f(-1)$

52. $f(x) = \dfrac{16}{x(x+4)}$; find $f(-8)$

53. $g(x) = |x+2| - 2$; find $g(-4)$

54. $g(x) = x - |x|$; find $g(5)$

Solve each equation by factoring.

55. $5x^4 + 10x^3 = 15x^2$ **56.** $4x^5 + 8x^4 = 32x^3$

57. $2x^{5/2} - 8x^{3/2} = 10x^{1/2}$

58. $3x^{5/2} + 3x^{3/2} = 18x^{1/2}$

Graph each function.

59. $f(x) = (x+1)^2 - 1$ **60.** $f(x) = (x-2)^2 - 4$

61. $f(x) = \begin{cases} 3x - 7 & \text{if } x \geq 2 \\ -x - 1 & \text{if } x < 2 \end{cases}$

(If you use a graphing calculator for Exercises 61 and 62, be sure to indicate any missing points.)

62. $f(x) = \begin{cases} 6 - 2x & \text{if } x > 2 \\ 2x - 1 & \text{if } x \leq 2 \end{cases}$

For each pair of functions $f(x)$ and $g(x)$, find
a. $f(g(x))$, **b.** $g(f(x))$.

63. $f(x) = x^2 + 1$; $g(x) = \dfrac{1}{x}$

64. $f(x) = \sqrt{x}$; $g(x) = 5x - 4$

65. $f(x) = \dfrac{x+1}{x-1}$; $g(x) = x^3$

66. $f(x) = |x|$; $g(x) = x + 2$

For each function, find and simplify the difference quotient $\dfrac{f(x+h) - f(x)}{h}$.

67. $f(x) = 2x^2 - 3x + 1$

68. $f(x) = \dfrac{5}{x}$

69. BUSINESS: Advertising Budget A company's advertising budget is $A(p) = 2p^{0.15}$, where p is the company's profit, and the profit is predicted to be $p(t) = 18 + 2t$, where t is the number of years from now. (Both A and p are in millions of dollars.) Express the advertising budget A as a function of t, and evaluate the function at $t = 4$.

70. a. Solve the equation $x^4 - 2x^3 - 3x^2 = 0$ by factoring.
b. Use a graphing calculator to graph $y = x^4 - 2x^3 - 3x^2$ and find the x-intercepts of the graph. Be sure that you understand why your answers to parts (a) and (b) agree.

71. a. Solve the equation $x^3 + 2x^2 - 3x = 0$ by factoring.
b. Use a graphing calculator to graph $y = x^3 + 2x^2 - 3x$ and find the x-intercepts of the graph. Be sure that you understand why your answers to parts (a) and (b) agree.

72. BUSINESS: Revenue The following table gives a company's annual revenue (in millions of dollars) from its overseas operations during its first 5 years.

Year	Revenue
1	2.0
2	1.8
3	1.9
4	2.1
5	2.8

a. Enter the table numbers into a graphing calculator and make a plot of the resulting points (Years on the x-axis and Revenue on the y-axis). What kind of curve do the points suggest?
b. Have your calculator fit such a curve to the data. Then enter the result in y_1, which gives a formula for the annual revenue each year. Plot the points together with the regression curve.
c. Use your formula to predict the revenue in years 6 and 7.

1.5 Exponential Functions

Graph each function.

73. $f(x) = 4^x$
74. $f(x) = \left(\frac{1}{4}\right)^x$

75. **PERSONAL FINANCE: Compound Interest** Find the value of $10,000 invested at 6% interest compounded quarterly for 5 years.

76. **PERSONAL FINANCE: Compound Interest** Find the value of $1,500 invested at 5% interest compounded daily for 10 years.

77. **PERSONAL FINANCE: Depreciation** A $21,000 car depreciates by 20% per year. Find the value after:
    a. 4 years.   b. 6 months.

78. **BUSINESS: Depreciation** A $70,000 ambulance depreciates by 30% per year. Find the value after:
    a. 5 years.   b. 6 months.

79. **BUSINESS: Continuous Compounding** A bond selling for $1000 has interest rate 4.8% compounded continuously. Find its value after ten years.

80. **BUSINESS: Continuous Compounding** A bond selling for $1,000,000 has interest rate 5.1% compounded continuously. Find its value after five years.

1.6 Logarithmic Functions

81. Find each logarithm *without* using a calculator.
    a. $\log 1000$ b. $\log \frac{1}{1000}$
    c. $\ln e^3$ d. $\ln \sqrt[4]{e}$

82. Find each logarithm *without* using a calculator.
    a. $\log \sqrt{10}$ b. $\log 10^8$
    c. $\ln \frac{1}{e}$ d. $\ln e^{3/2}$

Use the properties of natural logarithms to simplify each function.

83. $f(x) = \ln x^4 - \ln x^3 - \ln 1$
84. $f(x) = \ln e^{7x} - 5x - \ln e$

85. **PERSONAL FINANCE: Depreciation** A car depreciates by 22% per year. When will it be worth:
    a. half its original value?
    b. one quarter of its original value?

86. **BUSINESS: Depreciation** A corporate helicopter depreciates by 35% per year. When will it be worth:
    a. half its original value?
    b. one quarter of its original value?

87–88. **GENERAL: Fossils** In the following exercises, use the fact that the proportion of potassium-40 remaining after t million years is $e^{-0.00054t}$.

87. In 1984 in the Wind River Basin of Wyoming, scientists discovered a fossil of a small, three-toed horse, an ancestor of the modern horse. Estimate the age of this fossil if it contained 97.3% of its original potassium-40.

88. Estimate the age of a skull found in 1959 in Tanzania (dubbed "Nutcracker Man" because of its huge jawbone) that contained 99.9% of its original potassium-40.

Part I

Finite Mathematics

CHAPTER 2
MATHEMATICS OF FINANCE

CHAPTER 3
SYSTEMS OF EQUATIONS AND MATRICES

CHAPTER 4
LINEAR PROGRAMMING

CHAPTER 5
PROBABILITY

CHAPTER 6
STATISTICS

CHAPTER 7
MARKOV CHAINS

Chapters on Logic and Game Theory may be found on the Web at math.college.hmco.com; follow the links to this text's Website.

2 Mathematics of Finance

- 2.1 Simple Interest
- 2.2 Compound Interest
- 2.3 Annuities
- 2.4 Amortization

Application Preview

Musical Instruments as Investments*

No two guitars sound alike—good ones, anyway. Over the last 30 years, prices for high-quality acoustic instruments have soared. In 1993, a John D'Angelico New Yorker 18-inch-wide cutaway archtop ("jazz") guitar sold for $150,000, setting the record for the price of an acoustic fretted instrument not previously owned by a deceased superstar. Many pre-World War II instruments are valued not only for their craftsmanship but also for their distinctive sound. A mint condition Martin D-45 made in 1932–1942 that might have sold for $3500 in 1971 could be sold for $225,000 in 2003, while a Gibson 1932–1939 Mastertone Granada 5-string (replacement neck) banjo selling for just $1200 in 1971 would fetch $28,000 in 2003.

How can these price increases be compared to other investments such as stocks, bonds, or bank savings accounts? The dots on the graph show the typical selling prices over the 30-year period from 1971 to 2001 for a signed 1922 Gibson F-5 mandolin, together with a curve showing the increasing value of an investment earning 11% compound interest.

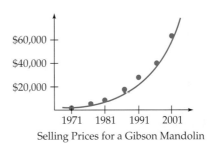

Selling Prices for a Gibson Mandolin

The dots closely match the curve, showing that this mandolin has increased in value by about 11% annually, better than many bond and stock investments. The values of fine musical instruments and other investments can be found by the methods of this section.

---

* Based on material originally provided by Stanley M. Jay and Larry Wexer of *mandolin bros. Ltd.* (Staten Island, New York), dealers in fine mandolins, guitars, and banjos.

101

2.1 SIMPLE INTEREST

Introduction

In the modern credit world, the old adage "time is money" has become a basic fact of economic life. When you open a savings account or take out a car loan, you directly experience the "time value of money." This chapter covers the basic financial properties of a loan between a lender and a borrower and the calculation of the interest and payments. We begin with simple interest.

Simple Interest

The *principal* of a loan is the amount of money borrowed from the lender, the *term* is the time the borrower has the money, and the *interest* is the additional money paid by the borrower for the use of the lender's money. The interest is called *simple interest* if it is calculated as a fixed percentage of the principal and is paid at the end of the term. The *interest rate* of the loan is the dollars of interest per $100 of principal per year (or "per annum"). It is usually stated as a percentage but is always written as a decimal in calculations. Simple interest is calculated as follows:

> **Simple Interest Formula**
>
> The interest I on a loan of P dollars at simple interest rate r for t years is
>
> $$I = Prt$$
>
> P = principal
> r = rate
> t = term

EXAMPLE 1

FINDING SIMPLE INTEREST

Find the interest on a loan of two million dollars at 7.2% simple interest for 4 months.

Solution

Writing the interest rate in decimal form and changing the term to years, we find that

$$I = 2{,}000{,}000 \cdot 0.072 \cdot \frac{4}{12} = 48{,}000 \qquad \begin{array}{l} I = Prt \text{ with} \\ P = 2{,}000{,}000, \\ r = 0.072, \text{ and } t = 4/12 \end{array}$$

The interest is $48,000.

Be Careful! When using a calculator, round off only your final answer. For example, if you use $t = 0.33$ in the previous example instead of $\frac{4}{12}$ or $\frac{1}{3}$, you would get the value $2{,}000{,}000 \cdot 0.072 \cdot 0.33 = \$47{,}520$, which is wrong by $480.

Practice Problem 1

Find the interest on a loan of $50,000 at 19.8% simple interest for 3 months. ➤ *Solution on page 108*

If you have values for any three unknowns in the interest formula $I = Prt$, you can solve for the fourth one.

EXAMPLE 2

FINDING THE SIMPLE INTEREST RATE

What is the interest rate of a loan charging $18 simple interest on a principal of $150 after 2 years?

Solution

We solve $I = Prt$ for interest rate r:

$$r = \frac{I}{Pt} \qquad I = Prt \text{ divided by } Pt$$

Then

$$r = \frac{18}{150 \cdot 2} = \frac{18}{300} = 0.06 \qquad \text{Substituting } I = 18, P = 150, \text{ and } t = 2$$

The interest rate is 6%.

Total Amount Due on a Loan

When the term of a loan is over, the borrower repays the principal and interest, so the total amount due is

$$\underbrace{P}_{\text{Principal}} + \underbrace{Prt}_{\text{Interest}} = \underbrace{P(1 + rt)}_{\text{Total amount}} \qquad \text{Factoring out the common term}$$

This gives the following formula:

Total Amount Due for Simple Interest

The total amount A due at the end of a loan of P dollars at simple interest rate r for t years is

$$A = P(1 + rt)$$

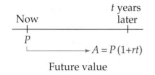
Future value

The amount due may also be regarded as the *accumulated value* of an investment or the *future value* of the principal. As before, this formula may be used as it is or it may be solved for any one of the other variables, as the next few examples will show.

EXAMPLE 3

FINDING THE TOTAL AMOUNT DUE

What is the total amount due on a loan of $3000 at 6% simple interest for 4 years?

Solution

$$A = 3000(1 + 0.06 \cdot 4)$$
$$= 3000 \cdot 1.24 = 3720$$

$A = P(1 + rt)$ with $P = 3000$, $r = 0.06$, and $t = 4$

The total due on the loan is $3720.

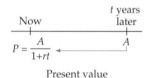
Present value

The $3000 loan in the above example grew to $3720 in 4 years. The $3720 is sometimes called the *future value* of the original $3000. Reversing our viewpoint, we say that the principal of $3000 is the *present value* of the later $3720. To find a formula for the present value, we solve the "total amount due" formula (on the previous page) for P by dividing by $1 + rt$:

$$P = \frac{A}{1 + rt}$$

Present value of the future amount A

EXAMPLE 4

FINDING A PRESENT VALUE

How much should be invested now at 8.6% simple interest if $10,000 is needed in 6 years?

Solution

The amount to invest now means the *present value* of $10,000 in 6 years. Using the above formula:

$$P = \frac{10{,}000}{1 + 0.086 \cdot 6} = \frac{10{,}000}{1.516} \approx 6596.31$$

$P = A/(1 + rt)$ with $A = 10{,}000$, $r = 0.086$, and $t = 6$

The amount required is $6596.31.

We emphasize that the *present value* is the amount that will grow to the required sum in the given time period (at the stated interest rate). For this reason, it gives the actual value *now* of an amount to be paid later.

Amounts like this, to be paid at some time in the future, occur in many situations from personal credit card balances to government bonds. Such "debts" are frequently bought and sold by banks and companies, and the present value gives the current value of such future payments. For example, at the interest rate stated in the preceding example, a payment of $10,000 to be received in 6 years is worth exactly $6596.31 now, and so should be bought or sold for this price.

Practice Problem 2

Find the present value of a "promissory note" that will pay $5000 in 4 years at 12% simple interest.   ➤ Solution on page 108

We found the present value by deriving a formula for it. We could instead have substituted the given numbers into the "total amount due" formula and then solved for P (which we may think of as standing for *principal* or *present value*). We will do the next example in this way, substituting numbers into the original formula and then solving for the remaining variable. You can use either method to solve these problems.

EXAMPLE 5

FINDING THE TERM OF A SIMPLE INTEREST LOAN

What is the term of a loan of $2000 at 4% simple interest if the amount due is $2400?

Solution

Substituting the given numbers for the appropriate variables in the "total amount due" formula (page 103) gives

$$2400 = 2000(1 + 0.04t) \qquad A = P(1 + rt) \text{ with } A = 2400, P = 2000, \text{ and } r = 0.04$$

$$1.2 = 1 + 0.04t \qquad \text{Dividing by 2000}$$

$$0.2 = 0.04t \qquad \text{Subtracting 1}$$

$$t = \frac{0.2}{0.04} = 5 \qquad \text{Dividing by 0.04 and reversing sides}$$

The term is 5 years.

Graphing Calculator Exploration

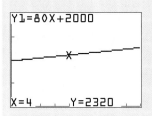

The formula $A = P(1 + rt)$ may be written as $A = (Pr)t + P$ so that it has the familiar $y = mx + b$ form of a straight line with slope Pr and y-intercept P (but with t and A instead of x and y).

a. Using the values $P = 2000$ and $P \cdot r = 2000 \cdot 0.04 = 80$ from the preceding example, graph the line $y_1 = 80x + 2000$ on the window [0, 8] by [0, 4000], so that y gives the amount of the loan for any term x.

b. Use TRACE to estimate the term x that gives $y = 2400$. How does your "graphical" answer compare to the answer for Example 5? Try ZOOMing IN to improve your estimate.

c. Add the horizontal line $y_2 = 2400$ to your graph and use INTERSECT to find the intersection point. Does the x-value of this point exactly match the answer for Example 5?

Discounted Loans and Effective Interest Rates

In a *discounted loan* the lender deducts the interest from the amount the borrower receives at the start. A discounted loan is better for the lender because getting the money early (so it can earn interest somewhere else) is always better than later. Since the borrower actually is receiving a smaller amount, we may recalculate the interest rate as a "standard" loan on this smaller amount. The resulting rate is called the *effective* simple interest rate of the loan.

EXAMPLE 6   FINDING THE EFFECTIVE SIMPLE INTEREST RATE

Determine the effective simple interest rate on a discounted loan of $1000 at 6% simple interest for 2 years.

Solution

The interest is $\$1000 \cdot 0.06 \cdot 2 = \$120$, so as a simple interest loan, the borrower receives only $P = 1000 - 120 = \$880$ and agrees to pay back $1000 at the end of 2 years. If we write r_s for the simple interest rate of this loan, the "total amount due" formula on page 103 gives

$$1000 = 880(1 + r_s \cdot 2) \qquad \begin{array}{l} A = P(1 + rt) \text{ with } A = 1000, \\ P = 880, \text{ and } t = 2 \end{array}$$

Solving for r_s,

$$r_s = \frac{1}{2}\left(\frac{1000}{880} - 1\right) \approx 0.0682 \qquad \text{Dividing by 880, subtracting 1, and dividing by 2}$$

The effective simple interest rate on this discounted loan is 6.82%.

Notice that the effective rate of 6.82% is significantly greater than the stated (or "nominal") rate of 6%. In general, the nominal rate in a discounted loan may be used to conceal the true cost to the borrower (and similarly, the true benefit to the lender). This is why effective rates are so important.

Using the same method as in Example 6, we can find a formula for the effective rate. For a discounted loan of amount A at interest rate r for term t, the borrower receives $A - Art = A(1 - rt)$ dollars. Therefore, the "total amount due" formula from page 103, but with the simple interest rate replaced by the effective rate r_s, gives

$$A = \underbrace{A(1 - rt)}_{\text{Principal}}(1 + r_s t) \qquad \begin{array}{l} A = P(1 + rt) \text{ with} \\ A(1 - rt) \text{ for } P \end{array}$$

Solving for r_s:

$$\frac{1}{1 - rt} = 1 + r_s t \qquad \begin{array}{l} \text{Canceling the } A\text{'s and} \\ \text{dividing by } (1 - rt) \end{array}$$

$$r_s t = \frac{1}{1 - rt} - 1 = \frac{1 - (1 - rt)}{1 - rt} = \frac{rt}{1 - rt} \qquad \begin{array}{l} \text{Reversing sides,} \\ \text{subtracting 1,} \\ \text{and simplifying} \end{array}$$

$$r_s = \frac{1}{t}\frac{rt}{1 - rt} = \frac{r}{1 - rt} \qquad \begin{array}{l} \text{Dividing by } t \text{ and} \\ \text{then canceling} \end{array}$$

This gives the following formula for r_s.

Effective Simple Interest Rate for a Discounted Loan

For a discounted loan at interest rate r for t years, the effective simple interest rate r_s is

$$r_s = \frac{r}{1 - rt}$$

Notice that the effective rate r_s will be larger than r (since the denominator is less than 1) and that it depends only on the *rate* and *term* of the loan, and not on its amount. For the loan in Example 6, our formula gives the answer that we found:

$$r_s = \frac{0.06}{1 - 0.06 \cdot 2} = \frac{0.06}{0.88} \approx 0.0682 \qquad r_s = \frac{r}{1-rt} \text{ with } r = 0.06 \text{ and } t = 2$$

Practice Problem 3 What is the effective simple interest rate of a discounted loan at 6% interest for 3 years? for 5 years? ➤ Solution below

2.1 Section Summary

The simple interest formula is

$$I = Prt$$

I = simple interest
P = principal
r = interest rate
t = term in years

We can solve for any one of the variables if the others are known. The total amount due at the end of the loan (principal plus interest) is

$$A = P(1 + rt) \qquad A = \text{total amount due}$$

This equation may be solved for any one of the variables, keeping in mind that the amount A at the end of the loan is the *future value* of the principal P, and P is the *present value* of the future amount A.

For a *discounted* loan, the interest is subtracted from the principal at the beginning of the loan. The effective simple interest rate r_s for a discounted loan at rate r for t years is

$$r_s = \frac{r}{1 - rt}$$

➤ **Solutions to Practice Problems**

1. $I = 50{,}000 \cdot 0.198 \cdot \frac{3}{12} = 2475.$
2. $P = \frac{5000}{1 + 4 \cdot 0.12} = \frac{5000}{1.48} \approx 3378.38$
3. $r_s = \frac{0.06}{1 - 0.06 \cdot 3} \approx 0.0732.$ The effective simple interest rate is 7.32%.

   For 5 years, $r_s = \frac{0.06}{1 - 0.06 \cdot 5} \approx 0.0857.$ The effective simple interest rate is 8.57%. Can you think of an intuitive reason for the effective rate to be higher if the term is longer? [*Hint:* Think of how much earlier the lender gets the money or of how much less the borrower really gets.]

2.1 Exercises

Find the simple interest on each loan.

1. $1500 at 7% for 10 years.
2. $2000 at 9% for 7 years.
3. $6000 at 6.5% for 8 years.
4. $4500 at 4.25% for 9 years.
5. $825 at 6.58% for 5 years 6 months.
6. $950 at 5.87% for 6 years 3 months.
7. $1280 at 4.8% for 3 months.
8. $5275 at 5.3% for 2 months.

Find the total amount due for each simple interest loan.

9. $1500 at 7% for 10 years.
10. $2000 at 9% for 7 years.
11. $6100 at 5.7% for 4 years 9 months.
12. $4500 at 6.3% for 3 years 6 months.
13. $3125 at 4.81% for 10 months.
14. $8775 at 13.11% for 7 months.

APPLIED EXERCISES

15. **Interest Rate** Find the interest rate on a loan charging $704 simple interest on a principal of $2750 after 4 years.

16. **Interest Rate** Find the interest rate on a loan charging $1127 simple interest on a principal of $4900 after 5 years.

17. **Principal** Find the principal of a loan at 8.4% if the simple interest after 5 years 6 months is $1155.

18. **Principal** Find the principal of a loan at 7.6% if the simple interest after 9 years 3 months is $2109.

19. **Term** Find the term of a loan of $175 at 9% if the simple interest is $63.

20. **Term** Find the term of a loan of $225 at 7% if the simple interest is $94.50.

21. **Present Value** How much should be invested now at 5.2% simple interest if $8670 is needed in 3 years?

22. **Present Value** How much should be invested now at 4.8% simple interest if $4530 is needed in 4 years 4 months?

ZERO COUPON BONDS A *zero coupon bond* pays only its *face value* on maturity (getting its name because it has no "coupons" for interest payments before that date). The *fair market price* is the present value of the face value at the current interest rate.

23. What is the fair market price of a $10,000 zero coupon bond due in 1 year if today's long-term simple interest rate is 5.81%?

24. What is the fair market price of a $15,000 zero coupon bond due in 1 year if today's long-term simple interest rate is 7.23%?

25. What is the fair market price of a $5000 zero coupon bond due in 2 years if today's long-term simple interest rate is 3.54%?

26. **How Interest Rates Affect Present Value** The present value formula $P = A/(1 + rt)$ can be viewed on your graphing calculator as a function y of the simple interest rate x by rewriting it as $y_1 = A/(1 + xt)$. Using $A = \$10{,}000$ and $t = 1$ year, graph this expression with window [0, 1] by [0, 12,000] to see the present value of a $10,000 zero coupon bond due in 1 year as a function of the interest rate. Use TRACE or VALUE to find the present value of the bond for interest rates of 4%, 5%, 6%, 7%, 8%, and 9%. Does a 1% increase in the interest rate always cause the same decrease in the present value of the bond?

27. **Term** What should be the term for a loan of $6500 at 7.3% simple interest if the lender wants to receive $9347 when the loan is paid off?

28. **Term** What should be the term for a loan of $5400 at 5.8% simple interest if the lender wants to receive $6966 when the loan is paid off?

29. **Term** How long will it take an investment at 8% simple interest to increase by 70%?

30. **Term** How long will it take an investment at 4.2% simple interest to increase by 26.6%?

DOUBLING TIME The *doubling time* of an investment is the number of years it takes for the value to double. This is the same as the number of years for the value to increase by 100%.

31. What is the doubling time of a 5% simple interest investment?

32. Show that the doubling time of a simple interest investment is $1/r$ where r is the simple interest rate.

33. **Effective Rate** What is the effective simple interest rate of a discounted loan at 4.6% interest for 3 years 6 months?

34. **Effective Rate** What is the effective simple interest rate of a discounted loan at 7.2% interest for 2 years 10 months?

35. **Discounted Loan** The Dewey, Cheetham, and Howe Loan Corporation offers discounted loans of $1000 at 5% for 20 years. Explain why their name is appropriate.

36. **How the Term Affects the Discounted Rate** The effective rate formula $r_s = r/(1 - rt)$ can be viewed on your graphing calculator as a function of the term t by rewriting it as $y_1 = r/(1 - rx)$. Using $r = 0.05$, graph this expression with window [0, 20] by [0, 1] to see the effective rate y of the discounted loan as a function of the term x. Use TRACE or VALUE to find the effective rate of the loan for terms of 5, 10, and 15 years. Does a 1-year increase in the term always cause the same increase in the effective rate?

37. **Bridge Loan** You have a buyer for your condominium, but the seller of your dream house wants to close now. In order to get enough money to go to the closing, you take out a 2-month bridge loan for $72,000 at 7.92% simple interest. You assume that you will close on the sale of your condominium before the 2 months are up. How much interest will you have to pay on the loan?

38. **Home Improvements** Because the contractor was short of cash when the job was finished, a plumber accepted as payment a promissory note for $2000 plus 18% simple interest due in 2 months. Needing cash himself, the plumber sold the note 1 month later for $2000 to a local loan agency. Find the simple interest rate the agency will earn on its investment.

BROKERAGE COMMISSIONS An Internet discount brokerage firm charges a commission of 10% on the first $20,000 plus 5% of the excess over $20,000 on each buy or sell transaction. Find the simple interest rate earned by each of the following investments after including the commissions paid to the brokerage firm.

39. Purchase 900 shares of American WebWide Education at $18.50 per share and sell them 4 months later at $26.75 per share.

40. Purchase 1500 shares of Well Care Deluxe at $11.90 per share and sell them 10 months later at $14.80 per share.

41. Purchase 800 shares of DucoFood Services at $28.70 per share and sell them 7 months later at $37.80 per share.

2.2 COMPOUND INTEREST

Introduction

When a simple interest loan is not paid off at the end of its term but is taken out as a new loan, the borrower owes the principal plus the interest. The interest on the new loan is called *compound interest* because it combines interest on both the original principal and on the unpaid interest. For example, if you borrow $8000 at 5% simple interest for 1 year, the formula on page 103 says that the amount due will be $8000(1 + 0.05) = $8400. If you do not pay it off, but borrow this sum for the next year, the amount due after 2 years will be $8000(1 + 0.05)(1 + 0.05) = $8000(1 + 0.05)^2 = $8820. After a third year, the amount due would become $8000(1 + 0.05)^3 = $9261. For each subsequent year, this amount would be multiplied by another (1 + 0.05), which is 1 plus the interest rate. Compounding clearly increases the amount of interest paid because of the interest on the interest.

Compound Interest

In general, the amount A due on a compound interest loan of P dollars at interest rate r per year compounded annually for t years is found by repeatedly multiplying the principal P by $(1 + r)$, once for each year:

$$A = \underbrace{P(1 + r)(1 + r) \cdots (1 + r)}_{t \text{ multiplications by } (1+r)} = P(1 + r)^t$$

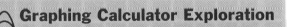

Graphing Calculator Exploration

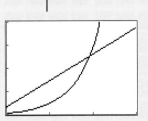

To see that compound interest eventually surpasses simple interest (even with a higher rate and principal), compare $500 invested at 4% compounded annually with $2500 invested at 8% simple interest.

a. Graph the simple interest amount $A = P(1 + rt)$ as $y_1 = 2500(1 + .08x)$ on the window [0, 150] by [0, 35,000] so that y_1 is the amount of the investment after x years.

b. Graph the compound interest amount $A = P(1 + r)^t$ as $y_2 = 500(1 + .04)^x$ on the same window.

(continues)

c. Use TRACE or INTERSECT to find when the compound interest amount equals the simple interest amount. What happens after this intersection point? Try your graphs with the larger windows [0, 300] by [0, 70,000] and by [0, 500,000].

Notice how the formulas for the amounts due under simple and compound interest differ:

Simple	$P(1 + rt)$
Compound	$P(1 + r)^t$

The formula for *compound* interest has t in the *exponent*, which is why it grows so much faster (eventually).

Banks always state *annual* interest rates, but the compounding can be done more than once a year. Some standard compounding periods are:

Compounding Frequency	Periods per Year
Annual	1
Semiannual	2
Quarterly	4
Monthly	12
Daily	365

Given the same annual interest rate, more frequent compounding is better since your money starts earning interest sooner. For example, if a bank offers 8% compounded quarterly, then the interest is calculated on a *quarterly* basis: each quarter you get 2%, (one quarter of the 8%), and this is done every *quarter*, so the exponent is the number of quarters, which is *four times* the number of years. In general, if the annual interest rate is r with m compoundings per year, then the interest rate per period is $\frac{r}{m}$ and the number of compounding periods in t years is mt. This leads to the following general formula for compound interest:

Compound Interest Formula

The amount A due on a loan of P dollars at yearly interest rate r compounded m times per year for t years is

$$A = P\left(1 + \frac{r}{m}\right)^{mt}$$

r = annual rate
m = periods per year
t = term (in years)

The stated yearly interest rate r is called the *nominal rate* of the loan.

EXAMPLE 1

FINDING THE AMOUNT DUE ON A COMPOUND INTEREST LOAN

Find the amount due on a loan of $1500 at 4.8% compounded monthly for 2 years.

Solution

The nominal rate of 4.8% expressed as a decimal is 0.048, and monthly compounding means that $m = 12$, so we have

$$A = 1500\left(1 + \frac{0.048}{12}\right)^{12 \cdot 2}$$

$A = P(1 + r/m)^{mt}$
with $P = 1500$, $r = 0.048$, $m = 12$, and $t = 2$

$$= 1500(1.004)^{24} \approx 1650.82$$

The amount due is $1650.82.

Practice Problem 1

Find the amount that is due on a loan of $3500 at 5.1% compounded quarterly for 3 years.
➤ Solution on page 119

A loan is really an amount that a bank *invests* in a borrower, so any question about a loan can be rephrased as a question about an *investment*. Example 1 could have asked for the value of an investment of $1500 that grows by 4.8% compounded monthly for 2 years—the answer would still be $1650.82. In this section we will speak interchangeably of *loans* and *investments*, since they are the same but from different viewpoints, depending whether you are the borrower or the lender.

In Example 1, the amount $1650.82 may be called the *future value* of the original $1500, and conversely, the $1500 is the *present value* of the later amount $1650.82. As before, the present value gives the value *today* of a payment that will be received at some time in the future. To find a formula for the present value, we solve the compound interest formula (on the previous page) for P, using the familiar $1/x^n = x^{-n}$ rule of exponents.

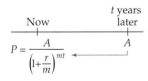

Future value

$$P = \frac{A}{\left(1 + \frac{r}{m}\right)^{mt}} = A\left(1 + \frac{r}{m}\right)^{-mt} \quad \text{Present value}$$

This formula should be intuitively reasonable: To find *future* value you multiply by $(1 + \frac{r}{m})^{mt}$, so to find *present* value you should *divide* by the same quantity.

EXAMPLE 2 FINDING THE PRESENT VALUE

How much should be invested now at 8.6% compounded weekly if $10,000 is needed in 6 years?

Solution

$$P = 10{,}000\left(1 + \frac{0.086}{52}\right)^{-52 \cdot 6} \approx 5971.58$$

$P = A(1 + r/m)^{-mt}$ with $A = 10{,}000$, $r = 0.086$, $m = 52$, and $t = 6$

The amount required is $5971.58.

Growth Times

We solved the preceding example by using the formula for present value. Instead, we could have substituted the given numbers into the compound interest formula and *then* solved for the remaining variable. We will do the next example in this way, finding a "growth time," the time for a loan to reach a given value. We will use the rule of logarithms $\log(x^n) = n \log x$ to "bring down the power." Logarithms with *any* base may be used in such calculations.

EXAMPLE 3 FINDING THE TERM OF A COMPOUND INTEREST LOAN

What is the term of a loan of $2000 at 6% compounded monthly that will have an amount due of $2400?

Solution

We substitute the given numbers into the compound interest formula (see page 112):

$$2400 = 2000\left(1 + \frac{0.06}{12}\right)^{12 \cdot t}$$

$A = P(1 + r/m)^{mt}$ with $A = 2400$, $P = 2000$, $r = 0.06$, and $m = 12$

$$1.2 = 1.005^{12t}$$

Dividing by 2000 and simplifying

$$\log 1.2 = \log 1.005^{12t}$$

Taking logarithms

$$\log 1.2 = 12t \log 1.005$$

Using $\log(x^n) = n \log x$ to bring down the power

$$12t = \frac{\log 1.2}{\log 1.005} \approx 36.555$$

Dividing by $\log 1.005$, reversing sides, and calculating logarithms

Months

This amount of time, $12t = 36.555$, is in *months* (t is years, but $12t$ is months). We round *up* to the nearest month (because a shorter term will not reach the needed amount), so the term needed is 37 months, or 3 years 1 month.

Hint: Instead of solving for t, if the compounding is monthly stop when you find $12t$ (the number of months), if quarterly stop when you find $4t$ (the number of quarters), if weekly stop when you find $52t$ (the number of weeks), and so on. Then round *up* to the next whole number of periods to ensure that the needed amount will actually be reached.

Notice that the actual amounts $2400 and $2000 did not matter in this calculation, only their *ratio* $\frac{2400}{2000} = 1.2$. Thus, a loan of $2000 grows to $2400 in the same amount of time that a loan of 20 *million* dollars would grow to 24 million dollars (at the stated interest rate) since the ratios are the same.

Therefore, in Example 3 we could have omitted the dollar amounts and simply asked for the term of a loan that would *multiply the principal by 1.2*. Furthermore, multiplying by 1.2 means *increasing* by 0.2, or 20%, so we could have asked for the term of a loan that would *increase the principal by 20%*—all three formulations are equivalent.

Be careful! Distinguish carefully between the *increase* and the *multiplier*. Increasing by 20% means *multiplying by 1.2* (the "1" keeps the original amount and the ".2" increases it by 20%). For example, to *increase* an amount by 35%, you would *multiply* it by 1.35; to increase an amount by 75%, you would multiply by 1.75; and to increase it by 100%, you would multiply by 2.

  What is the multiplier that corresponds to increasing the amount by 50%? ➤ Solution on page 119

EXAMPLE 4

FINDING THE TERM TO INCREASE THE PRINCIPAL

What is the term of a loan at 6% compounded weekly that will double the principal?

Solution

Doubling means multiplying the amount by 2. Using P for the unknown principal, we must solve

$$2P = P\left(1 + \frac{0.06}{52}\right)^{52 \cdot t}$$ P (for principal) on both sides

$$2 = (1 + 0.06/52)^{52t} \qquad \text{Canceling the } P\text{'s and simplifying}$$

$$\log 2 = 52t \log (1 + 0.06/52) \qquad \text{Taking logarithms and bringing down the power}$$

$$52t = \frac{\log 2}{\log (1 + 0.06/52)} \approx 601.1 \qquad \text{Dividing by } \log (1 + 0.06/52), \text{ reversing sides, and calculating logarithms}$$

Weeks

Rounding *up*, the term needed is 602 weeks, or 11 years 30 weeks.

Practice Problem 3 What is the term of a loan at 6% compounded quarterly that increases the principal by 50%? ▶ **Solution on page 119**

Rule of 72

The *doubling time* of an investment or a loan is the number of years it takes for the value to double. Doubling times are often estimated by using the *rule of 72*:

$$\left(\begin{array}{c}\text{Doubling}\\\text{time}\end{array}\right) \approx \frac{72}{r \times 100} \qquad \text{Divide 72 by the rate times 100}$$

The rule of 72, however, gives only an *approximation* for the doubling time, but it is often quite accurate. (A justification of this rule is given in Exercise 50.) For example, for a loan at 6% compounded quarterly the rule of 72 gives

$$\left(\begin{array}{c}\text{Doubling}\\\text{time}\end{array}\right) \approx \frac{72}{0.06 \times 100} = \frac{72}{6} = 12 \text{ years} \qquad \text{Rule of 72 estimate}$$

In Example 4 we found the correct doubling time to be 11 years 30 weeks, so the rule of 72 is not far off.

Effective Rates

Suppose that you are considering two certificates of deposit (CDs), one offering 6.2% compounded semiannually and the other offering 6.15% compounded monthly. Which should you choose, the one with the higher rate or the one with more frequent compounding? More generally, how can we compare any two different interest rates with different compoundings? We do so by calculating the actual percentage increase that each will generate in a year. This number is called the *annual percentage yield* or *APY*. More formally, the *APY* is also called the *effective rate of interest*, denoted r_e, which is the simple interest rate that will return the same amount on a one-year loan. Therefore, r_e can be found by solving

$$P(1 + r_e) = P\left(1 + \frac{r}{m}\right)^m \quad \text{Finding the "simple" rate that gives the "compound" rate for } t = 1$$

Canceling the P's and subtracting the 1 on the left gives the following.

Effective Rate for a Compound Interest Loan

For a compound interest loan at interest rate r compounded m times per year, the effective rate r_e is

$$r_e = \left(1 + \frac{r}{m}\right)^m - 1 \qquad r_e = APY$$

The effective rate can be interpreted as the interest earned by $1 for 1 year.

EXAMPLE 5 USING EFFECTIVE RATES TO COMPARE INVESTMENTS

A self-employed carpenter setting up a Keogh retirement plan must choose between a certificate of deposit (CD) with the First & Federal Bank at 6.2% compounded semiannually or a CD with the Chicago Nationswide Bank at 6.15% compounded monthly. Which is better?

Solution

The effective rate for 6.2% compounded semiannually (First & Federal) is

$$r_e = \left(1 + \frac{0.062}{2}\right)^2 - 1 \approx 0.0630 \qquad r_e \approx 6.30\%$$

For 6.15% compounded monthly (Chicago Nationswide), it is

$$r_e = \left(1 + \frac{0.0615}{12}\right)^{12} - 1 \approx 0.0633 \qquad r_e \approx 6.33\%$$

The CD at the Chicago Nationswide Bank is better.

Sometimes the interest rate is not given, and only the amounts at the beginning and end of the loan are known. How can we find the interest rate (called the *effective* rate of interest)?

The effective rate of return for an investment of P dollars that returns an amount A after t years can be found from the compound interest formula (page 112) with $m = 1$:

$$A = P(1 + r_e)^t$$

$A = P(1 + r/m)^{mt}$
with $m = 1$ and $r = r_e$

An example of this is a *zero coupon bond*, which pays its *face value* on maturity and sells now for a lower price. This kind of bond makes no payments before maturity and so has no interest "coupons."

EXAMPLE 6

EFFECTIVE RATE FOR A ZERO COUPON BOND

What is the effective rate of return of a $10,000 zero coupon bond maturing in 6 years and offered now for sale at $5500?

Solution

We solve

$10,000 = 5500(1 + r_e)^6$ $\quad A = P(1 + r_e)^t$ with $A = 10{,}000$, $P = 5500$, and $t = 6$

$\dfrac{10{,}000}{5500} = (1 + r_e)^6$ Dividing by 5500

$(100/55)^{1/6} = 1 + r_e$ Raising to power $\tfrac{1}{6}$

$r_e = (100/55)^{1/6} - 1 \approx 0.1048$ Solving for r_e

The effective rate is 10.48%.

2.2 Section Summary

The *compound interest formula* is

$$A = P\left(1 + \frac{r}{m}\right)^{mt}$$

A = amount due
P = principal
r = interest rate
m = compoundings per year
t = term in years

A is the *future value* of P, and P is the *present value* of A. Given values for four of the variables, we can solve for the fifth. Remember, however, to round the term *up* to the next whole number of compounding periods. The *doubling time*, the time for an amount to double $(A = 2P)$, can be approximated by the *rule of 72*: $t \approx 72/(r \times 100)$ years. Remember the difference between *increasing* by a percentage and *multiplying* by a number: Increasing by 50% means multiplying by 1.50.

The *effective rate* or *APY* of a loan is

$$r_e = \left(1 + \frac{r}{m}\right)^m - 1$$

r_e = effective rate of interest at rate r compounded m times each year

The effective rate of return for *any* investment of P dollars that returns A dollars after t years (such as a zero coupon bond) is found from the compound interest formula with $m = 1$, $A = P(1 + r_e)^t$.

> **Solutions to Practice Problems**

1. $A = 3500\left(1 + \frac{0.051}{4}\right)^{4 \cdot 3} \approx 4074.69$. The amount due is $4074.69.

2. 1.5

3. $1.5P = P\left(1 + \frac{0.06}{4}\right)^{4t}$ simplifies to $1.5 = (1 + 0.06/4)^{4t}$.

Taking logarithms: $\log 1.5 = \log(1 + 0.06/4)^{4t} = 4t \log(1 + 0.06/4)$

so $4t = \dfrac{\log 1.5}{\log(1 + 0.06/4)} \approx 27.2$ (quarters!). Rounding up gives 28 quarters, or 7 years.

2.2 Exercises

Determine the amount due on each compound interest loan.

1. $15,000 at 8% for 10 years if the interest is compounded:
   a. annually.　　**b.** quarterly.

2. $7500 at 6.5% for 8 years 6 months if the interest is compounded:
   a. annually.　　**b.** quarterly.

3. $17,500 at 8.5% for 6 years 9 months if the interest is compounded:
   a. annually.　　**b.** weekly.

4. $25,000 at 9% for 5 years if the interest is compounded:
   a. semiannually.　　**b.** monthly.

5. $12,000 at 7.5% for 4 years 6 months if the interest is compounded:
   a. semiannually.　　**b.** monthly.

6. $16,750 at 4.5% for 6 years 6 months if the interest is compounded:
   a. semiannually.　　**b.** monthly.

Calculate the present value of each compound interest loan.

7. $25,000 after 7 years at 12% if the interest is compounded:
   a. annually.　　**b.** quarterly.

8. $9500 after 4 years 6 months at 8.4% if the interest is compounded:
   a. annually.　　**b.** quarterly.

9. $19,500 after 6 years 9 months at 7.3% if the interest is compounded:
   a. annually.   b. weekly.

10. $16,500 after 5 years at 9% if the interest is compounded:
    a. semiannually.   b. monthly.

11. $11,500 after 4 years 3 months at 8.4% if the interest is compounded:
    a. semiannually.   b. monthly.

12. $2200 after 2 years 6 months at 7.5% if the interest is compounded:
    a. semiannually.   b. monthly.

Find the term of each compound interest loan.

13. 8.2% compounded quarterly to obtain $8400 from a principal of $2000.

14. 8.2% compounded annually to obtain $8400 from a principal of $2000.

15. 6.8% compounded quarterly to multiply the principal by 1.8.

16. 5.48% compounded daily to multiply the principal by 1.5.

17. 8.5% compounded monthly to increase the principal by 65%.

18. 8.5% compounded semiannually to increase the principal by 65%.

Use the "rule of 72" to estimate the doubling time (in years) for each interest rate, and then calculate it exactly.

19. 9% compounded annually.

20. 6% compounded quarterly.

21. 7.9% compounded weekly.

22. 9% compounded semiannually.

23. 6.1% compounded monthly.

24. 5.9% compounded daily.

Find the effective rate of each compound interest rate or investment.

25. 18% compounded monthly. [*Note:* This is a typical credit card interest rate, often stated as 1.5% per month.]

26. 4.3% compounded weekly.

27. 8.57% compounded semiannually.

28. 7.5% compounded quarterly.

29. A $50,000 zero coupon bond maturing in 8 years and selling now for $23,500.

30. A $10,000 zero coupon bond maturing in 12 years and selling now for $6400.

APPLIED EXERCISES

31. **Bond Funds** During the mid-1990s, the T. Rowe Price International Bond fund returned 10.43% compounded monthly. How much would a $5000 investment in this fund have been worth after 3 years?

32. **Mutual Funds** During the first half of the 1990s, the Twentieth Century Ultra aggressive growth mutual fund returned 19.83% compounded quarterly. How much would a $10,000 investment in this fund have been worth after 5 years?

33. **College Savings** How much would your parents have needed to set aside 17 years ago at 7.3% compounded weekly to give you $50,000 for college expenses today?

34. **College Savings** How much would your parents have needed to set aside 16 years ago at 6.7% compounded semiannually to give you $60,000 for college expenses today?

35. **Bond Funds** You have just received $125,000 from the estate of a long-lost rich uncle. If you invest all of your inheritance in a tax-free bond fund earning 6.9% compounded quarterly, how long do you have to wait to become a millionaire?

36. Home Buying You and your new spouse have decided to use all $6500 of your wedding present monies as a nest egg for a house down payment. Investing at 11.47% compounded monthly, how long must you wait to have enough to put 10% down on a $140,000 house?

37. Mutual Funds A $10,000 investment in the Fidelity Blue Chip Growth mutual fund in 1987 would have been worth $30,832 seven years later. What was the effective rate of this investment?

38. Rate Comparisons The First Federal Bank offers 4.7% compounded weekly passbook savings accounts while Consolidated Nationwide Savings offers 4.73% compounded quarterly. Which bank offers the better rate?

39. Rate Comparisons The Second Peoples National Bank offers a long-term certificate of deposit earning 6.43% compounded monthly. Your broker locates a $20,000 zero coupon bond rated AA by Standard & Poor's for $7965 and maturing in 14 years. Which investment will give the greater rate of return?

40. Horse Trading In March 1995, a descendant of an early Texas settler sent $100 to Sam Houston IV to make good on a $100 debt (possibly from the sale of a horse) owed for 160 years to Sam Houston, the hero of Texas independence. The check was donated to the Sam Houston Museum and the debt considered as settled. However, newspapers reported that a banker had estimated the accumulated interest on the loan to be $420 million. What yearly interest rate did the banker use in her calculations?

41. Lottery Winnings You have just won $100,000 from a lottery. If you invest all this amount in a tax-free money market fund earning 7% compounded weekly, how long do you have to wait to become a millionaire?

42. Becoming a Millionaire How much would you have to invest now at 6.5% interest compounded semiannually to have a million dollars in 40 years?

43. Rate Comparisons The People's State Bank offers 4.2% compounded quarterly, while Statewide Federal offers 4.1% compounded daily. Which bank offers the better rate?

44. Rate Comparisons The Southwestern Savings and Loan Bank offers a 10-year certificate of deposit earning 6.4% compounded monthly. Your broker offers you a $10,000 zero coupon bond costing $5325 and maturing in 10 years. Which gives the greater rate of return?

Explorations and Excursions

The following problems extend and augment the material presented in the text.

More About Compound Interest

45. Show that after two years, the (yearly) compound interest amount $A = P(1 + r)^2$ exceeds the simple interest amount $A = P(1 + 2r)$ by $P(r^2)$.

46. Show that after three years, the (yearly) compound interest amount $A = P(1 + r)^3$ exceeds the simple interest amount $A = P(1 + 3r)$ by $P(3r^2 + r^3)$.

47. Show that $P(1 + r/m)^{mt} = P(1 + r_e)^t$.

48. Show that the *effective m-compound rate* r_m for an investment compounded m times per year and returning A dollars from P dollars after t years is $r_m = m((A/P)^{1/mt} - 1)$. Then check this formula by showing that the rate r_m compounded m times for one year gives the effective rate $r_e = (A/P)^{1/t} - 1$.

More About the Rule of 72

49. Compare the "rule of 72" with the exact formula for the doubling time by graphing both as functions of the interest rate as follows. For interest compounded monthly, the number of years y for the loan to double at interest rate x is $y_1 = \log(2)/(12 \log(1 + x/12))$ while the "rule of 72" estimate is $y_2 = 72/(x \times 100)$. Graph both curves on the window [0, .5] by [0, 20] and explore them using the TRACE command. How closely do these curves match each other?

Note: Exercise 50 requires the *continuous* compounding formula from page 70.

50. a. Show that the doubling time for interest rate r compounded continuously is $\frac{1}{r} \ln 2$.

 b. How does the numerical value of $\ln 2$ explain the "rule of 72"?

2.3 ANNUITIES

Introduction

An *annuity* is a scheduled sequence of payments. Some annuities, such as retirement pensions, you receive the payments, while in others, such as car loans and home mortgages, you pay someone else. In this section we will consider an *ordinary annuity*, which is an annuity with equal payments made at the end of regular intervals, with the interest compounded at the end of each. For example, a car loan for 4 years with monthly payments of $200 at 6% compounded monthly is an ordinary annuity.

A First Example

While having a new car is nice, when the first payment of the above loan comes due at the end of the first month, you may wish that instead you were saving that $200 for yourself. If you were, then 4 years later your first $200 at 6% compounded monthly would have grown to

$$200\left(1 + \frac{0.06}{12}\right)^{47} = 200 \cdot 1.005^{47}$$

$$1 + 0.005 = 1.005$$

The exponent is 47 rather than 48 because the first payment was made at the end of the first month and so earns interest for only 47 months. Similarly, your second payment would grow to $200 \cdot 1.005^{46}$, your third to $200 \cdot 1.005^{45}$, and so on down to your last payment of 200 (at the end of the last month, so earning no interest).

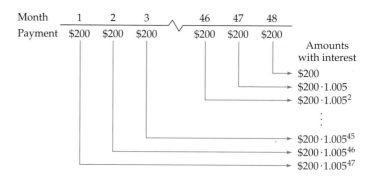

With patience and care we could evaluate and add up the numbers in the right side of this diagram, but there is an easier way. We will develop a formula for such sums and then use it to find the total amount of this annuity.

Geometric Series

The numbers in the right-hand side of the preceding diagram are found by successively multiplying by 1.005. In general, a sum of numbers each of which is a fixed multiple of the one before it is called a *geometric series*. The simplest geometric series is of the form

$$S = 1 + x + x^2 + \cdots + x^{n-1}$$

Multiplying by x gives the next number, and S is the sum of the series

If we multiply the sum by x, (which increases the exponent of each term) and subtract the original sum, we obtain

$$x \cdot S = \quad x + x^2 + \cdots + x^{n-1} + x^n \qquad x \text{ times } S$$
$$S = 1 + x + x^2 + \cdots + x^{n-1} \qquad S \text{ (lining up "like" terms)}$$

$$\underbrace{x \cdot S - S}_{S(x-1)} = -1 \qquad\qquad\qquad\qquad + x^n \qquad \text{Subtracting (most terms cancel)}$$

This last line can be written

$$S(x - 1) = x^n - 1 \qquad \text{Factoring on the left, reversing the order on the right}$$

so that

$$S = \frac{x^n - 1}{x - 1} \qquad \text{Dividing both sides by } x - 1 \text{ (for } x \neq 1\text{)}$$

Since S was defined as the sum $S = 1 + x + x^2 + \cdots + x^{n-1}$, this gives a formula for this sum:

$$1 + x + x^2 + \cdots + x^{n-1} = \frac{x^n - 1}{x - 1}$$

Multiplying both sides by a gives a formula for the sum of a geometric series:

Sum of a Geometric Series

$$a + ax + ax^2 + \cdots + ax^{n-1} = a\frac{x^n - 1}{x - 1} \qquad x \neq 1$$

For instance, the geometric series $3 + 6 + 12 + 24 + 48$, in which $a = 3$, $x = 2$, and the number of terms is $n = 5$, has the value

$$3 \cdot \frac{2^5 - 1}{2 - 1} = 93, \qquad a\frac{x^n - 1}{x - 1} \text{ with } a = 3, x = 2, \text{ and } n = 5$$

as you should check by adding up the five numbers.

Accumulated Amount Formula

For the payments at 6% compounded monthly for 4 years on page 122, we added terms of the form $200(1 + \frac{0.06}{12})^k$ with the exponent k taking values from 0 to $47 = 12 \cdot 4 - 1$. More generally, for payments of P dollars m times a year for t years at interest rate r, we would need to sum the geometric series

$$P + P\left(1 + \frac{r}{m}\right) + P\left(1 + \frac{r}{m}\right)^2 + \cdots + P\left(1 + \frac{r}{m}\right)^{mt-1}$$

Using the geometric series sum formula on the previous page, the sum of this series is

$$P\frac{\left(1 + \frac{r}{m}\right)^{mt} - 1}{\left(1 + \frac{r}{m}\right) - 1} \qquad a\frac{x^n - 1}{x - 1} \text{ with } a = P, \; x = (1 + \tfrac{r}{m}), \text{ and } n = mt$$

Simplifying the denominator to $\frac{r}{m}$ gives the following very useful formula.

Accumulated Amount of an Annuity

For an ordinary annuity with payments of P dollars m times a year for t years at interest rate r compounded at the end of each payment period, the accumulated amount (payments plus interest) will be

$$A = P\frac{(1 + r/m)^{mt} - 1}{r/m}$$

Although we will not use it, the archaic symbol $s_{\overline{n}|i}$ (pronounced "s sub n angle i") sometimes still occurs in business textbooks and tables to denote the value of $((1 + i)^n - 1)/i$. In this notation, the above formula takes the form $A = Ps_{\overline{n}|i}$ with $n = mt$ and $i = r/m$.

EXAMPLE 1

FINDING THE ACCUMULATED AMOUNT OF AN ANNUITY

Find the accumulated amount of the annuity of monthly payments for 4 years of $200 at 6% compounded monthly (the car loan annuity from page 122).

Solution

$$A = 200\frac{(1 + 0.06/12)^{12 \cdot 4} - 1}{0.06/12} \approx 10{,}819.57$$

$P\dfrac{(1 + r/m)^{mt} - 1}{r/m}$
with $P = 200$, $r = 0.06$, $m = 12$, and $t = 4$

Therefore, regular $200 payments at the end of each month for 4 years into a savings account earning 6% compounded monthly will total $10,819.57.

The amount of this annuity is significantly more than just the sum of 48 payments of $200, which would total $9600. The difference, $10{,}819.57 - 9600 = \$1219.57$, is the total interest earned by the payments.

Practice Problem 1 What is the final balance of a retirement account earning 5% interest compounded weekly if $40 is deposited at the end of every week for 35 years? How much of this final balance is interest?

▶ Solution on page 129

Sinking Funds

A *sinking fund* is a regular savings plan designed to provide a given amount after a certain number of years. For example, suppose you decide to save up for a $18,000 car instead of taking out a loan. What amount should you save each month to have the $18,000 in 3 years? If you just put your money in a shoe box, you will need to set aside $\$18{,}000 \div (12 \times 3) = \$500$ each month. It would make more sense to put your money into a savings account each month and let it earn compound interest. This is the same as setting up an annuity with a given accumulated amount and asking what the regular payment should be. Solving the accumulated amount formula (on the previous page) for P by dividing by the fraction following P gives the following formula.

Sinking Fund Payment

The payment P to make m times per year for t years at interest rate r compounded at each payment to accumulate amount A is

$$P = A\frac{r/m}{(1 + r/m)^{mt} - 1}$$

This formula may also be written $P = A/s_{\overline{n}|i}$ with $n = mt$ and $i = r/m$.

EXAMPLE 2

FINDING A SINKING FUND PAYMENT

What amount should be deposited at the end of each month for 3 years in a savings account earning 4.5% interest compounded monthly to accumulate $18,000 to buy a new car?

Solution

$$P = 18{,}000 \, \frac{0.045/12}{(1 + 0.045/12)^{12 \cdot 3} - 1} \approx 467.945$$

$P = A \dfrac{r/m}{(1 + r/m)^{mt} - 1}$
with $A = 18{,}000$,
$r = 0.045$, $m = 12$,
and $t = 3$

Rounding *up* to the next penny, $467.95 should be saved each month. (If we rounded down to $467.94, the savings account would accumulate slightly less than the amount needed.)

Notice that this amount is significantly lower than the $500 that would be needed *without* compound interest.

This example could also have been done using the accumulated amount formula (page 124) by substituting $A = 18{,}000$, $r = 0.045$, $m = 12$, and $t = 3$, and then solving for the remaining variable to find $P = \$467.95$.

Practice Problem 2

For wage earners paid every other week instead of monthly, Example 2 becomes: "What amount should be deposited biweekly for 3 years in a savings account earning 4.5% interest compounded biweekly to accumulate $18,000 for a new car?" Find this amount.

➤ Solution on page 129

How Long Will It Take?

Continuing with our sinking fund example, suppose you could save only $200 each month. How long would it take to accumulate the $18,000?

2.3 ANNUITIES

We solve the accumulated amount formula (page 124) for the number of periods mt.

$$A = P \frac{(1 + r/m)^{mt} - 1}{r/m}$$

$$\frac{A}{P}\frac{r}{m} + 1 = \left(1 + \frac{r}{m}\right)^{mt}$$

dividing by P, multiplying by r/m, and adding 1

$$mt \log\left(1 + \frac{r}{m}\right) = \log\left(\frac{A}{P}\frac{r}{m} + 1\right)$$

Switching sides and taking logs of both to bring down the exponent

Dividing by $\log(1 + r/m)$, we have:

Number of Periods

$$mt = \frac{\log\left(\dfrac{A}{P}\dfrac{r}{m} + 1\right)}{\log\left(1 + \dfrac{r}{m}\right)} \text{ periods}$$

As usual, we round *up* to the next whole number of compounding periods and then express the answer as years plus any extra periods. Notice that the time t depends only on the ratio $\frac{A}{P}$ between the accumulated amount A and the regular payment P.

EXAMPLE 3 FINDING THE TERM FOR A SINKING FUND

How long will it take to accumulate $18,000 by depositing $200 at the end of each month in a savings account earning 4.5% interest compounded monthly?

Solution

$$12t = \frac{\log\left(\dfrac{18{,}000}{200} \dfrac{0.045}{12} + 1\right)}{\log\left(1 + \dfrac{0.045}{12}\right)} \approx 77.7$$

$$mt = \frac{\log\left(\dfrac{A}{P}\dfrac{r}{m} + 1\right)}{\log\left(1 + \dfrac{r}{m}\right)}$$

with $A = 18{,}000$, $P = 200$, $r = 0.045$, and $m = 12$

The time is 77.7 months, which we round *up* to 78 months. Therefore, it will take 6 years 6 months to accumulate the $18,000.

This example could also have been done using the accumulated amount formula (page 124) by substituting $A = 18{,}000$, $P = 200$, $r = 0.045$, and $m = 12$, and then solving (using logarithms) for the remaining variable to find that $12t$ is 78 months.

Practice Problem 3

How long will it take to accumulate $18,000 by depositing $250 each month in a savings account earning 4.5% interest compounded monthly?
▶ Solution on next page

Graphing Calculator Exploration

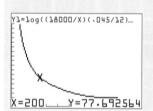

Depending on the size of the monthly deposit, the time needed to accumulate a given amount may vary greatly.

a. Using the values $r = 0.045$, $m = 12$, and $A = 18{,}000$ from Example 3, and replacing the payment amount P by x, graph $y_1 = \log((18000/x)(.045/12) + 1)/\log(1 + .045/12)$ on the window [0, 1000] by [0, 240]. The graph shows how y, the number of months needed to accumulate $18,000, decreases as the monthly payment x increases.

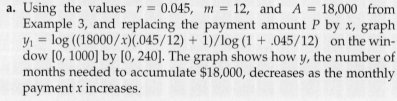

b. Use TRACE or VALUE to find y when $x = 200$ and when $x = 250$. How do your answers compare to the answers to Example 3 and Practice Problem 3?

c. Use TRACE or VALUE to find the years y needed when the monthly deposits x are $100, $150, $200, $250, and $300. Does each additional $50 reduce the time needed by the same amount? How does the graph show this?

Overview

In the previous section we found the eventual value of an investment, the initial amount necessary to reach that later amount, and the time it takes, just as we did in this section. What's the difference between the two sections? *There* we were considering investments of a *single* payment of P dollars, while *here* we are considering *multiple and regular payments* of P dollars.

2.3 Section Summary

For an *ordinary annuity* with regular payments of P dollars m times per year for t years at interest rate r compounded at the end of each payment, the accumulated amount A is

$$A = P\frac{(1 + r/m)^{mt} - 1}{r/m}$$

A = accumulated amount,
P = payment, r = interest rate,
m = times per year, t = years

To find how much of this amount is interest, subtract the total of all the payments, $P \cdot m \cdot t$, from it.

The annuity formula may be solved for the other variables. Solving for P finds the regular payment to make into a *sinking fund* to accumulate the amount A after t years:

$$P = A\frac{r/m}{(1 + r/m)^{mt} - 1} \qquad \text{Round } up$$

Solving for mt finds the number of periods required for the payments into a sinking fund to accumulate to a final amount A:

$$mt = \frac{\log\left(\frac{A}{P}\frac{r}{m} + 1\right)}{\log\left(1 + \frac{r}{m}\right)}$$

mt = number of periods
(other variables as above)

Round *up* to find the number of periods and then express your answer as a number of years plus any additional periods.

▶ Solutions to Practice Problems

1. $A = 40\dfrac{(1 + 0.05/52)^{52 \cdot 35} - 1}{0.05/52} \approx 197{,}590.27.$ The final balance is $197,590.27. Because the deposits total $40 \cdot 52 \cdot 35 = \$72{,}800$, the final balance contains $124,790.27 interest.

2. $P = 18{,}000\dfrac{0.045/26}{(1 + 0.045/26)^{26 \cdot 3} - 1} \approx 215.742.$ $215.75 should be deposited every other week. (Compare double this amount, $431.50, to the $467.95 monthly payment found in Example 2 on page 126.)

3. $12t = \dfrac{\log\left(\dfrac{18{,}000}{250}\dfrac{0.045}{12} + 1\right)}{\log\left(1 + \dfrac{0.045}{12}\right)} \approx 63.9$ months, which rounds up to 64 months. It will take 5 years 4 months.

2.3 Exercises

In the following ordinary annuities, the interest is compounded with each payment, and the payment is made at the end of the compounding period.

Find the accumulated amount of each annuity.

1. $1500 annually at 7% for 10 years.
2. $1750 annually at 8% for 7 years.
3. $2000 semiannually at 6% for 12 years.
4. $950 quarterly at 10.7% for 6 years.
5. $1000 monthly at 6.9% for 20 years.
6. $200 weekly at 11.3% for 5 years.

Find the required payment for each sinking fund.

7. Monthly deposits earning 5% to accumulate $5000 after 10 years.
8. Monthly deposits earning 4% to accumulate $6000 after 15 years.
9. Monthly deposits earning 11.7% to accumulate $14,000 after 5 years.
10. Quarterly deposits earning 6.4% to accumulate $50,000 after 20 years.
11. Yearly deposits earning 12.3% to accumulate $8500 after 12 years.
12. Weekly deposits earning 9.8% to accumulate $15,000 after 6 years.

Find the amount of time needed for each sinking fund to reach the given accumulated amount.

13. $1500 yearly at 8% to accumulate $100,000.
14. $2000 yearly at 9% to accumulate $125,000.
15. $800 semiannually at 6.6% to accumulate $80,000.
16. $500 quarterly at 8.2% to accumulate $19,500.
17. $235 monthly at 5.9% to accumulate $25,000.
18. $70 weekly at 10.9% to accumulate $7500.

APPLIED EXERCISES

19. **Retirement Savings** An individual retirement account, or IRA, earns tax-deferred interest and allows the owner to invest up to $3000 each year. Joe and Jill both will make IRA deposits for 30 years (from age 35 to 65) into stock mutual funds yielding 9.8%. Joe deposits $3000 once each year while Jill has $57.69 (which is 3000/52) withheld from her weekly paycheck and deposited automatically. How much will each have at age 65?

20. **Retirement Savings**
    a. Steve opens a retirement account yielding 9% and deposits $250 each month for the next 30 years (from age 35 to 65). How much will he have at age 65?
    b. Sue opens her own retirement account yielding 8% and deposits $250 each month for the 10 years from age 25 to 35. She then makes no further deposits but lets the accumulated amount earn compound interest for the next 30 years. How much will she have at age 65?
    c. Who has more money for retirement?

21. **Mutual Funds** How much must you invest each month in a mutual fund yielding 13.4% compounded monthly to become a millionaire in 10 years?

22. **Bond Funds** How much must you invest each quarter in a bond fund yielding 9.7% compounded quarterly to become a millionaire in 20 years?

23. **Lifetime Savings** The Oseola McCarty Scholarship Fund at the University of Southern Mississippi was established by a $150,000 gift from an 87-year-old woman who had dropped

out of sixth grade and worked for most of her life as a washerwoman. How much would she have had to save each week in a bank account earning 3.9% compounded weekly to have $150,000 after 75 years?

24. **Lifetime Savings** If each day your fairy godmother put $1 into a savings account at 8.37% compounded daily, how much would the account be worth when you were 65 years old?

25. **Home Buying** You and your new spouse each bring home $1500 each month after taxes and other payroll deductions. By living frugally, you intend to live on just one paycheck and save the other in a bond fund yielding 7.86% compounded monthly. How long will it take to have enough for a 20% down payment on a $165,000 condo in the city?

26. **Boat Buying** How long will it take to save $16,000 for a new motorboat if you deposit $375 each month into a money market fund yielding 5.73% compounded monthly?

Explorations and Excursions

The following problems extend and augment the material presented in the text.

More About Annuities

27. **Internal Rate of Return of an Annuity** Suppose that you make quarterly investments of $500 into a stock fund, and after 8 years the fund is worth $27,660. How can you find the "rate of return" on your money? If x is the (annual) rate of return, then you need to solve

$$500 \frac{(1 + x/4)^{4 \cdot 8} - 1}{x/4} = 27{,}660$$

Enter the left-hand side of this equation into your calculator as y_1 and GRAPH it on the window $[0, .2]$ by $[0, 30{,}000]$. Then use TRACE (or INTERSECT with $y_2 = 27{,}660$) to find the value of x where y_1 equals 27,660. The resulting value of x is called the *internal rate of return* of your investment.

28. **Stock Market** According to the *Wall Street Journal*, the $16\frac{1}{2}$ years from February 1966 to August 1982 was perhaps the most brutal period for stock market investors since the Great Depression in the 1930s. Ibbotson Associates estimates that if you had invested $100 each month into the stocks that make up the S&P 500 during this period, by the market low at mid-1982 your portfolio would have been worth almost $32,600 (with all dividends reinvested). Use the method from Exercise 27 (but now with *monthly* compounding) to find the internal rate of return for such an investment.

29. Solve any three of Exercises 1–6 as follows: Enter the accumulated amount formula (page 124) as $y = P((1 + r/m)^{mt} - 1)/(r/m)$. Then for each exercise, STORE the values for P, r, m, and t, and evaluate y to find the amount due.

More About Geometric Series

30. For each choice of values for a, x, and n, write out the geometric series

$$a + ax + ax^2 + \cdots + ax^{n-2} + ax^{n-1},$$

find the sum by adding up the terms, and then evaluate the corresponding $a \frac{x^n - 1}{x - 1}$ expression to check that both have the same value.

  a. $a = 2, \quad x = 3, \quad n = 5$.
  b. $a = 162, \quad x = \frac{1}{3}, \quad n = 5$.
  How do parts (a) and (b) differ?
  c. $a = 3, \quad x = 10, \quad n = 6$

31. *Requires a calculator with series operations.* Solve Exercise 30 as follows. Your calculator can LIST the numbers making up a geometric series as a *sequence* of values by using SEQUENCE. For the values of a, x, and n in Exercise 30a, list the numbers as the SEQUENCE $2(3^n)$ for n starting at 0, ending at 4, and increasing by 1. This list of numbers can be added using SUM. In this way, check each of the SUMs in Exercise 30. Then check that the geometric SEQUENCE $200(1 + 0.06/12)^n$ on page 122 for n from 0 to 47 SUMs to $10,819.57.

32. **Infinite Geometric Series** *Requires a calculator with series operations.* With the values $a = 1$ and $x = \frac{1}{2}$, the geometric series formula becomes

$$1 + \frac{1}{2} + \left(\frac{1}{2}\right)^2 + \cdots + \left(\frac{1}{2}\right)^{n-1} = \frac{\left(\frac{1}{2}\right)^n - 1}{\frac{1}{2} - 1}$$

a. Check this formula by evaluating both sides for $n = 5$, $n = 10$, and $n = 15$. As n gets larger, do these values get closer to 2?

b. Graph the SUM y of the first x terms of this geometric SEQUENCE on the window [0, 25] by [0, 3]. Use TRACE to explore the graph. As x gets larger, do these values get closer to 2?

An *infinite geometric series* is a sum of the form $a + ax + ax^2 + ax^3 + ax^4 + \cdots$, where the $+ \cdots$ indicates that the sum continues on to include every power of x. For $a = 1$ and $x = \frac{1}{2}$, we suspect from parts (a) and (b) that

$$1 + \tfrac{1}{2} + \left(\tfrac{1}{2}\right)^2 + \left(\tfrac{1}{2}\right)^3 + \left(\tfrac{1}{2}\right)^4 + \cdots = 2$$

c. Check this arithmetic: Because $\left(\tfrac{1}{2}\right)^n$ gets closer to 0 as n gets larger, the geometric series formula $\dfrac{\left(\tfrac{1}{2}\right)^n - 1}{\tfrac{1}{2} - 1}$ gets closer to $\dfrac{0 - 1}{\tfrac{1}{2} - 1}$ as n gets larger.

d. If $1 + \tfrac{1}{2} + \left(\tfrac{1}{2}\right)^2 + \left(\tfrac{1}{2}\right)^3 + \left(\tfrac{1}{2}\right)^4 + \cdots$ really is 2, it should behave just like 2. Show that

$$\tfrac{1}{2}\left(1 + \tfrac{1}{2} + \left(\tfrac{1}{2}\right)^2 + \left(\tfrac{1}{2}\right)^3 + \left(\tfrac{1}{2}\right)^4 + \cdots\right) = 1$$

by multiplying out the left side and using the fact that

$$1 + \tfrac{1}{2} + \left(\tfrac{1}{2}\right)^2 + \left(\tfrac{1}{2}\right)^3 + \left(\tfrac{1}{2}\right)^4 + \cdots = 2.$$

If $|x| < 1$, then x^n gets closer to 0 as n increases and we have the following formula for an *infinite* geometric series:

$$a + ax + ax^2 + \cdots = a \cdot \frac{1}{1 - x}$$

e. Find the value of

$$1 + \tfrac{1}{3} + \left(\tfrac{1}{3}\right)^2 + \left(\tfrac{1}{3}\right)^3 + \left(\tfrac{1}{3}\right)^4 + \cdots.$$

f. Find the value of

$$9 + 9\left(\tfrac{1}{10}\right) + 9\left(\tfrac{1}{10}\right)^2 + 9\left(\tfrac{1}{10}\right)^3 + 9\left(\tfrac{1}{10}\right)^4 + \cdots.$$

Can you write this geometric series as a decimal?

2.4 AMORTIZATION

Introduction

In this section we will find the *present value of an annuity*—the total value *now* of all the future payments. Such calculations are important for buying and selling annuities and for using them to pay off debts (amortization). For example, on pages 124–125 we found the accumulated amount for monthly payments of $200 for 4 years at 6% compounded monthly. But how much would a loan company give *right now* for that promise of future payments?

Present Value of an Annuity

We know that for an ordinary annuity with payments of P dollars m times per year for t years at interest rate r compounded at each payment, the accumulated amount A is

$$A = P\frac{(1 + r/m)^{mt} - 1}{r/m} \qquad \text{From page 124}$$

2.4 AMORTIZATION

On page 113 we saw that to find the present value of an amount, you simply divide by $(1 + r/m)^{mt}$. Therefore, the present value of this amount (denoted PV to distinguish it from P for payment) is

$$PV = P \frac{(1 + r/m)^{mt} - 1}{(r/m)(1 + r/m)^{mt}} \quad \text{Dividing by } (1 + r/m)^{mt}$$

$$= P \frac{1 - (1 + r/m)^{-mt}}{r/m} \quad \begin{array}{l}\text{Dividing numerator}\\\text{and denominator by}\\(1 + r/m)^{mt}\end{array}$$

This gives the formula:

Present Value of an Annuity

The present value PV of an ordinary annuity with payments of P dollars m times per year for t years at interest rate r compounded at each payment is

$$PV = P \frac{1 - (1 + r/m)^{-mt}}{r/m}$$

Although we will not use it, the archaic symbol $a_{n \rceil i}$ (pronounced "a sub n angle i") sometimes still occurs in business textbooks and tables to denote the value of $(1 - (1 + i)^{-n})/i$. In this notation, the above formula takes the form $PV = P a_{n \rceil i}$ with $n = mt$ and $i = r/m$.

EXAMPLE 1

FINDING THE PRESENT VALUE OF AN ANNUITY

What is the present value of a 6% car loan for 4 years with monthly payments of $200?

Solution

$$PV = 200 \frac{1 - (1 + 0.06/12)^{-12 \cdot 4}}{0.06/12} \approx 8516.06 \quad \begin{array}{l}PV = P \dfrac{1 - (1 + r/m)^{-mt}}{r/m}\\\text{with } P = 200, \; r = 0.06,\\m = 12, \text{ and } t = 4\end{array}$$

The present value is $8516.06.

In general, how can we interpret the present value of an annuity? It is the sum that must be deposited in a bank *now* (at the stated interest rate) to grow to exactly the amount of the annuity at the end of its term. For the loan in Example 1, in 4 years the present value will grow to $8516.06(1 + 0.06/12)^{12 \cdot 4} = \$10,819.57,$ which is exactly the sum of the annuity that we found on page 125. Therefore, the present value of $8516.06 is exactly the price for which the sequence of car payments should be sold, and the lender will not care which she receives.

Practice Problem 1

What is the present value of a 20-year retirement annuity paying $850 per month if the current long-term interest rate is 7.53%?

> Solution on page 140

Be Careful! The formulas for the *present value* and the *accumulated* or *future value* of an annuity (derived on page 124) look similar but are not the same, and should not be confused:

Present Value $\quad P \dfrac{1 - (1 + r/m)^{-mt}}{r/m}$

Future (or Accumulated) Value $\quad P \dfrac{(1 + r/m)^{mt} - 1}{r/m}$

As we saw on page 113, the future value formula divided by $(1 + r/m)^{mt}$ gives the present value formula.

Amortization

Suppose that you took out a loan to buy a house, and wanted to pay off the debt in a fixed number of payments, with the bank charging interest on the unpaid balance. A debt is *amortized* (or "killed off") if it is repaid by a regular sequence of payments.

Spreadsheet Exploration

The following spreadsheet* shows an "amortization table" for a debt of $100,000 to be paid off in five annual payments at 10% compounded yearly. Notice that the last payment is adjusted to *exactly* pay off the remaining debt and interest.

---

* See the Preface for how to obtain this and other Excel spreadsheets.

2.4 AMORTIZATION

	A	B	C	D	E	F	G
			=PMT(A5/(A11),A8*A11,A2)				
1	Debt		Payment				
2	100000		(26,379.75)				
3							
4	Annual Rate						
5	10%		Year	Payment	Interest	Debt Reduction	Outstanding Debt
6			0.00	0.00	0.00	0.00	100,000.00
7	Term in Years		1.00	(26,379.75)	10,000.00	(16,379.75)	83,620.25
8	5		2.00	(26,379.75)	8,362.03	(18,017.72)	65,602.53
9			3.00	(26,379.75)	6,560.25	(19,819.50)	45,783.03
10	Compoundings per Year		4.00	(26,379.75)	4,578.30	(21,801.45)	23,981.58
11	1		5.00	(26,379.74)	2,398.16	(23,981.58)	0.00
12							
13			Totals:	(131,898.74)	31,898.74	(100,000.00)	

The payment amount in cell C2 was found using the spreadsheet PMT function. The table shows how each payment is allocated to pay the interest and some of the debt, with the early payments paying more for interest than the later payments, which go almost entirely for debt.

How do we calculate the correct payments to amortize a debt? The payments form an annuity with present value PV equal to the debt D. To find the payment P to amortize the debt, we solve for P in the "present value of an annuity" formula on page 133 (dividing by the fraction following the P) and then replacing PV by D.

Amortization Payment

The payment P to make m times per year for t years at interest rate r compounded at each payment to amortize a debt of D dollars is

$$P = D \frac{r/m}{1 - (1 + r/m)^{-mt}}$$

This formula may also be written $P = D/a_{\overline{n}|i}$ with $n = mt$ and $i = r/m$.

EXAMPLE 2

FINDING THE PAYMENT TO AMORTIZE A DEBT

What monthly payment will amortize a $150,000 home mortgage at 8.6% in 30 years?

Solution

$$P = 150{,}000 \frac{0.086/12}{1 - (1 + 0.086/12)^{-12 \cdot 30}} \approx 1164.02$$

with $P = D \dfrac{r/m}{1 - (1 + r/m)^{-mt}}$, $D = 150{,}000$, $r = 0.086$, $m = 12$, and $t = 30$

The required payment is $1164.02 each month.

Notice that the borrower will pay a total of $1164.02 \cdot 12 \cdot 30 = $419,047.20 during the 30 years of the loan to pay off a $150,000 debt. The difference, $269,047.20, is the interest paid to the bank.

Practice Problem 2 What monthly payment will amortize a $150,000 home mortgage at 8.6% in 25 years? What is the total amount the borrower will pay?

➤ Solution on page 140

Graphing Calculator Exploration

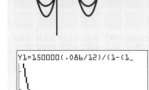

When amortizing a loan, a longer term means smaller payments, but the total amount the borrower pays is much larger.

a. To find the monthly payment y to amortize a debt over x years using the values $D = 150{,}000$, $r = 0.086$, and $m = 12$ from Practice Problem 2, graph the curve $y_1 = 150000(.086/12)/(1 - (1 + .086/12)^{-12x})$ on the window $[0, 50]$ by $[0, 5000]$.

b. Notice that as the term increases the payment decreases, at first rapidly but later more slowly. Use TRACE or VALUE to find the payments for mortgage terms of 15, 20, 25, and 30 years. How do your answers compare to the answers to Example 2 and Practice Problem 2?

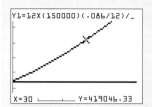

c. To find the total amount the borrower pays, multiply the payment by m and t (here replaced by x). Graph the new curve $y_1 = 12x(150000)(.086/12)/(1 - (1 + .086/12)^{-12x})$, along with the line $y_2 = 150000$ to represent the original debt, on the window [0, 50] by [0, 600,000] to find the total amount paid on a loan for x years.

d. Notice that as the term increases the total amount paid increases, well above the original debt of $150,000. Use TRACE or VALUE to find the total amount paid for mortgage terms of 15, 20, 25, and 30 years. Based on these two screens, does it make financial sense to extend the term of this loan beyond 30 years?

Unpaid Balance

How much does the borrower still owe partway through an agreed amortization schedule? Suppose you have made mortgage payments for 10 years on a 30-year loan and now you must move. How much should you pay to settle your debt? Your remaining payments form another annuity and are worth precisely the present value of this new annuity.

EXAMPLE 3

FINDING THE UNPAID BALANCE

What is the unpaid balance after 10 years of monthly payments on a 30-year mortgage of $150,000 at 8.6%?

Solution

First we must calculate the monthly payments. But for this problem, we know from Example 2 (on the previous page) that the monthly payment to amortize the original mortgage is $1164.02. The remaining payments thus form an annuity with $P = \$1164.02$, $r = 0.086$, $m = 12$, and $t = 20$ years. The amount still owed on the mortgage is the present value of this annuity, which we can calculate by the formula on page 133.

$$PV = 1164.02 \frac{1 - (1 + 0.086/12)^{-12 \cdot 20}}{0.086/12}$$

$$\approx 133{,}158.27$$

$PV = P \dfrac{1 - (1 + r/m)^{-mt}}{r/m}$

with $P = 1164.02$, $r = 0.086$, $m = 12$, and $t = 20$

The unpaid balance after 10 years of payments on this $150,000 mortgage is $133,158.27.

Notice that after the first 10 of 30 years, much less than one third of the mortgage has been paid. The early payments of any loan pay mostly interest and only a little of the principal.

As in this example, if you already know the payments, you use the same payments to calculate the unpaid balance. However, if you do *not* know the payments, you would first have to calculate them using the payment formula on page 135.

Practice Problem 3

How much is still owed on a 25-year mortgage of $150,000 at 8.6% after 13 years of monthly payments? [*Hint:* Use the result of Practice Problem 2 on page 136.] ➤ Solution on page 141

Equity

One often hears the term "equity" used with homes, such as a "home equity loan." *Equity* is defined as the *value of the home minus any unpaid mortgage balance,* and is a measure of the value that you (rather than the bank) have in your house. If the house with the mortgage in Example 3 now has a market value of $300,000, then the homeowner's equity after 10 years of payments is this value minus the unpaid balance: $300,000 - 133,158 = $166,842 (ignoring cents). This equity may be used to take out a new loan for a larger house.

Amortization Tables

The Spreadsheet Exploration on pages 134–135 showed an amortization table on a spreadsheet. Amortization tables can also be constructed on graphing calculators.

Graphing Calculator Exploration

The program* AMORTABL constructs an *amortization table* that shows how each payment is allocated to paying the interest and reducing the outstanding debt. To explore the table for the situation used in Examples 2 and 3, run this program by selecting it from the program menu and proceed as follows:

---

* See the Preface for information on how to obtain this and other graphing calculator programs.

2.4 AMORTIZATION

```
DEBT (DOLLARS)
   150000
RATE (DECIMAL)
    .086
NUMBER OF YEARS
     30
COMPOUNDINGS/YR
     12
```

```
     VIEW TABLE
1:FROM START
2:JUMP TO YEAR
3:AT END
4:FINISHED
```

a. Enter 150000 for the debt, .086 for the interest rate, 30 for the number of years, and 12 for the number of compoundings per year (because the payments are monthly).

b. After the table of values has been calculated by finding the payment rounded to the nearest penny and then applying each payment to the rounded interest due on the debt for that period and reducing the debt by the excess, you may choose which part of the table you wish to see.

c. The amortization table contains five columns: X is the number of the payment (negative X's or X's larger than the number of payments display zeros in the other columns), Y_1 is the payment, Y_2 is the interest part of the payment, Y_3 is the remaining part used to reduce the debt, and Y_4 is the debt remaining after this reduction. Use the arrow keys to scroll right and left through these columns.

X	Y_1	Y_2
0	0	0
1	1164	1075
2	1164	1074.4
3	1164	1073.7
4	1164	1073.1
5	1164	1072.4
6	1164	1071.8
X=0		

X	Y_3	Y_4
0	0	150000
1	89.02	149911
2	89.66	149821
3	90.3	149731
4	90.95	149640
5	91.6	149548
6	92.26	149456
Y_4=150000		

X	Y_1	Y_2
356	1164	40.8
357	1164	32.75
358	1164	24.64
359	1164	16.48
360	1160	8.25
361	0	0
362	419043	269043
X=361		

X	Y_3	Y_4
356	1123.2	4570
357	1131.3	3438.7
358	1139.4	2299.3
359	1147.5	1151.8
360	1151.8	0
361	0	0
362	150000	0
Y_4=0		

The second line after the end of the amortization (line 362) contains the total payments, interest, and debt reduction. Notice that the final payment is adjusted to correct for rounding errors and the debt is reduced to zero.

X	Y_3	Y_4
118	205.28	133573
119	206.75	133366
120	208.23	133158
121	209.72	132948
122	211.23	132737
123	212.74	132524
124	214.27	132310
Y_4=133157.56		

d. To use the table to find the remaining debt after a given number of years, "jump" to that position in the table, and scroll to the Y_4 column. After 10 years of payments, the table shows $133,157.56 remaining.

2.4 Section Summary

In this section we made extensive use of the formula for the accumulated amount of an annuity (page 124) that we derived in the previous section. The *present value* of an ordinary annuity compounded at each payment is

$$PV = P\frac{1 - (1 + r/m)^{-mt}}{r/m}$$

PV = present value,
P = payment, r = rate,
m = times per year, and
t = years

The payment to *amortize a debt* of D dollars is found by solving this formula for P.

$$P = D\frac{r/m}{1 - (1 + r/m)^{-mt}}$$

D = debt
(other variables are as before)

The amount still owed after making amortization payments for part of the term is the present value of a new annuity formed by the remaining payments and is found by the first formula.

This chapter has developed several formulas. The following diagram may help to organize some of them.

Single Payment	Simple Interest	$A = P(1 + rt)$	Section 2.1
	Compound Interest	$A = P(1 + r/m)^{mt}$	Section 2.2
Multiple Payments (Annuity)	Future Value	$A = P\frac{(1 + r/m)^{mt} - 1}{r/m}$	Section 2.3
	Present Value	$PV = P\frac{1 - (1 + r/m)^{-mt}}{r/m}$	Section 2.4

> **Solutions to Practice Problems**

1. $PV = 850\dfrac{1 - (1 + 0.0753/12)^{-12 \cdot 20}}{0.0753/12} \approx 105{,}272.468$.
   The present value is $105,272.47.

2. $P = 150{,}000 \dfrac{0.086/12}{1 - (1 + 0.086/12)^{-12 \cdot 25}} \approx 1217.97$.
   The required payment is $1217.97 each month. The borrower will pay a total of $1217.97 \cdot 12 \cdot 25 = \$365{,}391$.

3. The remaining payments form a 12-year annuity at 8.6% with monthly payments of $1217.97 (from Practice Problem 2). The present value of this annuity is

$$PV = 1217.97 \frac{1 - (1 + 0.086/12)^{-12 \cdot 12}}{0.086/12} \approx 109{,}174.18$$

so the amount still owed is $109,174.18. Notice that after more than half the term, less than half of the $150,000 loan is paid off.

2.4 Exercises

Calculate the present value of each annuity.

1. $15,000 annually at 7% for 10 years.
2. $18,000 annually at 6% for 15 years.
3. $10,000 semiannually at 11% for 7 years.
4. $3000 quarterly at 8% for 8 years.
5. $1400 monthly at 6.9% for 30 years.
6. $25 weekly at 7.7% for 6 years.

Determine the payment to amortize each debt.

7. Monthly payments on $100,000 at 5% for 25 years.
8. Monthly payments on $125,000 at 5.4% for 20 years.
9. Monthly payments on $6000 at 6% for 5 years.
10. Annual payments on $50,000 at 8.2% for 10 years.
11. Quarterly payments on $14,500 at 12.7% for 6 years.
12. Weekly payments on $2500 at 18% for 3 years.

Find the unpaid balance on each debt (you should already know the payments from Exercises 7–12).

13. After 6 years of monthly payments on $100,000 at 5% for 25 years.
14. After 15 years of monthly payments on $125,000 at 5.4% for 20 years.
15. After 2 years 11 months of monthly payments on $6000 at 6% for 5 years.
16. After 3 years of annual payments on $50,000 at 8.2% for 10 years.
17. After 2 years 3 months of quarterly payments on $14,500 at 12.7% for 6 years.
18. After 2 years of weekly payments on $2500 at 18% for 3 years.

APPLIED EXERCISES

19. **Contest Prizes** The super prize in a contest is $10 million. This prize will be paid out in equal yearly payments over the next 25 years. If the prize money is guaranteed by U.S. Treasury bonds yielding 7.9% and is placed into an escrow account when the contest is announced 1 year before the first payment, how much do the contest sponsors have to deposit in the escrow account?

20. **Grant Funding** In September 1994, the New York City public schools system was awarded a grant of $50 million over 5 years from the Annenberg Foundation. If the grant paid equal yearly amounts for the next 5 years and was financed at 6.8%, how much did this grant cost the Foundation when it was announced?

21. **NFL Contracts** When Michael Westbrook signed a 7-year, $18 million contract with the Washington Redskins, it included a $6.5 million signing bonus that was the largest for a wide receiver in NFL history. If the $18 million was paid out in equal quarterly payments for the 7 years and the current long-term interest rate was 6.85%, how much was the contract worth when it was signed?

22. **Escrow Accounts** On September 1, 1994, Judge Sam C. Pointer, Jr., of the Federal District Court in Birmingham approved a $4.25 billion settlement against the manufacturers of silicon breast implants to be paid out over the next 30 years. If this money was to be paid out in equal semiannual amounts from an escrow account earning 8.7% interest, how much would the manufacturers have to deposit into the account (6 months before the first payment) to meet the terms of the settlement?

23. **College Savings** When I graduated from high school last spring, my dear Aunt Sallie gave me a savings account passbook for an account earning 5.61%. She told me that starting next fall, there would be enough money for me to take out $10,000 every 6 months for the next 4 years to pay for my college expenses. I thought "Wow! Thanks! That's $80,000!" but she just smiled, shook her head "No," and told me to look in the passbook. How much was in the account when she gave it to me?

24. **Car Buying** A Cadillac dealer offers the following terms on a 2-year lease of a new Eldorado: either $1995 down and $339 per month for 24 months or $9599 down and nothing more to pay for 24 months. (a) If the current interest rate were 5.2%, which option would be cheaper? (b) Since the car dealer is an expert money manager, what must the current interest rate be if both offers are equally good for the dealer?

25. **Life Insurance** Just before his first attempt at bungee jumping, John decides to buy a life insurance policy. His annual income at age 30 is $35,000, so he figures he should get enough insurance to provide his wife and new baby with that amount each year for the next 35 years. If the long-term interest rate is 6.7%, what is the present value of John's future annual earnings? Rounding up to the next $50,000, how much life insurance should he buy?

26. **Life Insurance** A 60-year-old grandmother wants a life insurance policy that could replace her annual $55,000 earnings for the next 10 years. If the long-term interest rate is now 6.2%, how much life insurance (rounded up to the next $10,000) should she buy?

27. **Mortgages** Real estate prices are so high in Tokyo that some mortgages are written for 99 years in an attempt to keep down the monthly payments. (a) What is the monthly payment on a $500,000 mortgage at 8% for 99 years? (b) How much does this save the borrower each month compared to a 30-year mortgage at 8% for the same amount? (c) Which term results in the higher total payment?

28. **Apartment Rents** The Associates for International Research have estimated the following monthly rents for a two-bedroom apartment in major cities around the world.

Sydney	$1100	London	$2950
Cairo	$1200	Moscow	$5000
Buenos Aires	$1500	Shanghai	$6300
Paris	$2150	Tokyo	$7100
New York	$2300	Hong Kong	$7200

Assuming that the monthly rent represents the payment on the current market value of the apartment as a 30-year investment and that the long-term interest rate is 6.52%, estimate the current market value for an apartment in (a) Sydney, (b) Paris, (c) Moscow, and (d) Tokyo.

29. **Credit Cards** A MasterCard statement shows a balance of $560 at 13.9% compounded monthly. What monthly payment will pay off this debt in 1 year 8 months?

30. **Credit Cards** The MasterCard statement in Exercise 29 also states that the "minimum payment" is $15. Find, in two different ways, how long it will take to pay off this debt by making this minimum payment each month.

  a. To estimate the number of years required, STORE the values 15 for P, 0.139 for r, and

12 for m before graphing the present value formula $y = P(1 - (1 + r/m)^{-mx})/(r/m)$ on the window [0, 10] by [0, 1000]. Use TRACE or INTERSECT to estimate the number of years x that corresponds to the debt's present value $y = \$560$. Be sure to change this number of years into the number of months the payments must be made.

b. To estimate the number of years required, STORE the values 560 for D, 0.139 for r, and 12 for m before graphing the amortization payment formula

$$y = D(r/m)/(1 - (1 + r/m)^{-mx})$$

on the window [0, 10] by [0, 20]. Use TRACE or INTERSECT to estimate the number of years x that corresponds to the payment $y = \$15$. Be sure to change this number of years into the number of months the payments must be made.

c. Why are the answers from parts (a) and (b) exactly the same?

31. Credit Cards What monthly payment should you make on your Visa card to pay off a new $1575 stereo in 15 months if the interest rate is 14.8% compounded monthly?

32. Mortgages The *Town Gossip* local newspaper lost the libel suit filed by the village's former mayor. What quarterly payment will pay off the $150,000 judgment in 5 years at 11.6% compounded quarterly?

33. College Costs When Jill graduated from college, her loans with interest totaled $58,720. Because her last co-op employer invited her to stay on as a full-time employee, she was able to handle the monthly payments even though the interest rate was 9.4% and the bank expected everything to be paid back in 8 years. How much debt remained after 3 years of payments?

34. Car Buying Tom bought a new turbo Ford Thunderbird and financed it with a new-car loan of $18,000 at 13.78% for 4 years. After making the monthly payments for 7 months, he decided he couldn't study for his business courses at college and put in enough hours at his part-time job to keep up his grades, so he offered to give the car to his sister if she would take over the payments. What purchase price would she be paying for the car?

35. Automobile Financing A car dealer's newspaper ad offers "0% financing with guaranteed credit approval" and includes an example: "Finance $10,000 for 48 months—normal payment is $262.84 while 0% financing is $208.33—you save $2616." What is the "normal" finance rate? To estimate the answer, graph the payment formula

$$y = 10000(x/12)/(1 - (1 + x/12)^{-48})$$

on the window [0, .2] by [0, 500] to see the monthly payment y for a 4-year loan of $10,000 at interest rate x compounded monthly. Use TRACE or INTERSECT to find the rate x that gives a y-value near $262.84. (By the way, the small print at the bottom of the ad contains the statement "0% financing may affect selling price of car.")

36. Computer Financing A Radio Shack back-to-school sale ad offers a "complete PC system for $39 per month after no payments for 6 months." The ad explains that "if you do not pay the full amount of purchase [$1270] by the end of the deferred period," then "finance charges will accrue from the date of purchase and will be added to the purchase balance" and states that the APY is 22.55%.

a. Show that the nominal interest rate for this loan is 20.508%.

b. How much will the balance be after 6 months if you make no payments?

c. How long will it then take to pay off the debt by paying the $39 each month? To estimate the number of years required, STORE the values 39 for P, 0.20508 for r, and 12 for m before graphing the present value formula $y = P(1 - (1 + r/m)^{-mx})/(r/m)$ on the window [0, 20] by [0, 2000]. Use TRACE or INTERSECT to estimate the number of years x that corresponds to the present value y from part (b). Be sure to change this number of years into the number of months the payments must be made.

d. Using the advertised financing, what is the buyer's total cost?

37. Yearly Interest A home owner finances a new luxury car using a "home equity" loan so that she can deduct the interest each year when calculating her taxable income. If she plans to repay the $40,000 in equal monthly payments over 4 years and the interest rate is fixed at 9.35%, how much interest will she pay in the second year of the loan?

38. **Home Refinancing** A widow needs cash and decides to refinance the home she and her husband bought 20 years ago for $35,000 with a monthly payment, 30-year mortgage at 6.8% on 80% of the purchase price. The Home Sweet Home Mortgage Corporation has appraised her house at $125,000 and has agreed to lend her 70% of this value. How much cash will she have after paying off the old mortgage?

Explorations and Excursions

The following problems extend and augment the material presented in the text.

More About Amortization

39. Solve any three of Exercises 1–6 as follows: Enter the "present value of an annuity" formula as $y = P(1 - (1 + r/m)^{-mt})/(r/m)$. Then for each exercise, STORE the values for P, r, m, and t, and evaluate y to find the present value.

40. **Unpaid Balance Formula** Combine the "present value of an annuity" formula with the amortization payment formula to show that the unpaid balance after x years of payments made m times each year for t years on a debt of D dollars at interest rate r is

$$D(1 + r/m)^{mx} \frac{(1 + r/m)^{m(t-x)} - 1}{(1 + r/m)^{mt} - 1}$$

41. Solve any three of Exercises 7–12 as follows: Enter the amortization payment formula as $y = D(r/m)/(1 - (1 + r/m)^{-mt})$. Then for each exercise, STORE the values for $m, D, r,$ and t, and evaluate y to find the required payment.

42. **Present Value of an Annuity** Show that the present value of an annuity formula may be rewritten in the form

$$PV = P \frac{1 - (1 + i)^{-n}}{i}$$

where $i = r/m$ and $n = mt$.

43. Show that the present value of an annuity is the sum of the present values of the payments. [*Hint:* Use the notation of Exercise 42 and show that the present value of the first payment is $P(1 + i)^{-1}$, that of the second is $P(1 + i)^{-2}$, and so on for the n payments. Factor out the common terms from the sum and use the geometric series formula on page 123.]

44. Use the program AMORTABL to solve Exercise 13. How does the amortization table solution differ from your previous solution to Exercise 13? How much does the final payment in the table differ from the payments made during the rest of the table?

45. Use the program AMORTABL to solve Exercise 16. Be sure to enter "1" for the "compoundings per year." How does the amortization table solution differ from your previous solution to Exercise 16? How much does the final payment in the table differ from the payments made during the rest of the table?

Chapter Summary with Hints and Suggestions

Reading the text and doing the exercises in this chapter have helped you to master the following skills, which are listed by section (in case you need to review them) and are keyed to particular Review Exercises. Answers for all Review Exercises are given at the back of the book, and full solutions can be found in the Student Solutions Manual.

CHAPTER SUMMARY WITH HINTS AND SUGGESTIONS

2.1 Simple Interest

- Find the interest due on a simple interest loan. *(Review Exercises 1–4.)*

$$I = Prt$$

- Find the total amount due on a simple interest loan. *(Review Exercises 5–8.)*

$$A = P(1 + rt)$$

- Solve a simple interest situation for the interest, the interest rate, the principal, or the term. *(Review Exercises 9–13.)*

- Find a simple interest future or present value. *(Review Exercises 14–16.)*

$$A = P(1 + rt) \qquad P = \frac{A}{1 + rt}$$

- Find the effective simple interest rate of a discounted loan. *(Review Exercises 17–18.)*

$$r_s = \frac{r}{1 - rt}$$

- Solve an applied simple-interest problem. *(Review Exercises 19–21.)*

2.2 Compound Interest

- Determine the amount due on a compound interest loan. *(Review Exercises 22–26.)*

$$A = P(1 + r/m)^{mt}$$

- Find a future or present value. *(Review Exercises 27–31.)*

$$A = P(1 + r/m)^{mt} \qquad P = \frac{A}{(1 + r/m)^{mt}}$$

- Find the term needed for a given principal to grow to a future value. *(Review Exercises 32–36.)*

- Use the "rule of 72" to estimate a doubling time. *(Review Exercises 37–41.)*

$$\left(\begin{array}{c}\text{Doubling} \\ \text{time}\end{array}\right) \approx \frac{72}{r \times 100}$$

- Find the effective rate of a loan or investment. *(Review Exercises 42–46.)*

$$r_e = (1 + r/m)^m - 1$$

2.3 Annuities

- Find the accumulated amount of an ordinary annuity. *(Review Exercises 47–51.)*

$$A = P\frac{(1 + r/m)^{mt} - 1}{r/m}$$

- Calculate the regular payment to make into a sinking fund to accumulate a given amount. *(Review Exercises 52–56.)*

$$P = A\frac{r/m}{(1 + r/m)^{mt} - 1}$$

- Find the number of periods to make payments into a sinking fund to accumulate a given amount. *(Review Exercises 57–61.)*

$$mt = \frac{\log\left((A/P)(r/m) + 1\right)}{\log(1 + r/m)}$$

- Estimate the internal rate of return of regular investments using a graphing calculator. *(Review Exercises 62–66.)*

2.4 Amortization

- Find the present value of an ordinary annuity. *(Review Exercises 67–71.)*

$$PV = P\frac{1 - (1 + r/m)^{-mt}}{r/m}$$

- Find the regular payment to amortize a debt. *(Review Exercises 72–76.)*

$$P = D\frac{r/m}{1 - (1 + r/m)^{-mt}}$$

- Find the amount still owed after making amortization payments for part of the term. *(Review Exercises 77–81.)*

Hints and Suggestions

- (Overview) Compound interest is repeated simple interest with the interest added to the principal in each successive period. The *future value* is the amount the principal will become after all the interest is included. Reversing the point of view, the *present value* is the principal needed now that will grow to the final amount.

- The three basic amount formulas are: $P(1 + rt)$ for simple interest, $P(1 + r/m)^{mt}$ for compound interest, and $P((1 + r/m)^{mt} - 1)/(r/m)$ for an annuity (multiple payments). The other formulas all follow from these basic formulas.
- The rate is usually stated as a percent but is always used in decimal form for calculations: 6.7% in decimal form is 0.067.
- Round off only your final answer when using your calculator. Don't use the decimal 0.33 for the fraction $\frac{1}{3}$.
- Don't confuse a *percent increase* with the *multiplier. Increasing* an amount by 25% means *multiplying* it by 1.25.
- Round *up* to find the whole number of compoundings needed to reach a future value because a smaller number won't reach the stated goal.
- The "rule of 72" is only an approximation but is a helpful check that can catch "button pushing" errors when using your calculator.
- **Practice for Test:** Review Exercises 1, 5, 9, 13, 18, 24, 27, 32, 34, 40, 43, 47, 50, 54, 58, 68, 73, 78.

Review Exercises for Chapter 2

Practice test exercise numbers are in green.

Round all dollar amounts to the nearest penny, all times in years to two decimal places, and all terms to the appropriate number of compounding periods.

2.1 Simple Interest

Find the simple interest on each loan.

1. $1875 at 5.8% for 2 years.
2. $1150 at 9.2% for 6 months.
3. $8000 at $7\frac{1}{2}$% for 3 years 9 months.
4. $2385 at 11.3% for 1 year 3 months.

Find the total amount due on each simple interest loan.

5. $8900 at 5.9% for 2 years 6 months.
6. $1375 at 11.3% for 9 months.
7. $1795 at 6.38% for 3 years.
8. $3700 at 8.3% for 1 year 11 months.

Solve each problem.

9. Find the interest rate on a loan charging $272 simple interest on a principal of $2000 after 2 years.
10. Find the principal of a loan at 9.3% if the simple interest after 1 year 8 months is $279.
11. Find the term of a loan of $7600 at 9% if the simple interest is $1026.
12. What should be the term for a loan of $5500 at 8.7% simple interest if the lender wants to receive $7414 when the loan is paid off?
13. How long will it take an investment at 10% simple interest to increase by 50%?
14. How much should be invested now at 5.9% simple interest if $5208 is needed in 2 years 8 months?
15. What would be the fair market price of a $10,000 zero coupon bond due in 1 year if today's long-term simple interest rate is 6.45%?
16. What would be the fair market price of a $50,000 zero coupon bond due in 1 year if today's long-term simple interest rate is 5.78%?

17. What is the effective simple interest rate of a discounted loan at 12% interest for 6 years?

18. What is the effective simple interest rate of a discounted loan at 15% interest for 2 years?

19. **Treasury Bonds** *Forbes* magazine ranked John W. Kluge as the richest American, with a net worth of $5.2 billion, at the time of his divorce from Patricia Kluge. A newspaper reported that besides a 45-room Georgian mansion, Mrs. Kluge "... received $1 billion that invested conservatively in 30-year Treasury bonds would throw off $66 million a year." What was the interest rate when the article was written?

20. **Insurance Settlements** How large an insurance settlement does an accident victim need for a yearly income of $29,000 from long-term bond investments paying 5.8%?

21. **Furniture Buying** A furniture store offers a complete living room suite for $999 and will finance the entire price as a discounted loan at 10% for 1 year. What is the effective simple interest rate for this loan and how much will the buyer need to pay at the end of the year?

2.2 Compound Interest

Find the amount due on each compound interest loan.

22. $15,000 at 7.5% compounded quarterly for 10 years.

23. $65,000 at 8.25% compounded daily for 17 years.

24. **Mutual Funds** From 1970 to 1995, the AIM Charter Fund posted an average annual return of 15.55% including sales charges. How much would a $10,000 investment in this fund have been worth after 25 years?

25. **Mutual Funds** During the mid-1990s, the Brandywine long-term growth fund returned 19.6% compounded quarterly. How much would a $5000 investment in this fund have been worth after 4 years?

26. **National Debt** From 1835 to 1837, the United States not only was free of debt but actually had a surplus in the Treasury. On January 1, 1837, after $5 million had been set aside as a reserve fund, there remained $37,468,859. If that money had not been distributed to the 26 states but instead had been invested at 5% compounded monthly, would there have been enough by 1997 to pay off a $4.3 trillion national debt?

Find the present value of each compound interest loan.

27. $25,000 after 10 years at 9% compounded monthly.

28. $30,000 after 8 years at 6% compounded daily.

29. $175,000 after 16 years 6 months at 5.5% compounded quarterly.

30. **Certificates of Deposit** From 1968 to 1995, the average annual return on 6-month CDs was 7.79% compounded semiannually. How much would one have to have invested in 1968 in order to have $100,000 in 1995?

31. **Bond Funds** During the mid-1990s, the Northeast Investors Trust general-term bond fund returned 15.7% compounded monthly. How much would you need to invest in this fund to have $15,000 after 3 years 8 months?

Find the term of each compound interest loan.

32. 8.9% compounded quarterly to obtain $10,000 from a principal of $2000.

33. 11.5% compounded monthly to multiply the principal by 1.60.

34. How long will it take the value of an apartment building to increase by 80% if real estate prices are increasing 8% annually?

35. **Treasury Bonds** How long will you have to wait to become a millionaire if you invest all of your $850,000 lottery winnings in Treasury bonds paying 5.6% compounded semiannually?

36. **Art Appreciation** How long will it take for the value of an early Picasso pencil sketch to triple if its market value is increasing 6.5% annually?

For Exercises 37–41, use the "rule of 72" to estimate the doubling time (in years) for each interest rate and then calculate it exactly.

37. 8% compounded semiannually.
38. 6% compounded monthly.
39. 11.9% compounded weekly.
40. **Real Estate** How long will it take for the value of a house to double if real estate prices are increasing 12% each year?
41. **Index Funds** How long will it take an investment in a stock market index fund yielding 10% annually to double?
42. Find the effective rate of 13.25% compounded quarterly.
43. **Baseball Cards** A mint condition 1955 Topps "Sandy Koufax" baseball card selling for $500 in 1989 was worth $1350 in 1993. What is the effective rate of this price increase?
44. **Stocks** One of Wall Street's most cherished buys in the 1950s was Texas Instruments stock, which rose spectacularly from $72\frac{1}{8}$ to $214\frac{1}{4}$ in 18 months. What is the effective rate of this price increase?
45. **Baseball Teams** The Haas family bought the Oakland Athletics from Charles O. Finley for $12.7 million in 1980 and sold it in 1995 for $85 million. Neglecting all other expenses, what is the effective rate of return on their investment?
46. **Pianos** A properly maintained 90-year-old Steinway grand piano is now worth 13.6 times its initial purchase price. What is the effective rate of this increase?

2.3 Annuities

Find the accumulated amount of each annuity.

47. $1500 annually at 11% for 20 years.
48. $500 semiannually at 8% for 12 years.
49. **Home Buying** How much will you have for a vacation home if you save $25 each week for 16 years in a 5.5% passbook savings account?
50. **Retirement Savings** How much will an IRA stock fund earning 11.2% be worth if you deposit $180 each month for 35 years?
51. **Apartments** When André-François Raffray died at age 77 in Arles, France, he had been making $500 monthly payments to Jeanne Calment for 30 years for the right to take over her apartment when she died. Mrs. Calment, who at 120 was then the world's oldest person with the records to prove it, remarked that "In life, one sometimes makes bad deals." How much would Mr. Raffray's money be worth if instead he had deposited it into a 7.8% bond fund?

Find the required payment for each sinking fund.

52. Yearly deposits earning 7% to accumulate $16,000 after 8 years.
53. Quarterly deposits earning 4.9% to accumulate $40,000 after 25 years.
54. How much must you pay biweekly to pay off a 13.2% car loan of $19,500 in 5 years?
55. **Stock Funds** How much must you deposit each month into a stock fund earning 11.2% to accumulate $150,000 in 20 years?
56. **Home Buying** How much must you save each week in a 5.5% passbook savings account to have $26,000 for a vacation home in 16 years?

Find the number of years needed for each sinking fund to reach the given accumulated amount.

57. $1000 yearly at 10% to accumulate $100,000.
58. $250 quarterly at 8.1% to accumulate $75,000.
59. **Money Market Funds** How long must you save $275 each month in a 6.2% money market fund to accumulate $19,500 for a new car?
60. **Stock Funds** How long must you deposit $300 each month into a stock fund earning 11.2% to accumulate $150,000?
61. **Home Buying** How long must you save $25 each week in a 5.5% passbook savings account to have $26,000 for a vacation home?

Follow the method of Exercise 27 on page 131 to estimate the internal rate of return for each investment.

62. $150 quarterly to accumulate $10,000 after 10 years.

63. $225 monthly to accumulate $100,000 after 25 years.

64. $500 monthly to accumulate $1,000,000 after 30 years.

65. $1.25 weekly to accumulate $50,000 after 60 years.

66. $10 daily to accumulate $100,000 after 20 years.

2.4 Amortization

67. Find the present value of an annuity of $25,000 annually at 6% for 25 years.

68. **Football Contracts** Deion Sanders became the highest-paid defensive player in football when he signed a $25 million, 5-year contract with the Dallas Cowboys. If the $25 million was paid out in equal weekly payments over the 5 years and the current interest rate was 7.32%, how much was the contract worth when it was signed?

69. **Grant Funding** A foundation is funding a new program that will award twenty $250,000 grants each year for the next 8 years to promote art and music instruction in elementary schools. How much must be placed in a money market account paying 6.75% to fund this initiative?

70. **Contest Prizes** The grand prize in a lottery drawing is $14 million to be paid out in equal quarterly payments over the next 20 years. If the prize money is guaranteed by U.S. Treasury bonds yielding 7.6% and is placed into an escrow account when the lottery is announced 1 year before the first payment, how much do the lottery sponsors have to deposit in the escrow account?

71. **Home Buying** Before looking for their first house, the Jones family calculates that they could afford a $1250 monthly mortgage payment. How large a mortgage could they afford at 8.75% for 30 years?

72. Find the weekly payments to amortize a debt of $45,000 at 9.15% over 15 years.

73. **Credit Cards** A Discover card statement shows a balance of $945 at 15.7% compounded monthly. What monthly payment will pay off this debt in 2 years?

74. **Credit Cards** What monthly payment should you make on your Optima card to pay off $2784 of spring break vacation charges in 7 months if the interest rate is 18.9% compounded monthly?

75. **Home Buying** The Jones family has found a nice house in the suburbs. What is the monthly payment on their $150,000 mortgage at 8.75% for 30 years?

76. **Stock Funds** How much must you save each week in a stock fund yielding 10.1% to become a millionaire in 15 years?

Find the unpaid balance on each debt.

77. After 5 years of annual payments on $85,000 at 9.1% for 20 years.

78. After 3 years 8 months of monthly payments on $150,000 at 8.7% for 25 years.

79. **Car Buying** After 1 year 5 months of $297.92 monthly car payments on a 3-year loan at 4.6%, Karen wants to get rid of her sedan and get a red sports car. How much should she pay to settle her debt?

80. **Contract Buyout** The Middleville Central School Board made a mistake when they hired the new superintendent and agreed to a 6-year contract paying $125,000 per year. With current interest rates at 6.2%, how much should they offer to break the contract after the first year?

81. **Home Buying** After 4 years 9 months of monthly mortgage payments on their 30-year, $150,000 mortgage at 8.75%, the Jones family has to relocate to Atlanta and must sell their nice house in the suburbs. How much do they still owe on their mortgage?

3 Systems of Equations and Matrices

- 3.1 Systems of Two Linear Equations in Two Variables
- 3.2 Matrices and Linear Equations in Two Variables
- 3.3 Systems of Linear Equations and the Gauss–Jordan Method
- 3.4 Matrix Arithmetic
- 3.5 Inverse Matrices and Systems of Linear Equations
- 3.6 Introduction to Modeling: Leontief Models and Least Squares

Application Preview

A "Taxing" Problem

For a tax law to be acceptable, it must be perceived as fair by the taxpayers. When federal and state governments both try to tax the same income, abhorrence of "double jeopardy" suggests that the taxes paid to one entity should not be further taxed by the other.

Consider a simplified situation in which the federal tax is 20% of your taxable income after the state tax has first been deducted and the state tax is 4% of your taxable income after the federal tax has first been deducted. How can you find the correct taxes? There is a widespread misconception that since neither amount can be calculated first, the only way is by a process of successive approximation: First estimate the state tax, then subtract it from your income to calculate the federal tax, then readjust the state tax estimate based on the federal tax and begin again. The problem becomes even more complicated if city taxes are included. Perhaps as a result of this confusion, state income taxes do not allow the deduction of federal taxes although federal taxes *do* allow the deduction of state taxes. Contrary to this misconception, these taxes *can* be found exactly by the methods of this chapter (see pages 161–162), and the same methods can be used to solve many other problems.

3.1 SYSTEMS OF TWO LINEAR EQUATIONS IN TWO VARIABLES

Introduction

Neither of the statements "Bob and Sue together have $100" and "Bob has $20 less than Sue" tells us how much either has. But taken together, they force the conclusion that Bob has $40 and Sue has $60. Various methods of solving such simple problems have been used since antiquity, but many become unworkable as the problems become more complicated. This chapter provides a method that solves such problems and has the pleasing property that the method for complicated problems is an easy extension of the method for the simplest.

Systems of Equations

We begin with the simplest form of these problems.

Systems of Two Linear Equations in Two Variables

A system of two linear equations in two variables is any problem expressible in the form

$$\begin{cases} ax + by = h \\ cx + dy = k \end{cases}$$

where x and y are the variables and the constants a, b, c, d, h, k are such that at least one of the coefficients a, b, c, d is not zero.

Such equations are called *linear* because $ax + by = h$ (with at least one of a and b not equal to zero) is the same as the *general linear equation* (see page 12) whose graph is a line.

EXAMPLE 1

TWO EQUATIONS IN TWO VARIABLES

Express the statements "Bob and Sue together have $100" and "Bob has $20 less than Sue" as a system of two equations in two variables.

Solution

Let x represent the amount of money that Bob has and y represent the amount that Sue has. The statement that together they have $100 may be written "$x + y = 100$" and the statement that Bob has $20 less than Sue as "$x = y - 20$". Rearranging this second equation by subtracting y from both sides, we obtain the system of equations

$$\begin{cases} x + y = 100 \\ x - y = -20 \end{cases} \qquad \begin{array}{l} ax + by = h \text{ with} \\ a = 1, \ b = 1, \ h = 100 \\ cx + dy = k \text{ with} \\ c = 1, \ d = -1, \ k = -20 \end{array}$$

There are many other ways to express the two statements as a system of equations. For example, we could write the same equations but in the opposite order

$$\begin{cases} x - y = -20 \\ x + y = 100 \end{cases} \qquad \text{Switching the order of the equations}$$

or instead we could multiply one of the equations by 2

$$\begin{cases} 2x + 2y = 200 \\ x - y = -20 \end{cases} \qquad \begin{array}{l} x + y = 100 \\ \text{multiplied by 2} \end{array}$$

We could even add (or subtract) the equations and use the result to replace one of the original equations:

$$\begin{cases} x + y = 100 \\ 2x = 80 \end{cases} \qquad \begin{array}{l} x + y = 100 \\ \underline{x - y = -20} \\ 2x = 80 \end{array} \text{Adding}$$

It is an easy matter to check that each of these three systems has the same solution, $x = 40$ and $y = 60$. Notice that the second equation in the above system immediately gives $x = 40$, which is "half" of the solution. In fact, if we could write the system in the following way, we could immediately read off the *entire* solution:

$$\begin{cases} 1x + 0y = 40 \\ 0x + 1y = 60 \end{cases} \qquad \begin{array}{l} \longleftarrow \text{gives } x = 40 \\ \longleftarrow \text{gives } y = 60 \end{array}$$

This will be our goal—to simplify a system of equations until its solution is obvious (or until it is clear that there is no solution).

Practice Problem 1 Express the statements "a jar of pennies and nickels contains 80 coins" and "the coins in the jar are worth $1.60" as a system of two equations in two variables.   ➤ Solution on page 164

A *solution* of a system of equations in two variables is a pair of values for the variables that satisfy all the equations (such as $x = 40$, $y = 60$ for Example 1). The *solution set* is the collection of all solutions. *Solving* the system of equations means finding this solution set. In this section we will solve systems of linear equations by three different methods: *graphing, substitution,* and *elimination.* Each has its own advantages and disadvantages.

Graphical Representations of Equations

An equation of the form $ax + by = h$ (with at least one of a and b not zero) is the equation of a line written in *general form.* We begin by graphing linear equations.

$y = mx + b$

EXAMPLE 2 GRAPHING A LINEAR EQUATION IN TWO VARIABLES

Sketch the graph of each equation.

a. $2x + 3y = 12$ **b.** $2x - 3y = 0$
c. $3y = 12$ **d.** $2x = 12$

$-3y = -2x + 12$
$y = -2/3 x + 4$

Solution

a. To graph $2x + 3y = 12$ we find the intercepts by setting each of the variables in turn equal to zero:

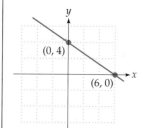

$\cdot\ 3y = 12$	$2x + 3y = 12$ with $x = 0$ Now divide by 3
$y = 4$ for the point (0, 4)	Since we began with $x = 0$
$2x = 12$	The original $2x + 3y = 12$ with $y = 0$
$x = 6$ for the point (6, 0)	Since we began with $y = 0$

The line through (0, 4) and (6, 0) is shown on the left.

b. To graph $2x - 3y = 0$ we begin by setting $x = 0$:

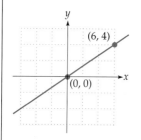

$-3y = 0$	$2x - 3y = 0$ with $x = 0$ Now divide by -3
$y = 0$ for the point (0, 0)	Since we began with $x = 0$

For another point we choose any other value of x.

$12 - 3y = 0$	$2x - 3y = 0$ with $x = 6$ Now subtract 12 and then divide by -3
$y = 4$ for the point (6, 4)	Since we began with $x = 6$

The line through (0, 0) and (6, 4) is shown on the left.

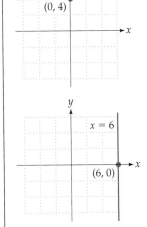

c. To graph $3y = 12$:

$$y = 4 \qquad \text{Dividing by 3}$$

$y = 4$ is a horizontal line. Its graph is shown on the left.

d. To graph $2x = 12$:

$$x = 6 \qquad \text{Dividing by 2}$$

$x = 6$ is a vertical line. Its graph is shown on the left.

The graph of a *pair* of equations $\begin{cases} ax + by = h \\ cx + dy = k \end{cases}$ may take three different forms: two lines intersecting at just one point (a *unique* solution), two lines that don't intersect (*no* solutions), or two lines that are the same (*infinitely many* solutions), as shown below. For lines that don't intersect, we say that the equations are *inconsistent*, and for two lines that are the *same* we call the equations *dependent*.

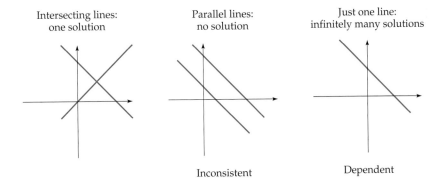

Intersecting lines: one solution

Parallel lines: no solution — Inconsistent

Just one line: infinitely many solutions — Dependent

The type and solution (if any) of a system of equations can be determined from the graph if it is drawn with sufficient accuracy.

EXAMPLE 3 SOLVING BY GRAPHING

Solve each system of equations by graphing the lines.

a. $\begin{cases} 2x + 3y = 12 \\ 2x - 3y = 0 \end{cases}$ b. $\begin{cases} 2x + 3y = 12 \\ 4x + 6y = 12 \end{cases}$ c. $\begin{cases} 2x + 3y = 12 \\ 4x + 6y = 24 \end{cases}$

Solution

The lines $2x + 3y = 12$ and $2x - 3y = 0$ were graphed in Example 2. The other lines may be graphed similarly.

a.
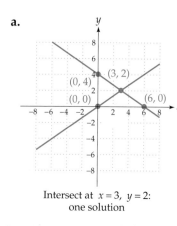

Intersect at $x = 3$, $y = 2$:
one solution

b.
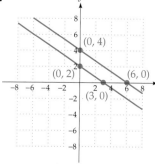

Lines are parallel:
no solution
(inconsistent)

c.
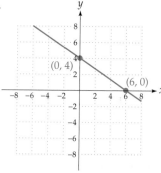

Just one line:
infinitely many solutions
(dependent)

In part (a), the unique solution $x = 3$, $y = 2$ is easily checked by substituting these values into the two equations:

$2 \cdot 3 + 3 \cdot 2 = 12$ $2x + 3y = 12$ with $x = 3$ and $y = 2$ (It checks: $12 = 12$)

$2 \cdot 3 - 3 \cdot 2 = 0$ $2x - 3y = 0$ with $x = 3$ and $y = 2$ (It checks: $0 = 0$)

In part (b) the lines do not intersect, so there is *no solution* (the equations are *inconsistent*).

In part (c) the lines are the *same* (the equations are *dependent*), so *every* point on the line $2x + 3y = 12$ is a solution.

For situations with infinitely many solutions, as in part (c), we may express *all* of these solutions at once as follows. We begin by solving either of the equations for one of the variables:

$$x = \frac{12 - 3y}{2} = 6 - \frac{3}{2}y \qquad \text{Solving } 2x + 3y = 12 \text{ for } x$$

and writing the solution as

$$\begin{cases} x = 6 - \frac{3}{2}t & \text{Replacing } y \text{ by } t \\ y = t & t \text{ may take any value} \end{cases}$$

where t is any number (called a *parameter*). For example, taking $t = 2$ gives $x = 3$, $y = 2$ and the point $(3, 2)$ is a solution because it satisfies both equations. All other solutions can be found by choosing other values for the parameter t.

Practice Problem 2 Find the solution corresponding to the parameter value $t = -2$ and verify that the resulting values satisfy the equations in Example 3c.

▶ Solution on page 164

Graphing Calculator Exploration

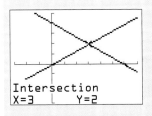

You can view a nonvertical line $ax + by = h$ by entering it as $y = (h - ax)/b$. To see the system $\begin{cases} 2x + 3y = 12 \\ 2x - 3y = 0 \end{cases}$ from Example 3a:

a. Enter the first equation as $y_1 = (12 - 2x)/3$ and the second as $y_2 = (-2x)/(-3)$. Graph them on the window $[-3, 7]$ by $[-3, 5]$.

b. Use TRACE or EVALUATE to check the x- and y-intercepts.

c. Use TRACE or INTERSECT to find the intersection point of these two lines.

Substitution Method

The graphical method is limited by the accuracy of your sketch. Our second way of solving a system of linear equations is called the *substitution method* and involves solving one of the equations for a variable and then substituting the result into the other equation.

EXAMPLE 4 **SOLVING BY THE SUBSTITUTION METHOD**

Solve $\begin{cases} x + 3y = 15 \\ 2x - 5y = 8 \end{cases}$ by the substitution method.

Solution

The first equation can easily be solved for x as $x = 15 - 3y$. We then substitute $15 - 3y$ for x in the second equation.

$$2(15 - 3y) - 5y = 8 \qquad \text{Replacing } x \text{ with } 15 - 3y$$
$$30 - 6y - 5y = 8 \qquad \text{Multiplying out}$$
$$-11y = -22 \qquad \text{Collecting like terms}$$
$$y = 2 \qquad \text{Dividing by } -11 \text{ gives the } y\text{-value}$$

Then

$$x = 15 - 3(2) = 15 - 6 = 9 \qquad x = 15 - 3y \text{ with } y = 2$$

The solution is $x = 9, \ y = 2$ as may be easily checked by substituting into the original equations.

The substitution method is particularly useful if one of the equations is easily solved for x or y.

Practice Problem 3   Solve $\begin{cases} 2x + y = 10 \\ x + 2y = 8 \end{cases}$ by the substitution method.

➤ Solution on page 164

What happens if we try to solve an *inconsistent* system, such as $\begin{cases} x + y = 10 \\ x + y = 20 \end{cases}$ (which is inconsistent since the same quantity can't equal both 10 and 20)? The first equation gives

$$x = 10 - y \qquad x + y = 10 \text{ solved for } x$$

Substituting this into the second equation

$$(10 - y) + y = 20 \qquad \begin{array}{l} x + y = 20 \text{ with } x \\ \text{replaced by } x = 10 - y \end{array}$$

which simplifies to

$$10 = 20 \qquad \text{A contradiction}$$

Interpret this result as follows: Looking for a value for y, we were led to the "impossible" equation $10 = 20$, meaning that *no* value of y works, so the original system has *no solutions* and is *inconsistent*. Whenever you find an "impossible" equation (saying that two different numbers are equal), the original system is inconsistent (has no solutions).

What happens if we try to solve a *dependent* system, such as $\begin{cases} x + y = 10 \\ 2x + 2y = 20 \end{cases}$ (which is dependent since one equation is twice the other)? The first equation gives

$$x = 10 - y \qquad\qquad x + y = 10 \text{ solved for } x$$

Substituting this into the second:

$$2(10 - y) + 2y = 20 \qquad\qquad \begin{array}{l}2x + 2y = 20 \text{ with } x \\ \text{replaced by } x = 10 - y\end{array}$$

After multiplying out and simplifying, we obtain:

$$20 = 20 \qquad\qquad \text{True but useless}$$

Interpret this result as follows: Looking for a value for y, we were led to the "useless" equation $20 = 20$ that is *always* true, meaning that *all* values of y work, so the original system has *infinitely many solutions* and is *dependent*. Whenever you find a "useless" equation (saying that a number equals itself), the original system is dependent.

Equivalent Systems of Equations

Two systems of equations are *equivalent* if they have the same solution. On page 153 we saw that we obtained equivalent systems by applying any of the following operations: switching the order of the equations, multiplying one equation by a (nonzero) constant, or adding or subtracting two equations. By combining the last two operations, we may even add a *multiple* of an equation to another. For example:

1. Switch the order of the equations. $\begin{cases} x + y = 100 \\ x - y = -20 \end{cases} \xrightarrow{\text{Switch equations}} \begin{cases} x - y = -20 \\ x + y = 100 \end{cases}$

2. Multiply or divide one of the equations by a nonzero number. $\begin{cases} x + y = 100 \\ x - y = -20 \end{cases} \xrightarrow{\substack{\text{Multiply second} \\ \text{equation by } -1}} \begin{cases} x + y = 100 \\ -x + y = 20 \end{cases}$

3. Add (or subtract) a multiple of one equation to (or from) the other. $\begin{cases} x + y = 100 \\ x - y = -20 \end{cases} \xrightarrow{\substack{\text{Add twice the first} \\ \text{to the second}}} \begin{cases} x + y = 100 \\ 3x + y = 180 \end{cases}$

These systems are all equivalent to each other because they all have exactly the same solutions.

Elimination Method

Our third way of solving equations, the *elimination* method, attempts to remove one variable from each and write them as an equivalent system of the form $\begin{cases} 1x + 0y = p \\ 0x + 1y = q \end{cases}$ so that the solution can be read off as $x = p$, $y = q$. If this cannot be done, the method will identify the system as inconsistent or dependent.

In these problems we may carry out subtractions in either order, *top line minus bottom* or *bottom line minus top*:

$$\begin{array}{c} 6 \\ \underline{15} \\ -9 \end{array} \leftarrow \text{top minus bottom} \quad \text{or} \quad \begin{array}{c} 6 \\ \underline{15} \\ 9 \end{array} \leftarrow \text{bottom minus top}$$
$$(6 - 15 = -9) \qquad\qquad (15 - 6 = 9)$$

EXAMPLE 5

SOLVING BY THE ELIMINATION METHOD

Solve $\begin{cases} x - 3y = 3 \\ 2x + 5y = 28 \end{cases}$ by the elimination method.

Solution

First we eliminate the x from the second equation by subtracting an appropriate multiple of the first (namely, *twice* the first in order to match the $2x$ in the second equation). Beginning with the original system:

$$\begin{cases} x - 3y = 3 \\ 2x + 5y = 28 \end{cases} \quad \times 2 \rightarrow \quad \begin{array}{l} 2x - 6y = 6 \\ \underline{2x + 5y = 28} \\ 11y = 22 \end{array} \leftarrow \text{bottom minus top}$$

$$\begin{cases} x - 3y = 3 \\ 11y = 22 \end{cases} \quad \text{Unchanged}$$

$$\begin{cases} x - 3y = 3 \\ y = 2 \end{cases} \quad \begin{array}{l} \text{Unchanged} \\ \text{Dividing } 11y = 22 \text{ by } 11 \end{array}$$

Next we eliminate the y from the first equation by subtracting an appropriate multiple of the second (namely, *three times* the second to match the $3y$ in the first).

$$\begin{cases} x - 3y = 3 \\ y = 2 \end{cases} \quad \begin{array}{c} \rightarrow \\ \times 3 \rightarrow \end{array} \quad \begin{array}{l} x - 3y = 3 \\ \underline{ 3y = 6} \\ x = 9 \end{array} \leftarrow \text{top plus bottom}$$

$$\begin{cases} x = 9 \\ y = 2 \end{cases} \quad \text{Unchanged} \quad \text{Same as} \quad \begin{cases} 1x + 0y = 9 \\ 0x + 1y = 2 \end{cases}$$

The original system has a unique solution $x = 9$, $y = 2$.

3.1 SYSTEMS OF TWO LINEAR EQUATIONS IN TWO VARIABLES

An inconsistent system of equations always leads to the impossible equation that zero equals a nonzero number. For instance, solving Example 3b by the elimination method could be done as follows:

$$\begin{cases} 2x + 3y = 12 \\ 4x + 6y = 12 \end{cases} \xrightarrow{\text{Divide second equation by 2}} \begin{cases} 2x + 3y = 12 \\ 2x + 3y = 6 \end{cases} \xrightarrow{\text{Subtract first from second}} \begin{cases} 2x + 3y = 12 \\ 0x + 0y = -6 \end{cases}$$

Since the last equation says $0 = -6$, which is contradictory, there is no solution and the equations are inconsistent.

A dependent system of equations always results in one equation becoming $0x + 0y = 0$. For instance, solving Example 3c by the elimination method could be done as follows:

$$\begin{cases} 2x + 3y = 12 \\ 4x + 6y = 24 \end{cases} \xrightarrow{\text{Divide second equation by 2}} \begin{cases} 2x + 3y = 12 \\ 2x + 3y = 12 \end{cases} \xrightarrow{\text{Subtract first from second}} \begin{cases} 2x + 3y = 12 \\ 0x + 0y = 0 \end{cases}$$

Since the last equation says $0 = 0$, which is uninformative but not contradictory, the solutions are all the points on the line $2x + 3y = 12$ and the system is dependent.

Practice Problem 4   Solve $\begin{cases} x + y = 100 \\ x - y = -20 \end{cases}$ by the elimination method.

➤ Solution on page 164

We conclude this section with the solution of the tax problem posed in the Application Preview on page 151.

EXAMPLE 6

A TAXING PROBLEM

What are the federal and state taxes on an income of $34,100 if the federal tax is 20% of the income after first deducting the state tax, and the state tax is 4% of the income after first deducting the federal tax?

Solution

Let x represent the federal tax and y represent the state tax. Then

$$\begin{cases} x = 0.20(34{,}100 - y) \\ y = 0.04(34{,}100 - x) \end{cases}$$ Federal tax is 20% after first subtracting the state tax
State tax is 4% after first subtracting the federal tax

$$\begin{cases} x = 6820 - 0.2y \\ y = 1364 - 0.04x \end{cases}$$ Multiplying out

Moving all variables to the left side of the equations gives the following system. To solve it, we eliminate the x from the second equation by subtracting 0.04 times the first equation (the 0.04 is chosen to match the $0.04x$ in the second equation):

$$\begin{cases} x + 0.2y = 6820 \\ 0.04x + y = 1364 \end{cases} \quad \times 0.04 \rightarrow \quad \begin{array}{l} 0.04x + 0.008y = 272.8 \\ \underline{0.04x + y = 1364} \\ 0.992y = 1091.2 \end{array} \leftarrow \text{bottom minus top}$$

$$\begin{cases} x + 0.2y = 6820 \\ 0.992y = 1091.2 \end{cases} \quad \text{Unchanged}$$

$$\begin{cases} x + 0.2y = 6820 \\ y = 1100 \end{cases} \quad \text{Dividing } 0.992y = 1091.2 \text{ by } 0.992$$

Now we eliminate the y from the first equation by subtracting 0.2 times the second (the 0.2 chosen to match the $0.2y$ in the first):

$$\begin{cases} x + 0.2y = 6820 \\ y = 1100 \end{cases} \quad \times 0.2 \rightarrow \quad \begin{array}{l} x + 0.2y = 6820 \\ \underline{0.2y = 220} \\ x = 6600 \end{array} \leftarrow \text{top minus bottom}$$

$$\begin{cases} x = 6600 \\ y = 1100 \end{cases} \quad \text{Unchanged}$$

The solution $x = 6600$, $y = 1100$ says that the federal tax is $6600 and the state tax is $1100. You should check that each of these taxes is the correct percentage of the $34,100 income after the other tax has been subtracted.

3.1 Section Summary

A *system of two linear equations in two variables* can be written in the form

$$\begin{cases} ax + by = h \\ cx + dy = k \end{cases}$$

and will be one of three types.

$\begin{cases} ax + by = h \\ cx + dy = k \end{cases}$	Graph		Number of Solutions
Has a unique solution	Lines intersect		One
Inconsistent	Parallel lines		None
Dependent	Just one line		Infinitely many

Solving by *graphing* is only as accurate as your sketch.

The *substitution method* solves one of the equations for a variable and uses this new expression to rewrite the other equation with just one variable. Solving this new equation and then finding the first variable solves the system of equations. This method is most useful when one of the coefficients of the variables (a, b, c, or d) is 1.

Equivalent systems of equations have the same solution and can be found by

1. Switching the order of the equations,
2. Multiplying or dividing one equation by a nonzero number, or
3. Replacing an equation by its sum or difference with a multiple of the other.

The *elimination method* solves the system by attempting to change it into an equivalent system of the form $\begin{cases} 1x + 0y = p \\ 0x + 1y = q \end{cases}$. If this can be done, the unique solution is $x = p$, $y = q$. If an equation of the form $0x + 0y = 1$ is obtained, there is *no solution*, and the system is *inconsistent*. If an equation of the form $0x + 0y = 0$ is obtained, there are *infinitely many solutions* and the system is *dependent*.

▶ Solutions to Practice Problems

1. Let x be the number of pennies in the jar and y be the number of nickels. Then the first statement may be expressed as $x + y = 80$ and the second as $x + 5y = 160$ (in cents, since each penny is worth 1¢ and each nickel is worth 5¢). The system of equations is $\begin{cases} x + y = 80 \\ x + 5y = 160 \end{cases}$

2. $\begin{cases} x = 6 - \frac{3}{2}(-2) = 9 \\ y = -2 \end{cases}$ Substituting these numbers into Example 3c:

$$\begin{cases} 2(9) + 3(-2) = 18 - 6 = 12 \checkmark \\ 4(9) + 6(-2) = 36 - 12 = 24 \checkmark \end{cases}$$

It checks!

3. Since the first equation can be solved for y as $y = 10 - 2x$, we can substitute $10 - 2x$ for y in the second equation: $x + 2(10 - 2x) = 8$. Multiplying out and collecting like terms, $x + 20 - 4x = 8$ so $-3x = -12$ giving $x = 4$. Substituting $x = 4$ into $y = 10 - 2x$ gives $y = 10 - 2(4) = 10 - 8 = 2$. The solution is $x = 4$, $y = 2$. There are several other ways of solving this problem by the substitution method, and all reach the same conclusion.

4. $\begin{cases} x + y = 100 \\ x - y = -20 \end{cases}$ $\xrightarrow{\text{Add second to first}}$ $\begin{cases} 2x + 0y = 80 \\ x - y = -20 \end{cases}$ $\xrightarrow{\text{Divide first by 2}}$

$\begin{cases} 1x + 0y = 40 \\ x - y = -20 \end{cases}$ $\xrightarrow{\text{Subtract first from second}}$ $\begin{cases} 1x + 0y = 40 \\ 0x - y = -60 \end{cases}$ $\xrightarrow{\text{Multiply second by } -1}$

$\begin{cases} 1x + 0y = 40 \\ 0x + 1y = 60 \end{cases}$

The solution is $x = 40$, $y = 60$. There are many other possible sequences of equivalent systems that solve this problem, and all reach the same conclusion.

3.1 Exercises

Represent each pair of statements as a system of two equations using the given definitions of x and y. Verify that the values given for x and y are a solution for the system of equations.

1. "The sum of two numbers is eighteen" and "the first number is two more than the second number." Let x be the first number and y be the second number. $x = 10$, $y = 8$

2. "The sum of two numbers is twenty-five" and "twice the first number added to the second number totals thirty-two." Let x be the first number and y be the second number. $x = 7$, $y = 18$

3. "Tom has $6 more than Alice" and "together, they have $40." Let x be the amount of money that Tom has and y be the amount Alice has. $x = 23$, $y = 17$

4. "Bill and Jessica together have $25" and "Jessica has $12." Let x be the amount of money that Bill has and y be the amount that Jessica has. $x = 13$, $y = 12$

5. "An envelope of $1 and $5 bills contains thirty bills" and "the money in the envelope is worth $70." Let x be the number of $1 bills and y be the number of $5 bills. $x = 20$, $y = 10$

6. "An envelope of $10 and $20 bills contains eight bills" and "the money in the envelope is worth $110." Let x be the number of $10 bills and y be the number of $20 bills. $x = 5$, $y = 3$

7. "A small theater sold tickets for all one hundred seats" and "the box office receipts of $650 came from adult tickets at $10 and child tickets at $5." Let x be the number of adult tickets sold and y be the number of child tickets sold. $x = 30$, $y = 70$

8. "A movie theater sold tickets for three hundred seats" and "the box office receipts of $2400 came from adult tickets at $9 and child tickets at $6." Let x be the number of adult tickets sold and y be the number of child tickets sold. $x = 200$, $y = 100$

9. "A corn and beet farmer planted 225 acres of crops" and "he planted twice as many acres of corn as acres of beets." Let x be the number of acres of corn and y be the number of acres of beets. $x = 150$, $y = 75$

10. "A stock and bond speculator invested $10,000 in the market" and "she invested three times as much in stocks as in bonds." Let x be the amount in stocks and y be the amount in bonds. $x = 7500$, $y = 2500$

Solve each system by graphing. If the solution is not unique, identify the system as "inconsistent" or "dependent."

You may use a graphing calculator if permitted by your instructor.

11. $\begin{cases} x + y = 6 \\ x - y = 2 \end{cases}$

12. $\begin{cases} -x + y = 2 \\ x + y = 4 \end{cases}$

13. $\begin{cases} 2x + y = 8 \\ x = 3 \end{cases}$

14. $\begin{cases} x + 2y = 10 \\ y = 4 \end{cases}$

15. $\begin{cases} x - y = 4 \\ -x + 2y = -6 \end{cases}$

16. $\begin{cases} 2x - y = 2 \\ x + 2y = 6 \end{cases}$

17. $\begin{cases} x + y = 10 \\ -x - y = 10 \end{cases}$

18. $\begin{cases} -2x + 4y = -16 \\ x - 2y = 4 \end{cases}$

19. $\begin{cases} x + y = 10 \\ -x - y = -10 \end{cases}$

20. $\begin{cases} 2x - 4y = 16 \\ -x + 2y = -8 \end{cases}$

Solve each system by the substitution method. If the solution is not unique, identify the system as "inconsistent" or "dependent."

21. $\begin{cases} x + 2y = 10 \\ y = 3 \end{cases}$

22. $\begin{cases} 2x + y = 8 \\ x = 2 \end{cases}$

23. $\begin{cases} 2x + y = 20 \\ x + y = 15 \end{cases}$

24. $\begin{cases} x + y = 12 \\ x + 2y = 14 \end{cases}$

25. $\begin{cases} 5x + 2y = 30 \\ 2x + y = 10 \end{cases}$

26. $\begin{cases} 2x + 3y = 25 \\ x - y = 5 \end{cases}$

27. $\begin{cases} 3x + 2y = 30 \\ x - y = -5 \end{cases}$

28. $\begin{cases} 2x + y = 20 \\ x + 3y = 15 \end{cases}$

29. $\begin{cases} -2x + 2y = -20 \\ x - y = 10 \end{cases}$

30. $\begin{cases} 3x - 2y = 30 \\ -6x + 4y = 30 \end{cases}$

Solve each system by the elimination method. If the solution is not unique, identify the system as "inconsistent" or "dependent."

31. $\begin{cases} x + y = 11 \\ 2x + 3y = 30 \end{cases}$

32. $\begin{cases} 3x + 2y = 30 \\ x + y = 13 \end{cases}$

33. $\begin{cases} 3x + y = 15 \\ x + 2y = 10 \end{cases}$

34. $\begin{cases} x + 3y = 30 \\ 2x + y = 10 \end{cases}$

35. $\begin{cases} x + 2y = 14 \\ 3x + 4y = 36 \end{cases}$

36. $\begin{cases} 2x + 3y = 30 \\ x - y = 10 \end{cases}$

37. $\begin{cases} 2x + 5y = 60 \\ 2x + 3y = 48 \end{cases}$ **38.** $\begin{cases} 4x + 3y = 36 \\ x + 3y = 18 \end{cases}$ **39.** $\begin{cases} 3x + 4y = -24 \\ 6x + 8y = 24 \end{cases}$ **40.** $\begin{cases} x - 4y = 20 \\ -2x + 8y = -40 \end{cases}$

APPLIED EXERCISES

Formulate each situation as a system of two linear equations in two variables. Be sure to state clearly the meaning of your x- and y-variables. Solve the system by the elimination method. Be sure to state your final answer in terms of the original question.

41. BUSINESS: Apartment Ownership A lawyer has found 60 investors for a limited partnership to purchase an inner-city apartment building, with each contributing either $5000 or $10,000. If the partnership raised $430,000, how many investors contributed $5000 and how many contributed $10,000?

42. BUSINESS: Violin Syndication Ninety supporters of a musical prodigy raised $1.2 million to purchase a violin by Peter Guarneri for her use. If each supporter contributed either $5000 or $20,000, how many contributed each amount?

43. GENERAL: Coins in a Jar A jar contains 60 nickels and dimes worth $4.30. How many of each are in the jar?

44. GENERAL: Bills in an Envelope An envelope found in a safe deposit box contains a total of ninety $5 and $20 bills worth $1200. How many of each are in the envelope?

45. PERSONAL FINANCE: Financial Planning A retired couple wish to invest their nest egg of $10,000 in a money market account paying 6% and in a stock mutual fund returning 11%. If their income tax and Social Security situation requires that they earn $1000 from these investments, how much should they invest in each?

46. PERSONAL FINANCE: Home Financing A young couple needs to borrow $168,000 to finance their first house. They could borrow the whole amount at 12% from their bank, but her father offers to lend them enough of the money at 5% (with the same terms as the bank loan) to reduce the overall interest rate to just 8%. How much does he lend them and how much do they borrow from the bank?

47. BUSINESS: Ice Hockey Concession Receipts The concession stand at an ice hockey rink had receipts of $7200 from selling a total of 3000 sodas and hot dogs. If each soda sold for $2 and each hot dog sold for $3, how many of each were sold?

48. BUSINESS: Baseball Tickets A college baseball game generated box office receipts of $4800 from 600 ticket sales. If general admission tickets were $12 and student tickets were half price, how many of each were sold?

49. PERSONAL FINANCE: Income Taxes Find the federal and state taxes on a taxable income of $49,900 if the federal tax is 10% of the taxable income after first deducting the state tax and the state tax is 2% of the taxable income after first deducting the federal tax.

50. GENERAL: Estate Division A will specifies that each of two sons receive one half of the $3 million estate after first deducting the other's share, and that any remainder is then to be given to their sister. How much does each son receive? How much is left for the sister?

51. ATHLETICS: Sports Nutrition The dietician at a sports training facility has determined that one of her athletes needs an additional 600 mg each of calcium and phosphorus daily. Two supplements are available containing the milligrams of calcium and phosphorus per tablet as given by the table. How many tablets of each supplement will provide the required calcium and phosphorus?

	Calcium	Phosphorus
Supplement A	150	100
Supplement B	120	120

52. BUSINESS: Bicycle Shop Management

A bicycle shop has $10,500 to spend on new bikes and 390 hours of assembly time to put them together. Each mountain bike costs $50 wholesale and takes two hours to assemble. Each racing bike costs $70 wholesale and takes two-and-one-half hours to assemble. How many of each can the shop buy and assemble to use all the available money and time?

Explorations and Excursions

The following problems extend and augment the material presented in the text.

More About Parameterizations

53. In Example 2a we sketched the line $2x + 3y = 12$, and after Example 3 we parameterized this line as $x = 6 - \frac{3}{2}t$, $y = t$. To graph these *parametric equations* on your calculator, change the MODE from FUNCtion to PARametric and enter the pair of equations $x_{1T} = 6 - (3/2)T$, $y_{1T} = T$. Set the WINDOW with Tmin $= -2$, Tmax $= 2$, Tstep $= 0.1$, Xmin $= -10$, Xmax $= 10$, Ymin $= -10$, and Ymax $= 10$, and watch as your calculator draws the graph. Change Tmin to -3, Tmax to 3, and GRAPH it again. Experiment with different values for Tmin and Tmax until you see how to get a "complete" picture of this line in this window.

54. Check that the line $2x - 5y = 8$ can be parameterized as $x = 4 + \frac{5}{2}t$, $y = t$. Set the WINDOW with Tmin $= -2$, Tmax $= 2$, Tstep $= 0.1$, Xmin $= -10$, Xmax $= 10$, Ymin $= -10$, and Ymax $= 10$. GRAPH these equations (be sure to first change the MODE from FUNCtion to PARametric) and use TRACE to explore this line segment. Experiment with different values for Tmin and Tmax until you see how to get a "complete" picture of this line in this window.

55. Find a parameterization for the line $x + 3y = 15$ and use your calculator to verify that your parametric equations determine a line with x-intercept $(15, 0)$ and y-intercept $(0, 5)$.

56. Use the parametric equations from Exercises 54 and 55 to GRAPH Example 4 on the WINDOW with Tmin $= -5$, Tmax $= 10$, Tstep $= 0.1$, Xmin $= -5$, Xmax $= 20$, Ymin $= -10$, and Ymax $= 10$. Use TRACE to explore both lines and verify that the intersection point is $(9, 2)$.

Round-off Errors

Round-off errors can completely misrepresent the true nature of a system of equations. (If you wish to remove fractions from a problem, multiply each equation through by the least common denominator instead of rounding off.)

57. Use the elimination method to show that the system $\begin{cases} x + \frac{1}{3}y = 39 \\ 2x + \frac{2}{3}y = 84 \end{cases}$ is inconsistent.

58. Rounding to one decimal place, the system in Exercise 57 becomes $\begin{cases} x + 0.3y = 39 \\ 2x + 0.7y = 84 \end{cases}$. Use the elimination method to show that this new system has the unique solution $x = 21$, $y = 60$.

59. Use the elimination method to show that the system $\begin{cases} x + \frac{2}{9}y = 9.79 \\ 4x + \frac{8}{9}y = 39.16 \end{cases}$ is dependent.

60. Rounding to two decimal places, the system in Exercise 59 becomes $\begin{cases} x + 0.22y = 9.79 \\ 4x + 0.89y = 39.16 \end{cases}$. Use the elimination method to show that this new system has the unique solution $x = 9.79$, $y = 0$.

61. Use the elimination method to show that the system $\begin{cases} 2.1x + \frac{1}{7}y = 157 \\ 3x + \frac{1}{5}y = 224 \end{cases}$ has the unique solution $x = 70$, $y = 70$. Rounding to two decimal places, this system becomes $\begin{cases} 2.10x + 0.14y = 157 \\ 3.00x + 0.20y = 224 \end{cases}$. Use the elimination method to show that this new system is inconsistent.

3.2 MATRICES AND LINEAR EQUATIONS IN TWO VARIABLES

Introduction

In this section we will use matrix notation to streamline the elimination method used in the previous section to solve systems of two linear equations in two variables. In the next section we shall extend this method to the solutions of systems of many linear equations in many variables.

Matrices

A *matrix* is a rectangular array of numbers called *elements*. This rectangular array has *rows* (with the first at the top, the second below the first, and so on) and *columns* (with the first on the left, the second to the right of the first, and so on). The *dimension* of a matrix with m rows and n columns is written $m \times n$ (rows × columns). Thus a 5×2 matrix is "tall and thin," while a 2×5 matrix is "short and wide." A *square matrix* has the same number of rows as columns. A *row matrix* has just one row, and a *column matrix* has just one column.

$$(1 \quad 2 \quad 3) \qquad \begin{pmatrix} 1 \\ 2 \\ 3 \end{pmatrix} \qquad \begin{pmatrix} 1 & 2 & 3 \\ 4 & 5 & 6 \\ 7 & 8 & 9 \end{pmatrix}$$

Row matrix   Column matrix   Square matrix
(dimension 1×3)   (dimension 3×1)   (dimension 3×3)

We name matrices with capital letters (A, B, C, . . .) and then the elements are named by the corresponding lowercase letter together with subscripts indicating the row and then the column. For instance, if

$$A = \begin{pmatrix} 1 & 2 & 3 & 4 \\ 5 & 6 & 7 & 8 \\ 9 & 10 & 11 & 12 \end{pmatrix}$$

then the dimension of A is 3×4 and $a_{2,3} = 7$ because 7 is the element in the second row and third column. This *double subscript* notation is sometimes used without the comma so that a_{23} means $a_{2,3}$. The elements on the *main diagonal* are those with the same row number as column number. For the matrix A above, the main diagonal consists of the elements $a_{1,1} = 1$, $a_{2,2} = 6$, and $a_{3,3} = 11$.

Augmented Matrices from Systems of Equations

An *augmented matrix* is a matrix created from two "smaller" matrices having the same number of rows by placing them beside each other and joining them into one "larger" matrix. The system of equations

3.2 MATRICES AND LINEAR EQUATIONS IN TWO VARIABLES

$\begin{cases} ax + by = h \\ cx + dy = k \end{cases}$ has a *coefficient matrix* $\begin{pmatrix} a & b \\ c & d \end{pmatrix}$ and a *constant term* *matrix* $\begin{pmatrix} h \\ k \end{pmatrix}$. Taken together, these form the augmented matrix $\begin{pmatrix} a & b & | & h \\ c & d & | & k \end{pmatrix}$, separating the original matrices by a vertical line. The augmented matrix represents the system of equations with the vertical bar representing the equals signs. The first column gives the coefficients of the x-variable, the second gives those of the y-variable, and the last column (after the bar) contains the constant terms after the equals signs.

Augmented Matrix of a System of Equations

The augmented matrix $\begin{pmatrix} a & b & | & h \\ c & d & | & k \end{pmatrix}$ represents the system of

equations $\begin{cases} ax + by = h \\ cx + dy = k \end{cases}$.

EXAMPLE 1

AUGMENTED MATRICES AND SYSTEMS OF EQUATIONS

a. Find the augmented matrix representing the system $\begin{cases} 2x + 3y = 24 \\ 4x + 5y = 60 \end{cases}$.

b. Find the system represented by the augmented matrix $\begin{pmatrix} 6 & 8 & | & 84 \\ 4 & 5 & | & 60 \end{pmatrix}$.

Solution

a. The system $\begin{cases} 2x + 3y = 24 \\ 4x + 5y = 60 \end{cases}$ is represented by $\begin{pmatrix} 2 & 3 & | & 24 \\ 4 & 5 & | & 60 \end{pmatrix}$.

b. The augmented matrix $\begin{pmatrix} 6 & 8 & | & 84 \\ 4 & 5 & | & 60 \end{pmatrix}$ represents $\begin{cases} 6x + 8y = 84 \\ 4x + 5y = 60 \end{cases}$.

Practice Problem 1

a. Find the augmented matrix representing the system $\begin{cases} 2x - y = 14 \\ x + 3y = 21 \end{cases}$.

b. Find the system represented by the augmented matrix $\begin{pmatrix} 3 & 2 & | & 35 \\ 1 & 3 & | & 21 \end{pmatrix}$.

➤ Solutions on page 177

Row Operations

We will solve a system of equations by operating on its augmented matrix, using the following *matrix row operations* to make the solution obvious. These row operations correspond to the steps that we used on page 159 to solve systems of equations, and result in a succession of *equivalent* matrices, each representing a system that has exactly the same solution as the original system.

> **Matrix Row Operations**
>
> 1. Switch any two rows.
> 2. Multiply or divide one of the rows by a nonzero number.
> 3. Replace a row by its sum or difference with a multiple of another row.

We may apply row operations to matrices of any size. We write short formulas for row operations using R for "row" and either a double arrow $\leftrightarrow$ for "is switched with" or an arrow $\rightarrow$ for "becomes." When we write the new matrix, the formula for each new row ends with an arrow $\rightarrow$ pointing to the row that was changed.

EXAMPLE 2

PERFORMING MATRIX ROW OPERATIONS

Carry out the indicated row operation on the given matrix.

a. $R_1 \leftrightarrow R_2$ on $\begin{pmatrix} 2 & 3 & | & 25 \\ 3 & 4 & | & 36 \end{pmatrix}$

b. $3R_2 \rightarrow R_2$ on $\begin{pmatrix} 2 & 3 & | & 30 \\ 1 & 1 & | & 13 \end{pmatrix}$

c. $R_1 - R_2 \rightarrow R_1$ on $\begin{pmatrix} 2 & 3 & | & 30 \\ 2 & -2 & | & 20 \end{pmatrix}$

Solution

a. $R_1 \leftrightarrow R_2$ says that row 1 and row 2 are switched:

$$\begin{matrix} R_2 \rightarrow \\ R_1 \rightarrow \end{matrix} \begin{pmatrix} 3 & 4 & | & 36 \\ 2 & 3 & | & 25 \end{pmatrix} \qquad \text{From} \quad \begin{pmatrix} 2 & 3 & | & 25 \\ 3 & 4 & | & 36 \end{pmatrix}$$

b. $3R_2 \rightarrow R_2$ says that 3 times row 2 becomes R_2:

$$\begin{matrix} \\ 3R_2 \rightarrow \end{matrix} \begin{pmatrix} 2 & 3 & | & 30 \\ 3 & 3 & | & 39 \end{pmatrix} \qquad \text{From} \quad \begin{pmatrix} 2 & 3 & | & 30 \\ 1 & 1 & | & 13 \end{pmatrix}$$

c. $R_1 - R_2 \to R_1$ says that row 1 minus row 2 becomes row 1:

$$R_1 - R_2 \to \begin{pmatrix} 0 & 5 & | & 10 \\ 2 & -2 & | & 20 \end{pmatrix} \qquad \text{From} \qquad \begin{pmatrix} 2 & 3 & | & 30 \\ 2 & -2 & | & 20 \end{pmatrix}$$

Practice Problem 2

a. Carry out $R_1 \leftrightarrow R_2$ on $\begin{pmatrix} 2 & 1 & | & 14 \\ 1 & -3 & | & 21 \end{pmatrix}$.

b. Carry out $R_1 + R_2 \to R_1$ on $\begin{pmatrix} 6 & 3 & | & 42 \\ 1 & -3 & | & 21 \end{pmatrix}$.

c. Carry out $\frac{1}{7}R_1 \to R_1$ on $\begin{pmatrix} 7 & 0 & | & 63 \\ 1 & -3 & | & 21 \end{pmatrix}$.

d. Is $0R_1 \to R_1$ a valid row operation?

e. Is $5R_1 \to R_2$ a valid row operation?

▶ Solutions on page 178

Solving Equations by Row Reduction

Two matrices are *equivalent* if one can be transformed into the other by a sequence of row operations. Since an augmented matrix represents a system of equations and the row operations correspond to the steps used to solve the system, we can solve the system by *row-reducing* the augmented matrix to an equivalent matrix that displays the solution. There are three possibilities. If we can obtain the form

$$\begin{pmatrix} 1 & 0 & | & p \\ 0 & 1 & | & q \end{pmatrix} \qquad \begin{cases} 1x + 0y = p \\ 0x + 1y = q \end{cases} \text{ or } \begin{cases} x = p \\ y = q \end{cases}$$

then the unique solution is $x = p$, $y = q$. However, if we obtain a row of zeros ending with a nonzero number:

$$0 \quad 0 \quad | \quad m \qquad\qquad 0x + 0y = m, \; m \neq 0$$

then the system is *inconsistent* and has *no solution* because the equation $0x + 0y = m$ makes the impossible claim that zero equals a nonzero number. On the other hand, if there is no such "inconsistent" row but there is a row consisting entirely of zeros:

$$0 \quad 0 \quad | \quad 0 \qquad\qquad 0x + 0y = 0$$

then the system is *dependent*, and there are *infinitely many solutions* because the equation $0x + 0y = 0$ is *always* true and represents no restriction at all. These important observations will be extended in the next section.

EXAMPLE 3

SOLVING EQUATIONS BY ROW REDUCTION

Solve $\begin{cases} x + 3y = 15 \\ 2x - 5y = 8 \end{cases}$ by row-reducing an augmented matrix.

Solution

The augmented matrix for this system is

$$\begin{pmatrix} 1 & 3 & | & 15 \\ 2 & -5 & | & 8 \end{pmatrix}$$

We hope to change the first column from $\begin{smallmatrix}1\\2\end{smallmatrix}$ to $\begin{smallmatrix}1\\0\end{smallmatrix}$ (which is the same as removing the x-variable from the second equation) and then the second column to $\begin{smallmatrix}0\\1\end{smallmatrix}$ (removing the y-variable from the first equation). Therefore we first want a 0 in the bottom-left corner. Subtracting twice the 1 in the top row from the 2 in the second row will achieve this 0, suggesting the row operation $R_2 - 2R_1 \to R_2$. We write "twice row 1" in small numbers above row 1 to make the subtraction of "row 2 minus twice row 1" easier:

$$\overset{2 \quad 6 \quad 30}{\begin{pmatrix} 1 & 3 & | & 15 \\ 2 & -5 & | & 8 \end{pmatrix}} \quad R_2 - 2R_1 \to R_2 \text{ gives}$$

$$R_2 - 2R_1 \to \begin{pmatrix} 1 & 3 & | & 15 \\ 0 & -11 & | & -22 \end{pmatrix}$$

The bottom row has a common factor of -11, which we divide out, obtaining the simpler matrix below. We then want a 0 where the 3 is. Subtracting three times the 1 in the second row from the 3 will achieve this, suggesting the row operation $R_1 - 3R_2 \to R_1$. To help with the subtraction we write "three times row 2" in small numbers under row 2:

$$\tfrac{1}{-11}R_2 \to \begin{pmatrix} 1 & 3 & | & 15 \\ 0 & 1 & | & 2 \end{pmatrix} \quad R_1 - 3R_2 \to R_1 \text{ gives}$$
$$ 0 \quad 3 \quad 6$$

$$R_1 - 3R_2 \to \begin{pmatrix} 1 & 0 & | & 9 \\ 0 & 1 & | & 2 \end{pmatrix} \quad \text{Equivalent to} \quad \begin{cases} 1x + 0y = 9 \\ 0x + 1y = 2 \end{cases}$$

The unique solution is $x = 9$, $y = 2$. (We solved this same system by the substitution method on pages 157–158 and found the same answer.)

Practice Problem 3 Solve $\begin{cases} 2x + y = 14 \\ x - 3y = 21 \end{cases}$ by row-reducing an augmented matrix.

➤ Solution on page 178

Graphing Calculator Exploration

Some graphing calculators can row-reduce matrices. If your calculator has a RREF command (for "reduced row-echelon form"), you can easily check your row reduction. For the system of equations in Example 3:

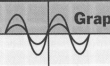

a. Enter $\begin{pmatrix} 1 & 3 & 15 \\ 2 & -5 & 8 \end{pmatrix}$ as matrix [A] using MATRX EDIT.

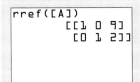

b. QUIT and select the RREF command from the MATRX MATH menu.

c. Apply RREF to the matrix [A].

Although this serves as a useful check of your answer, do not rely on it completely because the calculator sometimes returns an answer with round-off errors. Furthermore, this calculator command may not work for matrices with more rows than columns. We will be interested in such matrices in the next section.

EXAMPLE 4 — SOLVING EQUATIONS BY ROW REDUCTION

Solve $\begin{cases} 6x - 3y = 30 \\ -8x + 4y = -40 \end{cases}$ by row-reducing an augmented matrix.

Solution

The augmented matrix is

$$\begin{pmatrix} 6 & -3 & | & 30 \\ -8 & 4 & | & -40 \end{pmatrix}$$

When the numbers in a row have an obvious common factor, removing that factor can sometimes simplify the reduction.

$$\begin{matrix} \tfrac{1}{3}R_1 \to \\ \tfrac{1}{4}R_2 \to \end{matrix} \begin{pmatrix} 2 & -1 & | & 10 \\ -2 & 1 & | & -10 \end{pmatrix} \quad \text{(Want a 0 here)}$$

Adding the first row to the second, we get a zero at the bottom of the first column.

$$R_2 + R_1 \to \begin{pmatrix} 2 & -1 & 10 \\ 0 & 0 & 0 \end{pmatrix} \quad \text{(Want a 1 here)}$$

To finish, we divide the first row by 2 to make the row begin with a 1 on the left.

$$\tfrac{1}{2}R_1 \to \begin{pmatrix} 1 & -\tfrac{1}{2} & 5 \\ 0 & 0 & 0 \end{pmatrix} \quad \leftarrow \text{a zero row}$$

The zero row means that the system is *dependent*, so there are *infinitely many solutions*. The first row of this final matrix says that

$$x - \tfrac{1}{2}y = 5$$

or, solving for x,

$$x = 5 + \tfrac{1}{2}y$$

We may let y be *any* number t and then determine x from this equation. That is, the solutions may be parameterized as $x = 5 + \tfrac{1}{2}t$, $y = t$, where t is *any* number. The following table lists some of these solutions for various values of the parameter t.

3.2 MATRICES AND LINEAR EQUATIONS IN TWO VARIABLES

Parameterized solution

t	$x = 5 + \frac{1}{2}t$	$y = t$
-20	-5	-20
-10	0	-10
0	5	0
10	10	10
20	15	20

Evaluating x and y at t

There are many other sequences of row operations to reduce this augmented matrix and all reach the same conclusion.

Practice Problem 4

Verify that $x = -5$, $y = -20$, and $x = 15$, $y = 20$ from the preceding table solve $\begin{cases} 6x - 3y = 30 \\ -8x + 4y = -40 \end{cases}$.

➤ Solution on page 178

EXAMPLE 5

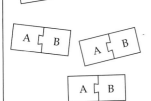

PRODUCTION MANAGEMENT

A worker in a plastics factory breaks apart sheets of component A and strips of component B and then snaps one of each together to make a finished item. If the worker can break off 20 A's or 30 B's per minute and snap together 10 pairs of A's and B's per minute, how should the worker's 440-minute workday be divided so as to complete as many items as possible with no unused pieces left over?

Solution

Let x be the number of minutes spent breaking apart sheets of component A and y be the number of minutes breaking apart strips of component B. The remainder of the worker's time, $440 - (x + y)$, will then be spent snapping A's and B's together. Because the worker needs as many A's as B's,

$$20x = 30y$$

Rate × time gives number finished

As many finished items as A's will be completed:

$$20x = 10(440 - (x + y))$$

Simplifies to $30x + 10y = 4400$

Rewriting these as a system of equations, the problem becomes

$$\begin{cases} 20x - 30y = 0 \\ 30x + 10y = 4400 \end{cases}$$

The augmented matrix is

$$\begin{pmatrix} 20 & -30 & | & 0 \\ 30 & 10 & | & 4400 \end{pmatrix}$$

We begin the row reduction by removing from both rows the common factor of 10.

$$\begin{array}{c} \frac{1}{10}R_1 \to \\ \frac{1}{10}R_2 \to \end{array} \begin{pmatrix} 2 & -3 & | & 0 \\ 3 & 1 & | & 440 \end{pmatrix} \quad \text{(Want a 1 here)}$$

To get a one in the upper left-hand corner:

$$R_2 - R_1 \to \begin{pmatrix} 1 & 4 & | & 440 \\ 3 & 1 & | & 440 \end{pmatrix} \quad \text{(Want a 0 here)}$$

To get a zero in the lower left-hand corner:

$$3R_1 - R_2 \to \begin{pmatrix} 1 & 4 & 440 \\ 0 & 11 & 880 \end{pmatrix}$$

Removing the common factor in row 2:

$$\tfrac{1}{11}R_2 \to \begin{pmatrix} 1 & 4 & 440 \\ 0 & 1 & 80 \end{pmatrix} \quad \text{(Want a 0 here)}$$

To get a zero above the 1 in the second row:

$$R_1 - 4R_2 \to \begin{pmatrix} 1 & 0 & 120 \\ 0 & 1 & 80 \end{pmatrix} \quad \begin{array}{l} x = 120 \\ y = 80 \end{array}$$

The solution is $x = 120$, $y = 80$. In terms of the original question, the worker should break apart sheets of component A for 120 minutes, break apart strips of component B for 80 minutes, and snap A's and B's together for the remaining 240 minutes.

If you have a graphing calculator with a RREF command, you may use it to check the result of the row reduction in Example 5.

3.2 Section Summary

The *augmented matrix* $\begin{pmatrix} a & b & | & h \\ c & d & | & k \end{pmatrix}$ represents the system of equations $\begin{cases} ax + by = h \\ cx + dy = k \end{cases}$. The rows correspond to the equations, and the first column corresponds to the x-variable, the second column to the y-variable, and the column after the bar to the constant terms of the equations.

The three basic row operations are

1. Switch any two rows.
2. Multiply or divide one of the rows by a nonzero number.
3. Replace a row by its sum or difference with a multiple of another row.

If these row operations lead to a row of zeros ending with a one, 0 0 | 1, then the system is *inconsistent* and has *no* solutions. If there is no such "inconsistent" row but there is a row of *all* zeros, 0 0 | 0, then the system is *dependent* and has *infinitely many* solutions. Otherwise, the system has a unique solution. In fact, any 2×3 augmented matrix is equivalent to exactly one of the five types below.

The augmented matrix $\begin{pmatrix} a & b & | & h \\ c & d & | & k \end{pmatrix}$ of $\begin{cases} ax + by = h \\ cx + dy = k \end{cases}$ is equivalent to exactly one of the following (where p and q stand for any numbers).

Matrix	Equations	Solution
$\begin{pmatrix} 1 & 0 & \| & p \\ 0 & 1 & \| & q \end{pmatrix}$	Unique solution	$x = p, \ y = q$
$\begin{pmatrix} 1 & p & \| & 0 \\ 0 & 0 & \| & 1 \end{pmatrix}$	Inconsistent	No solution
$\begin{pmatrix} 0 & 1 & \| & 0 \\ 0 & 0 & \| & 1 \end{pmatrix}$	Inconsistent	No solution
$\begin{pmatrix} 1 & p & \| & q \\ 0 & 0 & \| & 0 \end{pmatrix}$	Dependent	Infinitely many solutions: $x = q - pt, \ y = t$
$\begin{pmatrix} 0 & 1 & \| & p \\ 0 & 0 & \| & 0 \end{pmatrix}$	Dependent	Infinitely many solutions: $x = t, \ y = p$

▶ Solutions to Practice Problems

1. a. $\begin{pmatrix} 2 & -1 & | & 14 \\ 1 & 3 & | & 21 \end{pmatrix}$ b. $\begin{cases} 3x + 2y = 35 \\ x + 3y = 21 \end{cases}$

2. a. $\begin{matrix} R_2 \to \\ R_1 \to \end{matrix} \begin{pmatrix} 1 & -3 & | & 21 \\ 2 & 1 & | & 14 \end{pmatrix}$ **b.** $R_1 + R_2 \to \begin{pmatrix} 7 & 0 & | & 63 \\ 1 & -3 & | & 21 \end{pmatrix}$

c. $\tfrac{1}{7}R_1 \to \begin{pmatrix} 1 & 0 & | & 9 \\ 1 & -3 & | & 21 \end{pmatrix}$

d. No; you can multiply only by a *nonzero* number.

e. No; multiplying row 1 by 5 must still give row 1 (*not* row 2).

3. Starting with the augmented matrix $\begin{pmatrix} 2 & 1 & | & 14 \\ 1 & -3 & | & 21 \end{pmatrix}$, one possible sequence of row operations is

$$R_1 - R_2 \to \begin{pmatrix} 1 & 4 & | & -7 \\ 1 & -3 & | & 21 \end{pmatrix}$$

$$R_1 - R_2 \to \begin{pmatrix} 1 & 4 & | & -7 \\ 0 & 7 & | & -28 \end{pmatrix}$$

$$\tfrac{1}{7}R_2 \to \begin{pmatrix} 1 & 4 & | & -7 \\ 0 & 1 & | & -4 \end{pmatrix}$$

$$R_1 - 4R_2 \to \begin{pmatrix} 1 & 0 & | & 9 \\ 0 & 1 & | & -4 \end{pmatrix}$$

The unique solution is $x = 9$, $y = -4$. There are many other sequences of row operations to reduce this augmented matrix and all reach the same conclusion.

4. For $x = -5$, $y = -20$, the equations become

$$\begin{cases} 6(-5) - 3(-20) = -30 + 60 = 30 \checkmark \\ -8(-5) + 4(-20) = 40 - 80 = -40 \checkmark \end{cases}$$

For $x = 15$, $y = 20$, the equations become

$$\begin{cases} 6(15) - 3(20) = 90 - 60 = 30 \checkmark \\ -8(15) + 4(20) = -120 + 80 = -40 \checkmark \end{cases}$$

Both check!

3.2 Exercises

Find the dimension of each matrix and the value(s) of the specified element(s).

1. $\begin{pmatrix} 1 & 2 \\ 2 & 5 \\ 3 & 6 \end{pmatrix}$; $a_{1,1}, a_{3,2}$ **2.** $\begin{pmatrix} 6 & 3 \\ 5 & 2 \\ 4 & 1 \end{pmatrix}$; $a_{1,1}, a_{3,2}$ **4.** $\begin{pmatrix} 1 & -2 & -1 \\ -3 & 4 & 3 \\ 5 & -6 & -5 \\ -7 & 8 & 7 \end{pmatrix}$; $a_{2,2}, a_{4,3}$

3. $\begin{pmatrix} 1 & -1 & 2 & -2 \\ -3 & 3 & -4 & 4 \\ 5 & -5 & 6 & -6 \end{pmatrix}$; $a_{2,2}, a_{3,4}$ **5.** $\begin{pmatrix} 1 & 0 & 0 & 0 \\ 0 & 1 & 0 & 0 \\ 0 & 0 & 1 & 0 \\ 0 & 0 & 0 & 1 \end{pmatrix}$; $a_{1,1}, a_{2,2}, a_{3,3}, a_{4,4}, a_{1,4}$

6. $\begin{pmatrix} 1 & 0 & 0 \\ 0 & 1 & 0 \\ 0 & 0 & 1 \end{pmatrix}$; $a_{1,1}, a_{2,2}, a_{3,3}, a_{2,1}$

7. $(4\ 5\ 6\ 7)$; $a_{1,3}$ 8. $(2\ 3\ 4\ 5\ 6)$; $a_{1,4}$

9. $\begin{pmatrix} 9 \\ 8 \\ 7 \\ 6 \\ 5 \end{pmatrix}$; $a_{2,1}, a_{4,1}$ 10. $\begin{pmatrix} 2 \\ 8 \\ 3 \\ 7 \end{pmatrix}$; $a_{2,1}, a_{3,1}$

Find the augmented matrix representing the system of equations.

11. $\begin{cases} x + 2y = 2 \\ 3x + 4y = 12 \end{cases}$

12. $\begin{cases} 2x - y = 10 \\ -3x + 4y = 60 \end{cases}$

13. $\begin{cases} -4x + 3y = 84 \\ 5x - 2y = 70 \end{cases}$

14. $\begin{cases} -x + 2y = 2 \\ 2x - 3y = 6 \end{cases}$

15. $\begin{cases} 3x - 2y = 24 \\ x = 6 \end{cases}$

16. $\begin{cases} 4x - 3y = 24 \\ y = 8 \end{cases}$

17. $\begin{cases} 5x - 15y = 30 \\ -4x + 12y = 24 \end{cases}$

18. $\begin{cases} 12x - 4y = 36 \\ -15x + 5y = 45 \end{cases}$

19. $\begin{cases} x = 20 \\ y = 30 \end{cases}$

20. $\begin{cases} y = 18 \\ x = 12 \end{cases}$

Find the system of equations represented by the augmented matrix.

21. $\begin{pmatrix} 1 & 1 & | & 9 \\ 0 & 1 & | & 4 \end{pmatrix}$

22. $\begin{pmatrix} 1 & 0 & | & -3 \\ 1 & 1 & | & 5 \end{pmatrix}$

23. $\begin{pmatrix} -4 & 3 & | & -60 \\ 1 & -2 & | & 20 \end{pmatrix}$

24. $\begin{pmatrix} 3 & 4 & | & 24 \\ 1 & 2 & | & 6 \end{pmatrix}$

25. $\begin{pmatrix} 1 & -3 & | & -70 \\ 1 & 1 & | & 10 \end{pmatrix}$

26. $\begin{pmatrix} 1 & 1 & | & 5 \\ 2 & 3 & | & 7 \end{pmatrix}$

27. $\begin{pmatrix} 2 & 1 & | & 6 \\ 1 & 2 & | & -6 \end{pmatrix}$

28. $\begin{pmatrix} 3 & 1 & | & -24 \\ 1 & 3 & | & 24 \end{pmatrix}$

29. $\begin{pmatrix} 20 & -15 & | & 60 \\ -16 & 12 & | & -48 \end{pmatrix}$

30. $\begin{pmatrix} 20 & -15 & | & 60 \\ -16 & 12 & | & 48 \end{pmatrix}$

Carry out the row operation on the matrix.

31. $R_1 \leftrightarrow R_2$ on $\begin{pmatrix} 3 & 4 & | & 24 \\ 5 & 6 & | & 30 \end{pmatrix}$

32. $R_1 \leftrightarrow R_2$ on $\begin{pmatrix} 8 & 7 & | & 56 \\ 6 & 5 & | & 60 \end{pmatrix}$

33. $R_1 - R_2 \to R_1$ on $\begin{pmatrix} 8 & 7 & | & 56 \\ 6 & 5 & | & 60 \end{pmatrix}$

34. $2R_1 \to R_1$ on $\begin{pmatrix} 3 & 4 & | & 24 \\ 5 & 6 & | & 30 \end{pmatrix}$

35. $R_1 - 3R_2 \to R_1$ on $\begin{pmatrix} 5 & 6 & | & 30 \\ 1 & 2 & | & 18 \end{pmatrix}$

36. $R_1 - 3R_2 \to R_1$ on $\begin{pmatrix} 6 & 5 & | & 60 \\ 2 & 2 & | & -4 \end{pmatrix}$

37. $R_1 - R_2 \to R_2$ on $\begin{pmatrix} 6 & 6 & | & -12 \\ 6 & 5 & | & 60 \end{pmatrix}$

38. $R_1 - R_2 \to R_1$ on $\begin{pmatrix} 5 & 6 & | & 30 \\ 4 & 8 & | & 72 \end{pmatrix}$

39. $\frac{1}{8}R_2 \to R_2$ on $\begin{pmatrix} 1 & -2 & | & -42 \\ 0 & 8 & | & 120 \end{pmatrix}$

40. $\frac{1}{6}R_1 \to R_1$ on $\begin{pmatrix} 6 & 0 & | & 420 \\ 0 & 1 & | & -72 \end{pmatrix}$

Interpret each augmented matrix as the solution of a system of equations. State the solution or identify the system as "inconsistent" or "dependent."

41. $\begin{pmatrix} 1 & 0 & | & 7 \\ 0 & 1 & | & -3 \end{pmatrix}$

42. $\begin{pmatrix} 1 & 0 & | & -5 \\ 0 & 1 & | & 8 \end{pmatrix}$

43. $\begin{pmatrix} 1 & 1 & | & 0 \\ 0 & 0 & | & 1 \end{pmatrix}$

44. $\begin{pmatrix} 1 & -1 & | & 0 \\ 0 & 0 & | & 1 \end{pmatrix}$

45. $\begin{pmatrix} 0 & 1 & | & 0 \\ 0 & 0 & | & 1 \end{pmatrix}$

46. $\begin{pmatrix} 1 & 0 & | & 0 \\ 0 & 1 & | & 0 \end{pmatrix}$

47. $\begin{pmatrix} 1 & 2 & | & 3 \\ 0 & 0 & | & 0 \end{pmatrix}$

48. $\begin{pmatrix} 1 & -4 & | & 6 \\ 0 & 0 & | & 0 \end{pmatrix}$

49. $\begin{pmatrix} 0 & 1 & | & -3 \\ 0 & 0 & | & 0 \end{pmatrix}$

50. $\begin{pmatrix} 0 & 1 & | & 9 \\ 0 & 0 & | & 0 \end{pmatrix}$

Solve each system by row-reducing the corresponding augmented matrix. State the solution or identify the system as "inconsistent" or "dependent."

If you have a graphing calculator with a RREF command, use it to check your row reduction.

51. $\begin{cases} x + y = 5 \\ x = 3 \end{cases}$

52. $\begin{cases} x + y = 8 \\ y = 6 \end{cases}$

53. $\begin{cases} x + y = 4 \\ x - y = 2 \end{cases}$

54. $\begin{cases} x - 2y = 6 \\ x + y = 3 \end{cases}$

55. $\begin{cases} 2x + y = 4 \\ x + y = 3 \end{cases}$
56. $\begin{cases} x + y = 4 \\ 3x + y = 6 \end{cases}$
63. $\begin{cases} 6x + 2y = 18 \\ 5x + 2y = 10 \end{cases}$
64. $\begin{cases} 5x + 3y = 15 \\ 4x + 2y = 8 \end{cases}$

57. $\begin{cases} x + y = 5 \\ 2x + 3y = 12 \end{cases}$
58. $\begin{cases} 3x + y = 9 \\ 2x + y = 4 \end{cases}$
65. $\begin{cases} -3x + y = 3 \\ 5x - 2y = -10 \end{cases}$
66. $\begin{cases} 3x - y = 3 \\ -2x + y = 2 \end{cases}$

59. $\begin{cases} 2x + y = 20 \\ x + 3y = 15 \end{cases}$
60. $\begin{cases} x + y = 8 \\ 3x + 5y = 30 \end{cases}$
67. $\begin{cases} 2x - 6y = 18 \\ -3x + 9y = -27 \end{cases}$
68. $\begin{cases} 3x - 6y = 24 \\ -5x + 10y = -40 \end{cases}$

61. $\begin{cases} -x + 2y = 4 \\ x - 2y = 6 \end{cases}$
62. $\begin{cases} x - 3y = 12 \\ -x + 3y = 9 \end{cases}$
69. $\begin{cases} 4x + 7y = 56 \\ 2x + 3y = 30 \end{cases}$
70. $\begin{cases} 4x + 3y = 24 \\ 6x + 5y = 30 \end{cases}$

APPLIED EXERCISES

Express each situation as a system of two equations in two variables. Be sure to state clearly the meaning of your *x*- and *y*-variables. Solve the system by row-reducing the corresponding augmented matrix. State your final answer in terms of the original question.

 If you have a graphing calculator with a RREF command, use it to check your row reduction.

71. BUSINESS: Commodity Futures A corn and soybean commodities speculator invested $15,000 yesterday with twice as much in soybean futures as in corn futures. How much did she invest in each?

72. BUSINESS: Stamps and Coins A stamp and coin dealer spent $8000 at a numismatics auction last weekend. If he spent three times as much on coins as on stamps, how much did he spend on each?

73. GENERAL: Estate Division A will specifies that the older brother is to receive one half of the $12 million estate after first deducting the younger brother's share, the younger brother is to receive one third of the estate after first deducting the older brother's share, and the remainder is to be given to their sister. How much does each brother receive? How much is left over for their sister?

74. BUSINESS: Real Estate Taxes Find the state and city property taxes on an apartment building assessed at $833,000 if the state tax is 4% of the assessed value after first deducting the city tax and the city tax is 1% of the assessed value after first deducting the state tax.

75. BUSINESS: Racehorse Syndication The stud rights for a recent Triple-Crown winner were purchased by a syndicate for $3,675,000 and each member of the syndicate invested either $5000 or $25,000. If twice as many of the investors had invested $5000 but only half as many had invested $25,000, the syndicate would have raised only $3,150,000. How many investors contributed $5000 and how many contributed $25,000?

76. PERSONAL FINANCE: Financial Planning Last year, a retired couple's investments in a money market account yielding 5% and in a stock mutual fund yielding 14% paid a total of $16,800. If the stock mutual fund can return 17% this year and the money market account can continue to yield 5%, the couple will receive $17,850. How much is invested in each?

77. BIOMEDICAL: Geriatrics Nutrition The dietician at a senior care facility has decided to supplement the weekly menu with "NutraDrink" and "VitaPills," which contain calcium and vitamin C as listed in the table. If each resident needs an additional 300 mg of calcium and 240 mg of vitamin C per week, how many cans of NutraDrink and tablets of VitaPills should each resident be given per week?

	Calcium (mg)	Vitamin C (mg)
NutraDrink	25	20
VitaPills	20	16

78. GENERAL: Plant Fertilizer A garden field needs 308 pounds of potash and 330 pounds of nitrogen. Two brands of fertilizer, GrowRite and GreatGreen, are available and contain the amounts of potash and nitrogen per bag listed in the table. How many bags of each brand should be used to provide the required potash and nitrogen?

(Pounds per Bag)	GrowRite	GreatGreen
Potash	4	7
Nitrogen	5	6

79. SOCIAL SCIENCE: Political Advertising For the final days before the election, the campaign manager has a total of $36,000 to spend on TV and radio campaign advertisements. Each TV ad costs $3000 and is seen by 10,000 voters, while each radio ad costs $500 and is heard by 2000 voters. Ignoring repeated exposures to the same voter, how many TV and radio ads will contact 130,000 voters using the allocated funds?

80. MANAGEMENT SCIENCE: Consumer Preferences A marketing company wants to gather information on consumer preferences for laundry detergents using telephone interviews and direct mail questionnaires. Each attempted telephone interview costs $2.25 whether successful or not, and each mailed questionnaire costs $0.75 whether returned or not. The research director estimates that 21% of the telephone calls will result in usable interviews while only 7% of the questionnaires will be returned in usable condition. If the director's budget is $15,750 and she needs 1470 usable responses, how many telephone calls should she attempt and how many questionnaires should she mail out?

Explorations and Excursions

The following problems extend and augment the material presented in the text.

More about 2×3 *Row-Reduced Matrices*

The table on page 177 lists exactly five possible final matrices for the solution of a system of equations by reducing the corresponding augmented matrix. The following problems provide examples of each possibility.

Sketch each system of equations and then solve it by row-reducing the corresponding augmented matrix. State the solution or identify the system as "inconsistent" or "dependent."

If permitted by your instructor, graph each system on the window $[-30, 50]$ by $[-30, 50]$, use the RREF command to row-reduce the augmented matrix, and graph the parametric solution if the system is dependent.

81. $\begin{cases} x - 2y = -14 \\ 2x + 3y = 84 \end{cases}$

82. $\begin{cases} x + 2y = 56 \\ 2x - 3y = 42 \end{cases}$

83. $\begin{cases} 3x - 6y = 18 \\ -2x + 4y = 16 \end{cases}$

84. $\begin{cases} 4x - 6y = 36 \\ -2x + 3y = 30 \end{cases}$

85. $\begin{cases} 2y = 8 \\ 3y = 18 \end{cases}$

86. $\begin{cases} 5y = 20 \\ 3y = 30 \end{cases}$

87. $\begin{cases} 4x + 14y = 56 \\ 6x + 21y = 84 \end{cases}$

88. $\begin{cases} 6x + 14y = 42 \\ 9x + 21y = 63 \end{cases}$

89. $\begin{cases} 5y = 30 \\ 3y = 18 \end{cases}$

90. $\begin{cases} 2y = 18 \\ 3y = 27 \end{cases}$

Row Operations Are Reversible

Carry out the row operation on the augmented matrix $\begin{pmatrix} 3 & -10 & | & -65 \\ -4 & 13 & | & 84 \end{pmatrix}$ and then find a row operation to perform on your new matrix that will return it to the original $\begin{pmatrix} 3 & -10 & | & -65 \\ -4 & 13 & | & 84 \end{pmatrix}$.

91. $R_1 \leftrightarrow R_2$

92. $R_2 \leftrightarrow R_1$

93. $5R_1 \to R_1$

94. $3R_2 \to R_2$

95. $\frac{1}{4}R_2 \to R_2$

96. $\frac{1}{3}R_1 \to R_1$

97. $R_1 + R_2 \to R_1$

98. $R_2 + R_1 \to R_2$

99. $R_1 - R_2 \to R_1$

100. $R_2 - R_1 \to R_2$

3.3 SYSTEMS OF LINEAR EQUATIONS AND THE GAUSS–JORDAN METHOD

Introduction

In this section we use augmented matrices to solve larger systems of linear equations. We simply enlarge the augmented matrix to allow for more equations (rows) and more variables (columns) and then apply row operations to find an equivalent matrix that displays the solution.

Names for Many Variables

To deal with many variables, we now distinguish them by subscripts instead of different letters: x_1 ("x sub one"), x_2 ("x sub two"), and so on for as many as we need. For example:

$$\begin{cases} 3x_1 + 2x_2 + x_3 = 39 \\ 2x_1 + 3x_2 + x_3 = 34 \\ x_1 + 2x_2 + 3x_3 = 26 \end{cases}$$

We form the augmented matrix exactly as before, so for this system we have

$$\begin{matrix} x_1 & x_2 & x_3 & \end{matrix}$$
$$\begin{pmatrix} 3 & 2 & 1 & | & 39 \\ 2 & 3 & 1 & | & 34 \\ 1 & 2 & 3 & | & 26 \end{pmatrix}$$

← Variables corresponding to columns (last column represents constant terms)

← Each row represents an equation and the bar represents the equals sign

Row-Reduced Form

We continue to use row operations to "solve" the augmented matrix by finding an equivalent matrix that displays the solution. With our list of 2×3 matrices on page 177 as a guide, we make the following definition for matrices of any dimension. In this definition, a *zero row* is a row containing only zeros, and a *nonzero row* is a row with at least one nonzero element.

Row-Reduced Form

A matrix is in *row-reduced form* if it satisfies the following four conditions.

1. All zero rows are grouped below the nonzero rows.
2. The first nonzero element in each nonzero row is a 1. We call these particular 1s "leftmost 1s."

3.3 SYSTEMS OF LINEAR EQUATIONS AND THE GAUSS–JORDAN METHOD

3. If a column contains a leftmost 1, all other entries in the column are 0s.
4. Each leftmost 1 appears to the *right* of the leftmost 1 in the row above it.

The following matrices are in row-reduced form:

$$\begin{pmatrix} 0 & 1 & 2 & 0 & 1 \\ 0 & 0 & 0 & 1 & 4 \end{pmatrix}, \quad \begin{pmatrix} 1 & 2 & 3 & 0 \\ 0 & 0 & 0 & 1 \\ 0 & 0 & 0 & 0 \end{pmatrix}, \quad \text{and} \quad \begin{pmatrix} 1 & 0 & 0 & 0 & 1 \\ 0 & 1 & 0 & 0 & 2 \\ 0 & 0 & 1 & 0 & 3 \\ 0 & 0 & 0 & 1 & 4 \end{pmatrix}$$

Leftmost 1s — Zero row

Notice that the 1s in the upper right corners of the first and last matrices above are not *leftmost* 1s and do not need to have 0s below them. The following matrix is *not* in row-reduced form because although conditions (1), (2), and (3) are satisfied, condition (4) fails for the third row.

$$\begin{pmatrix} 1 & 0 & 0 & 2 \\ 0 & 0 & 1 & 4 \\ 0 & 1 & 0 & 6 \end{pmatrix}$$

Not row-reduced:

The 1 in the third row is not to the right of the 1 in the second row

Practice Problem 1 Find a row operation that will correct this defect and result in a matrix in row-reduced form. ➤ Solution on page 194

In many augmented matrices the row-reduced form will be particularly simple: to the left of the bar will be a square matrix with 1s down the main diagonal and 0s elsewhere, as shown below. In such cases the system has a *unique solution* and we may immediately read off the values of the variables from the numbers to the right of the bar (top to bottom):

$$\left(\begin{array}{ccc|c} 1 & 0 & 0 & 5 \\ 0 & 1 & 0 & -2 \\ 0 & 0 & 1 & 3 \end{array}\right) \text{ gives } \begin{cases} x_1 = 5 \\ x_2 = -2 \\ x_3 = 3 \end{cases} \text{ From } \begin{array}{l} 1x_1 + 0x_2 + 0x_3 = 5 \\ 0x_1 + 1x_2 + 0x_3 = -2 \\ 0x_1 + 0x_2 + 1x_3 = 3 \end{array}$$

This is the pattern of 1s and 0s that we seek, since it leads to a unique solution. If another pattern occurs, we will see how to interpret it as indicating either no solution or infinitely many solutions. We will row-reduce augmented matrices by following a procedure called the

For instance, you might get:

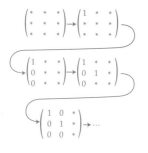

Gauss–Jordan Method

To row-reduce a matrix:

1. If any rows have leading 0s, switch the rows so that the ones with the *fewest* leading 0s are at the top, down to the ones with the *most* leading 0s at the bottom.
2. In the first row, find the leftmost nonzero entry and divide the row through by that number. This gives a *leftmost 1* in that row.
3. Add or subtract multiples of the first row to each other row to obtain 0s in the rest of the column above and below the leftmost 1 found in step 2.
4. Repeat steps 1, 2, and 3 but replacing "first row" by "second row" and then by "third row" and so on. Stop when you reach the bottom row or a row consisting entirely of zeros—when this happens the matrix is *row-reduced*.

Just as in the last section, there are three possibilities: a unique solution, no solution, or infinitely many solutions that can be parameterized.

EXAMPLE 1 SOLVING A SYSTEM USING THE GAUSS–JORDAN METHOD

Solve by the Gauss–Jordan method: $\begin{cases} 2x_1 + 4x_2 + 2x_3 = 2 \\ 3x_1 + 7x_2 + 3x_3 = 0 \\ x_1 + 2x_2 + 4x_3 = 7 \end{cases}$

Solution

Beginning with the augmented matrix, we carry out steps 1–4 of the Gauss–Jordan method.

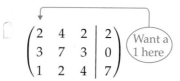

Step 1 concerns zero rows, and there are none. Step 2 says to divide row 1 by the first nonzero element, which is 2, to get a leftmost 1. The following row operation achieves this.

---

*Named after the mathematician Carl Friedrich Gauss (1777–1855) and the geodesist Wilhelm Jordan (1842–1899).

3.3 SYSTEMS OF LINEAR EQUATIONS AND THE GAUSS–JORDAN METHOD

$$\begin{array}{c}\frac{1}{2}R_1 \to \\ \\ \\ \end{array} \begin{pmatrix} 1 & 2 & 1 & | & 1 \\ 3 & 7 & 3 & | & 0 \\ 1 & 2 & 4 & | & 7 \end{pmatrix} \text{Want 0s here}$$

Step 3 says to subtract multiples of row 1 from the other rows to get 0s below the leftmost 1. The following two row operations do this.

$$\begin{array}{c}R_2 - 3R_1 \to \\ R_3 - R_1 \to \end{array} \begin{pmatrix} 1 & 2 & 1 & | & 1 \\ 0 & 1 & 0 & | & -3 \\ 0 & 0 & 3 & | & 6 \end{pmatrix} \text{Want a 0 here}$$

Row 2 already has a leftmost 1, so to achieve the 0 above it, in the next step we subtract twice row 2.

$$R_1 - 2R_2 \to \begin{pmatrix} 1 & 0 & 1 & | & 7 \\ 0 & 1 & 0 & | & -3 \\ 0 & 0 & 3 & | & 6 \end{pmatrix} \text{Want a 1 here}$$

Step 2 then says to divide row 3 by its first nonzero element, 3, to obtain a leftmost 1 as follows.

$$\frac{1}{3}R_3 \to \begin{pmatrix} 1 & 0 & 1 & | & 7 \\ 0 & 1 & 0 & | & -3 \\ 0 & 0 & 1 & | & 2 \end{pmatrix} \text{Want a 0 here}$$

The next step subtracts row 3 from row 1, achieving the final zero and completing the row-reduction of the augmented matrix.

$$R_1 - R_3 \to \begin{pmatrix} 1 & 0 & 0 & | & 5 \\ 0 & 1 & 0 & | & -3 \\ 0 & 0 & 1 & | & 2 \end{pmatrix}$$

The solution is $\begin{cases} x_1 = 5 \\ x_2 = -3. \\ x_3 = 2 \end{cases}$

Reading the numbers from the last column

Graphing Calculator Exploration

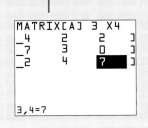

The program* ROWOPS carries out the arithmetic for the type of row operation you select from a menu. To have your calculator perform the row operations used in Example 1 (above), proceed as follows:

a. Enter the augmented matrix as matrix [A]. Because this matrix is too large to fit on the calculator screen, it will scroll from left to right and back again as you enter your numbers.

* See the Preface for information on how to obtain this and other programs.

```
 ROW OPERATIONS
1:  SWITCH
2:     MULTIPLY
3:     DIVIDE
4:  ADD TOGETHER
5:  SUBTRACT FROM
6:     ADD MULTIPLE
7:  QUIT
```

```
            [1 2 4 7]]
     DIVIDE
DIVIDE ROW 1
   BY 2
         [[1 2 1 1]
          [3 7 3 0]
          [1 2 4 7]]
```

b. Run the program ROWOPS. It will display the current values in the matrix [A] and you can use the arrows to scroll the screen to see the rest of it. Press ENTER to select the type of row operation you want to use.

c. Choose the type of row operation you want by using the arrows to move to its number and pressing ENTER, or just press the number. Enter the specific details for the operation and the program will carry out your request.

d. Press ENTER to select another row operation or to choose 7 and QUIT the program.

The program ROWOPS allows you to multiply rows by fractions and displays them in the usual 3/5 notation. The reduction of this matrix took six row operations to complete in Example 1. Can you find other ways to achieve the same result?

The system in Example 1 had a unique solution. If there is an "inconsistent" row, the system will have *no solutions*, just as in the previous section.

No Solutions

If a row-reduced matrix has a row of 0s ending in a 1 (such as 0 0 0 | 1), then the system of equations from which it came is *inconsistent* and has *no solutions*.

EXAMPLE 2

A SYSTEM WITH NO SOLUTION

Solve by the Gauss–Jordan method: $\begin{cases} 2x_1 - 4x_2 = 2 \\ -3x_1 + 6x_2 = 4 \end{cases}$.

Solution

The augmented matrix is

$$\begin{pmatrix} 2 & -4 & | & 2 \\ -3 & 6 & | & 4 \end{pmatrix}$$

Want a 1 here

There are no zero rows, so we go to step 2, dividing row 1 by the first nonzero element, 2, as follows.

$$\tfrac{1}{2}R_1 \rightarrow \begin{pmatrix} 1 & -2 & | & 1 \\ -3 & 6 & | & 4 \end{pmatrix} \quad \text{(Want a 0 here)}$$

Step 3 says to add or subtract a multiple of row 1 to row 2 to get a zero, so in the next matrix we add three times row 1.

$$R_2 + 3R_1 \rightarrow \begin{pmatrix} 1 & -2 & | & 1 \\ 0 & 0 & | & 7 \end{pmatrix} \quad \text{(Want a 1 here)}$$

Step 2 says to get a leading 1 in row 2, divide the row by 7 leading to the following matrix.

$$\tfrac{1}{7}R_2 \rightarrow \begin{pmatrix} 1 & -2 & | & 1 \\ 0 & 0 & | & 1 \end{pmatrix} \leftarrow \text{An "inconsistent" row}$$

Row 2 is of the form $0 \ 0 \ | \ 1$ saying that $0 = 1$, which means the system has *no solutions*.

The system is inconsistent and has no solutions.

Once we encountered the line $0 \ 0 \ | \ 7$ it was clear that by dividing by 7 we would obtain the "inconsistent" row $0 \ 0 \ | \ 1$. Therefore, in practice, whenever we encounter a row of 0s ending with any nonzero number after the bar we may stop at that point and declare that the system is inconsistent and has no solutions.

Practice Problem 2   Solve by the Gauss–Jordan method: $\begin{cases} 3x_1 - 6x_2 = 12 \\ -5x_1 + 10x_2 = -14 \end{cases}$

▶ **Solution on page 194**

The rule for infinitely many solutions is a little more complicated than in the case of two equations. It depends on the number of leftmost 1s compared to the number of variables.

> **Infinitely Many Solutions**
>
> For a row-reduced augmented matrix with no "inconsistent" rows, if the number of leading 1s is *less than* the number of variables, then the original system is *dependent* and has *infinitely many solutions*.

If there are infinitely many solutions, we may *parameterize* them as we did on pages 174–175. First we classify each variable as *determined* or *free* by looking at the column in the row-reduced augmented matrix corresponding to that variable:

The variable is *determined* if its column has a leftmost 1.

The variable is *free* if its column does *not* have a leftmost 1.

For example, given the following row-reduced matrix (with two columns to the left of the bar corresponding to two variables x_1 and x_2), the first variable x_1 is *determined* and the second variable x_2 is *free*.

x_1 is *determined* (its column has a leftmost 1)
x_2 is *free* (its column does not have a leftmost 1)

$$\begin{pmatrix} 1 & 2 & | & 3 \\ 0 & 0 & | & 0 \end{pmatrix}$$

$x_1 + 2x_2 = 3$
so $x_1 = 3 - 2x_2$

That is, x_2 is *free to take any value t*, while the value of x_1 is *determined* by the equation $x_1 = 3 - 2x_2$, giving the parameterized solution (replacing x_2 by t):

$$\begin{cases} x_1 = 3 - 2t \\ x_2 = t \end{cases}$$

From $x_1 = 3 - 2x_2$ with x_2 replaced by t

Free variables are sometimes called *independent* and determined variables are then called *dependent*.

EXAMPLE 3

A SYSTEM WITH INFINITELY MANY SOLUTIONS

Solve by the Gauss–Jordan method: $\begin{cases} 2x_1 + 6x_2 + 10x_3 = 8 \\ 3x_1 + 9x_2 + 15x_3 = 12. \\ 2x_1 + 5x_2 + 8x_3 = 7 \end{cases}$

Solution

The augmented matrix is

$$\begin{pmatrix} 2 & 6 & 10 & | & 8 \\ 3 & 9 & 15 & | & 12 \\ 2 & 5 & 8 & | & 7 \end{pmatrix}$$

Want a 1 here

Since there are no leading 0s, we go to step 2, which says to divide the first row by 2, as follows.

$\frac{1}{2}R_1 \to \begin{pmatrix} 1 & 3 & 5 & | & 4 \\ 3 & 9 & 15 & | & 12 \\ 2 & 5 & 8 & | & 7 \end{pmatrix}$

Want 0s here

To "zero-out" the rest of the column, in the next matrix we subtract appropriate multiples of the first row.

$\begin{matrix} \\ R_2 - 3R_1 \to \\ R_3 - 2R_1 \to \end{matrix} \begin{pmatrix} 1 & 3 & 5 & | & 4 \\ 0 & 0 & 0 & | & 0 \\ 0 & -1 & -2 & | & -1 \end{pmatrix}$

Has more leading 0s than

Since row 2 has more leading 0s than row 3, step 1 says to switch rows 2 and 3, as we do on the next page.

3.3 SYSTEMS OF LINEAR EQUATIONS AND THE GAUSS–JORDAN METHOD

$$\begin{array}{c} \\ R_3 \to \\ R_2 \to \end{array} \begin{pmatrix} 1 & 3 & 5 & | & 4 \\ 0 & -1 & -2 & | & -1 \\ 0 & 0 & 0 & | & 0 \end{pmatrix}$$

Want a 1 here

Now divide by -1 (which is equivalent to multiplying by -1) to get a leftmost 1 in the second row, as follows.

$$-1R_2 \to \begin{pmatrix} 1 & 3 & 5 & | & 4 \\ 0 & 1 & 2 & | & 1 \\ 0 & 0 & 0 & | & 0 \end{pmatrix}$$

Want a 0 here

Using the 1 to "zero-out" the rest of the column gives the matrix below.

$$R_1 - 3R_2 \to \begin{pmatrix} 1 & 0 & -1 & | & 1 \\ 0 & 1 & 2 & | & 1 \\ 0 & 0 & 0 & | & 0 \end{pmatrix}$$

Row-reduced!

The corresponding equations are $\begin{cases} x_1 \quad\;\; - x_3 = 1 \\ \quad x_2 + 2x_3 = 1 \end{cases}$.

Since the variables x_1 and x_2 are *determined* (their columns *do* have leftmost 1s) and the variable x_3 is *free* (column 3 does not have a leading 1), the free variable may take *any* value, $x_3 = t$, and we solve the other equations for the *determined* variables x_1 and x_2. The solution is:

$$\begin{cases} x_1 = 1 + t \\ x_2 = 1 - 2t \\ x_3 = t \end{cases}$$

From solving $x_1 - x_3 = 1$ for x_1
From solving $x_2 + 2x_3 = 1$ for x_2
and then replacing x_3 by t

We obtain all solutions by taking different values for t. For example:

$t = 3$ gives $\begin{cases} x_1 = 4 \\ x_2 = -5 \\ x_3 = 3 \end{cases}$ and $t = -2$ gives $\begin{cases} x_1 = -1 \\ x_2 = 5 \\ x_3 = -2 \end{cases}$.

It is easily checked that these are solutions to the original equations.

Practice Problem 3 Find the solution of a system of equations with augmented matrix equivalent to the row-reduced matrix

$$\begin{array}{cccc} x_1 & x_2 & x_3 & x_4 \end{array}$$
$$\begin{pmatrix} 1 & 2 & 0 & 0 & | & 0 \\ 0 & 0 & 1 & 0 & | & 3 \\ 0 & 0 & 0 & 1 & | & 4 \\ 0 & 0 & 0 & 0 & | & 0 \end{pmatrix}$$

➤ Solution on page 194

EXAMPLE 4

MANAGING PRODUCTION

A hand-thrown pottery shop manufactures plates, cups, and vases. Each plate requires 4 ounces of clay, 6 minutes of shaping, and 5 minutes of painting; each cup requires 4 ounces of clay, 5 minutes of shaping, and 3 minutes of painting; and each vase requires 3 ounces of clay, 4 minutes of shaping, and 4 minutes of painting. This week the shop has 165 pounds of clay, 59 hours of skilled labor for shaping, and 46 hours of skilled labor for painting. If the shop manager wishes to use all these resources fully, how many of each product should the shop produce?

Suggestion: List the information in a table.

	Plates	Cups	Vases
Clay	4	4	3
Shaping	6	5	4
Painting	5	3	4

Solution

Let x_1, x_2, and x_3 be the numbers of plates, cups, and vases produced. The number of ounces of clay required is then $4x_1 + 4x_2 + 3x_3$ and this must match the 165 pounds available:

$$4x_1 + 4x_2 + 3x_3 = 2640 \qquad \text{Use ounces on both sides of the equation}$$

Similarly, for the time in minutes required for shaping and painting,

$$6x_1 + 5x_2 + 4x_3 = 3540 \qquad \text{Use minutes on both sides of the equation}$$
$$5x_1 + 3x_2 + 4x_3 = 2760$$

Therefore, the augmented matrix is

$$\begin{pmatrix} 4 & 4 & 3 & | & 2640 \\ 6 & 5 & 4 & | & 3540 \\ 5 & 3 & 4 & | & 2760 \end{pmatrix}$$

We now solve this problem by row-reducing the augmented matrix using the Gauss–Jordan method.

$$\frac{1}{4}R_1 \to \begin{pmatrix} 1 & 1 & \frac{3}{4} & | & 660 \\ 6 & 5 & 4 & | & 3540 \\ 5 & 3 & 4 & | & 2760 \end{pmatrix} \qquad \text{Get a 1 in the first column of the first row}$$

$$\begin{matrix} \\ R_2 - 6R_1 \to \\ R_3 - 5R_1 \to \end{matrix} \begin{pmatrix} 1 & 1 & \frac{3}{4} & | & 660 \\ 0 & -1 & -\frac{1}{2} & | & -420 \\ 0 & -2 & \frac{1}{4} & | & -540 \end{pmatrix} \qquad \text{"Zero-out" the rest of the first column}$$

$$-R_2 \to \begin{pmatrix} 1 & 1 & \frac{3}{4} & | & 660 \\ 0 & 1 & \frac{1}{2} & | & 420 \\ 0 & -2 & \frac{1}{4} & | & -540 \end{pmatrix} \qquad \text{Get a 1 in the second column of the second row}$$

$$\begin{matrix} R_1 - R_2 \to \\ \\ R_3 + 2R_2 \to \end{matrix} \begin{pmatrix} 1 & 0 & \frac{1}{4} & | & 240 \\ 0 & 1 & \frac{1}{2} & | & 420 \\ 0 & 0 & \frac{5}{4} & | & 300 \end{pmatrix} \qquad \text{"Zero-out" the rest of the second column}$$

$$\tfrac{4}{5}R_3 \to \begin{pmatrix} 1 & 0 & \tfrac{1}{4} & | & 240 \\ 0 & 1 & \tfrac{1}{2} & | & 420 \\ 0 & 0 & 1 & | & 240 \end{pmatrix}$$

Get a 1 in the third column of the third row

$$\begin{array}{c} R_1 - \tfrac{1}{4}R_3 \to \\ R_2 - \tfrac{1}{2}R_3 \to \end{array} \begin{pmatrix} 1 & 0 & 0 & | & 180 \\ 0 & 1 & 0 & | & 300 \\ 0 & 0 & 1 & | & 240 \end{pmatrix}$$

"Zero-out" the rest of the third column

The system has a unique solution: $x_1 = 180$, $x_2 = 300$, $x_3 = 240$. You should check that these values satisfy the original equations and that the clay and skilled labor resources are fully used. In terms of the original question, the shop should produce 180 plates, 300 cups, and 240 vases this week.

EXAMPLE 5

MODELING A COMPUTER NETWORK

The office manager for the accounting division of a large company is writing a report to her supervisor about the demands placed on the computers serving her division. Besides meeting the needs of the accounting division, her four "file servers" are expected to accept and relay messages to and from other areas. The diagram shows the four file servers (I, II, III, IV), and the arrows and numbers indicate the data packets per minute passing from the senders to the receivers. Computers A through H are outside the accounting division. Connections passing an unknown number of data packets per minute are marked with variables (for instance, x_1 is the number of data packets per minute sent from server I to server II, while x_5 is the number leaving server IV to outside computer G). For the network to function, the number of data packets per minute arriving at each computer must match the number leaving it. How many data packets per minute must leave IV for G? Is it possible for each connection within the accounting division to carry no more than 1500 data packets per minute?

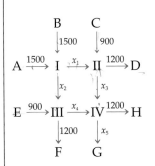

Solution

In terms of the notations from the diagram, the requirements that the data packets per minute arriving at each file server match the number leaving are represented by the equations

	"In" equals "out"
$1500 + 1500 = x_1 + x_2$	Server I
$x_1 + 900 = x_3 + 1200$	Server II
$900 + x_2 = x_4 + 1200$	Server III
$x_3 + x_4 = 1200 + x_5$	Server IV

Rewriting to put the variables on the left, we have the system

$$\begin{cases} x_1 + x_2 = 3000 \\ x_1 - x_3 = 300 \\ x_2 - x_4 = 300 \\ x_3 + x_4 - x_5 = 1200 \end{cases}$$

The following augmented matrix represents these equations, with its row-reduced form shown on the right. (You may wish to carry out the reduction—it can be done easily because of the many zeros and ones—or use your calculator's RREF command to check the reduction.)

$$\begin{pmatrix} 1 & 1 & 0 & 0 & 0 & | & 3000 \\ 1 & 0 & -1 & 0 & 0 & | & 300 \\ 0 & 1 & 0 & -1 & 0 & | & 300 \\ 0 & 0 & 1 & 1 & -1 & | & 1200 \end{pmatrix} \xrightarrow{\text{Leads to}} \begin{pmatrix} 1 & 0 & 0 & 1 & 0 & | & 2700 \\ 0 & 1 & 0 & -1 & 0 & | & 300 \\ 0 & 0 & 1 & 1 & 0 & | & 2400 \\ 0 & 0 & 0 & 0 & 1 & | & 1200 \end{pmatrix}$$

The variables x_1, x_2, x_3, and x_5 are *determined* (since their columns have leftmost 1s), and x_4 is a free variable (its column does *not* have a leftmost 1). We may parameterize the solution as follows:

$$\begin{aligned} x_1 &= 2700 - t \\ x_2 &= 300 + t \\ x_3 &= 2400 - t \\ x_4 &= t \qquad \text{The free variable} \\ x_5 &= 1200 \end{aligned}$$

To prevent a "back flow" along the connections, each variable must stay nonnegative, which means that $t \geq 0$ (since $x_4 = t$) and $t \leq 2400$ (since $x_3 = 2400 - t$). For example, the choice $t = 1200$ gives the solution

$$x_1 = 1500, \quad x_2 = 1500, \quad x_3 = 1200, \quad x_4 = 1200, \quad x_5 = 1200$$

In terms of the original questions, 1200 data packets per minute must leave computer IV for G, and the example with $t = 1200$ shows that it *is* possible to arrange the transmissions so that each connection within the accounting division carries no more than 1500 data packets per minute.

3.3 Section Summary

The solution of any system of linear equations can be found by representing it as an augmented matrix, applying row operations (page 170) using the Gauss–Jordan method (page 184) to find the equivalent row-reduced matrix (pages 182–183), and interpreting this final matrix as a statement about the original system of equations (pages 183 and 186–187). Each step in this method is a direct extension of the method for solving systems of two linear equations in two variables presented in the previous section.

For three equations in three unknowns with a unique solution, this would look like:

$$\begin{pmatrix} * & * & * & | & * \\ * & * & * & | & * \\ * & * & * & | & * \end{pmatrix} \xrightarrow{\text{Row-reduces to}} \begin{pmatrix} 1 & 0 & 0 & | & * \\ 0 & 1 & 0 & | & * \\ 0 & 0 & 1 & | & * \end{pmatrix}$$

$$\text{solution}$$

A row of zeros that ends in a nonzero number (such as $0\ 0\ 0\ |\ 1$) means that the system is *inconsistent* and has *no solutions*. If there is no such "inconsistent row" and if the number of leftmost 1s *is less* than the number of variables, then there are *infinitely many* solutions, which may be parameterized as on page 192.

A system of equations may have more than one free variable. The number of free variables will be equal to the total number of variables minus the number of leftmost 1s in the row-reduced form.

The three possibilities—unique solution, no solution, or infinitely many solutions—may be understood geometrically as follows. On page 155 we saw that equations in *two* variables represent *lines in the plane* and that two lines may intersect in a single point (a unique solution), not at all (parallel lines), or at infinitely many points (the same line). Analogously, equations in *three* variables represent *planes in three-dimensional space,* and three planes can intersect at a single point, not at all, or at infinitely many points, as shown below.

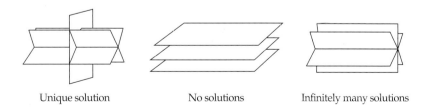

Unique solution   No solutions   Infinitely many solutions

> **Solutions to Practice Problems**

1. $R_2 \leftrightarrow R_3$

2.
$$\begin{pmatrix} 3 & -6 & | & 12 \\ -5 & 10 & | & -14 \end{pmatrix}$$

$\frac{1}{3}R_1 \to \begin{pmatrix} 1 & -2 & | & 4 \\ -5 & 10 & | & -14 \end{pmatrix}$

$R_2 + 5R_1 \to \begin{pmatrix} 1 & -2 & | & 4 \\ 0 & 0 & | & 6 \end{pmatrix}$

Inconsistent, so *no* solution.

3. $x_1, x_3,$ and x_4 are determined while x_2 is free. The solution is $\begin{cases} x_1 = -2t \\ x_2 = t \\ x_3 = 3 \\ x_4 = 4 \end{cases}$.

3.3 Exercises

Find the augmented matrix representing the system of equations.

1. $\begin{cases} x_1 + x_2 + x_3 = 4 \\ x_1 + 2x_2 + x_3 = 3 \\ x_1 + 2x_2 + 2x_3 = 5 \end{cases}$

2. $\begin{cases} x_1 + 5x_2 + 4x_3 = 6 \\ x_1 + x_2 + x_3 = 4 \\ 2x_1 + 3x_2 + 3x_3 = 9 \end{cases}$

3. $\begin{cases} 2x_1 - x_2 + 2x_3 = 11 \\ -x_1 + x_2 - 3x_3 = -12 \\ 2x_1 - 2x_2 + 7x_3 = 27 \end{cases}$

4. $\begin{cases} 4x_1 + 3x_2 - x_3 = 2 \\ 3x_1 + 3x_2 + 2x_3 = 9 \\ 2x_1 + x_2 - 3x_3 = -6 \end{cases}$

5. $\begin{cases} 2x_1 + x_2 + 5x_3 + 4x_4 + 5x_5 = 2 \\ x_1 + x_2 + 3x_3 + 3x_4 + 3x_5 = -1 \end{cases}$

6. $\begin{cases} 5x_1 + 2x_2 - 4x_3 + x_4 + 5x_5 = 7 \\ 3x_1 + x_2 - 3x_3 + x_4 + 3x_5 = 5 \end{cases}$

Find the system of equations represented by the augmented matrix.

7. $\begin{pmatrix} 4 & 3 & 2 & | & 11 \\ 3 & 3 & 1 & | & 6 \\ 1 & -2 & 3 & | & 13 \end{pmatrix}$

8. $\begin{pmatrix} 3 & -2 & 5 & | & 23 \\ -1 & 1 & -3 & | & -12 \\ 2 & -2 & 7 & | & 27 \end{pmatrix}$

9. $\begin{pmatrix} 2 & 1 & 1 & | & 7 \\ 2 & 2 & 1 & | & 6 \\ 3 & 3 & 2 & | & 10 \end{pmatrix}$

10. $\begin{pmatrix} 5 & 1 & 3 & | & 20 \\ 1 & 1 & 2 & | & 6 \\ 4 & 1 & 3 & | & 17 \end{pmatrix}$

11. $\begin{pmatrix} 8 & 3 & -2 & 19 & | & 15 \\ 3 & 1 & -1 & 7 & | & 6 \end{pmatrix}$

12. $\begin{pmatrix} 6 & -2 & -4 & -2 & | & 36 \\ 2 & -1 & -10 & 5 & | & 6 \end{pmatrix}$

Interpret each row-reduced matrix as the solution of a system of equations.

13. $\begin{pmatrix} 1 & 0 & 0 & | & 4 \\ 0 & 1 & 0 & | & 5 \\ 0 & 0 & 1 & | & -4 \end{pmatrix}$

14. $\begin{pmatrix} 1 & 0 & 0 & | & 0 \\ 0 & 1 & 0 & | & 0 \\ 0 & 0 & 1 & | & 1 \end{pmatrix}$

15. $\begin{pmatrix} 1 & 0 & 1 & | & 0 \\ 0 & 1 & 0 & | & 0 \\ 0 & 0 & 0 & | & 1 \end{pmatrix}$

16. $\begin{pmatrix} 1 & 1 & | & 0 \\ 0 & 0 & | & 1 \end{pmatrix}$

17. $\begin{pmatrix} 1 & 0 & -1 & | & -5 \\ 0 & 1 & 1 & | & 5 \\ 0 & 0 & 0 & | & 0 \end{pmatrix}$

18. $\begin{pmatrix} 1 & 1 & 0 & 0 & | & 2 \\ 0 & 0 & 1 & 0 & | & -1 \\ 0 & 0 & 0 & 1 & | & 3 \\ 0 & 0 & 0 & 0 & | & 0 \end{pmatrix}$

Use an appropriate row operation or sequence of row operations to find the equivalent row-reduced matrix.

If your instructor permits, you may carry out the calculations on a graphing calculator using the ROWOPS program.

19. $\begin{pmatrix} 0 & 1 & 0 & | & 2 \\ 1 & 0 & 0 & | & 1 \\ 0 & 0 & 1 & | & 3 \end{pmatrix}$

20. $\begin{pmatrix} 1 & 0 & 0 & | & 1 \\ 0 & 0 & 1 & | & 3 \\ 0 & 1 & 0 & | & 2 \end{pmatrix}$

21. $\begin{pmatrix} 1 & 0 & 1 & | & 4 \\ 0 & 1 & 0 & | & 2 \\ 0 & 0 & 1 & | & 3 \end{pmatrix}$

22. $\begin{pmatrix} 1 & 0 & 0 & | & 1 \\ 1 & 1 & 0 & | & 3 \\ 0 & 0 & 1 & | & 3 \end{pmatrix}$

23. $\begin{pmatrix} 2 & 4 & 0 & | & 6 \\ 0 & 0 & 1 & | & -3 \\ 0 & 0 & 0 & | & 0 \end{pmatrix}$

24. $\begin{pmatrix} 1 & 0 & 2 & | & 5 \\ 0 & 3 & -6 & | & 3 \\ 0 & 0 & 0 & | & 0 \end{pmatrix}$

Solve each system of equations by the Gauss–Jordan method. If the solution is not unique, identify the system as "dependent" or "inconsistent."

If you have a graphing calculator with a RREF command, use it to check your row reduction.

25. $\begin{cases} x_1 + x_2 + x_3 = 2 \\ x_1 + 2x_2 + 2x_3 = 3 \\ x_1 + 3x_2 + 2x_3 = 1 \end{cases}$

26. $\begin{cases} 2x_1 + 3x_2 + x_3 = 4 \\ 3x_1 + 3x_2 + 2x_3 = 12 \\ x_1 + 2x_2 + x_3 = 4 \end{cases}$

27. $\begin{cases} 2x_1 + 2x_2 - x_3 = -5 \\ -2x_1 - x_2 + x_3 = 3 \\ 3x_1 + 4x_2 - x_3 = -8 \end{cases}$

28. $\begin{cases} 3x_1 - 4x_2 + 2x_3 = -15 \\ -x_1 + 2x_2 - x_3 = 6 \\ 4x_1 - 3x_2 + 2x_3 = -16 \end{cases}$

29. $\begin{cases} 2x_1 + x_2 - 2x_3 + x_4 = 2 \\ x_1 + x_2 + 2x_3 + x_4 = 5 \\ x_1 + x_2 + x_3 + x_4 = 4 \\ 2x_1 + 2x_2 + 3x_3 + x_4 = 8 \end{cases}$

30. $\begin{cases} x_1 - x_2 + x_3 + 2x_4 = 7 \\ x_1 - 2x_2 + x_3 + 2x_4 = 8 \\ 2x_1 + 2x_2 + x_3 + 3x_4 = 7 \\ x_1 - x_2 + x_3 + x_4 = 6 \end{cases}$

31. $\begin{cases} 2x_1 + 3x_2 + x_3 = 4 \\ 3x_1 + 5x_2 + 2x_3 = 12 \\ x_1 + 2x_2 + x_3 = 3 \end{cases}$

32. $\begin{cases} x_1 + 3x_2 + 2x_3 = 6 \\ x_1 + 2x_2 + 2x_3 = 3 \\ 2x_1 + 5x_2 + 4x_3 = 8 \end{cases}$

33. $\begin{cases} 4x_1 + 3x_2 + 2x_3 = 24 \\ x_1 + x_2 + 3x_3 = 7 \\ 5x_1 + 4x_2 + 5x_3 = 31 \end{cases}$

34. $\begin{cases} 5x_1 - 7x_2 + 3x_3 = -9 \\ -x_1 + x_2 - x_3 = 1 \\ 4x_1 - 5x_2 + 3x_3 = -6 \end{cases}$

35. $\begin{cases} x_1 + x_2 + 2x_3 + x_4 = 2 \\ 2x_1 + 2x_2 + 3x_3 + 3x_4 = 9 \\ 2x_1 + x_2 + 2x_3 + 2x_4 = 7 \\ x_1 + x_2 + x_3 + x_4 = 4 \end{cases}$

36. $\begin{cases} 3x_1 + 7x_2 + 4x_3 + 4x_4 = -7 \\ 7x_1 + 7x_2 + 4x_3 + 7x_4 = 3 \\ 4x_1 + 3x_2 + 2x_3 + 3x_4 = 2 \\ 3x_1 + 2x_2 + x_3 + 3x_4 = 4 \end{cases}$

37. $\begin{cases} x_1 - x_2 + x_4 = 1 \\ 2x_1 - 2x_2 + 2x_4 = 2 \\ x_1 - x_2 - x_3 - x_4 = 1 \\ 2x_1 - 2x_2 - x_3 = 1 \end{cases}$

38. $\begin{cases} x_1 - x_3 + x_4 = 2 \\ x_2 + 2x_3 + x_4 = 0 \\ 2x_1 - 2x_2 - 6x_3 = 5 \\ x_1 + x_2 + x_3 + 2x_4 = 2 \end{cases}$

39. $\begin{cases} x_1 + x_2 + x_3 + 2x_4 = 3 \\ x_1 - x_3 + x_4 = 2 \\ x_1 + 2x_2 + 3x_3 + 3x_4 = 4 \\ x_2 + 2x_3 + x_4 = 1 \end{cases}$

40. $\begin{cases} x_1 + 2x_2 + x_3 - 2x_4 = -3 \\ 2x_1 + 4x_2 + 2x_3 - x_4 = 0 \\ x_1 + 2x_2 + x_3 + x_4 = 3 \\ x_1 + 2x_2 + x_3 = 1 \end{cases}$

APPLIED EXERCISES

Formulate each situation as a system of linear equations. Be sure to state clearly the meaning of each variable. Solve using the Gauss–Jordan method. State your final answer in terms of the original question.

If you have a graphing calculator with a RREF command, use it to check your row reduction.

41. GENERAL: Plant Fertilizer A backyard garden needs 35 pounds of potash, 68 pounds of nitrogen, and 25 pounds of phosphoric acid. Three brands of fertilizer, GrowRite, MiracleMix, and GreatGreen, are available and contain the amounts of potash, nitrogen, and phosphoric acid per bag listed in the table. How many bags of each brand should be used to provide the required potash, nitrogen, and phosphoric acid?

(Pounds per Bag)	GrowRite	MiracleMix	GreatGreen
Potash	4	6	7
Nitrogen	5	10	16
Phosphoric acid	3	4	5

42. ATHLETICS: Nutrition A student athlete has decided to "bulk up" before the wrestling season by supplementing his weekly diet with an additional 325 grams of protein, 185 grams of fiber, and 110 grams of fat. If a hamburger contains 20 grams of protein, 10 grams of fiber, and 5 grams of fat; a cheeseburger contains 25 grams of protein, 10 grams of fiber, and 5 grams of fat; and a "sloppy-joe" contains 20 grams of protein, 15 grams of fiber, and 10 grams of fat, how many of each should he eat this week to meet his goal?

43. BUSINESS: Oil Well Ownership An accountant manages a limited-partnership that exploits oil-depletion tax advantages for wealthy clients, and sells memberships in the partnership for $5000, $10,000 or $25,000. If the partnership has seven hundred members and an investment value of $6 million, how many memberships of each amount are there? What is the greatest possible number of $25,000 members?

44. PERSONAL FINANCE: Financial Planning An international investment banker wishes to invest $150,000 in U.S. and German stocks and bonds. Stocks are not as secure as bonds, so he plans to spread his investments so that he has three times as much in U.S. bonds as in U.S. stocks and twice as much in German bonds as in German stocks. Furthermore, he intends to invest just $45,000 in stocks altogether. How much does he invest in each?

45. PERSONAL FINANCE: Income Taxes
a. Find the federal, state, and city income taxes on a taxable income of $180,000 if the federal tax is 50% of the taxable income after first deducting the state and city taxes, the state tax is 17% of the taxable income after first deducting the federal and city taxes, and the city tax is 3% of the taxable income after first deducting the federal and state taxes. b. Although it appears that the taxpayer is facing a 70% nominal tax rate, use the actual taxes to determine the effective combined tax rate for this situation.

46. GENERAL: Estate Division As the old patriarch lay dying, he spoke to his four sons: "I leave you my fortune of 11,100 pieces of gold, which you must divide as I command. The oldest is to take one half, the second oldest is to take one third, the third oldest is to take one quarter, and the youngest is to take one fifth. But each of you must take your part from what remains after the others have taken theirs and then you must give any remainder to your mother and sister." The old woman was frantic: "How can there be any left for our daughter when you have given away more than everything?" "Don't worry, Mother," said the daughter, who had gotten an MBA while her brothers had been off fighting at Troy. How much did each son receive and how much was left over for the mother and daughter?

47. BUSINESS: Apparel Production a. A "limited edition" ladies fashion shop has a 240-yard supply of a silk fabric suitable for scarves, dresses, blouses, and skirts. Each scarf requires 1 yard of material, 3 minutes of cutting, and 5 minutes of sewing; each dress requires 3 yards of material, 14 minutes of cutting, and 40 minutes of sewing; each blouse requires 1.5 yards of material, 9 minutes of cutting, and 30 minutes of sewing; and each skirt requires 2 yards of material, 8 minutes of cutting, and 20 minutes of sewing. If the shop has 17 hours of skilled pattern cutter labor and 45 hours of skilled seamstress labor available, how many of each can be made? **b.** If 20 blouses and 10 skirts are made, how many scarves and dresses can be made?

48. BUSINESS: Agriculture Management A farmer grows wheat, barley, and oats on his 360-acre farm. The labor (for planting, tending, and harvesting) and capital (for seeds, fertilizers, and pesticides) requirements for each crop are given in the table. If the farmer can contract for 1050 days of migrant worker labor and has $18,600 set aside for capital expenses, how many acres of each crop can he grow?

(Per Acre)	Wheat	Barley	Oats
Days of labor	2	4	3
Capital expenses	$50	$40	$60

49. BUSINESS: Advertising The promotional director for a new movie has a total of $110,000 to spend on TV, radio, and newspaper advertisements in the metropolitan area. Each TV ad costs $2500 and is seen by 10,000 moviegoers, each radio ad costs $500 and is heard by 2000 moviegoers, and each newspaper ad costs $1000 and is read by 5000 moviegoers. Ignoring repeated exposures to the same person, how many of each ad will contact 500,000 moviegoers using the allocated funds?

50. MANAGEMENT SCIENCE: Mass Transit Part of a subway system is shown in the diagram. The numbers of subway cars per hour arriving at and leaving stations I, II, III, and IV are indicated by arrows with numbers or variables. In order for the subway to function, the numbers of cars arriving and leaving per hour must match at each station. To prevent collisions, the subway cars must travel in the directions indicated by the arrows, and so none of the variables may be negative. A group of concerned citizens has petitioned the city council to build an express service from station I to IV (indicated on the diagram as x_5) capable of handling 40 cars per hour. Is this a good idea or is it a waste of money?

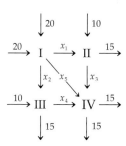

Explorations and Excursions

The following problems extend and augment the material presented in the text.

Row-Echelon Form

Row-echelon form has the same requirements as row-reduced form (pages 182–183) except that condition (3) is relaxed to require only that *the column* of each leftmost 1 contains zeros below it. Such a matrix is also called a *triangular matrix*, because the nonzero elements can only be on and above the main diagonal.

Find the system of equations represented by each row-echelon form matrix:

51. $\begin{pmatrix} 1 & 2 & | & 3 \\ 0 & 1 & | & 1 \end{pmatrix}$ **52.** $\begin{pmatrix} 1 & 3 & | & 9 \\ 0 & 1 & | & 2 \end{pmatrix}$

Unlike row-reduced matrices, there are many row-echelon form matrices equivalent to a given augmented matrix.

Row-reduce each group of matrices to show that they are all equivalent because they are all equivalent to the same row-reduced matrix.

53. $\begin{pmatrix} 1 & 2 & | & 3 \\ 2 & 3 & | & 5 \end{pmatrix}, \begin{pmatrix} 1 & 2 & | & 3 \\ 0 & 1 & | & 1 \end{pmatrix}, \begin{pmatrix} 1 & 3 & | & 4 \\ 0 & 1 & | & 1 \end{pmatrix}$

54. $\begin{pmatrix} 2 & 5 & | & 16 \\ 1 & 4 & | & 11 \end{pmatrix}, \begin{pmatrix} 1 & 3 & | & 9 \\ 0 & 1 & | & 2 \end{pmatrix}, \begin{pmatrix} 1 & -1 & | & 1 \\ 0 & 1 & | & 2 \end{pmatrix}$

Find an equivalent matrix in row-echelon form using the REF matrix command.

55. $\begin{pmatrix} 1 & 2 & | & 3 \\ 2 & 3 & | & 5 \end{pmatrix}$ **56.** $\begin{pmatrix} 1 & 2 & 2 & | & 7 \\ 1 & 1 & -1 & | & 0 \\ 2 & 2 & 1 & | & 6 \end{pmatrix}$

Back Substitution

If the augmented matrix for a system of equations is in row-echelon form, the last equation gives the value for the last variable. Substituting this value into the second-to-last equation, the value of the second-to-last variable is easy to find. Continuing upward through the equations, the values of all the remaining variables are easily found.

Solve each system of equations by back substitution.

57. $\begin{cases} x_1 + 2x_2 = 3 \\ x_2 = 1 \end{cases}$ **58.** $\begin{cases} x_1 + 3x_2 = 9 \\ x_2 = 2 \end{cases}$

59. $\begin{cases} x_1 + x_2 + x_3 = 4 \\ x_2 + x_3 = 3 \\ x_3 = 2 \end{cases}$ **60.** $\begin{cases} x_1 + 2x_2 + 2x_3 = 7 \\ x_2 + 2x_3 = 5 \\ x_3 = 2 \end{cases}$

61. $\begin{cases} x_1 + x_3 = 7 \\ x_2 + x_4 = 3 \\ x_3 = 4 \\ x_4 = 1 \end{cases}$ **62.** $\begin{cases} x_1 - x_2 + x_3 - x_4 = 4 \\ x_2 - x_3 + x_4 = -1 \\ x_3 - x_4 = 3 \\ x_4 = 1 \end{cases}$

Gauss–Jordan Elimination

Gauss–Jordan Elimination solves a system of equations by row-reducing the augmented matrix. It consists of two steps: First use row operations to find an equivalent row-echelon form matrix, and then use more row operations to back-substitute and complete the solution. Of course, if the row-echelon form matrix contains a row of zeros ending in a one, the corresponding system of equations is inconsistent, there can be no solution, and the method stops *without* attempting to carry out the back-substitution.

Solve each system of equations by Gauss–Jordan elimination.

63. $\begin{cases} x_1 + 3x_2 + x_3 = 2 \\ 2x_1 + 5x_2 + 2x_3 = 5 \\ x_1 + 2x_2 + 2x_3 = 5 \end{cases}$

64. $\begin{cases} 2x_1 + 2x_2 + x_3 + 3x_4 = 8 \\ x_1 + 2x_2 + x_3 + x_4 = 4 \\ 2x_1 + 3x_2 + x_3 + 2x_4 = 6 \\ 3x_1 + 2x_2 - x_3 + 2x_4 = 3 \end{cases}$

3.4 MATRIX ARITHMETIC

Introduction

Our focus thus far has been on the augmented matrix representation of a system of equations. This representation is "unnatural" in the sense that it somehow "contains" but does not show the equals signs from the equations. We shall now define the arithmetic of matrices so that we may write an entire system of equations as a single *matrix equation*.

In the next section, the solution of such a matrix equation will deepen our understanding of what it means to solve a system of equations.

Equality of Matrices

Two matrices are *equal* if they have the same dimension and if elements in corresponding locations are equal; that is, $A = B$ if $a_{i,j} = b_{i,j}$ for every row i and column j. For instance, the following matrices are equal:

$$\begin{pmatrix} 1 & 2 \\ 3 & 4 \end{pmatrix} = \begin{pmatrix} 1 & 1+1 \\ 4-1 & 2^2 \end{pmatrix}$$ Equal: same dimension and corresponding values are equal

The following matrices are *not* equal:

$$\begin{pmatrix} 1 & 2 \\ 3 & 4 \end{pmatrix} \neq \begin{pmatrix} 1 \\ 2 \\ 3 \\ 4 \end{pmatrix}$$ Not equal because their dimensions are different

$$\begin{pmatrix} 1 & 2 \\ 3 & 4 \end{pmatrix} \neq \begin{pmatrix} 1 & 2 \\ 4 & 3 \end{pmatrix}$$ Not equal because some corresponding values differ

Transpose of a Matrix

The *transpose* of a matrix A is the matrix A^t formed by turning each row into a column (or, equivalently, each column into a row). For example,

$$\begin{pmatrix} 1 & 2 & 3 \\ 4 & 5 & 6 \end{pmatrix}^t = \begin{pmatrix} 1 & 4 \\ 2 & 5 \\ 3 & 6 \end{pmatrix}$$ Transpose: First row becomes first column Second row becomes second column

Identity Matrix

An *identity matrix* is a square matrix with 1s on the main diagonal and 0s elsewhere. We write $I = I_n$ for the $n \times n$ identity matrix. The identity matrices I_1, I_2, I_3, and I_4 are written below.

$$I_1 = (1), \quad I_2 = \begin{pmatrix} 1 & 0 \\ 0 & 1 \end{pmatrix}, \quad I_3 = \begin{pmatrix} 1 & 0 & 0 \\ 0 & 1 & 0 \\ 0 & 0 & 1 \end{pmatrix}, \quad I_4 = \begin{pmatrix} 1 & 0 & 0 & 0 \\ 0 & 1 & 0 & 0 \\ 0 & 0 & 1 & 0 \\ 0 & 0 & 0 & 1 \end{pmatrix}$$

Scalar Multiplication

Scalar multiplication of a matrix simply means multiplying each element of a matrix by the same number. ("Scalar" is just another word for "number" and is used to distinguish a number from a matrix.) In particular, the *negative* of a matrix is $-A = -1 \cdot A$ and is found by multiplying each element by -1. For example, if

$$A = \begin{pmatrix} 1 & 2 & 3 \\ 4 & 5 & 6 \end{pmatrix}$$

then

$$3A = \begin{pmatrix} 3 & 6 & 9 \\ 12 & 15 & 18 \end{pmatrix} \quad \text{and} \quad -A = \begin{pmatrix} -1 & -2 & -3 \\ -4 & -5 & -6 \end{pmatrix}$$

> **Scalar Product**
>
> The product of a scalar (a number) times a matrix, kA, is the matrix A with each element multiplied by k.

Matrix Addition and Subtraction

For two matrices with the same dimensions, *matrix addition* means adding elements in corresponding locations.

$$\begin{pmatrix} 1 & 4 \\ 2 & 5 \\ 3 & 6 \end{pmatrix} + \begin{pmatrix} 11 & 12 \\ 10 & 9 \\ 7 & 8 \end{pmatrix} = \begin{pmatrix} 12 & 16 \\ 12 & 14 \\ 10 & 14 \end{pmatrix}$$

Matrices with *different* dimensions cannot be added.

$$\begin{pmatrix} 1 & 4 \\ 2 & 5 \\ 3 & 6 \end{pmatrix} + \begin{pmatrix} 7 & 8 & 9 \\ 10 & 11 & 12 \end{pmatrix} \quad \text{Is not possible}$$

Similarly, two matrices with the same dimensions can be *subtracted* by subtracting corresponding elements (or equivalently, adding the negative of the second matrix).

$$\begin{pmatrix} 11 & 12 \\ 10 & 9 \\ 7 & 8 \end{pmatrix} - \begin{pmatrix} 1 & 4 \\ 2 & 5 \\ 3 & 6 \end{pmatrix} = \begin{pmatrix} 10 & 8 \\ 8 & 4 \\ 4 & 2 \end{pmatrix}$$

> **Addition and Subtraction of Matrices**
>
> For two matrices A and B of the same dimensions, their sum $A + B$ and difference $A - B$ are found by adding or subtracting corresponding elements. Matrices of different dimensions *cannot* be added or subtracted.

A *zero matrix* has all elements equal to zero. We write 0 for the zero matrix of whatever size is appropriate for the situation. If two matrices are equal, their difference is a zero matrix: that is, if $A = B$, then $A - B = 0$ means that 0 is the zero matrix of the same dimension as A and B.

$$\begin{pmatrix} 1 & 2 \\ 3 & 4 \\ 5 & 6 \end{pmatrix} - \begin{pmatrix} 1 & 2 \\ 3 & 4 \\ 5 & 6 \end{pmatrix} = 0 \quad \text{means} \quad \begin{pmatrix} 1 & 2 \\ 3 & 4 \\ 5 & 6 \end{pmatrix} - \begin{pmatrix} 1 & 2 \\ 3 & 4 \\ 5 & 6 \end{pmatrix} = \begin{pmatrix} 0 & 0 \\ 0 & 0 \\ 0 & 0 \end{pmatrix}$$

Graphing Calculator Exploration

The basic matrix operations are found in the MATRX MATH menu. To have the calculator demonstrate the preceding concepts, proceed as follows:

```
MATRIX[A] 2 x3
[1    2    3   ]
[4    5    6   ]




2,3=6
```

a. Use MATRX EDIT to enter $\begin{pmatrix} 1 & 2 & 3 \\ 4 & 5 & 6 \end{pmatrix}$ in [A].

```
[A]
          [[1 2 3]
           [4 5 6]]
[A]^T
           [[1 4]
            [2 5]
            [3 6]]
```

b. Find the transpose of A.

```
3[A]
          [[3   6    9 ]
           [12  15  18]]

(1/2)[A]
        [[.5  1   1.5]
         [2   2.5  3   ]]
```

c. Find the scalar multiples $3A$ and $(1/2)A$. Check that attempting the division [A]/2 gives a "data type" error.

```
[A]+[A]
          [[2  4   6 ]
           [8  10  12]]

[A]-[A]
          [[0  0  0]
           [0  0  0]]
```

d. Find $A + A$ and $A - A$.

```
identity(4)
    [[1 0 0 0]
     [0 1 0 0]
     [0 0 1 0]
     [0 0 0 1]]
```

e. Find the 4 × 4 identity matrix.

EXAMPLE 1

USING MATRIX ARITHMETIC

A McBurger restaurant sells hamburgers for $1, cheeseburgers for $1.50, and fries for 75¢, while the BurgerQueen across the street sells hamburgers for $1.25, cheeseburgers for $1.75, and fries for 50¢. At McBurger the preparation costs are 30¢ per hamburger, 40¢ per cheeseburger, and 25¢ per order of fries, while at BurgerQueen the costs are 35¢ per hamburger, 45¢ per cheeseburger, and 20¢ per order of fries.

a. Represent the selling prices and preparation costs as matrices.
b. Using the matrices from (a), find the profit matrix for the items at the restaurants.
c. Using the matrix from (b), find the franchise fee charged on each item at each restaurant if this fee is 30% of the profit.

Solution

a. Let R be the revenue matrix of selling prices and let C be the cost matrix of preparation costs. We have 3 items from 2 restaurants, so we can choose the matrices to be either 2 × 3 or 3 × 2. We choose the matrices to be 2 × 3 with the rows corresponding to the different restaurants and the columns to the different menu items:

$$\begin{array}{c} \text{Hamburger} \quad \text{Cheeseburger} \quad \text{Fries} \\ \downarrow \qquad\qquad \downarrow \qquad\qquad \downarrow \\ \begin{array}{c}\text{McBurger} \rightarrow \\ \text{BurgerQueen} \rightarrow\end{array} \begin{pmatrix} \underline{} & \underline{} & \underline{} \\ \underline{} & \underline{} & \underline{} \end{pmatrix} \end{array}$$

From the given information, we obtain the revenue and cost matrices

$$R = \begin{pmatrix} 1.00 & 1.50 & 0.75 \\ 1.25 & 1.75 & 0.50 \end{pmatrix} \quad \text{and} \quad C = \begin{pmatrix} 0.30 & 0.40 & 0.25 \\ 0.35 & 0.45 & 0.20 \end{pmatrix}$$

b. Profit is revenue minus cost, so we find $P = R - C$ by matrix subtraction:

$$\overbrace{\begin{pmatrix} 1.00 & 1.50 & 0.75 \\ 1.25 & 1.75 & 0.50 \end{pmatrix}}^{R} - \overbrace{\begin{pmatrix} 0.30 & 0.40 & 0.25 \\ 0.35 & 0.45 & 0.20 \end{pmatrix}}^{C} = \overbrace{\begin{pmatrix} 0.70 & 1.10 & 0.50 \\ 0.90 & 1.30 & 0.30 \end{pmatrix}}^{P}$$

c. The franchise fee matrix is 30% of the profit matrix P, which we find by scalar multiplication:

$$0.30P = 0.30 \cdot \begin{pmatrix} 0.70 & 1.10 & 0.50 \\ 0.90 & 1.30 & 0.30 \end{pmatrix} = \begin{pmatrix} 0.21 & 0.33 & 0.15 \\ 0.27 & 0.39 & 0.09 \end{pmatrix}$$

Matrix Multiplication as Evaluation

To evaluate an expression like: $\quad 5x_1 + 6x_2 + 7x_3$

when x_1, x_2, and x_3 are equal to: $\quad 2, \quad 3, \text{ and } 4$

we simply multiply and add: $\quad 5 \cdot 2 + 6 \cdot 3 + 7 \cdot 4 = 56$

That is, evaluating linear expressions amounts to multiplying pairs of numbers and adding. With this kind of evaluation in mind, we define matrix multiplication as *multiplying in order the numbers from a row by the numbers from a column and adding the results*. For example, the matrix product of a row matrix by a column matrix is a 1×1 matrix:

$$(5 \quad 6 \quad 7) \cdot \begin{pmatrix} 2 \\ 3 \\ 4 \end{pmatrix} = (5 \cdot 2 + 6 \cdot 3 + 7 \cdot 4) = (56)$$

Of course, there must be exactly the same number of elements in the row as in the column. In general,

$$(a_1 \quad a_2 \quad \cdots \quad a_n) \cdot \begin{pmatrix} b_1 \\ b_2 \\ \vdots \\ b_n \end{pmatrix} = (a_1 b_1 + a_2 b_2 + \cdots + a_n b_n)$$

Practice Problem 1 Find $(1 \quad 2) \cdot \begin{pmatrix} 3 \\ 4 \end{pmatrix}$. ▶ Solution on page 209

If there are several rows on the left and several columns on the right, then we multiply *each row* times *each column*, with each answer placed in the product matrix at the row and column position from which it came. This means, for example, that when multiplying the *first row* of a matrix by the *second column* of another, place the result in the *first row* and *second column* of the answer. Note that we always take *rows* (from the left) times *columns* (on the right).

$$\begin{pmatrix} 3 & 2 & 1 \\ 2 & 0 & -2 \end{pmatrix} \begin{pmatrix} 2 & 7 \\ 3 & 6 \\ 4 & 5 \end{pmatrix} = \begin{pmatrix} (3\ 2\ 1)\begin{pmatrix}2\\3\\4\end{pmatrix} & (3\ 2\ 1)\begin{pmatrix}7\\6\\5\end{pmatrix} \\ (2\ 0\ -2)\begin{pmatrix}2\\3\\4\end{pmatrix} & (2\ 0\ -2)\begin{pmatrix}7\\6\\5\end{pmatrix} \end{pmatrix} = \begin{pmatrix} 16 & 38 \\ -4 & 4 \end{pmatrix}$$

First row times second column goes here, in first row, second column

Go from here

Try to multiply each row times column "in your head"

directly to here

In general, to multiply two matrices, the row-length of the first must match the column-length of the second, with the other numbers giving the dimension of the product:

$$\begin{matrix} A & \cdot & B \\ m \times p & & p \times n \end{matrix}$$

must match

$m \times n$

dimension of $A \cdot B$

Outside numbers give dimension
Inside numbers get "absorbed"

Matrix Multiplication

If A is an $m \times p$ matrix and B is a $p \times n$ matrix, the matrix product $A \cdot B$ is the $m \times n$ matrix whose element in row i and column j is the product of row i from A times column j from B. If the row-length of A does not equal the column-length of B, then the product $A \cdot B$ is not defined.

Even when both products $A \cdot B$ and $B \cdot A$ are defined, the different orders may give very different results. For example:

$$(5\ 6\ 7)\begin{pmatrix}2\\3\\4\end{pmatrix} = (5 \cdot 2 + 6 \cdot 3 + 7 \cdot 4) = (56)$$

Multiplying in one order gives a 1×1 matrix

$$\begin{pmatrix} 2 \\ 3 \\ 4 \end{pmatrix} (5 \ 6 \ 7) = \begin{pmatrix} 2 \cdot 5 & 2 \cdot 6 & 2 \cdot 7 \\ 3 \cdot 5 & 3 \cdot 6 & 3 \cdot 7 \\ 4 \cdot 5 & 4 \cdot 6 & 4 \cdot 7 \end{pmatrix} = \begin{pmatrix} 10 & 12 & 14 \\ 15 & 18 & 21 \\ 20 & 24 & 28 \end{pmatrix}$$

Multiplying in the other order gives a 3×3 matrix

That is:

> Matrix multiplication is not commutative.
>
> In general: $A \cdot B \neq B \cdot A$

Spreadsheet Exploration

The spreadsheet* below shows the sum and product of the matrices $A = \begin{pmatrix} 1 & 2 & 3 \\ 4 & -2 & 4 \\ 3 & 2 & 1 \end{pmatrix}$ and $B = \begin{pmatrix} 6 & 3 & 8 \\ 2 & 4 & 1 \\ 5 & 7 & 9 \end{pmatrix}$. The sum $A + B$ is found by adding the range of cells `a3:c5` (containing the entries of matrix A) to the range of cells `e3:g5` (containing the entries of matrix B, while the product $A \cdot B$ uses the matrix multiplication function `MMULT` applied to the two ranges of cells representing the matrices.

	A	B	C	D	E	F	G
1		Matrix A				Matrix B	
2							
3	1	2	3		6	3	8
4	4	-2	4		2	4	1
5	3	2	1		5	7	9
6							
7		Sum A+B				Product A*B	
8							
9	7	5	11		25	32	37
10	6	2	5		40	32	66
11	8	9	10		27	24	35
12							

If you modify this spreadsheet to calculate $B + A$ and $B \cdot A$, which answer will change and which will remain the same?

* See the Preface for how to obtain this and other Excel spreadsheets.

EXAMPLE 2

MATRIX MULTIPLICATION

Use the selling price ("revenue") matrix from Example 1 (pages 202–203) and matrix multiplication to find the total selling price for a meal of three hamburgers, a cheeseburger, and two fries at each of the restaurants.

Solution

The revenue matrix R from Example 1 gives the selling prices in rows, one row for each restaurant. If we write the three hamburgers, one cheeseburger, and two fries as a *column*, we can use matrix multiplication to find the total price at each restaurant.

$$\begin{pmatrix} 1.00 & 1.50 & 0.75 \\ 1.25 & 1.75 & 0.50 \end{pmatrix} \begin{pmatrix} 3 \\ 1 \\ 2 \end{pmatrix} = \begin{pmatrix} 1.00 \cdot 3 + 1.50 \cdot 1 + 0.75 \cdot 2 \\ 1.25 \cdot 3 + 1.75 \cdot 1 + 0.50 \cdot 2 \end{pmatrix} = \begin{pmatrix} 6.00 \\ 6.50 \end{pmatrix}$$

$\uparrow$ R (from Example 1) $\qquad\uparrow$ Order $\qquad\qquad\qquad\qquad\qquad\qquad\qquad\qquad\uparrow$ Prices

The total selling prices are $6.00 at McBurger and $6.50 at Burger-Queen.

Practice Problem 2

Modify Example 2 to find the total selling price for a meal of four hamburgers, two cheeseburgers, and five fries at each of the restaurants.

➤ Solution on page 209

Identity Matrices

Multiplication by an identity matrix (of the proper size) does not change the matrix:

$$\begin{pmatrix} 1 & 0 & 0 \\ 0 & 1 & 0 \\ 0 & 0 & 1 \end{pmatrix} \begin{pmatrix} 2 & 7 \\ 3 & 6 \\ 4 & 5 \end{pmatrix} = \begin{pmatrix} 2 & 7 \\ 3 & 6 \\ 4 & 5 \end{pmatrix} \qquad \text{Same matrix on the right as on the left}$$

The first row (1 0 0) picks out just the first value, the second (0 1 0) picks out the second value, and so on. (You should carefully verify this multiplication to see how the identity works.) To multiply by the identity matrix on the other side, we need to use an identity matrix of a different size:

$$\begin{pmatrix} 2 & 7 \\ 3 & 6 \\ 4 & 5 \end{pmatrix} \begin{pmatrix} 1 & 0 \\ 0 & 1 \end{pmatrix} = \begin{pmatrix} 2 & 7 \\ 3 & 6 \\ 4 & 5 \end{pmatrix} \qquad \text{Again, the matrix is duplicated}$$

Writing I for an identity matrix of the appropriate size, we have for any matrix A:

Multiplication by I

$$I \cdot A = A \cdot I = A$$

That is, the matrix I plays the role in matrix arithmetic that the number 1 plays in ordinary arithmetic—multiplying by it gives back exactly what you started with.

Matrix Multiplication and Systems of Equations

Because we defined matrix multiplication as evaluation of linear expressions (page 203), we may use it to write an entire system of equations as a single matrix equation. Consider the system of equations

$$\begin{cases} 2x_1 + 3x_2 + 3x_3 = 15 \\ 3x_1 + 4x_2 + 4x_3 = 22 \\ 3x_1 + 4x_2 + 5x_3 = 30 \end{cases}$$

The "coefficient" matrix A and the "constant term" matrix B are

$$A = \begin{pmatrix} 2 & 3 & 3 \\ 3 & 4 & 4 \\ 3 & 4 & 5 \end{pmatrix} \quad \text{and} \quad B = \begin{pmatrix} 15 \\ 22 \\ 30 \end{pmatrix}$$

Let X be the column matrix of the variables:

$$X = \begin{pmatrix} x_1 \\ x_2 \\ x_3 \end{pmatrix}$$

Then the matrix product AX is precisely the left sides of the original equations (as you should check):

$$\underbrace{\begin{pmatrix} 2 & 3 & 3 \\ 3 & 4 & 4 \\ 3 & 4 & 5 \end{pmatrix}}_{A} \underbrace{\begin{pmatrix} x_1 \\ x_2 \\ x_3 \end{pmatrix}}_{X} = \begin{pmatrix} 2x_1 + 3x_2 + 3x_3 \\ 3x_1 + 4x_2 + 4x_3 \\ 3x_1 + 4x_2 + 5x_3 \end{pmatrix}$$

The system of equations is the same as the matrix equation $AX = B$:

$$\begin{cases} 2x_1 + 3x_2 + 3x_3 = 15 \\ 3x_1 + 4x_2 + 4x_3 = 22 \\ 3x_1 + 4x_2 + 5x_3 = 30 \end{cases} \text{ is the same as } \underbrace{\begin{pmatrix} 2 & 3 & 3 \\ 3 & 4 & 4 \\ 3 & 4 & 5 \end{pmatrix}}_{A} \underbrace{\begin{pmatrix} x_1 \\ x_2 \\ x_3 \end{pmatrix}}_{X} = \underbrace{\begin{pmatrix} 15 \\ 22 \\ 30 \end{pmatrix}}_{B}$$

We can therefore represent the entire system of equations as just *one matrix equation* $AX = B$, as on the bottom of the previous page. This, at the very least, saves writing out the variables in each equation, a saving that is even greater for larger systems.

Matrix Multiplication and Row Operations

Because matrix multiplication of $I \cdot A$ picks out the elements of A in the correct order, switching the rows of I before multiplying will pick out the elements of A in a different order. For example:

I with rows 1 and 2 switched $\longrightarrow$
$$\begin{pmatrix} 0 & 1 & 0 \\ 1 & 0 & 0 \\ 0 & 0 & 1 \end{pmatrix} \begin{pmatrix} 2 & 7 \\ 3 & 6 \\ 4 & 5 \end{pmatrix} = \begin{pmatrix} 3 & 6 \\ 2 & 7 \\ 4 & 5 \end{pmatrix}$$
$\longleftarrow$ A with rows 1 and 2 switched

Replacing one of the 1s in I with a different value before multiplying will pick out the corresponding element of A that many times:

I with row 3 multiplied by 8 $\longrightarrow$
$$\begin{pmatrix} 1 & 0 & 0 \\ 0 & 1 & 0 \\ 0 & 0 & 8 \end{pmatrix} \begin{pmatrix} 2 & 7 \\ 3 & 6 \\ 4 & 5 \end{pmatrix} = \begin{pmatrix} 2 & 7 \\ 3 & 6 \\ 32 & 40 \end{pmatrix}$$
$\longleftarrow$ A with row 3 multiplied by 8

Replacing one of the 0s in I with another number before multiplying will add that multiple of the elements of another row to the current row of A:

I with twice row 3 added to row 1 $\longrightarrow$
$$\begin{pmatrix} 1 & 0 & 2 \\ 0 & 1 & 0 \\ 0 & 0 & 1 \end{pmatrix} \begin{pmatrix} 2 & 7 \\ 3 & 6 \\ 4 & 5 \end{pmatrix} = \begin{pmatrix} 10 & 17 \\ 3 & 6 \\ 4 & 5 \end{pmatrix}$$
$\longleftarrow$ A with twice row 3 added to row 1

Thus the row operations of switching two rows, multiplying a row by a constant, and adding a multiple of one row to another can be accomplished by matrix multiplications. Furthermore, to find the matrix that performs a given row operation, we need only apply the same row operation to the identity matrix (of the appropriate size).

> Any row operation can be accomplished by a matrix multiplication.

Furthermore, any *sequence* of row operations can be carried out by multiplying by a single matrix. This is because multiplying by several matrices on the left is equivalent to multiplying just once by the *product* of all of the matrices, which can also be found by applying the same sequence of row operations to the identity matrix. Therefore, if a matrix A is equivalent to the identity matrix, then there is a matrix R (corresponding to the row operations performed on the identity matrix) such that $R \cdot A = I$. This observation will be very important on page 214 of the next section.

3.4 Section Summary

Matrix arithmetic uses the following terms and operations.

Equal matrices	$A = B$	Same dimension and same values in same positions
Transpose of a matrix	A^t	Row and column positions reversed
Identity matrix	I	Square, 0s except for 1s on diagonal
Scalar multiple	kA	Multiply every element by k
Matrix addition	$A + B$	Add elements in same positions
Matrix subtraction	$A - B$	Subtract elements in same positions
Zero matrix	0	All elements are zeros
Matrix multiplication	$A \cdot B$	Rows on left times columns on right (row length of A = column length of B)

Any system of linear equations may be written as a matrix equation $AX = B$ where A is the coefficient matrix, X is the column matrix of variables, and B is the constant term matrix.

➤ Solutions to Practice Problems

1. $(1 \quad 2) \cdot \begin{pmatrix} 3 \\ 4 \end{pmatrix} = (1 \cdot 3 + 2 \cdot 4) = (11)$

2. $\begin{pmatrix} 1.00 & 1.50 & 0.75 \\ 1.25 & 1.75 & 0.50 \end{pmatrix} \begin{pmatrix} 4 \\ 2 \\ 5 \end{pmatrix} = \begin{pmatrix} 1.00 \cdot 4 + 1.50 \cdot 2 + 0.75 \cdot 5 \\ 1.25 \cdot 4 + 1.75 \cdot 2 + 0.50 \cdot 5 \end{pmatrix} = \begin{pmatrix} 10.75 \\ 11.00 \end{pmatrix}$

The prices are $10.75 at McBurger and $11 at BurgerQueen.

3.4 Exercises

Use the given matrices to find each expression.

$A = \begin{pmatrix} 1 & 2 & 3 \\ 4 & 5 & 6 \\ 7 & 8 & 9 \end{pmatrix} \quad B = \begin{pmatrix} 9 & 8 & 7 \\ 6 & 5 & 4 \\ 3 & 2 & 1 \end{pmatrix}$

$C = \begin{pmatrix} 1 & 6 & 8 \\ 4 & 2 & 7 \\ 9 & 5 & 3 \end{pmatrix}$

1. A^t
2. B^t
3. $3C$
4. $2A$
5. $-B$
6. $-C$
7. $A + C$
8. $B + C$
9. $C - (A + I)$
10. $(A + I) - B$

Find each matrix product.

11. $(1 \ -1 \ 1)\begin{pmatrix} 4 \\ 3 \\ 5 \end{pmatrix}$

12. $(1 \ 1 \ -1 \ -1)\begin{pmatrix} 3 \\ 2 \\ 4 \\ 1 \end{pmatrix}$

13. $\begin{pmatrix} 1 \\ 2 \end{pmatrix} \cdot (3 \ 4)$

14. $\begin{pmatrix} 2 \\ -1 \end{pmatrix} \cdot (4 \ 0)$

15. $\begin{pmatrix} 1 & 2 & 1 \\ 2 & 1 & 2 \end{pmatrix}\begin{pmatrix} 2 \\ -3 \\ 2 \end{pmatrix}$

16. $\begin{pmatrix} 1 & 3 & 1 \\ 3 & 1 & 3 \end{pmatrix}\begin{pmatrix} 4 \\ -2 \\ 4 \end{pmatrix}$

17. $\begin{pmatrix} 1 & 3 & 1 \\ 2 & 1 & 2 \end{pmatrix}\begin{pmatrix} 1 & 2 \\ 1 & 1 \\ 2 & 1 \end{pmatrix}$

18. $\begin{pmatrix} 1 & 2 & 1 \\ 3 & 1 & 3 \end{pmatrix}\begin{pmatrix} 2 & 3 \\ 3 & 1 \\ 2 & -1 \end{pmatrix}$

19. $\begin{pmatrix} 2 & 3 \\ 3 & 1 \\ 2 & -1 \end{pmatrix}\begin{pmatrix} 1 & 2 & 1 \\ 3 & 1 & 3 \end{pmatrix}$

20. $\begin{pmatrix} 1 & 2 \\ 1 & 1 \\ 2 & 1 \end{pmatrix}\begin{pmatrix} 1 & 3 & 1 \\ 2 & 1 & 2 \end{pmatrix}$

Use the given matrices to find each expression.
If your instructor permits, you may carry out the calculations on a graphing calculator.

$A = \begin{pmatrix} 2 & -1 & 1 \\ 1 & 0 & 1 \end{pmatrix} \quad B = \begin{pmatrix} 1 & 2 & 1 \\ 3 & -2 & 0 \end{pmatrix}$

$C = \begin{pmatrix} 2 & -1 & 2 \\ 1 & 1 & 1 \\ 0 & 2 & 1 \end{pmatrix}$

21. $A \cdot C$
22. $B \cdot C$
23. $C \cdot B^t$
24. $C \cdot A^t$
25. $(A - B) \cdot C$
26. $(B - A) \cdot C$
27. $A^t \cdot B + C$
28. $B^t \cdot A - C$
29. $B \cdot (C + I)$
30. $A \cdot (C - I)$

Rewrite each system of linear equations as a matrix equation $AX = B$.

31. $\begin{cases} x_1 + 5x_2 + 4x_3 = 6 \\ x_1 + x_2 + x_3 = 4 \\ 2x_1 + 3x_2 + 3x_3 = 9 \end{cases}$

32. $\begin{cases} x_1 + x_2 + x_3 = 4 \\ x_1 + 2x_2 + x_3 = 3 \\ x_1 + 2x_2 + 2x_3 = 5 \end{cases}$

33. $\begin{cases} 4x_1 + 3x_2 - x_3 = 2 \\ 3x_1 + 3x_2 + 2x_3 = 9 \\ 2x_1 + x_2 - 3x_3 = -6 \end{cases}$

34. $\begin{cases} 2x_1 - x_2 + 2x_3 = 11 \\ -x_1 + x_2 - 3x_3 = -12 \\ 2x_1 - 2x_2 + 7x_3 = 27 \end{cases}$

35. $\begin{cases} 5x_1 + 2x_2 - 4x_3 + x_4 + 5x_5 = 7 \\ 3x_1 + x_2 - 3x_3 + x_4 + 3x_5 = 5 \end{cases}$

36. $\begin{cases} 2x_1 + x_2 + 5x_3 + 4x_4 + 5x_5 = 2 \\ x_1 + x_2 + 3x_3 + 3x_4 + 3x_5 = -1 \end{cases}$

Rewrite each matrix equation $AX = B$ as a system of linear equations.

37. $\begin{pmatrix} 5 & 9 & 9 \\ 4 & 7 & 6 \\ 3 & 5 & 3 \\ 4 & 7 & 5 \end{pmatrix}\begin{pmatrix} x_1 \\ x_2 \\ x_3 \end{pmatrix} = \begin{pmatrix} 11 \\ 9 \\ 8 \\ 10 \end{pmatrix}$

38. $\begin{pmatrix} 6 & 3 & 5 \\ 1 & 2 & 2 \\ 4 & 3 & 4 \\ 5 & 1 & 3 \end{pmatrix}\begin{pmatrix} x_1 \\ x_2 \\ x_3 \end{pmatrix} = \begin{pmatrix} 8 \\ 1 \\ 5 \\ 7 \end{pmatrix}$

39. $\begin{pmatrix} 5 & 4 & 7 & 6 \\ 2 & 2 & 3 & 3 \\ 4 & 3 & 5 & 5 \\ 3 & 2 & 3 & 3 \end{pmatrix}\begin{pmatrix} x_1 \\ x_2 \\ x_3 \\ x_4 \end{pmatrix} = \begin{pmatrix} 18 \\ 9 \\ 16 \\ 11 \end{pmatrix}$

40. $\begin{pmatrix} 3 & 4 & 2 & 4 \\ 1 & 2 & 1 & 1 \\ 4 & 5 & 2 & 5 \\ 6 & 6 & 1 & 6 \end{pmatrix}\begin{pmatrix} x_1 \\ x_2 \\ x_3 \\ x_4 \end{pmatrix} = \begin{pmatrix} 12 \\ 4 \\ 14 \\ 15 \end{pmatrix}$

APPLIED EXERCISES

Formulate each situation in matrix form. Be sure to indicate the meaning of your rows and columns. Find the requested quantities using the appropriate matrix arithmetic.

41. **BUSINESS: Sales Commissions** A salesman at a furniture store sells mattresses manufactured by SlumberKing, DreamOn, and RestEasy. The selling prices for the three models of SlumberKing mattresses are $300 for the economy, $350 for the best, and $500 for the deluxe; DreamOn mattresses are $350 for the economy, $400 for the best, and $550 for the deluxe; and RestEasy mattresses are $400 for the economy, $500 for the best, and $700 for the deluxe. Represent these selling prices as a price matrix. Use this matrix to find the salesperson's commission matrix for these mattresses from these manufacturers if the commission is 15% of the selling price.

42. **BUSINESS: Sales Taxes** The Toys4U stores in Rockland and Martinville sell "Little Tykes" baseballs, bats, gloves, and caps. At the Rockland store, the retail prices are $1 per baseball, $8 per bat, $7 per glove, and $10 per cap while at the Martinville store, the prices are $1 per baseball, $9 per bat, $8 per glove, and $11 per cap. Represent these retail prices as a price matrix. Use this matrix to find the sales tax matrix for these items at these stores if the state sales tax is 5% of the retail price.

43. **BUSINESS: Automobile Sales** A car dealer sells sedans, station wagons, vans, and pickup trucks at sales lots in Oakdale and Roanoke. The "dealer markup" is the difference between the sticker price and the dealer invoice price. The dealer invoice prices at both locations are the same: $15,000 per sedan, $19,000 per wagon, $23,000 per van, and $25,000 per pickup. The sticker prices at the Oakdale lot are $18,900 per sedan, $22,900 per wagon, $26,900 per van, and $29,900 per pickup, while at the Roanoke lot the sticker prices are $19,900 per sedan, $21,900 per wagon, $27,900 per van, and $28,900 per pickup. Represent these prices as a dealer invoice matrix and a sticker price matrix. Use these matrices to find the dealer markup matrix for these vehicles at these sales lots.

44. **BUSINESS: Fuel Prices** An oil refinery in Louisiana produces gasoline, kerosene, and diesel fuel for sale at service stations in Tennessee, Alabama, and Florida. In Tennessee, the pump prices per gallon are $1.18 for gasoline, $0.87 for kerosene, and $1.09 for diesel fuel while the combined federal and state taxes per gallon are 52¢ for gasoline, 35¢ for kerosene, and 46¢ for diesel fuel. In Alabama, the pump prices per gallon are $1.15 for gasoline, $0.88 for kerosene, and $1.04 for diesel fuel, while the combined federal and state taxes per gallon are 48¢ for gasoline, 33¢ for kerosene, and 41¢ for diesel fuel. And in Florida, the pump prices per gallon are $1.27 for gasoline, $0.93 for kerosene, and $1.16 for diesel fuel while the combined federal and state taxes per gallon are 56¢ for gasoline, 41¢ for kerosene, and 48¢ for diesel fuel. Represent these prices and taxes as matrices. Use these matrices to find the pretax price matrix for these fuels in these states.

45. **MANAGEMENT SCIENCE: Overseas Manufacturing** A sports apparel company manufactures shorts, tee shirts, and caps in Costa Rica and Honduras for importation and sale in the United States. In Costa Rica, the labor costs per item are 75¢ per pair of shorts, 25¢ per tee shirt, and 45¢ per cap, while the costs of the necessary materials are $1.60 per pair of shorts, 95¢ per tee shirt, and $1.15 per cap. In Honduras, the labor costs per item are 80¢ per pair of shorts, 20¢ per tee shirt, and 55¢ per cap, while the costs of the necessary materials are $1.50 per pair of shorts, 80¢ per tee shirt, and $1.10 per cap. Represent these costs as a labor cost matrix and a materials cost matrix. Use these matrices to find the total cost matrix for these products in these countries.

46. **SOCIAL SCIENCE: Retirement Income** A study of retired Chicago municipal workers now living in the Ozark Plateau found that in Missouri the average monthly pension benefits

were $2700 for former police officers, $2500 for former mass transit workers, and $2800 for former firefighters, while in Arkansas the averages were $2750 for former police officers, $2300 for former mass transit workers, and $2900 for former firefighters. The average monthly Social Security benefits received by the same individuals in Missouri were $1100 for former police officers, $800 for former mass transit workers, and $1300 for former firefighters, while in Arkansas the averages were $1000 for former police officers, $900 for former mass transit workers, and $1200 for former firefighters. Represent these average monthly incomes as a Chicago pension matrix and a Social Security matrix. Use these matrices to find the average monthly retirement income matrix for these groups of retired employees in these states.

47. **GENERAL: Picnic Supplies** A supermarket advertises a "summer picnic sale" with 2-liter bottles of soda for 89¢, a large bottle of pickles for $1.29, packages of hot dogs for $2.39, and large bags of chips for $1.69. The Culbert family wants 12 sodas, 2 bottles of pickles, 3 packages of hot dogs, and 4 bags of chips. Represent the sale prices as a row matrix and the Culbert's shopping list as a column matrix. Use these matrices to find the total cost of these items at these prices.

48. **PERSONAL FINANCE: Part-Time Jobs** A college student makes money during the semester shelving books in the library at $8.50 per hour, tutoring freshmen at $30 per hour, and pumping gas on weekends at $6.75 per hour. Last week this student shelved books for 10 hours, tutored for 3 hours, and pumped gas for 16 hours. Represent the hourly pay rates as a row matrix and the hours worked as a column matrix. Use these matrices to find the total amount this student earned last week.

49. **MANAGEMENT SCIENCE: Furniture Production** A furniture company manufactures pine tables, chairs, and desks at factories in Wytheville and Andersen. Each table requires 2 hours of cutting and milling, 1 hour of assembly, and 2 hours of finishing; each chair requires 1.5 hours of cutting and milling, 1 hour of assembly, and 0.5 hours of finishing; and each desk requires 3 hours of cutting and milling, 2 hours of assembly, and 3 hours of finishing. At the Wytheville factory, the per-hour labor costs are $9 for cutting and milling work, $14 for assembly work, and $13 for finishing work, while at the Andersen factory, the per-hour labor costs are $10 for cutting and milling work, $13 for assembly work, and $12 for finishing work. Represent the table, chair, and desk labor requirements as a time matrix and the factory per-hour labor costs as a cost matrix. Use these matrices to find the production costs of these pieces of furniture at these factories.

50. **MANAGEMENT SCIENCE: Municipal Management** The Hollins County business manager has received a request from the police department for 3 new cars, 5 new motorcycles, and 1 van as well as a request from the rescue squad for 1 new car, 2 new vans, and 4 new ambulances. The local Ford dealer's prices are $20,000 per car, $8000 per motorcycle, $27,000 per van, and $52,000 per ambulance, while the local GM dealer's prices are $19,000 per car, $6000 per motorcycle, $29,000 per van, and $58,000 per ambulance. Represent the police and rescue requests as a vehicle matrix and the Ford and GM bids as a price matrix. Use these matrices to find the costs of these requests at these dealers.

Explorations and Excursions

The following problem extends and augments the material presented in the text.

51. Show that the solution of $\begin{cases} ax + by = h \\ cx + dy = k \end{cases}$ is

$$\begin{pmatrix} x \\ y \end{pmatrix} = \frac{1}{ad - bc} \begin{pmatrix} hd - bk \\ ak - hc \end{pmatrix}$$ for any values of $a, b, c, d, h,$ and k such that $ad - bc \neq 0$.

3.5 INVERSE MATRICES AND SYSTEMS OF LINEAR EQUATIONS

Introduction

We can write a system of linear equations as a single matrix equation $AX = B$. Just as we solved the simple equation $2x = 10$ by multiplying each side by $\frac{1}{2}$ (the inverse of 2) to obtain $1x = 5$, we will solve the matrix equation $AX = B$ by multiplying by an "inverse" matrix.

Inverse Matrices

The inverse of the number 2 is $\frac{1}{2}$ because $\frac{1}{2}$ times 2 (in either order) is 1. Similarly, the *inverse* of a matrix is defined as another matrix whose product with the first (in either order) gives I, the identity matrix. The inverse of the matrix A is denoted A^{-1} just as the inverse of 2 is $2^{-1} = \frac{1}{2}$.

> **Inverse Matrix**
>
> A square matrix A has an *inverse matrix*, denoted A^{-1}, if and only if
>
> $$A \cdot A^{-1} = I \quad \text{and} \quad A^{-1} \cdot A = I$$
>
> If the inverse A^{-1} exists, then A is *invertible*, and otherwise A is *singular*. The inverse A^{-1} will exist if and only if A is row-equivalent to the identity matrix I.

We read A^{-1} as "A inverse." The I in the equations stands for the identity matrix of the *same size* as A. Since the product in either order must give the identity matrix, if one matrix is the inverse of another, then the second is also the inverse of the first. Note that only square matrices can have inverses. Recall that a matrix being "row-equivalent to the identity matrix" means that it can be row-reduced to the identity by a sequence of row operations.

EXAMPLE 1

CHECKING INVERSE MATRICES

Verify that the inverse of $A = \begin{pmatrix} \frac{1}{2} & 3 \\ 1 & 5 \end{pmatrix}$ is $A^{-1} = \begin{pmatrix} -10 & 6 \\ 2 & -1 \end{pmatrix}$.

Solution

We must show that $A^{-1} \cdot A = I$ and $A \cdot A^{-1} = I$.

$$A^{-1} \cdot A = \begin{pmatrix} -10 & 6 \\ 2 & -1 \end{pmatrix} \begin{pmatrix} \frac{1}{2} & 3 \\ 1 & 5 \end{pmatrix} = \begin{pmatrix} 1 & 0 \\ 0 & 1 \end{pmatrix} = I$$

and

$$A \cdot A^{-1} = \begin{pmatrix} \frac{1}{2} & 3 \\ 1 & 5 \end{pmatrix} \begin{pmatrix} -10 & 6 \\ 2 & -1 \end{pmatrix} = \begin{pmatrix} 1 & 0 \\ 0 & 1 \end{pmatrix} = I$$

Multiplication in either order gives I

as required. (You should check the arithmetic in both multiplications.)

To verify that two matrices are inverses of each other, it is enough to multiply them in either order to get I (see Exercise 55 on page 226).

How to Find Inverse Matrices

Not all matrices have inverses. We saw on pages 208–209 that if a matrix A is equivalent to the identity matrix I, then the row reduction can be accomplished by multiplying A by the product R of the row operation matrices: $RA = I$, making R the inverse of A. Thus the inverse of an invertible matrix A can be found by adjoining an identity matrix, $(A \mid I)$, and row reducing to obtain $(I \mid A^{-1})$, with the inverse matrix on the right. This means we can calculate A^{-1} as follows:

> **Calculating an Inverse Matrix**
>
> If the square matrix A is invertible, then row-reducing the augmented matrix $(A \mid I)$ gives $(I \mid A^{-1})$, so that A^{-1} appears to the right of the bar.
>
> If row-reducing $(A \mid I)$ concludes with a matrix *other* than I on the left of the bar, then A is *singular* and does not have an inverse.

EXAMPLE 2 **CALCULATING AN INVERSE MATRIX**

Find the inverse of $A = \begin{pmatrix} 1 & 0 & 2 \\ 1 & 1 & 1 \\ 1 & 1 & 2 \end{pmatrix}$.

Solution

A is 3×3, so we augment it by I_3 and row-reduce $(A \mid I)$.

$$\begin{pmatrix} 1 & 0 & 2 & | & 1 & 0 & 0 \\ 1 & 1 & 1 & | & 0 & 1 & 0 \\ 1 & 1 & 2 & | & 0 & 0 & 1 \end{pmatrix} \quad (A \mid I)$$

$\underbrace{}_{A} \quad \underbrace{}_{I}$

There are many different sequences of row operations to reduce this augmented matrix, and all reach the same conclusion. One way is as follows:

$$R_3 - R_2 \to \begin{pmatrix} 1 & 0 & 2 & | & 1 & 0 & 0 \\ 1 & 1 & 1 & | & 0 & 1 & 0 \\ 0 & 0 & 1 & | & 0 & -1 & 1 \end{pmatrix}$$

$$\begin{array}{c} R_1 - 2R_3 \to \\ R_2 - R_3 \to \end{array} \begin{pmatrix} 1 & 0 & 0 & | & 1 & 2 & -2 \\ 1 & 1 & 0 & | & 0 & 2 & -1 \\ 0 & 0 & 1 & | & 0 & -1 & 1 \end{pmatrix}$$

$$R_2 - R_1 \to \begin{pmatrix} 1 & 0 & 0 & | & 1 & 2 & -2 \\ 0 & 1 & 0 & | & -1 & 0 & 1 \\ 0 & 0 & 1 & | & 0 & -1 & 1 \end{pmatrix} \quad \begin{array}{l} \text{Since } I \text{ is on} \\ \text{the left, } A^{-1} \\ \text{is on the right} \end{array}$$

$$\underbrace{\phantom{\begin{matrix}1 & 0 & 0\end{matrix}}}_{I} \quad \underbrace{\phantom{\begin{matrix}1 & 2 & -2\end{matrix}}}_{A^{-1}}$$

Therefore, $A^{-1} = \begin{pmatrix} 1 & 2 & -2 \\ -1 & 0 & 1 \\ 0 & -1 & 1 \end{pmatrix}$ is the inverse of $A = \begin{pmatrix} 1 & 0 & 2 \\ 1 & 1 & 1 \\ 1 & 1 & 2 \end{pmatrix}$.

Practice Problem 1 For the matrices A and A^{-1} in Example 2, verify that A^{-1} is the inverse of A by showing that $A^{-1}A = I$. ➤ Solution on page 222

Spreadsheet Exploration

The following spreadsheet* uses the matrix inverse function **MINVERSE** to calculate inverses of

$$\begin{pmatrix} 1 & 0 & 2 \\ 1 & 1 & 1 \\ 1 & 1 & 2 \end{pmatrix} \text{ and } \begin{pmatrix} 1 & -6 & 4 & 6 \\ 2 & 5 & 4 & 0 \\ 2 & 2 & 3 & 2 \\ 1 & 4 & 2 & -1 \end{pmatrix}.$$

*See the Preface for how to obtain this and other Excel spreadsheets.

	A	B	C	D	E	F	G	H	I
1		Matrix A					Inverse of A		
2									
3	1	0	2			1	2	-2	
4	1	1	1			-1	0	1	
5	1	1	2			0	-1	1	
6									
7	1	-6	4	6		-5	34	-8	-46
8	2	5	4	0		2	-15	4	20
9	2	2	3	2		-3.25E-17	2	-1	-2
10	1	4	2	-1		3	-22	6	29

However, the first answer is correct and the second is not. Spreadsheet programs sometimes produce round-off errors (such as the small number -3.25×10^{-17} instead of zero). The correct inverse of

$$\begin{pmatrix} 1 & -6 & 4 & 6 \\ 2 & 5 & 4 & 0 \\ 2 & 2 & 3 & 2 \\ 1 & 4 & 2 & -1 \end{pmatrix} \text{ is } \begin{pmatrix} -5 & 34 & -8 & -46 \\ 2 & -15 & 4 & 20 \\ 0 & 2 & -1 & -2 \\ 3 & -22 & 6 & 29 \end{pmatrix}$$

as you can easily check.

Be careful! Because apparent round-off errors might also indicate that A is invertible when it is in fact singular, *always* verify that the spreadsheet (or computer or calculator) answer does indeed satisfy $A^{-1}A = I$.

Solving $AX = B$ Using A^{-1}

Just as the equation $2x = 10$ is solved simply by multiplying both sides by 2^{-1} or $\frac{1}{2}$, we can solve $AX = B$ by multiplying both sides (on the left) by A^{-1}.

$AX = B$	Original equation
$A^{-1} \cdot A \cdot X = A^{-1} \cdot B$	Left-multiplying by A^{-1}
$I \cdot X = A^{-1} \cdot B$	Since $A^{-1}A = I$
$X = A^{-1}B$	Since $IX = X$

Notice that because matrix multiplication is *not* commutative, we must multiply both sides of $AX = B$ on the *left* by A^{-1}. (And besides, the product BA^{-1} is not defined because B is a column matrix and A^{-1} is a square matrix.)

3.5 INVERSE MATRICES AND SYSTEMS OF LINEAR EQUATIONS

> **Solving $AX = B$ Using A^{-1}**
>
> If A is invertible, the solution of $AX = B$ is
> $$X = A^{-1}B$$

If A is singular, then we must use the Gauss–Jordan method (from Section 3.3) and we will then find that there are either no solutions or infinitely many solutions.

EXAMPLE 3

SOLVING A SYSTEM USING AN INVERSE MATRIX

Use an inverse matrix to solve
$$\begin{cases} x_1 + 2x_3 = 22 \\ x_1 + x_2 + x_3 = 11 \\ x_1 + x_2 + 2x_3 = 20 \end{cases}.$$

Solution

Writing this system as $AX = B$, we have

$$\underbrace{\begin{pmatrix} 1 & 0 & 2 \\ 1 & 1 & 1 \\ 1 & 1 & 2 \end{pmatrix}}_{A} \underbrace{\begin{pmatrix} x_1 \\ x_2 \\ x_3 \end{pmatrix}}_{X} = \underbrace{\begin{pmatrix} 22 \\ 11 \\ 20 \end{pmatrix}}_{B} \qquad AX = B$$

Ordinarily, we would now find the inverse of matrix A by row-reducing $(A \mid I)$. However, this is exactly the matrix that we just row-reduced in Example 2, so we will simply use the A^{-1} found on page 215 and write the solution as $X = A^{-1}B$.

$$\underbrace{\begin{pmatrix} x_1 \\ x_2 \\ x_3 \end{pmatrix}}_{X} = \underbrace{\begin{pmatrix} 1 & 2 & -2 \\ -1 & 0 & 1 \\ 0 & -1 & 1 \end{pmatrix}}_{A^{-1}} \underbrace{\begin{pmatrix} 22 \\ 11 \\ 20 \end{pmatrix}}_{B} = \begin{pmatrix} 4 \\ -2 \\ 9 \end{pmatrix} \qquad \text{Using } A^{-1} \text{ from Example 2}$$

The solution is $x_1 = 4$, $x_2 = -2$, $x_3 = 9$.

We may check this solution by substituting into the original equations.

$$4 + 2 \cdot 9 = 22$$
$$4 - 2 + 9 = 11$$
$$4 - 2 + 2 \cdot 9 = 20 \qquad \text{It checks!}$$

Graphing Calculator Exploration

If A is invertible, you can solve $AX = B$ by entering the matrices A and B and then calculating $A^{-1}B$. To solve the problem from Example 3, enter $\begin{pmatrix} 1 & 0 & 2 \\ 1 & 1 & 1 \\ 1 & 1 & 2 \end{pmatrix}$ in [A], enter $\begin{pmatrix} 22 \\ 11 \\ 20 \end{pmatrix}$ in [B], and calculate $[A]^{-1}[B]$:

```
[A]
    [[1 0 2]
     [1 1 1]
     [1 1 2]]
```

```
[B]
    [[22]
     [11]
     [20]]
```

```
[A]-1[B]
    [[4 ]
     [-2]
     [9 ]]
```

Now choose three other integers, enter them in [B], and find $[A]^{-1}[B]$. Do these values satisfy your new equations?

Practice Problem 2 Solve $\begin{cases} x_1 + 2x_3 = -5 \\ x_1 + x_2 + x_3 = 10. \\ x_1 + x_2 + 2x_3 = 0 \end{cases}$

▶ Solution on page 222

[*Hint:* Write this system as $AX = B$ and notice that A is the same as in Example 3 but B is different, and so the solution can be found by matrix multiplication using the inverse found in Example 2.]

Solving $AX = B$ by finding the inverse and then using $X = A^{-1}B$ is only slightly more difficult than row-reducing the augmented matrix for the system of equations. However, if you need to solve $AX = B$ with the same A but with several different matrices B, using the inverse A^{-1} means that you only do the row reduction *once* and then the solution for any new matrix B is given by a simple matrix multiplication. Such problems, where A remains the same but B changes, occur frequently in applications.

Solving $AX = B$ for Many Different B's

If A is invertible, the solutions of the matrix equations

$$AX = B_1, \quad AX = B_2, \quad \ldots, \quad AX = B_n$$

may all be found by calculating A^{-1} once and then finding the solutions as the products

$$X = A^{-1}B_1, \quad X = A^{-1}B_2, \ldots, \quad X = A^{-1}B_n$$

EXAMPLE 4

JEWELRY PRODUCTION

An employee-owned jewelry company fabricates enameled gold rings, pendants, and bracelets. Each ring requires 3 grams of gold, 1 gram of enameling compound, and 2 hours of labor; each pendant requires 6 grams of gold, 2 grams of enameling compound, and 3 hours of labor; and each bracelet requires 8 grams of gold, 3 grams of enameling compound, and 2 hours of labor. Each of the five employee-owners works 160 hours each month, and the company has contracts guaranteeing the delivery of the grams of gold and enameling compound shown in the table on the first day of the months shown. How many rings, pendants, and bracelets should the company fabricate each month to use all the available materials and time?

	March	April	May	June
Gold	1720	2620	2460	2220
Enamel	600	960	900	800

Solution

Summarizing

	Rings	Pendants	Bracelets
Gold	3	6	8
Enamel	1	2	3
Labor	2	3	2

Let $x_1, x_2,$ and x_3 be the numbers of rings, pendants, and bracelets produced in one month. Then the required grams of gold, grams of enameling compound, and hours of labor are

Gold	$3x_1 + 6x_2 + 8x_3$
Enamel	$x_1 + 2x_2 + 3x_3$
Labor	$2x_1 + 3x_2 + 2x_3$

For the month of March, these quantities must match the amounts available:

$$\begin{cases} 3x_1 + 6x_2 + 8x_3 = 1720 \\ x_1 + 2x_2 + 3x_3 = 600 \\ 2x_1 + 3x_2 + 2x_3 = 800 \end{cases}$$

From the table and 5 workers at 160 hours each

The systems of equations for the other months follow in a similar manner, and we have four problems to solve:

	March	April
	$\begin{cases} 3x_1 + 6x_2 + 8x_3 = 1720 \\ x_1 + 2x_2 + 3x_3 = 600 \\ 2x_1 + 3x_2 + 2x_3 = 800 \end{cases}$	$\begin{cases} 3x_1 + 6x_2 + 8x_3 = 2620 \\ x_1 + 2x_2 + 3x_3 = 960 \\ 2x_1 + 3x_2 + 2x_3 = 800 \end{cases}$
	May	June
	$\begin{cases} 3x_1 + 6x_2 + 8x_3 = 2460 \\ x_1 + 2x_2 + 3x_3 = 900 \\ 2x_1 + 3x_2 + 2x_3 = 800 \end{cases}$	$\begin{cases} 3x_1 + 6x_2 + 8x_3 = 2220 \\ x_1 + 2x_2 + 3x_3 = 800 \\ 2x_1 + 3x_2 + 2x_3 = 800 \end{cases}$

The coefficient matrices for these four problems are the same, so if we find the inverse matrix for the March problem by row-reducing $(A \mid I)$, we will only have to carry out the row operations to reduce A once, and then the four problems can be solved with just four matrix multiplications.

We first find A^{-1} by row-reducing $(A \mid I)$. There are many different sequences of row operations to reduce this augmented matrix, and all reach the same conclusion. One possible way is as follows:

$$\begin{pmatrix} 3 & 6 & 8 & | & 1 & 0 & 0 \\ 1 & 2 & 3 & | & 0 & 1 & 0 \\ 2 & 3 & 2 & | & 0 & 0 & 1 \end{pmatrix} \quad (A \mid I)$$

$$\begin{array}{c} 3R_2 - R_1 \to \\ \\ 2R_2 - R_3 \to \end{array} \begin{pmatrix} 0 & 0 & 1 & | & -1 & 3 & 0 \\ 1 & 2 & 3 & | & 0 & 1 & 0 \\ 0 & 1 & 4 & | & 0 & 2 & -1 \end{pmatrix} \quad \text{Using 1 in the first column to zero-out the rest of that column}$$

$$\begin{array}{c} \\ R_2 - 2R_3 \to \\ \\ \end{array} \begin{pmatrix} 0 & 0 & 1 & | & -1 & 3 & 0 \\ 1 & 0 & -5 & | & 0 & -3 & 2 \\ 0 & 1 & 4 & | & 0 & 2 & -1 \end{pmatrix} \quad \text{Using 1 in the second column to zero-out the rest of that column}$$

$$\begin{array}{c} \\ R_2 + 5R_1 \to \\ R_3 - 4R_1 \to \end{array} \begin{pmatrix} 0 & 0 & 1 & | & -1 & 3 & 0 \\ 1 & 0 & 0 & | & -5 & 12 & 2 \\ 0 & 1 & 0 & | & 4 & -10 & -1 \end{pmatrix} \quad \text{Using 1 in third column to zero-out the rest of that column}$$

$$\begin{array}{c} R_2 \to \\ R_3 \to \\ R_1 \to \end{array} \begin{pmatrix} 1 & 0 & 0 & | & -5 & 12 & 2 \\ 0 & 1 & 0 & | & 4 & -10 & -1 \\ 0 & 0 & 1 & | & -1 & 3 & 0 \end{pmatrix} \quad \text{Switching rows to achieve the correct order}$$

Since the left side is I, the right side is the inverse matrix

$$A^{-1} = \begin{pmatrix} -5 & 12 & 2 \\ 4 & -10 & -1 \\ -1 & 3 & 0 \end{pmatrix}$$

3.5 INVERSE MATRICES AND SYSTEMS OF LINEAR EQUATIONS

The solution for each month is simply the product of this inverse matrix with the column matrix of the constant terms for that month.

Month	$X = A^{-1}B$	Solution
March	$\begin{pmatrix} x_1 \\ x_2 \\ x_3 \end{pmatrix} = \begin{pmatrix} -5 & 12 & 2 \\ 4 & -10 & -1 \\ -1 & 3 & 0 \end{pmatrix} \begin{pmatrix} 1720 \\ 600 \\ 800 \end{pmatrix} = \begin{pmatrix} 200 \\ 80 \\ 80 \end{pmatrix}$	200 rings 80 pendants 80 bracelets
April	$\begin{pmatrix} x_1 \\ x_2 \\ x_3 \end{pmatrix} = \begin{pmatrix} -5 & 12 & 2 \\ 4 & -10 & -1 \\ -1 & 3 & 0 \end{pmatrix} \begin{pmatrix} 2620 \\ 960 \\ 800 \end{pmatrix} = \begin{pmatrix} 20 \\ 80 \\ 260 \end{pmatrix}$	20 rings 80 pendants 260 bracelets
May	$\begin{pmatrix} x_1 \\ x_2 \\ x_3 \end{pmatrix} = \begin{pmatrix} -5 & 12 & 2 \\ 4 & -10 & -1 \\ -1 & 3 & 0 \end{pmatrix} \begin{pmatrix} 2460 \\ 900 \\ 800 \end{pmatrix} = \begin{pmatrix} 100 \\ 40 \\ 240 \end{pmatrix}$	100 rings 40 pendants 240 bracelets
June	$\begin{pmatrix} x_1 \\ x_2 \\ x_3 \end{pmatrix} = \begin{pmatrix} -5 & 12 & 2 \\ 4 & -10 & -1 \\ -1 & 3 & 0 \end{pmatrix} \begin{pmatrix} 2220 \\ 800 \\ 800 \end{pmatrix} = \begin{pmatrix} 100 \\ 80 \\ 180 \end{pmatrix}$	100 rings 80 pendants 180 bracelets

3.5 Section Summary

A square matrix A is *invertible* if there is an *inverse matrix* A^{-1} such that $A^{-1}A = I$ and $AA^{-1} = I$. A square matrix is *singular* if it is not invertible.

The inverse of an invertible matrix A may be found by row-reducing the augmented matrix $(A \mid I)$ to obtain $(I \mid A^{-1})$.

The solution of the matrix equation $AX = B$ is $X = A^{-1}B$, provided the matrix A is invertible.

The solutions to a succession of problems

$$AX = B_1, \quad AX = B_2, \quad \ldots, \quad AX = B_n$$

may be found from A^{-1} as

$$X = A^{-1}B_1, \quad X = A^{-1}B_2, \quad \ldots, \quad X = A^{-1}B_n$$

Solutions to Practice Problems

1. $A^{-1}A = \begin{pmatrix} 1 & 2 & -2 \\ -1 & 0 & 1 \\ 0 & -1 & 1 \end{pmatrix} \begin{pmatrix} 1 & 0 & 2 \\ 1 & 1 & 1 \\ 1 & 1 & 2 \end{pmatrix}$

$= \begin{pmatrix} 1+2-2 & 0+2-2 & 2+2-4 \\ -1+0+1 & 0+0+1 & -2+0+2 \\ 0-1+1 & 0-1+1 & 0-1+2 \end{pmatrix} = \begin{pmatrix} 1 & 0 & 0 \\ 0 & 1 & 0 \\ 0 & 0 & 1 \end{pmatrix} = I$

2. Using $X = A^{-1}B$:

$\begin{pmatrix} x_1 \\ x_2 \\ x_3 \end{pmatrix} = \begin{pmatrix} 1 & 2 & -2 \\ -1 & 0 & 1 \\ 0 & -1 & 1 \end{pmatrix} \begin{pmatrix} -5 \\ 10 \\ 0 \end{pmatrix} = \begin{pmatrix} 15 \\ 5 \\ -10 \end{pmatrix}$ so $\begin{cases} x_1 = 15 \\ x_2 = 5 \\ x_3 = -10 \end{cases}$

3.5 Exercises

Find each matrix product. Identify each pair of matrices as "a matrix and its inverse" or "not a matrix and its inverse."

1. $\begin{pmatrix} 1 & 2 \\ -1 & -1 \end{pmatrix}$ and $\begin{pmatrix} -1 & -2 \\ 1 & 1 \end{pmatrix}$

2. $\begin{pmatrix} 5 & 3 \\ -3 & -2 \end{pmatrix}$ and $\begin{pmatrix} 2 & 3 \\ -3 & -5 \end{pmatrix}$

3. $\begin{pmatrix} 1 & 1 & 0 \\ 2 & 1 & 1 \\ 1 & 0 & 0 \end{pmatrix}$ and $\begin{pmatrix} 0 & 0 & 1 \\ 1 & 0 & -1 \\ -1 & 1 & -1 \end{pmatrix}$

4. $\begin{pmatrix} 2 & 1 & -1 \\ -2 & 0 & 1 \\ -3 & -1 & 2 \end{pmatrix}$ and $\begin{pmatrix} 1 & -1 & 1 \\ 1 & 1 & 0 \\ 2 & -1 & 2 \end{pmatrix}$

5. $\begin{pmatrix} 4 & 6 & 3 \\ 3 & 4 & 1 \\ 5 & 7 & 3 \end{pmatrix}$ and $\begin{pmatrix} -5 & -3 & 6 \\ 4 & 3 & -5 \\ -1 & -2 & 2 \end{pmatrix}$

6. $\begin{pmatrix} 3 & 2 & 3 \\ 5 & 2 & 6 \\ 2 & 3 & 1 \end{pmatrix}$ and $\begin{pmatrix} 16 & -7 & -6 \\ -7 & 3 & 3 \\ -11 & 5 & 4 \end{pmatrix}$

7. $\begin{pmatrix} 10 & -4 & -7 \\ -7 & 3 & 5 \\ 4 & -1 & -3 \end{pmatrix}$ and $\begin{pmatrix} 4 & 5 & -1 \\ 1 & 2 & 1 \\ 5 & 6 & 2 \end{pmatrix}$

8. $\begin{pmatrix} 4 & 1 & 5 \\ -1 & 1 & -2 \\ 3 & 4 & 2 \end{pmatrix}$ and $\begin{pmatrix} 10 & 18 & -7 \\ -4 & -7 & -3 \\ -7 & -13 & 5 \end{pmatrix}$

For each matrix A, row-reduce $(A \mid I)$ to find the inverse matrix A^{-1} or to identify A as a singular matrix.

9. $\begin{pmatrix} 1 & 3 \\ 0 & 1 \end{pmatrix}$

10. $\begin{pmatrix} 1 & 4 \\ 0 & 1 \end{pmatrix}$

11. $\begin{pmatrix} 11 & 2 \\ 6 & 1 \end{pmatrix}$

12. $\begin{pmatrix} 5 & 7 \\ 2 & 3 \end{pmatrix}$

13. $\begin{pmatrix} 1 & 1 & 0 \\ 3 & 0 & 2 \\ 1 & 0 & 1 \end{pmatrix}$

14. $\begin{pmatrix} 1 & 0 & 1 \\ 1 & 1 & 0 \\ 5 & 0 & 4 \end{pmatrix}$

15. $\begin{pmatrix} 1 & 1 & 0 \\ 0 & 1 & 1 \\ 1 & 2 & 1 \end{pmatrix}$

16. $\begin{pmatrix} 1 & 0 & 1 \\ 2 & 1 & 3 \\ 0 & 1 & 1 \end{pmatrix}$

17. $\begin{pmatrix} 1 & 1 & 0 & 1 \\ 0 & 1 & 0 & 0 \\ 1 & 0 & 1 & 0 \\ 0 & 1 & 0 & 1 \end{pmatrix}$

18. $\begin{pmatrix} 1 & 0 & 0 & 1 \\ 0 & 1 & 1 & 1 \\ 2 & 0 & 0 & 1 \\ 0 & 1 & 0 & 1 \end{pmatrix}$

19. $\begin{pmatrix} 2 & 3 & 1 \\ 1 & 2 & 1 \\ 2 & 3 & 2 \end{pmatrix}$

20. $\begin{pmatrix} 1 & 2 & 2 \\ 1 & 1 & 1 \\ 1 & 3 & 2 \end{pmatrix}$

21. $\begin{pmatrix} 3 & -4 & 2 \\ -1 & 2 & -1 \\ 4 & -3 & 2 \end{pmatrix}$

22. $\begin{pmatrix} 2 & 2 & -1 \\ -2 & -1 & 1 \\ 3 & 4 & -1 \end{pmatrix}$

23. $\begin{pmatrix} 1 & 0 & -1 & 1 \\ 1 & 1 & 1 & 2 \\ 2 & -2 & -6 & 0 \\ 1 & 0 & 2 & 1 \end{pmatrix}$

24. $\begin{pmatrix} 1 & -1 & 0 & 1 \\ 2 & -2 & -1 & 0 \\ 1 & -1 & -1 & -1 \\ 2 & -2 & 0 & 2 \end{pmatrix}$

Rewrite each system of equations as a matrix equation $AX = B$ and use the inverse of A to find the solution. Be sure to check your solution in the original system of equations.

25. $\begin{cases} 11x_1 + 2x_2 = 9 \\ 6x_1 + x_2 = 5 \end{cases}$

26. $\begin{cases} 5x_1 + 7x_2 = 2 \\ 2x_1 + 3x_2 = 1 \end{cases}$

27. $\begin{cases} x_1 + x_2 = 2 \\ 3x_1 + 2x_3 = 5 \\ x_1 + x_3 = 2 \end{cases}$

28. $\begin{cases} x_1 + x_3 = 1 \\ x_1 + x_2 = 3 \\ 5x_1 + 4x_3 = 6 \end{cases}$

29. $\begin{cases} 2x_1 + 3x_2 + x_3 = 6 \\ x_1 + 2x_2 + x_3 = 4 \\ 2x_1 + 3x_2 + 2x_3 = 7 \end{cases}$

30. $\begin{cases} x_1 + 2x_2 + 2x_3 = 5 \\ x_1 + x_2 + x_3 = 3 \\ x_1 + 3x_2 + 2x_3 = 6 \end{cases}$

31. $\begin{cases} 3x_1 - 4x_2 + 2x_3 = 12 \\ -x_1 + 2x_2 - x_3 = -4 \\ 4x_1 - 3x_2 + 2x_3 = 15 \end{cases}$

32. $\begin{cases} 2x_1 + 2x_2 - x_3 = 1 \\ -2x_1 - x_2 + x_3 = 0 \\ 3x_1 + 4x_2 - x_3 = 2 \end{cases}$

33. $\begin{cases} x_1 - 2x_2 + x_3 + 2x_4 = 8 \\ x_1 - x_2 + x_3 + 2x_4 = 7 \\ 2x_1 + 2x_2 + x_3 + 3x_4 = 7 \\ x_1 - x_2 + x_3 + x_4 = 6 \end{cases}$

34. $\begin{cases} 2x_1 + x_2 - 2x_3 + x_4 = 2 \\ x_1 + x_2 + 2x_3 + x_4 = 5 \\ x_1 + x_2 + x_3 + x_4 = 4 \\ 2x_1 + 2x_2 + 3x_3 + x_4 = 8 \end{cases}$

35. $\begin{cases} x_1 - 2x_2 + x_3 + 2x_4 = 10 \\ x_1 - x_2 + x_3 + 2x_4 = 8 \\ 2x_1 + 2x_2 + x_3 + 3x_4 = 5 \\ x_1 - x_2 + x_3 + x_4 = 6 \end{cases}$

36. $\begin{cases} 2x_1 + x_2 - 2x_3 + x_4 = 8 \\ x_1 + x_2 + 2x_3 + x_4 = 3 \\ x_1 + x_2 + x_3 + x_4 = 4 \\ 2x_1 + 2x_2 + 3x_3 + x_4 = 5 \end{cases}$

37. $\begin{cases} x_1 + 2x_2 + x_3 = 11 \\ x_1 + 4x_2 + x_3 = 19 \\ 2x_1 + 2x_2 + x_3 = 13 \end{cases}$

38. $\begin{cases} 2x_1 + 3x_2 + x_3 = 4 \\ 3x_1 + 3x_2 + 2x_3 = 12 \\ x_1 + 2x_2 + x_3 = 4 \end{cases}$

39. $\begin{cases} 4x_1 + 6x_2 + 5x_3 + 9x_4 = 75 \\ 4x_1 + x_2 + 2x_3 + 4x_4 = 10 \\ x_1 + 4x_2 + 3x_3 + 6x_4 = 65 \\ 4x_1 + 3x_2 + 3x_3 + 4x_4 = 10 \end{cases}$

40. $\begin{cases} 3x_1 + 2x_2 + x_3 + 3x_4 = 10 \\ 4x_1 + 3x_2 + 3x_3 + 7x_4 = 15 \\ 6x_1 + 7x_2 + 3x_3 + 8x_4 = 50 \\ 5x_1 + 3x_2 + 3x_3 + 7x_4 = 10 \end{cases}$

APPLIED EXERCISES

Formulate each situation as a collection of systems of linear equations. Be sure to state clearly the meaning of each variable. Solve each collection using an inverse matrix. State your final answers in terms of the original questions.

If permitted by your instructor, you may use a graphing calculator to find the necessary inverse matrices and matrix products.

41. **BUSINESS: Movie Tickets** A five-screen multiplex cinema charges $10 for adults and $5 for children under twelve. The number of tickets sold for each of today's shows and the corresponding gross receipts are given in the table. How many tickets of each kind were sold for each film?

	Film No. 1	Film No. 2	Film No. 3	Film No. 4	Film No. 5
Tickets sold	500	400	450	500	600
Gross receipts	$3250	$3000	$3500	$4500	$6000

42. GENERAL: Summer Day Care An inner-city antipoverty foundation staffs summer day care sites serving children aged 6 to 12 at Hollis Avenue, Beaverton Boulevard, Gramson Park, and Riverside Street. Certified instructors earn $350 per week and supervise 8 children, and college student group leaders earn $250 per week and supervise 6 children. The number of children and the weekly payroll at each site are given in the table. How many instructors and group leaders work at each site?

	Children	Payroll
Hollis Avenue	92	$3900
Beaverton Boulevard	130	$5500
Gramson Park	124	$5300
Riverside Street	152	$6500

43. MANAGEMENT SCIENCE: Speculative Partnerships A broker offers limited partnerships in the amounts of $1000, $5000, and $10,000 in three organizations speculating in gold mines, oil wells, and modern art. The capitalization values and number of members in each are given in the table, together with the "altered value" of the capitalization should all of the $1000 memberships be replaced by the same number of $5000 memberships. How many memberships of each amount are in each partnership?

	Gold Mines	Oil Wells	Modern Art
Value	$2.0 million	$3.9 million	$2.5 million
Number of Members	500	700	600
"Altered Value"	$3.0 million	$4.5 million	$3.8 million

44. GENERAL: Plant Fertilizer The manager of a garden supply store has received the soil test results for the gardens of Mr. Smith, Mrs. Jones, Miss Roberts, and Mr. Wheeler. Using the size of each garden plot, the manager calculated the ounces of nutrients needed for each garden as listed in the first table. The store sells three brands of fertilizer, Great-Green, MiracleMix, and GrowRite, which contain the amounts of potash, nitrogen, and phosphoric acid per box listed in the second table. How many boxes of each fertilizer should be sold to each customer?

Ounces Needed	Mr. Smith	Mrs. Jones	Miss Roberts	Mr. Wheeler
Potash	26	31	25	25
Nitrogen	31	33	31	29
Phosphoric acid	19	20	19	18

Ounces per Box	GreatGreen	MiracleMix	GrowRite
Potash	4	5	2
Nitrogen	5	5	3
Phosphoric acid	3	3	2

45. BIOMEDICAL: Infant Nutrition A pediatric dietician at an inner-city foundling hospital needs to supplement each bottle of baby formula given to three infants in her ward with the units of vitamin A, vitamin D, calcium, and iron given in the first table. If four diet supplements are available with the nutrient content per drop given in the second table, how many drops of each supplement per bottle of formula should each infant receive?

Units Needed	Vitamin A	Vitamin D	Calcium	Iron
Billy	48	26	19	49
Susie	46	26	27	65
Jimmy	47	26	19	49

Units per Drop	Vitamin A	Vitamin D	Calcium	Iron
Supplement 1	5	3	0	1
Supplement 2	0	1	3	6
Supplement 3	4	2	3	7
Supplement 4	4	2	2	5

46. **PERSONAL FINANCE: International Investments** An international investment advisor recommends industrial stocks in Nigeria, Bolivia, Thailand, and Hungary, with four times as much in Thailand as in Bolivia and three times as much in Hungary as in Nigeria. Furthermore, he suggests that more than 20% but less than 25% of the investor's portfolio be Bolivian and Nigerian stocks. Acting on this advice, four investors commit the amounts given in the table. How much does each invest in each country?

	Total Investment	Bolivia and Nigeria
Mr. Croft	$100,000	$23,000
Mrs. Fredericks	$150,000	$33,000
Mr. Spencer	$90,000	$22,000
Ms. Winpeace	$135,000	$28,000

47. **PERSONAL FINANCE: Financial Planning** A retirement planning counselor recommends investing in a stock fund yielding 18%, a money market fund returning 6%, and a bond fund paying 8%, with twice as much in the bond and money market funds together as in the stock fund. Mr. and Mrs. Jordan have $300,000 to invest and need an annual return of $31,000; Mr. and Mrs. French have $234,900 to invest and need an annual return of $25,600; and Mr. and Mrs. Daimen have $270,000 to invest and need an annual return of $28,500. How much should each elderly couple place in each investment to receive their desired income?

48. **BUSINESS: Year-End Bonuses** At the end of each year, the owner of a small company gives himself, his salesman, and his secretary a bonus by dividing up whatever remains in the "office supplies" account. He takes one third for himself, gives one quarter to his salesman, and gives one fifth to his secretary, but each share is taken after the others have been given out. Any remaining money is then spent on the company's New Year's Eve dinner at the best restaurant in town. The ending balances in the office supplies account for four years are given in the table. How much did each person receive each year and how much was left over each year for the dinner party?

2000	2001	2002	2003
$2500	$2000	$3000	$2800

49. **MANAGEMENT SCIENCE: Mass Transit** The metropolitan area mass transit manager is revising the subway, bus, and jitney service to the suburbs of Brighton, Conway, Longwood, and Oakley. To meet federal clean air mandates, the mayor's office demands that twice as many electric subway cars as diesel buses and jitneys combined be used for each suburb. The transit workers' union contract requires that like the buses and jitneys, each subway car must have a driver/ticket taker whether or not that subway car is part of a longer train. The number of commuters from each suburb using mass transit and the number of transit workers assigned to each area are given in the table. If each subway car carries 70 commuters, each bus carries 60 commuters, and each jitney carries 10 commuters, how many of each vehicle must be assigned to each suburb?

	Commuters	Transit Workers
Brighton	11,500	180
Conway	9000	150
Longwood	9500	150
Oakley	10,250	165

50. **GENERAL: Community Food Pantry** A town-wide food drive to aid the interfaith ministries' food pantry for the needy was supported by the Boy Scouts, the Girl Scouts, and the Lions Club. Both dry food (in 10-ounce boxes) and canned food, in small (15-ounce) and large (40-ounce) sizes, were collected. The Boy Scouts collected 240 items weighing 355 pounds, of

which 315 pounds were canned goods; the Girl Scouts collected 240 items weighing 320 pounds, of which 260 pounds were canned goods; and the Lions Club collected 416 items weighing 565 pounds, of which 465 pounds were canned goods. How many boxes, small cans, and large cans did each group collect?

Explorations and Excursions

The following problems extend and augment the material presented in the text.

More About 2 × 2 Matrices

51. Show that the inverse of a 2×2 matrix $\begin{pmatrix} a & b \\ c & d \end{pmatrix}$ is $\dfrac{1}{ad - bc}\begin{pmatrix} d & -b \\ -c & a \end{pmatrix}$ provided that the *determinant* $ad - bc$ is not zero.

52. Solve $\begin{cases} ax + by = h \\ cx + dy = k \end{cases}$ for any values of $a, b, c, d, h,$ and k such that $ad - bc \neq 0$.

    [*Hint:* Use the preceding exercise and the formula $X = A^{-1}B$.]

More About Inverses

53. Show that the inverse of a product is the product of the inverses in the reverse order; that is, show that $(A \cdot B)^{-1} = B^{-1} \cdot A^{-1}$ for any invertible matrices A and B of the same dimension.

54. Show that the $n \times n$ identity matrix is unique; that is, given that $J \cdot A = A$ for every A, establish that $J = I$. [*Hint:* Consider $A = I$.]

55. Show that for an invertible matrix A, if $B \cdot A = I$, then $A \cdot B = I$ and $B = A^{-1}$.

56. Show that the inverse of a transposed matrix is the transpose of the inverse matrix; that is, show that $(A^t)^{-1} = (A^{-1})^t$ for any invertible matrix A.

Matrix Inverses and Geometric Series

Exercises 57–60 develop material that will be used in Chapter 7 on Markov Chains.

57. Show that the geometric series formula on page 123 can be written as
$$1 + x + x^2 + \cdots + x^{n-1} = \frac{1 - x^n}{1 - x}$$
and that the corresponding matrix equation is then
$$I + A + A^2 + \cdots + A^{n-1} = (I - A^n) \cdot (I - A)^{-1}$$
where A is a square matrix such that $I - A$ is invertible.

A square matrix with nonnegative entries is a *substochastic* matrix of size w (where $0 \leq w < 1$) if the sum of the entries in each row is no more than w.

58. Show that for square matrices A and B of the same dimension, if A is substochastic of size u and B is substochastic of size v, then $A \cdot B$ is substochastic of size $u \cdot v$.

59. Show that if A is a substochastic matrix of size w, then A^n is a substochastic matrix of size w^n for any positive integer n.

60. Use Exercises 59 and 57 to show that if A is a substochastic matrix of size w, then the inverse matrix $(I - A)^{-1}$ is the same as the infinite geometric series $I + A + A^2 + \cdots$.

3.6 INTRODUCTION TO MODELING: LEONTIEF MODELS AND LEAST SQUARES

Introduction

We conclude this chapter with two applications of matrix methods: the Leontief input–output model of an economy and least squares estimation. Because of the computational nature of these topics, you will need a graphing calculator or a computer.

Leontief "Open" Input–Output Models

Input–output analysis was developed by Wassily Leontief* to study the flow of goods and services among different sectors of an economy. In a "closed" model all the goods produced are used by the producers, while in an "open" model the economy produces more than is needed by the producers, with the extra output available to consumers. In the global economy of today, each country may be considered an open system. Given the relationships among the different sectors of an economy, we can calculate the extra production from the economic activity within each sector. And, conversely, we can determine the level of economic activity necessary within each sector to achieve a desired level of excess production. The Leontief "open" input–output economic model performs these two computational tasks by expressing the relation between the activity levels of the sectors and the extra production as a matrix equation.

We begin by considering the simple economy shown in the following diagram, consisting of a blacksmith (B), who makes nails, plows, and other tools; a carpenter (C), who builds barns and other useful buildings; and a farmer (F), who grows food.

The connections between the sectors of the economy are the values of materials needed to produce one dollar of output from each.

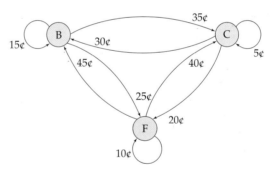

A diagram of a simple economy.

Each sector uses some of the output of the other sectors for its own production (for example, the carpenter uses nails from the blacksmith), and the arrows indicate the amount of output of one sector that is used by the other sector. To be precise, the number on an arrow from a sector indicates *the value of output from that sector needed to produce $1 of output of the target sector.* For example,

---

* Russian-born American economist (1906–1999), winner of the 1973 Nobel Memorial Prize for Economics.

means that 35¢ of the blacksmith's output is needed to produce $1 of the carpenter's output. Since each sector uses as input some of its own output (for example, the carpenter builds his own workshop), there are circular arrows from a sector to itself. These values of input materials from each sector needed to produce $1 of output from each sector are the connections between the sectors of the economy, and are the basic data needed to construct the Leontief model of the economy.

Let x_1, x_2, and x_3 be the values of the goods and services produced by the blacksmith, the carpenter, and the farmer, and let y_1, y_2, and y_3 be the values of the *excess* production from each that can be sold outside this economic system. The value produced by the blacksmith is the same as the values of the blacksmith's products used by the blacksmith, the carpenter, and the farmer, together with the amount sold outside this economy. Using the numbers from the diagram, we have

$$x_1 = \underbrace{0.15x_1}_{\substack{\text{Value of blacksmith's}\\\text{products used}\\\text{by the blacksmith}}} + \underbrace{0.35x_2}_{\substack{\text{Value of blacksmith's}\\\text{products used}\\\text{by the carpenter}}} + \underbrace{0.25x_3}_{\substack{\text{Value of blacksmith's}\\\text{products used}\\\text{by the farmer}}} + \underbrace{y_1}_{\substack{\text{Value of blacksmith's}\\\text{products sold}\\\text{to the "outside"}}}$$

Similarly, for the carpenter and the blacksmith,

$$x_2 = \underbrace{0.30x_1}_{\substack{\text{Value of carpenter's}\\\text{products used}\\\text{by the blacksmith}}} + \underbrace{0.05x_2}_{\substack{\text{Value of carpenter's}\\\text{products used}\\\text{by the carpenter}}} + \underbrace{0.20x_3}_{\substack{\text{Value of carpenter's}\\\text{products used}\\\text{by the farmer}}} + \underbrace{y_2}_{\substack{\text{Value of carpenter's}\\\text{products sold}\\\text{to the "outside"}}}$$

$$x_3 = \underbrace{0.45x_1}_{\substack{\text{Value of farmer's}\\\text{products used}\\\text{by the blacksmith}}} + \underbrace{0.40x_2}_{\substack{\text{Value of farmer's}\\\text{products used}\\\text{by the carpenter}}} + \underbrace{0.10x_3}_{\substack{\text{Value of farmer's}\\\text{products used}\\\text{by the farmer}}} + \underbrace{y_3}_{\substack{\text{Value of farmer's}\\\text{products sold}\\\text{to the "outside"}}}$$

In matrix form, this system of equations becomes

$$\begin{pmatrix} x_1 \\ x_2 \\ x_3 \end{pmatrix} = \begin{pmatrix} 0.15 & 0.35 & 0.25 \\ 0.30 & 0.05 & 0.20 \\ 0.45 & 0.40 & 0.10 \end{pmatrix} \begin{pmatrix} x_1 \\ x_2 \\ x_3 \end{pmatrix} + \begin{pmatrix} y_1 \\ y_2 \\ y_3 \end{pmatrix}$$

The numerical matrix in this equation was called the "interindustry matrix of technical coefficients" by Leontief and is now usually referred to as the "technology matrix" of the economy. The columns of this matrix are the input values for each sector as shown in the economy diagram. For example, the first column lists the amounts of the various outputs that are consumed by the blacksmith. In general:

3.6 INTRODUCTION TO MODELING: LEONTIEF MODELS AND LEAST SQUARES

Leontief "Open" Input–Output Model

A Leontief "open" input–output model is a matrix equation

$$X = AX + Y$$

where the column matrix X lists the values produced by each of the n economic sectors, the column matrix Y lists the values of the excess production of these same sectors, and the element in row i and column j of the technology matrix A represents the value of sector i goods and services needed to produce one dollar of sector j output.

Solving this matrix equation for the excess productions Y in terms of the sector production values X, we have

$$X - AX = Y$$

so that

$$Y = (I - A)X.$$

Alternatively, solving for the sector production values X required for given excess productions Y, we have

$$X = (I - A)^{-1}Y.$$

That is:

Excess and Sector Productions

In a Leontief "open" input–output model economy, the excess productions Y from given sector production values X is

$$Y = (I - A)X \qquad \text{Excess production}$$

The sector production values X necessary to provide required excess productions Y is

$$X = (I - A)^{-1}Y \qquad \text{Sector production}$$

Returning to our example economy, if a government survey of economic activity establishes that both the blacksmith and the farmer each produce \$200 of value each year and the carpenter produces \$160, so that X is

$$\begin{pmatrix} x_1 \\ x_2 \\ x_3 \end{pmatrix} = \begin{pmatrix} 200 \\ 160 \\ 200 \end{pmatrix} \qquad \begin{array}{l} \text{Blacksmith} \\ \text{Carpenter} \\ \text{Farmer} \end{array}$$

Then the amount of extra production is given by $Y = (I - A)X$:

$$\begin{pmatrix} y_1 \\ y_2 \\ y_3 \end{pmatrix} = \left[\begin{pmatrix} 1 & 0 & 0 \\ 0 & 1 & 0 \\ 0 & 0 & 1 \end{pmatrix} - \begin{pmatrix} 0.15 & 0.35 & 0.25 \\ 0.30 & 0.05 & 0.20 \\ 0.45 & 0.40 & 0.10 \end{pmatrix} \right] \begin{pmatrix} 200 \\ 160 \\ 200 \end{pmatrix} = \begin{pmatrix} 64 \\ 52 \\ 26 \end{pmatrix}$$

Therefore, the amount of extra production will be $64 from the blacksmith, $52 from the carpenter, and $26 from the farmer, for a total of $142.

How much will the production of each sector have to change in order to provide $151 of excess production, specifically $59 from the blacksmith, $48 from the carpenter, and $44 from the farmer? So Y is

$$\begin{pmatrix} y_1 \\ y_2 \\ y_3 \end{pmatrix} = \begin{pmatrix} 59 \\ 48 \\ 44 \end{pmatrix}$$

The formula $X = (I - A)^{-1}Y$ becomes (using a calculator)

$$\begin{pmatrix} x_1 \\ x_2 \\ x_3 \end{pmatrix} = \left[\begin{pmatrix} 1 & 0 & 0 \\ 0 & 1 & 0 \\ 0 & 0 & 1 \end{pmatrix} - \begin{pmatrix} 0.15 & 0.35 & 0.25 \\ 0.30 & 0.05 & 0.20 \\ 0.45 & 0.40 & 0.10 \end{pmatrix} \right]^{-1} \begin{pmatrix} 59 \\ 48 \\ 44 \end{pmatrix} = \begin{pmatrix} 200 \\ 160 \\ 220 \end{pmatrix}$$

Thus if the blacksmith and the carpenter continue to produce the same value as before but the farmer increases production by $20 to $220, the economy can generate an additional $9 in excess production (but notice that the respective sources of this excess change drastically).

Graphing Calculator Exploration

Leontief "open" input–output model calculations are simple to do on your calculator. To verify the above results for the blacksmith–carpenter–farmer economy example:

```
[A]
[[.15 .35 .25]
 [.3  .05 .2 ]
 [.45 .4  .1 ]]
```

a. Store the technology matrix A in [A].

```
[B]
        [[200]
        [160]
        [200]]
```

b. To find the excess productions Y for productions $X = \begin{pmatrix} 200 \\ 160 \\ 200 \end{pmatrix}$, store these values in [B].

```
(identity(3)-[A]
)*[B]
        [[64]
        [52]
        [26]]
```

c. Then use the MATRX MATH IDENTITY(3) command to find $Y = (I - A)X$.

```
(identity(3)-[A]
)⁻¹*[C]
        [[200]
        [160]
        [220]]
```

d. To find the sector productions X necessary to provide $Y = \begin{pmatrix} 59 \\ 48 \\ 44 \end{pmatrix}$ excess productions, store these values in [C] and use the $\boxed{x^{-1}}$ button to find $X = (I - A)^{-1}Y$.

There are many reasons to alter the production levels or the excess production levels of the sectors, such as raising the standard of living or supporting government programs through taxes. The Leontief "open" input–output model provides a tool to evaluate the effects of such possible changes.

Least Squares

The method of least squares was invented in 1794 by Carl Friedrich Gauss to find the best "compromise solution" to problems with inconsistent data. He became world famous as a scientist in 1801 when he used his method to predict when and where the asteroid Ceres could next be seen after it had been lost in the sun's glare on February 11, 1801 shortly after its discovery on January 1, 1801 by the Italian astronomer Piazzi. The sensation created by the accuracy of Gauss's prediction ultimately led to his appointment as astronomer at the Gottingen Observatory, assuring him the financial security that allowed him to pursue his many other ideas.

To understand how the least squares method is used, suppose we believe that some quantity y depends linearly on some other quantity x (that is, $y = mx + b$). Since "real world" measurements are not exact and always have errors, suppose we have data for x and y as follows:

232 CHAPTER 3 SYSTEMS OF EQUATIONS AND MATRICES

x	4	5	6	7
y	10	14	16	17

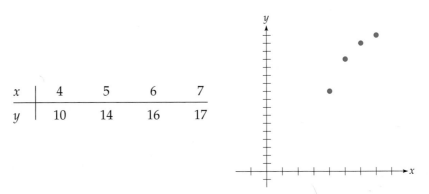

Clearly, these four data points are *not* collinear, so it is impossible to find a line $y = mx + b$ passing through all four of them. If we were to write the problems of finding m and b as a system of linear equations,

$$\begin{cases} 4m + b = 10 \\ 5m + b = 14 \\ 6m + b = 16 \\ 7m + b = 17 \end{cases} \longrightarrow \begin{pmatrix} 4 & 1 \\ 5 & 1 \\ 6 & 1 \\ 7 & 1 \end{pmatrix} \begin{pmatrix} m \\ b \end{pmatrix} = \begin{pmatrix} 10 \\ 14 \\ 16 \\ 17 \end{pmatrix} \qquad AX = B$$

then the system would be *inconsistent* since there is no solution. However, instead of an *exact* solution we can look for the best *compromise* line passing *closest* to the points. We multiply both sides of the above matrix equation by A^t.

$$\begin{pmatrix} 4 & 5 & 6 & 7 \\ 1 & 1 & 1 & 1 \end{pmatrix} \begin{pmatrix} 4 & 1 \\ 5 & 1 \\ 6 & 1 \\ 7 & 1 \end{pmatrix} \begin{pmatrix} m \\ b \end{pmatrix} = \begin{pmatrix} 4 & 5 & 6 & 7 \\ 1 & 1 & 1 & 1 \end{pmatrix} \begin{pmatrix} 10 \\ 14 \\ 16 \\ 17 \end{pmatrix} \qquad A^t A X = A^t B$$

Multiplying out yields

$$\begin{pmatrix} 126 & 22 \\ 22 & 4 \end{pmatrix} \begin{pmatrix} m \\ b \end{pmatrix} = \begin{pmatrix} 325 \\ 57 \end{pmatrix}$$

so

$$\begin{pmatrix} m \\ b \end{pmatrix} = \begin{pmatrix} 126 & 22 \\ 22 & 4 \end{pmatrix}^{-1} \begin{pmatrix} 325 \\ 57 \end{pmatrix} = \begin{pmatrix} \frac{1}{5} & -\frac{11}{10} \\ -\frac{11}{10} & \frac{63}{10} \end{pmatrix} \begin{pmatrix} 325 \\ 57 \end{pmatrix} = \begin{pmatrix} 2.3 \\ 1.6 \end{pmatrix}$$

These values, $m = 2.3$ and $b = 1.6$, give the line $y = 2.3x + 1.6$, which is called the "least squares" line, providing the best fit to the four points in that it minimizes the sum of the squared vertical distances.* The following graph shows the four points together with the least squares line.

---

*For a proof using calculus, see Section 13.4.

3.6 INTRODUCTION TO MODELING: LEONTIEF MODELS AND LEAST SQUARES

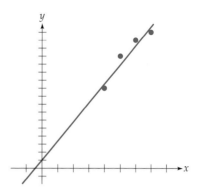

If the four points *did* lie on a line, then the m and b given by this procedure would be the *exact* values for the correct line. The least squares line is used widely in business and the sciences to predict future trends from current but imperfect data.

Least Squares Line

For the data points $(x_1, y_1), (x_2, y_2), \ldots, (x_n, y_n)$, form the matrices

$$A = \begin{pmatrix} x_1 & 1 \\ x_2 & 1 \\ \vdots & \vdots \\ x_n & 1 \end{pmatrix} \quad \text{and} \quad B = \begin{pmatrix} y_1 \\ y_2 \\ \vdots \\ y_n \end{pmatrix}.$$

The coefficients m and b for the least squares line $y = mx + b$ are then given by

$$\begin{pmatrix} m \\ b \end{pmatrix} = (A^t A)^{-1}(A^t B)$$

provided that the numbers $x_1, x_2, \ldots, x_n$ are distinct.

EXAMPLE 1 **FINDING A LEAST SQUARES LINE**

The numbers of paid subscribers to an internet service provider at the end of its first four years of operations are given in the following table. Find the least squares line for these data and use it to predict sales at the end of year 5.

Year	1	2	3	4
Subscribers (1000s)	9	13	14	18

Solution

Using $A = \begin{pmatrix} 1 & 1 \\ 2 & 1 \\ 3 & 1 \\ 4 & 1 \end{pmatrix}$ and $B = \begin{pmatrix} 9 \\ 13 \\ 14 \\ 18 \end{pmatrix}$ with the above formula, we find (see the following Graphing Calculator Exploration) that

$$\begin{pmatrix} m \\ b \end{pmatrix} = (A^tA)^{-1}(A^tB) = \begin{pmatrix} 2.8 \\ 6.5 \end{pmatrix}$$

Therefore, the least squares line is $y = 2.8x + 6.5$, as shown in the graph on the left. The predicted number of subscribers at the end of year 5 is found by substituting $x = 5$:

$$y = 2.8 \cdot 5 + 6.5 = 20.5 \quad \text{(in thousands)}$$

The predicted number of subscribers at the end of the fifth year is 20,500.

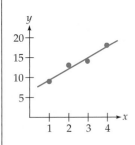

Graphing Calculator Exploration

The matrix calculations for the least squares line in the above example are easily carried out on a graphing calculator.

[A]
```
[[1 1]
 [2 1]
 [3 1]
 [4 1]]
```

a. Define a 4×2 matrix [A] containing the x-values.

[B]
```
[[9 ]
 [13]
 [14]
 [18]]
```

b. Define a 4×1 matrix [B] containing the y-values.

([A]^T[A])$^{-1}$([A]^T[B])
```
[[2.8]
 [6.5]]
```

c. Use the TRANSPOSE command and the $\boxed{x^{-1}}$ key to find $(A^tA)^{-1}(A^tB)$.

The least squares line can also be found by using the LinReg command on your graphing calculator.

3.6 Section Summary

A *Leontief "open" input–output model* is a matrix equation $X = AX + Y$, where the column matrix X lists the values produced by each of the n economic sectors, the column matrix Y lists the values of the excess production of these same sectors, and the element in row i and column j of the technology matrix A represents the value of sector i goods and services needed to produce one dollar of sector j output. Solving this model for the excess productions Y in terms of the sector production values X gives that $Y = (I - A)X$, while the sector production values X necessary to provide excess productions Y is $X = (I - A)^{-1}Y$.

The least squares line of a collection of data points can be found by the formula on page 233. It is widely used to analyze linear trends and predict future values from such trends.

3.6 Exercises (will be helpful.)

LEONTIEF "OPEN" INPUT–OUTPUT MODELS

Let A denote agriculture, C denote construction, E denote electronics, F denote fishing, H denote heavy industry, L denote light industry, M denote mining, R denote railroads, and T denote tourism.

Find the technology matrix for each economy diagram.

1., **2.**

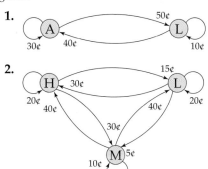

3.

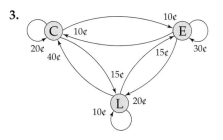

4.

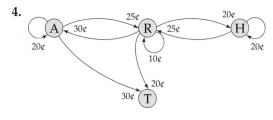

Draw an economy diagram for each technology matrix.

5. $\begin{pmatrix} 0.15 & 0.10 \\ 0.25 & 0.30 \end{pmatrix}$ for sectors M and H

6. $\begin{pmatrix} 0.10 & 0.30 & 0.10 \\ 0.20 & 0.40 & 0.20 \\ 0.10 & 0.30 & 0.10 \end{pmatrix}$ for sectors A, R, and M

7. $\begin{pmatrix} 0.20 & 0.30 & 0.10 \\ 0.50 & 0.20 & 0.20 \\ 0.30 & 0.10 & 0.20 \end{pmatrix}$ for sectors C, A, and L

8. $\begin{pmatrix} 0.10 & 0 & 0.20 \\ 0 & 0.20 & 0.10 \\ 0.10 & 0.40 & 0 \end{pmatrix}$ for sectors A, C, and R

Find the excess production Y of each economy with technology matrix A and economic activity level X.

9. $A = \begin{pmatrix} 0.20 & 0.30 \\ 0.35 & 0.25 \end{pmatrix}$ and $X = \begin{pmatrix} 130 \\ 110 \end{pmatrix}$

10. $A = \begin{pmatrix} 0.45 & 0.20 \\ 0.30 & 0.35 \end{pmatrix}$ and $X = \begin{pmatrix} 140 \\ 100 \end{pmatrix}$

11. $A = \begin{pmatrix} 0.05 & 0.15 & 0.20 \\ 0.15 & 0.05 & 0.15 \\ 0.10 & 0.10 & 0.05 \end{pmatrix}$ and $X = \begin{pmatrix} 150 \\ 170 \\ 140 \end{pmatrix}$

12. $A = \begin{pmatrix} 0.10 & 0.05 & 0.10 & 0.15 \\ 0.15 & 0.10 & 0.10 & 0.05 \\ 0 & 0.05 & 0.10 & 0 \\ 0.10 & 0 & 0.10 & 0.05 \end{pmatrix}$

and $X = \begin{pmatrix} 120 \\ 100 \\ 110 \\ 80 \end{pmatrix}$

Find, for each economy with technology matrix A, the economic activity level X necessary to generate excess production Y.

13. $A = \begin{pmatrix} 0.20 & 0.30 \\ 0.30 & 0.20 \end{pmatrix}$ and $Y = \begin{pmatrix} 84 \\ 51 \end{pmatrix}$

14. $A = \begin{pmatrix} 0.25 & 0.35 \\ 0.45 & 0.15 \end{pmatrix}$ and $Y = \begin{pmatrix} 33 \\ 57 \end{pmatrix}$

15. $A = \begin{pmatrix} 0.10 & 0.20 & 0 \\ 0 & 0.15 & 0.20 \\ 0.30 & 0.10 & 0.20 \end{pmatrix}$ and $Y = \begin{pmatrix} 60 \\ 31 \\ 50 \end{pmatrix}$

16. $A = \begin{pmatrix} 0.10 & 0.15 & 0.10 & 0.05 \\ 0.15 & 0.15 & 0.05 & 0.10 \\ 0 & 0.20 & 0.10 & 0.10 \\ 0.10 & 0 & 0.15 & 0.05 \end{pmatrix}$

and $Y = \begin{pmatrix} 60 \\ 55 \\ 60 \\ 70 \end{pmatrix}$

Represent each situation as a Leontief "open" input–output model by constructing an economy diagram and the corresponding technology matrix. Find the required excess production or level of economic activity and state your final answer in terms of the original question.

17. **ECONOMICS: Industrial Production**
The heavy and light industry sectors of the Birmingham economy depend on each other in the following way: Each dollar of production from the heavy industry sector requires $0.25 of heavy industry produce and $0.15 of light industry produce, while each dollar of production from the light industry sector requires $0.35 of heavy industry produce and $0.05 of light industry produce. The city leaders decide to increase production in order to generate tax revenues for a new sports stadium. How much must each type of industry produce to yield an excess production of $127 million of heavy industry produce and $221 million of light industry produce?

18. **ECONOMICS: County Production Planning**
An analysis of the mining, railroad, construction, and light industry sectors of the Hanover County economy revealed that each dollar produced by the mining sector requires $0.20 of mining products, $0.10 of railroad services, $0.10 of construction, and $0.10 of light industry products. Each dollar produced by the railroad sector requires $0.10 of mining products, $0.20 of railroad services, and $0.10 of light industry products. Each dollar produced by the construction sector requires $0.30 of mining products, $0.20 of construction, and $0.10 of light industry products. Each dollar produced by the light industry sector requires $0.10 of mining products, $0.20 of railroad services, and $0.10 of light industry products. If these

industries in Hanover County presently produce excess productions valued at $8 million from mining, $30 million from railroads, $32 million from construction, and $31 million from light industry, what is the current production level of each of these economic sectors?

19. **ECONOMICS: Developing Productivity** In an effort to raise the standard of living, the new government of a developing country wants to increase its excess production by stimulating the heavy industry sector of the national economy. An analysis of the relationships between its heavy industry, light industry, and railroad sectors found that the current production levels for these sectors are $100 million from heavy industry, $150 million from light industry, and $100 million from the railroads, with the technology matrix given in the table. Find the current excess production from these sectors of the economy. If the light industry and railroad productions remain the same, how much does this excess production increase with each $10 million increase in heavy industry production? What is the greatest heavy industry production level that this part of the national economy can tolerate?

From \ To	Heavy Industry	Light Industry	Railroads
Heavy industry	0.20	0.20	0.30
Light industry	0.50	0.20	0.20
Railroads	0.20	0.10	0.20

20. **ECONOMICS: Island Economy** An economist is studying the agriculture-, fishing-, and tourism-based economy of a small Pacific island nation. Although the country's excess production was $158 million last year, she has found that the economy actually produced $300 million, of which $142 million was consumed in the course of production. Her breakdown of each sector's production is given in the table. (The first line shows that the total value of agriculture production was $100 million, with $20 million consumed by the agriculture sector, $8 million consumed by the fishing sector, and $36 million consumed by the tourism sector, leaving an excess of $36 million.) If the same relations persist among the sectors of the economy, what will happen to the excess production next year if the fishing sector declines to $60 million while the other sectors produce the same amounts as before?

In Millions of $	Agriculture	Fishing	Tourism	Excess Production
Agriculture	$20	$8	$36	$36
Fishing	$10	$8	$36	$26
Tourism	$0	$0	$24	$96

LEAST SQUARES

Find the least squares best approximation line $y = mx + b$ for each collection of x and y data pairs.

21.
x	−1	0	1
y	24	36	42

22.
x	−1	0	1
y	8	8	14

23.
x	−1	0	1	2
y	120	90	70	30

24.
x	−1	0	1	3
y	761	656	691	551

25.
x	1	4	5	8	12
y	185	195	195	245	275

26.
x	1	4	5	7	8	11
y	50	80	110	140	170	200

Use the least squares best approximation line to make each prediction. Be sure to state your final answer in terms of the original question.

27. **BUSINESS: Real Estate** The new salesman at Abbott Associates, Real Estate Brokers, had

an impressive first three months with sales of $300,000, $480,000, and $600,000. If he can keep improving this much every month, how much will he sell in his fifth month?

28. BUSINESS: Commodity Futures A soybean speculator has been nervously watching the December delivery price per bushel drop from $6.59 on Monday to $6.41 on Tuesday and then to $6.29 on Wednesday. If this trend continues, what will the price be on Friday?

29. BEHAVIORAL SCIENCE: Price–Demand The market research division of a large candy manufacturer has test-marketed the new treat YummieCrunchies in five different markets at five different prices per 8-ounce bag to determine the relation between the selling price and the sales volume. The results of its study are presented in the table and are normalized to give the weekly sales per 20,000 consumers. How many weekly sales per 20,000 consumers can the manufacturer expect when it begins national distribution next week with an introductory price of 79¢ per 8-ounce bag?

Selling Price	Sales Volume
70¢	2050
55¢	2675
75¢	1975
90¢	1250
85¢	1425

30. MANAGEMENT SCIENCE: Advertising The accounts manager at television station WXXB claims that every dollar spent on commercials by local retailers generates $23 of sales. As proof, she shows prospective advertisers the following table of advertising budgets and gross receipts for five area companies that aired commercials last month. Is her claim justified?

Dollars Spent on WXXB Commercials	Gross Receipts
$8000	$245,000
$5000	$215,000
$2000	$125,000
$9000	$305,000
$6000	$230,000

Chapter Summary with Hints and Suggestions

Reading the text and doing the exercises in this chapter have helped you to master the following skills, which are listed by section (in case you need to review them) and are keyed to particular Review Exercises. Answers for all Review Exercises are given at the back of the book, and full solutions can be found in the Student Solutions Manual.

3.1 Systems of Two Linear Equations in Two Variables

- Represent a pair of statements as a system of two linear equations in two variables. *(Review Exercises 1–2.)*

$$\begin{cases} ax + by = h \\ cx + dy = k \end{cases}$$

- Solve a system of two linear equations in two variables by graphing. *(Review Exercises 3–4.)*

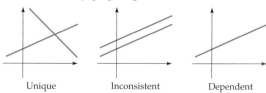

Unique   Inconsistent   Dependent

- Solve a system of two linear equations in two variables by the substitution method. *(Review Exercises 5–6.)*
- Solve a system of two linear equations in two variables by the elimination method. *(Review Exercises 7–8.)*
- Solve an applied problem by using a system of two equations in two variables and the elimination method. *(Review Exercises 9–10.)*

3.2 Matrices and Linear Equations in Two Variables

- Find the dimension of a matrix and identify a particular element using double subscript notation. *(Review Exercises 11–12.)*

$$A = \begin{pmatrix} a_{1,1} & \cdots & a_{1,n} \\ \vdots & & \\ a_{m,1} & \cdots & a_{m,n} \end{pmatrix}$$

- Carry out a given row operation on an augmented matrix. *(Review Exercises 13–14.)*

$$\begin{cases} ax + by = h \\ cx + dy = k \end{cases} \longleftrightarrow \left(\begin{array}{cc|c} a & b & h \\ c & d & k \end{array} \right)$$

- Solve a system of two linear equations in two variables by row-reducing an augmented matrix. *(Review Exercises 15–18.)*
- Solve an applied problem by using a system of two linear equations in two variable and row-reducing an augmented matrix. *(Review Exercises 19–20.)*

3.3 Systems of Linear Equations and the Gauss–Jordan Method

- Interpret a row-reduced matrix as the solution of a system of equations. If the solution is not unique, identify the system as "inconsistent" or "dependent." *(Review Exercises 21–22.)*

  0 0 $\cdots$ 0 | 1 means inconsistent.
  0 0 $\cdots$ 0 | 0 means dependent.

- Use the Gauss–Jordan method to find the equivalent row-reduced matrix. *(Review Exercises 23–24.)*

- Solve a system of equations by the Gauss–Jordan method. If the solution is not unique, identify the system as "inconsistent" or "dependent." *(Review Exercises 25–28.)*
- Formulate an application as a system of linear equations. Solve the system by the Gauss–Jordan method. State the final answer in terms of the original question. *(Review Exercises 29–30.)*

3.4 Matrix Arithmetic

- Find a matrix product or determine that the product is not defined. *(Review Exercises 31–32.)*

$$\underbrace{A \cdot B}_{m \times p \ \ p \times n} = \underbrace{C}_{m \times n}$$

- Find the value of a matrix expression involving scalar multiplication, matrix addition, subtraction, transposition, or multiplication. *(Review Exercises 33–34.)*

$$A \cdot I = I \cdot A = A \qquad A + 0 = 0 + A = A$$

- Rewrite a system of linear equations as a matrix equation $AX = B$. *(Review Exercises 35–36.)*
- Formulate an application in matrix form, and indicate the meaning of each row and column. Find the requested quantity using matrix arithmetic. *(Review Exercises 37–38.)*

3.5 Inverse Matrices and Systems of Linear Equations

- Find the product of a pair of matrices to identify the pair as "a matrix and its inverse" or "not a matrix and its inverse." *(Review Exercises 39–40.)*

$$AA^{-1} = A^{-1}A = I$$

- Find the inverse of a matrix or identify it as singular. *(Review Exercises 41–42.)*

$$(A|I) \to (I|A^{-1})$$

- Rewrite a system of equations as a matrix equation $AX = B$ and use the inverse of A to find the solution. *(Review Exercises 43–46.)*

$$X = A^{-1}B$$

- Formulate an application in terms of systems of linear equations. Solve the collection of systems of equations by finding the inverse of a common coefficient matrix and using matrix multiplications of this inverse times the various constant term matrices. State the final answer in terms of the original question. *(Review Exercises 47–48.)*

$$X = A^{-1}B_1, \ldots, X = A^{-1}B_n$$

3.6 Introduction to Modeling

Leontief "Open" Input–Output Models

- Find the technology matrix from an economy diagram. *(Review Exercises 49–50.)*
- Draw an economy diagram from a technology matrix. *(Review Exercises 51–52.)*
- Find the excess production of an economy with a given technology matrix and economic activity level. *(Review Exercises 53–54.)*

$$Y = (I - A)X$$

- Find, for an economy with a given technology matrix, the economic activity level necessary to generate a specified excess production. *(Review Exercises 55–56.)*

$$X = (I - A)^{-1}Y$$

- Represent an application as a Leontief "open" input–output model by constructing an economy diagram and the corresponding technology matrix. Find the required excess production or level of economic activity, and state the final answer in terms of the original question. *(Review Exercises 57–58.)*

Least Squares

- Find the least squares best approximation line $y = mx + b$ for a collection of x and y data pairs such that all the x-values are distinct. *(Review Exercises 59–62.)*

$$\begin{pmatrix} m \\ b \end{pmatrix} = (A^t A)^{-1}(A^t B)$$

- Use the least squares best approximation line to make a prediction in an application. *(Review Exercises 63–64.)*

Hints and Suggestions

- *(Overview)* Row operations on matrices are a generalization of the elimination method of finding the intersection of two lines, extending the technique to problems with many equations in many variables. Matrix arithmetic is similar to real number operations except that matrix multiplication is not commutative and many matrices do not have inverses. However, if a square matrix A *does* have an inverse, then the equation $AX = B$ can be solved as $X = A^{-1}B$, just as a real number linear equation can be solved by dividing. Matrices are used to represent and solve many large and complicated problems important to both science and society.

- Although not every system of linear equations has a solution, they can all be identified as "having a unique solution," "inconsistent," or "dependent" by row-reducing the corresponding augmented matrix.

- When setting up a word problem, look first for the questions "how many" or "how much" to help identify the variables. Be sure that finding values for your variables will answer the question stated in the problem. Use the rest of the given information to build equations describing facts about your variables.

- Many row-reduction problems require many steps to solve, so don't give up; keep improving the matrix until it meets all the requirements to be row-reduced. Make sure that each row operation you choose will move you toward your goal without undoing the parts you already have gotten the way you need.

- **Practice for Test:** Review Exercises 4, 6, 8, 11, 17, 20, 25, 30, 33, 35, 38, 41, 47, 49, 51, 55, 57, 59, and 63.

Review Exercises for Chapter 3

Practice test exercise numbers are in green.

3.1 Systems of Two Linear Equations in Two Variables

Represent each pair of statements as a system of two linear equations in two variables. Be sure to state clearly the meaning of your x- and y-variables.

1. "A small commuter airplane has thirty passengers" and "the ticket receipts of $3970 come from 30-day advance sale tickets at $79 and full fare tickets at $159."

2. "A cow and horse rancher has four hundred twenty animals" and "there are twice as many cows as horses."

Solve each system by graphing. If the solution is not unique, identify the system "inconsistent" or "dependent."

3. $\begin{cases} x + y = 18 \\ x - y = 8 \end{cases}$
4. $\begin{cases} 2x - 4y = 36 \\ -3x + 6y = -54 \end{cases}$

Solve each system by the substitution method. If the solution is not unique, identify the system as "inconsistent" or "dependent."

5. $\begin{cases} 5x - 2y = 10 \\ -2x + y = 2 \end{cases}$
6. $\begin{cases} -4x + 2y = 12 \\ 2x - y = 12 \end{cases}$

Solve each system by the elimination method. If the solution is not unique, identify the system as "inconsistent" or "dependent."

7. $\begin{cases} x + y = 12 \\ x + 3y = 18 \end{cases}$
8. $\begin{cases} 4x + 5y = 60 \\ 2x + 3y = 42 \end{cases}$

Formulate each situation as a system of equations. Be sure to state clearly the meaning of your x- and y-variables. Solve the system by the elimination method. State your final answer in terms of the original question.

9. **GENERAL: Garden Plants** A retired investment broker has rosebushes in his flower garden and tomato plants in his vegetable garden. He spends one hour each day tending his twenty-five plants. If each rosebush takes three minutes of care and each tomato plant takes two minutes, how many of each does he have?

10. **GENERAL: Fraternity Convention** Twenty-six members of the Alpha Alpha Alpha fraternity want to go to the national convention in Orlando and each has put in $10 for gas. If each car holds 5 people and uses $45 worth of gas and each van holds 8 people and uses $85 worth of gas, how many of each vehicle do they need for the trip?

3.2 Matrices and Linear Equations in Two Variables

Find the dimension of each matrix and the values of the specified elements.

11. $\begin{pmatrix} 8 & 3 & 4 \\ 1 & 5 & 9 \\ 6 & 7 & 2 \end{pmatrix}$; $a_{2,2}, a_{3,1}, a_{1,3}$

12. $\begin{pmatrix} 1 & 15 & 14 & 4 \\ 12 & 6 & 7 & 9 \\ 8 & 10 & 11 & 5 \\ 13 & 3 & 2 & 16 \end{pmatrix}$; $a_{2,3}, a_{3,2}, a_{4,1}$

Carry out the row operation on the matrix.

13. $3R_2 \rightarrow R_2$ on $\begin{pmatrix} 3 & 4 & | & 12 \\ 1 & 2 & | & 2 \end{pmatrix}$

14. $R_1 + R_2 \rightarrow R_1$ on $\begin{pmatrix} -1 & 2 & | & 2 \\ 2 & -3 & | & 6 \end{pmatrix}$

Solve each system by row-reducing an augmented matrix. If the solution is not unique, identify the system as "inconsistent" or "dependent."

15. $\begin{cases} x + 3y = 63 \\ 4x + 5y = 140 \end{cases}$
16. $\begin{cases} -3x + 4y = 60 \\ 2x - y = 10 \end{cases}$
17. $\begin{cases} 12x - 4y = 36 \\ -15x + 5y = 45 \end{cases}$
18. $\begin{cases} -16x + 12y = -48 \\ 20x - 15y = 60 \end{cases}$

Formulate each situation as a system of equations. Be sure to state clearly the meaning of your x- and y-variables. Solve the system by row-reducing the corresponding augmented matrix. State your final answer in terms of the original question.

19. BIOMEDICAL: Pharmaceuticals The pharmacist at the Charter Drug Shop filled 92 prescriptions today for antibiotics and cough suppressants. If there were 34 more prescriptions for antibiotics than for cough suppressants, how many prescriptions for each were filled?

20. BUSINESS: Classic Magazines A used book store offers grab bag packages of old *Life* and *The New Yorker* magazines from the 1940s containing 4 copies of *Life* and 3 copies of *The New Yorker* for $39 and larger bags of 12 copies of *Life* and 10 copies of *The New Yorker* for $122. What is the price of one copy of each old magazine?

3.3 Systems of Linear Equations and the Gauss–Jordan Method

Interpret each row-reduced matrix as the solution of a system of equations. If the solution is not unique, identify the system as "inconsistent" or "dependent."

21. $\begin{pmatrix} 1 & 0 & 0 & | & 3 \\ 0 & 1 & 0 & | & -3 \\ 0 & 0 & 1 & | & 6 \end{pmatrix}$ **22.** $\begin{pmatrix} 1 & 1 & 0 & | & 4 \\ 0 & 0 & 1 & | & 2 \\ 0 & 0 & 0 & | & 0 \end{pmatrix}$

Use the Gauss–Jordan method to find the equivalent row-reduced matrix.

23. $\begin{pmatrix} 0 & 1 & 1 & | & -1 \\ 1 & 0 & 1 & | & 6 \\ 2 & 0 & 1 & | & 10 \end{pmatrix}$ **24.** $\begin{pmatrix} 1 & 0 & 1 & 1 & | & 4 \\ 5 & 1 & 1 & 4 & | & 12 \\ 2 & 1 & 0 & 1 & | & 5 \\ 3 & 0 & 1 & 3 & | & 8 \end{pmatrix}$

Solve each system of equations by using the Gauss–Jordan method. If the solution is not unique, identify the system as "inconsistent" or "dependent." If permitted by your instructor, you may use a graphing calculator.

25. $\begin{cases} x_1 - 2x_3 = 2 \\ -x_1 + x_2 + 2x_3 = 1 \\ -x_1 + 2x_2 + 3x_3 + x_4 = 7 \\ x_1 - 2x_3 + x_4 = 4 \end{cases}$

26. $\begin{cases} x_1 - x_2 + x_4 = 3 \\ x_2 + x_3 = 3 \\ 2x_1 - x_2 + x_3 + 2x_4 = 9 \\ 2x_1 - x_2 + x_3 + x_4 = 5 \end{cases}$

27. $\begin{cases} x_1 + 2x_3 + x_4 = 3 \\ x_1 + x_2 + 3x_3 + x_4 = 2 \\ 3x_1 + 3x_2 + 9x_3 + 4x_4 = 7 \\ 2x_1 + 4x_3 + 2x_4 = 7 \end{cases}$

28. $\begin{cases} x_1 + x_2 + x_3 + x_4 = 2 \\ x_1 + x_2 + x_3 + x_4 = 1 \\ x_1 + x_2 + x_3 + x_4 = 0 \\ x_1 + x_2 + x_3 + x_4 = 3 \end{cases}$

Formulate each situation as a system of linear equations. Be sure to state clearly the meaning of each variable. Solve using the Gauss–Jordan method. State your final answer in terms of the original question. If permitted by your instructor, you may use a graphing calculator.

29. BUSINESS: Nursery Management The Nyack Nursery starts plants from seeds and sells potted plants to garden stores for resale. Each dahlia costs 16¢ to start and needs a 10¢ flowerpot and 5 ounces of soil; each chrysanthemum costs 11¢ to start and needs a 12¢ flowerpot and 6 ounces of soil; and each daisy costs 13¢ to start and needs an 8¢ flowerpot and 5 ounces of soil. If the nursery has $50 to spend on starting the plants, $48 to spend on flowerpots, and 153 pounds of potting soil, how many of each plant can it raise using all of the available resources?

30. GENERAL: Family Entertainment The Family Fun Center in Asheville offers the package specials listed in the table for an afternoon of fun at its go-kart track, miniature golf course, house of funny mirrors, and snack stand serving hot dogs and sodas. How much does one hot dog cost? How much does a go-kart ride cost if the mirror house is $1.50 and sodas are $1?

Go-kart	Golf	Mirror House	Hot Dog	Soda	Package Price
2	1	1	1	1	$15
3	2	'1	2	1	$23
4	2	2	3	2	$31

3.4 Matrix Arithmetic

Find each matrix product.

31. $\begin{pmatrix} 1 & 2 & 3 & 4 \\ 4 & 3 & 2 & 1 \end{pmatrix} \begin{pmatrix} 1 & 0 \\ 0 & -1 \\ 1 & 0 \\ 0 & -1 \end{pmatrix}$

32. $\begin{pmatrix} 1 & 0 \\ 0 & -1 \\ 1 & 0 \\ 0 & -1 \end{pmatrix} \begin{pmatrix} 1 & 2 & 3 & 4 \\ 4 & 3 & 2 & 1 \end{pmatrix}$

Use the given matrices to find each matrix expression.

$$A = \begin{pmatrix} -1 & 2 & -1 \\ 2 & -1 & 2 \end{pmatrix} \quad B = \begin{pmatrix} 3 & 6 & 8 \\ 7 & 5 & 4 \end{pmatrix}$$

$$C = \begin{pmatrix} 2 & 1 & 2 \\ 1 & -2 & 1 \\ 2 & 1 & 2 \end{pmatrix} \quad D = \begin{pmatrix} 5 & 8 \\ 7 & 6 \end{pmatrix}$$

33. $3D - A \cdot B^t + I$

34. $3I + A^t \cdot B - C$

Rewrite each system of linear equations as a matrix equation $AX = B$.

35. $\begin{cases} x_1 + 4x_2 + x_3 = 15 \\ 2x_1 + 8x_2 + 3x_3 = 26 \\ x_1 + 5x_2 + 2x_3 = 17 \end{cases}$

36. $\begin{cases} 2x_1 + 3x_2 - x_3 + x_4 = 20 \\ 5x_1 + 4x_2 + x_3 + 2x_4 = 35 \\ 2x_1 + x_2 + x_3 + x_4 = 12 \end{cases}$

Formulate each situation in matrix form. Be sure to indicate the meaning of your rows and columns. Find the requested quantities using the appropriate matrix arithmetic.

37. **GENERAL: Growing Grandchildren**
A proud grandmother's record book shows that her grandson Thomas is now 61 inches tall and weighs 90 pounds, while last year he was 58 inches tall and weighed 80 pounds; her grandson Richard is now 54 inches tall and weighs 75 pounds, while last year he was 52 inches tall and weighed 70 pounds; and her granddaughter Harriet is now 47 inches tall and weighs 60 pounds, while last year she was 46 inches tall and weighed 55 pounds. Represent these facts as a "this year" matrix and a "last year" matrix. Use these matrices to find how much each grandchild grew.

38. **MANAGEMENT SCIENCE: Spring Fashions**
The buyer for the ladies sportswear division of a large department store needs 200 jackets, 300 blouses, 250 skirts, and 175 pairs of slacks. The spring lines shown by both an East Coast designer and an Italian team are acceptable to her, but the East Coast designer wants $195 for each jacket, $85 for each blouse, $145 for each skirt, and $130 for each pair of slacks, while the Italian company wants $190 for each jacket, $90 for each blouse, $150 for each skirt, and $125 for each pair of slacks. Represent her needs as a column matrix and the prices as a price matrix. Use these matrices to find the cost of her order from each source.

3.5 Inverse Matrices and Systems of Linear Equations

Find each matrix product. Identify each pair of matrices as "a matrix and its inverse" or "not a matrix and its inverse."

39. $\begin{pmatrix} 1 & 2 & 3 \\ 1 & 1 & 1 \\ 0 & 1 & 3 \end{pmatrix}$ and $\begin{pmatrix} -2 & 3 & 1 \\ 3 & -3 & -2 \\ -1 & 1 & 1 \end{pmatrix}$

40. $\begin{pmatrix} -3 & 0 & 1 \\ 1 & 3 & 1 \\ -3 & 2 & 2 \end{pmatrix}$ and $\begin{pmatrix} -4 & -2 & 3 \\ 5 & 3 & -4 \\ -11 & -6 & 8 \end{pmatrix}$

Find the inverse of each matrix or identify it as singular.

41. $\begin{pmatrix} 1 & 1 & 0 \\ 0 & -3 & 1 \\ 2 & 3 & 0 \end{pmatrix}$

42. $\begin{pmatrix} 1 & 0 & 1 \\ 1 & 1 & 0 \\ 2 & 1 & 1 \end{pmatrix}$

Rewrite each system of equations as a matrix equation $AX = B$ and use the inverse of A to find the solution. Be sure to check the solution in the original system of equations.

43. $\begin{cases} 13x_1 + 4x_2 = 33 \\ 3x_1 + x_2 = 8 \end{cases}$

44. $\begin{cases} 3x_1 + 8x_2 = 25 \\ 2x_1 + 5x_2 = 16 \end{cases}$

45. $\begin{cases} 5x_1 + x_2 + 2x_3 = 11 \\ 2x_1 + 2x_2 + x_3 = 7 \\ 2x_1 + x_2 + x_3 = 5 \end{cases}$

244 CHAPTER 3 SYSTEMS OF EQUATIONS AND MATRICES

46. $\begin{cases} 3x_1 + 2x_2 + x_3 + 2x_4 = 7 \\ 2x_1 + 5x_2 + 2x_3 + 2x_4 = 10 \\ x_1 + 2x_2 + x_3 + x_4 = 4 \\ 2x_1 + 2x_2 + x_3 + 2x_4 = 6 \end{cases}$

Formulate each situation as a collection of systems of linear equations. Be sure to state clearly the meaning of each variable. Solve each collection using an inverse matrix. State your final answer in terms of the original question.

If permitted by your instructor, you may use a graphing calculator.

47. MANAGEMENT SCIENCE: Retail Displays A chain of furniture stores sells living room and bedroom suites and displays them in the store windows and inside on the showroom floor. A living room suite window display requires 3 square yards of window space and then the same furniture group is allowed 7 square yards of showroom space, while a bedroom suite window display requires 4 square yards of window space and then the same furniture group is allowed 9 square yards of showroom space. For furniture styles not given window space but shown only on the showroom floor, each living room suite is allowed 8 square yards and each bedroom suite is allowed 10 square yards. The setup costs per square yard are $50 for furniture shown in the window and on the showroom floor, while for furniture shown only on the showroom floor, $60 for living room suites and $70 for bedroom suites. The table lists the available space (in square yards) at the chain's stores in Kingman, Prescott, and Holbrook, together with the approved setup budgets and the amount of the total showroom space that must be given to furniture not displayed in the window. How many different living room and bedroom suites can be shown at each location?

	Window Space	Showroom Floor	Floor Space for Nonwindow Suites	Setup Budget
Kingman	18	131	90	$ 8,850
Prescott	21	182	134	$12,190
Holbrook	17	187	148	$12,680

48. PERSONAL FINANCE: Income Taxes Find the federal, state, and city income taxes on the taxable incomes of the individuals listed in the table if the federal income tax is 20% of the taxable income after first deducting the state and city taxes; the state income tax is 10% of the taxable income after first deducting the federal and city taxes; and the city income tax is 5% of the taxable income after first deducting the federal and state taxes.

	Mr. Dahlman	Mrs. Farrell	Ms. Mazlin	Mr. Seidner
Taxable income	$96,700	$48,350	$77,360	$145,050

3.6 Introduction to Modeling

Leontief "Open" Input–Output Models

Find the technology matrix for each economy diagram.

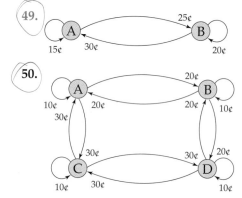

Draw an economy diagram for each technology matrix.

51. $\begin{pmatrix} 0.10 & 0 & 0.20 \\ 0.10 & 0.15 & 0 \\ 0 & 0.15 & 0.20 \end{pmatrix}$

52. $\begin{pmatrix} 0.15 & 0.10 & 0 & 0.15 \\ 0 & 0.10 & 0.20 & 0 \\ 0.15 & 0 & 0.10 & 0 \\ 0 & 0.10 & 0 & 0.15 \end{pmatrix}$

Find the excess production Y of each economy with technology matrix A and economic activity level X.

53. $A = \begin{pmatrix} 0.10 & 0.05 & 0.15 \\ 0.15 & 0.10 & 0.05 \\ 0.10 & 0.05 & 0.10 \end{pmatrix}$ and $X = \begin{pmatrix} 540 \\ 620 \\ 560 \end{pmatrix}$

54. $A = \begin{pmatrix} 0.15 & 0.15 & 0.10 & 0.05 \\ 0.10 & 0.10 & 0.05 & 0.10 \\ 0.10 & 0.05 & 0.10 & 0.15 \\ 0.05 & 0.10 & 0.15 & 0.10 \end{pmatrix}$

and $X = \begin{pmatrix} 280 \\ 200 \\ 260 \\ 240 \end{pmatrix}$

Find, for each economy with technology matrix A, the economic activity level X necessary to generate excess production Y.

55. $A = \begin{pmatrix} 0.10 & 0.35 \\ 0.40 & 0.15 \end{pmatrix}$ and $Y = \begin{pmatrix} 175 \\ 225 \end{pmatrix}$

56. $A = \begin{pmatrix} 0.10 & 0.10 & 0.20 \\ 0.30 & 0.20 & 0.10 \\ 0.10 & 0.30 & 0.20 \end{pmatrix}$ and $Y = \begin{pmatrix} 245 \\ 294 \\ 196 \end{pmatrix}$

Represent each situation as a Leontief "open" input–output model by constructing an economy diagram and the corresponding technology matrix. Find the required excess production or level of economic activity and state your final answer in terms of the original question.

57. BUSINESS: A 5&10 Problem Each of the four divisions of the Woolworth Corporation depends on the others in the following way: Each dollar of production from each division requires 10¢ of production from that division and 5¢ of production from each of the others. How much must each division produce to yield an excess production of $3 million from each?

58. SOCIAL SCIENCE: Energy Dependence As a result of treaty obligations, the domestic and foreign oil consumption of an industrialized nation is linked to the protection provided by its military forces. Each dollar of domestic oil produced requires 5¢ of domestic oil and 10¢ of military protection; each dollar of foreign oil requires 20¢ of military protection; and each dollar of military protection consumes 10¢ of domestic oil, 10¢ of foreign oil, and 5¢ of military protection. If the domestic oil production level is $940 million, the foreign production is $2000 million, and the military protection budget is $520 million, how much of each can be used elsewhere in the nation's economy?

Least Squares

Find the least squares best approximation line $y = mx + b$ for each collection of x and y data pairs.

59.
x	2	3	4
y	22	32	36

60.
x	3	5	6
y	238	364	462

61.
x	2	3	5	6
y	40	50	80	90

62.
x	10	11	12	13	14
y	140	150	180	200	210

Use the least squares best approximation line to make each prediction. Be sure to state your final answer in terms of the original question.

63. MANAGEMENT SCIENCE: Store Hours The owners of a "Mom and Pop" corner store have had their store open various numbers of hours on recent Mondays, the usual day off. The table shows the number of hours the store was open and the sales receipts for the day. How much could they expect to sell next Monday if they keep their store open for 12 hours?

Hours open	6	8	10	14
Total sales	$2230	$3035	$3770	$5065

64. MANAGEMENT SCIENCE: Production Accidents The supervisor of a factory assembly line gathered the data in the table over the last year to compare the number of minor accidents each month with the number of extra five-minute mini-breaks she allows during the day. How many minor accidents could she expect next month if she allows four mini-breaks during the day?

Mini-breaks	0	2	6	8
Minor accidents	21	16	8	3

4 Linear Programming

- 4.1 Linear Inequalities
- 4.2 Two-Variable Linear Programming Problems
- 4.3 The Simplex Method for Standard Maximum Problems
- 4.4 Standard Minimum Problems and Duality
- 4.5 Nonstandard Problems: The Dual Pivot Element and the Two-Stage Method*
- 4.6 Nonstandard Problems: Artificial Variables and the Big-M Method*

*These two sections provide *alternate* methods for solving the *same* types of problems. The choice of which one to cover is left to the instructor.

Application Preview

Managing an Investment Portfolio

The adage "never put all your eggs in one basket" particularly applies to investing money. There is no such thing as a low-risk yet high-return investment, so each investor must reconcile desire for high returns with fear of losing the principal. An "investment strategy" balances greed against safety by determining the investments to be made. Given the many different opportunities in today's global economy, the actual solution of this "portfolio management problem" can be very difficult.

Let us imagine a simplified situation in which there are only two possible investments: one with low risk and low return and the other with high risk and high return. Suppose you have $10,000 to invest and wish to protect yourself by investing at least as much in the low-risk investment as in the high-risk investment. We will express these restrictions as inequalities and then draw their graphs just as we drew the graphs of equations. The portfolio management problem is to find the combination of investments that gives the greatest possible return but still satisfies the restrictions. This will correspond to finding the point on our graph that maximizes the return subject to satisfying the inequalities. In this chapter we will develop ways to find this maximal point, first geometrically for simple problems and then numerically for larger problems. We will then use these methods to solve problems like the portfolio problem described above (see pages 320 and 336) as well as many other problems of practical importance.

4.1 LINEAR INEQUALITIES

Introduction

This chapter describes a method for solving a large and important class of problems called *linear programming problems*. A typical linear programming problem, such as the portfolio problem just described, asks for the maximum value of a linear function subject to a collection of linear inequalities. Such problems often arise in modern business, sometimes involving many variables, and are usually solved by a procedure known as the *simplex method*. Although the simplex method is a numerical algorithm, it is based on the geometry of linear inequalities such as those we studied on page 3. We will first consider linear programming problems in *two* variables since they can be solved graphically. In this section we will graph inequalities in two variables, restricting ourselves to inequalities using $\geq$ and $\leq$ signs since these are the ones that occur in practice.

Inequalities in Two Variables

To graph an inequality in two variables means to graph all of the points that satisfy the inequality. The *boundary* of the inequality is the corresponding *equality*, that is, an *equation*. For example, the boundary of the inequality $3x - 7y \leq 42$ is the equation $3x - 7y = 42$. The graph of the inequality is then one side or the other of the boundary, the correct side being determined by a *test point*. The correct side of the boundary is called the *feasible region* for the inequality.

EXAMPLE 1 GRAPHING A LINEAR INEQUALITY IN THE PLANE

Graph the linear inequality $2x + 3y \geq 12$.

Solution

First we graph its boundary, $2x + 3y = 12$, by plotting the intercepts. We find the y-intercept from $2x + 3y = 12$ with $x = 0$:

$$0 + 3y = 12 \quad \text{so} \quad y = 4 \qquad \text{For the point } (0, 4)$$

then the x-intercept from $2x + 3y = 12$ with $y = 0$:

$$2x + 0 = 12 \quad \text{so} \quad x = 6 \qquad \text{For the point } (6, 0)$$

Using these two points, we draw the boundary line shown on the left.

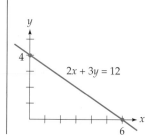

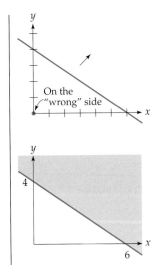

To determine the correct side we use the test point $(0, 0)$. (We may use any point *not on* the line.) Substituting $x = 0$, $y = 0$ into the inequality $2x + 3y \geq 12$ gives $0 \geq 12$, which is *false* (zero is *not* greater than twelve), meaning that the correct side of the line is the *opposite* side from the point $(0, 0)$, as indicated by the arrow in the graph.

Finally, we shade the correct side of the boundary to show the feasible region, as in the third graph.

The general procedure for graphing an inequality is as follows.

To Graph a Linear Inequality

1. Draw the boundary line (possibly using the intercepts).
2. Use a test point not on the line (possibly the origin) to determine the correct side. If this point satisfies the inequality, the point lies on the correct side; if not, the *other* side is the correct side.
3. Shade the correct side to show the feasible region.

Graphing Calculator Exploration

You can view any nonvertical boundary line $ax + by = h$ by entering it as $y = (h - ax)/b$. To see the boundary line $2x + 3y = 12$ from Example 1, enter it as $y_1 = (12 - 2x)/3$ and graph it on the window $[-5, 10]$ by $[-5, 10]$. Some calculators will also "shade" above or below the line provided you change the \ marker to the left of the y_1.

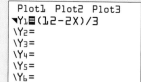
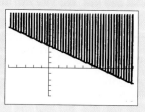

Practice Problem 1   Graph the linear inequality $3x - 5y \leq 60$.   ▶ Solution on page 257

Two or more inequalities joined by a brace { gives a *system* of inequalities, meaning that *each* of the inequalities must hold. To graph the feasible region for the system we graph each boundary, determine its correct side (which we mark with an arrow), and then shade the region that lies on the correct side of *all* of the lines. The system is *feasible* if there is *at least one point* in the feasible region, and *infeasible* if there are *no* points in the feasible region.

EXAMPLE 2

GRAPHING A SYSTEM OF INEQUALITIES

a. Graph the system $\begin{cases} x + 2y \leq 12 \\ x \geq 0 \\ y \geq 0 \end{cases}$

b. Is the system feasible?

c. What happens if the last inequality, $y \geq 0$, is replaced by $y \geq 7$?

Solution

a. We graph the boundary just as before, plotting the intercepts of $x + 2y = 12$: Substituting $x = 0$ gives $2y = 12$ or $y = 6$, for the point $(0, 6)$; substituting $y = 0$ into $x + 2y = 12$ gives $x = 12$ for the point $(12, 0)$. These two points give the line in the first of the following three graphs.

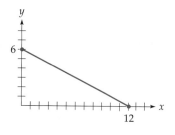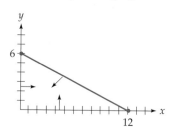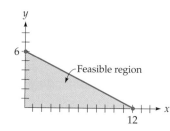

The point (0, 0) *does* satisfy $x + 2y \leq 12$ (since $0 + 0 \leq 12$), so the origin is on the *correct* side, giving the arrow in the second graph. We add the arrows on the axes because $x \geq 0$ means from the *y*-axis to the *right*, and $y \geq 0$ means from the *x*-axis *upwards*. From the three arrows we get the shaded region in the third graph.

b. Since there *are* points in the feasible region, the system *is* feasible.

c. Replacing $y \geq 0$ by $y \geq 7$ gives the graph shown on the left. There are no points on the correct side of all of the lines, so the system with $y \geq 7$ becomes *infeasible*.

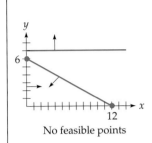

No feasible points

Practice Problem 2 Sketch the system $\begin{cases} x + 2y \leq 20 \\ x \geq 0 \\ y \geq 0 \end{cases}$ Is this system feasible?

➤ Solution on page 257

Vertices of Feasible Regions

The "corner points" of the feasible region are called *vertices*. More precisely:

> **Vertex**
>
> A *vertex* of a system of linear inequalities is an intersection point of two (or more) of the boundaries that satisfies all of the inequalities.

Each vertex may be found by solving for an intersection of two of the boundaries; if this point satisfies all of the inequalities, then it is a vertex of the region.

EXAMPLE 3 **FINDING THE VERTICES OF A FEASIBLE REGION**

Graph and find the vertices of the feasible region for $\begin{cases} x + 2y \leq 12 \\ x - y \geq 0 \\ y \geq 0 \end{cases}$

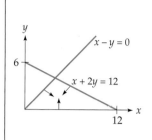

Solution

We graph the boundaries in the usual way, by plotting their intercepts. In Example 2 we found that the boundary $x + 2y = 12$ had intercepts $(0, 6)$ and $(12, 0)$. For the boundary $x - y = 0$ we use the intercept $(0, 0)$ and any other point satisfying $x - y = 0$, such as $(1, 1)$, leading to the graph on the left. The arrows are determined by test points just as in Examples 1 and 2. The inequality $y \geq 0$ means "upwards from the *x*-axis."

The only vertex that is not obvious from the graph is the intersection of $x + 2y = 12$ and $x - y = 0$. Subtracting:

$$x + 2y = 12$$
$$x - y = 0$$
$$\overline{3y = 12}\qquad \longleftarrow \text{Top minus bottom}$$

This last equation gives $y = 4$, which, when substituted into $x - y = 0$, gives $x - 4 = 0$ or $x = 4$, so the intersection point is $(4, 4)$. The feasible region is shown on the left, with vertices $(0, 0)$, $(4, 4)$, and $(12, 0)$.

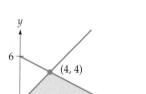

Notice that the point $(0, 6)$ is *not* a vertex of the feasible region since it is on the *wrong* side of the line $x - y = 0$ (it does not satisfy $x - y \geq 0$). Vertices of the feasible region must be on the correct side of *all* of the lines (that is, their coordinates must satisfy *all* of the inequalities).

Bounded and Unbounded Regions

A region is *bounded* if it is enclosed on all sides by boundary lines. For example, the feasible region in Example 3 is bounded. However, if in Example 3 the direction of the inequality $x + 2y \leq 12$ were reversed to become $x + 2y \geq 12$, the feasible region would be on the *other* side of the boundary $x + 2y = 12$, giving the *unbounded* region shown on the right below.

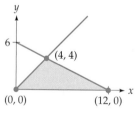
The bounded region from Example 3.

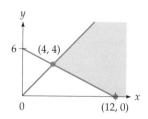
An unbounded region.

Practice Problem 3 Graph the feasible region $\begin{cases} x + 2y \geq 20 \\ x \geq 0 \\ y \geq 0 \end{cases}$ Is the region bounded?

[*Hint:* Modify your graph from Practice Problem 2.]

➤ Solution on page 258

Applications

Since most physical quantities are nonnegative, applied problems usually include the two *nonnegativity inequalities* $x \geq 0$ and $y \geq 0$, which mean that the feasible region will be in the first quadrant.

EXAMPLE 4   JEWELRY PRODUCTION

A small jewelry company prepares and mounts semiprecious stones. There are 10 lapidaries (who cut and polish the stones) and 12 jewelers (who mount the stones in gold settings). Each employee works 7 hours each day. Each tray of agates requires 5 hours of cutting and polishing and 4 hours of mounting, while each tray of onyxes requires 2 hours of cutting and polishing and 3 hours of mounting. How many trays of each stone can be processed each day?

Formulate this situation as a system of linear inequalities, sketch the feasible region, and find the vertices.

Solution

Summarizing

$\left. \begin{array}{l} \text{10 Lapidaries} \\ \text{12 Jewelers} \end{array} \right\} \times 7 \text{ hours}$

(hours)	Agates	Onyxes
Cut & Polish	5	2
Mount	4	3

Clearly, there are many different ways the company could operate each day: It could choose to do nothing and give the employees the day off, it could process just one kind of stone, or it could process some combination. But it could not exceed the amount of time the lapidaries could work (10 workers at 7 hours each is 70 work-hours) or the amount of time the jewelers could work (12 workers at 7 hours each is 84 work-hours). Nor could it process a negative number of trays.

Let

$$x = \begin{pmatrix} \text{Number of trays} \\ \text{of agates} \end{pmatrix} \quad \text{and} \quad y = \begin{pmatrix} \text{Number of trays} \\ \text{of onyxes} \end{pmatrix}$$

For the lapidaries,

$\underbrace{5x}_{\substack{x \text{ trays} \\ @ 5 \text{ hours} \\ \text{each}}} + \underbrace{2y}_{\substack{y \text{ trays} \\ @ 2 \text{ hours} \\ \text{each}}} \underbrace{\leq}_{\substack{\text{No} \\ \text{more} \\ \text{than}}} \underbrace{70}_{\substack{10 \text{ workers} \\ @ 7 \text{ hours} \\ \text{each}}} \qquad \begin{pmatrix} \text{Time} \\ \text{for} \\ \text{agates} \end{pmatrix} + \begin{pmatrix} \text{Time} \\ \text{for} \\ \text{onyxes} \end{pmatrix} \leq \begin{pmatrix} \text{Total} \\ \text{time for} \\ \text{lapidaries} \end{pmatrix}$

For the jewelers,

$$\underbrace{4x}_{\substack{x \text{ trays} \\ @ 4 \text{ hours} \\ \text{each}}} + \underbrace{3y}_{\substack{y \text{ trays} \\ @ 3 \text{ hours} \\ \text{each}}} \underbrace{\leq}_{\substack{\text{No} \\ \text{more} \\ \text{than}}} \underbrace{84}_{\substack{12 \text{ workers} \\ @ 7 \text{ hours} \\ \text{each}}} \qquad \begin{pmatrix} \text{Time} \\ \text{for} \\ \text{agates} \end{pmatrix} + \begin{pmatrix} \text{Time} \\ \text{for} \\ \text{onyxes} \end{pmatrix} \leq \begin{pmatrix} \text{Total} \\ \text{time for} \\ \text{jewelers} \end{pmatrix}$$

Combining these constraints with the nonnegativity conditions $x \geq 0$ and $y \geq 0$, we can represent the problem by the system of linear inequalities

$$\begin{cases} 5x + 2y \leq 70 \\ 4x + 3y \leq 84 \\ x \geq 0 \\ y \geq 0 \end{cases}$$

We can now proceed as usual. The intercepts of the boundary line $5x + 2y = 70$ are $(14, 0)$ and $(0, 35)$. The intercepts of the boundary line $4x + 3y = 84$ are $(21, 0)$ and $(0, 28)$. The nonnegativity conditions $x \geq 0$ and $y \geq 0$ place the feasible region in the first quadrant. Since the test point $(0, 0)$ satisfies the constraints, this region has the origin for one of its vertices.

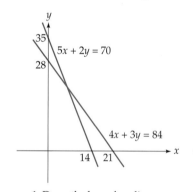

1. Draw the boundary lines

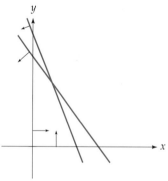

2. Mark the correct sides

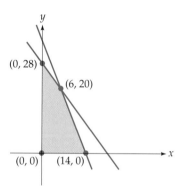

3. Shade the feasible region

Three of the four vertices of this bounded region are already known from the x- and y-intercepts of the boundary lines. The fourth is the intersection of $5x + 2y = 70$ and $4x + 3y = 84$:

$$\begin{array}{rlcrl} 5x + 2y &= 70 & \times 3 \rightarrow & 15x + 6y &= 210 \\ 4x + 3y &= 84 & \times 2 \rightarrow & 8x + 6y &= 168 \\ \hline & & & 7x &= 42 \end{array}$$ ← Top minus bottom

The solution to this last equation is $x = 6$ which, substituted into either equation, gives $y = 20$, for the intersection point $(6, 20)$. The feasible region has vertices $(0, 0)$, $(14, 0)$, $(6, 20)$, and $(0, 28)$, as shown on the right above.

Graphing Calculator Exploration

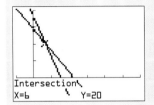

You can graph the boundary lines from Example 4 on your graphing calculator by entering them as $y_1 = (70 - 5x)/2$ and $y_2 = (84 - 4x)/3$.

a. Set the window to $[-10, 60]$ by $[-10, 40]$ and graph your lines. Remember that the nonnegativity conditions $x \geq 0$ and $y \geq 0$ put this region in the first quadrant.

b. You can use TRACE or INTERSECT to find the intersection point of the lines y_1 and y_2.

EXAMPLE 5

WASTE MANAGEMENT

The Marshall County trash incinerator in Norton burns 10 tons of trash per hour and co-generates 6 kilowatts (kW) of electricity, while the Wiseburg incinerator burns 5 tons per hour and co-generates 4 kilowatts. If the county needs to burn at least 70 tons of trash and co-generate at least 48 kilowatts of electricity each day, how many hours should each plant operate?

Formulate this situation as a system of linear inequalities, sketch the feasible region, and find the vertices.

Solution

Again, there are many different schedules of operating times that could burn all the trash and co-generate enough electricity. But there must be at least enough time to get the job done, and neither plant can operate a negative number of hours. Let

Summarizing

	Tons	kW
Norton	10	6
Wiseburg	5	4
	≥70	≥48

$$x = \begin{pmatrix} \text{Number of hours} \\ \text{Norton operates} \end{pmatrix} \quad \text{and} \quad y = \begin{pmatrix} \text{Number of hours} \\ \text{Wiseburg operates} \end{pmatrix}$$

For the amount of trash to be burned, we have the inequality

$$\underbrace{10x}_{\substack{x \text{ hours} \\ @ \ 10 \text{ tons} \\ \text{each}}} + \underbrace{5y}_{\substack{y \text{ hours} \\ @ \ 5 \text{ tons} \\ \text{each}}} \underbrace{\geq}_{\substack{\text{At} \\ \text{least}}} \underbrace{70}_{\substack{\text{Tons of} \\ \text{trash} \\ \text{to burn}}} \qquad \begin{pmatrix} \text{Tons} \\ \text{burned} \\ \text{at Norton} \end{pmatrix} + \begin{pmatrix} \text{Tons} \\ \text{burned at} \\ \text{Wiseburg} \end{pmatrix} \geq \begin{pmatrix} \text{Tons} \\ \text{to} \\ \text{burn} \end{pmatrix}$$

For the electricity to be produced,

$$\underbrace{6x}_{\substack{x \text{ hours} \\ @ \ 6 \text{ kWs} \\ \text{each}}} + \underbrace{4y}_{\substack{y \text{ hours} \\ @ \ 4 \text{ kWs} \\ \text{each}}} \underbrace{\geq}_{\substack{\text{At} \\ \text{least}}} \underbrace{84}_{\substack{\text{kWs} \\ \text{needed}}} \qquad \begin{pmatrix} \text{Norton} \\ \text{electricity} \end{pmatrix} + \begin{pmatrix} \text{Wiseburg} \\ \text{electricity} \end{pmatrix} \geq \begin{pmatrix} \text{Electricity} \\ \text{needed} \end{pmatrix}$$

Combining these constraints with the nonnegativity conditions $x \geq 0$ and $y \geq 0$, we can represent the problem by the system of linear inequalities

$$\begin{cases} 10x + 5y \geq 70 \\ 6x + 4y \geq 48 \\ x \geq 0 \\ y \geq 0 \end{cases}$$

The intercepts of $10x + 5y = 70$ are $(7, 0)$ and $(0, 14)$. The intercepts of $6x + 4y = 48$ are $(8, 0)$ and $(0, 12)$. The nonnegativity conditions place the feasible region in the first quadrant. Since the origin does not satisfy either of the first two inequalities, this region does *not* have the origin for one of its vertices.

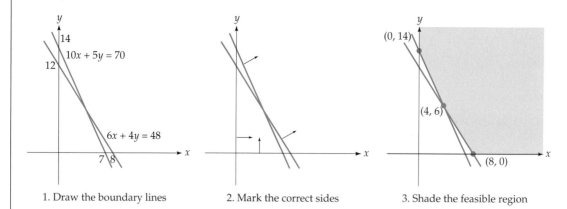

1. Draw the boundary lines
2. Mark the correct sides
3. Shade the feasible region

Two of the three vertices of this unbounded region are already known from the *x*- and *y*-intercepts of the boundary lines. The third is the intersection of $10x + 5y = 70$ and $6x + 4y = 48$, which, found in the usual way (omitting the details), is $(4, 6)$. The three vertices of this region are $(8, 0)$, $(4, 6)$, and $(0, 14)$, as shown in the third diagram above.

4.1 Section Summary

The *feasible region* of a system of linear inequalities $\begin{cases} ax + by \leq c \\ \vdots \end{cases}$ consists of all the (x, y) points on the correct sides of the *boundary lines* $\begin{cases} ax + by = c \\ \vdots \end{cases}$. We usually graph the boundary lines by plotting their

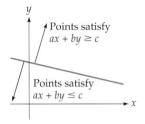

For *a*, *b*, and *c* all positive

intercepts. The correct side of the boundary can be found by trying a "test point" in the inequality [the origin $(0, 0)$ is usually the easiest to use, but any point *not* on the line will do].

The *vertices* are the "corner points" of the region and are the intersections of two (or more) boundary lines that satisfy the inequalities. These intersections can be found by solving the equations representing the lines.

A feasible region is *bounded* if it is enclosed on all sides by boundary lines, and *unbounded* if it is not.

Two special inequalities are the *nonnegativity conditions* $x \geq 0$ and $y \geq 0$, which place the feasible region in the first quadrant and frequently appear in word problems when the variables represent amounts of materials or other real objects.

> ### Solutions to Practice Problems

1. The boundary is the line $3x - 5y = 60$. From $x = 60/3 = 20$, the *x*-intercept is $(20, 0)$. From $y = 60/(-5) = -12$, the *y*-intercept is $(0, -12)$. Because $3 \cdot 0 - 5 \cdot 0$ is ≤ 60, the origin $(0, 0)$ is on the correct side of the boundary line.

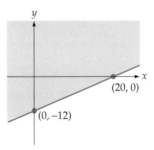

2. The boundary line is $x + 2y = 20$ with intercepts $(20, 0)$ and $(0, 10)$. The origin $(0, 0)$ is on the correct side of $x + 2y \leq 20$ (since $0 \leq 20$). $x \geq 0$ means "to the right of the *y*-axis" and $y \geq 0$ means "above the *x*-axis."

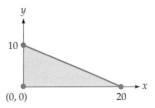

This system *is* feasible.

258 CHAPTER 4 LINEAR PROGRAMMING

3. The "reversed" inequality $x + 2y \geq 20$ means the other side of the boundary line, so the region is:

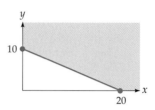

The region is *unbounded*.

4.1 Exercises

For each region, select the inequality it represents.

1. a. $5x + 8y \geq 40$
   b. $8x + 5y \geq 40$
   c. $5x + 8y \leq 40$
   d. $8x + 5y \leq 40$

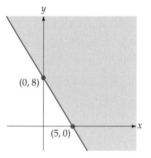

2. a. $7x + 6y \geq 42$
   b. $6x + 7y \geq 42$
   c. $7x + 6y \leq 42$
   d. $6x + 7y \leq 42$

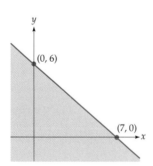

3. a. $-12x + 5y \geq -60$
   b. $5x - 12y \geq -60$
   c. $5x - 12y \leq 60$
   d. $10x + 24y \leq 60$

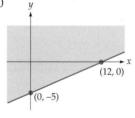

4. a. $5x - 3y \leq -15$
   b. $5x - 3y \geq -15$
   c. $-3x + 5y \leq 15$
   d. $-5x + 3y \leq 15$

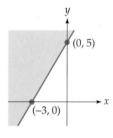

For each region, select the system of inequalities it represents.

5. a. $\begin{cases} 5x + 2y \leq 20 \\ y \geq 0 \end{cases}$
   b. $\begin{cases} 4x + 10y \leq 40 \\ x \leq 0 \end{cases}$
   c. $\begin{cases} 10x - 4y \leq 40 \\ y \leq 0 \end{cases}$
   d. $\begin{cases} -2x + 5y \geq -20 \\ y \geq 0 \end{cases}$

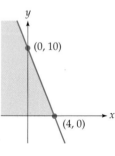

6. a. $\begin{cases} 5x - 3y \leq -30 \\ x \geq 0 \end{cases}$
   b. $\begin{cases} -6x + 10y \leq -60 \\ x \geq 0 \end{cases}$
   c. $\begin{cases} -5x + 3y \leq 30 \\ x \leq 0 \end{cases}$
   d. $\begin{cases} 5x + 3y \leq -30 \\ x \geq 0 \end{cases}$

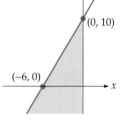

7. a. $\begin{cases} x - 3y \leq -60 \\ x \geq 0 \\ y \geq 0 \end{cases}$

b. $\begin{cases} 2x - 6y \geq -120 \\ x \geq 0 \\ y \geq 0 \end{cases}$

c. $\begin{cases} -60x + 20y \leq 240 \\ y \geq 0 \end{cases}$

d. $\begin{cases} 3x - y \geq -60 \\ x \leq 0 \\ y \geq 0 \end{cases}$

8. a. $\begin{cases} 2x + y \geq 16 \\ x \geq 0 \end{cases}$

b. $\begin{cases} 8x + 16y \leq 32 \\ y \leq 0 \\ x \geq 0 \end{cases}$

c. $\begin{cases} x + 2y \geq 16 \\ x \geq 0 \end{cases}$

d. $\begin{cases} 2x + y \geq -16 \\ y \geq 0 \end{cases}$

Sketch each system of inequalities. List all vertices and identify the region as "bounded" or "unbounded."

9. $\begin{cases} x + 2y \leq 40 \\ x \geq 0, \ y \geq 0 \end{cases}$

10. $\begin{cases} 3x + y \leq 90 \\ x \geq 0, \ y \geq 0 \end{cases}$

11. $\begin{cases} -2x + y \leq 10 \\ x \leq 10 \\ x \geq 0, \ y \geq 0 \end{cases}$

12. $\begin{cases} 2x - y \leq 20 \\ y \leq 40 \\ x \geq 0, \ y \geq 0 \end{cases}$

13. $\begin{cases} x + 2y \leq 8 \\ x + y \leq 6 \\ x \geq 0, \ y \geq 0 \end{cases}$

14. $\begin{cases} 2x + y \leq 10 \\ x + y \leq 8 \\ x \geq 0, \ y \geq 0 \end{cases}$

15. $\begin{cases} 5x + 2y \geq 20 \\ x \geq 0, \ y \geq 0 \end{cases}$

16. $\begin{cases} 4x + 5y \geq 20 \\ x \geq 0, \ y \geq 0 \end{cases}$

17. $\begin{cases} 4x + 3y \geq 24 \\ y \geq 4 \\ x \geq 0, \ y \geq 0 \end{cases}$

18. $\begin{cases} 3x + 4y \geq 12 \\ x \leq 8 \\ x \geq 0, \ y \geq 0 \end{cases}$

19. $\begin{cases} 3x + y \geq 12 \\ x + y \geq 8 \\ x \geq 0, \ y \geq 0 \end{cases}$

20. $\begin{cases} x + 3y \geq 15 \\ x + y \geq 9 \\ x \geq 0, \ y \geq 0 \end{cases}$

21. $\begin{cases} 2x + y \leq 80 \\ x + 3y \geq 30 \\ x \geq 0, \ y \geq 0 \end{cases}$

22. $\begin{cases} 3x + y \leq 90 \\ x + 2y \geq 20 \\ x \geq 0, \ y \geq 0 \end{cases}$

23. $\begin{cases} x - 3y \leq 15 \\ 2x + y \geq 30 \\ x \geq 0, \ y \geq 0 \end{cases}$

24. $\begin{cases} 3x + y \geq 30 \\ x - 2y \geq -60 \\ x \geq 0, \ y \geq 0 \end{cases}$

APPLIED EXERCISES

Formulate each situation as a system of inequalities, sketch the feasible region, and find the vertices. Be sure to state clearly the meaning of your x- and y-variables.

25. **BUSINESS: Livestock Management** A rancher raises goats and llamas on his 400-acre ranch. Each goat needs 2 acres of land and requires $100 of veterinary care per year, while each llama needs 5 acres of land and requires $80 of veterinary care per year. If the rancher can afford no more than $13,200 for veterinary care this year, how many of each animal can he raise?

26. **BUSINESS: Agriculture Management** A farmer grows wheat and barley on her 500-acre farm. Each acre of wheat requires 3 days of labor to plant, tend, and harvest, while each acre of barley requires 2 days of labor. If the farmer and her hired field hands can provide no more than 1200 days of labor this year, how many acres of each crop can she grow?

27. **BUSINESS: Production Planning** A boat company manufactures aluminum dinghies and rowboats. The amounts of metal work and painting needed for each are shown in the table, together with the number of hours of

skilled labor available for each task. How many of each kind of boat can the company manufacture?

	Dinghy	Rowboat	Labor Available
Metal work	2 hours	3 hours	120 hours
Painting	2 hours	2 hours	100 hours

28. **BUSINESS: Resource Allocation** A sailboat company manufactures fiberglass prams and yawls. The amounts of molding, painting, and finishing needed for each are shown in the table, together with the number of hours of skilled labor available for each task. How many of each kind of sailboat can the company manufacture?

	Pram	Yawl	Labor Available
Molding	3 hours	6 hours	150 hours
Painting	3 hours	2 hours	114 hours
Finishing	2 hours	6 hours	132 hours

29. **BIOMEDICAL: Nutrition** Joshua loves "junk food" but wants to stay within the recommended daily limits of 80 grams of fat and 2250 calories. If each serving of Sugar-Snaks contains 5 grams of fat and 125 calories and each bag of Gobbl'Ems contains 8 grams of fat and 250 calories, how much of each may he eat today?

30. **BIOMEDICAL: Diet Planning** Justin plays a lot of sports and wants to make sure he gets at least 70 grams of protein and 20 milligrams of iron each day. If each serving of Pro-Team Power Bars has 10 grams of protein and 2 milligrams of iron and each glass of Bulk-Up-Delight has 5 grams of protein and 2 milligrams of iron, how much of each should he eat each day?

31. **ENVIRONMENTAL SCIENCES: Pollution Control** A smelting company refines metals at two factories located in Ohio and in Pennsylvania. The smokestacks release both sulfur dioxide (which combines with water vapor to form "acid rain") and particulates (solid matter such as soot that can cause respiratory problems) at the rates shown in the table. The EPA has obtained an injunction against the company preventing it from releasing more than 64 pounds of sulfur dioxide and 60 pounds of particulates into the atmosphere each day. How many hours can the company operate these factories each day?

(Pounds Per Hour)	Sulfur Dioxide	Particulates
Ohio factory	4	5
Pennsylvania factory	4	3

32. **ENVIRONMENTAL SCIENCES: Pollution Control** A chemical company manufactures batteries at two factories located in Connecticut and in Alabama. The factories discharge both heavy metals (such as mercury and cadmium, which are very toxic) and nitric acid into the local river systems at the rates shown in the table. An environmental organization has obtained an injunction against the company preventing it from discharging more than 54 pounds of heavy metals and 60 pounds of nitric acid each day. How many hours can the company operate these factories each day?

(Pounds Per Hour)	Heavy Metals	Nitric Acid
Connecticut factory	6	4
Alabama factory	3	4

33. **PERSONAL FINANCE: Investment Strategy** An investment portfolio manager has $8 million to invest in stock and bond funds. If the amount invested in stocks can be no more than the amount invested in bonds, how much can be invested in each type of fund?

34. **PERSONAL FINANCE: Financial Planning** A retired couple want to invest their $20,000 life savings in bank certificates of deposit and Treasury bonds. If they want at least $5000 in each type of investment, how much can they invest in each?

4.2 TWO-VARIABLE LINEAR PROGRAMMING PROBLEMS

Introduction

In this section we will explain what a linear programming problem is and then use the geometry of feasible regions from the previous section to show that the solution of a linear programming problem occurs at a vertex of the region.

Linear Programming Problems

We begin by formulating an example of a linear programming problem, which we will then solve in Example 2.

EXAMPLE 1 **FARM MANAGEMENT**

A farmer grows soybeans and corn on his 300-acre farm. To maintain soil fertility, the farmer rotates the crops and always plants at least as many acres of soybeans as acres of corn. If each acre of soybeans yields a profit of $100 and each acre of corn yields a profit of $200, how many acres of each crop should the farmer plant to obtain the greatest possible profit?

Formulate this situation as a linear programming problem by identifying the variables, the objective function, and the constraints.

Solution

Since the question asks "how many acres of each crop," we let

$$x = \begin{pmatrix} \text{Number of} \\ \text{acres of soybeans} \end{pmatrix} \quad \text{and} \quad y = \begin{pmatrix} \text{Number of} \\ \text{acres of corn} \end{pmatrix}$$

The objective is to maximize the farmer's profit, and this profit is $P = 100x + 200y$ because the profits per acre are $100 for soybeans and $200 for corn. The 300-acre size of the farm leads to the constraint $x + y \leq 300$, while the crop rotation requirement that $x \geq y$ can be written as $x - y \geq 0$. Since x and y cannot be negative, we also have the nonnegativity constraints $x \geq 0$ and $y \geq 0$. This maximum linear programming problem may be written as

Maximize $P = 100x + 200y$ ← Objective function — Corn and soybean profits

Subject to $\begin{cases} x + y \leq 300 & \text{Size of farm} \\ x - y \geq 0 & \text{Crop rotation} \\ x \geq 0 \text{ and } y \geq 0 & \text{Nonnegativity} \end{cases}$ ← Constraints

We call $P = 100x + 200y$ the "objective function" because the objective of the entire procedure is to maximize it. The inequalities are called "constraints" because they constrain the profit by, for example, the limiting the crops to 300 acres.

In general, a *linear programming problem* consists of a linear *objective function* to be maximized or minimized subject to *constraints* in the form of a system of linear inequalities. Since an inequality with a $\geq$ can be changed into one with a $\leq$ by multiplying by -1 (for example, $x - y \geq 0$ can be written as $-x + y \leq 0$), every linear programming problem in two variables may be written in one of the following forms:

Linear Programming Problems

Maximize $P = Mx + Ny$

Subject to $\begin{cases} ax + by \leq c \\ \vdots \end{cases}$

or

Minimize $C = Mx + Ny$

Subject to $\begin{cases} ax + by \geq c \\ \vdots \end{cases}$

We call the objective function in the maximum problem P for profit and in the minimum problem C for cost, but problems with different goals are perfectly acceptable. Nonnegativity conditions should be included only when appropriate.

Fundamental Theorem of Linear Programming

A *solution* of a *maximum* linear programming problem is a point that satisfies the system of linear inequalities (a "feasible" point) and that gives the *largest* possible value of the objective function. Similarly, a solution of a *minimum* linear programming problem is a feasible point that gives the *smallest* possible value of the objective function. With either problem, if the constraints have no solutions (they are "infeasible") then the problem has no solution.

How can we solve such a problem? Consider the objective function $P = 100x + 200y$ from Example 1. If we solve this equation for y, we obtain:

$$200y = -100x + P \qquad \text{Switching sides}$$
$$y = -\tfrac{1}{2}x + \tfrac{1}{200}P \qquad \text{Dividing through by 200}$$

The $y = mx + b$ form of this equation shows that it represents a line of slope $-\tfrac{1}{2}$. Different values of P will give parallel "objective function lines," all with slope $-\tfrac{1}{2}$. In Example 2 we will see that the feasible region for Example 1 is the triangular region in the following graph. On this graph are drawn several of these objective function lines for various values of P. Higher lines correspond to larger value of P, so *maximizing* P means choosing the uppermost line that still intersects the feasible region.

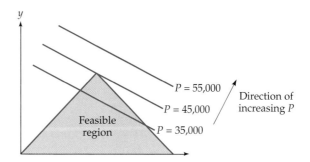

As the diagram shows, the uppermost line that intersects the region does so at a *vertex* of the region. (Any line with a higher value of P, such as $P = 55{,}000$, does not intersect the feasible region at all.) It should be clear that the situation will be the same for *any* linear objective function and *any* feasible region bounded by linear inequalities: *For a linear programming problem with a bounded region, the solution occurs at a vertex of the region.*

For an *unbounded* region the solution might not exist: The objective function lines might move in an unbounded direction as P increases, allowing higher and higher values of P, so there would never be a largest value. However, if increasing the value of P always means moving toward a boundary, then the solution *will* exist and again will occur at a vertex.

Similar reasoning applies to a *minimum* linear programming problem by finding the objective function line with the *minimum* value of P that intersects the feasible region. Combining these observations, we have established the following important result about the solution of a linear programming problem.

Fundamental Theorem of Linear Programming

If a linear programming problem has a solution, then it occurs at a *vertex* of the feasible region.

The Fundamental Theorem of Linear Programming gives us the following procedure to find a solution.

> **How to Solve a Linear Programming Problem**
>
> If the region is bounded, list the vertices and calculate the value of the objective function at each. The solution occurs at the vertex that gives the largest (or smallest) value.
>
> If the region is unbounded, first check whether the objective function "improves" in an unbounded direction. If it does, there is *no solution*; if it does not, list the vertices and calculate the value of the objective function at each; the solution occurs at the vertex that gives the largest (or smallest) value.

EXAMPLE 2

SOLUTION OF EXAMPLE 1

Solve the linear programming problem from Example 1 on page 261:

$$\text{Maximize} \quad P = 100x + 200y \quad \text{Objective function}$$

$$\text{Subject to} \quad \begin{cases} x + y \leq 300 \\ x - y \geq 0 \\ x \geq 0 \quad \text{and} \quad y \geq 0 \end{cases} \quad \text{Constraints}$$

Solution

We begin by sketching the region determined by the constraints. The nonnegativity conditions $x \geq 0$ and $y \geq 0$ place the region in the first quadrant. The boundary line $x + y = 300$ has intercepts $(300, 0)$ and $(0, 300)$, and the boundary line $x - y = 0$ passes through the origin $(0, 0)$ and the point $(1, 1)$. These points give the first of the following graphs. For test points, the origin satisfies the inequality $x + y \leq 300$, and the point $(300, 0)$ satisfies the inequality $x - y \geq 0$.

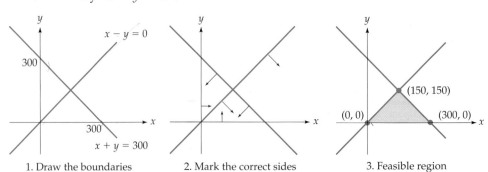

1. Draw the boundaries 2. Mark the correct sides 3. Feasible region

The region has three vertices, and two are already known from the x- and y-intercepts of the boundary lines. The third is the intersection of $x - y = 0$ and $x + y = 300$:

$$x + y = 300$$
$$x - y = 0$$
$$\overline{2x = 300} \quad \leftarrow \text{Adding}$$

The last equation gives $x = 150$, and from $x + y = 300$, we find $y = 150$, for the vertex $(150, 150)$. The vertices are $(0, 0)$, $(300, 0)$, and $(150, 150)$, as shown in the third graph.

Since the region is bounded, this problem *does* have a solution, and it occurs at a vertex. Evaluating the objective function at the vertices, we find the following values:

Vertex	Value of $P = 100x + 200y$	
(0, 0)	0	$0 = 100 \cdot 0 + 200 \cdot 0$
(300, 0)	30,000	$30{,}000 = 100 \cdot 300 + 200 \cdot 0$
(150, 150)	45,000 ← Largest	$45{,}000 = 100 \cdot 150 + 200 \cdot 150$

Since the largest value of the objective function occurs at the vertex $(150, 150)$, the solution of this problem is:

The maximum value of P is 45,000 at the vertex $(150, 150)$.

In terms of the original word problem in Example 1, the maximum profit is $45,000 when the farmer plants 150 acres of corn and 150 acres of soybeans.

Practice Problem 1 Solve the linear programming problem:

Maximize $P = 5x + 2y$

Subject to $\begin{cases} 3x + y \leq 60 \\ y \leq 36 \\ x \geq 0 \text{ and } y \geq 0 \end{cases}$

▶ Solution on page 271

EXAMPLE 3 A MINIMUM PROBLEM ON AN UNBOUNDED REGION

Solve the linear programming problem:

$$\text{Minimize} \quad C = 3x + 5y$$

$$\text{Subject to} \quad \begin{cases} 2x + y \geq 12 \\ x + y \geq 8 \\ x \geq 0 \text{ and } y \geq 0 \end{cases}$$

Solution

We begin by sketching the region determined by the constraints. The nonnegativity conditions $x \geq 0$ and $y \geq 0$ place the region in the first quadrant. The boundary line $2x + y = 12$ has intercepts $(6, 0)$ and $(0, 12)$, while the boundary line $x + y = 8$ has intercepts $(8, 0)$ and $(0, 8)$. Since the test point $(0, 0)$ does *not* satisfy $2x + y \geq 12$ and $x + y \geq 8$, the region is on the sides of these boundaries *away* from the origin.

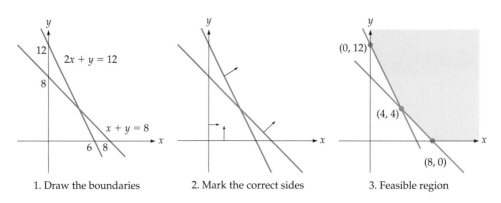

1. Draw the boundaries   2. Mark the correct sides   3. Feasible region

This unbounded region has three vertices, and two are already known from the *x*- and *y*-intercepts of the boundary lines. The third is the intersection of $2x + y = 12$ and $x + y = 8$, which we solve together:

$$\begin{array}{r} 2x + y = 12 \\ x + y = 8 \\ \hline x = 4 \end{array}$$ ← Top minus bottom

Substituting $x = 4$ into $x + y = 8$ gives $y = 4$, for the intersection point $(4, 4)$. The vertices are $(0, 12)$, $(4, 4)$, and $(8, 0)$, as shown in the third graph above.

The region is unbounded, so the solution may not exist. The third graph above shows that the region is unbounded in the *positive x direc-*

tion and also in the *positive y direction*, which means that we could increase x as much as we choose, and we could increase y as much as we choose. However, increasing either x or y will *increase* the objective function $C = 3x + 5y$ (since x and y are multiplied by *positive* numbers) which is the *opposite* of what we should do to minimize C. Therefore, the objective function does not "improve" in an unbounded direction, so the solution *does* exist. Evaluating the objective function at the vertices, we find the following values.

Vertex	C = 3x + 5y		
(8, 0)	24	← Smallest	$24 = 3 \cdot 8 + 5 \cdot 0$
(4, 4)	32		$32 = 3 \cdot 4 + 5 \cdot 4$
(0, 12)	60		$60 = 3 \cdot 0 + 5 \cdot 12$

Since the smallest value of the objective function occurs at the vertex (8, 0), the solution of this problem is:

The minimum value of C is 24 at the vertex (8, 0).

A linear programming problem with an unbounded feasible region may or may not have a solution. For example, in a maximum problem if the objective function can be made arbitrarily large by choosing x and y values far out in the unbounded direction, then it will not have a largest value, so the problem will have no solution. The general situation is described in the following box, where the phrase *the objective function "improves"* means that it *increases* for a *maximum* problem and that it *decreases* for a *minimum* problem.

Determining Whether a Solution Exists

If the feasible region is unbounded, the linear programming problem may not have a solution. First determine the direction(s) in which the feasible region is unbounded.

1. If the objective function *improves* when the x and y variables take values further out an unbounded direction, then the solution *does not exist*.
2. If the objective function does *not* improve when the x and y variables take values further out in any of the unbounded directions, then the solution *does* exist.

Practice Problem 2 Solve the linear programming problem:

$$\text{Maximize} \quad P = 5x + 11y$$
$$\text{Subject to} \quad \begin{cases} x + 3y \geq 60 \\ x \geq 0 \quad \text{and} \quad y \geq 0 \end{cases}$$

▶ Solution on page 271

EXAMPLE 4

A MANUFACTURING PROBLEM

A fully automated plastics factory produces two toys, a racing car and a jet airplane, in three stages: molding, painting, and packaging. After allowing for routine maintenance, the equipment for each stage can operate no more than 150 hours per week. Each batch of racing cars requires 6 hours of molding, 2.5 hours of painting, and 5 hours of packaging, while each batch of jet airplanes requires 3 hours of molding, 7.5 hours of painting, and 5 hours of packaging. If the profit per batch of toys is $120 for cars and $100 for airplanes, how many batches of each toy should be produced each week to obtain the greatest possible profit?

Solution

Summarizing (hours)	Car	Plane
Molding	6	3
Painting	2.5	7.5
Packaging	5	5
Profit	$120	$100

Let

$$x = \begin{pmatrix} \text{Number of} \\ \text{batches of cars} \end{pmatrix} \quad \text{and} \quad y = \begin{pmatrix} \text{Number of} \\ \text{batches of jets} \end{pmatrix}$$

made during the week. The profit is $P = 120x + 100y$ and the problem is to maximize P subject to the constraints that the molding, painting, and packaging processes can each take no more than 150 hours:

$$6x + 3y \leq 150 \quad \text{Molding time}$$
$$2.5x + 7.5y \leq 150 \quad \text{Painting time}$$
$$5x + 5y \leq 150 \quad \text{Packaging time}$$

and the nonnegativity conditions $x \geq 0$ and $y \geq 0$. Simplifying the constraints by removing common factors (dividing the molding constraint by 3, the painting constraint by 2.5, and the packaging constraint by 5), we can rewrite this problem as the linear programming problem

$$\text{Maximize} \quad P = 120x + 100y \quad \text{Objective function}$$

$$\text{Subject to} \quad \begin{cases} 2x + y \leq 50 \\ x + 3y \leq 60 \\ x + y \leq 30 \\ x \geq 0 \quad \text{and} \quad y \geq 0 \end{cases} \quad \text{Constraints}$$

Proceeding as usual, we graph the boundary lines by plotting their intercepts (the first graph), determine the correct sides by using the origin as a test point (the second graph), and finally shade the feasible region (the third graph).

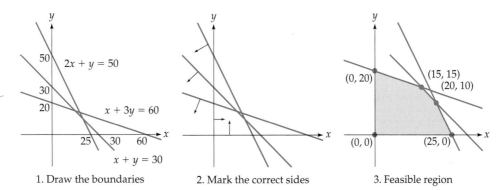

1. Draw the boundaries 2. Mark the correct sides 3. Feasible region

The third graph shows that the region has five vertices, three of which are already known from the x- and y-intercepts. The other two come from solving pairs of equations in the usual way, giving the following vertices:

$$
\begin{array}{ll}
2x + y = 50 & x + 3y = 60 \\
x + y = 30 & x + y = 30 \\
\hline
x = 20 & 2y = 30 \\
\text{Vertex: } (20, 10) & \text{Vertex: } (15, 15)
\end{array}
$$

[The intersection of $2x + y = 50$ and $x + 3y = 60$ is the point $(18, 14)$, but this is *not* a vertex of the region because it does not also satisfy $x + y \leq 30$: $18 + 14$ is not ≤ 30.] The vertices are $(0, 0)$, $(25, 0)$, $(20, 10)$, $(15, 15)$, and $(0, 20)$ as shown in the preceding diagram.

The region is bounded, so this problem *has* a solution, and it occurs at a vertex. Evaluating the objective function at the vertices, we find the following values:

Vertex	$P = 120x + 100y$	
(0, 0)	0	
(25, 0)	3000	
(20, 10)	3400	⟵ Largest
(15, 15)	3300	
(0, 20)	2000	

Since the largest value of the objective function occurs at the vertex $(20, 10)$, the solution is:

The maximum value is 3400 at the vertex $(20, 10)$.

> In terms of the original question, the maximum profit is $3400 when the factory produces 20 batches of racing car toys and 10 batches of jet airplane toys each week.

Be careful! Except in very simple problems, some boundary line intersections are *not* vertices of the feasible region because they violate at least one of the other constraints. Particularly when sketching regions by hand, you may want to find all the intersection points anyway and then verify which are feasible and which are not.

Extensions to Larger Problems

Our geometric method is limited to two variables and therefore to two products, which is unrealistic for a large company. In the next two sections we will develop a numerical method that applies to problems with *any* number of variables, and that also eliminates the need to find and check *all* of the vertices.

4.2 Section Summary

A *linear programming problem* consists of a linear *objective function* to be *maximized* or *minimized* subject to *constraints*, written as a system of linear inequalities. Every such problem can be written in the form

$$\text{Maximize } P = Mx + Ny \qquad \text{or} \qquad \text{Minimize } C = Mx + Ny$$
$$\text{Subject to } \begin{cases} ax + by \leq c \\ \vdots \end{cases} \qquad\qquad \text{Subject to } \begin{cases} ax + by \geq c \\ \vdots \end{cases}$$

where P and C are the objective functions and the braced systems are the constraints.

A *solution* is a feasible point that gives the largest (or smallest) possible value for the objective function. The *Fundamental Theorem of Linear Programming* says that if there is a solution, it occurs at a *vertex* of the feasible region. Thus the problem can be solved by graphing the feasible region of the constraints, evaluating the objective function at the vertices, and selecting the vertex with the optimal value. If the region is unbounded, you must check that a solution exists before selecting the "best" vertex.

Solutions to Practice Problems

1.

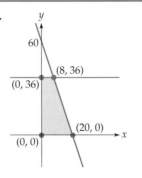

Vertex	$P = 5x + 2y$
(0, 0)	0
(20, 0)	100
(8, 36)	112
(0, 36)	72

The maximum value is 112 when $x = 8$ and $y = 36$.

2.

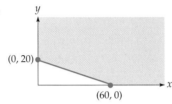

The region is unbounded in the positive x and positive y directions. Since the objective function $P = 5x + 11y$ increases ("improves") as x and y take values in those directions, *the solution does not exist.*

4.2 Exercises

For Exercises 1–5: Use the region below to find each maximum or minimum value. If such a value does not exist, explain why not.

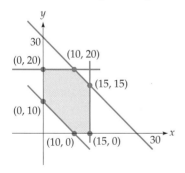

1. Maximum of $P = 2x + y$
2. Maximum of $P = x + 2y$
3. Minimum of $C = 3x + 4y$
4. Minimum of $C = 4x + 3y$
5. Maximum of $P = x - y$

For Exercises 6–10: Use the region below to find each maximum or minimum value. If such a value does not exist, explain why not.

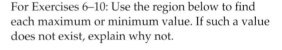
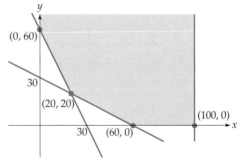

6. Minimum of $C = x + y$
7. Minimum of $C = 3x + y$
8. Maximum of $P = x + 3y$
9. Maximum of $P = 5x + 8y$
10. Maximum of $P = -x - y$

Solve each linear programming problem by sketching the region and labeling the vertices, deciding whether a solution exists, and then finding it if it does exist.

11. Maximize $P = 30x + 40y$

Subject to $\begin{cases} 2x + y \leq 16 \\ x + y \leq 10 \\ x \geq 0, \ y \geq 0 \end{cases}$

12. Maximize $P = 80x + 70y$

Subject to $\begin{cases} x + 2y \leq 18 \\ x + y \leq 10 \\ x \geq 0, \ y \geq 0 \end{cases}$

13. Minimize $C = 15x + 45y$

Subject to $\begin{cases} 2x + 5y \geq 20 \\ x \geq 0, \ y \geq 0 \end{cases}$

14. Minimize $C = 35x + 25y$

Subject to $\begin{cases} 5x + 3y \geq 60 \\ x \geq 0, \ y \geq 0 \end{cases}$

15. Maximize $P = 4x + 5y$

Subject to $\begin{cases} 2x + y \leq 50 \\ x + 3y \leq 75 \\ x \geq 0, \ y \geq 0 \end{cases}$

16. Maximize $P = 7x + 8y$

Subject to $\begin{cases} 3x + y \leq 90 \\ x + 2y \leq 60 \\ x \geq 0, \ y \geq 0 \end{cases}$

17. Minimize $C = 12x + 10y$

Subject to $\begin{cases} 4x + y \geq 40 \\ 2x + 3y \geq 60 \\ x \geq 0, \ y \geq 0 \end{cases}$

18. Minimize $C = 20x + 30y$

Subject to $\begin{cases} 3x + 2y \geq 120 \\ x + 4y \geq 80 \\ x \geq 0, \ y \geq 0 \end{cases}$

19. Maximize $P = 5x + 3y$

Subject to $\begin{cases} 2x + y \leq 90 \\ x + y \leq 50 \\ x + 2y \leq 90 \\ x \geq 0, \ y \geq 0 \end{cases}$

20. Maximize $P = 6x + 5y$

Subject to $\begin{cases} x + 2y \leq 96 \\ x + y \leq 54 \\ 2x + y \leq 96 \\ x \geq 0, \ y \geq 0 \end{cases}$

21. Maximize $P = 10x + 12y$

Subject to $\begin{cases} 3x + 2y \leq 180 \\ 4x + y \leq 120 \\ 3x + y \leq 105 \\ x \geq 0, \ y \geq 0 \end{cases}$

22. Maximize $P = 20x + 15y$

Subject to $\begin{cases} 2x + 3y \leq 60 \\ x + 4y \leq 40 \\ x + 3y \leq 33 \\ x \geq 0, \ y \geq 0 \end{cases}$

23. Minimize $C = 20x + 25y$

Subject to $\begin{cases} 3x + y \geq 60 \\ x + y \geq 42 \\ x + 3y \geq 60 \\ x \geq 0, \ y \geq 0 \end{cases}$

24. Minimize $C = 50x + 35y$

Subject to $\begin{cases} x + 3y \geq 72 \\ x + y \geq 48 \\ 3x + y \geq 72 \\ x \geq 0, \ y \geq 0 \end{cases}$

APPLIED EXERCISES

Formulate each situation as a linear programming problem by identifying the variables, the objective function, and the constraints. Be sure to state clearly the meaning of each variable. Determine whether a solution exists, and if it does, find it. State your final answer in terms of the original question.

25. **BUSINESS: Livestock Management** A rancher raises goats and llamas on his 400-acre ranch. Each goat needs 2 acres of land and requires $100 of veterinary care per year, while each llama needs 5 acres of land and requires $80 of veterinary care per year. The rancher can afford no more than $13,200 for veterinary care this year. If the expected profit is $60 for each goat and $90 for each llama, how many of each animal should he raise to obtain the greatest possible profit?

26. **BUSINESS: Agriculture Management** A farmer grows wheat and barley on her 500-acre farm. Each acre of wheat requires 3 days of labor to plant, tend, and harvest, while each acre of barley requires 2 days of labor. The farmer and her hired field hands can provide no more than 1200 days of labor this year. If the expected profit is $50 for each acre of wheat and $40 for each acre of barley, how many acres of each crop should she grow to obtain the greatest possible profit?

27. **BUSINESS: Resource Allocation** A sailboat company manufactures fiberglass prams and yawls. The amount of molding, painting, and finishing needed for each is shown in the table, together with the number of hours of skilled labor available for each task. If the expected profit is $150 for each pram and $180 for each yawl, how many of each kind of sailboat should the company manufacture to obtain the greatest possible profit?

	Pram	Yawl	Labor Available
Molding	3 hours	6 hours	150 hours
Painting	3 hours	2 hours	114 hours
Finishing	2 hours	6 hours	132 hours

28. **BUSINESS: Production Planning** A small jewelry company prepares and mounts semi-precious stones. There are 10 lapidaries (who cut and polish the stones) and 12 jewelers (who mount the stones in gold settings). Each employee works 7 hours each day. Each tray of agates requires 5 hours of cutting and polishing and 4 hours of mounting, while each tray of onyxes requires 2 hours of cutting and polishing and 3 hours of mounting. If the profit is $15 for each tray of agates and $10 for each tray of onyxes, how many trays of each stone should be processed each day to obtain the greatest possible profit?

29. **ENVIRONMENTAL SCIENCE: Waste Management** The Marshall County trash incinerator in Norton burns 10 tons of trash per hour and co-generates 6 kilowatts of electricity, while the Wiseburg incinerator burns 5 tons per hour and co-generates 4 kilowatts. The county needs to burn at least 70 tons of trash and co-generate at least 48 kilowatts of electricity every day. If the Norton incinerator costs $80 per hour to operate and the Wiseburg incinerator costs $50, how many hours should each incinerator operate each day with the least cost to the county?

30. **GENERAL: Disaster Relief** An international relief agency has been asked to provide medical support to a Caribbean island devastated by a recent hurricane. The agency estimates that it must be able to perform at least 50 major surgeries, 78 minor surgeries, and 130 outpatient services each day. The daily capacities of its portable field hospitals and clinics are given in the table, along with the daily operating costs. How many field hospitals and clinics should the agency airlift to the island to provide the needed help at the least daily cost?

	Major Surgeries	Minor Surgeries	Outpatient Services	Daily Cost
Field hospitals	5	3	2	$2000
Clinics	0	2	10	$500

31. **BIOMEDICAL: Nutrition** A pet store owner raises baby bunnies and feeds them leftover salad greens from a nearby restaurant. The weekly nutritional needs of each bunny are given in the table, as well as the nutrition provided by the salad greens and nutritional supplement drops. If the salad greens cost 2¢ per handful and the nutritional supplement costs 1¢ per drop, how many handfuls of greens and drops of supplement should each bunny receive weekly to minimize the owner's costs?

	Salad Greens (Per Handful)	Nutritional Supplement (Per Drop)	Minimum Weekly Requirement
Fiber (grams)	3	0	90
Vitamins (mg)	2	5	140
Minerals (mg)	1	6	84

32. **BIOMEDICAL: Diet Planning** The director of a school district's hot lunch program estimates that each lunch should contain at least 1000 calories and 15 grams of protein and no more than 30 grams of fat. The nutritional contents per ounce of the cafeteria's famous mystery foods X and Y are given in the table. If each ounce of X costs 5¢ and each ounce of Y costs 6¢, how much of each food should be served to meet the director's nutritional goals at the least cost?

	Calories	Protein	Fat
Mystery food X	200	1	3
Mystery food Y	100	3	3

33. **ENVIRONMENTAL SCIENCE: Land Reclamation** A state government included $1.2 million in its current appropriations bill to reclaim some of the land at a 3000-acre strip mine site. It will cost $800 per acre to return the land to productive grassland suitable for livestock and $500 per acre to return the land to forest suitable for commercial timber production. The appropriations bill also requires that at least 1800 acres be reclaimed immediately and then the income from leasing it to ranchers and/or paper companies be used to reclaim the rest of the site in the coming years. If long-term-use agreements yield $200 per acre of grassland and $150 per acre of forest, how many acres of each should be reclaimed this year to raise the greatest amount for next year's reclamation efforts?

34. **SOCIAL SCIENCE: Urban Renewal** A nonprofit urban development corporation has agreed to rebuild at least 24 city blocks in the south side of Megatropolis in which at least 15 blocks will be semidetached single family homes and at least 3 but not more than 10 blocks will be commercial buildings (retail stores and/or light industry). If it will cost $6 million to rebuild 1 block with homes and $7 million to rebuild 1 block for commercial use, how many blocks of each should be rebuilt to meet the goals at the least cost?

4.3 THE SIMPLEX METHOD FOR STANDARD MAXIMUM PROBLEMS

Introduction

Using the geometric method from the previous section as our guide, we shall explain the *simplex method* for solving linear programming problems. This method finds a *path of vertices* leading from an initial vertex to a solution without finding all of the intersection points of the boundary lines or even all of the vertices of the feasible region. We begin by defining a *standard maximum problem*, the type of problem that we will solve in this section. Other types of problems will be solved in succeeding sections.

Standard Maximum Problems

A standard maximum problem is a linear programming problem in which the objective function is to be maximized, the variables $x_1, x_2, \ldots, x_n$ are all nonnegative, and the other constraints have $\leq$ inequalities with nonnegative numbers on the right-hand sides.

Standard Maximum Problem

A problem is a *standard maximum problem* if it can be written:

Maximize an objective function

$$P = c_1 x_1 + c_2 x_2 + \cdots + c_n x_n$$

Subject to inequalities of the form

$$a_1 x_1 + a_2 x_2 + \cdots + a_n x_n \leq b \quad \text{with} \quad b \geq 0$$

and

$$x_1 \geq 0,\ x_2 \geq 0,\ \ldots,\ x_n \geq 0$$

Observe that if we substitute 0 for each variable, the inequalities are *all* satisfied, so for a standard maximum problem the origin *is* a vertex of the feasible region.

EXAMPLE 1

VERIFYING STANDARD MAXIMUM FORM

Is the following a standard maximum problem?

$$\text{Maximize}\quad P = 9x_1 + 16x_2$$

$$\text{Subject to}\quad \begin{cases} 3x_1 + 4x_2 \leq 24 \\ x_1 + 2x_2 \leq 10 \\ x_1 \geq 0\ \text{and}\ x_2 \geq 0 \end{cases}$$

Solution

The objective function is to be maximized, the variables are both nonnegative, and the other inequalities all have $\leq$ with nonnegative numbers on the right, so this *is* a standard maximum problem.

Practice Problem 1

Is the following a standard maximum problem?

Maximize $P = 7x_1 + 5x_2$

Subject to $\begin{cases} 2x_1 - 3x_2 \leq 24 \\ 5x_1 + 2x_2 \leq -4 \\ x_1 \geq 0 \text{ and } x_2 \geq 0 \end{cases}$

➤ Solution on page 291

Matrix Form of a Standard Maximum Problem

We may express a standard maximum problem in the matrix notation of the preceding chapter. Just as two matrices are *equal* if they have the same dimension and corresponding elements are equal, two matrices obey an *inequality* if they have the same dimension and corresponding elements obey that inequality. For example:

$$\begin{pmatrix} 1 & 2 \\ 3 & 4 \end{pmatrix} \leq \begin{pmatrix} 1 & 2 \\ 6 & 8 \end{pmatrix} \qquad \text{Matrices are the same size and corresponding elements obey } \leq$$

Using matrix inequalities, the problem written on the left below (from Example 1) may be expressed in the matrix form on the right:

Maximize $P = 9x_1 + 16x_2$

Subject to $\begin{cases} 3x_1 + 4x_2 \leq 24 \\ x_1 + 2x_2 \leq 10 \\ x_1 \geq 0 \\ x_2 \geq 0 \end{cases}$

Maximize $P = \begin{pmatrix} 9 & 16 \end{pmatrix} \begin{pmatrix} x_1 \\ x_2 \end{pmatrix}$

Subject to $\begin{cases} \begin{pmatrix} 3 & 4 \\ 1 & 2 \end{pmatrix} \begin{pmatrix} x_1 \\ x_2 \end{pmatrix} \leq \begin{pmatrix} 24 \\ 10 \end{pmatrix} \\ \begin{pmatrix} x_1 \\ x_2 \end{pmatrix} \geq 0 \end{cases}$

In this way, any standard maximum problem may be expressed in the following matrix form, using $X = \begin{pmatrix} x_1 \\ \vdots \\ x_n \end{pmatrix}$ for a column matrix of variables:

Standard Maximum Problem in Matrix Form*

Maximize $P = c^t X$

Subject to $\begin{cases} AX \leq b \\ X \geq 0 \end{cases}$ where $b \geq 0$

---

* Writing c^t for the row matrix of coefficients of the objective function is for historical reasons only. The mathematicians who developed these methods preferred *column* matrices, so row matrices were represented as *transposes* of column matrices, hence c^t.

The Initial Simplex Tableau

Equations are easier to solve than inequalities, so we first simplify the problem by changing the inequalities to equations by adding to each a *slack variable* that is nonnegative and represents the "amount not used." For example, in the inequality $4 \leq 7$ the two sides differ by 3, so adding 3 to the left side gives an *equation* $4 + 3 = 7$. Similarly, the inequality $x \leq 7$ can be rewritten as the *equation* $x + s = 7$ by adding a slack variable $s \geq 0$ to the left side to make up the difference between the sides. We do this for each inequality, using a different slack variable for each inequality.

EXAMPLE 2

INTRODUCING SLACK VARIABLES

Write the following inequalities (from Example 1) as equations with slack variables.

$$\begin{cases} 3x_1 + 4x_2 \leq 24 \\ x_1 + 2x_2 \leq 10 \end{cases}$$

Solution

We introduce a slack variable $s_1 \geq 0$ into the first inequality and another slack variable $s_2 \geq 0$ into the second to obtain the *equations* shown in the middle below.

$$\begin{cases} 3x_1 + 4x_2 \leq 24 \\ x_1 + 2x_2 \leq 10 \end{cases} \text{become} \begin{cases} 3x_1 + 4x_2 + s_1 = 24 \\ x_1 + 2x_2 + s_2 = 10 \end{cases} \text{or} \begin{cases} 3x_1 + 4x_2 + 1s_1 + 0s_2 = 24 \\ x_1 + 2x_2 + 0s_1 + 1s_2 = 10 \end{cases}$$

The version on the right above has *all* of the variables included, with the unnecessary ones multiplied by zeros.

Slack variables are sometimes called simply *slacks*.

We can write the objective function $P = 9x_1 + 16x_2$ (from Example 1) with all variables moved to the left of the equals sign, including the slacks multiplied by zeros:

$$P = 9x_1 + 16x_2 \quad \text{becomes} \quad P - 9x_1 - 16x_2 + 0s_1 + 0s_2 = 0$$

Since both the objective function and the constraints from Example 1 are now written as equations, with the variables in the same order, the entire problem can be summarized in a table, called the "simplex tableau" (using the French word for *table*, whose plural is *tableaux*). We state the general form and then give an example.

> **Initial Tableau**
>
> The initial tableau for the linear programming problem
>
> $$\text{Maximize } P = c^t X$$
> $$\text{Subject to } \begin{cases} AX \leq b \\ X \geq 0 \end{cases} \text{ is } \begin{array}{c|cc|c} & X & S & \\ \hline S & A & I & b \\ \hline P & -c^t & 0 & 0 \end{array}$$
>
> where I stands for the identity matrix and S stands for the slack variables.

EXAMPLE 3

WRITING A SIMPLEX TABLEAU

Write the initial simplex tableau for the following linear programming problem (from Example 1):

$$\text{Maximize } P = 9x_1 + 16x_2$$
$$\text{Subject to } \begin{cases} 3x_1 + 4x_2 \leq 24 \\ x_1 + 2x_2 \leq 10 \\ x_1 \geq 0, \text{ and } x_2 \geq 0 \end{cases}$$

Solution

We write the constraints as:

$$\begin{cases} 3x_1 + 4x_2 + 1s_1 + 0s_2 = 24 \\ x_1 + 2x_2 + 0s_1 + 1s_2 = 10 \end{cases} \quad \text{From Example 2}$$

and the objective function as:

$$P - 9x_1 - 16x_2 + 0s_1 + 0s_2 = 0 \quad \text{From } P = 9x_1 + 16x_2$$

For the initial simplex tableau we write the variables on the top, the constraint numbers in the middle, and the objective numbers on the bottom (all in the correct columns), and finally the slack variables on the left:

$$\begin{array}{c|cccc|c} & x_1 & x_2 & s_1 & s_2 & \\ \hline s_1 & 3 & 4 & 1 & 0 & 24 \\ s_2 & 1 & 2 & 0 & 1 & 10 \\ \hline P & -9 & -16 & 0 & 0 & 0 \end{array}$$

Column variables

Constraints

Objective function (signs changed)

Represents =

From the original problem:

Maximize $P = 9x_1 + 16x_2$

Subject to $\begin{cases} 3x_1 + 4x_2 \leq 24 \\ x_1 + 2x_2 \leq 10 \\ x_1 \geq 0, x_2 \geq 0 \end{cases}$

4.3 THE SIMPLEX METHOD FOR STANDARD MAXIMUM PROBLEMS

Practice Problem 2

Write the initial simplex tableau for the problem

Maximize $P = 3x_1 + 5x_2$

Subject to $\begin{cases} 2x_1 - 3x_2 \leq 24 \\ 5x_1 + 2x_2 \leq 20 \\ 3x_1 + 2x_2 \leq 12 \\ x_1 \geq 0 \text{ and } x_2 \geq 0 \end{cases}$

▶ Solution on page 291

Basic and Nonbasic Variables

In an initial simplex tableau like the one below (from Example 3), the slack variables s_1 and s_2 have two special properties.

	x_1	x_2	s_1	s_2	
s_1	3	4	1	0	24
s_2	1	2	0	1	10
P	-9	-16	0	0	0

1. Their columns contain all zeros except for exactly one 1, which occurs above the bottom row.
2. Each of the rows above the bottom row has exactly one of these special 1s.

These special variables determine a feasible point in the region without any further calculation: Just give the variable the value that is at the right-hand end of its row, and set all other variables equal to 0. For the above tableau, this means:

	x_1	x_2	s_1	s_2	
s_1	3	4	1	0	24 ← $s_1 = 24$
s_2	1	2	0	1	10 ← $s_2 = 10$
P	-9	-16	0	0	0	and $x_1 = x_2 = 0$ — Other variables set equal to 0

The general procedure for determining basic and nonbasic variables in a tableau is as follows.

Basic and Nonbasic Variables in a Simplex Tableau

Count the number of slack variables (or, equivalently, the number of constraints). This number, m, will be the number of basic variables throughout the solution.

1. For basic variables, choose any m variables whose columns have all zeros except for one 1 that must appear above the

bottom row and such that each of the rows above the bottom row has exactly one of these special 1s; assign to each basic variable the value at the right-hand end of the row containing this 1.

2. All *other* variables are *nonbasic* variables; assign to each the value 0.

3. At these values, the objective function equals the number in the bottom right corner.

Notice that statements 1 and 2 agree with the values we assigned to the variables in the tableau above the box. Furthermore, the objective function $P = 9x_1 + 16x_2$ evaluated at the assigned $x_1 = 0$ and $x_2 = 0$ is $P = 0$, the number in the bottom right corner of the tableau, agreeing with statement 3 in the box. Because of their importance, we list the basic variables on the left side of the tableau, each in the row corresponding to its special 1. We interpret the tableau as specifying a basic feasible point (a vertex) of the feasible region along with the value of the objective function at that point.

The Pivot Element

The simplex method begins by taking the slack variables as the basic variables and then successively finds better basic variables to increase the value of the objective function until the solution is found. The method mimics the geometric procedure that we used on pages 264–265 to find vertices, adding and subtracting equations as we did on page 172 to solve systems of equations, a process we now call *pivoting*. We will explain the process by referring to the problem in Example 1, but explaining how to carry out the steps on the simplex tableau since we use the tableau to represent the problem.

In the objective function from Example 1, $P = 9x_1 + 16x_2$, x_1 is multiplied by 9 and x_2 is multiplied by a larger number, 16, so increasing x_2 should do more to increase P than increasing x_1. In general, to increase *any* objective function, we should increase the variable with the *largest positive* coefficient. Since the bottom row of the tableau contains the *negatives* of these coefficients, this means choosing the column with the *smallest negative* number in the bottom row. This is the *pivot column*.

Pivot Column

The *pivot column* is the column with the smallest negative entry in the bottom row (omitting the right column). If there is a tie, choose the leftmost such column.

4.3 THE SIMPLEX METHOD FOR STANDARD MAXIMUM PROBLEMS

Be careful! The smallest negative number among 12, -8, and -5 is -8. (Students sometimes call this choosing *the "most negative" number*).

Having chosen the column and therefore the variable to increase, how much can we increase it without violating a constraint? The boundary lines for the constraints in Example 1 are on the left below. Recall from our geometric method that we cannot increase x_2 beyond the *intercepts* of these boundaries, which are calculated on the right below.

Boundary line	x_2-intercept	
$3x_1 + 4x_2 = 24$	$x_2 = \frac{24}{4} = 6$	← x_2 cannot exceed 6
$x_1 + 2x_2 = 10$	$x_2 = \frac{10}{2} = 5$	← x_2 cannot exceed 5

To stay below *two* numbers means staying below the *smaller* of them, so x_2 must not exceed the *smaller* of the two numbers, each number being the right-hand side of the inequality divided by the number in the pivot column. This gives the rule for choosing the *pivot row* from the simplex tableau:

Pivot Row

For each row (except the bottom), divide the rightmost entry by the pivot column entry (omitting any row with a zero or negative pivot column entry). The row with the smallest ratio is the *pivot row*. If there is a tie, choose the uppermost such row.

We ignore rows with zeros in the pivot column (because the corresponding inequality would not include that variable), and with negative numbers (because such an inequality would not restrict the variable from increasing).

Pivot Element

The *pivot element* is the entry in the pivot column and the pivot row.

The Pivot Operation

Once we have found the pivot element, we want to increase the selected variable as much as possible (up to the boundary of the feasible region) and to increase the objective function correspondingly. This is accomplished by the pivot operation, which is just the tableau version of the elimination method that we used for solving equations on pages 159–160.

Pivot Operation

1. Divide every entry in the pivot row by the pivot element to obtain a *new* pivot row, which will then have a 1 where the pivot element was. Replace the variable at the left-hand end of this row by the variable corresponding to the pivot column. The variables listed on the left will be the new *basic* variables.

2. Subtract multiples of the new pivot row from all other rows of the tableau to get zeros in the pivot column above and below the 1.

   Specifically, if R_p is the new pivot row found from step 1 and R is any *other* row, and if we denote the entry in that row and in the pivot column by p, then the row operation in step 2 for that row is $R - p \cdot R_p \rightarrow R$.

EXAMPLE 4 FINDING THE PIVOT ELEMENT AND PIVOTING

Find the pivot element in the following simplex tableau (from Example 3) and carry out the pivot operation.

	x_1	x_2	s_1	s_2	
s_1	3	4	1	0	24
s_2	1	2	0	1	10
P	-9	-16	0	0	0

Solution

The smallest negative number in the bottom row is -16, so the pivot column is column 2 (as shown below). To find the pivot row, we divide the last entry of each row by the pivot column entry for that row (skipping the bottom row and any row with a zero or a negative pivot column entry). We choose the row with the smallest nonnegative ratio, as shown by the calculations to the right of the following tableau. The pivot row is then row 2.

	x_1	x_2	s_1	s_2	
s_1	3	4	1	0	24
s_2	1	2	0	1	10
P	-9	-16	0	0	0

$\frac{24}{4} = 6$

$\frac{10}{2} = 5 \leftarrow$ Smallest ratio $\leftarrow$ Pivot row

Pivot column — Smallest negative

Pivot element (we want a 1 here)

4.3 THE SIMPLEX METHOD FOR STANDARD MAXIMUM PROBLEMS

Having found the pivot element, we carry out the pivot operation. The first step is to divide the pivot row by the pivot element, so we divide the second row by 2, which changes the pivot element to a 1. We also update the basis on the left—the pivot column variable x_2 replaces the pivot row variable s_2, so the new basic variables are s_1 and x_2.

		x_1	x_2	s_1	s_2	
	s_1	3	4	1	0	24
$-\frac{1}{2}R_2 \to$	x_2	$\frac{1}{2}$	1	0	$\frac{1}{2}$	5
	P	-9	-16	0	0	0

x_2 replaces s_2 in the basis

Want 0s here

The second step in the pivot operation is to subtract multiples of the *new* pivot row from the other rows to "zero out" the rest of the pivot column (the 4 and the -16). This is accomplished by the row operations listed on the left of the following tableau.

		x_1	x_2	s_1	s_2		
$R_1 - 4R_2 \to$	s_1	1	0	1	-2	4	$\leftarrow s_1 = 4$
	x_2	$\frac{1}{2}$	1	0	$\frac{1}{2}$	5	$\leftarrow x_2 = 5$
$R_3 + 16R_2 \to$	P	-1	0	0	8	80	$\leftarrow P = 80$

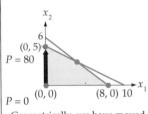

Geometrically, we have moved from the initial vertex (0, 0) to the vertex (0, 5), increasing P from 0 to 80.

This completes the pivot operation. The basic variables have the values shown on the right, with the other variables set equal to zero (in particular, $x_1 = 0$). Observe that when $x_1 = 0$ and $x_2 = 5$ the objective function $P = 9x_1 + 16x_2$ *does* take the value of 80 as given in the lower right corner of the tableau.

All of the steps we have just carried out—finding the pivot element, updating the basic variables on the left of the tableau, dividing to replace the pivot element by a 1, and zeroing out the rest of the column by subtracting multiples of the pivot row—constitute *one* pivot operation. You may want to carry out some of the calculations on scratch paper.

Notice that the tableau above still has a negative number (the -1) in the bottom row of the above tableau, meaning that there is more pivoting to be done.

EXAMPLE 5

PIVOTING A SECOND TIME

Perform the pivot operation on the following tableau (from the end of Example 4):

	x_1	x_2	s_1	s_2	
s_1	1	0	1	-2	4
x_2	$\frac{1}{2}$	1	0	$\frac{1}{2}$	5
P	-1	0	0	8	80

Solution

The -1 in the bottom row means that the pivot column is column 1. The calculations on the right below show that the pivot row is row 1.

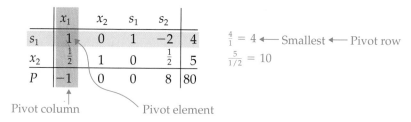

The pivot entry is already a 1, so no division is needed (but we update the basis by replacing s_1 by x_1). The two row operations listed on the left below complete the pivot operation.

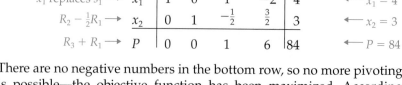

There are no negative numbers in the bottom row, so no more pivoting is possible—the objective function has been maximized. According to the box on pages 279–280, we take the values from this final tableau: P from the bottom right corner, the basic variables from the right side, with the nonbasic variables set to 0:

The maximum value of P is 84, when $x_1 = 4$ and $x_2 = 3$.

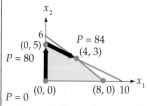

Geometrically, we have moved from (0, 5) to the vertex (4, 3), increasing P from 80 to 84.

Usually only the x-values are given in the solution, since only they were stated in the original problem. The nonbasic variables here take the values $s_1 = 0$ and $s_2 = 0$. In Example 7 we will see how to interpret the slack variables as "unused resources."

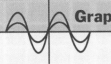

Graphing Calculator Exploration

The program* PIVOT carries out the pivot operation after you specify the pivot column and pivot row. To carry out the pivot operation with your calculator on the simplex tableau from Example 5, proceed as follows:

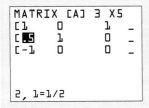

a. Enter the simplex tableau as one large matrix [A] having 3 rows and 5 columns. Because this matrix is too large to fit on the calculator screen, it will scroll from left to right and back again as you enter your numbers.

b. Run the program PIVOT. It will display the current tableau in matrix [A] and you can use the arrows to scroll the screen to see the rest of it. Press ENTER to enter the **Pivot Column** and then again to enter the **Pivot Row** of your pivot element.

c. After you ENTER the pivot row, the calculator will perform the pivot operation and display the new tableau. You can use the arrows to scroll the screen to see the rest of it. Press ENTER to exit the program.

If your problem requires several pivot operations, you can rerun the program with the new tableau by pressing ENTER again.

In general, we continue to pivot until no more pivoting is possible, that is, until we arrive at a *final tableau*.

How do we recognize and interpret a final tableau? Recall that choosing the smallest negative number in the bottom row was equivalent to choosing the variable that will increase the objective function most quickly. Therefore, if there is *no* negative number in the bottom row, then there is no variable that will increase the objective function, so P must be at its maximum value, and we read the solution from the tableau as we did in the preceding example. On the other hand, if there *is* a negative number in the bottom row (so the objective function

---

* See the Preface for information on how to obtain this and other programs.

can be increased) but there is *no* pivot row, then there are no constraints limiting the objective function, which can therefore be increased arbitrarily, meaning that there is *no solution*.

> **Interpreting a Final Tableau**
>
> 1. If all the numbers in the bottom row are positive or zero (so there is no pivot column), then the original problem *does* have a solution:
>   a. The (basic) variables listed on the left of the rows take the values at the right ends of those rows.
>   b. The (nonbasic) variables not listed on the left take the value 0.
>   c. The maximum value of the objective function is in the lower right corner of the tableau.
> 2. If there *is* a negative number in the bottom row but every other element in its column is negative or zero (so there is a pivot column but no pivot row), then the original problem has *no solution*.

EXAMPLE 6   INTERPRETING A FINAL TABLEAU

Interpret the following final tableaux:

a.

	x_1	x_2	s_1	s_2	
s_1	2	0	1	3	4
x_2	5	1	0	6	7
P	8	0	0	9	10

b.

	x_1	x_2	s_1	s_2	
s_1	-2	0	1	3	4
x_2	0	1	0	6	7
P	-8	0	0	9	10

Solution

a. The first tableau has no negative numbers in the bottom row, so the original problem *does* have a solution. The basic variables listed on the left, s_1 and x_2, take the values on the right, 4 and 7; the other variables, s_2 and x_1, take the value 0. The value of P is in the lower right corner:

The maximum value of P is 10, when $x_1 = 0$ and $x_2 = 7$.

b. The second tableau *does* have a negative number in the bottom row, -8, so there *is* a pivot column (column 1). However, there is no pivot *row* (the ratios are $\frac{4}{-2}$ which is negative and $\frac{7}{0}$ which is undefined). Therefore, there is *no solution*.

The Simplex Method

The simplex method proceeds by pivoting until a final tableau is reached, and then interpreting it.

> **Simplex Method**
>
> To solve a standard maximum problem (page 275) by the simplex method:
> 1. Construct the simplex tableau (see page 278).
> 2. Locate the pivot element (see pages 280–281) and go to step 3. If the tableau does not have a pivot element, go to step 4.
> 3. Perform the pivot operation (see page 282) using the pivot element and return to step 2.
> 4. If the final tableau does not have a pivot column (see page 280), then the solution occurs at the vertex where the (basic) variables listed on the left of the tableau take the values at the right-hand ends of those rows; the other (nonbasic) variables take the value 0. The maximum value of the objective function is in the bottom right corner of the tableau.
>   If the final tableau has a pivot *column* but no pivot *row* (see page 281), then the problem has *no solution*.

EXAMPLE 7 A MANUFACTURING PROBLEM WITH MANY VARIABLES

A pottery shop manufactures dinnerware in four different patterns by shaping the clay, decorating it, and then kiln-firing it. The numbers of hours required per place setting for the four designs are given in the following table, together with the expected profits. The shop employs two skilled workers to do the initial shaping and three artists to do the decorating, none of whom will work more than 40 hours each week. The kiln can be used no more than 55 hours per week. How many place settings of each pattern should be made this week to obtain the greatest possible profit?

	Classic	Modern	Art Deco	Floral
Shaping	2 hours	1 hour	4 hours	2 hours
Decorating	3 hours	1 hour	6 hours	4 hours
Kiln-firing	1 hour	1 hour	1 hour	1 hour
Expected profit	$10	$6	$9	$8

Solution

Let

$$x_1 = \begin{pmatrix} \text{Number of} \\ \text{Classic settings} \end{pmatrix}, \qquad x_2 = \begin{pmatrix} \text{Number of} \\ \text{Modern settings} \end{pmatrix},$$

$$x_3 = \begin{pmatrix} \text{Number of} \\ \text{Art Deco settings} \end{pmatrix}, \quad \text{and} \quad x_4 = \begin{pmatrix} \text{Number of} \\ \text{Floral settings} \end{pmatrix}$$

made during the week, so $x_1 \geq 0$, $x_2 \geq 0$, $x_3 \geq 0$, and $x_4 \geq 0$. From the last line in the table, the profit to be maximized is

$$P = 10x_1 + 6x_2 + 9x_3 + 8x_4.$$

The constraints on the time spent shaping, decorating, and firing come from the other numbers in the table, along with the time available for each:

$$2x_1 + x_2 + 4x_3 + 2x_4 \leq 80 \qquad \text{2 shapers @ 40 hours each}$$
$$3x_1 + x_2 + 6x_3 + 4x_4 \leq 120 \qquad \text{3 decorators @ 40 hours each}$$
$$x_1 + x_2 + x_3 + x_4 \leq 55 \qquad \text{Time kiln can be used}$$

The initial simplex tableau is

	x_1	x_2	x_3	x_4	s_1	s_2	s_3	
s_1	②	1	4	2	1	0	0	80
s_2	3	1	6	4	0	1	0	120
s_3	1	1	1	1	0	0	1	55
P	−10	−6	−9	−8	0	0	0	0

$\frac{80}{2} = 40 \longleftarrow$ Smallest ratio $\longleftarrow$ Pivot row
$\frac{120}{3} = 40$ (Tie for smallest ratio, so take top row)
$\frac{55}{1} = 55$

↑ Smallest negative — Pivot element

Pivot column ↗

When we pivot (on the 2 in column 1, row 1), the tableau becomes

	x_1	x_2	x_3	x_4	s_1	s_2	s_3	
x_1	1	1/2	2	1	1/2	0	0	40
s_2	0	−1/2	0	1	−3/2	1	0	0
s_3	0	①/2	−1	0	−1/2	0	1	15
P	0	−1	11	2	5	0	0	400

x_1 replaces s_1 in basis

$\frac{40}{1/2} = 80$
(Cannot use since $-\frac{1}{2}$ is < 0)
$\frac{15}{1/2} = 30 \longleftarrow$ Smallest ratio

Pivot row ↗

↑ Smallest negative — Pivot element

Pivot column ↗

Pivoting again (on the 1/2 in column 2, row 3), we reach the final tableau:

	x_1	x_2	x_3	x_4	s_1	s_2	s_3	
x_1	1	0	3	1	1	0	−1	25
s_2	0	0	−1	1	−2	1	1	15
x_2	0	1	−2	0	−1	0	2	30
P	0	0	9	2	4	0	2	430

x_2 replaces s_3 in the basis →

No negatives in bottom row so final tableau

The basic variables take the values at the right ends of their rows, the nonbasic variables are zero, and the objective function value appears in the bottom right corner. Therefore, the maximum value of P is 430 when $x_1 = 25$, $x_2 = 30$, $x_3 = 0$, and $x_4 = 0$. The fact that $s_2 = 15$ means that 15 hours of the available decorating time will not be needed.

Answer: The pottery shop should manufacture 25 place settings of Classic, 30 of Modern, and none of Art Deco and Floral.

How Good Is the Simplex Method?

Will the simplex method always find the solution? Even if it does find the solution, will it get to the answer after a reasonable number of tableaux?

In 1955, E. M. I. Beale published a linear programming problem with the awful property that the simplex method found the same sequence of tableaux over and over again. This repetition of tableaux is known as "cycling" because the tableaux repeat in a cyclic pattern. If a computer were solving this problem, it would never stop. Several examples of cycling and changes to the simplex method that prevent cycling are given in Exercises 42–46.

In 1972, V. Klee and G. J. Minty described a class of problems that take many more tableaux than their small size would suggest: With each new variable, the number of tableaux doubles. This means that a Klee–Minty problem with 2 variables takes 4 tableaux to solve, one with 3 variables takes 8 tableaux, one with 4 variables takes 16 tableaux, and so on. The 10-variable problem in this sequence will take 1024 tableaux with larger problems becoming even more unmanageable. Examples of these kinds of problems are given in Exercises 47–50.

Most "real world" computer programs for solving linear programming problems include subroutines to protect against cycling and

Klee–Minty situations. Stephen Smale has shown that "... the number of pivots required to solve a linear programming problem grows in proportion to the number of variables on average." Several other methods (by L. Khachiyan and N. Karmarkar) to solve linear programming problems guarantee that the number of steps required does not grow too fast for every problem. However, the simplex method continues to provide the fastest solutions to commercial problems.

4.3 Section Summary

The maximum linear programming problem

$$\text{Maximize } P = c^t X$$
$$\text{Subject to } \begin{cases} AX \leq b \\ X \geq 0 \end{cases}$$

is a *standard maximum problem* if $b \geq 0$. The *initial simplex tableau* represents this problem in the form

	X	S	
S	A	I	b
P	$-c^t$	0	0

The variables listed on the far left form a *basis*, and setting them equal to the values on the far right of their rows and setting the nonbasic variables equal to zero gives a *basic feasible point* that is a vertex of the region. The pivot column is the column with the smallest negative entry in the bottom row (omitting the right column), and the pivot row is the row with the smallest ratio between the last entry and the pivot column entry (omitting any row with a zero or negative pivot column entry, and the bottom row). The *pivot element* is the entry in the pivot column and pivot row.

The *pivot operation* divides the pivot row by the pivot element and then subtracts from each of the other rows the pivot column entry of that row times the new pivot row. The pivot column variable enters the basis, replacing the variable of the pivot row. The *simplex method* solves a standard maximum problem by pivoting until the tableau does not have a pivot element. If there is no pivot *column*, then the solution occurs at the vertex given by the basic variables, and the maximum value of the objective function appears in the bottom right corner. If there is a pivot column but no pivot *row*, then there is no solution.

The simplex method uses the pivot operation to make the bottom row "less negative" while keeping the right column nonnegative. That is, *the simplex method makes the tableau "more nearly optimal" while keep-*

4.3 THE SIMPLEX METHOD FOR STANDARD MAXIMUM PROBLEMS

ing it feasible. When the bottom row contains no negative entries, the tableau is *optimal* because the largest value of the objective function has been found.

▶ Solutions to Practice Problems

1. No. The -4 on the right of the second inequality violates the $b \geq 0$ condition in the box on page 275. (The -3 in the first inequality is not a violation: only the right-hand sides need to be nonnegative.)

2.

	x_1	x_2	s_1	s_2	s_3	
s_1	2	-3	1	0	0	24
s_2	5	2	0	1	0	20
s_3	3	2	0	0	1	12
P	-3	-5	0	0	0	0

4.3 Exercises

Construct the initial simplex tableau for each standard maximum problem.

1. Maximize $P = 8x_1 + 9x_2$

   Subject to $\begin{cases} 3x_1 + 2x_2 \leq 12 \\ 6x_1 + x_2 \leq 15 \\ x_1 \geq 0, \ x_2 \geq 0 \end{cases}$

2. Maximize $P = 10x_1 + 15x_2$

   Subject to $\begin{cases} 5x_1 + 2x_2 \leq 30 \\ 3x_1 + 4x_2 \leq 36 \\ x_1 \geq 0, \ x_2 \geq 0 \end{cases}$

3. Maximize $P = 13x_1 + 7x_2$

   Subject to $\begin{cases} 4x_1 + 3x_2 \leq 12 \\ 5x_1 + 2x_2 \leq 20 \\ x_1 + 6x_2 \leq 12 \\ x_1 \geq 0, \ x_2 \geq 0 \end{cases}$

4. Maximize $P = 8x_1 + 33x_2$

   Subject to $\begin{cases} 3x_1 + 2x_2 \leq 30 \\ x_1 + 4x_2 \leq 20 \\ 5x_1 + 6x_2 \leq 60 \\ x_1 \geq 0, \ x_2 \geq 0 \end{cases}$

5. Maximize $P = 5x_1 - 2x_2 + 10x_3 - 5x_4$

   Subject to $\begin{cases} 2x_1 + x_2 + x_3 + 3x_4 \leq 6 \\ x_1 + 4x_2 - 2x_3 + x_4 \leq 8 \\ x_1 \geq 0, \ x_2 \geq 0, \ x_3 \geq 0, \ x_4 \geq 0 \end{cases}$

6. Maximize $P = 5x_1 + 10x_2 - 30x_3 + 5x_4$

   Subject to $\begin{cases} 3x_1 + x_2 - 2x_3 + 4x_4 \leq 12 \\ 2x_1 - x_2 + 3x_3 + x_4 \leq 12 \\ x_1 \geq 0, \ x_2 \geq 0, \ x_3 \geq 0, \ x_4 \geq 0 \end{cases}$

7. Maximize $P = 10x_1 + 20x_2 + 15x_3$

   Subject to $\begin{cases} 8x_1 + x_2 + 4x_3 \leq 32 \\ 3x_1 + 5x_2 + 7x_3 \leq 30 \\ 6x_1 + 2x_2 + 9x_3 \leq 28 \\ x_1 \geq 0, \ x_2 \geq 0, \ x_3 \geq 0 \end{cases}$

8. Maximize $P = 10x_1 + 15x_2 + 5x_3$

   Subject to $\begin{cases} 7x_1 + 3x_2 + x_3 \leq 21 \\ 6x_1 + 5x_2 + 4x_3 \leq 20 \\ 8x_1 + 9x_2 + 2x_3 \leq 45 \\ x_1 \geq 0, \ x_2 \geq 0, \ x_3 \geq 0 \end{cases}$

For each simplex tableau, find the pivot element and carry out one complete pivot operation. If there is no pivot element, explain what the tableau shows about the solution of the original standard maximum problem.

9.
	x_1	x_2	s_1	s_2	
s_1	2	0	1	0	4
s_2	1	1	0	1	5
P	-7	-8	0	0	0

10.
	x_1	x_2	s_1	s_2	
s_1	1	-2	1	0	5
s_2	3	1	0	1	6
P	-4	-5	0	0	0

11.
	x_1	x_2	x_3	x_4	s_1	s_2	
s_1	4	2	6	2	1	0	12
s_2	3	1	2	1	0	1	8
P	-4	-5	6	-3	0	0	0

12.
	x_1	x_2	x_3	x_4	s_1	s_2	
s_1	5	3	1	3	1	0	30
s_2	6	2	2	4	0	1	16
P	-4	-6	-5	7	0	0	0

13.
	x_1	x_2	x_3	x_4	s_1	s_2	s_3	
s_1	0	0	2	1	1	1	-1	10
x_2	0	1	1	1	0	1	0	15
x_1	1	0	-2	0	0	-1	1	10
P	0	0	0	2	0	1	3	90

14.
	x_1	x_2	x_3	x_4	s_1	s_2	s_3	
x_2	0	1	-1	0	1	-1	0	20
x_4	1	0	1	1	0	1	0	40
s_3	1	0	2	0	-1	1	1	30
P	2	0	8	0	3	3	0	300

15.
	x_1	x_2	s_1	s_2	s_3	
s_1	0	0	1	-1	0	10
x_1	1	0	0	1	-1	10
x_2	0	1	0	1	0	20
P	0	0	0	25	-10	400

16.
	x_1	x_2	s_1	s_2	s_3	
s_1	0	0	1	0	-1	5
x_2	0	1	0	1	-1	10
x_1	1	0	0	1	0	20
P	0	0	0	35	-15	550

Solve each problem by the simplex method. (Exercises 17, 18, 25, and 26 can also be solved by the graphical method.)

 You may use the PIVOT program if permitted by your instructor.

17. Maximize $P = x_1 + 2x_2$

Subject to $\begin{cases} 3x_1 + x_2 \le 24 \\ x_1 + x_2 \le 14 \\ x_1 \ge 0, \; x_2 \ge 0 \end{cases}$

18. Maximize $P = x_1 + 3x_2$

Subject to $\begin{cases} 2x_1 + x_2 \le 24 \\ x_1 + x_2 \le 15 \\ x_1 \ge 0, \; x_2 \ge 0 \end{cases}$

19. Maximize $P = 30x_1 + 40x_2 + 15x_3$

Subject to $\begin{cases} 2x_1 + x_2 + 3x_3 \le 150 \\ 3x_1 + 2x_2 + x_3 \le 100 \\ x_1 \ge 0, \; x_2 \ge 0, \; x_3 \ge 0 \end{cases}$

20. Maximize $P = 30x_1 + 60x_2 + 15x_3$

Subject to $\begin{cases} 2x_1 + 3x_2 + x_3 \le 75 \\ 3x_1 + x_2 + 2x_3 \le 50 \\ x_1 \ge 0, \; x_2 \ge 0, \; x_3 \ge 0 \end{cases}$

21. Maximize $P = 6x_1 + 4x_2 + 5x_3$

Subject to $\begin{cases} x_1 - x_2 + x_3 \le 20 \\ x_1 + 2x_3 \le 10 \\ x_1 \ge 0, \; x_2 \ge 0, \; x_3 \ge 0 \end{cases}$

22. Maximize $P = 6x_1 + 5x_2 + 8x_3$

Subject to $\begin{cases} 2x_1 + x_3 \le 40 \\ x_1 - x_2 + x_3 \le 30 \\ x_1 \ge 0, \; x_2 \ge 0, \; x_3 \ge 0 \end{cases}$

23. Maximize $P = 4x_1 + 3x_2 - 5x_3 + 6x_4$

Subject to $\begin{cases} x_1 + x_2 + x_4 \le 60 \\ x_1 + x_3 + x_4 \le 40 \\ x_1 + x_2 + x_3 \le 50 \\ x_1 \ge 0, \; x_2 \ge 0, \; x_3 \ge 0, \; x_4 \ge 0 \end{cases}$

24. Maximize $P = 2x_1 + 3x_2 + 4x_3 + x_4$

Subject to $\begin{cases} x_1 + x_2 + x_4 \le 20 \\ x_1 + x_3 + x_4 \le 15 \\ x_1 + x_2 + x_3 \le 25 \\ x_1 \ge 0, \; x_2 \ge 0, \; x_3 \ge 0, \; x_4 \ge 0 \end{cases}$

25. Maximize $P = 4x_1 + 2x_2$

Subject to $\begin{cases} x_1 - x_2 \leq 1 \\ -x_1 + x_2 \leq 3 \\ x_1 + x_2 \leq 5 \\ x_1 \geq 0, \ x_2 \geq 0 \end{cases}$

26. Maximize $P = 12x_1 + 6x_2$

Subject to $\begin{cases} x_1 + x_2 \leq 8 \\ x_1 - x_2 \leq 2 \\ -x_1 + 2x_2 \leq 10 \\ x_1 \geq 0, \ x_2 \geq 0 \end{cases}$

27. Maximize $P = 40x_1 + 60x_2 + 50x_3$

Subject to $\begin{cases} 2x_1 + x_2 + 4x_3 \leq 400 \\ 4x_1 + 3x_2 + 2x_3 \leq 600 \\ x_1 \geq 0, \ x_2 \geq 0, \ x_3 \geq 0 \end{cases}$

28. Maximize $P = 140x_1 + 80x_2 + 100x_3$

Subject to $\begin{cases} 4x_1 + 2x_2 + x_3 \leq 100 \\ 2x_1 + 3x_2 + 3x_3 \leq 150 \\ x_1 \geq 0, \ x_2 \geq 0, \ x_3 \geq 0 \end{cases}$

29. Maximize $P = x_1 + 2x_2 + 3x_3$

Subject to $\begin{cases} x_1 + x_2 + x_3 \leq 15 \\ x_2 + x_3 \leq 10 \\ x_3 \leq 5 \\ x_1 \geq 0, \ x_2 \geq 0, \ x_3 \geq 0 \end{cases}$

30. Maximize $P = 15x_1 + 10x_2 + 5x_3$

Subject to $\begin{cases} x_1 + x_2 + x_3 \leq 3 \\ x_1 + x_2 \leq 2 \\ x_1 \leq 1 \\ x_1 \geq 0, \ x_2 \geq 0, \ x_3 \geq 0 \end{cases}$

APPLIED EXERCISES

Formulate each situation as a linear programming problem by identifying the variables, the objective function, and the constraints. Be sure to state clearly the meaning of each variable. Check that the problem is a standard maximum problem and then solve it by the simplex method. State your final answer in terms of the original question.

You may use the PIVOT program if permitted by your instructor.

31. BUSINESS: Production Planning An automotive parts shop rebuilds carburetors, fuel pumps, and alternators. The numbers of hours to rebuild and then inspect and pack each part are shown in the following table, together with the number of hours of skilled labor available for each task. If the profit is $12 for each carburetor, $14 for each fuel pump, and $10 for each alternator, how many of each should the shop rebuild to obtain the greatest possible profit?

	Rebuilding	Inspection & Packaging
Carburetor	5 hours	1 hour
Fuel Pump	4 hours	1 hour
Alternator	3 hours	0.5 hour
Labor Available	200 hours	45 hours

32. ENVIRONMENTAL SCIENCE: Pollution Control A storage yard at a coal-burning electric power plant can hold no more than 100,000 tons of coal. Two grades of coal are available: low-sulfur (1%) with an energy content of 20 million British thermal units (BTU) per ton and high-sulfur (2%) with an energy content of 30 million BTU per ton. If the next coal purchase may contain no more than 1400 tons of sulfur, how many tons of each type of coal should be purchased to obtain the most energy?

33. ENVIRONMENTAL SCIENCE: Recycling Management A volunteer recycle center accepts both used paper and empty glass bottles, which are then sorted and sold to a reprocessing company. The center has room to accept 800 crates of paper and glass each week and 50 hours of volunteer help to do the sorting. Each crate of paper products takes 5 minutes to sort and sells for 8¢, while each crate of bottles takes 3 minutes to sort and sells for 7¢. How many crates of each should the center accept each week to raise the most money for its ecology scholarship fund?

34. BUSINESS: Production Planning A small jewelry company prepares and mounts semiprecious stones. There are 20 lapidaries (who cut and polish the stones) and 24 jewelers (who mount the stones in gold settings). Each employee works 7 hours each day. Each tray of agates requires 5 hours of cutting and polishing and 4 hours to mount, each tray of onyxes requires 2 hours of cutting and polishing and 3 hours to mount, and each tray of garnets requires 6 hours of cutting and polishing and 3 hours to mount. If the profit is $15 for each tray of agates, $10 for each tray of onyxes, and $12 for each tray of garnets, how many trays of each stone should be processed each day to obtain the greatest possible profit?

35. BUSINESS: Production Planning Repeat the previous problem but with a profit of $13 per tray of garnets.

36. BUSINESS: Resource Allocation A furniture shop manufactures wooden desks, tables, and chairs. The numbers of hours to assemble and finish each piece are shown in the table, together with the numbers of hours of skilled labor available for each task. If the profit is $80 for each desk, $84 for each table, and $68 for each chair, how many of each should the shop manufacture to obtain the greatest possible profit?

	Assembly	Finishing
Desk	2 hours	2 hours
Table	1 hour	3 hours
Chair	2 hours	1 hour
Labor Available	200 hours	150 hours

37. SOCIAL SCIENCE: Political Advertising In the last few days before the election, a politician can afford to spend no more than $27,000 on TV advertisements and can arrange for no more than 10 ads. Each daytime ad costs $2000 and reaches 4000 viewers, each prime-time ad costs $3000 and reaches 5000 viewers, and each late-night ad costs $1000 and reaches 2000 viewers. Ignoring repeated viewings by the same person and assuming every viewer can vote, how many ads in each of the time periods will reach the most voters?

38. BUSINESS: Advertising The manager of a new mall may spend up to $18,000 on grand opening announcements in newspapers, on radio, and on TV. Each newspaper ad costs $300 and reaches 6000 readers, each one-minute radio commercial costs $800 and is heard by 10,000 listeners, and each 15-second TV spot costs $900 and is seen by 11,000 viewers. If there is time to arrange for no more than 5 newspaper ads and no more than 20 radio commercials and TV spots combined, how many of each should be placed to reach the largest number of potential customers? (Ignore multiple exposures to the same consumer.)

39. BUSINESS: Agriculture A farmer grows corn, peanuts, and soybeans on his 240-acre farm. To maintain soil fertility, the farmer rotates the crops and always plants at least as many acres of soybeans as the total acres of the other crops. Each acre of corn requires 2 days of labor and yields a profit of $150, each acre of peanuts requires 5 days of labor and yields a profit of $300, and each acre of soybeans requires 1 day of labor and yields a profit of $100. If the farmer and his family can put in at most 630 days of labor, how many acres of each crop should the farmer plant to obtain the greatest possible profit?

40. BUSINESS: Agriculture A farmer grows wheat, barley, and oats on her 500-acre farm. Each acre of wheat requires 3 days of labor (to plant, tend, and harvest) and costs $21 (for seed, fertilizer, and pesticides), each acre of barley requires 2 days of labor and costs $27, and each acre of oats requires 3 days of labor and costs $24. The farmer and her hired field hands can provide no more than 1200 days of labor this year, and she can afford to spend no more than $15,120. If the expected profit is $50 for each acre of wheat, $40 for each acre of barley, and $45 for each acre of oats, how many acres of each crop should she grow to obtain the greatest possible profit?

Explorations and Excursions

The following problems extend and augment the material presented in the text.

 The PIVOT program may be helpful (if permitted by your instructor).

More About the Pivot Operation

41. Show that pivoting on a tableau obeys the diagram

$$\begin{pmatrix} \vdots & \vdots \\ \cdots & c & q & \cdots \\ \cdots & p & r & \cdots \\ \vdots & \vdots \end{pmatrix} \longrightarrow \begin{pmatrix} \vdots & \vdots \\ \cdots & 0 & q - \dfrac{rc}{p} & \cdots \\ \cdots & 1 & r/p & \cdots \\ \vdots & \vdots \end{pmatrix}$$

where p is the pivot element, c is any other entry in the pivot column, r is any other entry in the pivot row, and q is any entry in the tableau not in the pivot row or pivot column. In words, this diagram defines the pivot operation in four steps: "(1) the pivot element becomes 1, (2) the rest of the pivot column becomes 0, (3) the rest of the pivot row is divided by the pivot element, and (4) every entry not in the pivot row or pivot column is decreased by the product of the pivot column entry in the same row with the pivot row entry in the same column divided by the pivot element." This form of the pivot operation makes it easy to check any particular number in a new tableau without repeating the row operations.

Cycling

42. Beale's Cycling Example Attempt to solve the following problem by the simplex method:

Maximize $P = (3/4 \ \ -20 \ \ 1/2 \ \ -6) \begin{pmatrix} x_1 \\ x_2 \\ x_3 \\ x_4 \end{pmatrix}$

Subject to
$\begin{cases} \begin{pmatrix} 1/4 & -8 & -1 & 9 \\ 1/2 & -12 & -1/2 & 3 \\ 0 & 0 & 1 & 0 \end{pmatrix} \begin{pmatrix} x_1 \\ x_2 \\ x_3 \\ x_4 \end{pmatrix} \leq \begin{pmatrix} 0 \\ 0 \\ 1 \end{pmatrix} \\ \text{and } \begin{pmatrix} x_1 \\ x_2 \\ x_3 \\ x_4 \end{pmatrix} \geq 0 \end{cases}$

You should return to the initial simplex tableau after pivoting at [column 1, row 1], [column 2, row 2], [column 3, row 1], [column 4, row 2], [column 5, row 1], and [column 6, row 2].

43. Rescaling Changing the size of the numbers in a problem without altering the feasible region can prevent cycling. Multiply the first constraint in Beale's cycling example (Exercise 42) by 4 to get the equivalent problem

Maximize $P = (3/4 \ \ -20 \ \ 1/2 \ \ -6) \begin{pmatrix} x_1 \\ x_2 \\ x_3 \\ x_4 \end{pmatrix}$

Subject to
$\begin{cases} \begin{pmatrix} 1 & -32 & -4 & 36 \\ 1/2 & -12 & -1/2 & 3 \\ 0 & 0 & 1 & 0 \end{pmatrix} \begin{pmatrix} x_1 \\ x_2 \\ x_3 \\ x_4 \end{pmatrix} \leq \begin{pmatrix} 0 \\ 0 \\ 1 \end{pmatrix} \\ \text{and } \begin{pmatrix} x_1 \\ x_2 \\ x_3 \\ x_4 \end{pmatrix} \geq 0 \end{cases}$

You should reach the final simplex tableau after pivoting at [column 1, row 1], [column 2, row 2], [column 3, row 1], [column 4, row 2], [column 1, row 3], and [column 5, row 2].

44. Bland's Rule Eliminate cycling by changing the choice of the pivot column to be *the leftmost column with a negative entry in the bottom row* (that is, ignore the size of the numbers and just look for the first negative). Use Bland's rule to solve Beale's cycling example (Exercise 42). You should reach the final simplex tableau after pivoting at [column 1, row 1], [column 2, row 2], [column 3, row 1], [column 4, row 2], [column 1, row 3], and [column 5, row 2].

45. A Nondeterministic Simplex Algorithm The simplex method (page 287) is a *deterministic* procedure to solve standard maximum problems because once the process is started, the rules *always* select the *same* sequence of pivot elements. If we allow *chance* to play a role in the selection of the pivot column or pivot row, the resulting solution method is nondeterministic in that the sequence of pivot elements in the solution is no longer determined but rather may change from one solution attempt to another. Let us change the rule on page 281 for the selection of the pivot row to state that *if there is a tie for the smallest ratio between two of the rows, the pivot row is to be selected by a coin flip.* Solve Beale's cycling example (Exercise 42) several times using this modified procedure. Do your solutions all use the same number of pivot operations?

46. **More Cycling Examples** A large class of cyclic examples was found by K. T. Marshall and J. W. Suurballe in 1969.

   a. Try to solve this typical Marshall–Suurballe problem by the simplex method.

$$\text{Maximize } P = (1 \quad -28 \quad -2 \quad -2)\begin{pmatrix} x_1 \\ x_2 \\ x_3 \\ x_4 \end{pmatrix}$$

Subject to
$$\begin{pmatrix} 1/2 & -18 & -2 & 4 \\ 1/4 & -5 & -1/2 & 1/2 \end{pmatrix}\begin{pmatrix} x_1 \\ x_2 \\ x_3 \\ x_4 \end{pmatrix} \leq \begin{pmatrix} 0 \\ 0 \end{pmatrix}$$

and $\begin{pmatrix} x_1 \\ x_2 \\ x_3 \\ x_4 \end{pmatrix} \geq 0$

You should return to the initial simplex tableau after pivoting at [column 1, row 1], [column 2, row 2], [column 3, row 1], [column 4, row 2], [column 5, row 1], and [column 6, row 2].

   b. Rescale the constraints of the problem in part (a) by multiplying the first by 2 and the second by 4. How many pivots does it now take to solve this rescaled problem?

   c. Solve the problem in part (a) using Bland's rule (see Exercise 44).
   d. Solve the problem in part (a) using the nondeterministic simplex algorithm given in Exercise 45.

Klee–Minty Problems

47. Solve the following two-variable problem by the simplex method. (Your fourth tableau should be the final tableau.)

$$\text{Maximize } P = (1 \quad 10)\begin{pmatrix} x_1 \\ x_2 \end{pmatrix}$$

Subject to
$$\begin{pmatrix} 0 & 1 \\ 1 & 20 \end{pmatrix}\begin{pmatrix} x_1 \\ x_2 \end{pmatrix} \leq \begin{pmatrix} 1 \\ 100 \end{pmatrix}$$

and $\begin{pmatrix} x_1 \\ x_2 \end{pmatrix} \geq 0$

48. Solve the following three-variable problem by the simplex method. (Your eighth tableau should be the final tableau.)

$$\text{Maximize } P = (1 \quad 10 \quad 100)\begin{pmatrix} x_1 \\ x_2 \\ x_3 \end{pmatrix}$$

Subject to
$$\begin{pmatrix} 0 & 0 & 1 \\ 0 & 1 & 20 \\ 1 & 20 & 200 \end{pmatrix}\begin{pmatrix} x_1 \\ x_2 \\ x_3 \end{pmatrix} \leq \begin{pmatrix} 1 \\ 100 \\ 10{,}000 \end{pmatrix}$$

and $\begin{pmatrix} x_1 \\ x_2 \\ x_3 \end{pmatrix} \geq 0$

49. Following the form of Exercises 47 and 48, write down a problem using four variables that will continue the pattern and require sixteen tableaux to solve.

50. Using the problems from Exercises 47–49, check that each of this family of Klee–Minty problems can be solved in one pivot by using Bland's rule (see Exercise 44) to choose the pivot column.

4.4 STANDARD MINIMUM PROBLEMS AND DUALITY

Introduction

In this section we will define and solve *standard minimum* linear programming problems. Rather than develop a new technique for minimum problems, we will find a relationship between minimum and maximum problems that allows us to solve one by solving the other.

Standard Minimum Problems

A *standard minimum problem* is a linear programming problem in which the objective function is to be minimized, all of the coefficients in the objective function are nonnegative, the variables $y_1, y_2, \ldots, y_n$ are all nonnegative, and the other constraints have $\geq$ inequalities.

> **Standard Minimum Problem**
>
> A problem is a *standard minimum problem* if it can be written:
>
> Minimize an objective function
>
> $C = b_1 y_1 + b_2 y_2 + \cdots + b_n y_n$ with *nonnegative* $b_1, b_2, \ldots, b_n$
>
> Subject to inequalities of the form
>
> $$a_1 y_1 + a_2 y_2 + \cdots + a_n y_n \geq c$$
>
> and
>
> $$y_1 \geq 0, \ y_2 \geq 0, \ldots, y_n \geq 0$$

We use C for the objective function to suggest cost, which is usually minimized, and y-variables to distinguish them from the x-variables used in maximum problems. Notice that in a minimum problem it is the *coefficients of the objective function* that must be nonnegative—the right-hand sides of the constraints may be of either sign.

EXAMPLE 1

VERIFYING STANDARD MINIMUM FORM

Is the following a standard minimum problem?

$$\text{Minimize} \quad C = 24 y_1 + 10 y_2$$

$$\text{Subject to} \quad \begin{cases} 3 y_1 + y_2 \geq 9 \\ 4 y_1 + 2 y_2 \geq 16 \\ y_1 \geq 0 \text{ and } y_2 \geq 0 \end{cases}$$

Solution

The objective function is to be minimized, the coefficients 24 and 10 in the objective function are nonnegative, the variables are nonnegative, and the other inequalities all have $\geq$ inequalities, so this *is* a standard minimum problem.

Practice Problem 1

Is the following a standard minimum problem?

Minimize $C = 24y_1 - 4y_2$

Subject to $\begin{cases} 2y_1 - 5y_2 \geq -10 \\ 3y_1 + 2y_2 \geq 6 \\ y_1 \geq 0 \quad \text{and} \quad y_2 \geq 0 \end{cases}$

▶ Solution on page 310

The Dual of a Standard Minimum Problem

Given a standard minimum problem, we want to create from it a particular maximum problem, called the *dual problem* of the original, which will help us to solve the original minimum problem. For this purpose, recall that the *transpose* of a matrix turns every row into a column, or equivalently, every column into a row (see page 199). Denoting a transpose by a superscripted *t*, we have, for example:

$$\begin{pmatrix} 1 & 2 & 3 \\ 4 & 5 & 6 \end{pmatrix}^t = \begin{pmatrix} 1 & 4 \\ 2 & 5 \\ 3 & 6 \end{pmatrix}$$

To create the dual, we summarize the minimum problem by writing its numbers in a matrix (with the objective function numbers in the bottom row), transpose the matrix, and then reinterpret the matrix as a maximum problem (with $\leq$ inequalities).

EXAMPLE 2

FINDING A DUAL PROBLEM

Find the dual of the following problem:

Minimize $C = 24y_1 + 10y_2$

Subject to $\begin{cases} 3y_1 + y_2 \geq 9 \\ 4y_1 + 2y_2 \geq 16 \\ y_1 \geq 0 \quad \text{and} \quad y_2 \geq 0 \end{cases}$

4.4 STANDARD MINIMUM PROBLEMS AND DUALITY

Solution

Beginning with the original problem on the left below, we summarize, transpose, and rewrite as a maximum problem as shown directly below the original.

Minimize $C = 24y_1 + 10y_2$

Subject to $\begin{cases} 3y_1 + y_2 \geq 9 \\ 4y_1 + 2y_2 \geq 16 \\ y_1 \geq 0 \text{ and } y_2 \geq 0 \end{cases}$

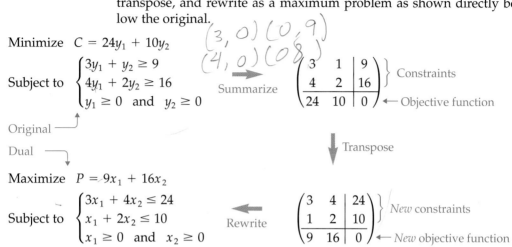

$\begin{pmatrix} 3 & 1 & | & 9 \\ 4 & 2 & | & 16 \\ 24 & 10 & | & 0 \end{pmatrix}$ } Constraints
← Objective function

Original ↗

Dual ↘

↓ Transpose

Maximize $P = 9x_1 + 16x_2$

Subject to $\begin{cases} 3x_1 + 4x_2 \leq 24 \\ x_1 + 2x_2 \leq 10 \\ x_1 \geq 0 \text{ and } x_2 \geq 0 \end{cases}$

← Rewrite

$\begin{pmatrix} 3 & 4 & | & 24 \\ 1 & 2 & | & 10 \\ 9 & 16 & | & 0 \end{pmatrix}$ } New constraints
← New objective function

Be careful! The matrices on the right are *not* tableaux so do not change the signs of the objective function numbers.

The dual of a standard minimum problem is a standard maximum problem. Since transposing the numbers a second time would just recover the original, we say that *either problem is the dual of the other.*

Practice Problem 2 Find the dual of the following problem:

Minimize $C = 24y_1 + 20y_2 + 12y_3$

Subject to $\begin{cases} 2y_1 + 5y_2 + 3y_3 \geq 3 \\ -3y_1 + 2y_2 + 2y_3 \geq 5 \\ y_1 \geq 0, \ y_2 \geq 0, \text{ and } y_3 \geq 0 \end{cases}$ ➤ Solution on page 310

Practice Problem 2 shows that a problem and its dual need not have the same number of variables or the same number of constraints.

The two problems in Example 2 use only two variables and so can be solved by the graphical method of Section 4.2: drawing graphs, shading regions, finding vertices, and determining which vertex maximizes or minimizes the objective function. The graphical solutions of these two problems are shown below, with the solution to each being where the two lines intersect.

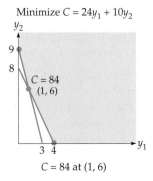

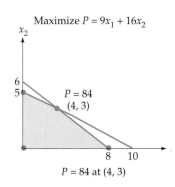

C = 84 at (1, 6)   P = 84 at (4, 3)

Notice that although the problems are quite different, the maximum value of P in one is the same as the minimum value of C in the dual (as you can easily check by substituting the intersection points into the objective functions). The fact that the minimum value of C in a standard minimum problem will *always* be the same as the maximum value of P in the dual maximum problem was a remarkable discovery, and is called the Duality Theorem. The following statement of the theorem also shows how to read the solution of the minimum problem from the final tableau of the dual problem. The Duality Theorem is proved in Exercise 37 at the end of this section.

Duality Theorem

If a maximum problem has a solution, then its final tableau also displays the solution to the dual minimum problem in its bottom row: first the slack variables, then the variables, and finally the objective function:

$$\begin{array}{|cccccc|c|}
\hline
x_1 & \cdots & x_n & s_1 & \cdots & s_m & \\
\vdots & & \vdots & \vdots & & \vdots & \\
\vdots & & \vdots & \vdots & & \vdots & \\
t_1 & \cdots & t_n & y_1 & \cdots & y_m & C \\
\hline
\end{array}$$

Slack variables y-variables ↑ Objective function

Even if this is not a final tableau, the values in the bottom row satisfy the constraints of the minimum problem with the exception of the nonnegativity conditions and yield the given value of the objective function.

4.4 STANDARD MINIMUM PROBLEMS AND DUALITY

The last sentence is not necessary for the purposes of this section but will be used in the following two sections. Note that for a minimum problem we use t's for the slack variables to distinguish them from the slacks in a maximum problem. Since dual problems are related by transposition, it is not surprising that the solution to the *minimum* problem appear in the last *row*, just as the solution to the *maximum* problem appears in the last *column*. If the maximum problem has no solution, then neither does the dual minimum problem.

For example, on pages 298–299 we began with the following minimum problem and found its dual.

$$\text{Minimize } C = 24y_1 + 10y_2$$

$$\text{Subject to } \begin{cases} 3y_1 + y_2 \geq 9 \\ 4y_1 + 2y_2 \geq 16 \\ y_1 \geq 0 \text{ and } y_2 \geq 0 \end{cases}$$

The dual maximum problem is the same one we solved in the previous section, obtaining the following final tableau on page 284:

	x_1	x_2	s_1	s_2	
x_1	1	0	1	-2	4
x_2	0	1	$-\frac{1}{2}$	$\frac{3}{2}$	3
P	0	0	1	6	84

Slacks y-variables   ↑ Objective function

From the bottom row of this tableau we may read off the solution to the original minimum problem:

The minimum value of C is 84, when $y_1 = 1$ and $y_2 = 6$.

The slacks, which we usually ignore, are $t_1 = 0$ and $t_2 = 0$

This is the same as the graphical solution shown on the previous page.

Practice Problem 3 Find the solution of the linear programming problem

$$\text{Minimize } C = 255y_1 + 435y_2 + 300y_3 + 465y_4$$

$$\text{Subject to } \begin{cases} 2y_1 + 3y_2 + 2y_3 + 4y_4 \geq 25 \\ y_1 + 3y_2 + 2y_3 + 2y_4 \geq 18 \\ 3y_1 + 4y_2 + 3y_3 + 2y_4 \geq 36 \\ y_1 \geq 0, \; y_2 \geq 0, \; y_3 \geq 0, \text{ and } y_4 \geq 0 \end{cases}$$

from the fact that the final tableau of its dual maximum problem is

	x_1	x_2	x_3	s_1	s_2	s_3	s_4	
x_3	0	0	1	0	−2	3	0	30
x_1	1	0	0	1	3	−5	0	60
x_2	0	1	0	−1	0	1	0	45
s_4	0	0	0	−2	−8	12	1	75
P	0	0	0	7	3	1	0	3390

▶ Solution on page 310

We may summarize the entire process as follows.

Solution of a Standard Minimum Problem

To solve a standard minimum problem:

1. Construct the dual maximum problem (pages 298–299).
2. Solve the dual maximum problem by the simplex method (page 287).
3. If there is a solution to the dual maximum problem, then there is a solution to the minimum problem: The values of the slacks, the variables, and the objective function appear (in that order) in the bottom row of the final tableau. If the maximum problem has no solution, then neither does the minimum problem.

EXAMPLE 3

SOLVING A STANDARD MINIMUM PROBLEM

$$\text{Minimize } C = 50y_1 + 60y_2$$

$$\text{Subject to } \begin{cases} y_1 + 4y_2 \geq 20 \\ 2y_1 + 3y_2 \geq 30 \\ y_1 - y_2 \geq 5 \\ y_1 \geq 0 \text{ and } y_2 \geq 0 \end{cases}$$

Solution

First we summarize, transpose, and write the dual:

$$\begin{pmatrix} 1 & 4 & | & 20 \\ 2 & 3 & | & 30 \\ 1 & -1 & | & 5 \\ 50 & 60 & | & 0 \end{pmatrix} \rightarrow \begin{pmatrix} 1 & 2 & 1 & | & 50 \\ 4 & 3 & -1 & | & 60 \\ 20 & 30 & 5 & | & 0 \end{pmatrix} \rightarrow$$

$$\text{Maximize } P = 20x_1 + 30x_2 + 5x_3$$

$$\text{Subject to } \begin{cases} x_1 + 2x_2 + x_3 \leq 50 \\ 4x_1 + 3x_2 - x_3 \leq 60 \\ x_1 \geq 0, \; x_2 \geq 0, \; x_3 \geq 0 \end{cases}$$

The initial simplex tableau for the dual maximum problem is

	x_1	x_2	x_3	s_1	s_2	
s_1	1	2	1	1	0	50
s_2	4	3	-1	0	1	60
P	-20	-30	-5	0	0	0

Bottom row states that
$t_1 = -20$, $t_2 = -30$, $t_3 = -5$,
$y_1 = 0$, $y_2 = 0$, and $C = 0$

To solve the dual maximum problem, we first pivot on the 3 in column 2 and row 2 to find the tableau

	x_1	x_2	x_3	s_1	s_2	
s_1	-5/3	0	5/3	1	-2/3	10
x_2	4/3	1	-1/3	0	1/3	20
P	20	0	-15	0	10	600

Bottom row states that
$t_1 = 20$, $t_2 = 0$, $t_3 = -15$,
$y_1 = 0$, $y_2 = 10$,
and $C = 600$

Then we pivot again on the 5/3 in column 3 and row 1 to reach the final tableau:

	x_1	x_2	x_3	s_1	s_2	
x_3	-1	0	1	3/5	-2/5	6
x_2	1	1	0	1/5	1/5	22
P	5	0	0	9	4	690

Bottom row states that
$t_1 = 5$, $t_2 = 0$, $t_3 = 0$,
$y_1 = 9$, $y_2 = 4$, and
$C = 690$

Answer: The minimum value is 690 when $y_1 = 9$ and $y_2 = 4$.

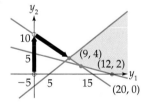

To compare this solution with the graphical method, we graph the feasible region of the minimum problem with y_1 on the horizontal axis and y_2 on the vertical axis. The path of nonfeasible points visited by the simplex tableaux of the dual maximum problem and leading to a feasible vertex is marked in bold.

The simplex method always begins at the origin and "pivots" to successively better vertices. Notice from the graph that in a minimum problem the origin is optimal (minimizing C to 0) but not feasible (not in the shaded region). This is in contrast to a maximum problem, where the origin is feasible but not optimal. That is, the simplex method *keeps the minimum problem optimal while making it feasible and keeps the maximum problem feasible while making it optimal.*

Shadow Prices and Marginal Values

The variables of the dual problem provide information about the original problem. Recall the dual minimum and maximum problems from pages 298–299 and 300, with their solutions written below them.

$$\text{Minimize } C = 24y_1 + 10y_2$$
$$\text{Subject to } \begin{cases} 3y_1 + y_2 \geq 9 \\ 4y_1 + 2y_2 \geq 16 \\ y_1 \geq 0 \text{ and } y_2 \geq 0 \end{cases}$$
$$C = 84 \text{ at } y_1 = 1, \; y_2 = 6$$

$$\text{Maximize } P = 9x_1 + 16x_2$$
$$\text{Subject to } \begin{cases} 3x_1 + 4x_2 \leq 24 \\ x_1 + 2x_2 \leq 10 \\ x_1 \geq 0 \text{ and } x_2 \geq 0 \end{cases}$$
$$P = 84 \text{ at } x_1 = 4, \; x_2 = 3$$

To illustrate their relationship as dual problems, suppose that the first problem involves minimizing the cost of using amounts y_1 and y_2 of two nutritional supplements that cost 24¢ and 10¢ per ounce (the objective function) subject to the daily requirements that you need at least 9 units of vitamins (the first constraint) and at least 16 units of minerals (the second constraint). The numbers of units of vitamins and minerals in each ounce of supplements A and B are given in the table on the left.

	A	B
Vitamins	3	4
Minerals	1	2

The solution says that 84¢ provides enough to satisfy both constraints, so it should be possible to see how much of this cost is for the vitamin constraint and how much is for the mineral constraint, leading to, in some sense, a *cost for the vitamins* and a *cost for the minerals*. Letting x_1 stand for this per-unit cost of vitamins and x_2 for the per-unit cost of minerals, we may interpret the two inequalities in the *dual* problem (on the right above) as saying that, for each supplement, the costs allocated to the ingredients cannot exceed the cost of the supplement.

$$3x_1 + 4x_2 \leq 24 \qquad \binom{\text{Vitamins}}{\text{in A}}\binom{\text{Cost of}}{\text{vitamins}} + \binom{\text{Minerals}}{\text{in A}}\binom{\text{Cost of}}{\text{minerals}} \leq \binom{\text{Cost}}{\text{of A}}$$

$$x_1 + 2x_2 \leq 10 \qquad \binom{\text{Vitamins}}{\text{in B}}\binom{\text{Cost of}}{\text{vitamins}} + \binom{\text{Minerals}}{\text{in B}}\binom{\text{Cost of}}{\text{minerals}} \leq \binom{\text{Cost}}{\text{of B}}$$

The values that we found, $x_1 = 4$ and $x_2 = 3$, then tell us that the cost of the vitamins is 4¢ per ounce and the cost of the minerals is 3¢ per ounce. These are called the *shadow prices* of vitamins and minerals. As an example of how such shadow prices are used, the 3¢ cost of minerals in this example means that if the mineral requirement were reduced by one unit (from 8 units to 7), the cost saving would be 3¢.

A similar analysis applied to the maximum problem on the right above, which we may interpret as maximizing the profit of a pottery company (the objective function) that sells x_1 bowls for $9 each and x_2 plates for $16, each made by "shapers" and "decorators," subject to the requirements that the shaping time cannot exceed 24 hours per day

	Shaping	Decorating
Bowls	3	1
Plates	4	2

(the first constraint) and the decorating time cannot exceed 10 hours per day (the second constraint), with the numbers of hours of shaping and decorating per bowl or plate given by the table on the left.

The profit is based on two processes, shaping and decorating, so it can be allocated between them. If we let y_1 stand for the *profit per hour based on shaping* and y_2 for the *profit per hour based on decorating*, then multiplying by the numbers in the table we would obtain constraints like those in the dual (minimum) problem on the left above. The solution $y_1 = 1$ then means that the shaping process contributes $1 per hour to the profit and $y_2 = 6$ means that the decorating process contributes $6 per hour to the profit. These numbers are called the *marginal values* of the shaping and decorating processes. For an example of how they are used, the $6 per hour contribution of decorating means that increasing the *decorating* time by 1 hour would raise the profit by $6. Further increases would be proportional (for example, an additional 5 hours would bring an additional $30) provided that none of the constraints are violated.

In general, for either a maximum or minimum problem, the *dual* variable with subscript i measures the contribution to the objective function (profit P or cost C) made by the quantity that is constrained by inequality number i, expressed in *objective units per constraint unit*.

Matrix Form

The relationships between a minimum problem and its dual maximum problem and the initial and final tableaux may be stated simply in matrix form. For a minimum problem, we use Y for a column of y-variables and A^t in the coefficients in the constraints (since we used A in the maximum problem).

Standard Minimum Problem   **Dual Maximum Problem**

Minimize $C = b^t Y$ $(b \geq 0)$ Maximize $P = c^t X$

Subject to $\begin{cases} A^t Y \geq c \\ Y \geq 0 \end{cases}$ Subject to $\begin{cases} AX \leq b \quad (b \geq 0) \\ X \geq 0 \end{cases}$

Notice that the *objective* numbers and the *right-hand sides of the constraints* become interchanged when writing the dual, so the requirement $b \geq 0$ applies to the objective numbers in the minimum problem and the right-hand sides of the constraints in the maximum problem. If the initial tableau for the maximum problem (on the left below) leads to a final tableau that solves the maximum problem (on the right below), then the minimum value is C with variables Y and slacks T taking values as shown in the bottom row.

Initial Tableau

$$\left(\begin{array}{c|cc|c} & X & S & \\ \hline S & A & I & b \\ \hline P & -c^t & 0 & 0 \end{array}\right)$$

If the solution exists

Final Tableau

$$\left(\begin{array}{c|cc|c} & X & S & \\ \hline & \vdots & \vdots & \vdots \\ \hline & T & Y & C \end{array}\right)$$

Mixed Constraints: A Transportation Problem

Although we have been careful to write our standard minimum problem with $\geq$ constraints, we do not mean to exclude the possibility of mixed constraints, some $\geq$ and some $\leq$. We simply multiply each $\leq$ inequality by -1 so it becomes a $\geq$ inequality. (Recall from page 297 that the minimum problem does not require that the right-hand sides of the inequalities be nonnegative.) The following example belongs to a general type of minimization problem with mixed constraints having the nice property that the pivot elements will all be 1s.

EXAMPLE 4

A TRANSPORTATION PROBLEM

A retail store chain has cartons of goods stored at warehouses in Maryland and Washington that must be distributed to its stores in Ohio and Louisiana. The cost to ship each carton from Maryland to Ohio is \$6, from Maryland to Louisiana is \$7, from Washington to Ohio is \$8, and from Washington to Louisiana is \$9. There are 300 cartons at the Maryland warehouse and 300 at the warehouse in Washington. If the Ohio stores need 200 cartons and the Louisiana stores need 300 cartons, how many cartons should be shipped from each warehouse to each state to incur the smallest shipping costs?

Solution

This is a minimization problem that requires four variables: one for the amount shipped from each warehouse to each state. Let

$$y_1 = \begin{pmatrix} \text{Cartons shipped} \\ \text{from MD to OH} \end{pmatrix}, \qquad y_2 = \begin{pmatrix} \text{Cartons shipped} \\ \text{from MD to LA} \end{pmatrix},$$

$$y_3 = \begin{pmatrix} \text{Cartons shipped} \\ \text{from WA to OH} \end{pmatrix}, \quad \text{and} \quad y_4 = \begin{pmatrix} \text{Cartons shipped} \\ \text{from WA to LA} \end{pmatrix}$$

4.4 STANDARD MINIMUM PROBLEMS AND DUALITY

Because each warehouse can ship no more than the number of cartons stored there, and each state must receive at least the required number of cartons, this problem is the linear programming problem

Minimize $C = 6y_1 + 7y_2 + 8y_3 + 9y_4$ Total shipping costs

Subject to
$$\begin{cases} y_1 + y_2 & \leq 300 & \text{Have 300 in MD} \\ y_3 + y_4 & \leq 300 & \text{Have 300 in WA} \\ y_1 + y_3 & \geq 200 & \text{Need 200 in OH} \\ y_2 + y_4 & \geq 300 & \text{Need 300 in LA} \\ y_1 \geq 0,\ y_2 \geq 0,\ y_3 \geq 0,\ y_4 \geq 0 & & \text{Nonnegativity} \end{cases}$$

We multiply the first two constraints by -1 to give them $\geq$ inequalities. This problem is then simply stated in matrix form:

$$\text{Minimize } C = \begin{pmatrix} 6 & 7 & 8 & 9 \end{pmatrix} \begin{pmatrix} y_1 \\ y_2 \\ y_3 \\ y_4 \end{pmatrix}$$

Subject to
$$\begin{cases} \begin{pmatrix} -1 & -1 & 0 & 0 \\ 0 & 0 & -1 & -1 \\ 1 & 0 & 1 & 0 \\ 0 & 1 & 0 & 1 \end{pmatrix} \begin{pmatrix} y_1 \\ y_2 \\ y_3 \\ y_4 \end{pmatrix} \geq \begin{pmatrix} -300 \\ -300 \\ 200 \\ 300 \end{pmatrix} \\ \text{and } \begin{pmatrix} y_1 \\ y_2 \\ y_3 \\ y_4 \end{pmatrix} \geq 0 \end{cases}$$

The dual maximum problem is

$$\text{Maximize } P = \begin{pmatrix} -300 & -300 & 200 & 300 \end{pmatrix} \begin{pmatrix} x_1 \\ x_2 \\ x_3 \\ x_4 \end{pmatrix}$$

Subject to
$$\begin{cases} \begin{pmatrix} -1 & 0 & 1 & 0 \\ -1 & 0 & 0 & 1 \\ 0 & -1 & 1 & 0 \\ 0 & -1 & 0 & 1 \end{pmatrix} \begin{pmatrix} x_1 \\ x_2 \\ x_3 \\ x_4 \end{pmatrix} \leq \begin{pmatrix} 6 \\ 7 \\ 8 \\ 9 \end{pmatrix} \\ \text{and } \begin{pmatrix} x_1 \\ x_2 \\ x_3 \\ x_4 \end{pmatrix} \geq 0 \end{cases}$$

The initial simplex tableau is

	x_1	x_2	x_3	x_4	s_1	s_2	s_3	s_4	
s_1	−1	0	1	0	1	0	0	0	6
s_2	−1	0	0	1	0	1	0	0	7
s_3	0	−1	1	0	0	0	1	0	8
s_4	0	−1	0	1	0	0	0	1	9
P	300	300	−200	−300	0	0	0	0	0

We pivot at [column 4, row 2], [column 3, row 1], and [column 1, row 3] to reach the final tableau:

	x_1	x_2	x_3	x_4	s_1	s_2	s_3	s_4	
x_3	0	−1	1	0	0	0	1	0	8
x_4	0	−1	0	1	−1	1	1	0	9
x_1	1	−1	0	0	−1	0	1	0	2
s_4	0	0	0	0	1	−1	−1	1	0
P	0	100	0	0	0	300	200	0	3700
					y_1	y_2	y_3	y_4	C

Bottom row states that $t_1 = 0$, $t_2 = 100$, $t_3 = 0$, $t_4 = 0$, $y_1 = 0$, $y_2 = 300$, $y_3 = 200$, $y_4 = 0$, and $C = 3700$

The Maryland warehouse should send nothing to Ohio and 300 cartons to Louisiana, while the Washington warehouse should send 200 cartons to Ohio and nothing to Louisiana to achieve the least shipping cost of $3700. The second slack shows that there will be 100 cartons unused in the Washington warehouse.

Spreadsheet Exploration

The spreadsheet* on the next page shows the solution of Example 4 using the Solver command. The entries in column E combine the coefficients and the current values of the variables to find the values that the Solver compares to the bounds and the goal of obtaining the minimum of the objective function.

---

* See the Preface for how to obtain this and other Excel spreadsheets.

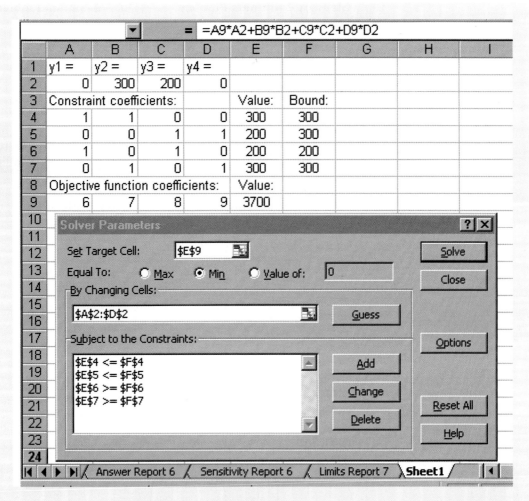

The answer, sensitivity, and limits reports generated by the Solver provide additional information about the solution. Do you see where the values in the bottom row and right column of the final tableau in the solution of Example 4 on the previous page appear in this spreadsheet solution?

4.4 Section Summary

We solve a standard minimum problem (page 297) by transposing to find its dual (pages 298–299), which we solve by the simplex method, and then interpret the solution according to the Duality Theorem (page 300):

Standard Minimum Problem

Minimize $C = b^t Y$

Subject to $\begin{cases} A^t Y \geq c \\ Y \geq 0 \end{cases}$

Dual Maximum Problem

Maximize $P = c^t X$

Subject to $\begin{cases} AX \leq b \\ X \geq 0 \end{cases}$

with $b \geq 0$ in both.

If the initial tableau for the dual problem (on the left below) leads to a tableau that solves the maximum problem (on the right), then the solution to the minimum problem may be read from the bottom row of the final tableau: the slack variables T, then the variables Y, followed by the minimum value C.

$$\begin{array}{c|cc|c} & X & S & \\ \hline S & A & I & b \\ P & -c^t & 0 & 0 \end{array} \quad \xrightarrow{\text{If the solution exists}} \quad \begin{array}{c|cc|c} & X & S & \\ \hline & \vdots & \vdots & \vdots \\ & T & Y & C \end{array}$$

Initial Tableau → Final Tableau

▶ **Solutions to Practice Problems**

1. No. The negative coefficient -4 in the objective function violates the condition in the box on page 297. (The -10 on the right-hand side of the first inequality is *not* a violation—only the objective coefficients need to be nonnegative.)

2. $\begin{pmatrix} 2 & 5 & 3 & | & 3 \\ -3 & 2 & 2 & | & 5 \\ 24 & 20 & 12 & | & 0 \end{pmatrix}^t \longrightarrow \begin{pmatrix} 2 & -3 & | & 24 \\ 5 & 2 & | & 20 \\ 3 & 2 & | & 12 \\ 3 & 5 & | & 0 \end{pmatrix}$

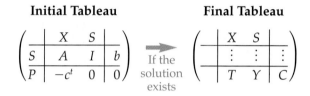

Notice that the original problem had three variables and two constraints, while the dual has two variables and three constraints. The numbers of variables and constraints will always interchange in this way because of the transposition.

3. The bottom row of the final tableau for the dual maximum problem displays the values for the slack variables $(t_1 = 0, \ t_2 = 0, \ t_3 = 0)$, the variables $(y_1 = 7, \ y_2 = 3, \ y_3 = 1, \ y_4 = 0)$, and the objective function $(C = 3390)$. The minimum value is 3390 when $y_1 = 7, \ y_2 = 3, \ y_3 = 1,$ and $y_4 = 0$.

4.4 Exercises

For each standard minimum problem construct the dual maximum problem and its initial simplex tableau.

1. Minimize $C = 84y_1 + 21y_2$

   Subject to $\begin{cases} 3y_1 + y_2 \geq 21 \\ 4y_1 - y_2 \geq 0 \\ y_1 \geq 0, \ y_2 \geq 0 \end{cases}$

2. Minimize $C = 7y_1 + 7y_2$

   Subject to $\begin{cases} 2y_1 + 3y_2 \geq 42 \\ 3y_1 + y_2 \geq 21 \\ y_1 \geq 0, \ y_2 \geq 0 \end{cases}$

3. Minimize $C = 60y_1 + 100y_2 + 300y_3$

   Subject to $\begin{cases} y_1 + 2y_2 + 3y_3 \geq 180 \\ 4y_1 + 5y_2 + 6y_3 \geq 120 \\ y_1 \geq 0, \ y_2 \geq 0, \ y_3 \geq 0 \end{cases}$

4. Minimize $C = 100y_1 + 60y_2 + 280y_3$

   Subject to $\begin{cases} 5y_1 + 9y_2 + 7y_3 \geq 315 \\ 2y_1 + 6y_2 + 4y_3 \geq 480 \\ y_1 \geq 0, \ y_2 \geq 0, \ y_3 \geq 0 \end{cases}$

5. Minimize $C = 3y_1 + 20y_2$

   Subject to $\begin{cases} 3y_1 + 2y_2 \geq 150 \\ y_1 + 4y_2 \geq 100 \\ 3y_1 + 4y_2 \geq 228 \\ y_1 \geq 0, \ y_2 \geq 0 \end{cases}$

6. Minimize $C = 60y_1 + 16y_2$

   Subject to $\begin{cases} y_1 + 4y_2 \geq 20 \\ 3y_1 + 2y_2 \geq 30 \\ 3y_1 + 4y_2 \geq 48 \\ y_1 \geq 0, \ y_2 \geq 0 \end{cases}$

7. Minimize $C = 15y_1 + 20y_2 + 5y_3$

   Subject to $\begin{cases} y_1 - y_2 - 2y_3 \leq 30 \\ y_1 + 2y_2 + y_3 \geq 30 \\ y_1 \geq 0, \ y_2 \geq 0, \ y_3 \geq 0 \end{cases}$

8. Minimize $C = 30y_1 + 40y_2 + 80y_3$

   Subject to $\begin{cases} -3y_1 + y_2 - y_3 \leq 60 \\ y_1 + y_2 + 2y_3 \geq 60 \\ y_1 \geq 0, \ y_2 \geq 0, \ y_3 \geq 0 \end{cases}$

Solve each standard minimum problem by finding the dual maximum problem and using the simplex method. (Exercises 9–16 can also be solved by the graphical method.)

You may use the PIVOT program if permitted by your instructor.

9. Minimize $C = 15y_1 + 10y_2$

   Subject to $\begin{cases} y_1 + 2y_2 \geq 20 \\ 3y_1 - y_2 \geq 60 \\ y_1 \geq 0, \ y_2 \geq 0 \end{cases}$

10. Minimize $C = 40y_1 + y_2$

    Subject to $\begin{cases} 5y_1 - y_2 \geq 5 \\ 4y_1 + y_2 \geq 4 \\ y_1 \geq 0, \ y_2 \geq 0 \end{cases}$

11. Minimize $C = 4y_1 + 3y_2$

    Subject to $\begin{cases} y_1 + y_2 \geq 15 \\ 3y_1 + y_2 \leq 60 \\ y_1 \geq 0, \ y_2 \geq 0 \end{cases}$

12. Minimize $C = 4y_1 + 5y_2$

    Subject to $\begin{cases} y_1 + y_2 \geq 20 \\ 2y_1 + y_2 \leq 50 \\ y_1 \geq 0, \ y_2 \geq 0 \end{cases}$

13. Minimize $C = 4y_1 + 5y_2$

    Subject to $\begin{cases} y_1 + y_2 \geq 10 \\ y_1 \geq 2 \\ y_2 \geq 3 \\ y_1 \geq 0, \ y_2 \geq 0 \end{cases}$

14. Minimize $C = 3y_1 + 2y_2$

Subject to $\begin{cases} y_1 + y_2 \geq 20 \\ y_1 \geq 4 \\ y_2 \geq 5 \\ y_1 \geq 0, \ y_2 \geq 0 \end{cases}$

15. Minimize $C = 2y_1 + y_2$

Subject to $\begin{cases} -y_1 + y_2 \leq 20 \\ y_1 - y_2 \leq 20 \\ y_1 + y_2 \geq 10 \\ y_1 \geq 0, \ y_2 \geq 0 \end{cases}$

16. Minimize $C = 2y_1 + 3y_2$

Subject to $\begin{cases} -y_1 + y_2 \leq 30 \\ y_1 - y_2 \leq 30 \\ y_1 + y_2 \geq 10 \\ y_1 \geq 0, \ y_2 \geq 0 \end{cases}$

17. Minimize $C = 10y_1 + 20y_2 + 10y_3$

Subject to $\begin{cases} -y_1 + y_2 + y_3 \geq 50 \\ y_1 + y_2 - y_3 \geq 30 \\ y_1 \geq 0, \ y_2 \geq 0, \ y_3 \geq 0 \end{cases}$

18. Minimize $C = 30y_1 + 50y_2 + 30y_3$

Subject to $\begin{cases} -y_1 + y_2 + y_3 \geq 10 \\ y_1 + y_2 - y_3 \geq 20 \\ y_1 \geq 0, \ y_2 \geq 0, \ y_3 \geq 0 \end{cases}$

19. Minimize $C = 30y_1 + 19y_2 + 30y_3$

Subject to $\begin{cases} 2y_1 + y_2 + y_3 \geq 6 \\ y_1 + y_2 + 2y_3 \geq 4 \\ y_1 \geq 0, \ y_2 \geq 0, \ y_3 \geq 0 \end{cases}$

20. Minimize $C = 60y_1 + 39y_2 + 60y_3$

Subject to $\begin{cases} 2y_1 + y_2 + y_3 \geq 6 \\ y_1 + y_2 + 2y_3 \geq 8 \\ y_1 \geq 0, \ y_2 \geq 0, \ y_3 \geq 0 \end{cases}$

21. Minimize $C = 20y_1 + 30y_2 + 40y_3$

Subject to $\begin{cases} y_1 - y_2 + y_3 \geq 15 \\ y_1 + y_2 + y_3 \geq 20 \\ y_1 - y_2 + y_3 \leq 10 \\ y_1 \geq 0, \ y_2 \geq 0, \ y_3 \geq 0 \end{cases}$

22. Minimize $C = 20y_1 + 50y_2 + 30y_3$

Subject to $\begin{cases} 2y_1 - y_2 + y_3 \leq 10 \\ y_1 + y_2 + y_3 \geq 30 \\ 2y_1 - y_2 + y_3 \geq 20 \\ y_1 \geq 0, \ y_2 \geq 0, \ y_3 \geq 0 \end{cases}$

23. Minimize $C = 132y_1 + 102y_2 + 60y_3$

Subject to $\begin{cases} 3y_1 + 2y_2 + y_3 \geq 48 \\ 4y_1 + 3y_2 + 2y_3 \geq 72 \\ 2y_1 + 2y_2 + y_3 \geq 42 \\ y_1 \geq 0, \ y_2 \geq 0, \ y_3 \geq 0 \end{cases}$

24. Minimize $C = 96y_1 + 144y_2 + 84y_3$

Subject to $\begin{cases} 2y_1 + 3y_2 + 2y_3 \geq 102 \\ 3y_1 + 4y_2 + 2y_3 \geq 132 \\ y_1 + 2y_2 + y_3 \geq 60 \\ y_1 \geq 0, \ y_2 \geq 0, \ y_3 \geq 0 \end{cases}$

APPLIED EXERCISES

Formulate each situation as a standard minimum linear programming problem by identifying the variables, the objective function, and the constraints. Be sure to state clearly the meaning of each variable. Solve it by finding the dual maximum problem and using the simplex method. State your final answer in terms of the original question.

 You may use the PIVOT program if permitted by your instructor.

25. BIOMEDICAL: Nutrition An athlete's training diet needs at least 44 more grams of carbohydrates, 12 more grams of fat, and 16 more grams of protein each day. A dietician recommends two food supplements, Bulk-Up

Bars (costing 48¢ each) and Power Drink (costing 45¢ per can), with nutritional contents (in grams) as given in the table. How much of each food supplement will provide the extra needed nutrition at the least cost?

	Bulk-Up Bar	Power Drink
Carbohydrates	4	3
Fat	1	1
Protein	2	1

26. **BIOMEDICAL: Diet Planning** The residents of an elder care facility need at least 156 more milligrams (mg) of calcium, 180 more micrograms (μg) of folate, and 66 more grams (g) of protein in their weekly diets. The staff chef has created a new fish entree and a salad that the residents should find appealing, and the nutritional contents of the new menu items are given in the accompanying table. If the fish entree costs 21¢ per ounce and the salad costs 15¢ per ounce, how many ounces of each should be added to every resident's weekly menu to provide the additional nutrition at the least cost?

(per ounce)	Fish Entree	Salad
Calcium (mg)	6	4
Folate (μg)	6	5
Protein (g)	2	2

27. **MANAGEMENT SCIENCE: Office Supplies** The office manager of a large accounting firm needs at least 315 more boxes of pens and 120 more boxes of pencils but has only $1500 left in his budget. Jack's Office Supplies has packages of 5 boxes of pens with 2 boxes of pencils on sale for $20, while John's Discount offers packages of 3 boxes of pens and 1 box of pencils for $11. How many packages from each store should he buy to restock the store room at the least cost?

28. **MANAGEMENT SCIENCE: Highway Construction** The project engineer for a highway construction company needs to cut through a small hill to make way for a new road. She estimates that at least 3500 cubic yards of dirt and at least 2400 cubic yards of crushed rock will have to be hauled away. A heavy-duty dump truck can haul either 10 cubic yards of dirt or 6 cubic yards of crushed rock, while a regular dump truck can haul either 5 cubic yards of dirt or 4 cubic yards of crushed rock. The dirt and crushed rock cannot be mixed, because they go to different dumping sites. The union contract demands that this job use at least 400 loads carried in heavy-duty dump trucks. If each heavy-duty dump truck load costs $90 and each regular dump truck load costs $50, how many truckloads with each type of truck and cargo will be needed to complete the job at the least cost?

29. **MANAGEMENT SCIENCE: Highway Construction** Repeat Exercise 28 with the additional requirement that there can be no more than 800 truckloads hauled from the job site in order to get the project finished on time.

30. **MANAGEMENT SCIENCE: Office Supplies** Repeat Problem 27 with the additional requirement that the office manager feels he must purchase at least $1200 worth of pens and pencils from Jack's Office Supplies because it gave him such a good deal on some filing cabinets last month.

31. **BUSINESS: Agriculture** A soil analysis of a farmer's field showed that he needs to apply at least 3000 pounds of nitrogen, 2400 pounds of phosphoric acid, and 2100 pounds of potash. Plant fertilizer is labeled with three numbers giving the percentages of nitrogen, phosphoric acid, and potash. The local farm supply store sells 15-30-15 Miracle Mix for 15¢ per pound and a 10-5-5 store brand for 8¢ per pound. How many pounds of each fertilizer should the farmer buy to meet the needs of the field at the least cost?

32. **GENERAL: Gardening** A weekend gardener's vegetable patch needs at least 10.2 pounds of nitrogen, 7.8 pounds of phosphoric acid, and 6.6 pounds of potash. Plant fertilizer is labeled with three numbers giving the percentages of nitrogen, phosphoric acid, and

potash. The local garden center sells 15-30-15 Miracle Mix for 16¢ per pound, 15-10-10 Grow Great for 13¢ per pound, and a 10-5-5 store brand for 8¢ per pound. How many pounds of each fertilizer should the gardener buy to meet the needs of the vegetable patch at the least cost?

33. MANAGEMENT SCIENCE: Transportation A retail store chain has cartons of goods stored at warehouses in Kentucky and Utah that must be distributed to its stores in Kansas, Texas, and Oregon. Each carton shipped from the Utah warehouse costs $2 whether it goes to Kansas, Texas, or Oregon. However, the cost to ship one carton from the Kentucky warehouse to Kansas is $2, to Texas is $4, and to Oregon is $5. There are 200 cartons at the Utah warehouse and 400 at the warehouse in Kentucky. If the Kansas stores need 200 cartons, the Texas stores need 300 cartons, and the Oregon stores need 100 cartons, how many cartons should be shipped from each warehouse to each state to incur the smallest shipping costs?

34. MANAGEMENT SCIENCE: Transportation A soda distributor has warehouses in Seaford and Centerville and needs to supply stores in Huntington and Towson. The costs of shipping one case of sodas are 8¢ from Seaford to Huntington, 5¢ from Seaford to Towson, 6¢ from Centerville to Huntington, and 4¢ from Centerville to Towson. There are 800 cases in the Seaford warehouse and 1200 in the Centerville warehouse. If the Huntington stores need at least 1000 cases and the Towson stores need at least 600 cases, how many cases should be shipped from each warehouse to each city to incur the smallest shipping costs?

Explorations and Excursions

The following problems extend and augment the material presented in the text.

Why $P \leq C$ for Dual Problems

35. *(Requires matrix algebra)* For the pair of dual problems

Maximize $P = c^t X$
Subject to $\begin{cases} AX \leq b \\ X \geq 0 \end{cases}$

Minimize $C = b^t Y$
Subject to $\begin{cases} A^t Y \geq c \\ Y \geq 0 \end{cases}$

show that $P \leq C$ for any feasible X and Y by justifying each $=$ and $\leq$ in the chain of statements

$$P = c^t \cdot X = X^t \cdot c \leq X^t \cdot A^t Y$$
$$= (X^t A^t) \cdot Y = (AX)^t \cdot Y \leq b^t \cdot Y = C$$

36. *(Requires matrix algebra)* Check the chain of matrix statements in Exercise 35 for the matrices

$$A = \begin{pmatrix} 3 & 4 \\ 1 & 2 \end{pmatrix}, \quad b = \begin{pmatrix} 24 \\ 10 \end{pmatrix}, \quad c = \begin{pmatrix} 9 \\ 16 \end{pmatrix}, \quad X = \begin{pmatrix} x_1 \\ x_2 \end{pmatrix},$$

and $Y = \begin{pmatrix} y_1 \\ y_2 \end{pmatrix}$ from Example 2 on pages 298–299.

A Proof of the Duality Theorem

37. *(Requires matrix algebra)* The following sequence of statements provides a proof of the Duality Theorem on page 300. Justify each statement to verify this proof.

a. If $mx + b$ is the same as $0x + v$ for every value of x, then $m = 0$ and $b = v$.

b. The dual problems in Exercise 35 may be rewritten as

Maximize $P = c^t X$
Subject to $\begin{cases} AX + S = b \\ X, S \geq 0 \end{cases}$

and

Minimize $C = b^t Y$
Subject to $\begin{cases} A^t Y - T = c \\ Y, T \geq 0 \end{cases}$

to clearly show both the variables (X and Y) and the slack variables (S and T).

c. The simplex tableau for the maximum problem (both the initial tableau and after *any* sequence of pivot operations) has numbers in the bottom row:

	x_1	$\cdots$	x_n	s_1	$\cdots$	s_m
	$\vdots$			$\vdots$		
P	w_1	$\cdots$	w_n	z_1	$\cdots$	z_m	V

To prove the Duality Theorem, we must show that these values can be used for the values of Y and T in the minimum problem.

That is, if we set $y_1 = z_1, \ldots, y_m = z_m$, and $t_1 = w_1, \ldots, t_n = w_n$, then the number V is the same as the value $C = b^t Y$ and the constraint $A^t Y - T = c$ is satisfied.

d. If we use the notations $W = \begin{pmatrix} w_1 \\ \vdots \\ w_n \end{pmatrix}$ and

$Z = \begin{pmatrix} z_1 \\ \vdots \\ z_m \end{pmatrix}$, the bottom row of the tableau

is the same as the matrix equation $P + W^t X + Z^t S = V$, and this equation holds for *any* values of X and S such that $AX + S = b$.

e. Since $P = c^t X$ and $S = b - AX$, we may rewrite the matrix equation $P + W^t X + Z^t S = V$ as

$$c^t X + W^t X + Z^t(b - AX) = V$$

f. That is, $X^t c + X^t W + b^t Z - (AX)^t Z = V$ and so

$$X^t \cdot (c + W - A^t Z) + b^t Z = V$$

Thus $(c + W - A^t Z) = 0$ and $b^t Z = V$. So the number V is the value of the minimum problem objective function $C = b^t Y$ when Y takes the value Z and this value for Y satisfies the constraint $A^t Y - T = c$ when the slack T takes the value W.

Complementary Slackness

38. Use the Duality Theorem for the simplex method (see page 300) to show that *at least* one of *every* pair of variables x_1 and t_1, x_2 and t_2, $\ldots$, x_n and t_n, y_1 and s_1, y_2 and s_2, $\ldots$, y_m and s_m is zero for the solution points of a linear programming problem and its dual. [*Hint:* What can you say about the bottoms of the basic variable columns in the simplex tableau of a maximum problem?]

39. Verify the Complementary Slackness Theorem (Exercise 38) for the final tableau given in Practice Problem 3 (see pages 301–302).

4.5 NONSTANDARD PROBLEMS: THE DUAL PIVOT ELEMENT AND THE TWO-STAGE METHOD

This section and the next are optional and present alternative methods for solving nonstandard linear programming problems. The choice of which one to cover is left to the instructor. The examples and exercises duplicate those in Section 4.6.

Introduction

In this section we extend the simplex method to maximum problems that include constraints with $\geq$ inequalities as well as $\leq$ inequalities. Because a $\geq$ inequality can be changed into a $\leq$ inequality by multiplying through by -1 (for example, changing $5 \geq -3$ into $-5 \leq 3$), this extension is equivalent to dropping the requirement that $b \geq 0$ in the constraints for the standard maximum problem on page 275.

The Dual Pivot Element

The origin is not a vertex for a nonstandard maximum problem because an inequality like $x_1 \geq 3$ separates the feasible region from the origin. The initial tableau begins at the origin so it is not feasible as well as not optimal. The question becomes: How can we move to a feasible point so that we can then work on making it optimal? On page 303 we saw that the pivot operations for a standard maximum problem make the dual minimum problem feasible. This gives the key to the answer to our question: Pivoting on "dual pivot elements" will make the problem feasible, and then pivoting on (regular) pivot elements will make it optimal.

Since dual problems interchange the roles of rows and columns, the following definition is just the definition of the pivot element with the roles of the rows and columns reversed, and we must take the *largest* ratio instead of the smallest because we are now dividing by negative numbers.

Dual Pivot Element

The *dual pivot row* is the row with the smallest negative entry in the rightmost column of the tableau (omitting the bottom row). If there is a tie, choose the uppermost such row.

The *dual pivot column* is the column with the largest ratio found by dividing the bottom entry by the dual pivot row entry, omitting any column with a zero or positive dual pivot row entry and omitting the rightmost column. If there is a tie, choose the leftmost such column.

The *dual pivot element* is the entry in the dual pivot row and the dual pivot column.

EXAMPLE 1

THE DUAL PIVOT ELEMENT

Find the dual pivot element in the following simplex tableau or explain why the tableau does not have a dual pivot element.

	x_1	x_2	x_3	x_4	s_1	s_2	s_3	
s_1	2	1	1	2	1	0	0	30
s_2	-1	-2	-1	-1	0	1	0	-5
s_3	-1	-1	-1	2	0	0	1	-10
P	-8	12	-10	14	0	0	0	0

Solution

The dual pivot row is row 3 because -10 is the smallest negative entry on the right (omitting the bottom row). The dual pivot column is column 3 because the ratio $\frac{-10}{-1} = 10$ is greater than the others (the remaining columns are not considered because their dual pivot row entries are zero or positive and the rightmost column is never considered).

	x_1	x_2	x_3	x_4	s_1	s_2	s_3	
s_1	2	1	1	2	1	0	0	30
s_2	−1	−2	−1	−1	0	1	0	−5
s_3	−1	−1	−1	2	0	0	1	−10
P	−8	12	−10	14	0	0	0	0

s_3 ← Smallest negative ← Dual pivot row

$\frac{-8}{-1} = 8 \quad \frac{12}{-1} = -12 \quad \frac{-10}{-1} = 10$ (Omit) (Omit) (Omit) (Omit)

↑ Largest ratio
↑ Dual pivot column

Dual pivot element

Choose the dual pivot row first then the dual pivot column

The dual pivot element is the −1 in row 3 and column 3.

Practice Problem 1 For each of the following simplex tableaux, find the dual pivot element or explain why the tableau does not have a dual pivot element.

a.

	x_1	x_2	x_3	x_4	s_1	s_2	s_3	
s_1	−1	−1	2	−2	1	0	0	−10
s_2	1	1	−1	−1	0	1	0	20
s_3	0	−1	3	2	0	0	1	−5
P	−5	−6	14	8	0	0	0	0

b.

	x_1	x_2	x_3	s_1	s_2	s_3	
s_1	1	5	1	1	0	0	25
s_2	−1	−2	−1	0	1	0	−15
s_3	2	1	4	0	0	1	−20
P	−6	10	−10	0	0	0	0

▶ Solutions on page 323

The Two-Stage Simplex Method

Pivoting on dual pivot elements will make the tableau feasible. After the tableau becomes feasible, pivoting on *regular* pivot elements will make it optimal, leading to the solution of the problem.

> **Two-Stage Simplex Method**
>
> To solve any maximum problem by the simplex method:
> 1. Write the constraints with $\leq$ inequalities (keeping the variables nonnegative), and construct the initial simplex tableau.
> 2. If at least one basic variable has a negative value in the rightmost column (the tableau is not feasible), go to step 3. If all the basic variables have nonnegative values in the rightmost column (the tableau is feasible), go to step 4.
> 3. Locate the dual pivot element (page 316), perform the pivot operation (page 282), and return to step 2. If the tableau has a dual pivot row but no dual pivot column, then there is *no solution* to the problem (the constraints are infeasible).
> 4. Locate the pivot element (pages 280–281), perform the pivot operation (page 282), and return to step 2. If the tableau does not have a pivot column, then the solution occurs at the vertex given by the basic variables, and the maximum value of the objective function appears in the bottom right corner of the tableau. If the tableau has a pivot column but no pivot row, then there is *no solution* to the problem.

EXAMPLE 2   THE TWO-STAGE SIMPLEX METHOD

Solve the following nonstandard linear programming problem by the two-stage simplex method.

$$\text{Maximize } P = 5x_1 + 7x_2$$

$$\text{Subject to } \begin{cases} x_1 + x_2 \leq 8 \\ x_1 + x_2 \geq 4 \\ 2x_1 + x_2 \geq 6 \\ x_1 \geq 0 \text{ and } x_2 \geq 0 \end{cases} \quad \begin{array}{l} \text{"Mixed constraints"} \\ \text{because they have} \\ \leq \text{ and } \geq \\ \text{inequalities} \end{array}$$

Solution

We change the second and third inequalities from $\geq$ to $\leq$ by multiplying by -1, obtaining $-x_1 - x_2 \leq -4$ and $-2x_1 - x_2 \leq -6$.

The initial simplex tableau is

	x_1	x_2	s_1	s_2	s_3	
s_1	1	1	1	0	0	8
s_2	−1	−1	0	1	0	−4
s_3	−2	−1	0	0	1	−6
P	−5	−7	0	0	0	0

$s_1 = 8$
$s_2 = -4$ Some basic variables are negative, so *not* feasible
$s_3 = -6$

Since s_2 and s_3 are negative, the tableau is not feasible and we look for a dual pivot element. The dual pivot row is row 3 (−6 is the smallest negative entry in the right column, omitting the bottom row). The dual pivot column is column 2 (the ratio $\frac{-7}{-1} = 7$ is greater than $\frac{-5}{-2} = 2.5$, and none of the other columns may be considered). The dual pivot element is the −1 in row 3 and column 2. Pivoting on this dual pivot element, the tableau becomes:

	x_1	x_2	s_1	s_2	s_3	
s_1	−1	0	1	0	1	2
s_2	1	0	0	1	−1	2
x_2	2	1	0	0	−1	6
P	9	0	0	0	−7	42

$s_1 = 2$
$s_2 = 2$ Feasible because *all* basic variables are nonnegative
$x_2 = 6$

The bottom row has a negative entry, so *not* optimal

(If some basic variables were still negative, we would look for another dual pivot element.) The tableau is now feasible but not optimal, so we look for a (regular) pivot element. The pivot column is column 5 (−7 is the only negative entry in the bottom row). The pivot row is row 1 (the other rows cannot be considered because their pivot column entries are negative). The pivot element is the 1 in column 5 and row 1. Pivoting on this pivot element, the tableau becomes:

	x_1	x_2	s_1	s_2	s_3	
s_3	−1	0	1	0	1	2
s_2	0	0	1	1	0	4
x_2	1	1	1	0	0	8
P	2	0	7	0	0	56

Feasible: all are nonnegative

Optimal: *all* are *nonnegative*

This is the final tableau because it is both feasible and optimal. The maximum is $P = 56$ when $x_1 = 0$ and $x_2 = 8$.

To compare this solution with the graphical method, we graph the feasible region with x_1 on the horizontal axis and x_2 on the vertical axis. The five vertices and the values of the objective function are given in the table. The path of vertices visited by the two-stage simplex method is marked in bold.

Vertex (x_1, x_2)	$P = 5x_1 + 7x_2$
(8, 0)	40
(4, 0)	20
(2, 2)	24
(0, 6)	42
(0, 8)	56

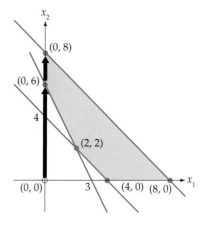

Notice that the first pivot reaches a feasible point and the second makes it optimal.

Practice Problem 2 Solve the following nonstandard linear programming problem by the two-stage simplex method.

Maximize $P = 4x_1 + x_2 + 3x_3$

Subject to $\begin{cases} x_1 + 2x_2 + x_3 \leq 50 \\ 2x_1 + x_2 + 2x_3 \geq 10 \\ x_1 \geq 0, \ x_2 \geq 0, \text{ and } x_3 \geq 0 \end{cases}$ ➤ Solution on page 323

EXAMPLE 3

MANAGING AN INVESTMENT PORTFOLIO

The manager of an $80 million mutual fund has three investment possibilities: secure government bonds yielding 4%, a blue chip growth stock paying 6%, and a biotechnology start-up company returning 12%. If the manager wants to invest at least $50 million in a combination of safe government bonds and blue chip stock, and at least as much in the bonds as the total in blue chip and biotechnology stocks, how much should be invested in each to obtain the greatest possible return?

4.5 NONSTANDARD PROBLEMS: THE DUAL PIVOT ELEMENT AND THE TWO-STAGE METHOD

Solution

Let

$$x_1 = \begin{pmatrix} \text{Amount in} \\ \text{government bonds} \end{pmatrix}$$

$$x_2 = \begin{pmatrix} \text{Amount in} \\ \text{blue chip stock} \end{pmatrix} \quad \text{In millions of dollars}$$

$$x_3 = \begin{pmatrix} \text{Amount in} \\ \text{biotechnology stock} \end{pmatrix}$$

The objective is to maximize the return

$$P = 0.04x_1 + 0.06x_2 + 0.12x_3 \quad \text{Rate} \times \text{amount for each investment}$$

subject to the restrictions

$$x_1 + x_2 + x_3 \leq 80 \quad \text{\$80 million available to invest}$$
$$x_1 + x_2 \geq 50 \quad \text{\$50 million in safe investments}$$
$$x_1 \geq x_2 + x_3 \quad \text{Bonds} \geq \text{total in stocks}$$
$$x_1 \geq 0, \; x_2 \geq 0, \; x_3 \geq 0 \quad \text{Nonnegativity}$$

Multiplying the second constraint by -1 to get $-x_1 - x_2 \leq -50$ and rewriting the third as $-x_1 + x_2 + x_3 \leq 0$, the initial simplex tableau is:

	x_1	x_2	x_3	s_1	s_2	s_3	
s_1	1	1	1	1	0	0	80
s_2	-1	-1	0	0	1	0	-50
s_3	-1	1	1	0	0	1	0
P	-0.04	-0.06	-0.12	0	0	0	0

$s_2 < 0$ ← Dual pivot row

Since this is not feasible (because $s_2 < 0$), we look for a dual pivot element. The dual pivot row is row 2 and the dual pivot column is column 2 (because the ratio $\frac{-0.06}{-1} = 0.06$ is greater than $\frac{-0.04}{-1} = 0.04$ and none of the other columns may be considered). Pivoting on the -1 in row 2 and column 2, the tableau becomes:

	x_1	x_2	x_3	s_1	s_2	s_3	
s_1	0	0	1	1	1	0	30
x_2	1	1	0	0	-1	0	50
s_3	-2	0	1	0	1	1	-50
P	0.02	0	-0.12	0	-0.06	0	3

$s_3 < 0$ ← Dual pivot row

This is still not feasible (because $s_3 < 0$). Pivoting on the dual pivot element in row 3 and column 1, the tableau becomes:

	x_1	x_2	x_3	s_1	s_2	s_3	
s_1	0	0	1	1	1	0	30
x_2	0	1	1/2	0	-1/2	1/2	25
x_1	1	0	-1/2	0	-1/2	-1/2	25
P	0	0	-0.11	0	-0.05	0.01	2.5

Feasible: all are ≥ 0

Not optimal: some are < 0

This is now feasible but not optimal (because the bottom row has some negative entries). Pivoting on the (regular) pivot element in column 3 and row 1, we obtain:

	x_1	x_2	x_3	s_1	s_2	s_3	
x_3	0	0	1	1	1	0	30
x_2	0	1	0	-1/2	-1	1/2	10
x_1	1	0	0	1/2	0	-1/2	40
P	0	0	0	0.11	0.06	0.01	5.8

Feasible: all are ≥ 0

Optimal: all are ≥ 0

This is the final tableau because it is both feasible and optimal.

The maximum is $P = 5.8$ when $x_1 = 40$, $x_2 = 10$, and $x_3 = 30$.

In terms of the original situation, the manager should invest $40 million in the government bonds, $10 million in the blue chip stock, and $30 million in the biotechnology start-up company for a maximum return of $5.8 million.

Any Linear Programming Problem Can Now Be Solved

The two-stage simplex method allows us to solve *any* maximum linear programming problem, no matter what mixture of $\leq$ and $\geq$ inequalities are present in the constraints, provided that the variables are nonnegative. A direct way to change a minimum problem into a maximum problem is described in Exercises 35–36. Should the initial formulation include equality constraints, Exercises 37–38 demonstrate two ways to change the problem into a form that we can solve. Exercises 39–41 explain how to reformulate a problem so that all the variables are nonnegative. Taken together, these techniques allow us to solve *any* linear programming problem.

4.5 Section Summary

A *nonstandard* maximum problem has both $\geq$ and $\leq$ constraints. Change the $\geq$ inequalities into $\leq$ form by multiplying by -1, and then construct the initial simplex tableau. In general, this tableau will have negative numbers in the rightmost column (making it not feasible) as well as negative numbers in the bottom row (making it not optimal).

The dual pivot *row* is the row with the smallest negative entry in the rightmost column of the tableau (omitting the bottom row). The dual pivot *column* is the column with the largest ratio of the bottom entry divided by the dual pivot row entry, omitting any column with a zero or positive dual pivot row entry and omitting the rightmost column.

The two-stage simplex method pivots on dual pivot elements until the tableau is feasible (the right-hand column is nonnegative) and then continues to pivot on (regular) pivot elements until the tableau is optimal (the bottom row is nonnegative). The problem has no solution if the dual pivot column or the (regular) pivot row does not exist.

▶ Solutions to Practice Problems

1. a. The dual pivot row is row 1 because -10 is the smallest negative entry in the rightmost column (omitting the bottom row). The dual pivot column is column 2 because the ratio $\frac{-6}{-1} = 6$ is greater than the ratio $\frac{-5}{-1} = 5$ for the first column and the ratio $\frac{8}{-2} = -4$ for the fourth column; the other columns may not be considered since their dual pivot row entries are zero or positive, and the rightmost column is never considered. The dual pivot element is the -1 in row 1 and column 2.

   b. The tableau does not have a dual pivot element. The dual pivot row is row 3 because the smallest negative entry in the rightmost column (omitting the bottom row) is -20. The other entries in row 3 are either zero or positive, so there is no dual pivot column. (This means that the constraints are infeasible. The third row represents the inequality $2x_1 + x_2 + 4x_3 \leq -20$ and this is impossible because the variables are nonnegative. The problem has no solution.)

2. Rewriting the second inequlity as $-2x_1 - x_2 - 2x_3 \leq -10$, the initial simplex tableau is

	x_1	x_2	x_3	s_1	s_2	
s_1	1	2	1	1	0	50
s_2	-2	-1	-2	0	1	-10
P	-4	-1	-3	0	0	0

This is not feasible because $s_2 = -10$. Pivoting on the dual pivot element in row 2 and column 1, the tableau becomes feasible (no negatives on the right-hand column):

	x_1	x_2	x_3	s_1	s_2	
s_1	0	3/2	0	1	1/2	45
x_1	1	1/2	1	0	-1/2	5
P	0	1	1	0	-2	20

Pivoting on the (regular) pivot element in column 5 and row 1, the tableau becomes optimal (no negatives in the bottom row):

	x_1	x_2	x_3	s_1	s_2	
s_2	0	3	0	2	1	90
x_1	1	2	1	1	0	50
P	0	7	1	4	0	200

This is the final tableau because it is both feasible and optimal. The maximum is $P = 200$ when $x_1 = 50$ and $x_2 = 0$.

4.5 Exercises

For each nonstandard linear programming problem, construct the initial simplex tableau and locate the dual pivot element. (You do not need to carry out the pivot operation.)

1. Maximize $P = 15x_1 + 20x_2 + 18x_3$

Subject to $\begin{cases} 3x_1 + 2x_2 + 8x_3 \leq 96 \\ 5x_1 + x_2 + 6x_3 \geq 30 \\ x_1 \geq 0, \ x_2 \geq 0, \ x_3 \geq 0 \end{cases}$

2. Maximize $P = 60x_1 + 90x_2 + 30x_3$

Subject to $\begin{cases} 4x_1 + 5x_2 + x_3 \geq 40 \\ 10x_1 + 18x_2 + 3x_3 \leq 150 \\ x_1 \geq 0, \ x_2 \geq 0, \ x_3 \geq 0 \end{cases}$

3. Maximize $P = 6x_1 + 4x_2 + 6x_3$

Subject to $\begin{cases} 2x_1 + x_2 + 3x_3 \geq 30 \\ x_1 + x_2 + 2x_3 \geq 20 \\ x_1 \geq 0, \ x_2 \geq 0, \ x_3 \geq 0 \end{cases}$

4. Maximize $P = 9x_1 + 5x_2 + 4x_3$

Subject to $\begin{cases} 3x_1 + x_2 + x_3 \geq 40 \\ 2x_1 + 2x_2 + x_3 \geq 30 \\ x_1 \geq 0, \ x_2 \geq 0, \ x_3 \geq 0 \end{cases}$

5. Maximize $P = 20x_1 + 30x_2 + 10x_3$

Subject to $\begin{cases} x_1 + x_2 + x_3 \geq 8 \\ x_1 + 2x_2 + 3x_3 \leq 30 \\ x_1 + 2x_2 + x_3 \leq 18 \\ x_1 \geq 0, \ x_2 \geq 0, \ x_3 \geq 0 \end{cases}$

6. Maximize $P = 24x_1 + 18x_2 + 30x_3$

Subject to $\begin{cases} x_1 + x_2 + x_3 \geq 4 \\ x_1 + 2x_2 + 3x_3 \leq 15 \\ x_1 + 2x_2 + x_3 \leq 9 \\ x_1 \geq 0, \ x_2 \geq 0, \ x_3 \geq 0 \end{cases}$

7. Maximize $P = 3x_1 + 2x_2 + 5x_3 + 4x_4$

Subject to $\begin{cases} x_1 + x_2 + x_3 + x_4 \geq 30 \\ 2x_1 + 3x_2 + 2x_3 + x_4 \geq 20 \\ 4x_1 + 2x_2 + x_3 + 2x_4 \leq 80 \\ x_1 \geq 0, \ x_2 \geq 0, \ x_3 \geq 0, \ x_4 \geq 0 \end{cases}$

8. Maximize $P = 6x_1 + 8x_2 + 3x_3 + 5x_4$

Subject to $\begin{cases} x_1 + x_2 + x_3 + x_4 \geq 70 \\ 3x_1 + 2x_2 + 4x_3 + 3x_4 \geq 60 \\ 4x_1 + x_2 + 2x_3 + 5x_4 \leq 100 \\ x_1 \geq 0, \ x_2 \geq 0, \ x_3 \geq 0, \ x_4 \geq 0 \end{cases}$

4.5 NONSTANDARD PROBLEMS: THE DUAL PIVOT ELEMENT AND THE TWO-STAGE METHOD

Solve each nonstandard linear programming problem by the two-stage simplex method. (Exercises 9, 10, 13, 14, 17, and 18 can also be solved by the graphical method.)

You may use the PIVOT program if permitted by your instructor.

9. Maximize $P = 4x_1 + x_2$

Subject to $\begin{cases} 3x_1 + 2x_2 \leq 120 \\ x_1 + x_2 \geq 50 \\ x_1 \geq 0, \ x_2 \geq 0 \end{cases}$

10. Maximize $P = x_1 + 4x_2$

Subject to $\begin{cases} 3x_1 + 2x_2 \leq 120 \\ 2x_1 + x_2 \geq 60 \\ x_1 \geq 0, \ x_2 \geq 0 \end{cases}$

11. Maximize $P = 30x_1 + 20x_2 + 28x_3$

Subject to $\begin{cases} 3x_1 + x_2 + 2x_3 \geq 30 \\ 4x_1 + x_2 + 3x_3 \leq 60 \\ x_1 \geq 0, \ x_2 \geq 0, \ x_3 \geq 0 \end{cases}$

12. Maximize $P = 18x_1 + 10x_2 + 20x_3$

Subject to $\begin{cases} 3x_1 + x_2 + 4x_3 \geq 30 \\ 2x_1 + x_2 + 5x_3 \leq 50 \\ x_1 \geq 0, \ x_2 \geq 0, \ x_3 \geq 0 \end{cases}$

13. Maximize $P = 4x_1 + 5x_2$

Subject to $\begin{cases} x_1 + 2x_2 \leq 12 \\ x_1 + x_2 \geq 15 \\ 2x_1 + x_2 \leq 12 \\ x_1 \geq 0, \ x_2 \geq 0 \end{cases}$

14. Maximize $P = 5x_1 + 4x_2$

Subject to $\begin{cases} 2x_1 + 3x_2 \leq 12 \\ x_1 + x_2 \geq 10 \\ 3x_1 + 2x_2 \leq 12 \\ x_1 \geq 0, \ x_2 \geq 0 \end{cases}$

15. Maximize $P = 15x_1 + 12x_2 + 18x_3$

Subject to $\begin{cases} 5x_1 + x_2 + 2x_3 \geq 30 \\ 2x_1 + x_2 + 3x_3 \leq 24 \\ x_1 \geq 0, \ x_2 \geq 0, \ x_3 \geq 0 \end{cases}$

16. Maximize $P = 4x_1 + 5x_2 + 6x_3$

Subject to $\begin{cases} 5x_1 + x_2 + 3x_3 \leq 30 \\ 2x_1 + x_2 + 2x_3 \geq 24 \\ x_1 \geq 0, \ x_2 \geq 0, \ x_3 \geq 0 \end{cases}$

17. Maximize $P = x_1 + x_2$

Subject to $\begin{cases} x_1 \leq 8 \\ x_2 \leq 5 \\ x_1 + 2x_2 \geq 6 \\ x_1 \geq 0, \ x_2 \geq 0 \end{cases}$

18. Maximize $P = x_1 + x_2$

Subject to $\begin{cases} x_1 \leq 6 \\ x_2 \leq 9 \\ 3x_1 + x_2 \geq 6 \\ x_1 \geq 0, \ x_2 \geq 0 \end{cases}$

19. Maximize $P = 4x_1 + 6x_2 + 12x_3 + 10x_4$

Subject to $\begin{cases} x_1 + x_2 + x_3 + x_4 \leq 60 \\ 2x_1 + x_2 + x_3 + 2x_4 \geq 10 \\ 2x_1 + 2x_2 + x_3 + x_4 \leq 100 \\ x_1 \geq 0, \ x_2 \geq 0, \ x_3 \geq 0, \ x_4 \geq 0 \end{cases}$

20. Maximize $P = 8x_1 - x_2 + 5x_3 + 14x_4$

Subject to $\begin{cases} x_1 + 2x_2 + 3x_4 \leq 55 \\ x_1 + x_2 + x_3 + 2x_4 \geq 25 \\ x_1 + 3x_3 + 3x_4 \leq 45 \\ x_1 \geq 0, \ x_2 \geq 0, \ x_3 \geq 0, \ x_4 \geq 0 \end{cases}$

21. Maximize $P = 5x_1 + 32x_2 + 3x_3 + 15x_4$

Subject to $\begin{cases} 5x_1 + 8x_2 + x_3 + 3x_4 \geq 18 \\ 2x_1 - x_2 + x_3 + x_4 \leq 10 \\ 3x_1 - 3x_2 + x_3 + 2x_4 \geq 6 \\ x_1 \geq 0, \ x_2 \geq 0, \ x_3 \geq 0, \ x_4 \geq 0 \end{cases}$

22. Maximize $P = 70x_1 + 12x_2 + 60x_3 + 20x_4$

Subject to $\begin{cases} 2x_1 + x_2 + x_3 + x_4 \leq 14 \\ 3x_1 + 3x_2 + x_3 + 2x_4 \leq 24 \\ 5x_1 + 3x_2 + 4x_3 + 2x_4 \geq 44 \\ x_1 \geq 0, \ x_2 \geq 0, \ x_3 \geq 0, \ x_4 \geq 0 \end{cases}$

23. Maximize $P = -x_1 + x_2 - 2x_3$

Subject to $\begin{cases} x_1 + x_2 + x_3 \leq 50 \\ x_1 + x_3 \geq 10 \\ x_2 + x_3 \geq 20 \\ x_1 \geq 0, \ x_2 \geq 0, \ x_3 \geq 0 \end{cases}$

24. Maximize $P = x_1 + 2x_2 + x_3$

Subject to $\begin{cases} x_1 + x_2 + x_3 \leq 30 \\ x_1 + x_2 \leq 25 \\ x_2 + x_3 \geq 15 \\ x_1 \geq 0, \ x_2 \geq 0, \ x_3 \geq 0 \end{cases}$

APPLIED EXERCISES

Express each situation as a nonstandard maximum problem. Be sure to state the meaning of each variable. Solve the problem by the two-stage simplex method. State your final answer in terms of the original question.

 You may use the PIVOT program if permitted by your instructor.

25. **PERSONAL FINANCE: Financial Planning** A retired couple want to invest their $20,000 life savings in bank certificates of deposit yielding 6% and Treasury bonds yielding 5%. If they want at least $5000 in each type of investment, how much should they invest in each to receive the greatest possible income?

26. **ENVIRONMENTAL SCIENCE: Recycling Management** A volunteer recycling center accepts both used paper and empty glass bottles, which it then sorts and sells to a reprocessing company. The center has room to accept a total of 800 crates of paper and glass each week and has 50 hours of volunteer help to do the sorting. Each crate of paper products takes 5 minutes to sort and sells for 8¢, while each crate of bottles takes 3 minutes to sort and sells for 7¢. To support the city's "grab that glass" recycle theme this week, the center wants to accept at least 600 crates of glass. How many crates of each should the center accept this week to raise the most money for its ecology scholarship fund?

27. **BUSINESS: Advertising** The manager of a new mall may spend up to $18,000 on "grand opening" announcements in newspapers, on radio, and on TV. Each newspaper ad costs $300 and reaches 5000 readers, each one-minute radio commercial costs $800 and is heard by 13,000 listeners, and each 15-second TV spot costs $900 and is seen by 15,000 viewers. If the manager wants at least 5 newspaper ads and at least 20 radio commercials and TV spots combined, how many of each should be placed to reach the largest number of potential customers? (Ignore multiple exposures to the same customer.)

28. **SOCIAL SCIENCE: Political Advertising** In the last few days before the election, a politician can afford to spend no more than $27,000 on TV advertisements and can arrange for no more than 10 ads. Each daytime ad costs $2000 and reaches 4000 viewers, each prime time ad costs $3000 and reaches 5000 viewers, and each late night ad costs $1000 and reaches 2000 viewers. To be sure to reach the widest variety of voters, the politician's advisors insist on at least 5 ads scheduled during daytime and late night combined. Ignoring repeated viewings by the same person and assuming every viewer can vote, how many ads in each of the time periods will reach the most voters?

29. **BUSINESS: Resource Allocation** A furniture shop manufactures wooden desks, tables, and chairs. The numbers of hours to assemble and finish each piece are shown in the table, together with the number of hours of skilled labor available for each task. To meet expected demand, a total of at least 30 desks and tables combined must be made. If the profit is $75 for each desk, $84 for each table, and $66 for each chair, how many of each should the company manufacture to obtain the greatest possible profit?

	Desk	Table	Chair	Labor Available
Assembly	2 hours	1 hour	2 hours	210 hours
Finishing	2 hours	3 hours	1 hour	150 hours

30. **BUSINESS: Production Planning** An automotive parts shop rebuilds carburetors, fuel pumps, and alternators. The numbers of hours to rebuild and then inspect and pack each part are shown in the table, together with the number of hours of skilled labor available for each task. At least 12 carburetors must be rebuilt. If the profit is $12 for each carburetor, $14 for each fuel pump, and $10 for each alternator, how many of each should the shop rebuild to obtain the greatest possible profit?

	Carburetor	Fuel Pump	Alternator	Labor Available
Rebuilding	5 hours	4 hours	3 hours	200 hours
Inspection & packaging	1 hour	1 hour	0.5 hour	45 hours

31. **BUSINESS: Agriculture** A farmer grows wheat, barley, and oats on her 500-acre farm. Each acre of wheat requires 3 days of labor (to plant, tend, and harvest) and costs $21 (for seed, fertilizer, and pesticides), each acre of barley requires 2 days of labor and costs $27, and each acre of oats requires 3 days of labor and costs $24. The farmer and her hired field hands can provide no more than 1200 days of labor this year. She wants to grow at least 100 acres of oats and can afford to spend no more than $15,120. If the profit is $50 for each acre of wheat, $40 for each acre of barley, and $45 for each acre of oats, how many acres of each crop should she grow to obtain the greatest possible profit?

32. **BUSINESS: Agriculture** A farmer grows corn, peanuts, and soybeans on his 240-acre farm. To maintain soil fertility, the farmer rotates the crops and always plants at least as many acres of soybeans as the total acres of the other crops. Because he has promised to sell some of his corn to a neighbor who raises cattle, he must plant at least 42 acres of corn. Each acre of corn requires 2 days of labor and yields a profit of $150, each acre of peanuts requires 5 days of labor and yields a profit of $300, and each acre of soybeans requires 1 day of labor and yields a profit of $100. If the farmer and his children can put in at most 630 days of labor, how many acres of each crop should the farmer plant to obtain the greatest possible profit?

33. **ENVIRONMENTAL SCIENCE: Pollution Control** A storage yard at a coal burning electric power plant can hold no more than 100,000 tons of coal. Two grades of coal are available: low sulfur (1%) with an energy content of 20 million BTU per ton and high sulfur (2%) with an energy content of 30 million BTU per ton. If existing contracts with the high-sulfur coal mine operator require that at least 50,000 tons of high sulfur coal be purchased, and if the next coal purchase may contain no more than 1400 tons of sulfur, how many tons of each type of coal should be purchased to obtain the most energy?

34. **BUSINESS: Production Planning** A small jewelry company prepares and mounts semi-precious stones. There are 20 lapidaries (who cut and polish the stones) and 24 jewelers (who mount the stones in gold settings). Each employee works 7 hours each day, 5 days each week. Each tray of agates requires 5 hours of cutting and polishing and 4 hours to mount, each tray of onyxes requires 2 hours of cutting and polishing and 3 hours to mount, and each tray of garnets requires 6 hours of cutting and polishing and 3 hours to mount. Furthermore, the company's owner has decided that the company will process at least 12 trays of agates each week. If the profit is $15 for each tray of agates, $10 for each tray of onyxes, and $13 for each tray of garnets, how many trays of each stone should be processed each week to obtain the greatest possible profit?

Explorations and Excursions

The following problems extend and augment the material presented in the text.

The PIVOT program may be helpful (if permitted by your instructor).

Minimum Problems and the Dual Pivot Element

35. a. Solve both of the following problems by the graphical method from Section 4.2. Is it true that the minimum of C is the same as -1 times the maximum of $P = -C$?

    Minimize $C = 3x + 4y$

    Subject to $\begin{cases} x + y \leq 10 \\ 2x + y \geq 8 \\ x \geq 0 \text{ and } y \geq 0 \end{cases}$

    Maximize $P = -3x - 4y$

    Subject to $\begin{cases} x + y \leq 10 \\ 2x + y \geq 8 \\ x \geq 0 \text{ and } y \geq 0 \end{cases}$

(continues)

b. Solve the following problem by finding the dual maximum problem and using the (regular) simplex method as we did in Section 4.4.

$$\text{Minimize } C = 3y_1 + 4y_2$$

$$\text{Subject to } \begin{cases} -y_1 - y_2 \geq -10 \\ 2y_1 + y_2 \geq 8 \\ y_1 \geq 0 \text{ and } y_2 \geq 0 \end{cases}$$

c. Solve the following nonstandard problem by the two-stage simplex method.

$$\text{Maximize } P = -3x_1 - 4x_2$$

$$\text{Subject to } \begin{cases} x_1 + x_2 \leq 10 \\ -2x_1 - x_2 \leq -8 \\ x_1 \geq 0 \text{ and } x_2 \geq 0 \end{cases}$$

d. Compare the tableaux and pivot elements in parts (b) and (c). Are the pivot elements the same numbers? Is the arithmetic to choose each pair of pivot elements the same? Is the "dual pivot element" just the "dual" of the (regular) pivot element?

36. Repeat the process of Exercise 35 to find "dual" solutions to the following problems.

$$\text{Minimize } C = 20y_1 + 18y_2$$

$$\text{Subject to } \begin{cases} 2y_1 + y_2 \geq 18 \\ y_1 + y_2 \geq 14 \\ 4y_1 + 3y_2 \geq 48 \\ y_1 \geq 0 \text{ and } y_2 \geq 0 \end{cases}$$

$$\text{Maximize } P = -20x_1 - 18x_2$$

$$\text{Subject to } \begin{cases} -2x_1 - x_2 \leq -18 \\ -x_1 - x_2 \leq -14 \\ -4x_1 - 3x_2 \leq -48 \\ x_1 \geq 0 \text{ and } x_2 \geq 0 \end{cases}$$

Equality Constraints

Equality constraints may either be rewritten as a pair of inequalities (one with $\leq$ and the other with $\geq$) or used to eliminate some of the variables from the problem.

37. Replace the equality constraint $2x_1 + x_2 = 6$ in the following problem by the two inequalities $2x_1 + x_2 \leq 6$ and $2x_1 + x_2 \geq 6$ and then solve it by the two-stage simplex method.

Check your answer by solving the original problem by the graphical method of Section 4.2.

$$\text{Maximize } P = 5x_1 + 3x_2$$

$$\text{Subject to } \begin{cases} x_1 + x_2 \leq 4 \\ 2x_1 + x_2 = 6 \\ x_1 \geq 0 \text{ and } x_2 \geq 0 \end{cases}$$

38. Solve the equality constraint $x_1 + x_3 = 20$ in the following problem for x_3 and then eliminate that variable from the objective function and the constraints (including the nonnegativity requirement that $x_3 \geq 0$). Solve the resulting two-variable problem by the two-stage simplex method. Be sure to state your solution in terms of the original problem.

$$\text{Maximize } P = 5x_1 + 6x_2 + 4x_3$$

$$\text{Subject to } \begin{cases} 2x_1 + x_2 + x_3 \leq 60 \\ x_1 + x_2 + x_3 \geq 40 \\ x_1 + x_3 = 20 \\ x_1 \geq 0, \ x_2 \geq 0, \text{ and } x_3 \geq 0 \end{cases}$$

More About Nonnegativity Conditions

The following problems show how to reformulate linear programming problems to have nonnegativity conditions on *all* the variables when the original versions do not.

39. Show that the first problem may be rewritten as the second by replacing the variable x_1 by a new variable u_1 where $u_1 = x_1 + 8$ and $u_1 \geq 0$. Solve the first problem by the graphical method from Section 4.2, solve the second by the simplex method from Section 4.3, and check that your answers agree.

$$\text{Maximize } P = 5x_1 + 3x_2$$

$$\text{Subject to } \begin{cases} 2x_1 + x_2 \leq 6 \\ x_1 + x_2 \leq 4 \\ x_1 \geq -8 \text{ and } x_2 \geq 0 \end{cases}$$

$$\text{Maximize } P = 5u_1 + 3x_2 - 40$$

$$\text{Subject to } \begin{cases} 2u_1 + x_2 \leq 22 \\ u_1 + x_2 \leq 12 \\ u_1 \geq 0 \text{ and } x_2 \geq 0 \end{cases}$$

40. Show that the first problem may be rewritten as the second by replacing the variable x_1 by a new variable u_1 where $u_1 = -x_1$ and $u_1 \geq 0$. Solve the first problem by the graphical method from Section 4.2, solve the second by the simplex method from Section 4.3, and check that your answers agree.

Maximize $P = 5x_1 + 3x_2$

Subject to $\begin{cases} 2x_1 + x_2 \leq 10 \\ -x_1 + x_2 \leq 4 \\ x_1 \leq 0 \text{ and } x_2 \geq 0 \end{cases}$

Maximize $P = -5u_1 + 3x_2$

Subject to $\begin{cases} -2u_1 + x_2 \leq 10 \\ u_1 + x_2 \leq 4 \\ u_1 \geq 0 \text{ and } x_2 \geq 0 \end{cases}$

Unrestricted Variables

41. A variable is *unrestricted* if it may take *any* value (positive, negative, or zero). Show that the first problem may be rewritten as the second by replacing the variable x_1 by two new variables u_1 and u_2 where $u_1 - u_2 = x_1$ and $u_1, u_2 \geq 0$. Solve the first problem by the graphical method from Section 4.2, solve the second by the simplex method from Section 4.3, and check that your answers agree.

Maximize $P = 3x_1 + 5x_2$

Subject to $\begin{cases} 2x_1 + x_2 \leq 6 \\ x_2 \leq 4 \\ x_1 \text{ unrestricted and } x_2 \geq 0 \end{cases}$

Maximize $P = 3u_1 - 3u_2 + 5x_2$

Subject to $\begin{cases} 2u_1 - 2u_2 + x_2 \leq 6 \\ x_2 \leq 4 \\ u_1 \geq 0, \ u_2 \geq 0, \text{ and } x_2 \geq 0 \end{cases}$

[*Hint:* Can any real number be written as the difference of two nonnegative numbers?]

4.6 NONSTANDARD PROBLEMS: ARTIFICIAL VARIABLES AND THE BIG-M METHOD

This and the previous section are optional and present alternative methods for solving nonstandard linear programming problems. The choice of which one to cover is left to the instructor. The examples and exercises duplicate those in Section 4.5.

Introduction

In this section we extend the simplex method to maximum problems with constraints involving $\geq$ and $=$ as well as the usual $\leq$. Because a basis for such a problem is not necessarily feasible or even immediately apparent, we introduce new "artificial" variables to provide an initial basis and then force the simplex method to seek a solution involving just the original variables.

Artificial Variables

The method requires that the right-hand side of each inequality be nonnegative. We multiply constraints by -1 as necessary to bring this about, remembering that such multiplication reverses the sense of the inequality (changing, for example $5 \geq -3$ into $-5 \leq 3$). We then introduce new variables into the constraints to make them into equations, but in different ways for the different types of constraints. Each of these new variables is required to be *nonnegative*.

- For $\leq$ constraints, we add slack variables in the usual way, so that, for example, $2x_1 + x_2 \leq 30$ becomes $2x_1 + x_2 + s_1 = 30$, with s_1 in the basis as $s_1 = 30$.

- For $\geq$ constraints, meaning that the left side may be *larger* than the right, we must *subtract* a slack variable* to obtain an equation, so that, for example, $x_1 + 2x_2 \geq 6$ becomes $x_1 + 2x_2 - s_2 = 6$. However, we cannot include s_2 in the basis because $-s_2 = 6$ means $s_2 = -6$, violating the nonnegativity condition of the variables. Therefore, we add an *artificial variable* a_1, giving $x_1 + 2x_2 - s_2 + a_1 = 6$. Such variables are called "artificial" because they have no interpretation in the original problem and are used only to obtain variables for the initial basis, in this case $a_1 = 6$.

- For *equality* constraints, which need no slack variables, and therefore do not give basic variables, we again introduce an artificial variable a_2 so that, for example, $x_1 + x_2 = 10$ becomes $x_1 + x_2 + a_2 = 10$, with $a_2 = 10$ in the basis.

The artificial variables a_1 and a_2 were added merely to create a feasible basis to begin with. However, since they have no real role in the problem, we must ensure that the simplex method removes them permanently from the basis. To do this, we subtract a large multiple M of each artificial variable from the objective function, so that, for example, $P = 8x_1 + 12x_2$ becomes $P = 8x_1 + 12x_2 - Ma_1 - Ma_2$. It is from this large number that the big-M method gets its name.

How do we know that there is a sufficiently large number M, and if so, how do we find it? Because a linear programming problem has a finite number of constraints, variables, and vertices, M need only be larger than the absolute values of a finite list of numbers, and so certainly exists. If the solution is to be calculated on a computer, a number near the largest possible number that the computer can represent will do (and if this doesn't suffice, the solution can't be calculated on that computer anyway). When calculating by hand, we will keep M as a *variable,* remembering that it stands for a large positive number.

The new variables gives us an "extended" maximum problem.

---

*Subtracted slack variables are sometimes called *surplus* variables.

Extended Problem

To form the extended problem for a nonstandard maximum problem:

1. Multiple constraints by -1 as necessary so that their right-hand sides are nonnegative.
2. In each $\leq$ constraint, add a slack variable.
3. In each $\geq$ constraint, subtract a slack variable and add an artificial variable.
4. In each $=$ constraint, add an artificial variable.
5. In the objective function, subtract M times each artificial variable.

Although it is not necessary, we will order constraints according to their type: first $\leq$, then $\geq$, and then $=$.

From the extended problem we write a *preliminary* tableau in the usual way, which we then modify to bring the artificial variables into the initial basis, giving the *initial* tableau.

EXAMPLE 1 CONSTRUCTING AN INITIAL TABLEAU

Construct the extended problem and write the preliminary and initial tableaux for the following nonstandard linear programming problem.

$$\text{Maximize } P = 8x_1 + 12x_2$$

$$\text{Subject to } \begin{cases} 2x_1 + x_2 \leq 30 \\ x_1 + 2x_2 \geq 6 \\ x_1 + x_2 = 10 \\ x_1 \geq 0 \text{ and } x_2 \geq 0 \end{cases} \quad \begin{array}{c}\text{"Mixed constraints"}\\ \leq, \geq, =\end{array}$$

Solution

The right-hand sides of the constraints are nonnegative, so we do not need to multiply by -1. Adding a slack variable to the $\leq$ constraint, subtracting a slack and adding an artificial variable to the $\geq$ constraint, and adding an artificial variable to the $=$ constraint, and finally subtracting M times the artificial variables from the objective function gives the extended problem:

Maximize $P = 8x_1 + 12x_2 - Ma_1 - Ma_2$ M is a large positive number

Subject to
$$\begin{cases} 2x_1 + x_2 + s_1 = 30 \\ x_1 + 2x_2 - s_2 + a_1 = 6 \\ x_1 + x_2 + a_2 = 10 \\ x_1 \geq 0, \ x_2 \geq 0, \ s_1 \geq 0, \\ s_2 \geq 0, \ a_1 \geq 0, \ a_2 \geq 0 \end{cases}$$

s_1 and s_2 are slack variables
a_1 and a_2 are artificial variables

All variables are nonnegative

From the extended problem we construct a *preliminary* tableau. Be sure you understand why the 1s and -1s are where they are, since later you may want to go directly from the original problem to the preliminary tableau (or even to the *initial* tableau given in the next step).

	x_1	x_2	s_1	s_2	a_1	a_2	
s_1	2	1	1	0	0	0	30
(a_1)	1	2	0	-1	1	0	6
(a_2)	1	1	0	0	0	1	10
P	-8	-12	0	0	M	M	0

"Preliminary" because the artificial variables a_1 and a_2 do not have 0s at the bottoms of their columns

To obtain zeros where the M's are we pivot on the 1s in their columns, which is equivalent to subtracting M times row 2 and M times row 3 from row 4, obtaining the following *initial tableau*, with the values of the basic variables written on the right.

	x_1	x_2	s_1	s_2	a_1	a_2		
s_1	2	1	1	0	0	0	30	$\leftarrow s_1 = 30$
a_1	1	2	0	-1	1	0	6	$\leftarrow a_1 = 6$
a_2	1	1	0	0	0	1	10	$\leftarrow a_2 = 10$
P	$-2M-8$	$-3M-12$	0	M	0	0	$-16M$	$\leftarrow P = -16M$

The basic solution on the right of the above tableau, with the remaining variables all set equal to zero, is *feasible* (all variables are nonnegative) but is far from optimal (P is negative). To solve the problem we would now pivot as usual on this tableau to try to make P optimal, in the process removing a_1 and a_2 from the basis.

> **Initial Tableau**
>
> The tableau written directly from the extended problem is the *preliminary* tableau. Pivoting on the 1s in the columns of the artificial variables gives the *initial* tableau.

4.6 NONSTANDARD PROBLEMS: ARTIFICIAL VARIABLES AND THE BIG-M METHOD

Practice Problem

For the following nonstandard maximum problem, write:
a. the extended problem
b. the preliminary tableau
c. the initial tableau.

Maximize $P = 5x_1 + 6x_2 - 3x_3$

Subject to $\begin{cases} x_1 - x_2 + 2x_3 \leq 10 \\ x_1 + 2x_2 - x_3 \geq 20 \\ x_1 \geq 0,\ x_2 \geq 0,\ x_3 \geq 0 \end{cases}$

▶ Solution on page 339

The Big-M Method

The big-M method solves a nonstandard maximum problem by introducing artificial variables to form the *extended* problem, writing the preliminary tableau to obtain the initial tableau that displays an initial basic feasible point, and then pivoting as usual to find the solution.

> **Big-M Method**
>
> To solve a nonstandard maximum problem by the big-M method:
>
> 1. Write the extended problem (page 331).
> 2. Construct the initial tableau for the extended problem (page 332).
> 3. Solve the extended problem by the simplex method (page 287).
> 4. If there is a solution to the extended problem in which every artificial variable is 0, then this is also the solution to the original problem. Otherwise, there is no solution to the original problem.

EXAMPLE 2

USING THE BIG-M METHOD

Solve the following nonstandard maximum problem by the big-M method.

Maximize $P = x_1 - x_2 + 3x_3$

Subject to $\begin{cases} x_1 + x_2 \leq 12 \\ x_2 + x_3 \geq 6 \\ x_1 + x_3 = 3 \\ x_1 \geq 0,\ x_2 \geq 0,\ x_3 \geq 0 \end{cases}$

Solution

The right-hand sides of the constraints are all nonnegative so we do not need to multiply any by -1. With slack and artificial variables the problem becomes:

Maximize $P = x_1 - x_2 + 3x_3 - Ma_1 - Ma_2$

Subject to
$$\begin{cases} x_1 + x_2 + s_1 = 12 \\ x_2 + x_3 - s_2 + a_1 = 6 \\ x_1 + x_3 + a_2 = 3 \\ x_1 \geq 0, \ x_2 \geq 0, \ x_3 \geq 0, \ s_1 \geq 0, \ s_2 \geq 0, \ a_1 \geq 0, \ a_2 \geq 0 \end{cases}$$

The *preliminary* tableau is

	x_1	x_2	x_3	s_1	s_2	a_1	a_2	
s_1	1	1	0	1	0	0	0	12
(a_1)	0	1	1	0	-1	1	0	6
(a_2)	1	0	1	0	0	0	1	3
P	-1	1	-3	0	0	M	M	0

"Preliminary" because the artificial variables a_1 and a_2 do not have 0s at the bottoms of their columns

To obtain the *initial* tableau we need 0s where the M's are, which we accomplish by pivoting on the 1s in their columns (column 6, row 2 and column 7, row 3), or equivalently, subtracting M times row 2 and M times row 3 from the bottom row.

	x_1	x_2	x_3	s_1	s_2	a_1	a_2	
s_1	1	1	0	1	0	0	0	12
a_1	0	1	1	0	-1	1	0	6
a_2	1	0	1	0	0	0	1	3
P	$-M-1$	$-M+1$	$-2M-3$	0	M	0	0	$-9M$

New pivot element

The pivot column is column 3 (because $-2M - 3$ is smaller than both $-M - 1$ and $-M + 1$ for large positive values of M) and the pivot row is row 3 (because $\frac{3}{1} = 3$ is smaller than $\frac{6}{1} = 6$ and we don't consider the first row because of the 0), and pivoting on the 1 in column 3 and row 3 gives:

	x_1	x_2	x_3	s_1	s_2	a_1	a_2	
s_1	1	1	0	1	0	0	0	12
a_1	-1	1	0	0	-1	1	-1	3
x_3	1	0	1	0	0	0	1	3
P	$M+2$	$-M+1$	0	0	M	0	$2M+3$	$-3M+9$

New pivot element

The bottom row was found by multiplying row 3 (the pivot row) by $2M + 3$ and adding it to the bottom row

We now pivot on column 2 (because $-M + 1$ is negative) and row 2 (because $\frac{3}{1} = 3$ is less than $\frac{12}{1} = 12$ and we don't consider the third row because of the 0), and pivoting on the 1 in column 2 and row 2 gives:

	x_1	x_2	x_3	s_1	s_2	a_1	a_2		
s_1	2	0	0	1	1	-1	1	9	← $s_1 = 9$
x_2	-1	1	0	0	-1	1	-1	3	← $x_2 = 3$
x_3	1	0	1	0	0	0	1	3	← $x_3 = 3$
P	3	0	0	0	1	$M-1$	$M+4$	6	← $P = 6$

This is the *final* tableau because there are no negative numbers in the bottom row. The solution of the extended problem is given on the right of the tableau, with the nonbasic variables $x_1, s_2, a_1,$ and a_2 all given the value zero. Therefore, the solution to the original problem is the same:

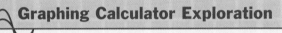

The maximum value of P is 6 when $x_1 = 0$, $x_2 = 3$, and $x_3 = 3$.

Graphing Calculator Exploration

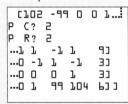

a. On calculators that perform only numerical calculations, the big-M method is carried out by using a particular large value for M. The TI-83 screen on the left shows the right-hand part of the final tableau in the preceding example, after pivoting on column 2 and row 2 using the program PIVOT (see page 285) with $M = 100$. The last column shows the solution we found, and the bottom row agrees with bottom row of the preceding tableau if $M = 100$.

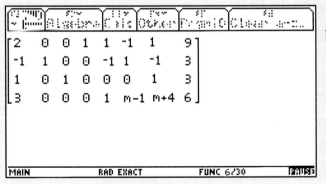

b. Some more advanced graphing calculators can perform symbolic as well as numerical calculations. Using the program* PIVOT for the TI-92 gives the same final tableau that we found by hand, with the last column again giving the solution that we found.

* See the Preface for information on how to obtain this and other programs.

EXAMPLE 3

MANAGING AN INVESTMENT PORTFOLIO

The manager of an $80 million mutual fund has three investment possibilities: secure government bonds yielding 4%, a blue chip growth stock paying 6%, and a biotechnology start-up company returning 12%. If the manager wants to invest at least $50 million in a combination of safe government bonds and blue chip stock, and at least as much in the bonds as the total in blue chip and biotechnology stocks, how much should be invested in each to obtain the greatest possible return?

Solution

Let

$$x_1 = \begin{pmatrix} \text{Amount in} \\ \text{government bonds} \end{pmatrix}$$

$$x_2 = \begin{pmatrix} \text{Amount in} \\ \text{blue chip stock} \end{pmatrix} \quad \text{In millions of dollars}$$

$$x_3 = \begin{pmatrix} \text{Amount in} \\ \text{biotechnology stock} \end{pmatrix}$$

The objective is to maximize the return

$$P = 0.04x_1 + 0.06x_2 + 0.12x_3 \qquad \text{Rate} \times \text{amount for each investment}$$

subject to the restrictions

$$x_1 + x_2 + x_3 \leq 80 \qquad \text{\$80 million available to invest}$$
$$x_1 + x_2 \geq 50 \qquad \text{\$50 million in safe investments}$$
$$x_1 \geq x_2 + x_3 \qquad \text{Bonds} \geq \text{total in stocks}$$
$$x_1 \geq 0, \quad x_2 \geq 0, \quad x_3 \geq 0 \qquad \text{Nonnegativity}$$

Rewriting the third constraint as $x_1 - x_2 - x_3 \geq 0$, we find that the second and third constraints require artificial variables. Pivoting twice on the preliminary tableau, the initial tableau is:

	x_1	x_2	x_3	s_1	s_2	s_3	a_1	a_2	
s_1	1	1	1	1	0	0	0	0	80
a_1	1	1	0	0	-1	0	1	0	50
a_2	1	-1	-1	0	0	-1	0	1	0
P	$-2M-0.04$	-0.06	$M-0.12$	0	M	M	0	0	$-50M$

Pivoting on the 1 in column 1 and row 3 yields

	x_1	x_2	x_3	s_1	s_2	s_3	a_1	a_2	
s_1	0	2	2	1	0	1	0	-1	80
a_1	0	2	1	0	-1	1	1	-1	50
x_1	1	-1	-1	0	0	-1	0	1	0
P	0	$-2M-0.10$	$-M-0.16$	0	M	$-M-0.04$	0	$-2M+0.04$	$-50M$

4.6 NONSTANDARD PROBLEMS: ARTIFICIAL VARIABLES AND THE BIG-M METHOD

Pivoting on the 2 in column 2 and row 2 gives:

	x_1	x_2	x_3	s_1	s_2	s_3	a_1	a_2	
s_1			1	1	1	0	-1	0	30
x_2	0	1	1/2	0	$-1/2$	1/2	1/2	$-1/2$	25
x_1	1	0	$-1/2$	0	$-1/2$	$-1/2$	1/2	1/2	25
P	0	0	-0.11	0	-0.05	0.01	$M + 0.05$	$M - 0.01$	2.5

There are now no artificial variables in the basis. Pivoting on the 1 in column 3 and row 1 brings us to the final tableau:

	x_1	x_2	x_3	s_1	s_2	s_3	a_1	a_2	
x_3	0	0	1	1	1	0	-1	0	30
x_2	0	1	0	$-1/2$	-1	1/2	1	$-1/2$	10
x_1	1	0	0	1/2	0	$-1/2$	0	1/2	40
P	0	0	0	0.11	0.06	0.01	$M - 0.06$	$M - 0.01$	5.8

Because the extended problem has a solution with the artificial variables all equal to zero, we have the solution of the original problem:

The maximum is $P = 5.8$ when $x_1 = 40$, $x_2 = 10$, and $x_3 = 30$.

In terms of the original situation, the manager should invest $40 million in the government bonds, $10 million in the blue chip stock, and $30 million in the biotechnology start-up company for a maximum return of $5.8 million.

Computer Exploration

Linear programming problems can also be solved by specialized computer programs. One of the best known, both in education and in industry, is LINDO (for **l**inear, **in**teractive, and **d**iscrete **o**ptimizer).* The following screen shows the LINDO 6.1 solution of Example 3. Note the similarity between the problem input format and our notation for linear programming problems.

---

*Further information about LINDO can be found at the Internet site *http://www.lindo.com* or obtained from LINDO Systems, Inc., Chicago, Illinois.

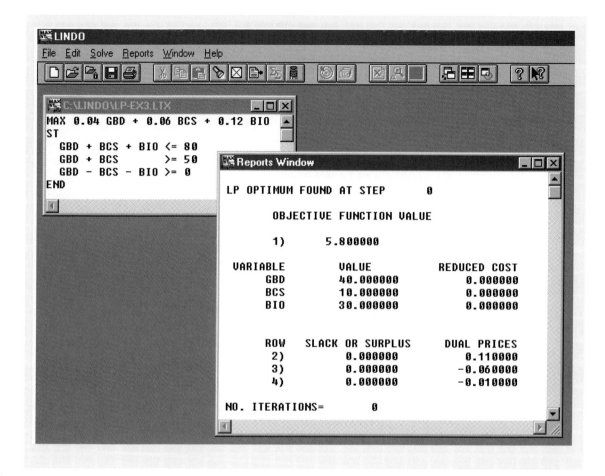

Any Linear Programming Problem Can Now Be Solved

The big-M method allows us to solve *any* maximum linear programming problem, no matter what mixture of ≤ and ≥ inequalities or = equalities are present in the constraints, provided that the variables are nonnegative. A direct way to change a minimum problem into a maximum problem is described in Exercises 35–36. Exercises 37–39 explain how to reformulate a problem so that all the variables are nonnegative. Taken together, these techniques allow us to solve *any* linear programming problem.

4.6 Section Summary

A nonstandard maximum problem may have any combination of ≤, ≥, and = constraints, which we write with nonnegative numbers on

4.6 NONSTANDARD PROBLEMS: ARTIFICIAL VARIABLES AND THE BIG-M METHOD

their right-hand sides. The big-M method (page 333) adds artificial variables to the $\geq$ and $=$ constraints and subtracts M times these variables from the objective function. The initial tableau for this extended problem then displays a basic feasible point including these variables, which the simplex method removes from the basis. If this extended problem has a solution in which every artificial variable is zero, then the original problem has the same solution; otherwise, the problem does not have a solution.

▶ **Solution to Practice Problem**

a. Maximize $P = 5x_1 + 6x_2 - 3x_3 - Ma_1$

Subject to
$$\begin{cases} x_1 - x_2 + 2x_3 + s_1 = 10 \\ x_1 + 2x_2 - x_3 - s_2 + a_1 = 20 \\ x_1 \geq 0, \ x_2 \geq 0, \ x_3 \geq 0 \\ s_1 \geq 0, \ s_2 \geq 0, \ a_1 \geq 0 \end{cases}$$

b.

	x_1	x_2	x_3	s_1	s_2	a_1	
s_1	1	-1	2	1	0	0	10
(a_1)	1	2	-1	0	-1	1	20
P	-5	-6	3	0	0	M	0

c.

	x_1	x_2	x_3	s_1	s_2	a_1	
s_1	1	-1	2	1	0	0	10
a_1	1	2	-1	0	-1	1	20
P	$-M-5$	$-2M-6$	$M+3$	0	M	0	$-20M$

4.6 Exercises

Construct the initial simplex tableau for each nonstandard linear programming problem.

1. Maximize $P = 15x_1 + 20x_2 + 18x_3$

Subject to $\begin{cases} 3x_1 + 2x_2 + 8x_3 \leq 96 \\ 5x_1 + x_2 + 6x_3 \geq 30 \\ x_1 \geq 0, \ x_2 \geq 0, \ x_3 \geq 0 \end{cases}$

2. Maximize $P = 60x_1 + 90x_2 + 30x_3$

Subject to $\begin{cases} 10x_1 + 18x_2 + 3x_3 \leq 150 \\ 4x_1 + 5x_2 + x_3 \geq 40 \\ x_1 \geq 0, \ x_2 \geq 0, \ x_3 \geq 0 \end{cases}$

3. Maximize $P = 6x_1 + 4x_2 + 6x_3$

Subject to $\begin{cases} 2x_1 + x_2 + 3x_3 \geq 30 \\ x_1 + x_2 + 2x_3 \geq 20 \\ x_1 \geq 0, \ x_2 \geq 0, \ x_3 \geq 0 \end{cases}$

4. Maximize $P = 9x_1 + 5x_2 + 4x_3$

Subject to $\begin{cases} 3x_1 + x_2 + x_3 \geq 40 \\ 2x_1 + 2x_2 + x_3 \geq 30 \\ x_1 \geq 0, \ x_2 \geq 0, \ x_3 \geq 0 \end{cases}$

5. Maximize $P = 20x_1 + 30x_2 + 10x_3$

Subject to $\begin{cases} x_1 + 2x_2 + 3x_3 \le 30 \\ x_1 + 2x_2 + x_3 \le 18 \\ x_1 + x_2 + x_3 \ge 8 \\ x_1 \ge 0, \ x_2 \ge 0, \ x_3 \ge 0 \end{cases}$

6. Maximize $P = 24x_1 + 18x_2 + 30x_3$

Subject to $\begin{cases} x_1 + 2x_2 + 3x_3 \le 15 \\ x_1 + 2x_2 + x_3 \le 9 \\ x_1 + x_2 + x_3 \ge 4 \\ x_1 \ge 0, \ x_2 \ge 0, \ x_3 \ge 0 \end{cases}$

7. Maximize $P = 3x_1 + 2x_2 + 5x_3 + 4x_4$

Subject to $\begin{cases} x_1 + x_2 + x_3 + x_4 \ge 30 \\ 2x_1 + 3x_2 + 2x_3 + x_4 \ge 20 \\ 4x_1 + 2x_2 + x_3 + 2x_4 = 80 \\ x_1 \ge 0, \ x_2 \ge 0, \ x_3 \ge 0, \ x_4 \ge 0 \end{cases}$

8. Maximize $P = 6x_1 + 8x_2 + 3x_3 + 5x_4$

Subject to $\begin{cases} x_1 + x_2 + x_3 + x_4 \ge 70 \\ 3x_1 + 2x_2 + 4x_3 + 3x_4 \ge 60 \\ 4x_1 + x_2 + 2x_3 + 5x_4 = 100 \\ x_1 \ge 0, \ x_2 \ge 0, \ x_3 \ge 0, \ x_4 \ge 0 \end{cases}$

Solve each nonstandard linear programming problem by the big-M method. (Problems 9, 10, 13, 14, 17, and 18 can also be solved by the graphical method.)

You may use the PIVOT program if permitted by your instructor.

9. Maximize $P = 4x_1 + x_2$

Subject to $\begin{cases} 3x_1 + 2x_2 \le 120 \\ x_1 + x_2 \ge 50 \\ x_1 \ge 0, \ x_2 \ge 0 \end{cases}$

10. Maximize $P = x_1 + 4x_2$

Subject to $\begin{cases} 3x_1 + 2x_2 \le 120 \\ 2x_1 + x_2 \ge 60 \\ x_1 \ge 0, \ x_2 \ge 0 \end{cases}$

11. Maximize $P = 30x_1 + 20x_2 + 28x_3$

Subject to $\begin{cases} 4x_1 + x_2 + 3x_3 \le 60 \\ 3x_1 + x_2 + 2x_3 \ge 30 \\ x_1 \ge 0, \ x_2 \ge 0, \ x_3 \ge 0 \end{cases}$

12. Maximize $P = 18x_1 + 10x_2 + 20x_3$

Subject to $\begin{cases} 2x_1 + x_2 + 5x_3 \le 50 \\ 3x_1 + x_2 + 4x_3 \ge 30 \\ x_1 \ge 0, \ x_2 \ge 0, \ x_3 \ge 0 \end{cases}$

13. Maximize $P = 4x_1 + 5x_2$

Subject to $\begin{cases} x_1 + 2x_2 \le 12 \\ 2x_1 + x_2 \le 12 \\ x_1 + x_2 \ge 15 \\ x_1 \ge 0, \ x_2 \ge 0 \end{cases}$

14. Maximize $P = 5x_1 + 4x_2$

Subject to $\begin{cases} 2x_1 + 3x_2 \le 12 \\ 3x_1 + 2x_2 \le 12 \\ x_1 + x_2 \ge 10 \\ x_1 \ge 0, \ x_2 \ge 0 \end{cases}$

15. Maximize $P = 15x_1 + 12x_2 + 18x_3$

Subject to $\begin{cases} 2x_1 + x_2 + 3x_3 \le 24 \\ 5x_1 + x_2 + 2x_3 \ge 30 \\ x_1 \ge 0, \ x_2 \ge 0, \ x_3 \ge 0 \end{cases}$

16. Maximize $P = 4x_1 + 5x_2 + 6x_3$

Subject to $\begin{cases} 5x_1 + x_2 + 3x_3 \le 30 \\ 2x_1 + x_2 + 2x_3 \ge 24 \\ x_1 \ge 0, \ x_2 \ge 0, \ x_3 \ge 0 \end{cases}$

17. Maximize $P = x_1 + x_2$

Subject to $\begin{cases} x_1 \le 8 \\ x_2 \le 5 \\ x_1 + 2x_2 \ge 6 \\ x_1 \ge 0, \ x_2 \ge 0 \end{cases}$

18. Maximize $P = x_1 + x_2$

Subject to $\begin{cases} x_1 \le 6 \\ x_2 \le 9 \\ 3x_1 + x_2 \ge 6 \\ x_1 \ge 0, \ x_2 \ge 0 \end{cases}$

19. Maximize $P = 4x_1 + 6x_2 + 12x_3 + 10x_4$

Subject to $\begin{cases} x_1 + x_2 + x_3 + x_4 \le 60 \\ 2x_1 + x_2 + x_3 + 2x_4 \ge 10 \\ 2x_1 + 2x_2 + x_3 + x_4 = 100 \\ x_1 \ge 0, \ x_2 \ge 0, \ x_3 \ge 0, \ x_4 \ge 0 \end{cases}$

20. Maximize $P = 8x_1 - x_2 + 5x_3 + 14x_4$

Subject to $\begin{cases} x_1 + 2x_2 + 3x_4 \leq 55 \\ x_1 + x_2 + x_3 + 2x_4 \geq 25 \\ x_1 + 3x_3 + 3x_4 = 45 \\ x_1 \geq 0, \; x_2 \geq 0, \; x_3 \geq 0, \; x_4 \geq 0 \end{cases}$

21. Maximize $P = 5x_1 + 32x_2 + 3x_3 + 15x_4$

Subject to $\begin{cases} 2x_1 - x_2 + x_3 + x_4 \leq 10 \\ 5x_1 + 8x_2 + x_3 + 3x_4 \geq 18 \\ 3x_1 - 3x_2 + x_3 + 2x_4 \geq 6 \\ x_1 \geq 0, \; x_2 \geq 0, \; x_3 \geq 0, \; x_4 \geq 0 \end{cases}$

22. Maximize $P = 70x_1 + 12x_2 + 60x_3 + 20x_4$

Subject to $\begin{cases} 2x_1 + x_2 + x_3 + x_4 \leq 14 \\ 3x_1 + 3x_2 + x_3 + 2x_4 \leq 24 \\ 5x_1 + 3x_2 + 4x_3 + 2x_4 \geq 44 \\ x_1 \geq 0, \; x_2 \geq 0, \; x_3 \geq 0, \; x_4 \geq 0 \end{cases}$

23. Maximize $P = -x_1 + x_2 - 2x_3$

Subject to $\begin{cases} x_1 + x_2 + x_3 \leq 50 \\ x_1 + x_3 \geq 10 \\ x_2 + x_3 \geq 20 \\ x_1 \geq 0, \; x_2 \geq 0, \; x_3 \geq 0 \end{cases}$

24. Maximize $P = x_1 + 2x_2 + x_3$

Subject to $\begin{cases} x_1 + x_2 + x_3 \leq 30 \\ x_1 + x_2 \leq 25 \\ x_2 + x_3 \geq 15 \\ x_1 \geq 0, \; x_2 \geq 0, \; x_3 \geq 0 \end{cases}$

APPLIED EXERCISES

Express each situation as a nonstandard maximum problem. Be sure to state the meaning of each variable. Solve the problem by the big-M method. State your final answer in terms of the original question.

 You may use the PIVOT program if permitted by your instructor.

25. PERSONAL FINANCE: Financial Planning A retired couple wants to invest their $20,000 life savings in bank certificates of deposit yielding 6% and Treasury bonds yielding 5%. If they want at least $5000 in each type of investment, how much should they invest in each to receive the greatest possible income?

26. ENVIRONMENTAL SCIENCE: Recycling Management A volunteer recycling center accepts both used paper and empty glass bottles which it then sorts and sells to a reprocessing company. The center has room to accept a total of 800 crates of paper and glass each week and has 50 hours of volunteer help to do the sorting. Each crate of paper products takes 5 minutes to sort and sells for 8¢, while each crate of bottles takes 3 minutes to sort and sells for 7¢. To support their city's "grab that glass" recycle theme this week, the center wants to accept at least 600 crates of glass. How many crates of each should the center accept this week to raise the most money for their ecology scholarship fund?

27. BUSINESS: Advertising The manager of a new mall may spend up to $18,000 on "grand opening" announcements in newspapers, on radio and on TV. Each newspaper ad costs $300 and reaches 5000 readers, each one-minute radio commercial costs $800 and is heard by 13,000 listeners, and each 15-second TV spot costs $900 and is seen by 15,000 viewers. If the manager wants at least 5 newspaper ads and at least 20 radio commercials and TV spots combined, how many of each should be placed to reach the largest number of potential customers? (Ignore multiple exposures to the same customer.)

28. SOCIAL SCIENCE: Political Advertising In the last few days before the election, a politician can afford to spend no more than $27,000 on TV advertisements and can arrange for no more than 10 ads. Each day time ad costs $2000 and reaches 4000 viewers, each prime time ad costs $3000 and reaches 5000 viewers, and each late night ad costs $1000 and reaches 2000 viewers. To be sure to reach the widest variety of voters, the politician's advisors insist on at least 5 ads scheduled during day time and late night combined. Ignoring repeated viewings by the same person and assuming every viewer can vote, how many ads in each of the time periods will reach the most voters?

29. **BUSINESS: Resource Allocation** A furniture shop manufactures wooden desks, tables and chairs. The numbers of hours to assemble and finish each piece are shown in the table together with the number of hours of skilled labor available for each task. To meet expected demand, a total of at least 30 desks and tables combined must be made. If the profit is $75 for each desk, $84 for each table, and $66 for each chair, how many of each should the company manufacture to obtain the greatest possible profit?

	Desk	Table	Chair	Labor Available
Assembly	2 hours	1 hour	2 hours	210 hours
Finishing	2 hours	3 hours	1 hours	150 hours

30. **BUSINESS: Production Planning** An automotive parts shop rebuilds carburetors, fuel pumps, and alternators. The numbers of hours to rebuild and then inspect and pack each part are shown in the table, together with the number of hours of skilled labor available for each task. At least 12 carburetors must be rebuilt. If the profit is $12 for each carburetor, $14 for each fuel pump, and $10 for each alternator, how many of each should the shop rebuild to obtain the greatest possible profit?

	Carburetor	Fuel Pump	Alternator	Labor Available
Rebuilding	5 hours	4 hours	3 hours	200 hours
Inspection & packaging	1 hour	1 hour	0.5 hour	45 hours

31. **BUSINESS: Agriculture** A farmer grows wheat, barley, and oats on her 500-acre farm. Each acre of wheat requires 3 days of labor (to plant, tend, and harvest) and costs $21 (for seed, fertilizer, and pesticides), each acre of barley requires 2 days of labor and costs $27, and each acre of oats requires 3 days of labor and costs $24. The farmer and her hired field hands can provide no more than 1200 days of labor this year. She wants to grow at least 100 acres of oats and can afford to spend no more than $15,120. If the profit is $50 for each acre of wheat, $40 for each acre of barley, and $45 for each acre of oats, how many acres of each crop should she grow to obtain the greatest possible profit?

32. **BUSINESS: Agriculture** A farmer grows corn, peanuts, and soybeans on his 240-acre farm. To maintain soil fertility, the farmer rotates the crops and always plants at least as many acres of soybeans as the total acres of the other crops. Because he has promised to sell some of his corn to a neighbor who raises cattle, he must plant at least 42 acres of corn. Each acre of corn requires 2 days of labor and yields a profit of $150, each acre of peanuts requires 5 days of labor and yields a profit of $300, and each acre of soybeans requires 1 day of labor and yields a profit of $100. If the farmer and his children can put in at most 630 days of labor, how many acres of each crop should the farmer plant to obtain the greatest possible profit?

33. **ENVIRONMENTAL SCIENCE: Pollution Control** A storage yard at a coal burning electric power plant can hold no more than 100,000 tons of coal. Two grades of coal are available: low sulfur (1%) with an energy content of 20 million BTU per ton and high sulfur (2%) with an energy content of 30 million BTU per ton. If existing contracts with the high-sulfur coal mine operator require that at least 50,000 tons of high-sulfur coal be purchased, and if the next coal purchase may contain no more than 1400 tons of sulfur, how many tons of each type of coal should be purchased to obtain the most energy?

34. **BUSINESS: Production Planning** A small jewelry company prepares and mounts semi-precious stones. There are 20 lapidaries (who cut and polish the stones) and 24 jewelers (who

mount the stones in gold settings). Each employee works 7 hours each day, 5 days each week. Each tray of agates requires 5 hours of cutting and polishing and 4 hours to mount, each tray of onyxes requires 2 hours of cutting and polishing and 3 hours to mount, and each tray of garnets requires 6 hours of cutting and polishing and 3 hours to mount. Furthermore, the company's owner has decided that they will process at least 12 trays of agates each week. If the profit is $15 for each tray of agates, $10 for each tray of onyxes, and $13 for each tray of garnets, how many trays of each stone should be processed each week to obtain the greatest possible profit?

Explorations and Excursions

The following problems extend and augment the material presented in the text.

The PIVOT program may be helpful (if permitted by your instructor).

Minimum Problems and the Big-M Method

35. a. Solve both of the following problems by the graphical method from Section 4.2. Is it true that the minimum of C is the same as -1 times the maximum of $P = -C$?

$$\text{Minimize} \quad C = 3x + 4y$$
$$\text{Subject to} \quad \begin{cases} x + y \leq 10 \\ 2x + y \geq 8 \\ x \geq 0 \text{ and } y \geq 0 \end{cases}$$

$$\text{Maximize} \quad P = -3x - 4y$$
$$\text{Subject to} \quad \begin{cases} x + y \leq 10 \\ 2x + y \geq 8 \\ x \geq 0 \text{ and } y \geq 0 \end{cases}$$

b. Solve the following problem by finding the dual maximum problem and using the (regular) simplex method as we did in Section 4.4.

$$\text{Minimize} \quad C = 3y_1 + 4y_2$$
$$\text{Subject to} \quad \begin{cases} -y_1 - y_2 \geq -10 \\ 2y_1 + y_2 \geq 8 \\ y_1 \geq 0 \text{ and } y_2 \geq 0 \end{cases}$$

c. Solve the following nonstandard problem by the big-M method.

$$\text{Maximize} \quad P = -3x_1 - 4x_2$$
$$\text{Subject to} \quad \begin{cases} x_1 + x_2 \leq 10 \\ 2x_1 + x_2 \geq 8 \\ x_1 \geq 0 \text{ and } x_2 \geq 0 \end{cases}$$

d. For each sequence of tableaux from parts (b) and (c), trace the basic points on the graphs from part (a). Which solution method is more efficient?

36. Repeat the process of Exercise 35 to compare the "dual" solution to the following minimum problem with the big-M solution of the corresponding maximum problem.

$$\text{Minimize} \quad C = 20y_1 + 18y_2$$
$$\text{Subject to} \quad \begin{cases} 2y_1 + y_2 \geq 18 \\ y_1 + y_2 \geq 14 \\ 4y_1 + 3y_2 \geq 48 \\ y_1 \geq 0 \text{ and } y_2 \geq 0 \end{cases}$$

$$\text{Maximize} \quad P = -20x_1 - 18x_2$$
$$\text{Subject to} \quad \begin{cases} 2x_1 + x_2 \geq 18 \\ x_1 + x_2 \geq 14 \\ 4x_1 + 3x_2 \geq 48 \\ x_1 \geq 0 \text{ and } x_2 \geq 0 \end{cases}$$

More About Nonnegativity Conditions

The following problems show how to reformulate linear programming problems to have nonnegativity conditions on *all* the variables when the original versions do not.

37. Show that the first problem may be rewritten as the second by replacing the variable x_1 by a new variable u_1 where $u_1 = x_1 + 8$ and $u_1 \geq 0$. Solve the first problem by the graphical method from Section 4.2, solve the second by the simplex method from Section 4.3, and check that your answers agree.

$$\text{Maximize} \quad P = 5x_1 + 3x_2$$
$$\text{Subject to} \quad \begin{cases} 2x_1 + x_2 \leq 6 \\ x_1 + x_2 \leq 4 \\ x_1 \geq -8 \text{ and } x_2 \geq 0 \end{cases}$$

Maximize $P = 5u_1 + 3x_2 - 40$

Subject to $\begin{cases} 2u_1 + x_2 \leq 22 \\ u_1 + x_2 \leq 12 \\ u_1 \geq 0 \text{ and } x_2 \geq 0 \end{cases}$

38. Show that the first problem may be rewritten as the second by replacing the variable x_1 by a new variable u_1 where $u_1 = -x_1$ and $u_1 \geq 0$. Solve the first problem by the graphical method from Section 4.2, solve the second by the simplex method from Section 4.3, and check that your answers agree.

Maximize $P = 5x_1 + 3x_2$

Subject to $\begin{cases} 2x_1 + x_2 \leq 10 \\ -x_1 + x_2 \leq 4 \\ x_1 \leq 0 \text{ and } x_2 \geq 0 \end{cases}$

Maximize $P = -5u_1 + 3x_2$

Subject to $\begin{cases} -2u_1 + x_2 \leq 10 \\ u_1 + x_2 \leq 4 \\ u_1 \geq 0 \text{ and } x_2 \geq 0 \end{cases}$

Unrestricted Variables

39. A variable is *unrestricted* if it may take *any* value (positive, negative, or zero). Show that the first problem may be rewritten as the second by replacing the variable x_1 by two new variables u_1 and u_2 where $u_1 - u_2 = x_1$ and $u_1, u_2 \geq 0$. Solve the first problem by the graphical method from Section 4.2, solve the second by the simplex method from Section 4.3, and check that your answers agree.

Maximize $P = 3x_1 + 5x_2$

Subject to $\begin{cases} 2x_1 + x_2 \leq 6 \\ x_2 \leq 4 \\ x_1 \text{ unrestricted and } x_2 \geq 0 \end{cases}$

Maximize $P = 3u_1 - 3u_2 + 5x_2$

Subject to $\begin{cases} 2u_1 - 2u_2 + x_2 \leq 6 \\ x_2 \leq 4 \\ u_1 \geq 0, \ u_2 \geq 0, \text{ and } x_2 \geq 0 \end{cases}$

[*Hint:* Can any real number be written as the difference of two nonnegative numbers?]

Chapter Summary with Hints and Suggestions

Reading the text and doing the exercises in this chapter have helped you to master the following skills, which are listed by section (in case you need to review them) and are keyed to particular Review Exercises. Answers for all Review Exercises are given at the back of the book, and full solutions can be found in the Student Solutions Manual.

4.1 Linear Inequalities

- Graph a linear inequality $ax + by \leq c$ by drawing the boundary line $ax + by = c$ from its intercepts, using a test point to choose the correct side, and shading in the feasible region. (*Review Exercises 1–2.*)

- Graph a system of linear inequalities $\begin{cases} ax + by \leq c \\ \vdots \end{cases}$ by drawing the boundary lines, finding the correct sides of the boundaries, and shading in the feasible region. List the vertices or "corners" of the region and identify the region as bounded or unbounded. (*Review Exercises 3–8.*)

- Formulate an applied situation as a system of linear inequalities and sketch the feasible region with the vertices. (*Review Exercises 9–10.*)

4.2 Two-Variable Linear Programming Problems

- Find the maximum or minimum of a linear function on a given region or explain why such a value does not exist. *(Review Exercises 11–14.)*

- Solve a two-variable linear programming problem by sketching the feasible region, determining whether a solution exists, and then finding it by evaluating the objective function at the vertices. *(Review Exercises 15–18.)*

- Solve an applied linear programming problem by identifying the variables, the objective function, and the constraints, and then using the vertices of the feasible region. *(Review Exercises 19–20.)*

4.3 The Simplex Method for Standard Maximum Problems

- Check that a maximum linear programming problem is a standard problem and construct the initial simplex tableau. *(Review Exercises 21–22.)*

$$\text{Maximize} \quad P = c^t X$$
$$\text{Subject to} \quad \begin{cases} AX \leq b \quad (b \geq 0) \\ X \geq 0 \end{cases}$$

	X	S	
S	A	I	b
P	$-c^t$	0	0

- Find the pivot element in a tableau and carry out one complete pivot operation. If there is no pivot element, explain what this means about the original problem. *(Review Exercises 23–24.)*

- Solve a standard maximum problem by the simplex method (construct the initial simplex tableau, pivot until the tableau does not have a pivot element, and interpret this final tableau). *(Review Exercises 25–28.)*

- Solve an applied linear programming problem by the simplex method. *(Review Exercises 29–30.)*

4.4 Standard Minimum Problems and Duality

- Check that a minimum linear programming problem is a standard problem, and construct the dual maximum problem and the initial simplex tableau. *(Review Exercises 31–34.)*

$$\text{Minimize} \quad C = b^t Y \qquad \text{Maximize} \quad P = c^t X$$
$$\text{Subject to} \quad \begin{cases} A^t Y \geq c \\ Y \geq 0 \end{cases} \qquad \text{Subject to} \quad \begin{cases} AX \leq b \\ X \geq 0 \end{cases}$$
$$\text{where} \ b \geq 0 \qquad\qquad \text{where} \ b \geq 0$$

- Solve a standard minimum problem by finding the dual maximum problem and using the simplex method. *(Review Exercises 35–38.)*

- Solve an applied minimum problem by finding the dual maximum problem and using the simplex method. *(Review Exercises 39–40.)*

4.5 Nonstandard Problems: The Dual Pivot Element and the Two-Stage Method

- For a nonstandard maximum problem, construct the initial simplex tableau and locate the dual pivot element. *(Review Exercises 41–42.)*

- Solve a nonstandard maximum problem by the two-stage simplex method. *(Review Exercises 43–48.)*

- Solve an applied nonstandard maximum problem by the two-stage simplex method. *(Review Exercises 49–50.)*

4.6 Nonstandard Problems: Artificial Variables and the Big-M Method

- Construct the initial simplex tableau for a nonstandard maximum problem by using artificial variables. *(Review Exercises 51–52.)*

- Solve a nonstandard maximum problem by the big-M method. *(Review Exercises 53–58.)*

- Solve an applied nonstandard maximum problem by the big-M method. *(Review Exercises 59–60.)*

Hints and Suggestions

- *(Overview)* A linear programming problem asks for the maximum or minimum of a linear objective function subject to constraints in the form of linear inequalities. If there is a solution to the problem, it occurs at a *vertex* of the feasible region determined by the constraints. A problem with two variables may be solved graphically, but one with more than two variables must be solved algebraically using the simplex method, duality (for a standard mini-mum problem), and either the two-stage simplex or the big-M method (for a nonstandard problem).

- While a problem with a *bounded* region always has a solution, a problem with an *unbounded* region may or may not have a solution (if it *does*, then the solution occurs at a vertex).

- When setting up a word problem, look first for the question "how many" or "how much" to help identify the variables. Be sure that finding values for your variables will answer the question stated in the problem. Use the rest of the given information to find the objective function and constraints. Include nonnegativity conditions for your variables as appropriate (for example, a farmer *cannot* raise a negative number of cows).

- A simplex tableau is *feasible* if it has no negative numbers in the rightmost *column* and is *optimal* if it has no negative numbers in the bottom *row*. If a simplex tableau is both feasible and optimal, it displays the solution of the problem.

- The pivot operation is not completed until *all* the rows of the tableau have been recalculated.

- If you are solving a standard maximum problem by the simplex method and you have a negative number in the rightmost column after pivoting, you've made an error in your choice of the pivot element and/or your arithmetic.

- Many simplex tableaux require several pivots to solve, so don't give up: Keep pivoting until the tableau does not have a pivot element.

- **Practice for Test:** Review Exercises 1, 7, 9, 11, 13, 16, 17, 20, 25, 29, 35, 39, and (41, 45, 50) or (51, 55, 60).

Review Exercises for Chapter 4

Practice test exercise numbers are in green.

4.1 Linear Inequalities

Graph each inequality.

1. $3x - 2y \leq 24$ 2. $x + y \geq -6$

Graph each system of inequalities. List all vertices and identify the region as "bounded" or "unbounded."

3. $\begin{cases} x + y \leq 20 \\ x \geq 0, \; y \geq 0 \end{cases}$ 4. $\begin{cases} x + 2y \geq 10 \\ x \geq 0, \; y \geq 0 \end{cases}$

5. $\begin{cases} 2x + y \leq 20 \\ x + y \leq 15 \\ x \geq 0, \; y \geq 0 \end{cases}$ 6. $\begin{cases} 2x + y \geq 12 \\ x + y \geq 8 \\ x \geq 0, \; y \geq 0 \end{cases}$

7. $\begin{cases} 3x + 2y \leq 24 \\ x + y \leq 10 \\ x \geq 2 \\ x \geq 0, \; y \geq 0 \end{cases}$ 8. $\begin{cases} x + y \leq 10 \\ -x + y \leq 4 \\ x \leq 9 \\ x \geq 0, \; y \geq 0 \end{cases}$

Formulate each application as a system of linear inequalities, sketch the feasible region, and find the vertices.

9. **GENERAL: Dog Training** A dog trainer works with Irish setters and Labrador retrievers for no more than 6 hours each day. Each training session with a "setter" takes half an hour and requires 8 dog treats, while each session with a "lab" takes three-quarters of an hour and also requires 8 dog treats. If the trainer has only 80 dog treats left, how many of each breed can he train today?

10. **GENERAL: Blending Bird Seed** A pet store owner blends custom bird seed by mixing SongBird and MeadowMix brands together. Each pound of SongBird contains 2 ounces of sunflower hearts and 3 ounces of crushed peanuts (along with a lot of cheap filler seed),

while each pound of MeadowMix contains 4 ounces of sunflower hearts and 2 ounces of crushed peanuts. If each bag of the custom blend is labeled "contains at least 104 ounces of sunflower hearts and 84 ounces of crushed peanuts," how many pounds of each brand must be put in each bag?

4.2 Two-Variable Linear Programming Problems

Use the region below to find each maximum or minimum value. If such a value does not exist, explain why not.

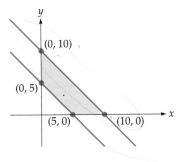

11. Maximum of $P = 3x + 4y$

12. Minimum of $C = 3x + 2y$

Use the region below to find each maximum or minimum value. If such a value does not exist, explain why not.

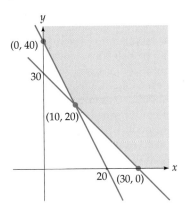

13. Maximum of $P = 4x + 3y$

14. Minimum of $C = 2x + 5y$

Solve each linear programming problem by sketching the region, labeling the vertices, checking whether a solution exists, and then finding it if it does.

15. Maximize $P = 20x + 15y$

Subject to $\begin{cases} x + 3y \le 18 \\ x + y \le 12 \\ x \ge 0, \ y \ge 0 \end{cases}$

16. Minimize $C = 30x + 40y$

Subject to $\begin{cases} 2x + y \ge 120 \\ x + 3y \ge 120 \\ x \ge 0, \ y \ge 0 \end{cases}$

17. Maximize $P = 20x + 30y$

Subject to $\begin{cases} 2x + y \le 24 \\ x + 4y \le 40 \\ x + y \le 13 \\ x \ge 0, \ y \ge 0 \end{cases}$

18. Minimize $C = 7x + 11y$

Subject to $\begin{cases} x + y \ge 20 \\ x - y \ge -10 \\ x \ge 0, \ y \ge 0 \end{cases}$

Formulate each application as a linear programming problem by identifying the variables, the objective function, and the constraints. Determine whether a solution exists, and if it does, find it using the vertices of the feasible region. State your final answer in terms of the original question.

19. BUSINESS: Production Planning A woodworking shop makes antique reproduction cases for wall and mantel clocks. The amount of cutting and finishing needed for each is shown in the table, together with the number of hours of skilled labor available for each task. If the profit for each case is $250 for wall clocks and $200 for mantel clocks, how many of each kind of clock case should the shop make to obtain the greatest possible profit?

	Wall Clock	Mantel Clock	Labor Available
Cutting	5 hours	3 hours	165 hours
Finishing	3 hours	4 hours	132 hours

20. **BIOMEDICAL: Nutrition** A large and active cat needs at least 70 grams of fat and 134.4 grams of protein each week. Each ounce of canned cat food contains 2 grams of fat and 3.2 grams of protein, while each ounce of dry food contains 2.5 grams of fat and 9.6 grams of protein. If canned cat food costs 5¢ per ounce and dry food costs 8¢ per ounce, how many ounces of each type of food will meet the cat's nutritional needs for the week at the least cost?

4.3 The Simplex Method for Standard Maximum Problems

Construct the initial simplex tableau for each standard maximum problem. (Do not solve the problem any further.)

21. Maximize $P = 4x_1 + 10x_2 + 9x_3$

    Subject to $\begin{cases} 5x_1 + 2x_2 + 3x_3 \leq 30 \\ 2x_1 + 3x_2 + 4x_3 \leq 24 \\ x_1 \geq 0, \ x_2 \geq 0, \ x_3 \geq 0 \end{cases}$

22. Maximize $P = 5x_1 + 7x_2$

    Subject to $\begin{cases} x_1 + x_2 \leq 10 \\ 2x_1 + x_2 \leq 14 \\ x_1 - x_2 \leq 4 \\ x_1 \geq 0, \ x_2 \geq 0 \end{cases}$

For each simplex tableau, find the pivot element and carry out one complete pivot operation. If there is no pivot element, explain what this means about the original problem.

23.

	x_1	x_2	x_3	s_1	s_2	s_3	
s_1	1	1	1	1	0	0	6
s_2	3	1	2	0	1	0	8
s_3	1	1	-3	0	0	1	6
P	-6	4	-8	0	0	0	0

24.

	x_1	x_2	s_1	s_2	s_3	
s_1	2	0	1	0	-2	8
s_2	6	0	0	1	3	42
x_2	1	1	0	0	1	10
P	5	0	0	0	15	150

Solve each problem by the simplex method.

25. Maximize $P = 7x_1 + 8x_2 + 7x_3$

    Subject to $\begin{cases} 2x_1 + x_2 + 4x_3 \leq 16 \\ x_1 + 2x_2 - x_3 \leq 30 \\ x_1 + x_2 + 3x_3 \leq 12 \\ x_1 \geq 0, \ x_2 \geq 0, \ x_3 \geq 0 \end{cases}$

26. Maximize $P = 8x_1 + 10x_2 + 9x_3$

    Subject to $\begin{cases} 3x_1 + 2x_2 + 3x_3 \leq 18 \\ 2x_1 + x_2 + 4x_3 \leq 14 \\ x_1 \geq 0, \ x_2 \geq 0, \ x_3 \geq 0 \end{cases}$

27. Maximize $P = 8x_1 + 7x_2 + 5x_3 + 7x_4$

    Subject to $\begin{cases} 3x_1 + 2x_2 + 4x_3 + 3x_4 \leq 12 \\ x_1 + x_2 + x_3 + x_4 \leq 5 \\ 3x_1 + x_2 + 5x_3 + 3x_4 \leq 15 \\ x_1 \geq 0, \ x_2 \geq 0, \ x_3 \geq 0, \ x_4 \geq 0 \end{cases}$

28. Maximize $P = 5x_1 + 4x_2 + 3x_3$

    Subject to $\begin{cases} x_1 + x_2 \leq 8 \\ x_2 + x_3 \leq 7 \\ x_1 + x_2 - x_3 \leq 6 \\ x_1 \geq 0, \ x_2 \geq 0, \ x_3 \geq 0 \end{cases}$

Formulate each application as a linear programming problem by identifying the variables, the objective function, and the constraints. Check that the problem is a standard maximum problem and then solve it by the simplex method. State your final answer in terms of the original question.

29. **BUSINESS: Book Publishing** A book publisher needs more copies of its latest best seller. By rescheduling several other projects over the next two weeks, it can arrange up to 300 hours of printing and 250 hours of binding. For each thousand copies, the trade edition requires 3 hours of printing and 2 hours of binding, the book club edition requires 2 hours of printing and 2 hours of binding, and the paperback edition requires 1 hour of printing and 1 hour of binding. If the profit per book is $7 for the trade edition, $6 for the book club edition, and $4 for the paperback, how many of each edition should be printed to obtain the greatest possible profit?

30. BUSINESS: Production Planning A fully automated plastics factory produces three toys—a racing car, a jet airplane, and a speed boat—in three stages: molding, painting, and packaging. After allowing for routine maintenance, the equipment can operate no more than 150 hours per week. Each batch of racing cars requires 6 hours of molding, 2.5 hours of painting, and 5 hours of packaging; each batch of jet airplanes requires 3 hours of molding, 7.5 hours of painting, and 5 hours of packaging; and each batch of speed boats requires 3 hours of molding, 5 hours of painting, and 5 hours of packaging. If the profits per batch of toys are $120 for cars, $100 for airplanes, and $110 for boats, how many batches of each toy should be produced each week to obtain the greatest possible profit?

4.4 Duality and Standard Minimum Problems

For each minimum linear programming problem check that it is a standard problem and then construct the dual maximum problem and the initial simplex tableau. (Do not solve the problem any further.)

31. Minimize $C = 10y_1 + 15y_2 + 20y_3$

Subject to $\begin{cases} 2y_1 - y_2 + y_3 \geq 40 \\ y_1 + y_2 + 3y_3 \geq 30 \\ y_1 \geq 0, \ y_2 \geq 0, \ y_3 \geq 0 \end{cases}$

32. Minimize $C = 30y_1 + 20y_2$

Subject to $\begin{cases} 5y_1 + 2y_2 \geq 210 \\ 3y_1 + 4y_2 \geq 252 \\ 5y_1 + 4y_2 \leq 380 \\ y_1 \geq 0, \ y_2 \geq 0 \end{cases}$

33. Minimize $C = 42y_1 + 36y_2$

Subject to $\begin{cases} 7y_1 + 4y_2 \geq 84 \\ y_1 + y_2 \geq 18 \\ y_1 \geq 0, \ y_2 \geq 0 \end{cases}$

34. Minimize $C = 130y_1 + 40y_2 + 98y_3$

Subject to $\begin{cases} 3y_1 + y_2 + 2y_3 \geq 51 \\ 4y_1 + y_2 + 2y_3 \geq 60 \\ 3y_1 + y_2 + 3y_3 \geq 57 \\ y_1 \geq 0, \ y_2 \geq 0, \ y_3 \geq 0 \end{cases}$

Solve each standard minimum problem by finding its dual and using the simplex method.

35. Minimize $C = 9y_1 + 24y_2 + 6y_3$

Subject to $\begin{cases} y_1 + 2y_2 \geq 18 \\ 3y_2 + y_3 \geq 15 \\ y_1 \geq 0, \ y_2 \geq 0, \ y_3 \geq 0 \end{cases}$

36. Minimize $C = 30y_1 + 36y_2$

Subject to $\begin{cases} 5y_1 + 3y_2 \geq 60 \\ y_1 + y_2 \geq 16 \\ y_2 \geq 5 \\ y_1 \geq 0, \ y_2 \geq 0 \end{cases}$

37. Minimize $C = 3y_1 + 4y_2 + 10y_3 + 8y_4$

Subject to $\begin{cases} -y_1 + y_2 + y_3 - 2y_4 \geq 10 \\ y_1 - y_2 - 2y_3 + y_4 \geq 20 \\ y_1 \geq 0, \ y_2 \geq 0, \ y_3 \geq 0, \ y_4 \geq 0 \end{cases}$

38. Minimize $C = 12y_1 + 5y_2 + 6y_3$

Subject to $\begin{cases} 3y_1 + y_2 + 3y_3 \geq 9 \\ 2y_1 + y_2 - 3y_3 \geq 12 \\ 4y_1 + 2y_2 + 2y_3 \geq 8 \\ y_1 \geq 0, \ y_2 \geq 0, \ y_3 \geq 0 \end{cases}$

Formulate each application as a standard minimum problem and then solve it by finding its dual and using the simplex method. State your final answer in terms of the original question.

39. GENERAL: School Supplies A student needs at least 3 new pens and 2 dozen ink cartridges. The bookstore sells single pens for 99¢, packages of 6 ink cartridges for 89¢, and a $1.19 "writer's combo" package containing 1 pen and 2 ink cartridges. How many of each will meet the student's needs at the least cost?

40. BUSINESS: Production Planning A pasta company is expanding its linguini production facility. Two machines are available: a small-capacity machine costing $5000 that produces 20 pounds per minute and needs 1 operator, and a large-capacity machine costing $6000 that produces 30 pounds per minute and needs 2 operators. The company wants to hire no more than 34 additional employees yet increase production by at least 600 pounds per minute. How many of each machine should the company buy to expand its production at the least cost?

4.5 Nonstandard Problems: The Dual Pivot Element and the Two-Stage Method

For each nonstandard maximum problem construct the initial simplex tableau and locate the dual pivot element. (Do not carry out the pivot operation.)

41. Maximize $P = 80x_1 + 30x_2$

Subject to $\begin{cases} 4x_1 + x_2 \le 40 \\ 2x_1 + 3x_2 \le 60 \\ x_1 + x_2 \ge 10 \\ x_1 \ge 0, \; x_2 \ge 0 \end{cases}$

42. Maximize $P = 4x_1 + 8x_2 + 6x_3 + 10x_4$

Subject to $\begin{cases} 2x_1 + 3x_2 - 3x_3 + x_4 \le 30 \\ 7x_1 - 5x_2 + x_3 + 5x_4 \ge 35 \\ x_1 \ge 0, \; x_2 \ge 0, \; x_3 \ge 0, \; x_4 \ge 0 \end{cases}$

Solve each nonstandard maximum problem by the two-stage simplex method.

43. Maximize $P = 3x_1 + 4x_2$

Subject to $\begin{cases} x_1 + x_2 \le 10 \\ x_1 + x_2 \ge 5 \\ x_1 \ge 0, \; x_2 \ge 0 \end{cases}$

44. Maximize $P = 8x_1 - 20x_2 - 18x_3$

Subject to $\begin{cases} x_1 + 3x_2 + 2x_3 \le 55 \\ x_1 + x_2 + x_3 \ge 25 \\ 2x_1 + 3x_2 + 3x_3 \le 70 \\ x_1 \ge 0, \; x_2 \ge 0, \; x_3 \ge 0 \end{cases}$

45. Maximize $P = 6x_1 - 9x_2$

Subject to $\begin{cases} 3x_1 + x_2 \le 24 \\ x_1 + 3x_2 \le 24 \\ x_1 + x_2 \ge 6 \\ x_1 \ge 0, \; x_2 \ge 0 \end{cases}$

46. Maximize $P = 12x_1 - 21x_2 + 2x_3$

Subject to $\begin{cases} 2x_1 - 3x_2 + x_3 \ge 20 \\ x_1 - x_2 + x_3 \le 25 \\ x_1 - 3x_2 \ge 5 \\ x_1 \ge 0, \; x_2 \ge 0, \; x_3 \ge 0 \end{cases}$

47. Maximize $P = 6x_1 + 7x_2$

Subject to $\begin{cases} x_1 \ge 2 \\ x_2 \ge 3 \\ x_1 \le 4 \\ x_2 \le 5 \\ x_1 \ge 0, \; x_2 \ge 0 \end{cases}$

48. Maximize $P = 8x_1 + 2x_2 + 10x_3$

Subject to $\begin{cases} 2x_1 + x_2 + x_3 \le 30 \\ x_1 + 2x_2 + x_3 \ge 5 \\ x_1 - x_2 + x_3 \ge 10 \\ x_1 \ge 0, \; x_2 \ge 0, \; x_3 \ge 0 \end{cases}$

Formulate each application as a nonstandard maximum problem and solve it by the two-stage simplex method. State your final answer in terms of the original question.

49. **MANAGEMENT SCIENCE: Lumber Production** A saw mill rough-cuts lumber and planes some of it to make finished-grade boards. The saw cuts 1 thousand board-feet per hour and can operate up to 10 hours each day, while the plane finishes 1 thousand board-feet per hour and can operate up to 8 hours each day. The profit per thousand board-feet is $120 for rough-cut lumber and $100 for finished-grade boards. If at least 4 thousand board-feet of finished-grade boards must be produced each day, how many board-feet of each type of lumber should be produced daily to obtain the greatest possible profit?

50. **PERSONAL FINANCE: Financial Planning** An international money manager wishes to invest no more than $150,000 in United States and Canadian stocks and bonds, with at least twice as much in bonds as in stocks. The current market yields are 10% on U.S. stocks, 5% on U.S. bonds, 11% on Canadian stocks, and 4% on Canadian bonds. If at least $30,000 must be invested in Canadian stocks, how much should be invested in each to obtain the greatest possible return?

4.6 Nonstandard Problems: Artificial Variables and the Big-M Method

Construct the initial simplex tableau for each nonstandard maximum problem.

51. Maximize $P = 80x_1 + 30x_2$
Subject to $\begin{cases} 4x_1 + x_2 \leq 40 \\ 2x_1 + 3x_2 \leq 60 \\ x_1 + x_2 \geq 10 \\ x_1 \geq 0, \ x_2 \geq 0 \end{cases}$

52. Maximize $P = 4x_1 + 8x_2 + 6x_3 + 10x_4$
Subject to $\begin{cases} 2x_1 + 3x_2 - 3x_3 + x_4 \leq 30 \\ 7x_1 - 5x_2 + x_3 + 5x_4 \geq 35 \\ x_1 \geq 0, \ x_2 \geq 0, \ x_3 \geq 0, \ x_4 \geq 0 \end{cases}$

Solve each nonstandard maximum problem by the big-M method.

53. Maximize $P = 3x_1 + 4x_2$
Subject to $\begin{cases} x_1 + x_2 \leq 10 \\ x_1 + x_2 \geq 5 \\ x_1 \geq 0, \ x_2 \geq 0 \end{cases}$

54. Maximize $P = 8x_1 - 20x_2 - 18x_3$
Subject to $\begin{cases} x_1 + 3x_2 + 2x_3 \leq 55 \\ 2x_1 + 3x_2 + 3x_3 \leq 70 \\ x_1 + x_2 + x_3 \geq 25 \\ x_1 \geq 0, \ x_2 \geq 0, \ x_3 \geq 0 \end{cases}$

55. Maximize $P = 6x_1 - 9x_2$
Subject to $\begin{cases} 3x_1 + x_2 \leq 24 \\ x_1 + 3x_2 \leq 24 \\ x_1 + x_2 \geq 6 \\ x_1 \geq 0, \ x_2 \geq 0 \end{cases}$

56. Maximize $P = 12x_1 - 21x_2 + 2x_3$
Subject to $\begin{cases} x_1 - x_2 + x_3 \leq 25 \\ x_1 - 3x_2 \geq 5 \\ 2x_1 - 3x_2 + x_3 \geq 20 \\ x_1 \geq 0, \ x_2 \geq 0, \ x_3 \geq 0 \end{cases}$

57. Maximize $P = 6x_1 + 7x_2$
Subject to $\begin{cases} x_1 \leq 4 \\ x_2 \leq 5 \\ x_1 \geq 2 \\ x_2 \geq 3 \\ x_1 \geq 0, \ x_2 \geq 0 \end{cases}$

58. Maximize $P = 8x_1 + 2x_2 + 10x_3$
Subject to $\begin{cases} 2x_1 + x_2 + x_3 \leq 30 \\ x_1 + 2x_2 + x_3 \geq 5 \\ x_1 - x_2 + x_3 \geq 10 \\ x_1 \geq 0, \ x_2 \geq 0, \ x_3 \geq 0 \end{cases}$

Formulate each application as a nonstandard maximum problem and solve it by the big-M method. State your final answer in terms of the original question.

59. MANAGEMENT SCIENCE: Lumber Production A saw mill rough-cuts lumber and planes some of it to make finished-grade boards. The saw cuts 1 thousand board-feet per hour and can operate up to 10 hours each day, while the plane finishes 1 thousand board-feet per hour and can operate up to 8 hours each day. The profit per thousand board-feet is $120 for rough-cut lumber and $100 for finished-grade boards. If at least 4 thousand board-feet of finished-grade boards must be produced each day, how many board-feet of each type of lumber should be produced daily to obtain the greatest possible profit?

60. PERSONAL FINANCE: Financial Planning An international money manager wishes to invest no more than $150,000 in United States and Canadian stocks and bonds, with at least twice as much in bonds as in stocks. The current market yields are 10% on U.S. stocks, 5% on U.S. bonds, 11% on Canadian stocks, and 4% on Canadian bonds. If at least $30,000 must be invested in Canadian stocks, how much should be invested in each to obtain the greatest possible return?

5 Probability

- 5.1 Sets, Counting, and Venn Diagrams
- 5.2 Permutations and Combinations
- 5.3 Probability Spaces
- 5.4 Conditional Probability and Independence
- 5.5 Bayes' Formula
- 5.6 Random Variables and Distributions

Application Preview

The Chevalier de Meré

Probability as a mathematical subject began when a French nobleman, Antoine Gombaude (whose title was the Chevalier de Meré), wanted to understand why he was losing money in his dice games. The Chevalier's first bet was that in four rolls of a die (the singular of *dice*) he could roll at least one 6, and he was winning so often that he had no more takers. The Chevalier then reasoned that in six times as many rolls he should get a *second* six, so he bet that in twenty-four rolls of *two* dice he could roll at least one *double*-6. Much to his surprise, he was losing these bets. Rather than just giving up, he had the good sense to seek the advice of one of the foremost mathematicians of the era, Blaise Pascal (1623–1662). Pascal corresponded with the other great French mathematician of the time, Pierre de Fermat (1601–1665), about the Chevalier's difficulties, and together they found that an analysis of the situation depended on a careful listing of all possible outcomes from throwing the dice. This analysis and the calculation of the corresponding probabilities is considered to be the beginning of the mathematical study of probability. We will analyze the Chevalier's bets on page 397 and see why he won one and lost the other.

5.1 SETS, COUNTING, AND VENN DIAGRAMS

Introduction

Probability is used frequently in everyday conversation, as in "you will probably get the job" or "it will probably rain today." In mathematics, however, we need to give a precise definition for probability, and we will see that it depends on a careful counting all of the possible events that can occur. With this in mind, we begin by reviewing sets and some ways of counting their members.

Sets and Set Operations

A *set* is any well-defined collection of objects (also called *elements* or *members*). By "well-defined" we mean you can tell whether an object is in the set or not. For example, we may speak of *the set of American citizens*, since there are specific conditions for being an American citizen. On the other hand, we cannot speak of *the set of all thin people*, since the word "thin" is not precisely defined.

We will often use diagrams to represent sets. We use a rectangle for the *universal set*, denoted U, which is the set of all of the elements that we are discussing. For example, depending on the question, the universal set might consist of all people, all customers of a company, all cars on the road, or any other collection of objects of interest. To represent sets within the universal set, we draw ellipses inside the rectangle containing the elements of the sets. Such drawings are called *Venn diagrams*.* There are several operations on sets, which we illustrate with Venn diagrams.

The *intersection* of sets A and B, denoted $A \cap B$, is the set of all elements that are in *both A and B*.

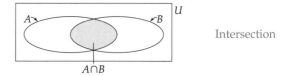

Intersection

Two sets A and B are *disjoint* if their intersection is empty (that is, if they have no elements in common).

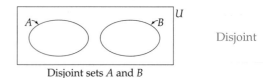

Disjoint sets A and B

---

* After the English logician John Venn (1843–1923), author of *Logic of Chance*.

The *union* of sets A and B, denoted $A \cup B$, is the set of all elements that are in *either A or B* (or both).

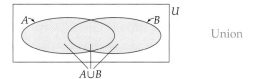

Union

The *complement* of a set A, denoted A^c, is the set of all elements *not* in A.

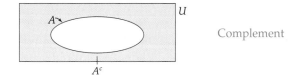

Complement

In general, the words "and," "or," and "not" translate into "intersection," "union," and "complement." We use the word "or" in the *inclusive* sense, meaning one possibility or the other *or both*. The complement of the universal set is the *empty* set, also called the *null* set, denoted $\emptyset$, which has no members. A *subset* of a set is a set (possibly empty) of elements from the original set. For a finite set A, we use the symbol $n(A)$ to mean the number of elements in A:

$$n(A) = \begin{pmatrix} \text{Number of} \\ \text{elements in } A \end{pmatrix}$$

For example, $n(\emptyset) = 0$ because the empty set has no elements. From the previous Venn diagram, it is clear that the elements of the set A^c are precisely the elements of the universal set U that are *not* in A. Therefore:

Complementary Principle of Counting

$n(A^c) = n(U) - n(A)$

The number of elements in the *complement* of a set is the number of elements in the universal set minus the number of elements in the original set

We will find this principle useful when it is easier to count the elements in the complement rather than the elements in the set itself.

Addition Principle for Counting

How can we count the number of elements in the *union* of two sets without counting them one by one? In the following Venn diagram there are $n(A) = 7$ elements (dots) in A and $n(B) = 5$ elements in B.

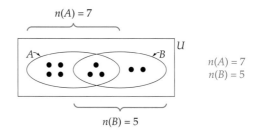

Altogether there are 9 elements in $A \cup B$, but adding $n(A) = 7$ and $n(B) = 5$ gives 12, not the correct total of 9 for the union. The diagram shows the trouble: Adding $n(A)$ and $n(B)$ counts the elements in the middle (in $A \cap B$) *twice*—once in A and a second time in B. To correct this double-counting, we must subtract the number of elements in $A \cap B$ from the total, obtaining the general rule:

Addition Principle of Counting

$$n(A \cup B) = n(A) + n(B) - n(A \cap B)$$

The number of elements in the *union* of two sets is the number of elements in one plus the number of elements in the other minus the number of elements in *both*

Of course, if A and B are disjoint, then $A \cap B$ is empty, so $n(A \cap B) = 0$, giving a simpler addition principle for disjoint sets:

$$n(A \cup B) = n(A) + n(B) \qquad \text{for } A, B \text{ disjoint}$$

CARS IN A PARKING LOT

A mall parking lot has 300 cars, of which 150 have alarm systems, 200 have sound systems, and 90 have both alarm and sound systems.

a. How many cars have an alarm system or a sound system?

b. How many have neither?

5.1 SETS, COUNTING, AND VENN DIAGRAMS

Solution

Let A be the set of cars with alarm systems and S be the set of cars with sound systems. Starting with the Venn diagram on the left, we enter the numbers successively, beginning with the intersection:

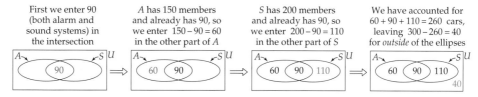

First we enter 90 (both alarm and sound systems) in the intersection

A has 150 members and already has 90, so we enter $150 - 90 = 60$ in the other part of A

S has 200 members and already has 90, so we enter $200 - 90 = 110$ in the other part of S

We have accounted for $60 + 90 + 110 = 260$ cars, leaving $300 - 260 = 40$ for *outside* of the ellipses

The 40 *outside* of the ellipses means that 40 cars have *neither* system. Answer:

260 cars have either an alarm system or a sound system. $60 + 90 + 110$
40 cars have neither an alarm system nor a sound system. $300 - 260$

Alternatively, since "or" means "union," we could find $n(A \cup S)$ by the Addition Principal of Counting, using $n(A) = 150$, $n(S) = 200$, and $n(A \cap S) = 90$:

$$n(A \cup S) = n(A) + n(S) - n(A \cap S) = 150 + 200 - 90 = 260$$

  150   200   90

Addition principle of counting

Again we see that 260 cars have an alarm or a sound system, so that $300 - 260 = 40$ have neither.

Hint: With Venn diagrams you usually begin at the "inside" and work "out."

Practice Problem 1   A survey of insurance coverage in 300 metropolitan businesses revealed that 150 offer their employees dental insurance, 150 offer vision coverage, and 100 offer both dental and vision coverage. How many of these businesses offer their employees dental or vision insurance? How many offer neither?   ▶ **Solution on page 361**

The Multiplication Principle for Counting

Suppose that you are choosing an outfit to wear, and you may choose any one of 3 shirts, S_1, S_2, or S_3, and either of 2 pairs of pants, P_1 or P_2. If the 3 shirts can be combined freely with the 2 pairs of pants, there are $3 \cdot 2 = 6$ different possible outfits, namely S_1P_1, S_1P_2, S_2P_1, S_2P_2, S_3P_1, and S_3P_2. The possible choices are shown in the *tree diagram* on the left, indicating a first choice of a shirt, then "branching"

358 CHAPTER 5 PROBABILITY

from each shirt to the second choice of a pair of pants, ending in 6 "leaves" at the bottom. More generally, we have the following *multiplication principle*, which will be very useful throughout this chapter.

> **Multiplication Principle for Counting**
>
> If two choices are to be made, and there are m possibilities for the first choice and n possibilities for the second choice, and if any first choice can be combined with any second choice, then the two choices together can be made in $m \cdot n$ ways.

The multiplication principle can also be proved by enumerating all of the possibilities, as in the following example.

EXAMPLE 2   COUNTING DIFFERENT PRODUCTS

A toy company makes red, green, blue, and yellow plastic cars, trucks, and planes. How many different kinds of toys do they make?

Solution

Let the set of colors be $C = \{\text{red, green, blue, yellow}\}$, and let the set of shapes be $S = \{\text{car, truck, plane}\}$. Each possible toy is described by its color and shape:

	Red	Green	Blue	Yellow
Car	(red, car)	(green, car)	(blue, car)	(yellow, car)
Truck	(red, truck)	(green, truck)	(blue, truck)	(yellow, truck)
Plane	(red, plane)	(green, plane)	(blue, plane)	(yellow, plane)

This rectangular table contains all possible combination of colors and shapes, and the number of boxes is found by multiplying length times width: There are $4 \cdot 3 = 12$ possible toys that can be made.

A convenient way of using the multiplication principle is to imagine making up a typical combination of the two choices. For instance, in the preceding example there were 4 choices for the color and 3 choices for the shape, giving

$4 \cdot 3 = 12$ possible toys

The Multiplication Principle for Counting generalizes to *more* than two choices.

> **Generalized Multiplication Principle for Counting**
>
> If k choices are to be made, and there are m_1 possibilities for the first choice, m_2 possibilities for the second choice, m_3 possibilities for the third choice, and so on down to m_k possibilities for the kth choice, and if the choices can be combined in any way, then the k choices can be made together in $m_1 \cdot m_2 \cdot \; \cdots \; \cdot m_k$ ways.

EXAMPLE 3

COUNTING PARKING PERMITS

A parking permit displays an identification code consisting of a letter (A to Z) followed by two digits (0 to 9). How many different permits can be issued?

Solution

Since each identification code consists of three symbols, we need to fill three blanks:

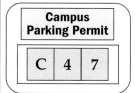

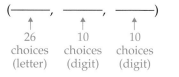

$26 \cdot 10 \cdot 10 = 2600$ ways

Therefore, 2600 different permits can be issued.

Practice Problem 2

A computer password consists of two letters (A to Z) followed by three digits (0 to 9). How many different passwords are there?

➤ Solution on page 362

The Number of Subsets of a Set

Given a set, how many subsets does it have? For example, the set $\{a, b\}$ has 4 subsets: $\{a\}$, $\{b\}$, $\{a,b\}$, and $\emptyset$. (We consider the empty set $\emptyset$ and the set itself to be subsets.) To form these subsets we had two choices: *include or exclude a* and *include or exclude b*. Two choices (for *a* and *b*) with each made in two possible ways (*include* or *exclude*) gives a total of $2 \cdot 2$ choices, for 4 subsets (as we saw).

$$(\underset{\underset{\text{(yes/no)}}{\underset{\uparrow}{\text{Include } a?}}}{\rule{2em}{0.4pt}}, \underset{\underset{\text{(yes/no)}}{\underset{\uparrow}{\text{Include } b?}}}{\rule{2em}{0.4pt}}) \qquad 2 \cdot 2 = 4 \text{ subsets}$$

In this same way we can count the number of subsets of *any* set. A set with *n* members means *n* choices: *include* or *exclude the first member, include* or *exclude the second member*, and so on. Making *n* choices, each in 2 possible ways (*include* or *exclude*) means a total of 2^n choices, for 2^n subsets.

Number of Subsets of a Set

A set with *n* elements has 2^n subsets.

EXAMPLE 4

FINDING THE SUBSETS OF A SET

List all subsets of the set $\{a,b,c\}$ and verify that there are $2^3 = 8$ of them.

Solution

The subsets are

$$\{a\}, \ \{b\}, \ \{c\}, \ \{a,b\}, \ \{a,c\}, \ \{b,c\}, \ \{a,b,c\}, \text{ and } \varnothing$$

Indeed, there are 8 subsets (including the set itself and the empty set).

EXAMPLE 5

COUNTING SUBSETS

A restaurant offers pizza with mushrooms, peppers, onions, pepperoni, and sausage. How many different types of pizza can be ordered?

Solution

The set of toppings has 5 members, so there are $2^5 = 32$ possible subsets, and so 32 different pizzas. (Do you see which of these subsets corresponds to "plain" pizza?)

5.1 Section Summary

For a set A, its *complement* A^c is the set of all elements *not* in A. For sets A and B their *intersection* $A \cap B$ is the set of elements that are in *both* A and B, and their *union* $A \cup B$ is the set of elements that are in *either A or B or both*. In the Venn diagrams below, U stands for the *universal* set of all elements being considered.

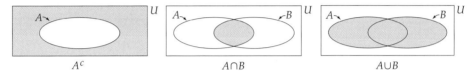

For a finite set A, $n(A)$ means the number of elements in A. The following formulas enable us to find the number of elements in one set by counting the elements in other sets:

$$n(A^c) = n(U) - n(A) \qquad \text{Complementary principle of counting}$$

$$n(A \cup B) = n(A) + n(B) - n(A \cap B) \qquad \text{Addition principle of counting}$$

The multiplication principle says that if two choices are to be made, and one can be made in m ways and the other can be made in n ways, and if the ways can be freely combined, then the two choices *together* can be made in $m \cdot n$ ways. The multiplication principle generalizes to more than two choices.

A set with n members has 2^n subsets.

▶ Solutions to Practice Problems

1. Let D be the set of businesses offering dental insurance, and let V be those offering vision insurance. Then, using a Venn diagram:

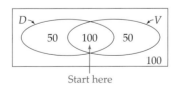

Start here

(continues)

Or, using the Addition Principle:

$$n(D \cup V) = n(D) + n(V) - n(D \cap V)$$
$$= 150 + 150 - 100 = 200$$

Two hundred businesses offer dental or vision insurance. One hundred offer neither.

2. Since $\underbrace{26 \cdot 26} \cdot \underbrace{10 \cdot 10 \cdot 10} = 676{,}000$, there are 676,000 different passwords.
   2 letters 3 digits

5.1 Exercises

Find each number using the following Venn diagram.

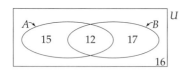

1. $n(A)$
2. $n(B)$
3. $n(U)$
4. $n(A \cap B)$
5. $n(A \cup B)$
6. $n(A^c)$
7. $n(B^c)$
8. $n(A \cap B^c)$
9. $n(A^c \cap B)$
10. $n(A \cup B^c)$

11. Given that $n(A) = 20$, $n(B) = 10$, $n(A \cap B) = 6$, and $n(U) = 40$, fill in the four regions in the Venn diagram.

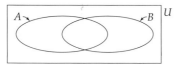

12. Given that $n(A) = 15$, $n(B) = 30$, $n(A \cap B) = 5$, and $n(U) = 65$, fill in the four regions in the Venn diagram.

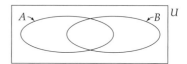

From your Venn diagram in Exercise 11 (on the left), find each number.

13. $n(A \cup B)$
15. $n(A^c)$
17. $n(B^c)$

From your Venn diagram in Exercise 12 (above), find each number.

14. $n(A \cup B)$
16. $n(A^c)$
18. $n(B^c)$

APPLIED EXERCISES (may be helpful.)

For Exercises 19–22, draw a tree diagram for each situation to find the number of possible pairs.

19. **GENERAL: Parking Permits** A parking permit sticker displays an identification code consisting of a letter (from A, B, and C) followed by a digit (from 1, 2, 3, and 4). How many different permits can be issued?

20. **GENERAL: Computer Passwords** A computer network password consists of two letters, the first being X, Y, or Z and the second

being A, B, C, D, or E. How many different passwords are there?

21. **GENERAL: Wardrobes** A clothing store sells windbreakers, ski jackets, and overcoats in your choice of red or blue. How many different kinds of coats do they sell?

22. **GENERAL: Dorm Chores** Ted, Bob, Fred, Jim, Bill, and Sam share a student dorm suite. How many ways can they choose who takes out the garbage and who sweeps the floor if they can choose the same person to do both? What if different men must do the chores?

23. **ATHLETICS: Lacrosse** The sophomore lacrosse team has 24 players, of whom 10 played defense last year, 12 played offense, and 5 played both defense and offense, while the rest of the players did not play last year. How many members of the team played last year?

24. **SOCIAL SCIENCE: Political Fund Raising** If 12,300 individuals contributed to the governor's first election campaign, 15,200 contributed to her second, and 7800 contributed to both, how many individuals made contributions?

25. **SOCIAL SCIENCE: Current Events** A survey of 1200 residents of New Orleans revealed that 760 get information on international events by watching television news programs, 530 from listening to radio news shows, and 290 from both television and radio, while the rest have no interest in international events. How many have no interest in international events?

26. **BUSINESS: International Marketing** Of the 30 freshman majoring in international marketing, 12 speak French, 8 speak German, and 5 speak both languages. How many do not speak either French or German?

27. **GENERAL: Parking Permits** Look back at Example 3 on page 359 and find the number of parking permits that can be issued if the letter O and the number 0 are omitted to avoid confusion.

28. **GENERAL: Computer Passwords** A computer network password is made up of letters A to Z and may be either four or five letters long. How many such passwords are there? Find the answer in two different ways:
    a. Find the number of four-letter passwords and add it to the number of five-letter passwords.
    b. Think of a four-letter password as really having length 5, where in the fifth position is a "blank" that can be thought of as one more member of an extended "alphabet" for the last position.

29. **GENERAL: Computer Passwords** How many eight-symbol computer passwords can be formed using the letters A to J and the digits 2 to 6?

30. **GENERAL: Bagels** A bagel shop offers 12 different kinds of bagels and 18 flavors of cream cheese. How many different orders for a bagel and cream cheese can customers place?

31. **GENERAL: Combination Locks** A suitcase combination lock has four wheels, each labeled with the digits 0 to 9. How many lock combinations are possible?

32. **GENERAL: Awards** A small community consists of 25 families, each consisting of two parents and three children. If one parent and one of his or her children are to be honored as "parent and child of the year," in how many different ways can the award be made?

33. **GENERAL: Committees** A college student governance committee is made up of 3 freshmen, 4 sophomores, 4 juniors, and 5 seniors. A subcommittee of 4 consisting of one person from each class is to go to a national convention. How many different subcommittees are there?

34. **GENERAL: Mix & Match Fashions** A designer has created a completely mixable line of five shirts, nine ties, and four pairs of slacks and advertises that "the possibilities are endless." Exactly how "endless" are the possible different outfits consisting of a shirt, a tie, and a pair of slacks?

35. **SOCIAL SCIENCE: False Testimony** A witness in a trial testified that he searched 20 cars, of which 12 were convertibles, 15 had out-of-state plates, and 5 were convertibles with

out-of-state plates. Explain why the witness was not telling the truth.

36. **SOCIAL SCIENCE: False Complaints** Your neighbor complained to the police that during the last month you played your music too loud on 22 days, your car blocked his driveway on 15 days, and you did both on 4 days. How would you prove to the police that your neighbor was not telling the truth?

37. **GENERAL: Cable TV Channels** A cable TV company offers regular service plus 6 premium channels that must be ordered separately. How many different options are there?

38. **GENERAL: Book Clubs** You join a book club that each month offers five books, of which any number (or none) can be ordered. How many possibilities are there? What if at least one book must be ordered?

39. **GENERAL: Classes** There are 10 students who have the prerequisites to take a math course. How many different possible classes are there? (Assume that a class must have at least one student.)

40. **GENERAL: Making Change** You have a penny, a nickel, a dime, a quarter, a half-dollar, and a dollar in your pocket. How many different amounts of money can you make?

Explorations and Excursions

The following problems extend and augment the material presented in the text.

De Morgan's Laws

De Morgan's Laws* show the relationship between the operations of complement, union, and intersection. De Morgan's Laws are:

$(A \cup B)^c = A^c \cap B^c$ De Morgan's First Law

$(A \cap B)^c = A^c \cup B^c$ De Morgan's Second Law

41. Verify De Morgan's First Law (above) for the sets A and B shown with their members in the following Venn diagram.

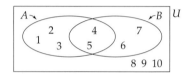

42. Verify De Morgan's Second Law (above Exercise 41) for the sets A and B shown with their members in the above Venn diagram.

43. Show De Morgan's First Law (above Exercise 41) by making Venn diagrams for $(A \cup B)^c$ and for $A^c \cap B^c$.

44. Show De Morgan's Second Law (above Exercise 41) by making Venn diagrams for $(A \cap B)^c$ and for $A^c \cup B^c$.

*After Augustus de Morgan (1806–1871), British mathematician and logician.

5.2 PERMUTATIONS AND COMBINATIONS

Introduction

In this section we will develop two very useful formulas for counting various types of choices, known as *permutations* and *combinations*, and then apply these formulas to a wide variety of problems. We begin by describing *factorial notation*.

Factorials

Products of successive integers from a number down to 1, such as $5 \cdot 4 \cdot 3 \cdot 2 \cdot 1$, are called *factorials*. We denote factorials by exclamation points, so the preceding product would be written 5! (read "5 factorial"). Formally:

Factorials

For any positive integer n,

$$n! = n(n-1) \cdot \cdots \cdot 1 \qquad \text{n factorial is the product of the integers from n down to 1}$$

$$0! = 1 \qquad \text{Zero factorial is 1}$$

The next example shows that some factorial expressions are most easily found by using cancellation *before* evaluating the factorials.

EXAMPLE 1

CALCULATING FACTORIALS

Find: **a.** $4!$ **b.** $\dfrac{7!}{6!}$ **c.** $\dfrac{6!}{3!}$ **d.** $\dfrac{100!}{99!}$

Solution

a. $4! = 4 \cdot 3 \cdot 2 \cdot 1 = 24$

b. $\dfrac{7!}{6!} = \dfrac{7 \cdot 6 \cdot 5 \cdot 4 \cdot 3 \cdot 2 \cdot 1}{6 \cdot 5 \cdot 4 \cdot 3 \cdot 2 \cdot 1} = \dfrac{7 \cdot \cancel{6 \cdot 5 \cdot 4 \cdot 3 \cdot 2 \cdot 1}}{\cancel{6 \cdot 5 \cdot 4 \cdot 3 \cdot 2 \cdot 1}} = 7$ When finding quotients of factorials, look for cancellation

c. $\dfrac{6!}{3!} = \dfrac{6 \cdot 5 \cdot 4 \cdot 3 \cdot 2 \cdot 1}{3 \cdot 2 \cdot 1} = \dfrac{6 \cdot 5 \cdot 4 \cdot \cancel{3 \cdot 2 \cdot 1}}{\cancel{3 \cdot 2 \cdot 1}} = 6 \cdot 5 \cdot 4 = 120$

d. $\dfrac{100!}{99!} = \dfrac{100 \cdot 99 \cdot \cdots \cdot 1}{99 \cdot \cdots \cdot 1} = 100$ Canceling $99 \cdot \cdots \cdot 1$

Parts (b) and (d) of the above example are instances of the formula $\dfrac{n!}{(n-1)!} = n$. For $n = 1$ this becomes $\dfrac{1!}{0!} = 1$, and multiplying each side by $0!$ gives $1 = 0!$, which is why we define $0!$ to be 1.

We will use factorials to count the number of ways that objects can be *ordered*.

Permutations

How many different orderings are there for the letters *a*, *b*, and *c*? We may list the orderings as *abc*, *acb*, *bac*, *bca*, *cab*, and *cba*, so there are 6. Each of these orderings is called a *permutation*. Instead of listing them all, we could observe that there are 3 ways of choosing the first letter, 2 ways of choosing the second (because one letter was "used up" in the first choice), and 1 way of choosing the last (whichever is left), so by the multiplication principle there are $3 \cdot 2 \cdot 1 = 6$ possible orderings, just as we found before. How many permutations are there of n distinct objects? By the same reasoning, the answer is $n(n-1) \cdot \cdots \cdot 1 = n!$.

EXAMPLE 2 COUNTING BATTING ORDERS

How many different batting orders are there for a 9-player baseball team?

Solution

The first batter can be any one of the 9 players, the second batter can be any one of the remaining 8, next can be any one of the remaining 7, and so on down to the last batter who will be the only one left. By the multiplication principle we should multiply together the numbers of ways of making each choice, so the number of batting orderings will be

$$9! = 9 \cdot 8 \cdot 7 \cdot 6 \cdot 5 \cdot 4 \cdot 3 \cdot 2 \cdot 1 = 362{,}880$$

Practice Problem 1

You have six different tasks to do today. In how many different orders can they be done? ▶ Solution on page 373

EXAMPLE 3 COUNTING ORDERINGS OF BOOKS

You have 3 math books, 4 history books, and 5 English books. In how many ways can these 12 books be arranged on the shelf if all books of the same subject are together?

Solution

There are 3! ways to arrange the math books among themselves, 4! ways to arrange the history books, and 5! ways to arrange the English books, so by the multiplication principle there are $3! \cdot 4! \cdot 5!$ ways if the order of the subjects is math then history then English. However, these three subjects can *themselves* be arranged in any of $3 \cdot 2 \cdot 1 = 3!$ orderings. For each of the 3! orderings of the subjects, the books can be arranged in $3! \cdot 4! \cdot 5!$ ways, so the total number of such arrangements is

$$3! \cdot 3! \cdot 4! \cdot 5! = 6 \cdot 6 \cdot 24 \cdot 120 = 103{,}680$$

To return to our baseball example, how many different orderings are there for just *the first 3 batters* on the 9-player team? Clearly any one of the 9 players can bat first, then any one of the remaining 8, then any one of the remaining 7, for a total of $9 \cdot 8 \cdot 7$ orderings (by the multiplication principle). In general, if we have n distinct objects and want to count all possible orderings of some r of them, there are n choices for the first, $n-1$ choices for the second, $n-2$ choices for the third, down to $n-r+1$ choices for the rth. (Notice that taking r of them means leaving $n-r$, so the last one kept is the preceding one, $n-r+1$.) Such orderings are called *permutations* and can be counted using the following formula.

Permutations*

The number of permutations (ordered arrangements) of n distinct objects taken r at a time is

$$_nP_r = \overbrace{n \cdot (n-1) \cdot \cdots \cdot (n-r+1)}^{r \text{ factors}} \quad \text{Product of } r \text{ numbers from } n \text{ down}$$

We define $_nP_0 = 1$ since there is exactly one way of taking zero objects from n, namely, taking nothing.

EXAMPLE 4

COUNTING NONSENSE WORDS

How many five letter "nonsense" words (that is, strings of letters without regard to meaning) can be made from the letters A to Z with no letter repeated?

Solution

Words are ordered arrangements of letters, so we use the permutation formula with $n = 26$ and $r = 5$.

$$_{26}P_5 = 26 \cdot 25 \cdot 24 \cdot 23 \cdot 22 = 7{,}893{,}600 \quad \text{Product of 5 numbers from 26 down}$$

There are 7,893,600 five-letter nonsense words with distinct letters.

---

*Alternative notations for $_nP_r$ are $P_{n,r}$, P_r^n, and $P(n, r)$.

(using factorials). Replacing the 8 and 3 by *any* positive integers, we obtain the following general formulas.

Combinations*

The number of combinations (*un*ordered arrangements) of n distinct objects taken r at a time is

$$_nC_r = \frac{n \cdot (n-1) \cdot \cdots \cdot (n-r+1)}{r \cdot (r-1) \cdot \cdots \cdot 1} \quad \begin{array}{l} \leftarrow r \text{ numbers beginning with } n \\ \leftarrow r! \end{array}$$

$$= \frac{n!}{r!(n-r)!} \qquad \text{In factorial form}$$

The second formula comes from the first by multiplying the numerator and denominator by $(n-r) \cdot \cdots \cdot 1 = (n-r)!$. We define $_nC_0 = 1$ since there is clearly one way to choose zero objects from n, namely, taking nothing.

Notice that 8 things taken 3 at a time is the same as 8 things taken 5 at a time:

$$_8C_3 = \frac{8 \cdot 7 \cdot 6}{3 \cdot 2 \cdot 1} = 56$$

$$_8C_5 = \frac{8 \cdot 7 \cdot 6 \cdot \cancel{5} \cdot \cancel{4}}{\cancel{5} \cdot \cancel{4} \cdot 3 \cdot 2 \cdot 1} = 56$$

Same value

There is a simple reason for this, which may be explained in terms of committees: Whenever you choose a committee of 3 from 8 people you are, in a sense, also selecting another committee, *the 5 left behind*. Any different choice of the 3 results in a different 5 being left behind. Therefore, from 8 people there must be exactly as many committees of 3 as committees of 5. This is also clear from the factorial formulas, since $_8C_3 = \frac{8!}{3!\,5!}$ and $_8C_5 = \frac{8!}{5!\,3!}$ differ only in the order of the denominators. As before, replacing 8 and 3 by *any* positive integers n and r shows that from n people there are exactly as many committees of r people as committees of $n-r$ people (in symbols: $_nC_r = {_nC_{n-r}}$). This relationship is useful in reducing calculation; for example, given $_8C_5$ it is easier to calculate it as $_8C_3$, as we saw in the calculations above this paragraph.

---

* Alternative notations for $_nC_r$ are $C_{n,r}$, C_r^n, $C(n, r)$, and $\binom{n}{r}$.

EXAMPLE 7

COUNTING PERMUTATIONS AND COMBINATIONS

A student club has 15 members.

a. How many ways can a president, vice president, and treasurer be chosen?

b. How many ways can a committee of three members be chosen?

Solution

Each question involves choosing three members from the club, but for the officers we want *ordered* arrangements (the order determines the offices: the president is listed first, the vice president second, and the treasurer third), while for the committee we want *unordered* arrangements. Thus part (a) asks for permutations, part (b) for combinations.

a. For the officers: $\quad _{15}P_3 = 15 \cdot 14 \cdot 13 = 2730 \quad$ 3 numbers from 15 down

b. For the committee: $\quad _{15}C_3 = \dfrac{15 \cdot 14 \cdot 13}{3 \cdot 2 \cdot 1} = 455 \quad$ Permutations divided by 3!

There are 2730 different ways of choosing the president, vice president, and treasurer, and 455 ways of choosing the committee.

Practice Problem 3

A college business major can also minor in computer science by taking any 7 courses from an approved list of 10 courses. How many different collections of courses will satisfy the requirements for the computer science minor? ➤ Solution on page 373

Graphing Calculator Exploration

Values of $_nC_r$ are easy to find if your calculator includes this command. The following screens show several values of $_nC_r$ for $n = 15$. Notice that $_nC_r$ gets larger and then smaller as r increases.

```
MATH NUM CPX PRB
1:rand
2:nPr
3:nCr
4:!
5:randInt(
6:randNorm(
7:randBin(
```

```
8 nCr 3
            56
8 nCr 4
            70
8 nCr 5
            56
```

A *standard deck* of 52 playing cards contains four *suits* (spades, hearts, diamonds, and clubs), each of which contains three *face cards* (jack, queen, and king), cards numbered 2 through 10, and an *ace*. A *hand* of cards is a selection of cards from the deck.

EXAMPLE 8

COUNTING 5-CARD HANDS

How many different 5-card hands are there?

Solution

Since the order in which the cards are dealt makes no difference to the final hand, we want the number of *combinations* of 52 cards taken 5 at a time.

$$_{52}C_5 = \frac{52 \cdot 51 \cdot 50 \cdot 49 \cdot 48}{5 \cdot 4 \cdot 3 \cdot 2 \cdot 1} = 2{,}598{,}960$$

There are 2,598,960 possible 5-card hands.

EXAMPLE 9

WHY FOUR OF A KIND BEATS A FULL HOUSE

In cards, "kind" means *value* (ace through king), so "four of a kind" means, for example, 4 aces or 4 sevens. Among all 5-card hands, how many have:

a. four of a kind?

b. three of a kind and two of a kind (called a *full house*)?

Solution

a. To get four of a kind, there are 13 ways to choose the "kind" (ace through king) and 48 ways to choose the fifth card (52 minus the 4 already chosen), giving:

$$13 \cdot 48 = 624 \qquad \text{Number of hands with } four of a kind$$

b. For a full house (three of a kind and two of a kind, such as 3 kings and 2 sevens), there are 13 ways to choose the first kind, and then 3 of the 4 cards of this kind can be chosen in $_4C_3 = {}_4C_1 = \frac{4}{1} = 4$ ways. For the "kind" in the two of a kind there are 12 choices (since one kind has been "used up"), and choosing 2 of these 4 cards can be done in $_4C_2 = \frac{4 \cdot 3}{2 \cdot 1} = 6$ ways, giving:

$$13 \cdot 4 \cdot 12 \cdot 6 = 3744$$

Number of hands with a *full house*

These numbers show why four of a kind beats a full house in poker: Four of a kind is much more rare, occurring only about a sixth as often as a full house.

5.2 Section Summary

Permutations are *ordered* choices; *combinations* are choices *without regard to order*. That is, in *permutations* a different order means a *different object* (*abc* and *bca* are different), but in combinations changing the order does *not* represent a new object (*abc* and *bca* are counted as the same). If order matters (as with letters in words or listings of people for president, vice president, and treasurer), use *permutations*; if order does not matter (as with committees or hands of cards), use *combinations*.

$$_nP_r = \overbrace{n \cdot (n-1) \cdot \cdots \cdot (n-r+1)}^{r \text{ factors}}$$

Permutations (orderings)

$$_nC_r = \frac{n \cdot (n-1) \cdot \cdots \cdot (n-r+1)}{r \cdot (r-1) \cdot \cdots \cdot 1} = \frac{n!}{r!(n-r)!}$$

Combinations (collections)

▶ Solutions to Practice Problems

1. There are $6! = 6 \cdot 5 \cdot 4 \cdot 3 \cdot 2 \cdot 1 = 720$ different orders.

2. Since the winners must be selected in order (first, second, and third place), there are $_{35}P_3 = 35 \cdot 34 \cdot 33 = 39{,}270$ different ways of choosing the winners.

3. Since the courses can be taken in any order, there are $_{10}C_7$ different ways. It is easier to calculate this as $_{10}C_3 = \frac{10 \cdot 9 \cdot 8}{3 \cdot 2 \cdot 1} = 120$ different ways to fulfill a minor in computer science.

5.2 Exercises

Calculate each factorial or quotient of factorials.

1. a. 2! b. 6!
2. a. 1! b. 5!
3. $\dfrac{10!}{7!}$
4. $\dfrac{12!}{10!}$
5. $\dfrac{200!}{198!}$
6. $\dfrac{1000!}{998!}$

Determine each number of permutations.

7. $_6P_3$
8. $_6P_2$
9. $_8P_1$
10. $_9P_8$
11. a. $_{13}P_4$ b. $_{13}P_5$ c. $_{13}P_6$
12. a. $_{11}P_5$ b. $_{11}P_6$ c. $_{11}P_7$
13. a. $_{10}P_1$ b. $_{10}P_5$ c. $_{10}P_{10}$
14. a. $_{12}P_1$ b. $_{12}P_6$ c. $_{12}P_{12}$

Find each number of combinations.

15. $_6C_2$
16. $_5C_3$
17. $_7C_3$
18. $_8C_1$
19. a. $_{11}C_4$ b. $_{11}C_5$ c. $_{11}C_6$ d. $_{11}C_7$
20. a. $_{13}C_5$ b. $_{13}C_6$ c. $_{13}C_7$ d. $_{13}C_8$
21. a. $_{12}C_1$ b. $_{12}C_2$ c. $_{12}C_6$ d. $_{12}C_{11}$ e. $_{12}C_{12}$
22. a. $_{10}C_1$ b. $_{10}C_2$ c. $_{10}C_5$ d. $_{10}C_9$ e. $_{10}C_{10}$

APPLIED EXERCISES

23. **SOCIAL SCIENCE: Election Ballots** How many different ways can the eight candidates for the school board election be listed on the ballot?

24. **MANAGEMENT SCIENCE: Employee Training** Six candidates in an employee training program are ranked according to the time it takes them to perform a task. Assuming that all of the times are different, how many different rankings are possible?

25. **BUSINESS: Commercial Art** A bank lobby wall has space to display 4 paintings by local artists. If 12 paintings are available, how many different ways can they be displayed on the wall?

26. **GENERAL: Postal Codes** How many five-digit ZIP codes have no repeated digits?

27. **GENERAL: Telephone Numbers** A telephone number consists of seven digits and the first digit cannot be a zero or a one. How many telephone numbers have no repeated digits?

28. **GENERAL: Horse Racing** How many different ways can the 14 horses in a race finish first, second, and third?

29. **GENERAL: Computer Passwords** A computer password is to consist of 4 alphanumeric characters with no repeats. (An alphanumeric character is a letter from A to Z or a digit from 0 to 9.) How many such passwords are there? How many are there if the letter O and the digit 0 are excluded to avoid confusion?

30. **GENERAL: Social Security Numbers** How many nine-digit Social Security numbers start with three odd digits, then have two even digits, and end with four distinct digits from 0 to 9?

31. **ATHLETICS: Team Sports** A league has 6 college teams, and each team must host each

of the others once in a season. How many games are played per season?

32. **GENERAL: Books** You have 2 science books, 3 language books, and 4 philosophy books. In how many ways can they be arranged on the shelf if all books of the same subject are to be together?

33. **ATHLETICS: Basketball Teams** A junior high girls' basketball team is to consist of 5 players. How many different teams can the manager select from a roster of 12 girls?

34. **GENERAL: Novels** By the end of an English class you are to have read 3 novels from a list of 21. How many choices are possible?

35. **GENERAL: Playing Cards** How many five-card hands from a standard deck will contain only spades? How many will contain only cards of the same suit?

36. **GENERAL: Playing Cards** How many five-card hands from a standard deck will contain only face cards? How many will contain no face cards?

37. **MANAGEMENT SCIENCE: Labor Negotiations** The union leadership and the management strike team have agreed to each select 4 representatives to enter into a "closed-room, around-the-clock" marathon in a final attempt to reach a settlement. If there are 10 members of the union leadership and 12 members of the management team, how many ways can the marathon negotiators be chosen?

38. **GENERAL: Lines and Circles** Ten distinct points have been marked on the circumference of a circle. How many lines can be drawn through pairs of these points?

39. **GENERAL: Test Taking** A test consists of 12 questions, from which each student chooses 10 to answer. How many choices are possible?

40. **GENERAL: Homework Grading** A sociology professor assigned 5 problems from Unit One, 10 problems from Unit Two, and 8 problems from Unit Three for next Wednesday and announced that she will grade only 2 of the assigned problems from Unit One, 4 of those from Unit Two, and 3 from Unit Three. How many different ways can she choose the problems that she will grade?

41. **SOCIAL SCIENCE: Senate Committees** The U.S. Senate consists of 100 members, two from each state. A committee of 5 members is to be chosen. How many such committees are possible? How many if the committee cannot have more than one senator from the same state?

42. **SOCIAL SCIENCE: Jury Verdicts** In how many ways can a civil jury of 6 members split 4 to 2 on a decision?

43. **ATHLETICS: Baseball Lineups** A sports broadcaster, irritated that a baseball manager kept changing his lineup, once said that the manager should "choose his most talented nine players, try them in all different batting orders, and then stick with the best one." To see whether this advice makes sense answer the following questions.
    a. Find the number of possible orderings for nine players (or see Example 2 on page 366).
    b. If the manager were to try each batting order for just 15 minutes, holding tryouts 8 hours a day 365 days a year, how long would it take to try them all?

44. **ATHLETICS: Lacrosse** A women's lacrosse team plays 12 games in a season, winning 7, losing 3, and tying 2. Show that the number of ways that this can occur is $(_{12}C_7)(_5C_3) = \frac{12!}{7!3!2!}$.

Explorations and Excursions

The following problems extend and augment the material presented in the text.

The Binomial Theorem

The numbers $_nC_r$ are often called *binomial coefficients* because of their connection with the *binomial theorem*:

Binomial Theorem

$$(x + y)^n = {_nC_0}x^n + {_nC_1}x^{n-1}y + {_nC_2}x^{n-2}y^2 + \cdots + {_nC_n}y^n$$

For any positive integer n

45. Prove the binomial theorem (above) by justifying the following equation.

$$\underbrace{(x + y)(x + y) \cdots (x + y)}_{n \text{ factors}} = x^n + {_nC_1}x^{n-1}y + {_nC_2}x^{n-2}y^2 + \cdots + y^n$$

- x^n: x from each factor
- ${_nC_1}x^{n-1}y$: y from one factor, x's from the others
- ${_nC_2}x^{n-2}y^2$: y's from two factors, x's from the other
- y^n: y from each factor

(Hint: $(x + y)(x + y) \cdots (x + y)$ is found by adding all products consisting of one term (either x or y) from each of the factors and combining similar terms: x^n comes from taking the x from each factor; $x^{n-1}y$ comes from taking y from one factor and x's from the others (explain why this can be done in ${_nC_1}$ ways); $x^{n-2}y^2$ takes y from two factors and x's from the others (explain why this can be done in ${_nC_2}$ ways); and so on. Then explain why the above equation is the same as the binomial theorem.)

46. Use the binomial theorem to expand $(x + y)^4$.

Pascal's Triangle

The coefficients of $(x + y)^n$ can also be found from *Pascal's triangle*,* in which each number is the sum of the two closest numbers in the row just above it and each row begins and ends with a 1.

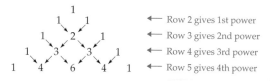

- Row 2 gives 1st power
- Row 3 gives 2nd power
- Row 4 gives 3rd power
- Row 5 gives 4th power

---

* Blaise Pascal, *Treatise on the Arithmetic Triangle*, (*Traité du Triangle Arithmétique*), published in 1665. "Pascal's" triangle dates back to the Chinese in about 1100, and certain parts of it to the Pythagoreans in 540 BC. For more information, see *Pascal's Arithmetical Triangle* by A.W.F. Edwards, Griffin, 1987.

47. Use the appropriate row of Pascal's triangle to expand $(x + y)^4$.

48. The connection between the binomial coefficients and Pascal's triangle is given by the formula ${_nC_k} + {_nC_{k-1}} = {_{n+1}C_k}$. Explain why this formula is the rule used to find the numbers in each row of Pascal's triangle and then show that it is correct in two ways:

  a. By algebra.

  b. By interpreting the two sides in terms of committees. [*Hint:* Think of forming committees of k people from n people plus yourself, and interpret the left-hand side as the number of committees excluding you plus the number of committees including you.]

Interpreting the Binomial Theorem in Terms of Committees

49. a. Show that evaluating the binomial theorem at $x = 1$ and $y = 1$ gives

$$2^n = {_nC_0} + {_nC_1} + {_nC_2} + \cdots + {_nC_n}.$$

  b. Interpret the above equation in terms of committees. [*Hint:* To interpret the left-hand side see page 360, and to interpret the right-hand side think of committees of different sizes.]

50. a. Show that evaluating the binomial theorem at $x = 1$ and $y = -1$ gives

$$0 = {_nC_0} - {_nC_1} + {_nC_2} - \cdots + (-1)^n {_nC_n}$$

  (with alternating signs).

  b. In the above equation, move the negative terms to the left-hand side and interpret the result in terms of committees with odd numbers of members compared to committees with even numbers of members.

5.3 PROBABILITY SPACES

Introduction

Some experiments always have the same outcome, while others involve "chance" or "random" effects that produce a variety of outcomes. In this section we will consider such "chance" experiments and show that, in spite of their random nature, we can draw many useful conclusions about their results. We restrict ourselves to experiments that have a finite number of possible outcomes.

Random Experiments and Sample Spaces

A *random* experiment is one that, when repeated under identical conditions, may produce different outcomes. Each repetition of the experiment is called a *trial*, and each result is an *outcome*. The set of all possible outcomes, exactly one of which must occur, is called the *sample space* for the experiment. The following table lists some simple random experiments and sample spaces.

Random Experiment	Sample Space	
a. Flip a coin	$\{H, T\}$	H means Heads, T means Tails
b. Flip a coin twice	$\{(H, H), (H, T), (T, H), (T, T)\}$	(H, T) means Heads then Tails
c. Choose a card from a deck and observe its suit	$\{\spadesuit, \heartsuit, \diamondsuit, \clubsuit\}$	
d. Roll a die	$\{1, 2, 3, 4, 5, 6\}$	

There may be more than one possible sample space for a given random experiment. For example, in the experiment of choosing a card (experiment (c) above) the sample space could be all 52 cards (the ace of spades to the king of clubs). In general, we will choose the *simplest* sample space that allows us to observe what we are interested in (which in experiment (c) is just the suit). The only restriction is that exactly one of the outcomes in the sample space must occur whenever the experiment is performed. In general, if the possible outcomes are represented as $e_1, e_2, \ldots, e_n$, then the sample space is the *set* of these possible outcomes $\{e_1, e_2, \ldots, e_n\}$.

Events

An *event* is a *subset* of the sample space.

EXAMPLE 1

ROLLING A DIE

For the experiment of rolling a die (experiment (d) on the previous page), represent the following events as subsets of the sample space $\{1, 2, 3, 4, 5, 6\}$:

a. Event E: *Rolling evens.*   **b.** Event F: *Rolling 5 or better.*

Solution

a. $E = \{2, 4, 6\}$ Rolling evens
b. $F = \{5, 6\}$ Rolling 5 or better

Practice Problem 1

For the experiment of flipping a coin twice (experiment (b) on the previous page), represent the event S: *same face* as a subset of the sample space $\{(H, H), (H, T), (T, H), (T, T)\}$. ▶ Solution on page 385

Probabilities of Possible Outcomes

Having identified the possible outcomes of an experiment, we now assign to each of them a probability from 0 (impossible) to 1 (certain). How do we assign probabilities? This depends on our knowledge of the experiment. For example, if there are three possible outcomes, and each seems equally likely, then we assign probability $\frac{1}{3}$ to each; if there are 10 outcomes that seem equally likely, then we assign probability $\frac{1}{10}$ to each. In general,

> **Equally Likely Outcomes**
>
> If each of the n possible outcomes in the sample space S is equally likely to occur, then we assign probability $\frac{1}{n}$ to each.

EXAMPLE 2

ASSIGNING EQUAL PROBABILITIES

a. A coin is said to be *fair* if "heads" and "tails" are equally likely. Therefore, for a fair coin we assign probabilities $P(H) = \frac{1}{2}$ and $P(T) = \frac{1}{2}$.

b. A die is said to be *fair* if each of its six faces is equally likely. Therefore, for a fair die we assign $P(1) = P(2) = P(3) = P(4) = P(5) = P(6) = \frac{1}{6}$.

c. A card randomly drawn from a standard deck is equally likely to be any one of the 52 cards. Therefore, the probability of drawing any particular card (such as the queen of hearts) is $\frac{1}{52}$.

EXAMPLE 3

ASSIGNING EQUAL PROBABILITY USING COMBINATIONS

A student club has 15 members. If a 3-member fund-raising committee is selected at random, what is the probability that the committee will consist of Bob, Sue, and Tim?

Solution

Since there are $_{15}C_3 = \frac{15 \cdot 14 \cdot 13}{3 \cdot 2 \cdot 1} = 455$ different 3-member committees that can be selected from the 15 members, and each committee is equally likely, the probability that any one particular committee is selected is $\frac{1}{455}$.

Not all outcomes are equally likely. For example, long-term weather records for an area might indicate that the probability that it will rain on any given day is $P(R) = \frac{1}{10}$ so that the probability of no rain is $P(N) = \frac{9}{10}$. Probabilities must be assigned on the basis of experience and knowledge of the basic experiment, but they are subject to two conditions: Each probability must be between 0 and 1 (inclusive), and they must add to 1.

EXAMPLE 4

ASSIGNING UNEQUAL PROBABILITIES

The arrow of the spinner on the right can point to any one of three regions labeled A, B, and C. If the probability of pointing to a region is proportional to its area, find the probability of pointing to each of the areas A, B, and C.

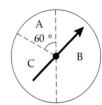

Solution

Let the sample space be $S = \{A, B, C\}$, representing the events that the spinner lands in regions A, B, or C. Clearly, region A is $\frac{60°}{360°} = \frac{1}{6}$ of the circle, region B is $\frac{1}{2}$ of the circle, and region C is the remaining $\frac{1}{3}$ of the circle. Therefore

$$P(A) = \tfrac{1}{6} \qquad P(B) = \tfrac{1}{2} \qquad P(C) = \tfrac{1}{3}$$

Practice Problem 2

Suppose that the spinner on the right is spun. What are the probabilities of the three outcomes?

▶ Solution on page 385

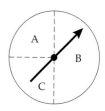

Probabilities of Events

We have assigned probabilities to events that consist of exactly one outcome. How do we find the probabilities of other events? We simply add up the probabilities of all of the outcomes in the event.

Probability Summation Formula

$$P(E) = \sum_{\text{All } e_i \text{ in } E} P(e_i)$$

The probability of an event is the sum of the probabilities of the possible outcomes in the event

Σ is the Greek capital letter sigma, which is equivalent to our capital S, and stands for "sum." From this formula, since the sample space S contains *all* of the outcomes,

$$P(S) = 1 \qquad \text{Something } must \text{ happen}$$

and, since the empty set contains *no* outcomes,

$$P(\varnothing) = 0 \qquad \text{"Nothing" cannot happen}$$

EXAMPLE 5

FINDING THE PROBABILITY OF AN EVENT

When rolling a pair of dice, what is the probability of rolling *doubles*?

Solution

For two dice, the sample space is

$S = \{$ (1, 1), (1, 2), (1, 3), (1, 4), (1, 5), (1, 6),
(2, 1), (2, 2), (2, 3), (2, 4), (2, 5), (2, 6),
(3, 1), (3, 2), (3, 3), (3, 4), (3, 5), (3, 6),
(4, 1), (4, 2), (4, 3), (4, 4), (4, 5), (4, 6),
(5, 1), (5, 2), (5, 3), (5, 4), (5, 5), (5, 6),
(6, 1), (6, 2), (6, 3), (6, 4), (6, 5), (6, 6) $\}$

Each of the 36 outcomes is equally likely, so each has probability $\frac{1}{36}$. Rolling doubles means rolling one of the outcomes (1, 1), (2, 2), (3, 3),

(4, 4), (5, 5), or (6, 6), each having probability $\frac{1}{36}$, so we add $\frac{1}{36}$ six times:

$$\underbrace{\frac{1}{36} + \frac{1}{36} + \cdots + \frac{1}{36}}_{6 \text{ times}} = \frac{6}{36} = \frac{1}{6} \qquad \text{Probability of rolling doubles}$$

In the preceding example, we could instead have taken the number of "favorable" outcomes (the 6 doubles) divided by the *total* number of possible outcomes (36) to get $\frac{6}{36} = \frac{1}{6}$, the same answer as before. Although this method is sometimes stated as a basic rule of probability, it is important to remember that it holds only when the outcomes are *equally likely*.

Probability for Equally-Likely Outcomes

$$P(E) = \frac{n(E)}{n(S)} \quad \begin{array}{l} \leftarrow \text{Number of outcomes "favorable to" (or in) } E \\ \leftarrow \text{Total number of outcomes} \end{array}$$

Probability That an Event Does *Not* Occur

An event E is a subset of the sample space, so the event that E does *not* occur, the event E^c, is the complement of E. Since E and E^c are disjoint and contain all outcomes, we must have $P(E) + P(E^c) = 1$. Solving this equation for $P(E^c)$ gives the probability of the complement:

Complementary Probability

$$P(E^c) = 1 - P(E) \qquad \begin{array}{l} \text{The probability that an event does } not \text{ occur} \\ \text{is 1 minus the probability that it } does \text{ occur} \end{array}$$

EXAMPLE 6

QUALITY CONTROL

A shipment of 100 memory chips contains 4 that are defective. The company that ordered them has a policy of testing 3 of the chips and, if any are defective, rejecting the entire shipment. What is the probability that the shipment is rejected?

Solution

The shipment will be rejected if among the three chosen chips there are 1, 2, or 3 defectives. It is much easier to find the probability of the *complementary* event, that the shipment is *accepted* (the number of defectives

5.3 Exercises

1. **SOCIAL SCIENCE: Committees** Find the sample space for a committee of two chosen from Alice, Bill, Carol, and David. Then find the sample space if both sexes must be represented. (*Use initials.*)

2. **SOCIAL SCIENCE: Committees** Find the sample space for a committee of two formed from Alice, Bill, Carol, David, and Edgar. Then find the sample space if the committee must include at least one woman. (*Use initials.*)

3. **GENERAL: Wardrobes** Find the sample space for choosing an outfit consisting of one of four shirts (S_1, S_2, S_3, or S_4) and one of two jackets (J_1 or J_2).

4. **GENERAL: Wardrobes** Find the sample space for choosing an outfit consisting of one of two shirts (S_1 or S_2), one of two jackets (J_1 or J_2), and one of two pairs of pants (P_1 or P_2).

5. **GENERAL: Marbles** A box contains three marbles, one red, one green, and one blue. A first marble is chosen, its color recorded, and then it is replaced in the box and a second marble is chosen, and its color is recorded. Find the sample space. Then find the sample space if the first marble is *not* replaced before the second is chosen.

6. **GENERAL: Coin Tossing** You toss a coin until you get the first head or until you have tossed it five times, whichever comes first. Find the sample space.

GENERAL: Dice A die is rolled, and $A = \{1, 3, 5\}$ (rolling an odd number), and $B = \{3, 6\}$ (rolling a three or a six). Specify each event as a subset of the sample space $S = \{1, 2, 3, 4, 5, 6\}$.

7. a. $A \cap B$ b. $A^c \cup B$

8. a. $A \cup B$ b. $A \cap B^c$

GENERAL: Dice Two dice are rolled. E is the event that the sum is even, F is the event of rolling at least one six, and G is the event that the sum is eight. List the outcomes for the following events.

9. a. $E \cap F$ b. $E^c \cap G$

10. a. $F \cap G$ b. $E^c \cap F \cap G$

Probabilities For the sample space $S = \{e_1, e_2, e_3, e_4, e_5\}$ with probabilities $P(e_1) = 0.10$, $P(e_2) = 0.30$, $P(e_3) = 0.40$, $P(e_4) = 0.05$, and $P(e_5) = 0.15$, and events $A = \{e_1, e_2, e_3\}$, $B = \{e_1, e_3, e_5\}$, and $C = \{e_2, e_4\}$, find each probability.

11. a. $P(A)$ b. $P(B \cup C)$

12. a. $P(B)$ b. $P(A \cup C)$

13. a. $P(A^c)$ b. $P(A^c \cap B)$

14. a. $P(B^c)$ b. $P(B^c \cap C)$

15. **GENERAL: Committees** If a committee of 3 is to be chosen at random from a class of 12 students, what is the probability of any particular committee being selected? What if the committee is to consist of a president, a vice president, and a treasurer?

16. **GENERAL: Committees** A committee of 3 is to be chosen at random from a class of 20 students. What is the probability that a particular committee will be selected? What if the 3 are to be a president, a vice president, and a treasurer?

17. **GENERAL: Marbles** One marble is selected at random from a box containing 6 red and 4 blue marbles, and its color noted, so the sample space is {R, B}. What probability should be assigned to each outcome?

GENERAL: Spinners Find the probability of the arrow landing in each numbered region. Assume that the probability of any region is proportional to its area and that areas that look the same size are the same size.

18.  19.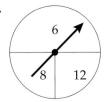

20. **GENERAL: Dartboards**
A circular dartboard has a radius of 12 inches with a circular bull's-eye of radius 2 inches at the center. You throw a dart and hit the dartboard. Assume that the probability of hitting any region is proportional to the area of the region.

    a. What is the probability of hitting the bull's-eye?
    b. What is the probability of hitting the rest of the dartboard?

21. **ATHLETICS: Baseball** A baseball hits the 12-foot-by-10-foot wall of your house. Assume that the probability of its hitting any region is proportional to the area of the region.

    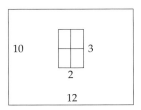

    a. What is the probability that it hits the 3-foot-by-2-foot window?
    b. What is the probability that it misses the window?

22. **GENERAL: Choosing Colors** One ball is to be chosen from a box containing red, blue, and green balls. If the probability that the chosen ball is red is $\frac{1}{5}$, and the probability that it is blue is $\frac{1}{3}$, what is the probability that it is green?

23. **SOCIAL SCIENCE: Political Contributions** In a town, 38% of the citizens contributed to the Republicans, 42% contributed to the Democrats, and 12% contributed to both. What percentage contributed to neither party?

24. **BUSINESS: Credit Cards** A store accepts both Visa cards and Mastercards. If 61% of its customers carry Visa cards, 52% carry Mastercards, and 28% carry both, what proportion carry a card that the store will accept?

25. **SOCIAL SCIENCE: Surveys** A college survey claimed that 63% of students took English composition, 48% took calculus, 15% took both, and 10% took neither. Show that these figures cannot be correct.

26. **GENERAL: Dice** You roll a pair of dice. Find the probability of:
    a. rolling a sum of 7.
    b. the first die show a higher number than the second.

27. **GENERAL: Marbles** A box contains 4 red and 8 green marbles. You reach in and remove 3 marbles all at once. Find the probability that these 3 marbles:
    a. are all red.
    b. are all of the same color.

28. **GENERAL: Lottery** In a lottery you choose 5 numbers out of 40. Then 6 numbers are announced, and you win something if you have 5 of the 6 numbers. What is the probability that you win something?*

29. **GENERAL: Guessing Numbers** Someone picks a number between 1 and 10 (inclusive) and you have three guesses. What is the probability that you will get it?

---

* One strategy for such lotteries is to avoid numbers that are dates or that are likely to be chosen by others. The probability of winning is the same for any number, and you want to avoid having to share the prize with others.

30. **GENERAL: Words** Among all 5-letter nonsense words (that is, without regard to meaning), what is the probability that a word has:
    a. no vowels?
    b. at least one vowel?

31. **GENERAL: Cards** If you are dealt 5 cards at random from an ordinary deck, what is the probability that your hand contains all four aces?

32. **GENERAL: Cards** If you are dealt 5 cards at random from an ordinary deck, what is the probability of being dealt a flush (all 5 cards of the same suit)?

33. **GENERAL: Elevator Stops** An elevator has 5 people and makes 7 stops. What is the probability that no two people get off on the same floor?

34. **BUSINESS: Defective Products** A carton of 24 CD players includes 4 that are missing a part. If you choose 4 at random, what is the probability that you get those 4?

35. **BUSINESS: Defective Products** A box of 100 screws contains 10 that are defective. If you choose 10 at random, what is the probability that none are defective?

36. **SOCIAL SCIENCE: Committees** A committee of 12 is to be formed from your class of 100 students. What is the probability that you and your best friend will be on the committee?

37. **SOCIAL SCIENCE: Senate Committees** The United States Senate consists of 100 members, 2 from each state. A committee of 8 senators is formed. What is the probability that it contains at least one senator from your state?

38. **GENERAL: Light Bulbs** Your house uses fifteen light bulbs, five in each of three different wattages, and you keep one spare bulb of each wattage. Two bulbs burn out. What is the probability that you have spares for both?

39. **GENERAL: Keys** You carry six keys in your pocket, two of which are for the two locks on your front door. You lose one key. What is the probability that you can get into your house through the front door?

40. **GENERAL: Shoes** You are rushing to leave on a trip, and you randomly grab four shoes from the five pairs in your closet. What is the probability that you take at least one pair?

41. **GENERAL: Monte Hall Problem** Contestants on Monte Hall's game show "Let's Make a Deal!" were asked to choose one of three doors to win the prize behind the door. Behind one door was a car and behind the others were goats. The contestant would choose a door, say door 1. Then Monte, who knew where the car was, would open one of the doors, say door 3, that had a goat and would then ask the contestant: "Do you want to change your choice to door 2?" Is it advantageous to switch? In 1990, Marilyn vos Savant wrote about this problem in her "Ask Marilyn" column in *Parade Magazine* and said that the contestant should switch. She subsequently received more than 10,000 letters, most of them disagreeing with her answer. Was she right?

42. **ATHLETICS: Batting Averages** Suppose that one baseball player (call him Babe) has a higher batting average than another (Ty) in the first half of the season, and also has the higher batting average in the second half of the season. Can Ty then have the higher batting average over the whole season? Consider the following batting averages (hits over at-bats) for the two players and complete the table by calculating hits over at-bats for each over the whole season.

	1st Half	2nd Half	Season
Babe	$\frac{60}{200} = 0.300$	$\frac{40}{100} = 0.400$	?
Ty	$\frac{29}{100} = 0.290$	$\frac{78}{200} = 0.390$	?

Interpreting these averages as probabilities, explain how the average of two higher probabilities turns out to be lower than the average of two lower probabilities. [*Hint:* Think about whether you are really averaging the averages or taking a *weighted average*, $\frac{2}{3}$ of one and $\frac{1}{3}$ of the other, and how that can make a difference.]

Explorations and Excursions

The following problems extend and augment the material presented in the text.

Odds

If E is an event with probability $P(E)$ such that $0 < P(E) < 1$, then "the *odds for* the event E" means the ratio $P(E) : P(E^c)$ [read: "$P(E)$ to $P(E^c)$"] while "the *odds against* the event E" means the ratio $P(E^c) : P(E)$ [read: "$P(E^c)$ to $P(E)$"]. Since odds are ratios, we multiply to clear the fractions, so that, for example, $\frac{2}{3} : \frac{1}{3}$ becomes $2 : 1$. For equally likely outcomes, the odds for an event can be interpreted as the ratio of the number of *favorable* outcomes to the number of *unfavorable* outcomes.

43. Show that the odds for heads in a flip of a coin are $1 : 1$.

44. Show that the odds for a two on a roll of a die are $1 : 5$.

45. Show that if $P(E) = \frac{n}{m}$ then the odds for E are $n : (m - n)$.

46. Show that if $P(E) = \frac{n}{m}$ then the odds against E are $(m - n) : n$.

47. Show that if the odds for E are $n : m$, then $P(E) = \frac{n}{n + m}$.

48. Show that if the odds against E are $n : m$, then $P(E) = \frac{m}{n + m}$.

49. ATHLETICS: Baseball A radio announcer says the odds that the Yankees will beat the Orioles are $7 : 4$. What does the announcer believe is the probability that the Yankees will beat the Orioles?

50. GENERAL: Horse Racing A bookie has changed his odds for Win-By-A-Neck from $3 : 14$ to $5 : 24$. Does he think the probability that this horse will win has gone up or down?

51. BUSINESS: Stocks A stock exchange announces that gainers beat losers by 7 to 2. What does this mean about the probability that a stock gained value?

5.4 CONDITIONAL PROBABILITY AND INDEPENDENCE

Introduction

We often ask about the probability of one event *given another*. For example, a card player may want to know the probability of being dealt a third ace given that the first two were aces. A smoker might ask about the probability of developing cancer given that one continues to smoke. Such questions involve *conditional probability*, which is the subject of this section. Conditional probability will then lead to a discussion of independence.

Conditional Probability

When two dice (always fair unless stated otherwise) are rolled, there are $6 \cdot 6 = 36$ possible outcomes in the sample space:

$$S = \{(1, 1), (1, 2), \ldots, (2, 1), (2, 2), \ldots, (6, 6)\}$$

See also page 380

EXAMPLE 1

FINDING A CONDITIONAL PROBABILITY

When rolling two dice, what is the probability that the first is a 3 given that the sum is 5?

Solution

Let B be the event that the sum is 5, so $B = \{(1, 4), (2, 3), (3, 2), (4,1)\}$. B contains four equally likely outcomes, and the first roll is a 3 in only one, (3, 2), so the probability that the first is a 3 (event A) given that the sum is 5 is one out of four, which we write as:

$$P(A \text{ given } B) = \frac{1}{4}$$

A: first is 3
B: sum is 5

In this example we found the conditional probability by restricting ourselves to a new *smaller* sample space corresponding to the "given" event B. The *conditional* probability was then the probability of A in B, that is, the probability of $A \cap B$ relative to the *restricted* sample space B. This leads to the following definition of conditional probability.

Conditional Probability

For events A and B with $P(B) > 0$, the conditional probability of A given B is

$$P(A \text{ given } B) = \frac{P(A \cap B)}{P(B)}$$

The probability of A given B is the probability of *both* divided by the probability of the *second*

We assume $P(B) > 0$ to avoid zero denominators. The conditional probability of A given B may also be written $P(A|B)$. If we show the events in a Venn diagram with area representing probability, then the conditional probability is the ratio of the intersection relative to the "given" area.

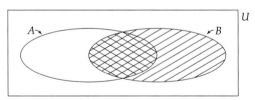

$$P(A \text{ given } B) = \frac{P(A \cap B)}{P(B)} = \frac{\text{Area} \boxtimes}{\text{Area} \boxtimes}$$

Area of the intersection over the area of B

We may check the result of Example 1 by using the above formula:

$$P(A \cap B) = \frac{1}{36}$$

The first is 3 *and* the sum is 5 means just (3, 2)

$$P(B) = \frac{4}{36} = \frac{1}{9}$$

The four outcomes were listed in Example 1

Then, according to the formula,

$$P(A \text{ given } B) = \frac{P(A \cap B)}{P(B)} = \frac{\frac{1}{36}}{\frac{1}{9}} = \frac{9}{36} = \frac{1}{4} \quad \text{Same answer as before}$$

The *unconditional* probability of getting a 3 on the first roll is $\frac{1}{6}$ (all six numbers are equally likely), so conditioning on the sum being 5 changed the probability from $\frac{1}{6}$ to $\frac{1}{4}$.

EXAMPLE 2

FINDING PROBABILITIES OF HANDS OF CARDS

You are playing cards with a friend, and each of you has been dealt 5 cards at random. If you have no face cards, what is the probability that your friend doesn't either?

Solution

Let A be the event that your friend has no face cards, and let B be the event that you have no face cards. The event $A \cap B$ means that *neither* of you has face cards, and since the deck has 12 face cards, this means that the first 10 cards dealt were from the 40 nonface cards.

$$P(A \cap B) = \frac{{}_{40}C_{10}}{{}_{52}C_{10}} \approx 0.0536 \quad \begin{array}{l}\leftarrow \text{Ways of choosing 10 from} \\ \text{the 40 nonface cards} \\ \leftarrow \text{Ways of choosing 10 from 52}\end{array}$$

Using a calculator

$$P(B) = \frac{{}_{40}C_5}{{}_{52}C_5} \approx 0.2532 \quad \begin{array}{l}\leftarrow \text{Ways of choosing 5 from the} \\ \text{40 nonface cards} \\ \leftarrow \text{Ways of choosing 5 from 52}\end{array}$$

Therefore,

$$P(A \text{ given } B) = \frac{P(A \cap B)}{P(B)} \approx \frac{0.0536}{0.2532} \approx 0.2117$$

The probability that your friend has no face cards given that you don't have any is about 21%.

Notice that the probability of having no face cards decreases from 25% to 21% when you include the information that you have none. Can you give an intuitive reason why this information should *decrease* the probability?

We could also solve this problem by looking directly at the restricted sample space: Given that you have 5 nonface cards, your friend's cards come from the remaining 47 cards, 12 of which are face cards and 35 of which are not. Therefore,

$$P(A \text{ given } B) = \frac{_{35}C_5}{_{47}C_5} \approx 0.2117$$

← 5 from the remaining 35 nonface cards
← 5 from the remaining 47

Same answer as before

We will see that this is true in general: You can solve conditional probability problems either by using the formula or by looking at the restricted sample space. You should use whichever way seems easier for that problem, but preferably use both ways to check that they agree.

The Product Rule for Probability

The conditional probability formula $P(A \text{ given } B) = \dfrac{P(A \cap B)}{P(B)}$ multiplied through by $P(B)$ gives $P(B) \cdot P(A \text{ given } B) = P(A \cap B)$. Reversing sides and interchanging A and B gives the following formula.

Product Rule for Probability

$$P(A \cap B) = P(A) \cdot P(B \text{ given } A)$$

The probability of $A \cap B$ is the probability of A times the probability of B *given A*

This rule is very useful when a conditional probability is known.

EXAMPLE 3

USING THE PRODUCT RULE

You need to be somewhere in half an hour, and your friend has borrowed your car. If your friend arrives soon (you give this a 50–50 chance), the probability that you will be on time is 90%. Otherwise, you will have to walk, with only a 60% chance of being on time. What is the probability that you arrive in your car and on time?

Solution

Much depends on whether your friend arrives soon, so that will be the "given" event, *A: Your friend arrives soon.* Then *B* is the other event, *B: You are on time.* The 90% is the probability of being on time *if (or given that)* your friend arrives soon, so $P(B \text{ given } A) = 0.90$. We are asked for the probability that you arrive in your car and on time, meaning that *both* event A and B occur. Therefore, by the Product Rule for Probabilities:

$$P(A \cap B) = P(A) \cdot P(B \text{ given } A)$$
$$= 0.5 \cdot 0.9 = 0.45$$

A: Friend arrives soon
B: You are on time

$\underbrace{P(A)}\ \underbrace{P(B \text{ given } A)}$

The probability that you arrive on time and in your car is 45%.

Suppose that instead we were asked for the probability of *arriving on time* (either by car or on foot). We will use a *tree diagram* to represent the different ways, similar to our use of tree diagrams to find different outcomes in Section 5.1.

EXAMPLE 4

USING A TREE DIAGRAM

For the situation in Example 3, find the probability that you arrive on time.

Solution

Again, much depends upon your friend's arrival, so we first "branch" on whether your friend arrives soon (event A, having probability 0.5) or not (event A^c, also having complementary probability 0.5). The second-stage branching represents whether you arrive on time (event B) or not (B^c), with the given (conditional) probabilities depending on the first-stage branching. Be sure you understand the correct placement of the probabilities given in the problem, 0.90 and 0.60, along with their complements 0.10 and 0.40. The probabilities at the end of *two* branchings are the *products* of the probabilities along those branches.

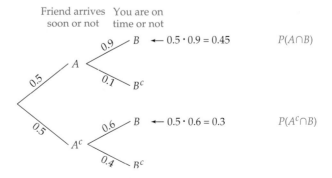

Arriving on time (by car or on foot) corresponds to the two branches ending in B. We multiply out those probabilities and add the results, 0.45 + 0.30 = 0.75. Answer:

> You have a 75% chance of arriving on time.

Notice that the uppermost branches give the 0.45 that we found in Example 3. In general, a tree diagram gives all possible probabilities that can be found from the given information.

Practice Problem 1

Use the tree diagram in Example 4 to answer the following questions.

a. What is the probability of arriving late and in your car?
b. What is the probability of arriving late? ▶ Solution on page 399

Notice several things about such tree diagrams:

- The branches from any point must have probabilities that add to 1.
- The probabilities at the right-hand end of a branch are found by multiplying along the branch.
- The probabilities written on the second-stage branches are *conditional* probabilities, so we are indeed using the Product Rule for Probability.
- There can be any number of branches per stage, and any number of stages.
- There is no need to multiply out all of the products, only the ones you need.

EXAMPLE 5

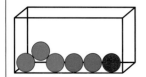

SAMPLING WITHOUT REPLACEMENT

A box contains three blue marbles, two green marbles, and one red marble. A marble is randomly chosen (and not replaced) and then a second marble is chosen. What is the probability that the second marble is green?

Solution

We branch on the color of the first marble, using B, G, and R for the events that it is blue, green, or red. Of the six marbles, 3 are blue, 2 are green, and 1 is red, so the probabilities on the first-stage branching are $\frac{3}{6} = \frac{1}{2}$, $\frac{2}{6} = \frac{1}{3}$, and $\frac{1}{6}$, as shown in the following diagram. For the second-stage branching, if a blue is chosen first, then the remaining 5 are 2 blue, 2 green, and 1 red, leading to the probabilities $\frac{2}{5}, \frac{2}{5}, \frac{1}{5}$ for the top part of the second stage choice. The other second-stage probabilities are calculated similarly, and depend on the color that was removed in the first stage.

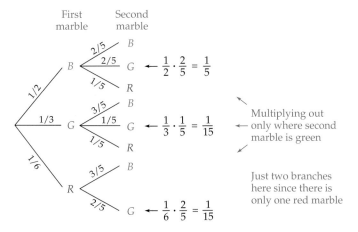

Multiplying out and adding up the branches that end in green:

The probability that the second marble is green is $\frac{1}{5} + \frac{1}{15} + \frac{1}{15} = \frac{5}{15} = \frac{1}{3}$.

Practice Problem 2 Use the diagram in Example 5 to find the probability that the second marble is blue. ▶ **Solution on page 399**

Independent Events

Roughly speaking, two events are said to be *independent* if one has nothing to do with the other, so that the occurrence of one has no bearing on the occurrence of the other. In terms of conditional probability, this means that for independent events A and B, we have $P(A \text{ given } B) = P(A)$. Using the definition of conditional probability, this equation becomes

$$\frac{P(A \cap B)}{P(B)} = P(A) \qquad \text{For } P(B) \neq 0$$

Multiplying each side by $P(B)$ gives the following equivalent condition, which we take as the *definition* of independence.

Independent Events

Events A and B are *independent* if

$$P(A \cap B) = P(A) \cdot P(B)$$

The probability of the intersection is the *product* of the probabilities

Events that are not independent are *dependent*.

Be careful! Independent does not mean the same thing as *disjoint*, or *mutually exclusive*. If events A and B have positive probabilities and are disjoint, then $P(A \cap B) = 0$ but $P(A) \cdot P(B)$ is positive, so they cannot be independent. Intuitively, disjointness is a very strong kind of *dependence*: The occurrence of one of two disjoint events guarantees the *non*occurrence of the other, so they cannot be independent.

The definition above gives a simple test for the independence of events A and B: Find $P(A \cap B)$ and find the product $P(A) \cdot P(B)$. If the results are *equal*, then A and B are *independent*; otherwise, A and B are *dependent*.

EXAMPLE 6

INDEPENDENT COIN TOSSES

A coin is tossed twice. Let A be the event that the first toss is heads, and let B be the event that the second toss is heads. Are the events A and B independent?

Solution

The sample space consists of four equally likely events:

$$S = \{(H, H), (H, T), (T, H), (T, T)\}$$

so

$$A = \{(H, H), (H, T)\}, \quad B = \{(H, H), (T, H)\}, \quad A \cap B = \{(H, H)\}$$

$$P(A) = \frac{2}{4} = \frac{1}{2}, \quad P(B) = \frac{2}{4} = \frac{1}{2}, \quad P(A \cap B) = \frac{1}{4}$$

Since $P(A) \cdot P(B) = \frac{1}{2} \cdot \frac{1}{2} = \frac{1}{4}$ and $P(A \cap B) = \frac{1}{4}$ are equal, the events A and B *are* independent.

If $P(A \cap B)$ and $P(A) \cdot P(B)$ had *not* been equal, the events would have been *dependent*. The fact that successive coin tosses are independent should come as no surprise, and this is sometimes expressed by saying that the coin has "no memory." Thus, each toss is a new experiment with "no influence from the past."

The concept of independence can be extended to more than two events.

Many Independent Events

A collection $E_1, E_2, \ldots, E_m$ of events is *independent* if any subcollection of them satisfies the multiplication formula:

$$P(E_i \cap E_j \cap \ldots \cap E_k) = P(E_i) \cdot P(E_j) \cdot \ldots \cdot P(E_k)$$

The probability of the intersection is the *product* of the probabilities

5.4 CONDITIONAL PROBABILITY AND INDEPENDENCE

In practice, the most important uses of independence do not involve *proving* it but *assuming* it. For example, just as successive coin tosses are independent of each other, different rolls of dice are independent of each other, so probabilities involving successive rolls can be found by multiplying together the probabilities for each roll. Using this idea, we can explain why the Chevalier de Meré won his first bet and lost his second.

EXAMPLE 7

WHY THE CHEVALIER DE MERÉ WON HIS FIRST BET

Recall from the Application Preview on page 353 that the Chevalier's first bet was that he could roll at least one six in four rolls of a die. Find the probability that he won this bet.

Solution

Instead of *at least one six* in four rolls it is easier to work with the complementary event, *no sixes in four rolls*, and subtract its probability from 1. Rolling a 6 has probability $\frac{1}{6}$, so *not* getting a six has probability $\frac{5}{6}$, and the probability that this will occur on four successive rolls is $\frac{5}{6} \cdot \frac{5}{6} \cdot \frac{5}{6} \cdot \frac{5}{6} = \left(\frac{5}{6}\right)^4$ (using independence). Subtracting this from 1 gives the probability of at least one six in four rolls:

$$1 - \left(\frac{5}{6}\right)^4 \approx 1 - 0.482 = 0.518$$

Therefore, he won his first bet about 52% of the time.

EXAMPLE 8

WHY THE CHEVALIER LOST HIS SECOND BET

The Chevalier's second bet was that he could roll at least one double-six in 24 rolls of two dice. Find the probability that he won this bet.

Solution

Again we begin with the complementary event. On one roll of two dice the probability of getting a double-six is $\frac{1}{36}$, so the probability of *not* getting a double-six is $\frac{35}{36}$, and the probability of this occurring on 24 successive tosses is $\left(\frac{35}{36}\right)^{24}$. Subtracting this from 1 gives the probability of at least one double-six in 24 rolls:

$$1 - \left(\frac{35}{36}\right)^{24} \approx 1 - 0.509 = 0.491$$

Therefore, he won his second bet only about 49% of the time.

Comparing the Chevalier's two bets, while both have probabilities close to $\frac{1}{2}$, he was right that in the long run the first is a winning bet and the second is a losing bet.

EXAMPLE 9

QUALITY CONTROL

A computer consists of a logic board, three memory chips, a screen, and an I/O (input-output) module. If these individual components work with probabilities 0.98, 0.99 (for each memory chip), 0.96, and 0.97 respectively, what percentage of the computers should be expected to work?

Solution

Assuming independence, we multiply the probabilities:

$$0.98 \cdot 0.99^3 \cdot 0.96 \cdot 0.97 \approx 0.885$$

About 89% of the assembled computers should be expected to work.

5.4 Section Summary

For events A and B, the conditional probability of A given B is

$$P(A \text{ given } B) = \frac{P(A \cap B)}{P(B)} \qquad \text{For } P(B) > 0$$

$P(A \text{ given } B)$ can also be found by restricting $P(A)$ to the sample space where B occurs. The above formula can be rewritten as:

$$P(A \cap B) = P(A) \cdot P(B \text{ given } A) \qquad \text{Product Rule for Probability}$$

This formula is useful in calculating probabilities along branches in *tree diagrams*.

Two events A and B are independent if

$$P(A \cap B) = P(A) \cdot P(B) \qquad \text{Probabilities multiply}$$

Be careful! When do you *add* probabilities and when do you *multiply* them? You *add* when the events are *disjoint* and you want *one or the other* to occur (subtracting the intersection if the events are not disjoint); you *multiply* if the events are *independent* and you want *both* to occur (using conditional probability if the events are not independent).

▶ Solutions to Practice Problems

1. **a.** From the branches corresponding to A and B^c, $0.50 \cdot 0.10 = 0.05$.

   b. From the branches ending in B^c, $0.05 + 0.50 \cdot 0.40 = 0.25$. [We could also have found this as 1 minus the probability of being on time, which we found in Example 4.]

2. Adding up the products along branches ending in blue:
$$\tfrac{1}{2} \cdot \tfrac{2}{5} + \tfrac{1}{3} \cdot \tfrac{3}{5} + \tfrac{1}{6} \cdot \tfrac{3}{5} = \tfrac{1}{2}.$$

5.4 Exercises

For Exercises 1–4, use the given values to find:
  a. $P(A$ given $B)$ **b.** $P(B$ given $A)$

1. $P(A) = 0.6$, $P(B) = 0.4$, $P(A \cap B) = 0.2$
2. $P(A) = 0.5$, $P(B) = 0.3$, $P(A \cap B) = 0.1$
3. $P(A) = 0.4$, $P(B) = 0.5$, $P(A \cup B) = 0.6$
4. $P(A) = 0.6$, $P(B) = 0.5$, $P(A \cup B) = 0.8$

5. **GENERAL: Marbles** A box contains 4 white, 2 red, and 4 black marbles. One marble is chosen at random, and it is not black. Find the probability that it is white.

6. **GENERAL: Cards** You select two cards at random from an ordinary deck. If the first card is a spade, what is the probability that the second card is a spade?

7. **BIOMEDICAL: Gender** Your friend has two children, and you know that at least one is a girl. What is the probability that both are girls? (Assume that girls and boys are equally likely.)

8. **BIOMEDICAL: Eye Color** Suppose that each of two children in a family has probability $\tfrac{1}{5}$ of having blue eyes, independently of each other. If at least one child has blue eyes, what is the probability that both have blue eyes?

9. **GENERAL: Cards** A deck contains three cards: One is red on both sides, one is blue on both sides, and the third is red on one side and blue on the other. One card is chosen at random from the deck, and the color on one side is observed. If this side is blue, what is the probability that the other side is blue?*

10. **BUSINESS: Credit Cards** Looking back at Exercise 24 on page 387, if a customer has at least one credit card, what is the probability that the customer has a Visa card?

11. **SOCIAL SCIENCE: Political Contributions** Looking back at Exercise 23 on page 387, if a person is a contributor, what is the probability that the person contributes to the Republican party?

12. **GENERAL: Cards** In the game of bridge, each of four players is dealt 13 cards. If a certain player has no aces, find the probability that that person's partner has:

    a. no aces   **b.** at least two aces

13. **GENERAL: Choosing Courses** You will take either a basket weaving course or a philosophy course, depending on what your advisor decides. You estimate that the probability of your getting an A in basket weaving is 0.95, while in philosophy it is 0.70. However, the

---

* Many people mistakenly believe that if one side is blue, then the probability that the other side is blue is $\tfrac{1}{2}$ since it can only be the blue–blue card or the blue–red card, and these are equally likely. The error comes from not realizing that the blue side is equally likely to be any one of the *three* blue sides. More than one probability student has made money by knowing the correct conditional probabilities and offering bets on the outcome.

chances of your advisor choosing the basket weaving course is only 20%, while there is an 80% chance of his putting you in the philosophy course. What is the probability of your ending up with an A?

14. **GENERAL: Multiple-Choice Tests** On a multiple-choice test you know the answers to 70% of the questions (and so get them right), and for the remaining 30% you choose randomly among the 5 answers. What percent of the answers should you expect to get right?

15. **GENERAL: Driving** Suppose that 70% of drivers are "careful" and 30% are "reckless." Suppose further that a careful driver has a 0.1 probability of being in an accident in a given year, while for a reckless driver the probability is 0.3. What is the probability that a randomly selected driver will have an accident within a year?

16. **MANAGEMENT SCIENCE: Quality Control** A computer manufacturer has assembly plants in three states. The Delaware plant produces 25% of the company's computers, the Michigan plant produces 35%, and the California plant produces the other 40%. The probabilities that a computer will pass inspection are 93% for the Delaware plant, 89% for the Michigan plant, and 94% for the California plant. What is the probability that a randomly selected computer from this company will pass inspection?

Independence For the experiment of tossing a coin twice, find whether events A and B are independent or dependent.

17. A: heads on the first toss
    B: different results on the two tosses

18. A: heads on the second toss
    B: the same results on both tosses

Independence For the experiment of rolling two dice, find whether events A and B are independent or dependent.

19. A: odd number on the first roll
    B: sum of the numbers is 4

20. A: even number on the first roll
    B: sum of the numbers is 10

21. **GENERAL: Dice** A pair of dice is rolled three times in succession. Find the probability that each of the rolls has a sum of seven.

22. **BUSINESS: Brand Loyalty** Suppose that each time that you buy a car, you choose between Ford and General Motors. Suppose that each time after the first, you stay with the same company with probability $\frac{2}{3}$ and switch with probability $\frac{1}{3}$. If you are equally likely to choose either company for your first car, what is the probability that your first and second choices will be Ford cars and your third and fourth choices will be General Motors cars?

23. **GENERAL: Class Attendance** Two students are registered for the same class and attend independently of each other, student A 90% of the time and student B 70% of the time. The teacher remembers that on a given day, at least one of them is in class. What is the probability that student A was in class that day?

24. **GENERAL: Class Attendance** For the students in Exercise 23, what is the probability that on any day:
    a. both will be in class?
    b. neither will be in class?
    c. at least one will be in class?

25. **GENERAL: Dice** Three dice are rolled. Find the probability of getting:
    a. all sixes
    b. all the same outcomes
    c. all different outcomes

26. **GENERAL: Fair Toss from an Unfair Coin** How can you make a fair game (one with probability $\frac{1}{2}$ of winning) from tossing an unfair coin? If a coin has probability p of heads (with p not necessarily $\frac{1}{2}$), show that for two tosses the events HT and TH have the same probability. Therefore, toss the coin twice: If you get HT call the result a "win," if you get TH call the result a "lose," and if you get HH or TT ignore those tosses and toss it twice more, repeating the procedure until you get HT or TH. "Win" and "lose" are equally likely, so each has probability $\frac{1}{2}$ of occurring first.

27. GENERAL: Dice When rolling two dice, find the probability of getting *a sum of 4 before a sum of 7*. [*Hint:* Let p be the probability of rolling a sum of 4 before a sum of 7. This event can occur in either of two ways: You roll the 4 on your first try, or your first try results in *neither* a 4 nor a 7, in which case you are back where you started and must subsequently roll a 4 before a 7 (again with probability p). This leads to the formula

$$p = P(\text{sum of 4}) + P(\text{sum of neither 4 nor 7}) \cdot p.$$

Find the two stated probabilities and then solve for p.]

28. GENERAL: Dice When rolling two dice, find the probability of rolling *a sum of 7 before a sum of 4*. [*Hint:* Modify the method outlined in Exercise 27.]

29. BIOMEDICAL: Longevity Based on *Life Tables*, in a population of 100,000 females, 89.8% will live to age 60 and 57.1% will live to age 80. Given that a woman is 60, what is the probability that she will live to be 80?

30. BIOMEDICAL: Longevity Based on *Life Tables*, in a population of 100,000 males, 81.5% will live to age 60 and 36.8% will live to age 80. Given that a man is 60, what is the probability that he will live to be 80?

31. MANAGEMENT SCIENCE: Reliability Travel between Seattle and Tacoma is sometimes impossible because of snow. The numbers p and q on possible road maps below are the probabilities that the roads are passable.

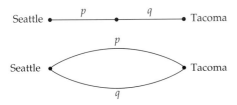

(In the language of *systems reliability* or *circuits*, the roads in the upper diagram are connected in *series* while in the lower diagram they are connected in *parallel*.)

a. For the first road map, what is the probability of being able to drive from Seattle to Tacoma?

b. For the second road map (with two possible routes), what is the probability of being able to drive from Seattle to Tacoma?

c. Evaluate your answers to parts (a) and (b) for $p = q = 0.5$ and then evaluate them for $p = q = 0.9$. Which is the more reliable highway design?

32. MANAGEMENT SCIENCE: Reliability (*continued*) Suppose that the parallel connection below has an unknown number of components (roadways), each operable (passable) with probability 80%. How many components are needed for the overall reliability to be at least 99.9%?

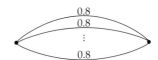

33. GENERAL: Shared Birthdays Assume that a person's birthday is equally likely to be on any of the 365 days of the year independently of every other person.

a. Show that the probability that a group of n people will all have different birthdays is

$$\frac{364}{365} \cdot \frac{363}{365} \cdot \frac{362}{365} \cdot \ldots \cdot \frac{366-n}{365}$$

b. Successively multiply these fractions together until you reach a number below 0.50, and thus find the first value of n for which this occurs.

c. Explain why this value of n has the following interpretation: For a group of this size, there is a greater than 50% probability that at least two people will share the same birthday.

Explorations and Excursions

The following problems extend and augment the material presented in the text.

More About Independence

34. If event A is such that $P(A) = 0$, and B is any other event, show that events A and B are independent.

35. If events A and B are independent, show that events A and B^c are independent.

36. If events A and B are independent, show that events A^c and B^c are independent. [*Hint:* Use the result of the previous exercise.]

37. In the experiment of tossing two coins, define events as follows:

    A: heads on first toss

    B: heads on second coin

    C: outcomes agree

    Show that A, B, and C are *pairwise* independent but do *not* satisfy

    $$P(A \cap B \cap C) = P(A) \cdot P(B) \cdot P(C)$$

    and so are not independent.

38. To see why the events in the preceding exercise are not independent, calculate $P(C)$ and also $P(C$ given $(A \cap B))$.

5.5 BAYES' FORMULA

Introduction

Sometimes we want to *reverse the order* in conditional probability. For example, the *probability of developing cancer given that one smokes* is different from the *probability of being a smoker given that one develops cancer*. The first might be of interest to a smoker and the second to a medical researcher. In this section we will develop *Bayes' formula** to reverse the order in conditional probability.

Bayes' Formula

In general, $P(A$ given $B)$ will not be the same as $P(B$ given $A)$.

Practice Problem 1   Explain the difference between

$$P(\text{you are rich given that you win the lottery})$$

and

$$P(\text{you win the lottery given that you are rich})$$

➤ Solution on page 406

Suppose that the sample space S is divided into two parts, S_1 and S_2, that are disjoint and whose union is the whole sample space $S_1 \cup S_2 = S$. These sets then partition any event A into two parts, the part in S_1 and

---

*Discovered by Thomas Bayes (1702–1761), an English nonconformist minister. Bayes was an amateur mathematician and a defender of the new subject of calculus when it was attacked as illogical and incorrect. His mathematical writings were not published during his lifetime; the one containing "Bayes' formula" was not published until three years after his death.

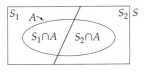

the part in S_2, as shown in the diagram on the left. The probability of A can then be found by adding the probabilities of the two parts:

$$P(A) = P(S_1 \cap A) + P(S_2 \cap A)$$

The definition of conditional probability, $P(S_1 \text{ given } A) = \dfrac{P(S_1 \cap A)}{P(A)}$ with the denominator replaced by the above expression gives:

$$P(S_1 \text{ given } A) = \frac{P(S_1 \cap A)}{P(S_1 \cap A) + P(S_2 \cap A)}$$

Using the Product Rule for Probability (page 392) in the numerator and twice in the denominator, we may write this in the following form:

$$P(S_1 \text{ given } A) = \frac{P(S_1) \cdot P(A \text{ given } S_1)}{P(S_1) \cdot P(A \text{ given } S_1) + P(S_2) \cdot P(A \text{ given } S_2)}$$

This formula generalizes by replacing S_1 and S_2 by *any* sets $S_1, S_2, \ldots, S_n$ that are pairwise disjoint and whose union is the entire sample space, $S_1 \cup S_2 \cup \cdots \cup S_n = S$:

Bayes' Formula

$$P(S_1 \text{ given } A) = \frac{P(S_1) \cdot P(A \text{ given } S_1)}{P(S_1) \cdot P(A \text{ given } S_1) + \cdots + P(S_n) \cdot P(A \text{ given } S_n)}$$

The sum in the denominator has n terms, one for each of the sets $S_1, S_2, \ldots, S_n$. Notice that the order in the conditional probability is reversed: S_1 *given* A on the left and A *given* S_1 on the right. Bayes' formula in this form is needed in only the most complicated cases. In fact we will use tree diagrams to carry out the calculations in Bayes' formula.

EXAMPLE 1

MEDICAL SCREENING

Suppose that a medical test for a disease is 90% accurate for both those who have the disease and for those who don't. Suppose, furthermore, that only 5% of the population has this disease. Given that a person tests positive for the disease (meaning that the test says that the person has the disease), what is the probability that the person actually has the disease? Would this test be useful for widespread medical screening?

Solution

The probabilities depend upon whether a person has the disease (event D, with probability 0.05) or not (event D^c, with probability

0.95), which represents the first-stage branching. The second-stage branching is according to whether the person tests positive (T) or not (T^c), with probabilities based on the test being 90% accurate.

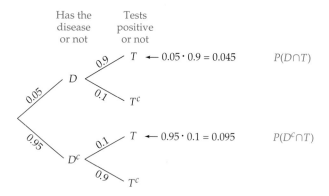

The probability that a randomly selected person has the disease given that the person tests positive is P(D given T), which, by the definition of conditional probability, is

$$P(D \text{ given } T) = \frac{P(D \cap T)}{P(T)}$$

For the numerator we take the branch though D and T (with probability 0.045) and for the denominator we add the two branches ending in T:

$$P(D \text{ given } T) = \frac{0.045}{0.045 + 0.095} = \frac{0.045}{0.14} \approx 0.32$$

Therefore, the probability that a person actually has the disease given that the person tests positive for it is only about 32%. In other words, 68% of those testing positive in fact do not have the disease and are being needlessly worried by a false diagnosis, so this test should *not* be used for widespread medical screening.

EXAMPLE 2 **ASSESSING VOTING PATTERNS**

Registered voters in Marin County are 45% Democratic, 30% Republican, and 25% Independent. In the last election for county supervisor, 70% of the Democrats voted, as did 80% of the Republicans and 90% of the Independents. What is the probability that a randomly selected voter in this election was a Democrat?

Solution

Since the information is stated in terms of the political parties, they represent the first-stage branching (D, R, or I), with the second-stage represented by voting or not voting (V or V^c). The probabilities come from the given values together with their complements.

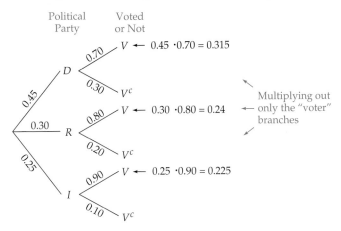

The probability that a randomly selected voter was a Democrat is P(D given V), which is

$$P(D \text{ given } V) = \frac{P(D \cap V)}{P(V)} \quad \text{Using the definition of conditional probability}$$

For the numerator $P(D \cap V)$ we take the branch through D and V (the uppermost branch), and for the denominator $P(V)$, we take the sum of all of the V branches:

$$\frac{0.24}{0.315 + 0.24 + 0.225} = \frac{0.315}{0.780} \approx 0.404$$

The probability that a voter in the election was a Democrat is about 40%.

Practice Problem 2 Using the diagram in Example 2, what is the probability that a randomly selected voter in the election was an Independent?

▶ Solution on page 406

Notice that as required for Bayes' formula, the three parties, Democrat, Republican, and Independent, do not overlap (the sets are *disjoint*) and include all voters (their probabilities *add to 1*). If the probabilities had not added to 1, we would have had to add in another group ("Others") with the remaining probability.

5.5 Section Summary

For pairwise disjoint sets $S_1, S_2, \ldots, S_n$ whose union is the entire sample space $S_1 \cup S_2 \cup \cdots \cup S_n = S$, Bayes' formula reverses the order in conditional probability:

$$P(S_1 \text{ given } A) = \frac{P(S_1) \cdot P(A \text{ given } S_1)}{P(S_1) \cdot P(A \text{ given } S_1) + \cdots + P(S_n) \cdot P(A \text{ given } S_n)}$$

In practice, we carry out this calculation using a tree diagram, where the numerator is the product along one branch and the denominator is the sum of several branches, using just the definition of conditional probability.

▶ Solutions to Practice Problems

1. P (you are rich given that you win the lottery) should be very high since you certainly will be rich if you win the lottery.

   P (you win the lottery given that you are rich) should be very small since your chances of winning the lottery are very small regardless of how much money you have.

2. Using the notation from the solution to Example 2, we have

   $$P(I \text{ given } V) = \frac{P(I \cap V)}{P(V)} = \frac{0.225}{0.315 + .024 + 0.225} = \frac{0.225}{0.780} \approx 0.288$$

   The probability that a voter in the last election was an Independent is about 29%.

5.5 Exercises

1. **SOCIAL SCIENCE: Voting** In a town, 60% of the citizens are Republicans and 40% are Democrats. In the last election 55% of the Republicans voted and 65% of the Democrats voted. If a voter is randomly selected, what is the probability that the person is a Republican?

2. **BIOMEDICAL: Colorblindness** An estimated 8% of men and 0.5% of women are colorblind. If a colorblind person is selected at random, what is the probability that the person is a man? (Assume that men and women occur in equal numbers.)

3. **BIOMEDICAL: Medical Testing** A new test is developed to test for a certain disease, giving "positive" or "negative" results to indicate that the person does or does not have the disease. For a person who actually *has* the disease, the test will give a positive result with probability 0.95 and a negative result with probability 0.05 (a so-called "false negative"). For a person who does *not* have the disease, the test will give a positive result with probability 0.05 (a "false positive") and a negative result with probability 0.95. Furthermore, only

one person in 1000 actually has this disease. If a randomly selected person is given the test and tests positive, what is the probability that the person actually has the disease?

4. **BIOMEDICAL: Home Pregnancy Testing** Suppose that 400 pregnant women take a home pregnancy test, and 397 of them test "positive" and the other 3 test "negative." Suppose also that 200 nonpregnant women take the test, and 184 of them test "negative" and the remaining 16 test "positive." What is the probability that a woman who tests positive is actually pregnant?

5. **MANAGEMENT SCIENCE: Manufacturing Defects** A computer chip factory has three machines, A, B, and C, for producing the memory chips. Machine A produces 50% of the factory's chips, machine B produces 30%, and machine C produces 20%. It is known that 3% of the chips produced by machine A are defective, as are 2% of chips produced by machine B and 1% of the chips from machine C. If a randomly selected chip from the factory's output is found to be defective, what is the probability that it was produced by machine B?

6. **MANAGEMENT SCIENCE: Manufacturing Defects** For the information in Exercise 5, if a randomly selected chip is found *not* to be defective, what is the probability that it came from machine B?

7. **SOCIAL SCIENCE: Airline Hijacking** The "hijacker profile" developed by the Federal Aviation Administration fits 90% of hijackers and only 0.05% of legitimate passengers. Based on historical data, assume that only 30 of 300 million passengers are hijackers. What is the probability that a person who fits the profile is actually a hijacker?

8. **SOCIAL SCIENCE: Polygraph Tests** Although seldom admissible in courts, polygraphs (sometimes called "lie detectors") are used by many businesses and branches of government. The Office of Technology Assessment has estimated that polygraphs have a "false negative" rate of 0.11 and a "false positive" rate of 0.20. Suppose that you are trying to find a single thief in your company of 100 employees. What is the probability that a person who fails a polygraph test is *not* the thief?

9. **BIOMEDICAL: Random Drug Testing** In 1986 the Reagan administration issued an executive order allowing agency heads to subject all employees to urine tests for drugs. Suppose that the test is 95% accurate both in identifying drug users and in clearing nonusers. Suppose also that 1% of employees use drugs. What is the probability that a person who tests positive is *not* a drug user?

10. **GENERAL: Grades** A probability class consists of 5 math majors, 3 science majors, and 2 whose major is "undecided." If the probability of earning an A is 90% for math majors, 85% for science majors, and 70% for "undecideds," what is the probability that a randomly selected A student is a math major?

11. **GENERAL: Grades** For the information in Exercise 10, what is the probability that a randomly selected A student is "undecided"?

12. **ATHLETICS: Olympic Athletes** Suppose that 10% of olympic athletes use steroids. A blood test is 95% correct in identifying drug users and has a 2% rate of false positives (that is, of *incorrectly* indicating steroid use). What is the probability that a champion cycler who tests positive uses steroids?

13. **BIOMEDICAL: Mammograms** Many doctors recommend that women in their forties have annual mammograms to detect breast cancer. Approximately 2% of women in their forties will develop breast cancer during that decade, and the early mammograms (no longer used) for that age group had a 30% rate of false positives and a 25% rate of false negatives. What is the probability that a woman who tested positive on this test had breast cancer?

14. **SOCIAL SCIENCE: Unemployment** A city has five districts, and 30% of its citizens live in district I, 25% in district II, 20% in district III, 15% in district IV, and 10% in district V. The unemployment rate in district I is 3%, while in

districts II, III, IV, and V it is 4%, 4%, 5%, and 6%, respectively. The local newspaper interviews a randomly selected unemployed resident. What is the probability that this resident lives in district I?

15. **SOCIAL SCIENCE: Unemployment** For the information in Exercise 14, what is the probability that this resident lives in district V?

16. **GENERAL: Automotive Advice** Your automobile mechanic says "Ninety percent of new cars I see in here for repairs were made in Detroit—practically none from Nashville. When you buy your next car, make sure it was made in Nashville." Is this good advice? Suppose that 10% of all new cars (whether from Detroit or Nashville) need repairs during their first year, so that all new cars are, in fact, equally good. Suppose also that 90% of all cars are manufactured in Detroit with the other 10% in Nashville. Show that your mechanic's observation was right, but his advice was not.

Explorations and Excursions

The following problem extends and augments the material presented in the text.

How to Ask an Embarrassing Question

17. Suppose that you want to determine the proportion of people who have a certain medical condition that they may not want to disclose if asked directly. Here is a way to find out while maintaining everyone's confidentiality: Each person in the survey is given a coin, and in a private room they toss the coin. If it comes up heads, they answer truthfully about the medical condition (Yes or No, written on a piece of paper), and if it is tails, they toss the coin again and write Yes or No depending on whether the second toss is heads or tails. Each person's written Yes or No discloses nothing about them since it is equally likely to be truthful or random. Given the resulting proportion $P(Y)$ of Yes answers, show how to find the actual proportion p of people who have the medical condition as follows.

a. Use a probability tree to find a formula for $P(Y)$ in terms of p.
b. Solve the resulting equation for p in terms of $P(Y)$.
c. If 62% of answers were Yes, what proportion of people actually have the condition?

5.6 RANDOM VARIABLES AND DISTRIBUTIONS

Introduction

Often a random experiment results in a *number*, such as the sum on the faces of two dice or your winnings in a lottery. Such numerical quantities that depend on chance events are called *random variables* and are the central objects of probability and statistics. In this section (and also in the following chapter) we will introduce some of the random variables that have proved most useful in applications.

Random Variables

A random variable is an assignment of a number to each outcome in the sample space. We will use capital letters such as X and Y for random variables.

EXAMPLE 1

DEFINING A RANDOM VARIABLE

A coin is tossed four times and

$$X = \begin{pmatrix} \text{Number} \\ \text{of heads} \end{pmatrix}$$

Find the possible values for X and the outcomes corresponding to each of its possible values.

Solution

The number of heads in four tosses can be 0, 1, 2, 3, or 4, so these are the possible values for X. The sample space consists of the 16 sequences of H's and T's shown on the left.

Since X is the number of heads, for any particular outcome we can find its value by counting H's. For example, (H, T, H, H) gives $X = 3$. The following table lists the possible values of X and the outcomes for which it takes those values (as you should check by "counting heads").

$$\begin{Bmatrix} (H,H,H,H), & (H,H,H,T), \\ (H,H,T,H), & (H,T,H,H), \\ (T,H,H,H), & (H,H,T,T), \\ (H,T,H,T), & (H,T,T,H), \\ (T,H,H,T), & (T,H,T,H), \\ (T,T,H,H), & (H,T,T,T), \\ (T,H,T,T), & (T,T,H,T), \\ (T,T,T,H), & (T,T,T,T) \end{Bmatrix}$$

Values of X	Outcomes
$X = 0$	(T, T, T, T)
$X = 1$	(H, T, T, T), (T, H, T, T), (T, T, H, T), (T, T, T, H)
$X = 2$	(H, H, T, T), (H, T, H, T), (H, T, T, H), (T, H, H, T), (T, H, T, H), (T, T, H, H)
$X = 3$	(H, H, H, T), (H, H, T, H), (H, T, H, H), (T, H, H, H)
$X = 4$	(H, H, H, H)

We can find the probability that X takes any particular value, such as $X = 2$, by adding up the probabilities of the outcomes corresponding to that value. The *probability distribution* of a random variable is the collection of these probabilities for its various values.

Random Variable

A *random variable* X is an assignment of a number to each element in the sample space. The *probability distribution* of the random variable X is the collection of all probabilities $P(X = x)$ for each possible value x.

EXAMPLE 2

FINDING A PROBABILITY DISTRIBUTION

A coin is tossed four times. Find and graph the probability distribution for

$$X = \begin{pmatrix} \text{Number} \\ \text{of heads} \end{pmatrix}$$

Solution

Using the table in Example 1, $X = 0$ occurs only for the outcome (T, T, T, T), which is one out of 16 equally likely outcomes, so $P(X = 0) = \frac{1}{16}$. The event $X = 1$ corresponds to 4 outcomes in the table, so $P(X = 1) = \frac{4}{16} = \frac{1}{4}$. The other probabilities $P(X = 2)$, $P(X = 3)$, and $P(X = 4)$ are similarly found by counting outcomes and dividing by 16, giving the probabilities in the following table. These probabilities are graphed on the left, the height of each bar being the probability that X takes the value at the bottom of the bar.

x	0	1	2	3	4
$P(X = x)$	$\frac{1}{16}$	$\frac{1}{4}$	$\frac{3}{8}$	$\frac{1}{4}$	$\frac{1}{16}$

Probability distribution for X

Observe that the *most likely* number of heads in 4 tosses is 2 and that the least likely are the extreme values 0 and 4, just as you might expect. The probabilities add to 1: $\frac{1}{16} + \frac{1}{4} + \frac{3}{8} + \frac{1}{4} + \frac{1}{16} = \frac{16}{16} = 1$, so *the area under the graph is 1* (since each bar has width 1).

Practice Problem 1 Your dog has a litter of four pups. Assuming that each pup is equally likely to be male or female, which is the more likely distribution: two of each sex or 3–1 split (3 of one sex and 1 of the other)? [*Hint:* Use Example 2.] ➤ Solution on page 418

EXAMPLE 3

RAFFLE WINNINGS

To raise money, a children's hospital sells four hundred raffle tickets for $100 each. First prize is a $3000 Florida vacation for two, second prize is a $1000 credit at the town's supermarket, and the five third prizes are $100 "dinner for two" gift certificates at a local restaurant. Find the probability distribution of the value of a raffle ticket.

Solution

If X represents the value of a ticket, then its possible values are $3000, 1000, 100,$ and $0,$ with probabilities

$P(X = 3000) = \frac{1}{400}$ — One $3000 prize among 400 tickets

$P(X = 1000) = \frac{1}{400}$ — One $1000 prize

$P(X = 100) = \frac{5}{400}$ — Five $100 prizes

$P(X = 0) = \frac{393}{400}$ — The other 393 tickets are worth nothing

Expected Value

On average, how much is each ticket worth? We divide the total value of all of the tickets, $3000 + 1000 + 100 \cdot 5 + 0 \cdot 393$, by the number of tickets, 400. The result can be written $3000 \cdot \frac{1}{400} + 1000 \cdot \frac{1}{400} + 100 \cdot \frac{5}{400} + 0 \cdot \frac{393}{400}$. Looking back at the above example, this is just the values of X multiplied by their probabilities and added. This leads to the following definition of the expected value of a random variable.

Expected Value

A random variable X taking values $x_1, x_2, \ldots, x_n$ with probabilities $p_1, p_2, \ldots, p_n$ has expected value:

$$E(X) = x_1 \cdot p_1 + x_2 \cdot p_2 + \cdots + x_n \cdot p_n$$

Sum of the possible values times their probabilities

The expected value is also called the *expectation* or the *mean* of the random variable (or of the probability distribution), and is sometimes denoted by μ (pronounced "mu," the Greek letter m). Using this definition, we may calculate the expected value of the raffle ticket in the preceding example by multiplying the possible values times their probabilities and adding:

$$E(X) = 3000 \cdot \tfrac{1}{400} + 1000 \cdot \tfrac{1}{400} + 100 \cdot \tfrac{5}{400} + 0 \cdot \tfrac{393}{400} = \$11.25$$

This expected value represents your average winnings per ticket. Since the tickets cost $100, you are on average losing $88.75 on each ticket. To put this in a more positive light, this latter figure represents your true generosity to the hospital.

Practice Problem 2 If the prizes in Example 3 were changed to one $3000 first prize, two $1000 second prizes, and ten $100 third prizes, what would be the expected value of a raffle ticket? ➤ Solution on page 418

EXAMPLE 4

EXPECTED DICE WINNINGS

You roll a die and win any *even* amount and lose any *odd* amount (in dollars). (That is, if you roll 2 you win $2, and if you roll 3 you lose $3.) What is the expected value of your winnings?

Solution

The expected value of $X = \begin{pmatrix} \text{Your} \\ \text{winnings} \end{pmatrix}$ is the sum of the values times their probabilities:

$$E(X) = -1 \cdot \tfrac{1}{6} + 2 \cdot \tfrac{1}{6} - 3 \cdot \tfrac{1}{6} + 4 \cdot \tfrac{1}{6} - 5 \cdot \tfrac{1}{6} + 6 \cdot \tfrac{1}{6} = \tfrac{3}{6} = \tfrac{1}{2}$$

Expected value is half a dollar, or 50¢

This answer of $\tfrac{1}{2}$ means that if you play the game many times, your *average winnings per play* will be about half a dollar. The expected value is positive, so this game is *favorable* to you. If the expected value were zero, the game would be *fair*, and if it were negative, the game would be *unfavorable* to you. The expected value can be interpreted as *the fair price for playing the game,* meaning that paying that amount before each play would make the game *fair*.

Note that if the values are equally likely, then the expected value is just the arithmetic average of the values. Generally, the probabilities will not all be equal, so the expected value will be a *weighted average*, weighting each value by its probability. If none of the probabilities are zero, the expected value will lie somewhere between the highest and lowest values of the random variable.

Practice Problem 3 Find the expected value of one roll of a die.

➤ Solution on page 418

EXAMPLE 5

CONCERT INSURANCE

A concert promoter is planning an outdoor concert and estimates a profit of $100,000 if there is no rain but only $10,000 if it rains. He can take out an insurance policy costing $15,000 that will pay $100,000 if it rains. The weather bureau estimate the chances of rain at 10%. Based on expected value, should he buy the insurance?

Solution

First we calculate the profit under both conditions.

	Without Insurance	With Insurance	
No rain	100,000	100,000 − 15,000 = 85,000	← Subtracting the cost of insurance
Rain	10,000	10,000 − 15,000 + 100,000 = 95,000	← and adding the insurance payment

The expected values are:

Without insurance: $\quad 100{,}000 \cdot \frac{9}{10} + 10{,}000 \cdot \frac{1}{10} = 91{,}000 \quad$ Each value times its probability

With insurance: $\quad\quad\quad 85{,}000 \cdot \frac{9}{10} + 95{,}000 \cdot \frac{1}{10} = 86{,}000$

The expected value *without* insurance is higher, so on this basis he should skip the insurance. (However, in view of the relatively small difference in expected profits, he might decide to take the insurance after all to reduce the risk.)

Binomial Distribution

Suppose a biased coin has probability p of coming up heads (and therefore probability $1 - p$ of tails) on any toss. Since successive tosses are independent, the probability of a particular succession of outcomes would be just the product of the probabilities for each toss. For example, the probability of the outcome *HTHHT* is $p \cdot (1-p) \cdot p \cdot p \cdot (1-p)$ which simplifies to $p^3(1-p)^2$. Clearly, any other particular ordering of 3 heads and 2 tails would have this same probability. How many such orderings are there? Choosing the three tosses on which heads will occur (so the other two tosses are automatically tails) can be done in exactly $_5C_3 = \frac{5 \cdot 4 \cdot 3}{3 \cdot 2 \cdot 1}$ ways (using the combinations formula from page 370). Therefore, the probability of exactly 3 heads in 5 tosses is $_5C_3 p^3(1-p)^2$.

We can generalize this result to any number n of tosses with k heads and $n - k$ tails to obtain the formula $_nC_k p^k(1-p)^{n-k}$. Furthermore, we spoke of heads and tails only for familiarity. The same ideas apply to any experiment that has just two possible outcomes, *success* and *failure*, occurring with probabilities p and $1 - p$, respectively, with successive repetitions being independent. Such repeated experiments with fixed probabilities are called *Bernoulli trials*, and the number of successes in Bernoulli trials is called a *binomial random variable*.[*]

---

[*] After the Swiss mathematician James Bernoulli (1654–1705), who first recognized the importance of such random variables. His main work was *Ars Conjectandi* (*Art of Conjecturing*).

Binomial Distribution

For independent repetitions of an experiment with probability p of success on each trial, the probability that in n trials the number X of successes will be k is

$$P(X = k) = {}_nC_k p^k (1-p)^{n-k} \quad \begin{array}{l} n = \text{number of trials} \\ k = \text{number of successes} \quad (0 \le k \le n) \\ p = \text{probability of success} \end{array}$$

X is called a *binomial random variable with parameters n and p*. Its expected value or mean is $E(X) = n \cdot p$.

The formula for the expected value is proven in Exercises 53–55, but may be made intuitively reasonable as follows: If a coin with probability of heads $\frac{1}{3}$ is tossed 12 times, you should expect about a third of them, $12 \cdot \frac{1}{3} = 4$, to be heads, and this is just the formula $E(X) = n \cdot p$. Recall that the number of combinations ${}_nC_k$ may be calculated from the formula ${}_nC_k = \frac{n \cdot (n-1) \cdots (n-k+1)}{k!} = \frac{n!}{k!(n-k)!}$. We may use the binomial distribution in any situation with independent trials having probability p of success, where the word "success" may mean whatever outcome we choose.

EXAMPLE 6   A BINOMIAL DISTRIBUTION

Let X be the number of heads in four tosses of a coin. Find the probability distribution and expected value of X.

Solution

Coin tossing is merely Bernoulli trials, so we want the probability distribution of a binomial random variable with $n = 4$ and $p = \frac{1}{2}$.

$P(X = 0) = 1 \cdot \left(\frac{1}{2}\right)^0 \cdot \left(\frac{1}{2}\right)^4 = \frac{1}{16}$ ${}_4C_0 p^0 (1-p)^4$

$P(X = 1) = 4 \cdot \left(\frac{1}{2}\right)^1 \cdot \left(\frac{1}{2}\right)^3 = \frac{4}{16} = \frac{1}{4}$ ${}_4C_1 p^1 (1-p)^3$

$P(X = 2) = 6 \cdot \left(\frac{1}{2}\right)^2 \cdot \left(\frac{1}{2}\right)^2 = \frac{6}{16} = \frac{3}{8}$ ${}_4C_2 p^2 (1-p)^2$

$P(X = 3) = 4 \cdot \left(\frac{1}{2}\right)^3 \cdot \left(\frac{1}{2}\right)^1 = \frac{4}{16} = \frac{1}{4}$ ${}_4C_3 p^3 (1-p)^1$

$P(X = 4) = 1 \cdot \left(\frac{1}{2}\right)^4 \cdot \left(\frac{1}{2}\right)^0 = \frac{1}{16}$ ${}_4C_4 p^4 (1-p)^0$

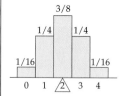

```
binompdf(4,.5)
(.0625 .25 .375...
Ans▶Frac
(1/16 1/4 3/8 1...
```

These five probabilities, graphed on the left, make up the probability distribution for X, the number of heads in four tosses. They could also have been found using a graphing calculator. For the expected value, we use the formula in the preceding box:

$$E(x) = n \cdot p = 4 \cdot \tfrac{1}{2} = 2$$

We found exactly this distribution in Example 2 (page 410). Here we found it using the binomial probability formula, and there we found it from basic principles. The results agree. The fact that the expected number of heads in four tosses of a fair coin is two should come as no surprise. This "mean" of 2 is the "balance point" of the probability distribution graph, as shown on the bottom left of the previous page.

Spreadsheet Exploration

The spreadsheet* below uses a "random number generator" to simulate 4 coin tosses repeated 400 times. The numbers of heads on each of these 4 tosses are recorded in cells `A21:T40` and the results are summarized in cells `G18:K18`. The graph below shows that the results of this experiment approximate the probability distribution found in Example 6.

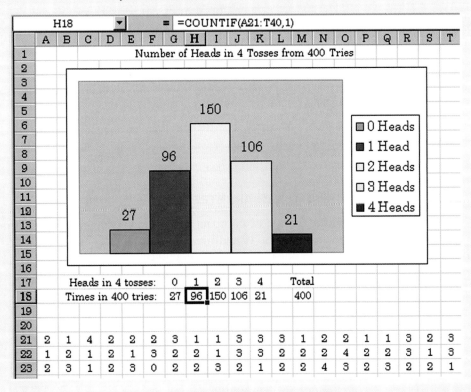

*See the Preface for information on how to obtain this and other Excel spreadsheets.

Practice Problem 4

Let X be the number of heads in six tosses of a coin. Find $P(X = 3)$ and $E(X)$.

▶ Solution on page 418

EXAMPLE 7

EMPLOYEE RETENTION

A restaurant manager estimates the probability that a newly hired waiter will still be working at the restaurant six months later is only 60%. For the five new waiters just hired, what is the probability that at least four of them will still be working at the restaurant in six months?

Solution

Assuming that the waiters decide independently of each other, their decisions make up five Bernoulli trials. Counting a waiter who stays as a "success," the number X of waiters who stay is a binomial random variable with $n = 5$ and $p = 0.6$. We want $P(X \geq 4)$, which means $P(X = 4 \text{ or } 5) = P(X = 4) + P(X = 5)$.

$$P(X = 4) + P(X = 5) = {}_5C_4(0.6)^4(0.4)^1 + {}_5C_5(0.6)^5(0.4)^0 \approx 0.337$$

$\underbrace{}_{5} \qquad \underbrace{}_{1}$

```
binompdf(5,.6,4)
+binompdf(5,.6,5
)
            .33696
```

The probability that at least four of the new waiters will stay for six months is only about 34%.

EXAMPLE 8

MONEY-BACK GUARANTEES

A manufacturer of compact disks (CD's) for computer storage sells them in packs of 10 with a "double your money back" guarantee if more than one CD is defective. If each CD is defective with a 1% probability independently of the others, what proportion of packages will require refunds?

Solution

If X is the number defective CD's in a package, then X is a binomial random variable with $n = 10$ and $p = 0.01$. Rather than calculating the probability of a refund (2 or more defectives), we find the complementary probability:

$$P(X = 0) + P(X = 1) = {}_{10}C_0(0.01)^0(0.99)^{10} + {}_{10}C_1(0.01)^1(0.99)^9 \approx 0.996$$

$\underbrace{}_{1} \qquad \underbrace{}_{10}$

```
binompdf(10,.01,
0)+binompdf(10,.
01,1)
         .9957337998
1-Ans
         .0042662002
```

Subtracting this answer from 1 (since it is the complementary probability) gives 0.004. Therefore, the company will have to provide refunds for only about 0.4% of its packs.

EXAMPLE 9

MULTIPLE-CHOICE TESTS

A multiple-choice test consists of 25 questions, each of which has 5 possible answers. What is the expected number of correct answers that would result from random guessing?

Solution

The 25 questions are 25 Bernoulli trials, and random guessing among the 5 possible answers for each question gives a probability $p = \frac{1}{5}$ of success for each. Therefore, the expected number of correct answers is

$$E(X) = 25 \cdot \tfrac{1}{5} = 5 \qquad\qquad E(X) = n \cdot p$$

This is why multiple-choice tests are often *rescaled* by only counting the number of correct answers above a certain expected number. In this case, the grade would be based on the number of correct answers beyond the first 5.

5.6 Section Summary

A *random variable* X is an assignment of a number to each outcome in a sample space. The *probability distribution* of X is the collection of probabilities $P(X = x)$ for each possible value x of X. The *expected value* or *mean* of X is

$$E(X) = x_1 \cdot P(X = x_1) + \cdots + x_n \cdot P(X = x_n)$$

Sometimes denoted μ

Bernoulli trials are independent repetitions of an experiment that results in *success* with probability p (and therefore *failure* with probability $1 - p$). The number X of successes in n Bernoulli trials is called a *binomial* random variable with parameters n and p and has probability distribution and mean

$$P(X = k) = {}_nC_k p^k (1-p)^{n-k} \qquad \text{For } k = 0, 1, \ldots, n$$
$$E(X) = n \cdot p$$

The following graphs show the binomial distribution for $n = 6$ and different values of p. Notice that the middle graph is symmetric, and that values away from $p = \frac{1}{2}$ skew the graph to one side or the other.

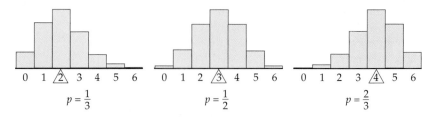

Under each graph the mean (calculated from $E(X) = n \cdot p$) is shown in a triangle to indicate that the distribution "balances" at that point.

▶ **Solutions to Practice Problems**

1. The sample space of all gender orders (such as F, M, M, F) is just like that for four coin tosses on page 410 but with different letters. Example 2 says that two of each gender has probability $\frac{3}{8}$ while a 3–1 split (which can occur in two ways) has probability $\frac{1}{4} + \frac{1}{4} = \frac{1}{2}$. Therefore, three of one sex and one of the other is more likely than two of each.

2. $E(X) = 3000 \cdot \frac{1}{400} + 1000 \cdot \frac{2}{400} + 100 \cdot \frac{10}{400} + 0 \cdot \frac{387}{400} = 15$.
   The expected value is $15.

3. $E(X) = 1 \cdot \frac{1}{6} + 2 \cdot \frac{1}{6} + 3 \cdot \frac{1}{6} + 4 \cdot \frac{1}{6} + 5 \cdot \frac{1}{6} + 6 \cdot \frac{1}{6} = \frac{21}{6} = 3.5$.
   This answer shows that the expected value may not be one of the possible values of the random variable.

4. $P(X = 3) = {}_6C_3 \left(\frac{1}{2}\right)^3 \left(\frac{1}{2}\right)^3 = \frac{6 \cdot 5 \cdot 4}{3 \cdot 2 \cdot 1} \left(\frac{1}{2}\right)^6 = 20 \cdot \frac{1}{64} = \frac{5}{16}$
   $E(x) = n \cdot p = 6 \cdot \frac{1}{2} = 3$

   ```
   binompdf(6,.5,3)
                  .3125
   Ans▶Frac
                  5/16
   ```

5.6 Exercises (will be helpful.)

1. **GENERAL: Coins** A coin is tossed three times, and X is the number of heads. Find and graph the probability distribution of X.

2. **GENERAL: Dice** Two dice are rolled, and X is the sum of the faces. Find and graph the probability distribution of X.

3. **GENERAL: Coins** You toss three coins and win $11 if they all agree (all heads or all tails), and otherwise you lose $1. Find and graph the probability distribution of your winnings.

4. **GENERAL: Marbles** Two marbles are chosen without replacement from a box containing

3 red and 5 green marbles. Let X be the number of green marbles chosen. Find and graph the probability distribution of X.

5. **GENERAL: Dice** A die is rolled, and you win $2 if it comes up odd, you lose $12 if it comes up 2, and you win $3 if it comes up 4 or 6. Find and graph the probability distribution of your winnings.

6. **GENERAL: Dice** A die is rolled and you win $8 if it comes up odd, you lose $15 if it comes up 2 or 4, and you win $6 if it comes up 6. Find and graph the probability distribution of your winnings.

7. **GENERAL: Dice** Two dice are rolled, and X is the larger of the two numbers that come up. Find and graph the probability distribution of X.

8. **GENERAL: Dice** Two dice are rolled, and X is the smaller of the two numbers that come up. Find and graph the probability distribution of X.

9. **Mean** Find the mean of the random variable in Exercise 1.

10. **Mean** Find the mean of the random variable in Exercise 2.

11. **Mean** Find the mean of the random variable in Exercise 3.

12. **Mean** Find the mean of the random variable in Exercise 4.

13. **Mean** Find the mean of the random variable in Exercise 5.

14. **Mean** Find the mean of the random variable in Exercise 6.

15. **Mean** Find the mean of the random variable in Exercise 7.

16. **Mean** Find the mean of the random variable in Exercise 8.

17. **GENERAL: Raffle Tickets** One thousand raffle tickets are sold, and there is one first prize worth $2000, one second prize worth $250, and 20 third prizes worth $50 each. Find the expected value of a ticket.

18. **BUSINESS: Insurance** A concert promoter is planning an outdoor concert and estimates a profit of $120,000 if there is no rain, but only $15,000 if it rains. She can take out an insurance policy costing $20,000 that will pay $90,000 if it rains. The weather bureau estimates the chances of rain at 20%. Based on expected value, should she take the insurance?

19. **BUSINESS: Insurance** A college is planning an outdoor fund raising event and predicts donations of $80,000 if there is no rain, but only $40,000 if it rains. They can take out an insurance policy costing $8,000 that will pay $40,000 if it rains. The weather bureau estimates the chances of rain at 15%. Based on expected value, should they take the insurance?

20. **Binomial Distribution** Find and graph the probability distribution of a binomial random variable with parameters $n = 4$ and $p = \frac{1}{5}$. Use the formula on page 414 to find its mean.

Binomial Distribution For a binomial random variable with the given parameters, find and graph its probability distribution, and use the formula on page 414 to find its mean.

21. $n = 20$ and $p = \frac{1}{2}$

22. $n = 25$ and $p = \frac{1}{2}$

23. $n = 20$ and $p = 0.8$

24. $n = 25$ and $p = 0.64$

25. **GENERAL: Coins** For 6 tosses of a coin, find the probability that the number of heads is between 2 and 4 (inclusive).

26. **GENERAL: Coins** What is the probability of getting exactly 5 heads in 10 tosses of a coin?

27. **GENERAL: Coins** What is the probability of getting exactly 4 heads in 8 tosses of a coin?

28. **GENERAL: Coins** For an unfair coin whose probability of heads is 0.7, what is the most likely number of heads in 10 tosses, and what is its probability? What is the *least* likely number of heads, and what is its probability?

29. **GENERAL: Coins** For an unfair coin whose probability of heads is $\frac{1}{3}$, what is the most likely number of heads in 6 tosses, and what is its probability? What is the *least* likely number of heads, and what is its probability?

30. **GENERAL: Dice** In rolling a die 10 times, what is the most likely number of sixes, and what is its probability? What is the mean number of sixes?

31. **GENERAL: Dice** In rolling a die 8 times, what is the most likely number of sixes, and what is its probability? What is the mean number of sixes?

32. **GENERAL: Testing** You know that one of three batteries is dead. If you test them one at a time until you find the defective one, what is the expected number of tests?

33. **BUSINESS: Sales** An automobile salesperson predicts that a customer will buy a $30,000 car with probability $\frac{1}{10}$, a $25,000 car with probability $\frac{1}{5}$, a $20,000 car with probability $\frac{3}{10}$, and otherwise buy nothing. What is the expected value of the sale?

34. **BUSINESS: Insurance** An insurance company estimates that on a typical policy it will have to pay out $10,000 with probability 0.05, $5000 with probability 0.1, and $1000 with probability 0.2. If the company wants to charge $200 more than the expected payout of the policy, what should it charge?

35. **BUSINESS: Product Quality** A company manufactures products of which 2% have hidden defects. If you buy 10 of them, then the number of defective ones in your purchase is a binomial random variable. Find its mean.

36. **SOCIAL SCIENCE: Multiple-Choice Testing** A multiple-choice test has 5 possible answers for each of 10 questions. For a student who guesses randomly:
    a. What is the mean of the number of right answers?
    b. What is the probability that this student gets 4 or more correct answers?

37. **SOCIAL SCIENCE: ESP** A person claims to have ESP (extrasensory perception) and calls 7 out of 10 tosses of a fair coin correctly. What is the probability of doing at least that well by guessing randomly?

38. **SOCIAL SCIENCE: Juries** Suppose that on a jury of 12 people it takes at least 9 votes to convict. If each person decides correctly with probability 0.9, what is the probability that a guilty person is convicted? What is the probability that an innocent person is found innocent? What assumptions are you making about independence?

39. **GENERAL: Communications** A one-digit message (0 or 1) is to be transmitted over a communications line that has a $\frac{1}{10}$ probability of changing the digit. Because of this, the message is to be transmitted in triplicate, 000 or 111, with *decoding by majority*, which means that the digit that occurs two or three times will be taken to be the message. What is the probability that the message is received correctly? How does this compare with the probability of correct reception when only *one* digit is transmitted?

40. **BIOMEDICAL: Family Distribution** Assuming that boys and girls are equally likely, what is the probability that a family of 6 children consists of 3 boys and 3 girls?

41. **GENERAL: Cards** A bridge hand consists of 13 cards. What is the probability that you get no aces in three consecutive hands?

42. **GENERAL: Target Practice** If the probability of hitting a target is $\frac{1}{4}$ and 5 shots are fired, what is the probability of hitting the target at least once?

43. **ATHLETICS: Winning a Series** Suppose that in a sports contest with no ties, the stronger team has probability $\frac{3}{5}$ of winning any particular game. Find the probability that the stronger team wins:
    a. A two-out-of-three series. [*Hint:* For ease of calculation, assume that all three games are played. What assumptions about independence are you making?]
    b. A three-out-of-five series.
    c. A four-out-of-seven series.

44. **ATHLETICS: Home Runs** On the basis of records from 1996 to 2002, the probability that Mark McGuire will hit a home run on any particular at-bat is $p_M = 0.110$. Model the number X of home runs by McGuire in a season by a binomial random variable with parameters

$n = 530$ (a typical number of at-bats in a season) and p_M. Find $P(X > 50)$, $P(X > 60)$, and $P(X > 70)$. Then do the same for Sammy Sosa, using his home-run-per-at-bat probability $p_S = 0.095$.

45. **SOCIAL SCIENCE: Multiple-Choice Tests** A multiple-choice test consists of 20 questions, each of which has 4 possible answers. What is the expected number of correct answers that would result from random guessing?

46. **SOCIAL SCIENCE: Multiple-Choice Tests** A multiple-choice test consists of 30 questions, each of which has 6 possible answers. What is the expected number of correct answers that would result from random guessing?

47. **GENERAL: Airline Safety** An airplane has four independent engines and each operates with probability 0.99. If it takes at least two engines to land safely, what is the probability of a safe landing? What if the two engines must be on opposite sides of the plane?

48. **ATHLETICS: Football** After scoring a touchdown, a football team has a choice of attempting to run or pass the ball into the end zone for two points (a *two-point conversion*) or kicking the ball through the goalposts for one point (a *field goal*). If the two-point conversions work 37% of the time and one-point field goals are successful 94% of the time,* which play has the better expected gain?

49. **GENERAL: Odd Man Out** Five people in an office decide who fetches (and pays for) mid-morning coffee in the following way. Each tosses a coin, all at the same time, and if everyone gets the same outcome except for one person (such as all tails except for one head), then that person gets coffee for everyone that day. If there is no "odd man out" on the first round, then they repeat the coin tosses as many times as necessary to choose someone.

a. What is the probability of getting an "odd man out" on one round of tosses?
b. What is the probability that it takes more than four rounds to choose an odd man out?
c. More than eight rounds?

50. **GENERAL: Expected Number of Trials Until a Success** Let T be the number of trials up to and including the first success in Bernoulli trials. We can find the expected number of trials, $E(T)$, by reasoning about what happens after the first trial: If the first trial is a success (this happens with probability p) then $T = 1$; if the first trial is a failure (this happens with probability $1 - p$), then we begin the wait for the first success all over again, with expected number of trials $E(T)$ beyond the first trial.*

a. Explain why the reasoning in the previous sentence leads to the equation
$$E(T) = 1 \cdot p + (1 + E(T)) \cdot (1 - p).$$
b. Solve this equation for $E(T)$ to find that $E(T) = \frac{1}{p}$.
c. What is the expected number of tosses up to and including the first 5 when rolling a die?
d. What is the expected number of tosses up to and including the first double-5 when rolling two dice?

51. **GENERAL: Coin Tossing** A coin is tossed n times. Given that there are exactly k heads, show that the probability of a head on any given toss is $\frac{k}{n}$.

52. **GENERAL: Coin Tossing** A coin is tossed n times where n is an even number. Let E be the event that the first toss is heads and F_k be the event that exactly k of the tosses are heads. For which values of k are E and F_k independent events?

Explorations and Excursions

The following exercises extend and augment the material presented in the text.

---

* According to data from the National Football League.

* Technically, this problem goes beyond the finite sample spaces we have discussed, since the first success might never occur. The reasoning, however, avoids this difficulty so that it causes no trouble.

The Mean of the Binomial Distribution

The following exercises prove the formulas $E(x) = np$ for binomial random variables with parameters n (number of trials) and p (probability of success on each trial). We will use the fact that adding together n binomial random variables, each with parameters 1 and p, gives a binomial random variable with parameters n and p (since adding up the number of "successes" in each of n Bernoulli trials gives the total number of successes in all n trials).

53. If X is binomial with $n = 1$, show that $E(X) = 1 \cdot p + 0 \cdot (1 - p) = p$.

54. Justify each numbered equals sign. For X and Y independent,

$$E(X + Y) \stackrel{1}{=} \sum_{\text{All } x,y} (x + y) \cdot P(X = x \text{ and } Y = y)$$

$$\stackrel{2}{=} \sum_{\text{All } x,y} (x + y) \cdot P(X = x) \cdot P(Y = y)$$

$$\stackrel{3}{=} \sum_{\text{All } x,y} x \cdot P(X = x) \cdot P(Y = y) + \sum_{\text{All } x,y} y \cdot P(X = x) \cdot P(Y = y)$$

$$\stackrel{4}{=} \left(\sum_{\text{All } x} x \cdot P(X = x)\right) \cdot \left(\sum_{\text{All } y} P(Y = y)\right) + \left(\sum_{\text{All } y} y \cdot P(Y = y)\right) \cdot \left(\sum_{\text{All } x} P(X = x)\right)$$

$$\stackrel{5}{=} \left(\sum_{\text{All } x} x \cdot P(X = x)\right) \cdot 1 + \left(\sum_{\text{All } y} y \cdot P(Y = y)\right) \cdot 1 = E(X) + E(Y)$$

That is, for independent random variables, $E(X + Y) = E(X) + E(Y)$ (in words, the expectation of the sum is the sum of the expectations). This result immediately generalizes to any number of independent random variables:

$$E(X_1 + X_2 + \cdots + X_n) = E(X_1) + E(X_2) + \cdots + E(X_n).$$

55. If X is a binomial random variable with parameters n and p, we may write $X = X_1 + X_2 + \cdots + X_n$, where $X_1, X_2, \ldots, X_n$ are independent binomial random variables each with parameters 1 and p. Justify:

$$E(X) = E(X_1 + X_2 + \cdots + X_n) = E(X_1) + E(X_2) + \cdots + E(X_n) = \underbrace{p + p + \cdots + p}_{n \text{ terms}} = np$$

Therefore, a binomial random variable with parameters n and p has mean $E(x) = np$.

Chapter Summary with Hints and Suggestions

Reading the text and doing the exercises in this chapter have helped you to master the following skills, which are listed by section (in case you need to review them) and are keyed to particular Review Exercises. Answers for all Review Exercises are given at the back of the book, and full solutions can be found in the Student Solutions Manual.

5.1 Sets, Counting, and Venn Diagrams

- Read and interpret a Venn diagram. *(Review Exercise 1.)*

- Use the complementary or the addition principles of counting or Venn diagrams to solve an applied problem. *(Review Exercise 2.)*

$$n(A^c) = n(U) - n(A)$$
$$n(A \cup B) = n(A) + n(B) - n(A \cap B)$$

- Use the multiplication principle of counting to solve an applied problem. *(Review Exercise 3.)*

5.2 Permutations and Combinations

- Calculate numbers of permutations and combinations. *(Review Exercise 4.)*

$$_nP_r = n \cdot (n-1) \cdot \cdots \cdot (n-r+1)$$
$$_nC_r = \frac{n(n-1) \cdot \cdots \cdot (n-r+1)}{r \cdot (r-1) \cdot \cdots \cdot 1} = \frac{n!}{r!(n-r)!}$$

- Use the permutation and combination formulas (possibly using 🖩) to solve an applied problem. *(Review Exercise 5–7.)*

5.3 Probability Spaces

- Find an appropriate sample space for a random experiment. *(Review Exercises 8–9.)*

- Describe an event in terms of the sample space. *(Review Exercise 10.)*

- Assign probabilities to outcomes in a sample space, and find probabilities of events. *(Review Exercises 11–15.)*

- Use the techniques of this section to find a probability in an applied problem. *(Review Exercises 16–24.)*

5.4 Conditional Probability and Independence

- Find a conditional probability using either the definition or a restricted sample space. *(Review Exercises 25–26.)*

$$P(A \text{ given } B) = \frac{P(A \cap B)}{P(B)}$$

- Use conditional probability to solve an applied problem. *(Review Exercises 27–29.)*

- Use a tree diagram to solve an applied problem. *(Review Exercises 30–32.)*

- Determine whether two events are independent or dependent. *(Review Exercise 33.)*

$$P(A \cap B) = P(A) \cdot P(B)$$

- Use independence to find a probability in an applied problem. *(Review Exercises 34–35.)*

5.5 Bayes' Formula

- Use Bayes' formula to solve an applied problem. *(Review Exercises 36–38.)*

5.6 Random Variables and Distributions

- Determine the outcomes that correspond to the values of a random variable. *(Review Exercise 39.)*

- Find and graph the probability distribution of a random variable, and find the mean. *(Review Exercises 40–44.)*

- For a binomial random variable, find and graph the probability distribution, and find its mean. *(Review Exercise 45.)*

$$P(X = k) = {_nC_k} p^k (1-p)^{n-k}$$
$$E(X) = n \cdot p$$

- Use the binomial probability distribution to solve an applied problem. *(Review Exercises 46–51.)*

Hints and Suggestions

- Much of probability depends on counting, and there are several principles to simplify counting large numbers of objects. Roughly speaking, the *complementary* principle says that you can count the *opposite* set and then subtract; the *addition* principle says that you can add numbers from two sets but you must then subtract what was double-counted; and the *multiplication* principle says that the number of two-part choices is the product of the number

- of first-part choices times the number of second-part choices.
- *Permutations* are arrangements where a different order means a different object (such as letters in a word or rankings of people). *Combinations* are arrangements where reordering does *not* make a new object (such as hands of cards or committees of people).
- A set of n objects has 2^n subsets (counting the "empty" subset and the set itself).
- Probabilities are assigned to possible outcomes of a random experiment so that each probability is between 0 and 1 (inclusive) and they add to 1. Equally likely outcomes should be assigned equal probabilities. Probabilities of more complicated events are found by adding up the probabilities of the outcomes in the event.
- Conditional probability is the relative probability of the intersection of the events compared to the probability of the *given* event. Conditional probabilities are sometimes more easily found from the restricted sample space than from the definition.
- A complicated probability can sometimes be found by a tree diagram, branching on some event on which others depend.
- Independent events "have nothing to do with each other," and the probability of both occurring is the *product* of their probabilities. This is not the same as *disjoint* events, where one precludes the other, and which therefore *cannot* be independent.
- Bayes' formula is useful for finding events of the form S_1 given A in terms of events of the form A given S_i where $S_1, \ldots, S_n$ are disjoint events whose union is the entire sample space.
- The mean of a random variable gives a *representative* or *typical* value.
- Bernoulli trials are repeated independent experiments with only two outcomes, *success* and *failure*. The number of successes in several Bernoulli trials is a *binomial* random variable.
- A graphing calculator that finds permutations, combinations, and binomial and other probabilities is very useful.
- **Practice for Test:** Review Exercises 1, 2, 3, 5, 7, 9, 10, 11, 16, 17, 18, 22, 23, 26, 30, 32, 35, 37, 40, 42, 43, 44, 47, 49, 51.

Review Exercises for Chapter 5

Practice test exercise numbers are in green.

5.1 Sets, Counting, and Venn Diagrams

1. **Venn Diagrams** For the Venn diagram below, find
   a. $n(A)$ b. $n(A \cup B)$ c. $n(B^c)$ d. $n(A^c \cap B)$

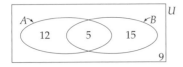

2. **GENERAL: Purchases** A survey of 1000 homeowners found that during the last year 230 bought an automobile, 340 bought a major appliance, and 540 bought neither. How many homeowners had bought both an automobile and a major appliance?

3. **GENERAL: Initials** Find how many three-letter monograms there are if:
   a. Repeated letters are allowed.
   b. Repeated letters are not allowed.

5.2 Permutations and Combinations

4. **Permutations and Combinations** Find:
   a. $_6P_3$ b. $_6C_3$

5. **GENERAL: Committees** How many 4-member committees can be formed from a club consisting of 20 students? What if the committee is to consist of a president, vice-president, secretary, and treasurer?

6. **GENERAL: Cards** How many 5-card hands are there that contain only spades?

7. **GENERAL: Cards** How many 5-card hands are there containing only cards 10 or higher? [*Hint:* Remember that aces may count high.]

5.3 Probability Spaces

8. **GENERAL: Wardrobes** Find the sample space for choosing an outfit consisting of one of two coats (C_1 or C_2), one of two scarves (S_1 or S_2), and one of two hats (H_1 or H_2).

9. **GENERAL: Marbles** A box contains three marbles: one blue, one yellow, and one red. A first marble is chosen, then it is replaced, and a second marble is chosen. Find the sample space. Then find the sample space if the first marble is *not* replaced before the second is chosen.

10. **Events** For the experiment of tossing a coin twice, describe each event as a subset of the sample space $\{(H, H), (H, T), (T, H), (T, T)\}$.
    a. At least one head.
    b. At most one head.
    c. Different faces.

11. **Probabilities of Events** Consider the sample space in Exercise 10.
    a. What probability should be assigned to each outcome?
    b. What is the probability of getting at least one head?
    c. What is the probability of getting different faces?

12. **GENERAL: Committees** A committee of 2 is to be chosen at random from a class of 15 students. What probability should be assigned to any particular committee? What if one of the 2 members is to be designated committee spokesperson?

GENERAL: Spinners Find the probability of the arrow landing in each numbered region. Assume that the probability of any region is proportional to its area, and assume that areas that look the same size are the same size.

13. 14.

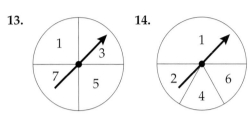

15. **GENERAL: Marbles** A marble is drawn at random from a box containing 3 red, 6 green, and 9 black marbles, and the color is noted. What is the probability of each outcome in the sample space $\{R, G, B\}$?

16. **BIOMEDICAL: Smoking and Weight** For a randomly selected person, the probability of being a smoker is 0.35, the probability of being overweight is 0.40, and the probability of being both a smoker and overweight is 0.20. What is the probability of being *neither* a smoker *nor* overweight?

17. **GENERAL: Dice** If you roll one die, find the probability of:
    a. Rolling at most 2.
    b. Rolling an even number.

18. **GENERAL: Lottery** In a lottery you choose 4 numbers out of 40. Then 5 numbers are announced, and you win something if you have 4 of the 5 numbers. What is the probability that you win something?

19. **GENERAL: Cards** You are dealt 5 cards at random from an ordinary deck. Find the probability of being dealt:
    a. All face cards.
    b. No face cards.

20. **GENERAL: Cards** You are dealt 13 cards at random from an ordinary deck. Find the probability of being dealt:
    a. No spades.
    b. No face cards.

21. **BUSINESS: Defective Products** You have 30 computer CDs, and 2 of them contain computer viruses. If you lend 5 to a friend, what is the probability that they are virus-free?

22. **BUSINESS: Defective Products** A store shelf has 50 light bulbs, of which 2 are defective. If you buy 4, what is the probability that you get both defective bulbs? Neither defective bulb?

23. **GENERAL: Committees** A committee of 3 is to be formed from your class of 30. What is the probability that both you and your best friend will be on it?

24. **BIOMEDICAL: Memory** You remember the names of 6 of the 10 people you just met. If you run into 2 of the 10 on the street, what is the probability that you will remember their names?

5.4 Conditional Probability and Independence

25. If $P(A) = 0.5$, $P(B) = 0.4$, and $P(A \cap B) = 0.3$, find
    a. $P(A$ given $B)$
    b. $P(B$ given $A)$

26. **GENERAL: Dice** If you roll two dice, what is the probability of at least one six given that the sum of the numbers is seven?

27. **BIOMEDICAL: Gender** If you know that a family of three children has at least one boy, what is the probability that all three are boys? Assume that boys and girls are equally likely.

28. **GENERAL: Cards** You draw three cards at random from a standard deck. If the first is an ace, what is the probability that the other two are also aces?

29. **GENERAL: Cards** If you are being dealt five cards and the first four are hearts, what is the probability that the fifth is also a heart?

30. **BUSINESS: Market Share** An automobile salesperson estimates that with a male customer she can make a sale with probability 0.3 and that with a female customer the probability is 0.4. If 80% of her customers are male and 20% female, what is the probability that for a randomly selected customer she can make a sale?

31. **GENERAL: Searches** An airplane is missing, and you estimate that with probability 0.6 it crashed in the mountains and with probability 0.4 it crashed in the valley. If it is in the mountains, a search there will find it with probability 0.8, whereas if it is in the valley, a search there will find it with probability 0.9. What is the probability that the plane will be found? (The complementary probabilities to the numbers 0.8 and 0.9 are called the *overlook probabilities* because they give the probability that it will *not* be found in a region if it is there.)

32. **BIOMEDICAL: Births** Twelve percent of all births are by cesarean (surgical) section, and of these births 96% survive. Overall, 98% of all babies survive delivery. For a randomly chosen mother who did not have a cesarean, what is the probability that her baby survives?

33. **GENERAL: Coins** For the experiment of tossing a coin twice, with sample space $\{(H, H), (H, T), (T, H), (T, T)\}$, are the following events independent or dependent?
    a. *At least one head* and *at most one head.*
    b. *Heads on first toss* and *same face on both tosses.*

34. **BUSINESS: Computer Malfunctions** An airline reservations computer breaks down with probability 0.01, at which time a second computer takes over, but it also has probability 0.01 of failing. Find the probability that the airlines will be able to take reservations.

35. **GENERAL: Coins** Five coins are tossed. Find the probability of getting
    a. All heads.
    b. All the same outcome.
    c. Alternating outcomes.

5.5 Bayes' Formula

36. **GENERAL: Coins** You have two coins in your pocket, one with two heads, and one with a head and a tail. You choose one at random and toss it, and it comes up heads. What

is the probability that it is the two-headed coin?

37. **MANAGEMENT SCIENCE: Manufacturing Defects** A soft drink bottler has three bottling machines: A, B, and C. Machine A bottles 25% of the company's output, machine B bottles 35%, and machine C bottles 40%. It is known that 4% of the bottles produced by machine A are defective, as are 3% of the bottles produced by machine B and 2% of the bottles from machine C. If a randomly selected bottle from the factory's output is found to be defective, what is the probability that it was produced by machine A?

38. **MANAGEMENT SCIENCE: Manufacturing Defects** For the information in Exercise 37, if a randomly selected bottle is *not* defective, what is the probability that it came from machine A?

5.6 Random Variables and Distributions

39. **GENERAL: Coins** Five coins are tossed, and X is the number of heads. List the outcomes (from the usual sample space) corresponding to each of the events $X = 5$, $X = 1$, and $X = 0$.

40. **GENERAL: Coins** You toss two coins. You win $34 if you get double heads, and otherwise you lose $2. Find and graph the probability distribution of your winnings.

41. **Mean** Find the mean of the random variable in Exercise 40.

42. **GENERAL: Marbles** Two marbles are chosen at random and without replacement from a box containing 5 red and 4 green marbles. If X is the number of green marbles chosen, find and graph the probability distribution of X.

43. **Mean** Find the mean of the random variable in Exercise 42.

44. **GENERAL: Raffle Tickets** Five hundred raffle tickets are sold for the following prizes: one first prize worth $1000, two second prizes worth $150, and 10 third prizes worth $25. Find the expected value of a ticket.

45. **Binomial Distribution** Find and graph the distribution of a binomial random variable with parameters $n = 4$ and $p = \frac{4}{5}$. Find its mean using the formula developed for a binomial random variable.

46. **GENERAL: Coins** For 8 tosses of a coin, find the probability that the number of heads is between 3 and 5 inclusive.

47. **BUSINESS: Sales** A television salesperson estimates that a customer will buy a $900 TV with probability $\frac{1}{10}$, a $500 TV with probability $\frac{1}{2}$, a $200 TV with probability $\frac{1}{10}$, and otherwise buy nothing. What is the expected value of the sale?

48. **GENERAL: Testing** You know that one of three keys will work in a lock. If you try them one at a time until you find the right one, what is the expected number of tries?

49. **BUSINESS: Product Quality** Of the products a company produces, 1% have hidden defects. If you buy a dozen of them, what is the mean of the number of defective ones in your purchase?

50. **BUSINESS: Product Quality** A company makes screws, and 98% of them are usable. If you buy a box of 100 screws, what is the probability of at least 97 being usable?

51. **GENERAL: Target Practice** You shoot at a target 5 times. If the probability of a hit on each shot is 75%, what is the probability that you hit it at least 4 times?

6 Statistics

6.1 Random Samples and Data Organization
6.2 Measures of Central Tendency
6.3 Measures of Variation
6.4 Normal Distributions and Binomial Approximation

Application Preview

Household Income

The following graph shows the percentages of U.S. households with 2001 incomes from $0 to $100,000 grouped in blocks of $5000 (the remaining 14% had incomes of $100,000 or more).

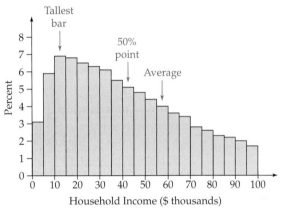

Source: Census Bureau

This is a *histogram*, which is just one of the graphical displays that we will use to exhibit data. There is a vast amount of information in this graph, and it is natural to ask whether there is some sort of "middle" income that might represent a "typical American household." There are several possible approaches. The tallest bar, representing the income block with the greatest number of households, has midpoint $12,500, but this is lower than the vast majority of households. The average of all incomes is $58,208, but if you look where this falls on the graph, it is much higher than most of the households (in fact, it's higher than 65% of them). The 50-percent point, with as many households above it as below it, is $42,228. Each of these figures has some claim to being a *representative* income, and in this chapter we will discuss the merits of each.

6.1 RANDOM SAMPLES AND DATA ORGANIZATION

Introduction

Often you have to make decisions based on incomplete information. For example, to find out how many Americans watch a television program you can't ask everyone, and to find out how long a light bulb lasts a company can't test all of its bulbs, particularly if it wants to have any left to sell. These and many other "real world" problems depend on taking samples and analyzing the data. Statistics is the branch of mathematics concerned with the collection, analysis, and interpretation of numerical data.

Random Samples

A *statistical population* is the collection of data that you are studying. Examples of statistical populations are the ages of citizens of the United States, the size of bank accounts in Illinois, and the brands of automobiles in California. Although in principle it is possible to measure the entire population, considerations of time and cost usually dictate that we measure only part of the population. For that purpose we use a *random sample* to represent the population.

Random Sample

A *random sample* is a selection of members of the population satisfying two requirements:

1. Every member of the population is equally likely to be included in the sample.
2. Every possible sample of the same size from the population is equally likely to be chosen.

A sample that does not meet these criteria is not representative and cannot be used to infer characteristics of the entire population. For example, a telephone survey to homes at 10 A.M. on a weekday would miss most employed people and so could not produce valid information about work skills. Random samples can be chosen using random number tables or by other methods, and we will assume in this chapter that the data we are analyzing are from a random sample.

Levels of Measurement

We shall consider only numerical data, which we classify into four types: *nominal, ordinal, interval,* and *ratio* data.

Nominal data (*nominal* means "in name only") means numbers used only to *identify* objects. For example, the numbers on the backs of football jerseys are nominal data, since they do not mean that "number 24" can run twice as fast or throw twice as far as "number 12."

Ordinal data (the word comes from "order") means numbers that can be arranged in order but differences between them are not meaningful. For example, you might rate the teachers you have had in college as "best" (number 1), "second best" (number 2), and so on. However, the difference between "1" and "2" may not be the same as the difference between "7" and "8."

Interval data means numbers that can be put in order *and* differences between them can be compared, but ratios cannot. For example, it makes sense to talk about the *difference* between two temperatures, like 40° and 80°, but it makes no sense to say that 80° is *twice* as hot as 40°, since the zero point in the Fahrenheit (or Celsius) system is arbitrary and does not mean *zero heat*.

Ratio data means numbers that can be put in order, whose differences can be compared, *and* whose *ratios* have meaning. Examples of ratio data are amounts of money and weights, since 80 of either really *is* twice 40. In general, with ratio data, the zero level *does* mean "none" of something.

When we refer to the *type* of a data collection, we mean the highest of these levels that applies, from nominal (lowest) to ratio (highest).

EXAMPLE 1

JUDGING LEVELS OF DATA MEASUREMENT

Identify the level of measurement of each data collection.

a. Ten vehicles passing through a turnpike toll booth were classified as 1 for a car, 2 for a bus, and 3 for a truck, giving data {1, 3, 2, 3, 1, 1, 1, 1, 3, 1}.

b. A rock concert enthusiast rates the last eight performances she attended on a scale from 1 (awful) to 10 (awesome) as {8, 5, 7, 9, 7, 10, 6, 9}.

c. The temperatures of five students at the college infirmary are {99.3, 102.1, 101.8, 101.5, 100.9} degrees Fahrenheit.

d. The weights of twelve members of the cross-country team are {124, 151, 132, 153, 142, 147, 120, 127, 154, 119, 118, 116} pounds.

Solution

a. *Nominal data:* The numbers are used only to identify the type of vehicle. The number 2 does not mean more or less than the number 3, so differences between these numbers have no meaning.

b. *Ordinal data:* The rankings put the performances in relative order, but no meaning can be given to differences between the rankings. The difference between ratings 5 and 6 may be more or less than the difference between ratings 9 and 10.

c. *Interval data:* The temperatures can be put in order from "lowest" to "highest," and differences between them can be compared. However, this is not ratio data because the zero point in temperature is arbitrary—101° is not 1% hotter than 100°.

d. *Ratio data:* The weights can be put in order, "10 pounds heavier" has meaning, *and* 200 pounds really is twice as heavy as 100 pounds.

Practice Problem 1 Students filling out a course evaluation were asked for the following information: (a) gender: 1—male, 2—female; (b) course rating: 1—poor, 2—acceptable, 3—good, 4—outstanding; and (c) current grade point average. Identify the level of measurement of the data gathered from each question. ➤ Solution on page 438

Bar Chart

A *bar chart* provides a visual summary of data with the number of times each value appears corresponding to the length of the bar for that value. The bars may be drawn vertically or horizontally, and they may have spaces between them.

EXAMPLE 2   **CONSTRUCTING A BAR CHART**

The colors of a random sample of thirty cars in the Country Corners Mall parking lot were recorded using: 1—blue, 2—green, 3—white,

4—yellow, 5—brown, and 6—black. Construct a bar chart for the data {3, 5, 2, 4, 1, 5, 1, 1, 1, 2, 4, 1, 6, 1, 6, 5, 6, 3, 1, 1, 1, 1, 1, 6, 3, 1, 2, 2, 4, 2}.

Solution

We first tally the frequency of each color and then sketch the bar chart.

Color	Tally	Frequency
1	⊬⊬⊬ ⊬⊬⊬ ‖	12
2	⊬⊬⊬	5
3	‖‖	3
4	‖‖	3
5	‖‖	3
6	‖‖‖	4

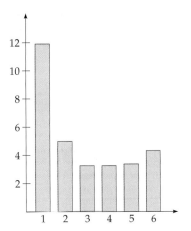

Notice that the most popular color by far is blue—about as popular as any three of the other colors combined. This observation is immediately obvious from the bar chart but is much more difficult to see from the list of data.

Spreadsheet Exploration

The following spreadsheet* shows the car colors and frequencies from Example 2, together with some of the many graphs that can be constructed from such data. The *pie chart* on the lower right is easily drawn by a computer, but it is difficult to draw accurately by hand because the size of the "slices" must be calculated first.

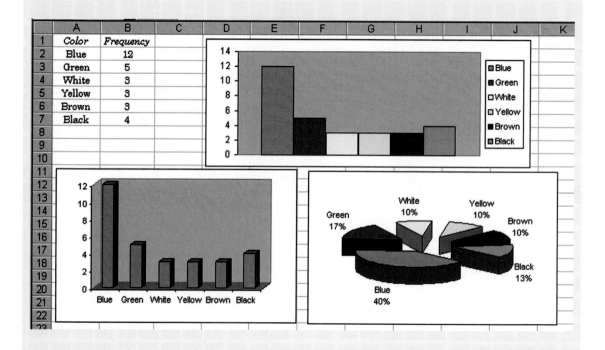

The pie chart is best for showing relative proportions because the total pie represents all of the data and the size of each slice shows its fraction of the whole.

Practice Problem 2   Construct a bar chart for the data given in Example 1(a) (page 431). Which type of vehicle was the most common? Which was the least common?   ➤ Solutions on page 438

---

* See the Preface for information on how to obtain this and other Excel spreadsheets.

Histogram

A *histogram* is a type of bar chart in which the sides of the bars touch and the *width* of the rectangles has meaning. The data is divided into *classes*, usually numbering from 5 to 15, one for each bar, and each class has the same *width*:

$$\begin{pmatrix}\text{Class} \\ \text{width}\end{pmatrix} \approx \frac{\begin{pmatrix}\text{Largest} \\ \text{data value}\end{pmatrix} - \begin{pmatrix}\text{Smallest} \\ \text{data value}\end{pmatrix}}{\begin{pmatrix}\text{Number} \\ \text{of classes}\end{pmatrix}}$$ Round the class width *up* so that all data will be covered

We then make a tally of the number of data values that fall into each class. Some histograms use the convention that any value on the boundary between two classes belongs to the upper class, while others are designed so that no data value falls on a class boundary.

EXAMPLE 3

CONSTRUCTING A HISTOGRAM

When FirstAmerica National Bank set up its website for online banking, they asked users to fill out a questionnaire. Those who said that they would use online banking at least twice a week listed their ages as follows: {34, 10, 53, 50, 80, 38, 39, 31, 52, 41, 46, 46, 41, 69, 73, 57, 40, 52, 47, 68, 22, 33, 51, 65, 23, 47, 64, 45, 26, 74}. Construct a histogram for the data and determine whether online banking is most popular with young, middle-aged, or elderly users.

Solution

Scanning the data, we see that the smallest and largest ages are 10 and 80. Therefore, if we choose 6 classes we obtain a class width of $\frac{80 - 10}{6} \approx 11.7$, which we round up to 12. Since we rounded up, we may choose our first class to start just below the lowest data point of 10, say at 9.5. We then add successively the class width of 12 to get the other class boundaries.

Class Boundaries	Tally	Frequency
9.5–21.5	|	1
21.5–33.5	ℍℍ	5
33.5–45.5	ℍℍ ||	7
45.5–57.5	ℍℍ ℍℍ	10
57.5–69.5	||||	4
69.5–81.5	|||	3

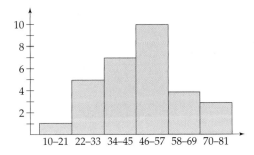

The histogram shows that FirstAmerica's website is most popular with middle-aged customers (in their 40's and 50's).

We may label the base of each bar with either the range of its data values or the class boundaries. Different numbers of bars or class boundaries will give different histograms, as the following Graphing Calculator Exploration shows.

Graphing Calculator Exploration

Histograms can be drawn on graphing calculators using the STAT PLOT command. To explore several histograms of the data from Example 3 (page 435), proceed as follows:

{34,10,53,50,80,38,39,31,52,41,4 6,46,41,69,73,57,40,52,47,68,22, 33,51,65,23,47,6 4,45,26,74}→L_1

a. Enter the data values as a list and store it in the list L_1.

b. Turn off the axes in the window FORMAT menu, turn on STAT PLOT 1, select the histogram icon with Xlist L_1, and set the WINDOW parameters Xmin = 9.5, Xmax = 81.5, Xscl = 12 to match the start, finish, and class width we used. GRAPH and then TRACE to explore your histogram.

  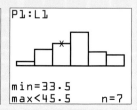

Of course, different choices of the class width, number of classes, and starting value will result in slightly different histograms. The following are several possibilities:

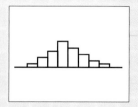 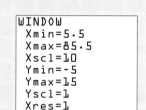 

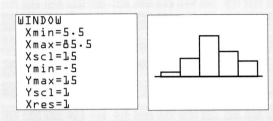

Notice that each histogram supports the conclusion that the website is most popular with those in the middle age range.

How do histograms compare to the graphs of probability distributions that we drew in Chapter 5? They are very similar, but in Chapter 5 the heights of the bars were *probabilities*, with the heights adding up to 1. Here the heights of the bars are *frequencies*, with the heights adding to equal the total number of data points.

6.1 Section Summary

A *statistical population* is the collection of data that you are studying. A *random sample* is a selection from the population such that every member is equally likely to be included and every possible sample of the same size is equally likely to be chosen.

There are four levels of data measurement:

Data	Uses
Nominal	Classify by name
Ordinal	Compare order (1st, 2nd, . . .)
Interval	Compare order and differences
Ratio	Compare order, differences, and ratios

Bar charts can be drawn for *any* type of data, with the length of the bars indicating the frequency of the values. Histograms involve dividing the data into adjacent intervals or *classes*, whose widths are calculated from the formula on page 435. Histograms are used only for interval and ratio data since only in these cases does the calculation of the class width have meaning.

Solutions to Practice Problems

1. a. Nominal b. Ordinal c. Ratio

2.

Vehicle	Tally	Frequency
1								6
2			1
3					3

1 = car
2 = bus
3 = truck

Cars were most common and buses were least common.

6.1 Exercises

Identify the level of data measurement in each situation.

1. **BUSINESS: Discount Brokers** A survey of ten randomly selected traders at a discount broker service records the numbers of transactions involving stocks, bonds, and mutual funds last Wednesday.

2. **BUSINESS: Travel Agencies** An airline surveys twenty randomly selected travel agencies in and around San Francisco and records the numbers of tickets sold at each last month to Hawaii, Alaska, Mexico, and New York.

3. **BUSINESS: Mortgage Banking** The home office of the First & National Bank sends questionnaires to two hundred randomly selected families who applied for mortgages last summer, asking for ratings (from "poor" to "outstanding") of the courtesy of the mortgage staff at the branch offices.

4. **BUSINESS: Shoe Sales** The national personnel manager for the Feets-R-Us shoe store franchise reviews the supervisor reports on thirty randomly selected sales employees and counts the number of "below average," "average," and "above average" performance evaluations.

5. **GENERAL: SAT Scores** The test results on the verbal SAT administered by the Educational Testing Service are numerical scores from 200 to 800. A survey records the verbal SAT scores of twelve randomly selected seniors at J. T. Marshall High School.

6. **SOCIAL SCIENCE: Midwest Aviation** A survey records the years of birth of fifty randomly selected commercial pilots residing in Missouri.

7. **SOCIAL SCIENCE: IQ Scores** An educational research team working on a grant from the National Institutes of Health records the IQ scores of eighty randomly selected fifth graders in the Baltimore County Public School system.

8. **BUSINESS: Quality Control** A major cereal manufacturer randomly selects thirty boxes of Corn Chex Crunches from the shipping warehouse and records the actual net weights.

9. **PERSONAL FINANCE: Disposable Income** The Chamber of Commerce tabulates the disposable family incomes of ninety randomly selected households in Thomsonville.

10. **BUSINESS: Financial Trends** The *Get Rich Slowly* investment newsletter reports the price-to-earnings ratios of sixty randomly selected technology stocks traded last month on the American Stock Exchange.

Construct a bar chart of the data in each situation.

11. **SOCIAL SCIENCE: Marital Status** A random sample of shoppers in the Kingston Valley Mall counted 39 "never married," 97 "married," 16 "widowed," and 8 "divorced" individuals.

12. **SOCIAL SCIENCE: Place of Birth** A random sample of skiers in Grand Junction last February counted 12 from Colorado, 27 from California, 19 from New York, 8 from Florida, and 6 from other states or nations.

13. **SOCIAL SCIENCE: State Legislators** A random sample of candidates running for election to the Nevada legislature counted 4 business executives, 7 real estate developers, 18 lawyers, 1 doctor, 3 certified public accountants, and 2 retirees.

14. **SOCIAL SCIENCE: Voter Concerns** A random sample of voters leaving the polls identified the most important issue in the Higginsville school board election as follows: 8 for teacher tenure, 54 for budget, 11 for after-school day care, 19 for the bond proposal for a new football stand at the high school, and 3 making "no comment."

For each situation construct a histogram and use it to answer the question. (For Exercises 19–25, histograms may vary.)

You may use a graphing calculator if your instructor permits it.

15. **BUSINESS: Shipping Weights** A random sample of twenty packages shipped from the Holiday Values Corporation mail order center found the following weights (in ounces). (Use 6 classes of width 8 starting at 12.5.)

29	21	60	23	39	24	33	49	37	17
13	27	56	21	30	43	34	24	18	13

Based on your histogram, which type of boxes are in greatest demand: lighter weight, midweight, or heavier weight?

16. **PERSONAL FINANCE: Blue Book Values** A random sample of thirty-two cars parked at Emmett's Field last June found the following "blue book" values (in thousands of dollars) (use 6 classes of width 4 starting at 2.5):

15	5	8	17	18	17	8	9
5	13	6	9	5	19	8	13
20	9	7	13	25	16	5	25
25	17	24	10	23	14	15	4

Based on your histogram, which two adjacent classes have the highest total number of cars?

17. **GENERAL: Websites** A random sample of twenty-eight student accounts at the campus Academic Computing Center showed the following number of website "hits" between 9 and 12 P.M. last Wednesday (use 8 classes of width 10 starting at 10.5):

38	29	16	35	69	63	13
58	11	58	34	73	45	57
70	16	45	37	39	88	69
47	30	40	12	47	20	61

Based on your histogram, which two classes have the highest number of hits?

18. **ATHLETICS: Football Players** A random sample of twenty-four defensive linemen in the East Coast Division II college conference found the following weights (in pounds) (use 8 classes of width 15 starting at 140.5):

173	171	211	169	258	213	143	236
191	163	192	235	196	208	157	221
174	200	195	198	228	152	188	201

Based on your histogram, are the weights evenly distributed or more concentrated in one part of the range?

19. **BUSINESS: Airline Tickets** A random sample of forty passengers traveling economy class from New York to Los Angeles last Thanksgiving weekend found the following one-way ticket prices (in dollars):

425	236	481	559	258	473	522	480
544	451	244	440	530	565	445	569
255	515	412	377	453	518	490	301
379	439	229	531	465	252	219	510
565	477	440	551	518	525	230	437

Based on your histogram, are more ticket prices grouped at the lower end, the middle, or the upper end of the range of costs?

20. **GENERAL: Insurance Claims** A random sample of thirty accident repair claims received in December at the western division office of the Some States Insurance Company were for the following dollar amounts:

5197	6012	1859	6207	3497	1097
4785	5950	3652	6279	3639	4553
6249	5999	6601	1193	5961	4625
5918	4289	2948	5043	1329	5483
968	5297	1423	5159	5527	4880

Based on your histogram, are more claims grouped at the lower end, in the middle, or at the upper end of the range?

21. **SOCIAL SCIENCE: Law Practice** A random sample of twenty-one attorneys hired within the past year at the U.S. Department of Justice found the following numbers of hours worked last week:

48	44	50	49	44	45	51
50	57	36	53	44	48	54
60	52	52	39	47	49	57

Based on your histogram, are more hours grouped at the lower end, the middle, or the upper end of the workweeks?

22. **GENERAL: Light Bulbs** A random sample of thirty-six 75-watt light bulbs found the following lifetimes (number of hours until burn out):

1070	1210	1280	1230	1340	1280
1200	1120	1310	1100	1170	1000
1290	1420	1190	1280	1160	1290
1250	1130	1180	1150	1110	1200
1230	1160	1240	1250	1230	1240
1000	1280	1180	1280	1270	1210

Based on your histogram, are the lifetimes roughly evenly distributed or concentrated in one part of the range?

23. **SOCIAL SCIENCE: Medical Insurance** A random sample of twenty private practice physicians in the Lehigh Valley found the following numbers of insurance plans accepted by each:

16	21	9	23	19	23	20	23	18	19
19	29	20	21	20	23	20	14	11	19

Based on your histogram, are the numbers of plans roughly equally distributed over the range or grouped in one area?

24. **BUSINESS: Dining Out** A random sample of forty dinner checks at the Lotus East Chinese Restaurant last Saturday evening showed the following total costs:

43	104	45	60	25	103	33	96
108	41	67	53	43	48	99	45
35	40	66	103	27	110	60	33
46	126	112	84	42	90	74	46
102	41	48	41	54	44	82	33

Based on your histogram, are the dinner checks grouped more at the lower or the upper end of the cost range?

25. **SOCIAL SCIENCE: Automobile Accidents** A random sample of twenty drivers involved in accidents in Columbia County during August showed the following ages:

23	16	75	18	19
74	17	21	73	27
61	18	72	18	44
16	75	29	17	24

Based on your histogram, are the accidents roughly evenly distributed across age groups? Give an explanation for the general shape of the histogram.

6.2 MEASURES OF CENTRAL TENDENCY

Introduction

We often hear of *averages,* from grade point averages to batting averages. Given a collection of data, we often use the average as a representative or typical value. In this section we will see that not all types of data are appropriately summarized by an average, and sometimes another kind of "typical" value is more representative. To avoid confusion with casual meanings of "average," these different kinds of typical values are called *measures of central tendency.*

Mode

The *mode* of a collection of data is the *most frequently occurring* value. If all of the values in a data set occur the same number of times, there is no mode, while if only a few of the values occur the same maximal number of times, there are several modes. The mode can be used with every level of data measurement, and it is the *only* one appropriate for *nominal* data (such as colors of cars, where "averages" have no meaning).

EXAMPLE 1

FINDING MODES

Find the mode for each collection of numbers.

a. {1, 2, 2, 2, 2, 3, 3, 4, 4, 4, 10}
b. {1, 2, 2, 2, 2, 3, 3, 4, 4, 4, 4}
c. {2, 2, 2, 3, 3, 3, 4, 4, 4}

Solution

a. Since 2 appears more often than any other value, the mode is 2.
b. Since both 2 and 4 occur four times and every other number occurs fewer times, the modes are 2 and 4.
c. Since each number occurs the same number of times, there is no mode.

The following are bar graphs for these collections of data. Notice that the presence of one number that is very different from the others, such as the 10 in the first graph, has no effect on the mode: If the 10 were replaced by 100 or even 1000, the mode would not change.

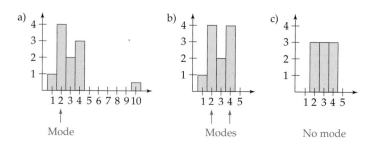

a) Mode b) Modes c) No mode

If a collection of data has just one mode (as in the first graph) it is said to be *unimodal,* and if it has two modes (as in the second) it is said to be *bimodal.*

Practice Problem 1 Find the mode of {17, 20, 20, 13, 10, 17, 10, 20}. ➤ Solution on page 448

Graphing Calculator Exploration

When finding the mode of a long list of values, it is often easier to first sort the list so that all the same values appear grouped within the list. For instance, to find the mode of {62, 56, 52, 53, 58, 64, 68, 67, 60, 61, 57, 58, 57, 63, 55}, proceed as follows:

a. Enter the data values as a list and STORE it in the list L_1 (or enter the numbers into L_1 directly using STAT and EDIT).

b. From the STAT or LIST menu, select SortA (to sort in *ascending* order). (You could also use SortD for descending order.)

c. Complete the ascending sort command by entering the name of the list of values, L_1.

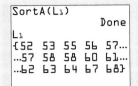

d. Examine the sorted list L_1 to identify the value(s) appearing most often. The modes of this collection of values are 57 and 58 because they appear twice and the others appear only once each.

Median

The *median* is the *middle* value of a list of data values when sorted in ascending or descending order. If there is an even number of data values, the median is the arithmetic average of the two middle values (their sum divided by 2). Since the median depends on arranging the values in order, the median is not appropriate for nominal data. For instance, it would be meaningless to find a "median color" of a car (see Example 2 on pages 432–433), since the assignment of numbers to colors was arbitrary. The median is appropriate for every level of data measurement except the nominal level.

EXAMPLE 2

FINDING MEDIANS

Find the median of each data set.

a. {1, 3, 4, 7, 11, 18, 39}
b. {1, 3, 4, 7, 11, 18, 1000}
c. {26, 16, 10, 6, 4, 2}
d. {62, 56, 52, 53, 58, 64, 68, 67, 60, 61, 57, 58, 57, 63, 55}

Solution

a. Since these seven numbers are arranged in ascending order, the median is the fourth number, 7, because three numbers are smaller and three are larger.

b. Just as in part (a), the median is 7. Notice that replacing the largest data value in the previous data set with a much larger value did not change the median.

c. Since these six values are arranged in descending order, the median is 8, the number halfway between the two middle values of 10 and 6. Thus the median need not be one of the data values.

d. Since these fifteen values are not in ascending or descending order, we first order them (as in the Graphing Calculator Exploration on the previous page):

$$\{52, 53, 55, 56, 57, 57, 58, 58, 60, 61, 62, 63, 64, 67, 68\}$$

The median of these is the middle value, 58, since there are seven data values no larger and seven values no smaller.

Practice Problem 2 Find the median of the values $\{4, 5, 5, 6, 7, 8\}$. ➤ Solution on page 448

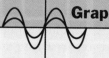

Graphing Calculator Exploration

The median of an unsorted list of values can be found with a single calculator command. To find the median of $\{62, 56, 52, 53, 58, 64, 68, 67, 60, 61, 57, 58, 57, 63, 55\}$, the unordered values from Example 2(d) above, proceed as follows:

Enter the data values as a list and store it in the list L_1. From the LIST MATH menu, select the median command and apply it to the list L_1.

```
{62,56,52,53,58,        NAMES  OPS  MATH       median(L1)
64,68,67,60,61,5         1:min(                          58
7,58,57,63,55}→L         2:max(
1                        3:mean(
                         4:median(
                         5:sum(
                         6:prod(
                         7↓stdDev(
```

Mean

The *mean* is the usual arithmetic average found by summing the values and dividing by the number of them:

Mean

The *mean* $\bar{x}$ of the n values $x_1, x_2, \ldots, x_n$ is

$$\bar{x} = \frac{1}{n}(x_1 + x_2 + \cdots + x_n)$$

The mean of a collection of numbers is equivalent to the mean of a random variable that is equally likely to be any of the numbers. The mean of a data set is sometimes called the *sample mean* to distinguish it from the mean of a probability distribution. Calculating the mean assumes that the distance between values is meaningful, so the mean is not an appropriate measure for ordinal or nominal data. For instance, it would be meaningless to find the mean of the concert ratings in Example 1(b) (page 431), from 1—awful to 10—awesome, because we do not know if a 6 and an 8 should average to a 7, since the intervals between them may not be the same. The mean is an appropriate measure of central tendency only for interval and ratio data.

EXAMPLE 3

FINDING MEANS

Find the mean of each data set.

a. {5, 16, 14, 1, 4, 20}

b. {5, 16, 14, 1, 4, 200}

```
mean({5,16,14,1,
4,20})
               10
mean({5,16,14,1,
4,200})
               40
```

Solution

a. $\bar{x} = \frac{1}{6}(5 + 16 + 14 + 1 + 4 + 20) = \frac{1}{6} \cdot 60 = 10$ The bar over x indicates *mean*

b. $\bar{x} = \frac{1}{6}(5 + 16 + 14 + 1 + 4 + 200) = \frac{1}{6} \cdot 240 = 40$

Notice that the data set in part (b) is the same as in part (a) except that the last number was changed from 20 to 200, with the result that the mean changed substantially. Thus, unlike the mode and the median, the mean is sensitive to extreme values.

The mean is where the histogram of the values would just balance.

From the data in Example 3

These calculations show that when finding the mean there is no need to arrange the numbers in any particular order and that the mean need *not* be one of the original data values.

Mean, Median, and Mode

We have three measures of central tendency, the mean, the median, and the mode. Each has advantages and disadvantages, and one may be more appropriate than the others as a "representative" value for a particular data set.

- The *mean* has the advantage of being easy to calculate, and takes all of the values into account, but it can be heavily influenced by extremes (see Example 3).
- The *median* is not influenced by extremes, and it can be used for ordinal data (that is, data that is only ranked, say from best to worst), but it is usually found by first sorting the data, which can be difficult for a large data set.
- The *mode* is easy to find, is not influenced by extremes, and can be used even for nominal data. However, a data set may not have a mode, or it may have several.

The following histograms show several possible relationships between the mean, the median, and the mode.

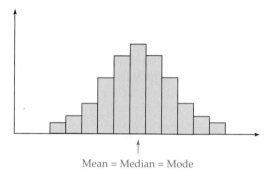

Mean = Median = Mode

For data that are symmetric about a single mode, the mean, the median, and the mode will all be the same.

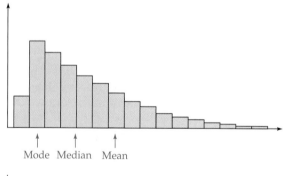

Mode Median Mean

For a *skewed* data set, the mode can occur at an early peak and the mean can be distorted by extreme values.

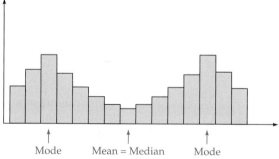

Mode   Mean = Median   Mode

For *bimodal* data, the modes may be quite far from the mean and median.

For an example of a histogram like that shown above, think of the graph of the number of automobile accidents for drivers of different ages: Teenage drivers have frequent accidents, then the number decreases with age, bottoming out in the 40s and 50s, then increasing again for elderly drivers with another peak in the top age bracket. That is, such a histogram would look like the one above but with the two peaks exactly at the ends (see Exercise 25 on page 440).

Practice Problem 3

For an extreme case of skewing, imagine that Bill Gates (the richest man in America, with an estimated wealth of $80 billion in 2001) lived in a town along with 7999 penniless people.

a. What would be the average (mean) wealth of people in the town?
b. What would be the median wealth?
c. What would be the mode?
d. Which of these seems the most "representative" measure of central tendency?

➤ Solution on page 448

EXAMPLE 4

BASEBALL SALARIES

In the 1995 baseball strike it was disclosed that the mean salary of major league baseball players was $1.2 million, while the median salary was only $500,000. What does this say about the distribution of baseball salaries?

Solution

If the *middle* salary is $500,000, then for the mean to be higher, there must be a few very large salaries. That is, a few baseball players must be receiving enormously high salaries to skew the mean that much. In such cases, the median is usually considered a more representative value for most players than the mean.

6.2 Section Summary

Although there are three measures of central tendency, not all of them can be used with every type of data. For nominal data only the mode is appropriate, for ordinal data either the mode or the median is appropriate, and for interval or ratio data all three are appropriate.

Measure of Central Tendency	Definition	Appropriate Data Levels
Mode	Most frequent	All levels
Median	Middle value	Ordinal, interval, and ratio
Mean	$\bar{x} = \dfrac{1}{n}(x_1 + x_2 + \cdots + x_n)$	Interval and ratio

The mean is sensitive to changes in the extremes, but the median and mode are not. The mean of a data set is sometimes called the *sample mean* to distinguish it from the mean of a probability distribution.

▶ Solutions to Practice Problems

1. Since 20 occurs three times and no other value occurs this often, the mode is 20.

2. Since these six values are arranged in ascending order, the median is 5.5, the number halfway between the two middle values of 5 and 6.

3. a. $\bar{x} = \dfrac{\$80 \text{ billion}}{8000} = \$10 \text{ million}$

   b. Median is $0.

   c. Mode is $0.

   d. The median and mode seem most representative.

6.2 Exercises (will be helpful throughout.)

Find the mode, median, and mean of each data set.

1. {6, 17, 12, 10, 15, 16, 5, 8, 9, 14, 9}
2. {10, 7, 9, 13, 11, 14, 15, 14, 8, 9}
3. {15, 14, 5, 7, 5, 14, 9, 7, 7, 14, 7, 14, 12}
4. {11, 19, 20, 11, 12, 18, 11, 20, 16, 18, 17, 10, 13, 11, 15}
5. {19, 11, 10, 19, 18, 12, 19, 10, 14, 12, 13, 17, 20, 19, 15}
6. {12, 5, 10, 6, 7, 6, 15, 7, 6, 5, 6, 5, 13, 10, 7, 13, 8, 11, 12}
7. {8, 15, 10, 14, 13, 14, 5, 10, 7, 9, 13, 14, 15, 14, 15, 19, 13, 12, 8}
8. {11, 8, 23, 21, 14, 16, 15, 11, 9, 22, 15, 11, 12, 11, 10, 15, 10, 25, 18, 6, 11}
9. {19, 9, 15, 8, 18, 15, 12, 22, 16, 19, 15, 19, 20, 24, 7, 14, 21, 19, 20, 5, 19}
10. {10, 13, 8, 15, 8, 12, 12, 12, 14, 8, 12, 11, 14, 9, 8, 14, 14, 14, 10, 10}

APPLIED EXERCISES

Find the mode, median, and mean for the data in each situation.

11. **GENERAL: Emergency Services** A random sample from the records of the Farmington Volunteer Ambulance Corps found the following twenty response times (in minutes) to 911 calls received during February:

    14 9 25 14 12 16 5 20 20 25
    20 24 12 7 19 12 6 13 24 18

12. **BUSINESS: Car Sales** A random sample of fifteen General Motors salesmen at Tri-State Dealers found the following earnings (in dollars) for the first week of September:

    1200 1200 1100 1300 900
    600   800   700   1200 500
    1000 1200 600   900   900

13. **ENVIRONMENTAL SCIENCE: Volunteer Recycling** A random sample of thirty volunteers helping sort newspapers, aluminum, and glass at the Parkerville Regional Recovery Center found the following hours contributed last week:

    10 7 13 11 7 11 10 5 9 14
    14 12 12 5 10 12 19 10 7 10
    9 21 13 10 7 18 10 15 10 12

14. **SOCIAL SCIENCE: Commuting Times** A random sample of twenty-four employees at the Grover–Smith Forge found the following morning commute times (in minutes) for last Monday:

    75 60 50 55 90 53 65 71
    33 81 60 74 71 75 83 53
    66 75 47 36 77 54 67 65

15. **SOCIAL SCIENCE: Unemployment** A random sample of nineteen newly unemployed workers in Wilmington last August reported the following times (in weeks) until they found new jobs:

    12 6 9 9 11 14 8 15 12 13
    9 8 9 7 9 5 6 13 15

16. **BIOMEDICAL: Cereal Killers** The calories per cup of several popular cold cereals, served with half a cup of milk and a peach or half a banana, are given below. Find the mode, the median, and the mean. Looking at the data, can you see why the mean is higher than the median?

Type	Calories
Banana Nut Crunch	360
Blueberry Morning	300
Cracklin Oat Bran	390
Fruit and Fibre	320
Great Grains	560
Mini Chex	360
Mother's Cream Oats	350
Müslix	420
Raspberry Granola	520
Raisin Bran Crunch	310
Raisin Nut Bran	390

17. **BIOMEDICAL: Chicken Sandwiches** *Consumer Reports* listed the calories for chicken sandwiches as follows. Find the mode, the median, and the mean. Looking at the data, can you see why the mean is higher than the median?

Type	Calories
Chick-fil-A	290
Wendy's Grilled Fillet	300
KFC Tender Roast	350
Arby's Grilled Deluxe	420
Wendy's Fillet (breaded)	430
McDonald's McGrill	450
KFC Original (breaded)	450
Burger King BK Broiler	500
McDonald's (breaded)	550
Burger King (breaded)	660
Boston Market	750

18. **SOCIAL SCIENCE: Raising IQs** When the mathematician Frederick Mosteller left Princeton for Harvard, he is reputed to have joked that he was "raising the average IQ of both places." How is this possible?

19. ATHLETICS: Basketball Salaries The salaries of the Los Angeles Lakers for the 2001–2002 season were as follows. Find the mode, the median, and the mean of the salaries. By looking at the data, can you see why the mean is so much higher than the median?

Player	Salary ($ million)
Kobe Bryant	11.3
Joe Crispin	0.3
Derek Fisher	3.0
Rick Fox	3.8
Devean George	0.8
Robert Horry	5.3
Lindsey Hunter	2.7
Mark Madsen	0.8
Jelani McCoy	0.5
Stanislav Medvedenkn	0.5
Shaquille O'Neal	21.4
Mitch Richmond	0.6
Brian Shaw	0.6
Samaki Walker	1.4

Source: HoopsWorld@Bskball.com

20. SOCIAL SCIENCE: New York Income New York State's mean income is the fourth highest in the nation, while its median income scores only 29th place, well below that of half of the states. What does this say about the incomes of New Yorkers?

21. PERSONAL FINANCE: Household Income The histogram in the Application Preview on page 429 shows the distribution of household incomes in the United States for the year 2001. The following is a random sample of these incomes in thousands of dollars: {38.2, 57.3, 102.1, 40.7, 47.9, 29.0, 38.1, 78.2, 33.6, 31.3, 112.8, 89.2, 67.9, 41.4, 39.6, 53.1, 113.3, 32.1, 39.7, 21.5}.

a. Find the mode, the median, and the mean.
b. Which is larger? Can you explain why in terms of the shape of the histogram on page 429?

Explorations and Excursions

The following problems extend and augment the material presented in the text.

Mean of Grouped Data

Sometimes the original data are not available and only the class intervals and frequencies are known. The mean of these grouped data may be estimated by assuming that each data value lies at the midpoint of its class. Therefore, we sum each midpoint value at the center of each class interval as many times as is indicated by the frequency and then divide that total by the sum of the frequencies.

22. For the grouped data

Class interval	14.5–19.5	19.5–24.5	24.5–29.5
Frequency	7	9	4

verify that the mean is

$$\bar{x} = \frac{17 \cdot 7 + 22 \cdot 9 + 27 \cdot 4}{7 + 9 + 4}$$

$$= \frac{119 + 198 + 108}{20} = \frac{425}{20} = 21\tfrac{1}{4}$$

23. For the grouped data

Class interval	6–12	13–19	20–26	27–33
Frequency	5	10	4	6

verify that the mean is

$$\bar{x} = \frac{9 \cdot 5 + 16 \cdot 10 + 23 \cdot 4 + 30 \cdot 6}{5 + 10 + 4 + 6}$$

$$= \frac{477}{25} = 19.08$$

Use the midpoints of the class intervals to estimate the mean of each set of grouped data.

24. BUSINESS: Truck Repairs The Wink-Quick package service maintains its own statewide fleet of delivery trucks. A random sample of the repair records found the following numbers of days last year that fifty trucks were not available for use because their repairs could not be completed overnight:

Number of days	0–2	3–5	6–8	9–11
Frequency	30	7	12	1

25. BUSINESS: Overtime Pay A random sample of payroll records for one hundred and ten craftspeople at the Old Fashion Vermont Country Furniture factory found the following amounts of overtime pay (in dollars) for the last week in November:

Overtime pay	0–19	20–39	40–59	60–79	80–99
Frequency	29	11	42	12	16

26. GENERAL: Charity Donations Donors at the Elktonburg Hospital Charity Ball are designated as "friends" for donations from $51 to $100, "benefactors" for $101 to $150, and "founders" for $151 to $200. A random sample of thirty hospital supporters found the following designations:

Designation	Friends	Benefactors	Founders
Frequency	9	6	15

6.3 MEASURES OF VARIATION

Introduction

It is helpful to summarize a given data set by a few numbers, as we did with random variables in the previous chapter. For a *representative* or *typical* value, we use one of the measures of central tendency (mode, median, or mean, whichever is most appropriate). But how do we measure whether the data are grouped tightly or spread widely about this central value? In this section we discuss three ways of describing the spread of the values. All three use differences among data values, so we must now restrict our discussion to data values at the interval or ratio level of measurement.

Range

The simplest measure of the spread of a data set is the *range*, which is the difference between the smallest and largest data values.

EXAMPLE 1

FINDING RANGES

Find the range of each data set.

a. {1, 5, 6, 7, 7, 7, 8, 15}
b. {1, 2, 10, 10, 12, 13, 15}

Solution

a. The largest value is 15 and the smallest is 1, so the range is $15 - 1 = 14$.

b. Again, the range is $15 - 1 = 14$. Notice that the range is *not* sensitive to the distribution of values between the two extremes.

Practice Problem 1 Find the range of the data values {15, 17, 27, 12, 30, 15, 10, 27, 25, 29}.

▶ Solution on page 459

Box-and-Whisker Plot

A *box-and-whisker plot* provides a quick way to visualize how data values are distributed between the largest and smallest by graphically displaying a *five-point summary* of the data. The *minimum* is the smallest value, the *maximum* is the largest, and we have already discussed the *median*. The *first quartile* is the median of the data values below the median (but not including the median), and the *third quartile* is the median of the data values above the median (but not including the median). The median may then be called the *second quartile*. The box extends from the first quartile to the third quartile with a vertical bar at the median. The whiskers extend from the box out to the minimum and maximum data values.

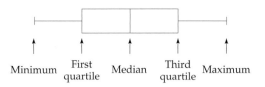

Notice that the distance between the extreme ends of the whiskers is the *range* of the data and that 50% of the data is contained in the box. The length of the box, from the first quartile to the third quartile, is sometimes called the *interquartile range*. In general, the quartiles will not be evenly spaced.

EXAMPLE 2

CONSTRUCTING A BOX-AND-WHISKER PLOT

Make a five-point summary and draw the box-and-whisker plot for the data values {20, 37, 65, 77, 78, 79, 81, 82, 83, 85, 87, 90}.

Solution

Since these values are already in order, the minimum is 20 and the maximum is 90. Since there are twelve values, the median is halfway between the sixth value (79) and the seventh (81), so the median is 80. The values below the median are the six values {20, 37, 65, 77, 78, 79}, and the first quartile is the median of these values (that is, the value halfway between 65 and 77), so the first quartile is 71. Similarly, the values above the median are the six values {81, 82, 83, 85, 87, 90}, and the third quartile is the median 84 because it is the value halfway between 83 and 85. The five-point summary of this data set and the corresponding box-and-whisker plot are shown below.

Minimum = 20
First quartile = 71
Median = 80
Third quartile = 84
Maximum = 90

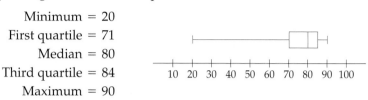

The box-and-whisker plot shows clearly that the bottom quarter of the values are spread rather thinly from 20 to 71, whereas the entire top half is concentrated between 80 and 90 (with a quarter of the values in the narrow range between 80 and 84). Such clustering is much easier to see from the box-and-whisker plot than from simply looking at the data.

Practice Problem 2

Make a five-point summary and draw the box-and-whisker plot for the data values {8, 10, 13, 14, 15, 16, 17, 19, 23, 24, 29}. From your plot, which quarter is the most tightly clustered? ▶ Solution on page 459

Graphing Calculator Exploration

Box-and-whisker plots can be drawn on graphing calculators by using the STAT PLOT command. To explore several box-and-whisker plots, including one of the data from Example 2 on the previous page, proceed as follows.

a. Turn off the axes in the window FORMAT menu and set the WINDOW parameters to Xmin = 0 and Xmax = 100.

b. Enter the data values:

{20, 37, 65, 77, 78, 79, 81, 82, 83, 85, 87, 90} as the list L_1,

{20, 31, 39, 41, 63, 80, 80, 80, 81, 87, 88, 90} as the list L_2, and

{20, 55, 58, 60, 60, 61, 63, 64, 65, 79, 81, 90} as the list L_3.

c. Turn on STAT PLOT 1 and select the box-and-whisker plot icon with Xlist L_1, and do the same for STAT PLOT 2 with Xlist L_2 and STAT PLOT 3 with Xlist L_3.

d. GRAPH and then TRACE to explore your box-and-whisker plots. The screen shows that the first quartile of the second list of values is 40. The first plot agrees with the one we drew in Example 2.

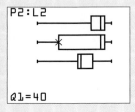

Interpreting Box-and-Whisker Plots

How do we interpret box-and-whisker plots? Consider an example.

DRAWING AND INTERPRETING BOX-AND-WHISKER PLOTS

The following five-point summaries are based on a survey of annual incomes (in thousands of dollars), for those with three different levels

of education: high school only, some college, and a bachelor's degree.*
Draw the box-and-whisker plot for each and interpret the results.

High school only: 9.6, 16.2, 23.0, 32.6, 50.1
Some college: 10.8, 19.1, 26.8, 37.5, 57.5
Bachelor's degree: 12.5, 24.9, 35.8, 51.1, 83.6

Solution

For each five-point summary (minimum, 1st quartile, median, 3rd quartile, and maximum) we plot the five points and draw the boxes and whiskers as follows.

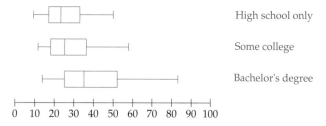

How do we interpret these plots? The first thing to look at is the median (the middle bar): Notice that it changes very little from "high school" to "some college," but then shifts significantly for "bachelor's degree," indicating that there is little income benefit from college unless you complete it. Then look at the boxes (the middle 50% of each group): As with the median, there is only a slight change from "high school" to "some college" and then a much larger change for "bachelor's degree." Notice that the minimum points (the left end of the left whisker) change very little (possibly because in any category some will leave employment), while the maximum increases much more dramatically (indicating a much greater earning potential for college graduates).

In each plot the right half of the box (the 25% above the mean) is wider than the left half (the 25% below the mean), and the right whisker is longer than the left whisker. This means that the distribution is more spread out to the right or *skewed to the right,* indicating that there are a few unusually high incomes. (If the data set were symmetric, the distances from the median to the quartiles and the extremes would be the same on either side.)

Clearly one can get much more information from comparing box-and-whisker plots than from just comparing medians.

* These and other data in this section are based on the *Current Population Survey* of the Bureau of Labor Statistics and the Bureau of the Census.

Sample Standard Deviation

A third way of measuring variation is the *sample standard deviation*, which estimates the typical variation of the data values from the (sample) mean of the data.

Sample Standard Deviation

The *sample standard deviation* of the n data values $x_1, x_2, \ldots, x_n$ is

$$s = \sqrt{\frac{(x_1 - \bar{x})^2 + \cdots + (x_n - \bar{x})^2}{n - 1}} \qquad \bar{x} = \text{mean}$$

The differences between the values and the mean are *squared*, $(x_i - \bar{x})^2$, so that positive differences don't cancel with negative ones. Since squaring gives an answer in *square units* (like "dollars squared"), we take the square root of the result to return to the original units. For n values we divide by $n - 1$. This is because the n numbers are related by the fact that their sum divided by n is the mean $\bar{x}$, so that any one value can be determined from the others together with the mean, leaving only $n - 1$ independent values in the sum. We call it the *sample* standard deviation to remind us of this.

EXAMPLE 4

CALCULATING A SAMPLE STANDARD DEVIATION

Find the sample standard deviation of the data values $\{9, 13, 16, 18\}$.

Solution

Since there are four values, $n = 4$. First we must find the mean:

$$\bar{x} = \tfrac{1}{4}(9 + 13 + 16 + 18) = \tfrac{1}{4}(56) = 14$$

Then

$$s = \sqrt{\frac{(9 - 14)^2 + (13 - 14)^2 + (16 - 14)^2 + (18 - 14)^2}{4 - 1}}$$

$$= \sqrt{\frac{25 + 1 + 4 + 16}{3}} = \sqrt{\frac{46}{3}} \approx 3.9 \qquad s = \sqrt{\frac{(x_1 - \bar{x})^2 + \cdots + (x_n - \bar{x})^2}{n - 1}}$$

The sample standard deviation s is (approximately) 3.9.

```
stdDev({9,13,16,
18})
           3.915780041
```

6.3 MEASURES OF VARIATION

Practice Problem 3 Find the sample standard deviation of the data values $\{6, 7, 12, 15\}$.

➤ Solution on page 459

Statistical Process Control: A Use of the Sample Variance

An important use of the sample standard deviation arises in industry. Any production process involves variation—not every item produced will be exactly the same. *Statistical process control* is a procedure for determining whether the variation is within reasonable limits. The method is based on the idea that if data follows the *normal distribution* (described in the next section), then 95% of the time the data will fall no more than two standard deviations away from the mean, and 99.7% of the time the data will fall no more than three standard deviations away from the mean. Generally, the actual mean and standard deviation are unknown, but in statistical process control they are replaced by *averages of sample means and of sample standard deviations*. The details differ from one implementation to another, but the following example is typical.

EXAMPLE 5 **STATISTICAL PROCESS CONTROL**

Ten samples of size 3 were taken from a manufacturing process, providing the data in the following table.

Sample number	1	2	3	4	5	6	7	8	9	10
Measurements	7	7	6	8	7	9	7	8	7	9
	8	6	7	7	7	6	7	7	6	7
	7	8	10	9	7	7	6	7	6	8

a. Make a control chart for the sample mean to test whether the process is under control using $k = 2$ standard deviations.
b. If a later sample is $\{8, 9, 10\}$, is the process under control?
c. If sometime later a sample is $\{9, 9, 10\}$, is the process still under control?

Solution

a. For each of the ten samples we find the mean and standard deviation, using the formulas on pages 444 and 456. The results are listed in the two new rows at the bottom of the table.

458 CHAPTER 6 STATISTICS

Sample number	1	2	3	4	5	6	7	8	9	10
Measurements	7	7	6	8	7	9	7	8	7	9
	8	6	7	7	7	6	7	7	6	7
	7	8	10	9	7	7	6	7	6	8
Sample means	7.3	7	7.7	8	7	7.3	6.7	7.3	6.3	8
Sample standard deviations	0.6	1	2.1	1	0	1.5	0.6	0.6	0.6	1

We then find the average of the ten sample means (called the *grand average*), written $\bar{\bar{x}}$, and the average of the ten sample standard deviations, written $\bar{s}$. The results are:

$$\bar{\bar{x}} = 7.26 \quad \text{and} \quad \bar{s} = 0.9 \qquad \text{For each, adding the 10 sample values and dividing by 10}$$

We make a *control chart*, with the *center line* at the grand average $\bar{\bar{x}}$, the upper control limit $k = 2$ standard deviations above the center line, and the lower control limit $k = 2$ standard deviations below the center line, as shown on the left.*

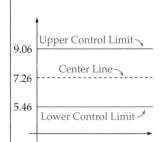

Upper control limit: $\quad 7.26 + 2 \cdot 0.9 = 9.06 \qquad \bar{\bar{x}} + k \cdot \bar{s}$
Lower control limit: $\quad 7.26 - 2 \cdot 0.9 = 5.46 \qquad \bar{\bar{x}} - k \cdot \bar{s}$

The process is considered to be *under control* if the sample mean lies between the control lines and *out of control* if not.

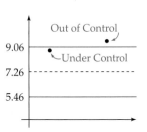

b. For the sample {8, 9, 10} the sample mean is $\bar{x} = \frac{8+9+10}{3} = 9$, which lies between the control lines, so the process is under control.

c. For the later sample {9, 9, 10} the sample mean is $\bar{x} = \frac{9+9+10}{3} = 9.3$, which is above the upper control limit, so the process is out of control. That is, after this sample the process should be stopped for repair.

One can also make analogous control tables for the sample standard deviation.

6.3 Section Summary

There are three ways to measure the spread of a data set. The simplest is the *range*, which is the largest data value minus the smallest. The

---

* The choice of k as 2 or 3 or some other value (often found from a *control chart table*) depends on the percentage of variation you choose to allow before declaring the process to be out of control. For further information, see *Introduction to Statistical Quality Control* by Douglas C. Montgomery (Wiley, 2001).

range gives no indication of the typical variation away from the mean, just the difference between the extremes.

A *box-and-whisker plot* graphically shows the range of each quarter of the data and can be drawn easily using a graphing calculator.

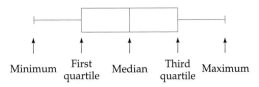

The *sample standard deviation* measures the typical spread of the data around the mean $\bar{x}$ as a single number and can be found easily using a calculator:

$$s = \sqrt{\frac{(x_1 - \bar{x})^2 + \cdots + (x_n - \bar{x})^2}{n - 1}} \qquad \bar{x} = \text{mean}$$

> **Solutions to Practice Problems**

1. The largest value is 30 and the smallest is 10, so the range is $30 - 10 = 20$.

2. The minimum is 8 and the maximum is 29.
   Since there are eleven values, the median is 16, the sixth value.
   From the first five values, {8, 10, 13, 14, 15}, the first quartile is 13, and from the last five values, {17, 19, 23, 24, 29}, the third quartile is 23.
   From this five-point summary, the box-and-whisker plot is

   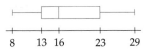

3. First, find the mean: $\bar{x} = \frac{1}{4}(6 + 7 + 12 + 15) = \frac{1}{4}(40) = 10$. Then

$$s = \sqrt{\frac{(6 - 10)^2 + (7 - 10)^2 + (12 - 10)^2 + (15 - 10)^2}{4 - 1}}$$

$$= \sqrt{\frac{16 + 9 + 4 + 25}{3}} = \sqrt{\frac{54}{3}} = \sqrt{18} \approx 4.2$$

6.3 Exercises (🧮 will be helpful throughout.)

Find the range of each data set.

1. {21, 44, 48, 52, 83}
2. {26, 35, 47, 59, 86, 92, 116}
3. {5, 25, 6, 9, 7, 7, 21, 19, 23, 16}
4. {2.0, 3.8, 5.2, 1.6, 4.3, 2.3, 5.4, 3.5, 2.5, 3.1}
5. {121, 115, 147, 163, 171, 116, 167, 147, 169, 131}

Make a five-point summary and draw the box-and-whisker plot for each data set.

6. {5, 8, 11, 13, 16, 20, 26}
7. {3, 10, 12, 14, 17, 21, 23}
8. {2, 3, 5, 9, 10, 12, 15, 17, 20}
9. {8, 9, 11, 12, 13, 14, 15, 19, 26}
10. {12, 8, 19, 17, 20, 8, 13, 14, 23, 5, 6, 9, 23, 12, 9}
11. {21, 13, 18, 14, 24, 8, 24, 24, 9, 12, 25, 20, 16, 17, 12}
12. {14, 17, 17, 13, 11, 6, 10, 21, 20, 18, 14, 11, 19, 14, 14}
13. {18, 7, 8, 18, 6, 15, 12, 15, 18, 24, 6, 18, 16, 17, 24}
14. {16, 9, 14, 12, 18, 15, 21, 19, 14, 18, 8, 4, 17, 15, 21, 14, 23, 13, 20, 17}
15. {19, 18, 7, 16, 5, 8, 9, 16, 18, 11, 20, 16, 17, 17, 8, 10, 15, 10, 8, 20}

Find the sample standard deviation of each data set.

16. {8, 13, 14, 17, 13}
17. {9, 2, 17, 13, 14}
18. {18, 11, 18, 1, 5, 19}
19. {22, 19, 11, 2, 20, 4}
20. {14, 11, 11, 9, 2, 13}

APPLIED EXERCISES

In Exercises 21–25, analyze the data by finding the range, the five-point summary, and the sample standard deviation and then drawing the box-and-whisker plot.

21. **BUSINESS: Stock Prices** Last Thursday was a very active trading day for shares of the new DiNextron technology stock. A random sample of twenty purchase transactions found the following share selling prices (in dollars):

    16 12 17 11 12 7 12 17 11 20
    10 11 14 17 16 14 25 9 17 5

22. **GENERAL: Kitchen Cabinets** A random sample of fifteen electric bills at Rick's Always Cooking cabinet factory found the following monthly electricity consumptions (in thousands of kilowatt hours):

    10 12 6 15 13
    15 12 9 14 6
    5 8 9 12 9

23. **BUSINESS: Airline Cancellations** A random sample of thirty airplane departures from Dallas International found the following numbers of "no shows" at the boarding gates:

    2 6 8 4 10 5 10 8 7 5
    10 10 9 6 11 9 6 6 10 8
    11 6 7 13 12 8 9 5 9 10

24. **SOCIAL SCIENCE: Malpractice Settlements** A random sample of twenty-four malpractice claims against a midwestern health maintenance organization (HMO) that were settled out of court found the following settlement amounts (in tens of thousands of dollars):

    16 22 16 18 29 22 30 18
    19 28 18 37 17 32 24 31
    19 16 16 6 14 15 8 18

25. **SOCIAL SCIENCE: Trial Schedules** A random sample from the clerk's files at the Marksburg District Court found the following jail waiting times (in weeks) between arraignment and trial (or plea bargain) for twenty individuals charged with felonies:

    11 35 15 33 20 21 2 21 23 24
    18 16 9 25 12 20 20 25 21 33

26. **SOCIAL SCIENCE: Income and Education** The following five-point summary is based on a survey of annual incomes (in thousands of dollars) for those whose education includes an advanced degree (beyond the bachelor's). Draw the box-and-whisker plot and interpret the results in comparison with the plots in Example 3 for people without an advanced degree.

    17.4, 30.2, 40.7, 54.4, 89.3

27. **SOCIAL SCIENCE: Income and Gender** The following five-point summaries are based on a survey of annual incomes (in thousands of dollars). The first summary is for men and

the second is for women. Draw the box-and-whisker plots for each. Interpret the results.

Men's incomes: 9.6, 18.4, 28.8, 40.6, 74.1
Women's income: 8.4, 15.0, 21.5, 31.7, 53.5

28. **ATHLETICS: Home Run Records** In 1998, a baseball record that had stood for 37 years, Roger Maris's record of 61 home runs in a season, was broken by Sammy Sosa (with 66) and by Mark McGwire (with 70). To compare their records, make a five-point summary and draw the box-and-whisker plot for the following lists of home runs per season for each player during their major league careers. Interpret the results.

Sammy Sosa: {4, 15, 10, 8, 33, 25, 36, 40, 36, 66, 63, 50, 64, 49}

Mark McGwire: {3, 49, 32, 33, 39, 22, 42, 9, 9, 39, 58, 70, 65, 32, 29}

Roger Maris: {14, 9, 19, 16, 39, 61, 33, 23, 26, 8, 13, 9, 5}

29. **GENERAL: Fuel Efficiency** The following tables list the fuel economy in miles per gallon for model year 2003 family sedans and for sport utility vehicles (SUVs). For each, make a five-point summary and draw the box-and-whisker plot. Give an interpretation of the results.

Family Sedans	mpg
Volkswagen Passat	23
Toyota Camry	22
Honda Accord	24
Nissan Maxima	21
Mazda 6	22
Volvo S40	22
Subaru Legacy	21
Mitsubishi Galant	22
Chevrolet Malibu	22
Ford Taurus	21
Mercury Sable	21
Hyundai XG350	19
Dodge Intrepid	19
Chrysler Sebring	21
Pontiac Grand Am	23
Saturn L200	23

SUVs	mpg
Audi Allroad	18
BMW X5	15
Honda Pilot	19
Accura MDX	18
Lexus RX300	18
Subaru Outback	21
Volvo XC70	18
Toyota Land Cruiser	14
Mercedes-Benz ML500	15
Chevrolet Suburban	13
GMC Yukon XL	13
Nissan Pathfinder	16
Buick Rendezvous	16
Dodge Durango	13
Jeep Grand Cherokee	16
Ford Excursion	10

Source: Consumer Reports

30. **MANAGEMENT SCIENCE: Process Control** Suppose that ten samples of size 3 taken from a manufacturing process provided the following data.

Sample	1	2	3	4	5	6	7	8	9	10
Measurements	4 1 4	4 3 5	3 4 7	5 4 6	4 5 4	6 5 4	4 4 3	5 4 4	4 5 3	6 4 6

a. Find the upper and lower control limits for the sample mean (use $k = 2$).
b. If a later sample were {5, 6, 7}, would the process be under control?
c. If a later sample were {7, 7, 6}, would the process be under control?

31. **MANAGEMENT SCIENCE: Process Control** Suppose that ten samples of size 3 taken from a manufacturing process provided the following data.

Sample	1	2	3	4	5	6	7	8	9	10
Measurements	3 4 3	3 2 4	2 3 5	4 3 5	3 4 3	5 4 3	3 3 2	4 3 3	3 2 2	5 3 4

a. Find the upper and lower control limits for the sample mean (use $k = 2$).
b. If a later sample were {4, 4, 6}, would the process be under control?
c. If a later sample were {5, 6, 5}, would the process be under control?

Explorations and Excursions

The following problems extend and augment the material presented in the text.

Calculating Standard Deviations

Although the formula on page 456 defines the sample standard deviation, there is a simpler way to calculate its value that is used in most computer programs and calculators.

32. Verify each step in the following transformation of the definition formula for the sample standard deviation

$$s = \sqrt{\frac{1}{n-1}\Sigma(x_k - \bar{x})^2}$$

The symbol Σ means sum, and the sum inside the square root can be written as follows:

$\Sigma (x_k - \bar{x})^2$
$= \Sigma (x_k^2 - 2x_k\bar{x} + \bar{x}^2)$
$= \left(\Sigma x_k^2\right) - 2\bar{x} \cdot \left(\Sigma x_k\right) + \bar{x}^2 \cdot \left(\Sigma 1\right)$
$= \left(\Sigma x_k^2\right) - 2 \cdot \left(\frac{1}{n} \cdot \Sigma x_k\right) \cdot \left(\Sigma x_k\right) + \left(\frac{1}{n} \cdot \Sigma x_k\right)^2 \cdot n$
$= \left(\Sigma x_k^2\right) - \frac{1}{n} \cdot \left(\Sigma x_k\right)^2$

Then

$$s = \sqrt{\frac{\left(\Sigma x_k^2\right) - \frac{1}{n}\left(\Sigma x_k\right)^2}{n - 1}}$$

That is, s can be found from the sum of the data values and the sum of their squares without calculating the mean $\bar{x}$.

33. Verify the following calculation of the sample standard deviation of the data {19, 18, 27, 5, 11}.

$\Sigma x_k = 19 + 18 + 27 + 5 + 11 = 80$

$\Sigma x_k^2 = 19^2 + 18^2 + 27^2 + 5^2 + 11^2 = 1560$

so

$s = \sqrt{\dfrac{1560 - \frac{1}{5}(80)^2}{5 - 1}} = \sqrt{70} \approx 8.37$

34. Verify the following calculation of the sample standard deviation of the data {11, 18, 29, 14, 24, 6}.

$\Sigma x_k = 102$ and $\Sigma x_k^2 = 2094$

so

$s = \sqrt{72} \approx 8.49$

35. Use the final formula from Exercise 32 to find the sample standard deviation of the data {9, 2, 17, 13, 14} and then compare your value with your answer to Exercise 17.

36. Use the final formula from Exercise 32 to find the sample standard deviation of the data {22, 19, 11, 2, 20, 4} and then compare your value with your answer to Exercise 19.

6.4 NORMAL DISTRIBUTIONS AND BINOMIAL APPROXIMATION

Introduction

The normal distribution, with its famous bell-shaped curve, is perhaps the most important of all probability distributions. The *central limit theorems* proved by Carl F. Gauss (1777–1855) and other mathematicians during the nineteenth century showed that many quantities, such as errors in measurement or means of random samples, follow the normal distribution. Today it is used to predict everything from sizes of newborn babies to stock market fluctuations to retirement benefits. After we examine some of its most important features, we will see how it is used in applications and examine its relation to the binomial distribution.

Discrete and Continuous Random Variables

The random variables we considered in Chapter 5 take values like 0, 1, 2, ... that are *separated* or *discrete*, and so are called *discrete random variables*. There are other random variables whose values can be *any number in an interval*, and are called *continuous random variables*.

Examples of continuous random variables are your (exact) weight at birth and the time until a light bulb burns out, since these will most likely not be whole numbers but may be any number (in some reasonable interval). For discrete random variables, we found probabilities of events by adding probabilities of outcomes, which can be interpreted geometrically as adding areas of bars of probability distributions that change in jumps (see page 410). Analogously, for continuous random variables, we will find probabilities by calculating areas under curves.

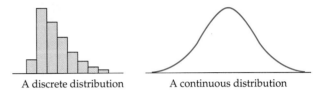

A discrete distribution	A continuous distribution

The most famous of all continuous distributions is the *normal distribution*, which we now discuss.

Normal Distribution

The normal distribution depends on two constants or *parameters*, its *mean* μ, which may be any number, its *standard deviation* σ (pronounced "sigma," the Greek letter "s"), which must be positive. The parameter σ measures the *spread* of the distribution away from its mean, and it is analogous to the sample standard deviation of the previous section. The square of the standard deviation is called the *variance*. The normal distribution is given by the following formula, although we will not use it explicitly in what follows.

Normal Probability Distribution

The normal probability distribution with mean μ and standard deviation σ is given by the curve

$$f(x) = \frac{1}{\sigma\sqrt{2\pi}} e^{-\frac{1}{2}\left(\frac{x-\mu}{\sigma}\right)^2} \qquad \text{for } -\infty < x < \infty$$

A random variable with this distribution is called a *normal random variable* with mean μ and standard deviation σ.

The curve has a central peak at μ and then falls back symmetrically on either side to approach the *x*-axis. Different values of μ change the location of the peak, as is shown below by three normal distributions with different means.

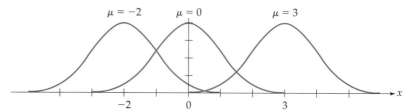
Three normal distributions with different means

The standard deviation σ measures how the values spread away from the mean. The distribution will be higher and more peaked if σ is small, and lower and more rounded if σ is large.

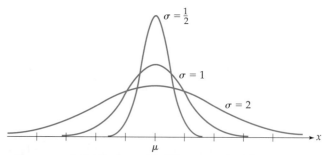
Three normal distributions with different standard deviations

As the distribution becomes "shorter" it also becomes "wider," and the area under the curve always stays at 1, although we will not prove this fact. The peaked shape of the normal distribution concentrates most of the probability (or area) near the center at μ. About 68% of the area under the normal curve is within one standard deviation of the mean, about 95% is within two standard deviations, and more than 99% is within three standard deviations. Notice also that the curve rises or falls most steeply at one standard deviation from the mean.

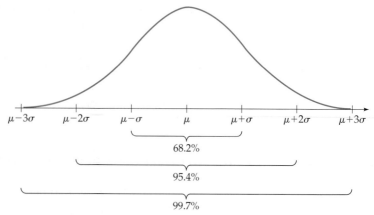
Area (probability) under the curve within 1, 2, and 3 standard deviations of the mean

6.4 NORMAL DISTRIBUTIONS AND BINOMIAL APPROXIMATION

The probability that a normal random variable has its value between any two given x-values corresponds to the *area* under the normal curve between those x-values. The probability is most easily found using a graphing calculator, but it can also be found using tables of values of the normal distribution.

> Readers who have graphing calculators that find normal probabilities should continue reading this section. Readers who do not have such calculators should now turn to page A-1 to read the appendix near the end of this book.

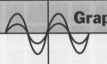 Graphing Calculator Exploration

Areas under normal distribution curves can be found easily on a graphing calculator. To find the area under the normal distribution curve with mean $\mu = 20$ and standard deviation $\sigma = 5$ from 10 (which is $\mu - 2\sigma$) to 30 (which is $\mu + 2\sigma$), proceed as follows:

```
DISTR DRAW
1:normalpdf(
2:normalcdf(
3:invNorm(
4:tpdf(
5:tcdf(
6:X²pdf(
7↓X²cdf(
```

a. From the DISTRIBUTION menu, select the normalcdf command to find the cumulative values of the normal distribution (that is, the area under the curve) between two given values.

```
normalcdf(10,30,
20,5)
          .954499876
```

b. Enter the (left) starting value and the (right) ending value as well as the mean and standard deviation. This area is the same as the probability that the normal random variable is between the two given values 10 and 30.

c. To *see* this area shaded under the normal curve, use the Shade-Norm command from the DISTRIBUTION DRAW menu with an appropriate window.

```
WINDOW
Xmin=0
Xmax=40
Xscl=5
Ymin=-.05
Ymax=.1
Yscl=1
Xres=1
```

```
DISTR DRAW
1:ShadeNorm(
2:Shade_t(
3:ShadeX²(
4:ShadeF(
```

```
ShadeNorm(10,30,
20,5)
```

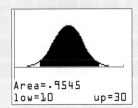
Area=.9545
low=10   up=30

EXAMPLE 1 FINDING A PROBABILITY FOR A NORMAL RANDOM VARIABLE

The heights of American men are approximately normally distributed with mean $\mu = 68.1$ inches and standard deviation $\sigma = 2.7$ inches.* Find the proportion of men who are between 5 feet 9 inches and 6 feet tall.

```
normalcdf(69,72,
68.1,2.7)
         .2951343633
```

Solution

Converting to inches, we want the probability of a man's height being between 69 inches and 72 inches. From the graphing calculator screen on the left, about 30% of American men are between 5 feet 9 inches and 6 feet tall.

Practice Problem 1 The weights of American women are approximately normally distributed with mean 134.7 pounds and standard deviation 30.4 pounds. Find the proportion of women who weigh between 130 and 150 pounds. ➤ Solution on page 472

z-Scores

We have seen that graphing calculators quickly calculate normal probabilities for *any* mean and standard deviation. The normal distribution with *mean 0 and standard deviation* 1 has special significance and is called the *standard* normal distribution (the word *standard* indicating mean 0 and standard deviation 1). The letter z is traditionally used for the variable in standard normal calculations. Since the mean is zero, the highest point on the curve is at $z = 0$, and the curve is symmetric about this value.

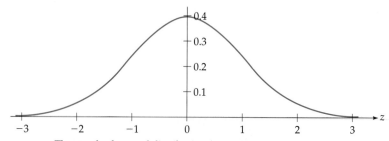

The standard normal distribution (mean 0 and standard deviation 1)

---

* These and other data in this section are from *Handbook of Human Factors and Ergonomics* by Gavriel Salvendy (Wiley-Interscience, 1997).

6.4 NORMAL DISTRIBUTIONS AND BINOMIAL APPROXIMATION

From the diagram on page 464 with $\mu = 0$ and $\sigma = 1$ we have the following facts about the standard normal distribution: More than

68% of its probability is between -1 and $+1$

95% of its probability is between -2 and $+2$

99% of its probability is between -3 and $+3$

To change any x-value for a normal distribution with mean μ and standard deviation σ into the corresponding z-value for the *standard normal distribution*, we subtract the mean and then divide by the standard deviation. The resulting number is called the *z-score* and this process is called *standardizing*.

z-Score

$$z = \frac{x - \mu}{\sigma}$$

Subtracting the mean μ and dividing by the standard deviation σ

The z-score gives the number of standard deviations the value is from the mean.

EXAMPLE 2

FINDING z-SCORES FOR HEIGHTS OF MEN

Using the mean and standard deviation for men's heights given in Example 1, find the z-scores corresponding to the following heights (in inches) and interpret the results:

a. 73.5   b. 62.7.

Solution

a. $z = \dfrac{73.5 - 68.1}{2.7} = \dfrac{5.4}{2.7} = 2$

Using $z = (x - \mu)/\sigma$ with $\mu = 68.1$ and $\sigma = 2.7$ from Example 1

Therefore, a height of 73.5 inches is two standard deviations *above* the mean.

b. $z = \dfrac{62.7 - 68.1}{2.7} = \dfrac{-5.4}{2.7} = -2$

Therefore, a height of 62.7 inches is two standard deviations *below* the mean.

Since the standard normal distribution has more than 95% of its probability between −2 and +2, these results mean that more than 95% of men have heights between 62.7 inches and 73.5 inches (5 feet 2.7 inches and 6 feet 1.5 inches).

Practice Problem 2 Using the mean and standard deviation for women's weights given in Practice Problem 1 on the previous page, find the z-scores for the following weights, and interpret your answers:

a. 165.1 lbs.   **b.** 104.3 lbs.   ▶ Solutions on page 472

The Normal and Binomial Distributions

As we saw on page 414, the binomial distribution has a kind of "bell" shape, with a peak at the expected value $\mu = np$ and falling on both sides to very small probabilities further away from the mean.

$n = 20, p = 0.4$ $n = 15, p = 0.5$ $n = 25, p = 0.7$

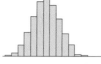 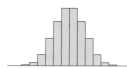 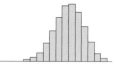

Several binomial distributions with different values for n and p

Since both the binomial and the normal distributions have similar "mound" shapes, could they be related? In the eighteenth century, Abraham de Moivre (1667–1754) and Pierre-Simon Laplace (1749–1827) discovered and proved that for any choice of p between 0 and 1, the binomial distribution *approaches* the normal distribution as n becomes large. This fundamental fact is known as the de Moivre–Laplace theorem.

Graphing Calculator Exploration

The program* SEENORML draws binomial distributions for $p = \frac{1}{2}$ with n from 2 to 30. Running this program shows how the "boxlike" binomial distributions change into "bell-shaped" curves as n

---

*See the Preface for information on how to obtain this and other graphing calculator programs.

increases. The window for each graph is different so that each fills as much as possible of the viewing area.

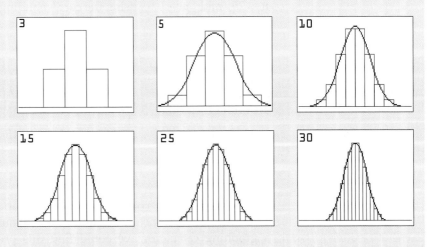

We may use the de Moivre–Laplace theorem to approximate binomial distributions by the normal distribution.

Normal Approximation to the Binomial

Let X be a binomial random variable with parameters n and p. If $np > 5$ and $n(1 - p) > 5$, then the distribution of X is approximately normal with mean $\mu = np$ and standard deviation $\sigma = \sqrt{np(1 - p)}$.

Approximating the *discrete* binomial distribution by the *continuous* normal distribution requires an adjustment: To have the same width for a "slice" of the normal distribution as for a slice of the binomial, we adopt the convention that the binomial probability $P(X = x)$ corresponds to the area under the normal distribution curve from $x - \frac{1}{2}$ to $x + \frac{1}{2}$. This is called the *continuous correction*.

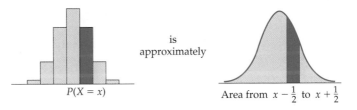

EXAMPLE 3

NORMAL APPROXIMATION OF A BINOMIAL PROBABILITY

Estimate $P(X = 12)$ for the binomial random variable X with $n = 25$ and $p = 0.6$ using the corresponding normal distribution.

Solution

Since

$$np = 25 \cdot 0.6 = 15 \quad \text{and} \quad n(1 - p) = 25 \cdot 0.4 = 10$$

are both greater than 5, we may use the normal distribution with

$$\mu = np = 25 \cdot 0.6 = 15$$

and

$$\sigma = \sqrt{np(1-p)} = \sqrt{25 \cdot 0.6 \cdot 0.4} = \sqrt{6}$$

By the continuous correction, we interpret the event $X = 12$ as $11\frac{1}{2} \le X < 12\frac{1}{2}$ to include numbers that would round to 12. Using a graphing calculator, we could find this probability as we did in Example 1 (page 466), but the following calculator commands also show the picture.

```
WINDOW
 Xmin=0
 Xmax=25
 Xscl=1
 Ymin=-.1
 Ymax=.25
 Yscl=1
 Xres=1
```

```
ShadeNorm(11.5,1
2.5,25*.6,√(25*.
6*.4))
```

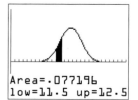

The required probability is approximately 0.077.

The normal approximation of 0.077 is indeed very close to the actual value calculated from the binomial probability formula (see page 414): $_{25}C_{12}(0.6)^{12}(1 - 0.6)^{13} \approx 0.076$.

EXAMPLE 4

MANAGEMENT MBA's

At a major Los Angeles accounting firm, 73% of the managers have MBA degrees. In a random sample of 40 of these managers, what is the probability that between 27 and 32 will have MBA's?

Solution

Because each manager either has or does not have an MBA, presumably independently of each other, the question asks for the probability that a binomial random variable X with $n = 40$ and $p = 0.73$ satisfies $27 \leq X \leq 32$. Since $np = 40 \cdot 0.73 = 29.2$ and $n(1 - p) = 40 \cdot 0.27 = 10.8$ are both greater than 5, this probability can be approximated as the area under the normal distribution curve with mean $\mu = np = 40 \cdot 0.73 = 29.2$ and standard deviation $\sigma = \sqrt{np(1 - p)} = \sqrt{40 \cdot 0.73 \cdot 0.27}$ from $26\frac{1}{2}$ to $32\frac{1}{2}$ (again using the continuous correction to include values that would round to between 27 and 32):

```
WINDOW
Xmin=0
Xmax=40
Xscl=5
Ymin=-.05
Ymax=.15
Yscl=1
Xres=1
```

```
ShadeNorm(26.5,3
2.5,40*.73,√(40*
.73*.27))
```

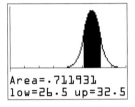

The probability is (about) 0.712, or about 71%.

The exact answer to the above problem, from summing the binomial distribution from 27 to 32, is 71.5%, so the normal approximation is very accurate.

Practice Problem 3 A brand of imported VCR is known to have defective tape rewind mechanisms in 8% of the units imported last April. If Jerry's Discount Electronics received a shipment of 80 of these VCR's, what is the probability that 10 or more are defective? ▶ Solution on next page

6.4 Section Summary

The graph of the normal distribution with mean μ and standard deviation σ is a bell-shaped curve that peaks at μ in the center and then falls back symmetrically on either side to approach the x-axis. The area under this curve between any two x-values is the probability that the normal random variable lies between these values. These areas and probabilities can be found easily with a graphing calculator or tables.

Any value x of a normal random variable with mean μ and standard deviation σ can be converted into a z-score using the formula

$$z = \frac{x - \mu}{\sigma}$$

This z-score indicates the number of standard deviations the x-value is away from the mean.

The normal distribution with mean $\mu = np$ and standard deviation $\sigma = \sqrt{np(1-p)}$ is a good approximation to the binomial distribution provided that both np and $n(1-p)$ are greater than 5.

> **Solutions to Practice Problems (for those using graphing calculators)**

```
normalcdf(130,15
0,134.7,30.4)
        .2540533807
```

1. About 25% of American women weigh between 130 and 150 pounds.

2. Using $\mu = 134.7$ and $\sigma = 30.4$ from Practice Problem 1:

   a. $z = \dfrac{165.1 - 134.7}{30.4} = \dfrac{30.4}{30.4} = 1$

   b. $z = \dfrac{104.3 - 134.7}{30.4} = \dfrac{-30.4}{30.4} = -1$

   Interpretation: More than 68% of women have weights between about 104 and 165 pounds.

```
normalcdf(9.5,80
.5,80*.08,√(80*.
08*.92))
        .1007041942
```

3. Since $np = 80 \cdot 0.08 = 6.4$ and $n(1-p) = 80 \cdot 0.92 = 73.6$ are both greater than 5, we may use the normal distribution with $\mu = np = 80 \cdot 0.08$ and $\sigma = \sqrt{np(1-p)} = \sqrt{80 \cdot 0.08 \cdot 0.92}$ to approximate $P(X \geq 10)$ as the area under the normal curve from 9.5 to 80.5 (corresponding to all 80 being defective). The probability of at least ten defective VCR's is (about) 0.10, or 10%.

6.4 Exercises

Let X be a normal random variable with mean $\mu = 12$ and standard deviation $\sigma = 3$. Find each probability as an area under the normal curve.

1. $P(9 \leq X \leq 15)$
2. $P(6 \leq X \leq 18)$
3. $P(3 \leq X \leq 21)$
4. $P(0 \leq X \leq 24)$
5. $P(12 \leq X \leq 15)$
6. $P(9 \leq X \leq 12)$
7. $P(15 \leq X \leq 16)$
8. $P(7 \leq X \leq 10)$
9. $P(6\tfrac{1}{2} \leq X \leq 11)$
10. $P(10\tfrac{1}{2} \leq X \leq 14)$

Find the z-score corresponding to each x-value.

11. $x = 6$ with $\mu = 4$ and $\sigma = 2$
12. $x = 20$ with $\mu = 15$ and $\sigma = 5$
13. $x = 10$ with $\mu = 10$ and $\sigma = 3$
14. $x = 6$ with $\mu = 6$ and $\sigma = 5$
15. $x = 15$ with $\mu = 20$ and $\sigma = 5$
16. $x = 6$ with $\mu = 8$ and $\sigma = 2$
17. $x = 15$ with $\mu = 20$ and $\sigma = 1$
18. $x = 6$ with $\mu = 8$ and $\sigma = 1$
19. $x = 130$ with $\mu = 100$ and $\sigma = 15$
20. $x = 90$ with $\mu = 100$ and $\sigma = 10$

Use the normal approximation to the binomial random variable X with $n = 20$ and $p = 0.7$ to find each probability.

21. $P(X = 10)$ **22.** $P(X = 12)$
23. $P(11 \leq X \leq 12)$ **24.** $P(8 \leq X \leq 10)$
25. $P(3 \leq X \leq 8)$ **26.** $P(13 \leq X \leq 19)$
27. $P(15 \leq X \leq 20)$ **28.** $P(2 \leq X \leq 7)$
29. $P(0 \leq X \leq 20)$ **30.** $P(14 \leq X \leq 20)$

APPLIED EXERCISES

Answer each question using an appropriate normal probability.

31. BIOMEDICAL: Weights Weights of men are approximately normally distributed with mean 163 pounds and standard deviation 28 pounds. If the minimum and maximum weights to be a volunteer fireman in Martinsville are 128 and 254, what proportion of men meet this qualification?

32. BUSINESS: Advertising The percentage of early afternoon television viewers who watch episodes of "As the World Spins" is a normal random variable with mean 22% and standard deviation 3%. The producers guarantee advertisers that the viewer percentage will be better than 20% or else the advertisers will receive free air time. What is the probability that the producers will have to give free air time after any particular episode?

33. BUSINESS: Airline Overbooking To avoid empty seats, airlines generally sell more tickets than there are seats. Suppose that the number of ticketed passengers who show up for a flight with 100 seats is a normal random variable with mean 97 and standard deviation 6. What is the probability that at least one person will have to be "bumped" from the flight?

34. GENERAL: SAT Scores The scores at Centerville High School on last year's mathematics SAT test were approximately normally distributed with mean 490 and standard deviation 140. What proportion of the scores were between 550 and 750?

35. BIOMEDICAL: Smoking In a large-scale study carried out in the 1960s of male smokers 35–45 years old, the number of cigarettes smoked daily was approximately normally distributed with mean 28 and standard deviation

10. What proportion of the smokers smoked between 30 and 40 (two packs!) each day?

36. GENERAL: Coin Tosses A fair coin is tossed fifty times. What is the probability that it comes up heads at least thirty times?

37. GENERAL: Actuarial Exams The probability of passing the first actuarial examination is 60%. If 940 college students take the exam in March, what is the probability that the number that pass is between 575 and 725?

38. BIOMEDICAL: Flu Shots Although getting a flu shot in October is a good way of planning ahead for the winter flu season, approximately 1% of those getting the shots develop side effects. If 800 students at Micheles College get flu shots, what is the probability that between 6 and 12 of them will develop side effects?

39. BUSINESS: Real Estate Nationwide, 40% of the sales force at the Greener Pastures real estate franchises are college graduates. If 200 of these sales personnel are chosen at random, what is the probability that between 85 and 100 of them are college graduates?

40. SOCIAL SCIENCE: Old Age If 3% of the population lives to age 90, what is the probability that between 10 and 15 of the 280 business majors at Henderson College will live to be that old?

Explorations and Excursions

The following exercises extend and augment the material presented in the text.

Chebychev's Theorem

(*Requires Section 5.6*) On page 411 we defined the *mean* or expected value of a random variable X as

$$\mu = E(X) = x_1 \cdot P(X = x_1) + \cdots + x_n \cdot P(X = x_n)$$

The standard deviation of a random variable X is defined as

$$\sigma = \sqrt{E(X-\mu)^2}$$
$$= \sqrt{(x_1-\mu)^2 \cdot P(X=x_1) + \cdots + (x_n-\mu)^2 \cdot P(X=x_n)}$$

The standard deviation measures the *spread* of the random variable away from its mean just as the sample standard deviation measures the spread of a data set away from its sample mean. The relationship between the standard deviation and the values of a random variable is given by *Chebychev's Theorem*:

> The probability that the values of a random variable lie within k standard deviations from its mean is at least $1 - \frac{1}{k^2}$.

For example, taking $k = 2$, Chebychev's Theorem says that the probability that the values of a random variable lie within 2 standard deviations from its mean is at least $1 - \frac{1}{2^2} = 1 - \frac{1}{4} = \frac{3}{4}$.

Use Chebychev's Theorem to find the probability that the values of a random variable lie within:

41. 3 standard deviations from its mean.

42. 4 standard deviations from its mean.

Suppose that a random variable has mean $\mu = 100$ and standard deviation $\sigma = 5$. Then, again taking $k = 2$, Chebychev's Theorem says that the probability that the values of the random variable lie between $\mu \pm 2\sigma = 100 \pm 2 \cdot 5$, that is, between 90 and 110, is at least $1 - \frac{1}{2^2} = \frac{3}{4}$.

43. State the result of Chebychev's Theorem with $k = 3$ for a random variable with $\mu = 100$ and $\sigma = 5$.

44. State the result of Chebychev's Theorem with $k = 4$ for a random variable with $\mu = 100$ and $\sigma = 5$.

While Chebychev's Theorem applies to all random variables, stronger results are available for *particular* random variables, such as normal random variables. For example, the probability that the values of a normal random variable with $\mu = 100$ and $\sigma = 5$ lie between 90 and 110 is greater than 95%, a stronger result than the $\frac{3}{4} = 75\%$ from Chebychev's Theorem.

45. For a normal random variable with $\mu = 100$ and $\sigma = 5$, find the probability that its values lies between 85 and 115. Compare the answer with that of Exercise 43, which gives the probability for any random variable.

46. For a normal random variable with $\mu = 100$ and $\sigma = 5$, find the probability that its values lie between 80 and 120. Compare the answer with that of Exercise 44, which gives the probability for any random variable.

Chapter Summary with Hints and Suggestions

Reading the text and doing the exercises in this chapter have helped you to master the following skills, which are listed by section (in case you need to review them) and are keyed to particular Review Exercises. Answers for all Review Exercises are given at the back of the book, and full solutions can be found in the Student Solutions Manual.

6.1 Random Samples and Data Organization

- Identify levels of data measurement (nominal, ordinal, interval, and ratio). *(Review Exercises 1–4.)*

- Construct a bar chart or a histogram for a given data set. *(Review Exercises 5–8.)*

6.2 Measures of Central Tendency

- Find the mode, median, and mean of a data set. *(Review Exercises 9–13.)*

  Mode: most frequent

  Median: middle

  Mean: $\bar{x} = \dfrac{x_1 + \cdots + x_n}{n}$

6.3 Measures of Variation

- Find the range of a data set.
  (Review Exercise 14.)

- Find the five-point summary of a data set and draw the box-and-whisker plot.
  (Review Exercise 15.)

- Find the sample standard deviation of a data set. (Review Exercise 16.)

$$s = \sqrt{\frac{(x_1 - \bar{x})^2 + \cdots + (x_n - \bar{x})^2}{n - 1}}$$

- Analyze the data in an application by finding the range, the five-point summary, and the sample standard deviation and then drawing the box-and-whisker plot.
  (Review Exercises 17–18.)

6.4 Normal Distributions and Binomial Approximation

- Find the probability that the normal random variable with mean μ and standard deviation σ is between two given values.
  (Review Exercises 19–20.)

- Find the z-score of an x-value of the normal random variable with mean μ and standard deviation σ. (Review Exercises 21–22.)

$$z = \frac{x - \mu}{\sigma}$$

- Use the normal approximation to the binomial random variable X to find the probability that X takes given values.
  (Review Exercises 23–24.)

- Solve an applied problem involving probabilities using the normal distribution.
  (Review Exercises 25–28.)

Hints and Suggestions

- (Overview) Properties of "many" can be inferred from just "some" only if the "some" are a random sample from the population. There are four different levels of data measurement: nominal, ordinal, interval, and ratio.

Data may be organized into bar charts or histograms. Measures of central tendency (mode, median, and mean) summarize all the data as one typical value, while measures of variation (range, box-and-whisker plot, and sample standard deviation) show how closely the values cluster about the center. The bell-shaped curve of the normal distribution applies to many situations, and z-scores give the number of standard deviations the values are from the mean. The binomial distribution is approximated by the normal distribution, provided both np and $n(1 - p)$ are sufficiently large.

- A bar chart may have its bars separated or touching. Histograms are for interval or ratio data and the bars touch each other on both sides.

- To find the mode or the median of data values, it is helpful to begin by sorting them in order.

- For measures of central tendency, with nominal ("in name only") data, only the mode is appropriate; with ordinal data, both the mode and the median are appropriate; with interval or ratio data, all three (mode, median, and mean) are appropriate.

- All three measures of variation are appropriate only for interval or ratio data.

- The mean is sensitive to changes in a few extreme values, while the mode and median are not.

- Box-and-whisker plots graphically show how the data are distributed among the four quartiles.

- Areas under the normal distribution curve are probabilities for the normal random variable.

- When using the normal distribution to approximate the binomial distribution, be sure to increase the largest value by $\frac{1}{2}$ and to decrease the smallest by $\frac{1}{2}$ (the continuous correction) to cover all the area represented by the boxes making up the binomial distribution.

- **Practice for Test:** Review Exercises 1, 3, 5, 7, 9, 17, 19, 21, 24, 26.

Review Exercises for Chapter 6

Practice test exercise numbers are in green.

(will be helpful throughout.)

6.1 Random Samples and Data Organization

Identify the level of data measurement in each situation.

1. **GENERAL: Chicago Commuters** A random sample of seventy Chicago area commuters counts the number traveling to work last Thursday by car, taxi, bus, and train.

2. **GENERAL: Gourmet Coffee** At the House of Java Coffee Emporium, a random sample of eighteen customers rates the new Turkish Sultan flavor on a scale from 1 ("I'll try something else") to 5 ("Love at first sip").

3. **GENERAL: Fourth of July** A random sample of five hundred suburban households records the backyard temperatures at 3 P.M. on Independence Day.

4. **ENVIRONMENTAL SCIENCE: Drinking Water** A random sample of eighty small municipalities (less than 50,000 residents) lists the lead content (in parts per billion) of the tap water.

Construct a bar chart of the data in each situation.

5. **GENERAL: Veterans of Foreign Wars** A random sample of forty members of the Hattiesburg V.F.W. identifies their branches of military service as 14 in the Army, 12 in the Navy, 6 in the Marines, and 8 in the Air Force.

6. **SOCIAL SCIENCE: Homicides** A random sample of twenty homicides investigated last year by the Macon County Coroner's Office lists the causes as 12 by handguns, 3 by firearms other than handguns, 2 by knives, 2 by blunt instruments, and 1 by use of hands.

Construct a histogram of the data in each situation.

7. **BUSINESS: Manufacturer's Rebates** A random sample of twenty requests for the $5 rebate offered last summer on two pairs of Ultra-Cool Shades sunglasses found that the checks were received by the consumers after waits of the following numbers of days:

45	53	60	34	60	37	52	65	44	49
54	51	61	43	49	63	53	53	56	55

8. **PERSONAL FINANCE: Checking Accounts** A random sample of thirty checking account balances on March 2 at the Lobsterman Trust Company branch in Bangor found the following amounts (rounded to the nearest dollar):

906	858	1168	1397	925	797
1036	1398	912	1059	931	815
698	787	711	1485	1048	937
1272	727	1339	1458	842	1264
1045	960	802	1204	1370	841

6.2 Measures of Central Tendency

Find the mode, median, and mean of each data set.

9. {14, 11, 15, 8, 5, 4, 10, 3, 12, 11, 6}

10. {5, 6, 15, 13, 15, 6, 11, 13, 13, 6, 13, 6, 8}

11. {12, 13, 13, 5, 9, 14, 13, 8, 9, 14}

12. **BUSINESS: Stock Earnings** The third-quarter earnings per share for twelve randomly selected textile stocks on the New York Stock Exchange were (in dollars):

12	8	8	11	4	7
8	7	5	13	10	9

13. **BUSINESS: Personal Calls** A random sample of the telephone records for fifteen office clerks at the southern regional office of the Marston Glassware Products Company

counted the following numbers of personal calls last week:

$$\begin{array}{ccccc} 4 & 15 & 23 & 26 & 2 \\ 22 & 25 & 4 & 27 & 18 \\ 30 & 28 & 11 & 20 & 6 \end{array}$$

6.3 Measures of Variation

14. Find the range of the data {38, 28, 32, 43, 41, 25, 35, 37, 43, 36}.

15. Find the five-point summary and draw the box-and-whisker plot for the data {4, 5, 7, 9, 10, 13, 14, 15, 20}.

16. Find the sample standard deviation of the data {5, 7, 9, 10, 13, 16}.

Analyze the data in each situation by finding the range, the five-point summary, and the sample standard deviation, and then drawing the box-and-whisker plot.

17. **BUSINESS: Bank Lines** A random sample of fifteen customers at the Middletown branch of the Peoples & First National Bank found the following waiting times (in minutes):

$$\begin{array}{ccccc} 5 & 6 & 6 & 5 & 3 \\ 5 & 8 & 3 & 9 & 6 \\ 7 & 9 & 5 & 9 & 6 \end{array}$$

18. **BUSINESS: Mortgages** A random sample of twenty mortgages filed last month at the Seaford Town Clerk office had the following values (in thousands of dollars):

$$\begin{array}{ccccc} 132 & 141 & 154 & 143 & 115 \\ 141 & 125 & 132 & 123 & 132 \\ 119 & 126 & 123 & 114 & 126 \\ 132 & 128 & 131 & 136 & 131 \end{array}$$

6.4 Normal Distributions and Binomial Approximation

Let X be a normal random variable with mean $\mu = 50$ and standard deviation $\sigma = 10$. Find each probability.

19. $P(30 \leq X \leq 60)$ 20. $P(55 \leq X \leq 75)$

Find the z-score corresponding to each x-value.

21. $x = 12$ with $\mu = 10$ and $\sigma = 1$
22. $x = 18$ with $\mu = 24$ and $\sigma = 3$

Use the normal approximation to the binomial random variable X with $n = 25$ and $p = 0.35$ to find each probability.

23. $P(X = 14)$ 24. $P(6 \leq X \leq 10)$

For exercises 25–28, answer each question using an appropriate normal distribution.

25. **BIOMEDICAL: Heights** Women's heights are approximately normally distributed with mean 63.2 inches and standard deviation 2.6 inches. Find the proportion of women with heights between 62 inches and 69 inches.

26. **GENERAL: Tire Wear** The life of Wear-Ever Super Tread automobile tires is normally distributed with mean 50,000 miles and standard deviation 5000 miles. What proportion of these tires last between 55,000 and 65,000 miles?

27. **GENERAL: Lost Luggage** During the Christmas rush, the chance that an airline will lose your suitcase increases to 1%. What is the probability that between 4 and 8 of the 690 suitcases checked in at Bellemeade Regional Airport this holiday season will be lost?

28. **BIOMEDICAL: Heart Disease** The leading cause of death in the United States is heart disease, which causes 32% of all deaths. In a random sample of 150 death certificates, what is the probability that between 50 and 60 list heart disease as the cause of death?

7 Markov Chains

- 7.1 States and Transitions
- 7.2 Regular Markov Chains
- 7.3 Absorbing Markov Chains

Application Preview

Weather in Sri Lanka*

The island nation of Sri Lanka, off the southeast coast of India, produces tea, coconuts, and rice. Only 16% of the land is arable and weather conditions vary from monsoon to drought, so accurate weather projections would do much to improve the country's economic output. Even if a month's total rainfall is adequate, crops can fail if the rain comes in a downpour after several dry weeks. A recent study modeled the weekly rainfall as a "Markov chain" (to be explained in this chapter), using 50 years of weather data to estimate the probabilities that a "wet" week will be followed by a "dry" week, a "wet" week will be followed by a "wet" week, and so on. The resulting probabilities were then used to simulate the rainfall in different areas over a year, with results such as the accompanying graph.

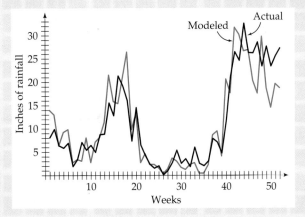

The model, which clearly tracks the actual rainfall quite well, can be used to predict the success of planting marginal areas with various crops whose growth rates and water requirements are known.

* Adapted in part from "On Development and Comparative Study of Two Markov Models of Rainfall in the Dry Zone of Sri Lanka" by B. V. R. Punyawardena and D. Kulasiri, Research Report 96/11, Centre for Computing and Biometrics, Lincoln University (Canterbury, New Zealand).

7.1 STATES AND TRANSITIONS

Introduction

When you read the morning newspaper or watch the evening TV news, much of the information is presented as the *change* since yesterday: The stock market is up, the Dodgers won again, and the weather is moderating. Sometimes the change is given in terms of probabilities ("There is a 70% probability that the rain will end tomorrow"). In this chapter we will explore the behavior of repeated changes that depend on probabilities.

States and Transitions

The Russell 2000 Index is an average of the prices of 2000 stocks traded on the New York Stock Exchange. Each trading day this index *gains*, *loses*, or remains *unchanged* compared with the previous day, and we can denote these three possible *states* by the letters G, L, and U. Watching the market for several successive trading days might result in a sequence of states such as

$$U, G, G, L, G, L, L, L, U, G, \ldots$$

Each *transition* from one day's state to the next begins from one of the three states G, L, and U and ends in one of them, so there are $3 \cdot 3 = 9$ possible transitions. From past stock market records, it is possible to estimate the probabilities of these transitions, and we can specify them by a *state-transition diagram* or, equivalently, by a *transition matrix*.

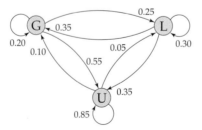

State-Transition Diagram   Transition Matrix

In the state-transition diagram, the number near the arrowhead gives the probability of the transition from the starting state to the ending state. For example, the transition from state G to state L (which we denote as $G \rightarrow L$) has probability 0.25, while the transition $U \rightarrow U$

$$\begin{array}{c} \text{Current State} \end{array} \begin{array}{c} \text{Next state} \\ \begin{array}{c} G L U \end{array} \\ \begin{array}{c} G \\ L \\ U \end{array} \begin{pmatrix} 0.20 & 0.25 & 0.55 \\ 0.35 & 0.30 & 0.35 \\ 0.10 & 0.05 & 0.85 \end{pmatrix} \end{array}$$

(that is, the market will remain unchanged) has probability 0.85. In the transition matrix, the number in a particular row and column gives the probability of the transition *from the row state to the column state*. We always list the row states from top to bottom in the same order as the column states from left to right. The matrix shows that the transition $U \to L$ has probability 0.05 and that the transition $L \to L$ (the market will continue to lose) has probability 0.30.

Practice Problem 1

What is the probability that the market will continue to gain?

▶ Solution on page 489

Notice that the sum of the entries in each row is 1, since any given state must be followed by one of the states G, L, or U. Such a matrix of probabilities is called a *transition matrix*.

Transition Matrix

A transition matrix is a square matrix such that the entries are nonnegative and the sum of each row is 1.

Practice Problem 2

Which of the following are transition matrices?

a. $\begin{pmatrix} 0.3 & 0.7 \\ 0.4 & 0.6 \end{pmatrix}$
b. $\begin{pmatrix} 0.8 & 0.1 \\ 0.6 & 0.4 \end{pmatrix}$
c. $\begin{pmatrix} 0.5 & 0.5 \\ 1.1 & -0.1 \end{pmatrix}$

▶ Solution on page 489

Markov Chains

Any collection of states (such as G, L, and U) together with the probabilities of passing from any state to any state (including from a state to itself) is called a *Markov chain*.*

Markov Chain

A *Markov chain* is a collection of *states* $S_1, S_2, \ldots, S_n$ together with a *transition matrix* T whose entry in row i and column j is the probability of the transition $S_i \to S_j$.

---

* After the Russian mathematician A. A. Markov (1856–1922) who first studied them.

A transition matrix alone may be considered a Markov chain by calling the states $S_1, S_2, \ldots, S_n$, corresponding to the rows (from top to bottom) and the columns (from left to right). Notice that the transition probabilities depend only on the present state and not on any earlier state. For this reason, a Markov chain is sometimes said to have *no memory*.

Types of Transition Matrices

What can we say about the effect of several successive transitions? Transition matrices may represent or contain several of the three basic types of transitions: *oscillating* (switching back and forth), *mixing* (moving among all possibilities), or *absorbing* (never leaving a state once it is reached).

EXAMPLE 1

TYPES OF TRANSITIONS

Characterize the transition corresponding to each transition matrix as "oscillating," "mixing," or "absorbing."

a. $\begin{pmatrix} 0 & 1 \\ 1 & 0 \end{pmatrix}$
b. $\begin{pmatrix} \frac{1}{3} & \frac{1}{3} & \frac{1}{3} \\ \frac{1}{3} & \frac{1}{3} & \frac{1}{3} \\ \frac{1}{3} & \frac{1}{3} & \frac{1}{3} \end{pmatrix}$
c. $\begin{pmatrix} 1 & 0 & 0 \\ \frac{1}{2} & \frac{1}{2} & 0 \\ \frac{1}{2} & 0 & \frac{1}{2} \end{pmatrix}$

Solution

We make a state-transition diagram for each transition matrix.

a. Since states S_1 and S_2 alternate, this is an *oscillating* transition. Repeating it several times would just flip back and forth between the two states.

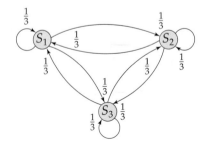

b. Since each state may lead to itself or to any of the other states, this is a *mixing* transition. Repeating it several times would give some combination of the states, and any mixture is possible.

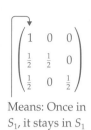

Means: Once in S_1, it stays in S_1

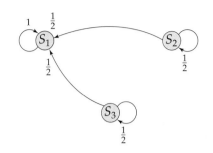

c. Since each of states S_2 and S_3 can reach state S_1 but then can never leave, this is an *absorbing* transition. Repeating it several times might lead to several occurrences of S_2 or S_3, but once the chain reached state S_1, it would remain there forever.

A state that can never be left once it is entered is called an *absorbing state*.

Practice Problem 3   Describe the action of the following transition matrix

and draw its transition diagram: $\begin{pmatrix} 1 & 0 & 0 \\ 0 & 1 & 0 \\ 0 & 0 & 1 \end{pmatrix}$.

▶ Solution on page 489

State Distribution Vectors

Given the transition matrix for a Markov chain together with information about its current state, what can we say about the chain after a transition? Consider an example.

EXAMPLE 2   **CALCULATING THE RESULT OF A TRANSITION**

$\begin{array}{c} \text{Current} \\ \text{state} \end{array} \begin{array}{c} \\ G \\ L \\ U \end{array} \begin{pmatrix} \text{Next state} \\ G \quad L \quad U \\ 0.20 \ 0.25 \ 0.55 \\ 0.35 \ 0.30 \ 0.35 \\ 0.10 \ 0.05 \ 0.85 \end{pmatrix}$

We return to the stock market situation with states G, L, and U (for *gaining*, *losing*, and *unchanged*) and the transition matrix shown on the left. If we have a portfolio of 100 stocks of which 50 gained, 20 lost, and 30 remained unchanged yesterday, how many may we expect to gain, lose, and remain unchanged in today's market?

Solution

We are given the numbers of stocks in the states G, L, and U, and the first column of the transition matrix gives the probabilities of transitions from these states into the state G, so multiplying and adding will give the number of gaining stocks expected today:

$(50 \ 20 \ 30) \begin{pmatrix} G \\ 0.20 \\ 0.35 \\ 0.10 \end{pmatrix} = 20$

$50 \cdot 0.20 + 20 \cdot 0.35 + 30 \cdot 0.10 = 20$

Gained $P(G \to G)$ (G) yesterday

Lost $P(L \to G)$ (L) yesterday

Unchanged $P(U \to G)$ (U) yesterday

Expected value is the sum of all the $n \cdot p$

Thus we can expect 20 of the stocks to gain today. Similarly, multiplying the given numbers by the probabilities in the second column (the probabilities of transitions into the state L) gives today's expected number of losers:

$$(50 \quad 20 \quad 30) \begin{pmatrix} 0.25 \\ 0.30 \\ 0.05 \end{pmatrix} = 20 \qquad \underbrace{50 \cdot 0.25}_{\substack{\text{Gained } P(G \to L) \\ \text{yesterday}}} + \underbrace{20 \cdot 0.30}_{\substack{\text{Lost } P(L \to L) \\ \text{yesterday}}} + \underbrace{30 \cdot 0.05}_{\substack{\text{Unchanged } P(U \to L) \\ \text{yesterday}}} = 20 \qquad \begin{array}{l}\text{Expected} \\ \text{number of} \\ \text{losing stocks}\end{array}$$

The remaining 60 stocks should remain unchanged today, and multiplying the given numbers by the *third* column of the transition matrix (the probabilities of transitions into the state U) gives exactly 60, as you should check.

Notice that these calculations are just the matrix product using the usual row and column multiplication described on page 203:

$$\underbrace{(50 \quad 20 \quad 30)}_{\text{Portfolio yesterday}} \cdot \underbrace{\begin{pmatrix} 0.20 & 0.25 & 0.55 \\ 0.35 & 0.30 & 0.35 \\ 0.10 & 0.05 & 0.85 \end{pmatrix}}_{\text{Transition matrix}} = \underbrace{(20 \quad 20 \quad 60)}_{\text{Portfolio today}}$$

Rephrasing our calculation in Example 2 in terms of probabilities for the 100 stocks in the portfolio, yesterday there were 50% gaining, 20% losing, and 30% unchanged, while after the transition to today, they became 20% gaining, 20% losing, and 60% unchanged. A row matrix consisting of such probabilities is called a *state distribution vector*, and multiplying it by the transition matrix gives the state distribution vector after one transition.

State Distribution Vector

A *state distribution vector* is a row matrix $D = (d_1 \; d_2 \; \ldots \; d_n)$ of nonnegative numbers whose sum is 1. If D represents the current probability distribution for a Markov chain with transition matrix T, then the state distribution vector one step later is $D \cdot T$.

Graphing Calculator Exploration

$$T = \begin{pmatrix} 0.20 & 0.25 & 0.55 \\ 0.35 & 0.30 & 0.35 \\ 0.10 & 0.05 & 0.85 \end{pmatrix}$$

To verify that the transition matrix from Example 2 (shown on the left) transforms the state distribution vector $D = (0.50 \quad 0.20 \quad 0.30)$ into $(0.20 \quad 0.20 \quad 0.60)$, enter T and D into matrices [A] and [B] and find their product as follows.

What happens if we repeat this calculation with the state distribution vector $D = (0.140 \quad 0.104 \quad 0.756)$?

The *k*th State Distribution Vector

What can we say about the future of a Markov chain given some information about the present? If we have an *initial* state distribution vector D_0 and transition matrix T, then we have just seen that the state distribution vector after the first transition is

$$D_1 = D_0 \cdot T \qquad \text{\textit{T} gives one transition}$$

After another transition we obtain

$$D_2 = (D_0 \cdot T) \cdot T = D_0 \cdot T^2 \qquad \text{Multiplying by \textit{T} again gives two transitions}$$

Multiplying again by T (that is, $D_0 \cdot T^3$) would give the third state distribution vector, and so on.

*k*th State Distribution Vector

$$D_k = D_0 \cdot T^k \qquad \text{Multiplying by } T^k \text{ gives the } k\text{th state distribution vector}$$

To put this another way, T^k gives the probabilities for k successive transitions, so the entry in row i and column j of T^k gives the probability of going from state S_i to state S_j in exactly k transitions.

EXAMPLE 3

$$T = \begin{pmatrix} 1 & 0 & 0 \\ \frac{1}{2} & \frac{1}{2} & 0 \\ \frac{1}{2} & 0 & \frac{1}{2} \end{pmatrix}$$

CALCULATING STATE DISTRIBUTION VECTORS

For the Markov chain with transition matrix on the left and initial state distribution vector $D_0 = \begin{pmatrix} \frac{1}{5} & \frac{2}{5} & \frac{2}{5} \end{pmatrix}$, calculate the next two state distribution vectors D_1 and D_2.

Solution

We obtain the next state distribution vector by multiplying the previous one by the transition matrix:

$$D_1 = D_0 \cdot T = \begin{pmatrix} \frac{1}{5} & \frac{2}{5} & \frac{2}{5} \end{pmatrix} \cdot \begin{pmatrix} 1 & 0 & 0 \\ \frac{1}{2} & \frac{1}{2} & 0 \\ \frac{1}{2} & 0 & \frac{1}{2} \end{pmatrix}$$

$$= \begin{pmatrix} \frac{1}{5} \cdot 1 + \frac{2}{5} \cdot \frac{1}{2} + \frac{2}{5} \cdot \frac{1}{2} & 0 + \frac{2}{5} \cdot \frac{1}{2} + 0 & 0 + 0 + \frac{2}{5} \cdot \frac{1}{2} \end{pmatrix} \quad \text{Row} \times \text{columns}$$

$$= \begin{pmatrix} \frac{3}{5} & \frac{1}{5} & \frac{1}{5} \end{pmatrix}$$

and

$$D_2 = D_1 \cdot T = \begin{pmatrix} \frac{3}{5} & \frac{1}{5} & \frac{1}{5} \end{pmatrix} \cdot \begin{pmatrix} 1 & 0 & 0 \\ \frac{1}{2} & \frac{1}{2} & 0 \\ \frac{1}{2} & 0 & \frac{1}{2} \end{pmatrix} = \begin{pmatrix} \frac{4}{5} & \frac{1}{10} & \frac{1}{10} \end{pmatrix} \quad \text{Omitting the details}$$

$D_0 = \begin{pmatrix} \frac{1}{5} & \frac{2}{5} & \frac{2}{5} \end{pmatrix}$
$D_2 = \begin{pmatrix} \frac{4}{5} & \frac{1}{10} & \frac{1}{10} \end{pmatrix}$

Comparing D_0 with D_2, we see that the first entry has increased from $\frac{1}{5}$ to $\frac{4}{5}$. That is, in only two transitions, the probability of being in state S_1 has increased from 20% to 80%. Since we saw in Example 1c on pages 482–483 that the state S_1 is absorbing for this transition matrix, the probability of being in S_1 should increase with each transition.

Graphing Calculator Exploration

For the absorbing Markov chain in Example 3, we may easily check that the probability of being in state S_1 increases with each transition. Entering T and D_0 into matrices [A] and [B] and calculating $D_k = D_0 \cdot T^k$ for successively higher values of k gives:

```
MATRIX[A]  3 x3        MATRIX[B]  1 x3        [B]*[A]
[[1    0    0  ]        [[.2   .4   .4 ]        [[.60  .20  .20]]   ←$D_1 = D_0 \cdot A$
 [.5  .5    0  ]                                Ans*[A]
 [.5   0   .5  ]]                               [[.80  .10  .10]]   ←$D_2 = D_0 \cdot A^2$
                                                Ans*[A]
                                                [[.90  .05  .05]]   ←$D_3 = D_0 \cdot A^3$
```

↑ Probability of being in S_1

Notice that the first entry of these state distribution vectors increases from 0.6 to 0.8 and then to 0.9, showing that the probability of being in state S_1 increases with each transition. These probabilities will continue to increase, even if we begin with a different initial distribution D_0.

Duration in a Given State

In general, how long can we expect a Markov chain to remain in its current state before moving to another state? The formula in the following box is proved on pages 492–493.

> **Expected Duration in a Given State**
>
> For a Markov chain in a state S_i, let p be the element in row i and column i of the transition matrix. If $p = 1$, then S_i is an absorbing state, so the Markov chain will remain in S_i forever. If $p < 1$, then the expected number of times that the chain will be in that state before moving to another state is
>
> $$E = \frac{1}{1-p}$$

EXAMPLE 4 FINDING AN EXPECTED TIME

The EasyDotCom Internet service provider classifies its residential customers by the number of connection hours used per week: H for high usage of more than 20 hours, M for moderate usage of between 5 and 20 hours, and I for infrequent usage of less than 5 hours. The company has found that the usage behavior of their subscribers can be modeled as a Markov chain with the transition matrix shown on the left.

If the Davidson household is presently a high user, how many weeks can it be expected to be a high user before changing to some other level of Internet use?

Solution

Since the probability that state H stays state H for one transition of this Markov chain is 0.75 (from the upper left entry in the transition matrix), the number of weeks (including the present week) that the Davidsons are expected to be heavy users before changing to some other level is

$$E = \frac{1}{1 - 0.75} = \frac{1}{0.25} = 4 \text{ weeks} \qquad E = \frac{1}{1-p} \text{ with } p = 0.75$$

Spreadsheet Exploration

The following spreadsheet* shows 500 observations of the Markov chain from Example 4 but labels the states 1 (yellow) for H, 2 (green) for M, and 3 (blue) for I. The transition matrix is stored in the upper left-hand corner, and the counters below it summarize the number of times each state occurred in this observation.

	A	B	C
1	0.75	0.2	0.05
2	0.2	0.7	0.1
3	0.25	0.2	0.55
4			
5	State #1 :		
6	232	0.464	
7			
8	State #2 :		
9	204	0.408	
10			
11	State #3 :		
12	64	0.128	

*See the Preface for how to obtain this and other Excel spreadsheets.

State H (yellow) sometimes appears just once and sometimes appears a dozen or more times in succession. Find the average length of these successions of H's, either by setting up your own modification of this spreadsheet or by counting from the screen image above. Is your answer consistent with our answer to Example 4?

7.1 Section Summary

A *transition matrix* is a square matrix of nonnegative entries with each row summing to 1. The entries are the transition probabilities between the states of the *Markov chain*. Transition matrices exhibit three basic transition behaviors:

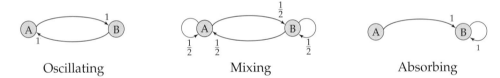

Oscillating   Mixing   Absorbing

An initial state distribution vector D_0 represents the probabilities of beginning in the various states, and the subsequent state distribution vectors $D_k = D_0 \cdot T^k$ are found by repeated multiplications by the transition matrix.

If the probability p that a Markov chain will stay in its present state for one transition is 1, then the chain will remain in that state forever. If it is less than 1, then the expected number of times it will be in this state before switching to some other state is $1/(1-p)$.

▶ Solutions to Practice Problems

1. The probability of the transition $G \to G$ is 0.20.

2. Only matrix (a). Matrix (b) fails because the first row does not sum to 1, and matrix (c) fails because one of the entries is negative.

3. The identity matrix sends each state to itself, so every state is absorbing.

7.1 Exercises

Construct a transition matrix for the state-transition diagram and identify the transition as "oscillating," "mixing," or "absorbing."

1.
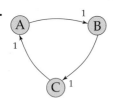

2.
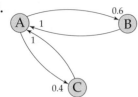

3.

4.
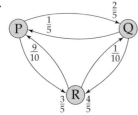

5.

6.
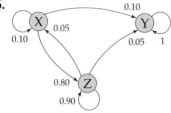

Construct a state-transition diagram for the transition matrix and identify the transition as "oscillating," "mixing," or "absorbing."

7. $\begin{pmatrix} 0 & 1 & 0 \\ 0.2 & 0 & 0.8 \\ 0 & 1 & 0 \end{pmatrix}$

8. $\begin{pmatrix} 0 & 1 & 0 & 0 \\ 0 & 0 & 0.3 & 0.7 \\ 1 & 0 & 0 & 0 \\ 1 & 0 & 0 & 0 \end{pmatrix}$

9. $\begin{pmatrix} 0.10 & 0.90 \\ 0.35 & 0.65 \end{pmatrix}$

10. $\begin{pmatrix} 0 & 1 & 0 & 0 \\ 0 & 0 & 0.3 & 0.7 \\ 0 & 1 & 0 & 0 \\ 1 & 0 & 0 & 0 \end{pmatrix}$

11. $\begin{pmatrix} 1 & 0 & 0 \\ 0.1 & 0.2 & 0.7 \\ 0.1 & 0.2 & 0.7 \end{pmatrix}$

12. $\begin{pmatrix} 1 & 0 & 0 & 0 \\ 0 & 0 & 0.3 & 0.7 \\ 0 & 1 & 0 & 0 \\ 0 & 0 & 0 & 1 \end{pmatrix}$

For each Markov chain transition matrix T and initial state distribution vector D_0, calculate the next two state distribution vectors D_1 and D_2.

13. $T = \begin{pmatrix} \frac{1}{4} & \frac{3}{4} \\ \frac{1}{2} & \frac{1}{2} \end{pmatrix}$, $D_0 = \begin{pmatrix} \frac{3}{5} & \frac{2}{5} \end{pmatrix}$

14. $T = \begin{pmatrix} \frac{1}{2} & \frac{1}{2} \\ \frac{1}{3} & \frac{2}{3} \end{pmatrix}$, $D_0 = \begin{pmatrix} \frac{2}{5} & \frac{3}{5} \end{pmatrix}$

15. $T = \begin{pmatrix} 0 & 0.2 & 0.8 \\ 0.6 & 0.4 & 0 \\ 0.4 & 0 & 0.6 \end{pmatrix}$,

$D_0 = (0.35 \quad 0.45 \quad 0.20)$

16. $T = \begin{pmatrix} 0.4 & 0.4 & 0.2 \\ 0.4 & 0 & 0.6 \\ 0.1 & 0.3 & 0.6 \end{pmatrix}$,

$D_0 = (0.25 \quad 0.60 \quad 0.15)$

For each Markov chain transition matrix T and present state, find the expected number of times the chain will be in that state before moving to some other state.

17. $T = \begin{pmatrix} 0.6 & 0.2 & 0.2 \\ 0.1 & 0.8 & 0.1 \\ 0.6 & 0.2 & 0.2 \end{pmatrix}$, presently in state S_2

18. $T = \begin{pmatrix} 0.8 & 0.2 & 0 \\ 0.3 & 0.6 & 0.1 \\ 0.1 & 0 & 0.9 \end{pmatrix}$, presently in state S_3

19. $T = \begin{pmatrix} 0.8 & 0.1 & 0.1 & 0 \\ 0.2 & 0.5 & 0.1 & 0.2 \\ 0 & 0 & 1 & 0 \\ 0 & 0.3 & 0.1 & 0.6 \end{pmatrix}$, presently in state S_4

20. $T = \begin{pmatrix} 1 & 0 & 0 & 0 \\ 0 & 0.20 & 0.70 & 0.10 \\ 0.05 & 0.10 & 0.85 & 0 \\ 0 & 0 & 0 & 1 \end{pmatrix}$, presently in state S_3

Represent each situation as a Markov chain by constructing a state-transition diagram and a transition matrix. Be sure to state your final answer in terms of the original question.

21. **GENERAL: Traffic Flow** A traffic light on Main Street operates on a one-minute timer. If it is green (to the traffic), at the end of the minute it will change to red only if a pedestrian has pressed a button at the crosswalk. The probability of a pedestrian pressing the button during any one-minute interval is 0.25. If the light is red, it turns back to green at the end of the minute. If the light is green now for the traffic, find the expected number of minutes that the light is green.

22. **SOCIAL SCIENCE: Learning Theory** The initial phase of a learning experiment is to establish that a rat will randomly wander in a maze prior to the introduction of food and other stimuli. The following maze has three connecting rooms, A, B, and C, and the location of the rat is observed at one-minute intervals. For both rooms A and B, the probability of remaining there is 50%, while the probability of remaining in room C is 75%. When the rat leaves a room, it is equally likely to choose either of the other two rooms. If the rat is in room B, find the expected number of minutes that the rat will remain there.

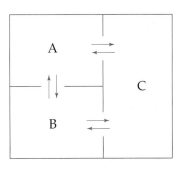

23. **GENERAL: Breakfast Habits** A survey of weekly breakfast eating habits found that although many people eat cereal, boredom and other reasons cause 5% to switch to something else (eggs, muffins, and so on) each week, while among those who are not eating cereal for breakfast, 15% will switch to cereal each week. If 70% of the residents of Cincinnati are eating cereal for breakfast this week, what percentage will be eating cereal next week? In two weeks?

24. **BUSINESS: Market Research** A telephone survey of households in the Atlanta metropolitan area found that 85% of those using a DVD player planned to rent movies or otherwise use it next month, while 10% of those not using a DVD player this month planned on using one next month. If 2% of the households currently use a DVD player, what percentage will be using a DVD player after one year?

25. **ATHLETICS: Retirement Sports** A survey of retirees in Arizona who golf or fish every day found that 60% of those who played golf one day switched to fishing the next and that 90% of those who went fishing switched to golf the next day. If 30% of these retirees are golfing today, what percentage will be fishing tomorrow? Golfing the day after tomorrow?

26. **MANAGEMENT SCIENCE: Emergency Medical Care** Authorization specialists at a national health maintenance organization spend twelve minutes to evaluate and process each request for emergency care, and their work loads are managed by a computerized distribution system. This system checks the work queue of each specialist every twelve minutes, transfers

finished work to the data storage center, and adds new requests to the queue, but never allows any queue to contain more than four requests (one "active" and three "pending review"). The number of new requests the system has to add to any particular queue is variable: 15% of the time there are none, 55% there is one, 25% there are two, and 5% there are three new requests. If a particular specialist does not have any "pending review" requests in her queue, how many minutes can we expect this situation to continue? [*Hint:* Define four states by the number of unfinished requests in the queue at the end of each twelve-minute block of time, just before the system checks it to remove finished work and add new work.]

27. **BUSINESS: Portfolio Management** An investment banker estimates the financial stability of mid-sized manufacturing companies as "secure," "doubtful," and "at risk." He has noticed that of the "secure" companies he follows, each year 5% decline to "doubtful" and the rest remain as they are; 10% of the "doubtful" companies improve to "secure," 5% decline to "at risk," and the rest remain as they are; and 5% of the "at risk" companies become bankrupt and never recover, 10% improve to "doubtful," and the rest remain as they are. If his current portfolio of investments is 80% "secure," 15% "doubtful," and 5% "at risk," what percentage will be "secure" in two years? (Assume that the present trends continue and that no changes are made to the portfolio.) In the long run, how many of the companies in the portfolio will become bankrupt?

28. **ATHLETICS: Baseball Players** A baseball fan tracks the careers of interesting players in the major and minor leagues and has found that from year to year, 80% of the major league players tracked stay in the majors, 10% drop down to the minor leagues, and 10% quit baseball or retire, while 20% of the minor league players tracked move up to the majors, 40% stay in the minors, and the rest quit or retire. If his current roster of interesting players is three-quarters major and one-quarter minor league players, what percentage of these players will have quit or retired in two years? What will happen to all the players in the long run?

29. **GENERAL: Working Mothers** A working mother of three teenagers expects them to help around the house and rates the laundry room each weekend as "empty," "manageable," or "you're all grounded!" She has noticed that if it is empty one weekend, then the following weekend with probability 0.4 it is empty, with probability 0.4 it is manageable, and with probability 0.2 it is so bad that "you're all grounded!" If it is manageable one weekend, then the following weekend with probability 0.5 it is empty and with probability 0.5 it is manageable. If it is so bad that she has to ground everybody, then the following weekend the laundry room is empty. If the laundry room is manageable this weekend, how many weekends will it be manageable before changing to either "you're all grounded!" or "empty"?

30. **ATHLETICS: Jogging Routes** Each morning a jogger runs either at the local high school track, around the neighborhood, or in the park. If she runs at the school one day, then the next day with probability 0.7 she runs there again and with probability 0.3 she switches to the park; if she runs around the neighborhood, then the next day with probability 0.9 she runs there again and with probability 0.1 she switches to the park; and if she runs in the park, then the next day she switches with probability 0.9 to the school and with probability 0.1 to around the neighborhood. If she runs around the neighborhood today, how many days can she be expected to run around the neighborhood before switching to the park?

Explorations and Excursions

The following problems extend and augment the material presented in the text.

Proof of the Formula for the Duration in a Given State

31. For a Markov chain in a nonabsorbing state S_i, let p be the element in row i and column i of the transition matrix T. Explain why the probability of the chain being in state S_i exactly n

times and then switching to some other state is $p^{n-1}(1-p)$.

32. Using the formula for expected value (page 411), explain why the expected number of times that the chain will be in state S_i is

$$E = 1 \cdot (1-p) + 2 \cdot p(1-p) + 3 \cdot p^2(1-p) + \cdots$$

33. Multiply out this series and combine similar terms to obtain

$$E = 1 + p^2 + p^3 + \cdots$$

34. Use the result of Exercise 32 on pages 131–132 to obtain $E = \dfrac{1}{1-p}$, which is the formula in the box on page 487.

7.2 REGULAR MARKOV CHAINS

Introduction

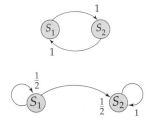

Everyone who takes their coffee or tea with cream and sugar knows that it makes no difference which is added first nor on which side of the cup it is done—a few stirs mixes everything uniformly. In this section we will study Markov chains that exhibit this sort of mixing—that is, ones that move among all of their states towards some sort of equilibrium.

Not all Markov chains exhibit this kind of mixing. The first chain on the left (from Example 1a on page 482) is an *oscillating* Markov chain, which bounces back and forth and never settles down to an equilibrium behavior. The second is an *absorbing* Markov chain, which eventually gets stuck in a single state rather than mixing around in all of them. In this section we begin by developing a condition on the transition matrix that guarantees mixing by avoiding oscillation and absorption, and then we show that the eventual equilibrium behavior of such a chain is independent of how it began.

Regular Markov Chains

Consider a Markov chain for which, in some number of moves, it is possible to go from any state to any state. Such a chain cannot rigidly oscillate nor be absorbing. Since we saw on page 485 that powers of the transition matrix give the probabilities of transitions in several moves, this means that some power of the transition matrix has all positive entries. Such chains are called *regular*.

> **Regular Markov Chains**
>
> A transition matrix is *regular* if some power of it has only positive entries. A Markov chain is regular if its transition matrix is regular.

EXAMPLE 1

We may determine whether a Markov chain is regular either from its transition matrix or its state-transition diagram.*

REGULAR MATRICES

Which of the following transition matrices are regular?

a. $A = \begin{pmatrix} \frac{1}{2} & \frac{1}{2} \\ 1 & 0 \end{pmatrix}$ b. $B = \begin{pmatrix} \frac{1}{2} & \frac{1}{2} \\ 0 & 1 \end{pmatrix}$ c. $C = \begin{pmatrix} 1 & 0 \\ 0 & 1 \end{pmatrix}$

Solution

We will first find the answer by calculating powers of the matrix and then check it from the state-transition diagram.

a. Since all entries of A^2 (calculated below) are positive, A is regular.

$$A^2 = \begin{pmatrix} \frac{1}{2} & \frac{1}{2} \\ 1 & 0 \end{pmatrix} \cdot \begin{pmatrix} \frac{1}{2} & \frac{1}{2} \\ 1 & 0 \end{pmatrix} = \begin{pmatrix} \frac{3}{4} & \frac{1}{4} \\ \frac{1}{2} & \frac{1}{2} \end{pmatrix}$$

The corresponding state-transition diagram shows that this is a mixing transition: Any state can be reached from every state in two steps. (Note that $S_2 \to S_2$ takes two steps.)

b. Since the lower left entry in B^k remains zero with each multiplication, B is not regular.

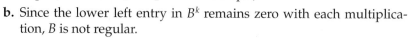

$B = \begin{pmatrix} \frac{1}{2} & \frac{1}{2} \\ 0 & 1 \end{pmatrix}$, $B^2 = \begin{pmatrix} \frac{1}{4} & \frac{3}{4} \\ 0 & 1 \end{pmatrix}$, $B^3 = \begin{pmatrix} \frac{1}{8} & \frac{7}{8} \\ 0 & 1 \end{pmatrix}$, ... Do you see why this entry will be 0 for *any* power of B?

The corresponding state-transition diagram shows that S_2 is an absorbing state for this transition, so S_1 is not reachable from S_2 in any number of steps.

c. Since every power of the identity matrix is still the identity matrix, $C = I$ is not regular.

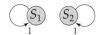

$C = \begin{pmatrix} 1 & 0 \\ 0 & 1 \end{pmatrix}$, $C^2 = \begin{pmatrix} 1 & 0 \\ 0 & 1 \end{pmatrix}$, $C^3 = \begin{pmatrix} 1 & 0 \\ 0 & 1 \end{pmatrix}$, ...

The state-transition diagram for C shows that both S_1 and S_2 are absorbing states.

Given a particular Markov chain, we can use the methods of the previous section to calculate the long-term distribution to see whether it is affected by the initial distribution.

---

* A transition matrix with only positive entries is called an *ergodic* matrix. Therefore, a transition matrix is *regular* if some power of it is ergodic. The term was coined by the physicist Ludwig Boltzmann (1844–1906) from the Greek words *erg* (work) and *ode* (path) when formulating his *ergodic hypothesis* to explain how the individual paths of the molecules in a gas lead to the uniform equilibrium that we experience.

EXAMPLE 2

CALCULATING A LONG-TERM DISTRIBUTION

A rental truck company has branches in Springfield, Tulsa, and Little Rock, and each rents trucks by the week. The trucks may be returned to any of the branches. The return location probabilities are shown in the following state-transition diagram. Find the distribution of the company's 180 trucks among the three cities after one year if the trucks are initially distributed as follows:

a. 36 in Springfield and 72 each in Tulsa and in Little Rock;

b. 90 in Springfield, 30 in Tulsa, and 60 in Little Rock;

c. all 180 in Tulsa.

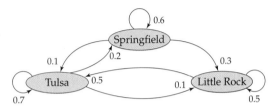

Solution

We will describe what happens in terms of the *proportion* of trucks in each city, with S_1, S_2, and S_3 standing for Springfield, Tulsa, and Little Rock, respectively. From the state-transition diagram, the transition matrix is

$$T = \begin{pmatrix} \text{Springfield} & \text{Tulsa} & \text{Little Rock} \\ 0.6 & 0.1 & 0.3 \\ 0.2 & 0.7 & 0.1 \\ 0 & 0.5 & 0.5 \end{pmatrix} \begin{matrix} \text{Springfield} \\ \text{Tulsa} \\ \text{Little Rock} \end{matrix}$$

Since T is regular (the state-transition diagram shows that it is possible to go from any state to any state in two steps), the distribution of trucks among the rental offices over successive weeks should show that they "mix around" in some predictable fashion rather than "bunching up" in one city. For each initial state distribution D_0, the distribution after one year is the result of 52 weekly transitions:

$$D_{52} = D_0 \cdot T^{52}$$

a. For the numbers of trucks given in part (a), the initial state distribution vector is

$$D_0 = \begin{pmatrix} \frac{36}{180} & \frac{72}{180} & \frac{72}{180} \end{pmatrix} = \begin{pmatrix} \frac{1}{5} & \frac{2}{5} & \frac{2}{5} \end{pmatrix} \quad \text{Each initial number divided by 180 trucks}$$

Then, after 52 weekly transitions, the distribution will be

$$D_{52} = \begin{pmatrix} \frac{1}{5} & \frac{2}{5} & \frac{2}{5} \end{pmatrix} \cdot \begin{pmatrix} 0.6 & 0.1 & 0.3 \\ 0.2 & 0.7 & 0.1 \\ 0 & 0.5 & 0.5 \end{pmatrix}^{52} = \begin{pmatrix} \frac{1}{4} & \frac{1}{2} & \frac{1}{4} \end{pmatrix} \quad \text{Using a calculator}$$

After one year there will be $\frac{1}{4} \cdot 180 = 45$ trucks in Springfield, $\frac{1}{2} \cdot 180 = 90$ trucks in Tulsa, and $\frac{1}{4} \cdot 180 = 45$ trucks in Little Rock.

b. The initial numbers of trucks in part (b) give

$$D_0 = \begin{pmatrix} \frac{90}{180} & \frac{30}{180} & \frac{60}{180} \end{pmatrix} = \begin{pmatrix} \frac{1}{2} & \frac{1}{6} & \frac{1}{3} \end{pmatrix}$$

so

$$D_{52} = \begin{pmatrix} \frac{1}{2} & \frac{1}{6} & \frac{1}{3} \end{pmatrix} \cdot \begin{pmatrix} 0.6 & 0.1 & 0.3 \\ 0.2 & 0.7 & 0.1 \\ 0 & 0.5 & 0.5 \end{pmatrix}^{52} = \begin{pmatrix} \frac{1}{4} & \frac{1}{2} & \frac{1}{4} \end{pmatrix} \quad \text{Using a calculator}$$

This is the same distribution found in part (a), so again there will be 45 trucks in Springfield, 90 trucks in Tulsa, and 45 trucks in Little Rock.

c. For the initial numbers in part (c), we have

$$D_0 = \begin{pmatrix} 0 & \frac{180}{180} & 0 \end{pmatrix} = \begin{pmatrix} 0 & 1 & 0 \end{pmatrix}$$

so

$$D_{52} = \begin{pmatrix} 0 & 1 & 0 \end{pmatrix} \cdot \begin{pmatrix} 0.6 & 0.1 & 0.3 \\ 0.2 & 0.7 & 0.1 \\ 0 & 0.5 & 0.5 \end{pmatrix}^{52} = \begin{pmatrix} \frac{1}{4} & \frac{1}{2} & \frac{1}{4} \end{pmatrix} \quad \text{Using a calculator}$$

Again we get the same distribution as in parts (a) and (b), so again there will be 45 trucks in Springfield, 90 trucks in Tulsa, and 45 trucks in Little Rock.

As we expected, the long-term distribution of the trucks did not depend on the initial distribution. Moreover, further mixing will not change the result:

$$\begin{pmatrix} \frac{1}{4} & \frac{1}{2} & \frac{1}{4} \end{pmatrix} \cdot \begin{pmatrix} 0.6 & 0.1 & 0.3 \\ 0.2 & 0.7 & 0.1 \\ 0 & 0.5 & 0.5 \end{pmatrix} = \begin{pmatrix} \frac{1}{4} & \frac{1}{2} & \frac{1}{4} \end{pmatrix} \quad \text{This distribution remains unchanged by another transition}$$

A distribution such as this that remains unchanged by a transition is called a *steady-state distribution*.

Steady-State Distribution

A state distribution vector D is a *steady-state distribution* for the Markov chain with transition matrix T if $D \cdot T = D$.

Practice Problem 1 Is $D = \begin{pmatrix} \frac{1}{2} & \frac{1}{2} \end{pmatrix}$ a steady-state distribution for the Markov chain with transition matrix $T = \begin{pmatrix} \frac{1}{4} & \frac{3}{4} \\ \frac{3}{4} & \frac{1}{4} \end{pmatrix}$?   ▶ Solution on page 501

The Fundamental Theorem of Regular Markov Chains

For the truck rental company in Example 2 we obtained the same long-run distribution regardless of the initial distribution. Can we obtain this long-run distribution directly from the transition matrix? The answer is "yes," and in two different ways.

For a regular Markov chain, calculating high powers T^k of the transition matrix T will result in rows that "settle down" to this steady-state distribution.

Graphing Calculator Exploration

To see whether the transition matrix for the truck rental company in Example 2 has the property that the rows of higher powers "settle down" to the steady-state distribution, we may calculate as follows.

```
MATRIX[A] 3 X3
[ .6    .1    .3   ]
[ .2    .7    .1   ]
[ 0     .5    .5   ]
```

a. Enter the matrix $T = \begin{pmatrix} 0.6 & 0.1 & 0.3 \\ 0.2 & 0.7 & 0.1 \\ 0 & 0.5 & 0.5 \end{pmatrix}$ in matrix [A].

b. Then calculate T^k for $k = 2, 4, 26,$ and 52 (a year of weekly transitions):

```
[A]²
[[ .38   .28   .34 ]
 [ .26   .56   .18 ]
 [ .1    .6    .3  ]]
```

```
[A]^4
[[ .2512   .4672  ....
 [ .2624   .4944  ....
 [ .224    .544   ....
```

```
[A]^26
[[ .2500000004  ....
 [ .2499999997  ....
 [ .2500000001  ....
```

```
[A]^52
[[ .25   .5   .25 ]
 [ .25   .5   .25 ]
 [ .25   .5   .25 ]]
```

Notice that the rows in successive screens all seem to be approaching the same distribution $\begin{pmatrix} \frac{1}{4} & \frac{1}{2} & \frac{1}{4} \end{pmatrix}$, with the last screen (for T^{52}) showing exactly this steady-state distribution, the same one that we found in Example 2.

Why does each column "settle down" to a single value as the matrix is raised to higher and higher powers? For a regular transition matrix, there is a power (call it n) such that all entries of T^n are positive. Consider the matrix product $T^n \cdot T^k$, which involves multiplying rows of T^n by columns of T^k. Each row of T^n consists of positive numbers adding to 1, so multiplying them by a column of T^k amounts to taking a *weighted average* of all of the column entries, which will bring the extremes of the column closer together. As k increases, the resulting column entries will therefore converge to a single "averaged" value (see page 412). This means that as we raise T to higher and higher powers, each column will tend towards a single value. This column value, being the long-run probability of entering into a state, must be the long-run probability of being in that state. Therefore, the rows must tend to the steady-state distribution, which is therefore unique. Note that this reasoning would fail if any of the elements of T^n were zero, which is why regularity is needed. This result is so important that it is called the Fundamental Theorem of Regular Markov Chains.

> **Fundamental Theorem of Regular Markov Chains**
>
> A regular Markov chain with transition matrix T has exactly one steady-state distribution D solving $D \cdot T = D$. Higher powers of the transition matrix T approximate arbitrarily closely a matrix each of whose rows is equal to D.

How to Solve $D \cdot T = D$

How can we find the steady-state distribution D that solves $D \cdot T = D$ without having to calculate arbitrarily high powers of the transition matrix? We begin by writing the equation with zero on the right:

$D \cdot T - D = 0$	From $D \cdot T = D$
$D \cdot (T - I) = 0$	Factoring (since $D \cdot I = D$)
$(T - I)^t \cdot D^t = 0$	Taking transposes since $(A \cdot B)^t = B^t \cdot A^t$
$(T^t - I) \cdot D^t = 0$	Since $(A - B)^t = A^t - B^t$ and $I^t = I$

Since D is a state distribution vector, the sum of the entries of D must be 1, which we can write as the matrix equation

$$(1 \cdots 1) \cdot D^t = 1 \qquad \text{Entries of } D \text{ add up to 1}$$

Combining these last two matrix equations into one large matrix equation, we need to solve

$$\begin{pmatrix} T^t - I \\ 1 \cdots 1 \end{pmatrix} \cdot D^t = \begin{pmatrix} 0 \\ 1 \end{pmatrix} \qquad \begin{pmatrix} 0 \\ 1 \end{pmatrix} \text{ means } \begin{pmatrix} 0 \\ \vdots \\ 0 \\ 1 \end{pmatrix}$$

7.2 REGULAR MARKOV CHAINS

We may now solve for D^t by row-reducing the augmented matrix $\begin{pmatrix} T^t - I & | & 0 \\ \hline 1 \cdots 1 & | & 1 \end{pmatrix}$ to obtain $\begin{pmatrix} I & | & D^t \\ \hline 0 & | & 0 \end{pmatrix}$, just as we did in Section 3 of Chapter 3 (see page 193).

> **Steady-State Distribution for a Regular Markov Chain**
>
> The steady-state distribution D for a *regular* Markov chain with transition matrix T may be found by row-reducing the augmented matrix
>
> $$\begin{pmatrix} T^t - I & | & 0 \\ \hline 1 \cdots 1 & | & 1 \end{pmatrix} \quad \text{to obtain} \quad \begin{pmatrix} I & | & D^t \\ \hline 0 & | & 0 \end{pmatrix}$$
>
> and then transposing all but the bottom entry in the last column.

EXAMPLE 3

$$T = \begin{pmatrix} 0.6 & 0.1 & 0.3 \\ 0.2 & 0.7 & 0.1 \\ 0 & 0.5 & 0.5 \end{pmatrix}$$

FINDING A STEADY-STATE DISTRIBUTION

Find the steady-state distribution for the transition matrix from the truck rental situation described in Example 2.

Solution

Following the procedure described above, we transpose T by changing rows into columns and then calculate $T^t - I$:

$$T^t - I = \begin{pmatrix} 0.6 & 0.2 & 0 \\ 0.1 & 0.7 & 0.5 \\ 0.3 & 0.1 & 0.5 \end{pmatrix} - \begin{pmatrix} 1 & 0 & 0 \\ 0 & 1 & 0 \\ 0 & 0 & 1 \end{pmatrix} = \begin{pmatrix} -0.4 & 0.2 & 0 \\ 0.1 & -0.3 & 0.5 \\ 0.3 & 0.1 & -0.5 \end{pmatrix}$$

We need to solve the matrix equation

$$\begin{pmatrix} -0.4 & 0.2 & 0 \\ 0.1 & -0.3 & 0.5 \\ 0.3 & 0.1 & -0.5 \\ 1 & 1 & 1 \end{pmatrix} \begin{pmatrix} d_1 \\ d_2 \\ d_3 \end{pmatrix} = \begin{pmatrix} 0 \\ 0 \\ 0 \\ 1 \end{pmatrix} \qquad \begin{pmatrix} T^t - I \\ 1 \cdots 1 \end{pmatrix} \cdot D^t = \begin{pmatrix} 0 \\ 1 \end{pmatrix}$$

We row-reduce the augmented matrix

$$\begin{pmatrix} -0.4 & 0.2 & 0 & | & 0 \\ 0.1 & -0.3 & 0.5 & | & 0 \\ 0.3 & 0.1 & -0.5 & | & 0 \\ \hline 1 & 1 & 1 & | & 1 \end{pmatrix} \quad \text{to obtain} \quad \begin{pmatrix} 1 & 0 & 0 & | & \frac{1}{4} \\ 0 & 1 & 0 & | & \frac{1}{2} \\ 0 & 0 & 1 & | & \frac{1}{4} \\ \hline 0 & 0 & 0 & | & 0 \end{pmatrix} \qquad \begin{pmatrix} T^t - I & | & 0 \\ \hline 1 \cdots 1 & | & 1 \end{pmatrix} \text{ reduces to } \begin{pmatrix} I & | & D^t \\ \hline 0 & | & 0 \end{pmatrix}$$

(omitting the details). The last column (omitting the bottom zero) is the transpose D^t of the steady-state distribution

$$D = \left(\tfrac{1}{4} \quad \tfrac{1}{2} \quad \tfrac{1}{4} \right)$$

This answer agrees with the fractions that we found in Example 2 (on pages 495–496) and also with the high powers of the transition matrix calculated in the Graphing Calculator Exploration on page 497.

Practice Problem 2 Use row-reduction to find the steady-state distribution of the Markov chain with transition matrix $T = \begin{pmatrix} 0 & 1 \\ \tfrac{1}{2} & \tfrac{1}{2} \end{pmatrix}$. ➤ Solution on next page

Graphing Calculator Exploration

We may solve Example 3 using a graphing calculator. The program* MARKOV displays powers of a transition matrix T, the augmented matrix $\left(\begin{array}{c|c} T^t - I & 0 \\ \hline 1 \cdots 1 & 1 \end{array} \right)$, and the row-reduced form showing the steady-state distribution. To find the steady-state distribution in Example 3, we proceed as follows:

a. Enter the T in matrix [A].

b. Run the program MARKOV and press ENTER several times to display T^k for $k = 1, 2, 3, 32,$ and 64, and then $\left(\begin{array}{c|c} T^t - I & 0 \\ \hline 1 \cdots 1 & 1 \end{array} \right)$ followed by its row-reduced form.

```
MATRIXCAJ 3 X3
[.6   .1   .3   ]
[.2   .7   .1   ]
[0    .5   .5   ]
```

```
[A]
[[3/5  1/10  3/10...
 [1/5  7/10  1/10...
 [0    1/2   1/2  ...
```

```
[A]^64
[[1/4  1/2  1/4]
 [1/4  1/2  1/4]
 [1/4  1/2  1/4]]
```

```
(A^T-I)X=0
AND (1..1)X=1 IS
[[-2/5  1/5   0   ...
 [1/10  -3/10 1/...
 [3/10  1/10  -1...
 [1     1     1  ...
```

```
SOLUTION
[[1  0  0  1/4]
 [0  1  0  1/2]
 [0  0  1  1/4]
 [0  0  0  0  ]]
```

*See the Preface for information on how to obtain this and other graphing calculator programs.

The steady state distribution $\begin{pmatrix} \frac{1}{4} & \frac{1}{2} & \frac{1}{4} \end{pmatrix}$ shown in the right-hand column agrees with the answer found earlier.

7.2 Section Summary

A transition matrix is *regular* if some power of it has all positive entries. The long-term behavior of a regular Markov chain with transition matrix T is the solution D of the *steady-state distribution* equation

$$D \cdot T = D$$

D may be found by row-reducing the augmented matrix

$$\left(\begin{array}{c|c} T^t - I & 0 \\ \hline 1 \cdots 1 & 1 \end{array} \right) \quad \text{to obtain} \quad \left(\begin{array}{c|c} I & D^t \\ \hline 0 & 0 \end{array} \right)$$

and then transposing the last column (omitting the bottom zero).

▶ Solutions to Practice Problems

1. $\begin{pmatrix} \frac{1}{2} & \frac{1}{2} \end{pmatrix} \cdot \begin{pmatrix} \frac{1}{4} & \frac{3}{4} \\ \frac{3}{4} & \frac{1}{4} \end{pmatrix} = \begin{pmatrix} \frac{1}{2} \cdot \frac{1}{4} + \frac{1}{2} \cdot \frac{3}{4} & \frac{1}{2} \cdot \frac{3}{4} + \frac{1}{2} \cdot \frac{1}{4} \end{pmatrix} = \begin{pmatrix} \frac{4}{8} & \frac{4}{8} \end{pmatrix} = \begin{pmatrix} \frac{1}{2} & \frac{1}{2} \end{pmatrix}$

   Therefore, $\begin{pmatrix} \frac{1}{2} & \frac{1}{2} \end{pmatrix}$ is a steady-state distribution for the transition matrix.

2. $T^t - I = \begin{pmatrix} 0 & 1 \\ \frac{1}{2} & \frac{1}{2} \end{pmatrix}^t - \begin{pmatrix} 1 & 0 \\ 0 & 1 \end{pmatrix} = \begin{pmatrix} 0 & \frac{1}{2} \\ 1 & \frac{1}{2} \end{pmatrix} - \begin{pmatrix} 1 & 0 \\ 0 & 1 \end{pmatrix} = \begin{pmatrix} -1 & \frac{1}{2} \\ 1 & -\frac{1}{2} \end{pmatrix}$

   so $\left(\begin{array}{c|c} T^t - I & 0 \\ \hline 1 \cdots 1 & 1 \end{array} \right) = \left(\begin{array}{cc|c} -1 & \frac{1}{2} & 0 \\ 1 & -\frac{1}{2} & 0 \\ \hline 1 & 1 & 1 \end{array} \right).$

   One of the many ways to row-reduce this matrix (all leading to the same result) is:

   $\begin{array}{c} \\ R_2 + R_1 \to \\ R_3 + R_1 \to \end{array} \begin{pmatrix} -1 & \frac{1}{2} & 0 \\ 0 & 0 & 0 \\ 0 & \frac{3}{2} & 1 \end{pmatrix} \begin{array}{c} -R_1 \to \\ R_3 \to \\ R_2 \to \end{array} \begin{pmatrix} 1 & -\frac{1}{2} & 0 \\ 0 & \frac{3}{2} & 1 \\ 0 & 0 & 0 \end{pmatrix} \Rightarrow \frac{2}{3} R_2 \to \begin{pmatrix} 1 & -\frac{1}{2} & 0 \\ 0 & 1 & \frac{2}{3} \\ 0 & 0 & 0 \end{pmatrix} \Rightarrow \begin{array}{c} R_1 + \frac{1}{2} R_2 \to \\ \\ \end{array} \begin{pmatrix} 1 & 0 & \frac{1}{3} \\ 0 & 1 & \frac{2}{3} \\ 0 & 0 & 0 \end{pmatrix}$

   The steady-distribution is $D = \begin{pmatrix} \frac{1}{3} & \frac{2}{3} \end{pmatrix}.$

7.2 Exercises

Identify each transition matrix as "regular" or "not regular."

1. $\begin{pmatrix} 0.15 & 0.85 \\ 0.40 & 0.60 \end{pmatrix}$

2. $\begin{pmatrix} 0 & 0 & 1 \\ 0 & 0 & 1 \\ 0.5 & 0.5 & 0 \end{pmatrix}$

3. $\begin{pmatrix} 0.75 & 0.25 \\ 1 & 0 \end{pmatrix}$

4. $\begin{pmatrix} 0 & 0.4 & 0.6 \\ 0 & 0.8 & 0.2 \\ 1 & 0 & 0 \end{pmatrix}$

5. $\begin{pmatrix} 0.75 & 0.25 \\ 0 & 1 \end{pmatrix}$

6. $\begin{pmatrix} 1 & 0 & 0 \\ 0 & 0.8 & 0.2 \\ 0 & 0.4 & 0.6 \end{pmatrix}$

7. $\begin{pmatrix} 0.7 & 0.3 & 0 \\ 0 & 0.3 & 0.7 \\ 0.7 & 0.3 & 0 \end{pmatrix}$

8. $\begin{pmatrix} 1 & 0 & 0 \\ 0 & 0 & 1 \\ 0 & 1 & 0 \end{pmatrix}$

18. $\begin{pmatrix} 0 & 0.5 & 0.5 \\ 1 & 0 & 0 \\ 0.5 & 0 & 0.5 \end{pmatrix}$, $S_2 \to S_2$

19. $\begin{pmatrix} 0 & 0 & 1 \\ 0.5 & 0 & 0.5 \\ 0 & 1 & 0 \end{pmatrix}$, $S_1 \to S_1$

20. $\begin{pmatrix} 0 & 0 & 0 & 1 \\ 0 & 0 & 1 & 0 \\ 0.5 & 0 & 0 & 0.5 \\ 0 & 1 & 0 & 0 \end{pmatrix}$, $S_4 \to S_1$

Represent each situation as a Markov chain by constructing a state-transition diagram and the corresponding transition matrix. Find the steady-state distribution and interpret it in terms of the original situation. Be sure to state your final answer in terms of the original question.

For each transition matrix T:
a. Find the steady-state distribution.
b. Calculate T^k for $k = 5, 10, 20,$ and 50 to verify that the rows of T^k approach the steady-state distribution.

9. $\begin{pmatrix} 0.4 & 0.6 \\ 0.6 & 0.4 \end{pmatrix}$

10. $\begin{pmatrix} 0.6 & 0.4 \\ 0.1 & 0.9 \end{pmatrix}$

11. $\begin{pmatrix} 0.55 & 0.45 \\ 0.30 & 0.70 \end{pmatrix}$

12. $\begin{pmatrix} 0.45 & 0.55 \\ 0.70 & 0.30 \end{pmatrix}$

13. $\begin{pmatrix} 0.1 & 0.8 & 0.1 \\ 0.2 & 0.7 & 0.1 \\ 0.4 & 0.5 & 0.1 \end{pmatrix}$

14. $\begin{pmatrix} 0.7 & 0.2 & 0.1 \\ 0.1 & 0.3 & 0.6 \\ 0.5 & 0.3 & 0.2 \end{pmatrix}$

15. $\begin{pmatrix} 0 & 0.6 & 0.4 \\ 0.1 & 0.9 & 0 \\ 0.3 & 0.5 & 0.2 \end{pmatrix}$

16. $\begin{pmatrix} 0 & 0.6 & 0.4 \\ 0.2 & 0.8 & 0 \\ 0 & 0.6 & 0.4 \end{pmatrix}$

For each regular transition matrix T:
a. Construct a state-transition diagram and find the smallest number k of transitions needed to move from the given state S_i to state S_j and the probability of that transition.
b. Use this k to verify that T^k has a positive transition probability in row i and column j equal to the probability found in (a).

17. $\begin{pmatrix} 0 & 0 & 1 \\ 0.5 & 0.5 & 0 \\ 0 & 1 & 0 \end{pmatrix}$, $S_2 \to S_3$

21. **SOCIAL SCIENCE: Voting Patterns** The students in the political science summer program at Edson State College are studying voting patterns in Marston County. Half of the students reviewed voter records at the County Clerk's Office and found that a person who voted in an election has an 80% chance of voting in the next election, while someone who did not vote in an election has a 30% chance of voting in the next election. The other half of the students conducted surveys door to door and at shopping centers and found that voter perceptions were somewhat different: 90% of those who claimed to have voted in the last election said they would vote in the next, while 40% of those who said they hadn't voted in the last election said they would vote in the next. If these findings are valid for predicting long-term trends, which survey is consistent with the national average of about 61% of eligible voters actually voting? (*Source:* Committee for the Study of the American Electorate)

22. **SOCIAL SCIENCE: Population Dynamics** Life is so good in Lucas, Marion, and Warren counties in upstate Maine that no one ever moves away from the tri-county region. However, each year 4% of the Lucas residents move to Marion

and 2% move to Warren; 2% of the Marion residents move to Lucas and 2% move to Warren; and 2% of the Warren residents move to Lucas and 1% move to Marion. After many years, how many reside in each county if the combined population of these three counties is 11,200?

23. **SOCIAL SCIENCE: Mass Transit** Commuters in the Pittsburgh metropolitan area either drive alone, join a car pool, or take the bus. Each month, of those who drive by themselves, 20% join a car pool, 30% switch to the bus, and the rest continue driving alone; of those who are in a car pool, 30% switch to driving alone, 20% switch to the bus, and the rest stay in a car pool; and of those who take the bus, 20% switch to driving alone, 30% join a car pool, and the rest continue taking the bus. After many months, how many of the three million commuters will be driving alone?

24. **GENERAL: Art Gallery Shows** The Harmon Gallery's showing of paintings by Clyberg, Stevensen, and Georgan attracted 1100 art dealers and collectors on its opening night. The security staff reported the following pattern of crowd movement every five minutes throughout the evening: Of the crowd in the Clyberg room, 10% stayed, 30% moved on to the Stevensen exhibit, and 60% went for more hors d'oeuvres. Of those in the Stevensen room, 30% stayed, 30% moved to the Clyberg exhibit, 20% moved to the Georgan exhibit, and 20% went for more hors d'oeuvres. Of those in the Georgan room, 20% stayed, 70% moved on to the Stevensen exhibit, and 10% went for more hors d'oeuvres. And of those in the central refreshments room, 40% stayed for more hors d'oeuvres, 10% went to the Clyberg exhibit, 20% went to the Stevensen exhibit, and 30% went to the Georgan exhibit. After several hours of milling about in this fashion, how many people were in each room at the gallery?

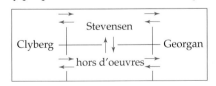

25. **BUSINESS: Market Share** The Peerless Products Corporation has decided to market a new toothpaste, DentiMint, designed to compete successfully with the market leaders: ConfiDent, SuperSmile, and MaxiWhite. Test marketing results from several cities indicate that each week, of those who used ConfiDent the previous week, 40% will buy it again, 20% will switch to SuperSmile, 30% will switch to MaxiWhite, and 10% will switch to DentiMint. Of those who used SuperSmile the previous week, 40% will buy it again, 20% will switch to ConfiDent, 10% will switch to MaxiWhite, and 30% will switch to DentiMint. Of those who used MaxiWhite the previous week, 40% will buy it again, 30% will switch to ConfiDent, 10% will switch to SuperSmile, and 20% will switch to DentiMint. And of those who used DentiMint the previous week, 40% will buy it again, 10% will switch to ConfiDent, 10% will switch to SuperSmile, and 40% will switch to MaxiWhite. If these buying patterns continue, what will the long-term market share be for DentiMint toothpaste?

26. **BUSINESS: Apple Harvests** The apple harvest in the Shenandoah Valley is rated as excellent, average, or poor. Following an excellent harvest, the chances of having an excellent, average, or poor harvest are 0.5, 0.3, and 0.2, respectively. Following an average harvest, the chances of having an excellent, average, or poor harvest are 0.3, 0.5, and 0.2, respectively. And following a poor harvest, the chances of having an excellent, average, or poor harvest are 0.6, 0.2, and 0.2, respectively. Assuming these trends continue, find the long-term chance for an excellent harvest.

27. **BUSINESS: Car Insurance** The records of an insurance broker in Boston classify the auto policy holders as preferred, satisfactory, poor, or in the assigned-risk pool. Each year 20% of the preferred policies are downgraded to satisfactory; 30% of the satisfactory policies are upgraded to preferred and another 20% are downgraded to poor; 60% of the poor policies are upgraded to satisfactory but 30% are placed in the assigned-risk pool; and just 20% of those in the assigned-risk pool are moved up to the poor classification. Assuming that these trends continue for many years, what percentage of the broker's auto policies are rated satisfactory or better?

28. **GENERAL: Rumor Accuracy** When repeating a rumor, gossips sometimes make mistakes and spread misinformation instead. Suppose the residents of a small town repeat a rumor that the mayor has been convicted of tax fraud at a trial held in the state capital, but that when they pass it on, 1% of the gossips reverse the result of whatever they were told. What is the long-term accuracy of the gossips' information?

29. **BUSINESS: Renters Who Move** The rental records at a Denver apartment building show that each year 78% of those who lived in the building the entire previous year will remain for the next year, while only 44% of those who moved in during the year will remain for the next year. If these observations represent a long-term trend, are they consistent with the U.S. Census statistic that about 34.3% of renters nationwide move each year?

30. **BUSINESS: Homeowners Who Move** The property transfer records at a real estate agency in the Salt Lake City suburbs show that each year 96% of those who owned their own house the entire previous year will remain for the next year, while only 46% of those who moved into their own house during the year will remain for the next year. If these observations represent a long-term trend, are they consistent with the U.S. Census statistic that about 8.9% of homeowners nationwide move each year?

7.3 ABSORBING MARKOV CHAINS

Introduction

A state is *absorbing* if, once you enter it, you cannot leave it. If a Markov chain has several absorbing states, will it always end up in one of them? If so, which one and how soon? Unlike regular Markov chains, where the initial distribution has no influence on the long-term outcome, with absorbing states the initial distribution *does* affect when and where the eventual absorption occurs. As in the regular case, we will see that much useful information is given by high powers of the transition matrix.

Absorbing Markov Chains

For an *absorbing state,* the transition probabilities are 1 from that state to itself and 0 to every other state. A Markov chain is *absorbing* if it has at least one absorbing state *and* if from every nonabsorbing state it is possible to reach an absorbing state (in some number of steps).

EXAMPLE 1 **FINDING WHETHER A MARKOV CHAIN IS ABSORBING**

Determine whether each Markov chain is absorbing.

a. b.

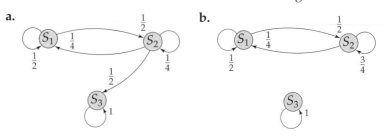

7.3 ABSORBING MARKOV CHAINS

Solution

a. This *is* an absorbing Markov chain since it has an absorbing state, S_3, which can be reached from S_1 in 2 steps and from S_2 in 1 step.

b. This is *not* an absorbing Markov chain. Although S_3 is an absorbing state, it cannot be reached from either S_1 or S_2 (the nonabsorbing states).

EXAMPLE 2 WETLANDS POLLUTION

The Southern Electric Company has filed an application with the Environmental Protection Agency to store heavy-metal waste (mostly cadmium and mercury) in a clay-lined storage pool at its manufacturing plant in Mayfield. An independent engineering consultant has certified that only 1% of the waste will leach out of the pool into the ground water each year. But a hydrologist working with the local chapter of People for a Cleaner Planet has pointed out that each year 2% of the contaminants in the ground water reach the Marlin Memorial Wetlands Conservation Area, where they remain indefinitely. Represent this situation by a transition diagram and matrix. Show that the wetlands are an absorbing state for the pollution.

Solution

The waste can be in one of three locations: at the plant (state P), in the ground water (state G), or in the wetlands (state W). The given information is shown in the following state-transition diagram:

$$T = \begin{array}{c} \text{Current} \\ \text{state} \end{array} \begin{array}{c} P \\ G \\ W \end{array} \overset{\begin{array}{ccc} \text{Next state} \\ P & G & W \end{array}}{\begin{pmatrix} 0.99 & 0.01 & 0 \\ 0 & 0.98 & 0.02 \\ 0 & 0 & 1 \end{pmatrix}}$$

Since these changes occur each year, we have a Markov chain with transition matrix T. W is an absorbing state because once the contamination reaches the Wetlands, it does not leave, and the other states P and G are nonabsorbing. The state-transition diagram shows that it is possible for some of the waste to move from the plant storage pool to the groundwater and on to the wetlands, so this *is* an absorbing Markov chain.

$$T^2 = \begin{pmatrix} 0.9801 & 0.0197 & 0.0002 \\ 0 & 0.9604 & 0.0396 \\ 0 & 0 & 1 \end{pmatrix}$$

We could also use the transition matrix to show that from each nonabsorbing state it is possible to reach the absorbing state. Squaring the transition matrix T gives the matrix on the left, and since the last column (corresponding to the wetlands) is all positive, it is possible to reach the wetlands from any other state in two steps.

How much of the heavy-metal waste will reach the Wetlands, and how soon is it expected to get there? Will *all* the contamination ultimately reach the wetlands? In Example 4 on pages 508–509 we will answer these important questions.

Standard Form

The transition matrix of an absorbing Markov chain is in *standard form* if it is written so that the absorbing states are listed *before* the nonabsorbing states. Since the absorbing states themselves may be in any order (as may the nonabsorbing states), the standard form is not necessarily unique.

> **Standard Form of an Absorbing Transition Matrix**
>
> An absorbing transition matrix is in *standard form* if the absorbing states appear before the nonabsorbing states, in which case it takes the form
>
> $$\begin{array}{c} \text{Absorbing } \{ \\ \text{Nonabsorbing } \{ \end{array} \left(\begin{array}{c|c} I & 0 \\ \hline R & Q \end{array} \right) \quad \begin{array}{l} I = \text{identity matrix} \\ 0 = \text{zero matrix} \end{array}$$
>
> R is the matrix of transition probabilities from the nonabsorbing states to the absorbing states and Q is the matrix of transition probabilities from nonabsorbing states to nonabsorbing states.

Any absorbing transition matrix can be rewritten in standard form by reordering the rows so the absorbing states are above the nonabsorbing states and then reordering the columns to have the same order as the rows.

EXAMPLE 3

REWRITING A TRANSITION MATRIX IN STANDARD FORM

Rewrite the following absorbing transition matrix in standard form and identify the matrices R and Q.

$$\begin{pmatrix} 0.80 & 0.03 & 0.10 & 0.02 & 0.05 \\ 0 & 1 & 0 & 0 & 0 \\ 0.10 & 0.07 & 0.20 & 0.03 & 0.60 \\ 0 & 0 & 0 & 1 & 0 \\ 0.05 & 0.01 & 0.10 & 0.04 & 0.80 \end{pmatrix}$$

Solution

We first identify the absorbing states. Since the row of an absorbing state consists of all zeros except one 1 in the *same column as the row*, the second and fourth rows represent absorbing states, which we will call A_1 and A_2. The other rows represent the nonabsorbing states, which we will call N_1, N_2 and N_3:

$$\begin{array}{c} \\ \text{Absorbing} \\ \text{rows (1's on} \\ \text{the main} \\ \text{diagonal)} \end{array} \begin{array}{c} N_1 \\ \rightarrow A_1 \\ N_2 \\ \rightarrow A_2 \\ N_3 \end{array} \begin{pmatrix} 0.80 & 0.03 & 0.10 & 0.02 & 0.05 \\ 0 & 1 & 0 & 0 & 0 \\ 0.10 & 0.07 & 0.20 & 0.03 & 0.60 \\ 0 & 0 & 0 & 1 & 0 \\ 0.05 & 0.01 & 0.10 & 0.04 & 0.80 \end{pmatrix}$$

Putting the rows in the order A_1, A_2, N_1, N_2, N_3, we have

$$\begin{array}{c} \\ A_1 \\ A_2 \\ N_1 \\ N_2 \\ N_3 \end{array} \begin{array}{ccccc} N_1 & A_1 & N_2 & A_2 & N_3 \end{array} \begin{pmatrix} 0 & 1 & 0 & 0 & 0 \\ 0 & 0 & 0 & 1 & 0 \\ 0.80 & 0.03 & 0.10 & 0.02 & 0.05 \\ 0.10 & 0.07 & 0.20 & 0.03 & 0.60 \\ 0.05 & 0.01 & 0.10 & 0.04 & 0.80 \end{pmatrix} \quad \text{First fix the rows}$$

and then putting the columns in the same order gives

$$\begin{array}{c} \\ A_1 \\ A_2 \\ N_1 \\ N_2 \\ N_3 \end{array} \begin{array}{ccccc} A_1 & A_2 & N_1 & N_2 & N_3 \end{array} \left(\begin{array}{cc|ccc} 1 & 0 & 0 & 0 & 0 \\ 0 & 1 & 0 & 0 & 0 \\ 0.03 & 0.02 & 0.80 & 0.10 & 0.05 \\ 0.07 & 0.03 & 0.10 & 0.20 & 0.60 \\ 0.01 & 0.04 & 0.05 & 0.10 & 0.80 \end{array} \right) \quad \text{Then fix the columns}$$

$$\underbrace{}_{R} \quad \underbrace{}_{Q}$$

R is the matrix of transition probabilities from the nonabsorbing states to the absorbing states, while Q is the matrix of transition probabilities from the nonabsorbing states to the nonabsorbing states:

$$R = \begin{array}{c} N_1 \\ N_2 \\ N_3 \end{array} \begin{pmatrix} A_1 & A_2 \\ 0.03 & 0.02 \\ 0.07 & 0.03 \\ 0.01 & 0.04 \end{pmatrix} \qquad Q = \begin{array}{c} N_1 \\ N_2 \\ N_3 \end{array} \begin{pmatrix} N_1 & N_2 & N_3 \\ 0.80 & 0.10 & 0.05 \\ 0.10 & 0.20 & 0.60 \\ 0.05 & 0.10 & 0.80 \end{pmatrix}$$

Practice Problem 1 Rewrite the transition matrix $\begin{pmatrix} & P & G & W \\ P & 0.99 & 0.01 & 0 \\ G & 0 & 0.98 & 0.02 \\ W & 0 & 0 & 1 \end{pmatrix}$ from Example 2 in standard form.

▶ Solution on page 513

Transition Times and Absorption Probabilities

Just as with regular Markov chains, the long-term behavior of an absorbing chain can be seen from high powers of the transition matrix. However, even more information about the ultimate behavior of the chain is provided by a theorem about the limiting form of these powers. The following results are proved in Exercises 41–50.

Expected Transition Times and Long-Term Absorption Probabilities

For an absorbing transition matrix in standard form

$$T = \left(\begin{array}{c|c} I & 0 \\ \hline R & Q \end{array} \right) \quad \begin{array}{l} \text{Absorbing states before} \\ \text{nonabsorbing states} \end{array}$$

high powers of T approach a matrix of the form

$$T^* = \left(\begin{array}{c|c} I & 0 \\ \hline F \cdot R & 0 \end{array} \right) \quad \begin{array}{l} T^* \text{ gives the long-term} \\ \text{transition probabilities} \end{array}$$

where the matrix F is given by $F = (I - Q)^{-1}$ and is called the *fundamental matrix* of the chain. It provides the following information:

1. The entry in row i and column j of F gives the expected number of times that the chain, if it begins in state N_i, will be in state N_j before being absorbed.

2. The sum of the entries in row i of F gives the expected number of times that the chain, if it begins in state N_i, will be in the nonabsorbing states before being absorbed.

3. The entry in row i and column j of the matrix $F \cdot R$ gives the probability that the chain, if it begins in state N_i, will be absorbed into the state A_j.

EXAMPLE 4 FINDING EXPECTED TIME UNTIL ABSORPTION

Find the expected number of years until the heavy-metal waste in the plant (state P) will reach the Wetlands Conservation Area (state W) for the Markov chain described in Example 2.

Solution

Writing the transition matrix from Example 2 in standard form as

$$T = \begin{array}{c} \\ W \\ G \\ P \end{array} \begin{array}{c} W \quad G \quad P \end{array} \left(\begin{array}{c|cc} 1 & 0 & 0 \\ \hline 0.02 & 0.98 & 0 \\ 0 & 0.01 & 0.99 \end{array} \right) \qquad \text{Listing first the absorbing state } W \text{ (for "wetlands")}$$

we have

$$Q = \begin{array}{c} \\ G \\ P \end{array} \begin{array}{c} G \quad P \end{array} \begin{pmatrix} 0.98 & 0 \\ 0.01 & 0.99 \end{pmatrix} \qquad \text{From } T = \left(\begin{array}{c|c} I & 0 \\ \hline R & Q \end{array} \right)$$

and

$$I - Q = \begin{pmatrix} 1 & 0 \\ 0 & 1 \end{pmatrix} - \begin{pmatrix} 0.98 & 0 \\ 0.01 & 0.99 \end{pmatrix} = \begin{pmatrix} 0.02 & 0 \\ -0.01 & 0.01 \end{pmatrix}$$

We calculate the fundamental matrix $F = (I - Q)^{-1}$ by row-reducing "$(A|I)$ to obtain $(I|A^{-1})$" (see page 214).

$$\begin{pmatrix} 0.02 & 0 & | & 1 & 0 \\ -0.01 & 0.01 & | & 0 & 1 \end{pmatrix} \qquad (A|I)$$

$$R_2 + \tfrac{1}{2}R_1 \rightarrow \begin{pmatrix} 0.02 & 0 & | & 1 & 0 \\ 0 & 0.01 & | & \tfrac{1}{2} & 1 \end{pmatrix}$$

$$\begin{array}{c} 50R_1 \rightarrow \\ 100R_2 \rightarrow \end{array} \begin{pmatrix} 1 & 0 & | & 50 & 0 \\ 0 & 1 & | & \underbrace{50 \quad 100}_{F} \end{pmatrix} \qquad (I|A^{-1})$$

The fundamental matrix F is the matrix on the right of the bar in the row-reduced matrix. From part (1) of the box on the previous page, its entries have the following interpretations.

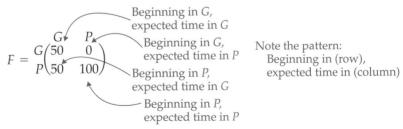

$$F = \begin{array}{c} \\ G \\ P \end{array} \begin{array}{c} G \quad P \end{array} \begin{pmatrix} 50 & 0 \\ 50 & 100 \end{pmatrix}$$

Note the pattern:
Beginning in (row),
expected time in (column)

From part (2) of the box on the previous page, the expected number of years until the heavy-metal waste in the plant (state P) is absorbed into the wetlands (the absorbing state W) is the sum of the numbers in row P: $50 + 100 = 150$ years.

Practice Problem 2

How long is it expected that the heavy-metal waste will spend in the plant storage pool (state P) before absorption? How long in the ground water (state G)? ➤ **Solution on page 513**

Graphing Calculator Exploration

(*Continuation of Example 3*) To find the probability that the absorbing Markov chain in Example 3 (pages 506–507) will be absorbed into state A_2 given that it started in state N_3, proceed as follows.

Enter the matrices Q and R from the solution of Example 3 into the calculator as [A] and [B], along with a 3×3 identity matrix in [I], as shown in the first three screens below.

```
MATRIXIAI 3 X3        MATRIX[B] 3 X2       MATRIX[I] 3 X3       ([I]-[A])^-1*[B]
[ .8  .1  .05 ]       [ .03  .02 ]         [ 1  0  0 ]          [[ .46  .54 ]
[ .1  .2  .6  ]       [ .07  .03 ]         [ 0  1  0 ]          [  .43  .57 ]
[ .05 .1  .8  ]       [ .01  .04 ]         [ 0  0  1 ]          [  .38  .62 ]]

3,3=.8                3,2=.04              3,3=1
```

Then calculate $F \cdot R$ (with $F = (I - Q)^{-1}$) as $([I] - [A])^{-1} * [B]$, as shown in the last screen above.

By part (3) of the result on page 508, the probability that the chain, beginning in state N_3, will be absorbed into state A_2 is the entry in row 3 and column 2 of $F \cdot R$, which is 0.62.

Practice Problem 3

From the matrix $F \cdot R$ in the preceding Graphing Calculator Exploration, what is the probability that the chain, if it begins in state N_1, will be absorbed into state A_2? ➤ **Solution on page 513**

Both the preceding Graphing Calculator Exploration and Practice Problem 3 found probabilities of being absorbed into state A_2, but beginning in different states. The fact that we obtained two different probabilities (0.62 and 0.54) shows that for an absorbing Markov chain, the initial state *does* affect the ultimate behavior. This is quite different from a *regular* Markov chain, where the initial state had no effect on the ultimate distribution.

EXAMPLE 5

FARM SIZE DISTRIBUTION*

An analysis of Census data for farms in the Grand Forks metropolitan area from 1950 to 2000 found the following probabilities for transitions between small (state S) family-owned farms of 50 to 200 acres, medium (state M) family-owned farms of 201 to 600 acres, large (state L) family-owned farms of 601 to 1000 acres, housing developments (state H), and corporate-owned agribusiness farms (state C). Notice that once a farm becomes a housing development or corporate-owned, it never returns to family-owned farm status again.

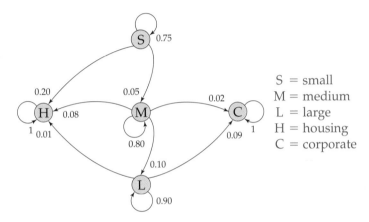

S = small
M = medium
L = large
H = housing
C = corporate

Assuming that these yearly transition patterns continue, what is the expected remaining lifetime of a medium family-owned farm (that is, the time until it becomes a housing development or owned by a corporation)? What is the probability that it will become a housing development?

Solution

We represent the situation as a Markov chain. Since H and C are absorbing states, one standard form for the transition matrix is

$$\begin{array}{c} \\ H \\ C \\ S \\ M \\ L \end{array} \begin{pmatrix} H & C & S & M & L \\ 1 & 0 & 0 & 0 & 0 \\ 0 & 1 & 0 & 0 & 0 \\ 0.20 & 0 & 0.75 & 0.05 & 0 \\ 0.08 & 0.02 & 0 & 0.80 & 0.10 \\ 0.01 & 0.09 & 0 & 0 & 0.90 \end{pmatrix}$$

---

*This example is based on ideas in "Projection of Farm Numbers for North Dakota with Markov Chains" by Ronald D. Krenz, *Agricultural Economics Research* **XVI**(3): 77–83.

So

$$R = \begin{matrix} & H & C \\ S & \\ M & \\ L & \end{matrix}\begin{pmatrix} 0.20 & 0 \\ 0.08 & 0.02 \\ 0.01 & 0.09 \end{pmatrix} \quad \text{and} \quad Q = \begin{matrix} & S & M & L \\ S & \\ M & \\ L & \end{matrix}\begin{pmatrix} 0.75 & 0.05 & 0 \\ 0 & 0.80 & 0.10 \\ 0 & 0 & 0.90 \end{pmatrix}$$

Using a calculator or pencil and paper, we find that the fundamental matrix $F = (I - Q)^{-1}$ is

$$\left(\begin{pmatrix} 1 & 0 & 0 \\ 0 & 1 & 0 \\ 0 & 0 & 1 \end{pmatrix} - \begin{pmatrix} 0.75 & 0.05 & 0 \\ 0 & 0.80 & 0.10 \\ 0 & 0 & 0.90 \end{pmatrix} \right)^{-1} = \underbrace{\begin{matrix} & S & M & L \\ S \\ M \\ L \end{matrix}\begin{pmatrix} 4 & 1 & 1 \\ 0 & 5 & 5 \\ 0 & 0 & 10 \end{pmatrix}}_{F} \quad \begin{array}{l} F \text{ gives} \\ \text{expected} \\ \text{times} \end{array}$$

and $F \cdot R$ is

$$\underbrace{\begin{matrix} S \\ M \\ L \end{matrix}\begin{pmatrix} 4 & 1 & 1 \\ 0 & 5 & 5 \\ 0 & 0 & 10 \end{pmatrix} \cdot \begin{matrix} & H & C \\ & \\ & \\ & \end{matrix}\begin{pmatrix} 0.20 & 0 \\ 0.08 & 0.02 \\ 0.01 & 0.09 \end{pmatrix}}_{F \cdot R} = \begin{matrix} & H & C \\ S \\ M \\ L \end{matrix}\begin{pmatrix} 0.89 & 0.11 \\ 0.45 & 0.55 \\ 0.10 & 0.90 \end{pmatrix} \quad \begin{array}{l} F \cdot R \text{ gives} \\ \text{absorption} \\ \text{probabilities} \\ \text{from } row \text{ to} \\ column \end{array}$$

The expected remaining lifetime of a medium family-owned farm (until it becomes housing or corporate-owned) is $0 + 5 + 5 = 10$ years, and the probability that it will become a housing development is 0.45, or 45%.

Practice Problem 4 From these matrices, what is the expected remaining lifetime of a *small* family-owned farm (until it becomes housing or corporate-owned), and what is the probability that it will become corporate-owned?

▶ Solution on next page

7.3 Section Summary

For an *absorbing* state, the transition probabilities are 1 from itself to itself and 0 to every other state. An *absorbing Markov chain* has at least one absorbing state and from each nonabsorbing state it is possible to reach some absorbing state (in some number of moves). The transition matrix is in *standard form* if the absorbing states are listed before the nonabsorbing states.

$$\begin{array}{c} \\ \text{Absorbing } \{ \\ \text{Nonabsorbing } \{ \end{array} \overbrace{\left(\begin{array}{c|c} I & 0 \\ \hline R & Q \end{array} \right)}^{\text{Absorbing Nonabsorbing}}$$

R gives the transition probabilities from the nonabsorbing states to the absorbing states, and Q gives the transition probabilities from the nonabsorbing states to the nonabsorbing states. Much information is given by the fundamental matrix $F = (Q - I)^{-1}$ and the product $F \cdot R$:

$$F = (Q - I)^{-1} = \begin{array}{c} N_1 \ldots \\ N_1 \\ \vdots \end{array} \!\!\! \left(\right) \qquad F \cdot R = \begin{array}{c} A_1 \ldots \\ N_1 \\ \vdots \end{array} \!\!\! \left(\right)$$

1. The entry in row i and column j of F gives the expected number of times that the chain, if it begins in state N_i, will be in state N_j before being absorbed.

2. The sum of the entries in row i of F gives the expected number of times that the chain, if it begins in state N_i, will be in the nonabsorbing states before being absorbed.

3. The entry in row i and column j of $F \cdot R$ gives the probability that the chain, if it begins in state N_i, will be absorbed into the state A_j.

▶ **Solutions to Practice Problems**

1. Both $\begin{array}{c} \\ W \\ P \\ G \end{array} \!\! \begin{array}{c} W \quad P \quad G \\ \left(\begin{array}{ccc} 1 & 0 & 0 \\ 0 & 0.99 & 0.01 \\ 0.02 & 0 & 0.98 \end{array} \right) \end{array}$ and $\begin{array}{c} \\ W \\ G \\ P \end{array} \!\! \begin{array}{c} W \quad G \quad P \\ \left(\begin{array}{ccc} 1 & 0 & 0 \\ 0.02 & 0.98 & 0 \\ 0 & 0.01 & 0.99 \end{array} \right) \end{array}$ are standard forms of the transition matrix.

2. The expected times are 100 years in the plant storage pool and then 50 years in the ground water.

3. 0.54 (from the entry in row 1 and column 2 of $F \cdot R$).

4. The expected lifetime is $4 + 1 + 1 = 6$ years, and the probability that it will become corporate-owned is 0.11, or 11%.

7.3 Exercises

For each state-transition diagram, identify the absorbing and the nonabsorbing states. For the given starting state, find the smallest number of steps required to reach an absorbing state, and the probability of that sequence of steps.

1. Starting in state S_1

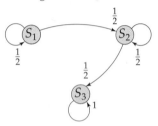

2. Starting in state S_2

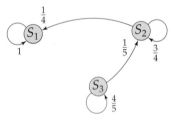

3. Starting in state S_4

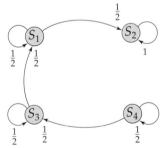

4. Starting in state S_4

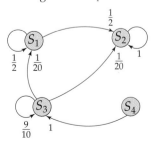

For each absorbing transition matrix T:

a. Draw a state-transition diagram, identify the absorbing and the nonabsorbing states, and find the smallest integer k such that after k transitions every nonabsorbing state can reach an absorbing state.

b. Verify your value for k by finding T^k and checking that every nonabsorbing state row has a positive probability in at least one absorbing state column.

5. $\begin{pmatrix} 1 & 0 & 0 \\ 0.20 & 0.55 & 0.25 \\ 0 & 0.25 & 0.75 \end{pmatrix}$

6. $\begin{pmatrix} 0.75 & 0 & 0 & 0.25 \\ 0 & 0.50 & 0.50 & 0 \\ 0 & 0 & 1 & 0 \\ 0 & 0.25 & 0.25 & 0.50 \end{pmatrix}$

7. $\begin{pmatrix} 0 & 1 & 0 & 0 \\ 0 & 1 & 0 & 0 \\ 0.2 & 0 & 0.6 & 0.2 \\ 0 & 0 & 0.2 & 0.8 \end{pmatrix}$

8. $\begin{pmatrix} 0.50 & 0.05 & 0.30 & 0.15 \\ 0 & 1 & 0 & 0 \\ 0.50 & 0 & 0.50 & 0 \\ 0 & 0 & 0 & 1 \end{pmatrix}$

Rewrite each absorbing transition matrix in standard form and identify the matrices R and Q.

9. $\begin{pmatrix} 0.88 & 0.02 & 0.10 \\ 0 & 1 & 0 \\ 0.04 & 0.16 & 0.80 \end{pmatrix}$

10. $\begin{pmatrix} 1 & 0 & 0 & 0 \\ 0.10 & 0.85 & 0.05 & 0 \\ 0.05 & 0.10 & 0.80 & 0.05 \\ 0 & 0 & 0 & 1 \end{pmatrix}$

11. $\begin{pmatrix} 0.80 & 0 & 0.20 & 0 & 0 \\ 0 & 1 & 0 & 0 & 0 \\ 0 & 0.20 & 0.50 & 0.30 & 0 \\ 0 & 0 & 0 & 1 & 0 \\ 0.20 & 0 & 0.05 & 0 & 0.75 \end{pmatrix}$

12. $\begin{pmatrix} 1 & 0 & 0 & 0 & 0 \\ 0 & 0.65 & 0 & 0.30 & 0.05 \\ 0 & 0 & 1 & 0 & 0 \\ 0.05 & 0.30 & 0.05 & 0.60 & 0 \\ 0 & 0 & 0 & 0 & 1 \end{pmatrix}$

20. $\begin{pmatrix} 1 & 0 & 0 & 0 & 0 \\ 0 & 1 & 0 & 0 & 0 \\ 0.04 & 0.16 & 0.60 & 0.10 & 0.10 \\ 0.06 & 0.04 & 0 & 0.90 & 0 \\ 0.08 & 0.02 & 0.20 & 0 & 0.70 \end{pmatrix}$

Find the fundamental matrix for each absorbing transition matrix in standard form. (You may use 🖩 if permitted by your instructor.)

13. $\begin{pmatrix} 1 & 0 & 0 \\ 0.4 & 0.4 & 0.2 \\ 0.2 & 0.2 & 0.6 \end{pmatrix}$

14. $\begin{pmatrix} 1 & 0 & 0 \\ 0.2 & 0.2 & 0.6 \\ 0.1 & 0.1 & 0.8 \end{pmatrix}$

15. $\begin{pmatrix} 1 & 0 & 0 & 0 \\ 0 & 1 & 0 & 0 \\ 0.10 & 0.10 & 0.20 & 0.60 \\ 0.05 & 0.15 & 0.20 & 0.60 \end{pmatrix}$

16. $\begin{pmatrix} 1 & 0 & 0 & 0 \\ 0 & 1 & 0 & 0 \\ 0.05 & 0 & 0.85 & 0.10 \\ 0.10 & 0.05 & 0.05 & 0.80 \end{pmatrix}$

17. $\begin{pmatrix} 1 & 0 & 0 & 0 \\ 0.1 & 0.2 & 0.1 & 0.6 \\ 0.1 & 0 & 0.9 & 0 \\ 0.2 & 0.2 & 0 & 0.6 \end{pmatrix}$

18. $\begin{pmatrix} 1 & 0 & 0 & 0 \\ 0.2 & 0.5 & 0.2 & 0.1 \\ 0.4 & 0 & 0.4 & 0.2 \\ 0.1 & 0 & 0.1 & 0.8 \end{pmatrix}$

19. $\begin{pmatrix} 1 & 0 & 0 & 0 & 0 \\ 0 & 1 & 0 & 0 & 0 \\ 0.10 & 0.10 & 0.20 & 0.10 & 0.50 \\ 0.05 & 0.05 & 0.10 & 0.80 & 0 \\ 0.20 & 0 & 0.20 & 0.10 & 0.50 \end{pmatrix}$

For Exercises 21–30, use the results of Exercises 13–20 as follows: For each given initial state, find the expected number of times in the given state before absorption, and find the probability of absorption in the given final state using the indicated transition matrix.

	Initial State	Expected Number of Times in State	Probability of Absorption into State	Use the Transition Matrix from Exercise
21.	N_1	N_1	A_1	13
22.	N_1	N_2	A_1	14
23.	N_2	N_1	A_2	15
24.	N_2	N_2	A_2	16
25.	N_2	N_2	A_1	17
26.	N_3	N_2	A_1	18
27.	N_1	N_3	A_1	19
28.	N_1	N_3	A_1	20
29.	N_2	any N_i	A_2	19
30.	N_2	any N_i	A_2	20

Represent each situation as an absorbing Markov chain by constructing a state-transition diagram and a transition matrix in standard form. Use the fundamental matrix to solve each problem. Be sure to state your final answers in terms of the original questions.

31. **BIOMEDICAL: Smallpox** To populations in the New World never before exposed to smallpox, its arrival with European explorers was devastating. Suppose the disease spread through a newly infected town in such a way that each month 20% of the uninfected population became sick, 15% of the sick recovered (and were therefore immune to further infections), and 35% of the sick perished.

a. How much of the town's original population survived?
b. For an infected person, what was the expected number of months for the disease to run its course?

32. **GENERAL: Random Walks and Gambler's Ruin** How long can two players gamble against each other before one wins all and the other is ruined? Suppose each player has $2 and they flip a coin with the loser paying $1 to the winner. Find the expected number of times the game will be played before it ends. [*Hint:* Assume the coin is fair, so the probability of winning or losing on each toss is $\frac{1}{2}$, and define five states for the game, each corresponding to one of the player's money: $0, $1, $2, $3, and $4. Which states are absorbing?]

33. **BIOMEDICAL: Geriatric Care** Each year at the Shady Oaks Assisted Living Facility, 10% of the independent residents are reclassified as requiring assistance, 2% die, and 8% are transferred to a nursing home, while 10% of those needing assistance die and 15% are transferred to a nursing home. What is the probability that a resident requiring assistance will ultimately be transferred to a nursing home?

34. **GENERAL: Inner-City Teachers** A study of teachers in the Chicago area found that every five years 20% of the inner-city teachers continued at their schools, 40% relocated to suburban schools, 20% accepted administrative jobs, and the remaining 20% either retired or quit; 10% of the suburban school teachers relocated to inner-city schools, 70% continued at their schools, 10% accepted administrative jobs, and the remaining 10% either retired or quit; and 10% of the teachers working in administrative positions returned to inner-city teaching, 20% relocated to suburban schools, 60% continued as administrators, and the remaining 10% either retired or quit. Assuming that these trends continue, find the expected number of years an inner-city school teacher will teach at an inner-city school before retiring or quitting.

35. **BUSINESS: Term Insurance** The term life insurance records of the All-County Insurance Company show that every five years those policies that were renewed are renewed again with probability 0.60, renewed with increased coverage with probability 0.20, closed with a death benefit payment with probability 0.07, and discontinued with probability 0.13, while those that had been renewed with increased coverage are renewed again with probability 0.30, renewed with increased coverage with probability 0.60, closed with a death benefit payment with probability 0.08, and discontinued with probability 0.02.
a. What is the expected number of years that a policy that has just been renewed will be in force?
b. What is the probability that a policy that is renewed with increased coverage will be closed with a death benefit payment?

36. **BUSINESS: Management Training** The Big-Beige-Box Computer Company has an in-house program of classroom work and apprenticeships that gives technicians entry to management careers. Each year 25% of those in the classroom move on to the apprenticeship program, 25% drop out, and the remainder continue in the classroom, while 40% of those in the apprenticeship program are promoted to supervisor positions, 10% drop out, and the remainder continue as apprentices. What percentage of those in the classroom will become supervisors?

37. **BIOMEDICAL: Dental Work** A dentist's records indicate that prior to extraction, a tooth needing no work remains that way for another year with probability 0.95, needs a filling with probability 0.03, needs a root canal with probability 0.01, and needs to be extracted with probability 0.01; a tooth with a filling one year needs no work the next with probability 0.90, needs a filling with probability 0.06, needs a root canal with probability 0.02, or needs to be extracted with probability 0.02; and (interestingly enough) a tooth with a root canal one year needs no work the next. Find the expected number of times a tooth needing a filling will need fillings before being extracted.

38. BIOMEDICAL: Genetics A rapidly mutating class of SMX viruses can be separated into distinct strains that are "successful," together with other "mutant" variations. After each year of replications, 90% of the successful strains are still successful, 5% have developed into mutants, and the remaining 5% have become extinct, while 20% of the previous mutants have further evolved to become successful strains, 40% are still mutants, and the remaining 40% have become extinct. What is the expected number of years until a mutant becomes extinct?

39. SOCIAL SCIENCE: Third-World Economic Development The Henderhohf Charitable Trust provides either business advice or low-interest loans to small companies in developing economies around the world. Each year 9% of the companies they advise become successful and leave the program, 65% continue receiving advice, 25% enter the loan program, and 1% go bankrupt, while 7% of the companies receiving loans become successful and leave the program, 15% need no further loans and return to receiving advice, 75% continue receiving loans, and 3% go bankrupt. Is the Trust's claim that "more than three-quarters of the companies they help become successful" justified?

40. GENERAL: College Graduation Rates As the students at Edson State College earn more credits, they progress from freshmen to sophomores to juniors to seniors and then graduate, unless they withdraw from college sometime along the way. If each year 64% of each level moves on to the next (including graduation), 20% remain at their current level, and 16% withdraw, what percentage of incoming freshmen will graduate?

Explorations and Excursions

The following problems extend and augment the material presented in the text.

Proof of the Long-Term Behavior of an Absorbing Markov Chain

The following exercises establish the results on page 508 concerning the long-term behavior of an absorbing Markov chain with transition matrix $T = \left(\begin{array}{c|c} I & 0 \\ \hline R & Q \end{array}\right)$ in standard form.

41. Show that

a. $T^2 = \left(\begin{array}{c|c} I & 0 \\ \hline R & Q \end{array}\right) \cdot \left(\begin{array}{c|c} I & 0 \\ \hline R & Q \end{array}\right)$

$= \left(\begin{array}{c|c} I & 0 \\ \hline R + Q \cdot R & Q^2 \end{array}\right)$

$= \left(\begin{array}{c|c} I & 0 \\ \hline (I + Q) \cdot R & Q^2 \end{array}\right)$

b. $T^3 = T^2 \cdot T = \left(\begin{array}{c|c} I & 0 \\ \hline R + Q \cdot R + Q^2 \cdot R & Q^3 \end{array}\right)$

$= \left(\begin{array}{c|c} I & 0 \\ \hline (I + Q + Q^2) \cdot R & Q^3 \end{array}\right)$

42. Verify Exercise 41 for

$$T = \begin{pmatrix} 1 & 0 & 0 & 0 \\ 0 & 1 & 0 & 0 \\ 0.15 & 0.05 & 0.60 & 0.20 \\ 0.03 & 0.07 & 0.20 & 0.70 \end{pmatrix}.$$

43. Extend Exercise 41 to show that

$$T^m = \left(\begin{array}{c|c} I & 0 \\ \hline (I + Q + \cdots + Q^{m-1}) \cdot R & Q^m \end{array}\right).$$

44. Verify Exercise 43 for

$$T = \begin{pmatrix} 1 & 0 & 0 & 0 & 0 \\ 0 & 1 & 0 & 0 & 0 \\ 0.02 & 0.08 & 0.60 & 0.20 & 0.10 \\ 0.07 & 0.03 & 0.10 & 0.70 & 0.10 \\ 0.15 & 0.05 & 0.20 & 0.40 & 0.20 \end{pmatrix} \text{ and } m = 4.$$

45. Use the fact that T is absorbing to explain why every row of

$$(I + Q + \cdots + Q^{m-1}) \cdot R$$

has at least one positive entry for sufficiently large values of m. Then show that

$$I + Q + \cdots + Q^{m-1} \to (I - Q)^{-1}$$

as m increases. [*Hint:* Use Exercise 60 on page 226.]

46. Verify Exercise 45 for $Q = \begin{pmatrix} 0.11 & 0.02 \\ 0.05 & 0.10 \end{pmatrix}$ by calculating the *finite* sum $I + Q + Q^2 + Q^3 + Q^4$ and checking that rounding this value to four decimal places matches the value of

$$(I - Q)^{-1} = \begin{pmatrix} 1.1250 & 0.0250 \\ 0.0625 & 1.1125 \end{pmatrix}$$

47. Verify Exercise 45 for

$$Q = \begin{pmatrix} 0.12 & 0.02 & 0.02 \\ 0.03 & 0.08 & 0.08 \\ 0.01 & 0.01 & 0.01 \end{pmatrix}$$ by calculating the

finite sum $I + Q + Q^2 + Q^3 + Q^4 + Q^5 + Q^6 + Q^7$ and checking that rounding this value to six decimal places matches the value of

$$(I - Q)^{-1} = \begin{pmatrix} 1.137500 & 0.025000 & 0.025000 \\ 0.038125 & 1.088750 & 0.088750 \\ 0.011875 & 0.011250 & 1.011250 \end{pmatrix}$$

48. Adapt Exercises 31–34 on pages 492–493 to show that the entry in row i and column j of the fundamental matrix $F = (I - Q)^{-1}$ is the expected number of times that the chain, if it begins in state N_i, will be in state N_j before being absorbed. This established the first result in the box on page 508.

49. Explain how the second result in the box on page 508 follows from Exercise 48.

50. Explain how the third result in the box on page 508 follows from Exercise 48 and the fact that the entry in row i and column j of the matrix R is the probability that state N_i moves to state A_j in one transition.

Chapter Summary with Hints and Suggestions

Reading the text and doing the exercises in this chapter have helped you to master the following skills, which are listed by section (in case you need to review them) and are keyed to particular Review Exercises. Answers for all Review Exercises are given at the back of the book, and full solutions can be found in the Student Solutions Manual.

7.1 States and Transitions

- Construct a transition matrix from a state-transition diagram and identify the transition as "oscillating," "mixing," or "absorbing." *(Review Exercises 1–2.)*

- Construct a state-transition diagram from a transition matrix and identify the transition as "oscillating," "mixing," or "absorbing." *(Review Exercises 3–4.)*

- Calculate the kth state distribution vector for a Markov chain with transition matrix T and initial distribution D_0. *(Review Exercises 5–6.)*

$$D_k = D_0 \cdot T^k$$

- Find the expected number of times a Markov chain will be in a given state before moving to some other state. *(Review Exercises 7–8.)*

$$E = \frac{1}{1 - p}$$

- Use a Markov chain to solve an applied problem. *(Review Exercises 9–10.)*

7.2 Regular Markov Chains

- Identify a transition matrix as "regular" or "not regular." *(Review Exercises 11–14.)*

- Find the steady-state distribution of a regular Markov chain by row-reduction, and by calculating powers of the transition matrix (using ▦). *(Review Exercises 15–16.)*

$$D \cdot T = D$$

$$\left(\begin{array}{c|c} T^t - I & 0 \\ \hline 1 \cdots 1 & 1 \end{array}\right) \longrightarrow \left(\begin{array}{c|c} I & D^t \\ \hline 0 & 0 \end{array}\right)$$

- Find the smallest positive number k of transitions needed to move from state S_i to state S_j in a regular Markov chain, and then verify this value (using ▦) by checking that T^k has a positive transition probability in row i and column j. *(Review Exercises 17–18.)*

- Use a regular Markov chain to solve an applied problem. *(Review Exercises 19–20.)*

7.3 Absorbing Markov Chains

- Identify the absorbing and nonabsorbing states in a state-transition diagram, and find the smallest number of transitions needed to move from a given nonabsorbing state to an absorbing state. *(Review Exercises 21–22.)*

- Identify absorbing and nonabsorbing states from a transition matrix, and find the smallest number of transitions needed to ensure that every nonabsorbing state can reach an absorbing state. *(Review Exercises 23–24.)*

- Rewrite an absorbing Markov chain transition matrix in standard form. *(Review Exercises 25–26.)*

$$T = \left(\begin{array}{c|c} I & 0 \\ \hline R & Q \end{array}\right)$$

- Find the fundamental matrix of an absorbing Markov chain. *(Review Exercises 27–30.)*

$$F = (I - Q)^{-1}$$

- Use the fundamental matrix to find the expected number of times in nonabsorbing states before absorption and the probability of absorption into a particular absorbing state. *(Review Exercises 31–38.)*

- Use an absorbing Markov chain to solve an applied problem. *(Review Exercises 39–40.)*

Hints and Suggestions

- *(Overview)* We discussed three general types of Markov chains: *oscillating, mixing,* and *absorbing.* Oscillating chains always move back and forth. Mixing (that is, regular) chains settle down to a steady-state that is independent of how they began. Absorbing chains end in absorption, with time until absorption and the final absorbing state depending on where they began.

- An entry in a particular row and column of a transition matrix gives the probability of moving *from* the *row* state *to* the *column* state.

- For a regular Markov chain, the steady-state distribution D can be found in two ways: by raising the transition matrix to a high power so the rows become D, or by row-reducing $\left(\begin{array}{c|c} T^t - I & 0 \\ \hline 1 \cdots 1 & 1 \end{array}\right)$ to obtain $\left(\begin{array}{c|c} I & D^t \\ \hline 0 & 0 \end{array}\right)$, where the superscript t means *transpose* (not a power).

- An *absorbing* Markov chain is not just a chain with an absorbing state. To be an absorbing Markov chain, it must be possible to move from each nonabsorbing state to some absorbing state (possibly in several moves).

- When writing an absorbing transition matrix in standard form, be sure to keep the state names with the correct rows and columns as you switch the absorbing rows to the top positions and then as you put the columns in the same order.

- For an absorbing Markov chain we could find the long-term behavior by raising the transition matrix to a high power, but we obtain much more information from the results in the box on page 508. These results depend on the *fundamental matrix* $F = (I - Q)^{-1}$. F gives expected times and $F \cdot R$ gives probabilities, in each case from the *row* state to the *column* state.

Review Exercises for Chapter 7

Practice test exercise numbers are in green.

7.1 States and Transitions

Construct a transition matrix T for the Markov chain represented by the state-transition diagram and identify the transition as "oscillating," "mixing," or "absorbing."

1.

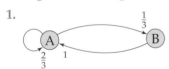

2.

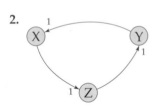

Construct a state-transition diagram for the Markov chain represented by the transition matrix and identify the transition as "oscillating," "mixing," or "absorbing."

3. $\begin{pmatrix} \frac{1}{3} & \frac{2}{3} \\ 0 & 1 \end{pmatrix}$

4. $\begin{pmatrix} 0 & \frac{2}{5} & \frac{3}{5} \\ 0 & \frac{4}{5} & \frac{1}{5} \\ \frac{1}{2} & \frac{1}{2} & 0 \end{pmatrix}$

For each Markov transition matrix T and initial state distribution vector D_0, calculate the next two state distribution vectors D_1 and D_2.

5. $T = \begin{pmatrix} 0.65 & 0.35 \\ 0.15 & 0.85 \end{pmatrix}$, $D_0 = (0.40 \quad 0.60)$

6. $T = \begin{pmatrix} 0 & 0.5 & 0.5 \\ 0.6 & 0.3 & 0.1 \\ 0.1 & 0.3 & 0.6 \end{pmatrix}$, $D_0 = (0 \quad 1 \quad 0)$

For each Markov transition matrix T and present state, find the expected number of times the chain will be in that state before moving to some other state.

7. $T = \begin{pmatrix} 0.95 & 0.05 \\ 0.45 & 0.55 \end{pmatrix}$, presently in state S_1

8. $T = \begin{pmatrix} 1 & 0 & 0 & 0 \\ 0.1 & 0.3 & 0 & 0.6 \\ 0.2 & 0 & 0.8 & 0 \\ 0.1 & 0.3 & 0 & 0.6 \end{pmatrix}$, presently in state S_3

Represent each situation as a Markov chain by constructing a state-transition diagram and the corresponding transition matrix. Be sure to state your final answer in terms of the original question.

9. **BUSINESS: Taxi Repairs** The Yellow-Top Cab Company has a large fleet of aging taxi cabs. Each day 11% break down and are removed from service, while 99% of those in the repair shop from yesterday are returned to service. What is the expected number of days a working cab will remain in service before breaking down?

10. **GENERAL: Two-Year Colleges** Each year 48% of the first-year students at the Marsten

Community College earn enough credits to become second-year students, 20% remain at the first-year level, and the remaining 32% withdraw. Also every year, 64% of the second-year students graduate, 20% remain at the second-year level, and the other 16% withdraw. If there are now 625 first-year and 250 second-year students, how many of these will be second-year students in two years?

7.2 Regular Markov Chains

Identify each transition matrix as "regular" or "not regular."

11. $\begin{pmatrix} 0.20 & 0.80 \\ 0.45 & 0.55 \end{pmatrix}$ **12.** $\begin{pmatrix} 0 & 1 \\ 1 & 0 \end{pmatrix}$

13. $\begin{pmatrix} 0 & 0.4 & 0.6 \\ 0 & 0.8 & 0.2 \\ 0.5 & 0 & 0.5 \end{pmatrix}$ **14.** $\begin{pmatrix} 0 & 0.2 & 0.8 \\ 0.1 & 0.2 & 0.7 \\ 0.3 & 0.2 & 0.5 \end{pmatrix}$

For each Markov transition matrix T:
a. Find the steady-state distribution.
b. Calculate T^k for $k = 5, 10, 20,$ and 50 to verify that the rows of T^k approach the steady-state distribution.

15. $\begin{pmatrix} 0.55 & 0.45 \\ 0.15 & 0.85 \end{pmatrix}$ **16.** $\begin{pmatrix} 0.3 & 0.6 & 0.1 \\ 0 & 0.4 & 0.6 \\ 0.3 & 0.6 & 0.1 \end{pmatrix}$

For each regular Markov transition matrix T:
a. Construct a state-transition diagram and find the smallest number k of steps needed to move from the given state S_i to state S_j and the probability of that sequence of steps.
b. Use this k to verify that T^k has a positive transition probability in row i and column j equal to the probability found in part (a).

17. $\begin{pmatrix} 0.4 & 0 & 0.6 \\ 0.3 & 0.7 & 0 \\ 0 & 0.3 & 0.7 \end{pmatrix}$, $S_1 \to S_2$

18. $\begin{pmatrix} 0.5 & 0.5 & 0 \\ 0.9 & 0 & 0.1 \\ 0.1 & 0 & 0.9 \end{pmatrix}$, $S_3 \to S_2$

Represent each situation as a Markov chain by constructing a state-transition diagram and a transition matrix. Find the steady-state distribution and interpret it in terms of the original situation. Be sure to state your final answer in terms of the original question.

19. BUSINESS: Customer Satisfaction Surveys of customer satisfaction with the repair service departments of dealers for a major car manufacturer show that of those rated "below average" one year, the next year 10% will remain that way, 70% will improve to "satisfactory," and the remaining 20% will improve to "excellent"; of those rated "satisfactory" one year, the next year 50% will remain that way, 20% will improve to "excellent," and the remaining 30% will slip to "below average"; and of those rated "excellent" one year, the next year 70% will remain that way and the remaining 30% will slip to "satisfactory." Assuming that this pattern has repeated for many years, how many of the manufacturer's 2660 dealers nationwide have service departments rated "excellent" by their customers?

20. GENERAL: Weather The old timers at Edna's Cafe in downtown Nora Springs will tell you the weather just keeps getting better and better for growing corn. In fact, if it was bad last year, there is a 20% chance it will now be terrific and a 60% chance it will be great this year; if last year was great, there is a 50% chance it will now be terrific and a 40% chance it will be great again this year; and if it was terrific last year, there is a 34% chance it will be that way again this year and a 54% chance it will be great this year. (Of course, they don't bother to mention that the weather might be bad this year because that could change their luck!) If they are right, how many of the next 25 years will have terrific weather for growing corn?

7.3 Absorbing Markov Chains

For each state-transition diagram, identify the absorbing and the nonabsorbing states. For the given starting state, find the smallest number of steps required to reach an absorbing state and the probability of that sequence of steps.

21. Starting in state S_1

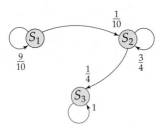

22. Starting in state S_4

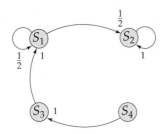

For each absorbing Markov transition matrix T:
a. Draw a state-transition diagram, identify the absorbing and the nonabsorbing states, and find the smallest integer k such that after k transitions, every nonabsorbing state can reach an absorbing state.
b. Verify your value for k by finding T^k and checking that every nonabsorbing state row has a positive probability in at least one absorbing state column.

23. $\begin{pmatrix} 0.6 & 0 & 0.2 & 0.2 \\ 0 & 1 & 0 & 0 \\ 0 & 0.5 & 0.5 & 0 \\ 0.5 & 0 & 0 & 0.5 \end{pmatrix}$

24. $\begin{pmatrix} 0.80 & 0 & 0 & 0.20 & 0 \\ 0 & 0.80 & 0.15 & 0 & 0.05 \\ 0 & 0 & 1 & 0 & 0 \\ 0.90 & 0.10 & 0 & 0 & 0 \\ 0 & 0 & 0 & 0 & 1 \end{pmatrix}$

Rewrite each absorbing Markov transition matrix in standard form and identify the matrices R and Q.

25. $\begin{pmatrix} 0.5 & 0.5 & 0 & 0 \\ 0 & 1 & 0 & 0 \\ 0.2 & 0 & 0.8 & 0 \\ 0 & 0 & 1 & 0 \end{pmatrix}$

26. $\begin{pmatrix} 0.6 & 0 & 0.3 & 0.1 \\ 0 & 1 & 0 & 0 \\ 0.1 & 0.1 & 0.8 & 0 \\ 0 & 0 & 0 & 1 \end{pmatrix}$

Find the fundamental matrix for each absorbing transition matrix in standard form. (You may use 🖩 if permitted by your instructor.)

27. $\left(\begin{array}{c|cc} 1 & 0 & 0 \\ \hline 0.3 & 0.3 & 0.4 \\ 0.1 & 0.1 & 0.8 \end{array} \right)$

28. $\left(\begin{array}{cc|cc} 1 & 0 & 0 & 0 \\ 0 & 1 & 0 & 0 \\ \hline 0.10 & 0 & 0.60 & 0.30 \\ 0.05 & 0.05 & 0.10 & 0.80 \end{array} \right)$

29. $\left(\begin{array}{c|ccc} 1 & 0 & 0 & 0 \\ \hline 0.1 & 0.7 & 0 & 0.2 \\ 0.1 & 0.1 & 0.8 & 0 \\ 0.2 & 0.4 & 0 & 0.4 \end{array} \right)$

30. $\left(\begin{array}{cc|ccc} 1 & 0 & 0 & 0 & 0 \\ 0 & 1 & 0 & 0 & 0 \\ \hline 0.05 & 0.15 & 0.20 & 0 & 0.60 \\ 0.10 & 0.10 & 0.20 & 0.50 & 0.10 \\ 0.10 & 0 & 0.10 & 0 & 0.80 \end{array} \right)$

For Exercises 31–38, use the results of Exercises 27–30 as follows: For each given initial state, find the expected number of times in the given state before absorption, and find the probability of absorption in the given final state using the indicated transition matrix.

Initial State	Expected Number of Times in State	Probability of Absorption into State	Use the Transition Matrix from Exercise
31. N_1	N_1	A_1	27
32. N_2	N_1	A_1	27
33. N_1	N_2	A_2	28
34. N_2	N_2	A_1	28
35. N_3	N_1	A_1	29
36. N_2	any N_i	A_1	29
37. N_1	N_2	A_2	30
38. N_3	any N_i	A_1	30

Represent each situation as an absorbing Markov chain by constructing a state-transition diagram and a transition matrix in standard form. Use the fundamental matrix to solve each problem. Be sure to state your final answers in terms of the original questions.

39. BUSINESS: Computer Obsolescence A personal computer (PC) market analyst has estimated that each year 60% of the PCs from the previous year are still manufactured, 20% have been upgraded to newer models, and the remainder have been discontinued, while 30% of the new-model PCs from the previous year are still manufactured, 60% undergo further upgrading, and the remainder are dis-continued. How many years is a new model expected to be manufactured before it is discontinued?

40. BUSINESS: Car Loans The Final Federal Credit Corporation classifies its car loans as paid in full (and thus closed out), current, late, overdue, or bad (in which case they are sold to a collection agency). Each month 10% of the current accounts are paid in full, 10% are late, and the rest remain current; 3% of the late accounts are paid in full, 30% improve to be current, 50% become overdue, 7% become bad, and the rest remain late; 2% of the overdue accounts are paid in full, 10% become current, 20% improve to be late, 18% become bad, and the rest remain overdue. What is the probability that an overdue account will (ultimately) be paid in full?

Part II

Calculus

CHAPTER 8
DERIVATIVES AND THEIR USES

CHAPTER 9
FURTHER APPLICATIONS OF DERIVATIVES

CHAPTER 10
EXPONENTIAL AND LOGARITHMIC FUNCTIONS

CHAPTER 11
INTEGRATION AND ITS APPLICATIONS

CHAPTER 12
INTEGRATION TECHNIQUES AND DIFFERENT EQUATIONS

CHAPTER 13
CALCULUS OF SEVERAL VARIABLES

8 Derivatives and Their Uses

- 8.1 Limits and Continuity
- 8.2 Rates of Change, Slopes, and Derivatives
- 8.3 Some Differentiation Formulas
- 8.4 The Product and Quotient Rules
- 8.5 Higher-Order Derivatives
- 8.6 The Chain Rule and the Generalized Power Rule
- 8.7 Nondifferentiable Functions

Application Preview

Temperature, Superconductivity, and Limits

It has long been known that there is a coldest possible temperature, called absolute zero, the temperature of an object if all heat could be removed from it. On the Fahrenheit temperature scale, absolute zero is 460 degrees below zero. On the "absolute" or "Kelvin" temperature scale (named after the nineteenth-century scientist Lord Kelvin), absolute zero temperature is assigned the value 0. Absolute zero is a temperature that can be *approached* but never actually *reached*. At temperatures approaching absolute zero, some metals become increasingly able to conduct electricity, with efficiencies approaching 100%, a state called *superconductivity*. The graph below gives the electrical conductivity of aluminum, showing the remarkable fact that as temperature decreases to absolute zero, aluminum becomes *superconducting*—its conductivity approaches a limit of 100% (see Exercise 81 on page 545).

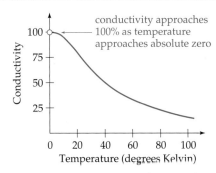

Conductivity of aluminum as a function of temperature

If $f(t)$ stands for the percent conductivity of aluminum at temperature t degrees Kelvin, then the fact that conductivity approaches 100 as temperature t decreases to zero may be written

$$\lim_{t \to 0^+} f(t) = 100 \qquad \text{Limit as } t \to 0^+ \text{ of conductivity is 100}$$

This is an example of *limits*, which are discussed in this section. Superconductivity has many commercial applications, from high-speed "mag-lev" trains that "float" on magnetic fields above the tracks to supercomputers and medical imaging devices.

8.1 LIMITS AND CONTINUITY

Introduction

In this chapter we begin the study of calculus and its applications. Quite simply, calculus is the study of rates of change. We will use calculus to analyze rates of inflation, rates of learning, rates of population growth, and rates of natural resource consumption. We begin by discussing two preliminary topics, limits and continuity, both of which will be treated intuitively rather than formally, and will be useful when we define the *derivative* in the next section.

Limits

The word "*limit*" is used in everyday conversation to describe the *ultimate* behavior of something, as in the "limit of one's endurance" or the "limit of one's patience." In mathematics, the word "limit" has a similar but more precise meaning.

The notation $x \to 3$ (read "x approaches 3") means that x takes values *arbitrarily close to 3 without ever reaching 3*. For example, the numbers 2.9, 2.99, 2.999, ... approach 3 (from the left), and the numbers 3.1, 3.01, 3.001, ... approach 3 (from the right). We can even let x approach 3 by alternating sides, taking values such as 2.9, 3.01, 2.999, 3.0001, ... We emphasize that $x \to 3$ means that x takes values closer and closer to 3 *but never equaling 3*.

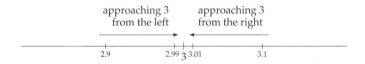

Given a function $f(x)$, if x approaching 3 causes the function to take values arbitrarily close to, say, 10, then we call 10 the limit of the function, and write

$$\lim_{x \to 3} f(x) = 10 \qquad \text{Limit of } f(x) \text{ as } x \text{ approaches 3 is 10}$$

The limit of a function may not exist (as we will see later), but if the limit *does* exist, it must be a *single* number. Limits can be defined formally,* but we will treat them intuitively, using the following definition.

> **Limit of a Function**
>
> $$\lim_{x \to c} f(x) \qquad \text{Limit of } f(x) \text{ as } x \text{ approaches } c$$
>
> is the number that $f(x)$ approaches as x approaches c $(x \neq c)$.

Finding Limits

Limits may be found from tables of values of $f(x)$ for x near c.

EXAMPLE 1

FINDING A LIMIT BY TABLES

Use tables to find $\lim_{x \to 3} (2x + 4)$. Limit of $2x + 4$ as x approaches 3

Solution

We choose *x*-values approaching 3, and calculate the resulting values of $f(x) = 2x + 4$. We make two tables, for $x < 3$ and for $x > 3$.

Numbers approaching 3 but *less* than 3:

x	$2x + 4$
2.9	9.8
2.99	9.98
2.999	9.998

Numbers approaching 3 but *greater* than 3:

x	$2x + 4$
3.1	10.2
3.01	10.02
3.001	10.002

Reading down these columns, the function values seem to approach 10.

Choosing *x*-values even closer to 3 (such as 2.9999 and 3.0001) would result in values of $2x + 4$ *even closer to 10, so the limit is 10*:

$$\lim_{x \to 3} (2x + 4) = 10$$

---

* Formally, limits are defined as follows: $\lim_{x \to c} f(x) = L$ if for every number $\varepsilon > 0$ there is a number $\delta > 0$ such that $|f(x) - L| < \varepsilon$ whenever $0 < |x - c| < \delta$.

This limit may be seen graphically: As $x \to 3$ (on the x-axis), $f(x)$ approaches 10 (on the y-axis).

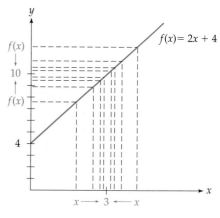

Graph showing $\lim_{x \to 3} (2x + 4) = 10$

The correct limit in Example 1 could have been found simply by *evaluating* the function at $x = 3$:

$$f(3) = 2 \cdot 3 + 4 = 10 \qquad \begin{array}{l} f(x) = 2x + 4 \text{ evaluated at } x = 3 \\ \text{gives the correct limit, 10} \end{array}$$

However, finding limits by this technique of *direct substitution* is not always possible, as the next example shows.

EXAMPLE 2

FINDING A LIMIT BY TABLES

Find $\lim_{x \to 0} (1 + x)^{1/x}$ correct to three decimal places.

Solution

The function values in the following tables were found using a calculator.

x	$(1 + x)^{1/x}$	x	$(1 + x)^{1/x}$
0.1	2.594	-0.1	2.868
0.01	2.705	-0.01	2.732
0.001	2.717	-0.001	2.720
0.0001	2.718	-0.0001	2.718

To use a graphing calculator to find these numbers, enter the function as $(1 + x)\wedge(1/x)$, and use the TABLE feature.

$$\lim_{x \to 0} (1 + x)^{1/x} \approx 2.718$$

From the agreement in the last columns of the tables

Practice Problem 1

Evaluate $(1 + x)^{1/x}$ at $x = 0.000001$ and $x = -0.000001$. Based on the results, give a better approximation for $\lim_{x \to 0} (1 + x)^{1/x}$.

➤ Solution on page 541

The actual value of this particular limit is the number $e \approx 2.71828$, which will be very important in our later work. Notice that $(1 + x)^{1/x}$ cannot be evaluated *at* $x = 0$ since the exponent would be $1/0$, which is undefined. This limit *requires* the limit process.

Which limits can be evaluated by direct substitution (as in Example 1) and which cannot (as in Example 2)? The answer comes from the following "Rules of Limits."

Rules of Limits

For any constants a and c, and any positive integer n:

1. $\lim_{x \to c} a = a$ — The limit of a constant is just the constant

2. $\lim_{x \to c} x^n = c^n$ — The limit of a power is the power of the limit

3. $\lim_{x \to c} \sqrt[n]{x} = \sqrt[n]{c}$ ($c > 0$ if n is even) — The limit of a root is the root of the limit

4. If $\lim_{x \to c} f(x)$ and $\lim_{x \to c} g(x)$ both exist, then

   a. $\lim_{x \to c} [f(x) + g(x)] = \lim_{x \to c} f(x) + \lim_{x \to c} g(x)$ — The limit of a sum is the sum of the limits

   b. $\lim_{x \to c} [f(x) - g(x)] = \lim_{x \to c} f(x) - \lim_{x \to c} g(x)$ — The limit of a difference is the difference of the limits

   c. $\lim_{x \to c} [f(x) \cdot g(x)] = [\lim_{x \to c} f(x)] \cdot [\lim_{x \to c} g(x)]$ — The limit of a product is the product of the limits

   d. $\lim_{x \to c} \dfrac{f(x)}{g(x)} = \dfrac{\lim_{x \to c} f(x)}{\lim_{x \to c} g(x)}$ (if $\lim_{x \to c} g(x) \neq 0$) — The limit of a quotient is the quotient of the limits

These rules, which may be proved from the definition of limit, can be summarized as follows.

Summary of Rules of Limits

For functions composed of additions, subtractions, multiplications, divisions, powers, and roots, limits may be evaluated by direct substitution, provided that the resulting expression is defined.

$$\lim_{x \to c} f(x) = f(c) \quad \text{Limit evaluated by direct substitution}$$

EXAMPLE 3

FINDING LIMITS BY DIRECT SUBSTITUTION

a. $\lim_{x \to 4} \sqrt{x} = \sqrt{4} = 2$ Direct substitution of $x = 4$ using Rule 3 or the Summary

b. $\lim_{x \to 6} \dfrac{x^2}{x+3} = \dfrac{6^2}{6+3} = \dfrac{36}{9} = 4$ Direct substitution of $x = 6$ (Rules 4, 2, and 1 or the Summary)

Practice Problem 2 Find $\lim_{x \to 3} (2x^2 - 4x + 1)$. ▶ Solution on page 541

If direct substitution into a quotient gives the undefined expression $\dfrac{0}{0}$, factoring, simplifying, and *then* using direct substitution may help.*

EXAMPLE 4

FINDING A LIMIT BY SIMPLIFYING

Find $\lim_{x \to 1} \dfrac{x^2 - 1}{x - 1}$.

Solution

Direct substitution of $x = 1$ into $\dfrac{x^2 - 1}{x - 1}$ gives $\dfrac{0}{0}$, which is undefined. But simplifying the fraction gives

$$\lim_{x \to 1} \dfrac{x^2 - 1}{x - 1} = \lim_{x \to 1} \dfrac{(x+1)(x-1)}{x - 1} = \lim_{x \to 1} \dfrac{(x+1)\cancel{(x-1)}}{\cancel{x - 1}} = \lim_{x \to 1} (x + 1) = 2$$

Factoring the numerator   Canceling the $(x - 1)$'s (since $x \neq 1$)   Now use direct substitution

Therefore, the limit *does* exist and equals 2.

---

* Recall from the definition on page 529 that the limit and the result from direct substitution are not always the same, so the limit may still exist even if direct substitution gives an undefined result.

Graphing Calculator Exploration

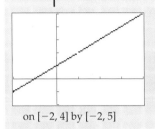

on [−2, 4] by [−2, 5]

The graph of $f(x) = \dfrac{x^2 - 1}{x - 1}$ (from Example 4) is shown on the left.

a. Can you explain why the graph appears to be a straight line?

b. From your knowledge of rational functions, should the graph really be a (complete) line? [*Hint:* See pages 50–51.] Can you see that a point is indeed missing from the graph?*

c. Use the graph on the left (where each tick mark is 1 unit) or enter the function and use TRACE on your calculator to verify that the limit is 2 as x approaches 1.

*To have your calculator show the point as missing, choose a window such that the x-value in question is midway between XMIN and XMAX, or see the Useful Hint in Graphing Calculator Terminology in the Preface.

Practice Problem 3 Find $\displaystyle\lim_{x \to 5} \dfrac{2x^2 - 10x}{x - 5}$ ➤ Solution on page 541

One-Sided Limits

In a *one-sided limit*, the variable approaches the number *from one side only*. For example, the limit as x approaches 3 *from the left*, denoted $x \to 3^-$, means the limit using only x-values to the *left* of 3, such as 2.9, 2.99, 2.999, ... The limit as x approaches 3 *from the right*, denoted $x \to 3^+$, means the limit using only x-values to the *right* of 3, such as 3.1, 3.01, 3.001, ... The limits from the left and from the right (sometimes called left and right limits) are exactly what we found in the two tables of Examples 1 and 2 (see pages 529–530). In the Application Preview on page 527, we used the notation $t \to 0^+$ to indicate that the temperature *decreased* to zero. In general:

Left and Right Limits

$\displaystyle\lim_{x \to c^-} f(x)$ means the limit of $f(x)$ Limit from the *left*

as $x \to c$ but with $x < c$

$\displaystyle\lim_{x \to c^+} f(x)$ means the limit of $f(x)$ Limit from the *right*

as $x \to c$ but with $x > c$

The Rules of Limits on page 531 also hold for one-sided limits. The limit that we defined earlier (page 529) is sometimes called the *two-sided* limit to distinguish it from one-sided limits. As in Examples 1 and 2, if the left and right limits both exist and have the same value, then the (two-sided) limit exists and has that same value.

$$\lim_{x \to c} f(x) = L \quad \text{if and only if } \textit{both} \text{ one-sided limits } \lim_{x \to c^-} f(x) \text{ and}$$

$$\lim_{x \to c^+} f(x) \text{ exist and equal the same number } L.$$

EXAMPLE 5

FINDING ONE-SIDED LIMITS

For the piecewise linear function $f(x) = \begin{cases} x + 1 & \text{if } x \leq 3 \\ 8 - 2x & \text{if } x > 3 \end{cases}$ graphed on the left, find the following limits or state that they do not exist.

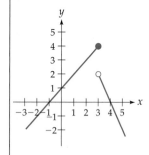

a. $\lim_{x \to 3^-} f(x)$ **b.** $\lim_{x \to 3^+} f(x)$ **c.** $\lim_{x \to 3} f(x)$

We give two solutions, one using the graph and one using the expression for the function. Both methods are important.

Solution (Using the graph)

a. $\lim_{x \to 3^-} f(x) = 4$

Approaching 3 *from the left* means using the line on the *left* of $x = 3$, which approaches height 4

b. $\lim_{x \to 3^+} f(x) = 2$

Approaching 3 *from the right* means using the line on the *right* of $x = 3$, which approaches height 2

c. $\lim_{x \to 3} f(x)$ does not exist

The two one-sided limits both exist, but they have different values (4 and 2), so *the limit does not exist*

Solution $\left(\text{Using } f(x) = \begin{cases} x + 1 & \text{if } x \leq 3 \\ 8 - 2x & \text{if } x > 3 \end{cases} \right)$

a. For $x \to 3^-$ we have $x < 3$, so $f(x)$ is given by the *upper* line of the function:

$$\lim_{x \to 3^-} f(x) = \lim_{x \to 3^-} (x + 1) = 3 + 1 = 4$$

b. For $x \to 3^+$ we have $x > 3$, so $f(x)$ is given by the *lower* line of the function:

$$\lim_{x \to 3^+} f(x) = \lim_{x \to 3^+} (8 - 2x) = 8 - 2 \cdot 3 = 2$$

c. $\lim_{x \to 3} f(x)$ does not exist since although the two one-sided limits exist, they are not equal.

Notice that the two methods found the same answers.

Practice Problem 4 Explain the difference among $x \to 3^-$, $x \to -3$, and $x \to -3^-$.

▶ Solution on page 541

Infinite Limits

We may use the symbols ∞ (infinity) and $-\infty$ (negative infinity) to indicate that the values of a function become arbitrarily large or arbitrarily small.* The following example shows some of the possibilities. The dashed lines on the graphs, where the function values approach ∞ or $-\infty$, are called *vertical asymptotes*.

EXAMPLE 6 **FINDING LIMITS INVOLVING $\pm \infty$**

For each function graphed below, use the limit notation with ∞ and $-\infty$ to describe its behavior as x approaches the vertical asymptote from the left, from the right, and from both sides.

a.

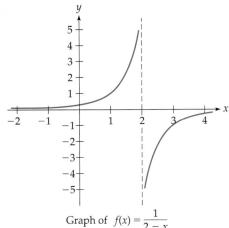

Graph of $f(x) = \dfrac{1}{2-x}$

b.

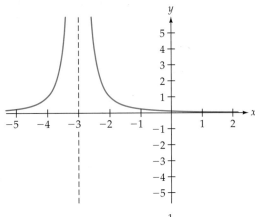

Graph of $f(x) = \dfrac{1}{(x+3)^2}$

*Limits as $x \to \infty$ and as $x \to -\infty$ will be discussed at the beginning of Section 12.3.

Solution

a. $\lim\limits_{x \to 2^-} f(x) = \infty$ The curve rises arbitrarily high as x approaches 2 from the left

$\lim\limits_{x \to 2^+} f(x) = -\infty$ The curve falls arbitrarily low as x approaches 2 from the right

$\lim\limits_{x \to 2} f(x)$ does not exist   The two one-sided limits do not agree

b. $\lim\limits_{x \to -3^-} f(x) = \infty$ The curve rises arbitrarily high as x approaches -3 from the left

$\lim\limits_{x \to -3^+} f(x) = \infty$ The curve rises arbitrarily high as x approaches -3 from the right

$\lim\limits_{x \to -3} f(x) = \infty$ The two one-sided limits agree

Be careful! To say that a limit exists means that the limit is a *number*, and since ∞ and $-\infty$ are not numbers, a statement such as $\lim\limits_{x \to c} f(x) = \infty$ means that *the limit does not exist*. The limit statement $\lim\limits_{x \to c} f(x) = \infty$ goes further to explain *why* the limit does not exist: the function values become too large to approach any limit.

Limits of Functions of Two Variables

Some limits involve two variables, with only one variable approaching a limit.

EXAMPLE 7

FINDING A LIMIT OF A FUNCTION OF TWO VARIABLES

Find $\lim\limits_{h \to 0} (x^2 + xh + h^2)$.

Solution

Only h is approaching zero, so x remains unchanged. Since the function involves only powers of h, we may evaluate the limit by direct substitution of $h = 0$:

$$\lim\limits_{h \to 0} (x^2 + xh + h^2) = x^2 + x \cdot 0 + 0^2 = x^2$$

Practice Problem 5 Find $\lim_{h \to 0} (3x^2 + 5xh + 1)$. ➤ Solution on page 541

Continuity

Intuitively, a function is said to be *continuous at c* if its graph passes through the point at $x = c$ *without a "hole" or a "jump."* For example, the first function below is *continuous* at c (it has no hole or jump at $x = c$), while the second and third are discontinuous at c.

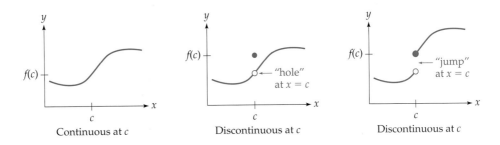

Continuous at c — Discontinuous at c ("hole" at x = c) — Discontinuous at c ("jump" at x = c)

In other words, a function is *continuous at c* if the curve *approaches the point at* $x = c$, which may be stated in terms of limits:

$$\lim_{x \to c} f(x) = f(c)$$

This equation means that the quantities on both sides must exist and be *equal*, which we make explicit as follows:

Continuity

A function f is continuous at c if the following three conditions hold:

1. $f(c)$ is defined — Function is *defined* at c
2. $\lim_{x \to c} f(x)$ exists — Left and right limits exist and agree
3. $\lim_{x \to c} f(x) = f(c)$ — Limit and value *at c* agree

f is *discontinuous* at c if one or more of these conditions *fails* to hold.

Condition 3, which is just the statement that the expressions in conditions 1 and 2 are equal to each other, may by itself be taken as the definition of continuity.

EXAMPLE 8　FINDING DISCONTINUITIES FROM A GRAPH

Each function below is *discontinuous at c* for the indicated reason.

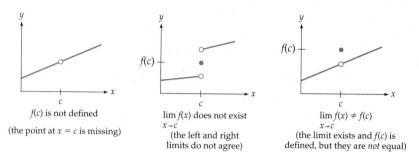

$f(c)$ is not defined
(the point at $x = c$ is missing)

$\lim_{x \to c} f(x)$ does not exist
(the left and right limits do not agree)

$\lim_{x \to c} f(x) \neq f(c)$
(the limit exists and $f(c)$ is defined, but they are *not* equal)

Practice Problem 6

For each graph below, determine whether the function is continuous at c. If it is *not* continuous, indicate the *first* of the three conditions in the definition of continuity (on the previous page) that is violated.

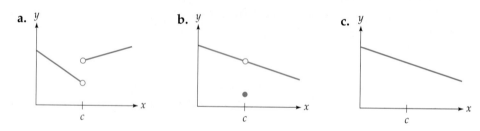

▶ Solutions on page 541

A function is continuous on an *open interval* (a, b) if it is continuous at each point of the interval. A function is continuous on a *closed interval* $[a, b]$ if it is continuous on the open interval (a, b) and has "one-sided continuity" at the endpoints: $\lim_{x \to a^+} f(x) = f(a)$ and $\lim_{x \to b^-} f(x) = f(b)$. A function that is continuous on the entire real line $(-\infty, \infty)$ is said to be *continuous everywhere*, or simply *continuous*.

Which Functions Are Continuous?

Which functions are continuous? *Linear* and *quadratic* functions are continuous, since their graphs are, respectively, straight lines and parabolas, with no holes or jumps. Similarly, *exponential* and *logarithmic* functions are continuous, as may be seen from graphs on page 84.

These and other continuous functions can be combined as follows to give other continuous functions.

> If functions f and g are continuous at c, then the following are also continuous at c:
>
> 1. $f \pm g$ — Sums and differences of continuous functions are continuous
> 2. $a \cdot f$ [for any constant a] — Constant multiples of continuous functions are continuous
> 3. $f \cdot g$ — Products of continuous functions are continuous
> 4. f/g [if $g(c) \neq 0$] — Quotients of continuous functions are continuous
> 5. $f(g(x))$ [for f continuous at $g(c)$] — Compositions of continuous functions are continuous

These statements, which can be proved from the Rules of Limits, show that the following types of functions are continuous:

> Every polynomial function is continuous.
>
> Every rational function is continuous except where the denominator is zero.

EXAMPLE 9 DETERMINING CONTINUITY

Determine whether each function is continuous.

a. $f(x) = x^3 - 3x^2 - x + 3$ b. $f(x) = \dfrac{1}{(x+1)^2}$ c. $f(x) = e^{x^2 - 1}$

Solution

The first function is continuous since it is a polynomial. The second (a rational function) is discontinuous at $x = -1$, where the denominator is zero. The rational function is continuous at all *other* values. The third function is continuous since it is the composition of the exponential function e^x and the polynomial $x^2 - 1$.

Calculator-drawn graphs of the functions in Example 9 are shown below. The polynomial (on the left) and the exponential function (on the right) are continuous, although you can't really tell from such graphs. The rational function (in the middle) exhibits the discontinuity at $x = -1$.

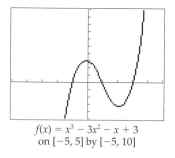

$f(x) = x^3 - 3x^2 - x + 3$
on $[-5, 5]$ by $[-5, 10]$

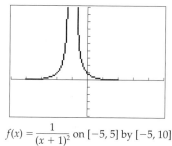

$f(x) = \dfrac{1}{(x + 1)^2}$ on $[-5, 5]$ by $[-5, 10]$

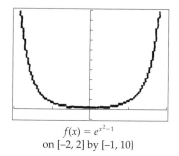

$f(x) = e^{x^2-1}$
on $[-2, 2]$ by $[-1, 10]$

8.1 Section Summary

The limit $\lim_{x \to c} f(x)$ is the number (if it exists) that $f(x)$ approaches as x approaches $c (x \neq c)$. Left and right limits are found by letting x approach c from one side only. If the left or right limit (or both) do not exist, or do not agree, then the limit does not exist. Limits can often be found from tables of function values for x near c, or by direct substitution of $x = c$ (provided that the function consists of additions, subtractions, multiplications, divisions, roots, and powers, and also provided that the resulting expression is defined). Sometimes it helps to simplify before direct substitution.

Continuity can be understood *geometrically* and *analytically*. Geometrically, a function is continuous at c if its graph has neither a hole nor a jump at $x = c$ (which is equivalent to the curve's passing through the point at c). Analytically, a function f is continuous at c if both sides of the following equation are defined and equal:

$$\lim_{x \to c} f(x) = f(c)$$

This equation may be interpreted as saying that continuity is equivalent to being able to *move the limit operation through the function:*

$$\lim_{x \to c} f(x) = f(\lim_{x \to c} x) = f(c)$$

8.1 LIMITS AND CONTINUITY

A polynomial is continuous everywhere, and a rational function is continuous everywhere except where the denominator is zero.

Solutions to Practice Problems

1. $\lim_{x \to 0}(1 + x)^{1/x} \approx 2.71828$

2. $\lim_{x \to 3}(2x^2 - 4x + 1) = 2 \cdot 3^2 - 4 \cdot 3 + 1$
   $= 18 - 12 + 1 = 7$

3. $\lim_{x \to 5} \dfrac{2x^2 - 10x}{x - 5} = \lim_{x \to 5} \dfrac{2x(x - 5)}{x - 5}$
   $= \lim_{x \to 5} \dfrac{2x(x - 5)}{x - 5} = \lim_{x \to 5} 2x = 10$

4. $x \to 3^-$ means: x approaches (positive) 3 *from the left*.
   $x \to -3$ means: x approaches -3 (the ordinary two-sided limit).
   $x \to -3^-$ means: x approaches -3 *from the left*.

5. $\lim_{h \to 0}(3x^2 + 5xh + 1) = 3x^2 + 5x \cdot 0 + 1$ Using direct substitution
   $= 3x^2 + 1$

6. **a.** Discontinuous, $f(c)$ is not defined
   b. Discontinuous, $\lim_{x \to c} f(x) \neq f(c)$
   c. Continuous

8.1 Exercises

Complete the tables and use them to find the given limit. Round calculations to three decimal places. A graphing calculator with a TABLE feature will be very helpful.

1. $\lim_{x \to 2}(5x - 7)$

x	5x − 7
1.9	
1.99	
1.999	

x	5x − 7
2.1	
2.01	
2.001	

2. $\lim_{x \to 4}(2x + 1)$

x	2x + 1
3.9	
3.99	
3.999	

x	2x + 1
4.1	
4.01	
4.001	

3. $\lim_{x \to 1} \dfrac{x^3 - 1}{x - 1}$

x	$\dfrac{x^3 - 1}{x - 1}$
0.9	
0.99	
0.999	

x	$\dfrac{x^3 - 1}{x - 1}$
1.1	
1.01	
1.001	

4. $\lim_{x \to 1} \dfrac{x^4 - 1}{x - 1}$

x	$\dfrac{x^4 - 1}{x - 1}$
0.9	
0.99	
0.999	

x	$\dfrac{x^4 - 1}{x - 1}$
1.1	
1.01	
1.001	

Find each limit by constructing tables similar to those in Exercises 1–4.

5. $\lim_{x \to 0} (1 + 2x)^{1/x}$

6. $\lim_{x \to 0} (1 - x)^{1/x}$

7. $\lim_{x \to 2} \dfrac{\dfrac{1}{x} - \dfrac{1}{2}}{x - 2}$

8. $\lim_{x \to 1} \dfrac{\sqrt{x} - 1}{x - 1}$

30. $\lim_{h \to 0} \dfrac{5x^4h - 9xh^2}{h}$

31. $\lim_{h \to 0} \dfrac{4x^2h + xh^2 - h^3}{h}$

32. $\lim_{h \to 0} \dfrac{x^2h - xh^2 + h^3}{h}$

Find each limit by graphing the function and using TRACE or TABLE to examine the graph near the indicated x-value.

9. $\lim_{x \to 1} \dfrac{\dfrac{1}{x} - 1}{1 - x}$
Use window [0, 2] by [0, 5].

10. $\lim_{x \to 1.5} \dfrac{2x^2 - 4.5}{x - 1.5}$
Use window [0, 3] by [0, 10].

11. $\lim_{x \to 4} \dfrac{x^{1.5} - 4x^{0.5}}{x^{1.5} - 2x}$

12. $\lim_{x \to 1} \dfrac{x - 1}{x - \sqrt{x}}$

[*Hint:* Choose a window whose x-values are centered at the limiting x-value.]

For each piecewise linear function $f(x)$ graphed in Exercises 33–36, find:

a. $\lim_{x \to 2^-} f(x)$ b. $\lim_{x \to 2^+} f(x)$ c. $\lim_{x \to 2} f(x)$

33.
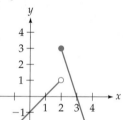

34.
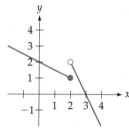

35.
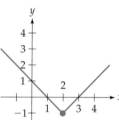

36.
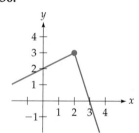

Find the following limits *without* using a graphing calculator or making tables.

13. $\lim_{x \to 3} (4x^2 - 10x + 2)$

14. $\lim_{x \to 7} \dfrac{x^2 - x}{2x - 7}$

15. $\lim_{x \to 5} \dfrac{3x^2 - 5x}{7x - 10}$

16. $\lim_{t \to 3} \sqrt[3]{t^2 + t - 4}$

17. $\lim_{x \to 3} \sqrt{2}$

18. $\lim_{q \to 9} \dfrac{8 + 2\sqrt{q}}{8 - 2\sqrt{q}}$

19. $\lim_{t \to 25} [(t + 5)t^{-1/2}]$

20. $\lim_{s \to 4} (s^{3/2} - 3s^{1/2})$

21. $\lim_{h \to 0} (5x^3 + 2x^2h - xh^2)$

22. $\lim_{h \to 0} (2x^2 + 4xh + h^2)$

23. $\lim_{x \to 2} \dfrac{x^2 - 4}{x - 2}$

24. $\lim_{x \to 1} \dfrac{x - 1}{x^2 + x - 2}$

25. $\lim_{x \to -3} \dfrac{x + 3}{x^2 + 8x + 15}$

26. $\lim_{x \to -4} \dfrac{x^2 + 9x + 20}{x + 4}$

27. $\lim_{x \to -1} \dfrac{3x^3 - 3x^2 - 6x}{x^2 + x}$

28. $\lim_{x \to 0} \dfrac{x^2 - x}{x^2 + x}$

29. $\lim_{h \to 0} \dfrac{2xh - 3h^2}{h}$

For each piecewise linear function, find:

a. $\lim_{x \to 4^-} f(x)$ b. $\lim_{x \to 4^+} f(x)$ c. $\lim_{x \to 4} f(x)$

37. $f(x) = \begin{cases} 3 - x & \text{if } x \leq 4 \\ 10 - 2x & \text{if } x > 4 \end{cases}$

38. $f(x) = \begin{cases} 5 - x & \text{if } x < 4 \\ 2x - 5 & \text{if } x \geq 4 \end{cases}$

39. $f(x) = \begin{cases} 2 - x & \text{if } x \leq 4 \\ x - 6 & \text{if } x > 4 \end{cases}$

40. $f(x) = \begin{cases} 2 - x & \text{if } x < 4 \\ 2x - 10 & \text{if } x \geq 4 \end{cases}$

For each function, find:

a. $\lim_{x \to 0^-} f(x)$ **b.** $\lim_{x \to 0^+} f(x)$ **c.** $\lim_{x \to 0} f(x)$

41. $f(x) = |x|$

42. $f(x) = -|x|$

43. $f(x) = \dfrac{|x|}{x}$

44. $f(x) = -\dfrac{|x|}{x}$

For each function, use the limit notation with ∞ and $-\infty$ to describe its behavior as x approaches the vertical asymptote from the left, from the right, and from both sides, as in Example 6 on pages 535–536.

45.

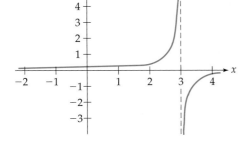

46.

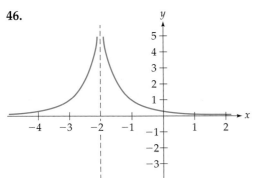

47.

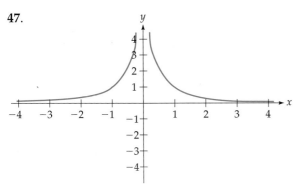

48.

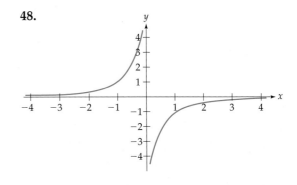

For each function, use the limit notation with ∞ and $-\infty$ to describe its behavior as x approaches the number where the function is undefined from the left, from the right, and from both sides, as in Example 6 on pages 535–536. (A table of values may be helpful.)

49. $f(x) = \dfrac{1}{x + 2}$

50. $f(x) = \dfrac{1}{x - 3}$

51. $f(x) = \dfrac{1}{(x - 3)^2}$

52. $f(x) = -\dfrac{1}{(x + 2)^2}$

Determine whether each of the following functions is continuous or discontinuous at c. If it is discontinuous, indicate the *first* of the three conditions in the definition of continuity (page 537) that is violated.

53.

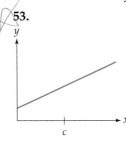

54.

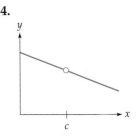

55.

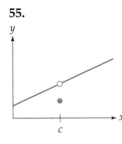

56.

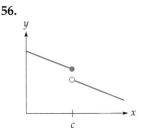

57.
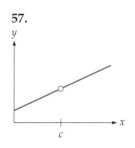

58.
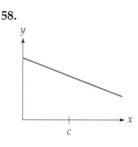

59.
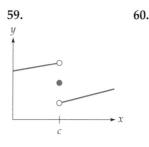

60.
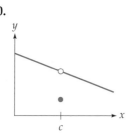

For each piecewise linear function:
a. Draw its graph (by hand or using a graphing calculator).
b. Find the limits as x approaches 3 from the left and from the right.
c. Is it continuous at $x = 3$? If not, indicate the first of the three conditions in the definition of continuity (page 537) that is violated.

61. $f(x) = \begin{cases} x & \text{if } x \leq 3 \\ 6 - x & \text{if } x > 3 \end{cases}$

62. $f(x) = \begin{cases} 5 - x & \text{if } x \leq 3 \\ x - 2 & \text{if } x > 3 \end{cases}$

63. $f(x) = \begin{cases} x & \text{if } x \leq 3 \\ 7 - x & \text{if } x > 3 \end{cases}$

64. $f(x) = \begin{cases} 5 - x & \text{if } x \leq 3 \\ x - 1 & \text{if } x > 3 \end{cases}$

Determine whether each function is continuous or discontinuous. If discontinuous, state where it is discontinuous.

65. $f(x) = 7x - 5$

66. $f(x) = 5x^3 - 6x^2 + 2x - 4$

67. $f(x) = \dfrac{x + 1}{x - 1}$

68. $f(x) = \dfrac{x^3}{(x + 7)(x - 2)}$

69. $f(x) = \dfrac{12}{5x^3 - 5x}$

70. $f(x) = \dfrac{x + 2}{x^4 - 3x^3 - 4x^2}$

71. $f(x) = \begin{cases} 3 - x & \text{if } x \leq 4 \\ 10 - 2x & \text{if } x > 4 \end{cases}$
[*Hint:* See Exercise 37.]

72. $f(x) = \begin{cases} 5 - x & \text{if } x < 4 \\ 2x - 5 & \text{if } x \geq 4 \end{cases}$
[*Hint:* See Exercise 38.]

73. $f(x) = \begin{cases} 2 - x & \text{if } x \leq 4 \\ x - 6 & \text{if } x > 4 \end{cases}$
[*Hint:* See Exercise 39.]

74. $f(x) = \begin{cases} 2 - x & \text{if } x < 4 \\ 2x - 10 & \text{if } x \geq 4 \end{cases}$
[*Hint:* See Exercise 40.]

75. $f(x) = |x|$
[*Hint:* See Exercise 41.]

76. $f(x) = \dfrac{|x|}{x}$
[*Hint:* See Exercise 43.]

77. By canceling the common factor, $\dfrac{(x - 1)(x + 2)}{x - 1}$ simplifies to $x + 2$. At $x = 1$, however, the function $\dfrac{(x - 1)(x + 2)}{x - 1}$ is *discontinuous* (since it is undefined where the denominator is zero), whereas $x + 2$ is *continuous*. Are these two functions, one obtained from the other by simplification, equal to each other? Explain.

APPLIED EXERCISES

78. GENERAL: Relativity According to Einstein's special theory of relativity, under certain conditions a 1-foot-long object moving with velocity v will appear to an observer to have length $\sqrt{1 - (v/c)^2}$, in which c is a constant equal to the speed of light. Find the limiting value of the apparent length as the velocity of the object approaches the speed of light by finding

$$\lim_{v \to c^-} \sqrt{1 - \left(\frac{v}{c}\right)^2}$$

79. **BUSINESS: Interest Compounded Continuously** If you deposit $1 into a bank account paying 10% interest compounded continuously (see pages 70–71), a year later its value will be

$$\lim_{x \to 0} \left(1 + \frac{x}{10}\right)^{1/x}$$

Find the limit by making a TABLE of values correct to two decimal places, thereby finding the value of the deposit in dollars and cents.

80. **BUSINESS: Interest Compounded Continuously** If you deposit $1 into a bank account paying 5% interest compounded continuously (see pages 70–71), a year later its value will be

$$\lim_{x \to 0} \left(1 + \frac{x}{20}\right)^{1/x}$$

Find the limit by making a TABLE of values correct to two decimal places, thereby finding the value of the deposit in dollars and cents.

81. **GENERAL: Superconductivity** The conductivity of aluminum at temperatures near absolute zero is approximated by the function $f(x) = \dfrac{100}{1 + .001x^2}$, which expresses the conductivity as a percent. Find the limit of this conductivity percent as the temperature x approaches 0 (absolute zero). (See the Application Preview on page 527.)

8.2 RATES OF CHANGE, SLOPES, AND DERIVATIVES

Introduction

In this section we will define the *derivative,* one of the two most important concepts in all of calculus, which measures the rate of change of a function or, equivalently, the slope of a curve.* We begin by discussing *rates of change*.

Average and Instantaneous Rate of Change

We often speak in terms of *rates of change* to express how one quantity changes with another. For example, in the morning the temperature might be "rising at the rate of 3 degrees per hour" and in the evening it might be "falling at the rate of 2 degrees per hour." In some situations, the rate of change of temperature is extremely important. For example, the manufacture of computer chips involves heating and cooling silicon very gradually. For simplicity, suppose that in one process the temperature of the silicon at time x hours is $f(x) = x^2$ degrees. We shall calculate the *average rate of change of temperature* over various time intervals, taking the change in temperature divided by the change in time. For example, at time 1 the temperature is $1^2 = 1$ degrees, and at time 3 the temperature is $3^2 = 9$ degrees, so the

---

*The second most important concept in calculus is the definite integral, which will be defined in Section 11.3.

temperature went up by $9 - 1 = 8$ degrees in 2 hours, for an average rate of:

$$\begin{pmatrix} \text{Average rate} \\ \text{of change} \\ \text{from 1 to 3} \end{pmatrix} = \frac{3^2 - 1^2}{2} = \frac{9 - 1}{2} = \frac{8}{2} = 4 \qquad \begin{array}{l}\text{Average rate} \\ \text{of change} \\ \text{over 2 hours}\end{array}$$

Similarly,

degrees per hour

$$\begin{pmatrix} \text{Average rate} \\ \text{of change} \\ \text{from 1 to 2} \end{pmatrix} = \frac{2^2 - 1^2}{1} = \frac{4 - 1}{1} = 3 \qquad \begin{array}{l}\text{Average rate} \\ \text{of change} \\ \text{over 1 hour}\end{array}$$

$$\begin{pmatrix} \text{Average rate} \\ \text{of change} \\ \text{from 1 to 1.5} \end{pmatrix} = \frac{1.5^2 - 1^2}{0.5} = \frac{2.25 - 1}{0.5} = \frac{1.25}{0.5} = 2.5 \qquad \begin{array}{l}\text{Average rate} \\ \text{of change} \\ \text{over 0.5 hour}\end{array}$$

$$\begin{pmatrix} \text{Average rate} \\ \text{of change} \\ \text{from 1 to 1.1} \end{pmatrix} = \frac{1.1^2 - 1^2}{0.1} = \frac{1.21 - 1}{0.1} = \frac{0.21}{0.1} = 2.1 \qquad \begin{array}{l}\text{Average rate} \\ \text{of change} \\ \text{over 0.1 hour}\end{array}$$

We see that rate of change of temperature over shorter and shorter time intervals is decreasing from 4 to 3 to 2.5 to 2.1, numbers that seem to be approaching 2 degrees per hour. To verify that this is indeed true, we generalize the process. We have been finding the average over a time interval from 1 to a slightly later time that we now call $1 + h$, where h is a small positive number. Notice that in each step we calculated the following expression for successively smaller values of h.

$$\begin{pmatrix} \text{Average rate} \\ \text{of change} \\ \text{from 1 to } 1 + h \end{pmatrix} = \frac{(1 + h)^2 - 1^2}{h} \qquad \begin{array}{l}\text{We used } h = 2, \ h = 1, \\ h = 0.5, \text{ and } h = 0.1\end{array}$$

If we now use our limit notation to let h approach zero, the amount of time will shrink to an instant, giving what is called the *instantaneous rate of change*:

$$\begin{pmatrix} \text{Instantaneous} \\ \text{rate of change} \\ \text{at time 1} \end{pmatrix} = \lim_{h \to 0} \frac{(1 + h)^2 - (1)^2}{h} \qquad \begin{array}{l}\text{Taking the limit as} \\ h \text{ approaches zero}\end{array}$$

For simplicity we have been using the function $f(x) = x^2$ and the time $x = 1$, but the same procedure applies to *any* function f and number x, leading to the following general definition:

Average and Instantaneous Rate of Change

The *average* rate of change of a function f between x and $x + h$ is defined as

$$\frac{f(x + h) - f(x)}{h} \qquad \begin{array}{l}\text{Difference quotient gives} \\ \text{the } \textit{average} \text{ rate of change}\end{array}$$

8.2 RATES OF CHANGE, SLOPES, AND DERIVATIVES

> The *instantaneous* rate of change of a function f at the number x is defined as
>
> $$\lim_{h \to 0} \frac{f(x+h) - f(x)}{h}$$
>
> Taking the limit makes it *instantaneous*

The fraction is just the difference quotient that we introduced on page 61; the numerator is the change in the *function* between two x-values, and the denominator is the change between the two x-values: $(x + h) - (x) = h$. We may use the second formula to check our guess that the average rate of change of temperature is indeed approaching 2 degrees per hour.

EXAMPLE 1

FINDING AN INSTANTANEOUS RATE OF CHANGE

Find the instantaneous rate of change of the temperature function $f(x) = x^2$ at time $x = 1$.

Solution

$\lim_{h \to 0} \dfrac{f(x+h) - f(x)}{h}$ Formula for the instantaneous rate of change of f at x

$= \lim_{h \to 0} \dfrac{f(1+h) - f(1)}{h}$ Substituting $x = 1$

$= \lim_{h \to 0} \dfrac{(1+h)^2 - (1)^2}{h}$ $f(x) = x^2$ gives $f(1+h) = (1+h)^2$ and $f(1) = 1^2$

$= \lim_{h \to 0} \dfrac{1 + 2h + h^2 - 1}{h}$ Expanding $(1+h)^2 = 1 + 2h + h^2$

$= \lim_{h \to 0} \dfrac{\cancel{1} + 2h + h^2 - \cancel{1}}{h}$ Simplifying

$= \lim_{h \to 0} \dfrac{h(2+h)}{h} = \lim_{h \to 0} \dfrac{\cancel{h}(2+h)}{\cancel{h}}$ Factoring out h and canceling (since $h \neq 0$)

$= \lim_{h \to 0} (2 + h) = 2$ Evaluating the limit by direct substitution

Since $f(x)$ gives the temperature at time x hours, this means that *the instantaneous rate of change of temperature at time $x = 1$ hour is 2 degrees per hour* (just as we had guessed earlier).

In this Example, f gave the temperature (degrees) at time x (hours), so the units of the rate of change were *degrees per hour*. In general, for a function f, the units of the instantaneous rate of change are *function units per x unit*.

Practice Problem 1 If $f(x)$ gives the population of a city in year x, what are the units of the instantaneous rate of change? ➤ **Solution on page 558**

Secant and Tangent Lines

We may "see" the average and instantaneous rates of change on the graph of a function. First, some terminology. A *secant line* to a curve is a line that passes through two points of the curve. A *tangent line* is a line that passes through a point of the curve and matches exactly the steepness of the curve at that point.*

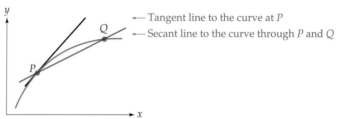

Tangent and secant lines to a curve

If the curve is the graph of a function f and the points P and Q have x-coordinates x and $x + h$, respectively (and so y-coordinates $f(x)$ and $f(x + h)$, respectively), then the graph of the curve and the secant line takes the form shown below.

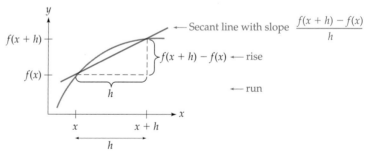

The slope $\dfrac{f(x + h) - f(x)}{h}$ of the secant line is the average rate of change of the function.

---

* The word "secant" comes from the Latin *secare*, "to cut," suggesting that the secant line "cuts" the curve at two points. The word "tangent" comes from the Latin *tangere*, "to touch," suggesting that the tangent line just "touches" the curve at one point.

Observe that the *slope* (rise over run) of the secant line is the difference quotient $\frac{f(x+h) - f(x)}{h}$, exactly the same as our earlier definition of the average rate of change of the function between x and $x + h$.

Furthermore, the two points where the secant line meets the curve are separated by a distance h along the x-axis. Letting $h \to 0$ forces the second point to approach the first, causing the secant line to approach the tangent line, the slope of which is then $\lim_{h \to 0} \frac{f(x+h) - f(x)}{h}$. But this is exactly the same as our earlier definition of the *instantaneous rate of change of the function at x*.

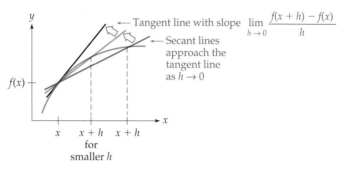

The slope $\lim_{h \to 0} \frac{f(x+h) - f(x)}{h}$ of the *tangent* line is the *instantaneous* rate of change of the function.

The last two diagrams have shown that the slope of the *secant* line is the *average* rate of change of the function, and the slope of the *tangent* lines is the *instantaneous* rate of change of the function. The fact that rates of change are related to slopes should come as no surprise: if a function has a large rate of change, its graph must be rising rapidly, giving it a large slope. Similarly, if a function has only a small rate of change, its graph will be rising slowly, giving it a small slope.

To summarize: For a function f and its graph:

$$\begin{pmatrix} \text{\textit{Average} rate} \\ \text{of change of } f \\ \text{between } x \text{ and } x+h \end{pmatrix} = \begin{pmatrix} \text{Slope of the \textit{secant}} \\ \text{line through the} \\ \text{points at } x \text{ and } x+h \end{pmatrix} = \frac{f(x+h) - f(x)}{h} \quad \text{Without the limit}$$

$$\begin{pmatrix} \textit{Instantaneous} \\ \text{rate of change} \\ \text{of } f \text{ at } x \end{pmatrix} = \begin{pmatrix} \text{Slope of the} \\ \textit{tangent} \text{ line} \\ \text{at } x \end{pmatrix} = \lim_{h \to 0} \frac{f(x+h) - f(x)}{h} \quad \text{With the limit}$$

In Example 1 we found the instantaneous rate of change of a function. We now use the same formula to find the slope of the tangent line to the graph of a function.

EXAMPLE 2

FINDING THE SLOPE OF A TANGENT LINE

Find the slope of the tangent line to $f(x) = \dfrac{1}{x}$ at $x = 2$.

Solution

$$\lim_{h \to 0} \frac{f(x+h) - f(x)}{h} \qquad \text{Formula for the slope of the tangent line at } x$$

$$= \lim_{h \to 0} \frac{f(2+h) - f(2)}{h} \qquad \text{Substituting } x = 2$$

$$= \lim_{h \to 0} \frac{\dfrac{1}{2+h} - \dfrac{1}{2}}{h} \qquad \text{Using the function } f(x) = \frac{1}{x}$$

$$= \lim_{h \to 0} \frac{1}{h}\left(\frac{1}{2+h} - \frac{1}{2}\right) \qquad \text{Since multiplying by } 1/h \text{ is equivalent to dividing by } h$$

$$= \lim_{h \to 0} \frac{1}{h} \cdot \frac{2 - (2+h)}{(2+h) \cdot 2} \qquad \text{Combining fractions using the common denominator } (2+h)\cdot 2$$

$$= \lim_{h \to 0} \frac{1}{h} \cdot \frac{-h}{(2+h) \cdot 2} \qquad \text{Simplifying the numerator}$$

$$= \lim_{h \to 0} \frac{1}{\cancel{h}} \cdot \frac{\cancel{-h}^{-1}}{(2+h) \cdot 2} = \lim_{h \to 0} \frac{-1}{(2+h) \cdot 2} \qquad \text{Cancelling}$$

$$= \frac{-1}{(2+0) \cdot 2} = \frac{-1}{4} = -\frac{1}{4} \qquad \text{Evaluating the limit by direct substitution of } h = 0$$

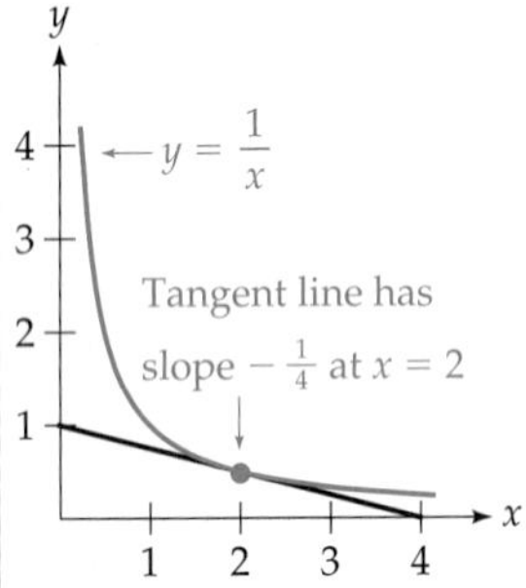

Therefore, the curve $y = \dfrac{1}{x}$ has slope $-\dfrac{1}{4}$ at $x = 2$, as shown in the graph on the left.

Graph: $y = \dfrac{1}{x}$, Tangent line has slope $-\dfrac{1}{4}$ at $x = 2$.

EXAMPLE 3

FINDING A TANGENT LINE

Use the result of the preceding example to find the equation of the tangent line to $f(x) = \dfrac{1}{x}$ at $x = 2$.

Solution

From Example 2 we know that the *slope* of the tangent line is $m = -\dfrac{1}{4}$. The point on the curve at $x = 2$ is $(2, \dfrac{1}{2})$, the y-coordinate coming from $y = f(2) = \dfrac{1}{2}$. The point-slope formula then gives:

$$y - \tfrac{1}{2} = -\tfrac{1}{4}(x - 2) \qquad \begin{array}{l} y - y_1 = m(x - x_1) \text{ with } m = -\tfrac{1}{4}, \\ x_1 = 2, \text{ and } y_1 = \tfrac{1}{2} \end{array}$$

$$y - \tfrac{1}{2} = -\tfrac{1}{4}x + \tfrac{1}{2} \qquad \text{Multiplying out}$$

$$\underbrace{y = 1 - \tfrac{1}{4}x}_{\text{Equation of the tangent line}} \qquad \text{Simplifying}$$

Graphing Calculator Exploration

On a graphing calculator, graph the curve $y = 1/x$ together with the line $y = 1 - x/4$ using the window [0, 4] by [0, 4] to see that the line is indeed the tangent line to the curve at $x = 2$. (For a way to have your graphing calculator find the *equation* for the tangent line, see Exercises 47–48 on page 560.)

The first of the following graphs shows the curve $y = \tfrac{1}{x}$ along with the tangent line that we found at $x = 2$. The next two graphs show the results of successively "zooming in" near the point of tangency, showing that the curve seems to straighten out and almost *become its own tangent line* on a smaller and smaller viewing window. For this reason, the tangent line is called *the best linear approximation to the curve near the point of tangency.*

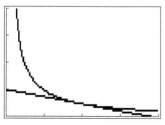

$y = 1/x$ and its tangent line at $x = 2$ on [0, 4] by [0, 4]

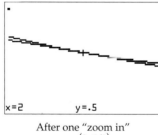

After one "zoom in" near $(2, 1/2)$

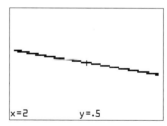

After a second "zoom in" centered at $(2, 1/2)$

Graphing Calculator Exploration

Use a graphing calculator to graph $y = x^3 - 2x^2 - 3x + 4$ (or any function of your choice). Then "zoom in" a few times around a point to see the curve straighten out and almost *become* its own tangent line.

The Derivative

In Examples 1 and 2 we found the instantaneous rate of change of a function and the slope of a tangent line of a curve *at a particular number or point*. It is much more efficient to carry out the same calculation but keeping x as a *variable*, obtaining a new function that gives the instantaneous rate of change or the slope of the tangent line at *any* value of x. This new function is denoted with a *prime*, $f'(x)$ (read: "f prime of x"), and is called *the derivative of f at x*.

> **Derivative**
>
> For a function f, the *derivative of f at x* is defined as
>
> $$f'(x) = \lim_{h \to 0} \frac{f(x+h) - f(x)}{h} \qquad \text{Limit of the difference quotient}$$
>
> (provided that the limit exists). The derivative $f'(x)$ gives the instantaneous rate of change of f at x and also the slope of the graph of f at x.

In general, the units of the derivative are *function units* per x unit.

EXAMPLE 4 FINDING THE DERIVATIVE OF A FUNCTION FROM THE DEFINITION

Find the derivative of $f(x) = x^2 - 7x + 150$.

Solution

$$f'(x) = \lim_{h \to 0} \frac{f(x+h) - f(x)}{h} \qquad \text{Definition of the derivative}$$

$$= \lim_{h \to 0} \frac{\overbrace{(x+h)^2 - 7(x+h) + 150}^{f(x+h)} - \overbrace{(x^2 - 7x + 150)}^{f(x)}}{h} \qquad \text{Using } f(x) = x^2 - 7x + 150$$

$$= \lim_{h \to 0} \frac{x^2 + 2xh + h^2 - 7x - 7h + 150 - x^2 + 7x - 150}{h} \qquad \text{Expanding and simplifying}$$

$$= \lim_{h \to 0} \frac{2xh + h^2 - 7h}{h} = \lim_{h \to 0} \frac{h(2x + h - 7)}{h} \qquad \text{Simplifying (since } h \neq 0\text{)}$$

$$= \lim_{h \to 0} (2x + h - 7) = 2x - 7 \qquad \text{Evaluating the limit by direct substitution}$$

Therefore, the derivative of $f(x) = x^2 - 7x + 150$ is $f'(x) = 2x - 7$.

The operation of calculating derivatives should be thought of as an operation on *functions,* taking one function [such as $f(x) = x^2 - 7x + 150$] and giving another [$f'(x) = 2x - 7$]. The resulting function is called "the derivative" because it is *derived* from the first, and the process of obtaining it is called "differentiation." If the derivative is defined at x, then the original function is said to be *differentiable* at x.

EXAMPLE 5

USING A DERIVATIVE IN AN APPLICATION

Refining crude oil into various products, such as gasoline, heating oil, and plastics, requires heating and cooling the oil at different rates. Suppose that the temperature of the oil at time x hours is $f(x) = x^2 - 7x + 150$ degrees Fahrenheit (for $0 \leq x \leq 8$). Find the instantaneous rate of change of temperature at times $x = 6$ hours and $x = 2$ hours and interpret the results.

Solution

The instantaneous rate of change means the derivative, so ordinarily we would now take the derivative of the temperature function $f(x) = x^2 - 7x + 150$. However, this is just what we did in the previous example, so we will use the result that the derivative is $f'(x) = 2x - 7$, evaluating this at $x = 6$ and at $x = 2$ and interpreting the results.

$$f'(6) = 2 \cdot 6 - 7 = 5 \qquad \text{Evaluating } f'(x) = 2x - 7 \text{ at } x = 6$$

Interpretation: After 6 hours, the temperature is increasing at the rate of 5 degrees per hour.

$$f'(2) = 2 \cdot 2 - 7 = -3 \qquad \text{Evaluating } f'(x) = 2x - 7 \text{ at } x = 2$$

Interpretation: After 2 hours, the temperature is *decreasing* at the rate of 3 degrees per hour.

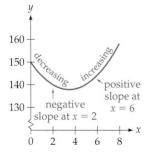

A derivative that is *positive* means that the original quantity is *increasing*, and a derivative that is *negative* means that the quantity is *decreasing*.

The graph of $f(x) = x^2 - 7x + 150$ on the left shows that the slope of the curve is indeed negative at $x = 2$ and positive at $x = 6$.

EXAMPLE 6

FINDING A DERIVATIVE FROM THE DEFINITION

Find the derivative of $f(x) = x^3$.

Solution

In our solution we will use the expansion $(x + h)^3 = x^3 + 3x^2h + 3xh^2 + h^3$ (found by multiplying together three copies of $(x + h)$).

$$f'(x) = \lim_{h \to 0} \frac{f(x + h) - f(x)}{h} \qquad \text{Definition of } f'(x)$$

$$= \lim_{h \to 0} \frac{(x + h)^3 - x^3}{h} \qquad \text{Using } f(x) = x^3$$

$$= \lim_{h \to 0} \frac{x^3 + 3x^2h + 3xh^2 + h^3 - x^3}{h} \qquad \text{Using the expansion of } (x + h)^3$$

$$= \lim_{h \to 0} \frac{x^3 + 3x^2h + 3xh^2 + h^3 - x^3}{h} = \lim_{h \to 0} \frac{h(3x^2 + 3xh + h^2)}{h} \qquad \text{Cancelling and then factoring out an } h$$

$$= \lim_{h \to 0} \frac{h(3x^2 + 3xh + h^2)}{h} = \lim_{h \to 0} 3x^2 + 3xh + h^2 \qquad \text{Cancelling again}$$

$$= 3x^2 \qquad \text{Evaluating the limit by direct substitution}$$

Therefore, the derivative of $f(x) = x^3$ is $f'(x) = 3x^2$.

Graphing Calculator Exploration

Some advanced graphing calculators have computer algebra systems that can simplify algebraic expressions. For example, the Texas Instruments *TI-89* calculator will find and simplify the difference quotient for the function $f(x) = x^3$ that we found in the preceding Example as follows:

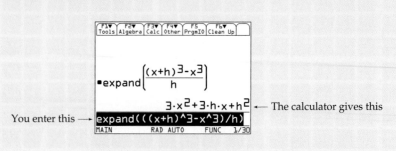

You enter this → `expand(((x+h)^3-x^3)/h)`

The calculator gives this ← $3 \cdot x^2 + 3 \cdot h \cdot x + h^2$

The calculator finds the (simplified) difference quotient as $3x^2 + 3xh + h^2$, and taking the limit (using direct substitution of $h = 0$) gives the derivative $f'(x) = 3x^2$, just as we found above.

Leibniz's Notation for the Derivative

Calculus was developed by Isaac Newton (1642–1727) and Gottfried Wilhelm Leibniz (1646–1716) in two different countries, so there naturally developed two different notations for the derivative. Newton denoted derivatives by a dot over the function, $\dot{f}$, a notation that has been largely replaced by our "prime" notation. Leibniz wrote the derivative of $f(x)$ by writing $\frac{d}{dx}$ in front of the function: $\frac{d}{dx}f(x)$. In Leibniz's notation, the fact that the derivative of x^3 is $3x^2$ is written

$$\frac{d}{dx}x^3 = 3x^2 \qquad \text{The derivative of } x^3 \text{ is } 3x^2$$

The following table shows equivalent expressions in the two notations.

Prime Notation		Leibniz's Notation	
$f'(x)$	=	$\frac{d}{dx}f(x)$	Prime and $\frac{d}{dx}$ both mean the derivative
y'	=	$\frac{dy}{dx}$	For y a function of x

Each notation has its own advantages, and we will use both.* Leibniz's notation comes from writing the definition of the derivative as:

$$\frac{dy}{dx} = \lim_{\Delta x \to 0} \frac{f(x + \Delta x) - f(x)}{\Delta x} \qquad \text{Definition of the derivative (page 552) with the change in } x \text{ written as } \Delta x$$

or

$$\frac{dy}{dx} = \lim_{\Delta x \to 0} \frac{\Delta y}{\Delta x} \qquad f(x + \Delta x) - f(x) \text{ is the change in } y, \text{ and so can be written } \Delta y$$

---

*Other notations for the derivative are $Df(x)$ and $D_x f(x)$, but we will not use them.

It is as if the limit turns Δ (read "Delta," the Greek letter D) into d, changing the $\frac{\Delta y}{\Delta x}$ into $\frac{dy}{dx}$. That is, Leibniz's notation reminds us that the derivative $\frac{dy}{dx}$ is the limit of the slope $\frac{\Delta y}{\Delta x}$.

Some functions are not *differentiable* (the derivative does not exist) at certain x-values. For example, the following diagram shows a function that has a "corner point" at $x = 1$. At this point the slope (and therefore the derivative) cannot be defined, so the function is not differentiable at $x = 1$. Other nondifferentiable functions will be discussed in Section 8.7.

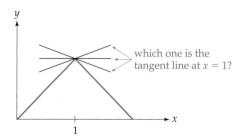

Since the tangent line cannot be uniquely defined at $x = 1$, the slope, and therefore the derivative, is undefined at $x = 1$.

The following diagram shows the geometric relationship between a function (upper graph) and its derivative (lower graph). Observe carefully how the *slope of f* is shown by the *sign of f'*.

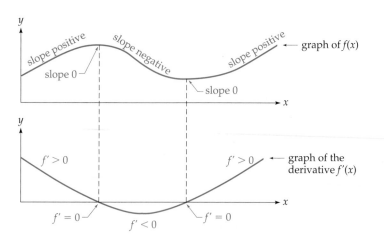

Practice Problem 2

The graph shows a function and its derivative. Which is the original funcion (#1 or #2) and which is its derivative? [*Hint:* Which curve has *slope* zero where the other has *value* zero?]

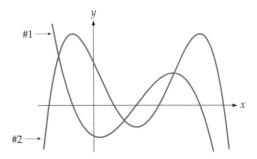

➤ **Solution on next page**

8.2 Section Summary

The derivative of the function f at x is

$$f'(x) = \lim_{h \to 0} \frac{f(x+h) - f(x)}{h} \qquad \text{Provided that the limit exists}$$

Some students remember the steps as **DESL** (pronounced "diesel"): write the **D**efinition, **E**xpress the numerator in terms of the function, **S**implify, and take the **L**imit.

The derivative $f'(x)$ gives both the *slope of the graph* of the function at x and the *instantaneous rate of change of the function* at x. In other words, "derivative," "slope," and "instantaneous rate of change" are merely the mathematical, the geometric, and the analytic versions of the same idea.

The derivative gives the rate of change *at a particular instant*, not an actual change over a period of time. Instantaneous rates of changes are like the speeds on an automobile speedometer—a reading of 50 mph at one moment does not mean that you will travel exactly 50 miles in the next hour, since the actual distance depends upon your speed during the entire hour. The derivative, however, may be interpreted as the *approximate* change resulting from a 1-unit increase in the independent variable. For example, if your speedometer reads 50 mph, then you may say that you will travel *about* 50 miles during the next hour, meaning that this will be true provided that your speed remains steady throughout the hour. (In Chapter 11 we will see how to calculate actual changes from rates of change that do not stay constant.)

> **Solutions to Practice Problems**
> 1. The units are people per year, measuring the rate of growth of the population.
> 2. #2 is the original function and #1 is its derivative.

8.2 Exercises

By imagining tangent lines at points P_1, P_2, and P_3, state whether the slopes are positive, zero, or negative at these points.

1.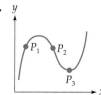

2.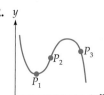

3.

4.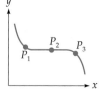

Use the tangent lines shown at points P_1 and P_2 to find the slopes of the curve at these points.

5.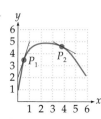

6.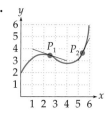

In Exercises 7 and 8, use the graph of each function $f(x)$ to make a rough sketch of the derivative $f'(x)$ showing where $f'(x)$ is positive, negative, and zero. (Omit scale on y-axis.)

7.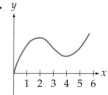

8.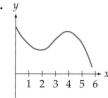

Find the average rate of change of the given function between the following pairs of x-values. [*Hint:* See page 546.]

a. $x = 1$ and $x = 3$
b. $x = 1$ and $x = 2$
c. $x = 1$ and $x = 1.5$
d. $x = 1$ and $x = 1.1$
e. $x = 1$ and $x = 1.01$
f. What number do your answers seem to be approaching?

9. $f(x) = x^2 + x$ 10. $f(x) = 2x^2 + 5$

Find the average rate of change of the given function between the following pairs of x-values. [*Hint:* See page 546.]

a. $x = 2$ and $x = 4$
b. $x = 2$ and $x = 3$
c. $x = 2$ and $x = 2.5$
d. $x = 2$ and $x = 2.1$
e. $x = 2$ and $x = 2.01$
f. What number do your answers seem to be approaching?

11. $f(x) = 2x^2 + x - 2$ 12. $f(x) = x^2 + 2x - 1$

Find the average rate of change of the given function between the following pairs of x-values. [Hint: See page 546.]

a. $x = 3$ and $x = 5$
b. $x = 3$ and $x = 4$
c. $x = 3$ and $x = 3.5$
d. $x = 3$ and $x = 3.1$
e. $x = 3$ and $x = 3.01$
f. What number do your answers seem to be approaching?

13. $f(x) = 5x + 1$ **14.** $f(x) = 7x - 2$

Find the average rate of change of the given function between the following pairs of x-values. [Hint: See page 546.]

a. $x = 4$ and $x = 6$
b. $x = 4$ and $x = 5$
c. $x = 4$ and $x = 4.5$
d. $x = 4$ and $x = 4.1$
e. $x = 4$ and $x = 4.01$
f. What number do your answers seem to be approaching?

15. $f(x) = \sqrt{x}$ **16.** $f(x) = \dfrac{4}{x}$

Use the formula on page 547 to find the instantaneous rate of change of the function at the given x-value. If you did the related problem in Exercises 9–16, compare your answers. [Hint: See Example 1.]

17. $f(x) = x^2 + x$ at $x = 1$
18. $f(x) = x^2 + 2x - 1$ at $x = 2$
19. $f(x) = 5x + 1$ at $x = 3$
20. $f(x) = \dfrac{4}{x}$ at $x = 4$

Use the formula on page 547 to find the slope of the tangent line to the curve at the given x-value. If you did the related problem in Exercises 9–16, compare your answers. [Hint: See Example 2.]

21. $f(x) = 2x^2 + x - 2$ at $x = 2$
22. $f(x) = 2x^2 + 5$ at $x = 1$
23. $f(x) = \sqrt{x}$ at $x = 4$
24. $f(x) = 7x - 2$ at $x = 3$

Find $f'(x)$ by using the definition of the derivative.

25. $f(x) = x^2 - 3x + 5$
26. $f(x) = 2x^2 - 5x + 1$
27. $f(x) = 1 - x^2$ **28.** $f(x) = \tfrac{1}{2}x^2 + 1$
29. $f(x) = 9x - 2$ **30.** $f(x) = -3x + 5$
31. $f(x) = \dfrac{x}{2}$ **32.** $f(x) = 0.01x + 0.05$
33. $f(x) = 4$ **34.** $f(x) = \pi$
35. $f(x) = ax^2 + bx + c$
(a, b, and c are constants)
36. $f(x) = (x + a)^2$
(a is a constant.) [Hint: First expand $(x + a)^2$.]
37. $f(x) = x^5$
[Hint: Use $(x + h)^5 = x^5 + 5x^4h + 10x^3h^2 + 10x^2h^3 + 5xh^4 + h^5$]
38. $f(x) = x^4$
[Hint: Use $(x + h)^4 = x^4 + 4x^3h + 6x^2h^2 + 4xh^3 + h^4$]
39. $f(x) = \dfrac{2}{x}$ **40.** $f(x) = \dfrac{1}{x^2}$
41. $f(x) = \sqrt{x}$ **42.** $f(x) = \dfrac{1}{\sqrt{x}}$
[Hint: Multiply the numerator and denominator of the difference quotient by $(\sqrt{x + h} + \sqrt{x})$ and then simplify.]
[Hint: Multiply the numerator and denominator of the difference quotient by $(\sqrt{x} + \sqrt{x + h})$ and then simplify.]
43. $f(x) = x^3 + x^2$ **44.** $f(x) = \dfrac{1}{2x}$

45. a. Find the equation for the tangent line to the curve $f(x) = x^2 - 3x + 5$ at $x = 2$, writing the equation in slope-intercept form. [Hint: Use your answer to Exercise 25.]
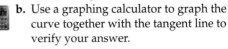
b. Use a graphing calculator to graph the curve together with the tangent line to verify your answer.

46. a. Find the equation for the tangent line to the curve $f(x) = 2x^2 - 5x + 1$ at $x = 2$, writing the equation in slope-intercept form.
[*Hint:* Use your answer to Exercise 26.]
  b. Use a graphing calculator to graph the curve together with the tangent line to verify your answer.

47. a. Graph the function $f(x) = x^2 - 3x + 5$ on the window $[-10, 10]$ by $[-10, 10]$. Then use the DRAW menu to graph the TANGENT line at $x = 2$. Your screen should also show the *equation* of the tangent line. (If you did Exercise 45, this equation for the tangent line should agree with the one you found there.)
  b. Add to your graph the tangent line at $x = 1$, and the tangent lines at any other x-values that you choose.

48. a. Graph the function $f(x) = 2x^2 - 5x + 1$ on the window $[-10, 10]$ by $[-10, 10]$. Then use the DRAW menu to graph the TANGENT line at $x = 2$. Your screen should also show the *equation* of the tangent line. (If you did Exercise 46, this equation for the tangent line should agree with the one you found there.)
  b. Add to your graph the tangent line at $x = 0$, and the tangent lines at any other x-values that you choose.

For each function in Exercises 49–54:
  a. Find $f'(x)$ using the definition of the derivative.
  b. Explain, by considering the original function, why the derivative is a constant.

49. $f(x) = 3x - 4$ **50.** $f(x) = 2x - 9$

51. $f(x) = 5$ **52.** $f(x) = 12$

53. $f(x) = mx + b$
(m and b are constants)

54. $f(x) = b$
(b is a constant)

APPLIED EXERCISES

55. BUSINESS: Temperature The temperature in an industrial pasteurization tank is $f(x) = x^2 - 8x + 110$ degrees centigrade after x minutes (for $0 \le x \le 12$).
  a. Find $f'(x)$ by using the definition of the derivative.
  b. Use your answer to part (a) to find the instantaneous rate of change of the temperature after 2 minutes. Be sure to interpret the sign of your answer.
  c. Use your answer to part (a) to find the instantaneous rate of change after 5 minutes.

56. GENERAL: Population The population of a town is $f(x) = 3x^2 - 12x + 200$ people after x weeks (for $0 \le x \le 20$).
  a. Find $f'(x)$ by using the definition of the derivative.
  b. Use your answer to part (a) to find the instantaneous rate of change of the population after 1 week. Be sure to interpret the sign of your answer. *(continues)*
  c. Use your answer to part (a) to find the instantaneous rate of change of the population after 5 weeks.

57. BEHAVIORAL SCIENCE: Learning Theory In a psychology experiment, a person could memorize x words in $f(x) = 2x^2 - x$ seconds (for $0 \le x \le 10$).
  a. Find $f'(x)$ by using the definition of the derivative.
  b. Evaluate your answer at $x = 5$ and interpret it as an instantaneous rate of change in the proper units.

58. BUSINESS: Advertising An automobile dealership finds that the number of cars that it sells on day x of an advertising campaign is $S(x) = -x^2 + 10x$ (for $0 \le x \le 7$).
  a. Find $S'(x)$ by using the definition of the derivative.
  b. Use your answer to part (a) to find the instantaneous rate of change on day $x = 3$. *(continues)*

c. Use your answer to part (a) to find the instantaneous rate of change on day $x = 6$.

Be sure to interpret the signs of your answers.

59. **BIOMEDICAL: Temperature** The temperature of a patient in a hospital on day x of an illness is given $T(x) = -x^2 + 5x + 100$ degrees Fahrenheit (for $1 < x < 5$).
   a. Find $T'(x)$ by using the definition of the derivative.
   b. Use your answer to part (a) to find the instantaneous rate of change of temperature on day 2.
   c. Use your answer to part (a) to find the instantaneous rate of change of temperature on day 3. *(continues)*

d. What do your answers to parts (b) and (c) tell you about the patient's health on those two days?

60. **BIOMEDICAL: Bacteria** The number of bacteria in a culture x hours after treatment with an antibiotic is given by
$f(x) = -x^2 + 12x + 1000$ (for $1 < x < 30$).
   a. Find $f'(x)$ by using the definition of the derivative.
   b. Use your answer to part (a) to find the instantaneous rate of change after 2 hours.
   c. Use your answer to part (a) to find the instantaneous rate of change after 20 hours.

Be sure to interpret the signs of your answers.

8.3 SOME DIFFERENTIATION FORMULAS

Introduction

In Section 8.2 we defined the *derivative* of a function and used it to calculate instantaneous rates of change and slopes. Even for a function as simple as $f(x) = x^2$, however, calculating the derivative from the definition was rather involved. Calculus would be of limited usefulness if all derivatives had to be calculated in this way.

In this section we will learn several *rules of differentiation* that will simplify finding the derivatives of many useful functions. The rules are derived from the definition of the derivative, which is why we studied the definition first. We will also learn another important use for differentiation: calculating "marginals" (marginal revenue, marginal cost, and marginal profit), which are used extensively in business and economics.

Derivative of a Constant

The first rule of differentiation shows how to differentiate a constant function.

For any constant c,
$$\frac{d}{dx} c = 0 \qquad \text{Derivative of a constant is zero}$$

EXAMPLE 1 DIFFERENTIATING A CONSTANT

$$\frac{d}{dx} 7 = 0$$

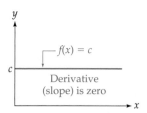

A constant function (a horizontal line) has derivative (slope) zero.

This rule is obvious geometrically. The graph of a constant function $f(x) = c$ is the horizontal line $y = c$. Since the slope of a horizontal line is zero, the derivative of $f(x) = c$ is zero.

This rule follows immediately from the definition of the derivative. The constant function $f(x) = c$ has the same value c for *any* value of x, so, in particular, $f(x + h) = c$ and $f(x) = c$. Substituting these into the definition of the derivative gives

$$f'(x) = \lim_{h \to 0} \frac{\overbrace{f(x+h)}^{c} - \overbrace{f(x)}^{c}}{h} = \lim_{h \to 0} \frac{c - c}{h} = \lim_{h \to 0} \frac{\overbrace{0}}{h} = \lim_{h \to 0} 0 = 0$$

Therefore, the derivative of a constant function $f(x) = c$ is $f'(x) = 0$.

Power Rule

One of the most useful differentiation formulas in all of calculus is called the *Power Rule*. It tells how to differentiate powers such as x^7 or x^{100}.

Power Rule

For any constant exponent n,

$$\frac{d}{dx} x^n = n \cdot x^{n-1}$$

To differentiate x^n, bring down the exponent as a mutiplier and then decrease the exponent by 1

A derivation of the Power Rule for positive integer exponents is given at the end of this section. We will use the Power Rule for *all* real numbers n, since more general proofs will be given later (see pages 589, 706–707, and 743–744).

EXAMPLE 2 **USING THE POWER RULE**

a. $\dfrac{d}{dx} x^7 = 7x^{7-1} = 7x^6$

    ↑ Bring down the exponent   ↑ Decrease the exponent by 1

b. $\dfrac{d}{dx} x^{100} = 100x^{100-1} = 100x^{99}$

c. $\dfrac{d}{dx} x^{-2} = -2x^{-2-1} = -2x^{-3}$ The Power Rule holds for negative exponents

d. $\dfrac{d}{dx} \sqrt{x} = \dfrac{d}{dx} x^{\frac{1}{2}} = \dfrac{1}{2} x^{\frac{1}{2}-1} = \dfrac{1}{2} x^{-\frac{1}{2}}$ And for fractional exponents

e. $\dfrac{d}{dx} x^1 = 1x^{1-1} = x^0 = 1$

This last result is used so frequently that it should be remembered separately.

$$\dfrac{d}{dx} x = 1 \qquad \text{The derivative of } x \text{ is 1}$$

From now on we will skip the middle step in these examples, differentiating powers in one step:

$$\dfrac{d}{dx} x^{50} = 50x^{49}$$

$$\dfrac{d}{dx} x^{2/3} = \dfrac{2}{3} x^{-1/3}$$

    ↖ $\tfrac{2}{3} - 1$

Practice Problem 1 Find

a. $\dfrac{d}{dx} x^2$ b. $\dfrac{d}{dx} x^{-5}$ c. $\dfrac{d}{dx} \sqrt[4]{x}$ ▶ Solutions on page 573

Constant Multiple Rule

The Power Rule shows how to differentiate a power such as x^3. The *Constant Multiple Rule* extends this result to functions such as $5x^3$, a constant *times* a function. Briefly, to differentiate a constant times a function, we simply "carry along" the constant and differentiate the function.

> **Constant Multiple Rule**
>
> For any constant c,
>
> $$\frac{d}{dx}[c \cdot f(x)] = c \cdot f'(x)$$
>
> The derivative of a constant times a function is the constant times the derivative of the function

(provided, of course, that the derivative $f'(x)$ exists). A derivation of this rule is given at the end of this section.

EXAMPLE 3 **USING THE CONSTANT MULTIPLE RULE**

a. $\dfrac{d}{dx} 5x^3 = 5 \cdot 3x^2 = 15x^2$ b. $\dfrac{d}{dx} 3x^{-4} = 3(-4)x^{-5} = -12x^{-5}$

Carry along the constant ⎯ Derivative of x^3

Again we will skip the middle step, bringing down the exponent and immediately multiplying it by the number in front of the x.

EXAMPLE 4 **CALCULATING DERIVATIVES MORE QUICKLY**

a. $\dfrac{d}{dx} 8x^{-1/2} = -4x^{-3/2}$ b. $\dfrac{d}{dx} 7x = 7 \cdot 1 = 7$

⎯ $8\left(-\tfrac{1}{2}\right)$ ⎯ Derivative of x

This last example, showing that the derivative of $7x$ is just 7, leads to a very useful general rule.

For any constant c,

$$\frac{d}{dx}(cx) = c \qquad \text{The derivative of a constant times } x \text{ is just the constant}$$

EXAMPLE 5

FINDING DERIVATIVES INVOLVING CONSTANTS

a. $\dfrac{d}{dx}(7x) = 7$ \qquad Using $\dfrac{d}{dx}(cx) = c$

b. $\dfrac{d}{dx} 7 = 0$ \qquad For a constant alone, the derivative is zero

c. $\dfrac{d}{dx}(7x^2) = 7 \cdot 2x = 14x$ \qquad But for a constant times a function, the derivative is the constant times the derivative of the function

Sum Rule

The *Sum Rule* extends differentiation to sums of functions. Briefly, to differentiate a *sum* of two functions, just differentiate the functions separately and add the results.

Sum Rule

$$\frac{d}{dx}[f(x) + g(x)] = f'(x) + g'(x) \qquad \text{The derivative of a sum is the sum of the derivatives}$$

(provided, of course, that both the derivatives $f'(x)$ and $g'(x)$ exist). A derivation of the Sum Rule is given at the end of this section. For example, the derivative of the sum

$$x^3 + x^5 \qquad \text{Sum of } x^3 \text{ and } x^5$$

is just the sum of the derivatives

$$3x^2 + 5x^4 \qquad \text{Differentiating each separately and adding}$$

A similar rule holds for the *difference* of two functions,

$$\frac{d}{dx}[f(x) + g(x)] = f'(x) + g'(x) \qquad \text{The derivative of a difference is the difference of the derivatives}$$

(provided that $f'(x)$ and $g'(x)$ exist). These two rules may be combined:

> **Sum-Difference Rule**
>
> $$\frac{d}{dx}[f(x) \pm g(x)] = f'(x) \pm g'(x)$$ Use both upper signs or both lower signs

Similar rules hold for sums and differences of any finite number of terms. Using these rules, we may differentiate any polynomial or, more generally, functions with variables raised to *any* constant powers.

EXAMPLE 6   USING THE SUM-DIFFERENCE RULE

a. $\dfrac{d}{dx}(x^3 - x^5) = 3x^2 - 5x^4$ Derivatives taken separately

b. $\dfrac{d}{dx}(5x^{-2} - 6x^{1/3} + 4) = -10x^{-3} - 2x^{-2/3}$ The constant 4 has derivative 0

Leibniz's Notation and Evaluation of Derivatives

Leibniz's derivative notation, $\dfrac{d}{dx}$, is often read "the derivative with respect to x" to emphasize that the independent variable is x. To differentiate a function of some *other* variable, the x in $\dfrac{d}{dx}$ is replaced by the other variable. For example:

Function	Derivative	
$f(t)$	$\dfrac{d}{dt}f(t)$	Use $\dfrac{d}{dt}$ for the derivative with respect to t
w^3	$\dfrac{d}{dw}w^3$	Use $\dfrac{d}{dw}$ for the derivative with respect to w

The following two notations both mean the derivative *evaluated* at $x = 2$.

$f'(2)$ — Derivative, Evaluated at $x = 2$

$\left.\dfrac{df}{dx}\right|_{x=2}$ — Derivative, Evaluated at $x = 2$ Bar | means "evaluated at"

Be careful! Both notations mean *first* differentiate and *then* evaluate.

EXAMPLE 7

EVALUATING A DERIVATIVE

If $f(x) = x^4$, find $f'(2)$.

Solution

$$f'(x) = 4x^3 \qquad \text{First differentiate}$$
$$f'(2) = 4 \cdot 2^3 = 4 \cdot 8 = 32 \qquad \text{Then evaluate}$$

Practice Problem 2   If $f(x) = x^3$, find $\left.\dfrac{df}{dx}\right|_{x=-1}$ ▶ Solution on page 573

Derivatives in Business and Economics: Marginals

There is another interpretation for the derivative, one that is particularly important in business and economics. Suppose that a company has calculated its revenue, cost, and profit functions, as defined below.

$$R(x) = \begin{pmatrix} \text{Total revenue (income)} \\ \text{from selling } x \text{ units} \end{pmatrix} \qquad \text{Revenue function}$$

$$C(x) = \begin{pmatrix} \text{Total cost of} \\ \text{producing } x \text{ units} \end{pmatrix} \qquad \text{Cost function}$$

$$P(x) = \begin{pmatrix} \text{Total profit from producing} \\ \text{and selling } x \text{ units} \end{pmatrix} \qquad \text{Profit function}$$

The term "marginal cost" means the additional cost of producing one more unit, $C(x+1) - C(x)$, which may be written $\dfrac{C(x+1) - C(x)}{1}$, which is just the difference quotient $\dfrac{C(x+h) - C(x)}{h}$ with $h = 1$. If many units are being produced, then $h = 1$ is a relatively small number compared with x, so this difference quotient may be approximated by its limit as $h \to 0$, that is by the *derivative* of the cost function. In view of this approximation, in calculus the marginal cost is *defined* to be the derivative of the cost function:

$$MC(x) = C'(x) \qquad \text{Marginal cost is the derivative of cost}$$

The marginal *revenue* function MR(x) and the marginal *profit* function MP(x) are similarly defined as the derivatives of the revenue and cost functions.

$$MR(x) = R'(x) \quad \text{Marginal revenue is the derivative of revenue}$$
$$MP(x) = P'(x) \quad \text{Marginal profit is the derivative of profit}$$

All of this can be summarized very briefly: "marginal" means "derivative of." We now have three interpretations for the derivative: *slopes*, *instantaneous rates of change*, and *marginals*.

EXAMPLE 8

FINDING AND INTERPRETING MARGINAL COST

A company manufactures cordless telephones and finds that its cost function (the total cost of manufacturing x telephones) is

$$C(x) = 400\sqrt{x} + 500 \quad \text{Cost function}$$

dollars, where x is the number of telephones produced.

a. Find the marginal cost function MC(x).
b. Find the marginal cost when 100 telephones have been produced, and interpret your answer.

Solution

a. The marginal cost function is the derivative of the cost function $C(x) = 400x^{1/2} + 500$:

$$MC(x) = 200x^{-1/2} = \frac{200}{\sqrt{x}} \quad \text{Derivative of C(x)}$$

b. To find the marginal cost when 100 telephones have been produced, we evaluate the marginal cost function at $x = 100$:

$$MC(100) = \frac{200}{\sqrt{100}} = \frac{200}{10} = \$20 \quad MC(x) = \frac{200}{\sqrt{x}} \text{ evaluated at } x = 100$$

Interpretation: When 100 telephones have been produced, the marginal cost is $20, meaning that to produce one more telephone costs about $20.

EXAMPLE 9   FINDING A LEARNING RATE

A psychology researcher finds that the number of names that a person can memorize in x minutes is approximately $f(x) = 6\sqrt[3]{x^2}$. Find the instantaneous rate of change of this function after 8 minutes and interpret your answer.

Solution

$$f(x) = 6x^{2/3} \qquad 6\sqrt[3]{x^2} \text{ in exponential form}$$

$$f'(x) = 6 \cdot \frac{2}{3} x^{-1/3} = 4x^{-1/3} \qquad \text{The instantaneous rate of change is } f'(x)$$

$$f'(8) = 4(8)^{-1/3} = 4\left(\frac{1}{\sqrt[3]{8}}\right) = 4\left(\frac{1}{2}\right) = 2 \qquad \text{Evaluating at } x = 8$$

Interpretation: After 8 minutes the person can memorize about two additional names per minute.

Functions as Single Objects

You may have noticed that calculus requires a more abstract point of view than precalculus mathematics. In earlier courses you looked at functions and graphs as collections of individual points, to be plotted one at a time. Now, however, we are operating on *whole functions* all at once (for example, differentiating the function x^3 to obtain the function $3x^2$). In calculus, the basic objects of interest are *functions*, and a function should be thought of as a *single* object.

This is in keeping with a trend toward increasing abstraction as you learn mathematics. You first studied single numbers, then points (pairs of numbers), then functions (collections of points), and now collections of functions (polynomials, differentiable functions, and so on). Each stage has been a generalization of the previous stage as you have reached higher levels of sophistication. This process of generalization or "chunking" of knowledge enables you to express ideas of wider applicability and power.

Derivatives on a Graphing Calculator

Graphing calculators have an operation called NDERIV (or something similar), standing for *numerical derivative*, which gives an *approximation* of the derivative of a function. Most do so by evaluating the *symmetric difference quotient*, $\dfrac{f(x + h) - f(x - h)}{2h}$ for a small value of h, such as

$h = 0.001$. The numerator represents the change in the function when x changes by $2h$ (from $x - h$ to $x + h$), and the denominator divides by this change in x. Geometrically, the symmetric difference quotient gives the slope of the secant line through two points on the curve h units on either side of the point at x. While NDERIV usually approximates the derivative quite closely, it sometimes gives erroneous results, as we will see in later sections. For this reason, using a graphing calculator effectively requires an understanding of both the calculus that underlies it and the technology that limits it.

Graphing Calculator Exploration

a. Use a graphing calculator to graph $y_1 = x^3 - x^2 - 6x + 3$ on $[-5, 5]$ by $[-10, 10]$.

b. Define y_2 as the derivative of y_1 (using NDERIV) and graph both functions.

c. Observe that where y_1 is horizontal, the *value* of y_2 is zero; where y_1 slopes *upward*, y_2 is *positive*; and where y_1 slopes *downward*, y_2 is *negative*. Would you be able to use these observations to identify which curve is the original function and which is the derivative?

d. Now check your answer to Example 9 as follows: Redefine y_1 as $y_1 = 6x^{2/3}$, reset the window to $[0, 10]$ by $[-10, 30]$, GRAPH y_1 and y_2, and EVALUATE y_2 at $x = 8$. Your answer should agree with that of Example 9.

8.3 Section Summary

Our development of calculus has followed two quite different lines—one technical (the *rules* of derivatives) and the other conceptual (the *meaning* of derivatives).

On the conceptual side, derivatives have three meanings:

- Instantaneous rates of change
- Slopes of curves
- Marginals

The fact that the derivative represents all three of these ideas simultaneously is one of the reasons that calculus is so useful.

On the technical side, although we have learned several differentiation rules, we really know how to differentiate only one kind of function, *x to a constant power:*

$$\frac{d}{dx} x^n = nx^{n-1}$$

The other rules,

$$\frac{d}{dx} [c \cdot f(x)] = c \cdot f'(x)$$

and

$$\frac{d}{dx} [f(x) \pm g(x)] = f'(x) \pm g'(x)$$

simply extend the Power Rule to sums, differences, and constant multiples of such powers. Therefore, any function to be differentiated must first be expressed in terms of powers. This is why we reviewed exponential notation so carefully in Chapter 1.

Verification of the Power Rule for Positive Integer Exponents

Multiplying $(x + h)$ times itself repeatedly gives

$$(x + h)^2 = x^2 + 2xh + h^2$$

$$(x + h)^3 = x^3 + 3x^2h + 3xh^2 + h^3$$

and in general, for any positive integer n,

$$(x + h)^n = x^n + nx^{n-1}h + \tfrac{1}{2}n(n-1)x^{n-2}h^2 + \cdots + nxh^{n-1} + h^n$$

$$= x^n + nx^{n-1}h + h^2[\tfrac{1}{2}n(n-1)x^{n-2} + \cdots + nxh^{n-3} + h^{n-2}] \quad \text{Factoring out } h^2$$

$$= x^n + nx^{n-1}h + h^2 \cdot P \quad \begin{array}{l} \text{\textit{P} stands for the polynomial} \\ \text{in the square bracket above} \end{array}$$

The resulting formula

$$(x + h)^n = x^n + nx^{n-1}h + h^2 \cdot P$$

will be useful in the following verification. To prove the Power Rule for any positive integer n, we use the definition of the derivative to differentiate $f(x) = x^n$.

$$f'(x) = \lim_{h \to 0} \frac{f(x+h) - f(x)}{h}$$ Definition of the derivative

$$= \lim_{h \to 0} \frac{(x+h)^n - x^n}{h}$$ Since $f(x+h) = (x+h)^n$ and $f(x) = x^n$

$$= \lim_{h \to 0} \frac{x^n + nx^{n-1}h + h^2 \cdot P - x^n}{h}$$ Expanding, using the formula derived earlier

$$= \lim_{h \to 0} \frac{nx^{n-1}h + h^2 \cdot P}{h}$$ Canceling the x^n and the $-x^n$

$$= \lim_{h \to 0} \frac{h(nx^{n-1} + h \cdot P)}{h}$$ Factoring out an h

$$= \lim_{h \to 0} (nx^{n-1} + h \cdot P)$$ Canceling the h (since $h \ne 0$)

$$= nx^{n-1}$$ Evaluating the limit by direct substitution

This shows that for any positive integer n, the derivative of x^n is nx^{n-1}.

Verification of the Constant Multiple Rule

For a constant c and a function f, let $g(x) = c \cdot f(x)$. If $f'(x)$ exists, we may calculate the derivative $g'(x)$ as follows:

$$g'(x) = \lim_{h \to 0} \frac{g(x+h) - g(x)}{h}$$ Definition of the derivative

$$= \lim_{h \to 0} \frac{c \cdot f(x+h) - c \cdot f(x)}{h}$$ Since $g(x+h) = c \cdot f(x+h)$ and $g(x) = c \cdot f(x)$

$$= \lim_{h \to 0} \frac{c \cdot [f(x+h) - f(x)]}{h}$$ Factoring out the c

$$= c \cdot \lim_{h \to 0} \frac{f(x+h) - f(x)}{h}$$ Taking c outside the limit leaves just the definition of the derivative $f'(x)$.

$$= c \cdot f'(x)$$ Constant Multiple Rule

This shows that the derivative of a constant times a function, $c \cdot f(x)$, is the constant times the derivative of the function, $c \cdot f'(x)$.

Verification of the Sum Rule

For two functions f and g, let their sum be $s = f + g$. If $f'(x)$ and $g'(x)$ exist, we may calculate $s'(x)$ as follows:

$$s'(x) = \lim_{h \to 0} \frac{s(x+h) - s(x)}{h}$$ Definition of the derivative

$$= \lim_{h \to 0} \frac{[f(x+h) + g(x+h)] - [f(x) + g(x)]}{h}$$ Since $s(x+h) = f(x+h) + g(x+h)$ and $s(x) = f(x) + g(x)$

$$= \lim_{h \to 0} \frac{f(x+h) + g(x+h) - f(x) - g(x)}{h}$$ Eliminating the brackets

$$= \lim_{h \to 0} \frac{f(x+h) - f(x) + g(x+h) - g(x)}{h}$$ Rearranging the numerator

$$= \lim_{h \to 0} \left[\frac{f(x+h) - f(x)}{h} + \frac{g(x+h) - g(x)}{h} \right]$$ Separating the fraction into two parts

$$= \underbrace{\lim_{h \to 0} \frac{f(x+h) - f(x)}{h}}_{f'(x)} + \underbrace{\lim_{h \to 0} \frac{g(x+h) - g(x)}{h}}_{g'(x)}$$ Using Limit Rule 4a on page 531

Recognizing the definition of the derivatives of f and g

$$= f'(x) + g'(x)$$ Sum Rule

This shows that the derivative of a sum $f(x) + g(x)$ is the sum of the derivatives $f'(x) + g'(x)$.

> **Solutions to Practice Problems**

1. a. $\dfrac{d}{dx} x^2 = 2x^{2-1} = 2x$

   b. $\dfrac{d}{dx} x^{-5} = -5x^{-5-1} = -5x^{-6}$

   c. $\dfrac{d}{dx} \sqrt[4]{x} = \dfrac{d}{dx} x^{1/4} = \dfrac{1}{4} x^{(1/4)-1} = \dfrac{1}{4} x^{-3/4}$

2. $\dfrac{df}{dx} = 3x^2$

   $\left.\dfrac{df}{dx}\right|_{x=-1} = 3(-1)^2 = 3$

8.3 Exercises

Find the derivative of each function.

1. $f(x) = x^4$
2. $f(x) = x^5$
3. $f(x) = x^{500}$
4. $f(x) = x^{1000}$
5. $f(x) = x^{1/2}$
6. $f(x) = x^{1/3}$
7. $g(x) = \frac{1}{2}x^4$
8. $f(x) = \frac{1}{3}x^9$
9. $g(w) = 6\sqrt[3]{w}$
10. $g(w) = 12\sqrt{w}$
11. $h(x) = \frac{3}{x^2}$
12. $h(x) = \frac{4}{x^3}$
13. $f(x) = 4x^2 - 3x + 2$
14. $f(x) = 3x^2 - 5x + 4$
15. $f(x) = \frac{1}{x^{1/2}}$
16. $f(x) = \frac{1}{x^{2/3}}$
17. $f(x) = \frac{6}{\sqrt[3]{x}}$
18. $f(x) = \frac{4}{\sqrt{x}}$
19. $f(r) = \pi r^2$
20. $f(r) = \frac{4}{3}\pi r^3$
21. $f(x) = \frac{1}{6}x^3 + \frac{1}{2}x^2 + x + 1$
22. $f(x) = \frac{1}{24}x^4 + \frac{1}{6}x^3 + \frac{1}{2}x^2 + x + 1$
23. $g(x) = \sqrt{x} - \frac{1}{x}$
24. $g(x) = \sqrt[3]{x} - \frac{1}{x}$
25. $h(x) = 6\sqrt[3]{x^2} - \frac{12}{\sqrt[3]{x}}$
26. $h(x) = 8\sqrt{x^3} - \frac{8}{\sqrt[4]{x}}$
27. $f(x) = \frac{10}{\sqrt{x}} - 9\sqrt[3]{x^5} + 17$
28. $f(x) = \frac{9}{\sqrt[3]{x}} - 16\sqrt{x^5} - 14$
29. $f(x) = \frac{x^2 + x^3}{x}$
30. $f(x) = x^2(x + 1)$

31. a. Find the derivative of $f(x) = 2$.
    b. Interpret your answer in terms of slope.
    c. Interpret your answer in terms of instantaneous rate of change.
32. a. Find the derivative of $f(x) = 3x$.
    b. Interpret your answer in terms of slope.
    c. Interpret your answer in terms of instantaneous rate of change.

Find the indicated derivatives.

33. If $f(x) = x^5$, find $f'(-2)$.
34. If $f(x) = x^4$, find $f'(-3)$.
35. If $f(x) = 6\sqrt[3]{x^2} - \frac{48}{\sqrt[3]{x}}$, find $f'(8)$.
36. If $f(x) = 12\sqrt[3]{x^2} + \frac{48}{\sqrt[3]{x}}$, find $f'(8)$.
37. If $f(x) = x^3$, find $\left.\frac{df}{dx}\right|_{x=-3}$
38. If $f(x) = x^4$, find $\left.\frac{df}{dx}\right|_{x=-2}$
39. If $f(x) = \frac{16}{\sqrt{x}} + 8\sqrt{x}$, find $\left.\frac{df}{dx}\right|_{x=4}$
40. If $f(x) = \frac{54}{\sqrt{x}} + 12\sqrt{x}$, find $\left.\frac{df}{dx}\right|_{x=9}$

41. Use a graphing calculator to verify that the derivative of a constant is zero, as follows. Define y_1 to be a constant (such as $y_1 = 5$) and then use NDERIV to define y_2 to be the derivative of y_1. Then graph the two functions together on an appropriate window and use TRACE to observe that the derivative y_2 is zero (graphed as a line along the x-axis), showing that the derivative of a constant is zero.

42. Use a graphing calculator to verify that the derivative of a linear function is a constant, as follows. Define y_1 to be a linear function (such as $y_1 = 3x - 4$) and then use NDERIV to define y_2 to be the derivative of y_1. Then graph the two functions together on an appropriate window and observe that the derivative y_2 is a constant (graphed as a horizontal line, such as $y_2 = 3$), verifying that the derivative of $y_1 = mx + b$ is $y_2 = m$.

APPLIED EXERCISES

43. BUSINESS: Marginal Profit An electronics company finds that its total profit from selling x computer chips is $P(x) = 0.02x^{3/2} - 3000$ dollars.
 a. Find the company's marginal profit function.
 b. Find the marginal profit when 10,000 units have been sold, and interpret your answer.

44. BUSINESS: Marginal Cost A steel mill finds that its cost function is

$$C(x) = 8000\sqrt{x} - 6000\sqrt[3]{x}$$

dollars, where x is the (daily) production of steel (in tons).

 a. Find the marginal cost function.
 b. Find the marginal cost when 64 tons of steel are produced.

45. BUSINESS: Marginal Profit (*43 continued*) Use a calculator to find the actual profit from the 10,001st computer chip, $P(10,001) - P(10,000)$, by evaluating the expression

$$\overbrace{[0.02(10,001)^{3/2} - 3000]}^{P(10,001)} - \overbrace{[0.02(10,000)^{3/2} - 3000]}^{P(10,000)}$$

Is your answer close to the answer of $3 found in Exercise 43(b)? Which way of finding the marginal profit was easier, using calculus (Exercise 43) or carrying out the calculation in this exercise?

46. BUSINESS: Marginal Cost (*44 continued*) Use a calculator to find the actual cost of the 65th ton of steel, $C(65) - C(64)$, by evaluating the expression:

$$\overbrace{(8000\sqrt{65} - 6000\sqrt[3]{65})}^{C(65)} - \overbrace{(8000\sqrt{64} - 6000\sqrt[3]{64})}^{C(64)}$$

Is your answer close to the answer of $375 found in Exercise 44(b)? Which way of finding the marginal cost was easier, using calculus (Exercise 44) or carrying out the calculation in this exercise?

47. GENERAL: Population A company that makes games for teenage children forecasts that the teenage population in the United States x years from now will be

$$P(x) = 12{,}000{,}000 - 12{,}000x + 600x^2 + 100x^3$$

Find the rate of change of the teenage population:

 a. x years from now.
 b. 1 year from now and interpret your answer.
 c. 10 years from now and interpret your answer.

48. BIOMEDICAL: Flu Epidemic The number of people newly infected on day t of a flu epidemic is $f(t) = 13t^2 - t^3$ (for $0 \le t \le 13$). Find the instantaneous rate of change of this number on

 a. day 5 and interpret your answer.
 b. day 10 and interpret your answer.

49. BUSINESS: Advertising It has been estimated that the number of people who will see a newspaper advertisement that has run for x consecutive days is of the form $N(x) = T - \frac{1}{2}T/x$ for $x \ge 1$, where T is the total readership of the newspaper. If a newspaper has a circulation of 400,000, an ad that runs for x days will be seen by

$$N(x) = 400{,}000 - \frac{200{,}000}{x}$$

people. Find how fast this number of potential customers is growing when this ad has run for 5 days.

50. ENVIRONMENTAL SCIENCE: Pollution An electrical generating plant burns high-sulfur oil, and the amount of sulfur dioxide pollution x miles downwind of the plant is $f(x) = 108x^{-2}$ parts per million (ppm). Find the instantaneous rate of change of the pollution level 2 miles from the source. Interpret your answer in the proper units.

51. BIOMEDICAL: Blood Flow Nitroglycerin is often prescribed to enlarge blood vessels that have become too constricted. If the cross-sectional area of a blood vessel t hours after nitroglycerin is administered is $A(t) = 0.01t^2$

square centimeters (for $1 \le t \le 5$), find the instantaneous rate of change of the cross-sectional area 4 hours after the administration of nitroglycerin.

52. **GENERAL: Hailstones** Hailstones are frozen raindrops that increase in size as long as the updrafts keep them in the clouds. The weight of a typical hailstone that remains in a cloud for t minutes is $W(t) = 0.05t^3$ ounces. Find the instantaneous rate of change of the weight after 2 minutes.

53. **PSYCHOLOGY: Learning Rates** A language school has found that its students can memorize $p(t) = 24\sqrt{t}$ phrases in t hours of class (for $1 \le t \le 10$). Find the instantaneous rate of change of this quantity after 4 hours of class.

54. **ENVIRONMENTAL SCIENCE: Water Quality** Downstream from a waste treatment plant the amount of dissolved oxygen in the water usually decreases for some distance (due to bacteria consuming the oxygen) and then increases (due to natural purification). A graph of the dissolved oxygen at various distances downstream looks like the curve below (known as the "oxygen sag"). The amount of dissolved oxygen is usually taken as a measure of the health of the river.

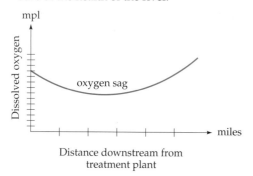

Distance downstream from treatment plant

Suppose that the amount of dissolved oxygen x miles downstream is $D(x) = 0.2x^2 - 2x + 10$ mpl (milligrams per liter) for $0 \le x \le 20$. Use this formula to find the instantaneous rate of change of the dissolved oxygen:

a. 1 mile downstream.
b. 10 miles downstream.

Interpret the signs of your answers.

55–56. **ECONOMICS: Marginal Utility** Generally, the more you have of something, the less valuable each additional unit becomes. For example, a dollar is less valuable to a millionaire than to a beggar. Economists define a person's "utility function" $U(x)$ for a product as the "perceived value" of having x units of that product. The *derivative* of $U(x)$ is called the *marginal utility function*, $MU(x) = U'(x)$. Suppose that a person's utility function for money is given by the function below. That is, $U(x)$ is the utility (perceived value) of x dollars.

a. Find the marginal utility function $MU(x)$.
b. Find $MU(1)$, the marginal utility of the first dollar.
c. Find $MU(1{,}000{,}000)$, the marginal utility of the millionth dollar.

55. $U(x) = 100\sqrt{x}$ 56. $U(x) = 12\sqrt[3]{x}$

57. **GENERAL: Smoking and Education** According to a recent study,* the probability that a smoker will quit smoking increases with the smoker's educational level. The probability (expressed as a percent) that a smoker with x years of education will quit is approximated by the equation $f(x) = 0.831x^2 - 18.1x + 137.3$ (for $10 \le x \le 16$).

a. Find $f(12)$ and $f'(12)$ and interpret these numbers. [*Hint*: $x = 12$ corresponds to a high school graduate.]
b. Find $f(16)$ and $f'(16)$ and interpret these numbers. [*Hint*: $x = 16$ corresponds to a college graduate.]

58. **BIOMEDICAL: Lung Cancer** Asbestos has been found to be a potent cause of lung cancer. According to one study of asbestos workers, the number of lung cancer cases in the group depended on the number t of years of exposure to asbestos according to the function $N(t) = 0.00437t^{3.2}$.

a. Graph this function on the window [0, 15] by [−10, 30].
b. Find $N(10)$ and $N'(10)$ and interpret these numbers.

---

*William Sander, "Schooling and Quitting Smoking," *The Review of Economics and Statistics* LXXVII(1):191–199, February 1995.

59–60. GENERAL: College Tuition The following tables give the annual college tuition costs (in dollars) for a year at a private (Exercise 59) or public (Exercise 60) college for the academic years ending in 1970–2000. To avoid large numbers, years are listed as years since 1970.

a. Enter these numbers into your graphing calculator and make a plot of the resulting points (Years Since 1970 on the x-axis and Tuition on the y-axis).
b. Have your calculator find the quadratic regression formula for these data. Then enter the result in function y_1. Plot the points together with the regression curve. Observe that the curve fits the points quite well.
c. Predict the tuition in the year 2010 by evaluating y_1 at $x = 40$ (years since 1970).
d. Define y_2 to be the derivative of y_1 (using NDERIV).
e. Predict the rate of change of tuition in the year 2010 by evaluating y_2 at $x = 40$. State your answer in the proper units.

59.

Years Since 1970	Tuition (private college)
0	1533
10	3130
20	8147
30	15,518

60.

Years Since 1970	Tuition (public college)
0	323
10	583
20	1356
30	3362

Sources: U.S. Department of Education, The College Board

8.4 THE PRODUCT AND QUOTIENT RULES

Introduction

In the previous section we learned how to differentiate the sum and difference of two functions—we simply take the sum or difference of the derivatives. In this section we learn how to differentiate the *product* and *quotient* of two functions. Unfortunately, we do not simply take the product or quotient of the derivatives. Matters are a little more complicated.

Product Rule

To differentiate the product of two functions, $f(x) \cdot g(x)$, we use the *Product Rule*.

Product Rule

$$\frac{d}{dx}[f(x) \cdot g(x)] = f'(x) \cdot g(x) + f(x) \cdot g'(x)$$

The derivative of a product is the derivative of the first times the second plus the first times the derivative of the second (provided, of course, that the derivatives $f'(x)$ and $g'(x)$ both exist). The formula is clearer if we write the functions simply as f and g.

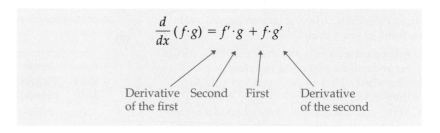

$$\frac{d}{dx}(f \cdot g) = \underbrace{f'}_{\text{Derivative of the first}} \cdot \underbrace{g}_{\text{Second}} + \underbrace{f}_{\text{First}} \cdot \underbrace{g'}_{\text{Derivative of the second}}$$

A derivation of the Product Rule is given at the end of this section.

EXAMPLE 1

USING THE PRODUCT RULE

Use the Product Rule to calculate $\frac{d}{dx}(x^3 \cdot x^5)$.

Solution

$$\frac{d}{dx}(x^3 \cdot x^5) = \underbrace{3x^2}_{\substack{\text{Derivative} \\ \text{of the first}}} \cdot \underbrace{x^5}_{\text{Second}} + \underbrace{x^3}_{\text{First}} \cdot \underbrace{5x^4}_{\substack{\text{Derivative} \\ \text{of the second}}} = 3x^7 + 5x^7 = 8x^7$$

We may check this answer by simplifying the original product, $x^3 \cdot x^5 = x^8$, and then differentiating:

$$\frac{d}{dx}\underbrace{(x^3 \cdot x^5)}_{x^8} = \frac{d}{dx}x^8 = 8x^7 \qquad \text{Agrees with above answer}$$

Notice that the derivative of a product is *not* the product of the derivatives: $(f \cdot g)' \neq f' \cdot g'$. For $x^3 \cdot x^5$ the product of the derivatives would be $3x^2 \cdot 5x^4 = 15x^6$, which is *not* the correct answer $8x^7$ that we found above. The Product Rule shows the correct way to differentiate a product.

EXAMPLE 2

USING THE PRODUCT RULE

Use the Product Rule to find $\frac{d}{dx}[(x^2 - x + 2)(x^3 + 3)]$.

Solution

$$\frac{d}{dx}[(x^2 - x + 2)(x^3 + 3)]$$

$$= \underbrace{(2x - 1)}_{\substack{\text{Derivative} \\ \text{of } x^2 - x + 2}}(x^3 + 3) + (x^2 - x + 2)\underbrace{(3x^2)}_{\substack{\text{Derivative} \\ \text{of } x^3 + 3}}$$

$$= 2x^4 + 6x - x^3 - 3 + 3x^4 - 3x^3 + 6x^2 \quad \text{Multiplying out}$$

$$= 5x^4 - 4x^3 + 6x^2 + 6x - 3 \quad \text{Simplifying}$$

Practice Problem 1   Use the Product Rule to find $\dfrac{d}{dx}[x^3(x^2 - x)]$.

▶ Solution on page 590

Quotient Rule

The *Quotient Rule* shows how to differentiate a quotient of two functions.

Quotient Rule

$$\frac{d}{dx}\left(\frac{f(x)}{g(x)}\right) = \frac{g(x) \cdot f'(x) - g'(x) \cdot f(x)}{[g(x)]^2}$$

← The bottom times the derivative of the top, minus the derivative of the bottom times the top
— The bottom squared

(provided that the derivatives $f'(x)$ and $g'(x)$ both exist and that $g(x) \neq 0$). A derivation of the Quotient Rule is given at the end of this section.

The Quotient Rule looks less formidable if we write the functions simply as f and g,

$$\frac{d}{dx}\left(\frac{f}{g}\right) = \frac{g \cdot f' - g' \cdot f}{g^2}$$

or even as

$$\frac{d}{dx}\left(\frac{\text{top}}{\text{bottom}}\right) = \frac{(\text{bottom}) \cdot \left(\dfrac{d}{dx}\text{top}\right) - \left(\dfrac{d}{dx}\text{bottom}\right) \cdot (\text{top})}{(\text{bottom})^2}$$

EXAMPLE 3 USING THE QUOTIENT RULE

Use the Quotient Rule to find $\dfrac{d}{dx}\left(\dfrac{x^9}{x^3}\right)$.

Solution

$$\dfrac{d}{dx}\left(\dfrac{x^9}{x^3}\right) = \dfrac{\overset{\text{Bottom}}{(x^3)}\overset{\text{Derivative of the top}}{(9x^8)} - \overset{\text{Derivative of the bottom}}{(3x^2)}\overset{\text{Top}}{(x^9)}}{\underset{\text{Bottom squared}}{(x^3)^2}} = \dfrac{9x^{11} - 3x^{11}}{x^6} = \dfrac{6x^{11}}{x^6} = 6x^5$$

We may check this answer by simplifying the original quotient and then differentiating:

$$\dfrac{d}{dx}\left(\underset{x^6}{\underbrace{\dfrac{x^9}{x^3}}}\right) = \dfrac{d}{dx}x^6 = 6x^5 \qquad \text{Agrees with above answer}$$

Notice that the derivative of a quotient is *not* the quotient of the derivatives:

$$\left(\dfrac{f}{g}\right)' \text{ is } not \text{ equal to } \dfrac{f'}{g'}$$

For the quotient $\dfrac{x^9}{x^3}$, taking the quotient of the derivatives would give $\dfrac{9x^8}{3x^2} = 3x^6$, which is *not* the correct answer $6x^5$ that we found above. The Quotient Rule shows the correct way to differentiate a quotient.

EXAMPLE 4 USING THE QUOTIENT RULE

Find $\dfrac{d}{dx}\left(\dfrac{x^2}{x+1}\right)$.

Solution The Quotient Rule gives

$$\frac{d}{dx}\left(\frac{x^2}{x+1}\right) = \frac{(x+1)(2x) - (1)(x^2)}{(x+1)^2} = \frac{2x^2 + 2x - x^2}{(x+1)^2} = \frac{x^2 + 2x}{(x+1)^2}$$

where the numerator components are: Bottom, Derivative of the top, Derivative of the bottom, Top, and the denominator is Bottom squared.

Practice Problem 2 Find $\dfrac{d}{dx}\left(\dfrac{2x^2}{x^2+1}\right)$.   ▶ Solution on page 590

Not every quotient requires the Quotient Rule. Some are simple enough to be differentiated by the Power Rule.

EXAMPLE 5 **DIFFERENTIATING A QUOTIENT BY THE POWER RULE**

Find the derivative of $y = \dfrac{5}{x^2}$

Solution

$$\frac{d}{dx}\left(\frac{5}{x^2}\right) = \frac{d}{dx}(5x^{-2}) = -10x^{-3} = -\frac{10}{x^3} \quad \text{Differentiated by the power rule}$$

In this example we rewrote the expression before and after the differentiation:

Begin	Rewrite	Differentiate	Rewrite
$y = \dfrac{5}{x^2}$	$y = 5x^{-2}$	$\dfrac{dy}{dx} = -10x^{-3}$	$\dfrac{dy}{dx} = -\dfrac{10}{x^3}$

This way is often much easier than using the Quotient Rule if the numerator or denominator is a constant.

EXAMPLE 6 FINDING THE COST OF CLEANER WATER

Practically every city must purify its drinking water and treat its wastewater. The cost of the treatment rises steeply for higher degrees of purity. If the cost of purifying a gallon of water to a purity of x percent is

$$C(x) = \frac{2}{100 - x} \quad \text{for } 80 < x < 100$$

dollars, find the rate of change of the purification costs when the purity is:

a. 90%   **b.** 98%

Solution

The rate of change of cost is the *derivative* of the cost function:

$$C'(x) = \frac{d}{dx}\left(\frac{2}{100-x}\right) = \frac{\overbrace{(100-x)(0)}^{\text{Derivative of 2}} - \overbrace{(-1)(2)}^{\text{Derivative of } 100-x}}{(100-x)^2} \quad \text{Differentiating by the Quotient Rule}$$

$$= \frac{0+2}{(100-x)^2} = \frac{2}{(100-x)^2} \quad \text{Simplifying (the derivative is undefined at } x = 100\text{)}$$

a. For 90% purity we evaluate at $x = 90$.

$$C'(90) = \frac{2}{(100-90)^2} = \frac{2}{10^2} = \frac{2}{100} = 0.02 \qquad C'(x) = \frac{2}{(100-x)^2} \text{ evaluated at } x = 90$$

Interpretation: At 90% purity, the rate of change of the cost is 0.02 dollar, meaning that the costs increase by about *2 cents for each additional percentage of purity.*

b. For 98% purity we evaluate $C'(x)$ at $x = 98$:

$$C'(98) = \frac{2}{(100-98)^2} = \frac{2}{2^2} = \frac{2}{4} = \frac{1}{2} = 0.50 \qquad C'(x) = \frac{2}{(100-x)^2} \text{ evaluated at } x = 98$$

Interpretation: At 98% purity, the rate of change of the cost is 0.50 dollar, meaning that the costs increase by about *50 cents for each additional percentage of purity.*

Notice that an extra percentage of purity above the 98% level is 25 times as costly as an extra percentage above the 90% purity level.

Graphing Calculator Exploration

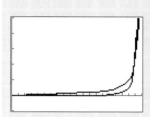

a. On a graphing calculator, enter the cost function from Example 6 as $y_1 = \dfrac{2}{(100-x)}$. Then use NDERIV to define y_2 to be the derivative of y_1.

b. Graph both y_1 and y_2 on the window [80, 100] by [−1, 5]. Your graph should resemble the one on the left (but you may have an additional "false" vertical line on the right).

c. Verify the results of Example 6 by evaluating y_2 at $x = 90$ and at $x = 98$.

d. Evaluate y_2 at $x = 100$, giving (supposedly) the derivative of y_1 at $x = 100$. However, in Example 6 we saw that the derivative of y_1 is *undefined* at $x = 100$. Your calculator is giving you a "false value" for the derivative, resulting from NDERIV's use of a symmetric difference quotient (see pages 569–570) and a positive value for h. Therefore, to use your calculator effectively, you must also understand calculus.

Marginal Average Cost

It is often useful to calculate not just the *total* cost of producing x units of some product, but also the *average cost per unit*, denoted $AC(x)$, which is found by dividing the total cost $C(x)$ by the number of units x.

$$AC(x) = \frac{C(x)}{x} \qquad \text{Average cost per unit is total cost divided by the number of units}$$

The derivative of the average cost function is called the *marginal average cost, MAC*.*

$$MAC(x) = \frac{d}{dx}\left[\frac{C(x)}{x}\right] \qquad \text{Marginal average cost is the derivative of average cost}$$

---

* The marginal average cost function is sometimes denoted $\overline{C}'(x)$, with similar notations used for marginal average revenue and marginal average profit.

Marginal average revenue *MAR*, and marginal average profit *MAP*, are defined similarly as the derivatives of average revenue per unit, $\frac{R(x)}{x}$, and average profit per unit, $\frac{P(x)}{x}$.

$$MAR(x) = \frac{d}{dx}\left[\frac{R(x)}{x}\right] \qquad \text{Marginal average revenue is the derivative of average revenue } \frac{R(x)}{x}$$

$$MAP(x) = \frac{d}{dx}\left[\frac{P(x)}{x}\right] \qquad \text{Marginal average profit is the derivative of average profit } \frac{P(x)}{x}$$

EXAMPLE 7

FINDING AND INTERPRETING MARGINAL AVERAGE COST

It costs a book publisher $12 to produce each book, and fixed costs are $1500. Therefore, the company's cost function is

$$C(x) = 12x + 1500 \qquad \text{Total cost of producing } x \text{ books}$$

a. Find the average cost function.
b. Find the marginal average cost function.
c. Find the marginal average cost at $x = 100$ and interpret your answer.

Solution

a. The average cost function is

$$AC(x) = \underbrace{\frac{12x + 1500}{x}}_{\substack{\text{Total cost divided} \\ \text{by number of units}}} = \underbrace{12 + \frac{1500}{x}}_{\text{Simplifying}} = \underbrace{12 + 1500x^{-1}}_{\text{In power form}}$$

b. The *marginal* average cost is the derivative of average cost. We could use the Quotient Rule on the first expression above, but it is easier to use the Power Rule on the last expression:

$$MAC(x) = \frac{d}{dx}(12 + 1500x^{-1}) = -1500x^{-2} = -\frac{1500}{x^2}$$

c. Evaluating at $x = 100$:

$$MAC(100) = -\frac{1500}{100^2} = -\frac{1500}{10{,}000} = -0.15 \qquad -\frac{1500}{x^2} \text{ at } x = 100$$

Interpretation: When 100 books have been produced, the average cost per book is decreasing (because of the negative sign) by about *15 cents per additional book produced.* This reflects the fact that while *total* costs rise when you produce more, the *average cost per unit* decreases, because of the economies of mass production.

Graphing Calculator Exploration

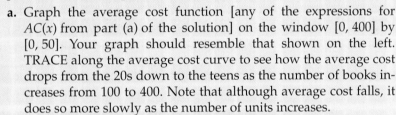

Use a graphing calculator to investigate further the effects of mass production in Example 7.

a. Graph the average cost function [any of the expressions for $AC(x)$ from part (a) of the solution] on the window [0, 400] by [0, 50]. Your graph should resemble that shown on the left. TRACE along the average cost curve to see how the average cost drops from the 20s down to the teens as the number of books increases from 100 to 400. Note that although average cost falls, it does so more slowly as the number of units increases.

b. To see exactly how rapidly the average cost declines, graph the *marginal* average cost function on the window [0, 400] by [−1, 1]. TRACE along this curve to see how the marginal average cost (which is negative since costs are decreasing) approaches zero as the number of units increases (the law of diminishing returns).

EXAMPLE 8

FINDING TIME SAVED BY SPEEDING

A certain mathematics professor drives 25 miles to his office every day, mostly on highways. If he drives at constant speed v miles per hour, his travel time (distance divided by speed) is

$$T(v) = \frac{25}{v}$$

hours. Find $T'(55)$ and interpret this number.

Solution Since $T(v) = \dfrac{25}{v}$ is a quotient, we could differentiate it by the Quotient Rule. However, it is easier to write $\dfrac{25}{v}$ as a *power*,

$$T(v) = 25v^{-1}$$

and differentiate using the Power Rule:

$$T'(v) = -25v^{-2}$$

This gives the rate of change of the travel time with respect to driving speed. $T'(v)$ is negative, showing that as speed increases, travel time *decreases*. Evaluating this at speed $v = 55$ gives

$$T'(55) = -25(55)^{-2} = \dfrac{-25}{(55)^2} \approx -0.00826 \qquad \text{Using a calculator}$$

This number, the rate of change of travel time with respect to driving speed, means that when driving at 55 miles per hour, you save only 0.00826 hour for each extra mile per hour of speed. Multiplying by 60 gives the saving in *minutes*:

$$(-0.00826)(60) \approx -0.50 = -\dfrac{1}{2}$$

That is, each extra mile per hour of speed saves only about half a minute, or 30 seconds. For example, speeding by 10 mph would save only about $\tfrac{1}{2} \cdot 10 = 5$ minutes. One must then decide whether this slight savings in time is worth the risk of an accident or a speeding ticket.

8.4 Section Summary

The following is a list of the differentiation formulas that we have learned so far. The letters c and n stand for constants, and f and g stand for differentiable functions of x.

$$\dfrac{d}{dx} c = 0$$

$$\dfrac{d}{dx} x^n = nx^{n-1} \qquad \text{special case: } \dfrac{d}{dx} x = 1$$

$$\dfrac{d}{dx} (c \cdot f) = c \cdot f' \qquad \text{special case: } \dfrac{d}{dx} (cx) = c$$

$$\frac{d}{dx}(f \pm g) = f' \pm g'$$

$$\frac{d}{dx}(f \cdot g) = f' \cdot g + f \cdot g'$$

$$\frac{d}{dx}\left(\frac{f}{g}\right) = \frac{g \cdot f' - g' \cdot f}{g^2} \qquad g \neq 0$$

These formulas are used extensively throughout calculus, and you are not yet ready to proceed to the next section until you have mastered them.

Verification of the Differentiation Formulas

We conclude this section with derivations of the Product and Quotient Rules, and the Power Rule in the case of *negative* integer exponents. First, however, we need to establish a preliminary result about an arbitrary function g:

If $g'(x)$ exists, then $\lim_{h \to 0} g(x + h) = g(x)$.

We begin with $\lim_{h \to 0} g(x + h)$ and show that it is equal to $g(x)$:

$$\lim_{h \to 0} g(x + h) = \lim_{h \to 0} [g(x + h) - g(x) + g(x)] \qquad \text{Subtracting and adding } g(x)$$

$$= \lim_{h \to 0} \left[\frac{g(x + h) - g(x)}{h} \cdot h + g(x) \right] \qquad \text{Dividing and multiplying by } h$$

$$= \lim_{h \to 0} \left[\frac{g(x + h) - g(x)}{h} \cdot h \right] + \lim_{h \to 0} g(x) \qquad \text{The limit of a sum is the sum of the limits}$$

$$= \underbrace{\lim_{h \to 0} \frac{g(x + h) - g(x)}{h}}_{g'(x)} \cdot \underbrace{\lim_{h \to 0} h}_{0} + g(x) \qquad \text{The limit of a product is the product of the limits}$$

$$= g'(x) \cdot 0 + g(x) \qquad \text{Since the first limit above is the definition of } g'(x) \text{ and the second limit is zero}$$

$$= g(x) \qquad \text{Simplifying}$$

This proves the result that if $g'(x)$ exists, then $\lim_{h \to 0} g(x + h) = g(x)$.

Replacing $x + h$ by a new variable y, this equation becomes

$$\lim_{y \to x} g(y) = g(x) \qquad y = x + h, \text{ so } h \to 0 \text{ implies } y \to x$$

According to the definition of continuity on page 537 (but with different letters), this equation means that the function g is *continuous* at x. Therefore, the result that we have shown can be stated simply:

If a function is *differentiable* at x, then it is *continuous* at x.

Or, even more briefly:

> Differentiability implies continuity.

Verification of the Product Rule

For two functions f and g, let their product be $p(x) = f(x) \cdot g(x)$. If $f'(x)$ and $g'(x)$ exist, we may calculate $p'(x)$ as follows.

$$p'(x) = \lim_{h \to 0} \frac{p(x+h) - p(x)}{h} \qquad \text{Definition of the derivative}$$

$$= \lim_{h \to 0} \frac{f(x+h)g(x+h) - f(x)g(x)}{h} \qquad \substack{p(x+h) = f(x+h) \cdot g(x+h) \\ \text{and } p(x) = f(x) \cdot g(x)}$$

$$= \lim_{h \to 0} \frac{f(x+h)g(x+h) - f(x)g(x+h) + f(x)g(x+h) - f(x)g(x)}{h} \qquad \substack{\text{Subtracting and adding} \\ f(x)g(x+h)}$$

$$= \lim_{h \to 0} \left[\frac{f(x+h)g(x+h) - f(x)g(x+h)}{h} + \frac{f(x)g(x+h) - f(x)g(x)}{h} \right] \qquad \substack{\text{Separating the fraction} \\ \text{into two parts}}$$

$$= \lim_{h \to 0} \frac{[f(x+h) - f(x)]g(x+h)}{h} + \lim_{h \to 0} \frac{f(x)[g(x+h) - g(x)]}{h} \qquad \substack{\text{Using Limit Rule 4a on} \\ \text{page 531 and factoring}}$$

$$= \lim_{h \to 0} \underbrace{\frac{[f(x+h) - f(x)]}{h}}_{f'(x)} \underbrace{\lim_{h \to 0} g(x+h)}_{g(x)} + \underbrace{f(x)}_{f(x)} \lim_{h \to 0} \underbrace{\frac{[g(x+h) - g(x)]}{h}}_{g'(x)} \qquad \substack{\text{Using Limit Rule 4c} \\ \text{on page 531}}$$

Recognizing the definitions of $f'(x)$ and $g'(x)$

$$= f'(x) \cdot g(x) + f(x) \cdot g'(x) \qquad \text{Product Rule}$$

Verification of the Quotient Rule

For two functions f and g with $g(x) \neq 0$, let the quotient be $q(x) = \dfrac{f(x)}{g(x)}$. If $f'(x)$ and $g'(x)$ exist, we may calculate $q'(x)$ as follows.

$$q'(x) = \lim_{h \to 0} \frac{q(x+h) - q(x)}{h} \qquad \text{Definition of the derivative}$$

$$= \lim_{h \to 0} \frac{\dfrac{f(x+h)}{g(x+h)} - \dfrac{f(x)}{g(x)}}{h} \qquad \substack{q(x+h) = \dfrac{f(x+h)}{g(x+h)} \\ \text{and } q(x) = \dfrac{f(x)}{g(x)}}$$

$$= \lim_{h \to 0} \frac{1}{h} \left[\frac{f(x+h)}{g(x+h)} - \frac{f(x)}{g(x)} \right]$$ Since dividing by h is equivalent to multiplying by $1/h$

$$= \lim_{h \to 0} \left[\frac{1}{h} \cdot \frac{g(x)f(x+h) - g(x+h)f(x)}{g(x+h)g(x)} \right]$$ Subtracting the fractions, using the common denominator $g(x+h)g(x)$

$$= \lim_{h \to 0} \left[\frac{1}{h} \cdot \frac{g(x)f(x+h) - g(x)f(x) - [g(x+h)f(x) - g(x)f(x)]}{g(x+h)g(x)} \right]$$ Subtracting and adding $g(x)f(x)$

$$= \lim_{h \to 0} \left[\frac{1}{g(x+h)g(x)} \cdot \frac{g(x)[f(x+h) - f(x)] - [g(x+h) - g(x)]f(x)}{h} \right]$$ Factoring in the numerator; switching the denominators

$$= \lim_{h \to 0} \left[\frac{1}{g(x+h)g(x)} \left(g(x) \lim_{h \to 0} \frac{f(x+h) - f(x)}{h} - \lim_{h \to 0} \frac{g(x+h) - g(x)}{h} f(x) \right) \right]$$ Using Limit Rules 4b and 4c on page 531

$$\underbrace{}_{\text{Approaches } g(x)} \quad \underbrace{}_{f'(x)} \quad \underbrace{}_{g'(x)}$$

$$= \frac{1}{[g(x)]^2} [g(x)f'(x) - g'(x)f(x)]$$ Using Limit Rules 1 and 4d on page 531

$$= \frac{g(x)f'(x) - g'(x)f(x)}{[g(x)]^2}$$ Quotient Rule

Verification of the Power Rule for Negative Integer Exponents

On pages 571–572 we proved the Power Rule for *positive* integer exponents. Using the Quotient Rule, we may now prove the Power Rule for *negative* integer exponents. Any negative integer n may be written as $n = -p$, where p is a *positive* integer. Then

$$\frac{d}{dx} x^n = \frac{d}{dx} \left(\frac{1}{x^p} \right)$$ Since $x^n = x^{-p} = \frac{1}{x^p}$

$$= \frac{x^p \cdot 0 - px^{p-1} \cdot 1}{x^{2p}}$$ Using the Quotient Rule, with $\frac{d}{dx} 1 = 0$ and $\frac{d}{dx} x^p = px^{p-1}$

$$= \frac{-px^{p-1}}{x^{2p}}$$ Simplifying

$$= -px^{p-1-2p} = -px^{-p-1}$$ Subtracting exponents and simplifying

$$\underbrace{}_{-p-1} \quad \underbrace{}_{n} \underbrace{}_{n-1}$$

$$= nx^{n-1}$$ Since $-p = n$ — Power Rule

This proves the Power Rule, $\frac{d}{dx} x^n = nx^{n-1}$, for negative integer exponents n.

Solutions to Practice Problems

1. $\dfrac{d}{dx}[x^3(x^2 - x)] = 3x^2(x^2 - x) + x^3(2x - 1)$
$= 3x^4 - 3x^3 + 2x^4 - x^3$
$= 5x^4 - 4x^3$

2. $\dfrac{d}{dx}\left(\dfrac{2x^2}{x^2 + 1}\right) = \dfrac{(x^2 + 1)4x - 2x \cdot 2x^2}{(x^2 + 1)^2}$
$= \dfrac{4x^3 + 4x - 4x^3}{(x^2 + 1)^2} = \dfrac{4x}{(x^2 + 1)^2}$

8.4 Exercises

Find the derivative of each function in two ways:
a. Using the *Product* Rule.
b. Multiplying out the function and using the *Power* Rule.
Your answers to parts (a) and (b) should agree.

1. $x^4 \cdot x^6$
2. $x^7 \cdot x^2$
3. $x^4(x^5 + 1)$
4. $x^5(x^4 + 1)$

Find the derivative of each function by using the Product Rule.

5. $f(x) = x^2(x^3 + 1)$
6. $f(x) = x^3(x^2 + 1)$
7. $f(x) = x(5x^2 - 1)$
8. $f(x) = 2x(x^4 + 1)$
9. $f(x) = (x^2 + 1)(x^2 - 1)$
10. $f(x) = (x^3 - 1)(x^3 + 1)$
11. $f(x) = (x^2 + x)(3x + 1)$
12. $f(x) = (x^2 + 2x)(2x + 1)$
13. $f(x) = (\sqrt{x} - 1)(\sqrt{x} + 1)$
14. $f(x) = (\sqrt{x} + 2)(\sqrt{x} - 2)$
15. $f(t) = 6t^{4/3}(3t^{2/3} + 1)$
16. $f(t) = 4t^{3/2}(2t^{1/2} - 1)$
17. $f(z) = (z^4 + z^2 + 1)(z^3 - z)$
18. $f(z) = (\sqrt[4]{z} + \sqrt{z})(\sqrt[4]{z} - \sqrt{z})$

Find the derivative of each function in two ways:
a. Using the *Quotient* rule.
b. Simplifying the original function and using the *Power* Rule.
Your answers to parts (a) and (b) should agree.

19. $\dfrac{x^8}{x^2}$
20. $\dfrac{x^9}{x^3}$
21. $\dfrac{1}{x^3}$
22. $\dfrac{1}{x^4}$

Find the derivative of each function by using the Quotient Rule.

23. $f(x) = \dfrac{x^4 + 1}{x^3}$
24. $f(x) = \dfrac{x^5 - 1}{x^2}$
25. $f(x) = \dfrac{x + 1}{x - 1}$
26. $f(x) = \dfrac{x - 1}{x + 1}$
27. $f(t) = \dfrac{t^2 - 1}{t^2 + 1}$
28. $f(t) = \dfrac{t^2 + 1}{t^2 - 1}$
29. $f(s) = \dfrac{s^3 - 1}{s + 1}$
30. $f(s) = \dfrac{s^3 + 1}{s - 1}$
31. $f(x) = \dfrac{x^4 + x^2 + 1}{x^2 + 1}$
32. $f(x) = \dfrac{x^5 + x^3 + x}{x^3 + x}$

Differentiate each function by rewriting before and after differentiating, as on page 581.

	Begin	Rewrite	Differentiate	Rewrite
33.	$y = \dfrac{3}{x}$			
34.	$y = \dfrac{x^2}{4}$			
35.	$y = \dfrac{3x^4}{8}$			
36.	$y = \dfrac{3}{2x^2}$			

37. PRODUCT RULE FOR THREE FUNCTIONS
Show that if f, g, and h are differentiable functions of x, then

$$\frac{d}{dx}(f \cdot g \cdot h) = f' \cdot g \cdot h + f \cdot g' \cdot h + f \cdot g \cdot h'$$

[*Hint:* Write the function as $f \cdot (g \cdot h)$ and apply the product rule twice.]

38. Derive the Quotient Rule from the Product Rule as follows.

a. Define the quotient to be a single function,
$$Q(x) = \frac{f(x)}{g(x)}.$$
b. Multiply both sides by $g(x)$ to obtain the equation $Q(x) \cdot g(x) = f(x)$.
c. Differentiate each side, using the Product Rule on the left side. *(continues)*

d. Solve the resulting formula for the derivative $Q'(x)$.
e. Replace $Q(x)$ by $\dfrac{f(x)}{g(x)}$ and show that the resulting formula for $Q'(x)$ is the same as the Quotient Rule.

Note that in this derivation when we differentiated $Q(x)$ we *assumed* that the derivative of the quotient exists, while in the derivation on pages 588–589 we *proved* that the derivative exists.

39. Find a formula for $\dfrac{d}{dx}[f(x)]^2$ by writing it as $\dfrac{d}{dx}[f(x)f(x)]$ and using the Product Rule. Be sure to simplify your answer.

40. Find a formula for $\dfrac{d}{dx}[f(x)]^{-1}$ by writing it as $\dfrac{d}{dx}\left[\dfrac{1}{f(x)}\right]$ and using the Quotient Rule. Be sure to simplify your answer.

Find the derivative of each function.

41. $(x^3 + 2)\dfrac{x^2 + 1}{x + 1}$ **42.** $(x^5 + 1)\dfrac{x^3 + 2}{x + 1}$

43. $\dfrac{(x^2 + 3)(x^3 + 1)}{x^2 + 2}$ **44.** $\dfrac{(x^3 + 2)(x^2 + 2)}{x^3 + 1}$

45. $\dfrac{\sqrt{x} - 1}{\sqrt{x} + 1}$ **46.** $\dfrac{\sqrt{x} + 1}{\sqrt{x} - 1}$

APPLIED EXERCISES

47. ECONOMICS: Marginal Average Revenue
Use the Quotient Rule to find a general expression for the marginal average revenue. That is, calculate $\dfrac{d}{dx}\left[\dfrac{R(x)}{x}\right]$ and simplify your answer.

48. ECONOMICS: Marginal Average Profit
Use the Quotient Rule to find a general expression for the marginal average profit. That is, calculate $\dfrac{d}{dx}\left[\dfrac{P(x)}{x}\right]$ and simplify your answer.

49. ENVIRONMENTAL SCIENCE: Water Purification If the cost of purifying a gallon of water to a purity of x percent is

$$C(x) = \frac{100}{100 - x} \text{ cents} \quad \text{for } 50 \leq x < 100$$

a. Find the instantaneous rate of change of the cost with respect to purity.
b. Evaluate this rate of change for a purity of 95% and interpret your answer.
c. Evaluate this rate of change for a purity of 98% and interpret your answer.

50. BUSINESS: Marginal Average Cost A toy company can produce plastic trucks at a cost of $8 each, while fixed costs are $1200 per day. Therefore, the company's cost function is $C(x) = 8x + 1200$.

  a. Find the average cost function
  $$AC(x) = \frac{C(x)}{x}.$$
  b. Find the marginal average cost function $MAC(x)$.
  c. Evaluate $MAC(x)$ at $x = 200$ and interpret your answer.

51. ENVIRONMENTAL SCIENCE: Water Purification (49 continued)

  a. Use a graphing calculator to graph the cost function $C(x)$ from Exercise 49 on the window [50, 100] by [0, 20]. TRACE along the curve to see how rapidly costs increase for purity (x-coordinate) increasing from 50 to near 100.
  b. To check your answers to Exercise 49, use the "dy/dx" or SLOPE feature of your calculator to find the slope of the cost curve at $x = 95$ and at $x = 98$. The resulting rates of change of the cost should agree with your answers to Exercise 49(b) and (c). Note that further purification becomes increasingly expensive at higher purity levels.

52. BUSINESS: Marginal Average Cost (50 continued)

  a. Graph the average cost function $AC(x)$ that you found in Exercise 50(a) on the window [0, 400] by [0, 50]. TRACE along the average cost curve to see how the average cost falls from the 20s down to the teens as the number of trucks increases. Note that although average cost falls, it does so more slowly as the number of units increases.
  b. To check your answer to Exercise 50, use the "dy/dx" or SLOPE feature of your calculator to find the slope of the average cost curve at $x = 200$. This slope gives the rate of change of the cost, which should agree with your answer to Exercise 50(c). Find the slope (rate of change) for other x-values to see that the rate of change of average cost tends toward zero (the law of diminishing returns).

53. BIOMEDICAL: Beverton-Holt Recruitment Curve Some organisms exhibit a density-dependent mortality from one generation to the next. Let $R > 1$ be the net reproductive rate (that is, the number of surviving offspring per parent), let $x > 0$ be the density of parents, and y be the density of surviving offspring. The Beverton-Holt recruitment curve is

$$y = \frac{Rx}{1 + \left(\frac{R-1}{K}\right)x}$$

where $K > 0$ is the *carrying capacity* of the organism's environment. Show that $\frac{dy}{dx} > 0$, and interpret this as a statement about the parents and the offspring.

54. BIOMEDICAL: Murrell's Rest Allowance Work-rest cycles for workers performing tasks that expend more than 5 kilocalories per minute (kcal/min) are often based on Murrell's formula

$$R(w) = \frac{w-5}{w-1.5} \quad \text{for } w \geq 5$$

for the number of minutes $R(w)$ of rest for each minute of work expending w kcal/min. Show that $R'(w) > 0$ for $w \geq 5$ and interpret this fact as a statement about the additional amount of rest required for more strenuous tasks.

55. BUSINESS: Marginal Average Profit A company's profit function is $P(x) = 12x - 1800$ dollars where x is the number sold.

  a. Find the average profit function $AP(x) = P(x)/x$.
  b. Find the marginal average profit function $MAP(x)$.
  c. Evaluate $MAP(x)$ at $x = 300$ and interpret your answer.

56. BUSINESS: Sales The number of bottles of whiskey that a store will sell in a month at a price of p dollars per bottle is

$$N(p) = \frac{2250}{p+7} \quad (p \geq 5)$$

Find the rate of change of this quantity when the price is $8 and interpret your answer.

57. GENERAL: Body Temperature If a person's temperature after x hours of strenuous exercise is $T(x) = x^3(4 - x^2) + 98.6$ degrees Fahrenheit (for $0 \leq x \leq 2$), find the rate of change of the temperature after 1 hour.

58. BUSINESS: CD Sales After x months, monthly sales of a compact disc are predicted to be $S(x) = x^2(8 - x^3)$ thousand (for $0 \leq x \leq 2$). Find the rate of change of the sales after 1 month.

59. GENERAL: Body Temperature *(57 continued)*
 a. Graph the temperature function $T(x)$ given in Exercise 57 on the window [0, 2] by [90, 110]. TRACE along the temperature curve to see how the temperature rises and then falls as time increases.
 b. To check your answer to Exercise 57, use the "dy/dx" or SLOPE feature of your calculator to find the slope (rate of change) of the curve at $x = 1$. Your answer should agree with your answer to Exercise 57.
 c. TRACE along the temperature curve to estimate the maximum temperature.

60. BUSINESS: CD Sales *(58 continued)*
 a. Graph the sales function $S(x)$ given in Exercise 58 on the window [0, 2] by [0, 12]. TRACE along the sales curve to see how the sales rise and then fall as x, the number of months, increases.
 b. To check your answer to Exercise 58, use the "dy/dx" or SLOPE feature of your calculator to find the slope (rate of change) of the curve at $x = 1$. Your answer should agree with your answer to Exercise 58.
 c. TRACE along the curve to estimate the maximum sales.

61. ECONOMICS: National Debt The following table gives the national debt (the amount of money that the federal government has borrowed from, and therefore owes to, its people), in millions of dollars, for the years 1970 to 2000. To avoid large numbers, years are listed in the table as years since 1970.

Years	Years Since 1970	National Debt (millions of dollars)
1970	0	370,100
1975	5	533,200
1980	10	907,700
1985	15	1,823,100
1990	20	3,233,300
1995	25	4,974,000
2000	30	5,674,200

Source: U.S. Treasury Department

 a. Enter these numbers into your graphing calculator and make a plot of the resulting points (Years Since 1970 on the x-axis and National Debt on the y-axis).
 b. Have your calculator find the quadratic regression formula for these data. Then enter the result in function y_1. Plot the points together with the regression curve. Observe that the curve fits the points quite well.
 c. The population of the United States (in millions) for these years is approximated by the linear function $y = 2.378x + 204.1$, where x is the number of years since 1970. Enter this function into your calculator in y_2.
 d. Use your calculator to define the function y_3 to be $y_1 \div y_2$, the national debt divided by the number of people, giving the "per capita national debt." Graph the function on the window [0, 40] by [0, 35,000] to show the per capita national debt for the years 1970 to 2010.
 e. TRACE along the curve y_3. The y-values represent the amount of money that the federal government owes to each one of its citizens in year x after 1970 (if the debt were equally distributed). Observe how rapidly the amount grows.
 f. Define function y_4 to be the derivative of y_3 (using the NDERIV feature of your calculator) and graph y_4 on the window [0, 40] by [0, 1500]. Use TABLE to evaluate both y_3 and y_4 at $x = 40$ and interpret your answers.

62–63. PITFALLS OF NDERIV ON A GRAPHING CALCULATOR

a. Find the derivative (by hand) of each function below, and observe that the derivative is undefined at $x = 0$.
b. Find the derivative of each function below by using NDERIV on a graphing calculator and evaluate the derivative at $x = 0$. If your calculator gives you an answer, this is a "false value" for the derivative, since in part (a) you showed that the derivative is undefined at $x = 0$. [For an explanation, see the Graphing Calculator Exploration part (d) on page 583.]

62. $y = \dfrac{1}{x}$

63. $y = \dfrac{1}{x^2}$

8.5 HIGHER-ORDER DERIVATIVES

Introduction

We have seen that from one function we can calculate a new function, the *derivative* of the original function. This new function, however, can itself be differentiated, giving what is called the *second derivative* of the original function. Differentiating again gives the *third derivative* of the original function, and so on. In this section we will calculate and interpret such higher-order derivatives.

Calculating Higher-Order Derivatives

EXAMPLE 1

FINDING HIGHER DERIVATIVES OF A POLYNOMIAL

From $f(x) = x^3 - 6x^2 + 2x - 7$ we may calculate

$$f'(x) = 3x^2 - 12x + 2 \qquad \text{"First" derivative of } f$$

Differentiating again gives

$$f''(x) = 6x - 12 \qquad \text{Second derivative of } f, \text{ read "} f \text{ double prime"}$$

and a third time:

$$f'''(x) = 6 \qquad \text{Third derivative of } f, \text{ read "} f \text{ triple prime"}$$

and a fourth time:

$$f''''(x) = 0 \qquad \text{Fourth derivative of } f, \text{ read "} f \text{ quadruple prime"}$$

All further derivatives of this function will, of course, be zero.

We also denote derivatives by replacing the primes by the number of differentiations in *parentheses*. For example, the fourth derivative may be denoted $f^{(4)}(x)$.

Practice Problem 1 If $f(x) = x^3 - x^2 + x - 1$, find:

a. $f'(x)$ b. $f''(x)$ c. $f'''(x)$ d. $f^{(4)}(x)$

▶ Solutions on page 604

While Example 1 showed that a polynomial can be differentiated "down to zero," the same is not true for all functions.

EXAMPLE 2

FINDING HIGHER DERIVATIVES OF A RATIONAL FUNCTION

Find the first five derivatives of $f(x) = \dfrac{1}{x}$.

Solution

$f(x) = x^{-1}$ $f(x)$ in power form

$f'(x) = -x^{-2}$ First derivative

$f''(x) = 2x^{-3}$ Second derivative

$f'''(x) = -6x^{-4}$ Third derivative

$f^{(4)}(x) = 24x^{-5}$ Fourth derivative

$f^{(5)}(x) = -120x^{-4}$ Fifth derivative

Clearly, we will never get to zero no matter how many times we differentiate.

Practice Problem 2 If $f(x) = 16x^{-1/2}$, find:

a. $f'(x)$ b. $f''(x)$

▶ Solutions on page 604

In Leibniz's notation, the second derivative $\dfrac{d}{dx}\dfrac{df}{dx}$ is written $\dfrac{d^2 f}{dx^2}$. The superscript goes after the d in the numerator and after the dx in the denominator. The following table shows equivalent statements in the two notations.

Prime Notation		Leibniz's Notation	
$f''(x)$	$=$	$\dfrac{d^2}{dx^2}f(x)$	Second derivative
y''	$=$	$\dfrac{d^2y}{dx^2}$	
$f'''(x)$	$=$	$\dfrac{d^3}{dx^3}f(x)$	Third derivative
y'''	$=$	$\dfrac{d^3y}{dx^3}$	
$f^{(n)}(x)$	$=$	$\dfrac{d^n}{dx^n}f(x)$	nth derivative
$y^{(n)}$	$=$	$\dfrac{d^ny}{dx^n}$	

Calculating higher derivatives merely requires repeated use of the same differentiation rules that we have been using.

EXAMPLE 3

FINDING A SECOND DERIVATIVE USING THE QUOTIENT RULE

Find $\dfrac{d^2}{dx^2}\left(\dfrac{x^2+1}{x}\right)$.

Solution

$$\dfrac{d}{dx}\left(\dfrac{x^2+1}{x}\right) = \dfrac{x(2x) - (x^2+1)}{x^2}$$ First derivative, using the Quotient Rule

$$= \dfrac{2x^2 - x^2 - 1}{x^2} = \dfrac{x^2 - 1}{x^2}$$ Simplifying

Differentiating this answer gives the *second* derivative of the original:

$$\dfrac{d}{dx}\left(\dfrac{x^2-1}{x^2}\right) = \dfrac{x^2(2x) - 2x(x^2-1)}{x^4}$$ Second derivative (derivative of the derivative)

$$= \dfrac{2x^3 - 2x^3 + 2x}{x^4} = \dfrac{2x}{x^4} = \dfrac{2}{x^3}$$ Simplifying

Answer: $\dfrac{d^2}{dx^2}\left(\dfrac{x^2+1}{x}\right) = \dfrac{2}{x^3}$

The function in this example was a quotient, so it was perhaps natural to use the Quotient Rule. It is easier, however, to simplify the original function first,

$$\frac{x^2+1}{x} = \frac{x^2}{x} + \frac{1}{x} = x + x^{-1}$$

and then differentiate by the Power Rule. So, the first derivative of $x + x^{-1}$ is $1 - x^{-2}$, and differentiating again gives $2x^{-3}$, agreeing with the answer found by the Quotient Rule. *Moral:* **Always simplify before differentiating.**

Practice Problem 3   Find $f''(x)$ if $f(x) = \dfrac{x+1}{x}$.   ▶ Solution on page 604

EXAMPLE 4

EVALUATING A SECOND DERIVATIVE

If $f(x) = \dfrac{1}{\sqrt{x}}$, find $f''\left(\dfrac{1}{4}\right)$.   *First differentiate, then evaluate*

Solution

$$f(x) = x^{-1/2} \qquad \text{$f(x)$ in power form}$$

$$f'(x) = -\frac{1}{2} x^{-3/2} \qquad \text{Differentiating once}$$

$$f''(x) = \frac{3}{4} x^{-5/2} \qquad \text{Differentiating again}$$

$$f''\left(\frac{1}{4}\right) = \frac{3}{4}\left(\frac{1}{4}\right)^{-5/2} = \frac{3}{4}(4)^{5/2} \qquad \text{Evaluating $f''(x)$ at $\frac{1}{4}$}$$

$$= \frac{3}{4}(\sqrt{4})^5 = \frac{3}{4}(2)^5 = \frac{3}{4}(32) = 24$$

Practice Problem 4   Find $\left.\dfrac{d^2}{dx^2}(x^4 + x^3 + 1)\right|_{x=-1}$.   ▶ Solution on page 604

Velocity and Acceleration

There is another important interpretation for the derivative, one that also gives a meaning to the *second* derivative. Imagine that you are driving along a straight road, and let s(t) stand for your distance (in miles) from your starting point after *t* hours of driving. Then the derivative s'(t) gives the instantaneous rate of change of distance with respect to time (miles per hour). However, "miles per hour" means speed or velocity, so the derivative of the *distance* function s(t) is just the *velocity* function v(t), giving your velocity at any time *t*.

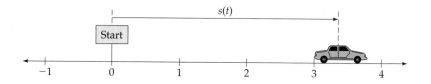

In general, for an object moving along a straight line, with distance measured from some fixed point, measured positively in one direction and negatively in the other (sometimes called "directed distance"),

$$\text{if} \quad s(t) = \begin{pmatrix} \text{Distance} \\ \text{at time } t \end{pmatrix}$$

$$\text{then} \quad s'(t) = \begin{pmatrix} \text{Velocity} \\ \text{at time } t \end{pmatrix}$$

Letting v(t) stand for the velocity at time *t*, we may state this simply as:

$$v(t) = s'(t) \qquad \text{The velocity function is the derivative of the distance function}$$

The units of velocity come directly from the distance and time units of s(t). For example, if distance is measured in feet and time in seconds, then the velocity is in *feet per second,* whereas if distance is in miles and time is in hours, then velocity is in *miles per hour.*

In everyday speech, the word "accelerating" means "speeding up." That is, acceleration means the rate of increase of speed, and since rates of increase are just derivatives, *acceleration is the derivative of velocity.* Since velocity is itself the derivative of distance, acceleration is the *second* derivative of distance. Letting a(t) stand for the acceleration at time *t*, we have:

Distance, Velocity, and Acceleration

$$s(t) = \begin{pmatrix} \text{Distance} \\ \text{at time } t \end{pmatrix}$$

$v(t) = s'(t)$ Velocity is the derivative of distance

$a(t) = v'(t) = s''(t)$ Acceleration is the derivative of velocity, and the second derivative of distance

Therefore, we now have an interpretation for the *second* derivative: If $s(t)$ represents distance, then the *first* derivative represents velocity, and the *second* derivative represents *acceleration*. (In physics, there is even an interpretation for the third derivative, which gives the rate of change of acceleration: It is called the "jerk," since it is related to motion being "jerky."*)

EXAMPLE 5

FINDING AND INTERPRETING VELOCITY AND ACCELERATION

A delivery truck is driving along a straight road, and after t hours its distance (in miles) east of its starting point is

$$s(t) = 24t^2 - 4t^3 \quad \text{for } 0 \leq t \leq 6$$

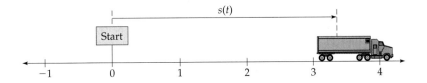

a. Find the velocity of the truck after 2 hours.
b. Find the velocity of the truck after 5 hours.
c. Find the acceleration of the truck after 1 hour.

Solution

a. To find velocity, we differentiate distance:

$v(t) = 48t - 12t^2$ Differentiating $s(t) = 24t^2 - 4t^3$

$v(2) = 48 \cdot 2 - 12 \cdot (2)^2$ Evaluating $v(t)$ at $t = 2$

$ = 96 - 48 = 48$ miles per hour Velocity after 2 hours

---

*See T. R. Sandlin, "The Jerk," *Physics Teacher* 28:36–40, January 1990.

b. At $t = 5$ hours:

$$v(5) = 48 \cdot 5 - 12 \cdot (5)^2 \qquad \text{Evaluating } v(t) \text{ at } t = 5$$
$$= 240 - 300 = -60 \text{ miles per hour} \quad \text{Velocity after 5 hours}$$

What does the negative sign mean? Since distances are measured *eastward* (according to the original problem), the "positive" direction is east, so a negative velocity means a *westward* velocity. Therefore, at time $t = 5$ the truck is driving *westward at 60 miles per hour* (that is, back toward its starting point).

c. The acceleration is

$$a(t) = 48 - 24t \qquad \text{Differentiating } v(t) = 48t - 12t^2$$
$$a(1) = 48 - 24 = 24 \qquad \text{Acceleration after 1 hour}$$

Therefore, after 1 hour the acceleration of the truck is 24 mi/hr² (it is speeding up).

Graphing Calculator Exploration

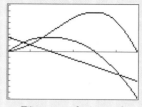

Distance, velocity, and acceleration on [0, 6] by [−150, 150]. Which is which?

Use a graphing calculator to graph the distance function $y_1 = 24x^2 - 4x^3$ (use x instead of t), the velocity function $y_2 = 48x - 12x^2$, and the acceleration function $y_3 = 48 - 24x$. (Alternatively, you could define y_2 and y_3 using NDERIV.) Your display should look like the one on the left. By looking at the graph, can you determine which curve represents distance, which represents velocity, and which represents acceleration? [*Hint:* Which curve gives the slope of which other curve?] Check your answer by using TRACE to identify functions 1, 2, and 3.

In Example 5, velocity was in miles per hour (mi/hr), so acceleration, the rate of change of velocity with respect to time, is in miles per hour *per hour*, written mi/hr². In general, the units of acceleration are distance/time².

Practice Problem 5 A helicopter rises vertically, and after t seconds its height above the ground is $s(t) = 6t^2 - t^3$ feet (for $0 \leq t \leq 6$).

a. Find its velocity after 2 seconds.
b. Find its velocity after 5 seconds.
c. Find its acceleration after 1 second.

[*Hint:* Distances are measured *upward,* so a negative velocity means downward.]

➤ Solutions on page 604

Other Interpretations of Second Derivatives

The second derivative has other meanings besides acceleration. In general, second derivatives measure how the rate of change is itself changing. That is, if the first derivative measures the rate of growth, then the second derivative tells whether the growth is speeding up or slowing down.

EXAMPLE 6 PREDICTING POPULATION GROWTH

Demographers predict that t years from now the population of a city will be:

$$P(t) = 2{,}000{,}000 + 28{,}800t^{1/3}$$

Find $P'(8)$ and $P''(8)$ and interpret these numbers.

Solution The derivative is

$$P'(t) = 9600t^{-2/3} \qquad \text{Derivative of } P(t)$$

so

$$P'(8) = 9600(8)^{-2/3} = 9600 \cdot \frac{1}{4} = 2400 \qquad \text{Evaluating at } t = 8$$

Interpretation: Eight years from now the population will be growing at the rate of 2400 people per year.

The second derivative is

$$P''(t) = -6400t^{-5/3} \qquad \text{Derivative of } P'(t) = 9600t^{-2/3}$$

$$P''(8) = -6400(8)^{-5/3} \qquad \text{Evaluating at } t = 8$$

$$= -6400 \cdot \frac{1}{32} = -200$$

The fact that the first derivative is positive and the second derivative (the rate of change of the derivative) is negative means that the growth is continuing but more slowly.

Interpretation: After 8 years, the growth rate is *decreasing* by about 200 people per year each year. In other words, in the following year the population will continue to grow, but at the slower rate of about $2400 - 200 = 2200$ people per year.

Graphing Calculator Exploration

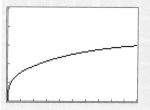

$y_1 = 2{,}000{,}000 + 28{,}800x^{1/3}$
on [0, 10] by
[2,000,000, 2,100,000]

Use a graphing calculator to graph the population function $y_1 = 2{,}000{,}000 + 28{,}800x^{1/3}$ (using x instead of t). Your graph should resemble the one on the left. Can you see from the graph that the first derivative is positive (sloping upward), and that the second derivative is negative (slope decreasing)? You may check these facts numerically using NDERIV.

ECONOMIC GROWTH SLOWED SHARPLY IN THIRD QUARTER

AN ANNUAL RATE OF 2.7%

Signs Point to Temporary Dip, but Strong Gains of Recent Years May Have Ended

By LOUIS UCHITELLE

The pace of economic activity slowed sharply in the summer quarter, the Commerce Department reported yester-

(Copyright © 2000 by the New York Times Co. Reprinted by permission.)

Statements about first and second derivatives occur frequently in everyday life, and may even make the front page of the newspaper. For example, the newspaper headline on the left (*New York Times*, October 28, 2000) says that the United States economy grew (the first derivative is positive) but more slowly (the second derivative is negative), following a curve like that shown below.

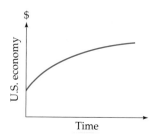

If y represents the size of the economy, then

$$\frac{dy}{dx} > 0 \quad \text{and} \quad \frac{d^2y}{dx^2} < 0$$

Practice Problem 6

The following headlines appeared recently in the *New York Times*. For each headline, sketch a curve representing the type of growth described and indicate the correct signs of the first and second derivatives.

a. Consumer Prices Rose in October at a Slower Rate
b. Households Still Shrinking, but Rate is Slower

➤ Solutions on page 605

Practice Problem 7 A recent mathematics publication* included the following statement: "In the fall of 1972 President Nixon announced that the rate of increase of inflation was decreasing. This was the first time a sitting president used the third derivative to advance his case for reelection." Explain why Nixon's announcement involved a third derivative.

➤ Solution on page 605

8.5 Section Summary

By simply repeating the process of differentiation we can calculate second, third, and higher derivatives. We also have another interpretation for the derivative, one that gives an interpretation for the second derivative as well. For distance measured along a straight line from some fixed point:

$$\text{If} \quad s(t) = \text{distance at time } t$$
$$\text{then} \quad s'(t) = \text{velocity at time } t$$
$$\text{and} \quad s''(t) = \text{acceleration at time } t.$$

Therefore, whenever you are driving along a straight road, your speedometer is the derivative of your odometer reading.

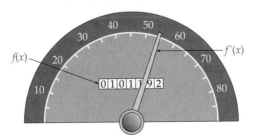

Velocity is the derivative of distance

Notices of the American Mathematical Society, vol. 43, no. 10 (1996), page 1108.

We now have *four* interpretations for the derivative: instantaneous rate of change, slope, marginals, and velocity. It has been said that science is at its best when it unifies, and the derivative, unifying these four different concepts, is one of the most important ideas in all of science. We also saw that the second derivative, which measures the rate of change of the rate of change, can show whether growth is speeding up or slowing down.

Remember, however, that derivatives measure just what an automobile speedometer measures: the velocity at a particular *instant*. Although this statement may be obvious for velocities, it is easy to forget when dealing with marginals. For example, suppose that the marginal cost for a product is $15 when 100 units have been produced [which may be written $C'(100) = 15$]. Therefore, costs are increasing at the rate of $15 per additional unit, but only at the instant when $x = 100$. Although this may be used to *estimate* future costs (*about* $15 for each additional unit), it does not mean that one additional unit will increase costs by exactly $15, two more by exactly $30, and so on, since the marginal rate usually changes as production increases. A marginal cost is only an *approximate* predictor of future costs.

▶ Solutions to Practice Problems

1. **a.** $f'(x) = 3x^2 - 2x + 1$
 b. $f''(x) = 6x - 2$
 c. $f'''(x) = 6$
 d. $f^{(4)}(x) = 0$

2. **a.** $f'(x) = -8x^{-3/2}$
 b. $f''(x) = 12x^{-5/2}$

3. $f(x) = \dfrac{x}{x} + \dfrac{1}{x} = 1 + x^{-1}$ Simplifying first

$f'(x) = -x^{-2}$

$f''(x) = 2x^{-3}$

4. $\dfrac{d}{dx}(x^4 + x^3 + 1) = 4x^3 + 3x^2$

$\dfrac{d^2}{dx^2}(x^4 + x^3 + 1) = \dfrac{d}{dx}(4x^3 + 3x^2) = 12x^2 + 6x$

$(12x^2 + 6x)\big|_{x=-1} = 12 - 6 = 6$

5. **a.** $v(t) = 12t - 3t^2$

$v(2) = 24 - 12 = 12 = 12$ ft/sec

 b. $v(5) = 60 - 75 = -15$ ft/sec or 15 ft/sec *downward*

 c. $a(t) = 12 - 6t$

$a(1) = 12 - 6 = 6$ ft/sec^2

6. a.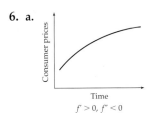
Consumer prices vs Time
$f' > 0, f'' < 0$

b.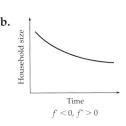
Household size vs Time
$f' < 0, f'' > 0$

7. Inflation is itself a derivative since it is the rate of change of the consumer price index. Therefore its growth rate is a second derivative, and the slowing of this growth would be a third derivative.

8.5 Exercises

For each function, find:
a. $f'(x)$ **b.** $f''(x)$ **c.** $f'''(x)$ **d.** $f^{(4)}(x)$

1. $f(x) = x^4 - 2x^3 - 3x^2 + 5x - 7$
2. $f(x) = x^4 - 3x^3 + 2x^2 - 8x + 4$
3. $f(x) = 1 + x + \frac{1}{2}x^2 + \frac{1}{6}x^3 + \frac{1}{24}x^4 + \frac{1}{120}x^5$
4. $f(x) = 1 + x + \frac{1}{2}x^2 + \frac{1}{6}x^3 + \frac{1}{24}x^4$
5. $f(x) = \sqrt{x^5}$
6. $f(x) = \sqrt{x^3}$

For each function, find **a.** $f''(x)$ and **b.** $f''(3)$.

7. $f(x) = \dfrac{x-1}{x}$
8. $f(x) = \dfrac{x+2}{x}$
9. $f(x) = \dfrac{x+1}{2x}$
10. $f(x) = \dfrac{x-2}{4x}$
11. $f(x) = \dfrac{1}{6x^2}$
12. $f(x) = \dfrac{1}{12x^3}$

Find the *second* derivative of each function.

13. $f(x) = (x^2 - 2)(x^2 + 3)$
14. $f(x) = (x^2 - 1)(x^2 + 2)$
15. $f(x) = \dfrac{27}{\sqrt[3]{x}}$
16. $f(x) = \dfrac{32}{\sqrt[4]{x}}$
17. $f(x) = \dfrac{x}{x-1}$
18. $f(x) = \dfrac{x}{x-2}$

Evaluate each expression.

19. $\dfrac{d^2}{dr^2}(\pi r^2)$
20. $\dfrac{d^3}{dr^3}\left(\dfrac{4}{3}\pi r^3\right)$

$f(x) = \dfrac{4}{3}\pi r^3$

21. $\dfrac{d^2}{dx^2} x^{10}\bigg|_{x=-1}$
22. $\dfrac{d^2}{dx^2} x^{11}\bigg|_{x=-1}$
23. $\dfrac{d^3}{dx^3} x^{10}\bigg|_{x=-1}$
24. $\dfrac{d^3}{dx^3} x^{11}\bigg|_{x=-1}$
25. $\dfrac{d^2}{dx^2} \sqrt{x^3}\bigg|_{x=1/16}$
26. $\dfrac{d^2}{dx^2} \sqrt[3]{x^4}\bigg|_{x=1/27}$

27. GENERAL: Velocity Each of the following three "stories," labeled **a, b,** and **c,** matches one of the velocity graphs, labeled (i), (ii), and (iii). For each story, choose the most appropriate graph.

a. I left my home and drove to meet a friend, but I got stopped for a speeding ticket. Afterward I drove on more slowly.

b. I started driving but then stopped to look at the map. Realizing that I was going the wrong way, I drove back the other way.

c. After driving for a while I got into some stop-and-go driving. Once past the tie-up I could speed up again.

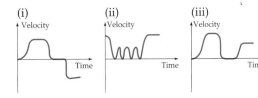

(i) Velocity vs Time   (ii) Velocity vs Time   (iii) Velocity vs Time

28. BUSINESS: Profit Each of the following three descriptions of a company's profit over time, labeled **a, b,** and **c,** matches one of the graphs, labeled (i), (ii), and (iii). For each description, choose the most appropriate graph.

a. Profits were growing increasingly rapidly.
b. Profits were declining but the rate of decline was slowing.
c. Profits were rising, but more and more slowly.

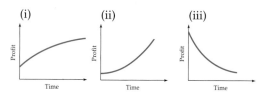

29. Find $\dfrac{d^{100}}{dx^{100}}(x^{99} - 4x^{98} + 3x^{50} + 6)$.

[*Hint:* No calculation is necessary. Think of what happens when an *n*th-degree polynomial is differentiated $n + 1$ times. For example, try differentiating x^3 four times.]

30. Find a general formula for $\dfrac{d^n}{dx^n}x^{-1}$.

[*Hint:* Calculate the first few derivatives and look for a pattern. You may use the "factorial" notation: $n! = n(n-1) \cdots 1$. For example, $3! = 3 \cdot 2 \cdot 1 = 6$.]

31. Verify the following formula for the *second* derivative of a product, where *f* and *g* are differentiable functions of *x*:

$$\dfrac{d^2}{dx^2}(f \cdot g) = f'' \cdot g + 2f' \cdot g' + f \cdot g''$$

[*Hint:* Use the Product Rule repeatedly.]

32. Verify the following formula for the *third* derivative of a product, where *f* and *g* are differentiable functions of *x*:

$$\dfrac{d^3}{dx^3}(f \cdot g) = f''' \cdot g + 3f'' \cdot g' + 3f' \cdot g'' + f \cdot g'''$$

[*Hint:* Differentiate the formula in Exercise 31 by the Product Rule.]

APPLIED EXERCISES

33. GENERAL: Velocity After *t* hours a freight train is $s(t) = 18t^2 - 2t^3$ miles due north of its starting point (for $0 \leq t \leq 9$).
  a. Find its velocity at time $t = 3$ hours.
  b. Find its velocity at time $t = 7$ hours.
  c. Find its acceleration at time $t = 1$ hour.

34. GENERAL: Velocity After *t* hours a passenger train is $s(t) = 24t^2 - 2t^3$ miles due west of its starting point (for $0 \leq t \leq 12$).
  a. Find its velocity at time $t = 4$ hours.
  b. Find its velocity at time $t = 10$ hours.
  c. Find its acceleration at time $t = 1$ hour.

35. GENERAL: Velocity A rocket can rise to a height of $h(t) = t^3 + 0.5t^2$ feet in *t* seconds. Find its velocity and acceleration 10 seconds after it is launched.

36. GENERAL: Velocity After *t* hours a car is a distance $s(t) = 60t + \dfrac{100}{t + 3}$ miles from its starting point. Find the velocity after 2 hours.

37. GENERAL: Impact Velocity A penny dropped from a building will fall a distance $s(t) = 16t^2$ feet in *t* seconds (neglecting air resistance).
  a. With what velocity will it hit the ground if it does so 5 seconds after it is dropped? (This is called the *impact velocity*.)
  b. Find the acceleration at any time *t*. (This number is called the *acceleration due to gravity*.)

38. GENERAL: Impact Velocity If a marble is dropped from the top of the Sears Tower in Chicago, its height above the ground *t* seconds after it is dropped will be $s(t) = 1454 - 16t^2$ feet (neglecting air resistance).
  a. How long will it take to reach the ground? [*Hint:* Find when the height equals zero.]
  b. Use your answer to part (a) to find the velocity with which it will strike the ground.

39. GENERAL: Maximum Height If a bullet from a 9-millimeter pistol is fired straight up from the ground, its height t seconds after it is fired will be $s(t) = -16t^2 + 1280t$ feet (neglecting air resistance) for $0 \le t \le 80$.

a. Find the velocity function.
b. Find the time t when the bullet will be at its maximum height. [*Hint*: At its maximum height the bullet is moving neither up nor down, and has velocity zero. Therefore, find the time when the velocity $v(t)$ equals zero.]
c. Find the maximum height the bullet will reach. [*Hint*: Use the time found in part (b) together with the height function $s(t)$.]

40. **BIOMEDICAL: Fever** The temperature of a patient t hours after taking a fever reducing medicine is $T(t) = 98 + 8/\sqrt{t}$ degrees Fahrenheit. Find $T(2)$, $T'(2)$, and $T''(2)$, and interpret these numbers.

41. **ECONOMICS: National Debt** The national debt of a South American country t years from now is predicted to be $D(t) = 65 + 9t^{4/3}$ billion dollars. Find $D'(8)$ and $D''(8)$ and interpret your answers.

42. **ENVIRONMENTAL SCIENCE: Earth Temperature** If the average temperature of the earth t years from now is predicted to be $T(t) = 65 - 4/t$ (for $t \ge 8$), find $T'(10)$ and $T''(10)$ and interpret your answers.

43–44. ENVIRONMENTAL SCIENCE: Sea Level The burning of fossil fuels (such as coal and oil) generates carbon dioxide, which traps heat in the atmosphere, thereby increasing the temperature of the earth (the "greenhouse effect").

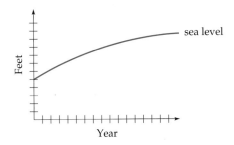

The higher temperature in turn melts the polar icecaps, raising the sea level. The precise results are very difficult to predict, but suppose the following:

43. In t years the average sea level on the East Coast of the United States will be $L(t) = 30 - 4t^{-1/2}$ feet (for $t > 2$). Find $L'(4)$ and $L''(4)$ and interpret your answers.

44. In t years the average sea level on the West Coast of the United States will be $L(t) = 35 - 8t^{-1/2}$ feet (for $t > 2$). Find $L'(4)$ and $L''(4)$ and interpret your answers.

45. **BUSINESS: Profit** The annual profit of the Digitronics company x years from now is predicted to be $P(x) = 5.27x^{0.3} - 0.463x^{1.52}$ million dollars (for $0 \le x \le 8$). Evaluate the profit function and its first and second derivatives at $x = 3$ and interpret your answers. [*Hint*: Enter the given function in y_1, define y_2 to be the derivative of y_1 (using NDERIV), and define y_3 to be the derivative of y_2. Then evaluate each at the stated x-value.]

46. **GENERAL: Population** The population of a country x years from now is predicted to be $P(x) = 3.17x^{1.3} + 0.192x^{0.74}$ million people (for $0 \le x \le 10$). Evaluate the population function and its first and second derivatives at $x = 3.75$ and interpret your answers. [See the hint in Exercise 45.]

47. **GENERAL: Windchill Index** The windchill index (revised in 2001) for a temperature of 32 degrees Fahrenheit and wind speed x miles per hour is

$$W(x) = 55.628 - 22.07x^{0.16}$$

a. Graph the windchill index on a graphing calculator using the window [0, 50] by [0, 40]. Then find the windchill index for wind speeds of $x = 15$ and $x = 30$ mph.
b. Notice from your graph that the windchill index has first derivative negative and second derivative positive. What does this mean about how successive 1-mph increases in wind speed affect the windchill index?
c. Verify your answer to part (b) by defining y_2 to be the derivative of y_1 (using NDERIV), evaluating it at $x = 15$ and $x = 30$, and interpreting your answers.

48. BIOMEDICAL: AIDS The cumulative number of cases of AIDS (acquired immunodeficiency syndrome) in the United States between 1981 and 2000 is given approximately by the function

$$f(x) = -0.0182x^4 + 0.526x^3 - 1.3x^2 + 1.3x + 5.4$$

in thousands of cases, where x is the number of years since 1980.

a. Graph this function on your graphing calculator on the window [1, 20] by [0, 800]. Notice that at some time in the 1990s the rate of growth began to slow.
b. Find when the AIDS epidemic began to slow. [*Hint:* Find where the second derivative of $f(x)$ is zero, and then convert the x-value to a year.]

(*Source:* Centers for Disease Control)

Find the second derivative of each function.

49. $(x^2 - x + 1)(x^3 - 1)$
50. $(x^3 + x - 1)(x^3 + 1)$
51. $\dfrac{x}{x^2 + 1}$
52. $\dfrac{x}{x^2 - 1}$
53. $\dfrac{2x - 1}{2x + 1}$
54. $\dfrac{3x + 1}{3x - 1}$

8.6 THE CHAIN RULE AND THE GENERALIZED POWER RULE

Introduction

In this section we will learn the last of the general rules of differentiation, the *Chain Rule* for differentiating composite functions. We will then prove a very useful special case of it, the *Generalized Power Rule* for differentiating powers of functions. We begin by reviewing composite functions.

Composite Functions

As we saw on page 56, *composite* functions are simply functions of functions: The composition of f with g evaluated at x is $f(g(x))$.

EXAMPLE 1   FINDING A COMPOSITE FUNCTION

For $f(x) = x^2$ and $g(x) = 4 - x$, find $f(g(x))$.

Solution

$$f(g(x)) = (4 - x)^2 \qquad \begin{array}{l} f(x) = x^2 \text{ with } x \\ \text{replaced by } g(x) = 4 - x \end{array}$$

Practice Problem 1   For the same $f(x) = x^2$ and $g(x) = 4 - x$, find $g(f(x))$.

➤ Solution on page 618

Graphing Calculator Exploration

Use a graphing calculator to verify that the compositions $f(g(x))$ and $g(f(x))$ above are different.

a. Enter $y_1 = x^2$ and $y_2 = 4 - x$.
b. Then define y_3 and y_4 to be the compositions in the two orders [on some calculators this is done by defining $y_3 = y_1(y_2)$ and $y_4 = y_2(y_1)$].
c. Graph y_3 and y_4 (but turn "off" y_1 and y_2) on the standard window and notice that the graphs are very different.

Besides building compositions out of simpler functions, we can also *de*compose functions into compositions of simpler functions.

EXAMPLE 2

DECOMPOSING A COMPOSITE FUNCTION

Find functions $f(x)$ and $g(x)$ such that $(x^2 + 1)^5$ is the composition $f(g(x))$.

Solution

Think of $(x^2 + 1)^5$ as an inside function $x^2 + 1$ followed by an outside operation ( )^5. We match the "inside" and "outside" parts of $(x^2 + 1)^5$ and $f(g(x))$.

$$\underbrace{(x^2 + 1)^5}_{\text{Inside function}} = f(\overbrace{g(x)}^{\text{Outside function}})$$

Therefore, $(x^2 + 1)^5$ can be written as $f(g(x))$ with $\begin{cases} f(x) = x^5 \\ g(x) = x^2 + 1 \end{cases}$

(Other answers are possible.)

Note that expressing a function as a composition involves thinking of the function in terms of "blocks," an inside block that starts a calculation and an outside block that completes it.

Practice Problem 2 Find $f(x)$ and $g(x)$ such that $\sqrt{x^5 - 7x + 1}$ is the composition $f(g(x))$.

➤ Solution on page 618

The Chain Rule

If we were asked to differentiate the function $(x^2 - 5x + 1)^{10}$, we could first multiply together ten copies of $x^2 - 5x + 1$ (certainly a time-consuming, tedious, and error-prone process), and then differentiate the resulting polynomial. There is, however, a much easier way, using the *Chain Rule*, which shows how to differentiate a composite function of the form $f(g(x))$.

Chain Rule

$$\frac{d}{dx} f(g(x)) = f'(g(x)) \cdot g'(x)$$

To differentiate $f(g(x))$, differentiate $f(x)$, then replace each x by $g(x)$, and finally multiply by the derivative of $g(x)$

(provided that the derivatives on the right-hand side of the equation exist). The name comes from thinking of compositions as "chains" of functions. A verification of the Chain Rule is given at the end of this section.

EXAMPLE 3   DIFFERENTIATING USING THE CHAIN RULE

Use the Chain Rule to find $\dfrac{d}{dx}(x^2 - 5x + 1)^{10}$.

Solution

$(x^2 - 5x + 1)^{10}$ is $f(g(x))$ with $\begin{cases} f(x) = x^{10} & \text{Outside function} \\ g(x) = x^2 - 5x + 1 & \text{Inside function} \end{cases}$

Since $f'(x) = 10x^9$, we have

$f'(g(x)) = 10(g(x))^9$ $f'(x) = 10x^9$ with x replaced by $g(x)$

$= 10(x^2 - 5x + 1)^9$ Using $g(x) = x^2 - 5x + 1$

Substituting this last expression into the Chain Rule gives:

$$\frac{d}{dx} f(g(x)) = f'(g(x)) \, g'(x) \qquad \text{Chain Rule}$$

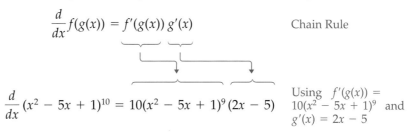

$$\frac{d}{dx}(x^2 - 5x + 1)^{10} = 10(x^2 - 5x + 1)^9 (2x - 5) \qquad \text{Using } f'(g(x)) = 10(x^2 - 5x + 1)^9 \text{ and } g'(x) = 2x - 5$$

This result says that to differentiate $(x^2 - 5x + 1)^{10}$, we bring down the exponent 10, reduce the exponent to 9 (steps familiar from the Power Rule), and finally multiply by the derivative of the inside function.

$$\frac{d}{dx}(x^2 - 5x + 1)^{10} = 10(x^2 - 5x + 1)^9(2x - 5)$$

- Inside function
- Bring down the power n
- Power $n - 1$
- Derivative of the inside function

Generalized Power Rule

Example 3 suggests a general rule for differentiating a function to a power.

Generalized Power Rule

$$\frac{d}{dx}[g(x)]^n = n \cdot [g(x)]^{n-1} \cdot g'(x)$$

To differentiate a function to a power, bring down the power as a multiplier, reduce the exponent by 1, and then multiply by the derivative of the inside function

(provided, of course, that the derivative $g'(x)$ exists). The Generalized Power Rule follows from the Chain Rule by reasoning similar to that of Example 3: The derivative of $f(x) = x^n$ is $f'(x) = nx^{n-1}$, so

$$\frac{d}{dx} f(g(x)) = f'(g(x)) \, g'(x) \qquad \text{Chain Rule}$$

gives

$$\frac{d}{dx}[g(x)]^n = n[g(x)]^{n-1} g'(x) \qquad \text{Generalized Power Rule}$$

EXAMPLE 4 — DIFFERENTIATING USING THE GENERALIZED POWER RULE

Find $\dfrac{d}{dx}\sqrt{x^4 - 3x^3 - 4}$.

Solution

$$\dfrac{d}{dx}\underbrace{(x^4 - 3x^3 - 4)}_{\text{Inside function}}{}^{1/2} = \underbrace{\dfrac{1}{2}}_{\substack{\text{Bring down} \\ \text{the } n}} (x^4 - 3x^3 - 4)^{\underbrace{-1/2}_{\substack{\text{Power} \\ n-1}}} \underbrace{(4x^3 - 9x^2)}_{\substack{\text{Derivative of} \\ \text{the inside function}}}$$

Think of the Generalized Power Rule "from the outside in." That is, first bring down the outer exponent and reduce the exponent by 1, and only then multiply by the derivative of the inside function.

Be careful! It is the *original function* (not the differentiated function) that is raised to the power $n - 1$. Only at the end do you multiply by the derivative of the inside function.

EXAMPLE 5 — SIMPLIFYING AND DIFFERENTIATING

Find $\dfrac{d}{dx}\left(\dfrac{1}{x^2+1}\right)^3$.

Solution Writing the function as $(x^2+1)^{-3}$ gives

$$\dfrac{d}{dx}\underbrace{(x^2+1)}_{\substack{\text{Inside} \\ \text{function}}}{}^{-3} = \underbrace{-3}_{\substack{\text{Bring down} \\ \text{the } n}}(x^2+1)^{\underbrace{-4}_{\substack{\text{Power} \\ n-1}}}\underbrace{(2x)}_{\substack{\text{Derivative of} \\ \text{the inside function}}} = -6x(x^2+1)^{-4} = -\dfrac{6x}{(x^2+1)^4}$$

Practice Problem 3 Find $\dfrac{d}{dx}(x^3 - x)^{-1/2}$. ▶ Solution on page 618

EXAMPLE 6

FINDING THE GROWTH RATE OF AN OIL SLICK

An oil tanker hits a reef, and after t days the radius of the oil slick is $r(t) = \sqrt{4t+1}$ miles. How fast is the radius of the oil slick expanding after 2 days?

Solution To find the rate of change of the radius, we differentiate:

$$\frac{d}{dt}(4t+1)^{1/2} = \frac{1}{2}(4t+1)^{-1/2}(4) = 2(4t+1)^{-1/2}$$

— Derivative of $4t+1$

At $t = 2$ this is

$$2(4 \cdot 2 + 1)^{-1/2} = 2 \cdot 9^{-1/2} = 2 \cdot \frac{1}{3} = \frac{2}{3}$$

Interpretation: After 2 days the radius of the oil slick is growing at the rate of $\frac{2}{3}$ of a mile per day.

Graphing Calculator Exploration

a. On a graphing calculator, enter the radius of the oil slick as $y_1 = \sqrt{4x+1}$. Then use NDERIV to define y_2 to be the derivative of y_1. Graph both y_1 and y_2 on the window [0, 6] by [0, 5].

b. Verify your answer to Example 6 by evaluating y_2 at $x = 2$. Do you get $\frac{2}{3}$?

c. Notice from your graph that the (derivative) function y_2 is *decreasing*, meaning that the radius is growing more *slowly*. Estimate when the radius will be growing by only $\frac{1}{2}$ mile per day. [*Hint:* One way is to use TRACE to follow along the curve y_2 to the x-value (number of days) when the y-coordinate is 0.5, zooming in if necessary. Your answer should be between 3 and 4 days. You can also use INTERSECT with $y_3 = 0.5$.

Some problems require the Generalized Power Rule in combination with another differentiation rule, such as the Product or Quotient Rule.

EXAMPLE 7 — DIFFERENTIATING USING TWO RULES

Find $\dfrac{d}{dx}[(5x-2)^4(9x+2)^7]$.

Solution Since this is a product of powers, $(5x-2)^4$ times $(9x+2)^7$, we use the Product Rule together with the Generalized Power Rule.

$$\dfrac{d}{dx}[(5x-2)^4(9x+2)^7] = \underbrace{4(5x-2)^3(5)}_{\text{Derivative of } (5x-2)^4}(9x+2)^7 + (5x-2)^4\underbrace{[7(9x+2)^6(9)]}_{\text{Derivative of } (9x+2)^7}$$

$$= 20(5x-2)^3(9x+2)^7 + 63(5x-2)^4(9x+2)^6 \quad \text{Simplifying}$$
$$\ \ \uparrow_{4\cdot 5} \uparrow_{7\cdot 9}$$

EXAMPLE 8 — DIFFERENTIATING USING TWO RULES

Find $\dfrac{d}{dx}\left(\dfrac{x}{x+1}\right)^4$.

Solution Since the function is a quotient raised to a power, we use the Quotient Rule together with the Generalized Power Rule. Working from the outside in, we obtain

$$\dfrac{d}{dx}\left(\dfrac{x}{x+1}\right)^4 = 4\left(\dfrac{x}{x+1}\right)^3 \underbrace{\dfrac{(x+1)(1)-(1)(x)}{(x+1)^2}}_{\text{Derivative of the inside function } \frac{x}{x+1}}$$

$$= 4\left(\dfrac{x}{x+1}\right)^3 \dfrac{x+1-x}{(x+1)^2} = 4\left(\dfrac{x}{x+1}\right)^3 \dfrac{1}{(x+1)^2} \quad \text{Simplifying}$$

$$= 4\dfrac{x^3}{(x+1)^3}\dfrac{1}{(x+1)^2} = \dfrac{4x^3}{(x+1)^5} \quad \text{Simplifying further}$$

EXAMPLE 9

DIFFERENTIATING USING A RULE TWICE

Find $\dfrac{d}{dz}[z^2 + (z^2 - 1)^3]^5$.

Solution Since this is a function to a power, where the inside function also contains a function to a power, we must use the Generalized Power Rule *twice*.

$$\dfrac{d}{dz}[z^2 + (z^2 - 1)^3]^5 = 5[z^2 + (z^2 - 1)^3]^4 \underbrace{[2z + 3(z^2 - 1)^2(2z)]}_{\text{Derivative of } z^2 + (z^2 - 1)^3}$$

$$= 5[z^2 + (z^2 - 1)^3]^4[2z + 6z(z^2 - 1)^2] \qquad \text{Simplifying}$$
$$ \uparrow$$
$$3 \cdot 2$$

Chain Rule in Leibniz's Notation

A composition may be written in two parts:

$$y = f(g(x)) \quad \text{is equivalent to} \quad y = f(u) \quad \text{and} \quad u = g(x)$$

The derivatives of these last two functions are:

$$\dfrac{dy}{du} = f'(u) \quad \text{and} \quad \dfrac{du}{dx} = g'(x)$$

The Chain Rule

$$\dfrac{d}{dx}\underbrace{f(g(x))}_{y} = \underbrace{f'(g(x))}_{} \cdot \underbrace{g'(x)}_{} \qquad \text{Chain Rule}$$

$$\underbrace{\phantom{\dfrac{d}{dx}f(g(x))}}_{\dfrac{dy}{dx}} \qquad \underbrace{}_{\dfrac{dy}{du}} \quad \underbrace{}_{\dfrac{du}{dx}}$$

with the indicated substitutions then becomes:

Chain Rule in Leibniz's Notation

For $y = f(u)$ with $u = g(x)$,

$$\dfrac{dy}{dx} = \dfrac{dy}{du} \cdot \dfrac{du}{dx}$$

In this form the Chain Rule is easy to remember, since it looks as if the *du* in the numerator and the denominator cancel:

$$\frac{dy}{dx} = \frac{dy}{du} \cdot \frac{du}{dx}$$

However, since derivatives are not really fractions (they are *limits* of fractions), this is only a convenient device for remembering the Chain Rule.

The Product Rule (page 577) showed that the derivative of a product is not the product of the derivatives. We now see where the product of the derivatives *does* appear: it appears in the Chain Rule, when differentiating composite functions. In other words, the product of the derivatives comes not from *products* but from *compositions* of functions.

A Simple Example of the Chain Rule

The derivation of the Chain Rule is rather technical, but we can show the basic idea in a simple example. Suppose that your company produces steel, and you want to calculate your company's total revenue in dollars per year. You would take the revenue from a ton of steel (dollars per ton) and multiply by your company's output (tons per year). In symbols:

$$\frac{\$}{\text{year}} = \frac{\$}{\text{ton}} \cdot \frac{\text{ton}}{\text{year}} \qquad \text{Note that ``ton'' cancels}$$

If we were to express these rates as derivatives, the equation above would become the Chain Rule.

8.6 Section Summary

To differentiate a composite function (a function of a function), we have the Chain Rule:

$$\frac{d}{dx} f(g(x)) = f'(g(x)) \cdot g'(x)$$

or, in Leibniz's notation, writing $y = f(g(x))$ as $y = f(u)$ and $u = g(x)$,

$$\frac{dy}{dx} = \frac{dy}{du} \cdot \frac{du}{dx} \qquad \text{The derivative of a composite function is the product of the derivatives}$$

8.6 THE CHAIN RULE AND THE GENERALIZED POWER RULE

To differentiate a function to a power, $[f(x)]^n$, we have the *Generalized Power Rule* (a special case of the Chain Rule when the "outer" function is a power):

$$\frac{d}{dx}[f(x)]^n = n \cdot [f(x)]^{n-1} \cdot f'(x)$$

For now, the Generalized Power Rule is more useful than the Chain Rule, but in Chapter 11 we will make important use of the Chain Rule.

Verification of the Chain Rule

Let $f(x)$ and $g(x)$ be differentiable functions. We define k by

$$k = g(x+h) - g(x)$$

or, equivalently,

$$g(x+h) = g(x) + k$$

Then

$$\lim_{h \to 0} \underbrace{[g(x+h) - g(x)]}_{k} = \lim_{h \to 0} k = 0$$

showing that $h \to 0$ implies $k \to 0$ (see pages 587–588). With these relations we may calculate the derivative of the composition $f(g(x))$.

$$\frac{d}{dx} f(g(x)) = \lim_{h \to 0} \frac{f(g(x+h)) - f(g(x))}{h} \qquad \text{Definition of the derivative of } f(g(x))$$

$$= \lim_{h \to 0} \left[\frac{f(g(x+h)) - f(g(x))}{g(x+h) - g(x)} \cdot \frac{g(x+h) - g(x)}{h} \right] \qquad \text{Dividing and multiplying by } g(x+h) - g(x)$$

$$= \lim_{h \to 0} \frac{f(g(x+h)) - f(g(x))}{g(x+h) - g(x)} \lim_{h \to 0} \frac{g(x+h) - g(x)}{h} \qquad \text{The limit of a product is the product of the limits (Limit Rule 4c on page 531)}$$

$$= \underbrace{\lim_{k \to 0} \frac{f(g(x) + k) - f(g(x))}{k}}_{f'(g(x))} \underbrace{\lim_{h \to 0} \frac{g(x+h) - g(x)}{h}}_{g'(x)} \qquad \text{Using the relations } g(x+h) = g(x) + k \text{ and } k = g(x+h) - g(x) \text{ and that } h \to 0 \text{ implies } k \to 0$$

$$= f'(g(x)) \cdot g'(x) \qquad \text{Chain Rule}$$

The last step comes from recognizing the first limit as the definition of the derivative f' at $g(x)$, and the second limit as the definition of the derivative $g'(x)$. This verifies the Chain Rule,

$$\frac{d}{dx} f(g(x)) = f'(g(x)) \cdot g'(x)$$

Solutions to Practice Problems

1. $g(f(x)) = 4 - f(x) = 4 - x^2$
2. $f(x) = \sqrt{x}, \ g(x) = x^5 - 7x + 1$
3. $\dfrac{d}{dx}(x^3 - x)^{-1/2} = -\dfrac{1}{2}(x^3 - x)^{-3/2}(3x^2 - 1)$

8.6 Exercises

Find functions f and g such that the given function is the composition $f(g(x))$.

1. $\sqrt{x^2 - 3x + 1}$
2. $(5x^2 - x + 2)^4$
3. $(x^2 - x)^{-3}$
4. $\dfrac{1}{x^2 + x}$
5. $\dfrac{x^3 + 1}{x^3 - 1}$
6. $\dfrac{\sqrt{x} - 1}{\sqrt{x} + 1}$
7. $\left(\dfrac{x + 1}{x - 1}\right)^4$
8. $\sqrt{\dfrac{x - 1}{x + 1}}$
9. $\sqrt{x^2 - 9} + 5$
10. $\sqrt[3]{x^3 + 8} - 5$

Use the Generalized Power Rule to find the derivative of each function.

11. $f(x) = (x^2 + 1)^3$
12. $f(x) = (x^3 + 1)^4$
13. $h(z) = (3z^2 - 5z + 2)^4$
14. $h(z) = (5z^2 + 3z - 1)^3$
15. $f(x) = \sqrt{x^4 - 5x + 1}$
16. $f(x) = \sqrt{x^6 + 3x - 1}$
17. $w(z) = \sqrt[3]{9z - 1}$
18. $w(z) = \sqrt[5]{10z - 4}$
19. $y = (4 - x^2)^4$
20. $y = (1 - x)^{50}$
21. $y = \left(\dfrac{1}{w^3 - 1}\right)^4$
22. $y = \left(\dfrac{1}{w^4 + 1}\right)^5$
23. $y = x^4 + (1 - x)^4$
24. $f(x) = (x^2 + 4)^3 - (x^2 + 4)^2$
25. $f(x) = \dfrac{1}{\sqrt[3]{(9x + 1)^2}}$
26. $f(x) = \dfrac{1}{\sqrt[3]{(3x - 1)^2}}$
27. $f(x) = [(x^2 + 1)^3 + x]^3$
28. $f(x) = [(x^3 + 1)^2 - x]^4$
29. $f(x) = 3x^2(2x + 1)^5$
30. $f(x) = 2x(x^3 - 1)^4$
31. $f(x) = (2x + 1)^3(2x - 1)^4$
32. $f(x) = (2x - 1)^3(2x + 1)^4$
33. $f(x) = \left(\dfrac{x + 1}{x - 1}\right)^3$
34. $f(x) = \left(\dfrac{x - 1}{x + 1}\right)^5$
35. $f(x) = x^2\sqrt{1 + x^2}$
36. $f(x) = x^2\sqrt{x^2 - 1}$
37. $f(x) = \sqrt{1 + \sqrt{x}}$
38. $f(x) = \sqrt[3]{1 + \sqrt[3]{x}}$

39. Find the derivative of $(x^2 + 1)^2$ in two ways:
   a. By the Generalized Power Rule.
   b. By "squaring out" the original expression and then differentiating.
   Your answers should agree.

40. Find the derivative of $\dfrac{1}{x^2}$ in three ways:
   a. By the Quotient Rule.
   b. By writing $\dfrac{1}{x^2}$ as $(x^2)^{-1}$ and using the Generalized Power Rule
   c. By writing $\dfrac{1}{x^2}$ as x^{-2} and using the (ordinary) Power Rule.
   Your answers should agree.

41. Find the derivative of $\dfrac{1}{3x+1}$ in two ways:

  a. By the Quotient Rule.
  b. By writing the function as $(3x+1)^{-1}$ and using the Generalized Power Rule.
  Your answers should agree. Which way was easier? Remember this for the future.

42. Find an expression for the derivative of the composition of three functions, $\dfrac{d}{dx} f(g(h(x)))$.
  [*Hint:* Use the Chain Rule twice.]

43. Suppose that $L(x)$ is a function such that $L'(x) = \dfrac{1}{x}$. Use the Chain Rule to show that the derivative of the composite function $L(g(x))$ is $\dfrac{d}{dx} L(g(x)) = \dfrac{g'(x)}{g(x)}$.

44. Suppose that $E(x)$ is a function such that $E'(x) = E(x)$. Use the Chain Rule to show that the derivative of the composite function $E(g(x))$ is $\dfrac{d}{dx} E(g(x)) = E(g(x)) \cdot g'(x)$.

Find the *second* derivative of each function.

45. $f(x) = (x^2 + 1)^{10}$ **46.** $f(x) = (x^3 - 1)^5$

APPLIED EXERCISES

47. BUSINESS: Cost A company's cost function is $C(x) = \sqrt{4x^2 + 900}$ dollars, where x is the number of units. Find the marginal cost function and evaluate it at $x = 20$.

48. BUSINESS: Cost *(continuation)* Graph the cost function $y_1 = \sqrt{4x^2 + 900}$ on the viewing window [0, 30] by [−10, 70]. Then use NDERIV to define y_2 as the derivative of y_1. Verify the answer to Exercise 47 by evaluating the marginal cost function y_2 at $x = 20$.

49. BUSINESS: Cost *(continuation)* Find the number x of units at which the marginal cost is 1.75. [*Hint:* TRACE along the marginal cost function y_2 to find where the y-coordinate is 1.75, giving your answer as the x-coordinate rounded to the nearest whole number.]

50. SOCIOLOGY: Educational Status A study* estimated how a person's social status (rated on a scale where 100 indicates the status of a college graduate) depended on years of education. Based on this study, with e years of education, a person's status is $S(e) = 0.22(e + 4)^{2.1}$. Find $S'(12)$ and interpret your answer.

*Robert L. Hamblin, "Mathematical Experimentation and Sociological Theory: A Critical Analysis," *Sociometry* 34:423–452, 1971.

51. SOCIOLOGY: Income Status A study* estimated how a person's social status (rated on a scale where 100 indicates the status of a college graduate) depended upon income. Based on this study, with an income of i thousand dollars, a person's status is $S(i) = 17.5(i - 1)^{0.53}$. Find $S'(25)$ and interpret your answer.

52. ECONOMICS: Compound Interest If $1000 is deposited in a bank paying r% interest compounded annually, 5 years later its value will be
$$V(r) = 1000(1 + 0.01r)^5 \text{ dollars}$$
Find $V'(6)$ and interpret your answer. [*Hint:* $r = 6$ corresponds to 6% interest.]

53. BIOMEDICAL: Drug Sensitivity The strength of a patient's reaction to a dose of x milligrams of a certain drug is $R(x) = 4x\sqrt{11 + 0.5x}$ for $0 \le x \le 140$. The derivative $R'(x)$ is called the *sensitivity* to the drug. Find $R'(50)$, the sensitivity to a dose of 50 mg.

54. GENERAL: Population The population of a city x years from now is predicted to be $P(x) = \sqrt[4]{x^2 + 1}$ million people for $1 \le x \le 5$. Find when the population will be growing at the rate of a quarter of a million people per year. [*Hint:* On a graphing calculator, enter the given population function in y_1, use NDERIV to define y_2 to be the derivative of y_1, and graph both on the window [1, 5] by [0, 3]. Then TRACE along y_2 to find the x-coordinate (rounded to the nearest tenth of a unit) at which the y-coordinate is 0.25. You may have to ZOOM IN to find the correct x-value.]

55. **BIOMEDICAL: Drug Sensitivity** (53 continued) For the reaction function given in Exercise 53, find the dose at which the sensitivity is 25. [*Hint:* On a graphing calculator, enter the reaction function $y_1 = 4x\sqrt{11 + 0.5x}$, and use NDERIV to define y_2 to be the derivative of y_1. Then graph y_2 on the window [0, 140] by [0, 50] and TRACE along y_2 to find the x-coordinate (rounded to the nearest whole number) at which the y-coordinate is 25. You may have to ZOOM IN to find the correct x-value.]

56. **BIOMEDICAL: Blood Flow** It follows from Poiseuille's Law that blood flowing through certain arteries will encounter a resistance of $R(x) = 0.25(1 + x)^4$, where x is the distance (in meters) from the heart. Find the instantaneous rate of change of the resistance at:
    a. 0 meters.   b. 1 meter.

57. **ENVIRONMENTAL SCIENCE: Pollution** The carbon monoxide level in a city is predicted to be $0.02x^{3/2} + 1$ ppm (parts per million), where x is the population in thousands. In t years the population of the city is predicted to be $x(t) = 12 + 2t$ thousand people. Therefore, in t years the carbon monoxide level will be
$$P(t) = 0.02(12 + 2t)^{3/2} + 1$$
ppm. Find $P'(2)$, the rate at which carbon monoxide pollution will be increasing in 2 years.

58. **PSYCHOLOGY: Learning** After p practice sessions, a subject could perform a task in $T(p) = 36(p + 1)^{-1/3}$ minutes for $0 \leq p \leq 10$. Find $T'(7)$ and interpret your answer.

59. **GENERAL: Greenhouse Gases and Global Warming** The burning of oil and other "fossil fuels" is increasing the concentration of "greenhouse gases" in the atmosphere. The chief greenhouse gas is carbon dioxide (CO_2), and the table below gives the atmospheric CO_2 concentration in parts per million (ppm) for every five years since 1970. To avoid large numbers, years are listed in the table as years since 1970.

	Years Since 1970	CO_2 (ppm)
1970	0	325.3
1975	5	331.0
1980	10	338.5
1985	15	345.7
1990	20	354.0
1995	25	362.0
2000	30	370.0

Source: Scripps Institute of Oceanography

a. Enter the table numbers into your graphing calculator and make a plot of the resulting points (Years Since 1970 on the x-axis and CO_2 on the y-axis).
b. Have your calculator find the linear regression formula for these data. Then enter the result as y_1, which gives a formula for CO_2 concentrations for each year. Plot the points together with the regression line. Observe that the line fits the points quite well.
c. Define $y_2 = 0.024x + 51.3$. This function is an estimate of the average global temperature y_2 (in degrees Fahrenheit) for any CO_2 concentration x (which is the "output" of function y_1). (This function was found similarly by another linear regression on temperature and CO_2 data.)
d. Define y_3 to be the *composition* of y_2 and y_1 [on some calculators, by defining $y_3 = y_2(y_1)$.] Therefore, y_3 gives the average global temperature for any value of x (Years Since 1970).
e. Turn "off" the point plots [part (a)] and the functions y_1 and y_2. Then graph y_3 on the window [0, 100] by [55, 65]. Find the slope of this linear function (using NDERIV or dy/dx), giving the predicted temperature rise per year.   (*continues*)

f. Use the slope (degrees per year) from part (e) to estimate how long it will take for temperatures to increase by 1.8 degrees. (It has been estimated that a 1.8-degree temperature increase could raise the sea level by 1 foot, inundating low-lying areas of the Mississippi delta and the food-producing river floodplains of Africa and Asia, and seriously disrupting world food supplies.)*

*The function y_1 that you found, describing how CO_2 levels increase with time, is widely accepted as accurate. The given function y_2 connecting temperature to CO_2 concentrations is less well established, since factors other than CO_2 affect temperature.

8.7 NONDIFFERENTIABLE FUNCTIONS

Introduction

In spite of all of the rules of differentiation, there are nondifferentiable functions—functions that cannot be differentiated at certain values. We begin this section by exhibiting such a function (the absolute value function) and showing that it is not differentiable at $x = 0$. We will then discuss general geometric conditions for a function to be nondifferentiable. Knowing where a function is not differentiable is important for understanding graphs and for interpreting answers from a graphing calculator.

Absolute Value Function

On page 52 we defined the absolute value function,

$$f(x) = |x| = \begin{cases} x & \text{if } x \geq 0 \\ -x & \text{if } x < 0 \end{cases}$$

Although the absolute value function is *defined* for *all* values of x, we will show that it is *not* differentiable at $x = 0$.

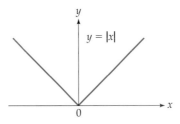

The graph of the absolute value function $f(x) = |x|$ has a "corner" at the origin.

EXAMPLE 1

SHOWING NONDIFFERENTIABILITY

Show that $f(x) = |x|$ is not differentiable at $x = 0$.

Solution We have no "rules" for differentiating the absolute value function, so we must use the definition of the derivative:

$$f'(x) = \lim_{h \to 0} \frac{f(x+h) - f(x)}{h}$$

provided that this limit exists. It is this provision, which until now we have steadfastly ignored, that will be important in this example. We will show that this limit, and hence the derivative, does not exist at $x = 0$.

For $x = 0$ the definition becomes

$$\lim_{h \to 0} \frac{f(0+h) - f(0)}{h} = \lim_{h \to 0} \frac{f(h) - f(0)}{h} = \lim_{h \to 0} \underbrace{\frac{|h| - |0|}{h}}_{\text{Using } f(x) = |x|} = \lim_{h \to 0} \frac{|h|}{h}$$

Now h can approach 0 through *positive* numbers such as 0.01, 0.001, 0.0001, ... (denoted $h \to 0^+$), or through *negative* numbers such as $-0.01, -0.001, -0.0001, \ldots$ (denoted $h \to 0^-$).

For *positive h*:

$$\lim_{h \to 0^+} \frac{|h|}{h} = \lim_{h \to 0^+} \frac{h}{h} = \lim_{h \to 0^+} 1 = 1$$

Since $|h| = h$ for $h > 0$ ⟶ $\frac{h}{h} = 1$

For *negative h*:

$$\lim_{h \to 0^-} \frac{|h|}{h} = \lim_{h \to 0^-} \frac{-h}{h} = \lim_{h \to 0^-} (-1) = -1$$

For $h < 0, |h| = -h$ (the negative sign makes the negative h positive) ⟶ $\frac{-h}{h} = -1$

The limit as $h \to 0$ must be a *single* number, the same regardless of how h approaches 0. Since the right and left limits do not agree (the first is $+1$ and the second is -1), the limit $\lim_{h \to 0} \frac{|h|}{h}$ does not exist, so the derivative does not exist. This is what we wanted to show—that the absolute value function is not differentiable at $x = 0$.

Geometric Explanation of Nondifferentiability

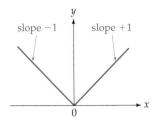

We can give a geometric and intuitive reason why the absolute value function is not differentiable at $x = 0$. Its graph consists of two straight lines with slopes $+1$ and -1 that meet in a corner at the origin. To the right of the origin the slope is $+1$ and to the left of the origin the slope is -1, but *at* the origin the two conflicting slopes make it impossible to define a *single* slope. Therefore, the slope (and hence the derivative) is undefined at $x = 0$.

Graphing Calculator Exploration

If your graphing calculator has an operation like ABS or something similar for the absolute value function, graph $y_1 = \text{ABS}(x)$ on the window $[-2, 2]$ by $[-1, 2]$. Use NDERIV to "find" the derivative of y_1 at $x = 0$. Your calculator may give a "false value" such as 0, resulting from its *approximating* the derivative by the symmetric difference quotient (see pages 569–570). The correct answer is that the derivative is *undefined* at $x = 0$, as we just showed. This is why it is important to understand nondifferentiability—your calculator may give a misleading answer.

Other Nondifferentiable Functions

For the same reason, at any "corner point" of a graph, where two different slopes conflict, the function will not be differentiable.

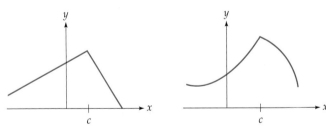

Each of the functions graphed here has a "corner point" at $x = c$, and so is not differentiable at $x = c$.

There are other reasons, besides a corner point, why a function may not be differentiable. If a curve has a vertical tangent line at a point, the slope will not be defined at that x-value, since the slope of a vertical line is undefined.

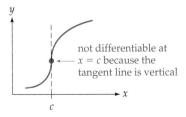

not differentiable at $x = c$ because the tangent line is vertical

We showed on pages 587–588 that if a function is differentiable, then it is continuous. Therefore, if a function is discontinuous (has a "jump") at some point, then it will not be differentiable at that x-value.

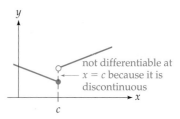
not differentiable at $x = c$ because it is discontinuous

Therefore, if a function f satisfies *any* of the following conditions:

1. f has a corner point at $x = c$,
2. f has a vertical tangent at $x = c$,
3. f is discontinuous at $x = c$,

then f will not be differentiable at c. In Chapter 9, when we use calculus for graphing, it will be important to remember these three conditions that make the derivative fail to exist.

Practice Problem

For the function graphed below, find the x-values at which the derivative is undefined.

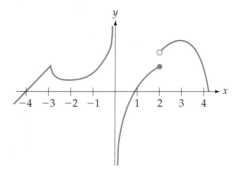

▶ Solution on next page

All differentiable functions are continuous (see pages 587–588), but *not* all continuous functions are differentiable—for example, $f(x) = |x|$. These facts are shown in the following diagram

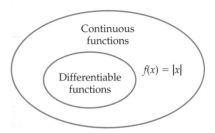

Spreadsheet Exploration

Another function that is not differentiable is $f(x) = x^{2/3}$. The following spreadsheet* calculates values of the difference quotient $\frac{f(x+h) - f(x)}{h}$ at $x = 0$ for this function. Since $f(0) = 0$, the difference quotient at $x = 0$ simplifies to:

$$\frac{f(x+h) - f(x)}{h} = \frac{f(0+h) - f(0)}{h} = \frac{f(h)}{h} = \frac{h^{2/3}}{h} = h^{-1/3}.$$

For example, cell **B5** evaluates $h^{-1/3}$ at $h = \frac{1}{1000}$ obtaining $\left(\frac{1}{1000}\right)^{-1/3} = 1000^{1/3} = \sqrt[3]{1000} = 10$. Column **B** evaluates this different quotient for the *positive* values of h in column **A**, while column **E** evaluates it for the corresponding negative values of h in column **D**.

	B5		=	=A5^(-1/3)	
	A	B	C	D	E
1	h	(f(0+h)-f(0))/h		h	(f(0+h)-f(0))/h
2	1.0000000	1.0000000		-1.0000000	-1.0000000
3	0.1000000	2.1544347		-0.1000000	-2.1544347
4	0.0100000	4.6415888		-0.0100000	-4.6415888
5	0.0010000	10.0000000		-0.0010000	-10.0000000
6	0.0001000	21.5443469		-0.0001000	-21.5443469
7	0.0000100	46.4158883		-0.0000100	-46.4158883
8	0.0000010	100.0000000		-0.0000010	-100.0000000
9	0.0000001	215.4434690		-0.0000001	-215.4434690

Notice that the values in column **B** are becoming arbitrarily large, while the values in column **E** are becoming arbitrarily small, so the difference quotient does not approach a limit as $h \to 0$. This shows that the derivative of $f(x) = x^{2/3}$ at 0 does not exist, so the function $f(x) = x^{2/3}$ is *not* differentiable at $x = 0$.

▶ **Solution to Practice Problem**

$x = -3$, $x = 0$, and $x = 2$

---

* See the Preface for how to obtain this and other Excel spreadsheets.

- Find the velocity and acceleration of a rocket. *(Review Exercise 60.)*

$$v(t) = s'(t) \qquad a(t) = v'(t) = s''(t)$$

- Find the maximum height of a projectile. *(Review Exercise 61.)*
- Find a company's profit function, then calculate and interpret the first and second derivatives. *(Review Exercise 62.)*

8.6 The Chain Rule and the Generalized Power Rule

- Find the derivative of a function using the Generalized Power Rule. *(Review Exercises 63–72.)*

$$\frac{d}{dx}f^n = n \cdot f^{n-1} \cdot f'$$

$$\frac{d}{dx}f(g(x)) = f'(g(x)) \cdot g'(x)$$

$$\frac{dy}{dx} = \frac{dy}{du} \cdot \frac{du}{dx}$$

- Find the derivative of a function using *two* differentiation rules. *(Review Exercises 73–84.)*
- Find the *second* derivative of a function using the Generalized Power Rule. *(Review Exercises 85–88.)*
- Find the derivative of a function in several different ways. *(Review Exercises 89–90.)*
- Use the Generalized Power Rule to find the derivative in an applied problem and interpret the answer. *(Review Exercises 91–92.)*
- Compare the profit from one unit to the marginal profit found by differentiation. *(Review Exercise 93.)*
- Find where the marginal profit equals a given number. *(Review Exercise 94.)*
- Use the Generalized Power Rule to solve an applied problem and interpret the answer. *(Review Exercises 95–96.)*

8.7 Nondifferentiable Functions

- See from a graph where the derivative is undefined. *(Review Exercises 97–100.)*

$$f' \text{ is undefined at } \begin{cases} \text{corner points} \\ \text{vertical tangents} \\ \text{discontinuities} \end{cases}$$

- Prove that a function is not differentiable at a given value. *(Review Exercises 101–102.)*

Hints and Suggestions

- (*Overview*) This chapter introduced one of the most important concepts in all of calculus, the *derivative*. First we defined it (using limits), then we developed several "rules of differentiation" to simplify its calculation.
- Remember the four interpretations of the derivative—*slopes, instantaneous rates of change, marginals,* and *velocities.*
- The *second* derivative gives the rate of change of the rate of change, and acceleration.
- Graphing calculators help to find limits, graph curves and their tangent lines, and calculate derivatives (using NDERIV) and second derivatives (using NDERIV twice). NDERIV, however, provides only an *approximation* to the derivative, and therefore sometimes gives a misleading result.
- The units of the derivative are important in applied problems. For example, if $f(x)$ gives the temperature in degrees at time x hours, then the derivative $f'(x)$ is in *degrees per hour*. In general, the units of the derivative $f'(x)$ are "*f*-units" per "*x*-unit."
- **Practice for Test:** Review Exercises 1, 2, 3, 4, 7, 9, 11, 15, 19, 20, 21, 25, 31, 37, 43, 47, 49, 57, 59, 62, 67, 71, 77, 83, 91, 94, 97.

Review Exercises for Chapter 8

Practice test exercise numbers are in green.

8.1 Limits and Continuity

For each limit, complete the limit table and find the limit (or state that the limit does not exist). Round calculations to three decimal places.

1. $\lim_{x \to 2} (4x + 2)$

x	4x + 2		x	4x + 2
1.9			2.1	
1.99			2.01	
1.999			2.001	

2. $\lim_{x \to 0} \dfrac{\sqrt{x+1} - 1}{x}$

x	$\dfrac{\sqrt{x+1}-1}{x}$
−0.1	
−0.01	
−0.001	

x	$\dfrac{\sqrt{x+1}-1}{x}$
0.1	
0.01	
0.001	

For each piecewise linear function, find:

a. $\lim_{x \to 5^-} f(x)$ b. $\lim_{x \to 5^+} f(x)$ c. $\lim_{x \to 5} f(x)$

3. $f(x) = \begin{cases} 2x - 7 & \text{if } x \leq 5 \\ 3 - x & \text{if } x > 5 \end{cases}$

4. $f(x) = \begin{cases} 4 - x & \text{if } x < 5 \\ 2x - 11 & \text{if } x \geq 5 \end{cases}$

Find the following limits (*without* using limit tables).

5. $\lim_{x \to 4} \sqrt{x^2 + x + 5}$

6. $\lim_{x \to 0} \pi$

7. $\lim_{s \to 16} \left(\dfrac{1}{2}s - s^{1/2}\right)$

8. $\lim_{r \to 8} \dfrac{r}{r^2 - 30\sqrt[3]{r}}$

9. $\lim_{x \to 1} \dfrac{x^2 - x}{x^2 - 1}$

10. $\lim_{x \to -1} \dfrac{3x^3 - 3x}{2x^2 + 2x}$

11. $\lim_{h \to 0} \dfrac{2x^2h - xh^2}{h}$

12. $\lim_{h \to 0} \dfrac{6xh^2 - x^2h}{h}$

For each function, state whether it is continuous or discontinuous. If it is discontinuous, state the values of x at which it is discontinuous.

13. $f(x) = 2x + 5$

14. $f(x) = x^2 - 1$

15. $f(x) = \dfrac{1}{x + 1}$

16. $f(x) = \dfrac{1}{x^2 + 1}$

17. $f(x) = \dfrac{x - 1}{x^2 + x}$

18. $f(x) = \dfrac{1}{|x| - 3}$

19. $f(x) = \begin{cases} 2x - 7 & \text{if } x \leq 5 \\ 3 - x & \text{if } x > 5 \end{cases}$

[*Hint:* See Exercise 3.]

20. $f(x) = \begin{cases} 4 - x & \text{if } x < 5 \\ 2x - 11 & \text{if } x \geq 5 \end{cases}$

[*Hint:* See Exercise 4.]

8.2 Rates of Change, Slopes, and Derivatives

Find the derivative of each function using the *definition* of the derivative (that is, as you did in Section 8.2).

21. $f(x) = 2x^2 + 3x - 1$

22. $f(x) = 3x^2 + 2x - 3$

(See instructions on previous page.)

23. $f(x) = \dfrac{3}{x}$ **24.** $f(x) = 4\sqrt{x}$

8.3 Some Differentiation Formulas

Find the derivative of each function.

25. $f(x) = 6\sqrt[3]{x^5} - \dfrac{4}{\sqrt{x}} + 1$

26. $f(x) = 4\sqrt{x^5} - \dfrac{6}{\sqrt[3]{x}} + 1$

Evaluate each expression.

27. If $f(x) = \dfrac{1}{x^2}$, find $f'\left(\dfrac{1}{2}\right)$.

28. If $f(x) = \dfrac{1}{x}$, find $f'\left(\dfrac{1}{3}\right)$.

29. If $f(x) = 12\sqrt[3]{x}$, find $f'(8)$.

30. If $f(x) = 6\sqrt[3]{x}$, find $f'(-8)$.

31. BUSINESS: Marginal Cost A company's cost function is $C(x) = 20 + 3x + \dfrac{54}{\sqrt{x}}$ dollars where x is the number of units (for $5 \le x \le 20$).
   a. Find the marginal cost function.
   b. Find the marginal cost at $x = 9$ and interpret your answer.

32. Learning Curves in Industry From pages 26–27 the learning curve for building Boeing 707 airplanes is $f(x) = 150x^{-0.322}$, where $f(x)$ is the time (in thousands of hours) that it took to build the xth Boeing 707. Find the instantaneous rate of change of this production time for the tenth plane, and interpret your answer.

33. GENERAL: Geometry The formula for the area of a circle is $A = \pi r^2$, where r is the radius of the circle and π is a constant.

   a. Show that the derivative of the area formula is $2\pi r$, the formula for the circumference of a circle.
   b. Give an explanation for this in terms of rates of change.

34. GENERAL: Geometry The formula for the volume of a sphere is $V = \tfrac{4}{3}\pi r^3$, where r is the radius of the sphere and π is a constant.

   a. Show that the derivative of the volume formula is $4\pi r^2$, the formula for the surface area of a sphere.
   b. Give an explanation for this in terms of rates of change.

8.4 The Product and Quotient Rules

Find the derivative of each function.

35. $f(x) = 2x(5x^3 + 3)$

36. $f(x) = x^2(3x^3 - 1)$

37. $f(x) = (x^2 + 5)(x^2 - 5)$

38. $f(x) = (x^2 + 3)(x^2 - 3)$

39. $y = (x^4 + x^2 + 1)(x^5 - x^3 + x)$

40. $y = (x^5 + x^3 + x)(x^4 - x^2 + 1)$

41. $y = \dfrac{x - 1}{x + 1}$ **42.** $y = \dfrac{x + 1}{x - 1}$

43. $y = \dfrac{x^5 + 1}{x^5 - 1}$ **44.** $y = \dfrac{x^6 - 1}{x^6 + 1}$

45. Find the derivative of $f(x) = \dfrac{2x + 1}{x}$ in three different ways, and check that the answers agree:
   a. By the Quotient Rule
   b. By writing the function in the form $f(x) = (2x + 1)(x^{-1})$ and using the Product Rule.
   c. By thinking of another way, which is the easiest of all.

46. BUSINESS: Sales The manager of an electronics store estimates that the number of cassette tapes that a store will sell at a price of x dollars is

$$S(x) = \frac{2250}{x+9}$$

Find the rate of change of this quantity when the price is $6 per tape, and interpret your answer.

47. BUSINESS: Marginal Average Profit A company's profit function is $P(x) = 6x - 200$ dollars, where x is the number of units.
   a. Find the average profit function.
   b. Find the marginal average profit function.
   c. Evaluate the marginal average profit function at $x = 10$ and interpret your answer.

48. BUSINESS: Marginal Average Cost A company's cost function is $C(x) = 5x + 100$ dollars, where x is the number of units.
   a. Find the average cost function.
   b. Find the marginal average cost function.
   c. Evaluate the marginal average cost function at $x = 20$ and interpret your answer.

8.5 Higher-Order Derivatives

Find the *second* derivative of each function.

49. $f(x) = 12\sqrt{x^3} - 9\sqrt[3]{x}$

50. $f(x) = 18\sqrt[3]{x^2} - 4\sqrt{x^3}$

51. $f(x) = \dfrac{1}{3x^2}$ **52.** $f(x) = \dfrac{1}{2x^3}$

Evaluate each expression.

53. If $f(x) = \dfrac{2}{x^3}$, find $f''(-1)$.

54. If $f(x) = \dfrac{3}{x^4}$, find $f''(-1)$.

55. $\left.\dfrac{d^2}{dx^2} x^6\right|_{x=-2}$ **56.** $\left.\dfrac{d^2}{dx^2} x^{-2}\right|_{x=-2}$

57. $\left.\dfrac{d^2}{dx^2} \sqrt{x^5}\right|_{x=16}$ **58.** $\left.\dfrac{d^2}{dx^2} \sqrt{x^7}\right|_{x=4}$

59. GENERAL: Population The population of a city t years from now is predicted to be $P(t) = 0.25t^3 - 3t^2 + 5t + 200$ thousand people. Find $P(10)$, $P'(10)$, and $P''(10)$ and interpret your answers.

60. GENERAL: Velocity A rocket rises $s(t) = 8t^{5/2}$ feet in t seconds. Find its velocity and acceleration after 25 seconds.

61. GENERAL: Velocity The fastest baseball pitch on record (thrown by Lynn Nolan Ryan of the California Angels on August 20, 1974) was clocked at 100.9 miles per hour (148 feet per second).
   a. If this pitch had been thrown straight up, its height after t seconds would have been $s(t) = -16t^2 + 148t + 5$ feet. Find the maximum height the ball would have reached.
   b. Verify your answer to part (a) by graphing the height function $y_1 = -16x^2 + 148x + 5$ on the window [0, 10] by [0, 400]. Then TRACE along the curve to find its highest point (or use the MAXIMUM feature of your calculator).

62. BUSINESS: Profit The table below gives a company's annual profit at the end of the year for years 1 through 6.

Year	Annual Profit (millions of dollars)
1	3.4
2	3.1
3	3.0
4	3.1
5	4.1
6	5.2

   a. Enter the data in the table into your graphing calculator and make a plot of the resulting points (Year on the x-axis and Annual Profit on the y-axis).
   b. Have your calculator find the quadratic regression formula for these data. Then enter the result in y_1, giving the annual profit for year x. Plot the points together with the regression curve (on an appropriate window). Observe that the curve fits the points quite well.
   c. Predict the annual profit at the end of year 7 by evaluating $y_1(7)$.   *(continues)*

d. Define y_2 to be the derivative of y_1 (using NDERIV). Find $y_2(7)$ and interpret this number.
e. Define y_3 to be the derivative of y_2 (so y_3 is the *second* derivative of y_1). Find and interpret $y_3(7)$.
f. Graph all three functions and the points on the window [0, 7] by [−1, 7]. Can you explain why y_3 appears to be a horizontal straight line?

8.6 The Chain Rule and the Generalized Power Rule

Find the derivative of each function.

63. $h(z) = (4z^2 - 3z + 1)^3$
64. $h(z) = (3z^2 - 5z - 1)^4$
65. $g(x) = (100 - x)^5$
66. $g(x) = (1000 - x)^4$
67. $f(x) = \sqrt{x^2 - x + 2}$
68. $f(x) = \sqrt{x^2 - 5x - 1}$
69. $w(z) = \sqrt[3]{6z - 1}$
70. $w(z) = \sqrt[3]{3z + 1}$
71. $h(x) = \dfrac{1}{\sqrt[5]{(5x + 1)^2}}$
72. $h(x) = \dfrac{1}{\sqrt[5]{(10x + 1)^3}}$
73. $g(x) = x^2(2x - 1)^4$
74. $g(x) = 5x(x^3 - 2)^4$
75. $y = x^3 \sqrt[3]{x^3 + 1}$
76. $y = x^4 \sqrt{x^2 + 1}$
77. $f(x) = [(2x^2 + 1)^4 + x^4]^3$
78. $f(x) = [(3x^2 - 1)^3 + x^3]^2$
79. $f(x) = \sqrt{(x^2 + 1)^4 - x^4}$
80. $f(x) = \sqrt{(x^3 + 1)^2 + x^2}$
81. $f(x) = (3x + 1)^4(4x + 1)^3$
82. $f(x) = (x^2 + 1)^3(x^2 - 1)^4$
83. $f(x) = \left(\dfrac{x + 5}{x}\right)^4$
84. $f(x) = \left(\dfrac{x + 4}{x}\right)^5$

Find the *second* derivative of each function.

85. $h(w) = (2w^2 - 4)^5$
86. $h(w) = (3w^2 + 1)^4$
87. $g(z) = z^3(z + 1)^3$
88. $g(z) = z^4(z + 1)^4$

89. Find the derivative of $(x^3 - 1)^2$ in two ways:

a. By the Generalized Power Rule.
b. By "squaring out" the original expression and then differentiating.

Your answers should agree.

90. Find the derivative of $g(x) = \dfrac{1}{x^3 + 1}$ in two ways:

a. By the Quotient Rule.
b. By the Generalized Power Rule.

Your answers should agree.

91. **BUSINESS: Marginal Profit** A company's profit from producing x tons of polyurethane is $P(x) = \sqrt{x^3 - 3x + 34}$ thousand dollars (for $0 \le x \le 10$). Find $P'(5)$ and interpret your answer.

92. **GENERAL: Compound Interest** If $500 is deposited in an account earning interest at r percent annually, after 3 years its value will be $V(r) = 500(1 + 0.01r)^3$ dollars. Find $V'(8)$ and interpret your answer.

93. **BUSINESS: Marginal Profit** (91 continued) Using a graphing calculator, graph the profit function from Exercise 91 on the window [0, 10] by [0, 30].

a. Evaluate the profit function at 4, 5, and 6 and calculate the following actual costs:

$P(5) - P(4)$ (actual cost of the fifth ton)

$P(6) - P(5)$ (actual cost of the sixth ton)

Compare these results with the "instantaneous" marginal profit at $x = 5$ found in Exercise 91.

b. Define another function to be the derivative of the previously entered profit function (using NDERIV) and graph both together. Find (to the nearest tenth of a unit) the x-value where the marginal profit reaches 4.

94. On a graphing calculator, graph the cost function $C(x) = \sqrt[3]{x^2 + 25x + 8}$ as y_1 on the window [0, 20] by [−2, 10]. Use NDERIV to define y_2 to be its derivative, graphing both. TRACE along the derivative to find the x-value (to the nearest unit) where the marginal cost is 0.25.

95. BIOMEDICAL: Blood Flow Blood flowing through an artery encounters a resistance of $R(x) = 0.25\,(0.01x + 1)^4$, where x is the distance (in centimeters) from the heart. Find the instantaneous rate of change of the resistance 100 centimeters from the heart.

96. GENERAL: Survival Rate Suppose that for a group of 10,000 people, the number who survive to age x is $N(x) = 1000\sqrt{100 - x}$. Find $N'(96)$ and interpret your answer.

8.7 Nondifferentiable Functions

For each function graphed below, find the values of x at which the derivative does not exist.

97.

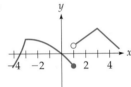

98.

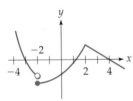

99.

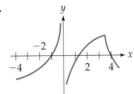

100.

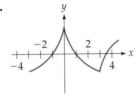

Use the definition of the derivative to show that the following functions are not differentiable at $x = 0$.

101. $f(x) = |5x|$ **102.** $f(x) = x^{3/5}$

9 Further Applications of Derivatives

- 9.1 Graphing Using the First Derivative
- 9.2 Graphing Using the First and Second Derivatives
- 9.3 Optimization
- 9.4 Further Applications of Optimization
- 9.5 Optimizing Lot Size and Harvest Size
- 9.6 Implicit Differentiation and Related Rates

Application Preview

Stevens' Law of Psychophysics

Suppose that you are given two weights and asked to judge how much heavier one is than the other. If one weight is actually *twice* as heavy as the other, most people will judge the heavier weight as being *less* than twice as heavy. This is one of the oldest problems in experimental psychology—how sensation (perceived weight) varies with stimulus (actual weight). Similar experiments can be performed for perceived brightness of a light compared with actual brightness, perceived effort compared with actual work, and so on. The results will vary somewhat from person to person, but the following diagram shows some typical stimulus-response curves (in arbitrary units).

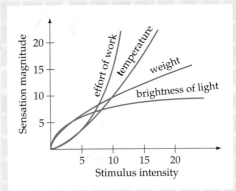

Notice, for example, that perceived effort increases more rapidly than actual work, which suggests that a 10% increase in an employee's work should be rewarded with a *greater* than 10% increase in pay.

Such stimulus-response curves were studied by the psychologist S. S. Stevens[*] at Harvard, who expressed them as power functions.

$$\text{Response} = a(\text{stimulus})^b$$

or

$$f(x) = ax^b \qquad \text{for constants } a \text{ and } b$$

In this chapter we will see that calculus can be very helpful for graphing such functions, such as showing whether they "curl upward" like the work and temperature curves or "curl downward" like the weight and brightness curves (see page 663).

---

[*] See S. S. Stevens, "On the Psychophysical Law," *Psychological Review* 64:153–181.

9.1 GRAPHING USING THE FIRST DERIVATIVE

Introduction

In this chapter we will put derivatives to two major uses: graphing and optimization. Graphing involves using calculus to find the most important points on a curve, then sketching the curve either by hand or by using a graphing calculator. Optimization means finding the largest or smallest values of a function (for example, maximizing profit or minimizing risk). We begin with graphing, since it will form the basis for optimization later in the chapter.

We saw in Chapter 8 that the derivative of a function gives the slope of its graph.

If $f' > 0$ on an interval, then f is *increasing* on that interval.

If $f' < 0$ on an interval, then f is *decreasing* on that interval.

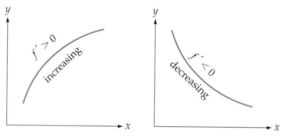

"Increasing" and "decreasing" on a graph mean rising and falling as you move *from left to right*, the same direction in which you are reading this book.

Relative Extreme Points and Critical Numbers

On a graph, a *relative maximum point* is a point that is at least as *high* as the neighboring points of the curve on either side, and a *relative minimum point* is a point that is at least as *low* as the neighboring points on either side.

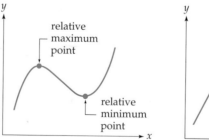
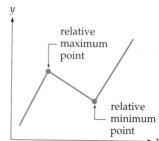

9.1 GRAPHING USING THE FIRST DERIVATIVE

The word "relative" means that although these points may not be the highest and lowest on the *entire* curve, they are the highest and lowest *relative to points nearby*. (Later we will use the terms "absolute maximum" and "absolute minimum" to mean the highest and lowest points on the entire curve.) A curve may have any number of relative maximum and minimum points (collectively, "relative extreme points"), even none. For a function f, the relative extreme points may be defined more formally in terms of the values of f.

> f has a *relative maximum value* at c if $f(c) \geq f(x)$ for all values of x near* c.
>
> f has a *relative minimum value* at c if $f(c) \leq f(x)$ for all values of x near* c.

In the first of the two graphs below, the relative extreme points occur where the slope is *zero* (where the tangent line is horizontal), and in the second graph they occur where the slope is *undefined* (at corner points). The x-coordinates of such points are called *critical numbers*.

Critical Number

A *critical number* of a function f is an x-value in the domain of f at which either

or
$$f'(x) = 0$$
$$f'(x) \text{ is undefined}$$

Derivative is zero or undefined

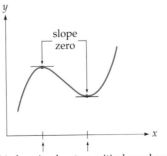

This function has two critical numbers (where the derivative is zero).

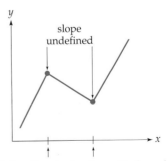

This function also has two critical numbers (but where the derivative is undefined).

---

* "Near c" means in some open interval containing c.

Graphing Functions

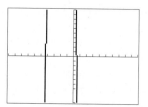

A "useless" graph of $f(x) = x^3 - 12x^2 - 60x + 36$ on the standard window $[-10, 10]$ by $[-10, 10]$

We graph a function by finding its critical numbers, making a "sign diagram" for the derivative to show the intervals of increase and decrease and the relative extreme points, and then drawing the curve on a graphing calculator or "by hand." Obtaining a reasonable graph even with a graphing calculator requires more than just pushing buttons, as shown in the graph of the function $f(x) = x^3 - 12x^2 - 60x + 36$ on the left. We will improve on this graph in the following example.

EXAMPLE 1

GRAPHING A FUNCTION

Graph the function $f(x) = x^3 - 12x^2 - 60x + 36$.

Solution

Step 1: Find critical numbers.

$f'(x) = 3x^2 - 24x - 60$ Derivative of $f(x) = x^3 - 12x^2 - 60x + 36$

$= 3(x^2 - 8x - 20) = 3(x - 10)(x + 2) = 0$ Factoring and setting equal to zero

Zero at $x = 10$ Zero at $x = -2$

The derivative is zero at $x = 10$ and at $x = -2$, and there are no numbers at which the derivative is undefined (it is a polynomial), so the critical numbers (CNs) are

$$\text{CN} \begin{cases} x = 10 \\ x = -2 \end{cases}$$

Both are in the domain of the orginal function

Step 2: *Make a sign diagram for the derivative f'.* A sign diagram for f' begins with a copy of the x-axis with the critical numbers written below it and the behavior of f' indicated above it.

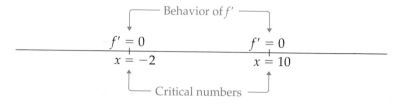

Since f' is continuous, it can change sign only at critical numbers, so f' must keep the same sign between consecutive critical numbers. We determine the sign of f' in each interval by choosing a "test point" in each interval and substituting it into f'. It is easiest to use the factored form: $f'(x) = 3(x - 10)(x + 2)$. For example, in the first interval, $f'(-3) = 3(-3 - 10)(-3 + 2) = 3(\text{negative})(\text{negative}) = (\text{positive})$. We indicate the sign of f' (the slope of f) by arrows: ↗ for positive slope, → for zero slope, and ↘ for negative slope.

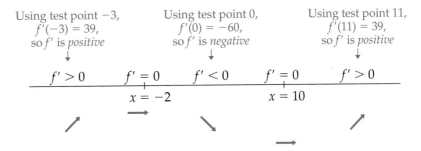

The sign diagram shows that to the left of $x = -2$ the function increases, then between $x = -2$ and $x = 10$ it decreases, and then to the right of $x = 10$ it increases again. Therefore, the open intervals of increase are $(-\infty, -2)$ and $(10, \infty)$, and the open interval of decrease is $(-2, 10)$.

Arrows ↗ → ↘ indicate a relative *maximum* point, and arrows ↘ → ↗ indicate a relative *minimum* point. We then list these points under the critical numbers.

$f' > 0$	$f' = 0$	$f' < 0$	$f' = 0$	$f' > 0$
	$x = -2$		$x = 10$	
	→			
↗	rel max	↘		↗
	$(-2, 100)$		→	
			rel min	
			$(10, -764)$	

The y-coordinates of the points were found by evaluating the *original* function at the x-coordinate: $f(-2) = 100$ and $f(10) = -764$. [*Hint:* Use 🖩 with TABLE or EVALUATE.]

Step 3: Sketch the graph. The arrows ↗ → ↘ → ↗ show the general shape of the curve: going up, level, down, level, and up again. The critical numbers show that we want to include x-values from before $x = -2$ to after $x = 10$, suggesting an interval such as $[-10, 20]$. The y-coordinates show that we want to go from above 100 to below -764, suggesting an interval of y-values such as $[-800, 200]$.

Using a graphing calculator, you then would graph on the window $[-10, 20]$ by $[-800, 200]$ (or some other reasonable window), as shown below.

By hand, you would plot the relative maximum point (with a "cap" ⌒) and the relative minimum point (with a "cup" ⌣), and then draw an "up-down-up" curve through them.

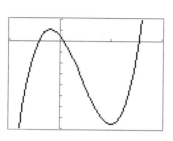
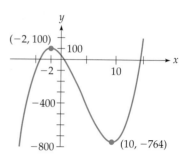

Two important observations:

1. Even with a graphing calculator, you still need calculus to find the relative extreme points that determine an appropriate window.
2. Given a calculator-drawn graph such as that on the left above, you should be able to make a hand-drawn graph such as that on the right, including numbers on the axes and coordinates of important points (using TRACE or TABLE if necessary).

Graphing Calculator Exploration

Do you see where the "useless" graph shown on page 638 fits into the "useful" graph on the left above? Check this by graphing $y_1 = x^3 - 12x^2 - 60x + 36$ first on the window $[-10, 20]$ by $[-800, 200]$ (obtaining the graph shown above) and then on the standard window $[-10, 10]$ by $[-10, 10]$.

Practice Problem 1 Find the critical numbers of $f(x) = x^3 - 12x + 8$.

$3x^2 - 12$

▶ Solution on page 645

In Example 1 there were no critical numbers at which the derivative was *undefined* (it was a polynomial). For an example of a function with a critical number where the derivative is *undefined*, think of the absolute value function $f(x) = |x|$. The graph has a "corner point"

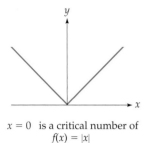

$x = 0$ is a critical number of $f(x) = |x|$

at the origin (as shown on the left), and so the derivative is undefined at the critical number $x = 0$.

First-Derivative Test for Relative Extreme Values

The graphical idea from the sign diagram, that ↗ → ↘ (up, level, and down) indicates a relative maximum and ↘ → ↗ (down, level, and up) indicates a relative minimum, can be stated more formally in terms of the derivative.

> **First-Derivative Test**
>
> If a function f has a critical number c, then at $x = c$ the function has
>
> a *relative maximum* if $f' > 0$ just before c and $f' < 0$ just after c.
>
> a *relative minimum* if $f' < 0$ just before c and $f' > 0$ just after c.

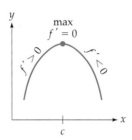

If f' is positive then negative, then f has a relative *maximum* at c.

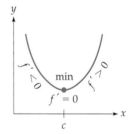

If f' is negative then positive, then f has a relative *minimum* at c.

If the derivative has the *same* sign on both sides of c, then the function has *neither* a relative maximum nor a relative minimum at $x = c$. [*Hint:* Use 📱 with TABLE or EVALUATE.]

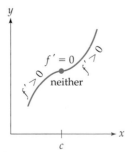

f' is positive on both sides, so f has *neither* at $x = c$.

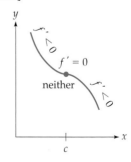

f' is negative on both sides, so f has *neither* at $x = c$.

The diagrams below show that the first-derivative test applies even at critical numbers where the derivative is *undefined* (abbreviated: f' und).

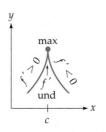

f' is positive then negative, so f has a relative *maximum* at $x = c$.

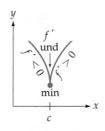

f' is negative then positive, so f has a relative *minimum* at $x = c$.

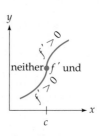

f' is positive on both sides, so f has *neither* at $x = c$.

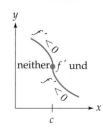

f' is negative on both sides, so f has *neither* at $x = c$.

EXAMPLE 2

GRAPHING A FUNCTION

Graph $f(x) = -x^4 + 4x^3 - 20$.

Solution

$$f'(x) = -4x^3 + 12x^2 \quad \text{Differentiating}$$
$$= -4x^2(x - 3) = 0 \quad \text{Factoring and setting equal to zero}$$

Critical numbers:

$$\text{CN} \begin{cases} x = 0 \\ x = 3 \end{cases} \quad \begin{array}{l} \text{From } -4x^2 = 0 \\ \text{From } (x - 3) = 0 \end{array}$$

We make a sign diagram for the derivative:

| $f' = 0$ | $f' = 0$ | Behavior of f' |
| $x = 0$ | $x = 3$ | Critical numbers |

We determine the sign of $f'(x) = -4x^2(x - 3)$ using test points in each interval, and then add arrows.

$f' > 0 \quad f' = 0 \quad f' > 0 \quad f' = 0 \quad f' < 0$
$\qquad\qquad x = 0 \qquad\qquad x = 3$

Finally, we interpret the arrows to describe the behavior of the functions, which we state at the bottom.

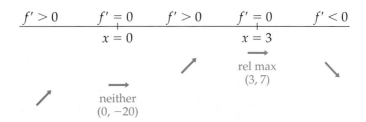

The open intervals of increase are $(-\infty, 0)$ and $(0, 3)$, and the open interval of decrease is $(3, \infty)$.

Using a graphing calculator, we would choose a window such as $[-2, 5]$ by $[-30, 10]$ to include the points $(0, -20)$ and $(3, 7)$, and graph the function.

By hand, we would plot $(0, -20)$ and $(3, 7)$, and join them by a curve that goes, according to the arrows, "up-level-up-level-down."

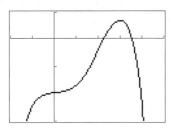
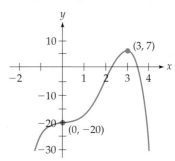

With a graphing calculator, an incomplete sign diagram may be enough to find an appropriate viewing window.

EXAMPLE 3

GRAPHING A RATIONAL FUNCTION

Graph the rational function $f(x) = \dfrac{1}{x^2 - 4x}$.

Solution

$$f(x) = \frac{1}{x^2 - 4x} = \frac{1}{x(x-4)} \qquad \text{Original function with factored denominator}$$

The denominator is zero at $x = 0$ and at $x = 4$, so the function is *undefined* at these numbers. The derivative is

$$f'(x) = \frac{(x^2 - 4x) \cdot 0 - (2x - 4) \cdot 1}{(x^2 - 4x)^2} \quad \text{Using the quotient rule on } \frac{1}{x^2 - 4x}$$

$$= \frac{-(2x - 4)}{(x^2 - 4x)^2} = \frac{-2(x - 2)}{(x^2 - 4x)^2} \quad \begin{array}{l}\text{Simplifying, and factoring}\\ \text{the numerator}\\ \text{Zero at } x = 2\end{array}$$

The derivative is zero at $x = 2$ and undefined (abbreviated: f' und) at $x = 0$ and at $x = 4$ (as was the original function). We list all of these on a sign diagram along with the signs of f' (from test points).

$$\begin{array}{cccccc} f' > 0 & f' \text{ und} & f' > 0 & f' = 0 & f' < 0 & f' \text{ und} & f' < 0 \\ \hline & + & & + & & + & \\ & x = 0 & & x = 2 & & x = 4 & \\ & & & (2, -\tfrac{1}{4}) & & & \end{array}$$

y-coordinate from
$f(x) = \dfrac{1}{x^2 - 4x}$ at $x = 2$

For the viewing window, we might choose x-values $[-2, 6]$ (to include the x-values in the sign diagram) and y-values $[-2, 2]$ (narrow, so that the $-\tfrac{1}{4}$ will be distinct from the x-axis). The graph is shown below (you may need to ignore some false vertical lines). A hand-drawn graph, which you should be able to make from the calculator graph, is next to it.

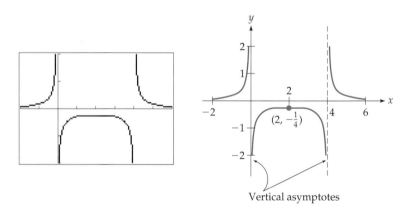

Vertical asymptotes

Notice in the previous example that where the function is undefined, the curve has a *vertical asymptote*. This is true of any (simplified) rational function, since the one-sided limits will be ∞ or $-\infty$ as x approaches one of these values.

Graphing Calculator Exploration

Graph the function $f(x) = \dfrac{1}{x^2 - 4x}$ on the standard window $[-10, 10]$ by $[-10, 10]$. Is the graph as clear as when graphed on the window we chose in Example 3? Choosing a good window is very important.

Practice Problem 2 Explain why, for a rational function, the function and its derivative will be undefined at the same x-values. Are such x-values critical numbers?   ➤ Solution on next page

9.1 Section Summary

To sketch a graph:

- Find the domain (by excluding any x-values at which the function is undefined).
- Find the critical numbers (where f' is zero or undefined, but where f is defined).
- List all of these on a sign diagram for the derivative, indicating the behavior of f' on each interval. Add arrows and relative extreme points.
- Finally, sketch the curve. If you use a graphing calculator, the sign diagram will suggest an appropriate window. If you graph by hand, your sign diagram will show you the shape of the curve.

If there are no relative extreme points, choose a few x-values, including any of special interest. On a graphing calculator, use these x-values to determine an x-interval, and use TABLE or TRACE to find a y-interval; then graph the function on the resulting window. By hand, plot the points corresponding to the chosen x-values and draw an appropriate curve through the points using your sign diagram.

➤ Solutions to Practice Problems

1. $f'(x) = 3x^2 - 12 = 3(x^2 - 4) = 3(x+2)(x-2)$

   CN $\begin{cases} x = -2 \\ x = 2 \end{cases}$

2. Because the denominator of f' will be the square of the denominator of f (from the Quotient Rule). Such x values are *not* critical numbers because they are not in the domain of the original function (see the definition on page 637).

9.1 EXERCISES

For the functions graphed in Exercises 1 and 2:
a. Find the intervals on which the derivative is positive.
b. Find the intervals on which the derivative is negative.

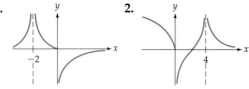

3. Which of the numbers 1, 2, 3, 4, 5, and 6 are critical numbers of the function graphed below?

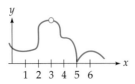

4. The first column in parts a through d shows the graphs of four functions, and the second column shows the graphs of their derivatives, but not necessarily in the same order. Write below each derivative the correct function from which it came.

a.

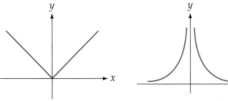

a) derivative of function: _____

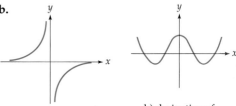

b) derivative of function: _____

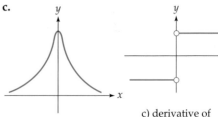

c) derivative of function: _____

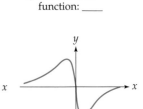

d) derivative of function: _____

Find the critical numbers of each function.

5. $f(x) = x^3 - 48x$
6. $f(x) = x^3 - 27x$
7. $f(x) = x^4 + 4x^3 - 8x^2 + 1$
8. $f(x) = x^4 + 4x^3 - 20x^2 - 12$
9. $f(x) = (2x - 6)^4$

10. $f(x) = (x^2 + 6x - 7)^2$
11. $f(x) = 3x + 5$
12. $f(x) = 4x - 12$
13. $f(x) = x^3 - 2x^2 + x + 11$
14. $f(x) = x^3 - x^2 + 15$

Sketch the graph of each function "by hand" after making a sign diagram for the derivative and finding all open intervals of increase and decrease.

15. $f(x) = x^4 + 4x^3 - 8x^2 + 64$
16. $f(x) = x^4 - 4x^3 - 8x^2 + 64$
17. $f(x) = -x^4 + 4x^3 - 4x^2 + 1$
18. $f(x) = -x^4 - 4x^3 - 4x^2 + 1$
19. $f(x) = 3x^4 - 8x^3 + 6x^2$
20. $f(x) = 3x^4 + 8x^3 + 6x^2$
21. $f(x) = (x - 1)^6$
22. $f(x) = (x - 1)^5$
23. $f(x) = (x^2 - 4)^2$
24. $f(x) = (x^2 - 2x - 8)^2$
25. $f(x) = x^2(x - 4)^2$
26. $f(x) = x(x - 4)^3$
27. $f(x) = x^2(x - 5)^3$
28. $f(x) = x^3(x - 5)^2$

Graph each function using a graphing calculator by first making a sign diagram for the derivative. For Exercises 29–32, also make a sketch from the screen, showing numbers on the axes and coordinates of relative extreme points. (Answers may vary depending on the window chosen.)

29. $f(x) = x^3 - 300x$ **30.** $f(x) = x^3 - 243x$
31. $f(x) = x^4 - 50x^2 - 25$
32. $f(x) = x^4 - 72x^2 - 4$
33. $f(x) = x^5 - 5x^4 + 5x^3 - 23$
34. $f(x) = x^5 + 5x^4 + 5x^3 - 2$
35. $f(x) = 0.01x^5 - 0.05x$
36. $f(x) = 0.02x^5 - 0.1x$

37. $f(x) = x^3 - 2x^2 + x + 11$
38. $f(x) = x^3 - x^2 + 12$
39. $f(x) = \sqrt{400 - x^2}$
40. $f(x) = \sqrt{2x - x^2}$

For each function, find all critical numbers and graph the function on an appropriate window using a graphing calculator. For Exercises 41 and 42, also make a sketch from the screen, showing numbers on the axes and coordinates of relative extreme points. You may need to ignore some false lines. (Answers may vary depending on the window chosen.)

41. $f(x) = \dfrac{1}{x^2 - 2x - 8}$

42. $f(x) = \dfrac{1}{x^2 - 2x - 3}$

43. $f(x) = \dfrac{8}{x^2 + 4}$ This curve is called the Witch of Agnesi.

44. $f(x) = \dfrac{4x}{x^2 + 4}$ This curve is called Newton's serpentine.

45. $f(x) = \dfrac{x^2}{x^2 + 1}$ **46.** $f(x) = f(x) = \dfrac{1}{x^2 + 1}$

47. $f(x) = \dfrac{x^2}{x - 3}$ **48.** $f(x) = \dfrac{x^2 - 3x - 1}{x - 5}$

49. $f(x) = \dfrac{10x^2}{x^2 - 5}$ **50.** $f(x) = \dfrac{x^2 + 4}{x}$

51. $f(x) = \dfrac{2x^2}{x^4 + 1}$ **52.** $f(x) = \dfrac{x^4 - 2x^2 + 1}{x^4 + 2}$

Graph each function on an appropriate window using a graphing calculator. (Answers may vary depending on the window chosen.)

53. $f(x) = \dfrac{1}{x - 5}$ **54.** $f(x) = \dfrac{1}{2 + x}$

55. $f(x) = \dfrac{1}{0.1 + 2^{-0.5x}}$ **56.** $f(x) = \dfrac{3x}{\sqrt{x^2 + 1}}$

57. Derive the formula $x = \dfrac{-b}{2a}$ for the x-coordinate of the vertex of parabola $y = ax^2 + bx + c$. [Hint: The slope is zero at the vertex, so finding the vertex means finding the critical number.]

58. Derive the formula $x = -b$ for the x-coordinate of the vertex of parabola $y = a(x+b)^2 + c$. [*Hint:* The slope is zero at the vertex, so finding the vertex means finding the critical number.]

APPLIED EXERCISES

59. BIOMEDICAL: Bacterial Growth A population of bacteria grows to size $p(x) = x^3 - 9x^2 + 24x + 10$ after x hours (for $x \geq 0$). Graph this population curve (based on, if you wish, a calculator graph), showing the coordinates of the relative extreme points.

60. BEHAVIORAL SCIENCE: Learning Curves A learning curve is a function $L(x)$ that gives the amount of time that a person requires to learn x pieces of information. Many learning curves take the form $L(x) = (x-a)^n + b$ (for $x \geq 0$), where a, b, and n are constants. Graph the learning curve $L(x) = (x-2)^3 + 8$ (based on, if you wish, a calculator graph), showing the coordinates of all corresponding points to critical numbers.

61. BUSINESS: Marginal and Average Cost A company's cost function is
$$C(x) = x^2 + 2x + 4$$
dollars, where x is the number of units.
   a. Enter the cost function in y_1 on a graphing calculator.
   b. Define y_2 to be the *marginal* cost function by defining y_2 to be the derivative of y_1 (using NDERIV).
   c. Define y_3 to be the company's *average* cost function, $AC(x) = \dfrac{C(x)}{x}$, by defining
   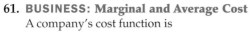
   $y_3 = \dfrac{y_1}{x}$.
   d. Turn off the function y_1 so that it will not be graphed, but graph the marginal cost function y_2 and the average cost function y_3 on the window [0, 10] by [0, 10]. Observe that the marginal cost function pierces the average cost function at its minimum point (use TRACE to see which curve is which function).
   e. To see that the final sentence of part (d) is true in general, change the coefficients in the cost function $C(x)$, or change the cost function to a cubic or some other function [so that $C(x)/x$ has a minimum]. Again turn off the cost function and graph the other two to see that the marginal cost function pierces the average cost function at its minimum. We will return to this observation later.

62. GENERAL: Airplane Flight Path A plane is to take off and reach a level cruising altitude of 5 miles after a horizontal distance of 100 miles, as shown in the diagram below. Find a polynomial flight path of the form $f(x) = ax^3 + bx^2 + cx + d$ by following steps i to iv to determine the constants a, b, c, and d.

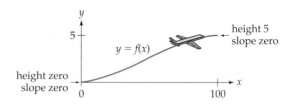

   i. Use the fact that the plane is on the ground at $x = 0$ [that is, $f(0) = 0$] to determine the value of d.
   ii. Use the fact that the path is horizontal at $x = 0$ [that is, $f'(0) = 0$] to determine the value of c.
   iii. Use the fact that at $x = 100$ the height is 5 and the path is horizontal to determine the values of a and b. State the function $f(x)$ that you have determined.
   iv. Use a graphing calculator to graph your function on the window [0, 100] by [0, 10] to verify its shape.

63. GENERAL: Drug Interception Suppose that the cost of a border patrol that intercepts x percent of the illegal drugs crossing a state border is
$$C(x) = \dfrac{600}{100 - x} \quad \text{million dollars (for } x < 100\text{).}$$

a. Graph this function on [0, 100] by [0, 100].
b. Observe that the curve is at first rather flat, but then rises increasingly steeply as x nears 100. Predict what the graph of the derivative would look like.
c. Check your prediction by defining y_2 to be the derivative of y_1 (using NDERIV) and graphing both y_1 and y_2.

64. **GENERAL: Aspirin** Clinical studies have shown that the analgesic (pain-relieving) effect of aspirin is approximately $f(x) = \dfrac{100x^2}{x^2 + 0.02}$ where $f(x)$ is the percentage of pain relief from x grams of aspirin.

a. Graph this "dose-response" curve for doses up to 1 gram, that is, on the window [0, 1] by [0, 100].
b. TRACE along the curve to see that the curve is very close to its maximum height of 100% or 1 by the time x reaches 0.65. This means that there is very little added effect in going above 650 milligrams, the amount of aspirin in two regular tablets, notwithstanding the aspirin companies' promotion of "extra strength" tablets.

[*Note:* Aspirin's dose-response curve is extremely unusual in that it levels off quite early, and, even more unusual, aspirin's effect in protecting against heart attacks even *decreases* as the dosage x increases above about 80 milligrams.]

9.2 GRAPHING USING THE FIRST AND SECOND DERIVATIVES

Introduction

In the previous section we used the first derivative to find the function's slope and relative extreme points and to draw its graph. In this section we will use the *second* derivative to find the "curl" or *concavity* of the curve, and to define the important concept of *inflection point*. The second derivative also gives us a very useful way to distinguish between maximum and minimum points of a curve.

Concavity and Inflection Points

A curve that curls upward is said to be *concave up*, and a curve that curls downward is said to be *concave down*. A point where the concavity *changes* (from up to down or down to up) is called an *inflection point*.

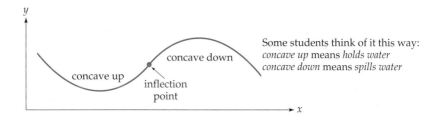

Some students think of it this way:
concave up means *holds water*
concave down means *spills water*

650 CHAPTER 9 FURTHER APPLICATIONS OF DERIVATIVES

Concavity shows how a curve *curls* or *bends* away from straightness.

A straight line (with any slope) has *no concavity*.

However, bending the two ends *upward* makes it *concave up*,

and bending the two ends *downward* makes it *concave down*.

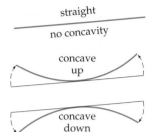

As these pictures show, a curve that is concave *up* lies *above* its tangent, whereas a curve that is concave *down* lies *below* its tangent (except, of course, at the point of tangency).

Practice Problem 1

For each of the following curves, label the parts that are concave up and the parts that are concave down. Then find all inflection points.

a.

b.

➤ Solutions on page 661

How can we use calculus to determine concavity? The key is the second derivative. The second derivative, being the derivative of the derivative, gives the rate of change of the slope, showing whether the slope is increasing or decreasing. That is, $f'' > 0$ means that the slope is increasing, and so the curve must be *concave up* (as in the diagram on the left below). Similarly, $f'' < 0$ means that the slope is decreasing, and so the curve must be *concave down* (as on the right below).

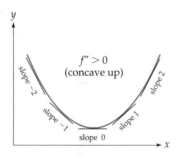

$f'' > 0$ means that the slope is increasing, so f is *concave up*.

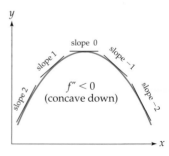

$f'' < 0$ means that the slope is decreasing, so f is *concave down*.

Since an inflection point is where the concavity changes, the second derivative must be negative on one side and positive on the other. Therefore, *at* an inflection point, f'' must be either zero or undefined. All of this may be summarized as follows.

Concavity and Inflection Points

$f'' > 0$ on an interval means that f is *concave up* (curls upward) on that interval.

$f'' < 0$ on an interval means that f is *concave down* (curls downward) on that interval.

An *inflection point* is where the concavity *changes* (f'' must be zero or undefined).

Graphing Calculator Exploration

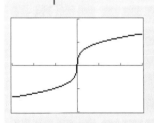

a. Use a graphing calculator to graph $y_1 = \sqrt[3]{x}$ on the window $[-3, 3]$ by $[-2, 2]$. Observe where the curve is concave up and where it is concave down.

b. Use NDERIV to define y_2 to be the derivative of y_1, and y_3 to be the derivative of y_2. Graph y_1 and y_3 (but turn off y_2 so that it will not be graphed).

c. Verify that y_3 (the second derivative) is positive where y_1 is concave up, and negative where y_1 is concave down.

d. Now change y_1 to $y_1 = \dfrac{x^2 + 2}{x^2 + 1}$ and observe that where this curve is concave up or down agrees with where y_3 is positive or negative. According to y_3, how many inflection points does y_1 have? Can you see them on y_1?

To find inflection points, we make a sign diagram for the *second* derivative to show where the concavity changes (where f'' changes sign). An example will make the method clear.

EXAMPLE 1

GRAPHING AND INTERPRETING A COMPANY'S ANNUAL PROFIT FUNCTION

A company's annual profit after x years is $f(x) = x^3 - 9x^2 + 24x$ million dollars (for $x \geq 0$). Graph this function, showing all relative extreme points and inflection points. Interpret the inflection points.

Solution

$$f'(x) = 3x^2 - 18x + 24 \qquad \text{Differentiating}$$
$$= 3(x^2 - 6x + 8) = 3(x-2)(x-4) \qquad \text{Factoring}$$

The critical numbers are $x = 2$ and $x = 4$, and the sign diagram for f' (found in the usual way) is

$f' > 0$	$f' = 0$	$f' < 0$	$f' = 0$	$f' > 0$
	$x = 2$		$x = 4$	
↗	rel max (2, 20)	↘	rel min (4, 16)	↗

To find the inflection points, we calculate the second derivative:

$$f''(x) = 6x - 18 = 6(x - 3) \qquad \text{Differentiating } f'(x) = 3x^2 - 18x + 24$$

This is zero at $x = 3$, which we enter on a sign diagram for the *second* derivative.

	$f'' = 0$		← Behavior of f''
	$x = 3$		← Where f'' is zero or undefined

We use test points to determine the sign of $f''(x) = 6(x - 3)$ on either side of $x = 3$, just as we did for the first derivative.

$f''(2) = 6(2 - 3) < 0 \qquad f''(4) = 6(4 - 3) > 0$

$f'' < 0$	$f'' = 0$	$f'' > 0$
	$x = 3$	
con dn		con up

IP (3, 18)

Concave down, concave up (so concavity *does* change)

IP means inflection point. The 18 comes from substituting $x = 3$ into $f(x) = x^3 - 9x^2 + 24x$

Using a graphing calculator, we would choose an *x*-interval such as [0, 6] (to include the *x*-values on the sign diagrams) and a *y*-interval such as [0, 30] (to include the origin and the *y*-values on the sign diagrams), and graph the function.

By hand, we would plot the relative maximum (⌢), minimum (⌣), and inflection point and sketch the curve according to the sign diagrams, being sure to show the concavity changing at the inflection point.

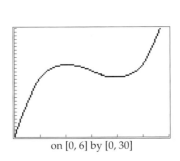
on [0, 6] by [0, 30]

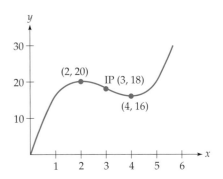

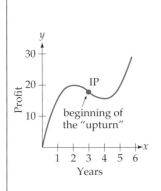

Interpretation of the inflection point: Observe what the graph shows— that the company's profit increased (up to year 2), then decreased (up to year 4), and then increased again. The inflection point at $x = 3$ is where the profit *first began to show signs of improvement.* It marks the end of the period of increasingly steep decline and the first sign of an "upturn," where a clever investor might begin to "buy in."

At an inflection point, the concavity (that is, the sign of f'') must *actually change.* For example, a second derivative sign diagram such as

$f'' < 0$	$f'' = 0$	$f'' < 0$
con dn	$x = 3$	con dn

sign of f'' (concavity) does *not* change

would mean that there is *not* an inflection point at $x = 3$, since the concavity is the same on both sides. For there to be an inflection point, the signs of f'' on the two sides *must be different.*

Practice Problem 2 For each curve on the following page, is there an inflection point? [*Hint:* Does the concavity change?]

Maximizing Profit

The famous economist John Maynard Keynes said, "The engine that drives Enterprise is Profit." Many management problems consist of maximizing profit, and require *constructing* the profit function before maximizing it. Such problems have three economic ingredients. The first is that profit is defined as *revenue minus cost*:

$$\text{Profit} = \text{Revenue} - \text{Cost}$$

The second ingredient is that revenue is *price times quantity*. For example, if a company sells 100 toasters for $25 each, the revenue will obviously be $25 \cdot 100 = \$2500$.

$$\text{Revenue} = \begin{pmatrix}\text{Unit}\\\text{price}\end{pmatrix} \cdot (\text{Quantity})$$

The third economic ingredient reflects the fact that, in general, price and quantity are inversely related: increasing the price decreases sales, whereas decreasing the price increases sales. To put this another way, "flooding the market" with a product drives the price down, whereas creating a shortage drives the price up. If the relationship between the price p and the quantity x that consumers will buy at that price is expressed as a function $p(x)$, it is called the *price function*.*

Price Function

$p(x)$ gives the price p at which consumers will buy exactly x units of the product.

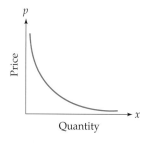

The price function $p(x)$ shows the inverse relation between price p and quantity x.

The price function, relating price and quantity, may be linear or curved (as shown on the left), but it will always be a *decreasing* function.

In actual practice, price functions are very difficult to determine, requiring extensive (and expensive) market research. In this section we will be given the price function. In the next section we will see how to do without price functions, at least in simple cases.

MAXIMIZING A COMPANY'S PROFIT

It costs the American Automobile Company $8000 to produce each automobile, and fixed costs (rent and other expenses that do not depend on the amount of production) are $20,000 per week. The company's price function is $p(x) = 22{,}000 - 70x$, where p is the price at which exactly x cars will be sold.

---

*We will use *lowercase p* for price and *capital P* for profit.

9.3 OPTIMIZATION

a. How many cars should be produced each week to maximize profit?
b. For what price should they be sold?
c. What is the company's maximum profit?

Solution **Revenue** is price times quantity, $R = p \cdot x$:

$$R = p \cdot x = \underbrace{(22{,}000 - 70x)}_{p(x)} x = 22{,}000x - 70x^2 \quad \text{Revenue function } R(x)$$

Replacing p by the price function $p = 22{,}000 - 70x$

Cost is the cost per car ($8000) times the number of cars (x) plus the fixed cost ($20,000):

$$C(x) = 8000x + 20{,}000 \quad \text{(Unit cost)} \cdot \text{(Quantity)} + \text{(Fixed costs)}$$

Profit is revenue minus cost:

$$P(x) = \underbrace{(22{,}000x - 70x^2)}_{R(x)} - \underbrace{(8000x + 20{,}000)}_{C(x)}$$

$$= -70x^2 + 14{,}00x - 20{,}000 \quad \text{Profit function (after simplification)}$$

a. We maximize the profit by setting its derivative equal to zero:

$$P'(x) = -140x + 14{,}000 = 0 \quad \text{Differentiating } P = -70x^2 + 14{,}000x - 20{,}000$$

$$-140x = -14{,}000 \quad \text{Solving}$$

$$x = \frac{-14{,}000}{-140} = 100 \quad \text{Only one critical number}$$

$$P''(x) = -140 \quad \text{From } P'(x) = -140x + 14{,}000$$

The second derivative is negative, so the profit is maximized at the critical number. (If the second derivative had involved x, we would have substituted the critical number $x = 100$.) Since x is the number of cars, the company should produce 100 cars per week (the time period stated in the problem).

b. The selling price p is found from the price function:

$$p = 22{,}000 - 70 \cdot 100 = \$15{,}000 \quad \begin{array}{l} p(x) = 22{,}000 - 70x \\ \text{evaluated at } x = 100 \end{array}$$

c. The maximum profit is found from the profit function:

$$P(100) = -70(100)^2 + 14{,}000(100) - 20{,}000 \quad \begin{array}{l} P(x) = -70x^2 + \\ 14{,}000x - 20{,}000 \\ \text{evaluated at } x = 100 \end{array}$$

$$= \$680{,}000$$

Finally, state the answer clearly in words.

The company should make 100 cars per week and sell them for $15,000 each. The maximum profit will be $680,000.

Actually, automobile dealers seem to prefer prices like $14,999 as if $1 makes a difference

Graphs of the Revenue, Cost, and Profit Functions

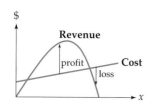

The graphs of the revenue and cost functions are shown on the left. At x-values where revenue is above cost, there is a profit, and where the cost is above the revenue, there is a loss.

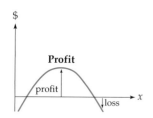

The height of the profit function at any x is the amount by which the revenue is above the cost in the graph on the left. Since profit equals revenue minus cost, we may differentiate each side of $P(x) = R(x) - C(x)$, obtaining

$$P'(x) = R'(x) - C'(x)$$

This shows that setting $P'(x) = 0$ (which we do to maximize profit) is equivalent to setting $R'(x) - C'(x) = 0$, which is equivalent to $R'(x) = C'(x)$. This last equation may be expressed in marginals, $MR = MC$, which is a classic economic criterion for maximum profit.

Classic Economic Criterion for Maximum Profit

At maximum profit:

$$\begin{pmatrix} \text{Marginal} \\ \text{revenue} \end{pmatrix} = \begin{pmatrix} \text{Marginal} \\ \text{cost} \end{pmatrix}$$

MAXIMIZING THE AREA OF AN ENCLOSURE

A farmer has 1000 feet of fence and wants to build a rectangular enclosure along a straight wall. If the side along the wall needs no fence, find the dimensions that make the enclosure as large as possible. Also find the maximum area.

9.3 OPTIMIZATION

Solution The largest enclosure means, of course, the largest area. We let variables stand for the length and width:

x = length (parallel to wall)
y = width (perpendicular to wall)

The problem becomes

$$\text{Maximize } A = xy \qquad \text{Area is length times width}$$

$$\text{subject to } x + 2y = 1000 \qquad \text{One } x \text{ side and two } y \text{ sides from 1000 feet of fence}$$

We must express the area $A = xy$ in terms of one variable. We use

$x = 1000 - 2y$ \qquad Solving $x + 2y = 1000$ for x

$A = xy = \underbrace{(1000 - 2y)}_{x} y = 1000y - 2y^2$ \qquad Substituting $x = 1000 - 2y$ into $A = xy$

$A' = 1000 - 4y = 0$ \qquad Maximizing $A = 1000y - 2y^2$ by setting the derivative equal to zero

$y = 250$ \qquad Solving $1000 - 4y = 0$ for y

Since $A'' = -4$, the second-derivative test shows that the area is indeed *maximized* when $y = 250$. The length x is

$x = 1000 - 2 \cdot 250 = 500$ \qquad Evaluating $x = 1000 - 2y$ at $y = 250$

Length (parallel to the wall) is 500 feet, width (perpendicular to the wall) is 250 feet, and area (length times width) is 125,000 square feet.

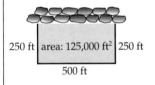

Spreadsheet Exploration

In the preceding example you might think that it does not matter how the fence is laid out as long as all 1000 feet are used. To see that the area enclosed really *does* change, and that the maximum occurs at $y = 250$, the following spreadsheet* calculates the area $A(y) = 1000y - 2y^2$ for y-values from 245 to 255. Notice that the area is largest for $y = 250$, and that for each change of 1 in the y-value the change in the area is smallest for widths closest to 250. This

---
* See the Preface for how to obtain this and other Excel spreadsheets.

verifies *numerically* that the derivative (rate of change) becomes zero as y approaches 250.

	A	B
	Width (perpendicular to wall)	Area Enclosed
1		
2		
3	245	124950
4	246	124968
5	247	124982
6	248	124992
7	249	124998
8	250	125000
9	251	124998
10	252	124992
11	253	124982
12	254	124968
13	255	124950

B8 = =1000*A8-2*A8^2

Based on this spreadsheet, how much area will the farmer lose if he mistakenly makes the width 249 or 251 feet instead of 250 (and therefore the length 502 or 498 feet)? Is this loss significant based on an area of 125,000 square feet? This is characteristic of maximization problems where the slope is zero—begin *near* the maximizing value is essentially as good as being *at* it.

EXAMPLE 5

MAXIMIZING THE VOLUME OF A BOX

An open-top box is to be made from a square sheet of metal 12 inches on each side by cutting a square from each corner and folding up the sides, as in the diagram below. Find the volume of the largest box that can be made in this way.

Square sheet

Corners removed

Side flaps folded up to make open-top box.

Solution Let $x =$ the length of the side of the square cut from each corner.

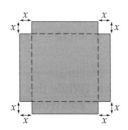

The 12" by 12" square with four x by x corners removed.

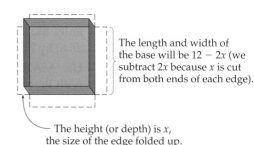

The length and width of the base will be $12 - 2x$ (we subtract $2x$ because x is cut from both ends of each edge).

The height (or depth) is x, the size of the edge folded up.

Therefore, the volume is

$$V(x) = (12 - 2x)(12 - 2x)x \qquad \text{(length)} \cdot \text{(width)} \cdot \text{(height)}$$

Since x is a length, $x > 0$, and since x inches are cut from *both* sides of each 12-inch edge, we must have $2x < 12$, so $x < 6$. The problem becomes

$$\begin{aligned}
\text{Maximize} \quad V(x) &= (12 - 2x)(12 - 2x)x & &\text{on } 0 < x < 6 \\
V(x) &= (144 - 48x + 4x^2)x & &\text{Multiplying out} \\
&= 4x^3 - 48x^2 + 144x & &\text{Multiplying out} \\
V'(x) &= 12x^2 - 96x + 144 & &\text{Differentiating} \\
&= 12(x^2 - 8x + 12) & &\text{Factoring} \\
&= 12(x - 2)(x - 6) & &\text{Factoring}
\end{aligned}$$

$$\text{CN} \begin{cases} x = 2 \\ x = 6 \end{cases} \quad \leftarrow \text{Not in the domain, so we eliminate it}$$

The second derivative is $V''(x) = 24x - 96$, which at $x = 2$ is

$$V''(2) = 48 - 96 < 0$$

Therefore, the volume is *maximized* at $x = 2$.

Maximum volume is 128 cubic inches. $\qquad$ From $V(x)$ evaluated at $x = 2$

APPLIED EXERCISES

21. **BIOMEDICAL: Pollen Count** The average pollen count in New York City on day x of the pollen season is $P(x) = 8x - 0.2x^2$ (for $0 < x < 40$). On which day is the pollen count highest?

22. **GENERAL: Fuel Economy** The fuel economy (in miles per gallon) of an average American compact car is $E(x) = -0.015x^2 + 1.14x + 8.3$, where x is the driving speed (in miles per hour, $20 \leq x \leq 60$). At what speed is fuel economy greatest?

23. **GENERAL: Fuel Economy** The fuel economy (in miles per gallon) of an average American midsized car is $E(x) = -0.01x^2 + 0.62x + 10.4$, where x is the driving speed (in miles per hour, $20 \leq x \leq 60$). At what speed is fuel economy greatest?

24. **GENERAL: Water Power** The proportion of a river's energy that can be obtained from an undershot waterwheel is $E(x) = 2x^3 - 4x^2 + 2x$, where x is the speed of the waterwheel relative to the speed of the river. Find the maximum value of this function on the interval [0, 1], thereby showing that only about 30% of a river's energy can be captured. Your answer should agree with the old millwright's rule that the speed of the wheel should be about one-third of the speed of the river.

25. **GENERAL: Timber Value** The value of a timber forest after t years is $V(t) = 480\sqrt{t} - 40t$ (for $0 \leq t \leq 50$). Find when its value is maximized.

26. **GENERAL: Longevity and Exercise** A recent study of the exercise habits of 17,000 Harvard alumni found that the death rate (deaths per 10,000 person-years) was approximately $R(x) = 5x^2 - 35x + 104$, where x is the weekly amount of exercise in thousands of calories $0 \leq x \leq 40$. Find the exercise level that minimizes the death rate.

27. **ENVIRONMENTAL SCIENCE: Pollution** Two chemical factories are discharging toxic waste into a large lake, and the pollution level at a point x miles from factory A toward factory B is $P(x) = 3x^2 - 72x + 576$ parts per million (for $0 \leq x \leq 50$). Find where the pollution is the least.

28. **BUSINESS: Maximum Profit** City Cycles Incorporated finds that it costs $70 to manufacture each bicycle, and fixed costs are $100 per day. The price function is $p(x) = 270 - 10x$, where p is the price (in dollars) at which exactly x bicycles will be sold. Find the quantity City Cycles should produce and the price it should charge to maximize profit. Also find the maximum profit.

29. **BUSINESS: Maximum Profit** Country Motorbikes Incorporated finds that it costs $200 to produce each motorbike, and that fixed costs are $1500 per day. The price function is $p(x) = 600 - 5x$, where p is the price (in dollars) at which exactly x motorbikes will be sold. Find the quantity Country Motorbikes should produce and the price it should charge to maximize profit. Also find the maximum profit.

30. **BUSINESS: Maximum Profit** A retired potter can produce china pitchers at a cost of $5 each. She estimates her price function to be $p = 17 - 0.5x$, where p is the price at which exactly x pitchers will be sold per week. Find the number of pitchers that she should produce and the price that she should charge in order to maximize profit. Also find the maximum profit.

31. **GENERAL: Parking Lot Design** A company wants to build a parking lot along the side of one of its buildings using 800 feet of fence. If the side along the building needs no fence, what are the dimensions of the largest possible parking lot?

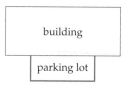

32. **GENERAL: Area** A farmer wants to make two identical rectangular enclosures along a straight river, as in the diagram shown below.

If he has 600 yards of fence, and if the sides along the river need no fence, what should be the dimensions of each enclosure if the total area is to be maximized?

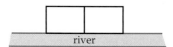

33. GENERAL: Area A farmer wants to make three identical rectangular enclosures along a straight river, as in the diagram shown below. If he has 1200 yards of fence, and if the sides along the river need no fence, what should be the dimensions of each enclosure if the total area is to be maximized?

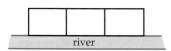

34. GENERAL: Area What is the area of the largest rectangle whose perimeter is 100 feet?

35. GENERAL: Package Design An open-top box is to be made from a square piece of cardboard that measures 18 inches by 18 inches by removing a square from each corner and folding up the sides. What are the dimensions and volume of the largest box that can be made in this way?

36. GENERAL: Gutter Design A long gutter is to be made from a 12-inch-wide strip of metal by folding up the two edges. How much of each edge should be folded up in order to maximize the capacity of the gutter? [*Hint*: Maximizing the capacity means maximizing the cross-sectional area, shown below.]

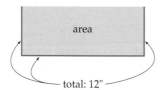

37. GENERAL: Maximizing a Product Find the two numbers whose sum is 50 and whose product is a maximum.

38. GENERAL: Maximizing Area Show that the largest rectangle with a given perimeter is a square.

39. BIOMEDICAL: Coughing When you cough, you are using a high-speed stream of air to clear your trachea (windpipe). During a cough your trachea contracts, forcing the air to move faster, but also increasing the friction. If a trachea contracts from a normal radius of 3 centimeters to a radius of r centimeters, the velocity of the airstream is $V(r) = c(3 - r)r^2$, where c is a positive constant depending on the length and the elasticity of the trachea. Find the radius r that maximizes this velocity. (X-ray pictures verify that the trachea does indeed contract to this radius.)

40. GENERAL: "Efishency" At what speed should a fish swim upstream so as to reach its destination with the least expenditure of energy? The energy depends on the friction of the fish through the water and on the duration of the trip. If the fish swims with velocity v, the energy has been found experimentally to be proportional to v^k (for constant $k > 2$) times the duration of the trip. A distance of s miles against a current of speed c requires time

$$\frac{s}{v - c}$$ (distance divided by speed). The energy required is then proportional to $\frac{v^k s}{v - c}$.

For $k = 3$, minimizing energy is equivalent to minimizing

$$E(v) = \frac{v^3}{v - c}$$

Find the speed v with which the fish should swim in order to minimize its expenditure $E(v)$. (Your answer will depend on c, the speed of the current.)

41. GENERAL: Athletic Fields A running track consists of a rectangle with a semicircle at each end, as shown below. If the perimeter is to be exactly 440 yards, find the dimensions (x and r) that maximize the area of the rectangle. [*Hint:* The perimeter is $2x + 2\pi r$.]

42. **GENERAL: Window Design** A Norman window consists of a rectangle topped by a semicircle, as shown below. If the perimeter is to be 18 feet, find the dimensions (x and r) that maximize the area of the window. [*Hint:* The perimeter is $2x + 2r + \pi r$.]

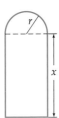

43–44. **GENERAL: Maximizing Capacity of a Computer Disk** Personal computers store information on disks by magnetically writing data on concentric circular "tracks" on the disk. The capacity of the disk depends on the radius x of the innermost track according to the formulas given below (in which x is in inches and y is in megabytes). Enter the formula into a graphing calculator and use MAXIMUM to find the inner radius x that maximizes the disk's capacity y. Also find the maximum capacity. (Computers actually use an inner radius slightly larger than the optimum value, which reduces the capacity slightly.)

43. $y = 0.988x(2.25 - x)$ for $0 \leq x \leq 2.25$ For 5.25-inch high-density double-sided disks)

44. $y = 1.06x(1.68 - x)$ for $0 \leq x \leq 1.68$ For 3.5-inch double-density double-sided disks)

45. **BIOMEDICAL: Bacterial Growth** A chemical reagent is introduced into a bacterial population, and t hours later the number of bacteria (in thousands) is $N(t) = 1000 + 15t^2 - t^3$ (for $0 \leq t \leq 15$).
    a. When will the population be the largest, and how large will it be?
    b. When will the population be growing at the fastest rate, and how fast? (What word applies to such a point?)

46. **GENERAL: Value of a Pulpwood Forest** The value of a pulpwood forest after growing for x years is predicted to be $V(t) = 400x^{0.4} - 40x$ thousand dollars (for $0 \leq x \leq 25$). Use a graphing calculator with MAXIMUM to find when the value will be maximized, and what the maximum value will be.

47–48. **GENERAL: Package Design** Use a graphing calculator (as explained on page 674) to find the side of the square removed and the volume of the box described in Example 5 (pages 672–673) if the square piece of metal is replaced by a:

47. 5 by 7 card (5 inches by 7 inches)

48. 6 by 8 card (6 inches by 8 inches)

You might try constructing such a box.

49. **BUSINESS: Maximum Revenue** A restaurant manager keeps records of the number of bottles of wine he sells per hour at different prices, and finds the following data.

Quantity Sold (x)	Price (p)
10	10
8	16
6	28

a. Enter these numbers into your graphing calculator and find the linear regression line giving price as a function of quantity sold, thereby finding the price function $p(x)$.
b. Find the company's revenue function. [*Hint:* Revenue is price times quantity.]
c. Find the quantity and price that maximize revenue.

50. **BUSINESS: Maximum Revenue** The manager of a campus bookstore keeps records of the number of baseball caps she sells per week as she varies the price, and finds the following data.

Quantity Sold (x)	Price (p)
15	14
12	17
9	23

a. Enter these numbers into your graphing calculator and find the linear regression line giving price as a function of quantity sold, thereby finding the price function $p(x)$.

b. Find the store's revenue function. [*Hint:* Revenue is price times quantity.]
c. Find the quantity and price that maximize revenue.

9.4 FURTHER APPLICATIONS OF OPTIMIZATION

Introduction

In this section we continue to solve optimization problems. In particular, we will see how to maximize a company's profit if we are not given the price function, provided that we are given information describing how price changes will affect sales. We will also see that sometimes x should be chosen as something *other* than the quantity sold.

EXAMPLE 1

FINDING PRICE AND QUANTITY FUNCTIONS

A store can sell 20 bicycles per week at a price of $400 each. The manager estimates that for each $10 price reduction she can sell two more bicycles per week. The bicycles cost the store $200 each. If x stands for *the number of $10 price reductions*, express the price p and the quantity q as functions of x.

Solution   Let

$$x = \text{the number of \$10 price reductions}$$

For example, $x = 4$ means that the price is reduced by $40 (four $10 price reductions). Therefore, in general, if there are x $10 price reductions from the original $400 price, then the price $p(x)$ is

$$p(x) = 400 - 10x \qquad \text{Price}$$

(Original price — Less x $10 price reductions)

The quantity sold $q(x)$ will be

$$q(x) = 20 + 2x \qquad \text{Quantity}$$

(Original quantity — Plus two for each price reduction)

We will return to this example and maximize the store's profit after a practice problem.

> **Practice Problem**
>
> A computer manufacturer can sell 1500 personal computers per month at a price of $3000 each. The manager estimates that for each $200 price reduction he will sell 300 more each month. If x stands for *the number of $200 price reductions*, express the price p and the quantity q as functions of x. ➤ **Solution on page 686**

EXAMPLE 2

MAXIMIZING PROFIT (*Continuation of Example 1*)

Using the information in Example 1, find the price of the bicycles and the quantity that maximize profit. Also find the maximum profit.

Solution In Example 1 we found

$$p(x) = 400 - 10x \qquad \text{Price}$$
$$q(x) = 20 + 2x \qquad \text{Quantity sold at that price}$$

Revenue is price times quantity, $p(x) \cdot q(x)$:

$$R(x) = (400 - 10x)(20 + 2x) \qquad p(x)q(x)$$
$$= 8000 + 600x - 20x^2 \qquad \text{Multiplying out and simplifying}$$

The cost function is unit cost times quantity:

$$C(x) = 200\underbrace{}(20 + 2x) = 4000 + 400x \qquad \text{If there were a fixed cost we would add it}$$

Unit cost / Quantity $q(x)$

Profit is revenue minus cost:

$$P(x) = \underbrace{(8000 + 600x - 20x^2)}_{R(x)} - \underbrace{(4000 + 400x)}_{C(x)}$$
$$= 4000 + 200x - 20x^2 \qquad \text{Simplifying}$$

We maximize profit by setting the derivative equal to zero:

$$200 - 40x = 0 \qquad \text{Differentiating } P = 4000 + 200x - 20x^2$$

The critical number is $x = 5$. The second derivative, $P''(x) = -40$, shows that the profit is *maximized* at $x = 5$. Since $x = 5$ is the number of $10 price reductions, the original price of $400 should be lowered by $50 ($10 five times), from $400 to $350. The quantity sold is found from the quantity function:

$$q(5) = 20 + 2 \cdot 5 = 30 \qquad q(x) = 20 + 2x \text{ at } x = 5$$

Finally, we state the answer clearly.

Sell the bicycles for $350 each.
Quantity sold: 30 per week.
Maximum profit: $4500.   From $P(x) = 4000 + 200x - 20x^2$ at $x = 5$

Exercise 25 will show how a graphing calculator enables you to modify the problem (such as changing the cost per bicycle) and then immediately recalculate the new answer.

Choosing Variables

Notice that in Example 2 we did not choose x to be the quantity sold, but instead to be *the number of $10 price reductions*. (Therefore, a negative x would have meant a price *increase*.) We chose this x because from it we could easily calculate both the new price and the new quantity. Other choices for x are also possible, but in situations where a price change will make one quantity rise and another fall, it is often easiest to choose x to be the *number of such changes*.

EXAMPLE 3 **MAXIMIZING HARVEST SIZE**

An orange grower finds that if he plants 80 orange trees per acre, each tree will yield 60 bushels of oranges. He estimates that for each additional tree that he plants per acre, the yield of each tree will decrease by 2 bushels. How many trees should he plant per acre to maximize his harvest?

Solution We take x equal to the number of "changes"—that is, let

$$x = \text{ the number of added trees per acre}$$

With x extra trees per acre,

Trees per acre: $80 + x$ Original 80 plus x more
Yield per tree: $60 - 2x$ Original yield less 2 per extra tree

Therefore, the total yield per acre will be

$$Y(x) = \underbrace{(60 - 2x)}_{\text{Yield per tree}}\underbrace{(80 + x)}_{\text{Tree per acre}} = 4800 - 100x - 2x^2$$

We maximize this by setting the derivative equal to zero:

$-100 - 4x = 0$ Differentiating $Y = 4800 - 100x - 2x^2$

$x = -25$ Negative!

The number of *added* trees is negative, meaning that the grower should plant 25 *fewer* trees per acre. The second derivative, $Y''(x) = -4$, shows that the yield is indeed maximized at $x = -25$. Therefore:

Plant 55 trees per acre. $80 - 25 = 55$

Earlier problems involved maximizing areas and volumes using only a fixed amount of material (such as a fixed length of fence). Instead, we could minimize the amount of materials for a fixed area or volume.

EXAMPLE 4

MINIMIZING PACKAGE MATERIALS

A moving company wishes to design an open-top box with a square base whose volume is exactly 32 cubic feet. Find the dimensions of the box requiring the least amount of materials.

Solution   The base is square, so we define

x = length of side of base
y = height

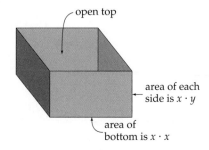

The volume (length·width·height) is $x \cdot x \cdot y$ or $x^2 y$, which (according to the problem) must equal 32 cubic feet:

$$x^2 y = 32$$

9.4 FURTHER APPLICATIONS OF OPTIMIZATION

The box consists of a bottom (area x^2) and four sides (each of area xy). Minimizing the amount of materials means minimizing the surface area of the bottom and four sides:

$$A = x^2 + 4xy \qquad \binom{\text{Area of}}{\text{bottom}} + \binom{\text{Area of}}{\text{four sides}}$$

As usual, we must express this area in terms of just *one* variable, so we use the volume requirement to express y in terms of x:

$$y = \frac{32}{x^2} \qquad \text{Solving } x^2y = 32 \text{ for } y$$

The area function becomes

$$A = x^2 + 4x\frac{32}{x^2} \qquad A = x^2 + 4xy \quad \text{with } y \text{ replaced by } \frac{32}{x^2}$$

$$= x^2 + \frac{128}{x} \qquad \text{Simplifying}$$

$$= x^2 + 128x^{-1} \qquad \text{Writing } \frac{1}{x} \text{ as } x^{-1}$$

We minimize this by finding the critical number:

$$A'(x) = 2x - 128x^{-2} \qquad \text{Differentiating } A = x^2 + 128x^{-1}$$

$$2x - \frac{128}{x^2} = 0 \qquad \text{Setting the derivative equal to zero}$$

$$2x^3 - 128 = 0 \qquad \text{Multiplying by } x^2 \text{ (since } x > 0\text{)}$$

$$x^3 = 64 \qquad \text{Adding 128 and then dividing by 2}$$

$$x = 4 \qquad \text{Taking cube roots}$$

The second derivative

$$A''(x) = 2 + 256x^{-3} = 2 + \frac{256}{x^3} \qquad \text{From } A'(x) = 2x - 128x^{-2}$$

is positive at $x = 4$, so the area is minimized. Therefore, the dimensions are

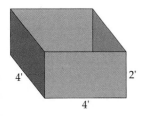

Base: 4 feet on each side

Height: 2 feet

Height from $y = 32/x^2$ at $x = 4$

Graphing Calculator Exploration

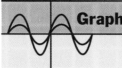

Use a graphing calculator to solve the previous example as follows:

a. Enter the area function to be minimized as $y_1 = x^2 + 128/x$.
b. Graph y_1 for x-values [0, 10], using TABLE or TRACE to find where y_1 seems to "bottom out" to determine an appropriate y-interval.
c. Graph the function on the window determined in part (b) and then use MINIMUM to find the minimum value. Your answer should agree with that found above. (Your calculator may give an inexact answer that needs to be rounded.)

Notice that either way required first finding the function to be minimized.

Minimizing the Cost of Materials

How would this problem have changed if the material for the bottom of the box had been more costly than the material for the sides? If, for example, the material for the sides cost $2 per square foot and the material for the base, needing greater strength, cost $4 per square foot, then instead of simply minimizing the surface area, we would minimize *total cost*:

$$\text{Cost} = \binom{\text{Area of}}{\text{bottom}}\binom{\text{Cost of bottom}}{\text{per square foot}} + \binom{\text{Area of}}{\text{sides}}\binom{\text{Cost of sides}}{\text{per square foot}}$$

Since the areas would be just as before, this cost would be

$$\text{Cost} = (x^2)(4) + (4xy)(2) = 4x^2 + 8xy$$

From here on we would proceed just as before, eliminating the y (using the volume relationship $x^2y = 32$) and then setting the derivative equal to zero.

Maximizing Tax Revenue

Governments raise money by collecting taxes. If a sales tax or an import tax is too high, trade will be discouraged and tax revenues will fall. If, on the other hand, the tax rate is too low, trade may flourish but tax revenues will again fall. Economists often want to determine the tax rate that maximizes revenue for the government. To do this, they must first predict the relationship between the tax on an item and the total sales of the item.

Suppose, for example, that the relationship between the tax rate t on an item and its total sales S is

$$S(t) = 9 - 20\sqrt{t} \qquad \begin{array}{l} t = \text{tax rate } (0 \leq t \leq 0.20) \\ S = \text{total sales (millions of dollars)} \end{array}$$

If the tax rate is $t = 0$ (0%), then the total sales will be

$$S(0) = 9 - 20\sqrt{0} = 9 \qquad \text{\$9 million}$$

If the tax rate is raised to $t = 0.16$ (16%), then sales will be

$$S(0.16) = 9 - 20\sqrt{0.16}$$
$$= 9 - (20)(0.4) = 9 - 8 = 1 \qquad \text{\$1 million}$$

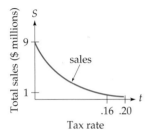

That is, raising the tax rate from 0% to 16% will discourage $8 million worth of sales. The graph of $S(t)$ on the left shows how total sales decrease as the tax rate increases. With such information (which may be found from historical data), one can find the tax rate that maximizes revenue.

EXAMPLE 5   MAXIMIZING TAX REVENUE

If economists predict that the relationship between the tax rate t on an item and the total sales S of that item (in millions of dollars) is

$$S(t) = 9 - 20\sqrt{t} \qquad \text{For } 0 \leq t \leq 0.20$$

find the tax rate that maximizes revenue to the government.

Solution

The government's revenue R is the tax rate t times the total sales $S(t) = 9 - 20\sqrt{t}$:

$$R(t) = t\underbrace{(9 - 20t^{1/2})}_{S(t)} = 9t - 20t^{3/2}$$

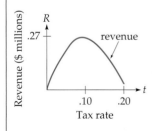

The graph of this function is shown on the left. To maximize it, we set its derivative equal to zero:

$9 - 30t^{1/2} = 0$	Derivative of $9t - 20t^{3/2}$
$9 = 30t^{1/2}$	Adding $30t^{1/2}$ to each side
$t^{1/2} = \dfrac{9}{30} = 0.3$	Switching sides and dividing by 30
$t = 0.09$	Squaring both sides

This gives a tax rate of $t = 9\%$. The second derivative,

$$R''(t) = -30 \cdot \frac{1}{2} t^{-1/2} = -\frac{15}{\sqrt{t}} \qquad \text{From } R' = 9 - 30t^{1/2}$$

is negative at $t = 0.09$, showing that the revenue is maximized. Therefore:

A tax rate of 9% maximizes revenue for the government.

Graphing Calculator Exploration

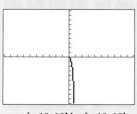

on $[-10, 10]$ by $[-10, 10]$

The graph of the function from Example 5, $y_1 = 9x - 20x^{3/2}$ (written in x instead of t for ease of entry) is shown on the left on the standard window $[-10, 10]$ by $[-10, 10]$. This might lead you to believe, erroneously, that the function is maximized at the endpoint $(0, 0)$.

a. Why does this graph not look like the graph at the bottom of the previous page? [*Hint:* Look at the scale.]

b. Can you find a window on which your graphing calculator will show a graph like the one on the left?

This example illustrates one of the pitfalls of graphing calculators—the part of the curve where the "action" takes place may be entirely hidden in one pixel. Calculus, on the other hand, will *always* find the critical value, no matter where it is, and then a graphing calculator can be used to confirm your answer by showing the graph on an appropriate window.

▶ **Solution to Practice Problem**

Price: $p(x) = 3000 - 200x$
Quantity: $q(x) = 1500 + 300x$

9.4 EXERCISES

1. **BUSINESS: Maximum Profit** An automobile dealer can sell 12 cars per day at a price of $15,000. He estimates that for each $300 price reduction he can sell two more cars per day. If each car costs him $12,000, and fixed costs are $1000, what price should he charge to maximize his profit? How many cars will he sell at this price? [*Hint:* Let x = the number of $300 price reductions.]

2. **BUSINESS: Maximum Profit** An automobile dealer can sell four cars per day at a price of $12,000. She estimates that for each $200 price reduction she can sell two more cars per day. If each car costs her $10,000, and her fixed costs are $1000, what price should she charge to maximize her profit? How many cars will she sell at this price? [*Hint:* Let x = the number of $200 price reductions.]

3. **BUSINESS: Maximum Revenue** An airline finds that if it prices a cross-country ticket at $200, it will sell 300 tickets per day. It estimates that each $10 price reduction will result in 30 more tickets sold per day. Find the ticket price (and the number of tickets sold) that will maximize the airline's revenue.

4. **ECONOMICS: Oil Prices** An oil-producing country can sell 1 million barrels of oil a day at a price of $25 per barrel. If each $1 price increase will result in a sales decrease of 50,000 barrels per day, what price will maximize the country's revenue? How many barrels will it sell at that price?

5. **BUSINESS: Maximum Revenue** Rent-A-Reck Incorporated finds that it can rent 60 cars if it charges $80 for a weekend. It estimates that for each $5 price increase it will rent three fewer cars. What price should it charge to maximize its revenue? How many cars will it rent at this price?

6. **GENERAL: Maximum Yield** A peach grower finds that if he plants 40 trees per acre, each tree will yield 60 bushels of peaches. He also estimates that for each additional tree that he plants per acre, the yield of each tree will decrease by 2 bushels. How many trees should he plant per acre to maximize his harvest?

7. **GENERAL: Maximum Yield** An apple grower finds that if she plants 20 trees per acre, each tree will yield 90 bushels of apples. She also estimates that for each additional tree that she plants per acre, the yield of each tree will decrease by 3 bushels. How many trees should she plant per acre to maximize her harvest?

8. **GENERAL: Fencing** A farmer has 1200 feet of fence and wishes to build two identical rectangular enclosures, as in the diagram. What should be the dimensions of each enclosure if the total area is to be a maximum?

9. **GENERAL: Minimum Materials** An open-top box with a square base is to have a volume of 4 cubic feet. Find the dimensions of the box that can be made with the smallest amount of material.

10. **GENERAL: Minimum Materials** An open-top box with a square base is to have a volume of 108 cubic inches. Find the dimensions of the box that can be made with the smallest amount of material.

11. **GENERAL: Largest Postal Package** The U.S. Postal Service will accept a package if its length plus its girth (the distance all the way around) does not exceed 84 inches. Find the dimensions and volume of the largest package with a square base that can be mailed.

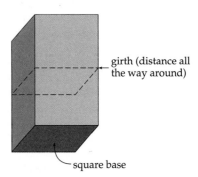

12. **GENERAL: Fencing** A homeowner wants to build, along his driveway, a garden surrounded by a fence. If the garden is to be 800 square feet, and the fence along the driveway costs $6 per foot whereas on the other three sides it costs only $2 per foot, find the

dimensions that will minimize the cost. Also find the minimum cost.

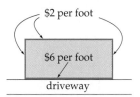

13. **GENERAL: Fencing** A homeowner wants to build, along her driveway, a garden surrounded by a fence. If the garden is to be 5000 square feet, and the fence along the driveway costs $6 per foot whereas on the other three sides it costs only $2 per foot, find the dimensions that will minimize the cost. Also find the minimum cost. (See the diagram above.)

14–15. ECONOMICS: Tax Revenue Suppose that the relationship between the tax rate t on imported shoes and the total sales S (in millions of dollars) is given by the function below. Find the tax rate t that maximizes revenue for the government.

14. $S(t) = 4 - 6\sqrt[3]{t}$ 15. $S(t) = 8 - 15\sqrt[3]{t}$

16. **BIOMEDICAL: Drug Concentration** If the amount of a drug in a person's blood after t hours is $f(t) = t/(t^2 + 9)$, when will the drug concentration be the greatest?

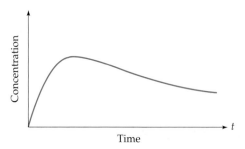

17. **GENERAL: Wine Storage** A case of vintage wine appreciates in value each year, but there is also an annual storage charge. The value of the wine after t years is $V(t) = 2000 + 96\sqrt{t} - 12t$ dollars (for $0 \le t \le 25$). Find the storage time that will maximize the value of the wine.

18. **GENERAL: Bus Shelter Design** A bus stop shelter, consisting of two square sides, a back, and a roof, as shown below, is to have volume 1024 cubic feet. What are the dimensions that require the least amount of materials?

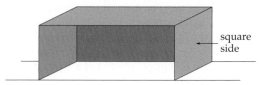

19. **GENERAL: Area** Show that the rectangle of fixed area whose perimeter is a minimum is a square.

20. **POLITICAL SCIENCE: Campaign Expenses** A politician estimates that by campaigning in a county for x days, she will gain $2x$ (thousand) votes, but her campaign expenses will be $5x^2 + 500$ dollars. She wants to campaign for the number of days that maximizes the number of votes per dollar, $f(x) = \dfrac{2x}{5x^2 + 500}$
For how many days should she campaign?

21. **GENERAL: Page Design** A page of 96 square inches is to have margins of 1 inch on either side and $1\frac{1}{2}$ inches at the top and bottom, as in the diagram. Find the dimensions of the page that maximize the print area.

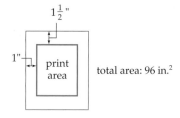

22. **BIOMEDICAL: Contagion** If an epidemic spreads through a town at a rate that is proportional to the number of infected people and to the number of uninfected people, then the rate is $R(x) = cx(p - x)$, where x is the number of infected people and c and p (the population) are positive constants. Show that the rate $R(x)$ is greatest when half of the population is infected.

23. **BIOMEDICAL: Contagion** If an epidemic spreads through a town at a rate that is proportional to the number of uninfected people

and to the square of the number of infected people, then the rate is $R(x) = cx^2(p - x)$, where x is the number of infected people and c and p (the population) are positive constants. Show that the rate $R(x)$ is greatest when two-thirds of the population is infected.

24. **BUSINESS: Maximizing Profit** An electronics store can sell 35 cellular telephones per week at a price of $200. The manager estimates that for each $20 price reduction she can sell 9 more per week. The telephones cost the store $100 each, and fixed costs are $700 per week.

    a. If x is the number of $20 price reductions, find the price $p(x)$ and enter it in y_1. Then enter the quantity function $q(x)$ in y_2.
    b. Make y_3 the revenue function by defining $y_3 = y_1 y_2$ (price times quantity).
    c. Make y_4 the cost function by defining y_4 as unit cost times y_2 plus fixed costs.
    d. Make y_5 the profit function by defining $y_5 = y_3 - y_4$ (revenue minus cost).
    e. Turn off y_1, y_2, y_3, and y_4 and graph the profit function y_5 for x-values $[-10, 10]$, using TABLE or TRACE to find an appropriate y-interval. Then use MAXIMUM to maximize it.
    f. Use EVALUATE to find the price and the quantity for this maximum profit.

25. **BUSINESS: Exploring a Profit Maximization Problem** Use a graphing calculator to further explore Example 2 (pages 680–681) as follows:

    a. Enter the price function $y_1 = 400 - 10x$ and the quantity function $y_2 = 20 + 2x$ into your graphing calculator.
    b. Make y_3 the revenue function by defining $y_3 = y_1 y_2$ (price times quantity).
    c. Make y_4 the cost function by defining $y_4 = 200 y_2$ (unit cost times quantity).
    d. Make y_5 the profit function by defining $y_5 = y_3 - y_4$ (revenue minus cost).
    e. Turn off y_1, y_2, y_3, and y_4 and graph the profit function y_5 on the window $[0, 10]$ by $[0, 10{,}000]$ and then use MAXIMUM to maximize it. Your answer should agree with that found in Example 2.

    Now change the problem!

    f. What if the store finds that it can buy the bicycles from another wholesaler for $150 instead of $200? In y_4, change the 200 to 150. Then graph the profit y_5 (you may have to turn off y_4 again) and maximize it. Find the new price and quantity by evaluating y_1 and y_2 (using EVALUATE) at the new x-value.
    g. What if cycling becomes more popular and the manager estimates that she can sell 30 instead of 20 bicycles per week at the original $400 price? Go back to y_2 and change 20 to 30 (keeping the change made earlier) and graph and find the price and quantity that maximize profit now.

    Notice how flexible this setup is for changing any of the numbers.

9.5 OPTIMIZING LOT SIZE AND HARVEST SIZE

Introduction

In this section we discuss two important applications of optimization, one economic and one ecological. The first concerns the most efficient way for a business to order merchandise (or for a manufacturer to produce merchandise), and the second concerns the preservation of animal populations that are harvested by people. Either of these applications may be read independently of the other.

Minimizing Inventory Costs

A business encounters two kinds of costs in maintaining inventory: storage costs (warehouse and insurance costs for merchandise not yet sold) and reorder costs (delivery and bookkeeping costs for each order). For example, if a furniture store expects to sell 250 sofas in a year, it could order all 250 at once (incurring high storage costs), or it could order them in many small lots, say 50 orders of five each, spaced throughout the year (incurring high reorder costs). Obviously, the best order size (or "lot" size) is the one that minimizes the total of storage plus reorder costs.

EXAMPLE 1

MINIMIZING INVENTORY COSTS

A furniture showroom expects to sell 250 sofas a year. Each sofa costs the store $300, and there is a fixed charge of $500 per order. If it costs $100 to store a sofa for a year, how large should each order be and how often should orders be placed to minimize inventory costs?

Solution Let

$$x = \text{lot size} \qquad x \text{ is the number of sofas in each order}$$

Storage Costs: If the sofas sell steadily throughout the year, and if the store reorders x more whenever the stock runs out, then its inventory during the year looks like the following graph.

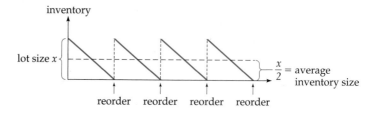

Notice that the inventory level varies from the lot size x down to zero, with an average inventory of $x/2$ sofas throughout the year. Because it costs $100 to store a sofa for a year, the total (annual) storage costs are

$$\begin{pmatrix}\text{Storage} \\ \text{costs}\end{pmatrix} = \begin{pmatrix}\text{Storage} \\ \text{per item}\end{pmatrix} \cdot \begin{pmatrix}\text{Average num-} \\ \text{ber of items}\end{pmatrix}$$

$$= 100 \cdot \frac{x}{2} = 50x$$

Reorder Costs: Each sofa costs $300, so an order of lot size x costs $300x$, plus the fixed order charge of $500:

$$\begin{pmatrix} \text{Cost} \\ \text{per order} \end{pmatrix} = 300x + 500$$

The yearly supply of 250 sofas, with x sofas in each order, requires $\dfrac{250}{x}$ orders. (For example, 250 sofas at 5 per order require $\dfrac{250}{5} = 50$ orders.) Therefore, the yearly reorder costs are

$$\begin{pmatrix} \text{Reorder} \\ \text{costs} \end{pmatrix} = \begin{pmatrix} \text{Cost} \\ \text{per order} \end{pmatrix} \cdot \begin{pmatrix} \text{Number} \\ \text{of orders} \end{pmatrix}$$

$$= (300x + 500) \cdot \left(\dfrac{250}{x} \right)$$

Total Cost: $C(x)$ is storage costs plus reorder costs:

$$C(x) = \begin{pmatrix} \text{Storage} \\ \text{costs} \end{pmatrix} + \begin{pmatrix} \text{Reorder} \\ \text{costs} \end{pmatrix}$$

$$= 100 \dfrac{x}{2} + (300x + 500) \left(\dfrac{250}{x} \right) \quad \text{Using the storage and reorder costs found earlier}$$

$$= 50x + 75{,}000 + 125{,}000 x^{-1} \quad \text{Simplifying}$$

To minimize $C(x)$, we differentiate:

$$C'(x) = 50 - 125{,}000 x^{-2} = 50 - \dfrac{125{,}000}{x^2} \quad \begin{array}{l} \text{Differentiating} \\ C = 50x + 75{,}000 + \\ 125{,}000 x^{-1} \end{array}$$

$$50 - \dfrac{125{,}000}{x^2} = 0 \quad \begin{array}{l} \text{Setting the derivative} \\ \text{equal to zero} \end{array}$$

$$50 x^2 = 125{,}000 \quad \begin{array}{l} \text{Multiplying by } x^2 \\ \text{and adding } 125{,}000 \\ \text{to each side} \end{array}$$

$$x^2 = \dfrac{125{,}000}{50} = 2500 \quad \begin{array}{l} \text{Dividing each} \\ \text{side by 50} \end{array}$$

$$x = 50 \quad \begin{array}{l} \text{Taking square roots} \\ (x > 0) \text{ gives} \\ \text{lot size 50} \end{array}$$

$$C''(x) = 25{,}000 x^{-3} = 250{,}000 \dfrac{1}{x^3} \quad \begin{array}{l} C'' \text{ is positive, so} \\ C \text{ is minimized} \\ \text{at } x = 50 \end{array}$$

At 50 sofas per order, the yearly 250 will require $\dfrac{250}{50} = 5$ orders. Therefore:

> Lot size is 50 sofas, with orders placed five times a year.

Reproduction Function

A reproduction function $f(p)$ gives the population a year from now if the current population is p.

For example, the reproduction function $f(p) = -\frac{1}{4}p^2 + 3p$ (where p and $f(p)$ are measured in thousands) means that if the population is now $p = 6$ (thousand), then a year from now the population will be

$$f(6) = -\frac{1}{4} 6^2 + 3 \cdot 6 = -9 + 18 = 9 \quad \text{(thousand)}$$

Therefore, during the year the population will increase from 6000 to 9000.

If, on the other hand, the present population is $p = 10$ (thousand), a year later the population will be

$$f(10) = -\frac{1}{4} \cdot 10^2 + 3 \cdot 10 = -25 + 30 = 5 \quad \text{(thousand)}$$

That is, during the year the population will decline from 10,000 to 5000 (perhaps because of inadequate food to support such a large population). In actual practice, reproduction functions are very difficult to calculate, but can sometimes be estimated by analyzing previous population and harvest data.*

Suppose that we have a reproduction function f and a current population of size p, which will therefore grow to size $f(p)$ next year. The *amount of growth* in the population during that year is

$$\begin{pmatrix} \text{Amount} \\ \text{of growth} \end{pmatrix} = f(p) - p$$

where $f(p)$ is Next year's population and p is Current population.

Harvesting this amount removes only the *growth*, returning the population to its former size p. The population will then repeat this growth, and taking the same harvest $f(p) - p$ will cause this situation to repeat itself year after year. The quantity $f(p) - p$ is called the sustainable yield.

Sustainable Yield

For reproduction function $f(p)$, the sustainable yield is

$$Y(p) = f(p) - p$$

---

*For more information, see J. Blower, L. Cook, and J. Bishop, *Estimating the Size of Animal Populations* (London: George Allen and Unwin Ltd., 1981).

We want the population size p that maximizes the sustainable yield $Y(p)$. To maximize $Y(p)$, we set its derivative equal to zero:

$Y'(p) = f'(p) - 1 = 0$ Derivative of $Y = f(p) - p$

$f'(p) = 1$ Solving for $f'(p)$

For a given reproduction function $f(p)$, we find the maximum sustainable yield by solving this equation (provided that the second-derivative test gives $Y''(p) = f''(p) < 0$).

Maximum Sustainable Yield

For reproduction function $f(p)$, the population p that results in the maximum sustainable yield is the solution to

$$f'(p) = 1$$

(provided that $f'(p) < 0$). The maximum sustainable yield is then

$$Y(p) = f(p) - p$$

Once we calculate the population p that gives the maximum sustainable yield, we wait until the population reaches this size and then harvest, year after year, an amount $Y(p)$.

Note that to find the maximum sustainable yield, we set the derivative $f'(p)$ equal to 1, not 0. This is because we are maximizing not the reproductive function $f(p)$ but rather the yield function $Y(p) = f(p) - p$.

EXAMPLE 3 **FINDING MAXIMUM SUSTAINABLE YIELD**

The reproduction function for the American lobster in an East Coast fishing area is $f(p) = -0.02p^2 + 2p$ (where p and $f(p)$ are in thousands). Find the population p that gives the maximum sustainable yield and find the size of the yield.

Solution We set the derivative of the reproduction function equal to 1:

$f'(p) = -0.04p + 2 = 1$ Differentiating $f(p) = -0.02p^2 + 2p$

$-0.04p = -1$ Subtracting 2 from each side

$p = \dfrac{-1}{-0.04} = 25$ Dividing by -0.04

The second derivative is $f''(p) = -0.04$, which is negative, showing that $p = 25$ (thousand) is the population that gives the maximum sustainable yield. The actual yield is found from the yield function $Y(p) = f(p) - p$:

$$Y(p) = \underbrace{-0.02p^2 + 2p}_{f(p)} - p = -0.02p^2 + p$$

$Y(25) = -0.02(25)^2 + 25$ Evaluating at $p = 25$
$= -12.5 + 25 = 12.5$ (thousand)

The population size for the maximum sustainable yield is 25,000, and the yield is 12,500 lobsters.   25,000 from $p = 25$

Graphing Calculator Exploration

Solve the previous example on a graphing calculator as follows:
a. Enter the reproduction function as $y_1 = -0.02x^2 + 2x$.
b. Define y_2 to be the derivative of y_1 (using NDERIV).
c. Define $y_3 = 1$.
d. Turn off y_1 and graph y_2 and y_3 on the window [0, 40] by [0, 2].
e. Use INTERSECT to find where y_2 and y_3 meet, thereby solving $f' = 1$. (You should find $x = 25$, as above.)
f. Find the yield by evaluating $y_1 - x$ at the x-value found in part (e).

In this problem, solving "by hand" was probably easier, but this graphing calculator method may be preferable if the reproduction function is more complicated.

9.5 EXERCISES

Lot Size

1. A supermarket expects to sell 4000 boxes of sugar in a year. Each box costs $2, and there is a fixed delivery charge of $20 per order. If it costs $1 to store a box for a year, what is the order size and how many times a year should the orders be placed to minimize inventory costs?

2. A supermarket expects to sell 5000 boxes of rice in a year. Each box costs $2, and there is a fixed delivery charge of $50 per order. If it costs $2 to store a box for a year, what is the order size and how many times a year should the orders be placed to minimize inventory costs?

3. A liquor warehouse expects to sell 10,000 bottles of scotch whiskey in a year. Each bottle costs $12, plus a fixed charge of $125 per order. If it costs $10 to store a bottle for a year, how many bottles should be ordered at a time and how many orders should the warehouse place in a year to minimize inventory costs?

4. A wine warehouse expects to sell 30,000 bottles of wine in a year. Each bottle costs $9, plus a fixed charge of $200 per order. If it costs $3 to store a bottle for a year, how many bottles should be ordered at a time and how many orders should the warehouse place in a year to minimize inventory costs?

5. An automobile dealer expects to sell 800 cars a year. The cars cost $9000 each plus a fixed charge of $1000 per delivery. If it costs $1000 to store a car for a year, find the order size and the number of orders that minimize inventory costs.

6. An automobile dealer expects to sell 400 cars a year. The cars cost $11,000 each plus a fixed charge of $500 per delivery. If it costs $1000 to store a car for a year, find the order size and the number of orders that minimize inventory costs.

Production Runs

7. A toy manufacturer estimates the demand for a game to be 2000 per year. Each game costs $3 to manufacture, plus setup costs of $500 for each production run. If a game can be stored for a year for a cost of $2, how many should be manufactured at a time and how many production runs should there be to minimize costs?

8. A toy manufacturer estimates the demand for a doll to be 10,000 per year. Each doll costs $5 to manufacture, plus setup costs of $800 for each production run. If it costs $4 to store a doll for a year, how many should be manufactured at a time and how many production runs should there be to minimize costs?

9. A producer of audio tapes estimates the yearly demand for a tape to be 1,000,000. It costs $800 to set up the machinery for the tape, plus $10 for each tape produced. If it costs the company $1 to store a tape for a year, how many should be produced at a time and how many production runs will be needed to minimize costs?

10. A compact disc manufacturer estimates the yearly demand for a CD to be 10,000. It costs $400 to set the machinery for the CD, plus $3 for each CD produced. If it costs the company $2 to store a CD for a year, how many should be burned at a time and how many production runs will be needed to minimize costs?

Maximum Sustainable Yield

11. Marine ecologists estimate the reproduction curve for swordfish in the Georges Bank fishing grounds to be $f(p) = -0.01p^2 + 5p$, where p and $f(p)$ are in hundreds. Find the population that gives the maximum sustainable yield, and the size of the yield.

12. The reproduction function for the Hudson Bay lynx is estimated to be $f(p) = -0.02p^2 + 5p$, where p and $f(p)$ are in thousands. Find the population that gives the maximum sustainable yield, and the size of the yield.

13. The reproduction function for the Antarctic blue whale is estimated to be $f(p) = -0.0004p^2 + 1.06p$, where p and $f(p)$ are in thousands. Find the population that gives the maximum sustainable yield, and the size of the yield.

14. The reproduction function for the Canadian snowshoe hare is estimated to be $f(p) = -0.025p^2 + 4p$, where p and $f(p)$ are in thousands. Find the population that gives the maximum sustainable yield, and the size of the yield.

15. A conservation commission estimates the reproduction function for rainbow trout in a

large lake to be $f(p) = 50\sqrt{p}$, where p and $f(p)$ are in thousands. Find the population that gives the maximum sustainable yield, and the size of the yield.

16. The reproduction function for oysters in a large bay is $f(p) = 30\sqrt[3]{p^2}$, where p and $f(p)$ are in pounds. Find the size of the population that gives the maximum sustainable yield, and the size of the yield.

17. Suppose that the reproduction function for Pacific salmon in a northwest fishing area is $f(p) = 24\sqrt[3]{p} - 9\sqrt{p}$ (for $0 \le p < 50$), where p and $f(p)$ are in thousands. Use a graphing calculator to find the population that gives the maximum sustainable yield, and the size of the yield. [*Hint:* Follow the steps in the Graphing Calculator Exploration on page 696 using this reproduction function and x-interval $[0, 50]$.]

18. BUSINESS: Exploring a Lot Size Problem Use a graphing calculator to explore Example 1 (pages 690–691) as follows:

a. Enter the total cost function, unsimplified, as $y_1 = 100(x/2) + (300x + 500)(250/x)$.

b. Graph y_1 on the window $[0, 200]$ by $[0, 150,000]$ and use MINIMUM to minimize it. Your answer should agree with the $x = 50$ found in Example 1.

c. Suppose that business improves, and the showroom expects to sell 350 per year instead of 250. In y_1, change the 250 to 350 and minimize it.

d. Suppose that a modest recession decreases sales (so change the 350 back to 250), and that inflation has driven the cost of storage up to \$125 (so change the 100 to 125), and minimize y_1.

9.6 IMPLICIT DIFFERENTIATION AND RELATED RATES

Introduction

A function written in the form $y = f(x)$ is said to be defined *explicitly*, meaning that y is defined by a rule or *formula $f(x)$ in x alone*. A function may instead be defined *implicitly*, meaning that y is defined by an *equation in x and y*, such as $x^2 + y^2 = 25$. In this section we will see how to differentiate such *implicit functions* when ordinary "explicit" differentiation is difficult or impossible. We will then use implicit differentiation to find rates of change.

Implicit Differentiation

The equation $x^2 + y^2 = 25$ defines a circle. While a circle is not the graph of a function (it violates the vertical line test, see page 34), the top half by itself defines a function, as does the bottom half by itself. To find these two functions, we solve $x^2 + y^2 = 25$ for y:

$y^2 = 25 - x^2$ Subtracting x^2 from each side of $x^2 + y^2 = 25$

$y = \pm\sqrt{25 - x^2}$ Plus or minus since when squared either one gives $25 - x^2$

The *positive* square root defines the top half of the circle (where y is positive), and the *negative* square root defines the bottom half (where y is negative). The equation $x^2 + y^2 = 25$ defines *both* functions at the same time.

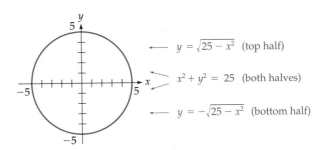

To find the slope anywhere on the circle, we could differentiate the "top" and "bottom" functions separately. However, it is easier to find both answers at once by differentiating *implicitly*, that is by differentiating both sides of the equation $x^2 + y^2 = 25$ with respect to x. Remember, however, that y is a *function* of x, so differentiating y^2 means differentiating a *function* squared, which requires the Generalized Power Rule:

$$\frac{d}{dx} y^n = n \cdot y^{n-1} \frac{dy}{dx}$$

EXAMPLE 1

DIFFERENTIATING IMPLICITLY

Use implicit differentiation to find $\dfrac{dy}{dx}$ when $x^2 + y^2 = 25$.

Solution

We differentiate both sides of the equation with respect to x:

$$\frac{d}{dx} x^2 + \frac{d}{dx} y^2 = \frac{d}{dx} 25 \qquad \text{Differentiating } x^2 + y^2 = 25$$

$$\downarrow \qquad \downarrow \qquad \downarrow$$

$$2x + 2y \frac{dy}{dx} = 0 \qquad \text{Using the Generalized Power Rule on } y^2$$

Solving for $\dfrac{dy}{dx}$:

$$2y \dfrac{dy}{dx} = -2x \qquad \text{Subtracting } 2x$$

$$\dfrac{dy}{dx} = -\dfrac{x}{y} \qquad \text{Canceling the 2's and dividing by } y$$

Therefore, $\dfrac{dy}{dx} = -\dfrac{x}{y}$ when x and y are related by $x^2 + y^2 = 25$.

Notice that the formula for $\dfrac{dy}{dx}$ involves both x and y. Implicit differentiation enables us to find derivatives that would otherwise be difficult or impossible to calculate, but at a "cost"—the result may depend on both x and y.

Remember that x and y play different roles: x is the *independent* variable, and y is a *function*. Therefore, we must include a $\dfrac{dy}{dx}$ (from the Generalized Power Rule) when differentiating y^n, but not when differentiating x^n since $\dfrac{dx}{dx} = 1$.

EXAMPLE 2

EVALUATING AN IMPLICIT DERIVATIVE *(Continuation of Example 1)*

Find the slope of the circle $x^2 + y^2 = 25$ at the points $(3, 4)$ and $(3, -4)$.

Solution

We simply evaluate the derivative $\dfrac{dy}{dx} = -\dfrac{x}{y}$ (found in Example 1) at the given points.

At $(3, 4)$: $\dfrac{dy}{dx} = -\dfrac{3}{4}$ $\leftarrow x$
$\leftarrow y$

At $(3, -4)$: $\dfrac{dy}{dx} = -\dfrac{3}{-4} = \dfrac{3}{4}$

Note that the negative sign in $\dfrac{dy}{dx} = -\dfrac{x}{y}$ gives the slope the correct sign: negative at $(3, 4)$ and positive at $(3, -4)$.

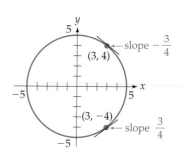

Graphing Calculator Exploration

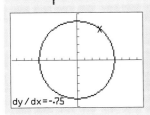

a. Graph the entire circle by graphing $y_1 = \sqrt{25 - x^2}$ and $y_2 = -\sqrt{25 - x^2}$. You may have to adjust the window to make the circle look "circular."

b. Verify the answer to Example 2 by finding the derivatives of y_1 and y_2 at $x = 3$.

c. Can you find the derivative at $x = 5$? Why not?

Derivatives should be evaluated only at points on the curve, so we evaluate $\frac{dy}{dx}$ only at *x*- and *y*-values *satisfying the original equation*. (It is easy to check that $x = 3$ and $y = \pm 4$ *do* satisfy $x^2 + y^2 = 25$.) Evaluating at a point not on the curve, such as (2, 3), would give a meaningless result.

The following are typical "pieces" that might appear in implicit differentiation problems.

EXAMPLE 3   FINDING DERIVATIVES—IMPLICIT AND EXPLICIT

a. $\dfrac{d}{dx} y^3 = 3y^2 \dfrac{dy}{dx}$ Differentiating y^3, so include $\dfrac{dy}{dx}$

b. $\dfrac{d}{dx} x^3 = 3x^2$ Differentiating x^3, so no $\dfrac{dx}{dx}$

c. $\dfrac{d}{dx}(x^3 y^5) = \underbrace{3x^2 \cdot y^5}_{\frac{d}{dx} x^3} + \underbrace{x^3 \cdot 5y^4 \dfrac{dy}{dx}}_{\frac{d}{dx} y^5}$ Using the Product Rule

$= 3x^2 y^5 + 5x^3 y^4 \dfrac{dy}{dx}$ Try to do problems such as this in one step, putting the constants in front from the start

Practice Problem

Find:

a. $\dfrac{d}{dx} x^4$ b. $\dfrac{d}{dx} y^2$ c. $\dfrac{d}{dx}(x^2 y^3)$ ➤ Solutions on page 707

In general, finding $\dfrac{dy}{dx}$ from an equation that defines y implicitly involves three steps:

1. Differentiate both sides of the equation *with respect to x*.
2. Collect all terms involving $\dfrac{dy}{dx}$ on one side, and all others on the other side.
3. Factor out the $\dfrac{dy}{dx}$ and solve for it by dividing.

EXAMPLE 4

FINDING AND EVALUATING AN IMPLICIT DERIVATIVE

For $y^4 + x^4 - 2x^2y^2 = 9$, find $\dfrac{dy}{dx}$ and evaluate it at $x = 2$, $y = 1$.

Solution

$$4y^3 \frac{dy}{dx} + 4x^3 - 4xy^2 - 4x^2y \frac{dy}{dx} = 0 \qquad \text{Differentiating with respect to } x, \text{ putting constants first}$$

$$4y^3 \frac{dy}{dx} - 4x^2y \frac{dy}{dx} = -4x^3 + 4xy^2 \qquad \text{Collecting } dy/dx \text{ terms on the left, others on the right}$$

$$(4y^3 - 4x^2y) \frac{dy}{dx} = -4x^3 + 4xy^2 \qquad \text{Factoring out } \frac{dy}{dx}$$

$$\frac{dy}{dx} = \frac{-4x^3 + 4xy^2}{4y^3 - 4x^2y} \qquad \text{Dividing by } 4y^3 - 4x^2y \text{ to solve for } dy/dx$$

$$= \frac{-x^3 + xy^2}{y^3 - x^2y} \qquad \text{Dividing by 4}$$

$$\frac{dy}{dx} = \frac{-(2)^3 + (2)(1)^2}{(1)^3 - (2)^2(1)} = \frac{-6}{-3} = 2 \qquad \text{Evaluating at } x = 2, y = 1$$

Note that in the example above the given point *is* on the curve, since $x = 2$ and $y = 1$ satisfy the original equation:

$$1^4 + 2^4 - 2 \cdot 2^2 \cdot 1 = 1 + 16 - 8 = 9$$

In economics, a *demand equation* is the relationship between the price p of an item and the quantity x that consumers will demand at that price. (All prices are in dollars unless otherwise stated.)

EXAMPLE 5

FINDING AND INTERPRETING AN IMPLICIT DERIVATIVE

For the demand equation $x = \sqrt{1900 - p^3}$, use implicit differentiation to find dp/dx. Then evaluate it at $x = 30$, $p = 10$ and interpret your answer.

Solution

$x^2 = 1900 - p^3$ Simplifying by squaring both sides of $x = \sqrt{1900 - p^3}$

$2x = -3p^2 \dfrac{dp}{dx}$ Differentiating both sides with respect to x

$\dfrac{dp}{dx} = -\dfrac{2x}{3p^2}$ Solving for $\dfrac{dp}{dx}$

$\dfrac{dp}{dx} = -\dfrac{60}{300} = -0.2$ Evaluating at $p = 10$ and $x = 30$

Interpretation: $dp/dx = -0.2$ says that the rate of change of price with respect to quantity is -0.2, so that increasing the quantity by 1 means decreasing the price by 0.20 (or 20 cents). Therefore, each 20-cent price decrease brings approximately one more sale (at the given values of x and p).

Our first step of squaring both sides of the original equation was not necessary, but it made the differentiation easier by avoiding the generalized Power Rule.

Notice that this particular demand function $x = \sqrt{1900 - p^3}$ can be solved *explicitly* for p:

$x^2 = 1900 - p^3$ Squaring

$p^3 = 1900 - x^2$ Adding p^3 and subtracting x^2

$p = (1900 - x^2)^{1/3}$ Taking cube roots

We can differentiate this *explicitly* with respect to x:

$p' = \dfrac{1}{3}(1900 - x^2)^{-2/3}(-2x)$ Using the Generalized Power Rule

$= -\dfrac{2}{3}x(1900 - x^2)^{-2/3}$ Simplifying

Evaluating at the given values of $x = 30$ and $p = 10$ gives

$p' = -\dfrac{2}{3} \cdot 30(1900 - 30^2)^{-2/3}$ Substituting $x = 30$

$= -20(1000)^{-2/3} = -\dfrac{20}{100} = -0.2$

This agrees with the answer by implicit differentiation. Which way was easier?

Related Rates

Sometimes *both* variables in an equation will be functions of a *third* variable, usually t for time. For example, for a seasonal product such as fur coats, the price p and weekly sales x will be related by a demand equation, and both price p and quantity x will depend on the time of year. Differentiating both sides of the demand equation with respect to time t will give an equation relating the derivatives dp/dt and dx/dt. Such "related rates" equations show how fast one quantity is changing relative to another. First, an "everyday" example.

EXAMPLE 6   FINDING RELATED RATES

A pebble thrown into a pond causes circular ripples to radiate outward. If the radius of the outer ripple is growing by 2 feet per second, how fast is the area of its circle growing at the moment when the radius is 10 feet?

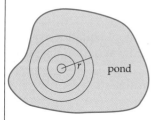

Solution The formula for the area of a circle is $A = \pi r^2$. Both the area A and the radius r of the circle increase with time, so both are functions of t. We are told that the radius is increasing by 2 feet per second ($dr/dt = 2$), and we want to know how fast the area is changing (dA/dt). To find the relationship between dA/dt and dr/dt, we differentiate both sides of $A = \pi r^2$ with respect to t.

$$\frac{dA}{dt} = 2\pi r \cdot \frac{dr}{dt} \qquad \text{Writing the 2 before the } \pi$$

$$\frac{dA}{dt} = 2\pi \cdot \underbrace{10}_{r} \cdot \underbrace{2}_{\frac{dr}{dt}} = \underbrace{40\pi \approx 125.6}_{\text{Using } \pi \approx 3.14} \qquad \text{Substituting } r = 10 \text{ and } \frac{dr}{dt} = 2$$

Therefore, at the moment when the radius is 10 feet, the area of the circle is growing at the rate of about 126 square feet per second.

We should be ready to interpret any *rate* as a derivative, just as we interpreted the radius growing by 2 feet per second as $dr/dt = 2$.

EXAMPLE 7

USING RELATED RATES TO FIND PROFIT GROWTH

A boat yard's total profit from selling x outboard motors is $P = -x^2 + 1000x - 2000$. If the outboards are selling at the rate of 20 per week, how fast is the profit changing when 400 motors have been sold?

Solution Profit P and quantity x both change with time, so both are functions of t. We differentiate both sides of $P = -x^2 + 1000x - 2000$ with respect to t and then substitute the given data.

$$\frac{dP}{dt} = -2x\frac{dx}{dt} + 1000\frac{dx}{dt} \quad \text{Differentiating with respect to } t$$

$$= \underbrace{-2(400)}_{x}\underbrace{20}_{dx/dt} + 1000\cdot\underbrace{20}_{dx/dt} \quad \begin{array}{l}\text{Substituting } x = 400 \text{ (number sold)}\\ \text{and } dx/dt = 20 \text{ (sales per week)}\end{array}$$

$$= -16{,}000 + 20{,}000 = 4000$$

Therefore, the company's profits are growing at the rate of $4000 per week.

EXAMPLE 8

USING RELATED RATES TO PREDICT POLLUTION

A study of urban pollution predicts that sulfur oxide emissions in a city will be $S = 2 + 20x + 0.1x^2$ tons, where x is the population (in thousands). The population of the city t years from now is expected to be $x = 800 + 20\sqrt{t}$ thousand people. Find how rapidly sulfur oxide pollution will be increasing 4 years from now.

Solution

Finding the rate of increase of pollution means finding $\dfrac{dS}{dt}$.

$$\frac{dS}{dt} = 20\frac{dx}{dt} + 0.2x\frac{dx}{dt} \quad \begin{array}{l} S = 2 + 20x + 0.1x^2 \\ \text{differentiated with respect to } t \\ (x \text{ is also a function of } t)\end{array}$$

We then find dx/dt from the other given equation:

$$\frac{dx}{dt} = 10t^{-1/2} \quad \begin{array}{l}x = 80 + 20t^{1/2} \text{ differentiated} \\ \text{with respect to } t\end{array}$$

$$= 10\cdot 4^{-1/2} = 10\cdot\frac{1}{2} = 5 \quad \begin{array}{l}\text{Substituting the given } t = 4 \\ \text{gives } dx/dt = 5\end{array}$$

$\dfrac{dS}{dt}$ then becomes

$$\underset{dx/dt}{\underbrace{\dfrac{dS}{dt} = 20 \cdot 5}} + 0.2 \underset{x}{\underbrace{(800 + 20\sqrt{4})}} \underset{dx/dt}{\underbrace{5}}$$

$$\dfrac{dS}{dt} = 20\dfrac{dx}{dt} + 0.2x\dfrac{dx}{dt}$$
with $dx/dt = 5$ and
$x = 800 + 20t^{1/2}$ at $t = 4$

$= 100 + 0.2(840)5 = 100 + 840 = 940$

Therefore, in 4 years the sulfur oxide emissions will be increasing at the rate of 940 tons per year.

Graphing Calculator Exploration

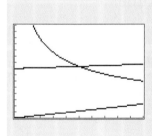

Verify the answer to the preceding example on a graphing calculator as follows:

a. Define $y_1 = 2 + 20x + 0.1x^2$ (the sulfur oxide function).
b. Define $y_2 = 800 + 20\sqrt{x}$ (the population function, using x for ease of entry).
c. Define $y_3 = y_1(y_2)$ (the composition of y_1 and y_2, giving pollution in year x).
d. Define y_4 to be the derivative of y_3 (using NDERIV, giving rate of change of pollution).
e. Graph these on the window [0, 10] by [0, 1500]. (Which function does *not* appear on the screen?)
f. Evaluate y_4 at $x = 4$ to verify the answer to Example 8.

Verification of the Power Rule for Rational Powers

On page 562 we stated the Power Rule for differentiation:

$$\dfrac{d}{dx}x^n = nx^{n-1}$$

Although we proved it only for *integer* powers, we have been using the Power Rule for *all* constant powers n. Using implicit differentiation, we may now prove the Power Rule for *rational* powers. (Recall that a rational number is of the form p/q, where p and q are integers with $q \neq 0$.) Let $y = x^n$ for a rational exponent $n = p/q$, and let x be a number at which $x^{p/q}$ is differentiable. Then

$y = x^n = x^{p/q}$ Since $n = p/q$

$y^q = x^p$ Raising each side to the power q

$qy^{q-1} \dfrac{dy}{dx} = px^{p-1}$ Differentiating each side implicitly with respect to x

$\dfrac{dy}{dx} = \dfrac{px^{p-1}}{qy^{q-1}}$ Dividing each side by qy^{q-1}

$= \dfrac{px^{p-1}}{q\left(x^{\frac{p}{q}}\right)^{q-1}} = \dfrac{p}{q} \cdot \dfrac{x^{p-1}}{x^{p-\frac{p}{q}}}$ Using $y = x^{p/q}$ and multiplying out the exponents in the denominator

$= \dfrac{p}{q} x^{\overbrace{p-1-\left(p-\frac{p}{q}\right)}^{-1+\frac{p}{q}}} = \dfrac{p}{q} x^{\frac{p}{q}-1} = nx^{n-1}$ Subtracting powers, simplifying, and replacing p/q by n (twice)

This is what we wanted to show, that the derivative of $y = x^n$ is $dy/dx = nx^{n-1}$ for any rational exponent $n = p/q$. This proves the Power Rule for rational exponents.

9.6 Section Summary

An equation in x and y may define one or more functions $y = f(x)$, which we may need to differentiate. Instead of solving the equation for y, which may be difficult or impossible, we can differentiate *implicitly*, differentiating both sides of the original equation with respect to x (writing a dy/dx or y' whenever we differentiate y) and solving for the derivative dy/dx. The derivative at any point of the curve may then be found by substituting the coordinates of that point.

Implicit differentiation is especially useful when several variables in an equation depend on an underlying variable, usually t for time. Differentiating the equation implicitly with respect to this underlying variable gives an equation involving the rates of change of the original variables. Numbers may then be substituted into this "related rate equation" to find a particular rate of change.

> **Solution to Practice Problem**

a. $\dfrac{d}{dx} x^4 = 4x^3$ b. $\dfrac{d}{dx} y^2 = 2y \dfrac{dy}{dx}$ c. $\dfrac{d}{dx}(x^2 y^3) = 2xy^3 + 3x^2 y^2 \dfrac{dy}{dx}$

9.6 EXERCISES

For each equation, use implicit differentiation to find dy/dx.

1. $y^3 - x^2 = 4$
2. $y^2 = x^4$
3. $x^3 = y^2 - 2$
4. $x^2 + y^2 = 1$
5. $y^4 - x^3 = 2x$
6. $y^2 = 4x + 1$
7. $(x + 1)^2 + (y + 1)^2 = 18$
8. $xy = 12$
9. $x^2 y = 8$
10. $x^2 y + xy^2 = 4$
11. $xy - x = 9$
12. $x^3 + 2xy^2 + y^3 = 1$
13. $x(y - 1)^2 = 6$
14. $(x - 1)(y - 1) = 25$
15. $y^3 - y^2 + y - 1 = x$
16. $x^2 + y^2 = xy + 4$
17. $\dfrac{1}{x} + \dfrac{1}{y} = 2$
18. $\sqrt[3]{x} + \sqrt[3]{y} = 2$
19. $x^3 = (y - 2)^2 + 1$
20. $\sqrt{xy} = x + 1$

For each equation, find dy/dx evaluated at the given values.

21. $y^2 - x^3 = 1$ at $x = 2, y = 3$
22. $x^2 + y^2 = 25$ at $x = -3, y = 4$
23. $y^2 = 6x - 5$ at $x = 1, y = -1$
24. $xy = 12$ at $x = 6, y = 2$
25. $x^2 y + y^2 x = 0$ at $x = -2, y = 2$
26. $y^2 + y + 1 = x$ at $x = 1, y = -1$
27. $x^2 + y^2 = xy + 7$ at $x = 3, y = 2$
28. $\sqrt[3]{x} + \sqrt[3]{y} = 3$ at $x = 1, y = 8$

For each demand equation, use implicit differentiation to find dp/dx.

29. $p^2 + p + 2x = 100$
30. $p^3 + p + 6x = 50$
31. $12p^2 + 4p + 1 = x$
32. $8p^2 + 2p + 100 = x$
33. $xp^3 = 36$
34. $xp^2 = 96$
35. $(p + 5)(x + 2) = 120$
36. $(p - 1)(x + 5) = 24$

APPLIED EXERCISES ON IMPLICIT DIFFERENTIATION

37. **BUSINESS: Demand Equation** A company's demand equation is $x = \sqrt{68 - p^2}$, where p is the price in dollars. Find dp/dx when $p = 2$ and interpret your answer.

38. **BUSINESS: Demand Equation** A company's demand equation is $x = \sqrt{650 - p^2}$, where p is the price in dollars. Find dp/dx when $p = 5$ and interpret your answer.

39. **BUSINESS: Sales** If a company spends r million dollars on research, its sales will be s million dollars, where r and s are related by $s^2 = r^3 - 55$.

   a. Find ds/dr by implicit differentiation and evaluate it at $r = 4$, $s = 3$. [*Hint*: Differentiate the equation with respect to r.]
   b. Find dr/ds by implicit differentiation and evaluate it at $r = 4$, $s = 3$. [*Hint*: Differentiate the original equation with respect to s.]
   c. Interpret your answers to parts (a) and (b) as rates of change.

40. **BIOMEDICAL: Muscle Contraction** When a muscle lifts a load, it does so according to the "fundamental equation of muscle contraction" $(L + m)(V + n) = k$, where L is the load that the muscle is lifting, V is the velocity of contraction of the muscle, and m, n, and k are constants. Use implicit differentiation to find dV/dL.

EXERCISES ON RELATED RATES

In each equation, x and y are functions of t. Differentiate with respect to t to find a relation between dx/dt and dy/dt.

41. $x^3 + y^2 = 1$
42. $x^5 - y^3 = 1$
43. $x^2 y = 80$
44. $xy^2 = 96$
45. $3x^2 - 7xy = 12$
46. $2x^3 - 5xy = 14$
47. $x^2 + xy = y^2$
48. $x^3 - xy = y^3$

APPLIED EXERCISES ON RELATED RATES

49. **GENERAL: Snowballs** A large snowball is melting so that its radius is decreasing at the rate of 2 inches per hour. How fast is the volume decreasing at the moment when the radius is 3 inches? [*Hint:* The volume of a sphere of radius r is $V = \frac{4}{3}\pi r^3$.]

50. **GENERAL: Hailstones** A hailstone (a small sphere of ice) is forming in the clouds so that its radius is growing at the rate of 1 millimeter per minute. How fast is its volume growing at the moment when the radius is 2 millimeters? [*Hint:* The volume of a sphere of radius r is $V = \frac{4}{3}\pi r^3$.]

51. **BIOMEDICAL: Tumors** The radius of a spherical tumor is growing by $\frac{1}{2}$ centimeter per week. Find how rapidly the volume is increasing at the moment when the radius is 4 centimeters. [*Hint:* The volume of a sphere of radius r is $V = \frac{4}{3}\pi r^3$.]

52. **BUSINESS: Profit** A company's profit from selling x units of an item is $P = 1000x - \frac{1}{2}x^2$ dollars. If sales are growing at the rate of 20 per day, find how rapidly profit is growing (in dollars per day) when 600 units have been sold.

53. **BUSINESS: Revenue** A company's revenue from selling x units of an item is given as $R = 1000x - x^2$ dollars. If sales are increasing at the rate of 80 per day, find how rapidly revenue is growing (in dollars per day) when 400 units have been sold.

54. **SOCIAL SCIENCE: Accidents** The number of traffic accidents per year in a city of population p is predicted to be $T = 0.002 p^{3/2}$. If the population is growing by 500 people a year, find the rate at which traffic accidents will be rising when the population is $p = 40{,}000$.

55. **SOCIAL SCIENCE: Welfare** The number of welfare cases in a city of population p is expected to be $W = 0.003 p^{4/3}$. If the population is growing by 1000 people per year, find the rate at which the number of welfare cases will be increasing when the population is $p = 1{,}000{,}000$.

56. GENERAL: Rockets A rocket fired straight up is being tracked by a radar station 3 miles from the launching pad. If the rocket is traveling at 2 miles per second, how fast is the distance between the rocket and the tracking station changing at the moment when the rocket is 4 miles up? [*Hint:* The distance D in the illustration satisfies $D^2 = 9 + y^2$. To find the value of D, solve $D^2 = 9 + 4^2$.]

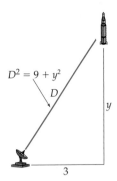

57. BIOMEDICAL: Poiseuille's Law Blood flowing through an artery flows faster in the center of the artery and more slowly near the sides (because of friction). The speed of the blood is $V = c(R^2 - r^2)$ millimeters (mm) per second, where R is the radius of the artery, r is the distance of the blood from the center of the artery, and c is a constant. Suppose that arteriosclerosis is narrowing the artery at the rate of $dR/dt = -0.01$ mm per year.

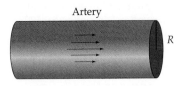

Find the rate at which blood flow is being reduced in an artery whose radius is $R = 0.05$ mm with $C = 500$. [*Hint:* Find dV/dt, considering r to be a constant. The units of dV/dt will be mm per second per year.]

58. IMPLICIT AND EXPLICIT DIFFERENTIATION The equation $x^2 + 4y^2 = 100$ describes an ellipse.
a. Use implicit differentiation to find its slope at the points $(8, 3)$ and $(8, -3)$.
b. Solve the equation for y, obtaining *two* functions, and differentiate both to find the slopes at $x = 8$. [Answers should agree with part (a).]
c. Use a graphing calculator to graph the two functions found in part (b) on an appropriate window. Then use NDERIV to find the derivatives at (or near) $x = 8$. [Answers should agree with parts (a) and (b).]

Notice that differentiating implicitly was easier than solving for y and then differentiating.

59. RELATED RATES: Speeding A traffic patrol helicopter is stationary a quarter of a mile directly above a highway, as shown in the diagram below. Its radar detects a car whose line-of-sight distance from the helicopter is half a mile and is increasing at the rate of 57 mph. Is the car exceeding the highway's speed limit of 60 mph?

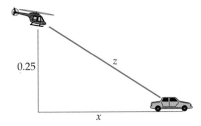

60. RELATED RATES: Speeding (*59 continued*)
In Exercise 59 you found that the car's speed was $\dfrac{(0.5)(57)}{\sqrt{(0.5)^2 - (0.25)^2}}$ mph. Enter this expression into a graphing calculator and then replace both occurrences of the line-of-sight distance 0.5 by 0.4 and calculate the new speed of the car. What if the line-of-sight distance were 0.3 mile?

Chapter Summary with Hints and Suggestions

Reading the text and doing the exercises in this chapter have helped you to master the following skills, which are listed by section (in case you need to review them) and are keyed to particular Review Exercises. Answers for all Review Exercises are given at the back of the book, and full solutions can be found in the Student Solutions Manual.

9.1 Graphing Using the First Derivative

9.2 Graphing Using the First and Second Derivatives

- Graph a polynomial, showing all relative extreme points and inflection points. *(Review Exercises 1–8.)*

  f' gives slope, f'' gives concavity

- Graph a fractional power function, showing all relative extreme points and inflection points. *(Review Exercises 9–12.)*

- Graph a rational function, showing all relative extreme points. *(Review Exercises 13–18.)*

9.3 Optimization

- Find the absolute extreme values of a given function on a given interval. *(Review Exercises 19–28.)*

- Show that the marginal cost function pierces the average cost function where the average cost is minimized. *(Review Exercises 29–30.)*

- Maximize the efficiency of a tugboat or a flying bird. *(Review Exercises 31, 38.)*

9.4 Further Applications of Optimization

- Solve a geometric optimization problem. *(Review Exercises 32–37.)*

- Maximize profit for a company. *(Review Exercise 39.)*

- Maximize revenue from an orchard. *(Review Exercise 40.)*

- Minimize the cost of a power cable. *(Review Exercise 41.)*

- Maximize tax revenue to the government. *(Review Exercise 42.)*

- Minimize the materials used in a container. *(Review Exercises 43–45.)*

9.5 Optimizing Lot Size and Harvest Size

- Find the lot size that minimizes production costs or inventory costs. *(Review Exercises 46–47.)*

- Find the population size that allows the maximum sustainable yield. *(Review Exercises 48–49.)*

9.6 Implicit Differentiation and Related Rates

- Find a derivative by implicit differentiation. *(Review Exercises 50–53.)*

- Find a derivative by implicit differentiation and evaluate it. *(Review Exercises 54–57.)*

- Solve a geometric related rates problem. *(Review Exercise 58.)*

- Use related rates to find the growth in profit or revenue. *(Review Exercises 59–60.)*

Hints and Suggestions

- Do not confuse *relative* and *absolute* extremes: a function may have several *relative* maximum points (high points compared to their neighbors), but it can have at most one *absolute* maximum value (the largest value of the function on its entire domain). *Relative* extremes are used in graphing, and *absolute* extremes are used in optimization.

- Graphing calculators can be very helpful for graphing functions. However, you must first find a window that shows the interesting parts of the curve (relative extreme points and inflection points), and that is where calculus is essential.

- We have two procedures for optimizing continuous functions. Both begin by finding all critical numbers of the function in the domain. Then:

  1. If the function has only one critical number, find the sign of the second derivative there: a *positive* sign means an absolute *minimum*, and a *negative* sign means an absolute *maximum*.

  2. If the interval is closed, the maximum and minimum values may be found by evaluating the function at all critical numbers in the interval and endpoints—the largest and smallest resulting values are the maximum and minimum values of the function.

  A good strategy is this: Find all critical numbers in the domain. If there is only one, use procedure 1 above. Otherwise, try to define the function on a closed interval and use procedure 2. If all else fails, make a sketch of the graph.

- Don't forget to use the second-derivative test in applied problems. Your employer will not be happy if you accidentally *minimize* your company's profits.

- In implicit differentiation problems, remember which is the *function* (usually y) and which is the independent variable (usually x).

- In related rate problems, begin by looking for an equation that relates the variables, and then differentiate it with respect to the underlying variable (usually t).

- **Practice for Test:** Review Exercises 3, 11, 21, 25, 33, 35, 39, 43, 45, 47, 53, 55.

Review Exercises for Chapter 9

Practice test exercise numbers are in green.

9.1 and 9.2 Graphing

Graph each function "by hand," showing all relative extreme points and inflection points.

1. $f(x) = x^3 - 3x^2 - 9x + 12$
2. $f(x) = x^3 + 3x^2 - 9x - 7$
3. $f(x) = x^4 - 4x^3 + 15$
4. $f(x) = x^4 + 4x^3 + 17$
5. $f(x) = x(x + 3)^2$
6. $f(x) = x(x - 6)^2$
7. $f(x) = x(x - 4)^3$
8. $f(x) = x(x + 4)^3$
9. $f(x) = \sqrt[7]{x^5} + 1$
10. $f(x) = \sqrt[7]{x^6} + 1$
11. $f(x) = \sqrt[7]{x^4} + 1$
12. $f(x) = \sqrt[7]{x^3} + 1$

Graph each function, showing all relative extreme points. If you use a graphing calculator, make a hand-drawn sketch showing all relative extreme points.

13. $f(x) = \dfrac{1}{x^2 - 6x}$
14. $f(x) = \dfrac{1}{x^2 + 4x}$
15. $f(x) = \dfrac{x^2}{x^4 + 1}$
16. $f(x) = \dfrac{2x^2}{1 + x^2}$
17. $f(x) = \dfrac{1 - 2x}{x^2}$
18. $f(x) = \dfrac{1 - x}{x^2}$

9.3 and 9.4 Optimization

Find the absolute extreme values of each function on the given interval.

19. $f(x) = 2x^3 - 6x$ on $[0, 5]$
20. $f(x) = 2x^3 - 24x$ on $[0, 5]$
21. $f(x) = x^4 - 4x^3 - 8x^2 + 64$ on $[-1, 5]$
22. $f(x) = x^4 - 4x^3 + 4x^2 + 1$ on $[0, 10]$
23. $h(x) = (x - 1)^{2/3}$ on $[0, 9]$
24. $f(x) = \sqrt{100 - x^2}$ on $[-10, 10]$
25. $g(w) = (w^2 - 4)^2$ on $[-3, 3]$
26. $g(x) = x(8 - x)$ on $[0, 8]$
27. $f(x) = \dfrac{x}{x^2 + 1}$ on $[-3, 3]$
28. $f(x) = \dfrac{x}{x^2 + 4}$ on $[-4, 4]$

29. **BUSINESS: Average and Marginal Cost** For the cost function $C(x) = 10{,}000 + x^2$, graph the marginal cost function $MC(x)$ and also the average cost function $AC(x) = C(x)/x$ on the same graph, showing that they intersect at the point where the average cost is minimized.

30. Prove that the marginal cost function intersects the average cost function at the point where the average cost is minimized (as shown in the following diagram) by justifying each numbered step in the following series of equations.
    At the x-value where the average cost function $AC(x) = \dfrac{C(x)}{x}$ is minimized, we must have
$$0 \stackrel{①}{=} \frac{d}{dx} AC(x) \stackrel{②}{=} \frac{xC'(x) - C(x)}{x^2}$$
$$\stackrel{③}{=} \frac{1}{x}\left[\frac{xC'(x) - C(x)}{x}\right]$$
$$\stackrel{④}{=} \frac{1}{x}\left[C'(x) - \frac{C(x)}{x}\right] \stackrel{⑤}{=} \frac{1}{x}[MC(x) - AC(x)].$$

⑥ Finally, explain why the equality of the first and last expressions in the series of equations proves the result.

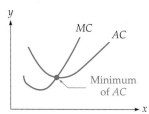

The marginal cost function pierces the average cost function at its minimum.

31. **GENERAL: Fuel Efficiency** At what speed should a tugboat travel upstream so as to use the least amount of fuel to reach its destination? If the tugboat's speed through the water is v, and if the speed of the current (relative to the land) is c, then the energy used is proportional to $E(v) = \dfrac{v^2}{v - c}$. Find the velocity v that minimizes the energy $E(v)$. Your answer will depend upon c, the speed of the current.

32. **GENERAL: Fencing** A homeowner wants to enclose three adjacent rectangular pens of equal size along a straight wall, as in the following diagram. If the side along the wall needs no fence, what is the largest total area that can be enclosed using only 240 feet of fence?

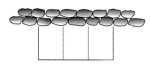

33. **GENERAL: Maximum Area** A homeowner wants to enclose three adjacent rectangular pens of equal size, as in the diagram below. What is the largest total area that can be enclosed using only 240 feet of fence?

34. **GENERAL: Unicorns** To celebrate the acquisition of Styria in 1261, Ottokar II sent hunters into the Bohemian woods to capture a unicorn. To display the unicorn at court, the king built a rectangular cage. The material for three sides of the cage cost 3 ducats per running cubit, while the fourth was to be gilded and cost 51 ducats per running cubit. In 1261 it was well known that a happy unicorn requires an area of 2025 square cubits. Find the dimensions that would keep the unicorn happy at the lowest cost.

35. **GENERAL: Box Design** An open-top box with a square base is to have a volume of exactly 500 cubic inches. Find the dimensions of the box that can be made with the smallest amount of materials.

36. **GENERAL: Packaging** Find the dimensions of the cylindrical tin can with volume 16π cubic inches that can be made from the least amount of tin. (*Note:* $16\pi \approx 50$ cubic inches.)

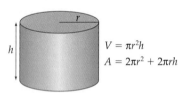

37. **GENERAL: Packaging** Find the dimensions of the open-top cylindrical tin can with volume 8π cubic inches that can be made from the least amount of tin. (*Note:* $8\pi \approx 25$ cubic inches.)

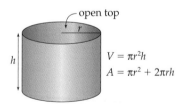

38. **GENERAL: Bird Flight** Let v be the flying speed of a bird, and let w be its weight. The power P that the bird must maintain during flight is $P = \dfrac{aw^2}{v} + bv^3$, where a and b are positive constants depending on the shape of the bird and the density of the air. Find the speed v that minimizes the power P.

39. **BUSINESS: Maximum Profit** A computer dealer can sell 12 personal computers per week at a price of $2000 each. He estimates that each $400 price decrease will result in three more sales per week. If the computers cost him $1200 each, what price should he charge to maximize his profit? How many will he sell at that price?

40. **GENERAL: Farming** A peach tree will yield 100 pounds of peaches now, which will sell for 40 cents a pound. Each week that the farmer waits will increase the yield by 10 pounds, but the selling price will decrease by 2 cents per pound. How long should the farmer wait to pick the fruit in order to maximize her revenue?

41. **GENERAL: Minimum Cost** A cable is to connect a power plant to an island that is 1 mile offshore and 3 miles downshore from the power plant. It costs $5000 per mile to lay a cable underwater and $3000 per mile to lay it underground. If the cost of laying the cable is to be minimized, find the distance x downshore from the island where the cable should meet the land.

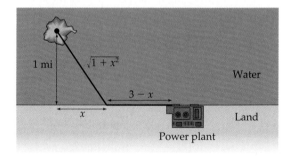

42. **ECONOMICS: Tax Revenue** Economists* have found that if cigarettes are taxed at rate t, then cigarette sales will be $S(t) = 64 - 51.26t$ (billion dollars annually).

---

*Michael Grossman, Gary Becker (winner of the 1992 Nobel Memorial Prize in Economics), Frank Chaloupka, and Kevin Murphy.

a. Use a graphing calculator to find the tax rate t that maximizes revenue to the government.
b. Multiply this tax rate times $2.02 (the average price of a pack of cigarettes) to find the actual tax per pack.

43. **GENERAL: Packaging** A 12-ounce soft drink can has volume 21.66 cubic inches. If the top and bottom are twice as thick as the sides, find the dimensions (radius and height) that minimize the amount of metal used in the can.

44. **GENERAL: Box Design** A standard 8.5- by 11-inch piece of paper can be made into a box with a lid by cutting x- by x-inch squares from two corners and x- by 5.5-inch rectangles from the other corners and then folding, as shown below. What value of x maximizes the volume of the box, and what is the maximum volume?

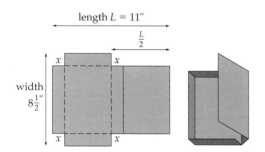

45. **GENERAL: Box Design** A standard 6- by 8-inch card can be made into a box with a lid by cutting x- by x-inch squares from two corners and x- by 4-inch rectangles from the other corners and then folding. (See the diagram above, but use length 8 and width 6.) What value of x maximizes the volume of the box, and what is the maximum volume?

9.5 Optimizing Lot Size and Harvest Size

46. **BUSINESS: Production Runs** A wallpaper company estimates the demand for a certain pattern to be 900 rolls per year. It costs $800 to set up the presses to print the pattern, plus $200 to print each roll. If the company can store a roll of wallpaper for a year at a cost of $4, how many rolls should it print at a time and how many printing runs will it need in a year to minimize production costs?

47. **BUSINESS: Lot Size** A motorcycle shop estimates that it will sell 500 motorbikes in a year. Each bike costs $300, plus a fixed charge of $500 per order. If it costs $200 to store a motorbike for a year, what is the order size and how many orders will be needed in a year to minimize inventory costs?

48. **MAXIMUM SUSTAINABLE YIELD** The reproduction function for the North American duck is estimated to be $f(p) = -0.02p^2 + 7p$, where p and $f(p)$ are measured in thousands. Find the size of the population that allows the maximum sustainable yield, and also find the size of the yield.

49. **MAXIMUM SUSTAINABLE YIELD** Ecologists estimate the reproduction function for striped bass in an East Coast fishing ground to be $f(p) = 60\sqrt{p}$, where p and $f(p)$ are measured in thousands. Find the size of the population that allows the maximum sustainable yield, and also the size of the yield.

9.6 Implicit Differentiation and Related Rates

For each equation, use implicit differentiation to find dy/dx.

50. $6x^2 + 8xy + y^2 = 100$
51. $8xy^2 - 8y = 1$
52. $2xy^2 - 3x^2y = 0$
53. $\sqrt{x} - \sqrt{y} = 10$

For each equation, find dy/dx evaluated at the given values.

54. $x + y = xy$ at $x = 2, y = 2$
55. $y^3 - y^2 - y = x$ at $x = 2, y = 2$
56. $xy^2 = 81$ at $x = 9, y = 3$
57. $x^2y^2 - xy = 2$ at $x = -1, y = 1$

58. GENERAL: Melting Ice A cube of ice is melting so that each edge is decreasing at the rate of 2 inches per hour. Find how fast the volume of the ice is decreasing at the moment when each edge is 10 inches long.

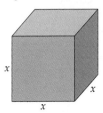

59. BUSINESS: Profit A company's profit from selling x units of a product is $P = 2x^2 - 20x$ dollars. If sales are growing at the rate of 30 per day, find the rate of change of profit when 40 units have been sold.

60. BUSINESS: Revenue A company finds that its revenue from selling x units of a product is $R = x^2 + 500x$ dollars. If sales are increasing at the rate of 50 per month, find the rate of change of revenue when 200 units have been sold.

61. BIOMEDICAL: Medication You swallow a spherical pill whose radius is 0.5 centimeter (cm), and it dissolves in your stomach so that its radius decreases at the rate of 0.1 cm per minute. Find the rate at which the volume is decreasing (the rate at which the medication is being made available to your system) when the radius is

   a. 0.5 cm
   b. 0.2 cm

Cumulative Review for Chapters 1, 8–9

The following exercises review some of the basic techniques that you learned in Chapters 1, 8, and 9. Answers to all of these cumulative review exercises are given in the answer section at the back of the book.

1. Find an equation for the line through the points $(-4, 3)$ and $(6, -2)$. Write your answer in the form $y = mx + b$.

2. Simplify $\left(\frac{4}{25}\right)^{-1/2}$.

3. Find, correct to three decimal places:
$$\lim_{x \to 0} (1 + 3x)^{1/x}$$

4. For the function $f(x) = \begin{cases} 4x - 8 & \text{if } x < 3 \\ 7 - 2x & \text{if } x \geq 3 \end{cases}$

   a. Draw its graph.
   b. Find $\lim_{x \to 3^-} f(x)$.
   c. Find $\lim_{x \to 3^+} f(x)$.
   d. Find $\lim_{x \to 3} f(x)$.
   e. Is $f(x)$ continuous or discontinuous, and if it is discontinuous, where?

5. Use the definition of the derivative, $f'(x) = \lim_{h \to 0} \frac{f(x + h) - f(x)}{h}$ to find the derivative of
$$f(x) = 2x^2 - 5x + 7.$$

6. Find the derivative of $f(x) = 8\sqrt{x^3} - \frac{3}{x^2} + 5$.

7. Find the derivative of $f(x) = (x^5 - 2)(x^4 + 2)$.

8. Find the derivative of $f(x) = \dfrac{2x - 5}{3x - 2}$.

9. The population of a city x years from now is predicted to be $P(x) = 3600x^{2/3} + 250{,}000$ people. Find $P'(8)$ and $P''(8)$ and interpret your answers.

10. Find $\dfrac{d}{dx}\sqrt{2x^2 - 5}$ and write your answer in radical form.

11. Find $\dfrac{d}{dx}[(3x+1)^4(4x+1)^3]$.

12. Find $\dfrac{d}{dx}\left(\dfrac{x-2}{x+2}\right)^3$ and simplify your answer.

13. Make sign diagrams for the first and second derivatives and draw the graph of the function $f(x) = x^3 - 12x^2 - 60x + 400$. Show on your graph all relative extreme points and inflection points.

14. Make sign diagrams for the first and second derivatives and draw the graph of the function $f(x) = \sqrt[3]{x^2 - 1}$. Show on your graph all relative extreme points and inflection points.

15. A homeowner wishes to use 600 feet of fence to enclose two identical adjacent pens, as in the diagram below. Find the largest total area that can be enclosed.

16. A store can sell 12 telephone answering machines per day at a price of $200 each. The manager estimates that for each $10 price reduction she can sell 2 more per day. The answering machines cost the store $80 each. Find the price and the quantity sold per day to maximize the company's profit.

17. For y defined implicitly by $x^3 + 9xy^2 + 3y = 43$, find $\dfrac{dy}{dx}$ and evaluate it at the point $(1, 2)$.

18. A large spherical balloon is being inflated at the rate of 32 cubic feet per minute. Find how fast the radius is increasing at the moment when the radius is 2 feet.

10 Exponential and Logarithmic Functions

- 10.1 Review of Exponential and Logarithmic Functions
- 10.2 Differentiation of Logarithmic and Exponential Functions
- 10.3 Two Applications to Economics: Relative Rates and Elasticity of Demand

Application Preview

Elasticity and Social Policy

It is common knowledge that if the price of an item rises, demand for it usually falls. But for a given price increase, will the drop in demand be large or small? To answer this question, economists have developed the concept of *elasticity of demand*. For a product selling at a particular price, the elasticity is a number $E \geq 0$ (calculated from a formula given later in this chapter) that measures the *responsiveness* of demand to price changes. For elasticities *greater* than 1, demand is said to be *elastic*, meaning that demand responds relatively *strongly* as price changes. For elasticities *less* than 1, demand is said to be *inelastic*, meaning that demand changes relatively *weakly* as price changes. To be specific, an elasticity $E = 2$ (elastic) means that a 1% increase in price would cause a drop of about 2% in demand, whereas an elasticity of $E = \frac{1}{2}$ (inelastic) means that a 1% increase in price would cause only about a $\frac{1}{2}$% decrease in demand.

Elasticity of demand has economic and social implications; consider two particular cases. For heroin sold in Detroit, elasticity has been estimated to be 0.267 (inelastic), meaning that demand for heroin will respond only slightly to price changes.* Therefore, a price increase will cause only a slight decrease in demand, and so the drug dealers' revenue, which is price times quantity, will actually *increase*. The increased revenues to drug dealers might also bring about an increase in crime to support the increased amount that addicts will have to pay. On the other hand, for marijuana sold on a college campus, elasticity has been estimated to be 1.013 (elastic), meaning that a small price increase will result in a larger decrease in demand, so that revenue to the drug dealers would fall.[†]

These different elasticities lead to different conclusions. In Detroit, police might decide to concentrate their efforts on heroin *users* rather than on dealers, since concentrating on dealers would cause a shortage, and the resulting price increases would paradoxically *increase* revenue to drug dealers and possibly increase crime. On college campuses, authorities might decide to concentrate on drug *suppliers*, since the resulting shortage would cause prices to rise, which, as we saw above, would *decrease* revenue to the dealers.

Elasticity of demand is defined in terms of derivatives of exponential and logarithmic functions, which are discussed in this chapter.

* Lester P. Silverman and Nancy L. Spruill, "Urban Crime and the Price of Heroin," *Journal of Urban Economics* 4:80–103.
† Charles T. Nisbit and Firouz Vakil, "Some Estimates of Price and Expenditure Elasticities of Demand for Marijuana Among U.C.L.A. Students," *Review of Economics and Statistics* 54:473–475.

10.1 REVIEW OF EXPONENTIAL AND LOGARITHMIC FUNCTIONS

Introduction

This section reviews certain aspects of exponential and logarithmic functions that will be useful in the next section when we differentiate these functions. It should serve as a review and continuation of, but not a replacement for, Sections 1.5 and 1.6 (pages 64–90) of Chapter 1, which you may want to look at before continuing.

Exponential Functions

A function with a variable in an exponent, such as $f(x) = 2^x$, is called an *exponential function*. The number being raised to the power is called the *base*, so $f(x) = 2^x$ has base 2. We may use any positive number as a base for an exponential function, and one of the most useful bases is the constant e defined on page 70.

$$e = \lim_{n \to \infty} \left(1 + \frac{1}{n}\right)^n = 2.71828 \ldots$$

The following graphs of exponential functions with bases $2, \frac{1}{2}$, and e can be obtained either from a graphing calculator or by plotting points.

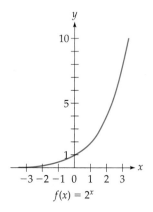
$f(x) = 2^x$

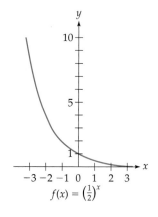
$f(x) = \left(\frac{1}{2}\right)^x$

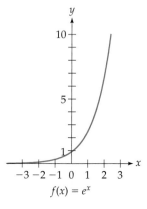
$f(x) = e^x$

Value Appreciation

Suppose that you make an investment of P dollars that increases in value ("appreciates") by 5% each year. After 1 year the value would be

$$\begin{pmatrix}\text{Value after} \\ \text{1 year}\end{pmatrix} = P + 0.05P = P(1 + 0.05) \quad \text{Factoring out the } P$$

10.1 REVIEW OF EXPONENTIAL AND LOGARITHMIC FUNCTIONS

Notice that increasing a quantity by 5% is the same as multiplying it by $(1 + 0.05)$. Therefore, to find the value after a second year we would multiply again by $(1 + 0.05)$, for a total of *two* multiplications by $(1 + 0.05)$, obtaining $P(1 + 0.05)^2$. For the value after t years we simply multiply by $(1 + 0.05)$ a total of t times:

$$\begin{pmatrix} \text{Value after} \\ t \text{ years} \end{pmatrix} = \overbrace{P(1 + 0.05) \cdot (1 + 0.05) \cdots \cdot (1 + 0.05)}^{t \text{ times}}$$

$$= P(1 + 0.05)^t$$

Clearly, the 5% can be replaced by *any* growth rate r, written in decimal form.

Value Appreciation

For an amount P that increases at rate r each year,

$$\begin{pmatrix} \text{Value after} \\ t \text{ years} \end{pmatrix} = P(1 + r)^t$$

This formula also follows directly from the compound interest formula (page 66) with $m = 1$ (for annual compounding).

EXAMPLE 1

FINDING THE VALUE OF A COIN COLLECTION

A coin collection appraised at $25,000 is expected to appreciate in value by 8.5% annually. Predict its value 6 years from now.

Solution

$25,000(1 + 0.085)^6$ $P(1 + r)^t$ with $P = 25,000$, $r = 0.085$, and $t = 6$

$= 25,000 \cdot 1.085^6 \approx 40,786.69$ Using a calculator

In 6 years the coin collection will be worth $40,787.

The same formula can be used to find *any* quantity that increases annually by a fixed percentage, such as populations or the spread of information. It can also be used for quantities that *decrease* by a fixed percentage by taking r to be *negative*.

Practice Problem 1 The 2002 population of St. Louis, Missouri, was 341,000 and decreasing at the rate of 1% per year. Predict the population in 2012. [*Hint:* Since the population is decreasing, take r to be *negative* 1%.] ➤ **Solution on page 730**

The formula $P(1 + r)^t$ involves raising the base $(1 + r)$ to a power t. In other applications, the base is e, the number defined on page 720.

EXAMPLE 2

PREDICTING THE POPULATION OF THE UNITED STATES

As the following graph shows, the population of the United States since 1900 has grown approximately exponentially, closely matching the curve $y = 80.25e^{0.0128x}$, where x is the number of years since 1900. Use this exponential function to predict the U.S. population in the year 2010.

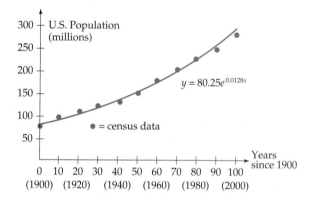

Solution

$$80.25e^{0.0128(110)} = 80.25e^{1.41} \approx 328$$

$80.25e^{0.0128x}$ with $x = 110$ (2010 is 110 years after 1900)

Using a calculator

In the year 2010 the population of the United States will be approximately 328 million (based on the last eleven censuses).

Behavioral Science: Learning Theory

One's ability to do a task generally improves with practice. Frequently, one's skill after t units of practice is given by a function of the form

$$S(t) = c(1 - e^{-kt})$$

where c and k are positive constants.

EXAMPLE 3

PREDICTING SKILL LEVELS

A secretarial school guarantees that after t weeks of training a person will type

$$S(t) = 100(1 - e^{-0.25t})$$

words per minute. If you hire a secretary who has been in the program for 3 weeks, how fast will he type?

Solution

We evaluate $S(t)$ at $t = 3$:

$$100(1 - e^{-0.25 \cdot 3}) = 100(1 - e^{-0.75}) \approx 53 \qquad \text{\small $100(1 - e^{-0.25t})$ with $t = 3$}$$

He will type approximately 53 words per minute.

Logarithmic Functions

On pages 78 and 83 we defined logarithms as exponents. The *common logarithm* of a number x, written $\log x$, is the exponent of 10 that gives that number.

Common Logarithms

$\log x = y$ is equivalent to $10^y = x$ $\qquad \log x$ means $\log_{10} x$

EXAMPLE 4

FINDING COMMON LOGARITHMS

a. $\log 1000 = 3$ $\qquad$ Since $10^3 = 1000$

b. $\log \dfrac{1}{10} = -1$ $\qquad$ Since $10^{-1} = \dfrac{1}{10}$

c. $\log \sqrt{10} = \dfrac{1}{2}$ $\qquad$ Since $10^{1/2} = \sqrt{10}$

Graphing Calculator Exploration

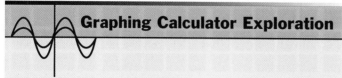

On a graphing calculator, common logarithms are easily found using the [LOG] key. Verify the common logarithms from Example 4 on your calculator.

Natural Logarithms

In calculus we work almost exclusively with *natural* (base *e*) logarithms. The natural logarithm of a number *x*, written ln *x*, is the power of *e* that gives that number.

Natural Logarithms

$\ln x = y$ is equivalent to $e^y = x$ ln *x* means $\log_e x$

EXAMPLE 5

FINDING NATURAL LOGARITHMS

a. $\ln e = 1$ Since $e^1 = e$
b. $\ln 1 = 0$ Since $e^0 = 1$
c. $\ln \sqrt{e} = \dfrac{1}{2}$ Since $e^{1/2} = \sqrt{e}$

Graphing Calculator Exploration

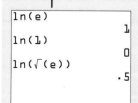

On a graphing calculator, natural logarithms are found using the [LN] key. Verify the natural logarithms from Example 5 on your calculator.

Practice Problem 2

Find (without using a calculator)

$$\ln \sqrt[3]{e}$$

▶ Solution on page 730

The following properties of natural logarithms were justified on pages 83 and 87–88. For positive numbers M and N and any number P:

Properties of Natural Logarithms

1. $\ln 1 = 0$	The natural log of 1 is 0 (since $e^0 = 1$)
2. $\ln e = 1$	The natural log of e is 1 (since $e^1 = e$)
3. $\ln e^x = x$	The natural log of e to a power is just the power (since $e^x = e^x$).
4. $e^{\ln x} = x$	e raised to the natural log of a number is just the number.
5. $\ln (M \cdot N) = \ln M + \ln N$	The natural log of a product is the sum of the logs
6. $\ln \left(\dfrac{1}{N}\right) = -\ln N$	The natural log of 1 over a number is minus the log of the number
7. $\ln \left(\dfrac{M}{N}\right) = \ln M - \ln N$	The natural log of a quotient is the difference of the logs
8. $\ln (M^P) = P \cdot \ln M$	The natural log of a number to a power is the power times the log

Properties 3 and 4 show that $y = \ln x$ and $y = e^x$ are *inverse functions*, so their graphs are reflections of each other in the diagonal line $y = x$.

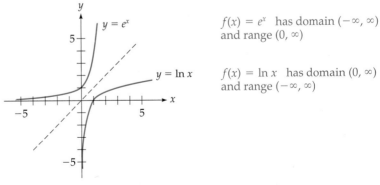

$f(x) = e^x$ has domain $(-\infty, \infty)$ and range $(0, \infty)$

$f(x) = \ln x$ has domain $(0, \infty)$ and range $(-\infty, \infty)$

$\ln x$ and e^x are inverse functions

The properties of natural logarithms are useful for simplifying functions.

EXAMPLE 6 — SIMPLIFYING A FUNCTION

$$f(x) = \ln(x^2) - \ln\sqrt{x}$$
$$= \ln(x^2) - \ln(x^{1/2}) \quad \text{Since } \sqrt{x} = x^{1/2}$$
$$= 2\ln x - \tfrac{1}{2}\ln x \quad \text{Using property 8 (twice)}$$
$$= \tfrac{3}{2}\ln x \quad \text{Combining}$$

EXAMPLE 7 — SIMPLIFYING A FUNCTION

$$f(x) = x + \ln(e^{-0.05x})$$
$$= x - 0.05x \quad \text{Using property 3}$$
$$= 0.95x \quad \text{Combining}$$

Drug Dosage

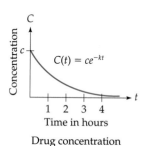

Drug concentration in the blood

The amount of a drug that remains in a person's bloodstream decreases exponentially with time. If the initial concentration is c (milligrams per milliliter of blood), the concentration t hours later will be

$$C(t) = ce^{-kt}$$

where the "absorption constant" k measures how rapidly the drug is absorbed.

Every medicine has a minimum concentration below which it is not effective. When the concentration falls to this level, another dose should be administered. If doses are administered regularly every T hours over a period of time, the concentration will look like the following graph:

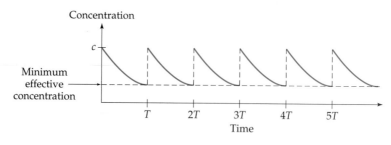

Drug concentration with repeated doses

10.1 REVIEW OF EXPONENTIAL AND LOGARITHMIC FUNCTIONS

The problem is to determine the time T at which the dose should be repeated so as to maintain an effective concentration.

EXAMPLE 8

CALCULATING DRUG DOSAGE

The absorption constant for penicillin is $k = 0.11$, and the minimum effective concentration is 2 mg/ml. If the original concentration is $c = 5$ mg/ml, find when another dose should be administered in order to maintain an effective concentration.

Solution The concentration formula $C(t) = ce^{-kt}$ with $c = 5$ and $k = 0.11$ is

$$C(t) = 5e^{-0.11t}$$

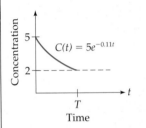

To find the time when this concentration reaches the minimum effective level of 2, we solve

$$5e^{-0.11t} = 2$$

$$e^{-0.11t} = 0.4 \qquad \text{Dividing by 5}$$

$$\underbrace{\ln e^{-0.11t}}_{-0.11t} = \ln 0.4 \qquad \text{Taking natural logs to bring down the exponent}$$

$$-0.11t = \ln 0.4$$

$$t = \frac{\ln 0.4}{-0.11} \approx \frac{-0.9163}{-0.11} \approx 8.3 \qquad \text{Solving for } t \text{ using a calculator}$$

The concentration will reach the minimum effective level in 8.3 hours, so the dose should be repeated approximately every 8 hours.

EXAMPLE 9

ESTIMATING LEARNING TIME (CONTINUATION OF EXAMPLE 3)

After t weeks of training, your secretary can type

$$S(t) = 100(1 - e^{-0.25t})$$

words per minute. How many weeks will he take to reach a speed of 80 words per minute?

Solution We solve for t in the following equation:

$$100(1 - e^{-0.25t}) = 80 \qquad \text{Setting } S(t) \text{ equal to 80}$$
$$1 - e^{-0.25t} = 0.80 \qquad \text{Dividing by 100}$$
$$-e^{-0.25t} = -0.20 \qquad \text{Subtracting 1}$$
$$e^{-0.25t} = 0.20 \qquad \text{Multiplying by } -1$$
$$-0.25t = \ln 0.20 \qquad \text{Taking natural logs}$$
$$t = \frac{\ln 0.20}{-0.25} \qquad \text{Solving for } t$$
$$\approx \frac{-1.6094}{-0.25} \approx 6.4 \qquad \text{Using a calculator}$$

He will reach a speed of 80 words per minute in about $6\frac{1}{2}$ weeks.

The last two examples have involved solving for t in an equation of the form $e^{at} = b$. Such equations occur frequently in applications, and are solved by taking logarithms of both sides to bring down the exponent. On a graphing calculator such equations can be solved by graphing the function (using x for ease of entry) together with a constant function and using INTERSECT.

Graphing Calculator Exploration

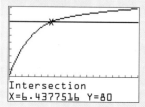

In Example 9, find when the typing speed $S(t)$ reaches 80 by entering $y_1 = 100(1 - e^{-0.25x})$ and $y_2 = 80$, graphing both (you should be able to find an appropriate window), and finding the value where they INTERSECT.

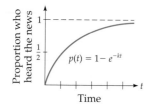

Social Science: Diffusion of Information by Mass Media

When a news bulletin is repeatedly broadcast over radio and television, the proportion of people who hear the bulletin within t hours is

$$p(t) = 1 - e^{-kt}$$

for some constant k.

EXAMPLE 10

PREDICTING THE SPREAD OF INFORMATION

A storm warning is broadcast, and the proportion of people who hear the bulletin within t hours of its first broadcast is $p(t) = 1 - e^{-0.30t}$. When will 75% of the people have heard the bulletin?

Solution

Equating the proportions gives $1 - e^{-0.30t} = 0.75$. Solving this equation as in the other examples (omitting the details) gives $t \approx 4.6$. Therefore, it takes about $4\frac{1}{2}$ hours for 75% of the people to hear the news.

10.1 Section Summary

A quantity P that increases at rate r per year for t years will grow to

$$P(1 + r)^t \qquad P = \text{initial amount}$$

For a quantity that *decreases*, r will be *negative*.

In calculus, the most useful base for exponential and logarithmic functions is the constant e, since it makes the differentiation formulas as simple as possible (as we will see in the next section).

$$e = \lim_{n \to \infty} \left(1 + \frac{1}{n}\right)^n = 2.71828\cdots$$

The natural logarithm of a number x, written $\ln x$, is the power of e that gives the number.

$$\ln x = y \quad \text{is equivalent to} \quad e^y = x \qquad \ln x \text{ means } \log_e x$$

Natural logarithms and exponentials are easily found on a calculator by using the $\boxed{\text{LN}}$ and $\boxed{\text{2nd}}$ keys. The properties of natural logarithms (see page 725) come directly from familiar properties of exponents, and are useful for solving for variables in exponents and for simplifying functions.

> **Solutions to Practice Problems**
>
> 1. $341{,}000(1 - 0.01)^{10} = 341{,}000 \cdot 0.99^{10} \approx 308{,}394$
>
>   In ten years the population will be about 308,400.
>
> 2. $\ln \sqrt[3]{e} = \ln e^{1/3} = \dfrac{1}{3}$

10.1 Exercises

Graph each function. If you are using [calculator], make a hand-drawn sketch from the screen.

1. $f(x) = 2^{x/2}$
2. $f(x) = \left(\tfrac{1}{2}\right)^{x/2}$
3. $f(x) = \ln(x + 1)$
4. $f(x) = 1 - \ln x$

[*Hint for Exercises 3 and 4:* You may need to sketch parts not shown on your graphing calculator.]

Evaluate each expression *without* using a calculator.

5. **a.** $e^{\ln 99}$ **b.** $\ln\left(\dfrac{1}{\sqrt{e}}\right)$ **c.** $\ln(e^{\ln 1})$

6. **a.** $\ln(e^{-500})$ **b.** $e^{\ln(e^2)}$ **c.** $\ln(e^{\ln e})$

Find each natural logarithm using a calculator and then raise e to the resulting power (include all decimal places shown on your screen) to check that you get back (approximately) the original number.

7. $\ln 2.75$
8. $\ln 5.85$

Use the properties of natural logarithms to simplify each function.

9. $f(x) = \ln(10x) - \ln 10$
10. $f(x) = \ln\left(\tfrac{x}{2}\right) + \ln 2 - \ln 1$
11. $f(x) = \ln\left(\tfrac{1}{x}\right) + \ln x + \ln e$
12. $f(x) = \ln(x^3) - \ln(x^2) - \ln x$
13. $f(x) = \dfrac{e^{\ln(3x^2)}}{x} - x^{\ln e}$
14. $f(x) = \dfrac{\ln(e^{4x^2})}{x} - e^{\ln x}$

APPLIED EXERCISES

15. **GENERAL: Art Appreciation** A Picasso etching, *The Frugal Repast*, bought for $250 in 1944, has been appreciating in value by 14% per year. Predict its value in the year 2010. (Round your answer to the nearest hundred dollars.)

16. **GENERAL: Art Appreciation** A Van Gogh painting, *Intérieur d'un Restaurant*, was auctioned in 1996 for $10,342,500. Assuming that it continues to appreciate in value by 10% per year, predict its value in 2016. (Round your answer to the nearest thousand dollars.)

17. **GENERAL: Asset Appreciation** A mint-condition 1910 Honus Wagner* baseball card was sold at auction in 1996 for $640,500, a record price for an item of sports memorabilia. Assuming that it continues to appreciate in value by 11% per year, predict its value in the year 2016. (Round your answer to the nearest hundred dollars.)

18. **GENERAL: Automobile Appreciation** A 1929 Packard 640 Custom Eight Model 341 Phaeton automobile sold in 2002 for $126,000.

---

* Honus Wagner, regarded as the greatest player of his time, played for the Pittsburgh Pirates. Baseball cards were then sold with cigarettes, and Wagner became even more famous when he refused to accept payment for the use of his picture as a protest against the use of tobacco.

If it continues to appreciate in value by 9%, predict its value in the year 2012. (Round your answer to the nearest hundred dollars.)

19. **GENERAL: Population** According to the United Nations Fund for Population Activities, the population of the world x years after the year 2000 will be $P(x) = 5.98e^{0.0125x}$ billion people (for $0 \leq x \leq 20$). Use this formula to predict the world population in 2015.

20. **GENERAL: Population** The most populous country today is China, with a (2000) population of 1.26 billion, increasing by 0.8% per year. The 2000 population of India was 1.02 billion, increasing by 1.7% per year. Assuming that these growth rates continue, which will be larger in 2025?

21–22. BIOMEDICAL: Drug Dosage If a dosage d of a drug is administered to a patient, the amount of the drug remaining in the tissues t hours later will be $f(t) = d \cdot e^{-kt}$ where k (the "absorption constant") depends on the drug.*

21. For the cardioregulator digoxin, the absorption constant is $k = 0.018$. For a dose of $d = 2$ mg, use the above formula to find the amount remaining in the tissues after:

    a. 24 hours   b. 48 hours

22. For the immunosupressent cyclosporine, the absorption constant is $k = 0.012$. For a dose of $d = 400$ mg, use the above formula to find the amount remaining in the tissues after:

    a. 24 hours   b. 48 hours

23. **BUSINESS: Advertising** After t days of advertisements for a new laundry detergent, the proportion of shoppers in a town who have seen the ad is $1 - e^{-0.04t}$. How long must the ad run to reach:

    a. 50% of the shoppers?
    b. 90% of the shoppers?
    c. Graph $p(x) = 1 - e^{-0.04x}$ on a graphing calculator (using an appropriate window) together with horizontal lines representing 50% and 90% and verify your answers to parts (a) and (b) using INTERSECT.

24. **BUSINESS: Advertising** After t days of advertisements for a new television program, the proportion of viewers in a city who have seen the ad is $1 - e^{-0.07t}$. How long must the ad run to reach:

    a. 60% of the viewers?
    b. 90% of the viewers?
    c. Graph $p(x) = 1 - e^{-0.07x}$ on a graphing calculator (using an appropriate window) together with horizontal lines representing 60% and 90% and verify your answers to parts (a) and (b) using INTERSECT.

25. **BEHAVIORAL SCIENCE: Memory** In a psychology experiment the proportion of students who could remember an eight-digit number correctly for t minutes was $e^{-0.08t}$.

    a. What proportion remembered the number correctly after 5 minutes?
    b. How long did it take for the proportion to fall to 20%?
    c. Graph $r(x) = e^{-0.08x}$ on a graphing calculator (using an appropriate window) together with a horizontal line representing 20% and verify your answer to part (b) using INTERSECT.

26. **BEHAVIORAL SCIENCE: Learning** In a psychology experiment, the proportion of information from a photograph that a student could remember after t minutes was $e^{-0.18t}$.

    a. What proportion was remembered after 2 minutes?
    b. How long did it take for the proportion to fall to 50%?
    c. Graph $r(x) = e^{-0.18x}$ on a graphing calculator (using an appropriate window) together with a horizontal line representing 50% and verify your answer to part (b) using INTERSECT.

27. **SOCIAL SCIENCE: Diffusion of Information by Mass Media** Election returns are broadcast in a town of 50,000 people, and the number of people who have heard the news within t hours is $50{,}000(1 - e^{-0.3t})$.

---

* For further details, see *Harrison's Principles of Internal Medicine*, 15th ed. (McGraw Hill, 2001).

a. How many people will have heard the news within 3 hours?
b. How long will it take for 40,000 people to hear the news?
c. Graph $p(x) = 50{,}000(1 - e^{-0.3x})$ on a graphing calculator (using an appropriate window) together with a horizontal line representing 40,000 and verify your answer to part (b) using INTERSECT.

28. SOCIAL SCIENCE: Diffusion of Information by Mass Media A bulletin about school closings is broadcast in a city of 200,000 people, and the number of people who have heard the bulletin within t hours is $200{,}000(1 - e^{-0.25t})$.

a. How many people will have heard the bulletin within 2 hours?
b. How long will it take for three quarters of the city to hear the bulletin?
c. Graph $p(x) = 200{,}000(1 - e^{-0.25x})$ on a graphing calculator (using an appropriate window) together with a horizontal line representing three quarters of the population and verify your answer to part (b) using INTERSECT.

29–30. BIOMEDICAL: Drug Dosage If the original concentration of a drug in a patient's bloodstream is c (milligrams per milliliter), then after t hours the concentration will be ce^{-kt}, where k is the absorption constant.

29. a. If the original concentration is 4 and the absorption constant is 0.25, when should the drug be readministered so that the concentration does not fall below the minimum effective concentration of 1.4?
b. Graph the concentration function on a graphing calculator (using an appropriate window) together with a horizontal line representing the minimum effective concentration and verify your answer to part (a) using INTERSECT.

30. a. If the original concentration is 5 and the absorption constant is 0.15, when should the drug be readministered so that the concentration does not fall below the minimum effective concentration of 2.7?
b. Graph the concentration function on a graphing calculator (using an appropriate window) together with a horizontal line representing the minimum effective concentration and verify your answer to part (a) using INTERSECT.

31. ENVIRONMENTAL SCIENCE: Radioactive Waste Hospitals use radioactive tracers in many medical tests. After the tracer is used, it must be stored as radioactive waste until its radioactivity has decreased enough to be disposed of as ordinary chemical waste. For the radioactive isotope of potassium, the proportion of radioactivity remaining after t days is $e^{-0.05t}$. How soon will the proportion of radioactivity decrease to 0.001 so that it can be disposed of as ordinary chemical waste?

32. ENVIRONMENTAL SCIENCE: Rain Forests It has been estimated[*] that the world's tropical rain forests are disappearing at the rate of 1.8% per year. If this rate continues, how soon will the rain forests be reduced to 50% of their present size? (Rain forests not only generate much of the oxygen that we breathe but also contain plants with unique medical properties, such as the rosy periwinkle, which has revolutionized the treatment of leukemia.)

33. BIOMEDICAL: Half-Life of a Drug The time required for the amount of a drug in one's body to decrease by half is called the "half-life" of the drug.

a. For a drug with absorption constant k, derive the following formula for its half-life:
$$(\text{Half-life}) = \frac{\ln 2}{k}$$

[*Hint:* Solve the equation $e^{-kt} = \frac{1}{2}$ for t and use the properties of logarithms.]

b. Find the half-life of the cardioregulator digoxin if its absorption constant is $k = 0.018$ and time is measured in hours.

---

[*] Paul R. Ehrlich and Edward O. Wilson, "Biodiversity Studies: Science and Policy," *Science* **253**:758–762.

34. SOCIAL SCIENCE: Cell Phone Usage Between 1990 and 2000 the number of cell phone subscribers worldwide increased by 52% annually. At this rate, what is the doubling time for cell phone subscribers? Express your answer in years and months.

35. ENVIRONMENTAL SCIENCE: Nuclear Waste More than half a century after the beginning of the nuclear age, not a single country has found a safe or permanent way to dispose of long-lived radioactive waste. Among the most hazardous radioactive waste is irradiated fuel from nuclear power plants, totaling 245,000 tons in 2000 and growing by 11.3% annually. At this rate, how long will it take for this amount to double? *(Source: Worldwatch Institute)*

36. SOCIAL SCIENCE: Education and Income According to a 1992 study,* each additional year of education increases one's income by 16%. How many additional years of schooling would then be required to double one's income?

* Orley Ashenfelter and Alan Krueger, "Estimate of the Economic Return to Schooling from a New Sample of Twins," *The American Economic Review*, **84**(5):1157–1173.

10.2 DIFFERENTIATION OF LOGARITHMIC AND EXPONENTIAL FUNCTIONS

Introduction

In the previous section we reviewed exponential and logarithmic functions, and in this section we differentiate these new functions and use their derivatives for graphing, optimization, and finding rates of change. We emphasize *natural* (base *e*) logs and exponentials, since most applications use these exclusively. Verifications of the differentiation rules are given at the end of the section.

Derivatives of Logarithmic Functions

The rule for differentiating the natural logarithm function is as follows:

Derivative of ln x

$$\frac{d}{dx} \ln x = \frac{1}{x}$$

The derivative of $\ln x$ is 1 over x

EXAMPLE 1 DIFFERENTIATING A LOGARITHMIC FUNCTION

Differentiate $f(x) = x^3 \ln x$.

Solution The function is a *product*, x^3 times $\ln x$, so we use the Product Rule.

$$\frac{d}{dx}(x^3 \ln x) = 3x^2 \ln x + x^3 \frac{1}{x} = 3x^2 \ln x + x^2$$

- Derivative of the first
- Second left alone
- First left alone
- Derivative of $\ln x$
- From $x^3 \frac{1}{x} = x^2$

Practice Problem 1 Differentiate $f(x) = \dfrac{\ln x}{x}$. ▶ Solution on page 745

The preceding rule, together with the Chain Rule, shows how to differentiate the natural logarithm of a *function*. For any differentiable function $f(x)$ that is positive:

Derivative of $\ln f(x)$

$$\frac{d}{dx} \ln f(x) = \frac{f'(x)}{f(x)}$$

The derivative of the natural log of a function is the derivative of the function over the function

Notice that the right-hand side does not involve logarithms at all.

EXAMPLE 2   **DIFFERENTIATING A LOGARITHMIC FUNCTION**

$$\frac{d}{dx} \ln(x^2 + 1) = \frac{2x}{x^2 + 1}$$

← Derivative of $x^2 + 1$
← Original function (without the ln)

As we observed, the answer does not involve logarithms.

Practice Problem 2   Find $\dfrac{d}{dx} \ln(x^3 - 5x + 1)$. ▶ Solution on page 745

EXAMPLE 3   DIFFERENTIATING A LOGARITHMIC FUNCTION

Find the derivative of $f(x) = \ln(x^4 - 1)^3$.

Solution   For this problem we need the rule for differentiating the natural logarithm of a function, together with the Generalized Power Rule [for differentiating $(x^4 - 1)^3$].

$$\frac{d}{dx} \ln(x^4 - 1)^3 = \frac{\frac{d}{dx}(x^4-1)^3}{(x^4-1)^3} \qquad \text{Using } \frac{d}{dx} \ln f = \frac{f'}{f}$$

$$= \frac{3(x^4-1)^2 \, 4x^3}{(x^4-1)^3} \qquad \text{Using the Generalized Power Rule}$$

$$= \frac{12x^3}{x^4 - 1} \qquad \text{Dividing top and bottom by } (x^4-1)^2$$

Alternative Solution   It is easier if we simplify first, using property 8 of logarithms (see the inside back cover) to bring down the exponent 3:

$$\ln(x^4-1)^3 = 3\ln(x^4-1) \qquad \text{Using } \ln(M^P) = P \ln M$$

Now we differentiate the simplified expression:

$$\frac{d}{dx} 3\ln(x^4-1) = 3\frac{4x^3}{x^4-1} = \frac{12x^3}{x^4-1} \qquad \text{Same answer as before}$$

Moral: Changing $\ln(\cdots)^n$ to $n \ln(\cdots)$ simplifies differentiation.

Derivatives of Exponential Functions

The rule for differentiating the exponential function e^x is as follows:

Derivative of e^x

$$\frac{d}{dx} e^x = e^x \qquad \text{The derivative of } e^x \text{ is simply } e^x$$

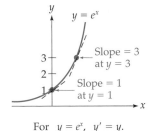

For $y = e^x$, $y' = y$.

This shows the rather surprising fact that e^x is its own derivative. Stated another way, the function e^x is unchanged by the operation of differentiation.

This rule can be interpreted graphically. If $y = e^x$, then $y' = e^x$, so that $y = y'$. This means that on the graph of $y = e^x$, the slope y' always equals the y-coordinate, as shown in the graph on the left.

Graphing Calculator Exploration

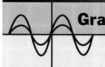

a. Define $y_1 = e^x$ and y_2 as the derivative of y_1 (using NDERIV) and graph them together on the window $[-3, 3]$ by $[-1, 10]$.
b. Why does the screen show only one curve?
c. Use TRACE to compare the values of y_1 and y_2 at some chosen x-value. Do the y-values agree *exactly*? If not, explain the slight discrepancy. [*Hint:* Is NDERIV really the derivative?]

EXAMPLE 4 **FINDING A DERIVATIVE INVOLVING e^x**

Find $\dfrac{d}{dx}\left(\dfrac{e^x}{x}\right)$.

Solution Since the function is a quotient, we use the Quotient Rule:

$$\frac{d}{dx}\left(\frac{e^x}{x}\right) = \frac{x \cdot e^x - 1 \cdot e^x}{x^2} = \frac{xe^x - e^x}{x^2}$$

(Derivative of e^x; Derivative of x)

EXAMPLE 5 **EVALUATING A DERIVATIVE INVOLVING e^x**

If $f(x) = x^2 e^x$, find $f'(1)$.

Solution

$$f'(x) = 2xe^x + x^2 e^x \qquad \text{Using the Product Rule on } x^2 \cdot e^x$$
$$f'(1) = 2(1)e^1 + (1)^2 e^1 \qquad \text{Substituting } x = 1$$
$$= 2e + e = 3e \qquad \text{Simplifying}$$

10.2 DIFFERENTIATION OF LOGARITHMIC AND EXPONENTIAL FUNCTIONS

In these problems we leave our answers in their "exact" forms, leaving e as e. Later, in applied problems, we will approximate our answers using $e \approx 2.718$ or a calculator.

Practice Problem 3 If $f(x) = xe^x$, find $f'(1)$. ▶ Solution on page 745

The rule for differentiating e^x, together with the Chain Rule, shows how to differentiate $e^{f(x)}$. For any differentiable function $f(x)$:

Derivative of $e^{f(x)}$

$$\frac{d}{dx} e^{f(x)} = e^{f(x)} \cdot f'(x)$$

The derivative of e to a function is e to the function times the derivative of the function

That is, to differentiate $e^{f(x)}$ we simply "copy" the original $e^{f(x)}$ and then multiply by the derivative of the exponent.

EXAMPLE 6 DIFFERENTIATING AN EXPONENTIAL FUNCTION

$$\frac{d}{dx} e^{x^4+1} = e^{x^4+1}(4x^3) = 4x^3 e^{x^4+1} \qquad \text{Reversing the order}$$

Copied — Derivative of the exponent

EXAMPLE 7 DIFFERENTIATING AN EXPONENTIAL FUNCTION

Find $\dfrac{d}{dx} e^{x^2/2}$.

Solution The exponent $x^2/2$ should first be rewritten as $\frac{1}{2}x^2$, a constant times x to a power, since then its derivative is easily seen to be x.

$$\frac{d}{dx} e^{x^2/2} = \frac{d}{dx} e^{\frac{1}{2}x^2} = e^{\frac{1}{2}x^2}(x) = xe^{\frac{1}{2}x^2} = xe^{x^2/2} \qquad \begin{array}{l}\text{Rewriting the exponent}\\ \text{in its original form}\end{array}$$

Derivative of the exponent

Practice Problem 4 Find $\dfrac{d}{dx} e^{1+x^3/3}$.

▶ Solution on page 745

The formulas for differentiating natural logarithmic and exponential functions are summarized as follows, with $f(x)$ written simply as f.

Logarithmic Formulas	Exponential Formulas	
$\dfrac{d}{dx} \ln x = \dfrac{1}{x}$	$\dfrac{d}{dx} e^x = e^x$	Top formulas apply only to $\ln x$ and e^x
$\dfrac{d}{dx} \ln f = \dfrac{f'}{f}$	$\dfrac{d}{dx} e^f = e^f \cdot f'$	Bottom formulas apply to ln and e of a *function*

e^x versus x^n

Notice that we do *not* take the derivative of e^x by the Power Rule,

$$\frac{d}{dx} x^n = nx^{n-1}$$

This is because the Power Rule applies to x^n, *a variable to a constant power*, while e^x is *a constant to a variable power*. The two types of functions are quite different, as their graphs show.

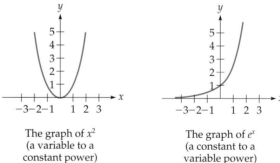

The graph of x^2 (a variable to a constant power)

The graph of e^x (a constant to a variable power)

Each type of function has its own differentiation formula.

$$\underbrace{\frac{d}{dx} x^n = nx^{n-1}}_{\text{For a variable } x \text{ to a constant power } n} \qquad \underbrace{\frac{d}{dx} e^x = e^x}_{\text{For the constant } e \text{ to a variable power } x}$$

10.2 DIFFERENTIATION OF LOGARITHMIC AND EXPONENTIAL FUNCTIONS

EXAMPLE 8

DIFFERENTIATING A LOGARITHMIC AND EXPONENTIAL FUNCTION

Find the derivative of $\ln(1 + e^x)$.

Solution

$$\frac{d}{dx}\ln(1 + e^x) = \frac{\frac{d}{dx}(1 + e^x)}{1 + e^x} = \frac{e^x}{1 + e^x}$$

$\underbrace{\qquad\qquad}$ Using $\frac{d}{dx}\ln f = \frac{f'}{f}$ $\qquad$ $\underbrace{\qquad\qquad}$ Working out the numerator

Functions of the form e^{kx} (for constant k) arise in many applications. The derivative of e^{kx} is as follows.

$$\frac{d}{dx}e^{kx} = e^{kx} \cdot k = ke^{kx} \qquad\qquad \text{Using } \frac{d}{dx}e^f = e^f \cdot f'$$

↑ Derivative of the exponent

This result is so useful that we record it as a separate formula.

Derivative of e^{kx}

$$\frac{d}{dx}e^{kx} = ke^{kx} \qquad\qquad \text{For any constant } k$$

This formula says that the rate of change (the derivative) of e^{kx} is proportional to itself. That is, the function $y = e^{kx}$ satisfies the *differential equation*

$$y' = ky$$

We noted this earlier when we observed that in exponential growth a quantity *grows in proportion to itself* (as in populations and savings accounts).

These differentiation formulas enable us to find instantaneous rates of change of logarithmic and exponential functions. In many applications the variable stands for time, so we use t instead of x.

EXAMPLE 9

FINDING A RATE OF IMPROVEMENT OF A SKILL

After t weeks of practice a pole vaulter can vault

$$H(t) = 14(1 - e^{-0.10t})$$

feet. Find the rate of change of the athlete's jumps after

a. 0 weeks (at the beginning of training)   **b.** 12 weeks

Solution

$$H(t) = 14(1 - e^{-0.10t}) = 14 - 14e^{-0.10t} \quad H(t) \text{ multiplied out}$$

We differentiate to find the rate of change

$$H'(t) = \underbrace{-14(-0.10)e^{-0.10t}}_{\text{Using } \frac{d}{dt}e^{kt} = ke^{kt}} = \underbrace{1.4e^{-0.10t}}_{\text{Simplifying}} \quad \begin{array}{l}\text{Differentiating}\\ 14 - 14e^{-0.10t}\end{array}$$

a. For the rate of change after 0 weeks:

$$H'(0) = 1.4e^{-0.10(0)} = 1.4e^0 = 1.4 \quad \begin{array}{l}H'(t) = 1.4e^{-0.10t}\\ \text{with } t = 0\end{array}$$

b. After 12 weeks:

$$H'(12) = 1.4e^{-0.10(12)}$$
$$= 1.4e^{-1.2} \approx 1.4(0.30) = 0.42 \quad \begin{array}{l}H'(t) = 1.4e^{-0.10t}\\ \text{with } t = 12\end{array}$$

Using a calculator

At first the vaults increased by 1.4 feet per week. After 12 weeks, the gain was only 0.42 foot (about 5 inches) per week.

This result is typical of learning a new skill: Early improvement is rapid, later improvement is slower.

Maximizing Consumer Expenditure

The amount of a commodity that consumers will buy depends on the price of the commodity. For a commodity whose price is p, let the consumer demand be given by a function $D(p)$. Multiplying the number of units $D(p)$ by the price p gives the total *consumer expenditure* for the commodity.

10.2 DIFFERENTIATION OF LOGARITHMIC AND EXPONENTIAL FUNCTIONS

Consumer Demand and Expenditure

Let $D(p)$ be the consumer demand at price p. Then the consumer expenditure is

$$E(p) = p \cdot D(p)$$

EXAMPLE 10 MAXIMIZING CONSUMER EXPENDITURE

If consumer demand for a commodity is $D(p) = 10{,}000e^{-0.02p}$ units per week, where p is the selling price, find the price that maximizes consumer expenditure.

Solution

Using the formula above for consumer expenditure

$$E(p) = p \cdot 10{,}000e^{-0.02p} = 10{,}000pe^{-0.02p} \qquad E(p) = p \cdot D(p)$$

To maximize $E(p)$ we differentiate:

$$E'(p) = \underbrace{10{,}000e^{-0.02p}}_{\text{Derivative of } 10{,}000p} + \underbrace{10{,}000p(-0.02)e^{-0.02p}}_{\text{Derivative of } e^{-0.02p}} \qquad \text{Using the Product Rule to differentiate } E(p) = 10{,}000p \cdot e^{-0.02p}$$

$$= 10{,}000e^{-0.02p} - 200pe^{-0.02p} \qquad \text{Simplifying}$$

$$= 200e^{-0.02p}(50 - p) \qquad \text{Factoring}$$

CN: $p = 50$ — Critical number from $(50 - p)$ (since e to a power is never zero)

We calculate E'' for the second-derivative test:

$$E''(p) = 200(-0.02)e^{-0.02p}(50 - p) + 200e^{-0.02p}(-1) \qquad \text{From } E'(p) = 200e^{-0.02p}(50 - p) \text{ using the Product Rule}$$

$$= -4e^{-0.02p}(50 - p) - 200e^{-0.02p} \qquad \text{Simplifying}$$

At the critical number $p = 50$,

$$E''(50) = -4e^{-0.02(50)}(50 - 50) - 200e^{-0.02(50)} \qquad \text{Substituting } p = 50$$

$$= -200e^{-1} = \frac{-200}{e} \qquad \text{Simplifying}$$

E'' is negative, so the expenditure $E(p)$ is maximized at $p = 50$:

Consumer expenditure is maximized at price $p = \$50$.

exponents. We begin by using one of the "inverse" properties of logarithms, $x = e^{\ln x}$, but with x replaced by x^n:

$$f(x) = x^n = e^{\ln x^n} = e^{n \ln x}$$
Using the property that logs bring down exponents

Differentiating:

$$f'(x) = e^{n \ln x}\left(n\frac{1}{x}\right)$$
Differentiating $f(x) = e^{n \ln x}$ by the $e^{f(x)}$ formula (page 737)

$\underbrace{\phantom{e^{n \ln x}}}_{x^{n-1}}$ $\underbrace{\phantom{(n\frac{1}{x})}}$ — Derivative of the exponent

$$= x^n \cdot \frac{1}{x} \cdot n = nx^{n-1}$$
Replacing $e^{n \ln x}$ by x^n, reordering, and combining x's

$\underbrace{}_{x^{n-1}}$

This is what we wanted to show—that the derivative of $f(x) = x^n$ is $f'(x) = nx^{n-1}$ for *all* exponents n. (The equation $x^n = e^{n \ln x}$ can be taken as a *definition* of x to a power. A definition of logarithms without recourse to exponents will be given in Exercise 75 on page 814).

Verification of the Differentiation Formulas

The formula for the derivative of the natural logarithm function comes from applying the definition of the derivative to $f(x) = \ln x$.

$$f'(x) = \lim_{h \to 0} \frac{f(x+h) - f(x)}{h} = \lim_{h \to 0} \frac{\ln(x+h) - \ln x}{h}$$
Definition of $f'(x)$

$$= \lim_{h \to 0} \frac{1}{h}[\ln(x+h) - \ln x]$$
Dividing by h is equivalent to multiplying by $1/h$

$$= \lim_{h \to 0} \frac{1}{h} \ln\left(\frac{x+h}{x}\right) = \lim_{h \to 0} \frac{1}{h} \ln\left(1 + \frac{h}{x}\right)$$
Using property 7 of logarithms (inside back cover), and simplifying

$$= \lim_{h \to 0} \frac{1}{x} \frac{x}{h} \ln\left(1 + \frac{h}{x}\right) = \lim_{h \to 0} \frac{1}{x} \ln\left(1 + \frac{h}{x}\right)^{x/h}$$
Dividing and multiplying by x, and then using property 8 of logarithms

$$= \lim_{n \to \infty} \frac{1}{x} \ln\left(1 + \frac{1}{n}\right)^n$$
Defining $n = x/h$, so that $h \to 0$ implies $n \to \infty$ (for $h > 0$)

$\underbrace{}$
Approaches e as $n \to \infty$

$$= \frac{1}{x} \ln e = \frac{1}{x}$$
Since $\ln e = 1$ (the same conclusion follows if $h < 0$)

This is the result that we wanted to show—that the derivative of the natural logarithm function $f(x) = \ln x$ is $f'(x) = \frac{1}{x}$.

10.2 DIFFERENTIATION OF LOGARITHMIC AND EXPONENTIAL FUNCTIONS

For the rule to differentiate the natural logarithm of a (positive) function, we begin with

$$\frac{d}{dx} f(g(x)) = f'(g(x))g'(x) \qquad \text{Chain Rule (from page 610) for differentiable functions } f \text{ and } g$$

$$\frac{d}{dx} \ln(g(x)) = \frac{1}{g(x)} g'(x) = \frac{g'(x)}{g(x)} \qquad \text{Taking } f(x) = \ln x, \text{ so } f'(x) = \frac{1}{x}$$

Replacing g by f, this is exactly the formula we wanted to show,

$$\frac{d}{dx} \ln f(x) = \frac{f'(x)}{f(x)}$$

To derive the rule for differentiating e^x we begin with

$$\ln e^x = x \qquad \text{Property 3 of natural logarithms}$$

and differentiate both sides:

$$\frac{\frac{d}{dx} e^x}{e^x} = 1 \qquad \text{Using } \frac{d}{dx} \ln f = \frac{f'}{f} \text{ on the left side}$$

Multiplying each side by e^x gives

$$\frac{d}{dx} e^x = e^x$$

This is the rule for differentiating e^x. This rule together with the Chain Rule gives the rule for differentiating $e^{f(x)}$, just as before.

▶ Solutions to Practice Problems

1. $f'(x) = \dfrac{d}{dx}\left(\dfrac{\ln x}{x}\right) = \dfrac{x\left(\frac{1}{x}\right) - 1 \cdot \ln x}{x^2}$ — Derivative of $\ln x$ / Derivative of x

   $= \dfrac{1 - \ln x}{x^2}$

2. $\dfrac{d}{dx} \ln(x^3 - 5x + 1) = \dfrac{3x^2 - 5}{x^3 - 5x + 1}$

3. $f'(x) = e^x + xe^x$

   $f'(1) = e^1 + 1e^1 = 2e$ (which is approximately $2 \cdot 2.718 = 5.436$)

4. $\dfrac{d}{dx} e^{1 + x^3/3} = \dfrac{d}{dx} e^{1 + \frac{1}{3}x^3} = e^{1 + \frac{1}{3}x^3}(x^2) = x^2 e^{1 + x^3/3}$

10.2 Exercises

Find the derivative of each function.

1. $f(x) = x^2 \ln x$
2. $f(x) = \dfrac{\ln x}{x^3}$
3. $f(x) = \ln x^2$
4. $f(x) = \ln(x^3 + 1)$
5. $f(x) = \ln \sqrt{x}$
6. $f(x) = \sqrt{\ln x}$
7. $f(x) = \ln(x^2 + 1)^3$
8. $f(x) = \ln(x^4 + 1)^2$
9. $f(x) = \ln(-x)$
10. $f(x) = \ln(5x)$
11. $f(x) = \dfrac{e^x}{x^2}$
12. $f(x) = x^3 e^x$
13. $f(x) = e^{x^3 + 2x}$
14. $f(x) = 2e^{7x}$
15. $f(x) = e^{x^3/3}$
16. $f(x) = \ln(e^x - 2x)$
17. $f(x) = x - e^{-x}$
18. $f(x) = x \ln x - x$
19. $f(x) = \ln e^{2x}$
20. $f(x) = \ln e^x$
21. $f(x) = e^{1+e^x}$
22. $f(x) = \ln(e^x + e^{-x})$
23. $f(x) = x^e$
24. $f(x) = ex$
25. $f(x) = e^3$
26. $f(x) = \sqrt{e}$
27. $f(x) = \ln(x^4 + 1) - 4e^{x/2} - x$
28. $f(x) = x^2 e^x - 2 \ln x + (x^2 + 1)^3$
29. $f(x) = x^2 \ln x - \frac{1}{2}x^2 + e^{x^2} + 5$
30. $f(x) = e^{-2x} - x \ln x + x - 7$

For each function, find the indicated expressions.

31. $f(x) = \dfrac{\ln x}{x^5}$, find a. $f'(x)$ b. $f'(1)$
32. $f(x) = x^4 \ln x$, find a. $f'(x)$ b. $f'(1)$
33. $f(x) = \ln(x^4 + 48)$, find a. $f'(x)$ b. $f'(2)$
34. $f(x) = x^2 \ln x - x^2$, find a. $f'(x)$ b. $f'(e)$
35. $f(x) = \ln(e^x - 3x)$, find a. $f'(x)$ b. $f'(0)$
36. $f(x) = \ln(e^x + e^{-x})$, find a. $f'(x)$ b. $f'(0)$

For each function:
a. Find $f'(x)$.
b. Evaluate the given expression and approximate it to three decimal places.

37. $f(x) = 5x \ln x$, find and approximate $f'(2)$.
38. $f(x) = e^{x^2/2}$, find and approximate $f'(2)$.
39. $f(x) = \dfrac{e^x}{x}$, find and approximate $f'(3)$.
40. $f(x) = \ln(e^x - 1)$, find and approximate $f'(3)$.

Find the *second* derivative of each function.

41. $f(x) = e^{-x^5/5}$
42. $f(x) = e^{-x^6/6}$

By calculating the first few derivatives, find a formula for the nth derivative of each function (k is a constant).

43. $f(x) = e^{kx}$
44. $f(x) = e^{-kx}$

Use your graphing calculator to graph each function on a window that includes all relative extreme points and inflection points, and give the coordinates of these points (rounded to two decimal places). [*Hint:* Use NDERIV once or twice with ZERO.] (Answers may vary depending on the graphing window chosen.)

45. $f(x) = e^{-2x^2}$
46. $f(x) = 1 - e^{-x^2/2}$
47. $f(x) = \ln(1 + x^2)$
48. $f(x) = e^x + e^{-x}$

Use your graphing calculator to graph each function on the indicated interval, and give the coordinates of all relative extreme points and inflection points (rounded to two decimal places). [*Hint:* Use NDERIV once or twice together with ZERO.] (Answers may vary depending on the graphing window chosen.)

49. $f(x) = \dfrac{x^2}{e^x}$ for $-1 \le x \le 8$
50. $f(x) = \dfrac{x}{e^x}$ for $-1 \le x \le 5$
51. $f(x) = x \ln |x|$ for $-2 \le x \le 2$

[Hint for Exercises 51–52: $|x|$ is sometimes entered as ABS (x).]

52. $f(x) = x^2 \ln |x|$ for $-2 \le x \le 2$

Use implicit differentiation to find dy/dx.

53. $y^2 - ye^x = 12$
54. $y^2 - x \ln y = 10$

APPLIED EXERCISES

55. GENERAL: Compound Interest A sum of $1000 at 5% interest compounded continuously will grow to $V(t) = 1000e^{0.05t}$ dollars in t years. Find the rate of growth after:

a. 0 years (the time of the original deposit).
b. 10 years.

[*Hint:* The rate of growth means the derivative.]

56. GENERAL: Depreciation A $10,000 automobile depreciates so that its value after t years is $V(t) = 10{,}000e^{-0.35t}$ dollars. Find the rate of change of its value:

a. when it is new ($t = 0$).
b. after 2 years.

[*Hint:* The rate of change means the derivative.]

57. GENERAL: Population The world population (in billions) is predicted to be $P(t) = 5.89e^{0.0175t}$, where t is the number of years after 2000. Find the rate of change of the population in the year 2010. [*Hint:* The rate of change means the derivative.]

58. BEHAVIORAL SCIENCE: Ebbinghaus Memory Model According to the Ebbinghaus model of memory, if one is shown a list of items, the percentage of items that one will remember t time units later is $P(t) = (100 - a)e^{-bt} + a$, where a and b are constants. For $a = 25$ and $b = 0.2$, this function becomes $P(t) = 75e^{-0.2t} + 25$. Find the rate of change of this percentage:

a. at the beginning of the test ($t = 0$).
b. after 3 time units.

[*Hint:* The rate of change means the derivative.]

59. BIOMEDICAL: Drug Dosage A patient receives an injection of 1.2 milligrams of a drug, and the amount remaining in the bloodstream t hours later is $A(t) = 1.2e^{-0.05t}$. Find the rate of change of this amount:

a. just after the injection (at time $t = 0$).
b. after 2 hours.

[*Hint:* The rate of change means the derivative.]

60. GENERAL: Temperature A covered cup of coffee at 200 degrees, if left in a 70-degree room, will cool to $T(t) = 70 + 130e^{-2.5t}$ degrees in t hours. Find the rate of change of the temperature:

a. at time $t = 0$.
b. after 1 hour.

61. BUSINESS: Sales The weekly sales (in thousands) of a new product are predicted to be $S(x) = 1000 - 900e^{-0.1x}$ after x weeks. Find the rate of change of sales after

a. 1 week.
b. 10 weeks.

62. SOCIAL SCIENCE: Diffusion of Information by Mass Media The number of people in a town of 50,000 who have heard an important news bulletin within t hours of its first broadcast is $N(t) = 50{,}000(1 - e^{-0.4t})$. Find the rate of change of the number of informed people:

a. at time $t = 0$.
b. after 8 hours.

63–64. ECONOMICS: Consumer Expenditure If consumer demand for a commodity is given by the function below (where p is the selling price in dollars), find the price that maximizes consumer expenditure.

63. $D(p) = 5000e^{-0.01p}$ **64.** $D(p) = 8000e^{-0.05p}$

65–66. BUSINESS: Maximizing Revenue Each of the following functions is a company's price function, where p is the price (in dollars) at which quantity x (in thousands) will be sold.

a. Find the revenue function $R(x)$. [*Hint:* Revenue is price times quantity, $p \cdot x$.]
b. Find the quantity and price that will maximize revenue.

65. $p = 400e^{-0.20x}$ **66.** $p = 4 - \ln x$

67. BIOMEDICAL: Population Growth The Gompertz growth curve models the size $N(t)$ of a population at time $t \geq 0$ as $N(t) = Ke^{-ae^{-bt}}$ where K, a, and b are positive constants. Show that $\dfrac{dN}{dt} = bN \ln\left(\dfrac{K}{N}\right)$ and interpret this derivative to make statements about the population growth when $N < K$ and when $N > K$.

68. BIOMEDICAL: Ricker Recruitment The population dynamics of many fish (such as salmon) can be described by the *Ricker curve* $y = axe^{-bx}$ for $x \geq 0$ where $a > 1$ and $b > 0$ are constants, x is the size of the parental stock, and y is the number of recruits (offspring). Determine the size of the parental stock that maximizes the number of recruits.

69. BIOMEDICAL: Reynolds Number An important characteristic of blood flow is the "Reynolds number." As the Reynolds number increases, blood flows less smoothly. For blood flowing through certain arteries, the Reynolds number is

$$R(r) = a \ln r - br$$

where a and b are positive constants and r is the radius of the artery. Find the radius r that maximizes the Reynolds number R. (Your answer will involve the constants a and b.)

70. BIOMEDICAL: Drug Concentration If a drug is injected intramuscularly, the concentration of the drug in the bloodstream after t hours will be

$$A(t) = \frac{c}{b-a}(e^{-at} - e^{-bt})$$

If the constants are $a = 0.4$, $b = 0.6$, and $c = 0.1$, find the time of maximum concentration.

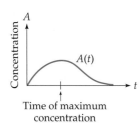

Time of maximum concentration

71. GENERAL: Temperature A mug of beer chilled to 40 degrees, if left in a 70-degree room, will warm to a temperature of $T(t) = 70 - 30e^{-3.5t}$ degrees in t hours. Enter this temperature function as y_1 (using x for ease of entry), define y_2 as its derivative (using NDERIV), and graph them on the window [0, 2] by [0, 80].

a. Evaluate y_1 and y_2 at $x = 0.25$ (using EVALUATE) and interpret your answers.
b. Evaluate y_1 and y_2 at $x = 1$ and interpret your answers.

72. SOCIAL SCIENCE: Diffusion of Information by Mass Media The number of people in a city of 200,000 who have heard a weather bulletin within t hours of its first broadcast is $N(t) = 200{,}000(1 - e^{-0.5t})$. Enter this function as y_1 (using x for ease of entry), define y_2 as its derivative (using NDERIV), and graph them on the window [0, 4] by [0, 200,000].

a. Evaluate y_1 and y_2 at $x = 0.5$ (using EVALUATE) and interpret your answers.
b. Evaluate y_1 and y_2 at $x = 3$ and interpret your answers.

73–74. ATHLETICS: World's Record 100-Meter Run In 1987 Carl Lewis set a new world's record of 9.93 seconds for the 100-meter run. The distance that he ran in the first x seconds was

$$11.274[x - 1.06(1 - e^{-x/1.06})] \text{ meters}$$

for $0 \leq x \leq 9.93$.* Enter this function as y_1, and define y_2 as its derivative (using NDERIV), so that y_2 gives the velocity after x seconds. Graph them on the window [0, 9.93] by [0, 100].

73. Trace along the velocity curve to verify that Lewis's maximum speed was about 11.27 meters per second. Find how quickly he reached a speed of 10 meters per second, which is 95% of his maximum speed.

74. Define y_3 as the derivative of y_2 (using NDERIV) so that y_3 gives the acceleration after x seconds, and graph y_2 and y_3 on the window [0, 9.93] by [0, 20]. Evaluate both y_2 and y_3 at $x = 0.1$ and also at $x = 9.93$ (using EVALUATE). Interpret your answers.

75. ECONOMICS: Consumer Expenditure If consumer demand for a commodity is given by $D(p) = 4000e^{-0.002p}$ (where p is the selling price in dollars), find the price that maximizes

---

* See W. G. Pritchard, "Mathematical Models of Running," *SIAM Review* **35**(3):359–379.

consumer expenditure. [*Hint:* Find the critical number of the expenditure function "by hand" and then graph it (using x for ease of entry) on an appropriate window to verify that it is maximized at this critical number.]

76. **BUSINESS: Maximizing Revenue** An electronics company finds that the price function for mobile telephones is $p(x) = 400e^{-0.005x}$, where p is the price in dollars at which quantity x (in thousands) will be sold. Find the quantity x that maximizes revenue. [*Hint:* Revenue is price times quantity, $p \cdot x$. Find the critical number "by hand" and then graph the revenue function on an appropriate window to verify that it is maximized at this critical number.]

77. **BIOMEDICAL: Drug Concentration** If a certain drug is injected intramuscularly, the concentration of the drug in the bloodstream after t hours will be $C(t) = 0.75(e^{-0.2t} - e^{-0.6t})$. Find the time of maximum concentration.

78. **ATHLETICS: How Fast Do Old Men Slow Down?** The fastest times for the marathon (26.2 miles) for male runners aged 35 to 80 are approximated by the function*

$$f(x) = \begin{cases} 106.2e^{0.0063x} & \text{if } x \leq 58.2 \\ 850.4e^{0.000614x^2 - 0.0652x} & \text{if } x > 58.2 \end{cases}$$

in minutes, where x is the age of the runner.

a. Graph this function on the window [35, 80] by [0, 240].
   [*Hint:* On some calculators, enter $y_1 = (106.2e^{0.0063x})(x \leq 58.2) + (850.4e^{0.000614x^2 - 0.0652x})(x > 58.2).$]
b. Find $f(35)$ and $f'(35)$ and interpret these numbers. [*Hint:* Use NDERIV or dy/dx.]
c. Find $f(80)$ and $f'(80)$ and interpret these numbers.

* Ray C. Fair, "How Fast Do Old Men Slow Down?," *Review of Economics and Statistics* LXXVI(1):103–118, February 1994.

EXPONENTIAL AND LOGARITHMIC FUNCTIONS TO OTHER BASES

The rules for differentiating exponential functions with (positive) base a are shown below:

Derivatives of a^x and $a^{f(x)}$

$$\frac{d}{dx} a^x = (\ln a)a^x$$

$$\frac{d}{dx} a^{f(x)} = (\ln a)a^{f(x)}f'(x) \qquad \text{For a differentiable function } f$$

For example,

$$\frac{d}{dx} 2^x = (\ln 2)2^x$$

$$\frac{d}{dx} 5^{3x^2+1} = (\ln 5)5^{3x^2+1}(6x) = 6(\ln 5)x \, 5^{3x^2+1}$$

Graphing Calculator Exploration

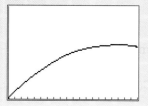

Enter the function from Practice Problem 1 into a graphing calculator as $y_1 = 300 + x^2$, and then "turn off" the function so that it will not graph. Then graph $y_2 = \dfrac{d}{dx} \ln y_1$ and $y_3 = \dfrac{y_1'}{y_1}$ (using NDERIV) together on the window [0, 20] by [0, 0.1]. Why do you get only one curve for the two functions?

Elasticity of Demand

Farmers are aware of the paradox that an abundant harvest usually brings *lower* total revenue than a poor harvest. The reason is simply that the larger quantities in an abundant harvest result in lower prices, which in turn cause increased demand, but the demand does *not* increase enough to compensate for the lower prices.

Revenue is price times quantity, and when one of these quantities increases, the other generally decreases. The question is whether the increase in one is enough to compensate for the decrease in the other. The concept of *elasticity of demand* was devised to answer this question.

Intuitively, elasticity of demand is a measure of how *responsive* demand is to price changes. Think of "elastic" as meaning "very responsive." If demand is elastic, a small price cut will bring a large increase in demand, so total revenue will rise. On the other hand, if demand is *in*elastic, a price cut will bring only a slight increase in demand, so total revenue will fall.

In general, *elastic* demand means that consumers will purchase *significantly* more or less in response to price changes. *Inelastic* demand means that consumers will buy only *slightly* more or less in response to price changes. (This is the cause of the farmers' difficulties: Demand for farm products is inelastic.)

Demand Functions

In general, if the price of an item rises, the demand will fall, and vice versa. If the relationship between the price p of an item and the quantity x that will be sold at that price can be expressed with x as a function of p, $x = D(p)$, that function is called the *demand function*.

10.3 TWO APPLICATIONS TO ECONOMICS: RELATIVE RATES AND ELASTICITY OF DEMAND

Demand Function

The demand function

$$x = D(p)$$

gives the quantity x of an item that will be demanded by consumers at price p.

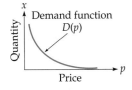

Law of downward-sloping demand

Since, in general, demand falls as prices rise, the slope of the demand function is negative, as shown on the left. This is known as the *law of downward-sloping demand*.

Calculating Elasticity of Demand

For a demand function $D(p)$, let us calculate the relative rate of change of demand divided by the relative rate of change of price. Using the derivative-of-the-logarithm formula,

$$\frac{\begin{pmatrix}\text{Relative rate of}\\\text{change of demand}\end{pmatrix}}{\begin{pmatrix}\text{Relative rate of}\\\text{change of price}\end{pmatrix}} = \frac{\dfrac{d}{dp}\ln D(p)}{\dfrac{d}{dp}\ln p} = \frac{\dfrac{D'(p)}{D(p)}}{\dfrac{1}{p}} = \underbrace{\frac{pD'(p)}{D(p)}}_{\text{Simplified}}$$

Because most demand functions are downward-sloping, the derivative $D'(p)$ is generally negative. Economists prefer to work with positive numbers, so the *elasticity of demand* is taken to be the negative of this quantity (in order to make it positive).*

Elasticity of Demand

For a demand function $D(p)$, the elasticity of demand is

$$E(p) = \frac{-p \cdot D'(p)}{D(p)}$$

Demand is *elastic* if $E(p) > 1$ and *inelastic* if $E(p) < 1$.

Elasticity, being composed of *relative* rates of change, does not depend on the units of the demand function. Therefore, elasticities can be compared between different products, and even between different countries.

---

*Some economists omit the negative sign.

EXAMPLE 2

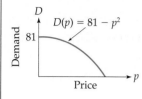

FINDING ELASTICITY OF DEMAND FOR COMMUTER BUS SERVICE

A bus line estimates the demand function for its daily commuter tickets to be $D(p) = 81 - p^2$ (in thousands of tickets), where p is the price in dollars ($0 \leq p \leq 9$). Find the elasticity of demand when the price is:

a. $3   **b.** $6

Solution

$$E(p) = \frac{-pD'(p)}{D(p)} \qquad \text{Definition of elasticity}$$

$$= \frac{-p(-2p)}{81 - p^2} \qquad \text{Substituting } D(p) = 81 - p^2 \text{ so } D'(p) = -2p$$

$$= \frac{2p^2}{81 - p^2} \qquad \text{Simplifying}$$

a. Evaluating at $p = 3$ gives

$$E(3) = \frac{2(3)^2}{81 - (3)^2} = \frac{18}{81 - 9} = \frac{18}{72} = \frac{1}{4} \qquad E(p) = \frac{2p^2}{81 - p^2} \text{ with } p = 3$$

Interpretation: The elasticity is less than 1, so demand for tickets is *inelastic* at a price of $3. This means that a small price change (up or down from this level) will cause only a *slight* change in demand. More precisely, elasticity of $\frac{1}{4}$ means that a 1% price change will cause only about a $\frac{1}{4}$% change in demand.

b. At the price of $6, the elasticity of demand is

$$E(6) = \frac{2(6)^2}{81 - (6)^2} = \frac{72}{81 - 36} = \frac{8}{5} = 1.6 \qquad E(p) = \frac{2p^2}{81 - p^2} \text{ with } p = 6$$

Interpretation: The elasticity is greater than 1, so demand is *elastic* at a price of $6. This means that a small change in price (up or down from this level) will cause a relatively *large* change in demand. In particular, an elasticity of 1.6 means that a price change of 1% will cause about a 1.6% change in demand.

The changes in demand are, of course, in the opposite direction from the changes in price. That is, if prices are *raised* by 1% (from the $6 level), demand will *fall* by 1.6%, whereas if prices are *lowered* by 1%, demand will *rise* by 1.6%. In the future we will assume that the *direc-*

tion of the change is clear, and say simply that a 1% change in price will cause about a 1.6% change in demand.

Practice Problem 2

For the demand function $D(p) = 90 - p$, find the elasticity of demand $E(p)$ and evaluate it at $p = 30$ and $p = 75$. (Be sure to complete this Practice Problem, as the results will be used shortly.)

▶ Solution on page 760

Graphing Calculator Exploration

For any demand function y_1 (written in terms of x), define y_2 as the elasticity function for y_1 by defining $y_2 = -x \cdot y_1'/y_1$ (using NDERIV). Try entering the demand function from Example 2 or Practice Problem 2 in y_1 and then evaluating y_2 at appropriate numbers to check the answers found there.

Using Elasticity to Increase Revenue

In Example 2 we found that at a price of $3, demand is inelastic ($E = \frac{1}{4} < 1$), and so demand responds only *weakly* to price changes. Therefore, to increase revenue the company should *raise* prices, since the higher prices will drive away only a relatively small number of customers. On the other hand, at a price of $6, demand is elastic ($E = 1.6 > 1$), and so demand is very responsive to price changes. In this case, to increase revenue the company should *lower* prices, since this will attract more than enough new customers to compensate for the price decrease. In general:

Using Elasticity to Increase Revenue

To increase revenue:

Raise prices if demand is *inelastic* ($E < 1$).

Lower prices if demand is *elastic* ($E > 1$).

This statement shows why elasticity of demand is important to any company that cuts prices in an attempt to boost revenue, or to any utility that raises prices in order to increase revenue. Elasticity shows whether the strategy will succeed or fail.

APPLIED EXERCISES ON ELASTICITY OF DEMAND

27. **Automobile Sales** An automobile dealer is selling cars at a price of $12,000. The demand function is $D(p) = 2(15 - 0.001p)^2$, where p is the price of a car. Should the dealer raise or lower the price to increase revenue?

28. **Liquor Sales** A liquor distributor wants to increase its revenues by discounting its best-selling liquor. If the demand function for this liquor is $D(p) = 60 - 3p$, where p is the price per bottle, and if the current price is $15, will the discount succeed?

29. **City Bus Revenues** The manager of a city bus line estimates the demand function to be $D(p) = 150{,}000\sqrt{1.75 - p}$, where p is the fare in dollars. The bus line currently charges a fare of $1.25, and it plans to raise the fare to increase its revenues. Will this strategy succeed?

30. **Newspaper Sales** The demand function for a newspaper is $D(p) = 80{,}000\sqrt{75 - p}$, where p is the price in cents. The publisher currently charges 50 cents, and it plans to raise the price to increase revenues. Will this strategy succeed?

31. **Electricity Rates** An electrical utility asks the Federal Regulatory Commission for permission to raise rates to increase revenues. The utility's demand function is

$$D(p) = \frac{120}{10 + p}$$

where p is the price (in cents) of a kilowatt-hour of electricity. If the utility currently charges 6 cents per kilowatt-hour, should the commission grant the request?

32. **Oil Prices** A Middle Eastern oil-producing country estimates that the demand for oil (in millions of barrels) is $D(p) = 28e^{-0.04p}$, where p is the price of a barrel of oil. To raise its revenues, should it raise or lower its price from its current level of $20 per barrel?

33. **Oil Prices** A European oil-producing country estimates that the demand for its oil (in millions of barrels) is $D(p) = 41e^{-0.06p}$, where p is the price of a barrel of oil. To raise its revenues, should it raise or lower its price from its current level of $20 per barrel?

34–35. **Liquor and Beer** The demand functions for distilled spirits and for beer are given below, where p is the retail price and $D(p)$ is the demand in gallons per capita.* For each demand function, find the elasticity of demand for any price p. [*Note:* You will find, in each case, that demand is inelastic. This means that taxation, which acts like a price increase, is an ineffective way of discouraging liquor consumption, but is an effective way of raising revenue.]

34. $D(p) = 3.509 p^{-0.859}$ (for distilled spirits)

35. $D(p) = 7.881 p^{-0.112}$ (for beer)

36. **Constant Elasticity**
   a. Show that for a demand function of the form $D(p) = \dfrac{c}{p^n}$, where c and n are positive constants, the elasticity is constant.
   b. What type of demand function has elasticity equal to 1 for every value of p?

37. **Linear Elasticity** Show that for a demand function of the form $D(p) = ae^{-cp}$, where a and c are positive constants, the elasticity of demand is $E(p) = cp$.

38–39. **Elasticity of Supply** A supply function $S(p)$ gives the total amount of a product that producers are willing to supply at a given price p. The *elasticity of supply* is defined as

$$E_S(p) = \frac{p \cdot S'(p)}{S(p)}$$

Elasticity of supply measures the relative increase in supply resulting from a small relative increase in price. It is less useful than elasticity of demand, however, since it is not related to total revenue.

---

* Stanley Ornstein and Dominique Hanssens, "Alcohol Control Laws and the Consumption of Distilled Spirits and Beer," *Journal of Consumer Research* 12:200–213. Variables in this study other than price have been ignored.

38. Use the preceding formula to find the elasticity of supply for a supply function of the form $S(p) = ae^{cp}$, where a and c are positive constants.

39. Use the preceding formula to find the elasticity of supply for a supply function of the form $S(p) = ap^n$, where a and n are positive constants.

40–41. Automobile Sales The following are demand functions for automobiles in a dealership, where p is the selling price.

a. Use the method described in the Graphing Calculator Exploration on page 757 to find the elasticity of demand at a price of $12,000.
b. Should the dealer raise or lower the price from this level to increase revenue?
c. Find the price at which elasticity equals 1. [*Hint:* Use INTERSECT.]

40. $D(p) = \sqrt[3]{20 - 0.001p}$

41. $D(p) = \dfrac{200}{8 + e^{0.0001p}}$

Chapter Summary with Hints and Suggestions

Reading the text and doing the exercises in this chapter have helped you to master the following skills, which are listed by section (in case you need to review them) and are keyed to particular Review Exercises. Answers for all Review Exercises are given at the back of the book, and full solutions can be found in the Student Solutions Manual.

10.1 Review of Exponential and Logarithmic Functions

- Find an appreciated value or another amount that has increased or decreased annually by a fixed percentage. *(Review Exercises 1–3.)*

  $P(1 + r)^t$ For decrease, r is negative

- Determine which of two drugs provides more medication. *(Review Exercise 4.)*

- Predict the world's largest city, computer memory capacity, or world cigarette production. *(Review Exercises 5–7.)*

- Predict how productivity improves with practice. *(Review Exercise 8.)*

- Use the properties of natural logarithms to simplify a function. *(Review Exercise 9.)*

  See the properties of natural logarithms on page 725

- Find and check a natural logarithm. *(Review Exercise 10.)*

- Predict the spread of information or the demand for oil. *(Review Exercises 11–12.)*

- Predict the time for a population or the demand for a product to increase. *(Review Exercises 13–14.)*

10.2 Differentiation of Logarithmic and Exponential Functions

- Find the derivative of a logarithmic or exponential function. *(Review Exercises 15–30.)*

  $$\dfrac{d}{dx} \ln x = \dfrac{1}{x} \qquad \dfrac{d}{dx} e^x = e^x \qquad \dfrac{d}{dx} \ln f = \dfrac{f'}{f}$$

  $$\dfrac{d}{dx} e^f = e^f \cdot f' \qquad \dfrac{d}{dx} e^{kx} = ke^{kx}$$

- Graph an exponential or a logarithmic function. *(Review Exercises 31–32.)*

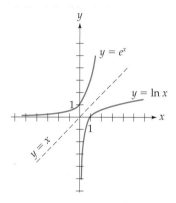

- Find the rate of change of sales, amount of medication, learning, temperature, or diffusion of information. *(Review Exercises 33–37.)*
- Find when a company maximizes its present value. *(Review Exercise 38.)*
- Maximize a company's revenue. *(Review Exercises 39–40.)*
- Maximize consumer expenditure for a product. *(Review Exercise 41.)*
- Graph an exponential or logarithmic function. *(Review Exercises 42–43.)*
- Maximize consumer expenditure or maximize revenue. *(Review Exercises 44–45.)*

10.3 Two Applications to Economics: Relative Rates and Elasticity of Demand

- Find the relative rate of change of a country's gross domestic product. *(Review Exercises 46–47.)*

$$\begin{pmatrix} \text{Relative} \\ \text{rate} \end{pmatrix} = \frac{d}{dt} \ln f = \frac{f'}{f}$$

- Find the elasticity of demand for a product, and its consequences. *(Review Exercises 48–50.)*

$$\begin{pmatrix} \text{Elasticity} \\ \text{of demand} \end{pmatrix} = \frac{-p \cdot D'(p)}{D(p)}$$

- Find the relative rate of change of population. *(Review Exercise 51.)*
- Find the elasticity of demand for a product, how it affects revenue, and what price gives unitary elasticity. *(Review Exercise 52.)*

Hints and Suggestions

- *(Overview)* Exponential and logarithmic functions should be thought of as just other types of functions, like polynomials, but having their own differentiation rules. In fact, in a sense they are more "natural" than polynomials because they give natural growth rates.

- A graphing calculator helps by drawing graphs of exponential and logarithmic functions, and finding intersection points (for example, where one population overtakes another). It is also useful for checking derivatives and graphically verifying maximum and minimum values.

- $P(1 + r)^t$ is the same as the compound interest formula (page 66) for annual compounding ($m = 1$) with annual interest rate or "growth rate" r. For *negative* values of r, it gives the result of annual *decreases* at rate r.

- e is just a number, approximately 2.718, calculated from the expression $\lim_{n \to \infty}(1 + \frac{1}{n})^n$. We use it as a base for exponential and logarithmic functions because e^x and $\ln x$ have much simpler differentiation rules than, say, 2^x and $\log x$.

- The Power Rule $\frac{d}{dx} x^n = nx^{n-1}$ is for differentiating *a variable to a constant power*, and $\frac{d}{dx} e^x = e^x$ is for differentiating *the constant e to a variable power*.

- **Practice for Test:** Review Exercises 2, 3, 8, 9, 11, 14, 15, 19, 21, 23, 27, 31, 33, 37, 41, 45, 47, 49, 51.

Review Exercises for Chapter 10

Practice test exercise numbers are in green.

10.1 Review of Exponential and Logarithmic Functions

1. **GENERAL: Asset Appreciation** A 1949 Scuderia Ferrari (166 MM Barchetta) automobile was sold at auction in 1996 for $1.65 million. If it continues to appreciate in value by 6% per year, find its value in the year 2016.

2. **GENERAL: Population** The population of Cleveland, Ohio, was 478,000 in 2000 and declining at the rate of 1.3% per year. Assuming that this trend continues, predict the population of Cleveland in 2015. (Round your answer to the nearest thousand people.)

3. **GENERAL: Wind Energy** Worldwide wind-generated electricity in 2001 was 24,900 MW, enough to power 14 million households, and is increasing annually by 30%. Assuming that this growth continues, how many households can be powered in 2011? (*Source*: Worldwatch Institute)

4. **BIOMEDICAL: Drug Concentration** If the concentration of a drug in a patient's bloodstream is c (milligrams per milliliter), then t hours later the concentration will be $C(t) = ce^{-kt}$, where k is a constant (the "elimination constant"). Two drugs are being compared: drug A with initial concentration $c = 2$ and elimination constant $k = 0.2$, and drug B with initial concentration $c = 3$ and elimination constant $k = 0.25$. Which drug will have the greater concentration 4 hours later?

5. **GENERAL: Population** The largest city in the world is Tokyo, with São Paulo (Brazil) smaller but growing faster. According to the Census Bureau, x years after 2000 the population of Tokyo will be $27.5e^{0.0034x}$ and the population of São Paulo will be $17.4e^{0.011x}$ (both in millions). Graph both functions on a calculator with the window [0, 100] by [0, 50]. When will São Paulo overtake Tokyo as the world's largest city (assuming that these rates continue to hold)?

6. **GENERAL: Moore's Law of Computer Memory** The amount of information that can be stored on a computer chip can be measured in megabits (a "bit" is a binary digit, 0 or 1, and a "megabit" is a million bits). The first 1-megabit chips became available in 1987, and 4-megabit chips became available in 1990. This quadrupling of capacity every 3 years is expected to continue, so that chip capacity will be $C(t) = 4^{t/3}$ megabits where t is the number of years after 1987. Use this formula (known as *Moore's Law*, after Gordon Moore, a founder of the Intel Corporation) to predict chip capacity in the year 2008. [*Hint:* What value of t corresponds to 2008?]

7. **BUSINESS: Cigarette Manufacturing** World cigarette production per person was 897 in 2002, and declining by 1% per year. Assuming that this trend continues, predict per-person cigarette production in 2017. (*Source*: U.S. Department of Agriculture)

8. **BUSINESS: Productivity** A factory worker assembles portable tape recorders, and after t days of practice his productivity is $30(1 - e^{-0.12t})$ tape recorders per hour. Find his productivity after:
    a. 5 days of practice   b. 15 days of practice
    c. Check your answers to parts (a) and (b) by graphing the productivity function on a graphing calculator on an appropriate window and then using TRACE or EVALUATE.

9. **GENERAL: Natural Logarithms** Simplify:
$$f(x) = \ln\left(\frac{1}{x}\right) + \ln(x^2) + \ln e$$

10. **GENERAL: Natural Logarithms**
    a. Use a calculator to find $\ln 9.3$.
    b. Raise e to the resulting power (include all decimal places shown) to verify that you return (approximately) to 9.3.

11. **SOCIAL SCIENCE: Diffusion of Information by Mass Media** In a city of a million people, news of election results broadcast over radio and television will reach $N(t) = 1{,}000{,}000(1 - e^{-0.3t})$ people within t hours. Find when the news will have reached 500,000 people.

12. **ECONOMICS: Oil Demand** The demand for oil in the United States is increasing by 3% per year. Assuming that this rate continues, how soon will demand increase by 50%?

13. **GENERAL: Population** During the years 1990–2000 the population of Colorado grew at the rate of 2.7% per year. Assuming that this growth continues, in how many years will the population double?

14. **BUSINESS: Memory Chip Demand** Demand for computer memory chips is increasing by 14% per year. Assuming that this growth continues, in how many years will demand double?

10.2 Differentiation of Logarithmic and Exponential Functions

Find the derivative of each function

15. $f(x) = \ln 2x$
16. $f(x) = \ln(x^2 - 1)^2$
17. $f(x) = \ln(1 - x)$
18. $f(x) = \ln \sqrt{x^2 + 1}$
19. $f(x) = \ln \sqrt[3]{x}$
20. $f(x) = \ln e^x$
21. $f(x) = \ln x^2$
22. $f(x) = x \ln x - x$
23. $f(x) = e^{-x^2}$
24. $f(x) = e^{1-x}$
25. $f(x) = \ln e^{x^2}$
26. $f(x) = e^{x^2 \ln x - x^2/2}$
27. $f(x) = 5x^2 + 2x \ln x + 1$
28. $f(x) = 2x^3 + 3x \ln x - 1$
29. $f(x) = 2x^3 - 3xe^{2x}$
30. $f(x) = 4x - 2x^2 e^{2x}$

Graph each function, showing all relative extreme points and inflection points.

31. $f(x) = \ln(x^2 + 4)$
32. $f(x) = 16e^{-x^2/8}$

33. **BUSINESS: Sales** The weekly sales (in thousands) of a new product after x weeks of advertising is $S(x) = 2000 - 1500e^{-0.1x}$. Find the rate of change of sales after:
    a. 1 week   b. 10 weeks.

34. **BIOMEDICAL: Drug Dosage** A patient receives an injection of 1.5 milligrams of a drug, and the amount remaining in the bloodstream t hours later is $A(t) = 1.5e^{-0.08t}$. Find the instantaneous rate of change of this amount:
    a. immediately after the injection (time $t = 0$).
    b. after 5 hours.

35. **BEHAVIORAL SCIENCE: Learning** In a test of short-term memory, the percent of subjects who remember an eight-digit number for at least t seconds is $P(t) = 100 - 200 \ln(t + 1)$. Find the rate of change of this percent after 5 seconds.

36. **GENERAL: Temperature** A thermos bottle that is chilled to 35 degrees Fahrenheit and then left in a 70-degree room will warm to a temperature of $T(t) = 70 - 35e^{-0.1t}$ degrees after t hours. Find the rate of change of the temperature:
    a. at time $t = 0$.   b. after 5 hours.

37. **SOCIAL SCIENCE: Diffusion of Information by Mass Media** The number of people in a town of 30,000 who have heard an important news bulletin within t hours of its first broadcast is $N(t) = 30{,}000(1 - e^{-0.3t})$. Find the instantaneous rate of change of the number of informed people after:
    a. 1 hour.   b. 8 hours.

38. **BUSINESS: Maximizing Present Value** A new company is growing so that its value t years from now will be $50t^2$ dollars. Therefore, its present value (at the rate of 8% compounded continuously) is $V(t) = 50t^2 e^{-0.08t}$ dollars (for $t > 0$). Find the number of years that maximizes the present value.

39–40. BUSINESS: Maximizing Revenue The function given below is a company's price function, where x is the quantity (in thousands) that will be sold at price p dollars.

a. Find the revenue function $R(x)$. [*Hint:* Revenue is price times quantity, $p \cdot x$.]
b. Find the quantity and price that will maximize revenue.

39. $p = 200e^{-0.25x}$ **40.** $p = 5 - \ln x$

41. ECONOMICS: Maximizing Consumer Expenditure Consumer demand for a commodity is estimated to be $D(p) = 25{,}000e^{-0.02p}$ units per month, where p is the selling price in dollars. Find the selling price that maximizes consumer expenditure.

42–43. Use your graphing calculator to graph each function on a window that includes all relative extreme points and inflection points, and give the coordinates of these points (rounded to two decimal places). [*Hint:* Use NDERIV once or twice with ZERO.] (Answers may vary depending on the graphing window chosen.)

42. $f(x) = \dfrac{x^4}{e^x}$ **43.** $f(x) = x^3 \ln |x|$

44. ECONOMICS: Consumer Expenditure If consumer demand for a commodity is $D(p) = 200e^{-0.0013p}$ (where p is the selling price in dollars), find the price that maximizes consumer expenditure.

45. BUSINESS: Maximizing Revenue A manufacturer finds that the price function for autofocus cameras is $p(x) = 20e^{-0.0077x}$, where p is the price in dollars at which quantity x (in thousands) will be sold. Find the quantity x that maximizes revenue.

10.3 Two Applications to Economics: Relative Rates and Elasticity of Demand

46–47. ECONOMICS: Relative Rate of Change The gross domestic product of a developing country is forecast to be $G(t) = 5 + 2e^{0.01t}$ million dollars t years from now. Find the relative rate of change:

46. 20 years from now. **47.** 10 years from now.

48. ECONOMICS: Elasticity of Demand A South American country exports coffee and estimates the demand function to be $D(p) = 63 - 2p^2$. If the country wants to raise revenues to improve its balance of payments, should it raise or lower the price from the present level of $3 per pound?

49. ECONOMICS: Elasticity of Demand A South African country exports gold and estimates the demand function to be $D(p) = 400\sqrt{600 - p}$. If the country wants to raise revenues to improve its balance of payments, should it raise or lower the price from the present level of $350 per ounce?

50. ECONOMICS: Elasticity of Demand The demand function for cigarettes is of the form $D(p) = 1.2p^{-0.44}$, where p is the price of a pack of cigarettes and $D(p)$ is the demand measured in packs per day per capita.* Find the elasticity of demand. [*Note:* You will find that demand is inelastic. This means that taxation, which acts like a price increase, is an ineffective way of discouraging smoking, but is an effective way of raising revenue.]

51. GENERAL: Relative Rate of Change The population of a city x years from now is projected to be $P(x) = 3.25 + 0.04x + 0.002x^3$ million people (for $0 \leq x \leq 10$). Find the relative rate of change 9 years from now.

52. GENERAL: Elasticity of Demand A boat dealer finds that the demand function for outboard motor boats near a large lake is $D(p) = 200 - 20p + p^2 - 0.03p^3$ (for $0 \leq p \leq 15$), where p is the selling price in thousands of dollars.

a. Use a graphing calculator to find the elasticity of demand at a price of $10,000. [*Hint:* What value of p corresponds to $10,000?]
b. Should the dealer raise or lower the price from this level to increase revenue?
c. Find the price at which elasticity equals 1.

---

*Jeffrey E. Harris, "Taxing Tar and Nicotine," *American Economic Review* **70**(3):300–311.

11 Integration and Its Applications

- 11.1 Antiderivatives and Indefinite Integrals
- 11.2 Integration Using Logarithmic and Exponential Functions
- 11.3 Definite Integrals and Areas
- 11.4 Further Applications of Definite Integrals: Average Value and Area Between Curves
- 11.5 Two Applications to Economics: Consumers' Surplus and Income Distribution
- 11.6 Integration by Substitution

Application Preview

Cigarette Smoking

Most cigarettes today have filters to absorb some of the toxic material or "tar" in the smoke before it is inhaled. The tobacco near the filter acts like an additional filter, absorbing tar until it is itself smoked, at which time it releases all of its accumulated toxins. A typical cigarette consists of 8 centimeters of tobacco followed by a filter.

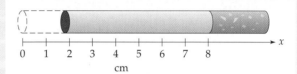

As the cigarette is smoked, tar is typically inhaled at the rate of $r(x) = 300e^{0.025x} - 240e^{0.02x}$ milligrams (mg) of tar per centimeter (cm) of tobacco, where x is the distance along the cigarette.* The amount of tar inhaled from any particular segment of the cigarette is the area under the graph of this function over that interval, which can be calculated by a process called *definite integration*, as explained in this chapter.

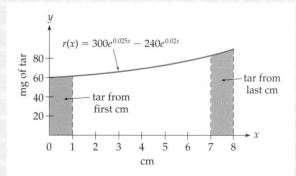

The results of integrating over the first and last centimeter of the cigarette are as follows (omitting the details—see Exercise 100 on page 816). The numbers represent milligrams of tar from the beginning and end of the cigarette.

---

*The actual rate depends upon the type of cigarette and the proportion of time that it is smoked rather than left to burn in the air. For further information, see Helen Marcus-Roberts and Maynard Thompson (eds), *Modules in Applied Mathematics*, vol 4, *Life Science Models*, Springer-Verlag, 1976, pp. 238–249.

Cigarette Smoking
(continued)

$$\begin{pmatrix} \text{Tar from} \\ \text{first cm} \end{pmatrix} = \int_0^1 (300e^{0.025x} - 240e^{0.2x})\, dx \approx 61 \qquad \text{Integrate from 0 to 1 for the first centimeter}$$

$$\begin{pmatrix} \text{Tar from} \\ \text{last cm} \end{pmatrix} = \int_7^8 (300e^{0.025x} - 240e^{0.02x})\, dx \approx 83 \qquad \text{Integrate from 7 to 8 for the last centimeter}$$

Notice that the last centimeter releases significantly more tar (about 36% more) than the first centimeter.

Moral: The Surgeon General has determined that smoking is hazardous to your health and the last puffs are 36% more hazardous than the first.

This is just one of the many applications of integration discussed in this chapter.

11.1 ANTIDERIVATIVES AND INDEFINITE INTEGRALS

Introduction

We have been studying differentiation and its uses. We now consider the reverse process, *anti*differentiation, which, for a given derivative, essentially recovers the original function. Antidifferentiation has many uses. For example, differentiation turns a cost function into a marginal cost function, and so antidifferentiation turns marginal cost back into cost. Later we will use antidifferentiation for other purposes, such as finding areas.

Antiderivatives and Indefinite Integrals

We begin with a simple example of antidifferentiation. Since the derivative of x^2 is $2x$, an *anti*derivative of $2x$ is x^2:

An antiderivative of $2x$ is x^2 Since the derivative of x^2 is $2x$

There are, however, other antiderivatives of $2x$. Each of the following is an antiderivative of $2x$:

$x^2 + 1$ $x^2 - 17$ $x^2 + e$ Since the derivative of each is $2x$

Clearly, we may add *any* constant to x^2 and the derivative will still be $2x$. Therefore, $x^2 + C$ is an antiderivative of $2x$ for *any* constant C.

Furthermore, it can be shown that there are no other antiderivatives of $2x$, and so the *most general* antiderivative of $2x$ is $x^2 + C$. The most general antiderivative of $2x$ is called the *indefinite integral* of $2x$, and is written with the $2x$ between an *integral sign* $\int$ and a dx:

$$\int 2x\, dx = x^2 + C$$

The indefinite integral of $2x$ is $x^2 + C$ (because the derivative of $x^2 + C$ is $2x$)

- Integral sign
- Integrand
- Arbitrary constant

The function to be integrated (here $2x$) is called the *integrand*. The dx reminds us that the variable of integration is x. The constant C is called an *arbitrary constant* because it may take any value, positive, negative, or zero.

An indefinite integral (sometimes called simply an *integral*) can always be checked by differentiation: the derivative of the answer must equal the integrand (as is the case with $x^2 + C$ and $2x$).

Indefinite Integral

$$\int f(x)\, dx = g(x) + C \qquad \text{The integral of } f(x) \text{ is } g(x) + C$$

if and only if

$$g'(x) = f(x) \qquad \text{the derivative of } g(x) \text{ is } f(x)$$

Integration Rules

There are several "rules" that simplify integration. The first, which is one of the most useful rules in all of calculus, shows how to integrate x to a constant power. It comes from "reversing" the power rule for differentiation (see page 562).

Power Rule for Integration

$$\int x^n\, dx = \frac{1}{n+1} x^{n+1} + C \qquad (n \neq -1)$$

To integrate x to a power, add 1 to the power and multiply by 1 over the new power

EXAMPLE 1

FINDING AN INDEFINITE INTEGRAL

$$\int x^3 \, dx = \frac{1}{4} x^4 + C \qquad \text{Using the Power Rule with } n = 3$$

with $n = 3$ and $n + 1 = 4$, so $\dfrac{1}{n+1} = \dfrac{1}{4}$.

Differentiating the answer $\frac{1}{4}x^4 + C$ immediately gives the integrand x^3, so the answer is correct.

Practice Problem 1

Find $\int x^2 \, dx$ and check your answer by differentiation.

➤ Solution on page 781

The proof of the Power Rule for Integration consists simply of differentiating the right-hand side.

$$\frac{d}{dx}\left(\frac{1}{n+1} x^{n+1} + C\right) = \frac{1}{n+1}(n+1)x^n = x^n$$

The power $n + 1$ brought down. The power decreased by 1. Simplified.

Since the derivative is the integrand x^n, the Power Rule for Integration is correct.

To integrate functions like $\sqrt{x}$ and $\dfrac{1}{x^2}$, we first express them as powers.

EXAMPLE 2

EXPRESSING AS A POWER BEFORE INTEGRATING

$$\int \sqrt{x} \, dx = \int x^{1/2} \, dx = \frac{2}{3} x^{3/2} + C \qquad \text{Using the Power Rule for Integration with } n = 1/2$$

with $n = \frac{1}{2}$, $n + 1 = \frac{1}{2} + 1 = \frac{3}{2}$, and $\dfrac{1}{n+1} = \dfrac{1}{3/2} = \dfrac{2}{3}$.

Note that in the answer, the multiple $\frac{2}{3}$ and the exponent $\frac{3}{2}$ are reciprocals of each other. Can you see, from the Power Rule, why this will always be so?

EXAMPLE 3

EXPRESSING AS A POWER BEFORE INTEGRATING

$$\int \frac{1}{x^2} dx = \int x^{-2} dx = \frac{1}{-1} x^{-1} + C = -x^{-1} + C$$

$n = -2$
$n + 1 = -1$

Simplified answer

Practice Problem 2 Find $\int \frac{dx}{x^3}$. [Hint: Equivalent to $\int \frac{1}{x^3} dx$.]

▶ Solution on page 781

EXAMPLE 4

INTEGRATING 1

$$\int 1 \, dx = \int x^0 \, dx = \frac{1}{1} x^1 + C = x + C \qquad \text{Using } x^0 = 1$$

$n = 0 \quad n + 1 = 1$

This result is so useful that it should be memorized.

$$\int 1 \, dx = x + C \qquad \text{The integral of 1 is } x + C$$

EXAMPLE 5

INTEGRATING WITH OTHER VARIABLES

a. $\int t^3 \, dt = \frac{1}{4} t^4 + C$

Using the Power Rule ($n = 3$) with dt since the variable is t ("integrating with respect to t")

b. $\int u^{-4} \, du = \frac{1}{-3} u^{-3} + C = -\frac{1}{3} u^{-3} + C$

Using the Power Rule ($n = -4$) with du since the variable is u ("integrating with respect to u")

Practice Problem 3 Find $\int z^{-1/2} \, dz$.   ▶ Solution on page 781

Notice what happens if we try to integrate x^{-1}:

$$\int x^{-1} \, dx = \frac{1}{0} x^0 + C \qquad \text{Undefined because of the } \tfrac{1}{0}$$

with $n = -1$ and $n + 1 = 0$.

The Power Rule for Integration fails for the exponent -1 because it leads to the undefined expression $\tfrac{1}{0}$. For this reason, the Power Rule for Integration includes the restriction "$n \neq -1$." It can integrate any power of x except x^{-1}.

The Constant Multiple and Sum Rules for differentiation (pages 564 and 565) lead immediately to analogous rules for simplifying integrals. The first rule says that the sum of two functions may be integrated one at a time.

Sum Rule for Integration

$$\int [f(x) + g(x)] \, dx = \int f(x) \, dx + \int g(x) \, dx \qquad \text{The integral of a sum is the sum of the integrals}$$

EXAMPLE 6   USING THE SUM RULE

$$\int (x^2 + x^3) \, dx = \int x^2 \, dx + \int x^3 \, dx \qquad \text{Using the Sum Rule to break the integral into two integrals}$$

$$= \frac{1}{3} x^3 + \frac{1}{4} x^4 + C \qquad \text{Using the Power Rule on each (one "+ C" is enough)}$$

The second rule says that a constant may be moved across the integral sign.

Constant Multiple Rule for Integration

For any constant k,

$$\int k \cdot f(x)\, dx = k \int f(x)\, dx$$

The integral of a constant times a function is the constant times the integral of the function

EXAMPLE 7

USING THE CONSTANT MULTIPLE RULE

$$\int 6x^2\, dx = 6 \int x^2\, dx$$

Using the Constant Multiple Rule to move the 6 across the integral sign

$$= 6 \cdot \frac{1}{3} x^3 + C$$

Using the Power Rule with $n = 2$

$$= 2x^3 + C$$

Simplifying

EXAMPLE 8

INTEGRATING A CONSTANT

$$\int 7\, dx = 7 \int 1\, dx = 7x + C$$

Moving the constant outside — The integral of 1 is x (plus C)

This leads to a very useful general rule. For any constant k,

Integral of a Constant

$$\int k\, dx = kx + C$$

The integral of a constant is the constant times x (plus C)

The Sum Rule can be extended to integrate the sum or difference of *any* number of terms, writing only one $+ C$ at the end (since any number of arbitrary constants can be added together to give just one).

EXAMPLE 9

INTEGRATING A SUM OF POWERS

$$\int (6x^2 - 3x^{-2} + 5)\, dx$$

$$= 6\int x^2\, dx - 3\int x^{-2}\, dx + 5\int 1\, dx \qquad \text{Breaking up the integral and moving constants outside}$$

$$= 6\cdot\frac{1}{3}x^3 - 3\cdot\frac{1}{-1}x^{-1} + 5x + C \qquad \text{Integrating each separately}$$

↑ From integrating the 1

$$= 2x^3 + 3x^{-1} + 5x + C \qquad \text{Simplifying}$$

EXAMPLE 10

REWRITING BEFORE INTEGRATING

$$\int \left(\frac{3\sqrt{x}}{2} - \frac{2}{\sqrt{x}}\right) dx = \int \left(\frac{3}{2}x^{1/2} - 2x^{-1/2}\right) dx \qquad \text{Writing as powers of } x$$

$$= \frac{3}{2}\cdot\frac{2}{3}x^{3/2} - 2\cdot\frac{2}{1}x^{1/2} + C \qquad \text{Integrating each term separately}$$

$$= x^{3/2} - 4x^{1/2} + C \qquad \text{Simplifying}$$

Practice Problem 4 Find $\int \left(\sqrt[3]{w} - \frac{4}{w^3}\right) dw$. ➤ Solution on page 781

Some integrals are so simple that they can be integrated "at sight."

EXAMPLE 11

INTEGRATING "AT SIGHT"

a. $\int 4x^3\, dx = x^4 + C$ By remembering that $4x^3$ is the derivative of x^4

b. $\int 7x^6\, dx = x^7 + C$ By remembering that $7x^6$ is the derivative of x^7

Practice Problem 5

Integrate "at sight" by noticing that each integrand is of the form nx^{n-1} and integrating to x^n without working through the Power Rule.

a. $\displaystyle\int 5x^4\, dx$ b. $\displaystyle\int 3x^2\, dx$

➤ Solutions on page 781

Algebraic Simplification of Integrals

Sometimes an integrand needs to be multiplied out or otherwise rewritten before it can be integrated.

EXAMPLE 12

EXPANDING BEFORE INTEGRATING

Find $\displaystyle\int x^2(x+6)^2\, dx$.

Solution

$$\int x^2(x+6)^2\, dx = \int x^2\underbrace{(x^2+12x+36)}_{(x+6)^2}\, dx \quad \text{"Squaring out" the }(x+6)^2$$

$$= \int (x^4+12x^3+36x^2)\, dx \quad \text{Multiplying out}$$

$$= \frac{1}{5}x^5 + 12\cdot\frac{1}{4}x^4 + 36\cdot\frac{1}{3}x^3 + C \quad \text{Integrating each term separately}$$

$$= \frac{1}{5}x^5 + 3x^4 + 12x^3 + C \quad \text{Simplifying}$$

Practice Problem 6

Find $\displaystyle\int \frac{6t^2 - t}{t}\, dt$.

[*Hint:* First simplify the integrand.]

➤ Solution on page 781

Since differentiation turns a cost function into a marginal cost function, integration turns a marginal cost function back into a cost function. To evaluate the constant, however, we need the fixed costs.

11.1 Exercises

Find each indefinite integral.

1. $\int x^4 \, dx$
2. $\int x^7 \, dx$
3. $\int x^{2/3} \, dx$
4. $\int x^{3/2} \, dx$
5. $\int \sqrt{u} \, du$
6. $\int \sqrt[3]{u} \, du$
7. $\int \dfrac{dw}{w^4}$
8. $\int \dfrac{dw}{w^2}$
9. $\int \dfrac{dz}{\sqrt{z}}$
10. $\int \dfrac{dz}{\sqrt[3]{z}}$
11. $\int 6x^5 \, dx$
12. $\int 9x^8 \, dx$
13. $\int (8x^3 - 3x^2 + 2) \, dx$
14. $\int (12x^3 + 3x^2 - 5) \, dx$
15. $\int \left(6\sqrt{x} + \dfrac{1}{\sqrt[3]{x}} \right) dx$
16. $\int \left(3\sqrt{x} + \dfrac{1}{\sqrt{x}} \right) dx$
17. $\int \left(16 \sqrt[3]{x^5} - \dfrac{16}{\sqrt[3]{x^5}} \right) dx$
18. $\int \left(14 \sqrt[4]{x^3} - \dfrac{3}{\sqrt[4]{x^3}} \right) dx$
19. $\int \left(10 \sqrt[3]{t^2} + \dfrac{1}{\sqrt[3]{t^2}} \right) dt$
20. $\int \left(21 \sqrt{t^5} + \dfrac{6}{\sqrt{t^5}} \right) dt$
21. $\int (x - 1)^2 \, dx$
22. $\int (x + 2)^2 \, dx$
23. $\int (1 + 10w) \sqrt{w} \, dw$
24. $\int (1 - 7w) \sqrt[3]{w} \, dw$
25. $\int \dfrac{6x^3 - 6x^2 + x}{x} \, dx$
26. $\int \dfrac{4x^4 + 4x^2 - x}{x} \, dx$
27. $\int (x - 2)(x + 4) \, dx$
28. $\int (x + 5)(x - 3) \, dx$
29. $\int (r - 1)(r + 1) \, dr$
30. $\int (3s + 1)(3s - 1) \, ds$
31. $\int \dfrac{x^2 - 1}{x + 1} \, dx$
32. $\int \dfrac{x^2 - 1}{x - 1} \, dx$
33. $\int (t + 1)^3 \, dt$
34. $\int (t - 1)^3 \, dt$

35. Find:
   a. $\int \dfrac{1}{x^3} \, dx$
   b. $\dfrac{\int 1 \, dx}{\int x^3 \, dx}$

   Notice that the answers are not the same, showing that you do not integrate a fraction by integrating the numerator and denominator separately.

36. Evaluate
   a. $\int x \, dx$
   b. $x \int 1 \, dx$

   Notice that the two answers are not the same, showing that a *variable* cannot be moved across the integral sign.

37. a. Verify that $\int x^2 \, dx = \frac{1}{3}x^3 + C$.
   b. Graph the five functions $\frac{1}{3}x^3 - 2, \frac{1}{3}x^3 - 1, \frac{1}{3}x^3, \frac{1}{3}x^3 + 1,$ and $\frac{1}{3}x^3 + 2$ (the solutions for five different values of C) on the graphing window [−3, 3] by [−5, 5]. Use TRACE to see how the constant shifts the curve vertically.
   c. Find the slopes (using NDERIV or dy/dx) of several of the curves at a particular x-value and check that in each case the slope is the square of the x-value. This verifies that the derivative of each curve is x^2, and so each is an integral of x^2.

38. a. Graph the five functions $\ln x - 2, \ln x - 1, \ln x, \ln x + 1,$ and $\ln x + 2$ on the graphing window [0, 4] by [−3, 3].

(continues)

b. Find the slope (using NDERIV or dy/dx) of several of the curves at a particular x-value and check that in each case the slope is the reciprocal of the x-value. This suggests that the derivative of each function is $1/x$.

c. Based on part (b), conjecture what is the indefinite integral of the function $1/x$ (for $x > 0$).

APPLIED EXERCISES

39. **BUSINESS: Cost** A company's marginal cost function is $MC = 20x^{3/2} - 15x^{2/3} + 1$, where x is the number of units, and fixed costs are $4000. Find the cost function.

40. **BUSINESS: Cost** A company's marginal cost function is $MC = 21x^{4/3} - 6x^{1/2} + 50$, where x is the number of units, and fixed costs are $3000. Find the cost function.

41. **BUSINESS: Revenue** A company's marginal revenue function is $MR = 12\sqrt[3]{x} + 3\sqrt{x}$, where x is the number of units. Find the revenue function. (Evaluate C so that revenue is zero when nothing is produced.)

42. **BUSINESS: Revenue** A company's marginal revenue function is $MR = 15\sqrt{x} + 4\sqrt[3]{x}$, where x is the number of units. Find the revenue function. (Evaluate C so that revenue is zero when nothing is produced.)

43. **GENERAL: Velocity** A Porsche 928 can accelerate from a standing start to a speed of $v(t) = -0.09t^2 + 8t$ feet per second after t seconds (for $0 \le t < 35$).

    a. Find a formula for the distance that it will travel from its starting point in the first t seconds. [*Hint:* Integrate velocity to find distance, and then use the fact that distance is 0 at time $t = 0$.]

    b. Use the formula that you found in part (a) to find the distance that the car will travel in the first 10 seconds.

44. **GENERAL: Velocity** A BMW 733i can accelerate from a standing start to a speed of $v(t) = -0.09t^2 + 6t$ feet per second after t seconds (for $0 \le t < 40$).

    a. Find a formula for the distance that it will travel from its starting point in the first t seconds. [*Hint:* Integrate velocity to find distance, and then use the fact that distance is 0 at time $t = 0$.]

    b. Use the formula that you found in part (a) to find the distance that the car will travel in the first 10 seconds.

45. **GENERAL: Learning** A person can memorize words at the rate of $3/\sqrt{t}$ words per minute.

    a. Find a formula for the total number of words that can be memorized in t minutes. [*Hint:* Evaluate C so that 0 words have been memorized at time $t = 0$.]

    b. Use the formula that you found in part (a) to find the total number of words that can be memorized in 25 minutes.

46. **BIOMEDICAL: Temperature** A patient's temperature is 108 degrees Fahrenheit and is changing at the rate of $t^2 - 4t$ degrees per hour, where t is the number of hours since taking a fever-reducing medication $(0 \le t \le 3)$.

    a. Find a formula for the patient's temperature after t hours. [*Hint:* Evaluate the constant C so that the temperature is 108 at time $t = 0$.]

    b. Use the formula that you found in part (a) to find the patient's temperature after 3 hours.

47. **ENVIRONMENTAL SCIENCE: Pollution** A chemical plant is adding pollution to a lake at the rate of $40\sqrt{t^3}$ tons per year, where t is the number of years that the plant has been in operation.

    a. Find a formula for the total amount of pollution that will enter the lake in the first t years of the plant's operation. [*Hint:* Evaluate C so that no pollution has been added at time $t = 0$.]

    (continues)

b. Use the formula that you found in part (a) to find how much pollution will enter the lake in the first 4 years of the plant's operation.
c. If all life in the lake will cease when 400 tons of pollution have entered the lake, will the lake "live" beyond 4 years?

48. **BUSINESS: Appreciation** A $20,000 art collection is increasing in value at the rate of $300\sqrt{t}$ dollars per year after t years.
   a. Find a formula for its value after t years. [*Hint:* Evaluate C so that its value at time $t = 0$ is $20,000.]
   b. Use the formula that you found in part (a) to find its value after 25 years.

REVIEW EXERCISE

This exercise will be important in the next section.

49. Find $\dfrac{d}{dx}\ln(-x)$.

11.2 INTEGRATION USING LOGARITHMIC AND EXPONENTIAL FUNCTIONS

Introduction

In the previous section we defined integration as the reverse of differentiation, and we introduced several integration formulas. In this section we develop integration formulas involving logarithmic and exponential functions. One of these formulas will answer a question that we could not answer earlier—namely, how to integrate x^{-1}, the only power not covered by the Power Rule.

The Integral $\int e^{ax}\, dx$

On page 739 we saw that to *differentiate* e^{ax}, we *multiply* by a to get ae^{ax}. Therefore, to *integrate* e^{ax}, the reverse process, we must *divide* by a. That is, for any $a \neq 0$:

$$\int e^{ax}\, dx = \frac{1}{a} e^{ax} + C$$

The integral of e to a constant times x is 1 over the constant times the original function (plus C)

EXAMPLE 1 INTEGRATING AN EXPONENTIAL FUNCTION

$$\int e^{2x}\, dx = \frac{1}{2} e^{2x} + C \qquad \text{Using the formula with } a = 2$$

$\uparrow_{a=2}\ \underbrace{\frac{1}{a}}\ \uparrow\text{Original function}$

As always, we may check the answer by differentiation.

$$\frac{d}{dx}\left(\frac{1}{2} e^{2x} + C\right) = \frac{1}{2} \cdot 2 \cdot e^{2x} = e^{2x} \qquad \text{Using } \frac{d}{dx} e^{ax} = a e^{ax}$$

The result is the integrand e^{2x}, so the integration is correct.

The proof of this rule consists simply of differentiating the right-hand side.

$$\frac{d}{dx}\left(\frac{1}{a} e^{ax} + C\right) = \frac{1}{a} a e^{ax} = e^{ax} \qquad \text{Using } \frac{d}{dx} e^{ax} = a e^{ax}$$

The result is the integrand, so the integration formula is correct.

EXAMPLE 2 INTEGRATING EXPONENTIAL FUNCTIONS

a. $\displaystyle\int e^{\frac{1}{2}x}\, dx = 2 e^{\frac{1}{2}x} + C \qquad\qquad a = \frac{1}{2} \ \text{ so }\ \frac{1}{a} = \frac{1}{\frac{1}{2}} = 2$

b. $\displaystyle\int 6 e^{-3x}\, dx = 6 \int e^{-3x}\, dx = 6\left(-\frac{1}{3}\right)\cdot e^{-3x} + C = -2 e^{-3x} + C$

$\qquad\qquad\qquad\qquad\quad \uparrow_{a=-3} \quad \uparrow_{\frac{1}{a}=-\frac{1}{3}}$

c. $\displaystyle\int e^{x}\, dx = 1 e^{x} + C = e^{x} + C \qquad\qquad a = 1,\ \text{ so }\ \frac{1}{a} = 1$

Each of these answers may be checked by differentiation. The last one says that e^x is the integral of itself, just as e^x is the derivative of itself.

$$\int e^x \, dx = e^x + C \qquad \text{The integral of } e^x \text{ is } e^x \text{ (plus C)}$$

Practice Problem 1 Find: **a.** $\displaystyle\int 12e^{4x} \, dx$ **b.** $\displaystyle\int e^{x/3} \, dx$

[*Hint:* For part (b), write $e^{x/3}$ as $e^{\frac{1}{3}x}$, then integrate, and rewrite again.]

➤ Solutions on page 794

When using these new integration formulas to solve applied problems, be careful to evaluate the constant C correctly. It will not always be equal to the initial value of the function.

EXAMPLE 3 FINDING TOTAL FLU CASES FROM THE RATE OF CHANGE

An influenza epidemic hits a large city and spreads at the rate of $12e^{0.2t}$ new cases per day, where t is the number of days since the epidemic began. The epidemic began with 4 cases.

a. Find a formula for the total number of flu cases in the first t days of the epidemic.

b. Use your formula to find the number of cases during the first 30 days.

Solution

a. To find the total number of cases, we integrate the growth rate $12e^{0.2t}$.

$$\underbrace{f(t)}_{\substack{\text{Total cases} \\ \text{in first } t \text{ days}}} = \int 12e^{0.2t} \, dt = 12 \int e^{0.2t} \, dt \qquad \text{Taking out the constant}$$

$$= 12 \cdot \frac{1}{\underset{\uparrow}{0.2}} e^{0.2t} + C = 60e^{0.2t} + C \qquad \text{Using the } \int e^{ax} dx \text{ formula}$$
$$\phantom{= 12 \cdot \frac{1}{0}}5$$

Evaluating $f(t)$ at $t = 0$ must give the initial number of cases:

$$f(0) = \underbrace{60e^{0.2(0)}}_{\text{Initial number of cases}} + C = 60 + C \qquad \begin{array}{l} f(t) = 60e^{0.2t} + C \text{ evaluated} \\ \text{at } t = 0 \text{ (the beginning of} \\ \text{the epidemic)} \end{array}$$

$\underbrace{\phantom{60e^{0.2(0)}}}_{e^0 = 1}$

This initial number must equal the given initial number, 4:

$$60 + C = 4 \qquad \begin{array}{l} \text{Initial number from the formula set} \\ \text{equal to the given initial number} \end{array}$$

$$C = -56 \qquad \text{Solving for } C$$

Replacing C by -56 gives the total number of flu cases within t days:

$$\underbrace{f(t)}_{\substack{\text{Number of cases} \\ \text{in first } t \text{ days}}} = \underbrace{60e^{0.2t} - 56}_{\substack{\text{Answer to} \\ \text{Part (a)}}} \qquad f(t) = 60e^{0.2t} + C \text{ with } C = -56$$

b. To find the number within 30 days, we evaluate at $t = 30$:

$$f(30) = 60e^{0.2(30)} - 56 \qquad f(t) = 60e^{0.2t} + C \text{ at } t = 30$$
$$= 60e^6 - 56 \approx 24{,}150 \qquad \text{Using a calculator}$$

Therefore, within 30 days the epidemic will have grown to more than 24,000 cases.

Evaluating the Constant C

Notice that we did *not* simply replace the constant C by the initial number of cases, 4. (This would have given the wrong initial number of cases.) Instead, we evaluated the function at the initial time $t = 0$ and set it equal to the initial number of cases:

$$\underbrace{60e^0 + C}_{\substack{f(t) \text{ evaluated} \\ \text{at } t = 0}} = \underbrace{4}_{\substack{\text{Given} \\ \text{initial value}}}$$

We then solved to find the correct value of C, $C = -56$, which we then substituted into the formula.

In general, to evaluate the constant C:
1. Evaluate the integral at the given number (usually $t = 0$) and set the result equal to the stated initial value.
2. Solve for C.
3. Write the answer with C replaced by its correct value.

The Integral $\int \dfrac{1}{x}\, dx$

The differentiation formula $\dfrac{d}{dx}\ln x = \dfrac{1}{x}$ can be read "backward" as an integration formula:

$$\int \frac{1}{x}\, dx = \ln x + C \qquad \text{The integral of 1 over } x \text{ is the natural log of } x \text{ (plus } C\text{)}$$

This formula, however, is restricted to $x > 0$, for only then is $\ln x$ defined. For $x < 0$ we can differentiate $\ln(-x)$, giving

$$\frac{d}{dx}\ln(-x) = \frac{-1}{-x} = \frac{1}{x} \qquad \text{Using } \frac{d}{dx}\ln f = \frac{f'}{f} \text{ and simplifying}$$

This result says that for $x < 0$, the integral of $1/x$ is $\ln(-x)$. The negative sign in $\ln(-x)$ serves only to make the already negative x positive, and this could be accomplished just as well with absolute value bars.

$$\int \frac{1}{x}\, dx = \ln|x| + C \qquad \text{The integral of 1 over } x \text{ is the natural logarithm of the absolute value of } x$$

This formula holds for negative *and* positive values of x, since in both cases $\ln|x|$ is defined. The integral can be written in three different ways, all of which have the same answer.

$$\int \frac{1}{x}\, dx = \int \frac{dx}{x} = \int x^{-1}\, dx = \ln|x| + C$$

EXAMPLE 4

INTEGRATING USING THE LN RULE

$$\int \frac{5}{2x}\, dx = \int \frac{5}{2}\frac{1}{x}\, dx = \frac{5}{2}\int \frac{1}{x}\, dx = \frac{5}{2}\ln|x| + C$$

                                    ↑   ⌣
                     Taking out   Using the
                   the constant ln formula

EXAMPLE 5

INTEGRATING NEGATIVE POWERS

$$\int (x^{-1} + x^{-2})\, dx = \int x^{-1}\, dx + \int x^{-2}\, dx = \ln|x| - x^{-1} + C$$

From the natural log formula ↑   From the Power Rule with $n = -2$

Practice Problem 2 Find $\int \dfrac{3}{4x}\, dx$.

▶ Solution on page 794

EXAMPLE 6

FINDING TOTAL SALES FROM THE SALES RATE

An electronics dealer estimates that during month t of a sale, a discontinued computer will sell at a rate of approximately $25/t$ per month, where $t = 1$ corresponds to the beginning of the sale, at which time none have been sold. Find a formula for the total number of computers that will be sold up to month t. Will the store's inventory of 64 computers be sold by month $t = 12$?

Solution

To find the total sales, we integrate the sales rate $\dfrac{25}{t}$:

$$\underbrace{S(t)}_{\text{Total sales in first } t \text{ months}} = \int \dfrac{25}{t}\, dt = 25 \int \dfrac{1}{t}\, dt = 25 \ln t + C \quad \text{Omitting absolute values, since } t > 0$$

To evaluate C, we evaluate at the given starting time $t = 1$:

$$\underbrace{S(1)}_{\text{Initial number of sales}} = 25 \underbrace{\ln 1}_{0} + C = C$$

The initial number of sales must be zero, giving $C = 0$, and substituting this into the sales function gives

$$\underbrace{S(t) = 25 \ln t}_{\text{Total number sold up to month } t} \qquad S(t) = 25 \ln t + C \text{ with } C = 0$$

To find the number sold up to month 12, we evaluate at $t = 12$:

$$S(12) = 25 \ln 12 \approx 62 \qquad \text{Using a calculator}$$

Therefore, all but two of the 64 computers will be sold by month 12.

In this example the initial time (the beginning of the sale) was given as $t = 1$ rather than the more usual $t = 0$. The initial time will be clear from the problem.

Consumption of Natural Resources

Just as the world population grows exponentially, so does the world's annual consumption of natural resources. We can estimate the total consumption at any time in the future by integrating the rate of consumption, and from this predict when the known reserves will be exhausted.

EXAMPLE 7

FINDING A FORMULA FOR TOTAL CONSUMPTION FROM THE RATE

The annual world consumption of tin is predicted* to be $0.24e^{0.01t}$ million metric tons per year, where t is the number of years since 2000. Find a formula for the total tin consumption within t years of 2000 and estimate when the known world reserves of 6 million metric tons will be exhausted.

Solution

To find the total consumption, we integrate the rate $0.24e^{0.01t}$.

$$\underbrace{C(t)}_{\substack{\text{Total tin consumed} \\ \text{in first } t \text{ years} \\ \text{after 2000}}} = \int 0.24 e^{0.01t}\, dt = 0.24 \int e^{0.01t}\, dt \qquad \text{Taking out the constant}$$

$$= 0.24 \underbrace{\frac{1}{0.01}}_{100} e^{0.01t} + C = 24 e^{0.01t} + C \qquad \text{Using the } \int e^{ax}\, dx \text{ formula}$$

---

*World Resources Institute and U.S. Geological Survey. Data for several exercises in this chapter come from these sources.

The total consumed in the first zero years must be zero, so $C(0) = 0$.

$$C(0) = \underbrace{24e^0}_{0} + \underbrace{C}_{\text{Must equal zero}} = 24 + C \qquad C(t) = 24e^{0.01t} + C \text{ with } t = 0$$

From $24 + C = 0$ we find $C = -24$. We substitute this into the formula for $C(t)$:

$$\underbrace{C(t) = 24e^{0.01t} - 24}_{\text{Formula for total consumption within first } t \text{ years since 2000}} \qquad C(t) = 24e^{0.01t} + C \text{ with } C = -24$$

To predict when the total world reserves of 6 million metric tons will be exhausted, we set this function equal to 6 and solve for t:

$24e^{0.01t} - 24 = 6$	$C(t)$ set equal to 6
$24e^{0.01t} = 30$	Adding 24
$e^{0.01t} = \dfrac{30}{24} = 1.25$	Dividing by 24
$\ln \underbrace{e^{0.01t}}_{0.01t} = \underbrace{\ln 1.25}_{0.223}$	Taking natural logs and using $\ln e^x = x$
$t \approx \dfrac{0.223}{0.01} = 22.3$	Solving $0.01t = 0.223$ for t

Therefore, the known world supply of tin will be exhausted in about 22 years after 2000, which means in about the year 2022 (assuming that consumption continues at the predicted rate).

Spreadsheet Exploration

Since integration is continuous summation, instead of integrating we could add up the annual tin consumption for each year to see when the total will reach the known reserves of 6 million metric tons. The following spreadsheet* shows this annual consumption

---

* See the Preface for how to obtain this and other Excel spreadsheets.

for each year (using the formula $0.24e^{0.01t}$ from the preceding example) in column B continuing into column F, with the cumulative totals shown in column C continuing into column G.

	A	B	C	D	E	F	G
		B2		= =0.24*EXP(0.01*$A2)			
1	Year	Annual Consumption	Cumulative Consumption		Year	Annual Consumption	Cumulative Consumption
2	1	0.24241204	0.24241204		16	0.28164261	4.18511691
3	2	0.24484832	0.48726036		17	0.28447316	4.46959007
4	3	0.24730909	0.73456945		18	0.28733217	4.75692224
5	4	0.24979459	0.98436404		19	0.29021990	5.04714215
6	5	0.25230506	1.23666910		20	0.29313666	5.34027881
7	6	0.25484077	1.49150987		21	0.29608273	5.63636154
8	7	0.25740196	1.74891183		22	0.29905842	5.93541996
9	8	0.25998890	2.00890073		23	0.30206400	6.23748396
10	9	0.26260183	2.27150256		24	0.30509980	6.54258376
11	10	0.26524102	2.53674358		25	0.30816610	6.85074986
12	11	0.26790674	2.80465032		26	0.31126322	7.16201308
13	12	0.27059924	3.07524956		27	0.31439147	7.47640454
14	13	0.27331881	3.34856837		28	0.31755115	7.79395570
15	14	0.27606571	3.62463408		29	0.32074260	8.11469830
16	15	0.27884022	3.90347430		30	0.32396611	8.43866441

Since the cumulative consumption is less than 6 million metric tons in year 22 and more in year 23, we conclude that the known world reserves of tin will be exhausted sometime in 2022, in agreement with Example 7.

The integration method of Example 7 has two advantages over addition: it is often easier (especially for a large number of years) and it includes changes in the amount *during* each year rather than just at the end of each year.

Incidentally, tin is used mainly for coating steel (a "tin" can is actually a steel can with a thin protective coating of tin to prevent rust). The predicted unavailability of tin is already causing major changes in the food-packaging industry.

Power Rule for Integration, Revisited

Our new integration formula shows how to integrate x^{-1}, the only power not covered by the Power Rule. Therefore, we can now write one "combined" formula for the integral $\int x^n \, dx$ for *any power n*.

11.2 INTEGRATION USING LOGARITHMIC AND EXPONENTIAL FUNCTIONS

Integrals of Powers of x

$$\int x^n \, dx = \begin{cases} \dfrac{1}{n+1} x^{n+1} + C & \text{if } n \neq -1 \\ \ln|x| + C & \text{if } n = -1 \end{cases}$$

Use the Power Rule if n is other than -1

Use the ln formula if n equals -1

It is a curious fact that every power of x integrates to another power of x, with the single exception of x^{-1}, which integrates to an entirely different kind of function, the natural logarithm.

11.2 Section Summary

We have three integration formulas:

$$\int x^n \, dx = \frac{1}{n+1} x^{n+1} + C \qquad n \neq -1$$

$$\int e^{ax} \, dx = \frac{1}{a} e^{ax} + C \qquad a \neq 0$$

$$\int \frac{1}{x} \, dx = \int \frac{dx}{x} = \int x^{-1} \, dx = \ln|x| + C$$

What is the difference between the bottom formula with and without the absolute value bars?

$$\int \frac{1}{x} \, dx = \ln|x| + C \qquad \text{versus} \qquad \int \frac{1}{x} \, dx = \ln x + C$$

Both are correct, but the second holds only for *positive* x-values (so that $\ln x$ is defined; see page 725), while the first holds for *negative* as well as positive x-values. We will sometimes use the second when we know that x is positive.

To evaluate C in an application we set the function (evaluated at the given number) equal to the stated initial value and solve for C.

> **Solutions to Practice Problems**

1. a. $\int 12e^{4x}\,dx = 12\int e^{4x}\,dx = 12\cdot\frac{1}{4}e^{4x} + C$
   $= 3e^{4x} + C$

   b. $\int e^{x/3}\,dx = \int e^{\frac{1}{3}x}\,dx = 3e^{\frac{1}{3}x} + C = 3e^{x/3} + C$

2. $\int \frac{3}{4x}\,dx = \frac{3}{4}\int \frac{1}{x}\,dx = \frac{3}{4}\ln|x| + C$

11.2 Exercises

Find each indefinite integral.

1. $\int e^{3x}\,dx$
2. $\int e^{4x}\,dx$
3. $\int e^{x/4}\,dx$
4. $\int e^{x/3}\,dx$
5. $\int e^{0.05x}\,dx$
6. $\int e^{0.02x}\,dx$
7. $\int e^{-2y}\,dy$
8. $\int e^{-3y}\,dy$
9. $\int e^{-0.5x}\,dx$
10. $\int e^{-0.4x}\,dx$
11. $\int 6e^{2x/3}\,dx$
12. $\int 24e^{-2u/3}\,du$
13. $\int -5x^{-1}\,dx$
14. $\int -\frac{1}{2}x^{-1}\,dx$
15. $\int \frac{3\,dx}{x}$
16. $\int \frac{dx}{2x}$
17. $\int \frac{3}{2v}\,dv$
18. $\int \frac{2}{3v}\,dv$
19. $\int \left(e^{3x} - \frac{3}{x}\right)dx$
20. $\int \left(e^{2x} - \frac{2}{x}\right)dx$
21. $\int (3e^{0.5t} - 2t^{-1})\,dt$
22. $\int (5e^{0.5t} - 4t^{-1})\,dt$
23. $\int (x^2 + x + 1 + x^{-1} + x^{-2})\,dx$
24. $\int (x^{-2} - x^{-1} + 1 - x + x^2)\,dx$
25. $\int (5e^{0.02t} - 2e^{0.01t})\,dt$
26. $\int (3e^{0.05t} - 2e^{0.04t})\,dt$

APPLIED EXERCISES

27–28. BIOMEDICAL: Epidemics A flu epidemic hits a college community, beginning with five cases on day $t = 0$. The rate of growth of the epidemic (new cases per day) is given by the following function $r(t)$, where t is the number of days since the epidemic began.

a. Find a formula for the total number of cases of flu in the first t days.
b. Use your answer to part (a) to find the total number of cases in the first 20 days.

27. $r(t) = 18e^{0.05t}$
28. $r(t) = 20e^{0.04t}$

29. BUSINESS: Sales In an effort to reduce its inventory, a music store runs a sale on its least popular compact discs (CDs). The sales rate (CDs sold per day) on day t of the sale is predicted to be $50/t$ (for $t \geq 1$), where $t = 1$ corresponds to the beginning of the sale, at which time none of the inventory of 200 CDs had been sold.

a. Find a formula for the total number of CDs sold up to day t.
b. Will the store have sold its inventory of 200 CDs by day $t = 30$?

30. BUSINESS: Sales In an effort to reduce its inventory, a book store runs a sale on its least popular mathematics books. The sales rate (books sold per day) on day t of the sale is predicted to be $60/t$ (for $t \geq 1$), where $t = 1$ corresponds to the beginning of the sale, at which time none of the inventory of 350 books had been sold.

a. Find a formula for the number of books sold up to day t.
b. Will the store have sold its inventory of 350 books by day $t = 30$?

31. GENERAL: Consumption of Natural Resources World consumption of silver is running at the rate of $17e^{0.02t}$ thousand metric tons per year, where t is measured in years and $t = 0$ corresponds to 2000.

a. Find a formula for the total amount of silver that will be consumed within t years of 2000.
b. When will the known world resources of 420 thousand metric tons of silver be exhausted? (Silver is used extensively in photography.)

32. GENERAL: Consumption of Natural Resources World consumption of copper is running at the rate of $15e^{0.04t}$ million metric tons per year, where t is measured in years and $t = 0$ corresponds to 2000.

a. Find a formula for the total amount of copper that will be used within t years of 2000.
b. When will the known world resources of 650 million metric tons of copper be exhausted?

33. GENERAL: Cost of Maintaining a Home The cost of maintaining a home generally increases as the home becomes older. Suppose that the rate of cost (dollars per year) for a home that is x years old is $200e^{0.4x}$.

a. Find a formula for the total maintenance cost during the first x years. (Total maintenance should be zero at $x = 0$.)
b. Use your answer to part (a) to find the total maintenance cost during the first 5 years.

34. BIOMEDICAL: Cell Growth A culture of bacteria is growing at the rate of $20e^{0.8t}$ cells per day, where t is the number of days since the culture was started. Suppose that the culture began with 50 cells.

a. Find a formula for the total number of cells in the culture after t days.
b. If the culture is to be stopped when the population reaches 500, when will this occur?

35. GENERAL: Freezing of Ice An ice cube tray filled with tap water is placed in the freezer, and the temperature of the water is changing at the rate of $-12e^{-0.2t}$ degrees Fahrenheit per hour after t hours. The original temperature of the tap water was 70 degrees.

a. Find a formula for the temperature of water that has been in the freezer for t hours.
b. When will the ice be ready? (Water freezes at 32 degrees.)

36. SOCIAL SCIENCE: Divorces The number of divorces per year in the United States is approximately $1.14e^{0.01t}$ million, where t is measured in years and $t = 0$ corresponds to 2000. Find a formula for the total number of divorces expected within t years of 2000.

37. BUSINESS: Total Savings A factory installs new equipment that is expected to generate savings at the rate of $800e^{-0.2t}$ dollars per year, where t is the number of years that the equipment has been in operation.

a. Find a formula for the total savings that the equipment will generate during its first t years.
b. If the equipment originally cost $2000, when will it "pay for itself"?

38. BUSINESS: Total Savings A company installs a new computer that is expected to generate savings at the rate of $20{,}000e^{-0.02t}$ dollars per year, where t is the number of years that the computer has been in operation.

a. Find a formula for the total savings that the computer will generate during its first t years.

b. If the computer originally cost $250,000, when will it "pay for itself"?

39. BUSINESS: Value of an Investment A real estate investment, originally worth $5000, grows continuously at the rate of $400e^{0.05t}$ dollars per year, where t is the number of years since the investment was made.

a. Find a formula for the value of the investment after t years.

b. Use your formula to find the value of the investment after 10 years.

40. BUSINESS: Value of an Investment A biotechnology investment, originally worth $20,000, grows continuously at the rate of $1000e^{0.10t}$ dollars per year, where t is the number of years since the investment was made.

a. Find a formula for the value of the investment after t years.

b. Use your formula to find the value of the investment after 7 years.

Find each indefinite integral [*Hint:* Use some algebra first.]

41. $\displaystyle\int \frac{(x+1)^2}{x}\,dx$

42. $\displaystyle\int \frac{(x-1)^2}{x}\,dx$

43. $\displaystyle\int \frac{(t-1)(t+3)}{t^2}\,dt$

44. $\displaystyle\int \frac{(t+2)(t-4)}{t^2}\,dt$

45. $\displaystyle\int \frac{(x-2)^3}{x}\,dx$

46. $\displaystyle\int \frac{(x+2)^3}{x}\,dx$

47. GENERAL: Consumption of Natural Resources World consumption of lead is running at the rate of $5.6e^{0.01t}$ million metric tons per year, where t is measured in years and $t = 0$ corresponds to 2000.

a. Find a formula for the total amount of lead that will be consumed within t years of 2000.

b. Use a graphing calculator to find when the world's known resources of 140 million metric tons of lead will be exhausted. [*Hint:* Use INTERSECT.] Lead has many uses, from batteries to shields against radioactivity.

48. GENERAL: Total Savings A homeowner installs a solar water heater that is expected to generate savings at the rate of $70e^{0.03t}$ dollars per year, where t is the number of years since it was installed.

a. Find a formula for the total savings within the first t years of operation.

b. Use a graphing calculator to find when the heater will "pay for itself" if it cost $800. [*Hint:* Use INTERSECT.]

11.3 DEFINITE INTEGRALS AND AREAS

Introduction

We begin this section by calculating areas under curves, leading to a definition of the *definite integral* of a function. The *Fundamental Theorem of Integral Calculus* then provides an easier way to calculate definite integrals using *indefinite* integrals. Finally, we will illustrate the wide variety of applications of definite integrals.

Area Under a Curve

The diagram below shows a continuous nonnegative function. We want to calculate the *area* under the curve and above the x-axis between the vertical lines $x = a$ and $x = b$.

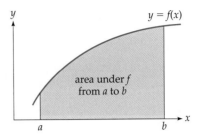

We begin by *approximating* the area by rectangles. In the first graph on the left below, the area under the curve is approximated by five rectangles with equal bases and with heights equal to the height of the curve at the left-hand edge of the rectangle. (These are called *left* rectangles.) Five rectangles, however, do not give a very accurate approximation for the area under the curve: They underestimate the actual area by the small white spaces just above the rectangles.

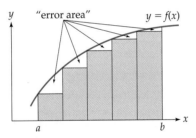
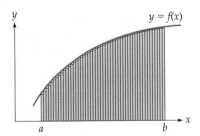

Area under f from a to b approximated by 5 rectangles

Area under f from a to b approximated by 50 rectangles

In the second graph, this same area is approximated by fifty rectangles, giving a much better approximation: the white "*error area*" between the curve and the rectangles is so small as to be almost invisible.

These diagrams suggest that more rectangles give a better approximation. In fact, the *exact* area under the curve is defined as the *limit* of the approximations as the number of rectangles *approaches infinity*. The following example shows how to carry out such an approximation for the area under a given curve, and afterward we will find the *exact* area by letting the number of rectangles approach infinity.

CHAPTER 11 INTEGRATION AND ITS APPLICATIONS

EXAMPLE 1

APPROXIMATING AREA BY RECTANGLES

Approximate the area under the curve $f(x) = x^2$ from 1 to 2 by five rectangles. Use rectangles with equal bases and with heights equal to the height of the curve at the left-hand edge of the rectangles.

Solution

For five rectangles, we divide the distance from $a = 1$ to $b = 2$ into five equal parts, so that each rectangle has width

$$\Delta x = \frac{2-1}{5} = \frac{1}{5} = 0.2 \qquad \text{For } n \text{ rectangles,} \quad \Delta x = \frac{b-a}{n}$$

Along the x-axis beginning at $x = 1$ we mark successive points with spacing $\Delta x = 0.2$, giving points 1, 1.2, 1.4, 1.6, 1.8, and 2, as shown below. Above each of the resulting subintervals we draw a rectangle whose height is the height of the curve at the left-hand edge of that rectangle. The curve is $y = x^2$, so these heights are the squares of the corresponding x-values.

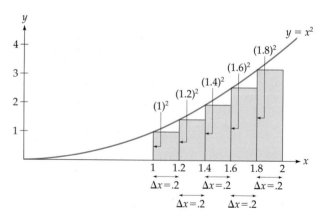

Area from 1 to 2 approximated by 5 rectangles

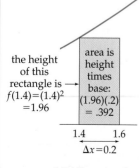

Middle rectangle

An enlarged view of the middle rectangle is shown on the left. The sum of the areas of the five rectangles, height times base $\Delta x = 0.2$ for each rectangle, is

| 1st rectangle | 2nd rectangle | 3rd rectangle | 4th rectangle | 5th rectangle | Adding height times base for each rectangle |

$(1)^2 \cdot (0.2) + (1.2)^2 \cdot (0.2) + (1.4)^2 \cdot (0.2) + (1.6)^2 \cdot (0.2) + (1.8)^2 \cdot (0.2)$

$= 0.2 + 0.288 + 0.392 + 0.512 + 0.648 = 2.04$ Multiplying out and summing.

Therefore, the area under the curve is approximately 2.04 square units.

As we saw earlier, using only five rectangles does not give a very accurate approximation for the true area under the curve. For greater accuracy we use more rectangles, calculating the area in the same way. The following table gives the "rectangular approximation" for the area under the curve in Example 1 for larger numbers of rectangles. The calculations were carried out on a graphing calculator using the program on page 812, rounding answers to three decimal places.

Number of Rectangles	Sum of Areas of Rectangles	
5	2.04	← Found in Example 1
10	2.185	The sum of the areas is approaching $2\frac{1}{3}$
100	2.318	
1,000	2.332	
10,000	2.333	

The areas in the right-hand column are approaching $2.333\ldots = 2\frac{1}{3}$, which is the *exact* area under the curve (as we will verify later). Therefore:

The area under the curve $f(x) = x^2$ from 1 to 2 is $2\frac{1}{3}$ square units.

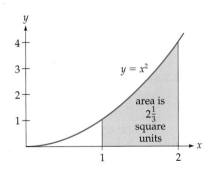

Areas are given in "square units," meaning that if the units on the graph are inches, feet, or some other units, then the area is in *square* inches, *square* feet, or, in general, some other *square* units.

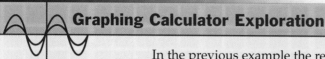

Graphing Calculator Exploration

In the previous example the rectangles touched the curve at the *left-hand* edge of the rectangle (and are called *left* rectangles). Similarly, we may use *midpoint* rectangles (which touch the curve at the *midpoint* of the rectangle), *right* rectangles (which touch the curve at their right-hand edge), or *random* rectangles (which touch the curve at a *random* x-value within the rectangle). The graphing calculator program* RIEMANN draws any of these types of rectangles for any function and calculates the sum of the areas, as shown below. The first screen shows five *midpoint* rectangles for $f(x) = x^2$ on the interval [1, 2]. The second screen shows the sum of the areas for various numbers of midpoint rectangles.

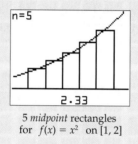

5 *midpoint* rectangles
for $f(x) = x^2$ on [1, 2]

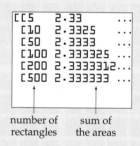

number of   sum of
rectangles   the areas

Notice from the second screen that the areas approach $2\frac{1}{3}$ very quickly, especially when compared with the table on the previous page for *left* rectangles. Do you see why midpoint rectangles are more accurate? [*Hint:* Compare the above graph for *midpoint* rectangles with the following graphs for *left* and *right* rectangles. Which type of rectangles seems to fit the curve most closely?]

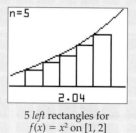

5 *left* rectangles for
$f(x) = x^2$ on [1, 2]

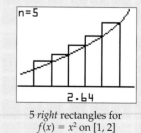

5 *right* rectangles for
$f(x) = x^2$ on [1, 2]

*See the Preface for how to obtain this and other programs.

Definite Integral

Approximating the area under a nonnegative function f by n rectangles means multiplying heights $f(x)$ by widths Δx and adding, obtaining

$$f(x_1) \cdot \Delta x + f(x_2) \cdot \Delta x + f(x_3) \cdot \Delta x + \cdots + f(x_n) \cdot \Delta x$$

The general procedure is as follows:

Area Under f from a to b Approximated by n Left Rectangles

1. Calculate the rectangle width $\Delta x = \dfrac{b-a}{n}$.

2. Find x-values $x_1, x_2, \ldots, x_n$ by successive additions of Δx beginning with $x_1 = a$.

3. Calculate the sum:

$$f(x_1) \cdot \Delta x + f(x_2) \cdot \Delta x + f(x_3) \cdot \Delta x + \cdots + f(x_n) \cdot \Delta x$$

The sum in step 3 is called a *Riemann sum*, after the great German mathematician Georg Bernhard Riemann (1826–1866).* The *limit* of the Riemann sum as the number n of rectangles approaches infinity gives the *area under the curve*, and is called the *definite integral of the function f from a to b*, written $\displaystyle\int_a^b f(x)\,dx$. Formally:

Definite Integral

Let f be a continuous function on an interval $[a, b]$. The definite integral of f from a to b is defined as

$$\int_a^b f(x)\,dx = \lim_{n \to \infty} [f(x_1) \cdot \Delta x + f(x_2) \cdot \Delta x + \cdots + f(x_n) \cdot \Delta x]$$

where $\Delta x = \dfrac{b-a}{n}$, and $x_1, x_2, \ldots, x_n$ are x-values beginning with $x_1 = a$ and obtained by successive additions of Δx. If f is nonnegative on $[a, b]$, then the definite integral gives the *area under the curve from a to b*.

---

* Actually, Riemann sums are slightly more general, allowing the subintervals to have different widths, with each width multiplied by the function evaluated at *any* x-value within that subinterval. For a continuous function, any Riemann sum will approach the same limiting value as the rectangles become arbitrarily narrow, so we may restrict our attention to the particular Riemann sums defined above (sometimes called "left Riemann sums").

The numbers *a* and *b* are called the *lower* and *upper limits of integration*.

Fundamental Theorem of Integral Calculus

A function followed by a vertical bar $\Big|_a^b$ with numbers *a* and *b* means evaluate the function at the *upper* number *b* and then subtract the evaluation at the *lower* number *a*.

$$F(x)\Big|_a^b = \underbrace{F(b)}_{\text{Evaluation at upper number}} - \underbrace{F(a)}_{\text{Evaluation at lower number}}$$

EXAMPLE 2 USING THE EVALUATION NOTATION

$$\underbrace{x^2\Big|_3^5}_{\text{Evaluation notation}} = \underbrace{(5)^2}_{\substack{x^2 \\ \text{at } x=5}} - \underbrace{(3)^2}_{\substack{x^2 \\ \text{at } x=3}} = \underbrace{25 - 9}_{\text{Simplifying}} = 16$$

Practice Problem 1

Evaluate $\sqrt{x}\,\Big|_4^{25}$ ▶ Solution on page 811

The following *Fundamental Theorem of Integral Calculus* shows how to evaluate definite integrals by using *indefinite* integrals. A geometric and intuitive justification of the theorem is given at the end of this section.

Fundamental Theorem of Integral Calculus

For a continuous function *f* on an interval [*a*, *b*],

$$\int_a^b f(x)\,dx = F(b) - F(a) \qquad \text{The right-hand side may be written } F(x)\Big|_a^b$$

where *F* is any antiderivative of *f*.

The theorem is "fundamental" in that it establishes a deep and unexpected connection between definite integrals (limits of Riemann sums) and antiderivatives. It says that definite integrals can be evaluated in two simple steps:

1. Find an *indefinite* integral of the function (omitting the + C).
2. *Evaluate* the result at b and *subtract* the evaluation at a.

EXAMPLE 3 FINDING A DEFINITE INTEGRAL BY THE FUNDAMENTAL THEOREM

Find $\int_1^2 x^2\,dx$.

Solution

Because x^2 is continuous on [1, 2], we can use the Fundamental Theorem.

$$\int_1^2 x^2\,dx = \underbrace{\frac{1}{3}x^3\Big|_1^2}_{\text{Integrating }x^2} = \underbrace{\frac{1}{3}2^3}_{\substack{\text{Evaluating}\\ \frac{1}{3}x^3\text{ at }x=2}} - \underbrace{\frac{1}{3}1^3}_{\substack{\text{And at}\\ x=1}} = \underbrace{\frac{8}{3} - \frac{1}{3}}_{\text{Subtracting}} = \frac{7}{3}$$

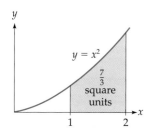

Since definite integrals of nonnegative functions give areas, this result means that the area under $f(x) = x^2$ from $x = 1$ to $x = 2$ is $\frac{7}{3}$ square units.

On pages 798–799 we found this same area by the much more laborious process of calculating Riemann sums and taking the limit as the number of rectangles approached infinity, and our answer here, $\frac{7}{3}$, agrees with the answer there, $2\frac{1}{3}$ square units. Whenever possible, we will calculate definite integrals and find areas in this much simpler way, using the Fundamental Theorem of Integral Calculus.

Graphing Calculator Exploration

a. Graph $y_1 = x^2$ on the graphing window [0, 2] by [−1, 4].

b. Verify the result of Example 3 by having your graphing calculator find the definite integral of x^2 from 1 to 2. [*Hint:* Use a command like FnInt or $\int f(x)\,dx$.]

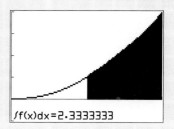

Practice Problem 2 Find the area under $y = x^3$ from 0 to 2 by evaluating $\int_0^2 x^3 \, dx$.

> Solution on page 811

EXAMPLE 4

FINDING THE AREA UNDER A CURVE

Find the area under $y = e^{2x}$ from $x = 0$ to $x = 1$.

Solution Because e^{2x} is nonnegative and continuous on $[0, 1]$, the area is given by the definite integral:

$$\int_0^1 e^{2x} \, dx = \underbrace{\left.\frac{1}{2} e^{2x} \right|_0^1}_{\substack{\text{Indefinite integral} \\ \text{(using the formula} \\ \text{for } \int e^{ax} \, dx)}} = \underbrace{\frac{1}{2} e^2 - \frac{1}{2} e^0}_{\substack{\text{Evaluating} \\ \text{at } x = 1 \text{ and} \\ \text{at } x = 0}} = \underbrace{\frac{1}{2} e^2 - \frac{1}{2}}_{\substack{\text{Simplifying} \\ \text{(using } e^0 = 1)}}$$

Therefore, the area is $\frac{1}{2} e^2 - \frac{1}{2}$ square units.

We leave the answer in this "exact" form (in terms of the number e). In an application we would use a calculator to approximate this answer as 3.19 square units.

EXAMPLE 5

FINDING THE AREA UNDER A CURVE

Find the area under $f(x) = \dfrac{1}{x}$ from $x = 1$ to $x = e$.

Solution Because $\dfrac{1}{x}$ is nonnegative and continuous on $[1, e]$, the area is:

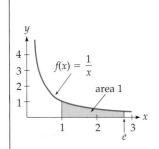

$$\int_1^e \frac{1}{x} \, dx = \left. \ln x \right|_1^e = \underbrace{\ln e}_{1} - \underbrace{\ln 1}_{0} = 1$$

Therefore, the area is 1 square unit.

From this example we can give an alternate definition of the number e: e is the number such that the definite integral of $1/x$ from 1 to e is 1.

Graphing Calculator Exploration

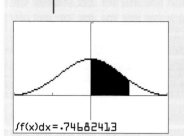

Use a graphing calculator to evaluate the definite integral

$$\int_0^1 e^{-x^2} \, dx.$$

(This definite integral, which is important in probability and statistics, cannot be evaluated by the Fundamental Theorem because the function e^{-x^2} has no simple antiderivative. Your calculator may find the answer by approximating the area under the curve by modified Riemann sums.)

Definite integrals have many of the properties of indefinite integrals. These properties follow from interpreting definite integrals as limits of Riemann sums.

Properties of Definite Integrals

$$\int_a^b c \cdot f(x) \, dx = c \int_a^b f(x) \, dx$$

A constant may be moved across the integral sign

$$\int_a^b [f(x) \pm g(x)] \, dx = \int_a^b f(x) \, dx \pm \int_a^b g(x) \, dx$$

Read both upper signs or both lower signs

The integral of a sum is the sum of the integrals (and similarly for differences)

EXAMPLE 6 USING THE PROPERTIES OF DEFINITE INTEGRALS

Find the area under $y = 24 - 6x^2$ from -1 to 1.

Solution

$$\int_{-1}^{1} (24 - 6x^2) \, dx = \left(24x - 6 \cdot \frac{1}{3} x^3 \right) \Big|_{-1}^{1}$$

$$= (24 - 2) - (-24 + 2) = 44$$

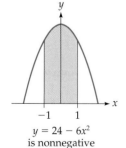

$y = 24 - 6x^2$ is nonnegative on $[-1, 1]$

Therefore, the area is 44 square units.

Total Cost of a Succession of Units

Given a marginal cost function, to find the *total* cost of producing, say, units 100 to 400, we could proceed as follows: *integrate* the marginal cost to find the total cost, *evaluate* at 400 to find the total cost up to unit 400, and *subtract* the evaluation at 100 to leave just the cost of units 100 to 400. However, these steps of integrating, evaluating, and subtracting are just the steps in evaluating a definite integral by the Fundamental Theorem. Therefore, the cost of a succession of units is equal to the definite integral of the marginal cost function.

Cost of a Succession of Units

For a marginal cost function $MC(x)$:

$$\begin{pmatrix} \text{Total cost of} \\ \text{units } a \text{ to } b \end{pmatrix} = \int_a^b MC(x)\, dx$$

EXAMPLE 7

FINDING THE COST OF A SUCCESSION OF UNITS

A company's marginal cost function is $MC(x) = \dfrac{75}{\sqrt{x}}$, where x is the number of units. Find the total cost of producing units 100 to 400.

Solution

$$\begin{pmatrix} \text{Total cost of} \\ \text{units } 100 \text{ to } 400 \end{pmatrix} = \int_{100}^{400} \frac{75}{\sqrt{x}}\, dx = 75 \int_{100}^{400} x^{-1/2}\, dx \quad \text{Integrating marginal cost}$$

$$= (75 \cdot 2 \cdot x^{1/2}) \Big|_{100}^{400}$$

$$= 150 \cdot (400)^{1/2} - 150 \cdot (100)^{1/2}$$

$$= 150 \cdot 20 - 150 \cdot 10 = 1500$$

The cost of producing units 100 to 400 is $1500.

Similarly, integrating *any* rate from a to b gives the *total accumulation* at that rate between a and b.

Total Accumulation at a Given Rate

$$\begin{pmatrix} \text{Total accumulation at} \\ \text{rate } f \text{ from } a \text{ to } b \end{pmatrix} = \int_a^b f(x)\, dx$$

The diagrams below illustrate this idea. In each case, the *curve* represents a *rate*, and the *area under the curve*, given by the definite integral, gives the *total accumulation* at that rate.

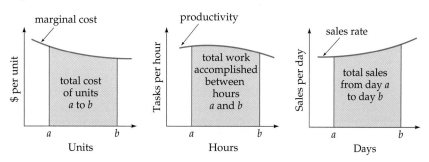

In repetitive tasks, a person's productivity usually increases with practice until it is slowed by monotony.

EXAMPLE 8

FINDING TOTAL PRODUCTIVITY FROM A RATE

A technician can test computer chips at the rate of $-3t^2 + 18t + 15$ chips per hour (for $0 \leq t \leq 6$), where t is the number of hours after 9:00 A.M. How many chips can be tested between 10:00 A.M. and 1:00 P.M.?

Solution

The total work accomplished is the integral of this rate from $t = 1$ (10 A.M.) to $t = 4$ (1 P.M.):

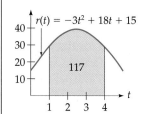

$$\int_1^4 (-3t^2 + 18t + 15)\, dt = \left(-t^3 + 18 \cdot \frac{1}{2} t^2 + 15t\right)\bigg|_1^4$$

$$= (-64 + 144 + 60) - (-1 + 9 + 15) = 117$$

That is, between 10 A.M. and 1 P.M., 117 chips can be tested.

Integration Notation

The symbol Σ (the Greek letter S) is used in mathematics to indicate a *sum*, and so the Riemann sum can be written $\sum_1^n f(x_k)\, \Delta x$. The fact that the Riemann sum approaches the definite integral can be expressed:

$$\sum_1^n f(x_k)\, \Delta x \longrightarrow \int_a^b f(x)\, dx$$

Σ becomes $\int$
Δ becomes d
as $n \to \infty$

That is, the n approaching infinity changes the Σ (a Greek S) into an integral sign $\int$ (a "stretched out" S), and the Δ (a Greek D) into a d. In other words, the integral notation reminds us that a definite integral represents a *sum of rectangles* of height $f(x)$ and width dx.

Verification of the Fundamental Theorem of Integral Calculus

The following is a geometric and intuitive justification of the Fundamental Theorem of Integral Calculus.

For a continuous and nonnegative function f on an interval $[a, b]$, we define a new function $A(x)$ as the *area under f from a to x*.

$$A(x) = \begin{pmatrix} \text{Area under } f \\ \text{from } a \text{ to } x \end{pmatrix}$$

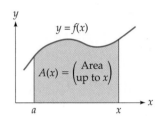

Therefore

$$A(b) = \begin{pmatrix} \text{Area under } f \\ \text{from } a \text{ to } b \end{pmatrix} \qquad \text{Replacing } x \text{ by } b \text{ gives the entire area from } a \text{ to } b$$

and

$$A(a) = 0 \qquad \text{The area "from } a \text{ to } a\text{" must be zero}$$

Subtracting these last two expressions, $A(b) - A(a)$, gives the total area from a to b [since $A(b)$ is the total area, and $A(a)$ is zero]. This same area can be expressed as the definite integral of f from a to b, leading to the equation

$$\int_a^b f(x)\, dx = A(b) - A(a) \qquad \text{Area under } f \text{ from } a \text{ to } b \text{ expressed in two ways}$$

If we can show that $A(x)$ is an antiderivative of $f(x)$, then the equation above will show that the definite integral can be found by an antiderivative evaluated at the upper and lower limits and subtracted, which will verify the Fundamental Theorem. To show that $A(x)$ is an antiderivative of $f(x)$, we show that $A'(x) = f(x)$, differentiating $A(x)$ by the definition of the derivative:

$$A'(x) = \lim_{h \to 0} \frac{A(x + h) - A(x)}{h}$$

In the numerator, $A(x + h)$ is the area under the curve up to $x + h$ and $A(x)$ is the area up to x, so subtracting them, $A(x + h) - A(x)$, leaves just the area from x to $x + h$, as shown below.

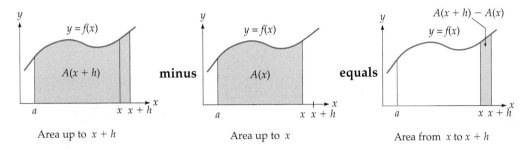

When h is small, this last area can be approximated by a rectangle of base h and height $f(x)$, where the approximation becomes exact as h approaches zero. Therefore, in the limit we may replace $A(x + h) - A(x)$ by the area of the rectangle, $h \cdot f(x)$:

$$A'(x) = \lim_{h \to 0} \frac{\overbrace{A(x + h) - A(x)}^{\approx h \cdot f(x)}}{h} = \lim_{h \to 0} \frac{h \cdot f(x)}{h} = f(x)$$

This equation says that $A'(x) = f(x)$, showing that $A(x)$ is an antiderivative of $f(x)$. This completes the verification of the Fundamental Theorem of Integral Calculus.

Functions Taking Positive and Negative Values

Riemann sums can be calculated for *any* continuous function, not just nonnegative functions. The following diagram illustrates a Riemann sum for a continuous function that takes positive *and* negative values.

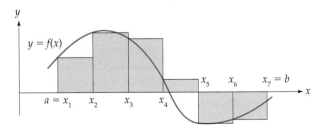

The two rightmost rectangles, where f is *negative*, will make a *negative* contribution to the sum. Taking the limit, the definite integral where the function lies *below* the x-axis will give the *negative* of the area between the curve and the x-axis. Therefore, the definite integral of such a function from a to b will give the *signed area* between the curve and the x-axis: the area *above* the axis minus the area *below* the axis, shown as A_{up} minus A_{down} in the diagram below.

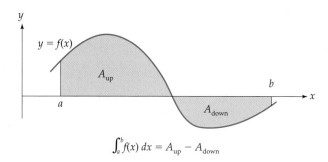

11.3 Section Summary

We began by approximating the area under a curve from a to b by Riemann sums (sums of rectangles). We then defined the *definite integral* $\int_a^b f(x)\,dx$ as the *limit* of the Riemann sum as the number of rectangles approaches infinity. The Fundamental Theorem of Integral Calculus showed how to evaluate definite integrals much more simply by evaluating *indefinite* integrals:

$$\int_a^b f(x)\,dx = F(x)\Big|_a^b \qquad \text{F is any antiderivative of f}$$

Distinguish carefully between definite and indefinite integrals: a *definite* integral is a *number*, whereas an *indefinite* integral is a function plus an arbitrary constant.

As for the *uses* of definite integrals, the definite integral of a nonnegative function gives the *area under the curve*:

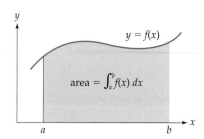

11.3 DEFINITE INTEGRALS AND AREAS

The definite integral of a *rate* gives the *total accumulation* at that rate:

$$\int_a^b f(x)\, dx = \begin{pmatrix} \text{Total accumulation at} \\ \text{rate } f \text{ from } a \text{ to } b \end{pmatrix}$$

> **Solutions to Practice Problems**

1. $\sqrt{x}\,\Big|_4^{25} = \sqrt{25} - \sqrt{4} = 5 - 2 = 3$

2. $\int_0^2 x^3\, dx = \tfrac{1}{4}x^4 \Big|_0^2 = \tfrac{1}{4}\cdot 16 - \tfrac{1}{4}\cdot 0 = 4$ square units

11.3 Exercises

Find the sum of the areas of the shaded rectangles under each graph. Round to two decimal places. [*Hint:* The width of each rectangle is the difference between the *x*-values at its base. The height of each rectangle is the height of the curve at the left edge of the rectangle.]

1.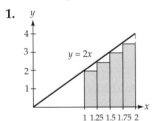

2.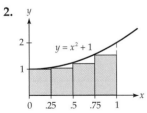

3.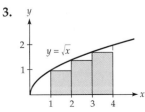

4.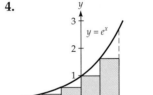

5.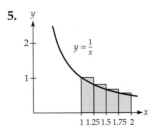

6.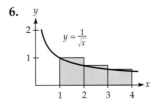

For each function:
i. *Approximate* the area under the curve from a to b by calculating a Riemann sum with the given number of rectangles. Use the method described in Example 1 on pages 798–799, rounding to three decimal places.
ii. Find the *exact* area under the curve from a to b by evaluating an appropriate definite integral using the Fundamental Theorem.

7. $f(x) = 2x$ from $a = 1$ to $b = 2$.
   For part (i), use 5 rectangles.
8. $f(x) = x^2 + 1$ from $a = 0$ to $b = 1$.
   For part (i), use 5 rectangles.
9. $f(x) = \sqrt{x}$ from $a = 1$ to $b = 4$.
   For part (i), use 6 rectangles.
10. $f(x) = e^x$ from $a = -1$ to $b = 1$.
    For part (i), use 8 rectangles.
11. $f(x) = \dfrac{1}{x}$ from $a = 1$ to $b = 2$.
    For part (i), use 10 rectangles.
12. $f(x) = \dfrac{1}{\sqrt{x}}$ from $a = 1$ to $b = 4$.
    For part(i), use 6 rectangles.

Use the graphing calculator program RIEMANN (see page 800), one of the programs given below, or a similar program to find the following Riemann sums.

i. Calculate the Riemann sum for each function for the following values of n: 10, 100, and 1000. Use left, right, or midpoint rectangles, making a table of the answers, rounded to three decimal places.
ii. Find the *exact* value of the area under the curve by evaluating an appropriate definite integral using the Fundamental Theorem. The values of the Riemann sums from part (i) should approach this number.

13. $f(x) = 2x$ from $a = 1$ to $b = 2$
14. $f(x) = x^2 + 1$ from $a = 0$ to $b = 1$
15. $f(x) = \sqrt{x}$ from $a = 1$ to $b = 4$
16. $f(x) = e^x$ from $a = -1$ to $b = 1$
17. $f(x) = \dfrac{1}{x}$ from $a = 1$ to $b = 2$
18. $f(x) = \dfrac{1}{\sqrt{x}}$ from $a = 1$ to $b = 4$

BASIC Program for (Left) Riemann Sums

(Lines 2, 3, 4, and 8 to be completed as indicated in small type)

```
Riemsum = 0
a =               fill in beginning x-value
b =               fill in ending x-value
n =               fill in number of rectangles
delta = (b - a)/n
x = a
FOR i = 1 to n
  Riemsum = Riemsum + (fill in the function)*delta
  x = x + delta
NEXT i
PRINT "RIEMANN SUM IS", Riemsum
END
```

TI-83 Program for (Left) Riemann Sums

(Enter the function in y_1 before executing)

```
0 → S
Disp "USES FUNCTION Y1"
Prompt A, B, N
(B-A)/N → D
A → X
For(I, 1, N)
  S + Y1*D → S
  X + D → X
End
Disp "RIEMANN SUM IS", S
```

Find the area under each curve between the given x-values. For Exercises 19–24, also make a sketch of the curve showing the region.

19. $f(x) = x^2$ from $x = 0$ to $x = 3$
20. $f(x) = x$ from $x = 0$ to $x = 4$
21. $f(x) = 4 - x$ from $x = 0$ to $x = 4$
22. $f(x) = 1 - x^2$ from $x = -1$ to $x = 1$
23. $f(x) = \dfrac{1}{x}$ from $x = 1$ to $x = 2$
24. $f(x) = e^x$ from $x = 0$ to $x = 1$
25. $f(x) = 8x^3$ from $x = 1$ to $x = 3$
26. $f(x) = 6x^2$ from $x = 2$ to $x = 3$
27. $f(x) = 6x^2 + 4x - 1$ from $x = 1$ to $x = 2$
28. $f(x) = 27 - 3x^2$ from $x = 1$ to $x = 3$
29. $f(x) = \dfrac{1}{\sqrt{x}}$ from $x = 4$ to $x = 9$
30. $f(x) = \dfrac{1}{x^2}$ from $x = 1$ to $x = 3$
31. $f(x) = 8 - 4\sqrt[3]{x}$ from $x = 0$ to $x = 8$
32. $f(x) = 9 - 3\sqrt{x}$ from $x = 0$ to $x = 9$
33. $f(x) = \dfrac{1}{x}$ from $x = 1$ to $x = 5$
34. $f(x) = \dfrac{1}{x}$ from $x = e$ to $x = e^3$
35. $f(x) = x^{-1} + x^2$ from $x = 1$ to $x = 2$
36. $f(x) = 6e^{2x}$ from $x = 0$ to $x = 2$
37. $f(x) = 2e^x$ from $x = 0$ to $x = \ln 3$
38. $f(x) = e^{-x}$ from $x = 0$ to $x = 1$
39. $f(x) = e^{x/2}$ from $x = 0$ to $x = 2$
40. $f(x) = e^{x/3}$ from $x = 0$ to $x = 3$

For each function:
a. Integrate ("by hand") to find the area under the curve between the given x-values.
b. Verify your answer to part (a) by having your calculator graph the function and find the area (using a command like FnInt or $\int f(x)\,dx$).

41. $f(x) = 12 - 3x^2$ from $x = 1$ to $x = 2$

42. $f(x) = 9x^2 - 6x + 1$ from $x = 1$ to $x = 2$
43. $f(x) = \dfrac{1}{x^3}$ from $x = 1$ to $x = 4$
44. $f(x) = \dfrac{1}{\sqrt[3]{x}}$ from $x = 8$ to $x = 27$
45. $f(x) = 2x + 1 + x^{-1}$ from $x = 1$ to $x = 2$
46. $f(x) = e^x$ from $x = 0$ to $x = 3$

Evaluate each definite integral.

47. $\displaystyle\int_0^1 (x^{99} + x^9 + 1)\,dx$
48. $\displaystyle\int_2^4 (1 + x^{-2})\,dx$
49. $\displaystyle\int_1^2 (6t^2 - 2t^{-2})\,dt$
50. $\displaystyle\int_{-2}^2 (3w^2 - 2w)\,dw$
51. $\displaystyle\int_1^4 \dfrac{1}{y^2}\,dy$
52. $\displaystyle\int_1^4 \dfrac{1}{\sqrt{z}}\,dz$
53. $\displaystyle\int_1^e \dfrac{dx}{x}$
54. $\displaystyle\int_1^{e^2} \dfrac{3}{x}\,dx$
55. $\displaystyle\int_1^3 (9x^2 + x^{-1})\,dx$
56. $\displaystyle\int_1^2 (x^{-1} - 4x^2)\,dx$
57. $\displaystyle\int_{-2}^{-1} 3x^{-1}\,dx$
58. $\displaystyle\int_{-3}^{-1} (1 + x^{-1})\,dx$
59. $\displaystyle\int_0^1 12e^{3x}\,dx$
60. $\displaystyle\int_0^2 3e^{x/2}\,dx$
61. $\displaystyle\int_{-1}^1 5e^{-x}\,dx$
62. $\displaystyle\int_0^1 (6x^2 - 4e^{2x})\,dx$
63. $\displaystyle\int_{\ln 2}^{\ln 3} e^x\,dx$
64. $\displaystyle\int_0^{\ln 5} e^x\,dx$
65. $\displaystyle\int_1^2 \dfrac{(x+1)^2}{x}\,dx$
66. $\displaystyle\int_1^2 \dfrac{(x+1)^2}{x^2}\,dx$

Use a graphing calculator to evaluate each definite integral, rounding answers to three decimal places. [Hint: Use a command like FnInt or $\int f(x)\,dx$.]

67. $\displaystyle\int_0^2 \dfrac{1}{x^2 + 1}\,dx$
68. $\displaystyle\int_{-1}^1 \sqrt{x^4 + 1}\,dx$
69. $\displaystyle\int_{-1}^1 e^{x^2}\,dx$
70. $\displaystyle\int_{-2}^2 e^{(-1/2)x^2}\,dx$
71. $\displaystyle\int_0^4 \sqrt{x}\,e^x\,dx$
72. $\displaystyle\int_1^4 x^x\,dx$

73. OMITTING THE C IN DEFINITE INTEGRALS

a. Evaluate the definite integral $\int_0^3 x^2 \, dx$.

b. Evaluate the same definite integral by completing the following calculation, in which the antiderivative includes a constant C.

$$\int_0^3 x^2 \, dx = \left(\frac{1}{3}x^3 + C\right)\bigg|_0^3 = \cdots$$

[The constant C should cancel out, giving the same answer as in part (a)].

c. Explain why the constant will cancel out of *any* definite integral. (We therefore omit the constant in definite integrals. However, be sure to keep the +C in *indefinite* integrals.)

74. Evaluate $\displaystyle\int_1^1 \frac{x^{43}e^{\sqrt[3]{x^2}} + x^{-39}}{\ln\sqrt[17]{x^3 + 11} + x^{199}} \, dx.$

[*Hint:* No work is necessary.]

75. Show that for any number $a > 0$,

$$\int_1^a \frac{1}{x} \, dx = \ln a$$

This equation is often used as a *definition* of natural logarithms, defining $\ln a$ as the area under the curve $y = 1/x$ between 1 and a.

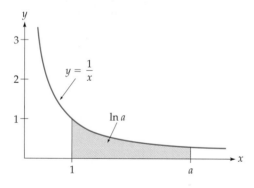

76. a. Try to evaluate the integral $\displaystyle\int_{-1}^1 \frac{1}{x^2} \, dx$.

b. Explain why the answer is negative in spite of the fact that the integrand is nonnegative, and so the integral should be positive. [*Hint:* In order to use the Fundamental Theorem, or even to *define* the definite integral, the integrand must be continuous on the interval. Is it?]

77–78. GEOMETRIC INTEGRATION Find $\displaystyle\int_0^4 f(x) \, dx$ for the function $f(x)$ graphed below.

[*Hint:* No calculus is necessary—just as we used integrals to find areas, we can use areas to find integrals. The curves shown are quarter circles.]

77.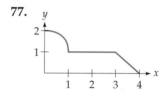

78.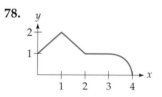

APPLIED EXERCISES

79. GENERAL: Electrical Consumption On a hot summer afternoon, a city's electricity consumption is $-3t^2 + 18t + 10$ units per hour, where t is the number of hours after noon ($0 \leq t \leq 6$). Find the total consumption of electricity between the hours of 1 and 5 P.M.

80. GENERAL: Weight An average child of age x years gains weight at the rate of $3.9x^{1/2}$ pounds per year (for $0 \leq x \leq 16$). Find the total weight gain from age 1 to age 9.

81–82. GENERAL: Repetitive Tasks After t hours of work, a bank clerk can process checks at the rate of r(t) checks per hour for the function r(t) given below. How many checks will the clerk process during the first three hours (time 0 to time 3)?

81. $r(t) = -t^2 + 90t + 5$

82. $r(t) = -t^2 + 60t + 9$

83–84. BUSINESS: Cost A company's marginal cost function is $MC(x)$ (given below), where x is the number of units. Find the total cost of the first hundred units ($x = 0$ to $x = 100$).

83. $MC(x) = 6e^{-0.02x}$ **84.** $MC(x) = 8e^{-0.01x}$

85. GENERAL: Price Increase The price of a double-dip ice cream cone is increasing at the rate of $18e^{0.08t}$ cents per year, where t is measured in years and $t = 0$ corresponds to 2000. Find the total change in price between the years 2000 and 2010.

86. BUSINESS: Sales An automobile dealer estimates that the newest model car will sell at the rate of $30/t$ cars per month, where t is measured in months and $t = 1$ corresponds to the beginning of January. Find the number of cars that will be sold from the beginning of January to the beginning of May.

87. BUSINESS: Tin Consumption World consumption of tin is running at the rate of $0.24e^{0.01t}$ million tons per year, where t is measured in years and $t = 0$ corresponds to the beginning of 2000. Find the total consumption of tin from the beginning of 2000 to the beginning of 2010.

88. SOCIOLOGY: Marriages There are approximately $2.2e^{0.01t}$ million marriages per year in the United States, where t is the number of years since 2000. Assuming that this rate continues, find the number of marriages from the year 2000 to the year 2010.

89. BEHAVIORAL SCIENCE: Learning A student can memorize words at the rate of $6e^{-t/5}$ words per minute after t minutes. Find the total number of words that the student can memorize in the first 10 minutes.

90. BIOMEDICAL: Epidemics An epidemic is spreading at the rate of $12e^{0.2t}$ new cases per day, where t is the number of days since the epidemic began. Find the total number of new cases in the first 10 days of the epidemic.

91. ECONOMICS: Pareto's Law The economist Vilfredo Pareto estimated that the number of people who have an income between A and B dollars ($A < B$) is given by a definite integral of the form

$$N = \int_A^B ax^{-b} \, dx \quad (b \neq 1)$$

where a and b are constants. Solve this integral.

92. BIOMEDICAL: Poiseuille's Law According to Poiseuille's law, the speed of blood in a blood vessel is given by $V = \dfrac{p}{4Lv}(R^2 - r^2)$, where R is the radius of the blood vessel, r is the distance of the blood from the center of the blood vessel, and p, L, and v are constants determined by the pressure and viscosity of the blood and the length of the vessel. The total blood flow is then given by

$$\begin{pmatrix}\text{Total} \\ \text{blood flow}\end{pmatrix} = \int_0^R 2\pi \frac{p}{4Lv}(R^2 - r^2)r \, dr$$

Find the total blood flow by finding this integral (p, L, v, and R are constants).

93–94. BUSINESS: Capital Value of an Asset The *capital value* of an asset (such as an oil well) that produces a continuous stream of income is the sum of the present value of all future earnings from the asset. Therefore, the capital value of an asset that produces income at the rate of $r(t)$ dollars per year (at a continuous interest rate i) is

$$\begin{pmatrix}\text{Capital} \\ \text{value}\end{pmatrix} = \int_0^T r(t)e^{-it} \, dt$$

where T is the expected life (in years) of the asset.

93. Use the formula in the preceding instructions to find the capital value (at interest rate $i = 0.06$) of an oil well that produces income at the constant rate of $r(t) = 240{,}000$ dollars per year for 10 years.

94. Use the formula in the preceding instructions to find the capital value (at interest rate $i = 0.05$) of a uranium mine that produces income at the rate of $r(t) = 560{,}000t^{1/2}$ dollars per year for 20 years.

95. GENERAL: Area
  a. Use your graphing calculator to find the area between 0 and 1 under the following curves: $y = x$, $y = x^2$, $y = x^3$, and $y = x^4$.
  b. Based on your answers to part (a), conjecture a formula for the area under $y = x^n$ between 0 and 1 for any value of $n > 0$.
  c. Prove your conjecture by evaluating an appropriate definite integral "by hand."

96. GENERAL: Dam Construction Ever since the Johnstown Dam burst in 1889, killing 2200 people, dam construction has become increasingly scientific.
  a. Estimate the amount of concrete needed to build the Snake River Dam by finding the area of the cross section shown in the following diagram and multiplying the result by the 574-foot length of the dam. All dimensions are in feet. [*Hint:* Find the cross-sectional area by integrating and using area formulas.]
  b. If a mixing truck carries 300 cubic feet of concrete, about how many truckloads would be needed to build the dam?

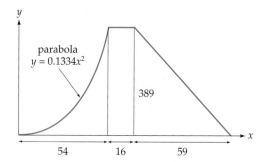

97. BIOMEDICAL: Drug Absorption An oral medication is absorbed into the bloodstream at the rate of $5e^{-0.04t}$ milligrams per minute, where t is the number of minutes since the medication was taken. Find the total amount of medication absorbed within the first 30 minutes.

98. BIOMEDICAL: Aortic Volume The rate of change of the volume of blood in the aorta t seconds after the beginning of the cardiac cycle is $-kP_0 e^{-mt}$ milliliters per second, where k, P_0, and m are constants (depending, respectively, on the elasticity of the aorta, the initial aortic pressure, and various characteristics of the cardiac cycle). Find the total change in volume from time 0 to time T (the end of the cardiac cycle). (Your answer will involve the constants k, P_0, m, and T.)

99. BUSINESS: Sales A dealer predicts that new cars will sell at the rate of $8xe^{-0.1x}$ sales per week in week x. Find the total sales in the first half year (week 0 to week 26).

100. GENERAL: Cigarette Smoking Reread, if necessary, the Application Preview on pages 769–770.
  a. Evaluate the definite integrals
  $$\int_0^1 (300e^{0.025x} - 240e^{0.02x})\, dx$$
  and
  $$\int_7^8 (300e^{0.025x} - 240e^{0.02x})\, dx$$
  to verify the answers given there for the amount of tar inhaled from the first and last centimeters of the cigarette.
  b. Evaluate the definite integral
  $$\int_0^8 (300e^{0.025x} - 240e^{0.02x})\, dx$$
  to find the amount of tar inhaled from smoking the entire cigarette.

101. GENERAL: Population A resort community swells at the rate of $100e^{0.4\sqrt{x}}$ new arrivals per day on day x of its "high season." Find the total number of arrivals in the first two weeks (day 0 to day 14).

102. GENERAL: Repetitive Tasks After t hours of work, a medical technician can carry out T-cell counts at the rate of $2x^2 e^{-x/4}$ tests per hour. How many tests will the technician process during the first eight hours (time 0 to time 8)?

11.4 FURTHER APPLICATIONS OF DEFINITE INTEGRALS: AVERAGE VALUE AND AREA BETWEEN CURVES

Introduction

In this section we will use definite integrals for two important purposes: finding average values of functions and finding areas between curves. Average values are used everywhere. Birth weights of babies are compared with average weights, and retirement benefits are determined by average income. Averages eliminate fluctuations, reducing a collection of numbers to a single "representative" number. Areas between curves are used to find quantities from trade deficits to lives saved by seat belts (see pages 830–831).

Average Value of a Function

The average value of n numbers is found by adding the numbers and dividing by n. For example,

$$\binom{\text{Average of}}{a, b, \text{ and } c} = \frac{1}{3}(a + b + c)$$

How can we find the average value of a *function* on an interval? For example, if a function gives the temperature over a 24-hour period, how can we calculate the *average temperature*? We could, of course, just take the temperature at every hour and then average these 24 values, but this would ignore the temperature at all of the intermediate times. Intuitively, the average should represent a "leveling off" of the curve to a uniform height, the dashed line shown on the left.

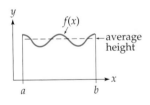

This leveling should use the "hills" to fill in the "valleys," maintaining the same total area under the curve. Therefore, the area under the line (a rectangle with base $(b - a)$ and height up to the line) must equal the area under the curve (the definite integral of the function).

$$\underbrace{(b - a)\binom{\text{Average}}{\text{height}}}_{\text{Area under line}} = \underbrace{\int_a^b f(x)\, dx}_{\text{Area under curve}} \quad \text{Equating the two areas}$$

Therefore:

$$\binom{\text{Average}}{\text{height}} = \frac{1}{b - a}\int_a^b f(x)\, dx \quad \text{Dividing by } (b - a)$$

This formula gives the average (or "mean") value of a continuous function on an interval.

> **Average Value of a Function**
>
> $$\begin{pmatrix}\text{Average value} \\ \text{of } f \text{ on } [a, b]\end{pmatrix} = \frac{1}{b-a}\int_a^b f(x)\, dx$$
>
> Average value is the definite integral of the function divided by the length of the interval

Finding the average value of a function by integrating and dividing by $b - a$ is analogous to averaging n numbers by adding and dividing by n (since integrals are continuous sums). A derivation of this formula by Riemann sums is given at the end of this section.

EXAMPLE 1

FINDING THE AVERAGE VALUE OF A FUNCTION

Find the average value of $f(x) = \sqrt{x}$ from $x = 0$ to $x = 9$.

Solution

$$\begin{pmatrix}\text{Average} \\ \text{value}\end{pmatrix} = \frac{1}{9-0}\int_0^9 \sqrt{x}\, dx = \frac{1}{9}\int_0^9 x^{1/2}\, dx \quad \text{Integral divided by the length of the interval}$$

$$= \frac{1}{9}\cdot\frac{2}{3}x^{3/2}\Big|_0^9 = \frac{2}{27}9^{3/2} - \frac{2}{27}0^{3/2} \quad \text{Integrating and evaluating}$$

$$= \frac{2}{27}(\sqrt{9})^3 - 0 = \frac{2}{27}\cdot 27 = 2 \quad \text{Simplifying}$$

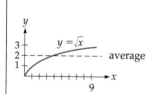

The average value of $f(x) = \sqrt{x}$ over the interval [0, 9] is 2.

EXAMPLE 2

FINDING AVERAGE POPULATION

The population of the United States is predicted to be $P(t) = 281e^{0.012t}$ million people, where t is the number of years since 2000. Predict the average population between the years 2010 and 2020.

Solution We integrate from $t = 10$ (year 2010) to $t = 20$ (year 2020).

$$\begin{pmatrix} \text{Average} \\ \text{value} \end{pmatrix} = \frac{1}{20-10} \int_{10}^{20} 281 e^{0.012t} \, dt \quad \text{Integral divided by the length of the interval}$$

$$= \frac{281}{10} \int_{10}^{20} e^{0.012t} \, dt$$

$$= 28.1 \frac{1}{0.012} e^{0.012t} \Big|_{10}^{20} \quad \text{Integrating by the } \int e^{ax} dx \text{ formula}$$

$$= 2342 e^{0.24} - 2342 e^{0.12} \approx 337 \quad \text{Evaluating, using a calculator}$$

The average population of the United States during the second decade of the twenty-first century will be about 337 million people.

Practice Problem 1 Find the average value of $f(x) = 3x^2$ from $x = 0$ to $x = 2$.

> Solution on page 826

Area Between Curves: Integrating "Upper Minus Lower"

We know that definite integrals give areas under curves. To calculate the area *between* two curves, we take the area under the *upper* curve and subtract the area under the *lower* curve.

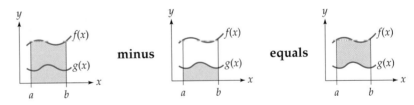

In terms of integrals:

$$\int_a^b f(x) \, dx \quad - \quad \int_a^b g(x) \, dx \quad = \quad \int_a^b [f(x) - g(x)] \, dx$$

↑ ↑
Upper Lower
curve curve

Therefore, the area between the curves can be written as a single integral:

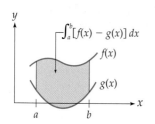

Area Between Curves

The area between two continuous curves $f(x) \geq g(x)$ on $[a, b]$ is

$$\begin{pmatrix} \text{Area between} \\ f \text{ and } g \text{ on } [a, b] \end{pmatrix} = \int_a^b [f(x) - g(x)]\, dx \qquad \text{Integrate "upper minus lower"}$$

Finding area by integrating "upper minus lower" works regardless of whether one or both curves dip below the x-axis.

EXAMPLE 3

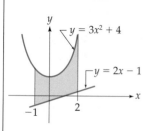

FINDING THE AREA BETWEEN CURVES

Find the area between $y = 3x^2 + 4$ and $y = 2x - 1$ from $x = -1$ to $x = 2$.

Solution The area is shown in the diagram on the left. (You may need to make a similar rough sketch for each problem to see which curve is "upper" and which is "lower.") We integrate "upper minus lower" between the given x-values.

$$\int_{-1}^{2} [\underbrace{(3x^2 + 4)}_{\text{Upper}} - \underbrace{(2x - 1)}_{\text{Lower}}]\, dx = \int_{-1}^{2} (3x^2 + 4 - 2x + 1)\, dx$$

$$= \int_{-1}^{2} (3x^2 - 2x + 5)\, dx \qquad \text{Simplifying}$$

$$= (x^3 - x^2 + 5x)\Big|_{-1}^{2} \qquad \text{Integrating}$$

$$= (8 - 4 + 10) - (-1 - 1 - 5) \qquad \text{Evaluating}$$

$$= 21 \text{ square units}$$

Practice Problem 2 Find the area between $y = 2x^2 + 1$ and $y = -x^2 - 1$ from $x = -1$ to $x = 1$.

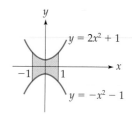

➤ Solution on page 827

EXAMPLE 4

FINDING SALES FROM EXTRA ADVERTISING

A company marketing high-definition television sets expects to sell them at the rate of $2e^{0.05t}$ thousand sets per month, where t is the number of months since they became available. However, with additional advertising using a sports celebrity, they should sell at the rate of $3e^{0.1t}$ thousand sets per month. How many additional sales would result from the celebrity endorsement during the first year?

Solution

We integrate the difference of the rates from month $t = 0$ (the beginning of the first year) to month $t = 12$ (the end of the year):

$$\int_0^{12} (3e^{0.1t} - 2e^{0.05t})\, dt = \left(3\underbrace{\frac{1}{0.1}}_{30} e^{0.1t} - 2 \underbrace{\frac{1}{0.05}}_{40} e^{0.05t} \right) \bigg|_0^{12}$$

$$= \underbrace{(30e^{1.2} - 40e^{0.6})}_{\text{Evaluating at } t=12} - \underbrace{(30e^0 - 40e^0)}_{\text{Evaluating at } t=0}$$

$$\approx \underbrace{26.7 - (-10)}_{\text{Using a calculator}} = 36.7 \qquad \text{In thousands}$$

The celebrity endorsement should result in 36.7 thousand, or 36,700 additional sales during the first year. (The profits from these additional sales must then be compared to the cost of the celebrity endorsement to decide whether it is worthwhile.)

Area Between Curves That Cross

At a point where two curves cross, the upper curve becomes the lower curve and the lower becomes the upper. The area between them must then be calculated by two (or more) integrals, upper minus lower on each interval.

EXAMPLE 5   FINDING THE AREA BETWEEN CURVES THAT CROSS

Find the area between the curves $y = 12 - 3x^2$ and $y = 4x + 5$ from $x = 0$ to $x = 3$.

Solution

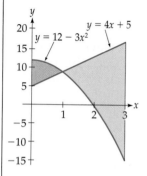

A sketch shows that the curves *do* cross. To find the intersection point, we set the functions equal to each other and solve.

$$4x + 5 = 12 - 3x^2 \qquad \text{Equating the two functions}$$
$$3x^2 + 4x - 7 = 0 \qquad \text{Combining all terms on the left}$$
$$(3x + 7)(x - 1) = 0 \qquad \text{Factoring}$$
$$x = 1 \qquad \begin{array}{l}\text{The other solution, } x = -7/3,\\ \text{is not in the interval } [0, 3]\end{array}$$

The curves cross at $x = 1$, so we must integrate separately over the intervals $[0, 1]$ and $[1, 3]$. The diagram shows which curve is upper and which is lower on each interval.

$$\int_0^1 [\underbrace{(12 - 3x^2)}_{\text{Upper}} - \underbrace{(4x + 5)}_{\text{Lower}}] \, dx + \int_1^3 [\underbrace{(4x + 5)}_{\text{Upper}} - \underbrace{(12 - 3x^2)}_{\text{Lower}}] \, dx \qquad \begin{array}{l}\text{Integrating upper}\\ \text{minus lower on}\\ \text{each interval}\end{array}$$

$$= \int_0^1 (-3x^2 - 4x + 7) \, dx + \int_1^3 (3x^2 + 4x - 7) \, dx \qquad \text{Simplifying the integrands}$$

$$= (-x^3 - 2x^2 + 7x)\Big|_0^1 + (x^3 + 2x^2 - 7x)\Big|_1^3 \qquad \text{Integrating and simplifying}$$

$$= -1 - 2 + 7 + 27 + 18 - 21 - (1 + 2 - 7) = 32 \qquad \text{Evaluating}$$

Therefore, the area between the curves is 32 square units.

Graphing Calculator Exploration

In Example 5 we used *two* integrals, since "upper" and "lower" switched at $x = 1$.

a. To see what happens if you integrate *without* regard to upper and lower, enter $y_1 = 12 - 3x^2$ and $y_2 = 4x + 5$ and graph them on the window [0, 3] by [−20, 20]. Have your calculator find the definite integral of $y_1 - y_2$ on the interval [0, 3]. (Use a command like FnInt or $\int f(x)\,dx$. You should get a negative answer, which cannot be correct for an area.)

b. Explain why the answer was negative. [*Hint:* Look at the graph.]

c. Finally, obtain the correct answer for the area by returning to the calculation in part (a) and integrating the *absolute value* of the difference, $|y_1 - y_2|$ [on some calculators, entered as ABS($y_1 - y_2$)].

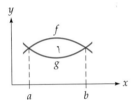

The *x*-values *a* and *b* are where the curves meet.

Areas Bounded by Curves

It is sometimes useful to find the *area bounded by two curves*, without being told the starting and ending *x*-values. In such problems the curves completely enclose an area, and the *x*-values for the upper and lower limits of integration are found by setting the functions equal to each other and solving.

EXAMPLE 6

FINDING AN AREA BOUNDED BY CURVES

Find the area bounded by the curves

$$y = 3x^2 - 12 \quad \text{and} \quad y = 12 - 3x^2$$

Solution The *x*-values for the upper and lower limits of integration are not given, so we find them by setting the functions equal to each other and solving.

$3x^2 - 12 = 12 - 3x^2$	Setting the functions equal to each other
$6x^2 - 24 = 0$	Combining everything on one side
$6(x^2 - 4) = 0$	Factoring
$6(x + 2)(x - 2) = 0$	Factoring further
$x = -2$ and $x = 2$	Solving

The smaller of these, $x = -2$, is the lower limit of integration and the larger, $x = 2$, is the upper limit. To determine which function is "upper" and which is "lower," we choose a "test value" between $x = -2$ and $x = 2$ ($x = 0$ will do), which we substitute into each function to see which is larger. Evaluating each of the original functions at the test point $x = 0$ yields

$$3x^2 - 12 = 3(0)^2 - 12 = -12 \quad \text{(Smaller)} \qquad \begin{array}{l} 3x^2 - 12 \\ \text{at } x = 0 \end{array}$$

$$12 - 3x^2 = 12 - 3(0)^2 = 12 \quad \text{(Larger)} \qquad \begin{array}{l} 12 - 3x^2 \\ \text{at } x = 0 \end{array}$$

Therefore, $y = 12 - 3x^2$ is the "upper" function (since it gives a higher y-value) and $y = 3x^2 - 12$ is the "lower" function. We then integrate upper minus lower between the x-values found earlier.

$$\int_{-2}^{2} \underbrace{[12 - 3x^2}_{\text{Upper}} - \underbrace{(3x^2 - 12)]}_{\text{Lower}} \, dx = \int_{-2}^{2} (24 - 6x^2) \, dx \qquad \text{Simplifying}$$

$$= (24x - 2x^3) \Big|_{-2}^{2} \qquad \text{Integrating}$$

$$= (48 - 16) - (-48 + 16) = 64 \qquad \text{Evaluating}$$

The area bounded by the two curves is 64 square units.

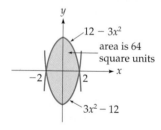

The two curves $y = 12 - 3x^2$ and $y = 3x^2 - 12$ are shown in the graph on the left. Notice that we were able to calculate the area between them without having to graph them.

Practice Problem 3 Find the area bounded by $y = 2x^2 - 1$ and $y = 2 - x^2$.

▶ Solution on page 827

For curves that intersect at *more* than two points, several integrals may be needed, integrating "upper minus lower" on each interval, as in Example 5. Test points in each interval will determine the upper and lower curves on that interval.

11.4 Section Summary

The average value of a continuous function over an interval is defined as the definite integral of the function divided by the length of the interval:

$$\begin{pmatrix} \text{Average value} \\ \text{of } f \text{ on } [a, b] \end{pmatrix} = \frac{1}{b-a} \int_a^b f(x)\, dx$$

To find the area between two curves:

1. If the x-values are not given, set the functions equal to each other and solve for the points of intersection.
2. Use a test point within each interval to determine which curve is "upper" and which is "lower."
3. Integrate "upper minus lower" on each interval.

If a curve lies *below* the x-axis, as in the following diagram, then the "upper" curve is the x-axis ($y = 0$) and the "lower" curve is $y = f(x)$, and so integrating "upper minus lower" results in integrating the *negative* of the function.

$$\int_a^b \underbrace{[0}_{\text{Upper}} - \underbrace{f(x)]}_{\text{Lower}}\, dx = \int_a^b [-f(x)]\, dx = -\int_a^b f(x)\, dx$$

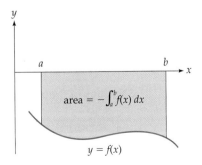

This case need not be remembered separately if you simply remember always to integrate "upper minus lower" over each interval.

Definite integration provides a very powerful method for finding areas, taking us far beyond the few formulas (for rectangles, triangles, and circles) that we knew before studying calculus.

Average Value Formula Derived from Riemann Sums

The formula for the average value of a function (page 818) can be derived using Riemann sums. For a continuous function f we could define a "sample average" by "sampling" the function at n points and averaging the results. From the interval $[a, b]$, we choose n numbers $x_1, x_2, \ldots, x_n$ from successive subintervals of length $\Delta x = \dfrac{b-a}{n}$, sum the resulting values of the function, and divide by n. This gives a "sample average" of the following form:

$$\frac{1}{n}[f(x_1) + f(x_2) + \cdots + f(x_n)] \qquad \text{Sum of } n \text{ values divided by } n$$

$$= \frac{1}{b-a} \cdot \underbrace{\frac{b-a}{n}}_{\Delta x}[f(x_1) + f(x_2) + \cdots + f(x_n)] \qquad \text{Dividing and multiplying by } b-a$$

$$= \frac{1}{b-a} \underbrace{[f(x_1) + f(x_2) + \cdots + f(x_n)] \cdot \Delta x}_{\text{This is a Riemann sum for } \int_a^b f(x)\, dx} \qquad \text{Moving } \Delta x = \dfrac{b-a}{n} \text{ to the right}$$

To get a more "representative" average, we increase the number n of sample points. Letting n approach infinity makes the above Riemann sum approach the definite integral $\int_a^b f(x)\, dx$, leading to the definition of the average value of a function:

$$\begin{pmatrix} \text{Average value} \\ \text{of } f \text{ on } [a, b] \end{pmatrix} = \frac{1}{b-a} \int_a^b f(x)\, dx$$

▶ **Solutions to Practice Problems**

1. $\dfrac{1}{2}\displaystyle\int_0^2 3x^2\, dx = \left.\dfrac{1}{2}\cdot x^3\right|_0^2 = \dfrac{1}{2}\cdot 2^3 - \dfrac{1}{2}\cdot 0^3 = \dfrac{1}{2}\cdot 8 = 4$

2. $\int_{-1}^{1} [(2x^2 + 1) - (-x^2 - 1)]\, dx$

$= \int_{-1}^{1} (2x^2 + 1 + x^2 + 1)\, dx$

$= \int_{-1}^{1} (3x^2 + 2)\, dx = (x^3 + 2x)\Big|_{-1}^{1}$

$= (1 + 2) - (-1 - 2) = 6$ square units

3. $2x^2 - 1 = 2 - x^2$

$3x^2 - 3 = 0$

$3(x^2 - 1) = 0$

$3(x + 1)(x - 1) = 0$

$x = 1$ and $x = -1$.

Test value $x = 0$ shows that $2 - x^2$ is "upper" and $2x^2 - 1$ is "lower."

$\int_{-1}^{1} [(2 - x^2) - (2x^2 - 1)]\, dx$

$= \int_{-1}^{1} (3 - 3x^2)\, dx = (3x - x^3)\Big|_{-1}^{1}$

$= (3 - 1) - (-3 + 1) = 4$ square units

11.4 Exercises

Average Value

Find the average value of each function over the given interval.

1. $f(x) = x^2$ on $[0, 3]$
2. $f(x) = x^3$ on $[0, 2]$
3. $f(x) = 3\sqrt{x}$ on $[0, 4]$
4. $f(x) = \sqrt[3]{x}$ on $[0, 8]$
5. $f(x) = \dfrac{1}{x^2}$ on $[1, 5]$
6. $f(x) = \dfrac{1}{x^2}$ on $[1, 3]$
7. $f(x) = 2x + 1$ on $[0, 4]$
8. $f(x) = 4x - 1$ on $[0, 10]$
9. $f(x) = 36 - x^2$ on $[-2, 2]$
10. $f(x) = 9 - x^2$ on $[-3, 3]$
11. $f(x) = 3$ on $[10, 50]$
12. $f(x) = 2$ on $[5, 100]$
13. $f(x) = e^{x/2}$ on $[0, 2]$
14. $f(x) = e^{-2x}$ on $[0, 1]$
15. $f(x) = \dfrac{1}{x}$ on $[1, 2]$
16. $f(x) = \dfrac{1}{x}$ on $[1, 10]$
17. $f(x) = x^n$ on $[0, 1]$, where n is a constant $(n > 0)$
18. $f(x) = e^{kx}$ on $[0, 1]$, where k is a constant $(k \neq 0)$
19. $f(x) = ax + b$ on $[0, 2]$, where a and b are constants

20. $f(x) = \dfrac{1}{x}$ on $[1, c]$, where c is a constant $(c > 1)$

21. $f(x) = e^{-x^4}$ on $[-1, 1]$

22. $f(x) = \sqrt{1 + x^4}$ on $[-2, 2]$

APPLIED EXERCISES ON AVERAGE VALUE

23–24. BUSINESS: Sales A store's sales on day x are given by the function $S(x)$ below. Find the average sales during the first 3 days (day 0 to day 3).

23. $S(x) = 200x + 6x^2$ 24. $S(x) = 400x + 3x^2$

25. **GENERAL: Temperature** The temperature at time t hours is $T(t) = -0.3t^2 + 4t + 60$ (for $0 \le t \le 12$). Find the average temperature between time 0 and time 10.

26. **BEHAVIORAL SCIENCE: Practice** After x practice sessions, a person can accomplish a task in $f(x) = 12x^{-1/2}$ minutes. Find the average time required from the end of session 1 to the end of session 9.

27. **ENVIRONMENTAL SCIENCE: Pollution** The amount of pollution in a lake x years after the closing of a chemical plant is $P(x) = 100/x$ tons (for $x \le 1$). Find the average amount of pollution between 1 and 10 years after the closing.

28. **GENERAL: Population** The population of the United States is predicted to be $P(t) = 281e^{0.012t}$ million, where t is the number of years after the year 2000. Find the average population between the years 2000 and 2050.

29. **BUSINESS: Compound Interest** A deposit of $1000 at 5% interest compounded continuously will grow to $V(t) = 1000e^{0.05t}$ dollars after t years. Find the average value during the first 40 years (that is, from time 0 to time 40).

30. **BIOMEDICAL: Bacteria** A colony of bacteria is of size $S(t) = 300e^{0.1t}$ after t hours. Find the average size during the first 12 hours (that is, from time 0 to time 12).

31. **BUSINESS: Profit** The MediGenics Corporation expects its profits to be $1.4e^{0.05x^2}$ million dollars per year x years from now. Find the company's average profit over the next decade (year 0 to year 10).

32. **GENERAL: Population** The population of a town is predicted to be $5.2e^{\sqrt{x}}$ thousand people x years from now. Find the average population over the next decade (year 0 to year 10).

AREA

33. Find the area between the curve $y = x^2 + 1$ and the line $y = 2x - 1$ (shown below) from $x = 0$ to $x = 3$.

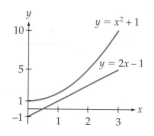

34. Find the area between the curve $y = x^2 + 3$ and the line $y = 2x$ (shown below) from $x = 0$ to $x = 3$.

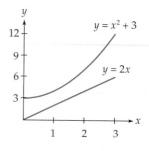

11.4 FURTHER APPLICATIONS OF DEFINITE INTEGRALS

35. Find the area between the curves $y = e^x$ and $y = e^{2x}$ (shown below) from $x = 0$ to $x = 2$. (Leave the answer in its exact form.)

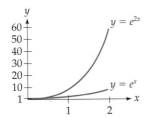

36. Find the area between the curves $y = e^x$ and $y = e^{-x}$ (shown in the following diagram) from $x = 0$ to $x = 1$. (Leave the answer in its exact form.)

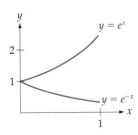

37. a. Sketch the parabola $y = x^2 + 4$ and the line $y = 2x + 1$ on the same graph.
  b. Find the area between them from $x = 0$ to $x = 3$.

38. a. Sketch the parabola $y = x^2 + 5$ and the line $y = 2x + 3$ on the same graph.
  b. Find the area between them from $x = 0$ to $x = 3$.

39. a. Sketch the parabola $y = 3x^2 - 3$ and the line $y = 2x + 5$ on the same graph.
  b. Find the area between them from $x = 0$ to $x = 3$.

40. a. Sketch the parabola $y = 3x^2 - 12$ and the line $y = 2x - 11$ on the same graph.
  b. Find the area between them from $x = 0$ to $x = 3$.

Find the area bounded by the given curves.

41. $y = x^2 - 1$ and $y = 2 - 2x^2$

42. $y = x^2 - 4$ and $y = 8 - 2x^2$

43. $y = 6x^2 - 10x - 8$ and $y = 3x^2 + 8x - 23$

44. $y = 3x^2 - x - 1$ and $y = 5x + 8$

45. $y = x^2$ and $y = x^3$

46. $y = x^3$ and $y = x^4$

47. $y = 7x^3 - 36x$ and $y = 3x^3 + 64x$

48. $y = x^n$ and $y = x^{n-1}$ (for $n > 1$)

49. $y = e^x$ and $y = x + 3$

[*Hint for Exercises 49–50:* Use INTERSECT to find the intersection points for the upper and lower limits of integration.]

50. $y = \ln x$ and $y = x - 2$

APPLIED EXERCISES ON AREA

51. GENERAL: Population The birthrate in Africa has increased from $17e^{0.02t}$ million births per year to $22e^{0.02t}$ million births per year, where t is the number of years since 2000. Find the total increase in population that will result from this higher birth rate between 2000 ($t = 0$) and 2020 ($t = 20$).

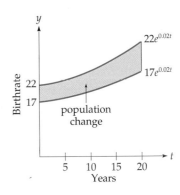

52. BUSINESS: Profit from Expansion A company expects profits of $60e^{0.02t}$ thousand dollars per month, but predicts that if it builds a new and larger factory, its profits will be $80e^{0.04t}$ thousand dollars per month, where t is the number of months from now. Find the extra profits resulting from the new factory during the first two years ($t = 0$ to $t = 24$). If the new factory will cost \$1,000,000, will this cost be paid off during the first two years?

53. BUSINESS: Net Savings A factory installs new machinery that saves $S(x) = 1200 - 20x$ dollars per year, where x is the number of years since installation. However, the cost of maintaining the new machinery is $C(x) = 100x$ dollars per year.

  a. Find the year x at which the maintenance cost $C(x)$ will equal the savings $S(x)$. (At this time, the new machinery should be replaced.)
  b. Find the accumulated net savings [savings $S(x)$ minus cost $C(x)$] during the period from $t = 0$ to the replacement time found in part (a).

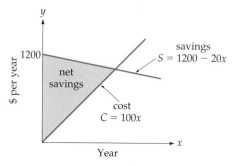

54. GENERAL: Design A graphic design consists of a white square containing the blue shape shown below. Find the area of the blue interior.

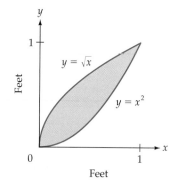

55. ECONOMICS: Balance of Trade A country's annual imports are $I(t) = 30e^{0.2t}$ and its exports are $E(t) = 25e^{0.1t}$, both in billions of dollars, where t is measured in years and $t = 0$ corresponds to the beginning of 2000. Find the country's accumulated trade deficit (imports minus exports) for the 10 years beginning with 2000.

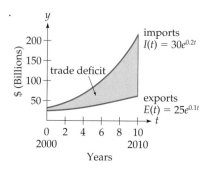

56. ECONOMICS: Cost of Labor Contracts An employer offers to pay workers at the rate of $30{,}000e^{0.04t}$ dollars per year, while the union demands payment at the rate of $30{,}000e^{0.08t}$ dollars per year, where $t = 0$ corresponds to the beginning of the contract. Find the accumulated difference in pay between these two rates over the 10-year life of the contract.

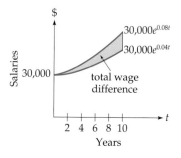

57–58. BUSINESS: Cumulative Profit A company finds that its marginal revenue function is $MR(x) = 700x^{-1}$ and its marginal cost function is $MC(x) = 500x^{-1}$ (both in thousands of dollars), where x is the number of units $(x > 1)$. Find the total profit from

57. $x = 100$ to $x = 200$

58. $x = 200$ to $x = 300$

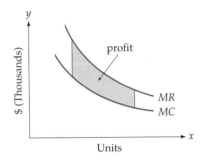

59. GENERAL: Lives Saved by Seat Belts Seat belt use in the United States has now risen to 66%, but nonusers still risk needless expense, injury, and death. The table below gives the number of automobile fatalities per year, and the predicted number of fatalities if everyone wore seat belts. To avoid large numbers, years are listed as years since 1996.

Years	Years Since 1996	Automobile Fatalities	Predicted Fatalities with 100% Seat Belt Use
1996	0	43,649	38,600
1997	1	43,458	38,400
1998	2	41,800	36,900
1999	3	41,300	36,500

Source: National Safety Council

a. Enter the first two columns of numbers into your graphing calculator and make a plot of the resulting points (Years Since 1996 on the x-axis and Fatalities on the y-axis).
b. Have your calculator find the linear regression formula for these data. Then enter the result as y_1, which gives a formula for fatalities each year. Plot the points together with the regression line.
c. Enter the last column of numbers into your calculator and make a plot of the resulting points (Years Since 1996 on the x-axis and Predicted Fatalities on the y-axis), keeping the earlier points too.
d. Have your calculator find the linear regression formula for the new points found in part (c). Then enter the result as y_2, which gives a formula for predicted fatalities each year. Plot both sets of points together with both regression lines.
e. Extend your viewing window to [0, 10] by [0, 50,000] to see what the lines predict for fatalities from 1996 to 2006. The area between these lines represents the lives that could be saved by seat belts.
f. Have your calculator find the area between these lines from 0 to 10, predicting the lives that would be saved by 100% seat-belt use between 1996 and 2006.

REVIEW EXERCISES

These exercises review material that will be helpful in Section 11.6.

Find the derivative of each function.

60. $e^{x^3 + 6x}$

61. $e^{x^2 + 5x}$

62. $\ln(x^3 + 6x)$

63. $\ln(x^2 + 5x)$

11.5 TWO APPLICATIONS TO ECONOMICS: CONSUMERS' SURPLUS AND INCOME DISTRIBUTION

Introduction

In this section we discuss several important economic concepts—consumers' surplus, producers' surplus, and the Gini index of income distribution—each of which is defined as the area between two curves.

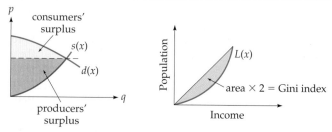

Consumers' Surplus

Imagine that you really liked pizza and were willing to pay $12 for a pizza pie. If, in fact, a pizza costs only $8, then you have, in some sense, "saved" $4, the $12 that you were willing to pay minus the $8 market price. If one were to add up this difference for all pizzas sold in a given period of time (the price that each consumer was willing to pay minus the price actually paid), the total savings would be called the *consumers' surplus* for that product. The consumers' surplus measures the benefit that consumers derive from an economy in which competition keeps prices low.

Demand Functions

Price and quantity are inversely related: if the price of an item rises, the quantity sold generally falls, and vice versa. Through market research, economists can determine the relationship between price and quantity for an item. This relationship can often be expressed as a *demand function* (or demand curve) $d(x)$, so called because it gives the price at which exactly x units will be demanded.

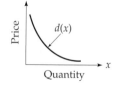

Demand Function

The demand function $d(x)$ for a product gives the price at which exactly x units will be sold.

$$d(x) = \begin{pmatrix} \text{Price when} \\ \text{demand is } x \end{pmatrix}$$

On page 668 we called $d(x)$ the *price function*.

Mathematical Definition of Consumers' Surplus

The demand curve gives the price that consumers are *willing* to pay, and the *market price* is what they *do* pay, so the amount by which the demand curve is above the market price measures the benefit or "surplus" to consumers. We add up all of these benefits by integrating, so the area between the demand curve and the market price line gives the *total benefit* that consumers derive from being able to buy at the market price. This total benefit (the shaded area in the diagram on the left) is called the *consumers' surplus*.

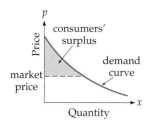

Consumers' Surplus

For a demand function $d(x)$ and demand level A, the market price B is the demand function evaluated at $x = A$, so that $B = d(A)$. The consumers' surplus is the area between the demand curve and the market price.

$$\begin{pmatrix} \text{Consumers'} \\ \text{surplus} \end{pmatrix} = \int_0^A [d(x) - B]\, dx$$

$\uparrow$ Demand function $\uparrow$ Market price

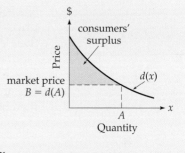

EXAMPLE 1

FINDING CONSUMERS' SURPLUS FOR ELECTRICITY

If the demand function for electricity is $d(x) = 1100 - 10x$ dollars (where x is in millions of kilowatt-hours, $0 \le x \le 100$), find the consumers' surplus at the demand level $x = 80$.

Solution The market price is the demand function $d(x)$ evaluated at $x = 80$.

$$\begin{pmatrix} \text{Market} \\ \text{price } B \end{pmatrix} = d(80) = 1100 - 10 \cdot 80 = 300 \qquad \begin{matrix} d(x) = 1100 - 10x \\ \text{at } x = 80 \end{matrix}$$

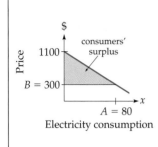

The consumers' surplus is the area between the demand curve and the market price line.

$$\begin{pmatrix} \text{Consumers'} \\ \text{surplus} \end{pmatrix} = \int_0^{80} \underbrace{(1100 - 10x}_{\text{Demand price}} - \underbrace{300)}_{\text{Market price}} dx$$

$$= \int_0^{80} (800 - 10x) \, dx = (800x - 5x^2) \Big|_0^{80}$$

$$= (64{,}000 - 32{,}000) - (0) = 32{,}000$$

Therefore, the consumers' surplus for electricity is $32,000.

How Consumers' Surplus Is Used

In Example 1, at demand level $x = 80$ the consumers' surplus was $32,000. If electricity usage were to increase to $x = 90$, the market price would then drop to $d(90) = 1100 - 10 \cdot 90 = 200$. We could then calculate the consumers' surplus at this higher demand level (and would find that the answer is $40,500). Therefore, a price decrease from $300 to $200 would mean that consumers would benefit by an additional $40,500 − $32,000 = $8500. This benefit would then be compared to the cost of a new generator to decide whether the expenditure would be worthwhile.

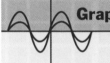

Graphing Calculator Exploration

a. Verify the solution to Example 1 on your graphing calculator by entering $y_1 = 1100 - 10x$ and then using FnInt or $\int f(x) \, dx$ to integrate $y_1(x) - y_1(80)$ from 0 to 80.

b. Find the consumers' surplus at demand level 90 by integrating $y_1(x) - y_1(90)$ from 0 to 90. [*Hint:* Simply return to the calculation of part (a) and replace 80 by 90. Your answer should agree with that in the preceding paragraph.]

Producers' Surplus

Just as the consumers' surplus measures the total benefit to consumers, the *producers' surplus* measures the total benefit that producers derive from being able to sell at the market price. Returning to our pizza example, if a pizza producer might just be willing to remain in business if the price of pizzas dropped to $5, the fact that pizzas can be sold for $8 gives the producer a "benefit" of $3. The sum of all such benefits is the *producers' surplus* for a product.

Supply Functions

Clearly, as the price of an item rises, so does the quantity that producers are willing to supply at that price. The relationship between the price of an item and the quantity that producers are willing to supply at that price can be expressed as a *supply function* (or supply curve) $s(x)$.

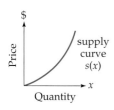

Supply Function

The supply function $s(x)$ for a product gives the price at which exactly x units will be supplied.

$$s(x) = \begin{pmatrix} \text{Price when} \\ \text{supply is } x \end{pmatrix}$$

Mathematical Definition of Producers' Surplus

As before, we integrate to find the total benefit, but now "upper" is the market price and "lower" is the supply curve $s(x)$.

Producers' Surplus

For a supply function $s(x)$ and demand level A, the market price is $B = s(A)$. The producers' surplus is the area between the market price and the supply curve.

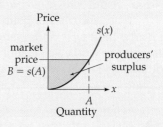

$$\begin{pmatrix} \text{Producers'} \\ \text{surplus} \end{pmatrix} = \int_0^A [B - s(x)]\, dx$$

↗ ↑
Market Supply
price function

EXAMPLE 2

FINDING PRODUCERS' SURPLUS

For the supply function $s(x) = 0.09x^2$ dollars and the demand level $x = 200$, find the producers' surplus.

Solution The market price is the supply function $s(x)$ evaluated at $x = 200$.

$$\begin{pmatrix} \text{Market} \\ \text{price } B \end{pmatrix} = s(200) = 0.09(200)^2 = 3600 \qquad \begin{array}{l} s(x) = 0.09x^2 \\ \text{at } x = 200 \end{array}$$

The producers' surplus is the area between the market price line and the supply curve.

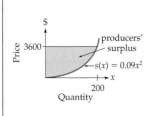

$$\begin{pmatrix} \text{Producers'} \\ \text{surplus} \end{pmatrix} = \int_0^{200} (\underbrace{3600}_{\substack{\text{Market} \\ \text{price}}} - \underbrace{0.09x^2}_{\substack{\text{Supply} \\ \text{function}}}) \, dx = (3600x - 0.03x^3) \Big|_0^{200}$$

$$= (720{,}000 - 240{,}000) - (0) = 480{,}000$$

Therefore, the producers' surplus is $480,000.

Consumers' Surplus and Producers' Surplus

The demand x at which the supply and demand curves intersect is called the *market demand*. The consumers' surplus and the producers' surplus can be shown together on the same graph. These two areas together give a numerical measure of the total benefit that consumers and producers derive from competition, showing that both consumers and producers benefit from an open market.

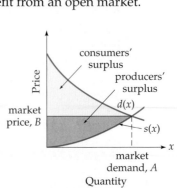

Gini Index of Income Distribution

In any society, some people make more money than others. To measure the "gap" between the rich and the poor, economists calculate the proportion of the total income that is earned by the lowest 20% of the population, and then the proportion that is earned by the lowest 40% of the population, and so on. This information (for the year 2000) is given in the table below (with percentages written as decimals), and is graphed on the right.

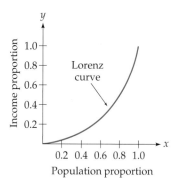

Proportion of Population	Proportion of Income
0.20	0.04
0.40	0.13
0.60	0.27
0.80	0.51
1.00	1.00

Source: U.S. Bureau of the Census

For example, the lowest 20% of the population earns only 4% of the total income, the lowest 40% earns only 13% of the total income, and so on. The curve is known as the *Lorenz curve* (after the American statistician Max Otto Lorenz).

Lorenz Curve

The Lorenz curve $L(x)$ gives the proportion of total income earned by the lowest proportion x of the population.

Gini Index

The Lorenz curve may be compared with two extreme cases of income distribution.

1. *Absolute equality of income* means that everyone earns exactly the same income, and so the lowest 10% of the population earns exactly 10% of the total income, the lowest 20% earns exactly 20% of the income, and so on. This gives the Lorenz curve $y = x$ shown on the following page.

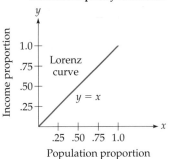

Absolute equality of income

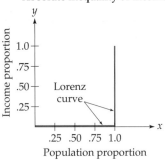

Absolute inequality of income

2. *Absolute inequality of income* means that nobody earns any income except one person, who earns all the income. This gives the Lorenz curve shown above on the right.

To measure how the actual distribution differs from absolute equality, we calculate the area between the actual distribution and the line of absolute equality $y = x$. Since this area can be at most $\frac{1}{2}$ (the area of the entire lower triangle), economists multiply the area by 2 to get a number between 0 (absolute equality) and 1 (absolute inequality). This measure is called the *Gini index*. Note that a higher Gini index means greater *in*equality (greater deviation from the line of absolute equality).

Gini Index

For a Lorenz curve $L(x)$, the Gini index is

$$\begin{pmatrix} \text{Gini} \\ \text{index} \end{pmatrix} = 2 \int_0^1 [x - L(x)]\, dx$$

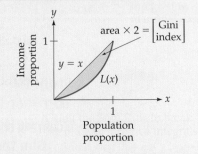

The Gini index varies from 0 (absolute equality) to 1 (absolute inequality).

EXAMPLE 3

FINDING THE GINI INDEX

The Lorenz curve for income distribution in the United States in 2000 was approximately $L(x) = x^{2.4}$. Find the Gini index.

Solution First we calculate the area between the curve of absolute equality $y = x$ and the Lorenz curve $y = x^{2.4}$.

$$\int_0^1 (x - x^{2.4})\, dx = \left(\frac{1}{2}x^2 - \frac{1}{3.4}x^{3.4}\right)\Big|_0^1$$

$$= \frac{1}{2} \cdot 1^2 - \frac{1}{3.4} \cdot 1^{3.4} - 0$$

$$\approx 0.5 - 0.294 = 0.206$$

Multiplying by 2 gives the Gini index.

$$\left(\begin{matrix}\text{Gini}\\\text{index}\end{matrix}\right) = 0.41 \qquad \text{Rounding to 2 decimal places.}$$

Practice Problem 1 In 1993, the Gini index for income was 0.39. Since 1993, has income distribution become more equal or less equal? ▸ **Solution on next page**

Graphing Calculator Exploration

In Example 3, how was the Lorenz function of the form x^n found? It was found by a method called "least squares" (discussed in Section 3.6 and more fully in Section 13.4), which minimizes the squared differences between the income proportions (from the table on page 837) and the curve x^n at the x-values 0.2, 0.4, 0.6, and 0.8. This amounts to minimizing the function

$$(0.2^x - 0.04)^2 + (0.4^x - 0.13)^2 + (0.6^x - 0.27)^2 + (0.8^x - 0.51)^2.$$

(Notice that here we are using x for the *exponent*.)

a. Graph this function on your calculator on the window $[0, 5]$ by $[-0.2, 1]$.

b. Find the x that minimizes the function. The resulting value is the exponent. Your answer should agree with the exponent in Example 3.

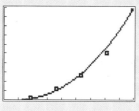

The graph on the left, on the window $[0, 1]$ by $[0, 1]$, shows that the function $x^{2.4}$ fits the points from the table on page 837 reasonably well. (Other types of functions besides x^n could also be used to fit these data.)

Chapter 11 Integration and its Applications

The Gini index for *total* wealth can be calculated similarly.

Practice Problem 2 The Lorenz curve for total wealth in the United States during 2000 was $L(x) = x^{6.9}$. Calculate the Gini index. ➤ Solution below

Practice Problem 2 shows that the Gini index for wealth is greater than the Gini index for income. That is, total wealth in the United States is distributed more unequally than total income. One reason for this is that we have an income tax but no "wealth" tax.

➤ Solutions to Practice Problems

1. Less equal

2. $\int_0^1 (x - x^{6.9})\, dx = \left(\frac{1}{2}x^2 - \frac{1}{7.9}x^{7.9}\right)\Big|_0^1 = \frac{1}{2} - \frac{1}{7.9} - 0 \approx 0.5 - 0.127 = 0.373$

Gini index for wealth is 0.75 (multiplying by 2 and rounding)

11.5 Exercises

For each demand function $d(x)$ and demand level x, find the consumers' surplus.

1. $d(x) = 4000 - 12x$, $x = 100$
2. $d(x) = 500 - x$, $x = 400$
3. $d(x) = 300 - \frac{1}{2}x$, $x = 200$
4. $d(x) = 200 - \frac{1}{2}x$, $x = 300$
5. $d(x) = 350 - 0.09x^2$, $x = 50$
6. $d(x) = 840 - 0.06x^2$, $x = 100$
7. $d(x) = 200e^{-0.01x}$, $x = 100$
8. $d(x) = 400e^{-0.02x}$, $x = 75$

For each supply function $s(x)$ and demand level x, find the producers' surplus.

9. $s(x) = 0.02x$, $x = 100$
10. $s(x) = 0.4x$, $x = 200$
11. $s(x) = 0.03x^2$, $x = 200$
12. $s(x) = 0.06x^2$, $x = 50$

For each demand function $d(x)$ and supply function $s(x)$:

a. Find the market demand (the positive value of x at which the demand function intersects the supply function).
b. Find the consumers' surplus at the market demand found in part (a).
c. Find the producers' surplus at the market demand found in part (a).

13. $d(x) = 300 - 0.4x$, $s(x) = 0.2x$
14. $d(x) = 120 - 0.16x$, $s(x) = 0.08x$
15. $d(x) = 300 - 0.03x^2$, $s(x) = 0.09x^2$
16. $d(x) = 360 - 0.03x^2$, $s(x) = 0.006x^2$
17. $d(x) = 300e^{-0.01x}$, $s(x) = 100 - 100e^{-0.02x}$
18. $d(x) = 400e^{-0.01x}$, $s(x) = 0.01x^{2.1}$

Find the Gini index for the given Lorenz curve.

19. $L(x) = x^{3.2}$ (the Lorenz curve for U.S. income in 1929)

20. $L(x) = x^3$ (the Lorenz curve for U.S. income in 1935)

21. $L(x) = x^{2.1}$ (the Lorenz curve for income in Sweden in 1990)

22. $L(x) = x^{15.3}$ (the Lorenz curve for wealth in Great Britain in 1990)

23. $L(x) = 0.4x + 0.6x^2$

24. $L(x) = 0.2x + 0.8x^3$

25. $L(x) = x^n$ (for $n > 1$)

26. $L(x) = \frac{1}{2}x + \frac{1}{2}x^n$ (for $n > 1$)

27. $L(x) = \dfrac{e^{x^2} - 1}{e - 1}$

28. $L(x) = 1 - \sqrt{1 - x}$

29. $L(x) = \dfrac{x + x^2 + x^3}{3}$

30. $L(x) = 0.62x^{7.15} + 0.38x^{9.47}$

31–32. The following tables give the distribution of family income in the United States: Exercise 31 is for the year 1977 and Exercise 32 is for 1989. Use the procedure described in the Graphing Calculator Exploration on page 411 to find the Lorenz function of the form x^n for the data. Then find the Gini index. If you do both problems, did family income become more concentrated or less concentrated from 1977 to 1989?

31.

Proportion (Lowest) of Families	Proportion of Income (1977)
0.20	0.06
0.40	0.18
0.60	0.34
0.80	0.57

Source: Congressional Budget Office

32.

Proportion (Lowest) of Families	Proportion of Income (1989)
0.20	0.04
0.40	0.14
0.60	0.29
0.80	0.51

Source: Congressional Budget Office

REVIEW EXERCISES

These exercises review material that will be helpful in the next section.

Find the derivative of each function.

33. $(x^5 - 3x^3 + x - 1)^4$ **34.** $(x^4 - 2x^2 - x + 1)^5$ **35.** $\ln(x^4 + 1)$ **36.** $\ln(x^3 - 1)$ **37.** e^{x^3} **38.** e^{x^4}

11.6 INTEGRATION BY SUBSTITUTION

Introduction

The Chain Rule (page 610) greatly expanded the range of functions that we could differentiate. In this section we will learn a similar technique for integration, called the *substitution method*, which will greatly expand the range of functions that we can integrate. First, however, we must define *differentials*.

Differentials

One of the notations for the derivative of a function $f(x)$ is df/dx. Although written as a fraction, df/dx was not defined as the quotient of two quantities df and dx, but as a single object, the *derivative*. We will now define df and dx separately (they are called *differentials*) so that their quotient $df \div dx$ is equal to the derivative df/dx. We begin with

$$\frac{df}{dx} = f' \qquad \text{Since } df/dx \text{ and } f' \text{ are both notations for the derivative}$$

$$df = f' \, dx \qquad \text{Multiplying each side by } dx$$

This leads to a definition for the differential df.

Differential

For a differentiable function $f(x)$, the differential df is

$$df = f'(x) \, dx \qquad df \text{ is the derivative times } dx$$

Note that df does *not* mean d times f. The dx is just the notation that appears at the end of integrals, arising from the Δx in the Riemann sum. The reason for finding differentials will be made clear shortly.

EXAMPLE 1   FINDING DIFFERENTIALS

Function $f(x)$	Differential df
$f(x) = x^2$	$df = 2x \, dx$
$f(x) = \ln x$	$df = \dfrac{1}{x} \, dx$
$f(x) = e^{x^2}$	$df = e^{x^2}(2x) \, dx$
$f(x) = x^4 - 5x + 2$	$df = \underbrace{(4x^3 - 5)}_{f'(x)} \, dx$

Each differential is the derivative of the function times dx

Practice Problem 1 For $f(x) = x^3 - 4x - 2$, find the differential df.

▶ Solution on page 852

The differential formula $df = f' \, dx$ is easy to remember because dividing both sides by dx gives

$$\frac{df}{dx} = f'$$

which simply says "the derivative equals the derivative." We may use other letters besides f and x.

EXAMPLE 2   CALCULATING DIFFERENTIALS IN OTHER VARIABLES

Function	Differential	
$u = x^3 + 1$	$du = 3x^2 \, dx$	Differentials end with d followed by the variable
$u = e^{2t} + 1$	$du = 2e^{2t} \, dt$	

Practice Problem 2

For $u = e^{-5t}$, find the differential du.   ▶ Solution on page 852

Substitution Method

Using differential notation, we can state three very useful integration formulas.

(A) $\displaystyle\int u^n \, du = \frac{1}{n+1} u^{n+1} + C \qquad n \neq -1$

(B) $\displaystyle\int e^u \, du = e^u + C$

(C) $\displaystyle\int \frac{du}{u} = \int \frac{1}{u} \, du = \int u^{-1} \, du = \ln|u| + C$

These formulas are easy to remember because they are exactly the formulas that we learned earlier (see pages 786 and 793) except that here we use the letter u to stand for a *function*. The du is the differential of the function. Each of these formulas may be justified by differentiating the right-hand side (see Exercises 61–63). A few examples will illustrate their use.

EXAMPLE 3

INTEGRATING BY SUBSTITUTION

Find $\int (x^2 + 1)^3 \, 2x \, dx$.

Solution

The integral involves a function to a power:

$$\int (x^2 + 1)^3 \, 2x \, dx \qquad (x^2 + 1)^3 \text{ is a function to a power}$$

as does formula (A): $\int u^n \, du \qquad u^n \text{ is a function to a power in } \int u^n \, du = \frac{1}{n+1} u^{n+1} + C$

To make the integral "fit" the formula we take $u = x^2 + 1$ and $n = 3$.

$$\int \underbrace{(x^2 + 1)^3}_{u^3} 2x \, dx \qquad u = x^2 + 1 \text{ and } n = 3$$

For $u = x^2 + 1$ the differential is $du = 2x \, dx$, which is exactly the remaining part of the integral. We then write the integral with each x-expression replaced by its equivalent u-expression.

$$\int \underbrace{(x^2 + 1)^3}_{u^3} \underbrace{2x \, dx}_{du} = \int \underbrace{u^3 \, du}_{\text{Written in terms of } u} \qquad \text{Using } u = x^2 + 1 \text{ and } du = 2x \, dx$$

The last integral we solve by formula (A):

$$\int u^3 \, du = \frac{1}{4} u^4 + C \qquad \int u^n \, du = \frac{1}{n+1} u^{n+1} + C$$

[formula (A)] with $n = 3$

Finally, we substitute back to the original variable x, using our relationship $u = x^2 + 1$, to get the answer:

$$\frac{1}{4}(x^2 + 1)^4 + C \qquad \frac{1}{4} u^4 + C \text{ with } u = x^2 + 1$$

The procedure is not as complicated as it might seem. All of these steps may be written together as follows.

$$\underbrace{\int (x^2 + 1)^3}_{u^3} \underbrace{2x\,dx}_{du} = \int u^3\,du = \frac{1}{4}u^4 + C = \frac{1}{4}(x^2 + 1)^4 + C$$

Choosing $u = x^2 + 1$ therefore $du = 2x\,dx$ | Substituting $u^3 = (x^2 + 1)^3$ $du = 2x\,dx$ | Integrating by formula (A) with $n = 3$ | Substituting back to x using $u = x^2 + 1$

We may check this answer by differentiation (using the Generalized Power Rule).

$$\frac{d}{dx}\underbrace{\left[\frac{1}{4}(x^2 + 1)^4 + C\right]}_{\text{Answer}} = \frac{1}{4} \cdot 4(x^2 + 1)^3 \underbrace{2x}_{\text{Derivative of the inside}} = \underbrace{(x^2 + 1)^3 \, 2x}_{\text{Integrand}}$$

Since the result of the differentiation agrees with the original integrand, the integration is correct.

Multiplying Inside and Outside by Constants

If the integral does not exactly match the form $\int u^n\,du$, we may sometimes still solve the integral by multiplying by constants.

EXAMPLE 4

INSERTING CONSTANTS BEFORE SUBSTITUTING

Find $\int (x^2 + 1)^3 x\,dx$.   Same as Example 3 but without the 2

Solution As before, we use formula (A) with $u = x^2 + 1$, which gives $du = 2x\,dx$. But the integral has only an $x\,dx$, not $2x\,dx$, which would allow us to substitute du.

$$\int \underbrace{(x^2 + 1)^3}_{u^3} \underbrace{x\,dx}_{\text{not }du = 2x\,dx \text{ because there is no 2}}$$

$u = x^2 + 1$, so $du = 2x\,dx$

Therefore, the integral is *not* in the form $\int u^3\,du$. (The integral must fit the formula exactly: *everything* in the integral must be accounted for either by the u^n or by the du.) However, we may multiply inside the integral by 2 as long as we compensate by also multiplying by $\frac{1}{2}$, and the $\frac{1}{2}$ may be written *outside* the integral (since constants may be moved across the integral sign), leading to the solution:

CHAPTER 11 INTEGRATION AND ITS APPLICATIONS

$$\frac{1}{2} \int \underbrace{(x^2 + 1)^3}_{u^3} \underbrace{2x \, dx}_{du} = \frac{1}{2} \int u^3 \, du = \frac{1}{2} \cdot \frac{1}{4} u^4 + C = \frac{1}{8}(x^2 + 1)^4 + C$$

Multiplying by $\frac{1}{2}$ and 2

$u = x^2 + 1$
$du = 2x \, dx$

Substituting

Integrating by formula (A)

Substituting back to x using $u = x^2 + 1$

This method of multiplying inside and outside by a constant is very useful, and may be used with the other substitution formulas as well.

EXAMPLE 5 USING OTHER SUBSTITUTION FORMULAS

Find $\int e^{x^3} x^2 \, dx$.

Solution

The integral involves e to a function:

$$\int e^{x^3} x^2 \, dx \qquad e^{x^3} \text{ is } e \text{ to a function}$$

as does formula (B) on page 843:

$$\int e^u \, du = e^u + C$$

Matching exponents of e gives $u = x^3$, and the differential of u is $du = 3x^2 \, dx$. The differential requires a 3, which is not in the integral, so we multiply inside by 3 and outside by $\frac{1}{3}$.

$$\int e^{x^3} x^2 \, dx = \frac{1}{3} \int \underbrace{e^{x^3}}_{e^u} \underbrace{3x^2 \, dx}_{du} = \frac{1}{3} \int e^u \, du = \frac{1}{3} e^u + C = \frac{1}{3} e^{x^3} + C$$

$u = x^3$
$du = 3x^2 \, dx$

Multiplying by $\frac{1}{3}$ and 3

Substituting

Integrating using formula (B)

Substituting back to x using $u = x^3$

Why did the 1/3 become part of the answer but the 3 disappeared? The 3 became part of the du ($du = 3x^2 \, dx$), which was then "used up" in the integration along with the integral sign.

EXAMPLE 6

RECOVERING COST FROM MARGINAL COST

A company's marginal cost function is $MC(x) = \dfrac{x^3}{x^4 + 1}$ and fixed costs are \$1000. Find the cost function.

Solution Cost is the integral of marginal cost.

$$C(x) = \int \frac{x^3 \, dx}{x^4 + 1}$$

The differential of the denominator is $4x^3 \, dx$, which except for the 4 is just the numerator. This suggests formula (C), $\int \dfrac{du}{u} = \ln |u| + C$ with $u = x^4 + 1$. We multiply inside by 4 (to complete the $du = 4x^3 \, dx$ in the numerator) and outside by $\tfrac{1}{4}$.

$$\int \frac{x^3 \, dx}{x^4 + 1} = \frac{1}{4} \int \frac{4x^3 \, dx}{x^4 + 1} = \frac{1}{4} \int \frac{du}{u} = \frac{1}{4} \ln |u| + C = \frac{1}{4} \ln(x^4 + 1) + C$$

$u = x^4 + 1$ Multiplying   Substituting   Integrating by   Substituting
$du = 4x^3 \, dx$ by 4 and $\tfrac{1}{4}$ formula (C)   back to x

We dropped the absolute value bars because $x^4 + 1$ is positive. To evaluate the constant C, we set the cost function (evaluated at $x = 0$) equal to the given fixed cost.

$$\frac{1}{4} \underbrace{\ln(1)}_{0} + C = 1000 \qquad \text{$\ln(x^4 + 1) + C$ at $x = 0$ set equal to 1000}$$

This gives $C = 1000$. Therefore, the company's cost function is

$$C(x) = \frac{1}{4} \ln(x^4 + 1) + 1000 \qquad \begin{array}{l} C(x) = \tfrac{1}{4}\ln(x^4 + 1) + C \\ \text{with } C = 1000 \end{array}$$

Which Formula to Use

The three formulas apply to three different types of integrals.

(A) $\displaystyle \int u^n \, du = \frac{1}{n + 1} u^{n+1} + C \quad (n \neq -1)$ Integrates a *function to a constant power* (except -1) times the differential of the function

Evaluating Definite Integrals by Substitution

Sometimes a definite integral requires a substitution. In such cases changing from x to u also requires changing the limits of integration from x-values to u-values, using the substitution formula for u.

EXAMPLE 9

EVALUATING A DEFINITE INTEGRAL BY SUBSTITUTION

Evaluate $\int_4^5 \dfrac{dx}{3-x}$.

Solution The differential of the denominator $3-x$ is $-1 \cdot dx$, which except for the -1 is just the numerator. This suggests formula (C), $\int \dfrac{du}{u} = \ln|u| + C$ with $u = 3 - x$ (from equating the denominators). We multiply inside and outside by -1.

$$\int_4^5 \frac{dx}{3-x} = -\int_4^5 \frac{-dx}{3-x} = -\int_{-1}^{-2} \frac{du}{u}$$

with $u = 3 - x$, $du = -dx$, $u = 3 - 4 = -1$, $u = 3 - 5 = -2$.

New upper and lower limits of integration for u are found by evaluating $u = 3 - x$ at the old x limits

$$= -\ln|u| \Big|_{-1}^{-2} = -\ln|-2| - (-\ln|-1|) = -\ln 2 + \ln 1 = -\ln 2$$

Graphing Calculator Exploration

Why is the answer to Example 9 *negative*? Graph $f(x) = \dfrac{1}{3-x}$ on the window $[3, 6]$ by $[-2, 2]$ to see why. Have your calculator find the definite integral from 4 to 5. How would you change the integral so that it gives the *area* between the curve and the x-axis?

11.6 INTEGRATION BY SUBSTITUTION

Definite integrals are used to find areas, total accumulations, and average values, and any of these uses may require a substitution.

EXAMPLE 10 FINDING TOTAL POLLUTION FROM A RATE

Pollution is being discharged into a lake at the rate of $r(t) = 400te^{t^2}$ tons per year, where t is the number of years since measurements began. Find the total amount of pollutant discharged into the lake during the first 2 years.

Solution The total accumulation is the definite integral from $t = 0$ (the beginning) to $t = 2$ (2 years later). Since the integral involves e to a function, we use the formula for $\int e^u \, du$ with $u = t^2$ (by equating exponents).

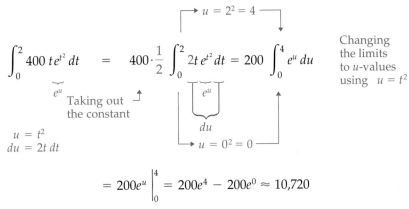

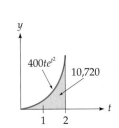

Therefore, during the first 2 years 10,720 tons of pollutant were discharged into the lake.

Notice that the du does not need to be all together, but can be in several separate pieces, as long as it is all *there*.

EXAMPLE 11 FINDING AVERAGE WATER DEPTH

After x months the water level in a newly built reservoir is $L(x) = 40x(x^2 + 9)^{-1/2}$ feet. Find the average depth during the first 4 months.

Solution The average value is the definite integral from $x = 0$ to $x = 4$ (the end of month 4) divided by the length of the interval.

$$\frac{1}{4}\int_0^4 40x(x^2+9)^{-1/2}\,dx = \frac{1}{4}\cdot 40\cdot\frac{1}{2}\int_0^4 2x(x^2+9)^{-1/2}\,dx = 5\int_9^{25} u^{-1/2}\,du$$

Changing the limits to u-values using $u = x^2 + 9$

$u = 4^2 + 9 = 25$
$u = 0^2 + 9 = 9$

$u = x^2 + 9$
$du = 2x\,dx$

$u^{-1/2}$
du

$L(x) = 40x(x^2+9)^{-1/2}$

average

Feet: 40, 30, 20, 10
Months: 1, 2, 3, 4

$$= 5\cdot 2u^{1/2}\Big|_9^{25} = 10(25)^{1/2} - 10(9)^{1/2} = 10\cdot 5 - 10\cdot 3 = 20$$

That is, the average depth of the reservoir over the last 4 months was 20 feet.

11.6 Section Summary

The three substitution formulas are listed on the inside back cover. Most of the work in using these formulas is making a problem "fit" the left-hand side of one of the formulas (choosing the u and adjusting constants to complete the du). Once a problem fits a left-hand side, the right-hand side immediately gives the answer (except for substituting back to the original variable).

Note that the du now plays a very important role: The du must be correct if the answer is to be correct. For example, the formula $\int e^u\,du = e^u + C$ should not be thought of as the formula for integrating e^u, but as the formula for integrating $e^u\,du$. The du is just as important as the e^u.

▶ Solutions to Practice Problems

1. $df = (3x^2 - 4)\,dx$
2. $du = -5e^{-5t}\,dt$
3. a. (B) b. (C) c. (A) d. (C)
4. Neither.
   a. Try formula (A) with $u = x^3 + 1$, so $du = 3x^2\,dx$. The problem has an x^3 for the differential instead of the needed x^2.
   b. Try formula (B) with $u = x^2$, so $du = 2x\,dx$. The problem does not have the x that is needed for the differential.

11.6 Exercises

Find each indefinite integral. [Integration formulas (A), (B), and (C) are on the inside back cover, numbered 5–7]

1. $\int (x^2 + 1)^9 \, 2x \, dx$
   [Hint: Use $u = x^2 + 1$ and formula 5.]

2. $\int (x^3 + 1)^4 \, 3x^2 \, dx$
   [Hint: Use $u = x^3 + 1$ and formula 5.]

3. $\int (x^2 + 1)^9 x \, dx$
   [Hint: Use $u = x^2 + 1$ and formula 5.]

4. $\int (x^3 + 1)^4 x^2 \, dx$
   [Hint: Use $u = x^3 + 1$ and formula 5.]

5. $\int e^{x^5} x^4 \, dx$
   [Hint: Use $u = x^5$ and formula 7.]

6. $\int e^{x^4} x^3 \, dx$
   [Hint: Use $u = x^4$ and formula 7.]

7. $\int \dfrac{x^5 \, dx}{x^6 + 1}$
   [Hint: Use $u = x^6 + 1$ and formula 6.]

8. $\int \dfrac{x^4 \, dx}{x^5 + 1}$
   [Hint: Use $u = x^5 + 1$ and formula 6.]

Show that each integral *cannot* be found by our substitution formulas.

9. $\int \sqrt{x^3 + 1} \, x \, dx$

10. $\int \sqrt{x^5 + 9} \, x^2 \, dx$

11. $\int e^{x^4} x^5 \, dx$

12. $\int e^{x^3} x^4 \, dx$

Find each indefinite integral by the substitution method or state that it cannot be found by our substitution formulas.

13. $\int (x^4 - 16)^5 x^3 \, dx$

14. $\int (x^5 - 25)^6 x^4 \, dx$

15. $\int e^{-x^2} x \, dx$

16. $\int e^{-x^4} x^3 \, dx$

17. $\int e^{3x} \, dx$

18. $\int e^{5x} \, dx$

19. $\int e^{x^2} x^2 \, dx$

20. $\int e^{x^3} x \, dx$

21. $\int \dfrac{dx}{1 + 5x}$

22. $\int \dfrac{dx}{1 + 3x}$

23. $\int (x^2 + 1)^9 5x \, dx$

24. $\int (x^2 - 4)^6 3x \, dx$

25. $\int \sqrt[4]{z^4 + 16} \, z^3 \, dz$

26. $\int \sqrt[3]{z^3 - 8} \, z^2 \, dz$

27. $\int \sqrt[4]{x^4 + 16} \, x^2 \, dx$

28. $\int \sqrt[3]{x^3 - 8} \, x \, dx$

29. $\int (2y^2 + 4y)^5 (y + 1) \, dy$

30. $\int (3y^2 - 6y)^3 (y - 1) \, dy$

31. $\int e^{x^2 + 2x + 5}(x + 1) \, dx$

32. $\int e^{x^3 - 3x + 7}(x^2 - 1) \, dx$

33. $\int \dfrac{x^3 + x^2}{3x^4 + 4x^3} \, dx$

34. $\int \dfrac{x^2 - x}{2x^3 - 3x^2} \, dx$

35. $\int \dfrac{x^3 + x^2}{(3x^4 + 4x^3)^2} \, dx$

36. $\int \dfrac{x^2 - x}{(2x^3 - 3x^2)^3} \, dx$

37. $\int \dfrac{x}{1 - x^2} \, dx$

38. $\int \dfrac{1}{1 - x} \, dx$

39. $\int (2x - 3)^7 \, dx$

40. $\int (5x + 9)^9 \, dx$

41. $\int \dfrac{e^{2x}}{e^{2x}+1}\,dx$
42. $\int \dfrac{e^{3x}}{e^{3x}-1}\,dx$
43. $\int \dfrac{\ln x}{x}\,dx$ [Hint: Let $u=\ln x$.]
44. $\int \dfrac{(\ln x)^2}{x}\,dx$ [Hint: Let $u=\ln x$.]
45. $\int \dfrac{e^{\sqrt{x}}}{\sqrt{x}}\,dx$ [Hint: Let $u=\sqrt{x}$.]
46. $\int \dfrac{e^{1/x}}{x^2}\,dx$ [Hint: Let $u=\dfrac{1}{x}$.]

Find each integral. [Hint: Try some algebra.]

47. $\int (x+1)x^2\,dx$
48. $\int (x+4)(x-2)\,dx$
49. $\int (x+1)^2 x^3\,dx$
50. $\int (x-1)^2 \sqrt{x}\,dx$

For each definite integral:
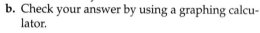
a. Evaluate it "by hand."
b. Check your answer by using a graphing calculator.

51. $\int_0^3 e^{x^2} x\,dx$
52. $\int_0^2 e^{x^3} x^2\,dx$
53. $\int_0^1 \dfrac{x}{x^2+1}\,dx$
54. $\int_2^3 \dfrac{x^2}{x^3-7}\,dx$
55. $\int_0^4 \sqrt{x^2+9}\,x\,dx$
56. $\int_0^3 \sqrt{x^2+16}\,x\,dx$
57. $\int_2^3 \dfrac{dx}{1-x}$
58. $\int_3^4 \dfrac{dx}{2-x}$
59. $\int_1^8 \dfrac{e^{\sqrt[3]{x}}}{\sqrt[3]{x^2}}\,dx$
60. $\int_1^4 \dfrac{e^{\sqrt{x}}}{\sqrt{x}}\,dx$

61. Prove the integration formula
$$\int u^n\,du = \dfrac{1}{n+1}u^{n+1}+C \quad (n\ne -1)$$
as follows.
a. Differentiate the right-hand side of the formula with respect to x (remembering that u is a function of x).
b. Verify that the result of part (a) agrees with the integrand in the formula (after replacing du in the formula by $u'\,dx$).

62. Prove the integration formula $\int e^u\,du = e^u + C$ by following the steps in Exercise 61.

63. Prove the integration formula $\int \dfrac{du}{u} = \ln u + C$ $(u>0)$ by following the steps in Exercise 61. (Absolute value bars come from applying the same argument to $-u$ for $u<0$.)

64. Find $\int (x+1)\,dx$:
a. by using the formula for $\int u^n\,du$ with $n=1$.
b. by dropping the parentheses and integrating directly.
c. Can you reconcile the two seemingly different answers? [Hint: Think of the arbitrary constant.]

APPLIED EXERCISES

65. **BUSINESS: Cost** A company's marginal cost function is $MC(x) = \dfrac{1}{2x+1}$ and its fixed costs are 50. Find the cost function.

66. **BUSINESS: Cost** A company's marginal cost function is $MC(x) = \dfrac{1}{\sqrt{2x+25}}$ and its fixed costs are 100. Find the cost function.

67. **GENERAL: Average Value** The population of a city is expected to be $P(x) = x(x^2+36)^{-1/2}$ million people after x years. Find the average population between year $x=0$ and year $x=8$.

68. **GENERAL: Area** Find the area between the curve $y = xe^{x^2}$ and the x-axis from $x=1$ to $x=3$. (Leave the answer in its exact form.)

69. BUSINESS: Average Sales A company's sales (in millions) during week x are given by $S(x) = \dfrac{1}{x+1}$. Find the average sales from week $x = 1$ to week $x = 4$.

70. BEHAVIORAL SCIENCE: Repeated Tasks A subject can perform a task at the rate of $\sqrt{2t+1}$ tasks per minute at time t minutes. Find the total number of tasks performed from time $t = 0$ to time $t = 12$.

71. BIOMEDICAL: Cholesterol An experimental drug lowers a patient's blood serum cholesterol at the rate of $t\sqrt{25-t^2}$ units per day, where t is the number of days since the drug was administered ($0 \le t \le 5$). Find the total change during the first 3 days.

72. BUSINESS: Total Sales During an automobile sale, cars are selling at the rate of $\dfrac{12}{x+1}$ cars per day, where x is the number of days since the sale began. How many cars will be sold during the first 7 days of the sale?

73. BUSINESS: Total Sales A real estate office is selling condominiums at the rate of $100e^{-x/4}$ per week after x weeks. How many condominiums will be sold during the first 8 weeks?

74. BUSINESS: Revenue An aircraft company estimates its marginal revenue function for helicopters to be $MR(x) = \sqrt{x^2 + 80x}(x + 40)$ thousand dollars, where x is the number of helicopters sold. Find the total revenue from the sale of the first 10 helicopters.

75–76. ENVIRONMENTAL SCIENCE: Pollution A factory is discharging pollution into a lake at the rate of $r(t)$ tons per year given below, where t is the number of years that the factory has been in operation. Find the total amount of pollution discharged during the first 3 years of operation.

75. $r(t) = \dfrac{t}{t^2 + 1}$ **76.** $r(t) = t\sqrt{t^2 + 16}$

Chapter Summary with Hints and Suggestions

Reading the text and doing the exercises in this chapter have helped you to master the following skills, which are listed by section (in case you need to review them) and are keyed to particular Review Exercises. Answers for all Review Exercises are given at the back of the book, and full solutions can be found in the Student Solutions Manual.

11.1 Antiderivatives and Indefinite Integrals

- Find an indefinite integral using the Power Rule. *(Review Exercises 1–8.)*

$$\int x^n \, dx = \frac{1}{n+1} x^{n+1} + C \quad (n \ne -1)$$

- Solve an applied problem involving integration. *(Review Exercises 9–10.)*

11.2 Integration Using Logarithmic and Exponential Functions

- Find an indefinite integral involving e^x or $\dfrac{1}{x}$. *(Review Exercises 11–18.)*

$$\int e^{ax} \, dx = \frac{1}{a} e^{ax} + C$$

$$\int \frac{1}{x} \, dx = \int x^{-1} \, dx = \ln|x| + C$$

- Solve an applied problem involving integration. *(Review Exercises 19–22.)*

11.3 Definite Integrals and Areas

- Evaluate a definite integral. *(Review Exercises 23–30.)*

$$\int_a^b f(x)\, dx = F(b) - F(a)$$

- Find the area under a curve. *(Review Exercises 31–36.)*

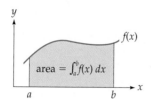

- Graph a function and find the area under it. *(Review Exercises 37–38.)*

- Solve an applied problem using definite integration. *(Review Exercises 39–44.)*

- Approximate the area under a curve by rectangles (Riemann sum). *(Review Exercises 45–46.)*

$$\int_a^b f(x)\, dx = \lim_{n \to \infty} [f(x_1) \cdot \Delta x + \cdots + f(x_n) \cdot \Delta x]$$

- Use a Riemann sum program to find Riemann sums. *(Review Exercises 47–48.)*

11.4 Further Applications of Definite Integrals: Average Value and Area Between Curves

- Find the area bounded by two curves. *(Review Exercises 49–56.)*

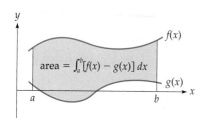

- Find the average value of a function on an interval. *(Review Exercises 57–60.)*

$$\left(\begin{array}{c}\text{Average of } f \\ \text{from } a \text{ to } b\end{array}\right) = \frac{1}{b - a} \int_a^b f(x)\, dx$$

- Solve an applied problem involving average value. *(Review Exercises 61–64.)*

- Solve an applied problem involving area between curves. *(Review Exercises 65–68.)*

11.5 Two Applications to Economics: Consumers' Surplus and Income Distribution

- Find the consumers' surplus for a product. *(Review Exercises 69–72.)*

$$\left(\begin{array}{c}\text{Consumers'} \\ \text{surplus}\end{array}\right) = \int_0^A [d(x) - B]\, dx$$

- Find the Gini index of income distribution. *(Review Exercises 73–76.)*

$$\left(\begin{array}{c}\text{Gini} \\ \text{index}\end{array}\right) = 2 \cdot \int_0^1 [x - L(x)]\, dx$$

11.6 Integration by Substitution

- Use a substitution to find an integral. *(Review Exercises 77–92.)*

$$\int u^n\, du = \frac{1}{n+1} u^{n+1} + C \quad (n \neq -1)$$

$$\int e^u\, du = e^u + C \qquad \int \frac{1}{u}\, du = \ln|u| + C$$

- Use a substitution to find a definite integral. *(Review Exercises 93–100.)*

- Use a substitution to find the area under a curve. *(Review Exercises 101–102.)*

- Use a substitution to find the average value of a function. *(Review Exercises 103–104.)*

- Use a substitution to solve an applied problem. *(Review Exercises 105–106.)*

Hints and Suggestions

- An *indefinite* integral is a function plus a constant, whereas a *definite* integral is a *number* (the signed area between the curve and the x-axis). The Fundamental Theorem of Integral Calculus shows how to evaluate the second in terms of the first.

- To integrate any power of x except x^{-1}, use the Power Rule; for x^{-1}, use the ln rule.

- *Indefinite* integrals have a $+C$. *Definite* integrals do not.

- Differentiation *breaks things down into parts*—for example, turning cost into marginal cost (cost per unit). Integration *combines back into a whole*—for example, combining all of the per-unit costs back into a total cost. In fact, the word "integrate" means "make whole."

- To find the area between two curves, integrate *upper* minus *lower*.

- For average values, don't forget the $\dfrac{1}{b-a}$.

- The substitution method can only "fix up" the *constant*; the variable part must already be correct (or else the function cannot be integrated by this technique).

- A graphing calculator helps by graphing functions, evaluating definite integrals, and finding areas under curves and total accumulations. With an appropriate program it can calculate Riemann sums with many rectangles.

- **Practice for Test:** Review Exercises 1, 5, 9, 15, 21, 23, 27, 31, 37, 39, 43, 49, 55, 57, 61, 69, 73, 81, 83, 93.

Review Exercises for Chapter 11

Practice test exercise numbers are in green.

11.1 Antiderivatives and Indefinite Integrals

Find each indefinite integral.

1. $\int (24x^2 - 8x + 1)\, dx$

2. $\int (12x^3 + 6x - 3)\, dx$

3. $\int (6\sqrt{x} - 5)\, dx$

4. $\int (8\sqrt[3]{x} - 2)\, dx$

5. $\int (10\sqrt[3]{x^2} - 4x)\, dx$

6. $\int (5\sqrt{x^3} - 6x)\, dx$

7. $\int (x+4)(x-4)\, dx$

8. $\int \dfrac{3x^3 + 2x^2 + 4x}{x}\, dx$

9. **BUSINESS: Cost** A company's marginal cost function is $MC(x) = x^{-1/2} + 4$, where x is the number of units. If fixed costs are 20,000, find the company's cost function.

10. **GENERAL: Population** The population of a town is now 40,000 and t years from now will be growing at the rate of $300\sqrt{t}$ people per year.
    a. Find a formula for the population of the town t years from now.
    b. Use your formula to find the population of the town 16 years from now.

11.2 Integration Using Logarithmic and Exponential Functions

Find each indefinite integral.

11. $\int e^{x/2}\, dx$

12. $\int e^{-2x}\, dx$

13. $\int 4x^{-1}\, dx$

14. $\int \dfrac{2}{x}\, dx$

15. $\int \left(6e^{3x} - \dfrac{6}{x}\right) dx$

16. $\int (x - x^{-1})\, dx$

17. $\int (9x^2 + 2x^{-1} + 6e^{3x})\, dx$

18. $\int \left(\dfrac{1}{x^2} + \dfrac{1}{x} + e^{-x}\right) dx$

19. **GENERAL: Consumption of Natural Resources** World consumption of aluminum is running at the rate of $50e^{0.05t}$ million tons per year, where t is the number of years since 2000.
   a. Find a formula for the total amount of aluminum consumed within t years of 2000.
   b. If consumption continues at this rate, when will the known resources of 7900 million tons of aluminum be exhausted?

20. **GENERAL: Total Savings** A homeowner installs a solar heating system, which is expected to generate savings at the rate of $200e^{0.1t}$ dollars per year, where t is the number of years since the system was installed.
   a. Find a formula for the total savings in the first t years.
   b. If the system originally cost $1500, when will it "pay for itself"?

21. **GENERAL: Consumption of Natural Resources** World consumption of zinc is running at the rate of $9e^{0.02t}$ million metric tons per year, where t is the number of years since 2000.
   a. Find a formula for the total amount of zinc consumed within t years of 2000.
   b. If consumption continues at this rate, when will the known resources of 430 million metric tons of zinc be exhausted? (Zinc is used to make protective coatings for iron and steel.)

22. **BUSINESS: Profit** A company's profit is growing at the rate of $200x^{-1}$ thousand dollars per month after x months, for $x \geq 1$.
   a. Find a formula for the total growth in the profit from month 1 to month x.
   b. When will the total growth in profit reach 600 thousand dollars?

11.3 Definite Integrals and Areas

Evaluate each definite integral.

23. $\int_1^9 \left(x - \dfrac{1}{\sqrt{x}}\right) dx$

24. $\int_2^5 (3x^2 - 4x + 5)\, dx$

25. $\int_1^{e^4} \dfrac{dx}{x}$

26. $\int_1^5 \dfrac{dx}{x}$

27. $\int_0^2 e^{-x}\, dx$

28. $\int_0^2 e^{x/2}\, dx$

29. $\int_0^{100} (e^{0.05x} - e^{0.01x})\, dx$

30. $\int_0^{10} (e^{0.04x} - e^{0.02x})\, dx$

For each function:
a. Find the area under the curve between the given x-values.
b. Verify your answer to part (a) by using a graphing calculator to find the area.

31. $f(x) = 6x^2 - 1$, $x = 1$ to $x = 2$
32. $f(x) = 9 - x^2$, $x = -3$ to $x = 3$
33. $f(x) = 12e^{2x}$, $x = 0$ to $x = 3$
34. $f(x) = e^{x/2}$, $x = 0$ to $x = 4$
35. $f(x) = \dfrac{1}{x}$, $x = 1$ to $x = 100$
36. $f(x) = x^{-1}$, $x = 1$ to $x = 1000$

Use a calculator to graph each function and find the area under it between the given x-values.

37. $f(x) = \dfrac{10}{x^4 + 1}$ from $x = -2$ to $x = 2$
38. $f(x) = e^{x^4}$ from $x = -1$ to $x = 1$

39. **GENERAL: Weight Gain** An average child of age t years gains weight at the rate of $1.7t^{1/2}$ kilograms per year. Find the total weight gain from age 1 to age 9.

40. **BEHAVIORAL SCIENCE: Learning** A student can memorize foreign vocabulary words

at the rate of $2/\sqrt[3]{t}$ words per minute, where t is the number of minutes since the studying began. Find the number of words that can be memorized from time $t = 1$ to time $t = 8$.

41. **ENVIRONMENTAL SCIENCE: Global Warming** The temperature of the earth is rising, as a result of the "greenhouse effect," in which carbon dioxide prevents the escape of heat from the atmosphere. If the temperature is rising at the rate of $0.15e^{0.1t}$ degrees per year, find the total rise in temperature over the next 10 years.

42. **BUSINESS: Cost** A company's marginal cost function is $MC(x) = x^{-1/2} + 4$ dollars, where x is the number of units. Find the total cost of the first 400 units (units $x = 0$ to $x = 400$).

43. **BUSINESS: Cost** A company's marginal cost function is $MC(x) = 22e^{-\sqrt{x}/5}$ dollars, where x is the number of units. Find the total cost of the first hundred units (units $x = 0$ to $x = 100$).

44. **BEHAVIORAL SCIENCE: Repetitive Tasks** A proofreader can read $15xe^{-0.25x}$ pages per hour, where x is the number of hours worked. Find the total number of pages that can be proofread in 8 hours.

Exercises on Riemann Sums

45. a. Approximate the area under the curve $f(x) = x^2$ from 0 to 2 using ten left rectangles with equal bases.
    b. Find the *exact* area under the curve between the given x-values by evaluating an appropriate definite integral using the Fundamental Theorem.

46. a. Approximate the area under the curve $f(x) = \sqrt{x}$ from 0 to 4 using ten left rectangles with equal bases. (Round calculations to three decimal places.)
    b. Find the *exact* area under the curve between the given x-values by evaluating an appropriate definite integral using the Fundamental Theorem.

47–48. Use the graphing calculator program RIEMANN (see page 800), one of the programs on page 812, or a similar program to find the following Riemann sums.

a. Calculate the Riemann sum for each function below for the following values of n: 10, 100, 1000. Use left, right, or midpoint rectangles, making a table of the answers, keeping three decimal places of accuracy.

b. Find the *exact* value of the area under the curve in each exercise below by evaluating an appropriate definite integral using the Fundamental Theorem. Your answers from part (a) should approach the number found from this integral.

47. $f(x) = e^x$ from $a = -2$ to $b = 2$

48. $f(x) = \dfrac{1}{x}$ from $a = 1$ to $b = 4$

11.4 Further Applications of Definite Integrals: Average Value and Area Between Curves

Find the area bounded by each pair of curves.

49. $y = x^2 + 3x$ and $y = 3x + 1$
50. $y = 12x - 3x^2$ and $y = 6x - 24$
51. $y = x^2$ and $y = x$
52. $y = x^4$ and $y = x$
53. $y = 4x^3$ and $y = 12x^2 - 8x$
54. $y = x^3 + x^2$ and $y = x^2 + x$
55. $y = e^x$ and $y = x + 5$
56. $y = \ln x$ and $y = \dfrac{x^2}{10}$

Find the average value of the function on the given interval.

57. $f(x) = \dfrac{1}{x}$ on $[1, 4]$
58. $f(x) = 6\sqrt{x}$ on $[1, 4]$
59. $f(x) = \sqrt{x^3 + 1}$ on $[0, 5]$
60. $f(x) = \ln(e^{x^2} + 10)$ on $[-2, 2]$

61. **GENERAL: Average Population** The population of the world, now more than 6 billion, is predicted to be $P(t) = 6e^{0.008t}$ billion, where t is the number of years after the year 2000. Find the average population between the years 2000 and 2100.

62. GENERAL: Compound Interest A deposit of $3000 in a bank paying 6% interest compounded continuously will grow to $V(t) = 3000e^{0.06t}$ dollars after t years. Find the average value during the first 20 years ($t = 0$ to $t = 20$).

63. GENERAL: Real Estate The value of a suburban plot of land being considered for rezoning is assessed at $4.3e^{0.01x^2}$ hundred thousand dollars x years from now. Find the average value over the next 10 years (year 0 to year 10).

64. GENERAL: Stock Price The price of a share of stock is expected to be $28e^{0.01x^{1.2}}$ dollars where x is the number of weeks from now. Find the average price over the next year (week 0 to week 52).

65. GENERAL: Art An artist wants to paint the interior of the shape shown below on the side of a building. How much area (in square meters) will the artist need to paint?

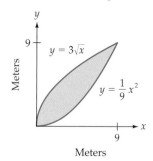

66. ECONOMICS: Balance of Trade A country's annual exports will be $E(t) = 40e^{0.2t}$ and its imports will be $I(t) = 20e^{0.1t}$ (both in billions of dollars per year), where t is the number of years from now. Find the accumulated trade surplus (exports minus imports) over the next 10 years. (See the graph below.)

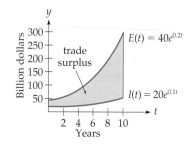

67. BUSINESS: Profit A company's annual revenue and annual costs are expected to be $R(t) = 50e^{0.08t}$ and $C(t) = 20e^{0.04t}$ million dollars per year, where t is the number of years from now. Find the cumulative profit over the next 8 years (year 0 to year 8).

68. GENERAL: Population China, with 21% of the world's population on only 7% of the world's arable land, has taken drastic measures to reduce its population, and has succeeded in reducing its birthrate (births per 1000 population) from 23.3 in 1987 to 16.2 in 2000. China's population is predicted to be $P(x) = 1260e^{0.009x}$ million people x years after 2000.

a. Multiply $P(x)$ by $\frac{23.3}{1000}$ to find the number of births per year at the old birthrate, and enter the result in your calculator as y_1. Then multiply $P(x)$ by $\frac{16.2}{1000}$ to find the number of births per year at the new birthrate, and enter the result as y_2.

b. Graph both y_1 and y_2 on the window [0, 10] by [0, 35], showing the difference in the number of births at the different birthrates during 2000–2010.

c. Find the area between the curves, thereby finding the decrease in the number of births during this period resulting from the lower birthrate.

11.5 Two Applications to Economics: Consumers' Surplus and Income Distribution

69–72. ECONOMICS: Consumers' Surplus For each demand function $d(x)$ and demand level x, find the consumers' surplus.

69. $d(x) = 8000 - 24x$, $x = 200$

70. $d(x) = 1800 - 0.03x^2$, $x = 200$

71. $d(x) = 300e^{-0.2\sqrt{x}}$, $x = 120$

72. $d(x) = \dfrac{100}{1 + \sqrt{x}}$, $x = 100$

73–76. ECONOMICS: Gini Index For each Lorenz curve, find the Gini index.

73. $L(x) = x^{3.5}$

74. $L(x) = x^{2.5}$

75. $L(x) = \dfrac{x}{2 - x}$

76. $L(x) = \dfrac{e^{x^4} - 1}{e - 1}$

11.6 Integration by Substitution

Find each integral or state that it cannot be evaluated by our substitution formulas.

77. $\int x^2 \sqrt[3]{x^3 - 1} \, dx$

78. $\int x^3 \sqrt{x^4 - 1} \, dx$

79. $\int x \sqrt[3]{x^3 - 1} \, dx$

80. $\int x^2 \sqrt{x^4 - 1} \, dx$

81. $\int \frac{dx}{9 - 3x}$

82. $\int \frac{dx}{1 - 2x}$

83. $\int \frac{dx}{(9 - 3x)^2}$

84. $\int \frac{dx}{(1 - 2x)^2}$

85. $\int \frac{x^2}{\sqrt[3]{8 + x^3}} \, dx$

86. $\int \frac{x}{\sqrt{9 + x^2}} \, dx$

87. $\int \frac{w + 3}{(w^2 + 6w - 1)^2} \, dw$

88. $\int \frac{t - 2}{(t^2 - 4t + 1)^2} \, dt$

89. $\int \frac{(1 + \sqrt{x})^2}{\sqrt{x}} \, dx$

90. $\int \frac{(1 + \sqrt[3]{x})^2}{\sqrt[3]{x^2}} \, dx$

91. $\int \frac{e^x}{e^x - 1} \, dx$

92. $\int \frac{1}{x \ln x} \, dx$

For each definite integral:
a. Evaluate it ("by hand") or state that it cannot be evaluated by our substitution formulas.
b. Verify your answer to part (a) by using a graphing calculator.

93. $\int_0^3 x \sqrt{x^2 + 16} \, dx$

94. $\int_0^4 \frac{dz}{\sqrt{2z + 1}}$

95. $\int_0^4 \frac{w}{\sqrt{25 - w^2}} \, dw$

96. $\int_1^2 \frac{x + 1}{(x^2 + 2x - 2)^2} \, dx$

97. $\int_3^9 \frac{dx}{x - 2}$

98. $\int_4^5 \frac{dx}{x - 6}$

99. $\int_0^1 x^3 e^{x^4} \, dx$

100. $\int_0^1 x^4 e^{x^5} \, dx$

Find the area under the given curve between the given x-values.

101. $y = \frac{x^2 + 6x}{\sqrt[3]{x^3 + 9x^2 + 17}}$ from $x = 1$ to $x = 3$

102. $y = \frac{x + 6}{\sqrt{x^2 + 12x + 4}}$ from $x = 0$ to $x = 3$

Find the average value of the function on the given interval.

103. $f(x) = xe^{-x^2}$ on $[0, 2]$

104. $f(x) = \frac{x}{x^2 - 3}$ on $[2, 4]$

105. **BUSINESS: Cost** A company's marginal cost function is $MC(x) = \frac{1}{\sqrt{2x + 9}}$ and fixed costs are 100. Find the cost function.

106. **BIOMEDICAL: Temperature** An experimental drug changes a patient's temperature at the rate of $\frac{3x^2}{x^3 + 1}$ degrees per milligram of the drug, where x is the amount of the drug administered. Find the total change in temperature resulting from the first 3 milligrams of the drug. [*Note:* The rate of change of temperature with respect to dosage is called the "drug sensitivity."]

12 Integration Techniques and Differential Equations

- 12.1 Integration by Parts
- 12.2 Integration Using Tables
- 12.3 Improper Integrals
- 12.4 Numerical Integration
- 12.5 Differential Equations
- 12.6 Further Applications of Differential Equations: Three Models of Growth

Application Preview

Improper Integrals and Eternal Recognition

Suppose that after you become rich and famous, you decide to commission a statue of yourself for your hometown. Your town, however, will accept this selfless gesture only if you pay for the perpetual upkeep of the statue by establishing a fund that will generate $2000 annually for every year in the future. Before deciding whether or not to accept this condition, you of course want to know how much it will cost. (Interestingly, we will see that a fund that will generate income forever does *not* require an infinite amount of money.) On page 869 we will find that to realize a yield of $2000 t years from now requires only its *present value* deposited now in a bank. At an interest rate of, say, 5% compounded continuously, $2000 in t years requires a deposit now of

$$\left(\begin{array}{c}\text{Present value of \$2000 at 5\%}\\ \text{compounded continuously}\end{array}\right) = 2000e^{-0.05t} \qquad \text{Amount times } e^{-0.05t}$$

Therefore, the size of the fund needed to generate $2000 annually *forever* is found by summing (integrating) this present value over the infinite time interval from zero to infinity (∞).

$$\left(\begin{array}{c}\text{Size of fund to yield an}\\ \text{annual \$2000 forever}\end{array}\right) = \int_0^\infty 2000e^{-0.05t}\, dt \qquad \text{Integrating out to infinity}$$

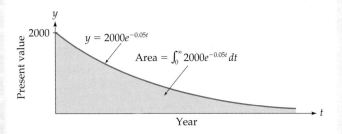

On page 894 we will learn how to evaluate such "improper integrals," finding that the value of this integral is $40,000, which shows that $40,000 deposited in a bank at 5% compounded continuously will generate $2000 every year *forever*. That is, $40,000 would pay for the perpetual upkeep of your statue, buying you (or at least your likeness) a kind of immortality.

Incidentally, to generate $2000 annually for only the first *hundred* years would require $\int_0^{100} 2000e^{-0.05t}\, dt \approx \$39{,}730$ (integrating from 0 to 100). This amount is only $270 less than the amount needed to generate the same sum *forever*. This small additional cost shows that the short term is expensive, but eternity is cheap.

12.1 INTEGRATION BY PARTS

Introduction

In this chapter we introduce further techniques for finding integrals: integration by parts, integration by tables, and numerical integration. We also discuss improper integrals (integrals over infinite intervals) and differential equations.

You may have felt that integration is "harder" than differentiation. One reason is that integration is an *inverse* process, carried out by *reversing* differentiation, and inverse processes are generally more difficult. Another reason is that, while we have the product and quotient rules for differentiating complicated expressions, there are no product and quotient rules for integration. The method of integration by parts, explained in this section, is in some sense a "product rule for integration" in that it comes from interpreting the product rule as an integration formula.

Integration by Parts

For two differentiable functions $u(x)$ and $v(x)$, hereafter denoted simply u and v, the Product Rule is

$$(uv)' = u'v + uv'$$

The derivative of a product is the derivative of the first times the second, plus the first times the derivative of the second

If we integrate both sides of this equation, integrating the left side "undoes" the differentiation.

$$uv = \int u'v\, dx + \int uv'\, dx$$

where $u'v\,dx = du$ and $uv'\,dx = dv$.

Using differential notation, $du = u'\, dx,\ dv = v'\, dx$

$$uv = \int v\, du + \int u\, dv$$

Formula above in differential notation

Solving this equation for the second integral $\int u\, dv$ gives

$$\int u\, dv = uv - \int v\, du$$

This formula is the basis for a technique called "integration by parts."

Integration by Parts

For differentiable functions u and v,

$$\int u\, dv = uv - \int v\, du$$

We use this formula to solve integrals by a "double substitution," substituting u for part of the given integral and dv for the rest, and then expressing the integral in the form $uv - \int v\, du$. The point is to choose the u and the dv so that the resulting integral $\int v\, du$ is *simpler* than the original integral $\int u\, dv$. A few examples will make the method clear.

EXAMPLE 1

INTEGRATING BY PARTS

Use integration by parts to find $\int xe^x\, dx$.

Solution

$\int \underbrace{x}_{u}\, \underbrace{e^x\, dx}_{dv}$ Original integral

We choose $u = x$ and $dv = e^x\, dx$

$\begin{bmatrix} u = x & dv = e^x\, dx \\ \downarrow & \downarrow \\ du = 1\, dx = dx & v = \int e^x\, dx = e^x \end{bmatrix}$

Differentiating u to find du and integrating dv to find v (omitting the C) (the arrows show which part leads to which other part)

$= \underbrace{x\, e^x}_{u\ \ v} - \int \underbrace{e^x\, dx}_{v\ \ du}$

Replacing the original integral $\int u\, dv$ by the right-hand side of the formula, $u \cdot v - \int v\, du$, with $u = x$, $v = e^x$, and $du = dx$

$= xe^x - e^x + C$

$\overset{\displaystyle \llcorner}{\text{From } \int e^x\, dx}$

Finding the new integral $\int e^x\, dx = e^x$ to give the final answer (with $+\, C$)

The procedure is not as complicated as it might seem. All the steps may be written together as follows:

$\int \underbrace{xe^x\, dx}_{u\ \ dv} = \underbrace{xe^x}_{uv} - \int \underbrace{e^x\, dx}_{v\ \ du} = xe^x - e^x + C$

$\overset{\displaystyle \llcorner}{\text{From } \int e^x\, dx}$

$\begin{bmatrix} u = x & dv = e^x\, dx \\ du = dx & v = \int e^x\, dx = e^x \end{bmatrix}$

We can check this answer by differentiation.

$$\frac{d}{dx}(xe^x - e^x + C) = \underbrace{e^x + xe^x}_{\text{Differentiating } xe^x \text{ by the Product Rule}} - e^x = xe^x$$

(Cancel)

Agrees with the original integrand, so the integration is correct

> **Remarks on the Integration by Parts Procedure**
>
> i. The differentials du and dv include the dx.
> ii. We omit the constant C when we integrate dv to get v because one $+ C$ at the end is enough.
> iii. The integration by parts formula does not give a "final answer," but rather expresses the given integral as $uv - \int v\, du$, a product $u \cdot v$ (already integrated), and a new integral $\int v\, du$. That is, integration by parts "exchanges" the original integral $\int u\, dv$ for another integral $\int v\, du$. The hope is that the second integral will be simpler than the first. In our example we "exchanged" $\int xe^x\, dx$ for the simpler $\int e^x\, dx$, which could be integrated immediately by formula 3 (inside back cover).
> iv. Integration by parts should be used only if the simpler formulas 1 through 7 (inside back cover) fail to solve the integral.

How to Choose the *u* and the *dv*

In Example 1, the choice of $\begin{cases} u = x \\ dv = e^x\, dx \end{cases}$ "exchanged" the original integral $\int xe^x\, dx$ for the simpler integral $\int e^x\, dx$. If we had instead chosen $\begin{cases} u = e^x \\ dv = x\, dx \end{cases}$ we would have "exchanged" the original integral for $\int x^2 e^x\, dx$ (as you may check), which is *more* difficult than the original (because of the x^2). Therefore, the first choice was the "right" choice in that it led to a solution. Generally, one choice for u and dv will be "best," and finding it may involve some trial and error. While there is no foolproof rule for finding the best u and dv, the following guidelines often help.

> **Guidelines for Choosing *u* and *dv***
>
> 1. Choose dv to be the most complicated part of the integral that can be integrated easily.
> 2. Choose u so that u' is simpler than u.

EXAMPLE 2 INTEGRATING BY PARTS

Find $\int x^2 \ln x \, dx$.

Solution None of the easier formulas (1 through 7 on the inside back cover) will solve the integral, as you may easily check. Therefore, we try integration by parts. The integrand is a product, x^2 times $\ln x$. The guidelines say to choose dv to be the most complicated part that can be easily integrated. We can integrate x^2 but not $\ln x$ (we know how to *differentiate* $\ln x$, but not how to *integrate* it), so we choose $dv = x^2 \, dx$, and therefore $u = \ln x$.

$$\int \underbrace{x^2 \ln x \, dx}_{\substack{u \\ dv}} = \underbrace{(\ln x)}_{u} \underbrace{\tfrac{1}{3} x^3}_{v} - \int \underbrace{\tfrac{1}{3} x^3}_{v} \underbrace{\tfrac{1}{x} dx}_{du} = \tfrac{1}{3} x^3 \ln x - \tfrac{1}{3} \int x^2 \, dx$$

Moving the $\tfrac{1}{3}$ outside and simplifying $x^3 \tfrac{1}{x}$ to x^2

$$\begin{bmatrix} u = \ln x & dv = x^2 \, dx \\ du = \dfrac{1}{x} dx & v = \int x^2 \, dx = \tfrac{1}{3} x^3 \end{bmatrix}$$

$$= \tfrac{1}{3} x^3 \ln x - \tfrac{1}{3} \tfrac{1}{3} x^3 + C = \tfrac{1}{3} x^3 \ln x - \tfrac{1}{9} x^3 + C$$

We may check this answer by differentiation.

$$\frac{d}{dx}\left(\tfrac{1}{3} x^3 \ln x - \tfrac{1}{9} x^3 + C\right) = \underbrace{x^2 \ln x + \tfrac{1}{3} x^3 \tfrac{1}{x}}_{\text{Differentiating } \tfrac{1}{3} x^3 \ln x \text{ by the Product Rule}} - \tfrac{1}{3} x^2$$

$$= x^2 \ln x + \tfrac{1}{3} x^2 - \tfrac{1}{3} x^2 = x^2 \ln x$$

From $x^3 \tfrac{1}{x}$ Cancel

Agrees with the original integrand, so the integration is correct

---

Practice Problem 1 Use integration by parts to find $\int x^3 \ln x \, dx$. ▶ Solution on page 872

CHAPTER 12 INTEGRATION TECHNIQUES AND DIFFERENTIAL EQUATIONS

Integration by parts is also useful because it simplifies integrating products of powers of linear functions.

EXAMPLE 3

INTEGRATING BY PARTS

Use integration by parts to find $\int (x-2)(x+4)^8\, dx$.

Solution The guidelines recommend that dv be the most complicated part that can be integrated. Both $x-2$ and $(x+4)^8$ can be integrated. For example,

$$\int (x+4)^8\, dx = \frac{1}{9}(x+4)^9 + C \qquad \text{By the substitution method with } u = x+4 \text{ (omitting the details)}$$

Since $(x+4)^8$ is more complicated than $(x-2)$, we take $dv = (x+4)^8\, dx$.

$$\int \underbrace{(x-2)}_{u}\underbrace{(x+4)^8\, dx}_{dv} = \underbrace{(x-2)}_{u}\underbrace{\frac{1}{9}(x+4)^9}_{v} - \int \underbrace{\frac{1}{9}(x+4)^9}_{v}\underbrace{dx}_{du}$$

$$\begin{bmatrix} u = x-2 & dv = (x+4)^8\, dx \\ du = dx & v = \int (x+4)^8\, dx \\ & = \frac{1}{9}(x+4)^9 \end{bmatrix}$$

$$= \frac{1}{9}(x-2)(x+4)^9 - \frac{1}{9}\int (x+4)^9\, dx$$

↑ Taking out the $\frac{1}{9}$

$$= \frac{1}{9}(x-2)(x+4)^9 - \underbrace{\frac{1}{9}\cdot\frac{1}{10}(x+4)^{10} + C}_{\text{Integrating by the substitution method}}$$

$$= \frac{1}{9}(x-2)(x+4)^9 - \frac{1}{90}(x+4)^{10} + C$$

Again we could check this answer by differentiation. (Do you see how the Product Rule, applied to the first part of this answer, will give a piece that will cancel with the derivative of the second part?)

Practice Problem 2 Use integration by parts to find $\int (x+1)(x-1)^3 \, dx$.

> Solution on page 873

Present Value of a Continuous Stream of Income

If a business or some other asset generates income continuously at the rate $C(t)$ dollars per year, where t is the number of years from now, then $C(t)$ is called a *continuous stream of income*.* On page 70 we saw that to find the value to which a sum will grow (the *future value* of the sum) under continuous compounding, we multiply it by e^{rt}, where r is the interest rate and t is the number of years. Just as with discrete interest (see page 113), to reverse the process and find the *present value* of a sum, the amount *now* that will later yield the stated sum, we would *divide* by e^{rt}, or equivalently, multiply by e^{-rt}. (We will refer to a rate with continuous compounding as a "continuous interest rate.") Therefore, the present value of the continuous stream $C(t)$ is found by multiplying by e^{-rt} and summing (integrating) over the time period.

Present Value of a Continuous Stream of Income

The present value of the continuous stream of income $C(t)$ dollars per year, where t is the number of years from now, for T years at continuous interest rate r is

$$\begin{pmatrix} \text{Present} \\ \text{value} \end{pmatrix} = \int_0^T C(t)e^{-rt} \, dt$$

EXAMPLE 4

FINDING THE PRESENT VALUE OF A CONTINUOUS STREAM OF INCOME

A business generates income at the rate of $2t$ million dollars per year, where t is the number of years from now. Find the present value of this continuous stream for the next 5 years at the continuous interest rate of 10%.

---

*$C(t)$ must be continuous, meaning that the income is being paid continuously rather than in "lump-sum" payments. However, even lump-sum payments can be approximated by a continuous stream if the payments are frequent enough or last long enough.

Solution

$$\begin{pmatrix} \text{Present} \\ \text{value} \end{pmatrix} = \int_0^5 2te^{-0.1t}\, dt \qquad \begin{array}{l} \text{Multiplying } C(t) = 2t \text{ by} \\ e^{-0.1t} \text{ (since } 10\% = 0.1\text{) and} \\ \text{integrating from 0 to 5 years} \end{array}$$

This is a *definite* integral, but we will ignore the limits of integration until after we have found the *indefinite* integral. None of the formulas 1 through 7 on the inside back cover will find this integral, so we try integration by parts with $u = 2t$ and $dv = e^{-0.1t}\, dt$. (Do you see why the guidelines on page 866 suggest this choice?)

$$\int \underbrace{2t}_{u}\, \underbrace{e^{-0.1t}\, dt}_{dv} \;=\; \underbrace{(2t)}_{u}\underbrace{(-10e^{-0.1t})}_{v} - \int \underbrace{(-10e^{-0.1t})}_{v}\, \underbrace{2\, dt}_{du}$$

$$\begin{bmatrix} u = 2t & dv = e^{-0.1t}\, dt \\ du = 2\, dt & v = \int e^{-0.1t}\, dt \\ & \quad = -10e^{-0.1t} \end{bmatrix}$$

$$= -20te^{-0.1t} + 20\int e^{-0.1t}\, dt$$
$$= -20te^{-0.1t} + 20(-10)e^{-0.1t} + C$$
$$= -20te^{-0.1t} - 200e^{-0.1t} + C$$

For the *definite* integral, we evaluate this from 0 to 5:

$$\left. (-20te^{-0.1t} - 200e^{-0.1t}) \right|_0^5 = (-20 \cdot 5 e^{-0.5} - 200e^{-0.5}) - (-20 \cdot 0 e^0 - 200e^0)$$

$$\underbrace{\phantom{(-20 \cdot 5 e^{-0.5} - 200e^{-0.5})}}_{\text{Evaluation at } t = 5} \quad \underbrace{}_{\text{Evaluation at } t = 0}$$

with the 0 and 200 labeled over the last two terms.

$$= -300e^{-0.5} + 200 \approx 18 \qquad \begin{array}{l} \text{In millions of} \\ \text{dollars (using} \\ \text{a calculator)} \end{array}$$

The present value of the stream of income over 5 years is approximately $18 million.

This answer means that $18 million at 10% interest compounded continuously would generate the continuous stream $C(t) = 2t$ million dollars for 5 years. This method is often used to determine the fair value of a business or some other asset, since it gives the present value of future income.

Graphing Calculator Exploration

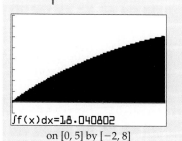

∫f(x)dx=18.040802
on [0, 5] by [−2, 8]

a. Verify the answer to Example 4 by graphing $2xe^{-0.1x}$ and finding the area under the curve from 0 to 5.

b. Can you explain why the curve increases less steeply farther to the right? [*Hint:* Think of the present value of money to be paid in the more distant future.]

Why is it necessary to learn integration by parts if integrals like the one in Example 4 can be evaluated on graphing calculators? One answer is that you should have a *variety* of ways to approach a problem—sometimes geometrically, sometimes analytically, and sometimes numerically (using a calculator). These various approaches mutually support one another; for example, you can solve a problem one way and check it another way. Furthermore, not all applications involve *definite* integrals. Example 4 might have asked for a *formula* for the present value up to any time t, and such formulas may be found by integrating "by hand" but not from most graphing calculators.

Remember that you should use integration by parts *only* if the "easier" formulas (1 through 7 on the inside back cover) fail to solve the integral.

Practice Problem 3   Which of the following integrals require integration by parts, and which can be found by the substitution formula $\int e^u \, du = e^u + C$? (Do not solve the integrals.)

a. $\int xe^x \, dx$ b. $\int xe^{x^2} \, dx$

➤ Solution on page 873

12.1 Section Summary

The integration by parts formula

$$\int u\, dv = uv - \int v\, du$$

is simply the integration version of the Product Rule. The guidelines on page 866 lead to the following suggestions for choosing u and dv in some commonly occurring integrals.

For Integrals of the Form:	Choose:	
$\int x^n e^{ax}\, dx$	$u = x^n$	$dv = e^{ax}\, dx$
$\int x^n \ln x\, dx$	$u = \ln x$	$dv = x^n\, dx$
$\int (x+a)(x+b)^n\, dx$	$u = x + a$	$dv = (x+b)^n\, dx$

Integration by parts can be useful in any situation involving integrals, such as recovering total cost from marginal cost, calculating areas, average values (the definite integral divided by the length of the interval), continuous accumulations, or present values of continuous income streams.

▶ Solutions to Practice Problems

1. $\displaystyle \int x^3 \ln x\, dx = (\ln x)\left(\frac{1}{4}x^4\right) - \int \frac{1}{4}x^4 \frac{1}{x}\, dx$

$$\left[\begin{array}{ll} u = \ln x & dv = x^3\, dx \\ du = \dfrac{1}{x}\, dx & v = \displaystyle\int x^3\, dx = \dfrac{1}{4}x^4 \end{array}\right]$$

$\displaystyle = (\ln x)\left(\frac{1}{4}x^4\right) - \frac{1}{4}\int x^3\, dx$

$\displaystyle = (\ln x)\left(\frac{1}{4}x^4\right) - \frac{1}{16}x^4 + C$

2. $\int (x+1)(x-1)^3\, dx$

$$\begin{bmatrix} u = x+1 & dv = (x-1)^3\, dx \\ du = dx & v = \int (x-1)^3\, dx = \tfrac{1}{4}(x-1)^4 \end{bmatrix}$$

$$= (x+1)\frac{1}{4}(x-1)^4 - \int \frac{1}{4}(x-1)^4\, dx$$

$$= \frac{1}{4}(x+1)(x-1)^4 - \frac{1}{4}\cdot\frac{1}{5}(x-1)^5 + C$$

$$= \frac{1}{4}(x+1)(x-1)^4 - \frac{1}{20}(x-1)^5 + C$$

3. **a.** Requires integration by parts

   b. Can be solved by the substitution $u = x^2$

12.1 Exercises

Integration by parts often involves finding integrals like the following when integrating dv to find v. Find the following integrals *without* using integration by parts (using formulas 1 through 7 on the inside back cover). Be ready to find similar integrals during the integration by parts procedure.

1. $\int e^{2x}\, dx$

2. $\int x^5\, dx$

3. $\int (x+2)\, dx$

4. $\int (x-1)\, dx$

5. $\int \sqrt{x}\, dx$

6. $\int e^{-0.5t}\, dt$

7. $\int (x+3)^4\, dx$

8. $\int (x-5)^6\, dx$

Use integration by parts to find each integral.

9. $\int xe^{2x}\, dx$

10. $\int xe^{3x}\, dx$

11. $\int x^5 \ln x\, dx$

12. $\int x^4 \ln x\, dx$

13. $\int (x+2)e^x\, dx$ [*Hint:* Take $u = x+2$.]

14. $\int (x-1)e^x\, dx$ [*Hint:* Take $u = x-1$.]

15. $\int \sqrt{x}\ln x\, dx$

16. $\int \sqrt[3]{x}\ln x\, dx$

17. $\int (x-3)(x+4)^5\, dx$

18. $\int (x+2)(x-5)^5\, dx$

19. $\int te^{-0.5t}\, dt$

20. $\int te^{-0.2t}\, dt$

21. $\int \frac{\ln t}{t^2}\, dt$

22. $\int \frac{\ln t}{\sqrt{t}}\, dt$

23. $\int s(2s+1)^4\, ds$

24. $\int \frac{x+1}{e^{3x}}\, dx$

25. $\int \frac{x}{e^{2x}}\, dx$

26. $\int \frac{\ln(x+1)}{\sqrt{x+1}}\, dx$

27. $\int \frac{x}{\sqrt{x+1}}\, dx$

28. $\int x\sqrt{x+1}\, dx$

29. $\int xe^{ax}\, dx$ $(a \neq 0)$

30. $\int (x+b)e^{ax}\,dx \quad (a \neq 0)$

31. $\int x^n \ln ax\,dx \quad (a \neq 0,\ n \neq -1)$

32. $\int (x+a)^n \ln(x+a)\,dx \quad (n \neq -1)$

33. $\int \ln x\,dx \quad$ [Hint: Take $u = \ln x,\ dv = dx$.]

34. $\int \ln x^2\,dx \quad$ [Hint: Take $u = \ln x^2,\ dv = dx$.]

35. $\int x^3 e^{x^2}\,dx \quad$ [Hint: Take $u = x^2,\ dv = xe^{x^2}$ and use a substitution to find v from dv.]

36. $\int x^3(x^2-1)^6\,dx \quad$ [Hint: Take $u = x^2$, $dv = x(x^2-1)^6\,dx$ and use a substitution to find v from dv.]

Find each integral by integration by parts or a substitution, as appropriate.

37. a. $\int xe^{x^2}\,dx$ b. $\int \dfrac{(\ln x)^3}{x}\,dx$

    c. $\int x^2 \ln 2x\,dx$ d. $\int \dfrac{e^x}{e^x+4}\,dx$

38. a. $\int \sqrt{\ln x}\,\dfrac{1}{x}\,dx$ b. $\int x^2 e^{x^3}\,dx$

    c. $\int x^7 \ln 3x\,dx$ d. $\int xe^{4x}\,dx$

Evaluate each definite integral using integration by parts. (Leave answers in exact form.)

39. $\int_0^2 xe^x\,dx$ 40. $\int_0^3 xe^x\,dx$

41. $\int_1^3 x^2 \ln x\,dx$ 42. $\int_1^2 x \ln x\,dx$

43. $\int_0^2 z(z-2)^4\,dz$ 44. $\int_0^4 z(z-4)^6\,dz$

45. $\int_0^{\ln 4} te^t\,dt$ 46. $\int_1^e \ln x\,dx$

Find in two different ways.

47. $\int x(x-2)^5\,dx$

    a. Use integration by parts.
    b. Use the substitution $u = x-2$ (so x is replaced by $u+2$) and then multiply out the integrand.

48. $\int x(x+4)^6\,dx$

    a. Use integration by parts.
    b. Use the substitution $u = x+4$ (so x is replaced by $u-4$) and then multiply out the integrand.

Derive each formula by using integration by parts on the left-hand side. (Assume $n > 0$.)

49. $\int x^n e^x\,dx = x^n e^x - n\int x^{n-1} e^x\,dx$

50. $\int (\ln x)^n\,dx = x(\ln x)^n - n\int (\ln x)^{n-1}\,dx$

51. Use the formula in Exercise 49 to find the integral $\int x^2 e^x\,dx$. [Hint: Apply the formula twice.]

52. Use the formula in Exercise 50 to find the integral $\int (\ln x)^2\,dx$. [Hint: Apply the formula twice.]

53. a. Find the integral $\int x^{-1}\,dx$ by integration by parts (using $u = x^{-1}$ and $dv = dx$), obtaining

$$\int x^{-1}\,dx = x^{-1}x - \int (-x^{-2})x\,dx$$

which gives

$$\int x^{-1}\,dx = 1 + \int x^{-1}\,dx$$

    b. Subtract the integral from both sides of this last equation, obtaining $0 = 1$. Explain this apparent contradiction.

54. We omit the constant of integration when we integrate dv to get v. Including the constant C

in this step simply replaces v by $v + C$, giving the formula

$$\int u\,dv = u(v + C) - \int (v + C)\,du$$

Multiplying out the parentheses and expanding the last integral into two gives

$$\int u\,dv = uv + Cu - \int v\,du - C\int du$$

Show that the second and fourth terms on the right cancel, giving the "old" integration by parts formula $\int u\,dv = uv - \int v\,du$. This shows that including the constant in the dv to v step gives the same formula. *One constant of integration at the end is enough.*

APPLIED EXERCISES

55. BUSINESS: Revenue If a company's marginal revenue function is $MR(x) = xe^{x/4}$, find the revenue function. [*Hint:* Evaluate the constant C so that revenue is 0 at $x = 0$.]

56. BUSINESS: Cost A company's marginal cost function is $MC(x) = xe^{-x/2}$ and fixed costs are 200. Find the cost function. [*Hint:* Evaluate the constant C so that the cost is 200 at $x = 0$.]

57. BUSINESS: Present Value of a Continuous Stream of Income An electronics company generates a continuous stream of income of $4t$ million dollars per year, where t is the number of years that the company has been in operation. Find the present value of this stream of income over the first 10 years at a continuous interest rate of 10%.

For Exercises 58–59:

a. Solve *without* using a graphing calculator.
b. Verify your answer to part (a) using a graphing calculator.

58. BUSINESS: Present Value of a Continuous Stream of Income An oil well generates a continuous stream of income of $60t$ thousand dollars per year, where t is the number of years that the rig has been in operation. Find the present value of this stream of income over the first 20 years at a continuous interest rate of 5%.

59. BIOMEDICAL: Drug Dosage A drug taken orally is absorbed into the bloodstream at the rate of $te^{-0.5t}$ milligrams per hour, where t is the number of hours since the drug was taken. Find the total amount of the drug absorbed during the first 5 hours.

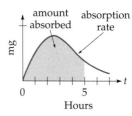

60. ENVIRONMENTAL SCIENCE: Pollution Contamination is leaking from an underground waste-disposal tank at the rate of $t \ln t$ thousand gallons per month, where t is the number of months since the leak began. Find the total leakage from the end of month 1 to the end of month 4.

61. GENERAL: Area Find the area under the curve $y = x \ln x$ and above the x-axis from $x = 1$ to $x = 2$.

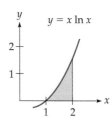

62. POLITICAL SCIENCE: Fund Raising A politician can raise campaign funds at the rate of 50$te^{-0.1t}$ thousand dollars per week during the first t weeks of a campaign. Find the average amount raised during the first 5 weeks.

63. BUSINESS: Product Recognition A company begins advertising a new product and finds that after t weeks the product is gaining customer recognition at the rate of $t^2 \ln t$ thousand customers per week (for $t \geq 1$). Find the total gain in recognition from the end of week 1 to the end of week 6.

64. GENERAL: Population The population of a town is increasing at the rate of $400te^{0.02t}$ people per year, where t is the number of years from now. Find the total gain in population during the next 5 years.

Repeated Integration by Parts

Sometimes an integral requires two or more integrations by parts. As an example, we apply integration by parts to the integral $\int x^2 e^x \, dx$.

$$\int \underbrace{x^2}_{u} \underbrace{e^x \, dx}_{dv} = \underbrace{x^2 e^x}_{u \; v} - \int \underbrace{e^x}_{v} \underbrace{2x \, dx}_{du} = x^2 e^x - 2 \int xe^x \, dx$$

$$\begin{bmatrix} u = x^2 & dv = e^x \, dx \\ du = 2x \, dx & v = \int e^x \, dx = e^x \end{bmatrix}$$

The new integral $\int xe^x \, dx$ is solved by a second integration by parts. Continuing with the previous solution, we choose new u and dv:

$$= x^2 e^x - 2 \left(\int \underbrace{xe^x \, dx}_{u \; dv} \right) \quad \begin{bmatrix} u = x & dv = e^x \, dx \\ du = dx & v = e^x \end{bmatrix}$$

$$= x^2 e^x - 2 \left(\underbrace{xe^x}_{uv} - \int \underbrace{e^x \, dx}_{v \; du} \right)$$

$$= x^2 e^x - 2(xe^x - e^x) + C$$

$$= x^2 e^x - 2xe^x + 2e^x + C$$

After reading the preceding explanation, find each integral by repeated integration by parts.

65. $\int x^2 e^{-x} \, dx$ **66.** $\int x^2 e^{2x} \, dx$

67. $\int (x+1)^2 e^x \, dx$ **68.** $\int (\ln x)^2 \, dx$

69. $\int x^2 (\ln x)^2 \, dx$ **70.** $\int x^3 e^x \, dx$

71–72. For each definite integral:

a. Evaluate it by integration by parts. (Give answer in its *exact* form.)

b. Verify your answer to part (a) using a graphing calculator.

71. $\int_0^2 x^2 e^x \, dx$ **72.** $\int_1^5 (\ln x)^2 \, dx$

Repeated Integration by Parts Using a Table

The solution to a repeated integration by parts problem can be organized in a table. As an example, we solve $\int x^2 e^{3x} \, dx$. We begin by choosing

$$u = x^2 \qquad dv = v' \, dx = e^{3x} \, dx$$

We then make a table consisting of the following three columns:

Alternating Signs	$u = x^2$ and Its Derivatives	$v' = e^{3x}$ and Its Antiderivatives
+	x^2	e^{3x}
−	$2x$	$\frac{1}{3} e^{3x}$ Using the formula
+	2	$\frac{1}{9} e^{3x}$ for $\int e^{ax} dx$
−	0	$\frac{1}{27} e^{3x}$

Stop when you get to 0

Finally, the solution is found by adding the *signed* products of the diagonals shown in the table:

$$\int x^2 e^{3x} \, dx = \frac{1}{3} x^2 e^{3x} - \frac{2}{9} xe^{3x} + \frac{2}{27} e^{3x} + C$$

After reading the preceding example, find each integral by repeated integration by parts using a table.

73. $\int x^2 e^{-x} \, dx$ **74.** $\int x^2 e^{2x} \, dx$

75. $\displaystyle\int x^3 e^{2x}\, dx$ **76.** $\displaystyle\int x^3 e^{-x}\, dx$ **77.** $\displaystyle\int (x-1)^3 e^{3x}\, dx$ **78.** $\displaystyle\int (x+1)^2 (x+2)^5\, dx$

12.2 INTEGRATION USING TABLES

Introduction

There are many techniques of integration, and only some of the most useful ones will be discussed in this book. Many of the advanced techniques lead to integration formulas, which can then be collected into a "table of integrals." In this section we will see how to find integrals by choosing an appropriate formula from such a table.*

On the inside back cover is a short table of integrals that we shall use. The formulas are grouped according to the type of integrand (for example, "Forms Involving $x^2 - a^2$"). Look at the table now (formulas 9 through 23) to see how it is organized.

Using Integral Tables

Given a particular integral, we first look for a formula that fits it exactly.

EXAMPLE 1   **INTEGRATING USING AN INTEGRAL TABLE**

Find $\displaystyle\int \frac{1}{x^2 - 4}\, dx$.

Solution The denominator $x^2 - 4$ is of the form $x^2 - a^2$ (with $a = 2$), so we look in the table of integrals under "Forms Involving $x^2 - a^2$." Formula 15,

$$\int \frac{1}{x^2 - a^2}\, dx = \frac{1}{2a} \ln\left|\frac{x-a}{x+a}\right| + C \qquad \text{Formula 15}$$

with $a = 2$ becomes our answer:

$$\int \frac{1}{x^2 - 4}\, dx = \frac{1}{4} \ln\left|\frac{x-2}{x+2}\right| + C \qquad \text{Formula 15 with } a = 2 \text{ substituted on both sides}$$

---

*Another way is to use a more advanced graphing calculator (see pages 883–884) or a computer software package such as Maple, Mathcad, or Mathematica.

Note that the expression $x^2 - a^2$ does not require that the last number be a "perfect square." For example, $x^2 - 3$ can be written $x^2 - a^2$ with $a = \sqrt{3}$.

Integral tables are useful in many applications, such as integrating a rate to find the total accumulation.

EXAMPLE 2

FINDING TOTAL SALES FROM THE SALES RATE

A company's sales rate is $\dfrac{x}{\sqrt{x+9}}$ sales per week after x weeks. Find a formula for the total sales after x weeks.

Solution To find the *total* sales $S(x)$ we integrate the *rate* of sales.

$$S(x) = \int \frac{x}{\sqrt{x+9}}\, dx$$

In the table on the inside back cover under "Forms Involving $\sqrt{ax+b}$" we find

$$\int \frac{x}{\sqrt{ax+b}}\, dx = \frac{2ax - 4b}{3a^2}\sqrt{ax+b} + C \qquad \text{Formula 13}$$

This formula with $a = 1$ and $b = 9$ gives the integral

$$S(x) = \int \frac{x}{\sqrt{x+9}}\, dx = \frac{2x - 36}{3}\sqrt{x+9} + C \qquad \text{Formula 13 with } a=1 \text{ and } b=9$$

$$= \left(\frac{2}{3}x - 12\right)\sqrt{x+9} + C \qquad \text{Simplifying}$$

To evaluate the constant C we use the fact that total sales at time $x = 0$ must be zero: $S(0) = 0$.

$$(-12)\sqrt{9} + C = 0 \qquad \begin{array}{l}(\tfrac{2}{3}x - 12)\sqrt{x+9} + C \\ \text{at } x = 0 \text{ set equal to zero}\end{array}$$

$$-36 + C = 0 \qquad \text{Simplifying}$$

Therefore, $C = 36$. Substituting this into $S(x)$ gives the formula for the total sales in the first x weeks.

$$S(x) = \left(\frac{2}{3}x - 12\right)\sqrt{x+9} + 36 \qquad \begin{array}{l} S(x) = (\tfrac{2}{3}x - 12)\sqrt{x+9} + C \\ \text{with } C = 36 \end{array}$$

EXAMPLE 3

GENETIC ENGINEERING

Under certain circumstances, the number of generations of bacteria needed to increase the frequency of a gene from 0.2 to 0.5 is

$$n = 2.5 \int_{0.2}^{0.5} \frac{1}{q^2(1-q)}\, dq$$

Find n (rounded to the nearest integer).

Solution Formula 12 (inside back cover) integrates a similar-looking fraction.

$$\int \frac{1}{x^2(ax+b)}\, dx = -\frac{1}{b}\left(\frac{1}{x} + \frac{a}{b}\ln\left|\frac{x}{ax+b}\right|\right) + C \qquad \text{Formula 12}$$

To make $(ax + b)$ into $(1 - x)$, we take $a = -1$ and $b = 1$, so the left-hand side of the formula becomes

$$\int \frac{1}{x^2(-x+1)}\, dx \quad \text{or} \quad \int \frac{1}{x^2(1-x)}\, dx \qquad \begin{array}{l}\text{From formula 12}\\ \text{with } a=-1,\ b=1\end{array}$$

Except for replacing x by q, this is the same as our integral. Therefore, the indefinite integral is found by formula 12 with $a = -1$ and $b = 1$ (which we express in the variable q).

$$\int \frac{1}{q^2(1-q)}\, dq = -\left(\frac{1}{q} - \ln\left|\frac{q}{1-q}\right|\right) + C \qquad \begin{array}{l}\text{Formula 12 with}\\ a=-1 \text{ and } b=1\end{array}$$

For the *definite* integral from 0.2 to 0.5, we evaluate and subtract.

$$-\left(\frac{1}{0.5} - \ln\left|\frac{0.5}{1-0.5}\right|\right) - \left[-\left(\frac{1}{0.2} - \ln\left|\frac{0.2}{1-0.2}\right|\right)\right]$$

$$\underbrace{\qquad\qquad\qquad\qquad}_{\text{Evaluation at } q=0.5} \qquad \underbrace{\qquad\qquad\qquad\qquad}_{\text{Evaluation at } q=0.2}$$

$$= -(2 - \underbrace{\ln 1}_{0}) + \left(5 - \underbrace{\ln \frac{0.2}{0.8}}_{\ln 0.25}\right) = -2 + 5 - \underbrace{\ln 0.25}_{\text{Using a calculator}} \approx 4.39$$

We multiply this by the 2.5 in front of the original integral.

$$(2.5)(4.39) \approx 10.98$$

Therefore, 11 generations are needed to raise the gene frequency from 0.2 to 0.5.

Graphing Calculator Exploration

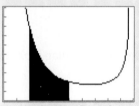

on [0, 1] by [0, 100]
(shaded from 0.2 to 0.5)

a. Verify the answer to Example 3 by finding the definite integral of
$$y = \frac{2.5}{x^2(1-x)}$$ from 0.2 to 0.5.

b. From the graph of the function shown on the left, which would require more generations: increasing the gene frequency from 0.1 to 0.2 or from 0.5 to 0.6?

c. Can you think of a reason for this? [*Hint:* Genes reproduce from other similar genes.]

Sometimes a substitution is needed to transform a formula to fit a given integral. In such cases both the x and the dx must be transformed. A few examples will make the method clear.

EXAMPLE 4 USING A TABLE WITH A SUBSTITUTION

Find $\int \dfrac{x}{\sqrt{x^4 + 1}}\, dx$.

Solution The table of integrals on the inside back cover has no formula involving x^4. However, $x^4 = (x^2)^2$, so a formula involving x^2, along with a substitution, might work. Formula 18 looks promising:

$$\int \frac{1}{\sqrt{x^2 \pm a^2}}\, dx = \ln\left|x + \sqrt{x^2 \pm a^2}\right| + C \quad \text{The } \pm \text{ means: use either the upper sign or the lower sign on both sides}$$

$$\int \frac{1}{\sqrt{x^2 + 1}}\, dx = \ln\left|x + \sqrt{x^2 + 1}\right| + C \quad \text{Formula 18 with } a = 1 \text{ and the upper sign}$$

With the substitution

$$x = z^2$$
$$dx = 2z\, dz \qquad \text{Differential of } x = z^2$$

this becomes

$$\int \frac{1}{\sqrt{z^4 + 1}} 2z\, dz = \ln\left|z^2 + \sqrt{z^4 + 1}\right| + C \qquad \begin{array}{l}\text{Above formula with}\\ x = z^2 \text{ and } dx = 2z\, dz\end{array}$$

Dividing by 2 and replacing z by x gives the integral that we wanted:

$$\int \frac{x}{\sqrt{x^4 + 1}}\, dx = \frac{1}{2}\ln\left(x^2 + \sqrt{x^4 + 1}\right) + C \qquad \begin{array}{l}\text{Dropping the absolute}\\ \text{value bars since}\\ x^2 + \sqrt{x^4 + 1} \text{ is positive}\end{array}$$

Given a particular integral, how do we choose a formula?

How To Choose a Formula

Find a formula that matches the *most complicated part* of the integral, making appropriate substitutions to change the formula into the given integral.

For instance, in Example 4 we matched the $\sqrt{x^4 + 1}$ in the given integral with the $\sqrt{x^2 \pm a^2}$ in the formula, and the rest of the integral followed from the differential.

Practice Problem Find $\int \dfrac{t}{9t^4 - 1}\, dt.$ [*Hint:* Use formula 15 with $x = 3t^2.$]

➤ Solution on page 884

EXAMPLE 5   USING A TABLE WITH A SUBSTITUTION

Find $\int \dfrac{e^{-2t}}{e^{-t} + 1}\, dt.$

Solution Looking in the table of integrals under "Forms Involving e^{ax} and $\ln x$," none of the formulas looks anything like this integral. However, replacing e^{-t} by x would make the denominator of our integral into $x + 1$, so formula 9 might help. This formula with $a = 1$ and $b = 1$ is

$$\int \frac{x}{x+1} dx = x - \ln|x+1| + C \qquad \text{Formula 9 with } a=1 \text{ and } b=1$$

With the substitution

$$x = e^{-t}$$
$$dx = -e^{-t} dt \qquad \text{Differential of } x = e^{-t}$$

formula 9 becomes

$$\int \frac{e^{-t}}{e^{-t}+1}(-e^{-t}) dt = e^{-t} - \ln|e^{-t}+1| + C$$

or

$$-\int \frac{e^{-2t}}{e^{-t}+1} dt = e^{-t} - \ln(e^{-t}+1) + C$$

Except for the negative sign, this is the given integral. Multiplying through by -1 gives the final answer.

$$\int \frac{e^{-2t}}{e^{-t}+1} dt = -e^{-t} + \ln(e^{-t}+1) + C$$

Reduction Formulas

Sometimes we must apply a formula several times to simplify an integral in stages.

EXAMPLE 6

USING A REDUCTION FORMULA

Find $\int x^3 e^{-x} dx$.

Solution In the integral table, we find formula 21.

$$\int x^n e^{ax} dx = \frac{1}{a} x^n e^{ax} - \frac{n}{a} \int x^{n-1} e^{ax} dx \qquad \text{Formula 21. With } n=3 \text{ and } a=-1, \text{ the left side fits our integral}$$

The right-hand side of this formula involves a new integral, but with a *lower* power of x. We will apply formula 21 several times, each time reducing the power of x until we eliminate it completely. Applying formula 21 with $n = 3$ and $a = -1$,

$$\int x^3 e^{-x}\, dx = -x^3 e^{-x} + 3 \int x^2 e^{-x}\, dx \qquad \text{After one application}$$

The power has been reduced

$$= -x^3 e^{-x} + 3\left(-x^2 e^{-x} + 2 \int x^1 e^{-x}\, dx\right) \qquad \begin{array}{l}\text{Applying formula 21} \\ \text{again (now with } n = 2) \\ \text{to the last integral above}\end{array}$$

$$= -x^3 e^{-x} - 3x^2 e^{-x} + 6 \int x^1 e^{-x}\, dx \qquad \text{Multiplying out}$$

$$= -x^3 e^{-x} - 3x^2 e^{-x} + 6\left(-xe^{-x} + \underbrace{\int x^0 e^{-x}\, dx}_{1}\right) \qquad \begin{array}{l}\text{Using formula 21 a third time} \\ \text{(now with } n = 1)\end{array}$$

$$= -x^3 e^{-x} - 3x^2 e^{-x} - 6xe^{-x} + 6 \int e^{-x}\, dx \qquad \begin{array}{l}\text{Now solve this last integral by the} \\ \text{formula } \int e^{ax}\, dx = \frac{1}{a} e^{ax} + C\end{array}$$

$$= -x^3 e^{-x} - 3x^2 e^{-x} - 6xe^{-x} - 6e^{-x} + C \qquad \begin{array}{l}\text{The solution, after three applications of formula 21}\end{array}$$

$$= -e^{-x}(x^3 + 3x^2 + 6x + 6) + C \qquad \text{Factoring}$$

We used formula 21 three times, reducing the x^3 in steps, first down to x^2, then to x^1, and finally to $x^0 = 1$, at which point we could solve the integral easily. If the power of x in the integral had been higher, more applications of formula 21 would have been necessary. Formulas such as 21 and 22 are called *reduction formulas*, since they express an integral in terms of a similar integral but with a smaller power of x.

Graphing Calculator Exploration

Some advanced graphing calculators can find *indefinite* integrals. For example, the Texas Instruments *TI-89* graphing calculator finds the integrals in Examples 5 and 6 as follows.

For Example 5

For Example 6

With some algebra, you can verify that these answers agree with those found in Examples 5 and 6, even though they do not look the same. (Notice that this calculator omits the arbitrary constant of integration.)

12.2 Section Summary

To find a formula that "fits" a given integral, we look for the formula whose left-hand side most closely matches the most complicated part of the integral. Then we choose constants (and possibly a substitution) to make the formula fit exactly. Although we have been using a very brief table, the technique is the same with a more extensive table. Many integral tables have been published, some of which are book-length, containing several thousand formulas.*

▶ Solution to Practice Problem

Formula 15 with the substitution $x = 3t^2$, $dx = 6t\,dt$, and $a = 1$ becomes

$$\int \frac{1}{9t^4 - 1} 6t\,dt = \frac{1}{2} \ln \left| \frac{3t^2 - 1}{3t^2 + 1} \right| + C$$

Dividing each side by 6 gives the answer:

$$\int \frac{1}{9t^4 - 1} dt = \frac{1}{12} \ln \left| \frac{3t^2 - 1}{3t^2 + 1} \right| + C$$

---

* A useful table of integrals containing more than 400 formulas is found in *CRC Standard Mathematical Tables and Formulae*, CRC Press, Boca Raton, Florida.

12.2 Exercises

For each integral, state the number of the integration formula (from the inside back cover) and the values of the constants a and b so that the formula fits the integral. (Do not evaluate the integral.)

1. $\int \dfrac{1}{x^2(5x-1)} dx$
2. $\int \dfrac{x}{2x-3} dx$
3. $\int \dfrac{1}{x\sqrt{-x+7}} dx$
4. $\int \dfrac{x}{\sqrt{-2x+1}} dx$
5. $\int \dfrac{x}{1-x} dx$
6. $\int \dfrac{1}{x\sqrt{1-4x}} dx$

Find each integral by using the integral table on the inside back cover.

7. $\int \dfrac{1}{9-x^2} dx$
   [Hint: Use formula 16 with $a = 3$.]

8. $\int \dfrac{1}{x^2-25} dx$
   [Hint: Use formula 15 with $a = 5$.]

9. $\int \dfrac{1}{x^2(2x+1)} dx$
   [Hint: Use formula 12 with $a = 2$, $b = 1$.]

10. $\int \dfrac{x}{x+2} dx$
    [Hint: Use formula 9 with $a = 1$, $b = 2$.]

11. $\int \dfrac{x}{1-x} dx$ [Hint: Use formula 9.]

12. $\int \dfrac{x}{\sqrt{1-x}} dx$ [Hint: Use formula 13.]

13. $\int \dfrac{1}{(2x+1)(x+1)} dx$

14. $\int \dfrac{x}{(x+1)(x+2)} dx$

15. $\int \sqrt{x^2-4} \, dx$

16. $\int \dfrac{1}{\sqrt{x^2-1}} dx$

17. $\int \dfrac{1}{z\sqrt{1-z^2}} dz$

18. $\int \dfrac{\sqrt{4+z^2}}{z} dz$

19. $\int x^3 e^{2x} dx$

20. $\int x^{99} \ln x \, dx$

21. $\int x^{-101} \ln x \, dx$

22. $\int (\ln x)^2 \, dx$

23. $\int \dfrac{1}{x(x+3)} dx$

24. $\int \dfrac{1}{x(x-3)} dx$

25. $\int \dfrac{z}{z^4-4} dz$

26. $\int \dfrac{z}{9-z^4} dz$

27. $\int \sqrt{9x^2+16} \, dx$

28. $\int \dfrac{1}{\sqrt{16x^2-9}} dx$

29. $\int \dfrac{1}{\sqrt{4-e^{2t}}} dt$

30. $\int \dfrac{e^t}{9-e^{2t}} dt$

31. $\int \dfrac{e^t}{e^{2t}-1} dt$

32. $\int \dfrac{e^{2t}}{1-e^t} dt$

33. $\int \dfrac{x^3}{\sqrt{x^8-1}} dx$

34. $\int x^2\sqrt{x^6+1} \, dx$

35. $\int \dfrac{1}{x\sqrt{x^3+1}} dx$

36. $\int \dfrac{\sqrt{1-x^6}}{x} dx$

37. $\int \dfrac{e^t}{(e^t-1)(e^t+1)} dt$

38. $\int \dfrac{e^{2t}}{(e^t-1)(e^t+1)} dt$

39. $\int x \, e^{x/2} \, dx$

40. $\int \dfrac{x}{e^x} dx$

41. $\int \dfrac{1}{e^{-x}+4} dx$

42. $\int \dfrac{1}{\sqrt{e^{-x}+4}} dx$

For each definite integral:
a. Evaluate it using the table of integrals on the inside back cover. (Leave answers in *exact* form.)
b. Use a graphing calculator to verify your answer to part (a).

43. $\int_4^5 \sqrt{x^2-16} \, dx$

44. $\int_0^4 \dfrac{1}{\sqrt{x^2+9}} dx$

45. $\int_2^3 \dfrac{1}{x^2-1} dx$

46. $\int_2^4 \dfrac{1}{1-x^2} dx$

47. $\int_3^5 \dfrac{\sqrt{25-x^2}}{x} dx$

48. $\int_3^4 \dfrac{1}{x\sqrt{25-x^2}} dx$

Find each integral by whatever means are necessary (either substitution or tables).

49. $\int \dfrac{1}{2x+6}\,dx$ **50.** $\int \dfrac{x}{x^2-4}\,dx$

51. $\int \dfrac{x}{2x+6}\,dx$ **52.** $\int \dfrac{1}{4-x^2}\,dx$

53. $\int x\sqrt{1-x^2}\,dx$ **54.** $\int \dfrac{x}{\sqrt{1-x^2}}\,dx$

55. $\int \dfrac{\sqrt{1-x^2}}{x}\,dx$ **56.** $\int \dfrac{1}{\sqrt{x^2-1}}\,dx$

57. $\int \dfrac{x-1}{(3x+1)(x+1)}\,dx$

58. $\int \dfrac{x-1}{x^2(x+1)}\,dx$ **59.** $\int \dfrac{x+1}{x\sqrt{1+x^2}}\,dx$

60. $\int \dfrac{x-1}{x\sqrt{x^2+4}}\,dx$

61. $\int \dfrac{x+1}{x-1}\,dx$ [*Hint:* After separating into two integrals, find one by a formula and the other by a substitution.]

62. $\int \dfrac{x+1}{\sqrt{x^2+1}}\,dx$ [*Hint:* After separating into two integrals, find one by a formula and the other by a substitution.]

Find each integral. [*Hint:* Separate each integral into two integrals, using the fact that the numerator is a sum or difference, and find the two integrals by two different formulas.]

APPLIED EXERCISES

63. BUSINESS: Total Sales A company's sales rate is $x^2 e^{-x}$ million sales per month after x months. Find a formula for the total sales in the first x months. [*Hint:* Integrate the sales rate to find the total sales and determine the constant C so that total sales are zero at time $x=0$.]

64. GENERAL: Population The population of a city is expected to grow at the rate of $x/\sqrt{x+9}$ thousand people per year after x years. Find the total change in population from year 0 to year 27.

65. BIOMEDICAL: Gene Frequency Under certain circumstances, the number of generations necessary to increase the frequency of a gene from 0.1 to 0.3 is

$$n = 3\int_{0.1}^{0.3} \dfrac{1}{q^2(1-q)}\,dq$$

Find n (rounded to the nearest integer).

66. BEHAVIORAL SCIENCE A subject in a psychology experiment gives responses at the rate of $t/\sqrt{t+1}$ correct answers per minute after t minutes.

a. Find the total number of correct responses from time $t=0$ to time $t=15$.

 b. Verify your answer to part (a) using a graphing calculator.

67. BUSINESS: Cost The marginal cost function for a computer chip manufacturer is $MC(x) = 1/\sqrt{x^2+1}$, and fixed costs are $\$2000$. Find the cost function.

68. SOCIAL SCIENCE: Employment An urban job placement center estimates that the number of residents seeking employment t years from now will be $t/(2t+4)$ million people.

a. Find the average number of job seekers during the period $t=0$ to $t=10$.

 b. Verify your answer to part (a) using a graphing calculator.

12.3 IMPROPER INTEGRALS

Introduction

In this section we define integrals over intervals that are infinite in length. Such "improper" integrals have many applications, such as in the Application Preview on page 863 at the beginning of this chapter.

Limits as x Approaches ± ∞

The notation $x \to \infty$ ("x approaches infinity") means that x takes on arbitrarily large values.

> **Limits Approaching ± ∞**
>
> $x \to \infty$ means: x takes values arbitrarily far to the *right* on the number line.
>
> $x \to -\infty$ means: x takes values arbitrarily far to the *left* on the number line.

Evaluating limits as x approaches positive or negative infinity is simply a matter of thinking about large and small numbers. The reciprocal of a large number is a small number. For example,

$$\frac{1}{1,000,000} = 0.000001 \qquad \text{One over a million is one one-millionth}$$

Similarly:

$$\lim_{x \to \infty} \frac{1}{x^2} = 0$$

$$\lim_{x \to \infty} \frac{1}{e^x} = \lim_{x \to \infty} e^{-x} = 0$$

As the denominator approaches infinity (with the numerator constant), the value approaches zero

These examples illustrate the following general rules.

$$\lim_{x \to \infty} \frac{1}{x^n} = 0 \quad (n > 0)$$ As x approaches infinity, 1 over x to a *positive* power approaches zero

$$\lim_{x \to \infty} e^{-ax} = 0 \quad (a > 0)$$ As x approaches infinity, e to a *negative* number times x approaches zero

EXAMPLE 1 EVALUATING LIMITS

a. $\lim_{b \to \infty} \dfrac{1}{b^2} = 0$ Using the first rule in the box above

b. $\lim_{b \to \infty} \left(3 - \dfrac{1}{b}\right) = 3$ Because the $\dfrac{1}{b}$ approaches zero

c. $\lim_{b \to \infty} (e^{-2b} - 5) = -5$ Because the e^{-2b} approaches zero

Similar rules hold for x approaching *negative* infinity.

$$\lim_{x \to -\infty} \frac{1}{x^n} = 0 \quad \text{(for integer } n > 0\text{)}$$ As x approaches *negative* infinity, 1 over x to a positive integer approaches zero

$$\lim_{x \to -\infty} e^{ax} = 0 \quad (a > 0)$$ As x approaches negative infinity, e to a positive number times x approaches zero (because the exponent is approaching $-\infty$)

Graphing Calculator Exploration

a. Define $y_1 = 1/x^2$ and $y_2 = e^{-x}$ and use the TABLE feature of your calculator to evaluate these functions at x-values like 1, 2, 3, 5, 10, 100, and 1000. Which of the limit rules above do the results verify? [*Note:* An answer such as 3E ⁻5 means $3 \cdot 10^{-5} = 0.00003$.]

b. Change y_2 to be $y_2 = e^x$ and use the TABLE to evaluate y_1 and y_2 at negative x-values like $-1, -2, -3, -5, -10,$ and -100. Which limit rules do the results verify?

Some quantities become arbitrarily large, and so do not have limits. (For a limit to exist, it must be *finite*.)

The following limits do not exist:

$\lim_{x \to \infty} x^n$	$(n > 0)$	As x approaches infinity, x to a positive power has no limit
$\lim_{x \to \infty} e^{ax}$	$(a > 0)$	As x approaches infinity, e to a positive number times x has no limit
$\lim_{x \to \infty} \ln x$		As x approaches infinity, the natural logarithm of x has no limit

---

EXAMPLE 2

FINDING WHETHER A LIMIT EXISTS

a. $\lim_{b \to \infty} b^3$ does not exist.   Because b^3 becomes arbitrarily large as b approaches infinity

b. $\lim_{b \to \infty} (\sqrt{b} - 1)$ does not exist.   Because $\sqrt{b}$ becomes arbitrarily large as b approaches infinity

---

Practice Problem 1

Evaluate the following limits (if they exist).

a. $\lim_{b \to \infty}\left(1 - \dfrac{1}{b}\right)$ b. $\lim_{b \to \infty}\left(\sqrt[3]{b} + 3\right)$

▶ Solutions on page 897

Improper Integrals

A definite integral over an interval of infinite length is an *improper integral*. As a first example, we evaluate the improper integral $\int_{1}^{\infty} \dfrac{1}{x^2}\, dx$, which gives the area under the curve $y = \dfrac{1}{x^2}$ from $x = 1$ extending arbitrarily far to the right.

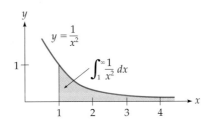

EXAMPLE 3

EVALUATING AN IMPROPER INTEGRAL

Evaluate $\int_{1}^{\infty} \frac{1}{x^2} dx$.

Solution To integrate to infinity, we first integrate over a *finite* interval, from 1 to some number b (think of b as some very large number), and then take the limit as b approaches ∞. First integrate from 1 to b.

$$\int_{1}^{b} \frac{1}{x^2} dx = \int_{1}^{b} x^{-2} dx = (-x^{-1})\Big|_{1}^{b} = \left(-\frac{1}{x}\right)\Big|_{1}^{b} \quad \text{Using the Power Rule}$$

$$= \underbrace{-\frac{1}{b}}_{\text{at } x = b} - \underbrace{\left(-\frac{1}{1}\right)}_{\text{at } x = 1} = -\frac{1}{b} + 1 \quad \text{Evaluating and simplifying}$$

Then take the limit of this answer as $b \to \infty$.

$$\lim_{b \to \infty} \left(-\frac{1}{b} + 1\right) = 1 \quad \text{Limit as } b \to \infty \text{ (the } \frac{1}{b} \text{ approaches zero)}$$

This gives the answer:

$$\int_{1}^{\infty} \frac{1}{x^2} dx = 1 \quad \text{Integral from 1 to } \infty \text{ equals 1}$$

Since the limit exists, we say that the improper integral is *convergent*. Geometrically, this procedure amounts to finding the area under the curve from 1 to some number b, shown on the left below, and then letting $b \to \infty$ to find the area arbitrarily far to the right.

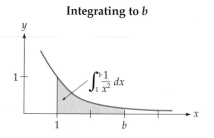
Integrating to b

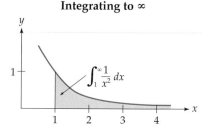
Integrating to ∞

Improper Integrals—Integrating to ∞

If f is continuous and nonnegative for $x \geq a$, we define

$$\int_a^\infty f(x)\,dx = \lim_{b \to \infty} \int_a^b f(x)\,dx$$

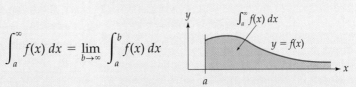

provided that the limit exists. The improper integral is said to be *convergent* if the limit exists, and *divergent* if the limit does not exist.

It is possible to define improper integrals for functions that take negative values, and even for discontinuous functions, but we shall not do so in this book, since most applications involve functions that are positive and continuous.

Practice Problem 2   Evaluate $\displaystyle\int_2^\infty \frac{1}{x^2}\,dx.$ ➤ Solution on page 897

EXAMPLE 4

FINDING WHETHER AN INTEGRAL DIVERGES

Evaluate $\displaystyle\int_1^\infty \frac{1}{\sqrt{x}}\,dx.$

Solution   Integrating up to b:

$$\int_1^b \frac{1}{\sqrt{x}}\,dx = \underbrace{\int_1^b x^{-1/2}\,dx \;=\; 2\cdot x^{1/2}\Big|_1^b}_{\text{Integrating by the Power Rule}} = \underbrace{2\sqrt{b} - 2\sqrt{1} = 2\sqrt{b} - 2}_{\text{Evaluating}}$$

Letting b approach infinity:

$$\lim_{b \to \infty} \left(2\sqrt{b} - 2\right) \text{ does not exist} \qquad \text{Because } \sqrt{b} \text{ becomes infinite as } b \to \infty$$

Therefore,

$$\int_1^\infty \frac{1}{\sqrt{x}}\,dx \text{ is } divergent \qquad \text{The integral cannot be evaluated}$$

Notice from Examples 3 and 4 that $\int_{1}^{\infty} \frac{1}{x^2} \, dx$ is convergent (its value is 1), whereas $\int_{1}^{\infty} \frac{1}{\sqrt{x}} \, dx$ is divergent (it does not have a finite value), as illustrated in the following diagrams.

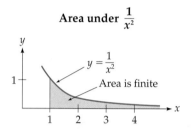

Area under $\frac{1}{x^2}$

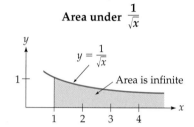

Area under $\frac{1}{\sqrt{x}}$

Intuitively, the areas differ because the curve $\frac{1}{x^2}$ lies much closer to the x-axis than does the curve $\frac{1}{\sqrt{x}}$ for large values of x, as shown above, and so has a smaller area under it.

Spreadsheet Exploration

The following spreadsheet* allows us to numerically "see" that one of these integrals converges while the other diverges. From Example 3 on page 890 we know that $\int_{1}^{\infty} \frac{1}{x^2} \, dx = \lim_{b \to \infty} \left(1 - \frac{1}{b}\right)$, while from Example 4 on page 891 we have $\int_{1}^{\infty} \frac{1}{\sqrt{x}} \, dx = \lim_{b \to \infty} (2\sqrt{b} - 2)$. The increasingly large values of b shown in column **A** are used to evaluate $1 - \frac{1}{b}$ in column **B** and $2\sqrt{b} - 2$ in column **C**.

* See the Preface for how to obtain this and other Excel spreadsheets.

	A	B	C
		=2*$A7^(1/2)-2	
	b	Integral of x^(-2)	Integral of x^(-1/2)
1	1	0	0.000
2	5	0.8	2.472
3	10	0.9	4.325
4	50	0.98	12.142
5	100	0.99	18.000
6	500	0.998	42.721
7	1000	0.999	61.246
8	5000	0.9998	139.421
9	10000	0.9999	198.000
10	50000	0.99998	445.214
11	100000	0.99999	630.456
12	500000	0.999998	1412.214

Is it clear which of these integrals (areas) is converging and which is diverging?

We may combine the two steps of integrating up to b and letting $b \to \infty$ into a single line. We show how to do this by evaluating the integral from Example 3 again, but more briefly.

$$\int_1^\infty x^{-2}\,dx = \lim_{b\to\infty}\int_1^b x^{-2}\,dx = \lim_{b\to\infty}(-x^{-1})\Big|_1^b = \lim_{b\to\infty}\left[-\frac{1}{b} - \left(-\frac{1}{1}\right)\right] = 1$$

Integrating; now use $x^{-1} = \dfrac{1}{x}$

Approaches 0

Permanent Endowments

Funds that generate steady income forever are called *permanent endowments*.

EXAMPLE 5

FINDING THE SIZE OF A PERMANENT ENDOWMENT

In the Application Preview on page 863 we found that the size of the fund necessary to generate $2000 annually forever (at 5% interest compounded continuously) is $\int_0^\infty 2000 e^{-0.05t}\, dt$. Find the size of this permanent endowment by evaluating the integral.

Solution

$$\int_0^\infty \underbrace{2000}_{\text{Annual income}} \underbrace{e^{-0.05t}}_{\substack{\text{Continuous} \\ \text{interest rate}}} dt \;=\; \lim_{b \to \infty} \left(\underbrace{2000}_{\text{Moved outside}} \int_0^b e^{-0.05t}\, dt \right)$$

$$= \lim_{b \to \infty} \left[2000(-20) e^{-0.05t} \Big|_0^b \right] \qquad \text{Integrating by } \int e^{ax}\, dx = \frac{1}{a} e^{ax}$$

$$= \lim_{b \to \infty} (\underbrace{-40{,}000 e^{-0.05b}}_{\text{Approaches } 0} + \underbrace{40{,}000 e^0}_{1}) = 40{,}000$$

Therefore, the size of the permanent endowment that will pay the $2000 annual maintenance forever is $40,000.

Permanent endowments are used to estimate the ultimate cost of anything that requires continuous long-term funding, from buildings to government agencies to toxic waste sites.

Finding Improper Integrals Using Substitutions

Solving an improper integral may require a substitution. In such cases we apply the substitution not only to the integrand but also to the differential and the upper and lower limits of integration.

EXAMPLE 6

FINDING AN IMPROPER INTEGRAL USING A SUBSTITUTION

Evaluate $\int_2^\infty \dfrac{x}{(x^2+1)^2}\, dx$.

Solution We use the substitution $u = x^2 + 1$, so $du = 2x\,dx$, requiring multiplication by 2 and by $\frac{1}{2}$. Notice how the substitution changes the limits.

$$\int_2^\infty \frac{x}{(x^2+1)^2}\,dx = \frac{1}{2}\int_2^\infty \frac{2x}{(x^2+1)^2}\,dx = \frac{1}{2}\int_5^\infty \frac{du}{u^2} = \frac{1}{2}\lim_{b\to\infty}\int_5^b u^{-2}\,du$$

as $x \to \infty$, $u = x^2 + 1 \to \infty$

$\begin{bmatrix} u = x^2 + 1 \\ du = 2x\,dx \end{bmatrix}$ $u = 2^2 + 1 = 5$

$$= \frac{1}{2}\lim_{b\to\infty}[-u^{-1}]\bigg|_5^b = \frac{1}{2}\lim_{b\to\infty}\left[-\frac{1}{b} - \left(-\frac{1}{5}\right)\right] = \frac{1}{2}\cdot\frac{1}{5} = \frac{1}{10}$$

Integrating   Approaches 0

To integrate over an interval that extends arbitrarily far to the *left*, we again integrate over a finite interval and then take the limit.

Improper Integrals — Integrating to $-\infty$

If f is continuous and nonnegative for $x \leq b$, we define

$$\int_{-\infty}^b f(x)\,dx = \lim_{a\to-\infty}\int_a^b f(x)\,dx$$

provided that the limit exists. The improper integral is *convergent* if the limit exists, and *divergent* if the limit does not exist.

To integrate over the *entire x*-axis, from $-\infty$ to ∞, we use two integrals, one from $-\infty$ to 0, and the other from 0 to ∞, and then add the results.

896 CHAPTER 12 INTEGRATION TECHNIQUES AND DIFFERENTIAL EQUATIONS

Improper Integrals—Integrating from $-\infty$ to ∞

If f is continuous and nonnegative for *all* values of x, we define

$$\int_{-\infty}^{\infty} f(x)\,dx = \lim_{a \to -\infty} \int_{a}^{0} f(x)\,dx + \lim_{b \to \infty} \int_{0}^{b} f(x)\,dx$$

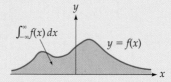

The improper integral is *convergent* if both limits exist, and *divergent* if either limit does not exist.

EXAMPLE 7

INTEGRATING TO $-\infty$

Evaluate $\int_{-\infty}^{3} 4e^{2x}\,dx$.

Solution

$$\int_{-\infty}^{3} 4e^{2x}\,dx = \lim_{a \to -\infty} \int_{a}^{3} 4e^{2x}\,dx$$

$$= \lim_{a \to -\infty} \left[4 \cdot \frac{1}{2} \cdot e^{2x} \Big|_{a}^{3} \right] \quad \text{Integrating}$$

$$= \lim_{a \to -\infty} \underbrace{(2e^{6} - 2e^{2a})}_{\text{Approaches 0 as } a \to -\infty} = 2e^{6}$$

Practice Problem 3 Evaluate the improper integral $\int_{-\infty}^{1} 12e^{3x}\,dx$. ➤ Solution on next page

---

12.3 Section Summary

A definite integral in which one or both limits of integration are infinite is called an "improper" integral. The improper integral of a continuous nonnegative function is defined as the *limit* of the integral over a finite interval. The integral is *convergent* if the limit exists, and *divergent* otherwise. This idea of dealing with the infinite by "dropping back to the finite and then taking the limit" is a standard technique in mathematics.

Several particular limits are helpful in evaluating improper integrals.

Approaching Infinity

$$\lim_{x \to \infty} \frac{1}{x^n} = 0 \quad (n > 0)$$

$$\lim_{x \to \infty} e^{-ax} = 0 \quad (a > 0)$$

Approaching Negative Infinity

$$\lim_{x \to -\infty} \frac{1}{x^n} = 0 \quad \text{(for integer } n > 0\text{)}$$

$$\lim_{x \to -\infty} e^{ax} = 0 \quad (a > 0)$$

Improper integrals give continuous sums over infinite intervals. For example, the total future output of an oil well can be found by integrating the production rate out to infinity, and the value of an asset that lasts indefinitely (such as land) can be found by integrating the present value of the future income out to infinity. Even when infinite duration is unrealistic, ∞ is used to represent "long-term behavior."

▶ Solutions to Practice Problems

1. a. $\lim_{b \to \infty} \left(1 - \frac{1}{b}\right) = 1 \quad$ because $\frac{1}{b}$ approaches 0

b. $\lim_{b \to \infty} \left(\sqrt[3]{b} + 3\right) \quad$ does not exist $\quad$ because $\sqrt[3]{b}$ gets arbitrarily large

2. $\int_2^b x^{-2}\, dx = (-x^{-1}) \Big|_2^b = -\frac{1}{b} - \left(-\frac{1}{2}\right) = \frac{1}{2} - \frac{1}{b}$

$\lim_{b \to \infty} \left(\frac{1}{2} - \frac{1}{b}\right) = \frac{1}{2}$

Therefore, $\int_2^\infty \frac{1}{x^2}\, dx = \frac{1}{2}$

3. $\int_{-\infty}^1 12e^{3x}\, dx = \lim_{a \to -\infty} \int_a^1 12e^{3x}\, dx = \lim_{a \to -\infty} \left[12 \cdot \frac{1}{3} e^{3x}\right]_a^1 = \lim_{a \to -\infty} (4e^3 - 4e^{3a}) = 4e^3$

12.3 Exercises

Evaluate each limit (or state that it does not exist).

1. $\lim\limits_{x \to \infty} \dfrac{1}{x^2}$

2. $\lim\limits_{b \to \infty} \left(\dfrac{1}{\sqrt{b}} - 8 \right)$

3. $\lim\limits_{b \to \infty} (1 - 2e^{-5b})$

4. $\lim\limits_{b \to \infty} (3e^{3b} - 4)$

5. $\lim\limits_{x \to \infty} (2 - e^{x/2})$

6. $\lim\limits_{x \to \infty} (1 - e^{-x/3})$

7. $\lim\limits_{b \to \infty} (3 + \ln b)$

8. $\lim\limits_{b \to \infty} (2 - \ln b^2)$

Evaluate each improper integral or state that it is divergent.

9. $\int_{1}^{\infty} \dfrac{1}{x^3}\, dx$

10. $\int_{1}^{\infty} \dfrac{1}{\sqrt[3]{x^4}}\, dx$

11. $\int_{2}^{\infty} 3x^{-4}\, dx$

12. $\int_{0}^{\infty} e^{-t}\, dt$

13. $\int_{2}^{\infty} \dfrac{1}{x}\, dx$

14. $\int_{1}^{\infty} \dfrac{1}{x^{0.99}}\, dx$

15. $\int_{1}^{\infty} \dfrac{1}{x^{1.01}}\, dx$

16. $\int_{10}^{\infty} e^{-x/5}\, dx$

17. $\int_{0}^{\infty} e^{-0.05t}\, dt$

18. $\int_{0}^{\infty} e^{0.01t}\, dt$

19. $\int_{5}^{\infty} \dfrac{1}{(x-4)^3}\, dx$

20. $\int_{0}^{\infty} \dfrac{x}{(x^2+1)^2}\, dx$

21. $\int_{0}^{\infty} \dfrac{x}{x^2+1}\, dx$

22. $\int_{0}^{\infty} \dfrac{x^2}{x^3+1}\, dx$

23. $\int_{0}^{\infty} x^2 e^{-x^3}\, dx$

24. $\int_{e}^{\infty} (\ln x)^{-2} \dfrac{1}{x}\, dx$

25. $\int_{-\infty}^{0} e^{3x}\, dx$

26. $\int_{-\infty}^{0} \dfrac{x^4}{(x^5-1)^2}\, dx$

27. $\int_{-\infty}^{1} \dfrac{1}{2-x}\, dx$

28. $\int_{-\infty}^{0} \dfrac{1}{1-x}\, dx$

29. $\int_{-\infty}^{\infty} \dfrac{e^x}{(1+e^x)^2}\, dx$

30. $\int_{-\infty}^{\infty} \dfrac{e^{-x}}{(1+e^{-x})^3}\, dx$

31. $\int_{-\infty}^{\infty} \dfrac{e^x}{1+e^x}\, dx$

32. $\int_{-\infty}^{\infty} \dfrac{e^{-x}}{1+e^{-x}}\, dx$

33. Use a graphing calculator to estimate the improper integrals $\int_{0}^{\infty} e^{\sqrt{x}}\, dx$ and $\int_{0}^{\infty} e^{-x^2}\, dx$ (if they converge) as follows:
   a. Define y_1 to be the definite integral (using FnInt) of $e^{\sqrt{x}}$ from 0 to x.
   b. Define y_2 to be the definite integral of e^{-x^2} from 0 to x.
   c. y_1 and y_2 then give the *areas* under these curves out to any number x. Make a TABLE of values of y_1 and y_2 for x-values such as 1, 10, 100, and 500. Which integral converges (and to what number, approximated to five decimal places) and which diverges?

34. Use a graphing calculator to estimate the improper integrals $\int_{0}^{\infty} \dfrac{1}{x^2+1}\, dx$ and $\int_{0}^{\infty} \dfrac{1}{\sqrt{x}+1}\, dx$ (if they converge) as follows:
   a. Define y_1 to be the definite integral (using FnInt) of $\dfrac{1}{x^2+1}$ from 0 to x.
   b. Define y_2 to be the definite integral of $\dfrac{1}{\sqrt{x}+1}$ from 0 to x.
   c. y_1 and y_2 then give the *areas* under these curves out to any number x. Make a TABLE of values of y_1 and y_2 for x-values such as 1, 10, 100, 500, and 10,000. Which integral converges (and to what number, approximated to five decimal places) and which diverges?

APPLIED EXERCISES

35. **GENERAL: Permanent Endowments** Find the size of the permanent endowment needed to generate an annual $12,000 forever at a continuous interest rate of 6%.

36. GENERAL: Permanent Endowments Show that the size of the permanent endowment needed to generate an annual C dollars forever at interest rate r compounded continuously is C/r dollars.

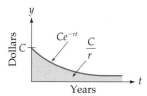

37. GENERAL: Permanent Endowments
   a. Find the size of the permanent endowment needed to generate an annual $1000 forever at a continuous interest rate of 10%.
   b. At this same interest rate, the size of the fund needed to generate an annual $1000 for precisely 100 years is $\int_0^{100} 1000 e^{-0.1t}\, dt$. Evaluate this integral (it is not an improper integral), approximating your answer using a calculator.
   c. Notice that the cost for the first 100 years is almost the same as the cost forever. This illustrates again the principle that in endowments, the short term is expensive, but eternity is cheap.

38–40. BUSINESS: Capital Value of an Asset
The capital value of an asset is defined as the present value of all future earnings. For an asset that may last indefinitely (such as real estate or a corporation), the capital value is

$$\binom{\text{Capital}}{\text{value}} = \int_0^\infty C(t) e^{-rt}\, dt$$

where $C(t)$ is the income per year and r is the continuous interest rate. Find the capital value of a piece of property that will generate an annual income of $C(t)$, for the function $C(t)$ given below, at a continuous interest rate of 5%.

38. $C(t) = 8000$ dollars

39. $C(t) = 50\sqrt{t}$ thousand dollars

40. $C(t) = 59\, t^{0.1}$ thousand dollars

41. BUSINESS: Oil Well Output An oil well is expected to produce oil at the rate of $50 e^{-0.05t}$ thousand barrels per month indefinitely, where t is the number of months that the well has been in operation. Find the total output over the lifetime of the well by integrating this rate from 0 to ∞. [*Note:* The owner will shut down the well when production falls too low, but it is convenient to estimate the total output as if production continued forever.]

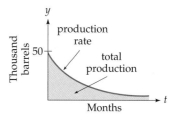

42. GENERAL: Duration of Telephone Calls Studies have shown that the proportion of telephone calls that last longer than t minutes is approximately $\int_t^\infty 0.3 e^{-0.3s}\, ds$. Use this formula to find the proportion of telephone calls that last longer than 4 minutes.

43. AREA Find the area between the curve $y = 1/x^{3/2}$ and the x-axis from $x = 1$ to ∞.

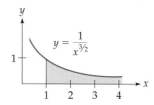

44. AREA Find the area between the curve $y = e^{-4x}$ and the x-axis from $x = 0$ to ∞.

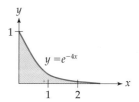

45. AREA Find the area between the curve $y = e^{-ax}$ (for $a > 0$) and the x-axis from $x = 0$ to ∞.

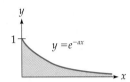

46. AREA Find the area between the curve $y = 1/x^n$ (for $n > 1$) and the x-axis from $x = 1$ to ∞.

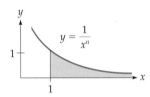

47. BEHAVIORAL SCIENCE: Mazes In a psychology experiment, rats were placed in a T-maze, and the proportion of rats who required more than t seconds to reach the end was $\int_{t}^{\infty} 0.05e^{-0.05s} \, ds$. Use this formula to find the proportion of rats who required more than 10 seconds.

48. SOCIOLOGY: Prison Terms If the proportion of prison terms that are longer than t years is given by the improper integral $\int_{t}^{\infty} 0.2e^{-0.2s} \, ds$, find the proportion of prison terms that are longer than 5 years.

49. BUSINESS: Product Reliability The proportion of light bulbs that last longer than t hours is predicted to be $\int_{t}^{\infty} 0.001e^{-0.001s} \, ds$. Use this formula to find the proportion of light bulbs that will last longer than 1200 hours.

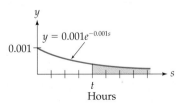

50. BUSINESS: Warranties When a company sells a product with a lifetime guarantee, the number of items returned for repair under the guarantee usually decreases with time. A company estimates that the annual rate of returns after t years will be $800e^{-0.2t}$. Find the total number of returns by summing (integrating) this rate from 0 to ∞.

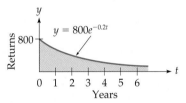

51. BUSINESS: Sales A publisher estimates that a book will sell at the rate of $16{,}000e^{-0.8t}$ books per year t years from now. Find the total number of books that will be sold by summing (integrating) this rate from 0 to ∞.

52. BIOMEDICAL: Drug Absorption To determine how much of a drug is absorbed into the body, researchers measure the difference between the dosage D and the amount of the drug excreted from the body. The total amount excreted is found by integrating the excretion rate $r(t)$ from 0 to ∞. Therefore, the amount of the drug absorbed by the body is

$$D - \int_{0}^{\infty} r(t) \, dt.$$

If the initial dose is $D = 200$ milligrams (mg), and the excretion rate is $r(t) = 40e^{-0.5t}$ mg per hour, find the amount of the drug absorbed by the body.

53–54. GENERAL: Permanent Endowments The formula for integrating the exponential function a^{bx} is $\int a^{bx} \, dx = \dfrac{1}{b \ln a} a^{bx} + C$ for constants $a > 0$ and b, as may be verified by using the differentiation formulas on page 749.

53. Use the formula above to find the size of the permanent endowment needed to generate an annual $2000 forever at 5% interest compounded annually.

[*Hint:* Find $\int_{0}^{\infty} 2000 \cdot 1.05^{-x} \, dx$.] Compare your answer with that found in Example 5 (page 894) for the same interest rate but compounded continuously.

54. Use the formula above to find the size of the permanent endowment needed to generate an annual $12,000 forever at 6% interest compounded annually.

[*Hint:* Find $\int_0^\infty 12{,}000 \cdot 1.06^{-x}\, dx$.] Compare your answer with that found in Exercise 35 (page 898) for the same interest rate but compounded continuously.

55. BUSINESS: Preferred Stock Since preferred stock can remain outstanding indefinitely, the present value per share is the limit of the present value of an annuity* paying that share's dividend D at interest rate r:

$$\begin{pmatrix}\text{Present}\\\text{value}\end{pmatrix} = \lim_{t\to\infty} D\left(\frac{1-(1+r)^{-t}}{r}\right)$$

Find this limit in terms of D and r.

56. BIOMEDICAL: Population Growth The *Gompertz growth curve* models the size $N(t)$ of a population at time $t \geq 0$ as

$$N(t) = Ke^{-ae^{-bt}}$$

where K, a, and b are positive constants. Find $\lim_{t\to\infty} N(t)$.

* See Section 2.4.

12.4 NUMERICAL INTEGRATION

Introduction

In spite of the many techniques of integration, there are still integrals that cannot be found by *any* method (as finite combinations of elementary functions). One example is the integral $\int e^{-x^2}\, dx$, which is closely related to the famous "bell-shaped curve" of probability and statistics. For *definite* integrals, however, it is always possible to *approximate* the actual value by interpreting it as the area under a curve, a process called *numerical integration*. We will discuss two of the most useful methods of numerical integration, based on approximating areas by *trapezoids* and by *parabolas* (known as "Simpson's Rule"). Both methods can be programmed on a calculator or computer. In fact, when you evaluate a definite integral on a graphing calculator using a command like FnInt, the calculator is using a numerical integration procedure similar to those described here. This section explains the mathematics behind such operations.

Rectangular Approximation and Riemann Sums

In Section 11.3 we approximated the area under a curve by rectangles, and we called the sum of the areas of the rectangles a *Riemann sum*.

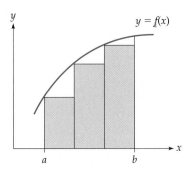

Area under $y = f(x)$ from a to b
approximated by three rectangles

However, these rectangles underestimate the true area under the curve by the white "error area" just above the rectangles. For greater accuracy we could increase the number of rectangles (as we did in Section 11.3), but this would involve more calculation and consequently more roundoff errors. Instead, we will replace the rectangles by shapes that fit the curve more closely.

Trapezoidal Approximation

We modify the approximating rectangles by allowing their tops to slant with the curve, as shown below, giving a much better "fit" to the curve.

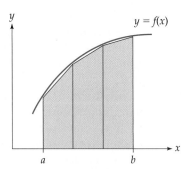

Area under $y = f(x)$ from a to b
approximated by three trapezoids

Such shapes, in which two sides are parallel, are called *trapezoids*. Clearly, approximating the area under the curve by trapezoids is more accurate than approximating it by rectangles: the white "error area" is much smaller.

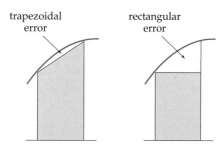

Area under a curve approximated by a
trapezoid (left) and by a rectangle (right).

The area of a trapezoid is the average of the two heights times the width.

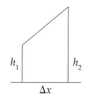

$$\begin{pmatrix} \text{Area of a} \\ \text{trapezoid} \end{pmatrix} = \underbrace{\frac{h_1 + h_2}{2}}_{\text{Average height}} \cdot \underbrace{\Delta x}_{\text{Width}}$$

If we use trapezoids that all have the same width Δx, we may add up all of the average heights first and then multiply by Δx. Furthermore, averaging the heights means dividing each height by 2 since $\frac{h_1 + h_2}{2} = \frac{h_1}{2} + \frac{h_2}{2}$. However, a side between two rectangles will be counted twice, once for the trapezoid on either side, thereby canceling the division by 2. Therefore, in adding up the heights, only the two outside heights, at a and b, should be divided by 2. This leads to the following procedure for trapezoidal approximation.

Trapezoidal Approximation

To approximate $\int_a^b f(x)\,dx$ by using n trapezoids:

1. Calculate $\Delta x = \dfrac{b - a}{n}$.   Trapezoid width

2. Find numbers $x_1, x_2, \ldots, x_{n+1}$ starting with $x_1 = a$ and successively adding Δx, ending with $x_{n+1} = b$.

3. The approximation for the integral is

$$\int_a^b f(x)\,dx \approx \left[\frac{1}{2}f(x_1) + f(x_2) + \cdots + f(x_n) + \frac{1}{2}f(x_{n+1})\right]\Delta x$$

This last formula calculates the total area of the n trapezoids shown below.

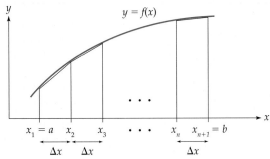

It is easiest to carry out the calculation in a table.

EXAMPLE 1

APPROXIMATING AN INTEGRAL USING TRAPEZOIDS

Approximate $\int_{1}^{2} x^2 \, dx$ using four trapezoids.

Solution The limits of integration are $a = 1$ and $b = 2$, and we are using $n = 4$ trapezoids. The method consists of the following six steps.

1. Calculate the trapezoid width $\Delta x = \dfrac{b - a}{n} = \dfrac{2 - 1}{4} = 0.25$.

2. List the x-values a through b with spacing Δx.

3. Apply $f(x)$ to each x-value.

x	$f(x) = x^2$	
Initial point a → 1	$(1)^2$ = $\cancel{1}$ 0.5	
add Δx → 1.25	$(1.25)^2 \approx 1.56$	4. Take half of first and last entries.
add Δx → 1.5	$(1.5)^2 = 2.25$	
add Δx → 1.75	$(1.75)^2 \approx 3.06$	5. Sum last column.
add Δx → Final point b 2.0	$(2)^2 = \cancel{4}$ 2	6. Multiply by Δx.
	9.37 · (0.25) ≈ 2.34	
	Final answer	

Therefore, the estimate using four trapezoids is $\int_{1}^{2} x^2 \, dx \approx 2.34$.

Earlier (see page 803) we evaluated this integral exactly, obtaining $\int_1^2 x^2 \, dx = \frac{7}{3} \approx 2.33$, and we can use this result to assess the accuracy of the trapezoidal method: our approximation of 2.34 is very accurate in spite of the fact that we used only four trapezoids. The relative error (the actual error, 0.01, divided by the actual value, 7/3) is $\frac{0.01}{7/3} \approx 0.004$. Expressed as a percentage, this is 0.4% (four tenths of one percent), which is also remarkably small. Notice also that the trapezoidal approximation of 2.34 using four trapezoids is far more accurate than the Riemann sum of 2.04 that we found on pages 798–799 using five (left) rectangles.

Error in Trapezoidal Approximation

The maximum error in trapezoidal approximation obeys the following formula.

Trapezoidal Error

For the trapezoidal approximation of $\int_a^b f(x) \, dx$ with n trapezoids,

$$(\text{Error}) \leq \frac{(b-a)^3}{12n^2} \max_{a \leq x \leq b} |f''(x)|$$

This formula is very difficult to use, because it involves maximizing the absolute value of the second derivative. We will not make further use of it except to observe that the n^2 in the denominator means that doubling the number of trapezoids reduces the maximum error by a factor of *four*.

Trapezoidal approximation is most easily carried out on a calculator or a computer, as shown in the following Graphing Calculator Exploration.

Graphing Calculator Exploration

The graphing calculator program* TRAPEZOD calculates the trapezoidal approximation of a definite integral and, for small values of n, draws the approximating trapezoids. The first screen

---

* See the Preface for how to obtain this and other programs.

below shows the approximation of $\int_1^2 x^2\, dx$ using four trapezoids, giving a value 2.34375 that agrees with the 2.34 found in the previous example. Even for only four trapezoids, the curve $y = x^2$ is almost indistinguishable from the tops of the trapezoids, the difference appearing only as a slight thickening of the line where they diverge slightly.

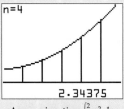

Approximating $\int_1^2 x^2\, dx$ by four trapezoids

For 500 trapezoids, the approximation is about $2\tfrac{1}{3}$

The screen on the right above shows the trapezoidal approximations as the number n of trapezoids increases from 5 to 500. The approximations do seem to approach the *exact* value $2\tfrac{1}{3}$ found in Example 1 by evaluating the definite integral.

Try using a command like FnInt to find the area under $y = x^2$ from $x = 1$ to $x = 2$. How do the answers compare? (FnInt uses a method similar to those described in this section, but is faster since it is designed into the calculator.)

The error formula given on the previous page is of limited usefulness on a computer since maximizing the second derivative is itself subject to error. What is done instead is to calculate the approximations for larger and larger values of n until successive approximations agree to the desired degree of accuracy. For example, the last two results in the above table give answers agreeing to five decimal places, indicating an error of less than 0.00001. While this procedure does not *guarantee* this accuracy, it is often accepted in practice.

Simpson's Rule (Parabolic Approximation)

For even greater accuracy, we could increase n (resulting in more calculations and more roundoff errors) or we could change the trapezoids, replacing the tops by *curves* chosen to fit the given curve more closely. Replacing the tops of the trapezoids by *parabolas* that pass through three points of the given curve leads to an even more accurate method of approximation, called Simpson's Rule.* The following diagram shows such a curve and its approximation by two parabolas.

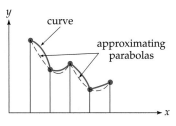

A curve approximated by two parabolas

Since each parabola spans two intervals, the number of intervals must be even. The area under the approximating parabolas is easily found by integration. The procedure is described as follows and is illustrated in Example 2. A justification of Simpson's Rule is given in Exercise 37.

Simpson's Rule (Parabolic Approximation)

To approximate $\int_a^b f(x)\,dx$ by Simpson's Rule using n intervals:

1. Calculate $\Delta x = \dfrac{b-a}{n}$. $\qquad n$ must be even

2. Find numbers $x_1, x_2, x_3, \ldots, x_{n+1}$ starting with $x_1 = a$ and successively adding Δx, ending with $x_{n+1} = b$.

3. The approximation for the integral is

$$\int_a^b f(x)\,dx \approx [f(x_1) + \underbrace{4f(x_2) + 2f(x_3) + \cdots + 4f(x_n)}_{\text{Alternating 4's and 2's}} + f(x_{n+1})]\frac{\Delta x}{3}$$

The function values are multiplied by "weights," which, written out by themselves, are

$$\underset{\underset{\text{Initial 1}}{\uparrow}}{1} \quad \underbrace{4 \quad 2 \quad 4 \quad 2 \quad 4 \quad \cdots \quad 4 \quad 2 \quad 4}_{\substack{\text{Alternating 4's and 2's} \\ \text{beginning and ending with 4}}} \quad \underset{\underset{\text{Final 1}}{\uparrow}}{1}$$

---

* Named after Thomas Simpson (1701–1761), an early user, but not the discoverer, of the formula.

EXAMPLE 2

APPROXIMATING AN INTEGRAL USING SIMPSON'S RULE

Approximate $\int_{3}^{5} \frac{1}{x} dx$ by Simpson's Rule with $n = 4$. $n = 4$ means 2 parabolas

Solution The method consists of the following six steps.

1. Calculate $\Delta x = \dfrac{b - a}{n} = \dfrac{5 - 3}{4} = 0.5$ n must be even

2. List x-values a through b with spacing Δx.
3. Apply $f(x)$ to each x-value.
4. Multiply by the weights to get

x	$f(x) = \dfrac{1}{x}$	weights		$f(x) \cdot$ weight
3	0.33333	1	← Initial 1	0.33333
3.5	0.28571	4		1.14284
4	0.25	2	Alternating 4's and 2's	0.50000
4.5	0.22222	4		0.88888
5	0.2	1	← Final 1	0.20000

a add Δx, add Δx, add Δx, add Δx, *b*

5. Sum last column. → 3.06505 · $\left(\dfrac{0.5}{3}\right) \approx 0.51084$

6. Multiply by $\dfrac{\Delta x}{3}$.   Final answer

Therefore, $\int_{3}^{5} \dfrac{1}{x} dx \approx 0.51084$.

This integral, too, can be found exactly. The answer is $\ln 5 - \ln 3 \approx 0.51083$, for an error of only 0.00001. The *relative* error (actual error divided by actual value) is

$$\dfrac{0.00001}{0.51083} \approx 0.00002 = 0.002\%$$

which is also extremely small.

IQ Distribution

Although it is increasingly clear that human intelligence cannot be measured by a single number, IQ tests are still widely used. (IQ stands for Intelligence Quotient, and is defined as mental age divided by chronological age, multiplied by 100.) The average American IQ is 100, and the distribution of IQs follows the famous "bell-shaped curve" so often used in statistics.*

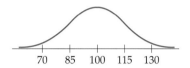

The proportion of Americans with IQs between two numbers A and B (with $A < B$) is given by the following integral:

$$\begin{pmatrix} \text{Proportion of Americans} \\ \text{with IQs between } A \text{ and } B \end{pmatrix} \approx \frac{1}{\sqrt{2\pi}} \int_{(A-100)/15}^{(B-100)/15} e^{-x^2/2} \, dx$$

For example, the proportion of Americans who have IQs between 115 and 145 is found by substituting $A = 115$ and $B = 145$ into the lower and upper limits in the formula above:

$$\begin{pmatrix} \text{Proportion of IQs} \\ \text{between 115 and 145} \end{pmatrix} = \frac{1}{\sqrt{2\pi}} \int_{1}^{3} e^{-x^2/2} \, dx$$

3 from $\dfrac{B - 100}{15} = \dfrac{145 - 100}{15} = \dfrac{45}{15} = 3$

1 from $\dfrac{A - 100}{15} = \dfrac{115 - 100}{15} = \dfrac{15}{15} = 1$

This integral cannot be found "by hand" (as a finite combination of elementary functions). It can be approximated by Simpson's Rule, just as in Example 2, but it is easier to carry out the calculation on a graphing calculator.

*For those familiar with Chapter 6, IQ scores are normally distributed with mean 100 and standard deviation 15.

Graphing Calculator Exploration

The graphing calculator program† SIMPSON uses Simpson's Rule to approximate definite integrals and, for small values of n, draws the approximating parabolas. The first screen below shows

†See the Preface for how to obtain this and other programs.

the approximation of $\dfrac{1}{\sqrt{2\pi}} \int_1^3 e^{-x^2/2}\,dx$ using $n = 4$ intervals (2 parabolas), giving approximately 0.157. Simpson's Rule is so accurate that the curve below is almost indistinguishable from the approximating parabolas, the difference again appearing only as a slight thickening of the line where they diverge slightly.

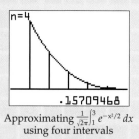

N	SMPSN APPRX
[[4	.15709468...
[10	.15730028...
[20	.15730504...
[50	.15730534...
[100	.15730535...
[200	.15730535...

Approximating $\dfrac{1}{\sqrt{2\pi}} \int_1^3 e^{-x^2/2}\,dx$ using four intervals

For 100 and 200 intervals the approximations agree to 8 decimal places

The screen on the right shows that Simpson's Rule converges extremely rapidly.

Try using a command like FnInt to find the area under $y = 1/\sqrt{2\pi}\, e^{-x^2/2}$ from $x = 1$ to $x = 3$. How do the answers compare?

Converting the answer 0.157 to a percentage, about 16% of all Americans have IQs in the range 115 to 145.

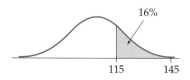

The Error in Simpson's Rule

The maximum error in Simpson's Rule obeys the following formula.

Error in Simpson's Rule

In approximating $\int_a^b f(x)\,dx$ by Simpson's Rule with n intervals,

$$(\text{Error}) \le \dfrac{(b-a)^5}{180 n^4} \max_{a \le x \le b} |f^{(4)}(x)|$$

This formula is difficult to use because it involves maximizing the absolute value of the fourth derivative. It does, however, show that Simpson's Rule is *exact* for cubics (third-degree polynomials), since cubics have zero fourth derivative, and that doubling the value of n reduces the maximum error by a factor of 16 (because of the n^4).

12.4 Section Summary

Some definite integrals cannot be evaluated exactly because it is impossible (or very difficult) to find an antiderivative. However, any definite integral can be *approximated* by numerical integration, and two of the most useful methods are trapezoidal approximation and Simpson's Rule (parabolic approximation). Each method involves choosing a number n of intervals and calculating a "weighted average" of function values at the endpoints of these intervals. Higher values of n generally give greater accuracy, but also involve more calculation (and therefore more roundoff errors). In practice, Simpson's Rule is generally the method of choice, since it gives greater accuracy for only slightly more effort. Trapezoidal approximation, however, has the advantage of having a simpler formula for its error. It is particularly appropriate to seek an approximation if the "exact" answer involves logarithms or exponentials, since these functions will probably be approximated anyway in the evaluation step.

It is curious that in Chapter 11 we used integrals to evaluate areas, and here we are using areas to evaluate integrals. This is typical of mathematics, in which any equivalence (such as definite integrals and areas) is exploited in both directions.

Programs for Trapezoidal Approximation and Simpson's Rule

On the following page are two programs for trapezoidal approximation and two for Simpson's Rule for the *TI-83* calculator and in BASIC. They may be adapted for other calculators or computers. The small print is explanation, not part of the program. These programs give only single numerical answers, whereas TRAPEZOD and SIMPSON draw graphs and make tables showing the approximations for several values of n.

TI-83 Program for Trapezoidal Approximation

(Enter the function in y_1 before executing.)

Disp "USES FUNCTION Y$_1$"
Prompt A, B, N
(B − A)/N→D
(Y$_1$(A) + Y$_1$(B))/2→S
A + D→X
For(I,1,N − 1)
S + Y$_1$→S
X + D→X
End
S*D→S
Disp "TRAP APPROX IS ", S

BASIC Program for Trapezoidal Approximation

(Lines 2, 3, 4, and 8 to be completed as indicated in small type)

approx = 0
a =   fill in beginning x-value
b =   fill in ending x-value
n =   fill in number of trapezoids
delta = (b − a)/n
x = a
FOR i = 1 TO n + 1
f = $\begin{pmatrix}\text{fill in the}\\\text{function}\end{pmatrix}$
IF i = 1 OR i = n + 1 THEN f = f/2
approx = approx + f
x = x + delta
NEXT i
approx = approx*delta
PRINT "TRAP APPROX IS", approx
END

TI-83 Program for Simpson's Rule

(Enter the function in y_1 before executing.)

Disp "USES FUNCTION Y$_1$"
Disp "N MUST BE EVEN"
Prompt A, B, N
(B − A)/N→D
Y$_1$(A) − Y$_1$(B)→S
A→X
For(I,1,N/2)
S + 4*Y$_1$(X + D) + 2*Y$_1$(X + 2D)→S
X + 2D→X
End
S*D/3→S
Disp "SIMP APPROX IS " , S

BASIC Program for Simpson's Rule

(Lines 1, 2, 3, and 9 to be completed as indicated in small type)

a =   fill in beginning x-value
b =   fill in ending x-value
n =   fill in number of intervals (even)
delta = (b − a)/n
approx = 0
x = a
FOR i = 1 TO n + 1
IF i = 1 OR i = n + 1 THEN weight = 1
approx = approx + $\begin{pmatrix}\text{fill in the}\\\text{function}\end{pmatrix}$* weight
x = x + delta
IF weight = 4 THEN weight = 2 ELSE weight = 4
NEXT i
approx = approx*delta/3
PRINT "SIMP APPROX IS", approx
END

12.4 Exercises

EXERCISES ON TRAPEZOIDAL APPROXIMATION

For each definite integral:
a. Approximate it "by hand," using trapezoidal approximation with $n = 4$ trapezoids. Round calculations to three decimal places.
b. Evaluate the integral exactly using antiderivatives, rounding to three decimal places.
c. Find the actual error (the difference between the actual value and the approximation).
d. Find the relative error (the actual error divided by the actual value, expressed as a percent).

1. $\int_1^3 x^2 \, dx$
2. $\int_1^2 x^3 \, dx$
3. $\int_2^4 \frac{1}{x} \, dx$
4. $\int_1^3 \frac{1}{x} \, dx$

Approximate each integral using trapezoidal approximation "by hand" with the given value of n. Round all calculations to three decimal places.

5. $\int_0^1 \sqrt{1 + x^2} \, dx, \quad n = 3$
6. $\int_0^1 \sqrt{1 + x^3} \, dx, \quad n = 3$
7. $\int_0^1 e^{-x^2} \, dx, \quad n = 4$
8. $\int_0^1 e^{x^2} \, dx, \quad n = 4$

Approximate each integral using the graphing calculator program TRAPEZOD (see pages 905–906) or one of the trapezoidal approximation programs on the previous page or a similar program. Use the following values for the numbers of intervals: 10, 50, 100, 200, 500. Then give an estimate for the value of the definite integral, keeping as many decimal places as the last two approximations agree (when rounded).

9. $\int_1^2 \sqrt{\ln x} \, dx$
10. $\int_0^1 \ln(x^2 + 1) \, dx$
11. $\int_{-1}^1 \sqrt{16 + 9x^2} \, dx$
12. $\int_{-1}^1 \sqrt{25 - 9x^2} \, dx$
13. $\int_{-1}^1 e^{x^2} \, dx$
14. $\int_0^4 \sqrt{1 + x^4} \, dx$

APPLIED EXERCISES ON TRAPEZOIDAL APPROXIMATION

15–16. GENERAL: IQs Use the formula on page 481 and TRAPEZOD or another trapezoidal approximation program (see the previous page) to find the proportion of Americans with IQs between the following two numbers. Use successively higher values of n until answers agree to four decimal places.

15. 100 and 130
16. 130 and 145

17–18. BUSINESS: Investment Growth An investment grows at a rate of $3.2 \, e^{\sqrt{t}}$ thousand dollars per year, where t is the number of years since the beginning of the investment. Use TRAPEZOD or another trapezoidal approximation program to estimate the total growth of the investment during the period stated below. Use successively higher values of n until answers agree to two decimal places.

17. In the first 2 years (year 0 to year 2)
18. In the first 3 years (year 0 to year 3)

EXERCISES ON SIMPSON'S RULE

Estimate each definite integral "by hand," using Simpson's Rule with $n = 4$. Round all calculations to three decimal places. Exercises 19–26 correspond to Exercises 1–8, in which the same integrals were estimated using trapezoids. If you did the corresponding exercise, compare your Simpson's Rule answer with your trapezoidal answer.

(See instructions on the previous page.)

19. $\int_1^3 x^2 \, dx$ 20. $\int_1^2 x^3 \, dx$

21. $\int_2^4 \frac{1}{x} \, dx$ 22. $\int_1^3 \frac{1}{x} \, dx$

23. $\int_0^1 \sqrt{1+x^2} \, dx$ 24. $\int_0^1 \sqrt{1+x^3} \, dx$

25. $\int_0^1 e^{-x^2} \, dx$ 26. $\int_0^1 e^{x^2} \, dx$

Approximate each integral using the graphing calculator program SIMPSON (see pages 909–910) or one of the Simpson's Rule approximation programs on page 912 or a similar program. Use the following values for the numbers of intervals: 10, 20, 50, 100, 200. Then give an estimate for the value of the definite integral, keeping as many decimal places as the last two approximations agree to (when rounded). Exercises 27–32 correspond to Exercises 9–14 in which the same integrals were estimated using trapezoids. If you did the corresponding exercise, compare your Simpson's Rule answer with your trapezoidal answer.

27. $\int_1^2 \sqrt{\ln x} \, dx$ 28. $\int_0^1 \ln(x^2+1) \, dx$

29. $\int_{-1}^1 \sqrt{16+9x^2} \, dx$ 30. $\int_{-1}^1 \sqrt{25-9x^2} \, dx$

31. $\int_{-1}^1 e^{x^2} \, dx$ 32. $\int_0^4 \sqrt{1+x^4} \, dx$

33–34. APPROXIMATION OF IMPROPER INTEGRALS For each improper integral:

a. Make it a "proper" integral by using the substitution $x = \frac{1}{t}$ and simplifying.

b. Approximate the proper integral using Simpson's Rule (either "by hand" or using a program) with $n = 4$ intervals, rounding your answer to three decimal places.

33. $\int_1^\infty \frac{1}{x^3+1} \, dx$ 34. $\int_1^\infty \frac{x}{x^3+1} \, dx$

APPLIED EXERCISES ON SIMPSON'S RULE

35. GENERAL: Suspension Bridges The cable of a suspension bridge hangs in a parabolic curve. The equation of the cable shown below is $y = \frac{x^2}{2000}$. Its length in feet is given by the integral.

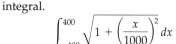

Approximate this integral using Simpson's Rule, using successively higher values of n until answers agree to the nearest whole number.

36. APPROXIMATION OF π The number π is the ratio of the circumference of a circle to its diameter (since $C = \pi D$). It can be shown that the following definite integral is equal to π.

$$\int_0^1 \frac{4}{x^2+1} \, dx = \pi$$

Find π by approximating this integral using a Simpson's Rule program, using successively higher values of n until answers agree to four decimal places.

JUSTIFICATION OF SIMPSON'S RULE

37. JUSTIFICATION OF SIMPSON'S RULE Justify Simpson's Rule by carrying out the following steps, which lead to the formula for Simpson's Rule (page 907) in a simple case.

i. Observe that if the three points shown in the following diagram lie on the parabola

$f(x) = ax^2 + bx + c$, then the following three equations hold:

$a(-d)^2 + b(-d) + c = y_1$ Since $f(-d) = y_1$
$a(0)^2 + b(0) + c = y_2$ Since $f(0) = y_2$
$a(d)^2 + b(d) + c = y_3$ Since $f(d) = y_3$

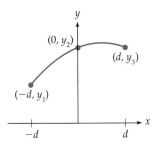

ii. Simplify these three equations to obtain

$$ad^2 - bd + c = y_1$$
$$c = y_2$$
$$ad^2 + bd + c = y_3$$

iii. Add the first and last equation plus 'four times the middle equation to obtain

$$2ad^2 + 6c = y_1 + 4y_2 + y_3$$

You will use this equation in step v.

iv. Evaluate the integral $\int_{-d}^{d} (ax^2 + bx + c)\, dx$ and simplify to show that the area under the parabola from $-d$ to d is

$$\text{Area} = \frac{2}{3}ad^3 + 2cd = \left(2ad^2 + 6c\right)\frac{d}{3}$$

v. Use the equation found in step iii to write this area as

$$\text{Area} = (y_1 + 4y_2 + y_3)\frac{d}{3}$$

vi. Use $y_1 = f(-d)$, $y_2 = f(0)$, $y_3 = f(d)$ and the fact that the spacing Δx is equal to d to rewrite this last equation as

$$\text{Area} = [f(-d) + 4f(0) + f(d)]\frac{\Delta x}{3}$$

This is exactly the formula for Simpson's Rule using one parabola ($n = 2$). For several parabolas placed next to each other, the function values *between* two neighboring parabolas are added twice (once for each side), and so should have weight 2. Therefore, successive function values are multiplied by the weights given on page 907:

1 4 2 4 2 4 $\cdots$ 4 1

12.5 DIFFERENTIAL EQUATIONS

Introduction

A differential equation is simply an equation involving derivatives. Practically any relationship involving rates of change can be expressed using a differential equation, and many differential equations can be solved by a technique called "separation of variables."

We will take y to be a function of x, sometimes writing it as $y(x)$.

We will write the derivative of y as either y' or $\frac{dy}{dx}$.

Differential Equation $y' = f(x)$

We have actually been solving differential equations since the beginning of Chapter 11. For example, the differential equation

$$y' = 2x$$

saying that the derivative of a function is $2x$, is solved simply by integrating:

$$y = \int 2x\, dx = x^2 + C \qquad y = x^2 + C \text{ satisfies } y' = 2x$$

In general, a differential equation of the form

$$y' = f(x)$$

is solved by integrating:

$$y = \int f(x)\, dx$$

Therefore, whenever we integrate a marginal cost function to find cost, or integrate a rate to find the total accumulation, we are solving a differential equation of the form $y' = f(x)$.

General and Particular Solutions

The solution of the differential equation $y' = 2x$ is $y = x^2 + C$, with an arbitrary constant C. We call $y = x^2 + C$ the *general solution* because taking all possible values of the constant C gives *all* solutions of the differential equation. If we take C to be a particular number, we get a *particular solution*. Some particular solutions of the differential equation $y' = 2x$ are

$y = x^2 + 2$ (taking $C = 2$)
$y = x^2$ (taking $C = 0$)
$y = x^2 - 2$ (taking $C = -2$)

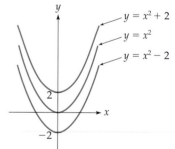

Three particular solutions from the general solution $y = x^2 + C$.

The different values of the arbitrary constant C give a "family" of curves, and the general solution $y = x^2 + C$ may be thought of as the entire family.

The solution of a differential equation is a *function*. The *general* solution contains an arbitrary constant, and a *particular* solution has this constant replaced by a particular number.

Verifying Solutions

Verifying that a function is a solution of a differential equation is simply a matter of calculating the necessary derivatives, substituting them into the differential equation, and checking that the two sides of the equation are equal.

EXAMPLE 1

VERIFYING A SOLUTION OF A DIFFERENTIAL EQUATION

Verify that $y = e^{2x} + e^{-x} - 1$ is a solution of the differential equation

$$y'' - y' - 2y = 2$$

Solution The differential equation involves y, y', and y'', so we calculate

$y = e^{2x} + e^{-x} - 1$	Given function
$y' = 2e^{2x} - e^{-x}$	Derivative
$y'' = 4e^{2x} + e^{-x}$	Second derivative

Then we substitute these into the differential equation.

$$y'' - y' - 2y = 2$$

$(4e^{2x} + e^{-x}) - (2e^{2x} - e^{-x}) - 2(e^{2x} + e^{-x} - 1) \stackrel{?}{=} 2$	$\stackrel{?}{=}$ means the equation may not be true
$4e^{2x} + e^{-x} - 2e^{2x} + e^{-x} - 2e^{2x} - 2e^{-x} + 2 \stackrel{?}{=} 2$	Expanding
$\cancel{4e^{2x}} + \cancel{e^{-x}} - \cancel{2e^{2x}} + \cancel{e^{-x}} - \cancel{2e^{2x}} - \cancel{2e^{-x}} + 2 \stackrel{?}{=} 2$	Canceling
$2 = 2$	It checks!

Since the equation is satisfied, the given function is indeed a solution of the differential equation.

If the two sides had not turned out to be exactly the same, the given function y would *not* have been a solution of the differential equation.

Practice Problem 1

Verify that $y = e^{-x} + e^{3x}$ is a solution of the differential equation

$$y'' - 2y' - 3y = 0$$

➤ Solution on page 929

The differential equations in Example 1 and Practice Problem 1 are called *second-order* differential equations because they involve second derivatives but no higher-order derivatives. From here on we will restrict our attention to *first-order* differential equations—that is, differential equations involving only the *first* derivative. Many first-order differential equations can be solved by a method called *separation of variables*.

Separation of Variables

A differential equation is said to be "separable" if the variables can be "separated" by moving every *x* and *dx* to one side of the equation and every *y* and *dy* to the other side. We may then solve the differential equation by integrating both sides. Several examples will make the method clear.

EXAMPLE 2   FINDING A GENERAL SOLUTION

Find the general solution of the differential equation $\dfrac{dy}{dx} = 2xy^2$.

Solution

$dy = 2xy^2 \, dx$	Multiplying both sides of the differential equation by *dx*
$\dfrac{dy}{y^2} = 2x \, dx$	Dividing each side by y^2 ($y \neq 0$); the variables are now separated: every *y* on one side, every *x* on the other
$y^{-2} \, dy = 2x \, dx$	In power form
$\displaystyle\int y^{-2} \, dy = \int 2x \, dx$	Integrating both sides
$-y^{-1} = x^2 + C$	Using the Power Rule (writing one *C* for both integrations)
$\dfrac{1}{y} = -x^2 - C$	Writing y^{-1} as $\dfrac{1}{y}$ and multiplying by -1
$y = \dfrac{1}{-x^2 - C}$	Taking reciprocals of both sides

This is the general solution of the differential equation. The solution may be left in this form, but if we replace the arbitrary constant *C* by

$-c$, another constant but with the opposite sign, this solution may be written

$$y = \frac{1}{-x^2 + c}$$

or, slightly shorter,

$$y = \frac{1}{c - x^2}$$

Reversing the order, giving the general solution of the differential equation

The differential equation $y' = 2xy^2$ has another solution: the function that is identically zero, $y(x) \equiv 0$. This is known as a *singular solution* and cannot be obtained from the general solution. We will not consider singular solutions further in this book except to say that $y = 0$ will generally be a solution whenever separating the variables results in dy being divided by a positive power of y, as in the previous example.

Separable Differential Equations

A first-order differential equation is *separable* if it can be written in the following form for some functions $f(x)$ and $g(y)$:

$$\frac{dy}{dx} = \frac{f(x)}{g(y)} \qquad \text{for } g(y) \neq 0$$

It is solved by separating variables and integrating:

$$\int g(y)\, dy = \int f(x)\, dx \qquad \text{Multiplying by } dx \text{ and } g(y) \text{ and integrating}$$

The preceding example asked for the *general* solution of a differential equation. Sometimes we will be given a differential equation together with some additional information that selects a *particular* solution from the general solution, information that determines the value of the arbitrary constant in the general solution. This additional information is called an *initial condition*.

EXAMPLE 3

FINDING A PARTICULAR SOLUTION

Solve the differential equation $y' = \dfrac{6x}{y^2}$ with the initial condition $y(1) = 2$.

Solution First we find the general solution by separating the variables.

$\dfrac{dy}{dx} = \dfrac{6x}{y^2}$ — Replacing y' by $\dfrac{dy}{dx}$

$y^2 \, dy = 6x \, dx$ — Multiplying both sides by dx and y^2 (the variables are separated)

$\displaystyle\int y^2 \, dy = \int 6x \, dx$ — Integrating both sides

$\dfrac{1}{3} y^3 = 3x^2 + C$ — Using the Power Rule

$y^3 = 9x^2 + \underbrace{3C}_{c}$ — Multiplying by 3

(3 times a constant is just another constant)

$y^3 = 9x^2 + c$ — Replacing $3C$ by c

$y = \sqrt[3]{9x^2 + c}$ — Taking cube roots

This is the general solution, with arbitrary constant c. The initial condition $y(1) = 2$ says that $y = 2$ when $x = 1$. We substitute these values into the general solution $y = \sqrt[3]{9x^2 + c}$ and solve for c.

$2 = \sqrt[3]{9 + c}$ — $y = \sqrt[3]{9x^2 + c}$ with $x = 1$ and $y = 2$

$8 = 9 + c$ — Cubing each side

$-1 = c$ — Solving for c gives $c = -1$

Therefore, we replace c by -1 in the general solution to obtain the particular solution:

$y = \sqrt[3]{9x^2 - 1}$ — $y = \sqrt[3]{9x^2 + c}$ with $c = -1$

This solution $y = \sqrt[3]{9x^2 - 1}$ with $x = 1$ gives $y = \sqrt[3]{9 - 1} = \sqrt[3]{8} = 2$, so the initial condition $y(1) = 2$ is indeed satisfied. We could also verify that the differential equation is satisfied by substituting this solution into it.

In general, solving a differential equation with an initial condition just means finding the general solution and then using the initial condition to evaluate the constant. Several solutions of the differential equation $y' = 6x/y^2$ are shown below, with the particular solution $y = \sqrt[3]{9x^2 - 1}$ shown in color.

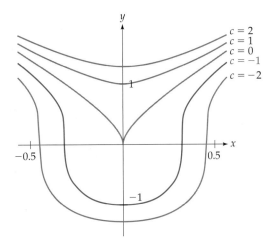

The solution $y = \sqrt[3]{9x^2 + c}$ for various values of c.

Graphing Calculator Exploration

A graphing calculator can help you to *see* what a differential equation is saying. The differential equation in the preceding example,

$$\frac{dy}{dx} = \frac{6x}{y^2} \qquad \text{Slope equals } \frac{6x}{y^2}$$

says that the slope at any point (x, y) in the plane is given by the formula $6x/y^2$. The graphing calculator program* SLOPEFLD draws a "slope field" consisting of many small line segments with

---

* See the Preface for how to obtain this and other programs. SLOPEFLD can draw the slope field of any differential equation of the form $y' = f(x, y)$.

the slopes specified by the differential equation. SLOPEFLD can also graph a particular solution. The slope field of the above differential equation is shown on the left below, and next to it is the same slope field with the particular solution found in the preceding example.

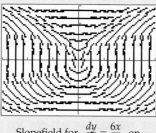

Slopefield for $\dfrac{dy}{dx} = \dfrac{6x}{y^2}$ on $[-0.6, 0.6]$ by $[-1.5, 1.5]$

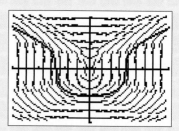

The same slopefield with the solution $y = \sqrt[3]{9x^2 - 1}$

Look back at the graph of the five solutions on the previous page and see how each of them follows the prescribed slopes in the above slope field. In fact, starting at *any* point in the plane you could draw a curve following the indicated slopes, thereby geometrically constructing a *solution curve*.

Practice Problem 2

Solve the differential equation and initial condition
$$\begin{cases} y' = \dfrac{6x^2}{y^4} \\ y(0) = 2 \end{cases}$$

▶ Solution on page 930

Recall that to solve for y in the logarithmic equation
$$\ln y = f(x)$$
we exponentiate both sides and simplify
$$y = e^{f(x)}$$
Using $e^{\ln y} = y$ on the left side

This idea will be useful in the next example.

EXAMPLE 4

FINDING A PARTICULAR SOLUTION

Solve the differential equation and initial condition $\begin{cases} \dfrac{dy}{dx} = xy \\ y(0) = 2 \end{cases}$

Solution

$$dy = xy\,dx \quad \text{Multiplying by } dx$$

$$\frac{dy}{y} = x\,dx \quad \text{Dividing by } y\ (y \neq 0). \text{ The variables are separated}$$

$$\int \frac{dy}{y} = \int x\,dx \quad \text{Integrating}$$

$$\ln y = \frac{1}{2}x^2 + C \quad \text{Evaluating the integrals (assuming } y > 0)$$

$$y = e^{\frac{1}{2}x^2 + C} \quad \text{Solving for } y \text{ by exponentiating}$$

$$y = e^{\frac{1}{2}x^2} \underbrace{e^C}_{c} \quad \text{e to a sum can be expressed as a product}$$

$$\phantom{y = e^{\frac{1}{2}x^2} e^C} \quad e^C \text{ is a positive constant } c$$

$$y = ce^{x^2/2} \quad \text{Replacing } e^C \text{ by } c \text{ and } e^{\frac{1}{2}x^2} \text{ by } e^{x^2/2}$$

This is the general solution. To satisfy the initial condition $y(0) = 2$, we substitute $y = 2$ and $x = 0$ and solve for c.

$$2 = ce^0 \quad y = ce^{x^2/2} \text{ with } y = 2 \text{ and } x = 0$$

$$2 = c \quad \text{Since } e^0 = 1$$

Substituting $c = 2$ into the general solution gives the particular solution

$$y = 2e^{x^2/2} \quad y = ce^{x^2/2} \text{ with } c = 2$$

We wrote the solution of the integral $\int \dfrac{dy}{y}$ as $\ln y$, without absolute value bars, thereby assuming that $y > 0$. *Keeping* the absolute value would have given $|y| = e^{\frac{1}{2}x^2 + C} = ce^{\frac{1}{2}x^2}$ so that $y = \pm ce^{\frac{1}{2}x^2}$ where c is a positive constant, showing that the general solution is $y = ce^{x^2/2}$, just as before, but now where c is *any* (positive or negative) constant. We will make similar simplifying assumptions in similar situations.

Note that the constant that was *added* in the integration step became a constant *multiplier* in the solution $y = ce^{x^2/2}$. In general, the constant may appear *anywhere* in the solution. The following diagram shows the solutions $y = ce^{x^2/2}$ for various values of c, with the particular solution $y = 2e^{x^2/2}$ shown in color.

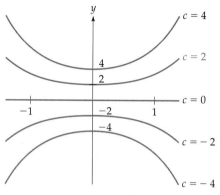

Solutions $y = ce^{x^2/2}$
for various values of c.

Graphing Calculator Exploration

The slope field for the differential equation in the preceding example, found by entering $y_1 = xy$, setting the window to $[-2, 2]$ by $[-6, 6]$ and running SLOPEFLD, is shown on the left below. Notice how the slopes are determined by the function $x \cdot y$: in the first quadrant they increase when either x or y increases, and they are positive in the first and third quadrants (where x and y have the same sign) and negative in the other quadrants.

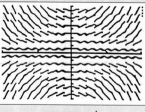

Slopefield for $\dfrac{dy}{dx} = xy$

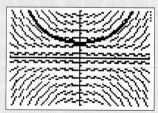

With the solution $y = 2e^{x^2/2}$

The screen on the right shows the same slope field with the particular solution found in the example. Do you see how the other solutions graphed just above this Graphing Calculator Exploration follow these slopes?

Practice Problem 3

Solve the differential equation and initial condition
$$\begin{cases} \dfrac{dy}{dx} = x^2 y \\ y(0) = 5 \end{cases}$$

➤ Solution on page 930

EXAMPLE 5

FINDING A GENERAL SOLUTION

Find the general solution of the differential equation
$$yy' - x = 0$$

Solution We are asked for the general solution rather than a particular solution.

$y \dfrac{dy}{dx} = x$	Replacing y' by dy/dx and moving the x to the other side
$y \, dy = x \, dx$	Multiplying by dx
$\displaystyle\int y \, dy = \int x \, dx$	Integrating
$\dfrac{1}{2}y^2 = \dfrac{1}{2}x^2 + C$	Using the Power Rule
$y^2 = x^2 + \underbrace{2C}_{c}$	Multiplying by 2

To solve for y we take the square root of each side. However, there are *two* square roots, one positive and one negative.

$$y = \sqrt{x^2 + c} \quad \text{and} \quad y = -\sqrt{x^2 + c}$$

These two solutions together are the general solution:

$$y = \pm\sqrt{x^2 + c}$$

Graphing Calculator Exploration

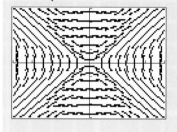

Use SLOPEFLD or a similar program to have your calculator draw the slope field for the differential equation in the previous example, $y' = x/y$, using the window $[-4, 4]$ by $[-4, 4]$. Try to draw a curve through the point $(0, 1)$ following the slopes. This would be the particular solution satisfying $y(0) = 1$.

For some differential equations, the integration step requires a substitution.

EXAMPLE 6   FINDING A PARTICULAR SOLUTION USING A SUBSTITUTION

Solve the differential equation and initial condition $\begin{cases} y' = xy - x \\ y(0) = 4 \end{cases}$

Solution

$\dfrac{dy}{dx} = xy - x$ Replacing y' by $\dfrac{dy}{dx}$

$\dfrac{dy}{dx} = x(y - 1)$ Factoring (to separate variables)

$\dfrac{dy}{y - 1} = x \, dx$ Dividing by $y - 1$ and multiplying by dx

$\displaystyle\int \dfrac{dy}{y - 1} = \int x \, dx$ Integrating

$\begin{bmatrix} u = y - 1 \\ du = dy \end{bmatrix}$ Using a substitution

$\displaystyle\int \dfrac{du}{u} = \int x \, dx$ Substituting

$\ln u = \dfrac{1}{2}x^2 + C$ Integrating (assuming $u > 0$)

$\ln(y - 1) = \dfrac{1}{2}x^2 + C$ Substituting back to y using $u = y - 1$

$y - 1 = e^{\frac{1}{2}x^2 + C}$	Solving for $y - 1$ by exponentiating
$y - 1 = \underbrace{e^{\frac{1}{2}x^2} \cdot e^C}_{c} = c \cdot e^{x^2/2}$	Replacing e^C by c and $e^{\frac{1}{2}x^2}$ by $e^{x^2/2}$
$y = c \cdot e^{x^2/2} + 1$	Adding 1 to each side

This is the general solution. To satisfy the initial condition $y(0) = 4$ we substitute $y = 4$ and $x = 0$.

$4 = ce^0 + 1$	$y = ce^{x^2/2} + 1$ with $y = 4$ and $x = 0$
$4 = c + 1$	Since $e^0 = 1$

This gives $c = 3$. Therefore, the particular solution is

$y = 3e^{x^2/2} + 1$	$y = ce^{x^2/2} + 1$ with $c = 3$

Accumulation of Wealth

The examples so far have *given* us a differential equation to solve. In this application we will first *derive* a differential equation to represent a situation and then solve it.

EXAMPLE 7

PREDICTING WEALTH

Suppose that you have saved $5000, and that you expect to save an additional $3000 during each year. If you deposit these savings in a bank account paying 5% interest compounded continuously, find a formula for your bank balance after t years.

Solution Let $y(t)$ stand for your bank balance (in thousands of dollars) after t years. Each year $y(t)$ grows by 3 (thousand dollars) plus 5% interest. This growth can be modeled by a differential equation:

Before continuing, be sure that you understand how this differential equation models the changes due to savings and interest.

We solve it by separating variables.

$$\frac{dy}{dt} = 3 + 0.05y \qquad \text{Replacing } y' \text{ by } \frac{dy}{dt}$$

$$\int \frac{dy}{3 + 0.05y} = \int dt \qquad \text{Dividing by } 3 + 0.05y, \text{ multiplying by } dt, \text{ and then integrating}$$

$$\begin{bmatrix} u = 3 + 0.05y \\ du = 0.05 dy \end{bmatrix} \qquad \text{Using a substitution}$$

$$20 \int \frac{du}{u} = \int dt \qquad \text{Substituting (the 20 comes from } \frac{1}{0.05} = 20\text{)}$$

$$20 \ln u = t + C \qquad \text{Integrating (assuming } u > 0\text{)}$$

$$\ln(3 + 0.05y) = 0.05t + \underbrace{0.05C}_{c} \qquad \text{Substituting } u = 3 + 0.05y \text{ and dividing by 20}$$

$$\ln(3 + 0.05y) = 0.05t + c \qquad \text{Replacing } 0.05C \text{ by } c$$

$$3 + 0.05y = e^{0.05t + c} = e^{0.05t} \underbrace{e^c}_{k} = ke^{0.05t} \qquad \text{Exponentiating and then simplifying constants}$$

$$0.05y = ke^{0.05t} - 3 \qquad \text{Subtracting 3}$$

$$y = \underbrace{20k}_{b}e^{0.05t} - 60 \qquad \text{Dividing by 0.05}$$

$$\text{(Always simplify arbitrary constants)}$$

$$y = be^{0.05t} - 60 \qquad \text{With arbitrary constant } b$$

You began at time $t = 0$ with 5 thousand dollars, which gives the initial condition $y(0) = 5$. We substitute $y = 5$ and $t = 0$.

$$5 = be^0 - 60 \qquad \begin{array}{l} y = be^{0.05t} - 60 \text{ with} \\ y = 5 \text{ and } t = 0 \end{array}$$

$$5 = b - 60 \qquad \text{Since } e^0 = 1$$

Therefore, $b = 65$, which we substitute into the general solution, obtaining a formula for your accumulated wealth after t years.

$$y = 65e^{0.05t} - 60 \qquad \begin{array}{l} y = be^{0.05t} - 60 \\ \text{with } b = 65 \end{array}$$

For example, to find your wealth after 10 years, we evaluate the solution of the differential equation at $t = 10$.

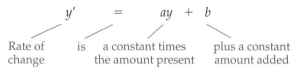

$y = 65e^{0.5} - 60 \approx 47.167 \qquad y = 65e^{0.05t} - 60 \text{ with } t = 10$

Therefore, after 10 years, you will have $47,167 in the bank.

12.5 Section Summary

We solved separable differential equations by separating the variables (moving every y to one side and every x to the other, with dx and dy in the numerators) and integrating both sides. The *general* solution of a differential equation includes an arbitrary constant, while a *particular* solution results from evaluating the constant, usually by applying an initial condition. The *slope field* of a differential equation allows you to construct geometrically a solution from any point by following the slopes away from the point.

We also saw how to *derive* a differential equation from information about how a quantity changes. For example, we can "read" the differential equation as follows.

$$\underbrace{y'}_{\text{Rate of change}} = \underbrace{ay}_{\substack{\text{is a constant times} \\ \text{the amount present}}} + \underbrace{b}_{\substack{\text{plus a constant} \\ \text{amount added}}}$$

The applied exercises show how differential equations lead to important formulas in a wide variety of fields.

▶ Solutions to Practice Problems

1. $y = e^{-x} + e^{3x}$
$y' = -e^{-x} + 3e^{3x}$
$y'' = e^{-x} + 9e^{3x}$

Substituting these expressions into the differential equation:

$(e^{-x} + 9e^{3x}) - 2(-e^{-x} + 3e^{3x}) - 3(e^{-x} + e^{3x}) \stackrel{?}{=} 0$

$e^{-x} + 9e^{3x} + 2e^{-x} - 6e^{3x} - 3e^{-x} - 3e^{3x} \stackrel{?}{=} 0$

$0 = 0 \qquad \text{It checks!}$

2. $\dfrac{dy}{dx} = \dfrac{6x^2}{y^4}$

$y^4\, dy = 6x^2\, dx$

$\displaystyle\int y^4\, dy = \int 6x^2\, dx$

$\tfrac{1}{5} y^5 = 2x^3 + C$

$y^5 = 10x^3 + 5C = 10x^3 + c$

$y = \sqrt[5]{10x^3 + c}$

The initial condition gives

$2 = \sqrt[5]{0 + c}$

$2 = \sqrt[5]{c}$

$32 = c$ Raising each side to the fifth power

Solution: $y = \sqrt[5]{10x^3 + 32}$

3. $\dfrac{dy}{dx} = x^2 y$

$\displaystyle\int \dfrac{dy}{y} = \int x^2\, dx$

$\ln y = \tfrac{1}{3} x^3 + C$

$y = e^{\tfrac{1}{3}x^3 + C} = e^{\tfrac{1}{3}x^3} e^C = c e^{x^3/3}$

The initial condition gives

$5 = c e^0 = c$

Solution: $y = 5 e^{x^3/3}$

12.5 Exercises

Verify that the function y satisfies the given differential equation.

1. $y = e^{2x} - 3e^x + 2$
$y'' - 3y' + 2y = 4$

2. $y = e^{5x} - 4e^x + 1$
$y'' - 6y' + 5y = 5$

3. $y = ke^{ax} - \dfrac{b}{a}$ (for constants a, b, and k)
$y' = ay + b$

4. $y = ax^2 + bx$ (for constants a and b)
$y' = \dfrac{y}{x} + ax$

Find the general solution of each differential equation or state that the differential equation is not separable. If the exercise says "and check," verify that your answer is a solution.

5. $y^2 y' = 4x$

6. $y^4 y' = 8x$

7. $y' = x + y$

8. $y' = xy - 1$

9. $y' = 6x^2 y$ and check

10. $y' = 12x^3 y$ and check

11. $y' = \dfrac{y}{x}$ and check

12. $y' = \dfrac{y^2}{x^2}$ and check

13. $yy' = 4x$
14. $yy' = 6x^2$
15. $y' = e^{xy}$
16. $y' = e^x + y$
17. $y' = 9x^2$
18. $y' = 6e^{-2x}$
19. $y' = \dfrac{x}{x^2 + 1}$
20. $y' = xy^2$
21. $y' = x^2 y$
22. $y' = \dfrac{x}{y}$
23. $y' = x^m y^n$ (for $m > 0$, $n \neq 1$)
24. $y' = x^m y$ (for $m > 0$)
25. $y' = 2\sqrt{y}$
26. $y' = 5 + y$
27. $xy' = x^2 + y^2$
28. $y' = \sqrt{x+y}$
29. $y' = xy + x$
30. $y' = x - 2xy$
31. $y' = ye^x - e^x$
32. $y' = ye^x - y$
33. $y' = ay^2$ (for constant $a > 0$)
34. $y' = axy$ (for constant a)
35. $y' = ay + b$ (for constants a and b)
36. $y' = (ay + b)^2$ (for constants $a \neq 0$ and b)

Solve each differential equation and initial condition and verify that your answer satisfies both the differential equation and the initial condition.

37. $\begin{cases} y^2 y' = 2x \\ y(0) = 2 \end{cases}$
38. $\begin{cases} y^4 y' = 3x^2 \\ y(0) = 1 \end{cases}$
39. $\begin{cases} y' = xy \\ y(0) = -1 \end{cases}$
40. $\begin{cases} y' = y^2 \\ y(2) = -1 \end{cases}$
41. $\begin{cases} y' = 2xy^2 \\ y(0) = 1 \end{cases}$
42. $\begin{cases} y' = 2xy^4 \\ y(0) = 1 \end{cases}$
43. $\begin{cases} y' = \dfrac{y}{x} \\ y(1) = 3 \end{cases}$
44. $\begin{cases} y' = \dfrac{2y}{x} \\ y(1) = 2 \end{cases}$
45. $\begin{cases} y' = 2\sqrt{y} \\ y(1) = 4 \end{cases}$
46. $\begin{cases} y' = \sqrt{y}\,e^x - \sqrt{y} \\ y(0) = 1 \end{cases}$
47. $\begin{cases} y' = y^2 e^x + y^2 \\ y(0) = 1 \end{cases}$
48. $\begin{cases} y' = xy - 5x \\ y(0) = 4 \end{cases}$
49. $\begin{cases} y' = ax^2 y \\ y(0) = 2 \end{cases}$ (for constant $a > 0$)
50. $\begin{cases} y' = axy \\ y(0) = 4 \end{cases}$ (for constant $a > 0$)

APPLIED EXERCISES

51. **BUSINESS: Elasticity** For a demand function $D(p)$, the elasticity of demand (see page 755) is defined as $E = \dfrac{-pD'}{D}$. Find demand functions $D(p)$ that have constant elasticity by solving the differential equation $\dfrac{-pD'}{D} = k$, where k is a constant.

52. **BIOMEDICAL: Cell Growth** A cell receives nutrients through its surface, and its surface area is proportional to the two-thirds power of its weight. Therefore, if $w(t)$ is the cell's weight at time t, then $w(t)$ satisfies $w' = aw^{2/3}$, where a is a positive constant. Solve this differential equation with the initial condition $w(0) = 1$ (initial weight 1 unit).

53–54. **BUSINESS: Continuous Annuities** An annuity is a fund into which one makes equal payments at regular intervals. If the fund earns interest at rate r compounded continuously, and deposits are made continuously at the rate of d dollars per year (a "continuous annuity"), then the value $y(t)$ of the fund after t years satisfies the differential equation $y' = d + ry$. (Do you see why?)

53. Solve the differential equation above for the continuous annuity $y(t)$ with deposit rate $d = \$1000$ and continuous interest rate $r = 0.05$, subject to the initial condition $y(0) = 0$ (zero initial value).

54. Solve the differential equation above for the continuous annuity $y(t)$, where d and r are unknown constants, subject to the initial condition $y(0) = 0$ (zero initial value).

55. GENERAL: Crime A medical examiner called to the scene of a murder will usually take the temperature of the body. A corpse cools at a rate proportional to the difference between its temperature and the temperature of the room. If $y(t)$ is the temperature (in degrees Fahrenheit) of the body t hours after the murder, and if the room temperature is 70°, then y satisfies

$$y' = -0.32(y - 70)$$

$y(0) = 98.6$ (body temperature initially 98.6°)

a. Solve this differential equation and initial condition.
b. Use your answer to part (a) to estimate how long ago the murder took place if the temperature of the body when it was discovered was 80°. [*Hint:* Find the value of t that makes your solution equal 80°.]

56. BIOMEDICAL: Glucose Levels Hospital patients are often given glucose (blood sugar) through a tube connected to a bottle suspended over their beds. Suppose that this "drip" supplies glucose at the rate of 25 mg per minute, and each minute 10% of the accumulated glucose is consumed by the body. Then the amount $y(t)$ of glucose (in excess of the normal level) in the body after t minutes satisfies

$$y' = 25 - 0.1y \quad \text{(Do you see why?)}$$

$y(0) = 0$ (zero excess glucose at $t = 0$)

Solve this differential equation and initial condition.

57. GENERAL: Friendships Suppose that you meet 30 new people each year, but each year you forget 20% of all of the people that you know. If $y(t)$ is the total number of people who you remember after t years, then y satisfies the differential equation $y' = 30 - 0.2y$. (Do you see why?) Solve this differential equation subject to the condition $y(0) = 0$ (you knew no one at birth).

58. ENVIRONMENTAL SCIENCE: Pollution For more than 75 years the Flexfast Rubber Company in Massachusetts discharged toxic toluene solvents into the ground at a rate of 5 tons per year. Each year approximately 10% of the accumulated pollutants evaporated into the air. If $y(t)$ is the total accumulation of pollution in the ground after t years, then y satisfies

$$y' = 5 - 0.1y \quad \text{(Do you see why?)}$$

$y(0) = 0$ (initial accumulation zero)

Solve this differential equation and initial condition to find a formula for the accumulated pollutant after t years.

59. GENERAL: Accumulation of Wealth Suppose that you now have $6000, you expect to save an additional $3000 during each year, and all of this is deposited in a bank paying 10% interest compounded continuously. Let $y(t)$ be your bank balance (in thousands of dollars) t years from now.

a. Write a differential equation that expresses the fact that your balance will grow by 3 (thousand dollars) and also by 10% of itself. [*Hint:* See Example 7.]
b. Write an initial condition to say that at time zero the balance is 6 (thousand dollars).
c. Solve your differential equation and initial condition.
d. Use your solution to find your bank balance $t = 25$ years from now.

60. GENERAL: Accumulation of Wealth Suppose that you now have $2000, you expect to save an additional $6000 during each year, and all of this is deposited in a bank paying 4% interest compounded continuously. Let $y(t)$ be your bank balance (in thousands of dollars) t years from now.

a. Write a differential equation that expresses the fact that your balance will grow by 6 (thousand dollars) and also by 4% of itself. [*Hint:* See Example 7.]
b. Write an initial condition to say that at time zero the balance is 2 (thousand dollars).
c. Solve your differential equation and initial condition.
d. Use your solution to find your bank balance $t = 20$ years from now.

61. BUSINESS: Sales Your company has developed a new product, and your marketing department has predicted how it will sell. Let $y(t)$

be the (monthly) sales of the product after t months.

a. Write a differential equation that says that the rate of growth of the sales will be four times the one-half power of the sales.
b. Write an initial condition that says that at time $t = 0$ sales were 10,000.
c. Solve this differential equation and initial value.
d. Use your solution to predict the sales at time $t = 12$ months.

62. BUSINESS: Sales Your company has developed a new product, and your marketing department has predicted how it will sell. Let $y(t)$ be the (monthly) sales of the product after t months.

a. Write a differential equation that says that the rate of growth of the sales will be six times the two-thirds power of the sales.
b. Write an initial condition that says that at time $t = 0$ sales were 1000.
c. Solve this differential equation and initial value.
d. Use your solution to predict the sales at time $t = 12$ months.

63. BIOMEDICAL: Bacterial Colony Let $y(t)$ be the size of a colony of bacteria after t hours.

a. Write a differential equation that says that the rate of growth of the colony is equal to eight times the three-fourths power of its present size.
b. Write an initial condition that says that at time zero the colony is of size 10,000.
c. Solve the differential equation and initial condition.
d. Use your solution to find the size of the colony at time $t = 6$ hours.

64. GENERAL: Value of a Building Let $y(t)$ be the value of a commercial building (in millions of dollars) after t years.

a. Write a differential equation that says that the rate of growth of the value of the building is equal to two times the one-half power of its present value.
b. Write an initial condition that says that at time zero the value of the building is 9 million dollars.

(continues)

c. Solve the differential equation and initial condition.
d. Use your solution to find the value of the building at time $t = 5$ years.

65. BIOMEDICAL: Heart Function In the *reservoir model*, the heart is viewed as a balloon that swells as it fills with blood (during a period called the *systole*), and then at time t_0 it shuts a valve and contracts to force the blood out (the *diastole*). Let $p(t)$ represent the pressure in the heart at time t.

a. During the diastole, which lasts from t_0 to time T, $p(t)$ satisfies the differential equation

$$\frac{dp}{dt} = -\frac{K}{R}p$$

Find the general solution $p(t)$ of this differential equation. (K and R are positive constants determined, respectively, by the strength of the heart and the resistance of the arteries. The differential equation states that as the heart contracts, the pressure decreases (dp/dt is negative) in proportion to itself.)

b. Find the particular solution that satisfies the condition $p(t_0) = p_0$. (p_0 is a constant representing the pressure at the transition time t_0.)

c. During the systole, which lasts from time 0 to time t_0, the pressure $p(t)$ satisfies the differential equation

$$\frac{dp}{dt} = KI_0 - \frac{K}{R}p$$

Find the general solution of this differential equation. (I_0 is a positive constant representing the constant rate of blood flow into the heart while it is expanding.) [*Hint*: Use the same u-substitution technique that was used in Example 7.]

d. Find the particular solution that satisfies the condition $p(t_0) = p_0$.

e. In parts (b) and (d) you found the formulas for the pressure $p(t)$ during the diastole ($t_0 \le t \le T$) and the systole ($0 \le t \le t_0$). Since the heart behaves in a cyclic fashion, these functions must satisfy $p(T) = p(0)$. Equate the solutions at these times (use the correct formula for each time) to derive the important relationship

$$R = \frac{p_0}{I_0} \frac{1 - e^{-KT/R}}{1 - e^{-Kt_0/R}}$$

66. BIOMEDICAL: Fick's Law Fick's Law governs the diffusion of a solute across a cell membrane. According to Fick's Law, the concentration $y(t)$ of the solute inside the cell at time t satisfies $\dfrac{dy}{dt} = \dfrac{kA}{V}(C_0 - y)$, where k is the diffusion constant, A is the area of the cell membrane, V is the volume of the cell, and C_0 is the concentration outside the cell.

a. Find the general solution of this differential equation. (Your solution will involve the constants k, A, V, and C_0.)
b. Find the particular solution that satisfies the initial condition $y(0) = y_0$, where y_0 is the initial concentration inside the cell.

The following exercises require the use of a slope field program.

For each differential equation and initial condition:

a. Use SLOPEFLD or a similar program to graph the slope field for the differential equation on the window $[-5, 5]$ by $[-5, 5]$.
b. Sketch the slope field on a piece of paper and draw a solution curve that follows the slopes and that passes through the point $(0, 2)$.

c. Solve the differential equation and initial condition.
d. Use SLOPEFLD or a similar program to graph the slope field and the solution that you found in part (c). How good was the sketch that you made in part (b) compared with the solution graphed in part (d)?

67. $\begin{cases} \dfrac{dy}{dx} = \dfrac{6x^2}{y^4} \\ y(0) = 2 \end{cases}$ **68.** $\begin{cases} \dfrac{dy}{dx} = \dfrac{x^2}{y^2} \\ y(0) = 2 \end{cases}$ **69.** $\begin{cases} \dfrac{dy}{dx} = \dfrac{4x}{y^3} \\ y(0) = 2 \end{cases}$

The following exercises require the use of a slope field program.

For each differential equation:

a. Use SLOPEFLD or a similar program to graph the slope field for the differential equation on the window $[-5, 5]$ by $[-5, 5]$.
b. Sketch the slope field on a piece of paper and draw a solution curve that follows the slopes and that passes through the given point.

70. $\dfrac{dy}{dx} = x - y^2$
point: $(0, 1)$

71. $\dfrac{dy}{dx} = \dfrac{x}{y^2 + 1}$
point: $(0, -1)$

72. $\dfrac{dy}{dx} = x \ln(y^2 + 1)$
point: $(0, -2)$

12.6 FURTHER APPLICATIONS OF DIFFERENTIAL EQUATIONS: THREE MODELS OF GROWTH

Introduction

This section continues our study of differential equations, but with a different approach. Instead of solving individual differential equations, we will solve three important classes of differential equations (for *unlimited*, *limited*, and *logistic* growth) and remember their solutions. This will enable us to solve many problems by identifying the appropriate differential equation and then immediately writing the solution. In this section we begin to *think in terms of differential equations*.

Proportion

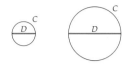

The circumference of a circle is proportional to its diameter.

We say that one quantity is *proportional to* another quantity if the first quantity is a *constant multiple* of the second. That is, y is proportional to x if $y = ax$ for some "proportionality constant" a. For example, the formula $C = \pi D$ for the circumference of a circle shows that the circumference C is proportional to the diameter D.

Unlimited Growth

In many situations the growth of a quantity is proportional to its present size. For example, a population of cells will grow in proportion to its present size, and a bank account earns interest in proportion to its current value. If a quantity y grows so that its rate of growth y' is proportional to its present size y, then y satisfies the differential equation

$$y' = ay$$

$\quad\;\;$ Rate of $\quad$ is propor- $\quad$ current
$\quad\;\;$ growth $\quad\quad$ tional to $\quad\;$ size

We solve this differential equation by separating variables.

$\dfrac{dy}{dt} = ay$ $\qquad\qquad$ Replacing y' by $\dfrac{dy}{dt}$

$\displaystyle\int \dfrac{dy}{y} = \int a\, dt$ $\qquad\;$ Dividing by y ($y \neq 0$), multiplying by dt, and integrating

$\ln y = at + C$ $\qquad\qquad$ Integrating ($y > 0$)

$y = e^{at+C} = e^{at}e^{C} = ce^{at}$ $\qquad$ Solving for y by exponentiating and then replacing e^C by c
$\qquad\qquad\;\;\uparrow_c$

At time $t = 0$ this becomes

$$y(0) = ce^0 = c \qquad\qquad y(t) = ce^{at} \text{ with } t = 0$$

Summarizing:

Unlimited Growth

The differential equation $\quad y' = ay$
with initial condition $\qquad\;\; y(0) = c$
is solved by $\qquad\qquad\qquad y = ce^{at}$

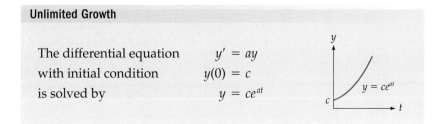

Such growth, where the rate of growth is proportional to the present size, is called *unlimited growth* because the solution y grows arbitrarily large. Given this result, whenever we encounter a differential equation of the form $y' = ay$, we can immediately write the solution $y = ce^{at}$, where c is the initial value.

EXAMPLE 1 PREDICTING ART APPRECIATION (UNLIMITED GROWTH)

An art collection, initially worth $25,000, continuously grows in value at the rate of 5% a year. Express this growth as a differential equation and find a formula for the value of the collection after t years. Then estimate the value of the art collection after 10 years.

Solution Growing continuously at the rate of 5% means that the value $y(t)$ grows by 5% *of itself*

$$y' = 0.05y$$

Rate of growth is 5% of the current value

This differential equation is of the form $y' = ay$ (unlimited growth) with $a = 0.05$ and initial value 25,000, so we may immediately write its solution.

$$y(t) = 25{,}000e^{0.05t} \qquad \begin{array}{l} y = ce^{at} \text{ with } a = 0.05 \\ \text{and } c = 25{,}000 \end{array}$$

This formula gives the value of the art collection after t years. To find the value after 10 years, we evaluate at $t = 10$:

$$y(10) = 25{,}000e^{0.05 \cdot 10} \qquad y(t) = 25{,}000e^{0.05t} \text{ with } t = 10$$
$$= 25{,}000e^{0.5} \approx 41{,}218 \qquad \text{Using a calculator}$$

In ten years the art collection will be worth $41,218.

The solution ce^{at} is the same as the continuous compounding formula Pe^{rt} (except for different letters). On page 73 we derived the formula Pe^{rt} rather laboriously, using the discrete interest formula $P(1 + r/m)^{mt}$, replacing m by rn, and taking the limit as $n \to \infty$. The present derivation, using differential equations, is much simpler and

Limited Growth

No real population can undergo unlimited growth for very long. Restrictions of food and space would soon slow its growth. If a quantity $y(t)$ cannot grow larger than a certain fixed maximum size M, and if its growth rate y' is proportional to how far it is from its upper limit, then y satisfies

$$\underbrace{y'}_{\text{Rate of growth}} = \underbrace{a}_{\substack{\text{is propor-}\\\text{tional to}}}\underbrace{(M - y)}_{\substack{\text{distance below}\\\text{upper bound } M}} \qquad a > 0$$

We solve this by separating variables.

$\dfrac{dy}{dt} = a(M - y)$ Replacing y' by $\dfrac{dy}{dt}$

$\displaystyle\int \dfrac{dy}{M - y} = \int a\, dt$ Dividing by $M - y$, multiplying by dt, and integrating

$\begin{bmatrix} u = M - y \\ du = -dy \end{bmatrix}$ Using a substitution

$-\displaystyle\int \dfrac{du}{u} = \int a\, dt$ Substituting

$-\ln u = at + C$ Integrating ($u > 0$)

$\ln(M - y) = -at - C$ Multiplying by -1 and replacing u by $(M - y)$

$M - y = e^{-at - C} = e^{-at}e^{-C} = ce^{-at}$ Solving for $M - y$ by exponentiating and simplifying

$y = M - ce^{-at}$ Subtracting M and multiplying by -1

We impose the initial conditon $y(0) = 0$ (size zero at time $t = 0$).

$0 = M - ce^0 = M - c$ $y = M - ce^{-at}$ with $y = 0$ and $t = 0$

Therefore, $c = M$, which gives the solution

$y = M - Me^{-at} = M(1 - e^{-at})$ $y = M - ce^{-at}$ with $c = M$

Summarizing:

> **Limited Growth**
>
> The differential equation
> $$y' = a(M - y)$$
> with initial condition
> $$y(0) = 0$$
> is solved by
> $$y = M(1 - e^{-at})$$

This type of growth, in which the rate of growth is proportional to the distance below an upper limit M, is called *limited growth* because the solution $y = M(1 - e^{-at})$ approaches the limit M as $t \to \infty$. Given this result, whenever we encounter a differential equation of the form

$$y' = a(M - y)$$

with initial value zero, we may immediately write the solution

$$y = M(1 - e^{-at})$$

Diffusion of Information by Mass Media

If a news bulletin is repeatedly broadcast over radio and television, the news spreads quickly at first, but later more slowly when most people have already heard it. Sociologists often assume that the rate at which news spreads is proportional to the number who have not yet heard the news. Let M be the population of a city, and let $y(t)$ be the number of people who have heard the news within t time units. Then y satisfies the differential equation

$$\underbrace{y'}_{\text{Rate of growth}} = \underbrace{a}_{\substack{\text{is propor-}\\\text{tional to}}} \underbrace{(M - y)}_{\substack{\text{number who have}\\\text{not heard the news}}}$$

We recognize this as the differential equation for limited growth, whose solution is $y = M(1 - e^{-at})$. It remains only to determine the values of the constants M and a.

EXAMPLE 2

PREDICTING SPREAD OF INFORMATION (LIMITED GROWTH)

An important news bulletin is broadcast to a town of 50,000 people, and after 2 hours 30,000 people have heard the news. Find a formula for the number of people who have heard the bulletin within t hours. Then find how many people will have heard the news within 6 hours.

Solution If $y(t)$ is the number of people who have heard the news within t hours, then y satisfies

$$y' = a(\underbrace{50{,}000 - y}_{\text{Number who have not heard the news}}) \qquad y' = a(M - y) \text{ with } M = 50{,}000$$

This is the differential equation for limited growth with $M = 50{,}000$, so the solution (from the preceding box) is

$$y = 50{,}000(1 - e^{-at}) \qquad y = M(1 - e^{-at}) \text{ with } M = 50{,}000$$

To find the value of the constant a, we use the given information that 30,000 people have heard the news within 2 hours.

$$30{,}000 = 50{,}000(1 - e^{-a \cdot 2}) \qquad \begin{array}{l} y = 50{,}000(1 - e^{-at}) \text{ with } y = 30{,}000 \\ \text{and } t = 2 \end{array}$$

$$0.6 = 1 - e^{-2a} \qquad \text{Dividing each side by } 50{,}000$$

$$0.4 = e^{-2a} \qquad \text{Subtracting 1 and then multiplying by } -1$$

$$-0.916 \approx -2a \qquad \begin{array}{l} \text{Taking natural logs (using } \ln e^x = x \\ \text{on the right)} \end{array}$$

$$a \approx 0.46 \qquad \text{Dividing by } -2$$

Therefore, the number of people who have heard the news within t hours is

$$y(t) = 50{,}000(1 - e^{-0.46t}) \qquad y = 50{,}000(1 - e^{-at}) \text{ with } a = 0.46$$

To find the number who have heard the news within 6 hours, we evaluate this solution at $t = 6$.

$$y(6) = 50{,}000(1 - e^{-(0.46)(6)}) = 50{,}000(1 - e^{-2.76}) \approx 46{,}835 \qquad \text{Using a calculator}$$

Within 6 hours about 46,800 people have heard the news.

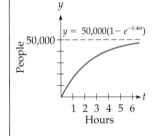

We found the value of the constant a by substituting the given data into the solution, simplifying, and taking logs. Similar steps will be required in many other problems, and in the future we shall omit the details.

Learning Theory

Psychologists have found that there seems to be an upper limit to the number of meaningless words that a person can memorize, and that memorizing becomes increasingly difficult approaching that bound. If M is this upper limit, and if $y(t)$ is the number of words that can be memorized in t minutes, then the situation is modeled by the differential equation for limited growth.

$$y' = a(M - y)$$

- y': Rate of increase
- a: is proportional to
- $(M - y)$: upper limit M minus number already memorized

This can be interpreted as saying that the rate at which new words can be memorized is proportional to the "unused memory capacity."

EXAMPLE 3   PREDICTING MEMORIZATION (LIMITED GROWTH)

Suppose that a person can memorize at most 100 meaningless words, and that after 15 minutes 10 words have been memorized. How long will it take to memorize 50 words?

Solution The number $y(t)$ of words that can be memorized in t minutes satisfies

$$y' = a(100 - y) \qquad \begin{array}{l} y' = a(M - y) \\ \text{with } M = 100 \end{array}$$

This is the differential equation for limited growth, so the solution is

$$y = 100(1 - e^{-at}) \qquad \begin{array}{l} y = M(1 - e^{-at}) \\ \text{with } M = 100 \end{array}$$

To evaluate the constant a, we use the given information that 10 words have been memorized in 15 minutes.

$$10 = 100(1 - e^{-a \cdot 15}) \qquad \begin{array}{l} y = 100(1 - e^{-at}) \text{ with} \\ y = 10 \text{ and } t = 15 \end{array}$$

Solving this equation for the constant a (omitting the details, which are the same as in Example 2) gives $a = 0.007$.

$$y(t) = 100(1 - e^{-0.007t}) \qquad \begin{array}{l} y = 100(1 - e^{-at}) \\ \text{with } a = 0.007 \end{array}$$

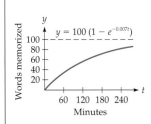

This solution gives the number of words that can be memorized in t minutes. To find how long it takes to memorize 50 words, we set this solution equal to 50 and solve for t.

$$50 = 100(1 - e^{-0.007t})$$

Solving for t (again the details are similar to those in Example 2) gives $t = 99$ minutes. Therefore, 50 words can be memorized in about 1 hour and 39 minutes.

Logistic Growth

Some quantities grow in proportion to both their present size *and* their distance from an upper limit M.

$$y' = ay(M - y)$$

Rate of growth / is propor- | tional to / present size \ upper limit M minus present size

This differential equation can be solved by separation of variables (see Exercise 55) to give the solution

$$y = \frac{M}{1 + ce^{-aMt}}$$

where c and a are positive constants. This function is called the *logistic* function, governing *logistic growth*.

Logistic Growth

The differential equation

$$y' = ay(M - y)$$

with initial condition

$$y(0) = \frac{M}{1 + c}$$

is solved by

$$y = \frac{M}{1 + ce^{-aMt}}$$

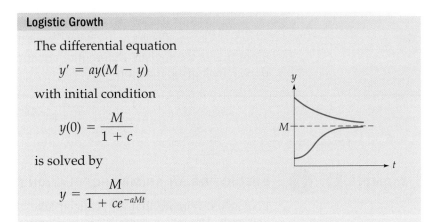

(The upper and lower curves in the box represent solutions whose initial values are, respectively, greater than or less than M.) As before, this result enables us to solve a differential equation "at sight," leaving only the evaluation of constants.

The lower curve in the box that rises to M is called a *sigmoidal* or *S-shaped curve*, and is used to model growth that begins slowly, then becomes more rapid, and finally slows again near the upper limit. Many different quantities grow according to sigmoidal curves.

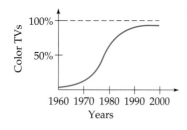

Percent of households owning a color television. (*Source*: Census Bureau).

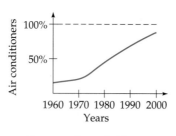

Percent of households with air conditioning. (*Source*: Worldwatch).

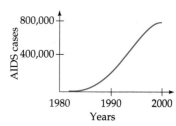

Total U.S. AIDS cases by year of diagnosis (*Source*: Centers for Disease Control).

Environmental Science

For an animal environment (such as a lake or a forest), the population that it can support will have an upper limit, called the *carrying capacity of the environment*. Ecologists often assume that an animal population grows in proportion to both its present size and its distance below the carrying capacity of the environment.

$$y' = ay(M - y)$$

Rate of growth is proportional to current size carrying capacity minus present size

Since this is the logistic differential equation, we know that the solution is the logistic function

$$y = \frac{M}{1 + ce^{-aMt}}$$

EXAMPLE 4

PREDICTING AN ANIMAL POPULATION (LOGISTIC GROWTH)

Ecologists estimate that an artificial lake can support a maximum of 2500 fish. The lake is initially stocked with 500 fish, and after 6 months the fish population is estimated to be 1500. Find a formula for the number of fish in the lake after t months, and estimate the fish population at the end of the first year.

Solution Letting $y(t)$ stand for the number of fish in the lake after t months, the situation is modeled by the logistic differential equation.

$$y' = ay(2500 - y) \qquad \begin{array}{l} y' = ay(M - y) \\ \text{with } M = 2500 \end{array}$$

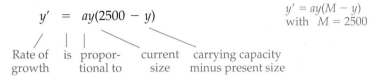

The solution is the logistic function with $M = 2500$.

$$y = \frac{2500}{1 + ce^{-a2500t}} \qquad y = \frac{M}{1 + ce^{-aMt}} \text{ with } M = 2500$$

$$= \frac{2500}{1 + ce^{-bt}} \qquad \text{Replacing } a \cdot 2500 \text{ by another constant } b$$

To evaluate the constants c and b, we use the fact that the lake was originally stocked with 500 fish.

$$500 = \frac{2500}{1 + ce^0} \qquad y = \frac{2500}{1 + ce^{-bt}} \text{ with } y = 500, t = 0$$

$$500 = \frac{2500}{1 + c} \qquad \text{Simplifying}$$

Solving this for c (omitting the details—the first step is to multiply both sides by $1 + c$) gives $c = 4$, so the logistic function becomes

$$y = \frac{2500}{1 + 4e^{-bt}} \qquad y = \frac{2500}{1 + ce^{-bt}} \text{ with } c = 4$$

To evaluate b, we substitute the information that the population is $y = 1500$ at $t = 6$.

$$1500 = \frac{2500}{1 + 4e^{-b \cdot 6}} \qquad y = \frac{2500}{1 + 4e^{-bt}} \text{ with } \begin{array}{l} y = 1500, \\ t = 6 \end{array}$$

Solving for b (again omitting the details—the first step is to multiply both sides by $1 + 4e^{-b \cdot 6}$) gives $b = 0.30$, so the logistic function becomes

$$y(t) = \frac{2500}{1 + 4e^{-0.3t}} \qquad y = \frac{2500}{1 + 4e^{-bt}} \text{ with } b = 0.3$$

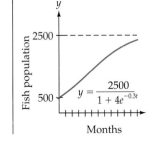

This is the formula for the population after t months. To find the population after a year, we evaluate at $t = 12$.

$$y(12) = \frac{2500}{1 + 4e^{-0.3(12)}} = \frac{2500}{1 + 4e^{-3.6}} \approx 2254 \quad \text{Using a calculator}$$

Therefore, the population at the end of the first year is 2254, which is about 90% of the carrying capacity of the lake.

Epidemics

Many epidemics spread at a rate proportional to both the number of people already infected (the "carriers") and also the number who have yet to catch the disease (the "susceptibles"). If $y(t)$ is the number of infected people at time t from a population of size M, then $y(t)$ satisfies the logistic differential equation

$$\underset{\substack{\text{Rate of}\\\text{growth}}}{y'} = \underset{\substack{\text{is propor-}\\\text{tional to}}}{a}\underset{\substack{\text{number}\\\text{infected}}}{y}\underset{\substack{\text{number}\\\text{susceptible}}}{(M - y)}$$

Therefore, the size of the infected population is given by the logistic function

$$y = \frac{M}{1 + ce^{-aMt}}$$

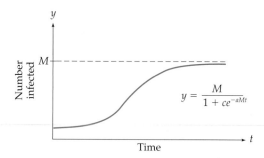

The constants are evaluated just as in Example 4, using the (initial) number of cases reported at time $t = 0$, and also the number of cases at some later time.

Spread of Rumors

Sociologists have found that rumors spread at a rate proportional to the number who have heard the rumor (the "informed") and the number who have not heard the rumor (the "uninformed"). Therefore, in a population of size M, the number $y(t)$ who have heard the rumor within t time units satisfies the logistic differential equation

$$y' = ay(M - y)$$

where y' is the Rate of growth, is proportional to, number informed, number uninformed.

The solution $y(t)$ is then the logistic function

$$y = \frac{M}{1 + ce^{-aMt}}$$

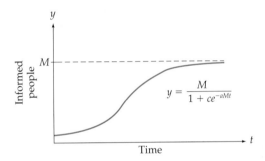

It remains only to evaluate the constants. The spread of a rumor is analogous to the spread of a disease, with an "informed" person being one who has been "infected."

Limited and Logistic Growth of Sales

The sales of a product whose total sales will approach an upper limit (market saturation) can be modeled by either the limited or the logistic equation. Which do you use when? For a product advertised over mass media, sales will at first grow rapidly, indicating a *limited* model. For a product becoming known only through "word of mouth," sales will at first grow slowly, indicating a *logistic* model.

Practice Problem

The graphs on the next page show the total sales through day t for two different products, A and B. Which of these products was advertised and which became known only by "word of mouth"? State an appropriate differential equation for each curve.

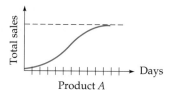
Product A

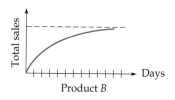
Product B

▶ Solution on next page

12.6 Section Summary

The unlimited, limited, and logistic growth models are summarized in the following table. If one of these differential equations governs a particular situation, we can write the solution immediately, evaluating the constants from the given data.

Three Models of Growth

Type	Differential Equation	Solution	Graph	Examples
Unlimited Growth is proportional to present size.	$y' = ay$	$y = ce^{at}$		Investments Bank accounts Unlimited populations
Limited Growth (starting at 0) is proportional to maximum size M minus present size.	$y' = a(M - y)$	$y = M(1 - e^{-at})$		Information spread by mass media Memorizing random information Total sales (advertised)
Logistic Growth is proportional to present size and to maximum size M minus present size.	$y' = ay(M - y)$	$y = \dfrac{M}{1 + ce^{-aMt}}$		Confined populations Epidemics Rumors Total sales (unadvertised)

To decide which (if any) of the three models applies in a given situation, think of whether the growth is proportional to *size*, to *unused capacity*, or to *both* (as shown in the chart). Notice that the differential equation gives much more insight into how the growth occurs than does the solution. This is what we meant at the beginning of the section by "thinking in terms of differential equations."

Graphing Calculator Exploration

Use program* SLOPEFLD or a similar program to graph the slope fields for the following differential equations (one at a time) on the window [0, 5] by [0, 5]:

a. $y' = y$
(unlimited)

b. $y' = 3 - y$
(limited)

c. $y' = y(3 - y)$
(logistic)

Do you see how the slope field gives a "picture" of the differential equation? Each graph is on the window [0, 5] by [0, 5].

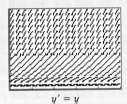
$y' = y$

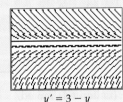
$y' = 3 - y$

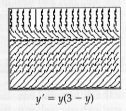
$y' = y(3 - y)$

*See the Preface for how to obtain this and other programs.

> ### Solution to Practice Problem

Product A, whose growth begins slowly, was not advertised, and the differential equation is logistic: $y' = ay(M - y)$. Product B, whose growth begins rapidly, was advertised, and the differential equation is limited: $y' = a(M - y)$.

12.6 Exercises

1. Verify that $y(t) = ce^{at}$ solves the differential equation for unlimited growth, $y' = ay$, with initial condition $y(0) = c$.

2. Verify that $y(t) = M(1 - e^{-at})$ solves the differential equation for limited growth, $y' = a(M - y)$, with initial condition $y(0) = 0$.

Determine the type of each differential equation: *unlimited* growth, *limited* growth, *logistic* growth, or *none* of these. (Do not solve, just identify the type.)

3. $y' = 0.02y$

4. $y' = 5(100 - y)$

5. $y' = 30(0.5 - y)$

6. $y' = 0.4y(0.01 - y)$

50. BIOMEDICAL: Glucose Levels Solve Exercise 56 on page 932 by factoring the right-hand side of the differential equation to write it in the form $y' = a(M - y)$.

OTHER GROWTH MODELS

Solve each differential equation by separation of variables.

51. GENERAL: Population If $y(t)$ is the size of a population at time t, then $\dfrac{y'}{y}$ is the population growth rate divided by the size of the population, and is called the *individual birthrate*. Suppose that the individual birthrate is proportional to the size of the population, $\dfrac{y'}{y} = ay$ for some constant a. Find a formula for the size of the population after t years.

52. GENERAL: Gompertz Curve Another differential equation that is used to model the growth of a population $y(t)$ is $y' = bye^{-at}$, where a and b are constants. Solve this differential equation.

53. GENERAL: Allometry Solve the differential equation of allometric growth: $y' = \dfrac{ay}{x}$ (where a is a constant). This differential equation governs the relative growth rates of different parts of the same animal.

54. GENERAL: Population Suppose that a population $y(t)$ in a certain environment grows in proportion to the *square* of the difference between the carrying capacity M and the present population, that is, $y' = a(M - y)^2$, where a is a constant. Solve this differential equation.

LOGISTIC GROWTH FUNCTION

55. Solve the logistic differential equation $y' = ay(M - y)$ as follows:

a. Separate variables to obtain
$$\frac{dy}{y(M - y)} = a\, dt$$

b. Integrate, using on the left-hand side the integration formula
$$\int \frac{dy}{y(M - y)} = \frac{1}{M}\ln\left(\frac{y}{M - y}\right)$$
(which may be checked by differentiation).

c. Exponentiate to solve for $\dfrac{y}{(M - y)}$ and then solve for y.

d. Show that the solution can be expressed as
$$y = \frac{M}{1 + ce^{-aMt}}$$

56. Find the inflection point of the logistic curve
$$f(x) = \frac{M}{1 + ce^{-aMx}}$$
and show that it occurs at midheight between $y = 0$ and the upper limit $y = M$ [*Hint:* Do you already know $f'(x)$?]

57. GENERAL: Raindrops *(Requires Slope Field Program)* Why do larger-sized raindrops fall faster than smaller ones? It depends on the resistance they encounter as they fall through the air. For large raindrops, the resistance to gravity's acceleration is proportional to the *square* of the velocity, whereas for small droplets, the resistance is proportional to the *first power* of the velocity. More precisely, their velocities obey the following differential equations, with each differential equation leading to a different *terminal velocity* for the raindrop:

i. $\dfrac{dv}{dt} = 32.2 - 0.1115v^2$ Downpour droplets, about 0.05 inch in diameter

ii. $\dfrac{dv}{dt} = 32.2 - 52.6v$ Drizzle droplets, about 0.003 inch in diameter

iii. $\dfrac{dv}{dt} = 32.2 - 5260v$ Fog droplets, about 0.0003 inch in diameter

(The 32.2 represents the force of gravity, and the other constant is determined experimentally.*)

a. Use a slope field program to graph the slope field of differential equation (i) on the window [0, 3] by [0, 20] (using x and y instead of t and v). From the slope field, must the solution curves rising from the bottom level off at a particular y-value? Estimate the value. This number is the terminal velocity (in feet per second) for a downpour droplet.

b. Do the same for differential equation (ii), but on the window [0, 0.1] by [0, 1]. What is the terminal velocity for a drizzle droplet?

c. Do the same for differential equation (iii), but on the window [0, 0.001] by [0, 0.01]. What is the terminal velocity for a fog droplet?

d. At this speed [from part (c)], how long would it take a fog droplet to fall 1 foot? This shows why fog clears so slowly.

*R. Gunn and Gilbert D. Kinzer, "The Terminal Velocity of Fall for Water Droplets in Stagnant Air," *Journal of Meteorology* **6**:243, 1949.

Chapter Summary with Hints and Suggestions

Reading the text and doing the exercises in this chapter have helped you to master the following skills, which are listed by section (in case you need to review them) and are keyed to particular Review Exercises. Answers for all Review Exercises are given at the back of the book, and full solutions can be found in the Student Solutions Manual.

12.1 Integration by Parts

- Find an integral using integration by parts. *(Review Exercises 1–14.)*

$$\int u\, dv = uv - \int v\, du$$

- Find an integral by whatever technique is necessary. *(Review Exercises 15–22.)*
- Solve an applied problem using integration by parts. *(Review Exercises 23–24.)*

$$\begin{pmatrix} \text{Present} \\ \text{value} \end{pmatrix} = \int_0^T C(t)e^{-rt}\, dt$$

$$\begin{pmatrix} \text{Total} \\ \text{accumulation} \end{pmatrix} = \int_0^T r(t)\, dt$$

12.2 Integration Using Tables

- Find an integral using a table of integrals. *(Review Exercises 25–36.)*
- Solve an applied problem using an integral table. *(Review Exercises 37–38.)*

12.3 Improper Integrals

- Evaluate an improper integral (if it is convergent). *(Review Exercises 39–56.)*

$$\int_a^\infty f(x)\, dx = \lim_{b \to \infty} \int_a^b f(x)\, dx$$

- Solve an applied problem involving an improper integral. *(Review Exercises 57–60.)*

- Predict whether an improper integral converges, and then check by evaluating it. *(Review Exercises 61–62.)*

12.4 Numerical Integration

- Approximate an integral using trapezoidal approximation "by hand." *(Review Exercises 63–68.)*

$$\int_a^b f(x)\,dx \approx \left[\frac{1}{2}f(x_1) + f(x_2) + \cdots + f(x_n) + \frac{1}{2}f(x_{n+1})\right] \cdot \Delta x$$

- Use a program to approximate an integral by trapezoidal approximation using a calculator. *(Review Exercises 69–74.)*

- Approximate an integral using Simpson's Rule "by hand." *(Review Exercises 75–80.)*

$$\int_a^b f(x)\,dx \approx [f(x_1) + 4f(x_2) + 2f(x_3) + \cdots + 4f(x_n) + f(x_{n+1})] \cdot \frac{\Delta x}{3}$$

- Use a program to approximate an integral by Simpson's Rule using a calculator. *(Review Exercises 81–86.)*

- Approximate an improper integral using trapezoidal approximation. *(Review Exercises 87–88.)*

12.5 Differential Equations

- Find the general solution of a differential equation by separation of variables. *(Review Exercises 89–98.)*

- Find a particular solution of a differential equation with an initial condition. *(Review Exercises 99–102.)*

- Solve an applied problem involving a differential equation. *(Review Exercises 103–106.)*

- Use a program to graph the slope field of a differential equation, and sketch the solution through a given point. *(Review Exercises 107–108.)*

12.6 Further Applications of Differential Equations: Three Models of Growth

- Choose an appropriate differential equation for an applied problem, and use it to solve the problem. *(Review Exercises 109–115.)*

Hints and Suggestions

- The unifying idea of this chapter is extensions of the concept of integration.

- Integration by parts takes one integral and gives another integral that, it is hoped, is simpler than the original integral. The formula is simply an integration version of the Product Rule. When using it, try to choose dv (including the dx) to be the most complicated part of the integrand that you can integrate, and, if possible, choose the u to be something that simplifies when differentiated.

- There are tables of integrals that are much longer than the one on the inside back cover. Longer tables, however, require much more time to search for the "right" formula. Other techniques (such as a substitution, integration by parts, or use of a formula more than once) may be used with an integral table.

- To find an integral, try the following methods. First try the "basic" formulas 1 through 4 on the inside back cover. Then try a substitution (formulas 5 through 7). If these methods fail, try integration by parts or an integral table. Remember that some integrals *cannot* be integrated (in terms of elementary functions). A *definite* integral can always be approximated by numerical methods.

- Before "evaluating" an improper integral, be sure that the integrand is defined over the interval, and that the integral is convergent. If the integral diverges, then it has no value and we simply state that it is divergent.

- Numerical integration involves approximating the area under a curve using geometric figures such as trapezoids or parabolas (Simpson's Rule). In practice, the calculations are usually carried out on a calculator or computer, but doing some "by hand" helps to make the method clear.

- A graphing calculator is very helpful for approximating definite integrals by trapezoidal approximation or Simpson's Rule for large values of n. Graphing calculators also have their own built-in numerical procedures for approximating integrals when you use FnInt.

- A differential equation is an equation involving derivatives (rates of change). Separation of variables involves separating the x's and y's to opposite sides of the equation and integrating both sides. Many useful differential equations can be solved by separation of variables, but many cannot. In fact, many differential equations cannot be solved by *any* method.

- For a differential equation, a solution involving an arbitrary constant is called a *general solution*, and a solution with the arbitrary constant replaced by a number is called a *particular solution*. The constant is usually evaluated by an *initial condition*, specifying the value of the solution at a particular point.

- A graphing calculator with a slope field program can show a "picture" of a differential equation of the form $dy/dx = f(x, y)$, drawing little slanted dashes with the correct slopes at many points of the plane. A solution can then be drawn through a given point following the indicated slopes.

- **Practice for Test:** Review Exercises 3, 5, 13, 25, 27, 33, 39, 41, 57, 63, 71, 79, 85, 89, 101, 103, 107, 111, 113, 115.

Review Exercises for Chapter 12

Practice test exercise numbers are in **green**.

12.1 Integration by Parts

Find each integral using integration by parts.

1. $\displaystyle\int xe^{2x}\, dx$

2. $\displaystyle\int xe^{-x}\, dx$

3. $\displaystyle\int x^8 \ln x\, dx$

4. $\displaystyle\int \sqrt[4]{x}\, \ln x\, dx$

5. $\displaystyle\int (x-2)(x+1)^5\, dx$

6. $\displaystyle\int (x+3)(x-1)^4\, dx$

7. $\displaystyle\int \frac{\ln t}{\sqrt{t}}\, dt$

8. $\displaystyle\int x^7 e^{x^4}\, dx$

9. $\displaystyle\int x^2 e^x\, dx$

10. $\displaystyle\int (\ln x)^2\, dx$

11. $\displaystyle\int x(x+a)^n\, dx$ (for constants a and $n > 0$)

12. $\displaystyle\int x(1-x)^n\, dx$ (for constant $n > 0$)

13. $\displaystyle\int_0^5 xe^x\, dx$

14. $\displaystyle\int_1^e x \ln x\, dx$

Find each integral by a substitution or by integration by parts, as appropriate.

15. $\displaystyle\int \frac{dx}{1-x}$

16. $\displaystyle\int xe^{-x^2}\, dx$

17. $\displaystyle\int x^3 \ln 2x\, dx$

18. $\displaystyle\int \frac{dx}{(1-x)^2}$

19. $\displaystyle\int \frac{\ln x}{x}\, dx$

20. $\displaystyle\int \frac{e^{2x}}{e^{2x}+1}\, dx$

21. $\displaystyle\int \frac{e^{\sqrt{x}}}{\sqrt{x}}\, dx$

22. $\displaystyle\int (e^{2x}+1)^3 e^{2x}\, dx$

23. **BUSINESS: Present Value of a Continuous Stream of Income** A company generates a continuous stream of income of $25t$ million dollars per year, where t is the number of years that the company has been in operation.

   a. Find the present value of this stream for the first 10 years at 5% interest compounded

(*continues*)

continuously. (Do not use a graphing calculator.)

b. Verify your answer to part (a) using FnInt on a graphing calculator.

24. ENVIRONMENTAL SCIENCE: Pollution
Radioactive waste is leaking out of cement storage vessels at the rate of $te^{0.2t}$ hundred gallons per month, where t is the number of months since the leak began.
a. Find the total leakage during the first 3 months. (Do not use a graphing calculator.)
b. Verify your answer to part (a) using FnInt on a graphing calculator

12.2 Integration Using Tables

Use the integral table on the inside back cover to find each integral.

25. $\displaystyle\int \frac{1}{25-x^2}\, dx$ 26. $\displaystyle\int \frac{1}{x^2-4}\, dx$

27. $\displaystyle\int \frac{x}{(x-1)(x-2)}\, dx$

28. $\displaystyle\int \frac{1}{(x-1)(x-2)}\, dx$

29. $\displaystyle\int \frac{1}{x\sqrt{x+1}}\, dx$ 30. $\displaystyle\int \frac{x}{\sqrt{x+1}}\, dx$

31. $\displaystyle\int \frac{1}{\sqrt{x^2+9}}\, dx$ 32. $\displaystyle\int \frac{1}{\sqrt{x^2+16}}\, dx$

33. $\displaystyle\int \frac{z^3}{\sqrt{z^2+1}}\, dz$ 34. $\displaystyle\int \frac{e^{2t}}{e^t+2}\, dt$

35. $\displaystyle\int x^2 e^{2x}\, dx$ 36. $\displaystyle\int (\ln x)^4\, dx$

37. BUSINESS: Cost A company's marginal cost function is $MC(x) = \dfrac{1}{(2x+1)(x+1)}$ and fixed costs are 1000 (all in dollars). Find the company's cost function.

38. GENERAL: Population The population of a town is growing at the rate of $\sqrt{t^2+1600}$ people per year, where t is the number of years from now.

a. Find the total increase in population during the first 30 years. (Do not use a graphing calculator.)

b. Verify your answer to part (a) using FnInt on a graphing calculator.

12.3 Improper Integrals

Find the value of each improper integral or state that it is divergent.

39. $\displaystyle\int_1^\infty \frac{1}{x^5}\, dx$ 40. $\displaystyle\int_1^\infty \frac{1}{x^6}\, dx$ 41. $\displaystyle\int_1^\infty \frac{1}{\sqrt[5]{x}}\, dx$

42. $\displaystyle\int_1^\infty \frac{1}{\sqrt[6]{x}}\, dx$ 43. $\displaystyle\int_0^\infty e^{-2x}\, dx$

44. $\displaystyle\int_4^\infty e^{-0.5x}\, dx$ 45. $\displaystyle\int_0^\infty e^{2x}\, dx$

46. $\displaystyle\int_4^\infty e^{0.5x}\, dx$ 47. $\displaystyle\int_0^\infty e^{-t/5}\, dt$

48. $\displaystyle\int_{100}^\infty e^{-t/10}\, dt$ 49. $\displaystyle\int_0^\infty \frac{x^3}{(x^4+1)^2}\, dx$

50. $\displaystyle\int_0^\infty \frac{x^4}{(x^5+1)^2}\, dx$ 51. $\displaystyle\int_{-\infty}^0 e^{2t}\, dt$

52. $\displaystyle\int_{-\infty}^0 e^{4t}\, dt$ 53. $\displaystyle\int_{-\infty}^4 \frac{1}{(5-x)^2}\, dx$

54. $\displaystyle\int_{-\infty}^8 \frac{1}{(9-x)^2}\, dx$ 55. $\displaystyle\int_{-\infty}^\infty \frac{e^{-x}}{(1+e^{-x})^4}\, dx$

56. $\displaystyle\int_{-\infty}^\infty \frac{e^{-x}}{(1+e^{-x})^3}\, dx$

57. GENERAL: Permanent Endowments Find the size of the permanent endowment needed to generate an annual $6000 forever at an interest rate of 10% compounded continuously.

58. GENERAL: Automobile Age Insurance records indicate that the proportion of cars on the road that are more than x years old is approximated by the integral $\int_x^\infty 0.21e^{-0.21t}\, dt$. Find the proportion of cars that are more than 5 years old.

59. BUSINESS: Book Sales A publisher estimates that the demand for a certain book will be $12e^{-0.05t}$ thousand copies per year, where t is the number of years since the book's publication. Find the total number of books that will be sold from the publication date onward.

60. GENERAL: Resource Consumption If the rate of consumption of a certain mineral is $300e^{-0.04t}$ million tons per year (where t is the number of years from now), find the total amount of the mineral that will be consumed from now on.

For each improper integral, use a graphing calculator to evaluate it (or to show that it diverges) as follows:
a. Define y_1 to be the definite integral (using FnInt) of the integrand from 1 to x.
b. Make a TABLE of values of y_1 for x-values such as 1, 10, 100, and 1000. Does the integral converge (and if so, to what number) or does it diverge?
c. Verify your answers to part (b) by evaluating the improper integral "by hand."

61. $\displaystyle\int_1^\infty \frac{1}{x^3}\,dx$ **62.** $\displaystyle\int_1^\infty \frac{1}{\sqrt[3]{x}}\,dx$

12.4 Numerical Integration

Estimate each integral using trapezoidal approximation with the given value of n. (Round all calculations to three decimal places.)

63. $\displaystyle\int_0^1 \sqrt{1+x^4}\,dx, \quad n=3$

64. $\displaystyle\int_0^1 \sqrt{1+x^5}\,dx, \quad n=3$

65. $\displaystyle\int_0^1 e^{x^2/2}\,dx, \quad n=4$

66. $\displaystyle\int_0^1 e^{-x^2/2}\,dx, \quad n=4$

67. $\displaystyle\int_{-1}^1 \ln(1+x^2)\,dx, \quad n=4$

68. $\displaystyle\int_{-1}^1 \ln(x^3+2)\,dx, \quad n=4$

Use a trapezoidal approximation program such as TRAPEZOD to approximate each integral. Use successively higher values of n until the results agree to three decimal places (rounded).

69. $\displaystyle\int_0^1 \sqrt{1+x^4}\,dx$ **70.** $\displaystyle\int_0^1 \sqrt{1+x^5}\,dx$

71. $\displaystyle\int_0^1 e^{x^2/2}\,dx$ **72.** $\displaystyle\int_0^1 e^{-x^2/2}\,dx$

73. $\displaystyle\int_{-1}^1 \ln(1+x^2)\,dx$ **74.** $\displaystyle\int_{-1}^1 \ln(x^3+2)\,dx$

Estimate each integral using Simpson's Rule (parabolic approximation) with the given value of n. (Round all calculations to four decimal places.)

75. $\displaystyle\int_0^1 \sqrt{1+x^4}\,dx, \quad n=4$

76. $\displaystyle\int_0^1 \sqrt{1+x^5}\,dx, \quad n=4$

77. $\displaystyle\int_0^1 e^{x^2/2}\,dx, \quad n=4$

78. $\displaystyle\int_0^1 e^{-x^2/2}\,dx, \quad n=4$

79. $\displaystyle\int_{-1}^1 \ln(1+x^2)\,dx, \quad n=4$

80. $\displaystyle\int_{-1}^1 \ln(x^3+2)\,dx, \quad n=4$

Use a Simpson's Rule approximation program such as SIMPSON to approximate each integral. Use successively higher values of n until the rounded results agree to six decimal places.

81. $\displaystyle\int_0^1 \sqrt{1+x^4}\,dx$ **82.** $\displaystyle\int_0^1 \sqrt{1+x^5}\,dx$

83. $\displaystyle\int_0^1 e^{x^2/2}\,dx$ **84.** $\displaystyle\int_0^1 e^{-x^2/2}\,dx$

85. $\displaystyle\int_{-1}^1 \ln(1+x^2)\,dx$ **86.** $\displaystyle\int_{-1}^1 \ln(x^3+2)\,dx$

For each improper integral:
a. Make it a "proper" integral by using the substitution $x=\dfrac{1}{t}$ and simplifying.

(continues)

b. Approximate the proper integral using trapezoidal approximation with $n = 4$. Keep three decimal places.

87. $\int_1^\infty \dfrac{1}{x^2+1}\,dx$ 88. $\int_1^\infty \dfrac{x^2}{x^4+1}\,dx$

12.5 Differential Equations

Find the general solution of each differential equation.

89. $y^2 y' = x^2$ 90. $y' = x^2 y$ 91. $y' = \dfrac{x^3}{x^4+1}$

92. $y' = xe^{-x^2}$ 93. $y' = y^2$ 94. $y' = y^3$

95. $y' = 1 - y$ 96. $y' = \dfrac{1}{y}$

97. $y' = xy - y$ 98. $y' = x^2 + x^2 y$

Solve each differential equation and initial condition.

99. $y^2 y' = 3x^2$ 100. $y' = \dfrac{y}{x^2}$
    $y(0) = 1$ $y(1) = 1$

101. $y' = \dfrac{y}{x^3}$ 102. $y' = \sqrt[3]{y}$
     $y(1) = 1$ $y(1) = 0$

103. **GENERAL: Accumulation of Wealth** Suppose that you now have $10,000 and that you expect to save an additional $4000 during each year, and all of this is deposited in a bank paying 5% interest compounded continuously. Let $y(t)$ be your bank balance (in thousands of dollars) after t years.
   a. Write a differential equation and initial condition to model your bank balance.
   b. Solve your differential equation and initial condition.
   c. Use your solution to find your bank balance after 10 years.

104. **ENVIRONMENTAL SCIENCE: Pollution** A town discharges 4 tons of pollutant into a lake annually, and each year bacterial action removes 25% of the accumulated pollution.
   a. Write a differential equation and initial condition for the amount of pollution in the lake.
   b. Solve your differential equation to find a formula for the amount of pollution in the lake after t years.

105. **GENERAL: Fever Thermometers** How long should you keep a thermometer in your mouth to take your temperature? *Newton's Law of Cooling* says that the thermometer reading rises at a rate proportional to the difference between your actual temperature and the present reading. For a fever of 106 degrees Fahrenheit, the thermometer reading $y(t)$ after t minutes in your mouth satisfies $y' = 2.3(106 - y)$ with $y(0) = 70$ (initially at room temperature). (The constant 2.3 is typical for household thermometers.) Solve this differential equation and initial condition.

106. **GENERAL: Fever Thermometers** (Continued) Use your solution $y(t)$ in Exercise 105 to calculate $y(1)$, $y(2)$, and $y(3)$, the thermometer readings after 1, 2, and 3 minutes. Do you see why 3 minutes is the usually recommended time for keeping the thermometer in your mouth?

The following exercises require the use of a slope field program.

For each differential equation and initial condition:
   a. Use SLOPEFLD or a similar program to graph the slope field for the differential equation on the window $[-5, 5]$ by $[-5, 5]$.
   b. Sketch the slope field on a piece of paper and draw a solution curve that follows the slopes and that passes through the point $(0, -2)$.
   c. Solve the differential equation and initial condition.
   d. Use SLOPEFLD or a similar program to graph the slope field and the solution that you found in part (c). How good was the sketch that you made in part (b) compared with the solution graphed in part (d)?

107. $\begin{cases} \dfrac{dy}{dx} = \dfrac{x^2}{y^2} \\ y(0) = -2 \end{cases}$ 108. $\begin{cases} \dfrac{dy}{dx} = \dfrac{x}{y^2} \\ y(0) = -2 \end{cases}$

12.6 Further Applications of Differential Equations: Three Models of Growth

For each situation, write an appropriate differential equation (unlimited, limited, or logistic). Then find its solution and solve the problem.

109. **GENERAL: Postage Stamps** On average, the price of a first-class postage stamp grows continuously at the rate of about 7% per year. If in 2002 the price was 37 cents, estimate the price in the year 2010.

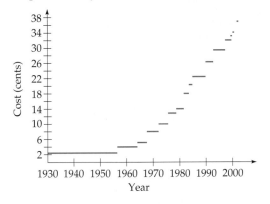

110. **ECONOMICS: Computer Expenditure** The amount spent by Americans on computers and related hardware is increasing continuously at the rate of 12% per year. If in 2000 the amount was $16.8 billion, estimate the amount in the year 2010.

111. **BIOMEDICAL: Epidemics** A virus spreads through a university community of 8000 people at a rate proportional to both the number already infected and the number not yet infected. If it begins with 10 cases and grows in a week to 150 cases, estimate the size of the epidemic after 2 weeks.

112. **SOCIAL SCIENCE: Rumors** A rumor spreads through a school of 500 students at a rate proportional to both the number who have heard and the number who have not heard the rumor. If the rumor began with 2 students and within a day had spread to 75, how many students will have heard the rumor within 2 days?

113. **BUSINESS: Total Sales** A manufacturer estimates that he can sell a maximum of 10,000 videocassette recorders in a city. His total sales grow at a rate proportional to how far they are below this upper limit. If after 7 months the total sales are 3000, find a formula for the total sales after t months. Then use your answer to estimate the total sales at the end of the first year.

114. **GENERAL: Learning** Suppose that the maximum rate at which a mail carrier can sort letters is 60 letters per minute, and that she learns at a rate proportional to how far she is below this upper limit. If after 2 weeks on the route she can sort 25 letters per minute, how many weeks will it take her to sort 50 letters per minute?

115. **BUSINESS: Advertising** A new product is advertised extensively on television to a city of 500,000 people, and the number of people who have seen the ads increases at a rate proportional to the number who have not yet seen them. If within 2 weeks 200,000 people have seen the ads, how long must the product be advertised to reach 400,000 people?

13 Calculus of Several Variables

- 13.1 Functions of Several Variables
- 13.2 Partial Derivatives
- 13.3 Optimizing Functions of Several Variables
- 13.4 Least Squares
- 13.5 Lagrange Multipliers and Constrained Optimization
- 13.6 Total Differentials and Approximate Changes
- 13.7 Multiple Integrals

Application Preview

Safe Cars, Unsafe Streets

Since 1970, improvements in automobile design and increased use of seat belts have lowered the number of automobile-related fatalities in the United States. On the other hand, during this same period the greater availability of handguns, together with other socioeconomic causes, has increased the number of gunshot fatalities. The data are shown in the following table and graph.

	Years Since 1970	Automobile Fatalities	Gunshot Fatalities
1970	0	55,800	26,800
1975	5	47,100	33,400
1980	10	53,200	32,900
1985	15	45,800	31,800
1990	20	47,600	36,000
1995	25	43,900	37,900
2000	30	41,300	32,200

Source: National Center for Health Statistics

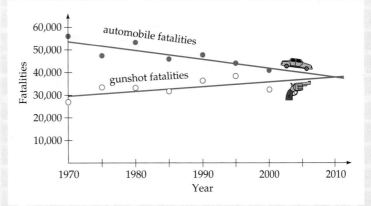

Two lines have been fit to these data points by a technique called "least squares." The lines show that at some time guns will overtake automobiles as the leading cause of accidental death in America. (In Exercise 31 on page 1013 you will find the precise year.)

In this chapter we will discuss the least squares technique for fitting lines or curves to data points, as well as other useful applications of functions of several variables.

13.1 FUNCTIONS OF SEVERAL VARIABLES

Introduction

Many quantities depend on *several* variables. For example, the "wind-chill" factor announced by the weather bureau during the winter depends on two variables: temperature and wind speed. The cost of a telephone call depends on *three* variables: distance, duration, and time of day.

In this chapter we define functions of two or more variables and learn how to differentiate and integrate them. We use derivatives for calculating rates of change and optimizing functions, and integrals for finding volumes, continuous sums, and average values.

Graphing calculators will be less useful in this chapter because of their small screens and limited computing power, but computer-drawn pictures of three-dimensional surfaces will be very useful.

Functions of Two Variables

A function f that depends on *two* variables, x and y, is written $f(x, y)$ (read: f of x and y). The *domain* of the function is the set of all ordered pairs (x, y) for which it is defined. The *range* is the set of all resulting values of the function. Formally:

Function of Two Variables

A function f of two variables is a rule such that to each ordered pair (x, y) in the domain of f there corresponds one and only one number $f(x, y)$.

If the domain is not stated, it will always be taken to be the largest set of ordered pairs for which the function is defined (the "natural domain").

EXAMPLE 1

FINDING THE DOMAIN OF A FUNCTION

For $f(x, y) = \dfrac{\sqrt{x}}{y^2}$, find   **a.** the domain   **b.** $f(9, -1)$

Solution

a. $\{(x, y) \mid x \geq 0, y \neq 0\}$ For $\sqrt{x}/y^2$, x cannot be negative (because of the $\sqrt{}$), and y cannot be zero

b. $f(9, -1) = \dfrac{\sqrt{9}}{(-1)^2} = \dfrac{3}{1} = 3$ $f(x, y) = \sqrt{x}/y^2$ with $x = 9$ and $y = -1$

EXAMPLE 2   **FINDING THE DOMAIN OF A FUNCTION INVOLVING LOGARITHMS AND EXPONENTIALS**

For $g(u, v) = e^{uv} - \ln u$, find   **a.** the domain   **b.** $g(1, 2)$

Solution

a. $\{(u, v) \mid u > 0\}$ u must be positive so that its logarithm is defined

b. $g(1, 2) = e^{1 \cdot 2} - \underbrace{\ln 1}_{0} = e^2 - 0 = e^2$ $g(u, v) = e^{uv} - \ln u$ with $u = 1$ and $v = 2$

Practice Problem 1   For $f(x, y) = \dfrac{\ln x}{e^{\sqrt{y}}}$, find   **a.** the domain   **b.** $f(e, 4)$

➤ Solutions on page 970

Functions of two variables are used in many applications.

EXAMPLE 3   **FINDING A COMPANY'S COST FUNCTION**

A company manufactures three-speed and ten-speed bicycles. It costs $100 to make each three-speed bicycle, it costs $150 to make each ten-speed bicycle, and fixed costs are $2500. Find the cost function, and use it to find the cost of producing 15 three-speed bicycles and 20 ten-speed bicycles.

Solution   Let:

x = the number of three-speed bicycles
y = the number of ten-speed bicycles

The cost function is

$$C(x, y) = \underset{\substack{\uparrow \\ \text{Unit} \\ \text{cost}}}{100x} + \underset{\substack{\uparrow \\ \text{Quan-} \\ \text{tity}}}{} \underset{\substack{\uparrow \\ \text{Unit} \\ \text{cost}}}{150y} + \underset{\substack{\uparrow \\ \text{Quan-} \\ \text{tity}}}{} \underset{\substack{\uparrow \\ \text{Fixed} \\ \text{costs}}}{2500}$$

The cost of producing 15 three-speed bicycles and 20 ten-speed bicycles is found by evaluating $C(x, y)$ at $x = 15$ and $y = 20$:

$$C(15, 20) = 100 \cdot 15 + 150 \cdot 20 + 2500$$
$$= 1500 + 3000 + 2500 = 7000$$

Producing 15 three-speed and 20 ten-speed bicycles costs $7000.

The variables x and y in the preceding example stand for numbers of bicycles, and so should take only integer values. Instead, however, we will allow x and y to be "continuous" variables, and round to integers at the end if necessary.

Some other "everyday" examples of functions of two variables are as follows:

$A(l, w) = lw$ Area of a rectangle of length l and width w

$f(w, v) = kwv^2$ Length of the skid marks for a car of weight w and velocity v skidding to a stop (k is a constant depending on the road surface)

Cobb–Douglas Production Functions

A function used to model the output of a company or a nation is called a *production function*, and the most famous is the Cobb–Douglas production function*

$$P(L, K) = aL^b K^{1-b} \qquad \text{For constants } a > 0 \text{ and } 0 < b < 1$$

This function expresses the total production P as a function of L, the number of units of labor, and K, the number of units of capital. (Labor

---

* First used by Charles Cobb and Paul Douglas in a landmark study of the American economy published in 1928.

is measured in work-hours, and capital means *invested* capital, including the cost of buildings, equipment, and raw materials.)

EXAMPLE 4

EVALUATING A COBB–DOUGLAS PRODUCTION FUNCTION

Cobb and Douglas modeled the output of the American economy by the function $P(L, K) = L^{0.75}K^{0.25}$. Find $P(150, 220)$.

Solution

$$P(150, 220) = (150)^{0.75}(220)^{0.25}$$ $P(L, K) = L^{0.75}K^{0.25}$ with $L = 150$ and $K = 220$

$$\approx (42.9)(3.85) \approx 165$$ Using a calculator

That is, 150 units of labor and 220 units of capital should result in approximately 165 units of production.

Graphing Calculator Exploration

The windchill index (revised in 2001), announced by the weather bureau during the winter to measure the combined effect of wind and cold, is calculated from the formula below, where x is wind speed (miles per hour) and y is temperature (degrees Fahrenheit).

$$W(x, y) = 35.74 + 0.6215y - 35.75x^{0.16} + 0.4275yx^{0.16}$$

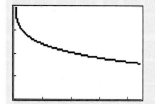

a. Enter this function into your graphing calculator but with y (temperature) replaced by 32 so that it becomes a function of one variable, x (wind speed). Then graph the function on the window [0, 45] by [0, 40]. The graph shows how the perceived temperature decreases as wind speed increases.

b. Find the wind speed that makes it feel like 18 degrees (that is, find x where $y = 18$).

c. Notice that the graph drops more steeply for low wind speeds than for high wind speeds. What does this mean about the effect of an extra 5 miles per hour of wind on a calm day as opposed to a windy day? (Exercises 36 and 37 continue this analysis.)

Functions of Three or More Variables

Functions of three (or more) variables are defined analogously. Some examples are:

$V(l, w, h) = lwh$ — Volume of a rectangular solid of length l, width w, and height h

$W(P, r, n) = Pe^{rn}$ — Worth of P dollars invested at a continuous interest rate r for n years

$f(w, x, y, z) = \dfrac{w + x + y + z}{4}$ — Average of four numbers

---

EXAMPLE 5

FINDING THE DOMAIN OF A FUNCTION

For $f(x, y, z) = \dfrac{\sqrt{x}}{y} + \ln \dfrac{1}{z}$, find **a.** the domain **b.** $f(4, -1, 1)$

Solution

a. In $f(x, y, z) = \dfrac{\sqrt{x}}{y} + \ln \dfrac{1}{z}$ we must have $x \geq 0$ (because of the square root), $y \neq 0$ (since it is a denominator), and $z > 0$ (so that $1/z$ has a logarithm). Therefore, the domain is

$$\{ (x, y, z) \mid x \geq 0, y \neq 0, z > 0 \}$$

b. $f(4, -1, 1) = \dfrac{\sqrt{4}}{-1} + \ln \dfrac{1}{1} = \dfrac{2}{-1} + \underbrace{\ln 1}_{0} = -2$

EXAMPLE 6

FINDING THE VOLUME AND AREA OF A DIVIDED BOX

An open-top box is to have a center divider, as shown in the diagram. Find formulas for the volume V of the box and for the total amount of material M needed to construct the box.

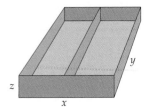

Solution The volume is length times width times height.

$$V = xyz$$

The box consists of a bottom, a front and back, two sides, and a divider, whose areas are shown in the diagram. Therefore, the total amount of material (the area) is

$$M = \underset{\substack{| \\ \text{Bottom}}}{xy} + \underset{\substack{| \\ \text{Back} \\ \text{and front}}}{2xz} + \underset{\substack{| \\ \text{Sides and} \\ \text{divider}}}{3yz}$$

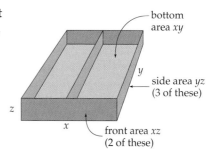

bottom area xy

side area yz (3 of these)

front area xz (2 of these)

Practice Problem 2 Find a formula for the total amount of material M needed to construct an open-top box with three parallel dividers. Use the variables shown in the diagram.

▶ **Solution on page 970**

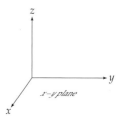

Graph of a Function of Two Variables

Graphing a function of two variables requires a three-dimensional coordinate system. We draw three perpendicular axes as shown on the right.* We will usually draw only the positive half of each axis, although each axis extends infinitely far in the negative direction as well. The plane at the base is called the *x-y plane*.

A point in a three-dimensional coordinate system is specified by three coordinates, giving its distances from the origin in the x, y, and z directions. For example, the point

$$(2, 3, 4)$$
$$\underset{\substack{| \\ x\text{-coordinate}}}{\uparrow} \underset{\substack{| \\ y\text{-coordinate}}}{\uparrow} \underset{\substack{| \\ z\text{-coordinate}}}{\uparrow}$$

is plotted by starting at the origin, moving 2 units in the x direction, 3 units in the y direction, and then 4 units in the (vertical) z direction.

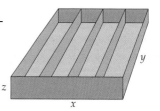

The three-dimensional ("right-handed") coordinate system

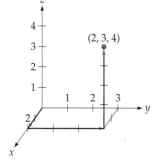

The point (2, 3, 4)

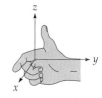

*This is called a "right-handed" coordinate system because the x, y, and z axes correspond to the first two fingers and thumb of the right hand.

To graph a function of two variables we choose values for x and y, calculate z-values from $z = f(x, y)$, and plot the points (x, y, z).

EXAMPLE 7

GRAPHING A FUNCTION OF TWO VARIABLES

To graph $f(x, y) = 18 - x^2 - y^2$, we set z equal to the function.

$$z = 18 - x^2 - y^2 \qquad \text{z replaces $f(x, y)$}$$

Then we choose values for x and y. Choosing $x = 1$ and $y = 2$ gives

$$z = 18 - 1^2 - 2^2 = 13 \qquad \text{$z = 18 - x^2 - y^2$ with $x = 1$ and $y = 2$}$$

for the point

$$(1, 2, 13) \qquad \text{The chosen $x = 1$, $y = 2$, and the calculated z}$$

Choosing $x = 2$ and $y = 3$ gives

$$z = 18 - 2^2 - 3^2 = 5 \qquad \text{$z = 18 - x^2 - y^2$ with $x = 2$ and $y = 3$}$$

for the point

$$(2, 3, 5) \qquad \text{The chosen $x = 2$, $y = 3$, and the calculated z}$$

These points (1, 2, 13) and (2, 3, 5) are plotted on the graph on the left below. The completed graph of the function is shown on the right.

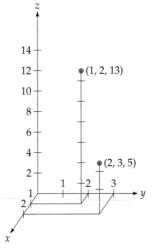

The points (1, 2, 13) and (2, 3, 5) of the function $f(x, y) = 18 - x^2 - y^2$

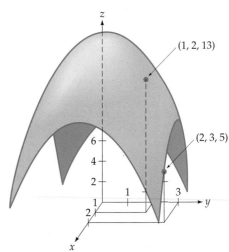

The graph of $f(x, y) = 18 - x^2 - y^2$

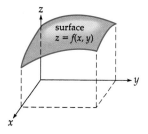

The graph of a function of two variables is a surface whose height above the point (x, y) in the x-y plane is $z = f(x, y)$.

In general, the graph of a function of *two* variables is a *surface* above or below the x-y plane, just as the graph of a function of *one* variable is a *curve* above or below the x-axis.

Graphing functions of two variables involves drawing three-dimensional graphs, which is very difficult. Graphing functions of *more* than two variables requires *more* than three dimensions, and is impossible. For this reason we will not graph functions of several variables. We will, however, often speak of a function of two variables as representing a *surface* in three-dimensional space.

Spreadsheet Exploration

The following spreadsheet* graph of $f(x, y) = 18 - x^2 - y^2$ from the previous example is a chart showing the values of the function that were calculated for values of x and y between -5 and 5.

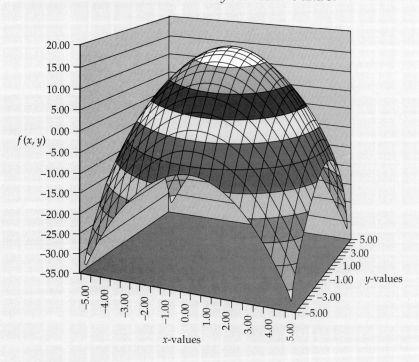

*See the Preface for how to obtain this and other Excel spreadsheets.

968 CHAPTER 13 CALCULUS OF SEVERAL VARIABLES

Just as with functions of one variable, useful graphs can be constructed only if we know the values of the variables corresponding to points of interest. As we have already seen, these values are easily found using calculus.

Relative Extreme Points and Saddle Points

Certain points on a surface of such a graph are of special importance.

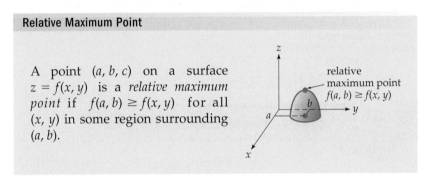

Relative Maximum Point

A point (a, b, c) on a surface $z = f(x, y)$ is a *relative maximum point* if $f(a, b) \geq f(x, y)$ for all (x, y) in some region surrounding (a, b).

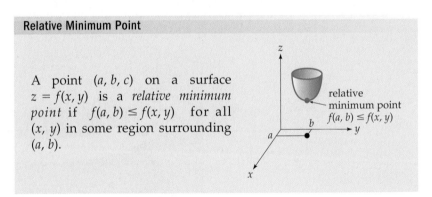

Relative Minimum Point

A point (a, b, c) on a surface $z = f(x, y)$ is a *relative minimum point* if $f(a, b) \leq f(x, y)$ for all (x, y) in some region surrounding (a, b).

As before, the term *relative extreme point* means a point that is either a relative maximum or a relative minimum point. A surface can have any number of relative extreme points, even none.

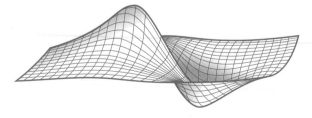

A surface with two relative extreme points: one relative maximum and one relative minimum.

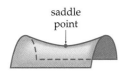

saddle point

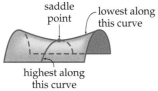

saddle point — lowest along this curve

highest along this curve

The point shown on the left is called a *saddle point* (so named because the diagram resembles a saddle).

A saddle point is a point that is the highest point along one curve of the surface and the lowest point along another curve. A saddle point is *not* a relative extreme point.

If we think of a surface $z = f(x, y)$ as a landscape, then relative maximum and minimum points correspond to "hilltops" and "valley bottoms," and a saddle point corresponds to a "mountain pass" between two peaks.

Gallery of Surfaces

The following are the graphs of a few functions of two variables.

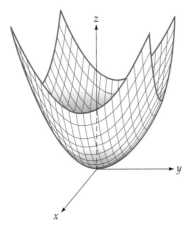

The surface $f(x, y) = x^2 + y^2$ has a relative minimum point at the origin.

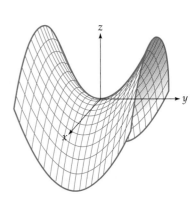

The surface $f(x, y) = y^2 - x^2$ has a saddle point at the origin.

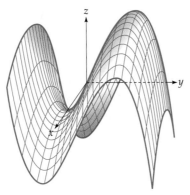

The surface $f(x, y) = 12y + 6x - x^2 - y^3$ has a saddle point and a relative maximum point.

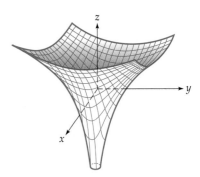

The surface $f(x, y) = \ln(x^2 + y^2)$ has no relative extreme points. It is undefined at $(0, 0)$.

13.1 Section Summary

Just as for a function of one variable, a *function of several variables* gives exactly one value for each point in its domain. For a function of *two* variables, denoted $f(x, y)$, the values $z = f(x, y)$ determine a *surface* above or below the *x-y* plane. The surface may have *relative maximum points* and *relative minimum points* ("hilltops" and "valley bottoms") or even *saddle points* (high points along one curve and low points along another), as shown on the previous pages.

> **Solutions to Practice Problems**
>
> 1. a. $\{(x, y) \mid x > 0, y \geq 0\}$
>
>   b. $f(e, 4) = \dfrac{\ln e}{e^{\sqrt{4}}} = \dfrac{1}{e^2} = e^{-2}$
>
> 2. $M = xy + 2xz + 5yz$

13.1 Exercises

For each function, find the domain.

1. $f(x, y) = \dfrac{1}{xy}$

2. $f(x, y) = \dfrac{\sqrt{x}}{\sqrt{y}}$

3. $f(x, y) = \dfrac{1}{x - y}$

4. $f(x, y) = \dfrac{\sqrt[3]{x}}{\sqrt[3]{y}}$

5. $f(x, y) = \dfrac{\ln x}{y}$

6. $f(x, y) = \dfrac{x}{\ln y}$

7. $f(x, y, z) = \dfrac{e^{1/y} \ln z}{x}$

8. $f(x, y, z) = \dfrac{\sqrt{x} \ln y}{z}$

For each function, evaluate the given expression.

9. $f(x, y) = \sqrt{99 - x^2 - y^2}$, find $f(3, -9)$

10. $f(x, y) = \sqrt{75 - x^2 - y^2}$, find $f(5, -1)$

11. $g(x, y) = \ln(x^2 + y^4)$, find $g(0, e)$

12. $g(x, y) = \ln(x^3 - y^2)$, find $g(e, 0)$

13. $w(u, v) = \dfrac{1 + 2u + 3v}{uv}$, find $w(-1, 1)$

14. $w(u, v) = \dfrac{2u + 4u}{v - u}$, find $w(1, -1)$

15. $h(x, y) = e^{xy + y^2 - 2}$, find $h(1, -2)$

16. $h(x, y) = e^{x^2 - xy - 4}$, find $h(1, -2)$

17. $f(x, y) = xe^y - ye^x$, find $f(1, -1)$

18. $f(x, y) = xe^y + ye^x$, find $f(-1, 1)$

19. $f(x, y, z) = xe^y + ye^z + ze^x$, find $f(1, -1, 1)$

20. $f(x, y, z) = xe^y + ye^z + ze^x$, find $f(-1, 1, -1)$

21. $f(x, y, z) = z \ln \sqrt{xy}$, find $f(-1, -1, 5)$

22. $f(x, y, z) = z\sqrt{x} \ln y$, find $f(4, e, -1)$

APPLIED EXERCISES

23. BUSINESS: Stock Yield The *yield* of a stock is defined as $Y(d, p) = \dfrac{d}{p}$ where d is the dividend per share and p is the price of a share of stock. Find the yield of a stock that sells for $140 and offers a dividend of $2.20.

24. BUSINESS: Price-Earnings Ratio The price-earnings ratio of a stock is defined as $R(P, E) = \dfrac{P}{E}$ where P is the price of a share of stock and E is its earnings. Find the price-earnings ratio of a stock that is selling for $140 with earnings of $1.70.

25. GENERAL: Scuba Diving The maximum duration of a scuba dive (in minutes) can be estimated from the formula

$$T(v, d) = \dfrac{33v}{d + 33}$$

where v is the volume of air (at sea-level pressure) in the tank and d is the depth of the dive. Find $T(90, 33)$.

26. SOCIAL SCIENCE: Cephalic Index Anthropologists define the *cephalic index* to distinguish the head shapes of different people. For a head of width W and length L (measured from above), the cephalic index is

$$C(W, L) = 100 \dfrac{W}{L}$$

Calculate the cephalic index for a head of width 8 inches and length 10 inches.

27. ECONOMICS: Cobb–Douglas Functions A company's production is estimated to be $P(L, K) = 2L^{0.6}K^{0.4}$. Find $P(320, 150)$.

28. BIOMEDICAL: Body Area The surface area (in square feet) of a person of weight w pounds and height h feet is approximated by the function $A(w, h) = 0.55\, w^{0.425} h^{0.725}$. Use this function to estimate the surface area of a person who weighs 160 pounds and who is 6 feet tall. (Such estimates are important in certain medical procedures.)

29. ECONOMICS: Cobb–Douglas Functions Show that the Cobb–Douglas production function $P(L, K) = aL^b K^{1-b}$ satisfies the equation $P(2L, 2K) = 2 \cdot P(L, K)$. This shows that doubling the amounts of labor and capital doubles production, a property called *returns to scale*.

30. ECONOMICS: Cobb–Douglas Functions Show that the Cobb–Douglas function production $P(L, K) = aL^b K^{1-b}$ with $0 < b < 1$ satisfies

$P(2L, K) < 2P(L, K)$ and $P(L, 2K) < 2P(L, K)$

This shows that doubling the amounts of either labor or capital alone results in *less* than double production, a property called *diminishing returns*.

31. GENERAL: Telephone Calls For two cities with populations x and y that are d miles apart, the number of telephone calls per hour between them can be estimated by the function of three variables

$$f(x, y, d) = \dfrac{3xy}{d^{2.4}}$$

(This is called the *gravity model*.) Use the gravity model to estimate the number of calls between two cities of populations 40,000 and 60,000 that are 600 miles apart.

32. ENVIRONMENTAL SCIENCE: Tag and Recapture Estimates Ecologists estimate the size of animal populations by capturing and tagging a few animals, and then releasing them. After the first group has mixed with the population, a second group of animals is captured, and the number of tagged animals in this group is counted. If originally T animals were tagged, and the second group is of size S and contains t tagged animals, then the population is estimated by the function of three variables

$$P(T, S, t) = \dfrac{TS}{t}$$

Estimate the size of a deer population if 100 deer were tagged, and then a second group of 250 contained 20 tagged deer.

CHAPTER 13 CALCULUS OF SEVERAL VARIABLES

$$\frac{\partial}{\partial x} f(x, y) = \begin{pmatrix} \text{Derivative of } f \text{ with respect} \\ \text{to } x \text{ with } y \text{ held constant} \end{pmatrix}$$

$$\frac{\partial}{\partial x} f(x, y) = \begin{pmatrix} \text{Derivative of } f \text{ with respect} \\ \text{to } y \text{ with } x \text{ held constant} \end{pmatrix}$$

EXAMPLE 1   FINDING A PARTIAL DERIVATIVE WITH RESPECT TO x

Find $\dfrac{\partial}{\partial x} x^3 y^4$.

Solution The $\partial/\partial x$ means differentiate with respect to x, holding y (and therefore y^4) constant. We therefore differentiate the x^3 and carry along the "constant" y^4:

$$\frac{\partial}{\partial x} x^3 y^4 = 3x^2 y^4$$

$\partial/\partial x$ means differentiate with respect to x, treating y (and therefore y^4) as a constant

Derivative of x^3; Carry along the "constant" y^4

EXAMPLE 2   FINDING A PARTIAL WITH RESPECT TO y

Find $\dfrac{\partial}{\partial y} x^3 y^4$.

Solution

$$\frac{\partial}{\partial y} x^3 y^4 = x^3 4 y^3 = 4 x^3 y^3$$

Carry along the "constant" x^3; Derivative of y^4; Writing the 4 first

$\partial/\partial y$ means differentiate with respect to y, treating x (and therefore x^3) as a constant

Practice Problem 1   Find **a.** $\dfrac{\partial}{\partial x} x^4 y^2$ **b.** $\dfrac{\partial}{\partial y} x^4 y^2$ ➤ Solutions on page 984

13.2 PARTIAL DERIVATIVES

EXAMPLE 3

FINDING A PARTIAL DERIVATIVE

Find $\dfrac{\partial}{\partial x} y^4$.

Solution

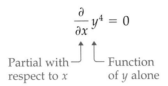

$$\dfrac{\partial}{\partial x} y^4 = 0$$

The derivative of a constant is zero (since $\partial/\partial x$ means hold y constant)

Partial with respect to x — Function of y alone

Practice Problem 2 Find $\dfrac{\partial}{\partial y} x^2$. ▶ Solution on page 985

EXAMPLE 4

FINDING A PARTIAL OF A POLYNOMIAL IN TWO VARIABLES

Find $\dfrac{\partial}{\partial x}(2x^4 - 3x^3y^3 - y^2 + 4x + 1)$.

Solution

$$\dfrac{\partial}{\partial x}(2x^4 - 3x^3y^3 - y^2 + 4x + 1) = 8x^3 - 9x^2y^3 + 4$$

Differentiating with respect to x, so each y is held constant

$$\dfrac{\partial}{\partial x} y^2 = 0$$

Practice Problem 3 Find $\dfrac{\partial}{\partial y}(2x^4 - 3x^3y^3 - y^2 + 4x + 1)$. ▶ Solution on page 985

Subscript Notation for Partial Derivatives

Partial derivatives are often denoted by subscripts: a subscript x means the partial with respect to x, and a subscript y means the partial with respect to y.*

---

* Sometimes subscripts 1 and 2 are used to indicate partial derivatives with respect to the first and second variables: $f_1(x, y)$ means $f_x(x, y)$ and $f_2(x, y)$ means $f_y(x, y)$. We will not use this notation in this book.

$$f_x(x, y) = \frac{\partial}{\partial x} f(x, y) \qquad f_x \text{ means the partial of } f \text{ with respect to } x$$

$$f_y(x, y) = \frac{\partial}{\partial y} f(x, y) \qquad f_y \text{ means the partial of } f \text{ with respect to } y$$

EXAMPLE 5

USING SUBSCRIPT NOTATION

Find $f_x(x, y)$ if $f(x, y) = 5x^4 - 2x^2y^3 - 4y^2$.

Solution

$$f_x(x, y) = 20x^3 - 4xy^3 \qquad \text{Differentiating with respect to } x, \text{ holding } y \text{ constant}$$

EXAMPLE 6

FINDING A PARTIAL INVOLVING LOGS AND EXPONENTIALS

Find both partials of $f = e^x \ln y$.

Solution

$$f_x = e^x \ln y \qquad \text{The derivative of } e^x \text{ is } e^x \text{ (times the "constant" } \ln y)$$

$$f_y = e^x \frac{1}{y} \qquad \text{The derivative of } \ln y \text{ is } \frac{1}{y} \text{ (times the "constant" } e^x)$$

EXAMPLE 7

FINDING A PARTIAL OF A FUNCTION TO A POWER

Find f_y if $f = (xy^2 + 1)^4$.

Solution

$$f_y = 4(xy^2 + 1)^3 (x2y)$$

↑ Partial of the inside with respect to y

Using the Generalized Power Rule (the derivative of f^n is $nf^{n-1}f'$, but with f' meaning a *partial*)

$$= 8xy(xy^2 + 1)^3 \qquad \text{Simplifying}$$

EXAMPLE 8 — FINDING A PARTIAL OF A QUOTIENT

Find $\dfrac{\partial g}{\partial x}$ if $g = \dfrac{xy}{x^2 + y^2}$.

Solution

Partial of the top with respect to x
Partial of the bottom with respect to x

$$\frac{\partial g}{\partial x} = \frac{(x^2 + y^2)y - 2x \cdot xy}{(x^2 + y^2)^2} \qquad \text{Using the Quotient Rule}$$

Bottom squared

$$= \frac{x^2 y + y^3 - 2x^2 y}{(x^2 + y^2)^2} = \frac{y^3 - x^2 y}{(x^2 + y^2)^2} \qquad \text{Simplifying}$$

EXAMPLE 9 — FINDING A PARTIAL OF THE LOGARITHM OF A FUNCTION

Find $f_x(x, y)$ if $f(x, y) = \ln(x^2 + y^2)$.

Solution

Partial of the bottom with respect to x

$$f_x(x, y) = \frac{2x}{x^2 + y^2} \qquad \text{Derivative of } \ln f \text{ is } \frac{f'}{f}$$

An expression such as $f_x(2, 5)$, which involves both differentiation and evaluation, means *first differentiate and then evaluate.**

EXAMPLE 10 — EVALUATING A PARTIAL DERIVATIVE

Find $f_y(1, 3)$ if $f(x, y) = e^{x^2 + y^2}$.

---

* $f_x(2, 5)$ may be written $\dfrac{\partial f}{\partial x}(2, 5)$ or $\dfrac{\partial f}{\partial x}\bigg|_{(2, 5)}$, again meaning first differentiate, then evaluate.

Solution

$$f_y(x, y) = e^{x^2+y^2}(2y)$$

Derivative of e^f is $e^f \cdot f'$

Partial of the exponent with respect to y

$$f_y(1, 3) = e^{1^2+3^2}(2 \cdot 3)$$

Evaluating at $x = 1$ and $y = 3$

$$= 6e^{10}$$

Simplifying

Practice Problem 4 Find $f_y(1, 2)$ if $f(x, y) = e^{x^3+y^3}$. ➤ Solution on page 985

Partial Derivatives in Three or More Variables

Partial derivatives in three or more variables are defined similarly. That is, the partial derivative of $f(x, y, z)$ with respect to any one variable is the "ordinary" derivative with respect to that variable, holding all other variables constant.

EXAMPLE 11   **FINDING A PARTIAL OF A FUNCTION OF THREE VARIABLES**

$$\frac{\partial}{\partial x}(x^3 y^4 z^5) = 3x^2 y^4 z^5$$

$\partial/\partial x$ means differentiate with respect to x, holding y and z constant

Hold constant — Derivative of x^3

Practice Problem 5 Find $\dfrac{\partial}{\partial y}(x^3 y^4 z^5)$. ➤ Solution on page 985

EXAMPLE 12   **EVALUATING A PARTIAL IN THREE VARIABLES**

Find $f_z(1, 1, 1)$ if $f(x, y, z) = e^{x^2+y^2+z^2}$.

Solution

$$f_z(x, y, z) = e^{x^2+y^2+z^2}(2z)$$

Partial with respect to z

$$= 2z e^{x^2+y^2+z^2}$$

Writing the $2z$ first

$$f_z(1, 1, 1) = 2 \cdot 1 \cdot e^{1^2+1^2+1^2} = 2e^3$$

Evaluating

Interpreting Partial Derivatives as Rates of Change

Since partials are just "ordinary" derivatives with the other variable(s) held constant, they give *instantaneous rates of change* with respect to one variable at a time.

Partials as Rates of Change

$$f_x(x, y) = \begin{pmatrix} \text{Instantaneous rate of change of } f \text{ with} \\ \text{respect to } x \text{ when } y \text{ is held constant} \end{pmatrix}$$

$$f_y(x, y) = \begin{pmatrix} \text{Instantaneous rate of change of } f \text{ with} \\ \text{respect to } y \text{ when } x \text{ is held constant} \end{pmatrix}$$

This is why they are called *partial* derivatives: not all the variables are changed at once, only a "partial" change is made.

Cobb–Douglas Production Functions

Recall that a Cobb–Douglas production function $P(L, K) = aL^b K^{1-b}$ expresses production P as a function of L (units of labor) and K (units of capital). Each partial therefore gives the rate of increase of production with respect to one of these variables while the other is held constant.

EXAMPLE 13

INTERPRETING THE PARTIALS OF A COBB–DOUGLAS PRODUCTION FUNCTION

Find and interpret $P_L(120, 200)$ and $P_K(120, 200)$ for the Cobb–Douglas function $P(L, K) = 20L^{0.6}K^{0.4}$.

Solution

$$P_L = 12L^{-0.4}K^{0.4}$$

Partial with respect to L (the 12 is 20 times 0.6)

$$P_L(120, 200) = 12(120)^{-0.4}(200)^{0.4} \approx 14.7$$

Substituting $L = 120$ and $K = 200$, and evaluating using a calculator

Interpretation: $P_L = 14.7$ means that production increases by about 14.7 units for each additional unit of labor (when $L = 120$ and $K = 200$). This is called the *marginal productivity of labor*.

$$P_K = 8L^{0.6}K^{-0.6}$$

Partial with respect to K (the 8 is 20 times 0.4)

$$P_K(120, 200) = 8(120)^{0.6}(200)^{-0.6} \approx 5.9$$

Substituting $L = 120$ and $K = 200$, and evaluating using a calculator

Interpretation: $P_K = 5.9$ means that production increases by about 5.9 units for each additional unit of capital (when $L = 120$ and $K = 200$). This is called the *marginal productivity of capital*.

These numbers show that to increase production, additional units of labor are more than twice as effective as additional units of capital (at the levels $L = 120$ and $K = 200$).

Partial derivatives give the marginals for one product at a time.

Interpreting Partials as Marginals

Let $C(x, y)$ be the (total) cost function for x units of product A and y units of product B. Then

$$C_x(x, y) = \begin{pmatrix} \text{Marginal cost function for product A when} \\ \text{production of product B is held constant} \end{pmatrix}$$

$$C_y(x, y) = \begin{pmatrix} \text{Marginal cost function for product B when} \\ \text{production of product A is held constant} \end{pmatrix}$$

Similar statements hold, of course, for revenue and profit functions: the partials give the marginals for one variable at a time when the other variables are held constant.

EXAMPLE 14 **INTERPRETING PARTIALS OF A PROFIT FUNCTION**

A company's profit from producing x radios and y televisions per day is $P(x, y) = 4x^{3/2} + 6y^{3/2} + xy$. Find the marginal profit functions. Then find and interpret $P_y(25, 36)$.

Solution

$$P_x(x, y) = 6x^{1/2} + y$$ 	Marginal profit for radios when television production is held constant

$$P_y(x, y) = 9y^{1/2} + x$$ 	Marginal profit for televisions when radio production is held constant

$$P_y(25, 36) = 9\underbrace{(36)^{1/2}}_{6} + 25 = 79$$ 	Evaluating P_y at $x = 25$ and $y = 36$

Interpretation: Profit increases by about $79 per additional television (when producing 25 radios and 36 televisions per day).

Interpreting Partials Geometrically

A function $f(x, y)$ represents a surface in three-dimensional space, and the partial derivatives are the slopes along the surface in different directions: $\partial f / \partial x$ gives the slope of the surface "in the x-direction," and $\partial f / \partial y$ gives the slope of the surface "in the y-direction" at the point (x, y).

To put this colloquially, if you were on the surface $z = f(x, y)$, then $\partial f / \partial x$ would be the steepness of the surface *in the x-direction*, and $\partial f / \partial y$ would be the steepness of the surface *in the y-direction* from the point (x, y).

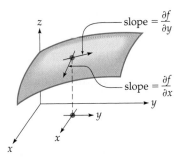

Partial derivatives are slopes.

For example, the following graph shows gridlines in the x direction (roughly up and down the page, with the positive x direction being down) and in the y direction (roughly across the page, with the positive y direction being to the right).

982 CHAPTER 13 CALCULUS OF SEVERAL VARIABLES

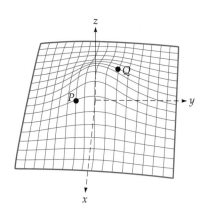

Therefore, at the point P on the graph the partials would have the following signs:

$$\frac{\partial f}{\partial x} < 0 \qquad \text{Walking from } P \text{ in the positive } x \text{ direction would mean walking } downhill$$

$$\frac{\partial f}{\partial y} > 0 \qquad \text{Walking from } P \text{ in the positive } y \text{ direction would mean walking } uphill$$

Practice Problem 6 On the preceding graph at the point Q,

a. Is $\dfrac{\partial f}{\partial x}$ positive or negative?

b. Is $\dfrac{\partial f}{\partial y}$ positive or negative?

[*Hint:* Leaving Q in those directions, would you be walking uphill or downhill?] ➤ Solutions on page 985

Higher-Order Partial Derivatives

We can differentiate a function more than once to obtain *higher-order* partials.

Second-Order Partials

Subscript Notation	∂ Notation	In Words
f_{xx}	$\dfrac{\partial^2}{\partial x^2} f$	Differentiate twice with respect to x
f_{yy}	$\dfrac{\partial^2}{\partial y^2} f$	Differentiate twice with respect to y

13.2 PARTIAL DERIVATIVES

Subscript Notation	∂ Notation	In Words
f_{xy}	$\dfrac{\partial^2}{\partial y\, \partial x} f$	Differentiate first with respect to x, then with respect to y
f_{yx}	$\dfrac{\partial^2}{\partial x\, \partial y} f$	Differentiate first with respect to y, then with respect to x

Each notation means differentiate first with respect to the variable *closest* to f.

Calculating a "second partial" such as f_{xy} is a two-step process: first calculate f_x, and then differentiate the *result* with respect to y.

EXAMPLE 15

FINDING SECOND-ORDER PARTIALS

Find all four second-order partials of $f(x, y) = x^4 + 2x^2y^2 + x^3y + y^4$.

Solution First we calculate

$$f_x = 4x^3 + 4xy^2 + 3x^2y \qquad \text{Partial of } f \text{ with respect to } x$$

Then from this we find f_{xx} and f_{xy}:

$$f_{xx} = 12x^2 + 4y^2 + 6xy \qquad \text{Differentiating } f_x = 4x^3 + 4xy^2 + 3x^2y \text{ with respect to } x$$

$$f_{xy} = 8xy + 3x^2 \qquad \text{Differentiating } f_x = 4x^3 + 4xy^2 + 3x^2y \text{ with respect to } y$$

Now, returning to the original function $f = x^4 + 2x^2y^2 + x^3y + y^4$, we calculate

$$f_y = 4x^2y + x^3 + 4y^3 \qquad \text{Partial of } f \text{ with respect to } y$$

Then, from this,

$$f_{yy} = 4x^2 + 12y^2 \qquad \text{Differentiating } f_y = 4x^2y + x^3 + 4y^3 \text{ with respect to } y$$

$$f_{yx} = 8xy + 3x^2 \qquad \text{Differentiating } f_y = 4x^2y + x^3 + 4y^3 \text{ with respect to } x$$

Notice that these "mixed partials" are equal:

$$\left. \begin{array}{l} f_{xy} = 8xy + 3x^2 \\ f_{yx} = 8xy + 3x^2 \end{array} \right\} \text{Equal}$$

For each function, calculate the third-order partials
a. f_{xxy}, b. f_{xyx}, and c. f_{yxx}.

33. $f(x, y) = x^4 y^3 - e^{2x}$ 34. $f(x, y) = x^3 y^4 - e^{2y}$

For each function of three variables, find the partials a. f_x, b. f_y, and c. f_z.

35. $f = xy^2 z^3$

36. $f = x^2 y^3 z^4$

37. $f = (x^2 + y^2 + z^2)^4$ 38. $f = (xyz + 1)^3$

39. $f = e^{x^2 + y^2 + z^2}$

40. $f = \ln(x^2 - y^3 + z^4)$

For each function, evaluate the stated partial.

41. $f = 3x^2 y - 2xz^2$, find $f_x(2, -1, 1)$

42. $f = 2yz^3 - 3x^2 z$, find $f_z(2, -1, 1)$

43. $f = e^{x^2 + 2y^2 + 3z^2}$, find $f_y(-1, 1, -1)$

44. $f = e^{2x^3 + 3y^3 + 4z^3}$, find $f_y(1, -1, 1)$

APPLIED EXERCISES

45–46. BUSINESS: Marginal Profit An electronics company's profit $P(x, y)$ from making x tape decks and y CD players per day is given below.

a. Find the marginal profit function for tape decks.
b. Evaluate your answer to part (a) at $x = 200$ and $y = 300$ and interpret the result.
c. Find the marginal profit function for CD players.
d. Evaluate your answer to part (c) at $x = 200$ and $y = 100$ and interpret the result.

45. $P(x, y) = 2x^2 - 3xy + 3y^2 + 150x + 75y + 200$
46. $P(x, y) = 3x^2 - 4xy + 4y^2 + 80x + 100y + 200$

47–48. BUSINESS: Cobb–Douglas Production Functions A company's production is given by the Cobb–Douglas function $P(L, K)$ below, where L is the number of units of labor and K is the number of units of capital.

a. Find $P_L(27, 125)$ and interpret this number.
b. Find $P_K(27, 125)$ and interpret this number.
c. From your answers to parts (a) and (b), which will increase production more: an additional unit of labor or an additional unit of capital?

47. $P(L, K) = 270 L^{1/3} K^{2/3}$

48. $P(L, K) = 225 L^{2/3} K^{1/3}$

49. BUSINESS: Sales A store's TV sales depend on x, the price of the televisions, and y, the amount spent on advertising, according to the function $S(x, y) = 200 - 0.1x + 0.2y^2$. Find and interpret the marginals S_x and S_y.

50. ECONOMICS: Value of an MBA A 1973 study found that a businessperson with a master's degree in business administration (MBA) earned an average salary of $S(x, y) = 10{,}990 + 1120x + 873y$ dollars, where x is the number of years of work experience before the MBA, and y is the number of years of work experience after the MBA. Find and interpret the marginals S_x and S_y.

51. SOCIOLOGY: Status A study found that a person's status in a community depends on the person's income and education according to the function $S(x, y) = 7x^{1/3} y^{1/2}$, where x is income (in thousands of dollars) and y is years of education beyond high school.

a. Find $S_x(27, 4)$ and interpret this number.
b. Find $S_y(27, 4)$ and interpret this number.

52. BIOMEDICAL: Blood Flow The resistance of blood flowing through an artery of radius r and length L (both in centimeters) is $R(r, L) = 0.08 L r^{-4}$.

a. Find $R_r(0.5, 4)$ and interpret this number.
b. Find $R_L(0.5, 4)$ and interpret this number.

53. GENERAL: Highway Safety The length in feet of the skid marks from a truck of weight w (tons) traveling at velocity v (miles per hour) skidding to a stop on a dry road is $S(w, v) = 0.027 wv^2$.

a. Find $S_w(4, 60)$ and interpret this number.
b. Find $S_v(4, 60)$ and interpret this number.

54. GENERAL: Windchill Temperature The windchill temperature (revised in 2001) announced by the weather bureau during the cold weather measures how cold it "feels" for a given temperature and wind speed. The formula is $C(t, w) = 35.74 + 0.6215t - 35.75w^{0.16} + 0.4275tw^{0.16}$ where t is the temperature (in degrees Fahrenheit) and w is the wind speed (in miles per hour). Find and interpret $C_w(30, 20)$.

COMPETITIVE AND COMPLEMENTARY COMMODITIES

55. ECONOMICS: Competitive Commodities Certain commodities (such as butter and margarine) are called "competitive" or "substitute" commodities because one can substitute for the other. If $B(b, m)$ gives the daily sales of butter as a function of b, the price of butter, and m, the price of margarine:
  a. Give an interpretation of $B_b(b, m)$.
  b. Would you expect $B_b(b, m)$ to be positive or negative? Explain.
  c. Give an interpretation of $B_m(b, m)$.
  d. Would you expect $B_m(b, m)$ to be positive or negative? Explain.

56. ECONOMICS: Complementary Commodities Certain commodities (such as washing machines and clothes dryers) are called "complementary" commodities because they are often used together. If $D(d, w)$ gives the monthly sales of dryers as a function of d, the price of dryers, and w, the price of washers:
  a. Give an interpretation of $D_d(d, w)$.
  b. Would you expect $D_d(d, w)$ to be positive or negative? Explain.
  c. Give an interpretation of $D_w(d, w)$.
  d. Would you expect $D_w(d, w)$ to be positive or negative? Explain.

13.3 OPTIMIZING FUNCTIONS OF SEVERAL VARIABLES

Introduction

The graph of a function of two variables is a *surface*, with relative maximum and minimum points ("hilltops" and "valley bottoms") and saddle points ("mountain passes"). In this section we will see how to maximize and minimize such functions by finding critical points and using a two-variable version of the second-derivative test.

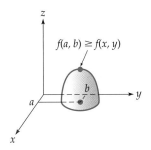

f has a relative *maximum* value at (a, b).

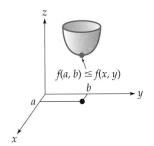

f has a relative *minimum* value at (a, b).

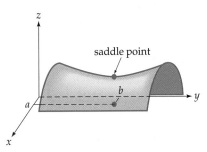

f has a saddle point at (a, b) (neither a maximum nor a minimum).

For simplicity, we will consider only functions whose first and second partials are defined everywhere. Such optimization techniques have many applications, such as maximizing profit for a company that makes several products.

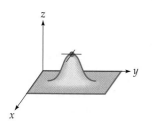

A critical point
(both partials are zero)

Critical Points

At the very top of a smooth hill, the slope or steepness in any direction must be zero. That is, a straight stick would balance horizontally at the top. Since the partials f_x and f_y are the slopes in the x and y directions, these partials must both be zero at a relative maximum point, and similarly for a relative minimum point or a saddle point. A point (a, b) at which both partials are zero is called a *critical point* of the function.*

Critical Point

(a, b) is a critical point of $f(x, y)$ if

$$f_x(a, b) = 0 \quad \text{and} \quad f_y(a, b) = 0 \qquad \text{Both first partials are zero}$$

Relative maximum and minimum values can occur only at critical points.

EXAMPLE 1

FINDING CRITICAL POINTS

Find all critical points of

$$f(x, y) = 3x^2 + y^2 + 3xy + 3x + y + 6$$

Solution We want all points at which both partials are zero.

$$f_x = 6x + 3y + 3 \quad \text{and} \quad f_y = 2y + 3x + 1 \qquad \text{Partials}$$

$$\begin{aligned} 6x + 3y + 3 &= 0 \\ 3x + 2y + 1 &= 0 \end{aligned} \quad \begin{array}{l} \text{Partials} \\ \text{set equal to zero} \\ \text{Reordered so the } x\text{- and } y\text{-terms line up} \end{array}$$

To solve these equations simultaneously, we multiply the second by -2 so that the x-terms drop out when we add.

---

*We use the term "critical point" for the *pair* of values (a, b) at which the two partials are zero. The corresponding point on the graph has *three* coordinates $(a, b, f(a, b))$.

$$6x + 3y + 3 = 0 \qquad \text{First equation}$$
$$-6x - 4y - 2 = 0 \qquad \text{Second equation times } -2$$
$$-y + 1 = 0 \qquad \text{Adding (} x \text{ drops out)}$$
$$y = 1 \qquad \text{From solving } -y + 1 = 0$$

Substituting $y = 1$ into either equation gives x:

$$6x + 3 + 3 = 0 \qquad \text{Substituting } y = 1 \text{ into } 6x + 3y + 3 = 0$$
$$6x = -6 \qquad \text{Simplifying}$$
$$x = -1 \qquad \text{Solving}$$

These x- and y-values give one critical point.

$$\text{CP: } (-1, 1) \qquad \text{From } x = -1, \; y = 1$$

Second-Derivative Test for Functions $f(x, y)$: The D-Test

To determine whether $f(x, y)$ has a relative maximum, a relative minimum, or a saddle point at a critical point, we use the following D-test, which is a generalization of the second-derivative test on page 658.

D-Test

If (a, b) is a critical point of the function $f(x, y)$, then for D defined by

$$D = f_{xx}(a, b) \cdot f_{yy}(a, b) - [f_{xy}(a, b)]^2$$

More briefly,
$$D = f_{xx}f_{yy} - (f_{xy})^2$$

i. f has a relative *maximum* at (a, b) if $D > 0$ and $f_{xx}(a, b) < 0$.
ii. f has a relative *minimum* at (a, b) if $D > 0$ and $f_{xx}(a, b) > 0$.

Different signs for $f_{xx}(a, b)$

iii. f has a *saddle point* at (a, b) if $D < 0$.

The following observations may help in understanding the D-test.

1. The D-test is used only *after* finding the critical points. The test is then applied to each critical point, one at a time.
2. $D > 0$ appears only in parts i and ii, and so guarantees that the function has either a relative maximum or a relative minimum. Then all that remains to be done is to use the "old"

second-derivative test (checking the sign of f_{xx}) to determine which one (maximum or minimum) occurs.

3. $D < 0$ means a saddle point, regardless of the sign of f_{xx}. (A saddle point is neither a maximum nor a minimum.)

4. $D = 0$ means that the D-test is inconclusive—the function may have a maximum, a minimum, or a saddle point at the critical point.

EXAMPLE 2 FINDING RELATIVE EXTREME VALUES OF A POLYNOMIAL

Find the relative extreme values of

$$f(x, y) = 3x^2 + y^2 + 3xy + 3x + y + 6$$

Solution We find critical points by setting the two partials equal to zero and solving. But we did this for the same function in Example 1, finding one critical point, $(-1, 1)$. For the D-test, we calculate the second partials.

$$f_{xx} = 6 \qquad \text{From } f_x = 6x + 3y + 3$$
$$f_{yy} = 2 \qquad \text{From } f_y = 2y + 3x + 1$$
$$f_{xy} = 3 \qquad \text{From } f_x = 6x + 3y + 3$$

Calculating D:

$$D = 6 \cdot 2 - (3)^2 = 12 - 9 = 3 \qquad D = f_{xx}f_{yy} - (f_{xy})^2$$

Positive

D is positive and f_{xx} is positive (since $f_{xx} = 6$), so f has a *relative minimum* (part ii of the D-test) at the critical point $(-1, 1)$. (If f_{xx} had been negative, there would have been a relative maximum.) The relative minimum *value* is found by evaluating $f(x, y)$ at $(-1, 1)$.

$$f(-1, 1) = 3 + 1 - 3 - 3 + 1 + 6 \qquad \begin{array}{l} f = 3x^2 + y^2 + 3xy + 3x + y + 6 \\ \text{evaluated at } x = -1, \ y = 1 \end{array}$$
$$= 5$$

Relative minimum value: $f = 5$ at $x = -1, \ y = 1$.

EXAMPLE 3 FINDING RELATIVE EXTREME VALUES OF AN EXPONENTIAL FUNCTION

Find the relative extreme values of $f(x, y) = e^{x^2 - y^2}$.

Solution

$$f_x = e^{x^2-y^2}(2x)$$
$$f_y = e^{x^2-y^2}(-2y)$$ Partials

$$e^{x^2-y^2}(2x) = 0$$
$$e^{x^2-y^2}(-2y) = 0$$ Partials set equal to zero

CP: (0, 0) Since $\begin{cases} e^{x^2-y^2}(2x) = 0 \\ e^{x^2-y^2}(-2y) = 0 \end{cases}$ only at $x = 0, y = 0$

For the *D*-test we calculate the second partials:

$$f_{xx} = e^{x^2-y^2}(2x)(2x) + e^{x^2-y^2}(2)$$ From $f_x = e^{x^2-y^2}(2x)$ using the Product Rule
$$= 4x^2 e^{x^2-y^2} + 2e^{x^2-y^2}$$ Simplifying

$$f_{yy} = e^{x^2-y^2}(-2y)(-2y) + e^{x^2-y^2}(-2)$$ From $f_y = e^{x^2-y^2}(-2y)$ using the Product Rule
$$= 4y^2 e^{x^2-y^2} - 2e^{x^2-y^2}$$ Simplifying

$$f_{xy} = e^{x^2-y^2}(2x)(-2y) = -4xy e^{x^2-y^2}$$ From $f_x = e^{x^2-y^2}(2x)$ treating x as a constant

Evaluating at the critical point (0, 0):

$f_{xx}(0, 0) = 0e^0 + 2e^0 = 0 + 2 = 2$ $f_{xx} = 4x^2 e^{x^2-y^2} + 2e^{x^2-y^2}$ at (0, 0)
$f_{yy}(0, 0) = 0e^0 - 2e^0 = 0 - 2 = -2$ $f_{yy} = 4y^2 e^{x^2-y^2} - 2e^{x^2-y^2}$ at (0, 0)
$f_{xy}(0, 0) = 0e^0 = 0$ $f_{xy} = -4xy e^{x^2-y^2}$ at (0, 0)

Therefore *D* is

$$D = (2)(-2) - (0)^2 = -4 - 0 = -4 \qquad D = f_{xx}f_{yy} - (f_{xy})^2$$
     /   \   \
   f_{xx} f_{yy} f_{xy} Negative

Since *D* is negative, the function has a *saddle point* (part iii of the *D*-test) at the critical point (0, 0).

    f has no relative extreme values
    (it has a saddle point at $x = 0, y = 0$).

Maximizing Profit

If a company makes too few of its products, the resulting lost sales will lower the company's profits. On the other hand, making too many will "flood the market" and depress prices, again resulting in lower profits. Therefore, any realistic profit function must have a maximum at some "intermediate" point (x, y), which therefore must be a *relative* maximum point. Many applied problems are solved in this way: by knowing that the *absolute* extreme values (the highest and lowest values on the entire domain) must exist, and finding them as *relative* extreme points.

Suppose that a company produces two products, A and B, and that the two price functions are

$$p(x) = \begin{pmatrix} \text{Price at which exactly } x \text{ units} \\ \text{of product } A \text{ will be sold} \end{pmatrix}$$

and

$$q(y) = \begin{pmatrix} \text{Price at which exactly } y \text{ units} \\ \text{of product } B \text{ will be sold} \end{pmatrix}$$

If $C(x, y)$ is the (total) cost function, then the company's profit will be

$$\underbrace{P(x, y)}_{\text{Profit}} = \underbrace{p(x) \cdot x}_{\substack{\text{Price times} \\ \text{quantity for} \\ \text{product } A}} + \underbrace{q(y) \cdot y}_{\substack{\text{Price times} \\ \text{quantity for} \\ \text{product } B}} - \underbrace{C(x, y)}_{\text{Cost}} \quad \begin{array}{l} \text{Revenue for each} \\ \text{product (price times} \\ \text{quantity) minus the} \\ \text{cost function} \end{array}$$

EXAMPLE 4 MAXIMIZING PROFIT FOR A COMPANY

Universal Motors makes compact and midsized cars. The price function for compacts is $p = 17 - 2x$ (for $0 \le x \le 8$), and the price function for midsized cars is $q = 20 - y$ (for $0 \le y \le 20$), both in thousands of dollars, where x and y are, respectively, the numbers of compact and midsized cars produced per hour. If the company's cost function is

$$C(x, y) = 15x + 16y - 2xy + 5$$

thousand dollars, find how many of each car should be produced and the prices that should be charged in order to maximize profit. Also find the maximum profit.

Solution The profit function is

$$P(x, y) = \underbrace{(17 - 2x)x}_{\substack{\text{Price Quantity} \\ \text{For compacts}}} + \underbrace{(20 - y)y}_{\substack{\text{Price Quantity} \\ \text{For midsized}}} - \underbrace{(15x + 16y - 2xy + 5)}_{\text{Cost}}$$

$$= 17x - 2x^2 + 20y - y^2 - 15x - 16y + 2xy - 5 \quad \text{Multiplying out}$$
$$= -2x^2 - y^2 + 2xy + 2x + 4y - 5 \quad \text{Simplifying}$$

We maximize $P(x, y)$ in the usual way:

$$P_x = -4x + 2y + 2$$
$$P_y = -2y + 2x + 4$$
$$\left.\right\} \text{Partials}$$

$$-4x + 2y + 2 = 0 \quad \text{Partials set equal to zero}$$
$$\underline{2x - 2y + 4 = 0} \quad \text{Rearranged to line up } x\text{'s and } y\text{'s}$$
$$-2x \quad\quad + 6 = 0 \quad \text{Adding (the } y\text{'s cancel)}$$
$$x = 3 \quad \text{From solving } -2x + 6 = 0$$
$$y = 5 \quad \text{From substituting } x = 3 \text{ into either equation (omitting the details)}$$

These two values give one critical point.

$$\text{CP: } (3, 5)$$

For the D-test we calculate the second partials:

$$P_{xx} = -4 \quad P_{xy} = 2 \quad P_{yy} = -2 \quad \text{From } P_x = -4x + 2y + 2 \text{ and } P_y = 2y + 2x + 4$$

$$D = (-4)(-2) - (2)^2 = 4 \quad\quad D = P_{xx}P_{yy} - (P_{xy})^2$$

D is positive and $P_{xx} = -4$ is negative, so profit is indeed *maximized* at $x = 3$ and $y = 5$. To find the prices, we evaluate the price functions:

$$p = 17 - 2 \cdot 3 = 11 \quad \text{(thousand dollars)} \quad \begin{array}{l} p = 17 - 2x \\ \text{evaluated at } x = 3 \end{array}$$

$$q = 20 - 5 = 15 \quad \text{(thousand dollars)} \quad \begin{array}{l} q = 20 - y \\ \text{evaluated at } y = 5 \end{array}$$

The profit comes from the profit function:

$$P(3, 5) = \underbrace{-2 \cdot 3^2 - 5^2 + 2 \cdot 3 \cdot 5 + 2 \cdot 3 + 4 \cdot 5 - 5}_{\substack{P = -2x^2 - y^2 + 2xy + 2x + 4y - 3 \\ \text{evaluated at } x = 3, \ y = 5}}$$

$$= 8 \text{ (thousand dollars)}$$

Profit is maximized when the company produces 3 compacts per hour, selling them for $11,000 each, and 5 midsized cars per hour, selling them for $15,000 each. The maximum profit will be $8000 per hour.

The D-test ensures that you have *maximized* profit rather than minimized it.*

Some functions have *more* than one critical point.

---

* How are the price and cost functions in such problems found? Price functions may be constructed by the methods used on pages 679–680 or by more sophisticated techniques of market research. Cost functions may be found simply as the sum of the unit cost times the number of units for each product, or by regression techniques based on the least squares method described in the following section for constructing functions from data.

EXAMPLE 5

FINDING RELATIVE EXTREME VALUES

Find the relative extreme values of
$$f(x, y) = x^2 + y^3 - 6x - 12y$$

Solution

$$2x - 6 = 0 \qquad f_x = 0$$
$$3y^2 - 12 = 0 \qquad f_y = 0$$

The first gives

$$x = 3 \qquad \text{Solving } 2x - 6 = 0$$

and the second gives

$$3y^2 = 12 \qquad \text{Adding 12 to each side of } 3y^2 - 12 = 0$$
$$y^2 = 4 \qquad \text{Dividing by 3}$$
$$y = \pm 2 \qquad \text{Taking square roots}$$

From $x = 3$ and $y = \pm 2$ we get *two* critical points:

$$\text{CP: } (3, 2) \quad \text{and} \quad (3, -2)$$

Calculating the second partials and substituting them into D:

$$D = (2)(6y) - (0)^2 = 12y \qquad \begin{array}{l} D = (f_{xx})(f_{yy}) - (f_{xy})^2 \text{ with} \\ f_{xx} = 2, \ f_{yy} = 6y, \ f_{xy} = 0 \end{array}$$

We apply the D-test to the critical points one at a time.

At $(3, 2)$: $\quad D = 12 \cdot 2 > 0 \qquad \qquad \begin{array}{l} D = 12y \text{ evaluated} \\ \text{at } (3, 2) \end{array}$

$\qquad \qquad f_{xx} = 2 > 0$

$\qquad \qquad$ *relative minimum* at $x = 3, \ y = 2 \qquad$ Since D and f_{xx} are both positive

At $(3, -2)$: $\quad D = 12 \cdot (-2) < 0 \qquad \qquad \begin{array}{l} D = 12y \text{ evaluated} \\ \text{at } (3, -2) \end{array}$

$\qquad \qquad$ *saddle point* at $x = 3, \ y = -2 \qquad$ Since D is negative

Answer:

Relative minimum value: $f = -25$ at $\begin{cases} x = 3 & f = -25 \text{ from} \\ y = 2 & f = x^2 + y^3 - 6x - 12y \\ & \text{at } (3, 2) \end{cases}$

(saddle point at $x = 3, \ y = -2$)

Competition and Collusion

In 1938, the French economist Antoine Cournot published the following comparison of a monopoly (a market with only one supplier) and a duopoly (a market with two suppliers).

13.3 OPTIMIZING FUNCTIONS OF SEVERAL VARIABLES

Monopoly. Imagine that you are selling spring water from your own spring (or any product whose cost of production is negligible). Since you are the only supplier in town, you have a "monopoly." Suppose that your price function is $p = 6 - 0.01x$, where p is the price in dollars at which you will sell precisely x gallons ($0 \leq x \leq 600$). Your revenue is then

$$R(x) = (6 - 0.01x)x = 6x - 0.01x^2 \qquad \text{Price } (6 - 0.01x) \text{ times quantity } x$$

You maximize revenue by setting its derivative equal to zero:

$$R'(x) = 6 - 0.02x = 0$$

$$x = \frac{6}{0.02} = 300 \qquad \text{Solving } 6 - 0.02x = 0$$

Therefore, you should sell 300 gallons per day. (The second-derivative test will verify that revenue is maximized.) The price will be

$$p = 6 - 0.01 \cdot 300 = 6 - 3 = 3 \qquad \begin{array}{l} p = 6 - 0.01x \\ \text{evaluated at } x = 300 \end{array}$$

or $3 dollars per gallon. Your maximum revenue will be

$$R(300) = 6 \cdot 300 - 0.01(300)^2 = \$900 \qquad \begin{array}{l} R(x) = 6x - 0.01x^2 \\ \text{evaluated at } x = 300 \end{array}$$

Duopoly. Suppose now that your neighbor opens a competing spring water business. (A market such as this with two suppliers is called a "duopoly.") Now both of you must share the same market. If your neighbor sells y gallons per day (and you sell x), you must both sell at price

$$p = 6 - 0.01(x + y) = 6 - 0.01x - 0.01y \qquad \begin{array}{l} \text{Price function } p = 6 - 0.01x \\ \text{with } x \text{ replaced by the} \\ \text{combined quantity } x + y \end{array}$$

Each of you calculates revenue as price times quantity:

$$\left(\begin{array}{c}\text{Your}\\ \text{revenue}\end{array}\right) = p \cdot x = (6 - 0.01x - 0.01y)x = 6x - 0.01x^2 - 0.01xy$$

$$\left(\begin{array}{c}\text{Neighbor's}\\ \text{revenue}\end{array}\right) = p \cdot y = (6 - 0.01x - 0.01y)y = 6y - 0.01xy - 0.01y^2$$

You each want to maximize revenue, so you set the partials equal to zero:

$$6 - 0.02x - 0.01y = 0 \qquad \begin{array}{l} \text{Partial of } 6x - 0.01x^2 - 0.01xy \\ \text{with respect to } x, \text{ set equal to zero} \end{array}$$

$$6 - 0.01x - 0.02y = 0 \qquad \begin{array}{l} \text{Partial of } 6y - 0.01xy - 0.01y^2 \\ \text{with respect to } y, \text{ set equal to zero} \end{array}$$

23. BUSINESS: Price Discrimination An automobile manufacturer sells cars in America and Europe, charging different prices in the two markets. The price function for cars sold in America is $p = 20 - 0.2x$ thousand dollars (for $0 \leq x \leq 100$), and the price function for cars sold in Europe is $q = 16 - 0.1y$ thousand dollars (for $0 \leq y \leq 160$), where x and y are the numbers of cars sold per day in America and Europe, respectively. The company's cost function is

$$C = 20 + 4(x + y) \quad \text{thousand dollars}$$

a. Find the company's profit function. [*Hint:* Profit is revenue from America plus revenue from Europe minus costs, where each revenue is price times quantity.]
b. Find how many cars should be sold in each market to maximize profit. Also find the price for each market.

24. BIOMEDICAL: Drug Dosage In a laboratory test the combined antibiotic effect of x milligrams of medicine A and y milligrams of medicine B is given by the function

$$f(x, y) = xy - 2x^2 - y^2 + 110x + 60y$$

(for $0 \leq x \leq 55$, $0 \leq y \leq 60$). Find the amounts of the two medicines that maximize the antibiotic effect.

25. PSYCHOLOGY: Practice and Rest A subject in a psychology experiment who practices a skill for x hours and then rests for y hours achieves a test score of $f(x, y) = xy - x^2 - y^2 + 11x - 4y + 120$ (for $0 \leq x \leq 10$, $0 \leq y \leq 4$). Find the numbers of hours of practice and rest that maximize the subject's score.

26. SOCIOLOGY: Absenteeism The number of office workers near a beach resort who call in "sick" on a warm summer day is

$$f(x, y) = xy - x^2 - y^2 + 110x + 50y - 5200$$

where x is the air temperature ($70 \leq x \leq 100$) and y is the water temperature ($60 \leq y \leq 80$). Find the air and water temperatures that maximize the number of absentees.

27–28. ECONOMICS: Competition and Collusion Compare the outputs of a monopoly and a duopoly by repeating the analysis on pages 994–996 for the following price function. That is:

a. For a monopoly, calculate the quantity x that maximizes your revenue. Also calculate the price p and the revenue R.
b. For the duopoly, calculate the quantities x and y that maximize revenue for each duopolist. Calculate the price p and the two revenues.
c. Are more goods produced under a monopoly or a duopoly?
d. Is the price lower under a monopoly or a duopoly?

27. $p = 12 - 0.005x$ dollars $\quad (0 \leq x \leq 2400)$

28. $p = a - bx$ dollars (for positive numbers a and b with $0 \leq x \leq a/b$)

29. BUSINESS: Maximum Profit An automobile dealer can sell 8 sedans per day at a price of $20,000 and 4 SUVs (sport utility vehicles) per day at a price of $25,000. She estimates that for each $400 decrease in price of the sedans she can sell two more per day, and for each $600 decrease in price for the SUVs she can sell one more. If each sedan costs her $16,800 and each SUV costs her $19,000, and fixed costs are $1000 per day, what price should she charge for the sedans and the SUVs to maximize profit? How many of each type will she sell at these prices? [*Hint:* Let x be the number of $400 price decreases for sedans and y be the number of $600 price decreases for SUVs, and use the method of Examples 1 and 2 on pages 679–681 for each type of car.]

30. BUSINESS: Maximum Revenue An airline flying to a Midwest destination can sell 20 coach-class tickets per day at a price of $250 and six business-class tickets per day at a price of $750. It finds that for each $10 decrease in the price of the coach ticket, it will sell four more per day, and for each $50 decrease in the business-class price, it will sell two more per day. What prices should the airline charge for the coach- and business-class tickets to maximize revenue? How many of each type will be sold at these prices? [*Hint:* Let x be the number of $10 price decreases for coach tickets and y be the number of $50 price decreases for business-class tickets, and use the method of Examples 1 and 2 on pages 679–681 for each type of ticket.]

OPTIMIZING FUNCTIONS OF THREE VARIABLES

31. BUSINESS: Price Discrimination An automobile manufacturer sells cars in America, Europe, and Asia, charging a different price in each of the three markets. The price function for cars sold in America is $p = 20 - 0.2x$ (for $0 \le x \le 100$), the price function for cars sold in Europe is $q = 16 - 0.1y$ (for $0 \le y \le 160$), and the price function for cars sold in Asia is $r = 12 - 0.1z$ (for $0 \le z \le 120$), all in thousands of dollars, where x, y, and z are the numbers of cars sold in America, Europe, and Asia, respectively. The company's cost function is $C = 22 + 4(x + y + z)$ thousand dollars.

a. Find the company's profit function $P(x, y, z)$. [*Hint:* The profit will be revenue from America plus revenue from Europe plus revenue from Asia minus costs, where each revenue is price times quantity.]
b. Find how many cars should be sold in each market to maximize profit. [*Hint:* Set the three partials P_x, P_y, and P_z equal to zero and solve. Assuming that the maximum exists, it must occur at this point.]

32. ECONOMICS: Competition and Collusion Suppose that in the discussion of competition and collusion (pages 994–996), *two* of your neighbors began selling spring water. Use the price function $p = 36 - 0.01x$ (for $0 \le x \le 3600$) and repeat the analysis, but now comparing a monopoly with competition among *three* suppliers (a "triopoly"). That is:

a. For a monopoly, calculate the quantity x that maximizes your revenue. Also calculate the price p and the revenue R.
b. For a triopoly, find the quantities x, y, and z for the three suppliers that maximize revenue for each. Also calculate the price p and the three revenues. [*Hint:* Find the three revenue functions, one for each supplier, and maxmimize each with respect to that supplier's variable.]
c. Are more goods produced under a monopoly or under a triopoly?
d. Is the price lower under a monopoly or under a triopoly?

EXERCISES WITH MORE THAN ONE CRITICAL POINT

Find the relative extreme values of each function.

33. $f(x, y) = x^3 + y^3 - 3xy$
34. $f(x, y) = x^5 + y^5 - 5xy$
35. $f(x, y) = 12xy - x^3 - 6y^2$
36. $f(x, y) = 6xy - x^3 - 3y^2$
37. $f(x, y) = 2x^4 + y^2 - 12xy$
38. $f(x, y) = 16xy - x^4 - 2y^2$

13.4 LEAST SQUARES

Introduction

You may have wondered how the mathematical models in this book were developed. For example, how are the constants a and b in the Cobb–Douglas production function $P = aL^bK^{1-b}$ determined for a particular company or nation? The problem of finding the function that fits a collection of data can be viewed geometrically as the problem of fitting a curve to a collection of points. The simplest case of this problem is fitting a straight line to a collection of points, and the most widely used method for doing this is called *least squares*. Least squares is the method that graphing calculators use to find regression lines and curves. Even if you continue to use your calculator to carry out the calculations, this section is important in that it explains what you are actually doing.

Least squares lines are used extensively in forecasting and for detecting underlying trends in data.

A First Example

The graph on the left shows a company's annual sales (in millions) over a 3-year period. How can we fit a straight line to these three points? Clearly, these points do not lie exactly on a line, and so rather than an "exact" fit, we want the line $y = ax + b$ that fits these three points most closely.

Let d_1, d_2, and d_3 stand for the vertical distances between the three points and the line $y = ax + b$. The line that minimizes the sum of the squares of these vertical deviations is called the *least squares line* or the *regression line*.* Squaring the deviations ensures that none are negative, so that a deviation below the line does not "cancel" one above the line.

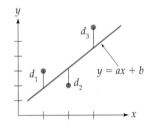

EXAMPLE 1   MINIMIZING THE SQUARED DEVIATIONS

The table below gives a company's sales (in millions) in year x. Find the least squares line for the data. Then use the line to predict sales in year 4.

x	y
1	10
2	12
3	25

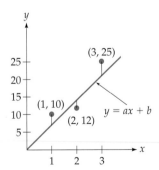

---

* The word "regression" comes from an early use of this technique to determine whether unusually tall parents have unusually tall children. It seems that tall parents do have tall offspring, but not quite as tall, with successive generations exhibiting a "regression" toward the average height of the population.

13.4 LEAST SQUARES

Solution The vertical deviations are found by calculating the heights of the line $y = ax + b$ at each x-value minus the y-values from the table. These differences are then squared and summed.

$$S = (a \cdot 1 + b - 10)^2 + (a \cdot 2 + b - 12)^2 + (a \cdot 3 + b - 25)^2$$

- $(a \cdot 1 + b)$: Height of the line $y = ax + b$ at $x = 1$
- 10: y-value of the point at $x = 1$
- $(a \cdot 2 + b)$: Height of the line $y = ax + b$ at $x = 2$
- 12: y-value of the point at $x = 2$
- $(a \cdot 3 + b)$: Height of the line $y = ax + b$ at $x = 3$
- 25: y-value of the point at $x = 3$

This sum S depends on a and b, the numbers that determine the line $y = ax + b$. To minimize S, we set its partials with respect to a and b equal to zero:

$$\frac{\partial S}{\partial a} = 2(a + b - 10) + 2(2a + b - 12) \cdot 2 + 2(3a + b - 25) \cdot 3 \quad \text{Differentiating each part of } S \text{ by the Generalized Power Rule}$$

$$= 2a + 2b - 20 + 8a + 4b - 48 + 18a + 6b - 150 \quad \text{Multiplying out}$$

$$= 28a + 12b - 218 \quad \text{Combining terms}$$

$$\frac{\partial S}{\partial b} = 2(a + b - 10) + 2(2a + b - 12) + 2(3a + b - 25) \quad \text{Differentiating each part of } S \text{ by the Generalized Power Rule}$$

$$= 2a + 2b - 20 + 4a + 2b - 24 + 6a + 2b - 50 \quad \text{Multiplying out}$$

$$= 12a + 6b - 94 \quad \text{Combining terms}$$

We set the two partials equal to zero and solve:

$$\left. \begin{array}{l} 28a + 12b - 218 = 0 \\ 12a + 6b - 94 = 0 \end{array} \right\} \quad \text{Partials set equal to zero}$$

$$28a + 12b - 218 = 0 \quad \text{First equation}$$

$$\underline{-24a - 12b + 188 = 0} \quad \text{Second multiplied by } -2$$

$$4a \quad\quad - 30 = 0 \quad \text{Adding (the } b\text{'s drop out)}$$

$$a = \frac{30}{4} = 7.5 \quad \text{Solving } 4a - 30 = 0$$

$$b = \frac{4}{6} \approx 0.67 \quad \text{From substituting } a = 7.5 \text{ into } 12a + 6b - 94 = 0 \text{ and solving for } b$$

These values for a and b give the least squares line. (The *D*-test would show that S has indeed been minimized.)

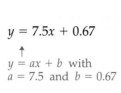
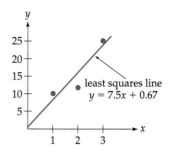

$y = 7.5x + 0.67$

$y = ax + b$ with $a = 7.5$ and $b = 0.67$

least squares line $y = 7.5x + 0.67$

To predict the sales in year 4, we evaluate the least squares line at $x = 4$:

$y = 7.5(4) + 0.67 = 30.67$ $y = 7.5x + 0.67$ evaluated at $x = 4$

Prediction: 30.67 million sales in year 4.

The slope of the line is 7.5, meaning that the linear trend in the company's sales is a growth of 7.5 million sales per year.

Graphing Calculator Exploration

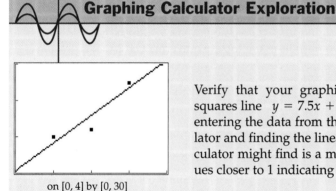

on [0, 4] by [0, 30]

Verify that your graphing calculator gives the same least squares line $y = 7.5x + 0.67$ that we found in Example 1 by entering the data from the table on page 1000 into your calculator and finding the linear regression line. The r that your calculator might find is a measure of "goodness of fit," with values closer to 1 indicating a better fit.

Least Squares Line for *n* Points

Example 1 used only three points, which is too few for most realistic applications. If we carry out the same steps for n points, we would obtain the following formulas (in which Σ stands for sum).

Least Squares Line

For data

x	y
x_1	y_1
x_2	y_2
.	.
.	.
.	.
x_n	y_n

calculate

$$a = \frac{n\Sigma xy - (\Sigma x)(\Sigma y)}{n\Sigma x^2 - (\Sigma x)^2}$$

$$b = \frac{1}{n}(\Sigma y - a\Sigma x)$$

n = number of data points
Σx = sum of x's
Σy = sum of y's
Σxy = sum of products $x \cdot y$
Σx^2 = sum of squares of x's

The least squares line is then $y = ax + b$

From now on we will find least squares lines by using these formulas, a derivation of which is given at the end of this section.

EXAMPLE 2

FINDING THE LEAST SQUARES LINE

A 1955 study compared cigarette smoking with the mortality rate for lung cancer in several countries. Find the least squares line that fits these data. Then use the line to predict lung cancer deaths if per capita cigarette consumption is 600 cigarettes per month (a pack a day).

	Cigarette Consumption (per capita)	Lung Cancer Deaths (per million)
Norway	250	90
Sweden	300	120
Denmark	350	170
Canada	500	150

Solution The procedure for calculating a and b consists of six steps, beginning with the following table.

1. List the x- and y-values 2. Multiply $x \cdot y$ 3. Square each x

x	y	xy	x^2
250	90	22,500	62,500
300	120	36,000	90,000
350	170	59,500	122,500
500	150	75,000	250,000
1400	530	193,000	525,000
‖	‖	‖	‖
Σx	Σy	Σxy	Σx^2

← 4. Sum each column

5. Calculate a and b using the formulas on the preceding page.

n = number of points

$$a = \frac{(4)(193{,}000) - (1400)(530)}{(4)(525{,}000) - 1400^2} \qquad a = \frac{n\,\Sigma xy - (\Sigma x)(\Sigma y)}{n\,\Sigma x^2 - (\Sigma x)^2}$$

$$= \frac{30{,}000}{140{,}000} \approx 0.21$$

$$b = \frac{1}{4}[530 - 0.21(1400)] = 59 \qquad b = \frac{1}{n}(\Sigma y - a\,\Sigma x)$$

6. The least squares line is $y = ax + b$ with the a and b values above

$$y = 0.21x + 59 \qquad y = ax + b \text{ with } a = 0.21 \text{ and } b = 59$$

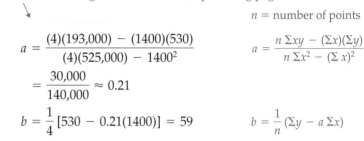

This line is graphed on the left, and shows that lung cancer mortality increases with cigarette smoking.

To predict the lung cancer deaths if per capita cigarette consumption reaches 600, we evaluate the least squares line at $x = 600$.

$$y = 0.21 \cdot 600 + 59 = 185 \qquad y = 0.21x + 59 \text{ with } x = 600$$

Predicted annual mortality: 185 deaths per million.

Criticism of Least Squares

Least squares is the most widely used method for fitting lines to points, but it does have one weakness: the vertical deviations from the line are squared, so one large deviation, when squared, can have an unexpectedly large influence on the line. For example, the graph on the right shows four points and their least squares line. The line fits the points quite closely.

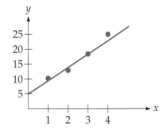

The second graph adds a fifth point, one quite out of line with the others, and shows the least squares line for the five points. The added point has an enormous effect on the line, causing it to slope downward even though all of the other points suggest an upward slope. In actual applications, one should inspect the points for such "outliers" before calculating the least squares line.

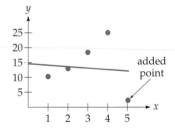

Fitting Exponential Curves by Least Squares

It is not always appropriate to fit a straight line to a set of points. Sometimes a collection of points will suggest a *curve* rather than a line, such as one of the following exponential curves.

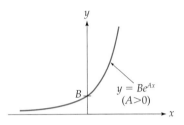
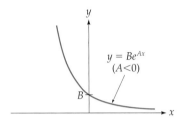

Least squares can be used to fit an exponential curve of the form $y = Be^{Ax}$ to a collection of points as follows. Taking natural logs of both sides of $y = Be^{Ax}$ gives

$$\ln y = \ln(Be^{Ax}) = \ln B + \ln e^{Ax} = \ln B + Ax \qquad \text{Using the properties of natural logarithms}$$

If we introduce a new variable $Y = \ln y$, and also let $b = \ln B$, then $\ln y = \ln B + Ax$ becomes

$$Y = b + Ax \qquad \text{Or, equivalently, } Y = Ax + b$$

Therefore, we fit a straight line to the *logarithms* of the y-values. (Now we are minimizing not the squared deviations, but the squared deviations of the logarithms.) The procedure consists of the eight steps shown in the following example.

EXAMPLE 3

FITTING AN EXPONENTIAL CURVE BY LEAST SQUARES

The world population* since 1850 is shown in the table below. Fit an exponential curve to these data and predict the world population in the year 2050.

Year	Population (billions)
1850	1.26
1900	1.65
1950	2.52
2000	6.08

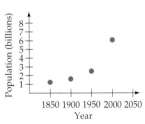

* *Source:* U.S. Census Bureau

23.

x	y
0	1
1	2
2	5
3	10

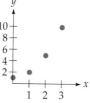

24.

x	y
0	2
1	4
2	7
3	15

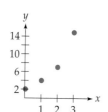

25.

x	y
−1	20
0	18
1	15
3	4
5	1

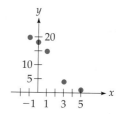

26.

x	y
−2	20
0	12
2	9
4	6
5	5

APPLIED EXERCISES ON FITTING EXPONENTIAL CURVES

27. POLITICAL SCIENCE: Cost of a Congressional Victory The following table shows average amounts spent by winners of seats in the House of Representatives in presidential election years. Fit an exponential curve to these data and use it to predict the cost of a House seat in the years 2004 and 2008.

	x	Cost ($1000)
1992	1	544
1996	2	764
2000	3	833

Source: Center for Responsive Politics

28. GENERAL: Drunk Driving The following table shows how a driver's blood-alcohol level (% grams per dekaliter) affects the probability of being in a collision. A collision factor of 3 means that the probability of a collision is 3 times as large as normal. Fit an exponential curve to the data. Then use your curve to estimate the collision factor for a blood-alcohol level of 15.

Blood-Alcohol Level	Collision Factor
0	1
6	1.1
8	3
10	6

29. BUSINESS: Super Bowl Advertising The following table shows the cost of television advertising during the Super Bowl in *thousands of dollars per second*. Fit an exponential curve to the data and then use it to predict the cost in the year 2004.

	x	Cost ($1000 per Second)
1997	1	40
1998	2	40.3
1999	3	53.3
2000	4	73.3
2001	5	80

Source: Nielsen Media Research

30. GENERAL: Stamp Prices The following table shows the cost of a first-class postage stamp in past years. Fit an exponential curve to these data and use your answer to predict the cost of a stamp in the year 2020.

	x	Postage (cents)
1940	1	3
1960	2	4
1980	3	15
2000	4	34

31. GENERAL: Safe Cars, Unsafe Streets Reread the Application Preview on page 959 at the beginning of this chapter and carry out the following steps to find when gunshot fatalities will overtake automobile fatalities in the United States.

a. Enter into your calculator the data from the table on page 959, *Years since 1970* into list L_1, *Automobile Fatalities* into L_2, and *Gunshot Fatalities* into L_3.

b. Have your calculator find the linear regression formula for L_1 and L_2 and enter it into function y_1, giving an approximate linear formula for the number of gunshot fatalities x years after 1970. *(continues)*

c. Have your calculator find the linear regression formula for L_1 and L_3 and enter it into function y_2, giving an approximate linear formula for the number of automobile fatalities x years after 1970. Graph all of the data and both functions on the window [0, 50] by [0, 60000].

d. Have your calculator find where the curves intersect, and interpret the x-value (years after 1970) as the year when gunshot fatalities will overtake automobile fatalities.

32. GENERAL: Cost of College Education The following table shows the cost of one year of tuition at a private four-year college. Fit an exponential curve to the data and then use it to predict the cost for the academic year 2005–2006.

	x	Cost ($1000)
1975–76	1	2.3
1980–81	2	3.6
1985–86	3	5.4
1990–91	4	9.4
1995–96	5	12.4
2000–01	6	16.3

Source: The College Board

13.5 LAGRANGE MULTIPLIERS AND CONSTRAINED OPTIMIZATION

Introduction

In Section 13.3 we optimized functions of several variables. Some problems, however, involve maximizing or minimizing a function subject to a *constraint*. For example, a company might want to maximize production subject to the constraint of staying within its budget, or a soft drink distributor might want to design the least expensive aluminum can subject to the constraint that it hold exactly 12 ounces of soda. In this section we solve such "constrained optimization" problems by the method of Lagrange multipliers, invented by the French mathematician Joseph Louis Lagrange (1736–1813).

1026 CHAPTER 13 CALCULUS OF SEVERAL VARIABLES

APPLIED EXERCISES

Solve each using Lagrange multipliers. (The stated extreme values *do* exist.)

25. **GENERAL: Fences** A parking lot, divided into two equal parts, is to be constructed against a building, as shown in the diagram. Only 6000 feet of fence are to be used, and the side along the building needs no fence.

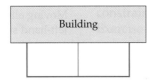

a. What are the dimensions of the largest area that can be so enclosed?
b. Evaluate and give an interpretation for $|\lambda|$.

26. **GENERAL: Fences** Three adjacent rectangular lots are to be fenced in, as shown in the diagram, using 12,000 feet of fence. What is the largest total area that can be so enclosed?

27–28. **GENERAL: Container Design** A cylindrical tank without a top is to be constructed with the least amount of material (bottom plus side area). Find the dimensions if the volume is to be:

27. 160 cubic feet
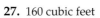
28. 120 cubic feet
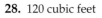

29. **GENERAL: Postal Regulations** The U.S. Postal Service will accept a package if its length plus its girth is not more than 84 inches. Find the dimensions and volume of the largest package with a square end that can be mailed. (See the diagram in the next column.)

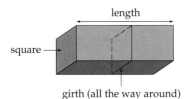

30. **GENERAL: Postal Regulations** Solve Exercise 29, but now for a package with a round end, so that the package is a cylinder rather than a rectangular solid. Compare the volume with that of Exercise 29.

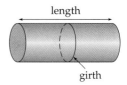

31. **BUSINESS: Maximum Production** A company's output is given by the Cobb–Douglas production function $P = 200L^{3/4}K^{1/4}$, where L and K are the numbers of units of labor and capital. Each unit of labor costs $50 and each unit of capital costs $100, and $8000 is available to pay for labor and capital.

a. How many units of labor and of capital should be used to maximize production?
b. Evaluate and give an interpretation for $|\lambda|$.

32. **BUSINESS: Production Possibilities** A company manufactures two products, in quantities x and y. Because of limited materals and capital, the quantities produced must satisfy the equation $2x^2 + 5y^2 = 32{,}500$. (This curve is called a *production possibilities curve.*) If the company's profit function is $P = 4x + 5y$ dollars, how many of each product should be made to maximize profit? Also find the maximum profit.

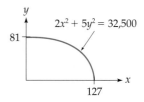

33. **GENERAL: Package Design** A metal box with a square base is to have a volume of 45 cubic inches. If the top and bottom cost 50 cents per square inch and the sides cost 30 cents per square inch, find the dimensions that minimize the cost. [*Hint:* The cost of the box is the area of each part (top, bottom, and sides) times the cost per square inch for that part. Minimize this subject to the volume constraint.]

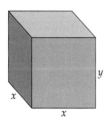

34. **GENERAL: Building Design** A one-story building is to have 8000 square feet of floor space. The front of the building is to be made of brick, which costs $120 per linear foot, and the back and sides are to be made of cinder-block, which costs only $80 per linear foot.

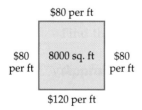

a. Find the length and width that minimize the cost of the building. [*Hint:* The cost of the building is the length of the front, back, and sides, each times the cost per foot for that part. Minimize this subject to the area constraint.]
b. Evaluate and give an interpretation for $|\lambda|$.

Functions of Three Variables

(The stated extreme values *do* exist.)

35. Minimize $f(x, y, z) = x^2 + y^2 + z^2$
    subject to $2x + y - z = 12$.

36. Minimize $f(x, y, z) = x^2 + y^2 + z^2$
    subject to $x - y + 2z = 6$.

37. Maximize $f(x, y, z) = x + y + z$
    subject to $x^2 + y^2 + z^2 = 12$.

38. Maximize $f(x, y, z) = xyz$
    subject to $x^2 + y^2 + z^2 = 12$.

39. **GENERAL: Building Design** A one-story storage building is to have a volume of 250,000 cubic feet. The roof costs $32 per square foot, the walls $10 per square foot, and the floor $8 per square foot. Find the dimensions that minimize the cost of the building.

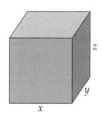

40. **GENERAL: Container Design** An open-top box with two parallel partitions, as in the diagram, is to have volume 64 cubic inches. Find the dimensions that require the least amount of material.

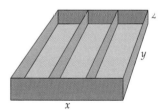

13.6 TOTAL DIFFERENTIALS AND APPROXIMATE CHANGES

Introduction

In this section we define the *total differential* of a function of several variables, and use it to approximate the change in the function resulting from changes in the independent variables. In addition to giving

However, for *dependent* variables, "Δ" and "d" have different meanings: Δ indicates the *actual* change, and d indicates the total differential.

For some functions, calculating the actual change Δf can be complicated. The total differential provides a simple *linear approximation* for the actual change. The partials f_x and f_y give the rate of change of f per unit change in x and y, respectively. Changing x by Δx units changes f by approximately $f_x \cdot \Delta x$ units, and changing y by Δy units changes f by approximately $f_y \cdot \Delta y$ units. Therefore, changing *both* x and y should change f by approximately the *sum* of these changes. In symbols:

> **Differential Approximation Formula**
>
> $$\underbrace{f(x + \Delta x, y + \Delta y) - f(x, y)}_{\text{Change in } f} \approx \underbrace{f_x \cdot \Delta x + f_y \cdot \Delta y}_{\substack{\text{Total differential of } f \\ (\text{since } \Delta x = dx \text{ and } \Delta y = dy)}}$$

Written more compactly:

$$\Delta f \approx df \qquad \begin{array}{l} \text{Since } \Delta f = f(x + \Delta x, y + \Delta y) - f(x, y) \\ \text{and } df = f_x(x, y) \cdot dx + f_y(x, y) \cdot dy \end{array}$$

The partials in the first formula are evaluated at (x, y), and the approximation improves as Δx and Δy approach zero. (See pages 1033–1034 for a geometric explanation of the approximation $\Delta f \approx df$.)

EXAMPLE 3

APPROXIMATING AN ACTUAL CHANGE BY A DIFFERENTIAL

For $f(x, y) = x^2 + 4xy + y^3$ and values $x = 3$, $y = 2$, $\Delta x = 0.2$, and $\Delta y = -0.1$, find: **a.** Δf **b.** df

Solution

a. From the given values,

$$x + \Delta x = 3 + 0.2 = 3.2 \quad \text{and} \quad y + \Delta y = 2 - 0.1 = 1.9$$

The change Δf is

$\Delta f = f(3.2, 1.9) - f(3, 2)$ \qquad $\Delta f = f(x + \Delta x, y + \Delta y) - f(x, y)$

$= \underbrace{3.2^2 + 4 \cdot (3.2) \cdot (1.9) + 1.9^3}_{f(3.2, 1.9)} - \underbrace{(3^2 + 4 \cdot 3 \cdot 2 + 2^3)}_{f(3, 2)}$ \qquad Using $f(x, y) = x^2 + 4xy + y^3$

$= 10.24 + 24.32 + 6.859 - (9 + 24 + 8)$ \qquad Evaluating

$= 41.419 - 41 = 0.419$ \qquad Change in f is $\Delta f = 0.419$

b. The total differential df is

$$df = \underbrace{(2x + 4y)}_{f_x(x,y)} \cdot dx + \underbrace{(4x + 3y^2)}_{f_y(x,y)} \cdot dy \quad \text{Partials of } x^2 + 4xy + y^3 \text{ times } dx \text{ and } dy$$

$$= (2 \cdot 3 + 4 \cdot 2) \cdot (0.2) + (4 \cdot 3 + 3 \cdot 2^2) \cdot (-0.1) \quad \begin{array}{l}\text{Evaluating at } x = 3,\\ y = 2, \ dx = 0.2,\\ \text{and } dy = -0.1\end{array}$$

$$= 2.8 - 2.4 = 0.4 \quad \text{Total differential is } df = 0.4$$

We found $\begin{cases} \Delta f = 0.419 \\ df = 0.4 \end{cases}$. The total differential $df = 0.4$ provides a reasonably accurate approximation for the actual change $\Delta f = 0.419$. The approximation would be even more accurate for smaller values of Δx and Δy.

Why should we bother calculating an *approximation* df when with a calculator we can easily find the *exact* change Δf? The total differential df has the advantage of *linearity*: If we were to *double* the changes Δx and Δy in the independent variables, then the differential would also double, from 0.4 to 0.8; if we were to *halve* the changes Δx and Δy, then the differential would also be halved, from 0.4 to 0.2. The *actual* change Δf admits no such simple modification—whenever Δx and Δy change it has to be recalculated "from scratch" using the formula $\Delta f = f(x + \Delta x, y + \Delta y) - f(x, y)$. The linearity of the total differential df is an advantage when you are trying to understand how various changes in x and y would affect the value of f.

EXAMPLE 4

ESTIMATING ADDITIONAL PROFIT

The American Farm Machinery Company finds that if it manufactures x economy tractors and y heavy-duty tractors per month, then its profit (in thousands of dollars) will be $P(x, y) = 3x^{4/3} + 0.05xy + 4y$. If the company now manufactures 125 economy tractors and 100 heavy-duty tractors per month, find an approximation for the additional profit from manufacturing 3 more economy tractors and 2 more heavy-duty tractors per month.

Solution We want an estimate for the change in the profit $P(x, y)$ when production increases above the levels $x = 125$ and $y = 100$ by amounts $\Delta x = 3$ and $\Delta y = 2$. The partials are

$$P_x = 4x^{1/3} + 0.05y \qquad P_y = 0.05x + 4 \qquad \begin{array}{l}\text{Partials of}\\ P = 3x^{4/3} + 0.05xy + 4y\\ \text{with respect to } x \text{ and } y\end{array}$$

The total differential is

$$dP = (4x^{1/3} + 0.05y) \cdot dx + (0.05x + 4) \cdot dy \qquad dP = P_x \cdot dx + P_y \cdot dy$$

$$= (4 \cdot 125^{1/3} + 0.05 \cdot 100) \cdot 3 + (0.05 \cdot 125 + 4) \cdot 2 \qquad \begin{array}{l}\text{Substituting}\\ x = 125,\ y = 100,\\ \Delta x = 3,\ \text{and}\ \Delta y = 2\end{array}$$

$$\underbrace{\sqrt[3]{125} = 5}$$

$$= (20 + 5) \cdot 3 + (6.25 + 4) \cdot 2 = 75 + 20.5 = 95.5 \quad \begin{array}{l}\text{In thousands}\\ \text{of dollars}\end{array}$$

Therefore, producing 3 more economy tractors and 2 more heavy-duty tractors will generate about $95,500 in additional profit.

The actual change in the profit function, found by applying the Δf formula on page 601, is $\Delta P = P(125 + 3, 100 + 2) - P(125, 100) \approx 96.04$, so the approximation of 95.5 is indeed quite accurate.

The *linearity* of the total differential means that if sometime later the company wanted to estimate the additional profit from *doubling* these changes—making 6 more economy tractors and 4 more heavy-duty tractors—they would need only to double the estimate of $95,500 to immediately obtain $191,000.

Estimating Errors

No physical measurement can ever be made with perfect accuracy. If you can estimate the maximum error in a measurement, then you can use differentials to estimate the resulting error in a calculation. The measurement errors may be expressed as *percentage* errors or *actual* numbers (as in Example 6, to be discussed shortly). The following example estimates the percentage error in calculating the volume of a cylinder. Such calculations have applications from predicting variations in soft-drink cans to ensuring safety margins for artificial arteries.

EXAMPLE 5 **ESTIMATING THE ERROR IN CALCULATING VOLUME**

A cylinder is measured to have radius r and height h, but these measurements may be in error by up to 1%. Estimate the resulting percentage error in calculating the volume of the cylinder.

13.6 TOTAL DIFFERENTIALS AND APPROXIMATE CHANGES

Solution The height and radius being in error by 1% means that

$$\Delta r = 0.01r$$
$$\Delta h = 0.01h$$

For each, the change is 1% of the value

The volume of a cylinder is $V = \pi r^2 h$, and the total differential is

$$dV = 2\pi rh \cdot dr + \pi r^2 \cdot dh \qquad \text{Partials of } V = \pi r^2 h \text{ times } dr \text{ and } dh$$

$$= 2\pi rh \cdot 0.01r + \pi r^2 \cdot 0.01h \qquad \text{Substituting } dr = \Delta r = 0.01r \text{ and } dh = \Delta r = 0.01h$$

$$= \pi r^2 h \underbrace{(2 \cdot 0.01 + 0.01)}_{0.03} \qquad \text{Factoring out } \pi r^2 h$$

$$= 0.03 \pi r^2 h = 0.03 \cdot V \qquad \text{Change is 3\% of volume}$$

Volume V of the cylinder

This result, $dV = 0.03 \cdot V$, means that the volume may be in error by as much as 3% if the radius and height are "off" by 1%.

Geometric Visualization of *df* and Δ*f*

A function $f(x, y)$ represents a *surface* in three-dimensional space, and the change Δ*f* represents the change in height when moving from one point to another along the surface, as shown below.

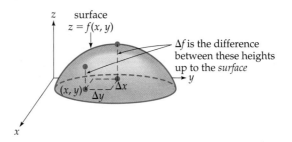

Δ*f* is the difference between these heights up to the *surface*

The total differential *df* represents the change in height when moving from one point to another along the *plane* that best fits the surface at the first point, as shown on the following page. This plane is called the *tangent plane* to the surface at the point.

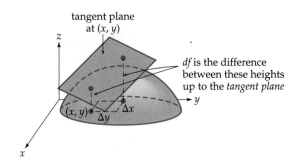

Since the tangent plane fits the surface closely near the point (x, y), changes df in height along the *tangent plane* should be very close to changes Δf in height along the *surface*, which is why the total differential df closely approximates the actual change Δf for small changes in x and y.

Total Differential of a Function of Three Variables

The total differential may be generalized to apply to functions of three (or more) variables, multiplying each partial by "d" of that variable and adding.

Total Differential of $f(x, y, z)$

For a function $f(x, y, z)$, the total differential df is

$$df = f_x(x, y, z) \cdot dx + f_y(x, y, z) \cdot dy + f_z(x, y, z) \cdot dz \qquad \text{Partials times } dx, dy, \text{ and } dz$$

The total differential of a function $f(x, y, z)$ gives an estimate for the change Δf when the variables are changed by amounts Δx, Δy, and Δz:

$$\underbrace{f(x + \Delta x, y + \Delta y, z + \Delta z) - f(x, y, z)}_{\text{Change in } f} \approx \underbrace{\frac{\partial f}{\partial x} \cdot \Delta x + \frac{\partial f}{\partial y} \cdot \Delta y + \frac{\partial f}{\partial z} \cdot \Delta z}_{\substack{\text{Total differential of } f \\ (\text{since } \Delta x = dx, \\ \Delta y = dy, \Delta z = dz)}}$$

or, more briefly:

$$\Delta f \approx df$$

The partials in the formula above are evaluated at (x, y, z), and the approximation is increasingly accurate for smaller values of Δx, Δy, and Δz.

EXAMPLE 6

ESTIMATING THE ERROR IN CALCULATING VOLUME

A rectangular box is measured to be 30 inches long, 24 inches wide, and 10 inches high. If the maximum errors in measuring the length, width, and height of the box are, respectively, 0.3, 0.2, and 0.1 inch, estimate the maximum error in calculating its volume.

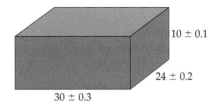

Solution The volume of the box is length times width times height, $V = x \cdot y \cdot z$. We want to estimate the change in volume resulting from changing the dimensions $x = 30$, $y = 24$, and $z = 10$ by the amounts $\Delta x = 0.3$, $\Delta y = 0.2$, and $\Delta z = 0.1$. The total differential is

$$df = \overbrace{y \cdot z}^{V_x} \cdot dx + \overbrace{x \cdot z}^{V_y} \cdot dy + \overbrace{x \cdot y}^{V_z} \cdot dz \qquad \text{Partials of } V = x \cdot y \cdot z \text{ times } dx, dy, \text{ and } dz$$

$$= 24 \cdot 10 \cdot 0.3 + 30 \cdot 10 \cdot 0.2 + 30 \cdot 24 \cdot 0.1 \qquad \text{Substituting } x = 30,\ y = 24,\ z = 10,\ dx = 0.3,\ dy = 0.2,\ \text{and}\ dz = 0.1$$

$$= 72 + 60 + 72 = 204 \qquad \text{Multiplying out and adding}$$

That is, the maximum error in calculating the volume is approximately 204 cubic inches.

While an error of 204 cubic inches may seem large, the *relative* error (the error divided by the volume, expressed as a percentage) is only

$$\frac{204}{30 \cdot 24 \cdot 10} = \frac{204}{7200} \approx 0.028 = 2.8\% \qquad \text{Relative error is 2.8\%}$$

Justification of the Interpretation of the Lagrange Multiplier λ

In the preceding section we used Lagrange multipliers to optimize an objective function $f(x, y)$ subject to a constraint $g(x, y) = 0$. We did so by setting the partials of $F(x, y, \lambda) = f(x, y) + \lambda g(x, y)$ equal to zero, and we interpreted the absolute value of the "Lagrange multiplier" $|\lambda|$ as *the number of additional objective units per additional constraint unit*. This interpretation may be justified as follows. Setting the partials of F with respect to x and y equal to zero gives

$$\begin{cases} f_x + \lambda g_x = 0 \\ f_y + \lambda g_y = 0 \end{cases} \quad \text{or, equivalently,} \quad \begin{cases} f_x = -\lambda g_x \\ f_y = -\lambda g_y \end{cases}$$

If we increase x and y by amounts Δx and Δy, the resulting change in the objective function $f(x, y)$ can be approximated by the total differential:

$$\underbrace{f(x + \Delta x, y + \Delta y) - f(x, y)}_{\Delta f} \approx f_x \Delta x + f_y \Delta y \qquad \Delta f \approx df$$

$$= -\lambda g_x \Delta x - \lambda g_y \Delta y \qquad \begin{array}{l}\text{Substituting}\\ f_x = -\lambda g_x \text{ and}\\ f_y = -\lambda g_y\end{array}$$

$$= -\lambda \underbrace{(g_x \Delta x + g_y \Delta y)}_{dg} \qquad \begin{array}{l}\text{Factoring out } -\lambda\\ \text{leaves the total}\\ \text{differential of } g\end{array}$$

$$\approx -\lambda \cdot \Delta g \qquad \begin{array}{l}\text{Replacing } dg \text{ by } \Delta g\\ \text{(since } dg \approx \Delta g\text{)}\end{array}$$

These equations, read from beginning to end, say that $\Delta f \approx -\lambda \cdot \Delta g$, or

$$\frac{\Delta f}{\Delta g} \approx -\lambda \qquad\qquad \text{Dividing by } \Delta g$$

The left-hand side of this is the ratio of the change in the objective function f to the change in the constraint function g. Taking absolute values and letting Δx and Δy approach zero (to make the approximation exact) shows that $|\lambda|$ is *the number of objective units per additional constraint unit*, as stated on page 1021.

13.6 Section Summary

For a function $f(x, y)$, the total differential df is defined as the partials multiplied by dx and dy and added:

$$df = f_x(x, y) \cdot dx + f_y(x, y) \cdot dy \qquad\qquad df = \frac{\partial f}{\partial x} dx + \frac{\partial f}{\partial y} dy$$

The actual change in the function when x and y change by amounts Δx and Δy is

$$\Delta f = f(x + \Delta x, y + \Delta y) - f(x, y)$$

For small values of $\Delta x = dx$ and $\Delta y = dy$, the actual change Δf can be approximated by the total differential df:

$$f(x + \Delta x, y + \Delta y) - f(x, y) \approx f_x(x, y) \cdot dx + f_y(x, y) \cdot dy \qquad \Delta f \approx df$$

The actual change Δf may depend on Δx and Δy in very complicated ways, but the total differential df is *linear* in $\Delta x = dx$ and $\Delta y = dy$, and is therefore easier to calculate.

▶ Solution to Practice Problem

$g_x = 2xe^{5y} \qquad g_y = x^2 5e^{5y}$ Partials

$dg = 2xe^{5y} \cdot dx + 5x^2 e^{5y} \cdot dy$ Total differential

13.6 Exercises

Find the total differential of each function.

1. $f(x, y) = x^2 y^3$
2. $f(x, y) = x^4 y^{-1}$
3. $f(x, y) = 6x^{1/2} y^{1/3} + 8$
4. $f(x, y) = 100 x^{0.05} y^{0.02} - 7$
5. $g(x, y) = \dfrac{x}{y}$
6. $g(x, y) = \dfrac{x}{x + y}$
7. $g(x, y) = (x - y)^{-1}$
8. $g(x, y) = \sqrt{x^2 + y^2}$
9. $z = \ln(x^3 - y^2)$
10. $z = x^2 \ln y$
11. $z = xe^{2y}$
12. $z = e^{3x - 2y}$
13. $w = 2x^3 + xy + y^2$
14. $w = 3x^2 - xy^{-1} + y^3$
15. $f(x, y, z) = 2x^2 y^3 z^4$
16. $f(x, y, z) = xy + yz + xz$
17. $f(x, y, z) = \ln(xyz)$
18. $f(x, y, z) = \ln(x^2 + y^2 + z^2)$
19. $f(x, y, z) = e^{xyz}$
20. $f(x, y, z) = e^{x^2 + y^2 + z^2}$

For the given function and values, find:
   a. Δf b. df

21. $f(x, y) = x^2 + xy + y^3$, $x = 4$, $\Delta x = dx = 0.2$, $y = 2$, $\Delta y = dy = -0.1$
22. $f(x, y) = x^3 + xy + y^3$, $x = 5$, $\Delta x = dx = 0.01$, $y = 3$, $\Delta y = dy = -0.01$
23. $f(x, y) = e^x + xy + \ln y$, $x = 0$, $\Delta x = dx = 0.05$, $y = 1$, $\Delta y = dy = 0.01$
24. $f(x, y) = \ln(x^2 + y^2)$, $x = 6$, $\Delta x = dx = 0.1$, $y = 8$, $\Delta y = dy = 0.2$
25. $f(x, y, z) = xy + z^2$, $x = 3$, $\Delta x = dx = 0.03$, $y = 2$, $\Delta y = dy = 0.02$, $z = 1$, $\Delta z = dz = 0.01$
26. $f(x, y, z) = x^2 + y^2 + z^2$, $x = 3$, $\Delta x = dx = 0.1$, $y = 4$, $\Delta y = dy = 0.1$, $z = 5$, $\Delta z = dz = 0.1$

APPLIED EXERCISES

Use total differentials to solve the following exercises.

27. GENERAL: Measurement Errors A rectangle is measured to be 150 feet by 100 feet, but each measurement may be "off" by half a foot. Estimate the error in calculating the area. Then estimate the error in calculating the area if each measurement is "off" by one foot.

28. GENERAL: Telephone Calls For two cities with populations x and y (in thousands) that are 500 miles apart, the number of telephone calls per day between them can be modeled by the function $12xy$. For two cities with populations 40 thousand and 60 thousand, estimate the number of additional telephone calls if each city grows by 1 thousand people. Then estimate the number of additional calls if instead each city were to grow by only 500 people.

29–30. BUSINESS: Profit An electronics company's profit in dollars from making x tape decks and y CD players per day is given by the following profit function $P(x, y)$. If the company currently produces 200 tape decks and 300 CD players, estimate the extra profit that would result from producing five more tape decks and four more CD players.

29. $P(x, y) = 2x^2 - 3xy + 3y^2$

30. $P(x, y) = 3x^2 - 4xy + 4y^2$

31. GENERAL: Highway Safety The emergency stopping distance in feet for a truck of weight w tons traveling at v miles per hour on a dry road is $S = 0.027wv^2$. For a truck that weighs 4 tons and is usually driven at 60 miles per hour, estimate the extra stopping distance if it has an extra half ton of load and is traveling 5 miles per hour faster than usual.

32. GENERAL: Scuba Diving The maximum duration (in minutes) of a scuba dive can be estimated by the formula $T = \dfrac{33v}{d + 33}$, where v is the volume of air in the tank (in cubic feet at sea-level pressure) and d is the depth (in feet) of the dive. For values $v = 100$ and $d = 67$, estimate the change in duration if an extra 20 cubic feet of air is added and the dive is 10 feet deeper.

33. GENERAL: Relative Error in Calculating Area A rectangle is measured to have length x and width y, but each measurement may be in error by 1%. Estimate the percentage error in calculating the area.

34. GENERAL: Relative Error in Calculating Volume A rectangular solid is measured to have length x, width y, and height z, but each measurement may be in error by 1%. Estimate the percentage error in calculating the volume.

35. BIOMEDICAL: Cardiac Output Medical researchers calculate the quantity of blood pumped through the lungs (in liters per minute) by the formula $C = \dfrac{x}{y - z}$, where x is the amount of oxygen absorbed by the lungs (in milliliters per minute), and y and z are, respectively, the concentrations of oxygen in the blood just after and just before passing through the lungs (in milliliters of oxygen per liter of blood). Typical measurements are $x = 250$, $y = 160$, and $z = 150$. Estimate the error in calculating the cardiac output C if each measurement may be "off" by 5 units.

36. GENERAL: Windchill The windchill index (revised in 2001) announced during the winter by the weather bureau measures how cold it "feels" for a given temperature t (in degrees Fahrenheit) and wind speed w (in miles per hour). It is calculated by the formula $C(t, w) = 35.74 + 0.6215t - 35.75w^{0.16} + 0.4275tw^{0.16}$. If the temperature is 30 degrees and the wind speed is 10 miles per hour, estimate the change in the windchill temperature if the wind speed increases by 4 miles per hour and the temperature drops by 5 degrees.

37. THE SLOPE OF $f(x, y) = c$ On page 1024 we used the fact that the slope in the x-y plane of the curve defined by $f(x, y) = c$ (for constant c) is given by the formula $-\dfrac{f_x}{f_y}$.

Verify this formula by justifying the following five steps.

a. If $f(x, y) = c$ can be solved explicitly for a function $y = F(x)$, then we may write $f(x, F(x)) = c$.
Justify: $f(x + \Delta x, F(x + \Delta x)) - f(x, F(x)) = 0$.

b. Justify: $f(x + \Delta x, F(x + \Delta x) - F(x) + F(x)) - f(x, F(x)) = 0$.

c. Defining ΔF by $\Delta F = F(x + \Delta x) - F(x)$, we may write the equation above as
$$f(x + \Delta x, \Delta F + F(x)) - f(x, F(x)) = 0$$
Then, writing F for $F(x)$, this becomes
$$f(x + \Delta x, F + \Delta F) - f(x, F) = 0$$
Justify: $f_x \Delta x + f_y \Delta F \approx 0$.

d. Justify: $\dfrac{\Delta F}{\Delta x} \approx -\dfrac{f_x}{f_y}$.

e. Justify: $\dfrac{dF}{dx} = -\dfrac{f_x}{f_y}$.

This shows that the slope of $F(x)$, and therefore the slope of $f(x, y) = c$, is $-\dfrac{f_x}{f_y}$.

38. **BIOMEDICAL: Blood Vessels** A section of an artery is measured to have length 12 centimeters and diameter 0.8 centimeter. If each measurement may be off by 0.1 centimeter, find the volume with an estimate of the error.

13.7 MULTIPLE INTEGRALS

Introduction

This section discusses *integration* of functions of several variables. We define the *double integral* of a function by considering the volume under a surface $z = f(x, y)$. We then evaluate double integrals by "iterated" (repeated) single integrations. Finally, we use double integrals to calculate volumes, average values, and total accumulations. We restrict our attention to continuous functions (surfaces that have no holes or breaks), since most functions encountered in applications satisfy this restriction.

Rectangular Regions, Volumes, and Double Integrals

The points (x, y) in the plane with x taking values between numbers a and b and y taking values between numbers c and d determine a *rectangular region R*.

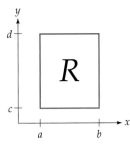

Rectangular region $R = \{(x,y) | a \le x \le b, c \le y \le d\}$

The graph below shows a nonnegative function $f(x, y)$ defined on a rectangular region R. We want to find the *volume* under the surface f and above the region R.

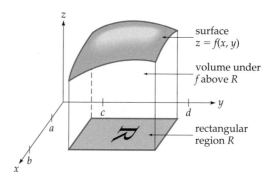

Volume under the surface $z = f(x, y)$ lying above a rectangular region R

We begin by *approximating* the volume by rectangular solids ("boxes") extending from R up to the surface. We divide R into small rectangles by drawing lines parallel to the x- and y-axes with spacing Δx and Δy, as shown below.* On each of these small rectangles we erect a rectangular solid with height $f(x_i, y_j)$, the height of the surface at some point (x_i, y_j) in the base rectangle.

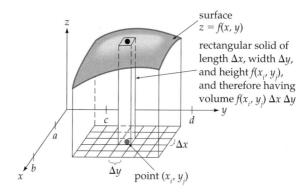

Volume under f over R showing one of the rectangular solids

The volume of the rectangular solid is $f(x_i, y_j) \cdot \Delta x \cdot \Delta y$ (height times length times width), and the sum of the volumes of all such rectangular

---

* Technically, the parallel lines need not have equal spacing. However, we will be letting the spacings approach zero, and the final results are the same for equal or unequal spacing.

solids approximates the volume under f above R:

$$\begin{pmatrix} \text{Volume under} \\ f \text{ above } R \end{pmatrix} \approx \sum f(x_i, y_j) \cdot \Delta x \cdot \Delta y \qquad \text{Σ means sum over all base rectangles}$$

The *limit* of this sum as both Δx and Δy approach zero (so that the base rectangles become smaller and more numerous) gives the *exact* volume, and is called the *double integral of f over R*, denoted $\iint_R f(x, y) \, dx \, dy$.

Double Integrals

The double integral of a continuous function $f(x, y)$ on a rectangular region R is

$$\iint_R f(x, y) \, dx \, dy = \lim_{\Delta x, \Delta y \to 0} \sum f(x_i, y_j) \cdot \Delta x \cdot \Delta y \qquad \text{The sum is over all rectangles in R, each containing one (x_i, y_j)}$$

If $f(x, y)$ is nonnegative on R, then the double integral gives the volume under f over R.

Iterated Integrals

Evaluating double integrals from the definition is difficult. Fortunately, double integrals can be evaluated by two separate "single" integrations, integrating with respect to one variable at a time while holding the other variable constant. (This is analogous to partial differentiation with respect to one variable, holding the other variable constant.) Such repeated integrals are called *iterated* integrals ("iterated" means "repeated"). A proof that double integrals can be evaluated as iterated integrals can be found in a more theoretical calculus book.

EXAMPLE 1

EVALUATING AN ITERATED INTEGRAL

Evaluate the iterated integral $\int_0^1 \int_0^2 (3x^2 + 6xy^2) \, dx \, dy$.

Solution The two separate integrations will be clearer if we use parentheses:

$$\int_0^1 \left(\int_0^2 (3x^2 + 6xy^2) \, dx \right) dy \qquad \text{An inner x-integral and an outer y-integral}$$

CHAPTER 13 CALCULUS OF SEVERAL VARIABLES

The inner integral gives

$$\int_0^2 (3x^2 + 6xy^2)\, dx = \left(x^3 + 6 \cdot \frac{1}{2} x^2 y^2\right)\Big|_{x=0}^{x=2}$$

- dx means integrate with respect to x holding y constant
- Integral of $3x^2$
- Integral of x
- Held constant

$$= 2^3 + 3 \cdot 2^2 y^2 - 0 = 8 + 12 y^2$$

- Evaluated at $x = 2$
- And at $x = 0$
- Simplified

We now apply the outer y-integral to this result:

$$\int_0^1 (8 + 12 y^2)\, dy = (8y + 4 y^3)\Big|_{y=0}^{y=1} = 8 + 4 - 0 = 12$$

- Result of the inner integral
- dy means integrate with respect to y
- $12 \cdot \frac{1}{3}$
- Evaluated at $y = 1$
- And at $y = 0$
- Final answer

Therefore:

$$\int_0^1 \int_0^2 (3x^2 + 6xy^2)\, dx\, dy = 12$$

The iterated integral equals 12

Always solve an iterated integral "from the inside out."

$$\int_0^1 \left(\int_0^2 (3x^2 + 6xy^2)\, dx \right) dy$$

- Limits for y
- Limits for x
- First integrate with respect to x
- Then with respect to y

In Example 1 we integrated first with respect to x and then with respect to y. The next example shows that switching the order of integration gives the same answer, provided that we also switch the x and y limits of integration. That is,

$$\int_0^1 \int_0^2 (3x^2 + 6xy^2)\, dx\, dy$$

is equal to

$$\int_0^2 \int_0^1 (3x^2 + 6xy^2)\, dy\, dx$$

EXAMPLE 2

REVERSING THE ORDER OF INTEGRATION

Evaluate $\int_0^2 \int_0^1 (3x^2 + 6xy^2)\, dy\, dx.$ Same as Example 1, but with the order of integration reversed

Solution First we evaluate the inner *y*-integral:

$$\int_0^1 (3x^2 + 6xy^2)\, dy = (3x^2 y + 2xy^3)\Big|_{y=0}^{y=1} = 3x^2 + 2x - 0$$

- Integrate with respect to *y*
- *x* held constant
- $\frac{1}{3} \cdot 6$
- Evaluated at $y=1$
- And at $y=0$

$$= 3x^2 + 2x$$

Then we apply the outer *x*-integral to this expression:

$$\int_0^2 (3x^2 + 2x)\, dx = (x^3 + x^2)\Big|_{x=0}^{x=2} = 8 + 4 - 0 = 12$$

- From inner integration
- Evaluated at $x=2$
- And at $x=0$
- Final answer

Therefore:

$$\int_0^2 \int_0^1 (3x^2 + 6xy^2)\, dy\, dx = 12$$

Notice that Examples 1 and 2 (in which the order of integration was reversed) gave the same answer, 12. Reversing the order of integration *always* gives the same answer, provided that the function is continuous.

Reversing the Order of Integration

For a continuous $f(x, y)$

$$\int_c^d \int_a^b f(x, y)\, dx\, dy$$

is equal to

$$\int_c^d \int_a^b f(x, y)\, dx\, dy$$

Double Integrals and Volumes

Earlier, we defined double integrals over rectangular regions. Double integrals can be evaluated by *either* of two iterated integrals (integrating in either order). The limits of integration are taken directly from the region R.

Evaluating Double Integrals

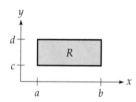

The double integral $\iint_R f(x, y)\, dx\, dy$

over the region $R = \{(x, y) \mid a \leq x \leq b, c \leq y \leq d\}$

can be evaluated by finding *either* of the iterated integrals

$$\int_c^d \int_a^b f(x, y)\, dx\, dy \quad \text{or} \quad \int_a^b \int_c^d f(x, y)\, dy\, dx$$

The limits of integration come from the definition of the rectangle R.

EXAMPLE 3

EVALUATING A DOUBLE INTEGRAL

Evaluate $\iint_R y^2 e^{-x}\, dx\, dy$ where $R = \{(x, y) \mid 0 \leq x \leq 2, -1 \leq y \leq 1\}$.

Solution This double integral can be evaluated by finding either of the iterated integrals

$$\int_{-1}^{1} \int_0^2 y^2 e^{-x}\, dx\, dy \quad \text{or} \quad \int_0^2 \int_{-1}^{1} y^2 e^{-x}\, dy\, dx \qquad \text{Limits of integration come from } R$$

We find the second one, beginning with the inner integral.

$$\int_{-1}^{1} y^2 e^{-x}\, dy = \left(\frac{1}{3} y^3 e^{-x}\right)\Big|_{y=-1}^{y=1} = \frac{1}{3} e^{-x} - \left(\frac{1}{3}(-1)e^{-x}\right)$$

Integrated — Held constant — Evaluated at $y = 1$ — And at $y = -1$

$$= \frac{1}{3} e^{-x} + \frac{1}{3} e^{-x} = \frac{2}{3} e^{-x}$$

Then we integrate this with respect to x:

$$\int_0^2 \frac{2}{3} e^{-x} \, dx = -\frac{2}{3} e^{-x} \Big|_{x=0}^{x=2} = -\frac{2}{3} e^{-2} - \left(-\frac{2}{3} e^0\right) = -\frac{2}{3} e^{-2} + \frac{2}{3}$$

Practice Problem

Show that evaluating this same double integral by the iterated integral in the *other* order gives the same answer. That is, evaluate the iterated integral

$$\int_{-1}^{1} \int_0^2 y^2 e^{-x} \, dx \, dy$$

➤ Solution on page 1050

The volume under a surface can be found by evaluating a double integral (since this is how double integrals were defined).

> **Volume by Double Integrals**
>
> For a nonnegative continuous function $f(x, y)$, the volume under the surface $z = f(x, y)$ and above a rectangular region R in the x-y plane is
>
> $$\begin{pmatrix} \text{Volume under} \\ f \text{ above } R \end{pmatrix} = \iint_R f(x, y) \, dx \, dy$$
>
>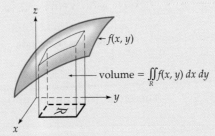
>
> If the surface lies *below* the x-y plane, this integral gives the *negative* of the volume.

EXAMPLE 4

FINDING THE VOLUME UNDER A SURFACE

A modernistic tent with closed sides is constructed according to the function shown below. To design ventilation and heating systems, it is necessary to know the volume under the tent. Find the volume under the tent on the indicated rectangle.

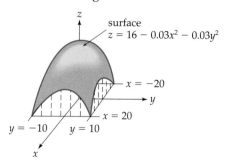

Solution

The volume is the integral of the function over the rectangle R:

$$\int_{-10}^{10} \int_{-20}^{20} (16 - 0.03x^2 - 0.03y^2) \, dx \, dy \qquad \text{Limits of integration come from } R$$

The inner integral is

$$\int_{-20}^{20} (16 - 0.03x^2 - 0.03y^2) \, dx = (16x - 0.01x^3 - 0.03y^2 x) \Big|_{x=-20}^{x=20}$$

$$= \underbrace{320 - 80 - 0.6y^2}_{\text{Evaluated at } x = 20} - \underbrace{(-320 + 80 + 0.6y^2)}_{\text{Evaluated at } x = -20} = 480 - 1.2y^2$$

Integrating this with respect to y:

$$\int_{-10}^{10} (480 - 1.2y^2) \, dy = (480y - 0.4y^3) \Big|_{-10}^{10} \qquad \text{From integrating } 1.2y^2$$

$$= 4800 - 400 - (-4800 + 400) = 8800$$

Therefore, the volume under the tent is 8800 cubic feet.

Average Value

On page 818 the average value of a function of *one* variable over an interval was defined as the definite integral of the function divided by the length of the interval. For similar reasons, the average value of a function $f(x, y)$ of *two* variables over a region is defined as the *double integral* divided by the *area* of the region.

Average Value

$$\begin{pmatrix} \text{Average value} \\ \text{of } f \text{ over } R \end{pmatrix} = \frac{1}{\text{Area of } R} \iint_R f(x, y)\, dx\, dy$$

Double integral over the region divided by the area of the region

For a rectangular region R, the area of R is simply length times width.

EXAMPLE 5

FINDING THE AVERAGE TEMPERATURE OVER A REGION

The temperature x miles east and y miles north of a weather station is $T(x, y) = 60 + 2x - 4y$ degrees. Find the average temperature over the rectangular region R extending 2 miles north and south from the station and 5 miles east (as shown in the following diagram).

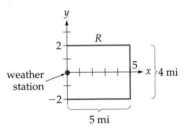

Solution The area of the region R is $4 \cdot 5 = 20$ square miles (length times width). The average temperature is the double integral divided by 20:

$$\text{Average} = \frac{1}{20} \int_{-2}^{2} \int_{0}^{5} (60 + 2x - 4y)\, dx\, dy$$

The inner integral is

$$\int_{0}^{5} (60 + 2x - 4y)\, dx = (60x + x^2 - 4yx)\Big|_{x=0}^{x=5}$$

$$= 300 + 25 - 20y - 0 = 325 - 20y$$

Evaluated at $x = 5$

The (outer) integral of this expression is

$$\int_{-2}^{2} (325 - 20y)\, dy = (325y - 10y^2)\Big|_{y=-2}^{y=2} = 650 - 40 - (-650 - 40)$$

$$= 610 + 690 = 1300$$

Finally, for the average we divide by 20 (the area of the region):

$$\frac{1300}{20} = 65$$

The average temperature over the region is 65 degrees.

Integrating over More General Regions

We can also integrate over regions R that are bounded by curves, provided that the curves as well as the function being integrated are continuous.

> **Double Integrals over Regions Between Curves**
>
> Let R be the region bounded by a lower curve $y = g(x)$ and an upper curve $y = h(x)$ from $x = a$ to $x = b$, as shown below. Then the double integral of $f(x, y)$ over R is
>
> $$\iint_R f(x, y)\, dx\, dy = \int_a^b \int_{g(x)}^{h(x)} f(x, y)\, dy\, dx$$
>
>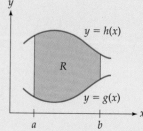
>
> If f is nonnegative, this double integral gives the volume under the surface $f(x, y)$ above R.

EXAMPLE 6

FINDING THE VOLUME UNDER A SURFACE

Find the volume under the surface $f(x, y) = 12xy$ and above the region R shown on the following page.

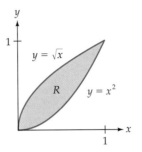

Solution The region R is bounded by the upper curve $h(x) = \sqrt{x}$ and the lower curve $g(x) = x^2$ from $x = 0$ to $x = 1$. From the boxed definition on the previous page, the volume is given by the following integral:

$$\text{Volume} = \int_0^1 \int_{x^2}^{\sqrt{x}} 12xy \, dy \, dx$$

We solve the inner integral first:

$$\int_{x^2}^{\sqrt{x}} 12xy \, dy = 12x \cdot \frac{1}{2} y^2 \Big|_{y=x^2}^{y=\sqrt{x}} = 6xy^2 \Big|_{y=x^2}^{y=\sqrt{x}}$$

$$= \underbrace{6x(\sqrt{x})^2}_{\substack{\text{Evaluating} \\ \text{at } y = \sqrt{x}}} - \underbrace{6x(x^2)^2}_{\substack{\text{Evaluating} \\ \text{at } y = x^2}} = \underbrace{6x^2 - 6x^5}_{\text{Simplified}}$$

Now we integrate the result with respect to x:

$$\int_0^1 (6x^2 - 6x^5) \, dx = \left(6 \cdot \frac{1}{3} x^3 - x^6\right) \Big|_0^1$$

$$= (2x^3 - x^6) \Big|_0^1 = 2 - 1 - (0) = 1$$

Therefore, the volume under the surface and above the region R is 1 cubic unit.

13.7 Section Summary

The double integral of a nonnegative function $f(x, y)$ over a region R gives the *volume* under the surface above R. (This is analogous to defining the definite integral of a function $f(x)$ of *one* variable as the *area* under the curve.)

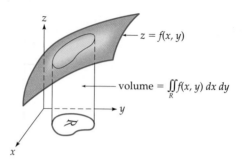

We evaluate a double integral by evaluating an iterated (repeated) integral:

$$\iint_R f(x, y)\, dx\, dy \quad \text{over} \quad R = \{(x, y) \mid a \le x \le b, c \le y \le d\}$$

is found by evaluating either of the two iterated integrals

$$\int_c^d \int_a^b f(x, y)\, dx\, dy \quad \text{or} \quad \int_a^b \int_c^d f(x, y)\, dy\, dx \qquad \text{Integrating in either order}$$

Note the distinction: *double* integrals are written with R (which must be specified) under the double integral, and *iterated* integrals are written with upper and lower *limits* on each integral sign.

Double integrals do more than just find volumes; they give *continuous sums* (as illustrated in Exercises 43 and 44), and when divided by the area of the region they give the *average value of the function over the region*.

Exercises 47–50 discuss "triple" integrals of functions $f(x, y, z)$ of three variables. Triple integrals are evaluated by *iterated* integrals (but *three* of them), integrating successively with respect to one variable at a time, holding all others constant.

▶ Solution to Practice Problem

$$\int_0^2 y^2 e^{-x}\, dx = -y^2 e^{-x}\Big|_{x=0}^{x=2} = -y^2 e^{-2} + y^2 e^0$$

$$= -y^2 e^{-2} + y^2$$

$$\int_{-1}^1 (-y^2 e^{-2} + y^2)\, dy = \left(-\frac{1}{3} y^3 e^{-2} + \frac{1}{3} y^3\right)\Big|_{-1}^1$$

$$= -\frac{1}{3} e^{-2} + \frac{1}{3} - \left(\frac{1}{3} e^{-2} - \frac{1}{3}\right) = -\frac{2}{3} e^{-2} + \frac{2}{3} \qquad \text{(same as before)}$$

13.7 Exercises

Evaluate each (single) integral.

1. $\int_1^{x^2} 8xy^3 \, dy$
2. $\int_1^{y^2} 10x^4 \, dx$
3. $\int_{-y}^{y} 9x^2 y \, dx$
4. $\int_{-x}^{x} 6xy^2 \, dy$
5. $\int_0^x (6y - x) \, dy$
6. $\int_0^y (4x - y) \, dx$

Evaluate each iterated integral.

7. $\int_0^2 \int_0^1 4xy \, dx \, dy$
8. $\int_0^2 \int_0^1 8xy \, dx \, dy$
9. $\int_0^2 \int_0^1 x \, dy \, dx$
10. $\int_0^4 \int_0^3 y \, dx \, dy$
11. $\int_0^1 \int_0^2 x^3 y^7 \, dx \, dy$
12. $\int_0^1 \int_0^3 x^8 y^2 \, dy \, dx$
13. $\int_1^3 \int_0^2 (x + y) \, dy \, dx$
14. $\int_1^2 \int_0^4 (x - y) \, dx \, dy$
15. $\int_{-1}^1 \int_0^3 (x^2 - 2y^2) \, dx \, dy$
16. $\int_{-1}^1 \int_0^3 (2x^2 + y^2) \, dy \, dx$
17. $\int_{-3}^3 \int_0^3 y^2 e^{-x} \, dy \, dx$
18. $\int_{-2}^2 \int_0^2 xe^{-y} \, dx \, dy$
19. $\int_{-2}^2 \int_{-1}^1 ye^{xy} \, dx \, dy$
20. $\int_{-1}^1 \int_{-1}^1 xe^{xy} \, dy \, dx$
21. $\int_0^2 \int_x^1 12xy \, dy \, dx$
22. $\int_0^1 \int_y^1 4xy \, dx \, dy$
23. $\int_3^5 \int_0^y (2x - y) \, dx \, dy$
24. $\int_2^4 \int_0^x (x - 2y) \, dy \, dx$
25. $\int_{-3}^3 \int_0^{4x} (y - x) \, dy \, dx$
26. $\int_{-1}^1 \int_0^{2y} (x + y) \, dx \, dy$
27. $\int_0^1 \int_{-y}^{y} (x + y^2) \, dx \, dy$
28. $\int_0^2 \int_{-x}^{x} (x^2 - y) \, dy \, dx$

For each double integral:
a. Write the *two* iterated integrals that are equal to it.
b. Evaluate *both* iterated integrals (the answers should agree).

29. $\iint_R 3xy^2 \, dx \, dy$
   with $R = \{(x, y) \mid 0 \le x \le 2, 1 \le y \le 3\}$

30. $\iint_R 6x^2 y \, dx \, dy$
   with $R = \{(x, y) \mid 0 \le x \le 1, 1 \le y \le 2\}$

31. $\iint_R ye^x \, dx \, dy$
   with $R = \{(x, y) \mid -1 \le x \le 1, 0 \le y \le 2\}$

32. $\iint_R xe^y \, dx \, dy$
   with $R = \{(x, y) \mid 0 \le x \le 1, -2 \le y \le 2\}$

Use integration to find the volume under each surface $f(x, y)$ above the region R.

33. $f(x, y) = x + y$
    $R = \{(x, y) \mid 0 \le x \le 2, 0 \le y \le 2\}$

34. $f(x, y) = 8 - x - y$
    $R = \{(x, y) \mid 0 \le x \le 4, 0 \le y \le 4\}$

35. $f(x, y) = 2 - x^2 - y^2$
    $R = \{(x, y) \mid 0 \le x \le 1, 0 \le y \le 1\}$

36. $f(x, y) = x^2 + y^2$
    $R = \{(x, y) \mid 0 \le x \le 2, 0 \le y \le 2\}$

Use integration to find the volume under each surface $f(x, y)$ above the region R.

37. $f(x, y) = 2xy$ over the region shown in the graph.

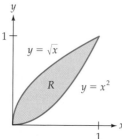

38. $f(x, y) = 3xy^2$ over the region shown in the graph.

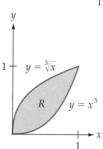

39. $f(x, y) = e^y$ over the region shown in the graph.

40. $f(x, y) = e^y$ over the region shown in the graph.

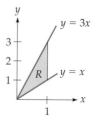

APPLIED EXERCISES

41. GENERAL: Average Temperature The temperature x miles east and y miles north of a weather station is given by the function $f(x, y) = 48 + 4x - 2y$. Find the average temperature over the region R shown below.

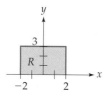

42. ENVIRONMENTAL SCIENCE: Average Air Pollution The air pollution near a chemical refinery is $f(x, y) = 20 + 6x^2y$ parts per million (ppm), where x and y are the numbers of miles east and north of the refinery. Find the average pollution level for the region R shown below.

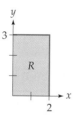

43. GENERAL: Total Population of a Region The population density (people per square mile) x miles east and y miles north of the center of a city is $P(x, y) = 12{,}000e^{x-y}$. Find the total population of the region R shown in the following diagram. [*Hint:* Integrate the population density over the region R. This is an example of a double integral as a *sum*, giving a *total* population over a region.]

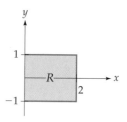

44. BUSINESS: Value of Mineral Deposit The value of an offshore mineral deposit x miles east and y miles north of a certain point is $f(x, y) = 4x + 6y^2$ million dollars per square mile. Find the total value of the tract shown on next page. [*Hint:* Integrate the function over the region R. This is an example of a double integral as a *sum*, giving a *total* value over a region.]

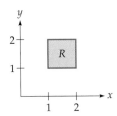

45–46. GENERAL: Volume of a Building To estimate heating and air conditioning costs, it is necessary to know the volume of a building.

45. A conference center has a curved roof whose height is $f(x, y) = 40 - 0.006x^2 + 0.003y^2$. The building sits on a rectangle extending from $x = -50$ to $x = 50$ and $y = -100$ to $y = 100$. Use integration to find the volume of the building. (All dimensions are in feet.)

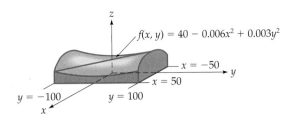

46. An airplane hangar has a curved roof whose height is $f(x, y) = 40 - 0.03x^2$. The building sits on a rectangle extending from $x = -20$ to $x = 20$ and $y = -100$ to $y = 100$. Use integration to find the volume of the building. (All dimensions are in feet.)

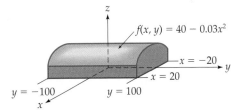

Triple Integrals

Evaluate each triple iterated integral. [*Hint:* Integrate with respect to one variable at a time, treating the other variables as constants, working from the inside out.]

47. $\displaystyle\int_1^2 \int_0^3 \int_0^1 (2x + 4y - z^2)\, dx\, dy\, dz$

48. $\displaystyle\int_1^2 \int_0^3 \int_0^2 (6x - 2y + z^2)\, dx\, dy\, dz$

49. $\displaystyle\int_1^2 \int_0^2 \int_0^1 2xy^2z^3\, dx\, dy\, dz$

50. $\displaystyle\int_1^3 \int_0^1 \int_0^2 12x^3y^2z\, dx\, dy\, dz$

Chapter Summary with Hints and Suggestions

Reading the text and doing the exercises in this chapter have helped you to master the following skills, which are listed by section (in case you need to review them) and are keyed to particular Review Exercises. Answers for all Review Exercises are given at the back of the book, and full solutions can be found in the Student Solutions Manual.

13.1 Functions of Several Variables

- Find the domain of a function of two variables. (*Review Exercises 1–4.*)

13.2 Partial Derivatives

- Find the first and second partials of a function of two variables. (*Review Exercises 5–12.*)

$$\frac{\partial}{\partial x}f(x,y) = f_x(x,y) = \lim_{h\to 0}\frac{f(x+h,y)-f(x,y)}{h}$$

$$\frac{\partial}{\partial y}f(x,y) = f_y(x,y) = \lim_{h\to 0}\frac{f(x,y+h)-f(x,y)}{h}$$

- Evaluate the first partials of a function of two variables. *(Review Exercises 13–16.)*

- Solve an applied problem involving partials, and interpret the answer. *(Review Exercises 17–18.)*

$$\frac{\partial}{\partial x}f(x,y) = \begin{pmatrix}\text{Rate of change of } f \text{ with respect}\\ \text{to } x \text{ when } y \text{ is held constant}\end{pmatrix}$$

$$\frac{\partial}{\partial y}f(x,y) = \begin{pmatrix}\text{Rate of change of } f \text{ with respect}\\ \text{to } y \text{ when } x \text{ is held constant}\end{pmatrix}$$

13.3 Optimizing Functions of Several Variables

- Find the relative extreme values of a function. *(Review Exercises 19–30.)*

$$\begin{cases} f_x = 0 \\ f_y = 0 \end{cases} \qquad D = f_{xx}f_{yy} - (f_{xy})^2$$

- Solve an applied problem by optimizing a function of two variables. *(Review Exercises 31–32.)*

13.4 Least Squares

- Find a least squares line "by hand." *(Review Exercises 33–34.)*

- Find the least squares line for actual data, and use it to make a prediction. *(Review Exercises 35–36.)*

13.5 Lagrange Multipliers and Constrained Optimization

- Solve a constrained maximum or minimum problem using Lagrange multipliers. *(Review Exercises 37–42.)*

$$F(x,y,\lambda) = f(x,y) + \lambda g(x,y) \qquad \begin{cases} F_x = 0 \\ F_y = 0 \\ F_\lambda = 0 \end{cases}$$

- Find *both* extreme values in a constrained optimization problem using Lagrange multipliers. *(Review Exercises 43–44.)*

- Solve an applied problem using Lagrange multipliers. *(Review Exercises 45–48.)*

13.6 Total Differentials and Approximate Changes

- Find the total differential of a function. *(Review Exercises 49–54.)*

$$df = \frac{\partial f}{\partial x}dx + \frac{\partial f}{\partial y}dy$$

- Solve an applied problem by using the total differential to estimate an actual change. *(Review Exercise 55.)*

$$\Delta f = f(x + \Delta x, y + \Delta y) - f(x,y) \qquad \Delta f \approx df$$

- Estimate the relative error in an area calculation. *(Review Exercise 56.)*

13.7 Multiple Integrals

- Evaluate an iterated integral. *(Review Exercises 57–60.)*

$$\int_c^d \int_a^b f(x,y)\,dx\,dy = \int_a^b \int_c^d f(x,y)\,dy\,dx$$

- Find the volume under a surface above a region using a double integral. *(Review Exercises 61–64.)*

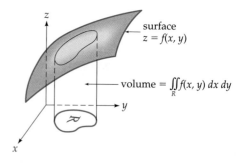

- Find the average population over a region using a double integral. *(Review Exercise 65.)*

- Find the total value of a region using a double integral. *(Review Exercise 66.)*

Hints and Suggestions

- The graph of a function of two variables is represented by a *surface* above (or below) the *x-y* plane. Such three-dimensional graphs are difficult to draw "by hand" but can be shown on some graphing calculators and computer screens. Such surfaces have maximum and minimum points (just as with functions of one variable), but also a new phenomenon, *saddle points* (see page 987).

- The graph of a function of *three* or more variables would require *four* or more dimensions, and so cannot be drawn.

- When finding partial derivatives, remember which is the variable of differentiation and then treat the other variable as a constant. Other than this, the differentiation formulas are the same.

- Partials give instantaneous rates of change with respect to one variable while the other is held constant. They also give *marginals* with respect to one product while production of the other is held constant.

- The *D*-test (page 989) applies only to critical points, where the first partials are zero. These critical points must be found first, and then the *D*-test is applied to each, one at a time.

- Least squares is carried out automatically by a graphing calculator or computer when it finds the linear regression line.

- Solving a constrained optimization problem by Lagrange multipliers is often easier than eliminating one of the variables, as was done in Sections 9.3 and 9.4.

- Multiple integrals give volume under a function if the function is nonnegative.

- When finding the average value of a function of two variables, don't forget to divide by the area of the region over which you are integrating.

- **Practice for Test:** Review Exercises 3, 7, 15, 17, 19, 33, 41, 45, 49, 51, 57, 65.

Review Exercises for Chapter 13

Practice test exercise numbers are in **green**.

13.1 Functions of Several Variables

For each function, state the domain.

1. $f(x, y) = \dfrac{\sqrt{x}}{\sqrt[3]{y}}$
2. $f(x, y) = \dfrac{\sqrt[3]{x}}{\sqrt{y}}$
3. $f(x, y) = e^{1/x} \ln y$
4. $f(x, y) = \dfrac{\ln y}{x}$

13.2 Partial Derivatives

For each function f, calculate
a. f_x, **b.** f_y, **c.** f_{xy}, and **d.** f_{yx}.

5. $f(x, y) = 2x^5 - 3x^2y^3 + y^4 - 3x + 2y + 7$
6. $f(x, y) = 3x^4 + 5x^3y^2 - y^6 - 6x + y - 9$
7. $f(x, y) = 18x^{2/3}y^{1/3}$
8. $f(x, y) = \ln(x^2 + y^3)$
9. $f(x, y) = e^{x^3 - 2y^3}$
10. $f(x, y) = 3x^2 e^{-5y}$
11. $f(x, y) = ye^{-x} - x \ln y$
12. $f(x, y) = x^2 e^y + y \ln x$

For each function, calculate
a. $f_x(1, -1)$ and **b.** $f_y(1, -1)$.

13. $f(x, y) = \dfrac{x + y}{x - y}$
14. $f(x, y) = \dfrac{x}{x^2 + y^2}$
15. $f(x, y) = (x^3 + y^2)^3$
16. $f(x, y) = (2xy - 1)^4$

17. **BUSINESS: Marginal Productivity** A company's production is given by the Cobb–Douglas function $P(L, K) = 160L^{3/4}K^{1/4}$, where L is the number of units of labor and K is the number of units of capital.
    a. Find $P_L(81, 16)$ and interpret this number.

(continues)

b. Find $P_K(81, 16)$ and interpret this number.
c. From your answers to parts (a) and (b), which will increase production more, an additional unit of labor or an additional unit of capital?

18. **BUSINESS: Advertising** A clothing designer's sales S depend on x, the amount spent on television advertising, and y, the amount spent on print advertising (both in thousands of dollars), according to the function

$$S(x, y) = 60x^2 + 90y^2 - 6xy + 200$$

Find $S_x(2, 3)$ and $S_y(2, 3)$, and interpret these numbers.

13.3 Optimizing Functions of Several Variables

For each function, find all relative extreme values.

19. $f(x, y) = 2x^2 - 2xy + y^2 - 4x + 6y - 3$
20. $f(x, y) = x^2 - 2xy + 2y^2 - 6x + 4y + 2$
21. $f(x, y) = 2xy - x^2 - 5y^2 + 2x - 10y + 3$
22. $f(x, y) = 2xy - 5x^2 - y^2 + 10x - 2y + 1$
23. $f(x, y) = 2xy + 6x - y + 1$
24. $f(x, y) = 4xy - 4x + 2y - 4$
25. $f(x, y) = e^{-(x^2+y^2)}$ 26. $f(x, y) = e^{2(x^2+y^2)}$
27. $f(x, y) = \ln(5x^2 + 2y^2 + 1)$
28. $f(x, y) = \ln(4x^2 + 3y^2 + 10)$
29. $f(x, y) = x^3 - y^2 - 12x - 6y$
30. $f(x, y) = y^2 - x^3 + 12x - 4y$

31. **BUSINESS: Maximum Profit** A boatyard builds 18-foot and 22-foot sailboats. Each 18-foot boat costs $3000 to build, each 22-foot boat costs $5000 to build, and the company's fixed costs are $6000. The price function for the 18-foot boats is $p = 7000 - 20x$, and that for the 22-foot boats is $q = 8000 - 30y$ (both in dollars), where x and y are the numbers of 18-foot and 22-foot boats, respectively.

a. Find the company's cost function $C(x, y)$.
b. Find the company's revenue function $R(x, y)$.   *(continues)*

c. Find the company's profit function $P(x, y)$.
d. Find the quantities and prices that maximize profit. Also find the maximum profit.

32. **BUSINESS: Price Discrimination** A company sells farm equipment in America and Europe, charging different prices in the two markets. The price function for harvesters sold in America is $p = 80 - 0.2x$, and the price function for harvesters sold in Europe is $q = 64 - 0.1y$ (both in thousands of dollars), where x and y are the numbers sold per day in America and Europe, respectively. The company's cost function is $C = 100 + 12(x + y)$ thousand dollars.

a. Find the company's profit function.
b. Find how many harvesters should be sold in each market to maximize profit. Also find the price for each market.

13.4 Least Squares

For each exercise:
a. Find the least squares line "by hand."
b. Check your answer using a graphing calculator.

33.

x	y
1	-1
3	6
4	6
5	10

34.

x	y
1	7
2	4
4	2
5	-1

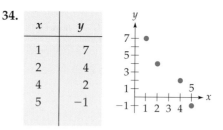

35. **GENERAL: The Aging of America** The population of Americans who are over 65 years old is growing faster than the population at large. Find the least squares line for the follow-

ing data for the over-65 population, and use it to predict the size of the over-65 population in the year 2010 ($x = 6$).

x	y = Population (millions)
1960	1	16.7
1970	2	20.1
1980	3	25.5
1990	4	31.2
2000	5	34.7

36. ECONOMICS: Unemployment The unemployment rate in the United States for several years is given in the following table. Find the least squares line for these data and use it to predict the average unemployment rate for the year 2010 ($x = 7$).

x	y = Percent Unemployed
1980	1	7.1
1985	2	7.2
1990	3	5.6
1995	4	5.6
2000	5	4.1

13.5 Lagrange Multipliers and Constrained Optimization

Use Lagrange multipliers to optimize each function subject to the given constraint. (The stated extreme values *do* exist.)

37. Maximize $f(x, y) = 6x^2 - y^2 + 4$
subject to $3x + y = 12$.

38. Maximize $f(x, y) = 4xy - x^2 - y^2$
subject to $x + 2y = 26$.

39. Minimize $f(x, y) = 2x^2 + 3y^2 - 2xy$
subject to $2x + y = 18$.

40. Minimize $f(x, y) = 12xy - 1$
subject to $y - x = 6$.

41. Minimize $f(x, y) = e^{x^2 + y^2}$
subject to $x + 2y = 15$.

42. Maximize $f(x, y) = e^{-x^2 - y^2}$
subject to $2x + y = 5$.

Use Lagrange multipliers to find the maximum *and* minimum values of each function subject to the given constraint. (Both extreme values *do* exist.)

43. $f(x, y) = 6x - 18y$
subject to $x^2 + y^2 = 40$

44. $f(x, y) = 4xy$
subject to $x^2 + y^2 = 32$

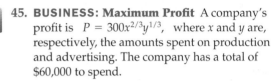

45. BUSINESS: Maximum Profit A company's profit is $P = 300x^{2/3}y^{1/3}$, where x and y are, respectively, the amounts spent on production and advertising. The company has a total of $60,000 to spend.

a. Use Lagrange multipliers to find the amounts for production and advertising that maximize profit.

b. Evaluate and give an interpretation for $|\lambda|$.

46. BIOMEDICAL: Nutrition A nursing home uses two vitamin supplements, and the nutritional value of x ounces of the first together with y ounces of the second is $4x + 2xy + 8y$. The first costs $2 per ounce, the second costs $1 per ounce, and the nursing home can spend only $8 per patient per day.

a. Use Lagrange multipliers to find how much of each supplement should be used to maximize the nutritional value subject to the budget constraint.

b. Evaluate and give an interpretation for $|\lambda|$.

47. ECONOMICS: Least Cost Rule A company's production is given by the Cobb–Douglas function $P = 60L^{2/3}K^{1/3}$, where L and K are the numbers of units of labor and capital. Each unit of labor costs $25 and each unit of capital costs $100. The company wants to produce exactly 1920 units.

a. Find the numbers of units of labor and capital that meet the production requirements at the lowest cost.

b. Find the marginal productivity of labor and the marginal productivity of capital. [*Hint:* This means the partials of P with respect to L and K.]

(*continues*)

c. Show that at the values found in part (a), the following relationship holds:

$$\frac{\text{Marginal productivity of labor}}{\text{Marginal productivity of capital}} = \frac{\text{Price of labor}}{\text{Price of capital}}$$

This is called the "least cost rule."

48. **GENERAL: Container Design** An open-top box with a square base and two perpendicular dividers, as shown in the diagram, is to have a volume of 576 cubic inches. Use Lagrange multipliers to find the dimensions that require the least amount of material.

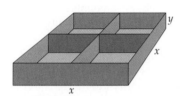

13.6 Total Differentials and Approximate Changes

Find the total differential of each function.

49. $f(x, y) = 3x^2 + 2xy + y^2$

50. $f(x, y) = x^2 + xy - 3y^2$

51. $g(x, y) = \ln(xy)$

52. $g(x, y) = \ln(x^3 + y^3)$

53. $z = e^{x-y}$

54. $z = e^{xy}$

55. **BUSINESS: Sales** A clothing designer's sales S depend on x, the amount spent on television advertising, and y, the amount spent on print advertising (all in thousands of dollars) according to the formula $S(x, y) = 60x^2 - 6xy + 90y^2 + 200$. If the company now spends 2 thousand dollars on television advertising and 3 thousand dollars on print advertising, use the total differential to estimate the change in sales if television advertising is increased by $500 and print advertising is decreased by $500. [*Hint:* Δx and Δy must be in thousands of dollars, just as x and y are.] Then estimate the change in sales if each of the changes in advertising were halved.

56. **GENERAL: Relative Error in Calculating Area** A triangular piece of real estate is measured to have length x feet and altitude y feet, but each measurement may be in error by 1%. Estimate the percentage error in calculating the area by using a total differential.

13.7 Multiple Integrals

Evaluate each iterated integral.

57. $\displaystyle\int_0^4 \int_{-1}^1 2xe^{2y} \, dy \, dx$

58. $\displaystyle\int_{-1}^1 \int_0^3 (x^2 - 4y^2) \, dx \, dy$

59. $\displaystyle\int_{-1}^1 \int_{-y}^y (x + y) \, dx \, dy$

60. $\displaystyle\int_{-2}^2 \int_{-x}^x (x + y) \, dy \, dx$

Find the volume under the surface $f(x, y)$ above the region R.

61. $f(x, y) = 8 - x - y$
    $R = \{(x, y) \mid 0 \le x \le 2, 0 \le y \le 4\}$

62. $f(x, y) = 6 - x - y$
    $R = \{(x, y) \mid 0 \le x \le 4, 0 \le y \le 2\}$

63. $f(x, y) = 12xy^3$ over the region shown in the graph.

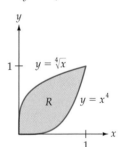

64. $f(x, y) = 15xy^4$ over the region shown in the graph.

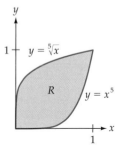

65. GENERAL: Average Population of a Region
The population per square mile x miles east and y miles north of the center of a city is $P(x, y) = 12{,}000 + 100x - 200y$. Find the *average* population over the region shown below.

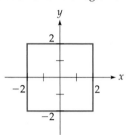

66. GENERAL: Total Value of a Region
The value of land x blocks east and y blocks north of the center of a town is $V(x, y) = 40 - 4x - 2y$ hundred thousand dollars per block. Find the *total* value of the parcel of land shown in the graph above.

Cumulative Review for Chapters 1, 8–13

The following exercises review some of the basic techniques that you learned in Chapters 1 and 8–13. Answers to all of these cumulative review exercises are given in the answer section at the back of the book. A graphing calculator is suggested but not required.

1. Draw the graph of the function
$$f(x) = \begin{cases} 2x - 1 & \text{if } x \leq 3 \\ 7 - x & \text{if } x > 3 \end{cases}$$

2. Simplify: $\left(\dfrac{1}{8}\right)^{-2/3}$

3. Use the definition of the derivative
$$f'(x) = \lim_{h \to 0} \frac{f(x + h) - f(x)}{h}$$
to find the derivative of $f(x) = \dfrac{1}{x}$.

4. If $f(x) = 12\sqrt[3]{x^2} - 4$, find $f'(8)$.

5. A camera store finds that if it sells disposable cameras at a price of p dollars each, it will sell $S(p) = \dfrac{800}{p + 8}$ of them per week. Find $S'(12)$ and interpret the answer.

6. Find $\dfrac{d}{dx}[x^2 + (2x + 1)^4]^3$.

7. Make sign diagrams for the first and second derivatives and graph the function $f(x) = x^3 + 9x^2 - 48x - 148$, showing all relative extreme points and inflection points.

8. Make a sign diagram for the first derivative and graph the function $f(x) = \dfrac{1}{x^2 - 4x}$, showing all asymptotes and relative extreme points.

9. A homeowner wants to use 80 feet of fence to make a rectangular enclosure along an existing stone wall. If the side along the existing wall needs no fence, what are the dimensions of the enclosure that has the largest possible area?

10. An open-top box with a square base is to have a volume of 108 cubic feet. Find the dimensions of the box of this type that can be made with the least amount of material.

11. A spherical balloon is being inflated at the constant rate of 128 cubic feet per minute. Find

how fast the radius is increasing at the moment when the radius is 4 feet.

12. A sum of $1000 is deposited in a bank account paying 8% interest. Find the value of the account after 3 years if the interest is compounded:
   a. quarterly   b. continuously

13. In t years the population of a county is predicted to be $P(t) = 12{,}000e^{0.02t}$, and the population of a neighboring county is predicted to be $Q(t) = 9000e^{0.04t}$. In how many years will the second county overtake the first in population?

14. A sum is deposited in a bank account paying 6% interest compounded monthly. How many years will it take for the value to increase by 50%?

15. Make sign diagrams for the first and second derivatives and graph the function $f(x) = e^{-x^2/2}$, showing all relative extreme points and inflection points.

16. Find $\int (12x^2 - 4x + 1)\, dx$.

17. Pollution is being discharged into a lake at the rate of $18e^{0.02t}$ million gallons per year, where t is the number of years from now. Find a formula for the total amount of pollution that will be discharged into the lake during the next t years.

18. Find the area bounded by $y = 20 - x^2$ and $y = 8 - 4x$.

19. Find the average value of $f(x) = 12\sqrt{x}$ over the interval $[0, 4]$.

20. Find: a. $\int \dfrac{x^2\, dx}{x^3 + 1}$ b. $\int \dfrac{e^{\sqrt{x}}\, dx}{\sqrt{x}}$

21. Find by integration by parts: $\int xe^{4x}\, dx$.

22. Use the integral table on the inside back cover to find $\int \dfrac{\sqrt{4 - x^2}}{x}\, dx$. Check your answer by differentiating and then simplifying.

23. Evaluate $\int_1^\infty \dfrac{1}{x^3}\, dx$.

24. Use trapezoidal approximation with $n = 4$ trapezoids to approximate $\int_0^1 \sqrt{x^2 + 1}\, dx$. (If you use a graphing calculator, compare your answer with the value obtained by using FnInt.)

25. Use Simpson's Rule with $n = 4$ to approximate $\int_0^1 \sqrt{x^2 + 1}\, dx$. (If you use a graphing calculator, compare your answer with the value obtained by using FnInt.)

26. a. Find the general solution to the differential equation $y' = x^3 y$.
    b. Then find the particular solution that satisfies $y(0) = 2$.

27. Find the first partial derivatives of the function $f(x, y) = x \ln y + y e^{2x}$.

28. Find the relative extreme values of $f(x, y) = 2x^2 - 2xy + y^2 + 4x - 6y + 12$.

29. Find the least squares line for the following points:

x	y
1	−3
3	1
5	3
7	8

30. Use Lagrange multipliers to find the minimum value of $f(x, y) = 3x^2 + 2y^2 - 2xy$ subject to the constraint $x + 2y = 18$.

31. Find the total differential of $f(x, y) = 2x^2 + xy - 3y^2 + 4$.

32. Find the volume under the surface $f(x, y) = 12 - x - 2y$ over the rectangle
$R = \{(x, y) \mid 0 \le x \le 2, 0 \le y \le 3\}$.

Appendix

NORMAL PROBABILITIES USING TABLES

This appendix shows how to find normal probabilities from tables. It replaces pages 465–471 (through Practice Problem 3) of Chapter 6 (Statistics) for readers who do not have graphing calculators that calculate normal probabilities.

z-Scores

The normal distribution with *mean* 0 and *standard deviation* 1 has special significance and is called the *standard* normal distribution (the word *standard* indicating mean 0 and standard deviation 1). The letter z is traditionally used for the variable in standard normal calculations. Since the mean is zero, the highest point on the curve is at $z = 0$, and the curve is symmetric about this value.

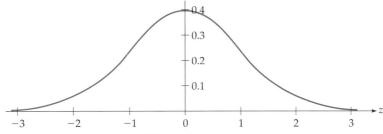

The standard normal distribution (mean 0 and standard deviation 1)

From the diagram on page 464 with $\mu = 0$ and $\sigma = 1$ we have the following facts about the standard normal distribution: More than

        68% of its probability is between -1 and $+1$

        95% of its probability is between -2 and $+2$

        99% of its probability is between -3 and $+3$

To change any x-value for a normal distribution with mean μ and standard deviation σ into the corresponding z-value for the *standard normal distribution*, we subtract the mean and then divide by the standard deviation. The resulting number is called the *z-score* and this process is called *standardizing*.

z-Score

$$z = \frac{x - \mu}{\sigma}$$

Subtracting the mean μ and dividing by the standard deviation σ

The z-score gives the number of standard deviations the value is from the mean.

EXAMPLE 1

FINDING z-SCORES FOR HEIGHTS OF MEN

The heights of American men are approximately normally distributed with mean $\mu = 68.1$ inches and standard deviation $\sigma = 2.7$ inches.* Find the z-scores corresponding to the following heights (in inches) and interpret the results:

a. 73.5 **b.** 62.7

Solution

a. $z = \dfrac{73.5 - 68.1}{2.7} = \dfrac{5.4}{2.7} = 2$ Using $z = (x - \mu)/\sigma$ with $\mu = 68.1$ and $\sigma = 2.7$

Therefore, a height of 73.5 inches is two standard deviations *above* the mean.

b. $z = \dfrac{62.7 - 68.1}{2.7} = \dfrac{-5.4}{2.7} = -2$

Therefore, a height of 62.7 inches is two standard deviations *below* the mean.

Since the standard normal distribution has more than 95% of its probability between -2 and $+2$, these results mean that more than 95% of men have heights between 62.7 inches and 73.5 inches (5 feet 2.7 inches and 6 feet 1.5 inches).

---

* These and other data in the appendix are from *Handbook of Human Factors and Ergonomics* by Gavriel Salvendy (Wiley).

Practice Problem 1 The weights of American women are approximately normally distributed with mean 134.7 pounds and standard deviation 30.4 pounds. Find the z-scores for the following weights, and interpret your answers:

a. 165.1 lbs.   **b.** 104.3 lbs.   ▶ Solutions on page A-8

On page A-10 of this appendix is a brief table for the normal distribution, giving the probability that a standard normal random variable takes a value between 0 and any given positive number. For example, to find $P(0 \leq Z \leq 1.24)$ (in words, the probability that a standard normal random variable has a value between 0 and 1.24), we locate in the table the row headed **1.2** and the column headed **0.04** (the second decimal place), and $P(0 \leq Z \leq 1.24)$ is the number where this row and column intersect, **0.3925**.

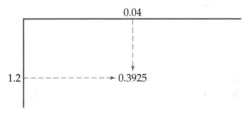

In the table on page A-10, the row headed **1.2** and the column headed **0.04** intersect at the table value **0.3925**

Therefore,

$$P(0 \leq Z \leq 1.24) \approx 0.3925$$

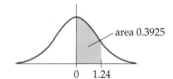

area 0.3925

Probabilities for other intervals can be found by adding and subtracting such areas and using the symmetry of the normal curve, as the following examples illustrate.

EXAMPLE 2

FINDING A PROBABILITY FOR A NORMAL RANDOM VARIABLE

Using the mean and standard deviation for the heights of men given in Example 1, find the proportion of men who are between 5 feet 9 inches and 6 feet tall.

Solution

Converting to inches, we want the probability of a man's height being between 69 inches and 72 inches. We must convert these numbers to the corresponding z-scores for a *standard* normal random variable by subtracting the mean and dividing by the standard deviation:

$$x = 69 \text{ corresponds to} \quad z = \frac{69 - 68.1}{2.7} \approx 0.33 \quad \text{Using } z = \frac{x - \mu}{\sigma}$$

$$x = 72 \text{ corresponds to} \quad z = \frac{72 - 68.1}{2.7} \approx 1.44 \quad \text{with } \mu = 68.1 \text{ and } \sigma = 2.7$$

Using these values, we then want $P(0.33 \leq Z \leq 1.44)$, which is equivalent to the shaded area in the first of the following graphs, which is equal to the difference between the next two areas on the right. These two areas (or probabilities) are found from the table on page A-10, with the calculations shown below.

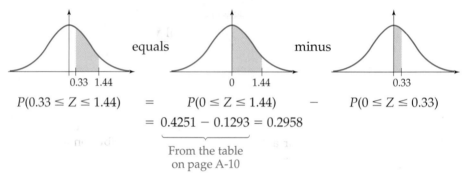

$$P(0.33 \leq Z \leq 1.44) \quad = \quad P(0 \leq Z \leq 1.44) \quad - \quad P(0 \leq Z \leq 0.33)$$
$$= 0.4251 - 0.1293 = 0.2958$$

From the table on page A-10

Therefore, about 30% of American men are between 5 feet 9 inches and 6 feet tall.

Practice Problem 2 Using the mean and standard deviation for weights of women given in Practice Problem 1, find the proportion of women who weigh between 130 and 150 pounds. ▶ Solution on page A-8

The Normal and Binomial Distributions

As we saw on page 414, the binomial distribution has a kind of "bell" shape, with a peak at the expected value $\mu = np$ and falling on both sides to very small probabilities further away from the mean.

$n = 20, p = 0.4$ \qquad $n = 15, p = 0.5$ \qquad $n = 25, p = 0.7$

Several binomial distributions with different values for n and p

Since both the binomial and the normal distributions have similar "mound" shapes, could they be related? In the eighteenth century, Abraham de Moivre (1667–1754) and Pierre-Simon Laplace (1749–1827) discovered and proved that for any choice of p between 0 and 1, the binomial distribution *approaches* the normal distribution as n becomes large. This fundamental fact is known as the *de Moivre–Laplace theorem*.

We may use the de Moivre–Laplace theorem to approximate binomial distributions by the normal distribution.

Normal Approximation to the Binomial

Let X be a binomial random variable with parameters n and p. If $np > 5$ and $n(1 - p) > 5$, then the distribution of X is approximately normal with mean $\mu = np$ and standard deviation $\sigma = \sqrt{np(1 - p)}$.

Approximating the *discrete* binomial distribution by the *continuous* normal distribution requires an adjustment: To have the same width for a "slice" of the normal distribution as for the binomial, we adopt the convention that the binomial probability $P(X = x)$ corresponds to the area under the normal distribution curve from $x - \frac{1}{2}$ to $x + \frac{1}{2}$. This is called the *continuous correction*.

$P(X = x)$ is approximately Area from $x - \frac{1}{2}$ to $x + \frac{1}{2}$

EXAMPLE 3

NORMAL APPROXIMATION OF A BINOMIAL PROBABILITY

Estimate $P(X = 12)$ for a binomial random variable X with $n = 25$ and $p = 0.6$ using the corresponding normal distribution.

Solution

Since

$$np = 25 \cdot 0.6 = 15 \quad \text{and} \quad n(1 - p) = 25 \cdot 0.4 = 10$$

are both greater than 5, we may use the normal distribution with

$$\mu = np = 25 \cdot 0.6 = 15$$

and

$$\sigma = \sqrt{np(1-p)} = \sqrt{25 \cdot 0.6 \cdot 0.4} = \sqrt{6} \approx 2.45$$

By the continuous correction we interpret the event $X = 12$ as $11.5 \leq X < 12.5$ to include numbers that would round to 12. Converting these x-values into z-scores, we find that

$$x = 11.5 \text{ corresponds to } z = \frac{11.5 - 15}{2.45} \approx -1.43 \quad \text{Using } z = \frac{x - \mu}{\sigma}$$

$$x = 12.5 \text{ corresponds to } z = \frac{12.5 - 15}{2.45} \approx -1.02 \quad \text{with } \mu = 15 \text{ and } \sigma = 2.45$$

The probability $P(-1.43 \leq Z \leq -1.02)$ is represented below by the shaded area in the first graph, which, by symmetry, is equivalent to the second graph, which in turn is equivalent to the difference between the third and fourth graphs. The calculation with the probabilities found from the normal table is shown below.

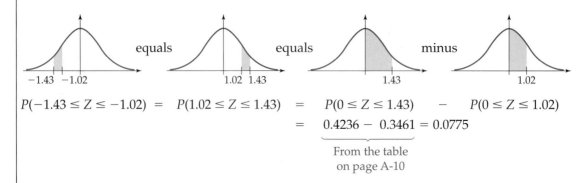

$P(-1.43 \leq Z \leq -1.02) = P(1.02 \leq Z \leq 1.43) = P(0 \leq Z \leq 1.43) - P(0 \leq Z \leq 1.02)$

$= 0.4236 - 0.3461 = 0.0775$

From the table on page A-10

The required probability is approximately 0.078.

The normal approximation of 0.0775 is indeed very close to the actual value calculated from the binomial probability formula (see page 414): $_{25}C_{12}(0.6)^{12}(1 - 0.6)^{13} \approx 0.076$.

EXAMPLE 4 MANAGEMENT MBAs

At a major Los Angeles accounting firm, 73% of the managers have MBA degrees. In a random sample of 40 of these managers, what is the probability that between 27 and 32 will have MBAs?

NORMAL PROBABILITIES USING TABLES A-7

Solution

Because each manager either has or does not have an MBA, presumably independently of each other, the question asks for the probability that a binomial random variable X with $n = 40$ and $p = 0.73$ satisfies $27 \leq X \leq 32$. Since both $np = 40 \cdot 0.73 = 29.2$ and $n(1-p) = 40 \cdot 0.27 = 10.8$ are greater than 5, this probability can be approximated as the area under the normal distribution curve with mean $\mu = np = 40 \cdot 0.73 = 29.2$ and standard deviation $\sigma = \sqrt{np(1-p)} = \sqrt{40 \cdot 0.73 \cdot 0.27} \approx 2.81$ from $26\frac{1}{2}$ to $32\frac{1}{2}$ (again using the continuous correction to include values that would round to between 27 and 32). We convert the x-values to z-scores:

$x = 26.5$ corresponds to $z = \dfrac{26.5 - 29.2}{2.81} \approx -0.96$

$x = 32.5$ corresponds to $z = \dfrac{32.5 - 29.2}{2.81} \approx 1.17$

Using $z = \dfrac{x - \mu}{\sigma}$ with $\mu = 29.2$ and $\sigma = 2.81$

The probability $P(-0.96 \leq Z \leq 1.17)$ is represented by the shaded area in the first graph, which is equivalent to the sum of the next two areas, with the calculation shown below.

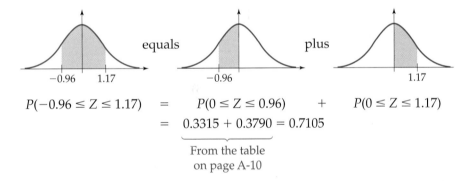

$P(-0.96 \leq Z \leq 1.17) \quad = \quad P(0 \leq Z \leq 0.96) \quad + \quad P(0 \leq Z \leq 1.17)$
$\quad = \quad 0.3315 + 0.3790 = 0.7105$

From the table on page A-10

The probability that the number of MBAs will be between 27 and 32 is 0.711, or about 71%.

The exact answer to the preceding problem, from summing the binomial distribution from 27 to 32, is 71.5%, so the normal approximation is very accurate.

Practice Problem 3 A brand of imported VCR is known to have defective tape rewind mechanisms in 8% of the units imported last April. If Jerry's Discount Electronics received a shipment of 80 of these VCR's, what is the probability that 10 or more are defective? ▶ Solution on page A-9

A-8 APPENDIX

> After completing Practice Problem 3, you should return to page 471 to read the Section Summary and do the Exercises in Section 6.4.

▶ Solutions to Practice Problems

1. Using $\mu = 134.7$ and $\sigma = 30.4$:

   a. $z = \dfrac{165.1 - 134.7}{30.4} = \dfrac{30.4}{30.4} = 1$

   b. $z = \dfrac{104.3 - 134.7}{30.4} = \dfrac{-30.4}{30.4} = -1$

   Interpretation: More than 68% of women have weights between about 104 and 165 pounds.

2. We first change the weights into z-scores using $z = \dfrac{x - \mu}{\sigma}$ with $\mu = 134.7$ and $\sigma = 30.4$:

   $x = 130$ corresponds to $z = \dfrac{130 - 134.7}{30.4} \approx -0.15$

   $x = 150$ corresponds to $z = \dfrac{150 - 134.7}{30.4} \approx 0.50$

   Using these values, we then want $P(-0.15 \leq Z \leq 0.50)$, which is equivalent to the shaded area shown in the graph on the left below, which is equal to the *sum* of the following two shaded areas. The two areas are found from the table on page A-10, with the calculation shown below.

   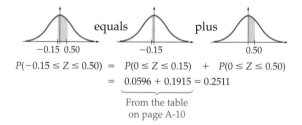

   $P(-0.15 \leq Z \leq 0.50) = P(0 \leq Z \leq 0.15) + P(0 \leq Z \leq 0.50)$
   $= \underbrace{0.0596 + 0.1915}_{\text{From the table on page A-10}} = 0.2511$

   Therefore, about 25% of American women weigh between 130 and 150 pounds.

3. Since $np = 80 \cdot 0.08 = 6.4$ and $n(1-p) = 80 \cdot 0.92 = 73.6$ are each greater than 5, we may use the normal distribution with $\mu = np = 80 \cdot 0.08 \approx 6.4$ and $\sigma = \sqrt{np(1-p)} = \sqrt{80 \cdot 0.08 \cdot 0.92} \approx 2.43$ to approximate $P(X \geq 10)$ as the area under the normal curve from 9.5 to 80.5 (corresponding to all 80 being defective, and again including rounding). Converting the x-values into z-scores using $z = \dfrac{x - \mu}{\sigma}$ with $\mu = 6.4$ and $\sigma = 2.43$:

$x = 9.5$ corresponds to $z = \dfrac{9.5 - 6.4}{2.43} \approx 1.28$

$x = 80.5$ corresponds to $z = \dfrac{80.5 - 6.4}{2.43} \approx 30.49$

The probability $P(1.28 \leq Z \leq 30.49)$ is represented by the shaded area on the left, which is equivalent to the difference between the two areas on the right with the calculation shown below.

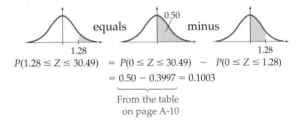

$P(1.28 \leq Z \leq 30.49) = P(0 \leq Z \leq 30.49) - P(0 \leq Z \leq 1.28)$
$= 0.50 - 0.3997 = 0.1003$

From the table on page A-10

The probability of at least ten defective VCR's is (about) 0.10, or 10%.

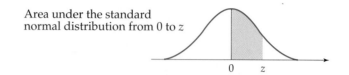

Area under the standard normal distribution from 0 to z

x	0.00	0.01	0.02	0.03	0.04	0.05	0.06	0.07	0.08	0.09
0.0	0.0000	0.0040	0.0080	0.0120	0.0160	0.0199	0.0239	0.0279	0.0319	0.0359
0.1	0.0398	0.0438	0.0478	0.0517	0.0557	0.0596	0.0636	0.0675	0.0714	0.0754
0.2	0.0793	0.0832	0.0871	0.0910	0.0948	0.0987	0.1026	0.1064	0.1103	0.1141
0.3	0.1179	0.1217	0.1255	0.1293	0.1331	0.1368	0.1406	0.1443	0.1480	0.1517
0.4	0.1554	0.1591	0.1628	0.1664	0.1700	0.1736	0.1772	0.1808	0.1844	0.1879
0.5	0.1915	0.1950	0.1985	0.2019	0.2054	0.2088	0.2123	0.2157	0.2190	0.2224
0.6	0.2258	0.2291	0.2324	0.2357	0.2389	0.2422	0.2454	0.2486	0.2518	0.2549
0.7	0.2580	0.2612	0.2642	0.2673	0.2704	0.2734	0.2764	0.2794	0.2823	0.2852
0.8	0.2881	0.2910	0.2939	0.2967	0.2996	0.3023	0.3051	0.3078	0.3106	0.3133
0.9	0.3159	0.3186	0.3212	0.3238	0.3264	0.3289	0.3315	0.3340	0.3365	0.3389
1.0	0.3413	0.3438	0.3461	0.3485	0.3508	0.3531	0.3554	0.3577	0.3599	0.3621
1.1	0.3643	0.3665	0.3686	0.3708	0.3729	0.3749	0.3770	0.3790	0.3810	0.3820
1.2	0.3849	0.3869	0.3888	0.3907	0.3925	0.3944	0.3962	0.3980	0.3997	0.4015
1.3	0.4032	0.4049	0.4066	0.4082	0.4099	0.4115	0.4131	0.4147	0.4162	0.4177
1.4	0.4192	0.4207	0.4222	0.4236	0.4251	0.4265	0.4279	0.4292	0.4306	0.4319
1.5	0.4332	0.4345	0.4357	0.4370	0.4382	0.4394	0.4406	0.4418	0.4429	0.4441
1.6	0.4452	0.4463	0.4474	0.4484	0.4495	0.4505	0.4515	0.4525	0.4535	0.4545
1.7	0.4554	0.4564	0.4573	0.4582	0.4591	0.4599	0.4608	0.4616	0.4625	0.4633
1.8	0.4641	0.4649	0.4656	0.4664	0.4671	0.4678	0.4686	0.4693	0.4699	0.4706
1.9	0.4713	0.4719	0.4726	0.4732	0.4738	0.4744	0.4750	0.4756	0.4761	0.4767
2.0	0.4772	0.4778	0.4783	0.4788	0.4793	0.4798	0.4803	0.4808	0.4812	0.4817
2.1	0.4821	0.4826	0.4830	0.4834	0.4838	0.4842	0.4846	0.4850	0.4854	0.4857
2.2	0.4861	0.4864	0.4868	0.4871	0.4875	0.4878	0.4881	0.4884	0.4887	0.4890
2.3	0.4893	0.4896	0.4898	0.4901	0.4904	0.4906	0.4909	0.4911	0.4913	0.4916
2.4	0.4918	0.4920	0.4922	0.4925	0.4927	0.4929	0.4931	0.4932	0.4934	0.4936
2.5	0.4938	0.4940	0.4941	0.4943	0.4945	0.4946	0.4948	0.4949	0.4951	0.4952
2.6	0.4953	0.4955	0.4956	0.4957	0.4959	0.4960	0.4961	0.4962	0.4963	0.4964
2.7	0.4965	0.4966	0.4967	0.4968	0.4969	0.4970	0.4971	0.4972	0.4973	0.4974
2.8	0.4974	0.4975	0.4976	0.4977	0.4977	0.4978	0.4979	0.4979	0.4980	0.4981
2.9	0.4981	0.4982	0.4982	0.4983	0.4984	0.4984	0.4985	0.4985	0.4986	0.4986
3.0	0.4987	0.4987	0.4987	0.4988	0.4988	0.4989	0.4989	0.4989	0.4990	0.4990
3.1	0.4990	0.4991	0.4991	0.4991	0.4992	0.4992	0.4992	0.4992	0.4993	0.4993
3.2	0.4993	0.4993	0.4994	0.4994	0.4994	0.4994	0.4994	0.4995	0.4995	0.4995
3.3	0.4995	0.4995	0.4995	0.4996	0.4996	0.4996	0.4996	0.4996	0.4996	0.4997
3.4	0.4997	0.4997	0.4997	0.4997	0.4997	0.4997	0.4997	0.4997	0.4997	0.4998
3.5	0.4998	0.4998	0.4998	0.4998	0.4998	0.4998	0.4998	0.4998	0.4998	0.4998

Answers to Selected Exercises

Exercises 1.1 page 14

1. $\{x \mid 0 \leq x < 6\}$ 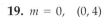 **3.** $\{x \mid x \leq 2\}$
5. a. Increase by 15 units **b.** Decrease by 10 units **7.** $m = -2$ **9.** $m = \frac{1}{3}$ **11.** $m = 0$
13. Slope is undefined
15. $m = 3$, $(0, -4)$ **17.** $m = -\frac{1}{2}$, $(0, 0)$ **19.** $m = 0$, $(0, 4)$

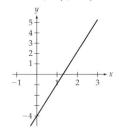 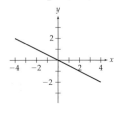 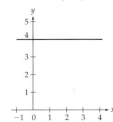

21. Slope and y-intercept do not exist.   **23.** $m = \frac{2}{3}$, $(0, -4)$ **25.** $m = -1$, $(0, 0)$

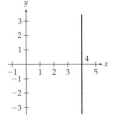 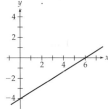 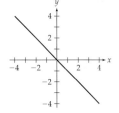

27. $m = 1$, $(0, 0)$ **29.** $m = \frac{1}{3}$, $(0, \frac{2}{3})$ **31.** $m = \frac{2}{3}$, $(0, -1)$

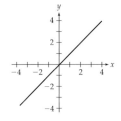 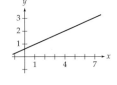 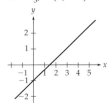

33.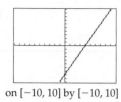
on [−10, 10] by [−10, 10]

35.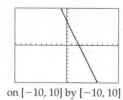
on [−10, 10] by [−10, 10]

37.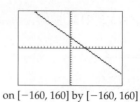
on [−160, 160] by [−160, 160]

39. $y = -2.25x + 3$ **41.** $y = 5x + 3$ **43.** $y = -4$ **45.** $x = 1.5$ **47.** $y = -2x + 13$
49. $y = -1$ **51.** $y = -2x + 1$ **53.** $y = \frac{3}{2}x - 2$
55. $y = -x + 5$, $y = -x - 5$, $y = x + 5$, $y = x - 5$
57. Substituting $(0, b)$ into $y - y_1 = m(x - x_1)$ gives $y - b = m(x - 0)$, or $y = mx + b$.
59. $(-b/m, 0)$, $m \neq 0$ **61. a.**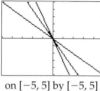
on [−5, 5] by [−5, 5]
b.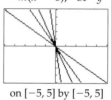
on [−5, 5] by [−5, 5]

63. Low: [0, 8); average: [8, 20); high: [20, 40); critical [40, ∞)
65. a. 3 minutes 38.28 seconds **b.** In about 2033
67. a. $y = 4x + 2$ **b.** $10 million **c.** $22 million **69. a.** $y = \frac{9}{5}x + 32$ **b.** 68°
71. a. $V = 50{,}000 - 2200t$ **b.** $39,000 **c.**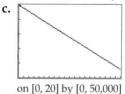
on [0, 20] by [0, 50,000]

73. Smaller populations increase towards the carrying capacity
75. a.  **b.** Men: 28.7 years; women: 27.2 years **c.** Men: 30 years; women: 28.9 years
77. a. 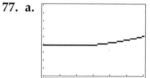 **b.** About 40% (from 39.8) **c.** About 60% (from 60.4)

79. b.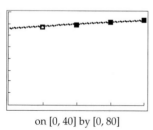
on [0, 40] by [0, 80]

$y_1 = 0.203x + 65.92$ **c.** 79.1 years

Exercises 1.2 page 28

1. 64 **3.** $\frac{1}{16}$ **5.** 8 **7.** $\frac{8}{5}$ **9.** $\frac{1}{32}$ **11.** $\frac{8}{27}$ **13.** 1 **15.** $\frac{4}{9}$ **17.** 5 **19.** 125 **21.** 8
23. 4 **25.** -32 **27.** $\frac{125}{216}$ **29.** $\frac{9}{25}$ **31.** $\frac{1}{4}$ **33.** $\frac{1}{2}$ **35.** $\frac{1}{8}$ **37.** $\frac{1}{4}$ **39.** $-\frac{1}{2}$ **41.** $\frac{1}{4}$
43. $\frac{4}{5}$ **45.** $\frac{64}{125}$ **47.** -243 **49.** 2.14 **51.** 274.37 **53.** -128 **55.** 6.25 **57.** 0.5
59. 0.4 **61.** 0.977 (rounded) **63.** 2.720 (rounded) **65.** x^{10} **67.** z^{27} **69.** x^8 **71.** w^5
73. y^5/x **75.** $27y^4$ **77.** $u^2v^2w^2$ **79.** 25.6 feet **81.** Costs will be multiplied by about 2.3.
83. 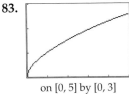 Capacity can be multiplied by about 3.2.

on [0, 5] by [0, 3]

85. 125 beats per minute **87.** Heart rate decreases more slowly as body weight increases.

on [0, 200] by [0, 150]

89. About 42.6 thousand work-hours, or 42,600 work-hours, rounded to the nearest hundred hours.
91. a. About 32 times more ground motion **b.** About 8 times more ground motion
93. About 312 mph
95. $x \approx 18.2$. Therefore, the land area must be increased by a factor of more than 18 to double the number of species.

on [0, 100] by [0, 4]

97. b. $y_1 = 3261x^{-0.267}$ **c.** 1147 work-hours

on $[-2, 32]$ by [1,000, 3,500]

Exercises 1.3 page 45

1. Yes **3.** No **5.** No **7.** No **9.** Domain: $\{x \mid x \leq 0 \text{ or } x \geq 1\}$; Range: $\{y \mid y \geq -1\}$
11. a. $f(10) = 3$ **b.** $\{x \mid x \geq 1\}$ **c.** $\{y \mid y \geq 0\}$
13. a. $h(-5) = -1$ **b.** $\{z \mid z \neq -4\}$ **c.** $\{y \mid y \neq 0\}$
15. a. $h(81) = 3$ **b.** $\{x \mid x \geq 0\}$ **c.** $\{y \mid y \geq 0\}$ **17. a.** $f(-8) = 4$ **b.** $\mathbb{R}$ **c.** $\{y \mid y \geq 0\}$
19. a. $f(0) = 2$ **b.** $\{x \mid -2 \leq x \leq 2\}$ **c.** $\{y \mid 0 \leq y \leq 2\}$
21. a. $f(-25) = 5$ **b.** $\{x \mid x \leq 0\}$ **c.** $\{y \mid y \geq 0\}$

23. 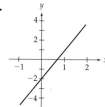   25. 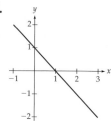   27. 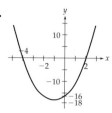   29.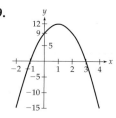

31. a. (20, 100)   b.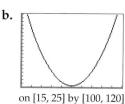

on [15, 25] by [100, 120]

33. a. (−40, −200)   b.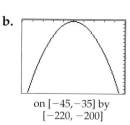

on [−45,−35] by [−220, −200]

35. $x = 7$, $x = -1$ 37. $x = 3$, $x = -5$ 39. $x = 4$, $x = 5$ 41. $x = 0$, $x = 10$
43. $x = 5$, $x = -5$ 45. $x = -3$ 47. $x = 1$, $x = 2$ 49. No solutions   51. No solutions
53. $x = -4$, $x = 5$ 55. $x = 4$, $x = 5$ 57. $x = -3$ 59. No (real) solutions
61. $x = 1.14$, $x = -2.64$ 63. a. The slopes are all 2, but the y-intercepts differ   b. $y = 2x - 8$
65. $C(x) = 4x + 20$ 67. $P(x) = 15x + 500$ 69. a. 17.7 lbs/in.²   b. 15,765 lbs/in.²
71. 132 feet   73. a. 400 cells   b. 5200 cells   75. About 208 mph   77. 2.92 seconds
79. a. Break even at 40 units and at 200 units.
    b. Profit maximized at 120 units. Max profit is $12,800 per week.
81. a. Break even at 20 units and at 80 units.
    b. Profit maximized at 50 units. Max profit is $1800 per day.

83. $v = \dfrac{c}{w + a} - b$ 85. b. 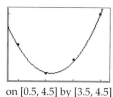   c. $y_1(5) = 4.575$, so about 4.6 million units

on [0.5, 4.5] by [3.5, 4.5]

Exercises 1.4 page 60

1. Domain: $\{x \mid x > 0 \text{ or } x < -4\}$; Range: $\{y \mid y > 0 \text{ or } y < -2\}$
3. a. $f(-3) = 1$ b. $\{x \mid x \neq -4\}$ c. $\{y \mid y \neq 0\}$
5. a. $f(-1) = -\tfrac{1}{2}$ b. $\{x \mid x \neq 1\}$ c. $\{y \mid y \leq 0 \text{ or } y \geq 4\}$
7. a. $f(2) = 1$ b. $\{x \mid x \neq 0, x \neq -4\}$ c. $\{y \mid y > 0 \text{ or } y \leq -3\}$
9. a. $g(-5) = 3$ b. $\mathbb{R}$ c. $\{y \mid y \geq 0\}$ 11. $x = 0$, $x = -3$, $x = 1$
13. $x = 0$, $x = 2$, $x = -2$ 15. $x = 0$, $x = 3$ 17. $x = 0$, $x = 5$ 19. $x = 0$, $x = 3$
21. $x = -2$, $x = 0$, $x = 4$ 23. $x = -1$, $x = 0$, $x = 3$ 25. $x = 0$, $x = 3$ 27. $x = -5$, $x = 0$
29. $x = 0$, $x = 1$ 31. $x \approx -1.79$, $x = 0$, $x \approx 2.79$

33. 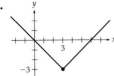 **35.** 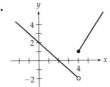 **37.**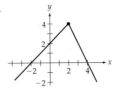

39. Polynomial   **41.** Piecewise linear   **43.** Polynomial   **45.** Rational   **47.** Piecewise linear
49. Polynomial   **51.** None (not a polynomial because of the fractional exponent)
53. a. y_3 **b.** y_1 **c.** 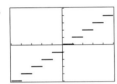   **d.** $(10, 1000)$.

55. a. 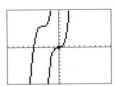 $y = \text{INT}(x)$ on $[-5, 5]$ by $[-5, 5]$ Note that each line segment in this graph includes its left endpoint but excludes its right endpoint, so it should be drawn like •——○.

b. Domain: $\mathbb{R}$; Range: $\{..., -3, -2, -1, 0, 1, 2, 3, ...\}$, that is, the set of all integers.

57. a. $(7x-1)^5$ **b.** $7x^5 - 1$ **59. a.** $\dfrac{1}{x^2+1}$ **b.** $\left(\dfrac{1}{x}\right)^2 + 1$

61. a. $(\sqrt{x}-1)^3 - (\sqrt{x}-1)^2$ **b.** $\sqrt{x^3 - x^2 - 1}$ **63. a.** $\dfrac{(x^2-x)^3 - 1}{(x^2-x)^3 + 1}$ **b.** $\left(\dfrac{x^3-1}{x^3+1}\right)^2 - \dfrac{x^3-1}{x^3+1}$

65. a. $f(g(x)) = acx + ad + b$ **b.** Yes   **67.** $10x + 5h$ or $5(2x+h)$ **69.** $4x + 2h - 5$

71. $14x + 7h - 3$ **73.** $3x^2 + 3xh + h^2$ **75.** $\dfrac{-2}{(x+h)x}$

77. $\dfrac{-2x-h}{x^2(x+h)^2}$ or $\dfrac{-2x-h}{x^2(x^2+2xh+h^2)}$ or $\dfrac{-2x-h}{x^4 + 2x^3h + h^2x^2}$

79. Shifted left 3 units and up 6 units

81. a. $300   **b.** $500   **c.** $2000   **d.**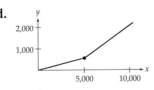

B-6 CHAPTER 1

83. a. $f(\tfrac{2}{3}) = 7$, $f(1\tfrac{1}{3}) = 14$, $f(4) = 29$, and $f(10) = 53$ **b.**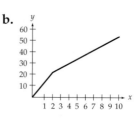

85. $R(v(t)) = 2(60 + 3t)^{0.3}$, $R(v(10)) \approx 7.714$ million dollars
87. a. 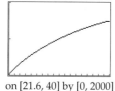 **b.** About $x = 27.9$ mpg

on [21.6, 40] by [0, 2000]

Exercises 1.5 page 74

1. a. 7.389 **b.** 0.135 **c.** 1.649 **3. a.** e^3 **b.** e^2 **c.** e^5
5. [graph] **7.** [graph] **9.** 5.697 (rounded)

11. a. e^x **b.** e^x **c.** e^x **d.** e^x **e.** e^x will exceed any power of x for large enough values of x.
13. a. $2143.59 **b.** $2203.76 **c.** $2225.54 **15. a.** $2196 **b.** $2096
17. Yes, the trust fund would be worth $1,000,508.52.
19. 9.9% compounded continuously (yielding about $1.1041) is better than 10% compounded quarterly (yielding about $1.1038). While the difference is small for $1 for one year, it would be more significant for larger sums and longer times.
21. Yes, it will be worth $1000.18 **23.** $3105.41 **25. a.** $3570 (rounded) **b.** $16,125 (rounded)
27. 7.0 billion **29. a.** 0.53 (the chances are better than 50–50) **b.** 0.70 (quite likely)
31. a. 0.267 or 26.7% **b.** 0.012 or 1.2% **33. a.** 1.3 milligrams **b.** 0.84 milligram **35.** 208
37. a. About 153 degrees **b.** About 123 degrees **39.** 38 **41.** By about 25%
43. a. 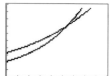 **b.** In about 2060 (from $x \approx 60.2$)

Exercises 1.6 page 89

1. a. 5 **b.** −2 **c.** $\tfrac{1}{2}$ **3. a.** 5 **b.** −1 **c.** $\tfrac{1}{3}$ **5. a.** 0 **b.** 1 **c.** $\tfrac{2}{3}$
7. a. 2 **b.** −1 **c.** $\tfrac{1}{2}$ **9. a.** 1.348 **b.** 3.105 **11.** $\ln x$ **13.** $2\ln x$ or $\ln x^2$ **15.** $\ln x$

17. $3x$ **19.** $7x$ **21.** 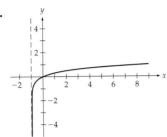   **23.** Domain: $\{x|x > 1 \text{ or } x < -1\}$
Range: $\mathbb{R}$

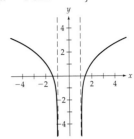

25. a. In about 1.9 years   **b.** In about 3.9 years   **27.** In about 2.5 years
29. About 31,400 years   **31.** About 1.7 million years   **33.** In 8 years
35. 228 million years   **37. a.** $\log_b b^x = x$, property 8 of logarithms   **b.** Follows directly from (a)
   c. Using the change-of-base formula, cancellation

Chapter 1 Review Exercises page 93

1. $\{x | 2 < x \leq 5\}$ **2.** $\{x | -2 \leq x < 0\}$
3. $\{x | x \geq 100\}$ **4.** $\{x | x \leq 6\}$
5. Hurricane: $[74, \infty)$; storm: $[55, 74)$; gale: $[38, 55)$; small craft warning: $[21, 38)$
6. a. $(0, \infty)$ **b.** $(-\infty, 0)$ **c.** $[0, \infty)$ **d.** $(-\infty, 0]$ **7.** $y = 2x - 5$ **8.** $y = -3x + 3$
9. $x = 2$ **10.** $y = 3$ **11.** $y = -2x + 1$ **12.** $y = 3x - 5$ **13.** $y = 2x - 1$ **14.** $y = -\frac{1}{2}x + 1$
15. a. $V = 25{,}000 - 3000t$ **b.** $13{,}000$ **16. a.** $V = 78{,}000 - 5000t$ **b.** $38{,}000$
17. b.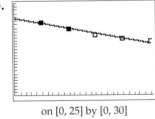

The regression line fits the data well.
c. 13.7 million tons in the year 2010 [from $y_1(35)$]

on [0, 25] by [0, 30]

18. b.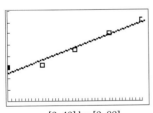

The regression line fits the data reasonably well.
c. 81 to 1 in the year 2010 [from $y_1(50) = 81.4$]

on [0, 40] by [0, 80]

19. 36   **20.** $\frac{3}{4}$ **21.** 8   **22.** 10   **23.** $\frac{1}{27}$ **24.** $\frac{1}{1000}$ **25.** $\frac{9}{4}$ **26.** $\frac{64}{27}$ **27.** 13.97   **28.** 112.32
29. a. $f(11) = 2$ **b.** $\{x | x \geq 7\}$ **c.** $\{y | y \geq 0\}$ **30. a.** $g(-1) = \frac{1}{2}$ **b.** $\{t | t \neq -3\}$ **c.** $\{y | y \neq 0\}$
31. a. $h(16) = \frac{1}{8}$ **b.** $\{w | w > 0\}$ **c.** $\{y | y > 0\}$ **32. a.** $w(8) = \frac{1}{16}$ **b.** $\{z | z \neq 0\}$ **c.** $\{y | y > 0\}$
33. Yes   **34.** No

35. 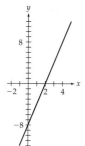   **36.** 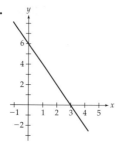   **37.** 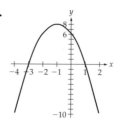   **38.**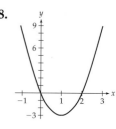

39. $x = 0$, $x = -3$ **40.** $x = 5$, $x = -1$ **41.** $x = -2$, $x = 1$ **42.** $x = 1$, $x = -1$

43. a. Vertex: $(5, -50)$ **b.** 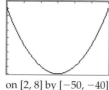 on $[2, 8]$ by $[-50, -40]$ **44. a.** Vertex: $(-7, -64)$ **b.** 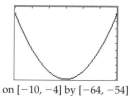 on $[-10, -4]$ by $[-64, -54]$

45. $C(x) = 45 + 0.12x$ **46.** $I(t) = 800t$ **47.** $T(x) = 70 - \dfrac{x}{300}$

48. $C(t) = 27 + 0.58t$; In about 5 years from 2000, so in 2005.

49. a. Break even at 15 and 65 units.   **b.** Profit maximized at 40 units. Max profit is $1250 per week.

50. a. Break even at 150 and 450 units   **b.** Profit maximized at 300 units. Max profit is $67,500 per month.

51. a. $f(-1) = 1$ **b.** $\{x \mid x \neq 0 \text{ and } x \neq 2\}$ **c.** $\{y \mid y > 0 \text{ or } y \leq -3\}$

52. a. $f(-8) = \frac{1}{2}$ **b.** $\{x \mid x \neq 0 \text{ and } x \neq -4\}$ **c.** $\{y \mid y > 0 \text{ or } y \leq -4\}$

53. a. $g(-4) = 0$ **b.** $\mathbb{R}$ **c.** $\{y \mid y \geq -2\}$ **54. a.** $g(-5) = -10$ **b.** $\mathbb{R}$ **c.** $\{y \mid y \leq 0\}$

55. $x = 0$, $x = 1$, $x = -3$ **56.** $x = 0$, $x = 2$, $x = -4$ **57.** $x = 0$, $x = 5$ **58.** $x = 0$, $x = 2$

59. 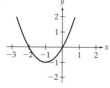   **60.** 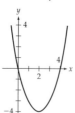   **61.** 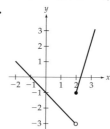   **62.**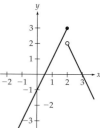

63. a. $f(g(x)) = \left(\dfrac{1}{x}\right)^2 + 1 = \dfrac{1}{x^2} + 1$ **b.** $g(f(x)) = \dfrac{1}{x^2 + 1}$

64. a. $f(g(x)) = \sqrt{5x - 4}$ **b.** $g(f(x)) = 5\sqrt{x} - 4$

65. a. $f(g(x)) = \dfrac{x^3 + 1}{x^3 - 1}$ **b.** $g(f(x)) = \left(\dfrac{x + 1}{x - 1}\right)^3$ **66. a.** $f(g(x)) = |x + 2|$ **b.** $g(f(x)) = |x| + 2$

67. $4x + 2h - 3$ **68.** $\dfrac{-5}{(x + h)x}$ **69.** $A(p(t)) = 2(18 + 2t)^{0.15}$, $A(p(4)) \approx \$3.26$ million

70. a. $x = -1$, $x = 0$, $x = 3$
b.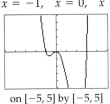
on $[-5, 5]$ by $[-5, 5]$

71. a. $x = -3$, $x = 0$, $x = 1$
b.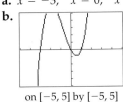
on $[-5, 5]$ by $[-5, 5]$

72. a. The points suggest a parabolic (quadratic) curve.
b.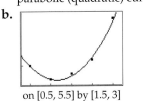
on $[0.5, 5.5]$ by $[1.5, 3]$
c. $3.6 million, $4.8 million

73.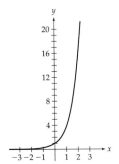

74.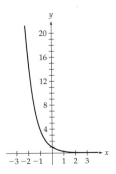

75. $13,468.55   **76.** $2473   **77. a.** $8602 (rounded)   **b.** $18,783 (rounded)
78. a. $11,765 (rounded)   **b.** $58,566 (rounded)   **79.** $1616.07   **80.** $1,290,461.62
81. a. 3   **b.** -3 **c.** 3   **d.** $\frac{1}{4}$ **82. a.** $\frac{1}{2}$ **b.** 8   **c.** -1 **d.** $\frac{3}{2}$ **83.** $f(x) = \ln x$
84. $f(x) = 2x - 1$ **85. a.** In about 2.8 years   **b.** In about 5.6 years
86. a. In about 1.6 years   **b.** In about 3.2 years   **87.** About 50.7 million years
88. About 1.85 million years

Exercises 2.1 page 109

1. $1050   **3.** $3120   **5.** $298.57   **7.** $15.36   **9.** $2550   **11.** $7751.58   **13.** $3250.26
15. 6.4%   **17.** $2500   **19.** 4 years   **21.** $7500   **23.** $9450.90   **25.** $4669.41   **27.** 6 years
29. 8 years 9 months   **31.** 20 years   **33.** 5.48%
35. The $1000 is the same as the interest, so you would be agreeing to pay $1000 of interest on a loan of $0.
37. $950.40   **39.** 58.25%   **41.** 17.89%

Exercises 2.2 page 119

1. a. $32,383.87   **b.** $33,120.59   **3. a.** $30,352.10   **b.** $31,046.39
5. a. $16,713.76   **b.** $16,799.62   **7. a.** $11,308.73   **b.** $10,926.92
9. a. $12,119.55   **b.** $11,917.53   **11. a.** $8106.29   **b.** $8057.40
13. 17 years and 9 months   **15.** 8 years and 9 months   **17.** 5 years and 11 months
19. About 8 years; 8.04 rounds up to 9 years   **21.** About 9.11 years; 8 years and 41 weeks
23. About 11.80 years; 11 years 5 months   **25.** 19.56%   **27.** 8.75%   **29.** 9.90%
31. $6827.69   **33.** $14,467.34   **35.** 30 years and 6 months   **37.** 17.45%
39. The bond gives the greater return (6.80%; the CD returns only 6.62%).   **41.** 32 years and 48 weeks

B-10 CHAPTER 2

43. People's State Bank offers the higher effective yield (4.27%; Statewide Federal's effective rate is 4.18%).
45. $P(1+r)^2 - P(1+2r) = P((1+r)^2 - (1+2r)) = P(1 + 2r + r^2 - 1 - 2r) = P(r^2)$
47. $P(1+r_e)^t = P(1 + (1+r/m)^m - 1)^t = P((1+r/m)^m)^t = P(1+r/m)^{mt}$
49. The two curves match each other very closely.

Exercises 2.3 page 130

1. $20,724.67 **3.** $68,852.94 **5.** $514,647.37 **7.** $32.20 **9.** $172.81 **11.** $345.85
13. 24 years **15.** 22 years and 6 months **17.** 7 years and 2 months
19. Joe will have $475,172.10, and Jill will have $546,822.42. **21.** $4001.27 **23.** $6.39
25. 1 year and 9 months **27.** 13.2% (from $x \approx 0.1319$)

Exercises 2.4 page 141

1. $105,353.72 **3.** $95,896.48 **5.** $212,572.08 **7.** $584.59 **9.** $116 **11.** $872.41
13. $85,934.02 **15.** $2719.69 **17.** $10,284.16 **19.** $4,306,638 **21.** $14,204,378 **23.** $70,777.95
25. $468,407.35; John should buy $500,000 of life insurance.
27. a. $3334.58   **b.** It saves $334.24 each month.   **c.** The longer mortgage costs an extra $2,640,705.84.
29. $31.53   **31.** $115.66   **33.** $41,640.29   **35.** 11.9%   **37.** $2531.31
45. The amortization table uses an annual payment of $7518.83 and the unpaid balance after 3 years is $38,879.49, a difference of 2¢ compared to Exercise 16. The final payment is $7518.87, a correction of 4¢ from the others in the table.

Chapter 2 Review Exercises page 146

1. $217.50   **2.** $52.90   **3.** $2250   **4.** $336.88   **5.** $10,212.75   **6.** $1491.53   **7.** $2138.56
8. $4288.61   **9.** 6.8%   **10.** $1800   **11.** 1 year and 6 months   **12.** 4 years   **13.** 5 years
14. $4500   **15.** $9394.08   **16.** $47,267.91   **17.** 42.86%   **18.** 21.43%   **19.** 6.6%   **20.** $500,000
21. 11.1%; $1110   **22.** $31,535.24   **23.** $264,205.92   **24.** $370,893.09   **25.** $10,749.24
26. No; it would only be worth $110 billion   **27.** $10,198.43   **28.** $18,564.23   **29.** $71,056.01
30. $12,702.50   **31.** $8466.50   **32.** 18 years and 6 months   **33.** 4 years and 2 months   **34.** 8 years
35. 3 years   **36.** 18 years   **37.** 9 years; 9 years (from 8.84)   **38.** 12 years; 11 years and 7 months
39. 6 years; 5 years and 44 weeks   **40.** 6 years; 7 years (from 6.12)   **41.** 7.2 years; 8 years (from 7.27)
42. 13.92%   **43.** 28.19%   **44.** 106.64%   **45.** 13.51%   **46.** 2.94%   **47.** $96,304.25
48. $19,541.30   **49.** $33,322.04   **50.** $935,211.12   **51.** $715,609.58   **52.** $1559.49   **53.** $205.98
54. $106.28   **55.** $168.76   **56.** $19.51   **57.** 26 years   **58.** 24 years and 6 months
59. 5 years and 1 month   **60.** 15 years and 7 months   **61.** 13 years and 6 months   **62.** 9.80%
63. 2.97%   **64.** 9.42%   **65.** 6.58%   **66.** 3.00% (with 365 days per year)   **67.** $319,583.90
68. $20,923,409   **69.** $30,147,922   **70.** $6,773,662.45   **71.** $158,891   **72.** $106.12   **73.** $46.14

74. $423.17 **75.** $1180.05
76. $548.25; do not confuse paying off a current debt with accumulating money in the future.
77. $75,148.14   **78.** $142,742.00   **79.** $5449.20   **80.** $523,692.95   **81.** $143,927.90

Exercises 3.1 page 164

1. $\begin{cases} x + y = 18 \\ x - y = 2 \end{cases}$ **3.** $\begin{cases} x - y = 6 \\ x + y = 40 \end{cases}$ **5.** $\begin{cases} x + y = 30 \\ x + 5y = 70 \end{cases}$ **7.** $\begin{cases} x + y = 100 \\ 10x + 5y = 650 \end{cases}$ **9.** $\begin{cases} x + y = 225 \\ x - 2y = 0 \end{cases}$

11. The solution is $x = 4$, $y = 2$.   **13.** The solution is $x = 3$, $y = 2$.

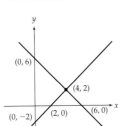 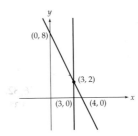

15. The solution is $x = 2$, $y = -2$.   **17.** There is no solution. The equations are inconsistent.

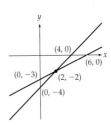 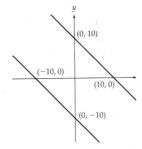

19. There are infinitely many solutions that may be parameterized as $x = 10 - t$, $y = t$. The system is dependent.

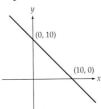

21. $x = 4$, $y = 3$.   **23.** $x = 5$, $y = 10$.   **25.** $x = 10$, $y = -10$.   **27.** $x = 4$, $y = 9$.
29. There are infinitely many solutions that may be parameterized as $x = 10 + t$, $y = t$. The system is dependent.
31. $x = 3$, $y = 8$.   **33.** $x = 4$, $y = 3$.   **35.** $x = 8$, $y = 3$.   **37.** $x = 15$, $y = 6$.
39. There is no solution. The system is inconsistent.

41. 34 investors contributed $5000 and 26 contributed $10,000.
43. There are 34 nickels and 26 dimes in the jar.
45. The retired couple should invest $2000 in the money market account and $8000 in the stock mutual fund.
47. The concession stand sold 1800 sodas and 1200 hot dogs.
49. The federal tax is $4900 and the state tax is $900.
51. The required calcium and phosphorus can be provided by just 5 tablets of supplement B (with none of supplement A) each day.

Exercises 3.2 page 178

1. $3 \times 2; 1; 6$ **3.** $3 \times 4; 3; -6$ **5.** $4 \times 4; 1; 1; 1; 1; 0$ **7.** $1 \times 4; 6$ **9.** $5 \times 1; 8; 6$

11. $\begin{pmatrix} 1 & 2 & | & 2 \\ 3 & 4 & | & 12 \end{pmatrix}$ **13.** $\begin{pmatrix} -4 & 3 & | & 84 \\ 5 & -2 & | & 70 \end{pmatrix}$ **15.** $\begin{pmatrix} 3 & -2 & | & 24 \\ 1 & 0 & | & 6 \end{pmatrix}$ **17.** $\begin{pmatrix} 5 & -15 & | & 30 \\ -4 & 12 & | & 24 \end{pmatrix}$

19. $\begin{pmatrix} 1 & 0 & | & 20 \\ 0 & 1 & | & 30 \end{pmatrix}$ **21.** $\begin{cases} x + y = 9 \\ y = 4 \end{cases}$ **23.** $\begin{cases} -4x + 3y = -60 \\ x - 2y = 20 \end{cases}$ **25.** $\begin{cases} x - 3y = -70 \\ x + y = 10 \end{cases}$

27. $\begin{cases} 2x + y = 6 \\ x + 2y = -6 \end{cases}$ **29.** $\begin{cases} 20x - 15y = 60 \\ -16x + 12y = -48 \end{cases}$ **31.** $\begin{matrix} R_2 \to \\ R_1 \to \end{matrix} \begin{pmatrix} 5 & 6 & | & 30 \\ 3 & 4 & | & 24 \end{pmatrix}$

33. $R_1 - R_2 \to \begin{pmatrix} 2 & 2 & | & -4 \\ 6 & 5 & | & 60 \end{pmatrix}$ **35.** $R_1 - 3R_2 \to \begin{pmatrix} 2 & 0 & | & -24 \\ 1 & 2 & | & 18 \end{pmatrix}$

37. $R_1 - R_2 \to \begin{pmatrix} 6 & 6 & | & -12 \\ 0 & 1 & | & -72 \end{pmatrix}$ **39.** $\frac{1}{8}R_2 \to \begin{pmatrix} 1 & -2 & | & -42 \\ 0 & 1 & | & 15 \end{pmatrix}$

41. $x = 7, y = -3$.
43. No solution. The system is inconsistent.
45. No solution. The system is inconsistent.
47. $x = 3 - 2t, y = t$ (infinitely many solutions). The system is dependent.
49. $x = t, y = -3$ (infinitely many solutions). The system is dependent.
51. $x = 3, y = 2$.   **53.** $x = 3, y = 1$.   **55.** $x = 1, y = 2$.   **57.** $x = 3, y = 2$.
59. $x = 9, y = 2$.   **61.** No solution. The system is inconsistent.   **63.** $x = 8, y = -15$.
65. $x = 4, y = 15$.   **67.** $x = 9 + 3t, y = t$ (infinitely many solutions). The system is dependent.
69. $x = 21, y = -4$.
71. The commodities speculator invested $10,000 in soybean futures and $5000 in corn futures.
73. The older brother receives $4.8 million and the younger brother receives $2.4 million, leaving $4.8 million for their sister.
75. 175 members invested $5000 and 112 members invested $25,000.
77. The dietician probably wants positive whole numbers for the solution, so the possibilities are

Cans of NutraDrink:	12	8	4	0
Tablets of VitaPills:	0	5	10	15

79. The campaign manager should use 7 TV ads and 30 radio ads.

ANSWERS TO SELECTED EXERCISES   **B-13**

81. $x = 18$, $y = 16$.

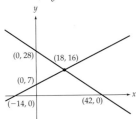

83. No solution. The system is inconsistent.

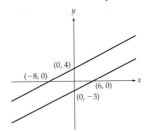

85. No solution. The system is inconsistent.

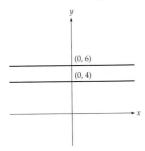

87. $x = 14 - \frac{7}{2}t$, $y = t$ (infinitely many solutions). The system is dependent.

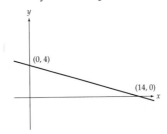

89. $x = t$, $y = 6$ (infinitely many solutions). The system is dependent.

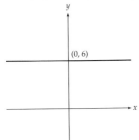

91. $\begin{pmatrix} 3 & -10 & | & -65 \\ -4 & 13 & | & 84 \end{pmatrix} \xrightarrow[R_1 \to]{R_2 \to} \begin{pmatrix} -4 & 13 & | & 84 \\ 3 & -10 & | & -65 \end{pmatrix} \xrightarrow[R_1 \to]{R_2 \to} \begin{pmatrix} 3 & -10 & | & -65 \\ -4 & 13 & | & 84 \end{pmatrix}$

93. $\begin{pmatrix} 3 & -10 & | & -65 \\ -4 & 13 & | & 84 \end{pmatrix} \xrightarrow{5R_1 \to} \begin{pmatrix} 15 & -50 & | & -325 \\ -4 & 13 & | & 84 \end{pmatrix} \xrightarrow{\frac{1}{5}R_1 \to} \begin{pmatrix} 3 & -10 & | & -65 \\ -4 & 13 & | & 84 \end{pmatrix}$

95. $\begin{pmatrix} 3 & -10 & | & -65 \\ -4 & 13 & | & 84 \end{pmatrix} \xrightarrow{\frac{1}{4}R_2 \to} \begin{pmatrix} 3 & -10 & | & -65 \\ -1 & 13/4 & | & 21 \end{pmatrix} \xrightarrow{4R_2 \to} \begin{pmatrix} 3 & -10 & | & -65 \\ -4 & 13 & | & 84 \end{pmatrix}$

97. $\begin{pmatrix} 3 & -10 & | & -65 \\ -4 & 13 & | & 84 \end{pmatrix} \xrightarrow{R_1 + R_2 \to} \begin{pmatrix} -1 & 3 & | & 19 \\ -4 & 13 & | & 84 \end{pmatrix} \xrightarrow{R_1 - R_2 \to} \begin{pmatrix} 3 & -10 & | & -65 \\ -4 & 13 & | & 84 \end{pmatrix}$

99. $\begin{pmatrix} 3 & -10 & | & -65 \\ -4 & 13 & | & 84 \end{pmatrix} \xrightarrow{R_1 - R_2 \to} \begin{pmatrix} 7 & -23 & | & -149 \\ -4 & 13 & | & 84 \end{pmatrix} \xrightarrow{R_1 + R_2 \to} \begin{pmatrix} 3 & -10 & | & -65 \\ -4 & 13 & | & 84 \end{pmatrix}$

Exercises 3.3 page 194

1. $\begin{pmatrix} 1 & 1 & 1 & | & 4 \\ 1 & 2 & 1 & | & 3 \\ 1 & 2 & 2 & | & 5 \end{pmatrix}$ **3.** $\begin{pmatrix} 2 & -1 & 2 & | & 11 \\ -1 & 1 & -3 & | & -12 \\ 2 & -2 & 7 & | & 27 \end{pmatrix}$ **5.** $\begin{pmatrix} 2 & 1 & 5 & 4 & 5 & | & 2 \\ 1 & 1 & 3 & 3 & 3 & | & -1 \end{pmatrix}$

7. $\begin{cases} 4x_1 + 3x_2 + 2x_3 = 11 \\ 3x_1 + 3x_2 + x_3 = 6 \\ x_1 - 2x_2 + 3x_3 = 13 \end{cases}$ **9.** $\begin{cases} 2x_1 + x_2 + x_3 = 7 \\ 2x_1 + 2x_2 + x_3 = 6 \\ 3x_1 + 3x_2 + 2x_3 = 10 \end{cases}$ **11.** $\begin{cases} 8x_1 + 3x_2 - 2x_3 + 19x_4 = 15 \\ 3x_1 + x_2 - x_3 + 7x_4 = 6 \end{cases}$

13. $x_1 = 4, \; x_2 = 5, \; x_3 = -4.$

15. The system is inconsistent, so no solution.

17. The system is dependent, with solution $x_1 = -5 + t, \; x_2 = 5 - t, \; x_3 = t.$

19. $\begin{pmatrix} 1 & 0 & 0 & | & 1 \\ 0 & 1 & 0 & | & 2 \\ 0 & 0 & 1 & | & 3 \end{pmatrix}$ **21.** $\begin{pmatrix} 1 & 0 & 0 & | & 1 \\ 0 & 1 & 0 & | & 2 \\ 0 & 0 & 1 & | & 3 \end{pmatrix}$ **23.** $\begin{pmatrix} 1 & 2 & 0 & | & 3 \\ 0 & 0 & 1 & | & -3 \\ 0 & 0 & 0 & | & 0 \end{pmatrix}$

25. $x_1 = 1, \; x_2 = -2, \; x_3 = 3.$

27. $x_1 = 1, \; x_2 = -2, \; x_3 = 3.$

29. $x_1 = 1, \; x_2 = 1, \; x_3 = 1, \; x_4 = 1.$

31. The system is inconsistent, with no solution.

33. The system is dependent, with solution $x_1 = 3 + 7t, \; x_2 = 4 - 10t, \; x_3 = t.$

35. $x_1 = 2, \; x_2 = 1, \; x_3 = -2, \; x_4 = 3.$

37. The system is dependent and inconsistent, so *no* solution.

39. The system is dependent, with solution $x_1 = 2 + t_1 - t_2, \; x_2 = 1 - 2t_1 - t_2, \; x_3 = t_1, \; x_4 = t_2.$

41. The gardener should use 2 bags of GrowRite, 1 bag of MiracleMix, and 3 bags of GreatGreen.

43. Let m be the number of $25,000 members in the partnership. Then there are $200 + 3m$ members at $5000 and $500 - 4m$ members at $10,000. Because the number of each can never be negative, there can be no more than 125 of the $25,000 members in the partnership.

45. a. The federal tax is $80,510, the state tax is $16,490, and the city tax is $2490.
  b. The effective combined tax rate is (about) 55.3%.

47. a. Letting B be the number of blouses and S be the number of skirts, the shop can make $60 + \tfrac{6}{5}B - \tfrac{4}{5}S$ scarves and $60 - \tfrac{9}{10}B - \tfrac{2}{5}S$ dresses (where none of these quantities are negative).
  b. 76 scarves and 38 dresses.

49. Letting R be the number of radio ads, the number of TV ads is $20 - \tfrac{1}{5}R$, and the number of newspaper ads is 60. The promotional director may choose R to be 0, 5, 10, 15, 20, ..., 85, 90, 95, or 100 because these are the only values for R that will result in a whole number of TV ads, while ensuring that neither value is negative.

51. $\begin{cases} x_1 + 2x_2 = 3 \\ x_2 = 1 \end{cases}$ **53.** All are equivalent to $\begin{pmatrix} 1 & 0 & | & 1 \\ 0 & 1 & | & 1 \end{pmatrix}$ and so to each other.

55. $\begin{pmatrix} 1 & 0 & | & 1 \\ 0 & 1 & | & 1 \end{pmatrix}$ **57.** $x_1 = 1, \; x_2 = 1.$ **59.** $x_1 = 1, \; x_2 = 1, \; x_3 = 2.$

61. $x_1 = 3, \; x_2 = 2, \; x_3 = 4, \; x_4 = 1.$ **63.** $x_1 = 3, \; x_2 = -1, \; x_3 = 2.$

Exercises 3.4 page 209

1. $\begin{pmatrix} 1 & 4 & 7 \\ 2 & 5 & 8 \\ 3 & 6 & 9 \end{pmatrix}$ **3.** $\begin{pmatrix} 3 & 18 & 24 \\ 12 & 6 & 21 \\ 27 & 15 & 9 \end{pmatrix}$ **5.** $\begin{pmatrix} -9 & -8 & -7 \\ -6 & -5 & -4 \\ -3 & -2 & -1 \end{pmatrix}$ **7.** $\begin{pmatrix} 2 & 8 & 11 \\ 8 & 7 & 13 \\ 16 & 13 & 12 \end{pmatrix}$

ANSWERS TO SELECTED EXERCISES B-15

9. $\begin{pmatrix} -1 & 4 & 5 \\ 0 & -4 & 1 \\ 2 & -3 & -7 \end{pmatrix}$ **11.** (6)   **13.** $\begin{pmatrix} 3 & 4 \\ 6 & 8 \end{pmatrix}$ **15.** $\begin{pmatrix} -2 \\ 5 \end{pmatrix}$ **17.** $\begin{pmatrix} 6 & 6 \\ 7 & 7 \end{pmatrix}$ **19.** $\begin{pmatrix} 11 & 7 & 11 \\ 6 & 7 & 6 \\ -1 & 3 & -1 \end{pmatrix}$

21. $\begin{pmatrix} 3 & -1 & 4 \\ 2 & 1 & 3 \end{pmatrix}$ **23.** $\begin{pmatrix} 2 & 8 \\ 4 & 1 \\ 5 & -4 \end{pmatrix}$ **25.** $\begin{pmatrix} -1 & -4 & -1 \\ -2 & 6 & -1 \end{pmatrix}$ **27.** $\begin{pmatrix} 7 & 1 & 4 \\ 0 & -1 & 0 \\ 4 & 2 & 2 \end{pmatrix}$

29. $\begin{pmatrix} 5 & 5 & 6 \\ 7 & -7 & 4 \end{pmatrix}$ **31.** $\begin{pmatrix} 1 & 5 & 4 \\ 1 & 1 & 1 \\ 2 & 3 & 3 \end{pmatrix} \begin{pmatrix} x_1 \\ x_2 \\ x_3 \end{pmatrix} = \begin{pmatrix} 6 \\ 4 \\ 9 \end{pmatrix}$ **33.** $\begin{pmatrix} 4 & 3 & -1 \\ 3 & 3 & 2 \\ 2 & 1 & -3 \end{pmatrix} \begin{pmatrix} x_1 \\ x_2 \\ x_3 \end{pmatrix} = \begin{pmatrix} 2 \\ 9 \\ -6 \end{pmatrix}$

35. $\begin{pmatrix} 5 & 2 & -4 & 1 & 5 \\ 3 & 1 & -3 & 1 & 3 \end{pmatrix} \begin{pmatrix} x_1 \\ x_2 \\ x_3 \\ x_4 \\ x_5 \end{pmatrix} = \begin{pmatrix} 7 \\ 5 \end{pmatrix}$ **37.** $\begin{cases} 5x_1 + 9x_2 + 9x_3 = 11 \\ 4x_1 + 7x_2 + 6x_3 = 9 \\ 3x_1 + 5x_2 + 3x_3 = 8 \\ 4x_1 + 7x_2 + 5x_3 = 10 \end{cases}$ **39.** $\begin{cases} 5x_1 + 4x_2 + 7x_3 + 6x_4 = 18 \\ 2x_1 + 2x_2 + 3x_3 + 3x_4 = 9 \\ 4x_1 + 3x_2 + 5x_3 + 5x_4 = 16 \\ 3x_1 + 2x_2 + 3x_3 + 3x_4 = 11 \end{cases}$

41. Let P be the "price" matrix of selling prices and C be the "commission" matrix so that $C = 0.15P$. Choosing P to be a 3×3 matrix with the rows representing the manufacturers SlumberKing, DreamOn, and RestEasy in that order and the columns representing the models "economy," "best," and "deluxe" in that order, $P = \begin{pmatrix} 300 & 350 & 500 \\ 350 & 400 & 550 \\ 400 & 500 & 700 \end{pmatrix}$ and $C = \begin{pmatrix} 45.00 & 52.50 & 75.00 \\ 52.50 & 60.00 & 82.50 \\ 60.00 & 75.00 & 105.00 \end{pmatrix}$.

43. Let D be the "dealer invoice" matrix and S be the "sticker price" matrix so that the "markup" matrix M is $M = S - D$. Choosing D to be a 2×4 matrix with the rows representing the sales lots in Oakdale and Roanoke in that order and the columns representing the vehicle models "sedan," "station wagon," "van," and "pickup truck" in that order, $D = \begin{pmatrix} 15{,}000 & 19{,}000 & 23{,}000 & 25{,}000 \\ 15{,}000 & 19{,}000 & 23{,}000 & 25{,}000 \end{pmatrix}$ and $S = \begin{pmatrix} 18{,}900 & 22{,}900 & 26{,}900 & 29{,}900 \\ 19{,}900 & 21{,}900 & 27{,}900 & 28{,}900 \end{pmatrix}$ so then $M = \begin{pmatrix} 3900 & 3900 & 3900 & 4900 \\ 4900 & 2900 & 4900 & 3900 \end{pmatrix}$.

45. Let L be the "labor costs" matrix and M be the "materials cost" matrix so that the "total cost" matrix T is $T = L + M$. Choosing L to be a 2×3 matrix of values in pennies with the rows representing the countries Costa Rica and Honduras in that order and the columns representing the apparel items "shorts," "tee-shirts," and "caps" in that order, $L = \begin{pmatrix} 75 & 25 & 45 \\ 80 & 20 & 55 \end{pmatrix}$ and $M = \begin{pmatrix} 160 & 95 & 115 \\ 150 & 80 & 110 \end{pmatrix}$ so then $T = \begin{pmatrix} 235 & 120 & 160 \\ 230 & 100 & 165 \end{pmatrix}$.

47. Let P be the "sale price" row matrix with values in dollars and N be the "number of items" column matrix so that the "total cost" matrix C is $C = P \cdot N$. Choosing the columns of P and the rows of N to represent the items "bottles of soda," "bottles of pickles," "packages of hot dogs," and "bags of chips" in that order, $P = (0.89 \quad 1.29 \quad 2.39 \quad 1.69)$ and $N = \begin{pmatrix} 12 \\ 2 \\ 3 \\ 4 \end{pmatrix}$ so then $C = (27.19)$. The total cost of these items at these prices is $27.19.

49. Let T be the "time" matrix and L be the "labor hourly cost" matrix so that the "production cost" matrix C is $C = T \cdot L$. Choosing T to be a 3×3 matrix with the rows representing the furniture items "table," "chair," and "desk" in that order and the columns representing the manufacturing steps "cutting and milling," "assembly," and "finishing" in that order, $T = \begin{pmatrix} 2 & 1 & 2 \\ 1.5 & 1 & 0.5 \\ 3 & 2 & 3 \end{pmatrix}$, and choosing L to be a 3×2 matrix with the rows representing the manufacturing steps "cutting and milling," "assembly," and "finishing" in that order and the columns representing the factory locations "Wytheville" and "Andersen" in that order, $L = \begin{pmatrix} 9 & 10 \\ 14 & 13 \\ 13 & 12 \end{pmatrix}$, so then $C = \begin{pmatrix} 58 & 57 \\ 34 & 34 \\ 94 & 92 \end{pmatrix}$. For these choices of T and L, the rows of C represent the furniture items "table," "chair," and "desk" in that order, and the columns represent the factory locations "Wytheville" and "Andersen" in that order.

51. Begin with the augmented matrix $\left(\begin{array}{cc|c} a & b & h \\ c & d & k \end{array} \right)$ and apply the "symbolic" row operations $dR_1 - bR_2 \to R_1$, $(ad - bc)R_2 \to R_2$, $R_2 - cR_1 \to R_2$, and $\frac{1}{d}R_2 \to R_2$ to obtain $\left(\begin{array}{cc|c} ad - bc & 0 & hd - bk \\ 0 & ad - bc & ak - hc \end{array} \right)$. Dividing through by $ad - bc$ gives the identity matrix on the left of the bar with the desired solution on the right.

Exercises 3.5 page 222

1. $\begin{pmatrix} 1 & 2 \\ -1 & -1 \end{pmatrix} \begin{pmatrix} -1 & -2 \\ 1 & 1 \end{pmatrix} = \begin{pmatrix} 1 & 0 \\ 0 & 1 \end{pmatrix}$ so this *is* a matrix and its inverse.

3. $\begin{pmatrix} 1 & 1 & 0 \\ 2 & 1 & 1 \\ 1 & 0 & 0 \end{pmatrix} \begin{pmatrix} 0 & 0 & 1 \\ 1 & 0 & -1 \\ -1 & 1 & -1 \end{pmatrix} = \begin{pmatrix} 1 & 0 & 0 \\ 0 & 1 & 0 \\ 0 & 0 & 1 \end{pmatrix}$ so this *is* a matrix and its inverse.

5. $\begin{pmatrix} 4 & 6 & 3 \\ 3 & 4 & 1 \\ 5 & 7 & 3 \end{pmatrix} \begin{pmatrix} -5 & -3 & 6 \\ 4 & 3 & -5 \\ -1 & -2 & 2 \end{pmatrix} = \begin{pmatrix} 1 & 0 & 0 \\ 0 & 1 & 0 \\ 0 & 0 & 1 \end{pmatrix}$ so this *is* a matrix and its inverse.

7. $\begin{pmatrix} 10 & -4 & -7 \\ -7 & 3 & 5 \\ 4 & -1 & -3 \end{pmatrix} \begin{pmatrix} 4 & 5 & -1 \\ 1 & 2 & 1 \\ 5 & 6 & 2 \end{pmatrix} = \begin{pmatrix} 1 & 0 & -28 \\ 0 & 1 & 20 \\ 0 & 0 & -11 \end{pmatrix}$ so this is *not* a matrix and its inverse.

9. $\begin{pmatrix} 1 & -3 \\ 0 & 1 \end{pmatrix}$ **11.** $\begin{pmatrix} -1 & 2 \\ 6 & -11 \end{pmatrix}$ **13.** $\begin{pmatrix} 0 & 1 & -2 \\ 1 & -1 & 2 \\ 0 & -1 & 3 \end{pmatrix}$ **15.** Singular matrix

17. $\begin{pmatrix} 1 & 0 & 0 & -1 \\ 0 & 1 & 0 & 0 \\ -1 & 0 & 1 & 1 \\ 0 & -1 & 0 & 1 \end{pmatrix}$ **19.** $\begin{pmatrix} 1 & -3 & 1 \\ 0 & 2 & -1 \\ -1 & 0 & 1 \end{pmatrix}$ **21.** $\begin{pmatrix} 1 & 2 & 0 \\ -2 & -2 & 1 \\ -5 & -7 & 2 \end{pmatrix}$ **23.** Singular matrix

25. $\begin{pmatrix} x_1 \\ x_2 \end{pmatrix} = \begin{pmatrix} -1 & 2 \\ 6 & -11 \end{pmatrix} \begin{pmatrix} 9 \\ 5 \end{pmatrix} = \begin{pmatrix} 1 \\ -1 \end{pmatrix}$ so $\begin{cases} x_1 = 1 \\ x_2 = -1 \end{cases}$

27. $\begin{pmatrix} x_1 \\ x_2 \\ x_3 \end{pmatrix} = \begin{pmatrix} 0 & 1 & -2 \\ 1 & -1 & 2 \\ 0 & -1 & 3 \end{pmatrix} \begin{pmatrix} 2 \\ 5 \\ 2 \end{pmatrix} = \begin{pmatrix} 1 \\ 1 \\ 1 \end{pmatrix}$ so $\begin{cases} x_1 = 1 \\ x_2 = 1 \\ x_3 = 1 \end{cases}$

29. $x_1 = 1$, $x_2 = 1$, $x_3 = 1$.   31. $x_1 = 4$, $x_2 = -1$, $x_3 = -2$.
33. $x_1 = 2$, $x_2 = -1$, $x_3 = 2$, $x_4 = 1$.   35. $x_1 = 1$, $x_2 = -2$, $x_3 = 1$, $x_4 = 2$.
37. $x_1 = 2$, $x_2 = 4$, $x_3 = 1$.   39. $x_1 = -5$, $x_2 = 10$, $x_3 = -20$, $x_4 = 15$.
41. The multiplex sold 150 adult tickets and 350 child tickets for Film No. 1; 200 adult tickets and 200 child tickets for Film No. 2; 250 adult tickets and 200 child tickets for Film No. 3; 400 adult tickets and 100 child tickets for Film No. 4; and 600 adult tickets and no child tickets for Film No. 5.
43. The Gold Mines partnership consists of 250 members at $1000, 150 members at $5000, and 100 members at $10,000; the Oil Wells partnership consists of 150 members at $1000, 350 members at $5000, and 200 members at $10,000; and the Modern Art partnership consists of 325 members at $1000, 115 members at $5000, and 160 members at $10,000.
45. Billy needs 4 drops of Supplement 1, no drops of Supplement 2, 5 drops of Supplement 3, and 2 drops of Supplement 4; Susie needs 2 drops of Supplement 1, 2 drops of Supplement 2, 3 drops of Supplement 3, and 6 drops of Supplement 4; and Jimmy needs 3 drops of Supplement 1, 1 drop of Supplement 2, no drops of Supplement 3, and 8 drops of Supplement 4.
47. Mr. and Mrs. Jordan should invest $100,000 in the stock fund, $150,000 in the money market fund, and $50,000 in the bond fund; Mr. and Mrs. French should invest $78,300 in the stock fund, $51,100 in the money market fund, and $105,500 in the bond fund; and Mrs. Daimen should invest $90,000 in the stock fund, $105,000 in the money market fund, and $75,000 in the bond fund.
49. The mass transit manager should assign 120 subway cars, 50 buses, and 10 jitneys to Brighton; 100 subway cars, 30 buses, and 20 jitneys to Conway; 100 subway cars, 40 buses, and 10 jitneys to Longwood; and 110 subway cars, 40 buses, and 15 jitneys to Oakley.
51. $\frac{1}{ad-bc}\begin{pmatrix} d & -b \\ -c & a \end{pmatrix}\begin{pmatrix} a & b \\ c & d \end{pmatrix} = \frac{1}{ad-bc}\begin{pmatrix} da-bc & db-bd \\ -ca+ac & -cb+ad \end{pmatrix} = \frac{1}{ad-bc}\begin{pmatrix} ad-bc & 0 \\ 0 & ad-bc \end{pmatrix} = \begin{pmatrix} 1 & 0 \\ 0 & 1 \end{pmatrix}$
53. $(A \cdot B)(B^{-1} \cdot A^{-1}) = A\underbrace{BB^{-1}}_{I}A^{-1} = AIA^{-1} = AA^{-1} = I$ and similarly for $(B^{-1} \cdot A^{-1})(A \cdot B)$
55. If $B \cdot A = I$, then multiplying on the right by A^{-1} gives $B \cdot A \cdot A^{-1} = I \cdot A^{-1}$ so that $B \cdot I = A^{-1}$ and $B = A^{-1}$. Then $A \cdot B = A \cdot A^{-1} = I$.
57. The formula in the box with $a = 1$, reversing the order on the left, and multiplying numerator and denominator on the right by -1 gives the required equation, which can be written
$1 + x + x^2 + \cdots + x^{n-1} = (1 - x^n)(1 - x)^{-1}$. The corresponding matrix equation is then
$1 + A + A^2 + \cdots + A^{n-1} = (1 - A^n)(1 - A)^{-1}$.
59. Using the result of Exercise 58:
$A^3 = A \cdot A^2$ is substochastic of order $w \cdot w^2 = w^3$
$A^4 = A \cdot A^3$ is substochastic of order $w \cdot w^3 = w^4$
$\vdots$
$A^n = A \cdot A^{n-1}$ is substochastic of order $w \cdot w^{n-1} = w^n$

Exercises 3.6 page 235

1. $\begin{pmatrix} 0.30 & 0.50 \\ 0.40 & 0.10 \end{pmatrix}$ for sectors A and L   3. $\begin{pmatrix} 0.20 & 0.10 & 0.15 \\ 0.10 & 0.30 & 0.20 \\ 0.40 & 0.15 & 0.10 \end{pmatrix}$ for sectors C, E, and L

5.

7.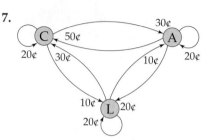

9. $Y = \begin{pmatrix} 71 \\ 37 \end{pmatrix}$ **11.** $Y = \begin{pmatrix} 89 \\ 118 \\ 101 \end{pmatrix}$ **13.** $X = \begin{pmatrix} 150 \\ 120 \end{pmatrix}$ **15.** $X = \begin{pmatrix} 80 \\ 60 \\ 100 \end{pmatrix}$

17. Heavy industry production must be $300 million and light industry production must be $280 million.

19. The current excess production from these sectors is $115 million (made up of $20 million from heavy industry, $50 million from light industry, and $45 million from the railroads). Each $10 million increase in heavy industry production raises the excess production by $1 million (but because the heavy industry production consumes both light industry and railroad production, this increase is composed of an $8 million increase in heavy industry production together with decreases of $5 million from light industry and $2 million from the railroads). When the heavy industry production level reaches $200 million, all light industry excess production will be consumed by the heavy industry sector, and no further expansion will be possible without expanding the light industry production.

21. $y = 9x + 34$ **23.** $y = -29x + 92$ **25.** $y = 9x + 165$

27. His sales in the fifth month may be expected to be $910,000.

29. At 79¢ per 8-ounce bag, the manufacturer can expect 1711 sales per 20,000 customers.

Chapter 3 Review Exercises page 241

1. Let x be the number of 30-day advance sale tickets and let y be the number of full-fare tickets:
$$\begin{cases} x + y = 30 \\ 79x + 159y = 3970 \end{cases}$$

2. Let x be the number of cows and let y be the number of horses: $\begin{cases} x + y = 420 \\ x - 2y = 0 \end{cases}$

3. $x = 13$, $y = 5$.

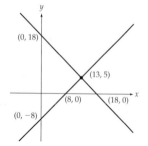

4. There are infinitely many solutions that may be parameterized as $x = 18 + 2t$, $y = t$. The system is dependent.

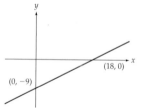

5. $x = 14$, $y = 30$.   **6.** No solution. The system is inconsistent.
7. $x = 9$, $y = 3$.   **8.** $x = -15$, $y = 24$.

ANSWERS TO SELECTED EXERCISES   B-19

9. He has 10 rosebushes and 15 tomato plants in his garden.

10. The fraternity should use 2 cars and 2 vans for the trip to Orlando.   11. 3×3; 5; 6; 4

12. 4×4; 7; 10; 13   13. $3R_2 \to \begin{pmatrix} 3 & 4 & | & 12 \\ 3 & 6 & | & 6 \end{pmatrix}$ 14. $R_1 + R_2 \to \begin{pmatrix} 1 & -1 & | & 8 \\ 2 & -3 & | & 6 \end{pmatrix}$

15. $x = 15$, $y = 16$.   16. $x = 20$, $y = 30$.   17. No solution. The system is inconsistent.

18. $x = 3 + \frac{3}{4}t$, $y = t$ (infinitely many solutions). The system is dependent.

19. The pharmacist filled 63 prescriptions for antibiotics and 29 prescriptions for cough suppressants.

20. Each old *Life* magazine is $6 and each old *The New Yorker* is $5.

21. $x_1 = 3$, $x_2 = -3$, $x_3 = 6$.

22. The system is dependent, with solution $x_1 = 4 - t$, $x_2 = t$, $x_3 = 2$.

23. $\begin{pmatrix} 1 & 0 & 0 & | & 4 \\ 0 & 1 & 0 & | & -3 \\ 0 & 0 & 1 & | & 2 \end{pmatrix}$ 24. $\begin{pmatrix} 1 & 0 & 0 & 1 & | & 0 \\ 0 & 1 & 0 & -1 & | & 0 \\ 0 & 0 & 1 & 0 & | & 0 \\ 0 & 0 & 0 & 0 & | & 1 \end{pmatrix}$

25. $x_1 = 4$, $x_2 = 3$, $x_3 = 1$, $x_4 = 2$.

26. The system is dependent, with solution $x_1 = 2 - t$, $x_2 = 3 - t$, $x_3 = t$, $x_4 = 4$.

27. The system is inconsistent, so there is no solution.

28. The system is inconsistent, so there is no solution.

29. The Nyack Nursery should raise 48 dahlias, 328 chrysanthemums, and 48 daisies.

30. Each hot dog costs $1. If the mirror house is $1.50 and a soda is $1, then a go-kart ride costs $4.50.

31. $\begin{pmatrix} 4 & -6 \\ 6 & -4 \end{pmatrix}$ 32. $\begin{pmatrix} 1 & 2 & 3 & 4 \\ -4 & -3 & -2 & -1 \\ 1 & 2 & 3 & 4 \\ -4 & -3 & -2 & -1 \end{pmatrix}$ 33. $\begin{pmatrix} 15 & 25 \\ 5 & 2 \end{pmatrix}$ 34. $\begin{pmatrix} 12 & 3 & -2 \\ -2 & 12 & 11 \\ 9 & 3 & 1 \end{pmatrix}$

35. $\begin{pmatrix} 1 & 4 & 1 \\ 2 & 8 & 3 \\ 1 & 5 & 2 \end{pmatrix} \begin{pmatrix} x_1 \\ x_2 \\ x_3 \end{pmatrix} = \begin{pmatrix} 15 \\ 26 \\ 17 \end{pmatrix}$ 36. $\begin{pmatrix} 2 & 3 & -1 & 1 \\ 5 & 4 & 1 & 2 \\ 2 & 1 & 1 & 1 \end{pmatrix} \begin{pmatrix} x_1 \\ x_2 \\ x_3 \\ x_4 \end{pmatrix} = \begin{pmatrix} 20 \\ 35 \\ 12 \end{pmatrix}$

37. Let T be the "this year" matrix and L be the "last year" matrix so that the "growth" matrix G is $G = T - L$. Choosing T to be a 3×2 matrix with the rows representing the grandchildren Thomas, Richard, and Harriet in that order and the columns representing their heights and weights in that order,

$$T = \begin{pmatrix} 61 & 90 \\ 54 & 75 \\ 47 & 60 \end{pmatrix} \text{ and } L = \begin{pmatrix} 58 & 80 \\ 52 & 70 \\ 46 & 55 \end{pmatrix} \text{ so then } G = \begin{pmatrix} 3 & 10 \\ 2 & 5 \\ 1 & 5 \end{pmatrix}.$$

38. Let N be the "number of items needed" column matrix and P be the "price" matrix so that the "cost of her order" matrix C is $C = P \cdot N$. Choosing the rows of N to represent the items "jacket," "blouse," "skirt," and "slacks" in that order, N is the 4×1 matrix $N = \begin{pmatrix} 200 \\ 300 \\ 250 \\ 175 \end{pmatrix}$. Then the columns of P must also represent the items "jacket," "blouse," "skirt," and "slacks" in that order, so P is the 2×4 matrix

$P = \begin{pmatrix} 195 & 85 & 145 & 130 \\ 190 & 90 & 150 & 125 \end{pmatrix}$ and the rows represent the prices of the East Coast designer and the Italian team in the order. Then $C = \begin{pmatrix} 123{,}500 \\ 124{,}375 \end{pmatrix}$. The cost of her order from the East Coast designer is \$123,500, and from the Italian team it is \$124,375.

39. $\begin{pmatrix} 1 & 2 & 3 \\ 1 & 1 & 1 \\ 0 & 1 & 3 \end{pmatrix} \begin{pmatrix} -2 & 3 & 1 \\ 3 & -3 & -2 \\ -1 & 1 & 1 \end{pmatrix} = \begin{pmatrix} 1 & 0 & 0 \\ 0 & 1 & 0 \\ 0 & 0 & 1 \end{pmatrix}$ so this *is* a matrix and its inverse.

40. $\begin{pmatrix} -3 & 0 & 1 \\ 1 & 3 & 1 \\ -3 & 2 & 2 \end{pmatrix} \begin{pmatrix} -4 & -2 & 3 \\ 5 & 3 & -4 \\ -11 & -6 & 8 \end{pmatrix} = \begin{pmatrix} 1 & 0 & -1 \\ 0 & 1 & -1 \\ 0 & 0 & -1 \end{pmatrix}$ so this is *not* a matrix and its inverse.

41. $\begin{pmatrix} 3 & 0 & -1 \\ -2 & 0 & 1 \\ -6 & 1 & 3 \end{pmatrix}$ 42. Singular matrix 43. $\begin{pmatrix} x_1 \\ x_2 \end{pmatrix} = \begin{pmatrix} 1 & -4 \\ -3 & 13 \end{pmatrix} \begin{pmatrix} 33 \\ 8 \end{pmatrix} = \begin{pmatrix} 1 \\ 5 \end{pmatrix}$ so $\begin{cases} x_1 = 1 \\ x_2 = 5 \end{cases}$

44. $\begin{pmatrix} x_1 \\ x_2 \end{pmatrix} = \begin{pmatrix} -5 & 8 \\ 2 & -3 \end{pmatrix} \begin{pmatrix} 25 \\ 16 \end{pmatrix} = \begin{pmatrix} 3 \\ 2 \end{pmatrix}$ so $\begin{cases} x_1 = 3 \\ x_2 = 2 \end{cases}$

45. $\begin{pmatrix} x_1 \\ x_2 \\ x_3 \end{pmatrix} = \begin{pmatrix} 1 & 1 & -3 \\ 0 & 1 & -1 \\ -2 & -3 & 8 \end{pmatrix} \begin{pmatrix} 11 \\ 7 \\ 5 \end{pmatrix} = \begin{pmatrix} 3 \\ 2 \\ -3 \end{pmatrix}$ so $\begin{cases} x_1 = 3 \\ x_2 = 2 \\ x_3 = -3 \end{cases}$

46. $\begin{pmatrix} x_1 \\ x_2 \\ x_3 \\ x_4 \end{pmatrix} = \begin{pmatrix} 1 & 0 & 0 & -1 \\ 0 & 1 & -2 & 0 \\ 0 & -2 & 6 & -1 \\ -1 & 0 & -1 & 2 \end{pmatrix} \begin{pmatrix} 7 \\ 10 \\ 4 \\ 6 \end{pmatrix} = \begin{pmatrix} 1 \\ 2 \\ -2 \\ 1 \end{pmatrix}$ so $\begin{cases} x_1 = 1 \\ x_2 = 2 \\ x_3 = -2 \\ x_4 = 1 \end{cases}$

47. The Kingman store can display 7 living room suites (2 in the window and 5 more only on the showroom floor) and 8 bedroom suites (3 in the window and 5 more only on the showroom floor); the Prescott store can display 11 living room suites (3 in the window and 8 more only on the showroom floor) and 10 bedroom suites (3 in the window and 7 more only on the showroom floor); and the Holbrook store can display 9 living room suites (3 in the window and 6 more only on the showroom floor) and 12 bedroom suites (2 in the window and 10 more only on the showroom floor).

48. Mr. Dahlman's taxes are \$17,100 (federal), \$7600 (state), and \$3600 (city); Mrs. Farrell's taxes are \$8550 (federal), \$3800 (state), and \$1800 (city); Ms. Mazlin's taxes are \$13,680 (federal), \$6080 (state), and \$2880 (city); and Mr. Seidner's taxes are \$25,650 (federal), \$11,400 (state), and \$5400 (city).

49. $\begin{pmatrix} 0.15 & 0.25 \\ 0.30 & 0.20 \end{pmatrix}$ for sectors A and B 50. $\begin{pmatrix} 0.10 & 0.20 & 0.30 & 0 \\ 0.20 & 0.10 & 0 & 0.20 \\ 0.30 & 0 & 0.10 & 0.30 \\ 0 & 0.20 & 0.30 & 0.10 \end{pmatrix}$ for sectors A, B, C, and D

51. 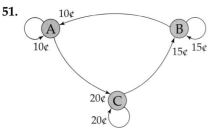 **52.**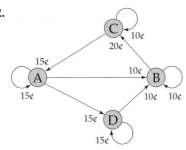

53. $Y = \begin{pmatrix} 371 \\ 449 \\ 419 \end{pmatrix}$ **54.** $Y = \begin{pmatrix} 170 \\ 115 \\ 160 \\ 143 \end{pmatrix}$ **55.** $X = \begin{pmatrix} 364 \\ 436 \end{pmatrix}$ **56.** $X = \begin{pmatrix} 457 \\ 605 \\ 529 \end{pmatrix}$

57. Each division must produce $4 million.
58. The other sectors of the economy can use $841 million of domestic oil and $1948 million of foreign oil, but all the military protection budget is consumed in the production of this output.
59. $y = 7x + 9$ **60.** $y = 73x + 14$ **61.** $y = 13x + 13$ **62.** $y = 19x - 52$
63. They can expect sales of $4405.   **64.** She can expect 12 minor accidents.

Exercises 4.1 page 258

1. b **3.** c **5.** a **7.** b
9. The vertices are (0, 0), (40, 0), and (0, 20). The region is bounded.   **11.** The vertices are (0, 0), (10, 0), (10, 30), and (0, 10). The region is bounded.

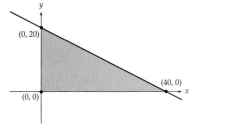 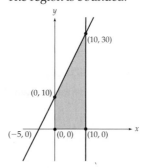

13. The vertices are (0, 0), (6, 0), (4, 2), and (0, 4). The region is bounded.   **15.** The vertices are (4, 0) and (0, 10). The region is unbounded.

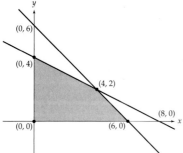 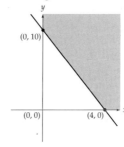

17. The vertices are (3, 4) and (0, 8).
The region is unbounded.

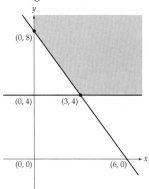

19. The vertices are (8, 0), (2, 6), and (0, 12).
The region is unbounded.

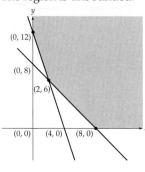

21. The vertices are (30, 0), (40, 0), (0, 80), and (0, 10).
The region is bounded.

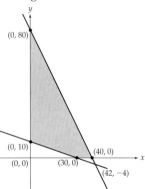

23. The vertices are (15, 0) and (0, 30).
The region is unbounded.

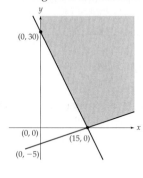

25. Let x be the number of goats and let y be the number of llamas.

$$\begin{cases} 2x + 5y \le 400 & \text{(Land)} \\ 100x + 80y \le 13{,}200 & \text{(Money)} \\ x \ge 0, \ y \ge 0 & \text{(Nonnegativity)} \end{cases}$$

The vertices are (0, 0), (132, 0), (100, 40), and (0, 80).

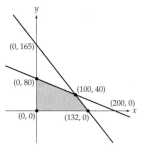

27. Let x be the number of dinghies and let y be the number of rowboats.

$$\begin{cases} 2x + 3y \le 120 & \text{(Metal work)} \\ 2x + 2y \le 100 & \text{(Painting)} \\ x \ge 0, \ y \ge 0 & \text{(Nonnegativity)} \end{cases}$$

The vertices are (0, 0), (50, 0), (30, 20), and (0, 40).

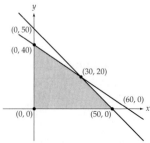

29. Let x be the number of servings of SugarSnaks and let y be the number of bags of Gobbl'Ems.

$$\begin{cases} 5x + 8y \leq 80 & \text{(Fat)} \\ 125x + 250y \leq 2250 & \text{(Calories)} \\ x \geq 0, \; y \geq 0 & \text{(Nonnegativity)} \end{cases}$$

The vertices are (0, 0), (16, 0), (8, 5), and (0, 9).

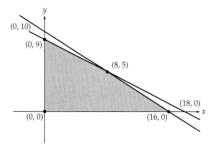

31. Let x be the number of hours the Ohio factory operates and let y be the number of hours the Pennsylvania factory operates.

$$\begin{cases} 4x + 4y \leq 64 & \text{(Sulfur dioxide)} \\ 5x + 3y \leq 60 & \text{(Particulates)} \\ x \geq 0, \; y \geq 0 & \text{(Nonnegativity)} \end{cases}$$

The vertices are (0, 0), (12, 0), (6, 10), and (0, 16).

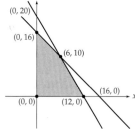

33. Let x be the amount of money (in millions of dollars) invested in stock funds and let y be the amount of money (in millions of dollars) invested in bond funds.

$$\begin{cases} x + y \leq 8 & \text{(Money)} \\ x \leq y & \text{(Limit risk)} \\ x \geq 0, \; y \geq 0 & \text{(Nonnegativity)} \end{cases}$$

The vertices are (0, 0), (4, 4), and (0, 8).

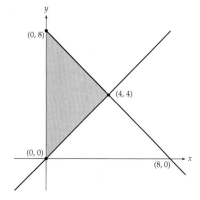

Exercises 4.2 page 271

1. The maximum is 45 (when $x = 15$, $y = 15$).   **3.** The minimum is 30 (when $x = 10$, $y = 0$).
5. The maximum is 15 (when $x = 15$, $y = 0$).   **7.** The minimum is 60 (when $x = 0$, $y = 60$).
9. There is no maximum because the region is unbounded in the positive y direction and P increases for increasing y values.
11. The maximum is 400 (when $x = 0$, $y = 10$).   **13.** The minimum is 150 (when $x = 10$, $y = 0$).
15. The maximum is 160 (when $x = 15$, $y = 20$).   **17.** The minimum is 232 (when $x = 6$, $y = 16$).
19. The maximum is 230 (when $x = 40$, $y = 10$).   **21.** The maximum is 1080 (when $x = 0$, $y = 90$).

23. The minimum is 885 (when $x = 33$, $y = 9$).
25. The rancher should raise 100 goats and 40 llamas to obtain the greatest possible profit of $9600.
27. The company should manufacture 32 prams and 9 yawls to obtain the greatest possible profit of $6420.
29. The county should operate the Norton incinerator 4 hours each day and the Wiseburg incinerator 6 hours each day to obtain the least cost of $620.
31. Each bunny should receive 30 handfuls of greens and 16 drops of supplement each week to minimize the owner's costs at 76¢.
33. No grassland and 2400 acres of forest should be reclaimed this year to raise the greatest possible amount of $360,000 in long-term leases for use in next year's reclamation efforts.

Exercises 4.3 page 291

1.

	x_1	x_2	s_1	s_2	
s_1	3	2	1	0	12
s_2	6	1	0	1	15
P	-8	-9	0	0	0

3.

	x_1	x_2	s_1	s_2	s_3	
s_1	4	3	1	0	0	12
s_2	5	2	0	1	0	20
s_3	1	6	0	0	1	12
P	-13	-7	0	0	0	0

5.

	x_1	x_2	x_3	x_4	s_1	s_2	
s_1	2	1	1	3	1	0	6
s_2	1	4	-2	1	0	1	8
P	-5	2	-10	5	0	0	0

7.

	x_1	x_2	x_3	s_1	s_2	s_3	
s_1	8	1	4	1	0	0	32
s_2	3	5	7	0	1	0	30
s_3	6	2	9	0	0	1	28
P	-10	-20	-15	0	0	0	0

9. The smallest negative entry in the bottom row is -8, so the pivot column is column 2. The first row may not be considered for the pivot row since the pivot column entry is 0. The pivot row is row 2 and the pivot element is the 1 in column 2 and row 2. Since the pivot element is already 1, the pivot operation consists of one row operation:

		x_1	x_2	s_1	s_2	
	s_1	2	0	1	0	4
	x_2	1	1	0	1	5
$R_3 + 8R_2 \to$	P	1	0	0	8	40

11. The smallest negative entry in the bottom row is -5, so the pivot column is column 2. The ratios are $\frac{12}{2} = 6$ and $\frac{8}{1} = 8$, so the pivot row is row 1. The pivot element is the 2 in column 2 and row 1 of the simplex tableau. The first step of the pivot operation is the row operation $\frac{1}{2}R_1 \to R_1$ to replace the pivot element with a 1. Then:

		x_1	x_2	x_3	x_4	s_1	s_2	
	x_2	2	1	3	1	1/2	0	6
$R_2 - R_1 \to$	s_2	1	0	-1	0	$-1/2$	1	2
$R_3 + 5R_1 \to$	P	6	0	21	2	5/2	0	30

13. No pivot column because there are no negative entries in the bottom row. The solution has been found: The maximum is 90 when $x_1 = 10$, $x_2 = 15$, $x_3 = 0$, and $x_4 = 0$ (and $s_1 = 10$, $s_2 = 0$, $s_3 = 0$).

15. The smallest negative entry in the bottom row is -10 so the pivot column is column 5. Since all of the entries in the pivot column are zero or negative, there is no pivot row. There is no maximum value.
17. The maximum is 28 when $x_1 = 0$, $x_2 = 14$.
19. The maximum is 2000 when $x_1 = 0$, $x_2 = 50$, $x_3 = 0$.
21. There is no maximum (the second pivot column is column 2 but there is no pivot row).
23. The maximum is 300 when $x_1 = 0$, $x_2 = 20$, $x_3 = 0$, $x_4 = 40$.
25. The maximum is 16 when $x_1 = 3$, $x_2 = 2$.
27. The maximum is 12,600 when $x_1 = 0$, $x_2 = 160$, $x_3 = 60$.
29. The maximum is 30 when $x_1 = 5$, $x_2 = 5$, $x_3 = 5$.
31. The shop should rebuild no carburetors, 35 fuel pumps, and 20 alternators for a greatest possible profit of $690.
33. The recycling center should accept 300 crates of paper products and 500 crates of glass bottles to raise the greatest possible amount of $59 each week.
35. The company should process no agates, 49 trays of onyxes, and 7 trays of garnets each day to obtain the greatest possible profit of $581.
37. The politician should run 3 daytime ads, 7 prime time ads, and no late night ads to reach 47,000 voters, the most possible.
39. The farmer should plant 30 acres of corn, 90 acres of peanuts, and 120 acres of soybeans for the greatest possible profit of $43,500.

43. The final tableau is

	x_1	x_2	x_3	x_4	s_1	s_2	s_3	
x_3	0	0	1	0	0	0	1	1
s_1	0	-8	0	30	1	-2	3	3
x_1	1	-24	0	6	0	2	1	1
P	0	2	0	21/2	0	3/2	5/4	5/4

The maximum is $\frac{5}{4}$ when $x_1 = 1$, $x_2 = 0$, $x_3 = 1$, $x_4 = 0$.

45. No.   **47.** The maximum is 100 when $x_1 = 100$, $x_2 = 0$.

49. Maximize $P = \begin{pmatrix} 1 & 10 & 100 & 1000 \end{pmatrix} \begin{pmatrix} x_1 \\ x_2 \\ x_3 \\ x_4 \end{pmatrix}$

Subject to $\begin{pmatrix} 0 & 0 & 0 & 1 \\ 0 & 0 & 1 & 20 \\ 0 & 1 & 20 & 200 \\ 1 & 20 & 200 & 2000 \end{pmatrix} \begin{pmatrix} x_1 \\ x_2 \\ x_3 \\ x_4 \end{pmatrix} \leq \begin{pmatrix} 1 \\ 100 \\ 10{,}000 \\ 1{,}000{,}000 \end{pmatrix}$ and $\begin{pmatrix} x_1 \\ x_2 \\ x_3 \\ x_4 \end{pmatrix} \geq 0$.

The maximum is 1,000,000 when $x_1 = 1{,}000{,}000$, $x_2 = 0$, $x_3 = 0$, $x_4 = 0$.

Exercises 4.4 page 311

1. The dual problem is maximize $P = 21x_1$ subject to $\begin{cases} 3x_1 + 4x_2 \leq 84 \\ x_1 - x_2 \leq 21 \\ x_1 \geq 0, \ x_2 \geq 0 \end{cases}$

with initial simplex tableau

	x_1	x_2	s_1	s_2	
s_1	3	4	1	0	84
s_2	1	-1	0	1	21
P	-21	0	0	0	0

3. The dual problem is maximize $P = 180x_1 + 120x_2$ subject to $\begin{cases} x_1 + 4x_2 \le 60 \\ 2x_1 + 5x_2 \le 100 \\ 3x_1 + 6x_2 \le 300 \\ x_1 \ge 0, x_2 \ge 0 \end{cases}$

with initial simplex tableau

	x_1	x_2	s_1	s_2	s_3	
s_1	1	4	1	0	0	60
s_2	2	5	0	1	0	100
s_3	3	6	0	0	1	300
P	-180	-120	0	0	0	0

5. The dual problem is maximize $P = 150x_1 + 100x_2 + 228x_3$ subject to $\begin{cases} 3x_1 + x_2 + 3x_3 \le 3 \\ 2x_1 + 4x_2 + 4x_3 \le 20 \\ x_1 \ge 0, x_2 \ge 0, x_3 \ge 0 \end{cases}$

with initial simplex tableau

	x_1	x_2	x_3	s_1	s_2	
s_1	3	1	3	1	0	3
s_2	2	4	4	0	1	20
P	-150	-100	-228	0	0	0

7. The dual problem is maximize $P = -30x_1 + 30x_2$ subject to $\begin{cases} -x_1 + x_2 \le 15 \\ x_1 + 2x_2 \le 20 \\ 2x_1 + x_2 \le 5 \\ x_1 \ge 0, x_2 \ge 0 \end{cases}$

with initial simplex tableau

	x_1	x_2	s_1	s_2	s_3	
s_1	-1	1	1	0	0	15
s_2	1	2	0	1	0	20
s_3	2	1	0	0	1	5
P	30	-30	0	0	0	0

9. The minimum is 300 when $y_1 = 20$, $y_2 = 0$.   11. The minimum is 45 when $y_1 = 0$, $y_2 = 15$.
13. The minimum is 43 when $y_1 = 7$, $y_2 = 3$.   15. The minimum is 10 when $y_1 = 0$, $y_2 = 10$.
17. The minimum is 900 when $y_1 = 0$, $y_2 = 40$, $y_3 = 10$.
19. The minimum is 98 when $y_1 = 2$, $y_2 = 2$, $y_3 = 0$.   21. There is no minimum.
23. The minimum is 2376 when $y_1 = 6$, $y_2 = 12$, $y_3 = 6$.
25. The athlete should use 8 Bulk-Up Bars and 4 cans of Power Drink to receive the needed fat and protein at the least possible cost of $5.64.
27. The office manager should buy 45 packages from Jack's Office Supplies and 30 packages from John's Discount to restock the store room at the least possible cost of $1230.
29. The project engineer should use 350 heavy-duty dump truck loads of dirt, 300 heavy-duty dump truck loads of crushed rock, no regular dump truck loads of dirt, and 150 regular dump truck loads of crushed rock to get the project finished on time at the least possible cost of $66,000.
31. The farmer should buy 8000 pounds of Miracle Mix and 18,000 pounds of the store brand to meet the fertilizer needs of his field at the least possible cost of $2640.
33. The Kentucky warehouse should ship 200 cartons to Kansas, 200 cartons to Texas, and none to Oregon, and the Utah warehouse should ship none to Kansas, 100 cartons to Texas, and 100 cartons to Oregon to incur the smallest possible shipping cost of $1600.

39. The final tableau shows that the solutions of the minimum and maximum problems are both 3390 with the variables taking the following values.

Minimum problem	$y_1 = 7$	$y_2 = 3$	$y_3 = 1$	$y_4 = 0$	$t_1 = 0$	$t_2 = 0$	$t_3 = 0$
Maximum problem	$s_1 = 0$	$s_2 = 0$	$s_3 = 0$	$s_4 = 75$	$x_1 = 60$	$x_2 = 45$	$x_3 = 30$

Thus one of each pair x_1 and t_1, x_2 and t_2, x_3 and t_3, s_1 and y_1, s_2 and y_2, s_3 and y_3, and s_4 and y_4 is zero.

Exercises 4.5 page 324

1.

	x_1	x_2	x_3	s_1	s_2	
s_1	3	2	8	1	0	96
s_2	−5	−1	−6	0	1	−30
P	−15	−20	−18	0	0	0

The dual pivot element is the −1 in row 2 and column 2.

3.

	x_1	x_2	x_3	s_1	s_2	
s_1	−2	−1	−3	1	0	−30
s_2	−1	−1	−2	0	1	−20
P	−6	−4	−6	0	0	0

The dual pivot element is the −1 in row 1 and column 2.

5.

	x_1	x_2	x_3	s_1	s_2	s_3	
s_1	−1	−1	−1	1	0	0	−8
s_2	1	2	3	0	1	0	30
s_3	1	2	1	0	0	1	18
P	−20	−30	−10	0	0	0	0

The dual pivot element is the −1 in row 1 and column 2.

7.

	x_1	x_2	x_3	x_4	s_1	s_2	s_3	
s_1	−1	−1	−1	−1	1	0	0	−30
s_2	−2	−3	−2	−1	0	1	0	−20
s_3	4	2	1	2	0	0	1	80
P	−3	−2	−5	−4	0	0	0	0

The dual pivot element is the −1 in row 1 and column 3.

9. The maximum is 110 when $x_1 = 20$, $x_2 = 30$.
11. The maximum is 1200 when $x_1 = 0$, $x_2 = 60$, $x_3 = 0$.
13. There is no maximum. **15.** The maximum is 270 when $x_1 = 2$, $x_2 = 20$, $x_3 = 0$.
17. The maximum is 13 when $x_1 = 8$, $x_2 = 5$.
19. The maximum is 720 when $x_1 = 0$, $x_2 = 0$, $x_3 = 60$, $x_4 = 0$.
21. The maximum is 808 when $x_1 = 0$, $x_2 = 14$, $x_3 = 0$, $x_4 = 24$.
23. The maximum is 30 when $x_1 = 10$, $x_2 = 40$, $x_3 = 0$.
25. The retired couple should invest $15,000 in certificates of deposit and $5000 in treasury bonds.
27. The manager should place 5 newspaper ads, 15 radio commercials, and 5 TV spots to reach 295,000 potential customers, the greatest possible number.
29. The furniture shop should manufacture 20 desks, 10 tables, and 80 chairs to obtain the greatest possible profit of $7620.
31. The farmer should grow 100 acres of wheat, 300 acres of barley, and 100 acres of oats to obtain the greatest possible profit of $21,500.
33. The electric power plant should purchase 40,000 tons of low-sulfur coal and 50,000 tons of high-sulfur coal to obtain the most energy of 2,300,000 million BTUs.
35. a. The minimum of C is 12 when $x = 4$, $y = 0$, while the maximum of P is −12, also when $x = 4$, $y = 0$. Yes.

b. The simplex tableaux for the dual of this problem are

	x_1	x_2	s_1	s_2	
s_1	−1	2	1	0	3
s_2	−1	1	0	1	4
P	10	−8	0	0	0

and then

	x_1	x_2	s_1	s_2	
x_2	−1/2	1	1/2	0	3/2
s_2	−1/2	0	−1/2	1	5/2
P	6	0	4	0	12

c. The simplex tableaux for this nonstandard problem are

	x_1	x_2	s_1	s_2	
s_1	1	1	1	0	10
s_2	−2	−1	0	1	−8
P	3	4	0	0	0

and then

	x_1	x_2	s_1	s_2	
s_1	0	1/2	1	1/2	6
x_1	1	1/2	0	−1/2	4
P	0	5/2	0	3/2	−12

d. Yes; yes and yes.

37. The maximum is 16 when $x_1 = 2$, $x_2 = 2$. **39.** The maximum is 16 when $x_1 = 2$, $x_2 = 2$.
41. The maximum is 23 when $x_1 = 1$, $x_2 = 4$.

Exercises 4.6 page 339

1.

	x_1	x_2	x_3	s_1	s_2	a_1	
s_1	3	2	8	1	0	0	96
a_1	5	1	6	0	−1	1	30
P	−5M − 15	−M − 20	−6M − 18	0	M	0	−30M

3.

	x_1	x_2	x_3	s_1	s_2	a_1	a_2	
a_1	2	1	3	−1	0	1	0	30
a_2	1	1	2	0	−1	0	1	20
P	−3M − 6	−2M − 4	−5M − 6	M	M	0	0	−50M

5.

	x_1	x_2	x_3	s_1	s_2	s_3	a_1	
s_1	1	2	3	1	0	0	0	30
s_2	1	2	1	0	1	0	0	18
a_1	1	1	1	0	0	−1	1	8
P	−M − 20	−M − 30	−M − 10	0	0	M	0	−8M

7.

	x_1	x_2	x_3	x_4	s_1	s_2	a_1	a_2	a_3	
a_1	1	1	1	1	−1	0	1	0	0	30
a_2	2	3	2	1	0	−1	0	1	0	20
a_3	4	2	1	2	0	0	0	0	1	80
P	−7M − 3	−6M − 2	−4M − 5	−4M − 4	M	M	0	0	0	−130M

9. The maximum is 110 when $x_1 = 20$, $x_2 = 30$.
11. The maximum is 1200 when $x_1 = 0$, $x_2 = 60$, $x_3 = 0$.
13. There is no maximum. **15.** The maximum is 270 when $x_1 = 2$, $x_2 = 20$, $x_3 = 0$.
17. The maximum is 13 when $x_1 = 8$, $x_2 = 5$.
19. The maximum is 480 when $x_1 = 0$, $x_2 = 40$, $x_3 = 20$, $x_4 = 0$.
21. The maximum is 808 when $x_1 = 0$, $x_2 = 14$, $x_3 = 0$, $x_4 = 24$.
23. The maximum is 30 when $x_1 = 10$, $x_2 = 40$, $x_3 = 0$.

25. The retired couple should invest $15,000 in certificates of deposit and $5000 in Treasury bonds.
27. The manager should place 5 newspaper ads, 15 radio commercials, and 5 TV spots to reach 295,000 potential customers, the greatest possible number.
29. The furniture shop should manufacture 20 desks, 10 tables, and 80 chairs to obtain the greatest possible profit of $7620.
31. The farmer should grow 100 acres of wheat, 300 acres of barley, and 100 acres of oats to obtain the greatest possible profit of $21,500.
33. The electric power plant should purchase 40,000 tons of low-sulfur coal and 50,000 tons of high-sulfur coal to obtain the most energy of 2,300,000 million BTUs.
35. a. The minimum of C is 12 when $x = 4$, $y = 0$, while the maximum of P is -12, also when $x = 4$, $y = 0$. Yes.
37. The maximum is 16 when $x_1 = 2$, $x_2 = 2$.   **39.** The maximum is 23 when $x_1 = 1$, $x_2 = 4$.

Chapter 4 Review Exercises page 346

1. 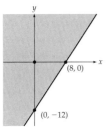   **2.**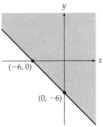

3. The vertices are $(0, 0)$, $(20, 0)$, and $(0, 20)$. The region is bounded.

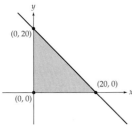

4. The vertices are $(10, 0)$ and $(0, 5)$. The region is unbounded.

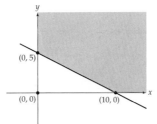

5. The vertices are $(0, 0)$, $(10, 0)$, $(5, 10)$, and $(0, 15)$. The region is bounded.

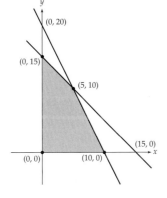

6. The vertices are $(8, 0)$, $(4, 4)$, and $(0, 12)$. The region is unbounded.

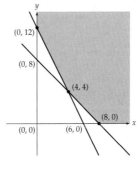

7. The vertices are (2, 0), (8, 0), (4, 6), and (2, 8). The region is bounded.

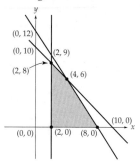

8. The vertices are (0, 0), (9, 0), (9, 1), (3, 7), and (0, 4). The region is bounded.

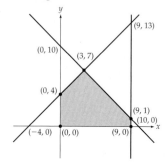

9. Let x be the number of Irish setters and let y be the number of Labrador retrievers.

$$\begin{cases} \frac{1}{2}x + \frac{3}{4}y \leq 6 & \text{(Time in hours)} \\ 8x + 8y \leq 80 & \text{(Dog treats)} \\ x \geq 0, \ y \geq 0 & \text{(Nonnegativity)} \end{cases}$$

The vertices are (0, 0), (10, 0), (6, 4), and (0, 8).

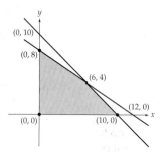

10. Let x be the number of pounds of SongBird brand bird seed and let y be the number of pounds of MeadowMix brand bird seed.

$$\begin{cases} 2x + 4y \geq 104 & \text{(Sunflower hearts)} \\ 3x + 2y \geq 84 & \text{(Crushed peanuts)} \\ x \geq 0, \ y \geq 0 & \text{(Nonnegativity)} \end{cases}$$

The vertices are (52, 0), (16, 18), and (0, 42).

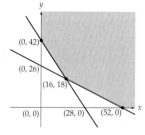

11. The maximum is 40 when $x = 0$, $y = 10$. **12.** The minimum is 10 when $x = 0$, $y = 5$.
13. There is no maximum because the region is unbounded to the right and upward, and P increases in these directions.
14. The minimum is 60 when $x = 30$, $y = 0$. **15.** The maximum is 240 when $x = 12$, $y = 0$.
16. The minimum is 2400 when $x = 48$, $y = 24$. **17.** The maximum is 350 when $x = 4$, $y = 9$.
18. The minimum is 140 when $x = 20$, $y = 0$.
19. The shop should make 24 wall clock cases and 15 mantle clock cases to obtain the greatest possible profit of $9000.
20. The cat's nutritional needs can be met by 30 ounces of canned food and 4 ounces of dry food at a least cost of $1.82.

21.

	x_1	x_2	x_3	s_1	s_2	
s_1	5	2	3	1	0	30
s_2	2	3	4	0	1	24
P	-4	-10	-9	0	0	0

22.

	x_1	x_2	s_1	s_2	s_3	
s_1	1	1	1	0	0	10
s_2	2	1	0	1	0	14
s_3	1	-1	0	0	1	4
P	-5	-7	0	0	0	0

23. The smallest negative entry in the bottom row is -8, so the pivot column is column 3. The ratios are $\frac{6}{1} = 6$, $\frac{8}{2} = 4$, and (omit), so the pivot row is row 2. The pivot element is the 2 in column 3 and row 2 of the simplex tableau. The first step of the pivot operation is the row operation $\frac{1}{2}R_2 \to R_2$ to replace the pivot element with a 1. Then:

	x_1	x_2	x_3	s_1	s_2	s_3	
$R_1 - R_2 \to s_1$	-1/2	1/2	0	1	-1/2	0	2
x_3	3/2	1/2	1	0	1/2	0	4
$R_3 + 3R_2 \to s_3$	11/2	5/2	0	0	3/2	1	18
$R_4 + 8R_2 \to P$	6	8	0	0	4	0	32

24. No pivot column because there are no negative entries in the bottom row. The solution has been found: The maximum is 150 when $x_1 = 0$ and $x_2 = 10$ (and $s_1 = 8$, $s_2 = 42$, and $s_3 = 0$).

25. The maximum is 96 when $x_1 = 0$, $x_2 = 12$, $x_3 = 0$.

26. The maximum is 90 when $x_1 = 0$, $x_2 = 9$, $x_3 = 0$.

27. The maximum is 37 when $x_1 = 2$, $x_2 = 3$, $x_3 = 0$, $x_4 = 0$.

28. The maximum is 61 when $x_1 = 8$, $x_2 = 0$, $x_3 = 7$.

29. The publisher should print no trade edition copies, no book club edition copies, and 250,000 paperback edition copies for the greatest possible profit of $1,000,000.

30. The plastics factory should produce 20 batches of toy racing cars, no toy jet airplanes, and 10 batches of toy speed boats for the greatest possible profit of $3500.

31. It is a standard problem because $\begin{pmatrix} 10 \\ 15 \\ 20 \end{pmatrix} \geq 0$. The dual problem is

$$\text{Maximize } P = 40x_1 + 30x_2$$

$$\text{Subject to } \begin{cases} 2x_1 + x_2 \leq 10 \\ -x_1 + x_2 \leq 15 \\ x_1 + 3x_2 \leq 20 \\ x_1 \geq 0 \text{ and } x_2 \geq 0 \end{cases}$$

The initial simplex tableau is:

	x_1	x_2	s_1	s_2	s_3	
s_1	2	1	1	0	0	10
s_2	-1	1	0	1	0	15
s_3	1	3	0	0	1	20
P	-40	-30	0	0	0	0

32. It is a standard problem because $\begin{pmatrix} 30 \\ 20 \end{pmatrix} \geq 0$. The dual problem is

Maximize $P = 210x_1 + 252x_2 - 380x_3$

Subject to $\begin{cases} 5x_1 + 3x_2 - 5x_3 \leq 30 \\ 2x_1 + 4x_2 - 4x_3 \leq 20 \\ x_1 \geq 0, \ x_2 \geq 0, \ \text{and} \ x_3 \geq 0 \end{cases}$

The initial simplex tableau is:

	x_1	x_2	x_3	s_1	s_2	
s_1	5	3	−5	1	0	30
s_2	2	4	−4	0	1	20
P	−210	−252	380	0	0	0

33. It is a standard problem because $\begin{pmatrix} 42 \\ 36 \end{pmatrix} \geq 0$. The dual problem is

Maximize $P = 84x_1 + 18x_2$

Subject to $\begin{cases} 7x_1 + x_2 \leq 42 \\ 4x_1 + x_2 \leq 36 \\ x_1 \geq 0 \ \text{and} \ x_2 \geq 0 \end{cases}$

The initial simplex tableau is:

	x_1	x_2	s_1	s_2	
s_1	7	1	1	0	42
s_2	4	1	0	1	36
P	−84	−18	0	0	0

34. It is a standard problem because $\begin{pmatrix} 130 \\ 40 \\ 98 \end{pmatrix} \geq 0$. The dual problem is

Maximize $P = 51x_1 + 60x_2 + 57x_3$

Subject to $\begin{cases} 3x_1 + 4x_2 + 3x_3 \leq 130 \\ x_1 + x_2 + x_3 \leq 40 \\ 2x_1 + 2x_2 + 3x_3 \leq 98 \\ x_1 \geq 0, \ x_2 \geq 0, \ \text{and} \ x_3 \geq 0 \end{cases}$

The initial simplex tableau is:

	x_1	x_2	x_3	s_1	s_2	s_3	
s_1	3	4	3	1	0	0	130
s_2	1	1	1	0	1	0	40
s_3	2	2	3	0	0	1	98
P	−51	−60	−57	0	0	0	0

35. The minimum is 192 when $y_1 = 8$, $y_2 = 5$, $y_3 = 0$.
36. The minimum is 510 when $y_1 = 11$, $y_2 = 5$. **37.** There is no solution.
38. The minimum is 60 when $y_1 = 0$, $y_2 = 12$, $y_3 = 0$.
39. The student should buy no single pens, 3 packages of ink cartridges, and 3 "writer's combo" packages to spend the least possible amount of $6.24.

40. The pasta company should purchase 18 small-capacity machines and 8 large-capacity machines to expand its linguini production at the least possible cost of $138,000.

41.

	x_1	x_2	s_1	s_2	s_3	
s_1	4	1	1	0	0	40
s_2	2	3	0	1	0	60
s_3	−1	−1	0	0	1	−10
P	−80	−30	0	0	0	0

The dual pivot element is the −1 in row 3 and column 1.

42.

	x_1	x_2	x_3	x_4	s_1	s_2	
s_1	2	3	−3	1	1	0	30
s_2	−7	5	−1	−5	0	1	−35
P	−4	−8	−6	−10	0	0	0

The dual pivot element is the −1 in row 2 and column 3.

43. The maximum is 40 when $x_1 = 0$, $x_2 = 10$.
44. The maximum is 280 when $x_1 = 35$, $x_2 = 0$, $x_3 = 0$.
45. The maximum is 48 when $x_1 = 8$, $x_2 = 0$.
46. The maximum is 300 when $x_1 = 25$, $x_2 = 0$, $x_3 = 0$.
47. The maximum is 59 when $x_1 = 4$, $x_2 = 5$.
48. The maximum is 300 when $x_1 = 0$, $x_2 = 0$, $x_3 = 30$.
49. The sawmill should produce 6 thousand board-feet of rough-cut lumber and 4 thousand board-feet of finished-grade boards each day to obtain the greatest possible profit of $1120.
50. The money manager should invest $100,000 in U.S. bonds and $50,000 in Canadian stocks (and nothing in U.S. stocks or in Canadian bonds) to obtain the greatest possible return of $10,500.

51.

	x_1	x_2	s_1	s_2	s_3	a_1	
s_1	4	1	1	0	0	0	40
s_2	2	3	0	1	0	0	60
a_1	1	1	0	0	−1	1	10
P	−M − 80	−M − 30	0	0	M	0	−10M

52.

	x_1	x_2	x_3	x_4	s_1	s_2	a_1	
s_1	2	3	−3	1	1	0	0	30
a_1	7	−5	1	5	0	−1	1	35
P	−7M − 4	5M − 8	−M − 6	−5M − 10	0	M	0	−35M

53. The maximum is 40 when $x_1 = 0$, $x_2 = 10$.
54. The maximum is 280 when $x_1 = 35$, $x_2 = 0$, $x_3 = 0$.
55. The maximum is 48 when $x_1 = 8$, $x_2 = 0$.
56. The maximum is 300 when $x_1 = 25$, $x_2 = 0$, $x_3 = 0$.
57. The maximum is 59 when $x_1 = 4$, $x_2 = 5$.
58. The maximum is 300 when $x_1 = 0$, $x_2 = 0$, $x_3 = 30$.
59. The sawmill should produce 6 thousand board-feet of rough-cut lumber and 4 thousand board-feet of finished-grade boards each day to obtain the greatest possible profit of $1120.

60. The money manager should invest $100,000 in U.S. bonds and $50,000 in Canadian stocks (and nothing in U.S. stocks and in Canadian bonds) to obtain the greatest possible return of $10,500.

Exercises 5.1 page 362

1. 27   **3.** 60   **5.** 44   **7.** 31   **9.** 17
11. 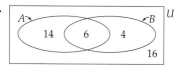   **13.** 24   **15.** 20   **17.** 30

19. 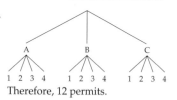   **21.**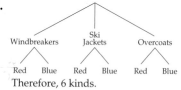

Therefore, 12 permits.   Therefore, 6 kinds.

23. 17   **25.** 200   **27.** $25 \cdot 9 \cdot 9 = 2025$ **29.** $15^8 = 2,562,890,625$ **31.** $10^4 = 10,000$
33. $3 \cdot 4 \cdot 4 \cdot 5 = 240$
35. According to the Addition Principle, the number of cars that were either convertibles or had out-of-state plates is $12 + 15 - 5 = 22$, which is impossible since he searched only 20 cars.
37. $2^6 = 64$ **39.** $2^{10} - 1 = 1024 - 1 = 1023$
41. The members of $(A \cup B)^c$ are $8, 9,$ and $10,$ and these are also the members of $A^c \cap B^c$.
43.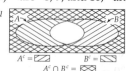

Note that the ▨ shadings agree.

Exercises 5.2 page 374

1. a. 2   **b.** 720   **3.** 720   **5.** 39,800   **7.** 120   **9.** 8
11. a. 17,160   **b.** 154,440   **c.** 1,235,520   **13. a.** 10   **b.** 30,240   **c.** 3,628,800
15. 15   **17.** 35   **19. a.** 330   **b.** 462   **c.** 462   **d.** 330
21. a. 12   **b.** 66   **c.** 924   **d.** 12   **e.** 1
23. $8! = 40,320$ **25.** $_{12}P_4 = 11,880$ **27.** $8 \cdot {}_9P_6 = 483,840$
29. $36 \cdot 35 \cdot 34 \cdot 33 = 1,413,720$; $34 \cdot 33 \cdot 32 \cdot 31 = 1,113,024$ **31.** $_6P_2 = 30$ **33.** $_{12}C_5 = 792$
35. $_{13}C_5 = 1287$; $4 \cdot 1287 = 5148$ **37.** $({}_{10}C_4)({}_{12}C_4) = 103,950$ **39.** $_{12}C_{10} = {}_{12}C_2 = 66$
41. $_{100}C_5 = 75,287,520$; $\frac{100 \cdot 98 \cdot 96 \cdot 94 \cdot 92}{5!} = 67,800,320$ **43. a.** 362,880   **b.** Over 31 years
45. Choosing one factor from n factors can be done in $_nC_1$ ways, choosing two factors from n can be done in $_nC_2$ ways, and so on. The left-hand sides of the equations are equal, and since $_nC_0 = {}_nC_n = 1$, so are the right-hand sides.
47. Using the fifth row of Pascal's triangle: $(x + y)^4 = x^4 + 4x^3y + 6x^2y^2 + 4xy^3 + y^4$.
49. a. The left-hand side of the binomial theorem becomes $(1 + 1)^n = 2^n$ and the right-hand side, after omitting the 1s, becomes $_nC_0 + {}_nC_1 + {}_nC_2 + \cdots + {}_nC_n$.

b. The left-hand side is the total number of committees that can be made from n people (including an "empty" committee and the committee of all n), and the right-hand side gives the number of committees of size 0, size 1, size 2, and so on up to size n, which must add up to the *total* number of committees.

Exercises 5.3 page 386

1. $\{(A, B), (A, C), (A, D), (B, C), (B, D), (C, D)\}$; $\{(A, B), (A, D), (B, C), (C, D)\}$
3. $\{(S_1, J_1), (S_1, J_2), (S_2, J_1), (S_2, J_2), (S_3, J_1), (S_3, J_2), (S_4, J_1), (S_4, J_2)\}$
5. $\{(R, R), (R, G), (R, B), (G, R), (G, G), (G, B), (B, R), (B, G), (B, B)\}$;
$\{(R, G), (R, B), (G, R), (G, B), (B, R), (B, G)\}$
7. a. $\{3\}$ **b.** $\{2, 3, 4, 6\}$ **9. a.** $\{(6, 2), (6, 4), (6, 6), (2, 6), (4, 6)\}$ **b.** $\varnothing$ **11. a.** 0.80   **b.** 1
13. a. 0.20   **b.** 0.15   **15.** $1/(_{12}C_3) = \frac{1}{220}$, $1/(_{12}P_3) = \frac{1}{1320}$ **17.** $P(R) = \frac{6}{10} = \frac{3}{5}$, $P(B) = \frac{4}{10} = \frac{2}{5}$
19. $P(6) = \frac{1}{2}$, $P(8) = \frac{1}{4}$, $P(12) = \frac{1}{4}$ **21. a.** $\frac{1}{20}$ **b.** $\frac{19}{20}$ **23.** 32%
25. $63\% + 48\% - 15\% + 10\% = 106\%$ when it should add to 100%.
27. a. $\frac{_4C_3}{_{12}C_3} = \frac{1}{55}$ **b.** $\frac{_4C_3}{_{12}C_3} + \frac{_8C_3}{_{12}C_3} = \frac{3}{11}$ **29.** $1 - \frac{_9C_3}{_{10}C_3} = \frac{3}{10}$ **31.** $\frac{_{48}C_1}{_{52}C_5} \approx 0.000018$
33. $\frac{7 \cdot 6 \cdot 5 \cdot 4 \cdot 3}{7^5} \approx 0.15$ **35.** $\frac{_{90}C_{10}}{_{100}C_{10}} \approx 0.33$ **37.** $1 - \frac{_{98}C_8}{_{100}C_8} \approx 0.15$ **39.** $\frac{_4C_1}{_6C_1} = \frac{2}{3}$
41. Yes   **43.** $P(H) = \frac{1}{2}$ and $P(T) = \frac{1}{2}$ give $\frac{1}{2} : \frac{1}{2}$ or $1 : 1$.
45. $P(E^c) = \frac{m-n}{m}$ so the odds are $\frac{n}{m} : \frac{m-n}{m}$, or $n : (m-n)$.
47. Since the probabilities must add to 1, we divide the odds by $n + m$, so $P(E) = \frac{n}{n+m}$.
49. $\frac{7}{11}$ **51.** $\frac{7}{9}$

Exercises 5.4 page 399

1. a. $\frac{0.2}{0.4} = \frac{1}{2}$ **b.** $\frac{0.2}{0.6} = \frac{1}{3}$ **3.** Both parts require first finding $P(A \cap B) = 0.3$.   **a.** $\frac{3}{5}$ **b.** $\frac{3}{4}$
5. $\frac{2}{3}$ **7.** $\frac{1}{2}$ **9.** $\frac{1}{3}$ **11.** $\frac{0.38}{0.68} \approx 0.56$ **13.** $0.95 \cdot 0.20 + 0.70 \cdot 0.80 = 0.75$ **15.** 0.16
17. Independent $[P(A) \cdot P(B) = \frac{1}{2} \cdot \frac{1}{2}$ and $P(A \cap B) = \frac{1}{4}]$
19. Dependent $[P(A) \cdot P(B) = \frac{1}{2} \cdot \frac{3}{36} = \frac{1}{24}$ but $P(A \cap B) = \frac{2}{36} = \frac{1}{18}]$
21. $\left(\frac{1}{6}\right)^3 = \frac{1}{216}$ **23.** $\frac{0.9}{1 - 0.1 \cdot 0.3} \approx 0.928$, so about 93%   **25. a.** $\left(\frac{1}{6}\right)^3 = \frac{1}{216}$ **b.** $\frac{6}{216} = \frac{1}{36}$ **c.** $\frac{35}{36}$
27. $\frac{1}{3}$ **29.** 0.636   **31. a.** pq **b.** $1 - (1 - p)(1 - q)$ **c.** 0.25 and 0.75, 0.81 and 0.99, parallel
33. b. 23   **35.** $P(A \cap B^c) = P(A) - P(A \cap B) = P(A) - P(A) \cdot P(B) = P(A)[1 - P(B)] = P(A) \cdot P(B^c)$
37. $P(A \cap B) = \frac{1}{4} = \frac{1}{2} \cdot \frac{1}{2} = P(A) \cdot P(B)$
$P(A \cap C) = \frac{1}{4} = \frac{1}{2} \cdot \frac{1}{2} = P(A) \cdot P(C)$
$P(B \cap C) = \frac{1}{4} = \frac{1}{2} \cdot \frac{1}{2} = P(B) \cdot P(C)$
$P(A \cap B \cap C) = \frac{1}{4}$ but $P(A) \cdot P(B) \cdot P(C) = \frac{1}{2} \cdot \frac{1}{2} \cdot \frac{1}{2} = \frac{1}{8}$

Exercises 5.5 page 406

1. $\frac{0.55 \cdot 0.60}{0.55 \cdot 0.60 + 0.65 \cdot 0.40} \approx 0.559$, so about 56%   **3.** $\frac{0.95 \cdot 0.001}{0.95 \cdot 0.001 + 0.05 \cdot 0.999} \approx 0.019$, so about 2%
5. $\frac{0.02 \cdot 0.3}{0.03 \cdot 0.5 + 0.02 \cdot 0.3 + 0.01 \cdot 0.2} \approx 0.261$, so about 26%
7. $\dfrac{0.90 \cdot \frac{30}{300,000,000}}{0.90 \cdot \frac{30}{300,000,000} + 0.0005 \cdot \frac{299,999,970}{300,000,000}} \approx 0.00018$, so about 0.02%

9. $\frac{0.05 \cdot 0.99}{0.05 \cdot 0.99 + 0.95 \cdot 0.01} \approx 0.839$, so about 84% **11.** $\frac{0.70 \cdot 0.2}{0.90 \cdot 0.5 + 0.85 \cdot 0.3 + 0.70 \cdot 0.2} \approx 0.166$, so about 17%
13. $\frac{0.75 \cdot 0.02}{0.75 \cdot 0.02 + 0.3 \cdot 0.98} \approx 0.049$, so about 5% **15.** $\frac{0.006}{0.009 + 0.01 + 0.008 + 0.0075 + 0.006} \approx 0.148$, so about 15%
17. a. $P(Y) = p \cdot \frac{1}{2} + p \cdot \frac{1}{4} + (1-p) \cdot \frac{1}{4}$ **b.** $p = 2P(Y) - \frac{1}{2}$ **c.** 74%

Exercises 5.6 page 418

1.
x	0	1	2	3
$P(X = x)$	$\frac{1}{8}$	$\frac{3}{8}$	$\frac{3}{8}$	$\frac{1}{8}$

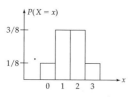

3.
x	-1	11
$P(X = x)$	$\frac{3}{4}$	$\frac{1}{4}$

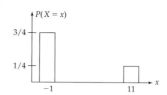

5.
x	-12	2	3
$P(X = x)$	$\frac{1}{6}$	$\frac{1}{2}$	$\frac{1}{3}$

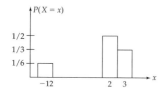

7.
x	1	2	3	4	5	6
$P(X = x)$	$\frac{1}{36}$	$\frac{1}{12}$	$\frac{5}{36}$	$\frac{7}{36}$	$\frac{1}{4}$	$\frac{11}{36}$

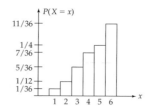

9. $E(X) = 0 \cdot \frac{1}{8} + 1 \cdot \frac{3}{8} + 2 \cdot \frac{3}{8} + 3 \cdot \frac{1}{8} = \frac{3}{2}$ **11.** $E(X) = 2$
13. $E(X) = 0$ **15.** $E(X) = \frac{161}{36} \approx 4.5$ **17.** $E(X) = 3\frac{1}{4}$ **19.** No ($72,000 versus $74,000)
21. $E(X) = 20 \cdot \frac{1}{2} = 10$ **23.** $E(X) = 16$

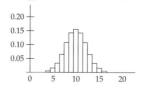
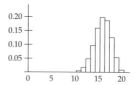

25. 0.78125, or about 78% **27.** $_8C_4(\frac{1}{2})^8 = \frac{70}{256} \approx 0.273$, or about 27%
29. Most likely: 2 heads, with probability 0.329 (approx.)
Least likely: 6 heads, with probability 0.0014 (approx.)
31. Most likely: 1 six, with probability 0.372 (approx.), $E(X) = \frac{4}{3}$ **33.** $14,000 **35.** $E(X) = 0.2$
37. $P(7 \le X \le 10) \approx 0.172$ **39.** 0.972 (compared to 0.90 for transmitting a single digit)
41. $(_{48}C_{13}/_{52}C_{13})^3 \approx 0.028$, or about 3%
43. a. $_3C_2(\frac{3}{5})^2(\frac{2}{5})^1 + _3C_3(\frac{3}{5})^3(\frac{2}{5})^0 \approx 0.648$, or about 65% **b.** 0.682, or about 68% **c.** 0.710, or about 71%
45. 5 **47.** 0.999996, 0.9998
49. a. $\frac{5}{16}$ **b.** $(\frac{11}{16})^4 \approx 0.223$ or about 22% **c.** $(\frac{11}{16})^8 \approx 0.05$ or about 5%
53. By the definition of expected value.
55. The expected value of a sum is the sum of the expected values (from Exercise 54) and the formula for the expected value of a binomial random variable with parameters 1 and p (from Exercise 53).

Chapter 5 Review Exercises page 424

1. a. 17 **b.** 32 **c.** 21 **d.** 15 **2.** 110 **3. a.** $26^3 = 17{,}576$ **b.** $26 \cdot 25 \cdot 24 = 15{,}600$
4. a. 120 **b.** 20 **5.** $_{20}C_4 = 4845$, $_{20}P_4 = 116{,}280$ **6.** $_{13}C_5 = 1287$ **7.** $_{20}C_5 = 15{,}504$
8. $\{(C_1, S_1, H_1), (C_1, S_1, H_2), (C_1, S_2, H_1), (C_1, S_2, H_2), (C_2, S_1, H_1), (C_2, S_1, H_2), (C_2, S_2, H_1), (C_2, S_2, H_2)\}$
9. a. $\{(B, B), (B, Y), (B, R), (Y, B), (Y, Y), (Y, R), (R, B), (R, Y), (R, R)\}$
  b. $\{(B, Y), (B, R), (Y, B), (Y, R), (R, B), (R, Y)\}$
10. a. $\{(H, H), (H, T), (T, H)\}$ **b.** $\{(H, T), (T, H), (T, T)\}$ **c.** $\{(H, T), (T, H)\}$
11. a. $\frac{1}{4}$ **b.** $\frac{3}{4}$ **c.** $\frac{1}{2}$ **12.** $1/{_{15}C_2} = \frac{1}{105}$, $1/{_{15}P_2} = \frac{1}{210}$ **13.** $P(1) = P(3) = P(5) = P(7) = \frac{1}{4}$
14. $P(1) = \frac{1}{2}$, $P(2) = P(4) = P(6) = \frac{1}{6}$ **15.** $P(R) = \frac{1}{6}$, $P(G) = \frac{1}{3}$, $P(B) = \frac{1}{2}$ **16.** 0.45
17. a. $\frac{1}{3}$ **b.** $\frac{1}{2}$ **18.** $_5C_4/{_{40}C_5} \approx 0.00000760$
19. a. $_{12}C_5/{_{52}C_5} \approx 0.000305$, or about 0.03% **b.** $_{40}C_5/{_{52}C_5} \approx 0.253$, or about 25%
20. a. $_{39}C_{13}/{_{52}C_{13}} \approx 0.0128$, or about 1% **b.** $_{40}C_{13}/{_{52}C_{13}} \approx 0.0189$, or about 2%
21. $_{28}C_5/{_{30}C_5} \approx 0.690$, or about 69%
22. $_{48}C_2/{_{50}C_4} \approx 0.0049$, or about 0.5%; $_{48}C_4/{_{50}C_4} \approx 0.8449$, or about 84.5%
23. $_{28}C_1/{_{30}C_3} = \frac{1}{145} \approx 0.00690$, or about 0.7%
24. $_6C_2/{_{10}C_2} = \frac{1}{3}$ **25. a.** $P(A \text{ given } B) = \frac{3}{4}$ **b.** $P(B \text{ given } A) = \frac{3}{5}$ **26.** $\frac{1}{3}$ **27.** $\frac{1}{7}$
28. 0.00235, or about 0.2% **29.** $\frac{3}{16}$ **30.** 0.32 **31.** $0.8 \cdot 0.6 + 0.9 \cdot 0.4 = 0.84$ **32.** 0.983 (approx.)
33. a. No $\left(\frac{3}{4} \cdot \frac{3}{4} \neq \frac{1}{2}\right)$ **b.** Yes $\left(\frac{1}{2} \cdot \frac{1}{2} = \frac{1}{4}\right)$ **34.** 0.9999 **35. a.** $\frac{1}{32}$ **b.** $\frac{1}{16}$ **c.** $\frac{1}{16}$
36. $\frac{2}{3}$ **37.** $\frac{0.04 \cdot 0.25}{0.04 \cdot 0.25 + 0.03 \cdot 0.35 + 0.02 \cdot 0.40} \approx 0.351$, or about 35% **38.** 0.247, or about 25%

39.

Value of X	Outcomes
X = 5	(H, H, H, H, H)
X = 1	(H, T, T, T, T), (T, H, T, T, T), (T, T, H, T, T), (T, T, T, H, T), (T, T, T, T, H)
X = 0	(T, T, T, T, T)

40.

x	34	−2
P(X = x)	$\frac{1}{4}$	$\frac{3}{4}$

41. $E(X) = 7$ **42.**

x	0	1	2
P(X = x)	$\frac{5}{18}$	$\frac{5}{9}$	$\frac{1}{6}$

43. $E(X) = \frac{8}{9}$ **44.** $E(X) = \$3.10$

45.

x	0	1	2	3	4
P(X = x)	$\frac{1}{625}$	$\frac{16}{625}$	$\frac{96}{625}$	$\frac{256}{625}$	$\frac{256}{625}$

$E(X) = \frac{16}{5} = 3\frac{1}{5}$

46. 0.711 (approx.), or about 71% **47.** $360 **48.** 2 **49.** $E(X) = 12 \cdot 0.01 = 0.12$
50. 0.859, or about 86% **51.** 0.633, or about 63%

Exercises 6.1 page 438

1. Nominal **3.** Ordinal **5.** Ordinal **7.** Interval **9.** Ratio

11. 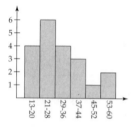 1 = Never married
2 = Married
3 = Widowed
4 = Divorced

13. 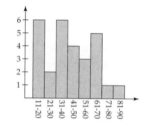 1 = Business executives
2 = Developers
3 = Lawyers
4 = Doctors
5 = CPAs
6 = Retirees

15. Using 6 classes of width 8 beginning at 12.5:

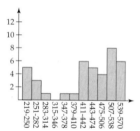

The lighter weight boxes are in greatest demand.

17. Using 8 classes of width 10 beginning at 10.5:

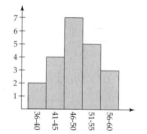

The highest number of hits were in classes 11–20 and 31–40.

19. Using 11 classes of width 32 beginning at 218.5:

More ticket prices are grouped at the upper end.

21. Using 5 classes of width 5 beginning at 35.5:

More are grouped in the middle.

23. Using 7 classes of width 3 beginning at 8.5:

They are grouped in the middle.

25. Using 10 classes of width 6 beginning at 15.5:

Not evenly distributed: teenagers and the elderly have more accidents.

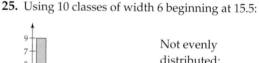

Exercises 6.2 page 448

1. Mode = 9, median = 10, $\bar{x} = 11$
3. Modes = 7 and 14 (bimodal), median = 9, $\bar{x} = 10$
5. Mode = 19, median = 15, $\bar{x} = 15.2$
7. Mode = 14, median = 13, $\bar{x} = 12$
9. Mode = 19, median = 18, $\bar{x} = 16$

11. Modes = 12 and 20 (bimodal), median = 15, $\bar{x}$ = 15.75, each in minutes
13. Mode = 10, median = 10, $\bar{x}$ = 11.1, each in hours   15. Mode = 9, median = 9, $\bar{x}$ = 10, each in weeks
17. The mode is 450, the median is 450, and the mean is 468. The unusually high 750 skews the mean higher.
19. The modes are 0.5, 0.6 and 0.8, the median is 1.1, and the mean is 3.8 (rounded, all in millions of dollars). The two very high salaries skew the mean.
21. There is no mode, the median is 41.05, and the mean is 55.35 (all in thousands of dollars). The relatively few extremely high salaries of the wealthiest people skew the mean higher.
25. $\bar{x} \approx \$44.95$

Exercises 6.3 page 459

1. 62   3. 20   5. 56   7. Minimum = 3
First quartile = 10
Median = 14
Third quartile = 21
Maximum = 23

9. Minimum = 8
First quartile = 10
Median = 13
Third quartile = 17
Maximum = 26

11. Minimum = 8
First quartile = 12
Median = 17
Third quartile = 24
Maximum = 25

13. Minimum = 6
First quartile = 8
Median = 16
Third quartile = 18
Maximum = 24

15. Minimum = 5
First quartile = 8.5
Median = 15.5
Third quartile = 17.5
Maximum = 20

17. $s \approx 5.79$ (from $\bar{x} = 11$)   19. $s \approx 8.63$ (from $\bar{x} = 13$)

21. Range = 20
Minimum = 5
First quartile = 11
Median = 13
Third quartile = 17
Maximum = 25
$s \approx 4.64$
} All in dollars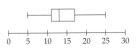

23. Range = 11
Minimum = 2
First quartile = 6
Median = 8
Third quartile = 10
Maximum = 13
$s \approx 2.55$
} All in numbers of people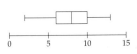

25. Range = 33
Minimum = 2
First quartile = 15.5
Median = 20.5
Third quartile = 24.5
Maximum = 35
$s \approx 8.19$
} All in weeks

27.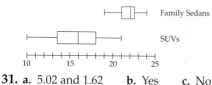

Men
Women

Gender inequality increases with income

Thousands of dollars

29. Family Sedans $\begin{cases} \text{Minimum} = 19 \\ \text{First quartile} = 21 \\ \text{Median} = 22 \\ \text{Second quartile} = 22.5 \\ \text{Maximum} = 24 \end{cases}$ SUVs $\begin{cases} \text{Minimum} = 10 \\ \text{First quartile} = 13.5 \\ \text{Median} = 16 \\ \text{Second quartile} = 18 \\ \text{Maximum} = 21 \end{cases}$

Family Sedans
SUVs

The worst family sedan is better than three-quarters of the SUVs.

31. a. 5.02 and 1.62   **b.** Yes   **c.** No

Exercises 6.4 page 472

(*Note:* Answers may differ depending on rounding and on whether graphing calculators or tables were used.)
 1. 0.6827   **3.** 0.9973   **5.** 0.3413   **7.** 0.0674   **9.** 0.3361   **11.** $z = 1$ **13.** $z = 0$
15. $z = -1$ **17.** $z = -5$ **19.** $z = 2$
21. 0.0298 [from $\mu = 14$, $\sigma \approx 2.05$, after checking that $np > 5$ and $n(1 - p) > 5$]
23. 0.1883 [from $\mu = 14$, $\sigma \approx 2.05$, after checking that $np > 5$ and $n(1 - p) > 5$]
25. 0.0036 [from $\mu = 14$, $\sigma \approx 2.05$, after checking that $np > 5$ and $n(1 - p) > 5$]
27. 0.4029 [from $\mu = 14$, $\sigma \approx 2.05$, after checking that $np > 5$ and $n(1 - p) > 5$]
29. 0.9992 [from $\mu = 14$, $\sigma \approx 2.05$, after checking that $np > 5$ and $n(1 - p) > 5$]
31. About 0.89, or 89%
33. About 0.25, or 25% using $P(x \geq 101)$. [Alt. answers: 28% for $P(x > 100.5)$ or 31% for $P(x > 100)$]
35. About 0.31, or 31%
37. About 0.24, or 24% [from $\mu = 564$, $\sigma \approx 15$, after checking that $np > 5$ and $n(1 - p) > 5$]
39. About 0.26, or 26% [from $\mu = 80$, $\sigma \approx 6.93$, after checking that $np > 5$ and $n(1 - p) > 5$]
41. $\frac{8}{9} \approx 89\%$
43. The probability that the values of the random variable lie between 85 and 115 is at least $\frac{8}{9} \approx 89\%$.
45. 99.7%. This is a stronger result than the probability of $\frac{8}{9} \approx 89\%$ found in Exercise 43.

Chapter 6 Review Exercises page 476

1. Nominal   **2.** Ordinal   **3.** Interval   **4.** Ratio

5.

1 = Army
2 = Navy
3 = Marines
4 = Air Force

6.

1 = Handguns
2 = Other firearms
3 = Knives
4 = Blunt instruments
5 = Hands

7. Using 8 classes of width 4 beginning at 33.5:

8. Using 10 classes of width 79 beginning at 697.5:

9. Mode = 11, median = 10, $\bar{x} = 9$ **10.** Modes = 6 and 13 (bimodal), median = 11, $\bar{x} = 10$
11. Mode = 13, median = 12.5, $\bar{x} = 11$ **12.** Mode = 8, median = 8, $\bar{x} = 8.50$ (all in dollars)
13. Mode = 4, median = 20, $\bar{x} = 17.4$ (all in numbers of calls)   **14.** 18
15. Minimum = 4   **16.** $s = 4$
First quartile = 6
Median = 10
Third quartile = 14.5
Maximum = 20

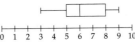

17. Range = 6
Minimum = 3
First quartile = 5
Median = 6   All in minutes
Third quartile = 8
Maximum = 9
$s \approx 1.96$

18. Range = 40
Minimum = 114
First quartile = 124   All in
Median = 131   thousands
Third quartile = 134   of dollars
Maximum = 154
$s \approx 9.74$

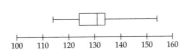

19. 0.8186 ⎫ Answers may
20. 0.3023 ⎭ differ slightly
21. $z = 2$ **22.** $z = -2$
23. 0.015 [from $\mu = 8.75$, $\sigma \approx 2.38$, after checking that $np > 5$ and $n(1 - p) > 5$]
24. 0.68 [from $\mu = 8.75$, $\sigma \approx 2.38$, after checking that $np > 5$ and $n(1 - p) > 5$]
25. About 0.66, or 66%   **26.** About 0.16, or 16%
27. About 0.63, or 63% [from $\mu = 6.9$, $\sigma \approx 2.61$, after checking that $np > 5$ and $n(1 - p) > 5$]
28. About 0.38, or 38% [from $\mu = 48$, $\sigma \approx 5.71$, after checking that $np > 5$ and $n(1 - p) > 5$]

Exercises 7.1 page 490

1. $\begin{array}{c} \\ A \\ B \\ C \end{array} \begin{pmatrix} A & B & C \\ 0 & 1 & 0 \\ 0 & 0 & 1 \\ 1 & 0 & 0 \end{pmatrix}$, oscillating   **3.** $\begin{array}{c} \\ D \\ E \end{array} \begin{pmatrix} D & E \\ \frac{2}{5} & \frac{3}{5} \\ 1 & 0 \end{pmatrix}$, mixing   **5.** $\begin{array}{c} \\ F \\ G \end{array} \begin{pmatrix} F & G \\ \frac{2}{5} & \frac{3}{5} \\ 0 & 1 \end{pmatrix}$, absorbing

7. Oscillating   **9.** Mixing

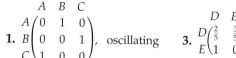

11. Absorbing

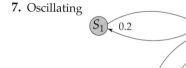

13. $D_1 = \begin{pmatrix} \frac{7}{20} & \frac{13}{20} \end{pmatrix}$, $D_2 = \begin{pmatrix} \frac{33}{80} & \frac{47}{80} \end{pmatrix}$

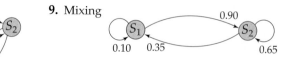

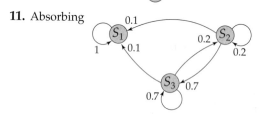

15. $D_1 = (0.35 \quad 0.25 \quad 0.40)$, $D_2 = (0.31 \quad 0.17 \quad 0.52)$ **17.** 5   **19.** $2\frac{1}{2}$
21. The expected number of minutes that a green light for traffic on Main Street will be green before changing to red is 4 minutes.
23. Next week, 71% will be eating cereal; the week after, 71.8% will be eating cereal.
25. Tomorrow, 25% will be fishing; the day after, 52.5% will be golfing.
27. In two years, approximately 75.4% of the portfolio will be "secure." Because "bankrupt" is an absorbing state for all the other states, ultimately every company in the portfolio will become bankrupt.
29. The expected number of weekends that the laundry room will be manageable, given that it is now manageable, is 2 weekends.
31. $\underbrace{p \cdots \cdots p}_{\text{Stays in } S_i \text{ and then}} \cdot \underbrace{(1 - p)}_{\text{leaves } S_i} = p^{n-1}(1 - p)$
for $n - 1$ transitions

33. $E = 1 \cdot (1 - p) + 2 \cdot p(1 - p) + 3 \cdot p^2(1 - p) + \cdots = \underbrace{(1 - p)}_{+p} + \underbrace{(2p - 2p^2)}_{+p^2} + \underbrace{(3p^2 - 3p^3)}_{+p^3} + \cdots$

$= 1 + p + p^2 + p^3 + \cdots$

Exercises 7.2 page 502

1. Regular **3.** Regular **5.** Not regular **7.** Regular

9. $D = \left(\frac{1}{2} \; \frac{1}{2}\right)$, $T^{20} = \begin{pmatrix} 0.5 & 0.5 \\ 0.5 & 0.5 \end{pmatrix}$ **11.** $D = \left(\frac{2}{5} \; \frac{3}{5}\right)$, $T^{20} = \begin{pmatrix} 0.4 & 0.6 \\ 0.4 & 0.6 \end{pmatrix}$

13. $D = \left(\frac{1}{5} \; \frac{7}{10} \; \frac{1}{10}\right)$, $T^{20} = \begin{pmatrix} 0.2 & 0.7 & 0.1 \\ 0.2 & 0.7 & 0.1 \\ 0.2 & 0.7 & 0.1 \end{pmatrix}$ **15.** $D = \left(\frac{1}{10} \; \frac{17}{20} \; \frac{1}{20}\right)$, $T^{50} = \begin{pmatrix} 0.10 & 0.85 & 0.05 \\ 0.10 & 0.85 & 0.05 \\ 0.10 & 0.85 & 0.05 \end{pmatrix}$

17.  $k = 2$, with probability 0.5

19. 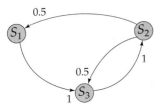 $k = 3$, with probability 0.5

21. The voter records lead to a steady-state distribution with 60% of the population voting, which is consistent with the national average (the surveys lead to a steady-state distribution with 80% of the population voting, which is not consistent with the national average).

23. One million of three million commuters will be driving alone.

25. The long-term market share for DentiMint toothpaste will be 24%.

27. 45% + 30% = 75% of the broker's auto policies are rated satisfactory or better.

29. Yes; a long-term moving probability of $\frac{1}{3}$ is consistent with the national average of about 34.3%.

Exercises 7.3 page 514

1. S_3 is absorbing; S_1 and S_2 are nonabsorbing. 2 steps, with probability $\frac{1}{2} \cdot \frac{1}{2} = \frac{1}{4}$

3. S_2 is absorbing; S_1, S_3, and S_4 are nonabsorbing. 3 steps, with probability $\frac{1}{2} \cdot \frac{1}{2} \cdot \frac{1}{2} = \frac{1}{8}$

5. a. 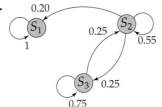 S_1 is absorbing; S_2 and S_3 are nonabsorbing. $k = 2$

b. $T^2 = \begin{pmatrix} 1 & 0 & 0 \\ 0.31 & 0.365 & 0.325 \\ 0.05 & 0.325 & 0.625 \end{pmatrix}$

7. a.

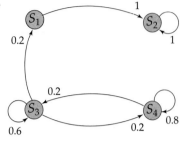

S_2 is absorbing; S_1, S_3, and S_4 are nonabsorbing. $k = 3$

b. $T^3 = \begin{pmatrix} 0 & 1 & 0 & 0 \\ 0 & 1 & 0 & 0 \\ 0.08 & 0.32 & 0.296 & 0.304 \\ 0.056 & 0.04 & 0.304 & 0.6 \end{pmatrix}$

9. $\begin{array}{c} \\ S_2 \\ S_1 \\ S_3 \end{array} \begin{pmatrix} S_2 & S_1 & S_3 \\ 1 & 0 & 0 \\ 0.02 & 0.88 & 0.10 \\ 0.16 & 0.04 & 0.80 \end{pmatrix}$ Other answers are possible.

$\underbrace{}_{R} \underbrace{}_{Q}$

11. $\begin{array}{c} S_2 \\ S_4 \\ S_3 \\ S_1 \\ S_5 \end{array} \begin{pmatrix} S_2 & S_4 & S_3 & S_1 & S_5 \\ 1 & 0 & 0 & 0 & 0 \\ 0 & 1 & 0 & 0 & 0 \\ 0.20 & 0.30 & 0.50 & 0 & 0 \\ 0 & 0 & 0.20 & 0.80 & 0 \\ 0 & 0 & 0.05 & 0.20 & 0.75 \end{pmatrix}$ Other answers are possible.

$\underbrace{}_{R} \underbrace{}_{Q}$

13. $\begin{pmatrix} 2 & 1 \\ 1 & 3 \end{pmatrix}$ **15.** $\begin{pmatrix} 2 & 3 \\ 1 & 4 \end{pmatrix}$ **17.** $\begin{pmatrix} 2 & 2 & 3 \\ 0 & 10 & 0 \\ 1 & 1 & 4 \end{pmatrix}$ **19.** $\begin{pmatrix} 2 & 2 & 2 \\ 1 & 6 & 1 \\ 1 & 2 & 3 \end{pmatrix}$

21. 2; 1 **23.** 1; 0.70 **25.** 10; 1 **27.** 2; 0.70 **29.** $1 + 6 + 1 = 8; 0.40$

31. a. 30% of the town's original population survived.
b. The expected time for the disease to run its course was 2 months.

33. With 0.60 probability, a resident requiring assistance will ultimately be transferred to a nursing home.

35. a. A policy that has just been renewed can be expected to continue for $4 + 2 = 6$ periods, so $6 \cdot 5 = 30$ years.
b. A policy that is renewed with increased coverage will close with a death benefit payment with probability 0.53.

37. A tooth needing a filling can be expected to need 4 fillings before being extracted.

39. Since 80% of the companies receiving advice become successful and 76% of those receiving loans become successful, the trust's claim is justified.

Chapter 7 Review Exercises page 520

1. $\begin{array}{c} \begin{array}{cc}A & B\end{array} \\ \begin{array}{c}A\\B\end{array}\begin{pmatrix} \frac{2}{3} & \frac{1}{3} \\ 1 & 0 \end{pmatrix} \end{array}$, mixing

2. $\begin{array}{c} \begin{array}{ccc}X & Y & Z\end{array} \\ \begin{array}{c}X\\Y\\Z\end{array}\begin{pmatrix} 0 & 0 & 1 \\ 1 & 0 & 0 \\ 0 & 1 & 0 \end{pmatrix} \end{array}$, oscillating

3. Absorbing

4. Mixing

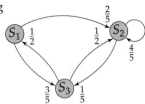

5. $D_1 = (0.35 \quad 0.65)$, $D_2 = (0.325 \quad 0.675)$
6. $D_1 = (0.60 \quad 0.30 \quad 0.10)$, $D_2 = (0.19 \quad 0.42 \quad 0.39)$
7. 20
8. 5
9. A working cab can be expected to remain in service for 100 days before breaking down.
10. Of these 875 students, 130 will be at the second-year level after 2 years.
11. Regular
12. Not regular
13. Regular
14. Regular
15. $D = (\frac{1}{4} \quad \frac{3}{4})$, $T^{50} = \begin{pmatrix} 0.25 & 0.75 \\ 0.25 & 0.75 \end{pmatrix}$
16. $D = (\frac{3}{20} \quad \frac{1}{2} \quad \frac{7}{20})$, $T^{20} = \begin{pmatrix} 0.15 & 0.50 & 0.35 \\ 0.15 & 0.50 & 0.35 \\ 0.15 & 0.50 & 0.35 \end{pmatrix}$

17.

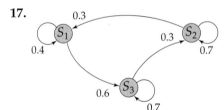

2 steps, with probability $0.6 \cdot 0.3 = 0.18$

18.

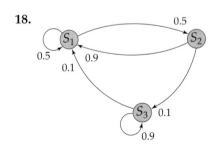

2 steps, with probability $0.1 \cdot 0.5 = 0.05$

19. Of the manufacturer's 2660 dealers, $0.40 \cdot 2660 = 1064$ have service departments rated "excellent" by their customers.
20. Out of the next 25 years, 10 will have terrific weather for growing corn, if the old timers are right.
21. S_3 is absorbing; S_1 and S_2 are nonabsorbing. 2 steps, with probability $\frac{1}{10} \cdot \frac{1}{4} = \frac{1}{40}$
22. S_2 is absorbing; S_1, S_3, and S_4 are nonabsorbing. 3 steps, with probability $1 \cdot 1 \cdot \frac{1}{2} = \frac{1}{2}$

23. a.

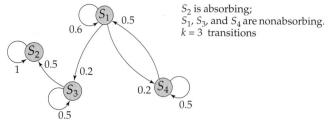

S_2 is absorbing; S_1, S_3, and S_4 are nonabsorbing. $k = 3$ transitions

b. $T^3 = \begin{pmatrix} 0.386 & 0.21 & 0.202 & 0.202 \\ 0 & 1 & 0 & 0 \\ 0 & 0.875 & 0.125 & 0 \\ 0.505 & 0.05 & 0.16 & 0.285 \end{pmatrix}$

24. a.

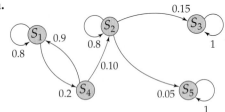

S_3 and S_5 are absorbing; S_1, S_2, and S_4 are nonabsorbing. $k = 3$ transitions

b. $T^3 = \begin{pmatrix} 0.8 & 0.032 & 0.003 & 0.164 & 0.001 \\ 0 & 0.512 & 0.366 & 0 & 0.122 \\ 0 & 0 & 1 & 0 & 0 \\ 0.738 & 0.082 & 0.027 & 0.144 & 0.009 \\ 0 & 0 & 0 & 0 & 1 \end{pmatrix}$

25.

$\begin{array}{c|c|ccc} & S_2 & S_1 & S_3 & S_4 \\ \hline S_2 & 1 & 0 & 0 & 0 \\ S_1 & 0.5 & 0.5 & 0 & 0 \\ S_3 & 0 & 0.2 & 0.8 & 0 \\ S_4 & 0 & 0 & 1 & 0 \end{array}$
$\qquad \underbrace{}_{R} \underbrace{}_{Q}$

Other answers are possible.

26.

$\begin{array}{c|cc|cc} & S_2 & S_4 & S_3 & S_1 \\ \hline S_2 & 1 & 0 & 0 & 0 \\ S_4 & 0 & 1 & 0 & 0 \\ S_3 & 0.1 & 0 & 0.8 & 0.1 \\ S_1 & 0 & 0.1 & 0.3 & 0.6 \end{array}$
$\qquad \underbrace{}_{R} \underbrace{}_{Q}$

Other answers are possible.

27. $\begin{pmatrix} 2 & 4 \\ 1 & 7 \end{pmatrix}$ **28.** $\begin{pmatrix} 4 & 6 \\ 2 & 8 \end{pmatrix}$ **29.** $\begin{pmatrix} 6 & 0 & 2 \\ 3 & 5 & 1 \\ 4 & 0 & 3 \end{pmatrix}$ **30.** $\begin{pmatrix} 2 & 0 & 6 \\ 1 & 2 & 4 \\ 1 & 0 & 8 \end{pmatrix}$ **31.** 2; 1   **32.** 1; 1

33. 6; 0.3   **34.** 8; 0.6   **35.** 4; 1   **36.** $3 + 5 + 1 = 9$; 1   **37.** 0; 0.3   **38.** $1 + 0 + 8 = 9$; 0.85

39. A new model PC can be expected to be manufactured for $3 + 4 = 7$ years before it is discontinued.

40. With probability 0.39 an overdue account will (ultimately) be paid in full.

Exercises 8.1 page 541

1.

x	5x − 7	x	5x − 7
1.9	2.5	2.1	3.5
1.99	2.95	2.01	3.05
1.999	2.995	2.001	3.005

$\lim_{x \to 2}(5x - 7) = 3$

3.

x	$\dfrac{x^3 - 1}{x - 1}$	x	$\dfrac{x^3 - 1}{x - 1}$
0.9	2.71	1.1	3.31
0.99	2.97	1.01	3.03
0.999	2.997	1.001	3.003

$\lim_{x \to 1} \dfrac{x^3 - 1}{x - 1} = 3$

5. 7.389 (rounded) **7.** −0.25 **9.** 1 **11.** 2 **13.** 8 **15.** 2 **17.** $\sqrt{2}$ **19.** 6 **21.** $5x^3$
23. 4 **25.** $\frac{1}{2}$ **27.** −9 **29.** 2x **31.** $4x^2$ **33. a.** 1 **b.** 3 **c.** Does not exist
35. a. −1 **b.** −1 **c.** −1 **37. a.** −1 **b.** 2 **c.** Does not exist
39. a. −2 **b.** −2 **c.** −2 **41. a.** 0 **b.** 0 **c.** 0
43. a. −1 **b.** 1 **c.** Does not exist **45.** $\lim_{x \to 3^-} f(x) = \infty$, $\lim_{x \to 3^+} f(x) = -\infty$, $\lim_{x \to 3} f(x)$ does not exist
47. $\lim_{x \to 0^-} f(x) = \infty$, $\lim_{x \to 0^+} f(x) = \infty$, $\lim_{x \to 0} f(x) = \infty$
49. $\lim_{x \to -2^-} f(x) = -\infty$, $\lim_{x \to -2^+} f(x) = \infty$, $\lim_{x \to -2} f(x)$ does not exist
51. $\lim_{x \to 3^-} f(x) = \infty$, $\lim_{x \to 3^+} f(x) = \infty$, $\lim_{x \to 3} f(x) = \infty$ **53.** Continuous **55.** Discontinuous, (3) is violated
57. Discontinuous, (1) is violated **59.** Discontinuous, (2) is violated
61. a. 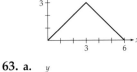 **b.** $\lim_{x \to 3^-} f(x) = 3$, $\lim_{x \to 3^+} f(x) = 3$ **c.** Yes

63. a. 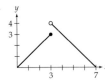 **b.** $\lim_{x \to 3^-} f(x) = 3$, $\lim_{x \to 3^+} f(x) = 4$ **c.** No, (2) is violated **65.** Continuous

67. Discontinuous at $x = 1$ **69.** Discontinuous at $x = -1$, $x = 0$, and $x = 1$
71. Discontinuous at $x = 4$ **73.** Continuous **75.** Continuous
77. The two functions are *not* equal to each other, since at $x = 1$ one is defined and the other is not (see pages 50–51).
79. 1.11 (dollars) **81.** 100

Exercises 8.2 page 558

1. At P_1: positive slope
   At P_2: negative slope
   At P_3: zero slope

3. At P_1: positive slope
   At P_2: negative slope
   At P_3: zero slope

5. At P_1: slope is 3
   At P_2: slope is $-\frac{1}{2}$

7. Your graph should look roughly like the following:

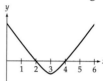

9. a. 5   b. 4   c. 3.5   d. 3.1   e. 3.01   f. 3
11. a. 13   b. 11   c. 10   d. 9.2   e. 9.02   f. 9
13. a. 5   b. 5   c. 5   d. 5   e. 5   f. 5
15. a. 0.2247   b. 0.2361   c. 0.2426   d. 0.2485   e. 0.2498   f. 0.25
17. 3   19. 5   21. 9   23. $\frac{1}{4}$ 25. $f'(x) = 2x - 3$ 27. $f'(x) = -2x$ 29. $f'(x) = 9$
31. $f'(x) = \frac{1}{2}$ 33. $f'(x) = 0$ 35. $f'(x) = 2ax + b$ 37. $f'(x) = 5x^4$ 39. $f'(x) = \frac{-2}{x^2}$
41. $f'(x) = \frac{1}{2\sqrt{x}}$ 43. $f'(x) = 3x^2 + 2x$
45. a. $y = x + 1$ b.   47. a. 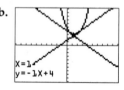   b.
   on viewing window $[-10, 10]$ by $[-10, 10]$

49. a. $f'(x) = 3$ b. The graph of $f(x) = 3x - 4$ is a straight line with slope 3.
51. a. $f'(x) = 0$ b. The graph of $f(x) = 5$ is a horizontal straight line with slope 0.
53. a. $f'(x) = m$ b. The graph of $f(x) = mx + b$ is a straight line with slope m.
55. a. $f'(x) = 2x - 8$ b. Decreasing at the rate of 4 degrees per minute (since $f'(2) = -4$)
    c. Increasing at the rate of 2 degrees per minute (since $f'(5) = 2$)
57. a. $f'(x) = 4x - 1$
    b. When 5 words have been memorized, the memorization time is increasing at the rate of 19 seconds per word
59. a. $T'(x) = -2x + 5$ b. Increasing at the rate of 1 degree per day
    c. Decreasing at the rate of 1 degree per day   d. Deteriorating on day 2, improving on day 3

Exercises 8.3 page 574

1. $4x^3$ 3. $500x^{499}$ 5. $\frac{1}{2}x^{-1/2}$ 7. $2x^3$ 9. $2w^{-2/3}$ 11. $-6x^{-3}$ 13. $8x - 3$
15. $-\frac{1}{2}x^{-3/2} = -\frac{1}{2\sqrt{x^3}}$ 17. $-2x^{-4/3} = -\frac{2}{\sqrt[3]{x^4}}$ 19. $2\pi r$ 21. $\frac{1}{2}x^2 + x + 1$ 23. $\frac{1}{2}x^{-1/2} + x^{-2}$
25. $4x^{-1/3} + 4x^{-4/3}$ 27. $-5x^{-3/2} - 15x^{2/3}$ 29. $1 + 2x$
31. a. $f'(x) = 0$ b. The graph of the constant function $f(x) = 2$ is a horizontal line and therefore has slope 0.   c. Since $f(x)$ is a constant function, its rate of change is zero.

33. 80   **35.** 3   **37.** 27   **39.** 1
41. For $y_1 = 5$ and window $[-10, 10]$ by $[-10, 10]$, your calculator screen should look like the following:

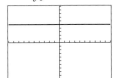

43. a. $MP(x) = 0.03x^{1/2}$ **b.** $MP(10,000) = 3$. *Interpretation:* After 10,000 chips, the profit on each additional chip is about \$3.   **45.** 3.00007, which is close to \$3.
47. a. $P'(x) = -12,000 + 1200x + 300x^2$ **b.** Decreasing by about 10,500 per year
   c. Increasing by about 30,000 per year
49. Increasing by about 8000 people per additional day
51. Increasing by about 0.08 square centimeter per hour
53. Increasing by about 6 phrases per hour   **55. a.** $MU(x) = 50x^{-1/2}$ **b.** 50   **c.** 0.05
57. a. $f(12) \approx 40$. *Interpretation:* The probability of a high school graduate quitting smoking is about 40%.
   $f'(12) \approx 1.8$. *Interpretation:* The probability of quitting increases by about 1.8% for each additional year of education.
   b. $f(16) \approx 60$. *Interpretation:* The probability of a college graduate quitting smoking is about 60%.
   $f'(16) \approx 8.5$. *Interpretation:* The probability of quitting increases by about 8.5% for each additional year of education.
59. c. \$26,043   **e.** Tuition increasing at the rate of \$1191 per year

Exercises 8.4 page 590

1. $10x^9$ **3.** $9x^8 + 4x^3$ **5.** $5x^4 + 2x$ **7.** $15x^2 - 1$ **9.** $4x^3$ **11.** $9x^2 + 8x + 1$ **13.** 1
15. $36t + 8t^{1/3}$ (after simplification)   **17.** $7z^6 - 1$ (after simplification)   **19.** $6x^5$ **21.** $-\dfrac{3}{x^4}$ or $-3x^{-4}$
23. $\dfrac{x^4 - 3}{x^4}$ **25.** $-\dfrac{2}{(x-1)^2}$ **27.** $\dfrac{4t}{(t^2+1)^2}$ **29.** $\dfrac{2s^3 + 3s^2 + 1}{(s+1)^2}$ (after simplification)
31. $\dfrac{2x^5 + 4x^3}{(x^2+1)^2}$ (after simplification)   **33.** $y = 3x^{-1}$, $\dfrac{dy}{dx} = -3x^{-2}$, $\dfrac{dy}{dx} = -\dfrac{3}{x^2}$
35. $y = \dfrac{3}{8}x^4$, $\dfrac{dy}{dx} = \dfrac{3}{2}x^3$, $\dfrac{dy}{dx} = \dfrac{3x^3}{2}$ **37.** $\dfrac{d}{dx}(fgh) = f'(gh) + f(gh)' = f'gh + fg'h + fgh'$ **39.** $2f(x)f'(x)$
41. $3x^2 \cdot \dfrac{x^2+1}{x+1} + (x^3+2) \cdot \dfrac{x^2+2x-1}{(x+1)^2}$ **43.** $\dfrac{3x^6 + 13x^4 + 18x^2 - 2x}{(x^2+2)^2}$ **45.** $\dfrac{x^{-1/2}}{(x^{1/2}+1)^2} = \dfrac{1}{\sqrt{x}(\sqrt{x}+1)^2}$
47. $MAR(x) = \dfrac{xR'(x) - R(x)}{x^2}$
49. a. $C'(x) = \dfrac{100}{(100-x)^2}$ **b.** Increasing by 4¢ per additional percentage of purity
   c. Increasing by 25¢ per additional percentage of purity.
51. b. Rates of change are 4 and 25
53. $\dfrac{dy}{dx} = \dfrac{R}{\left(1 + \dfrac{R-1}{K}x\right)^2} > 0$; As the population increases, the number of offspring increases.

55. a. $AP(x) = \dfrac{12x - 1800}{x} = 12 - \dfrac{1800}{x} = 12 - 1800x^{-1}$

b. $MAP(x) = \dfrac{1800}{x^2}$ or $1800x^{-2}$

c. $MAP(300) = \dfrac{2}{100}$, so average profit is increasing by 2¢ per additional unit.

57. Increasing at the rate of 7 degrees per hour   **59. b.** 7   **c.** About 104.5 degrees

61. b. 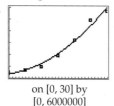   **d.** 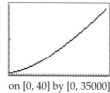   **f.**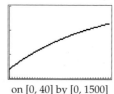

on [0, 30] by [0, 6000000]   on [0, 40] by [0, 35000]   on [0, 40] by [0, 1500]

f. (continued)
$y_3(40) \approx 33{,}325$, so in 2010 the per capita national debt should be \$33,325.
$y_4(40) \approx 1219$, so in 2010 the per capita national debt should be growing by \$1219 per year.

63. a. $y' = -2x^{-3} = -\dfrac{2}{x^3}$ (which is undefined at $x = 0$)

b. Your calculator should give "Error" but may, incorrectly, give "0."

Exercises 8.5 page 605

1. a. $4x^3 - 6x^2 - 6x + 5$ **b.** $12x^2 - 12x - 6$ **c.** $24x - 12$ **d.** 24

3. a. $1 + x + \dfrac{1}{2}x^2 + \dfrac{1}{6}x^3 + \dfrac{1}{24}x^4$ **b.** $1 + x + \dfrac{1}{2}x^2 + \dfrac{1}{6}x^3$ **c.** $1 + x + \dfrac{1}{2}x^2$ **d.** $1 + x$

5. a. $\dfrac{5}{2}x^{3/2}$ **b.** $\dfrac{15}{4}x^{1/2}$ **c.** $\dfrac{15}{8}x^{-1/2}$ **d.** $-\dfrac{15}{16}x^{-3/2}$ **7. a.** $-\dfrac{2}{x^3}$ or $-2x^{-3}$ **b.** $-\dfrac{2}{27}$

9. a. $\dfrac{1}{x^3} = x^{-3}$ **b.** $\dfrac{1}{27}$ **11. a.** x^{-4} **b.** $\dfrac{1}{81}$ **13.** $12x^2 + 2$ **15.** $12x^{-7/3}$

17. $\dfrac{2x-2}{(x^2-2x+1)^2} = \dfrac{2}{(x-1)^3}$ **19.** 2π **21.** 90   **23.** -720 **25.** 3

27. a. iii   **b.** i   **c.** ii   **29.** 0

31. $\dfrac{d^2}{dx^2}(fg) = \dfrac{d}{dx}(f'g + fg') = f''g + f'g' + f'g' + fg'' = f''g + 2f'g' + fg''$

33. a. 54 mph   **b.** -42 mph or 42 mph south   **c.** 24 mi/hr²   **35.** 310 ft/sec, 61 ft/sec²

37. a. 160 ft/sec   **b.** 32 ft/sec²   **39. a.** $-32t + 1280$ **b.** 40 seconds   **c.** 25,600 feet

41. $D'(8) = 24$: After 8 years the debt is growing by \$24 billion per year.
$D''(8) = 1$: After 8 years the debt will be growing increasingly rapidly, with the rate of growth growing by about \$1 billion per year per year.

43. $L'(4) = \tfrac{1}{4}$: After 4 years the sea level will be rising by $\tfrac{1}{4}$ foot per year.
$L''(4) = -\tfrac{3}{32}$: After 4 years the rate of growth will be slowing by about $\tfrac{3}{32}$ foot per year per year.

45. $P(3) \approx 4.87$, $P'(3) \approx -0.51$, $P''(3) \approx -0.39$
Interpretation: In 3 years the profit will be about \$4.87 million, decreasing at the rate of \$0.51 million per year, and the decline of profit will be accelerating.

47. a. 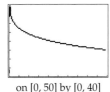 Approximately 22° and 18°

on [0, 50] by [0, 40]

b. Each successive 1-mph increase in wind speed lowers the windchill index, but less so as wind speed rises.

c. $y_2(15) \approx -0.4°$ and $y_2(30) \approx -0.2°$. *Interpretation:* At a wind speed of 15 mph, each additional mph decreases the windchill index by about 0.4°, whereas at a wind speed of 30 mph, each additional mph decreases the windchill index by only about 0.2°.

49. $20x^3 - 12x^2 + 6x - 2$ **51.** $\dfrac{2x^5 - 4x^3 - 6x}{(x^2 + 1)^4} = \dfrac{2x^3 - 6x}{(x^2 + 1)^3}$ **53.** $\dfrac{-32x - 16}{(4x^2 + 4x + 1)^2} = \dfrac{-16}{(2x + 1)^3}$

Exercises 8.6 page 618

Note: For Exercises 1 through 9, there are other possible correct answers.

1. $f(x) = \sqrt{x}, g(x) = x^2 - 3x + 1$ **3.** $f(x) = x^{-3}, g(x) = x^2 - x$ **5.** $f(x) = \dfrac{x+1}{x-1}, g(x) = x^3$

7. $f(x) = x^4, g(x) = \dfrac{x+1}{x-1}$ **9.** $f(x) = \sqrt{x} + 5, g(x) = x^2 - 9$ **11.** $6x(x^2 + 1)^2$

13. $4(3z^2 - 5z + 2)^3(6z - 5)$ **15.** $\tfrac{1}{2}(x^4 - 5x + 1)^{-1/2}(4x^3 - 5)$ **17.** $\tfrac{1}{3}(9z - 1)^{-2/3}(9) = 3(9z - 1)^{-2/3}$

19. $-8x(4 - x^2)^3$ **21.** $-12w^2(w^3 - 1)^{-5}$ **23.** $4x^3 - 4(1 - x)^3$

25. $-\tfrac{2}{3}(9x + 1)^{-5/3}(9) = -6(9x + 1)^{-5/3}$ **27.** $3[(x^2 + 1)^3 + x]^2[6x(x^2 + 1)^2 + 1]$

29. $6x(2x + 1)^5 + 30x^2(2x + 1)^4 = 6x(2x + 1)^4(7x + 1)$

31. $6(2x + 1)^2(2x - 1)^4 + 8(2x + 1)^3(2x - 1)^3 = 2(2x + 1)^2(2x - 1)^3(14x + 1)$

33. $-6\dfrac{(x + 1)^2}{(x - 1)^4}$ **35.** $2x(1 + x^2)^{1/2} + x^3(1 + x^2)^{-1/2}$ **37.** $\dfrac{1}{4}x^{-1/2}(1 + x^{1/2})^{-1/2}$

39. a. $4x(x^2 + 1)$ **b.** $4x^3 + 4x$ **41. a.** $-\dfrac{3}{(3x + 1)^2}$ **b.** $-3(3x + 1)^{-2}$

43. $\dfrac{d}{dx}L(g(x)) = L'(g(x))\,g'(x) = \dfrac{1}{g(x)}g'(x) = \dfrac{g'(x)}{g(x)}$ **45.** $20(x^2 + 1)^9 + 360x^2(x^2 + 1)^8$

47. $MC(x) = 4x(4x^2 + 900)^{-1/2}$ $MC(20) = \tfrac{8}{5} = 1.60$ **49.** $x = 27$

51. $S'(25) \approx 2.1$. *Interpretation:* At income level $25,000, status increases by about 2.1 units for each additional $1000 of income.

53. $R'(50) = 32\tfrac{1}{3}$ **55.** 26 mg   **57.** $P'(2) = 0.24$; pollution is increasing by about 0.24 ppm per year.

59. b. 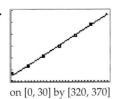 **e.** Slope ≈ 0.036 **f.** $\dfrac{1.8}{0.036} \approx 50$ years

on [0, 30] by [320, 370]

Exercises 8.7 page 626

1. $-2, 0,$ and 2 3. -3 and 3

5. $\lim\limits_{h \to 0} \dfrac{f(x + h) - f(x)}{h}$ simplifies to $\lim\limits_{h \to 0} \dfrac{|2h|}{h}$, which gives $\begin{cases} 2 & \text{for } h > 0 \\ -2 & \text{for } h < 0' \end{cases}$ so the limit (and therefore the derivative at $x = 0$) does not exist.

7. $\lim\limits_{h \to 0} \dfrac{f(x + h) - f(x)}{h}$ simplifies to $\lim\limits_{h \to 0} \dfrac{h^{2/5}}{h} = \lim\limits_{h \to 0} \dfrac{1}{h^{3/5}}$, which does not exist. Therefore, the derivative at $x = 0$ does not exist.

9. If you got a numerical answer, it is wrong, since the function is undefined at $x = 0$, so the derivative at $x = 0$ does not exist. (For an explanation, see the Graphing Calculator Exploration on page 623.)

11. a. The formula comes from substituting the given function and x-value into the definition of the derivative and simplifying.
    b. 3.16, 31.6, 316 (rounded)   c. No, No   d.

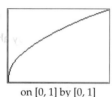

on $[0, 1]$ by $[0, 1]$

Chapter 8 Review Exercises page 629

1.

x	4x + 2	x	4x + 2
1.9	9.6	2.1	10.4
1.99	9.96	2.01	10.04
1.999	9.996	2.001	10.004

$\lim\limits_{x \to 2} (4x + 2) = 10$

2.

x	$\dfrac{\sqrt{x + 1} - 1}{x}$	x	$\dfrac{\sqrt{x + 1} - 1}{x}$
-0.1	0.513	0.1	0.488
-0.01	0.501	0.01	0.499
-0.001	0.500	0.001	0.500

$\lim\limits_{x \to 0} \dfrac{\sqrt{x + 1} - 1}{x} = 0.5$

3. a. 3   b. -2 c. Does not exist   4. a. -1 b. -1 c. -1 5. 5   6. π 7. 4
8. 2   9. $\tfrac{1}{2}$ 10. -3 11. $2x^2$ 12. $-x^2$ 13. Continuous   14. Continuous
15. Discontinuous at $x = -1$ 16. Continuous   17. Discontinuous at $x = 0$ and at $x = -1$
18. Discontinuous at $x = 3$ and at $x = -3$ 19. Discontinuous at $x = 5$ 20. Continuous
21. $4x + 3$ 22. $6x + 2$ 23. $-\dfrac{3}{x^2} = -3x^{-2}$ 24. $\dfrac{2}{\sqrt{x}}$ 25. $10x^{2/3} + 2x^{-3/2}$
26. $10x^{3/2} + 2x^{-4/3}$ 27. -16 28. -9 29. 1   30. $\tfrac{1}{2}$
31. a. $C'(x) = 3 - 27x^{-3/2}$ b. 2. *Interpretation:* Costs are increasing by about $2 per additional unit.
32. $f'(10) = -2.3$ (thousand hours). *Interpretation:* After 10 planes, the construction time is decreasing by about 2300 hours for each additional plane built.

33. a. $\dfrac{dA}{dr} = 2\pi r$ **b.** As the radius increases, the area "grows by a circumference."

34. a. $V' = \frac{4}{3}\pi r^2 \cdot 3 = 4\pi r^2$ **b.** As radius increases, volume "grows by a surface area."

35. $40x^3 + 6$ (after simplification)   **36.** $15x^4 - 2x$ **37.** $2x(x^2 - 5) + (x^2 + 5)2x = 4x^3$ **38.** $4x^3$

39. $(4x^3 + 2x)(x^5 - x^3 + x) + (x^4 + x^2 + 1)(5x^4 - 3x^2 + 1) = 9x^8 + 5x^4 + 1$

40. $(5x^4 + 3x^2 + 1)(x^4 - x^2 + 1) + (x^5 + x^3 + x)(4x^3 - 2x) = 9x^8 + 5x^4 + 1$ **41.** $\dfrac{2}{(x+1)^2}$

42. $\dfrac{-2}{(x-1)^2}$ **43.** $\dfrac{-10x^4}{(x^5-1)^2}$ **44.** $\dfrac{12x^5}{(x^6+1)^2}$

45. a. $-\dfrac{1}{x^2}$ **b.** $-x^{-2}$ (after simplification)   **c.** By simplifying to $f(x) = 2 + x^{-1}$, the derivative is $-x^{-2}$.

46. $S'(6) = -10$, so at a price of $6 each, sales will decrease by about 10 for each dollar increase in price.

47. a. $AP(x) = \dfrac{6x - 200}{x}$ **b.** $MAP(x) = \dfrac{200}{x^2}$

   c. $MAP(10) = 2$, so average profit is increasing by about $2 per additional unit.

48. a. $AC(x) = \dfrac{5x + 100}{x}$ **b.** $MAC(x) = \dfrac{-100}{x^2}$

   c. $MAC(20) = -0.25$, so average cost is decreasing by about 25¢ per additional unit.

49. $9x^{-1/2} + 2x^{-5/3}$ **50.** $-4x^{-4/3} - 3x^{-1/2}$ **51.** $2x^{-4}$ **52.** $6x^{-5}$ **53.** -24 **54.** 60

55. 480 **56.** $\frac{3}{8}$ **57.** 15 **58.** 70

59. $P(10) = 200$, $P'(10) = 20$, $P''(10) = 9$. *Interpretation:* 10 years from now the population will be 200 thousand, growing at the rate of 20 thousand per year, and the growth will be accelerating.

60. Velocity 2500 ft/sec; acceleration 150 ft/sec²   **61. a.** 347.25 feet

62. b. 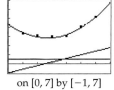   **c.** $y_1(7) = 6.76$; at the end of year 7 the annual profit will be about $6.76 million.

on [0.5, 6.5] by [2.5, 5.5]

   d. $y_2(7) \approx 1.77$. *Interpretation:* At the end of the seventh year, the annual profit will be growing by about $1.77 million per year.

   e. $y_3(7) \approx 0.41$. *Interpretation:* will be growing increasingly fast, with the rate of growth increasing by about $0.41 million per year per year.   **f.** y_1 is a quadratic function, and so its first derivative (y_2) is linear, and its second derivative (y_3) is constant, and the graph of a constant is a horizontal line.

on [0, 7] by [−1, 7]

63. $3(4z^2 - 3z + 1)^2(8z - 3)$ **64** $4(3z^2 - 5z - 1)^3(6z - 5)$ **65.** $-5(100 - x)^4$ **66.** $-4(1000 - x)^3$

67. $\frac{1}{2}(x^2 - x + 2)^{-1/2}(2x - 1)$ **68.** $\frac{1}{2}(x^2 - 5x - 1)^{-1/2}(2x - 5)$ **69.** $2(6z - 1)^{-2/3}$ **70.** $(3z + 1)^{-2/3}$

71. $-2(5x + 1)^{-7/5}$ **72.** $-6(10x + 1)^{-8/5}$ **73.** $2x(2x - 1)^4 + 8x^2(2x - 1)^3 = 2x(2x - 1)^3(6x - 1)$

74. $5(x^3 - 2)^4 + 60x^3(x^3 - 2)^3 = 5(x^3 - 2)^3(13x^3 - 2)$ **75.** $3x^2(x^3 + 1)^{1/3} + x^5(x^3 + 1)^{-2/3}$

76. $4x^3(x^2 + 1)^{1/2} + x^5(x^2 + 1)^{-1/2}$ 77. $3[(2x^2 + 1)^4 + x^4]^2[16x(2x^2 + 1)^3 + 4x^3]$
78. $2[(3x^2 - 1)^3 + x^3][18x(3x^2 - 1)^2 + 3x^2]$ 79. $\frac{1}{2}[(x^2 + 1)^4 - x^4]^{-1/2}[8x(x^2 + 1)^3 - 4x^3]$
80. $[(x^3 + 1)^2 + x^2]^{-1/2}[3x^2(x^3 + 1) + x]$ 81. $12(3x + 1)^3(4x + 1)^3 + 12(3x + 1)^4(4x + 1)^2$
82. $6x(x^2 + 1)^2(x^2 - 1)^4 + 8x(x^2 + 1)^3(x^2 - 1)^3$ 83. $\frac{-20}{x^2}\left(\frac{x + 5}{x}\right)^3 = \frac{-20(x + 5)^3}{x^5}$
84. $-20\left(\frac{x + 4}{x}\right)^4 \frac{1}{x^2} = -20\frac{(x + 4)^4}{x^6}$ 85. $20(2w^2 - 4)^4 + 320w^2(2w^2 - 4)^3$
86. $24(3w^2 + 1)^3 + 432w^2(3w^2 + 1)^2$ 87. $6z(z + 1)^3 + 18z^2(z + 1)^2 + 6z^3(z + 1)$
88. $12z^2(z + 1)^4 + 32z^3(z + 1)^3 + 12z^4(z + 1)^2$
89. a. $6x^2(x^3 - 1)$ b. $6x^5 - 6x^2$ 90. a. $\frac{-3x^2}{(x^3 + 1)^2}$ b. $-(x^3 + 1)^{-2}(3x^2)$
91. $P'(5) = 3$. *Interpretation:* When producing 5 tons, profit increases by about 3 thousand dollars for each additional ton.
92. $V'(8) = 17.496$. *Interpretation:* Value increased by about $17.50 for each additional percentage of interest.
93. a. $P(5) - P(4) \approx 2.73$, $P(6) - P(5) \approx 3.23$, both of which are near 3   b. At about $x = 7.6$
94. $x \approx 16$ 95. 0.08
96. $N'(96) = -250$. *Interpretation:* At age 96, the number of survivors is decreasing by about 250 people per year.
97. $x = -3, x = 1, x = 3$ 98. $x = 2, x = -2$ 99. $x = 0, x = 3.5$ 100. $x = 0, x = 3$
101. $\lim_{h \to 0}\frac{f(x + h) - f(x)}{h}$ simplifies to $\lim_{h \to 0}\frac{|5h|}{h}$, which gives $\begin{cases} 5 & \text{for } h > 0 \\ -5 & \text{for } h < 0 \end{cases}$, and so the limit (and therefore the derivative at $x = 0$) does not exist.
102. $\lim_{h \to 0}\frac{f(x + h) - f(x)}{h}$ simplifies to $\lim_{h \to 0}\frac{h^{3/5}}{h} = \lim_{h \to 0}\frac{1}{h^{2/5}}$, which does not exist. Therefore, the derivative at $x = 0$ does not exist.

Exercises 9.1 page 646

1. a. $(-\infty, -2)$ and $(0, \infty)$ b. $(-2, 0)$ 3. All but 3 (where the function is undefined)   5. 4 and -4
7. 0, -4, and 1   9. 3   11. No CNs   13. 1 and $\frac{1}{3}$
15.

$f' < 0$	$f' = 0$	$f' > 0$	$f' = 0$	$f' < 0$	$f' = 0$	$f' > 0$
	$x = -4$		$x = 0$		$x = 1$	
↘		↗		↘		↗
	rel min $(-4, -64)$		rel max $(0, 64)$		rel min $(1, 61)$	

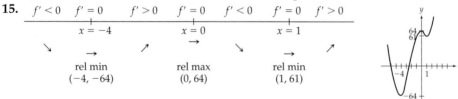

Open intervals of increase: $(-4, 0)$ and $(1, \infty)$; Open intervals of decrease: $(-\infty, -4)$ and $(0, 1)$

17.

$f' > 0$	$f' = 0$	$f' < 0$	$f' = 0$	$f' > 0$	$f' = 0$	$f' < 0$
	$x = 0$		$x = 1$		$x = 2$	
↗		↘		↗		↘
	rel max $(0, 1)$		rel min $(1, 0)$		rel max $(2, 1)$	

Open intervals of increase: $(-\infty, 0)$ and $(1, 2)$; Open intervals of decrease: $(0, 1)$ and $(2, \infty)$

19.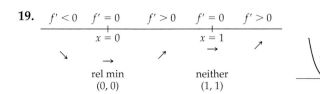

Open intervals of increase: $(0, 1)$ and $(1, \infty)$; Open interval of decrease: $(-\infty, 0)$

21.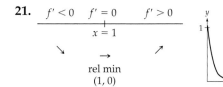

Open interval of increase: $(1, \infty)$; Open interval of decrease: $(-\infty, 1)$

23.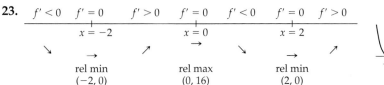

Open intervals of increase: $(-2, 0)$ and $(2, \infty)$; Open intervals of decrease: $(-\infty, -2)$ and $(0, 2)$

25.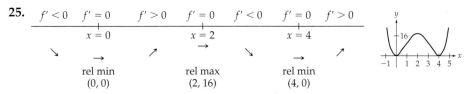

Open intervals of increase: $(0, 2)$ and $(4, \infty)$; Open intervals of decrease: $(-\infty, 0)$ and $(2, 4)$

27.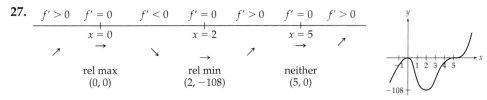

Open intervals of increase: $(-\infty, 0)$, $(2, 5)$, and $(5, \infty)$; Open interval of decrease: $(0, 2)$

29.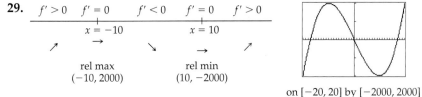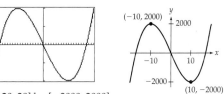

on $[-20, 20]$ by $[-2000, 2000]$

31.

```
  f' < 0    f' = 0   f' > 0   f' = 0   f' < 0   f' = 0   f' > 0
 ─────────────┼──────────────┼──────────────┼──────────────
           x = −5          x = 0          x = 5
    ↓          →     ↗            ↘           →     ↗
            rel min          rel max         rel min
           (−5, −650)       (0, −25)        (5, −650)
```

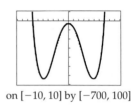

on [−10, 10] by [−700, 100]

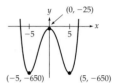

33.

```
  f' > 0   f' = 0   f' > 0   f' = 0   f' < 0   f' = 0   f' > 0
 ─────────────┼──────────────┼──────────────┼──────────────
           x = 0           x = 1          x = 3
    ↗          →     ↗            ↘           →     ↗
           neither          rel max         rel min
          (0, −23)         (1, −22)        (3, −50)
```

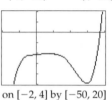

on [−2, 4] by [−50, 20]

35.

```
  f' > 0   f' = 0   f' < 0   f' = 0   f' > 0
 ─────────────┼──────────────┼──────────────
          x = −1           x = 1
    ↗          →     ↘            →     ↗
          rel max          rel min
         (−1, 0.04)       (1, −0.04)
```

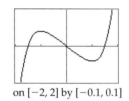

on [−2, 2] by [−0.1, 0.1]

37.

```
  f' > 0   f' = 0   f' < 0   f' = 0   f' > 0
 ─────────────┼──────────────┼──────────────
          x = 1/3           x = 1
    ↗          →     ↘            →     ↗
          rel max          rel min
         (1/3, 11.15)      (1, 11)
```

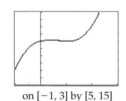

on [−1, 3] by [5, 15]

39.

```
  f' und    f' > 0   f' = 0   f' < 0   f' und
 ─────────────┼──────────────┼──────────────
  x = −20           x = 0           x = 20
             ↗            ↘
                    rel max
                    (0, 20)
```

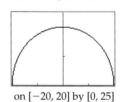

on [−20, 20] by [0, 25]

41. Critical number: 1

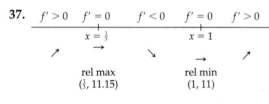

on [−4, 6] by [−2, 2]

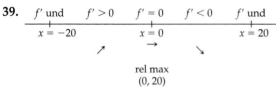

(1, −1/9)

43. Critical number: 0

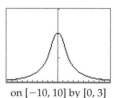

on [−10, 10] by [0, 3]

45. Critical number: 0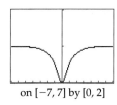
on [−7, 7] by [0, 2]

47. Critical numbers: 0 and 6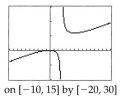
on [−10, 15] by [−20, 30]

49. Critical number: 0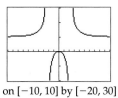
on [−10, 10] by [−20, 30]

51. Critical numbers: −1, 0, and 1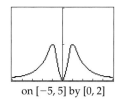
on [−5, 5] by [0, 2]

53.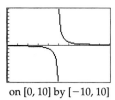
on [0, 10] by [−10, 10]

55.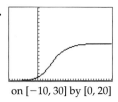
on [−10, 30] by [0, 20]

57. $f'(x) = 2ax + b = 0$ at $x = \dfrac{-b}{2a}$

59.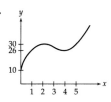

61. d.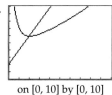
on [0, 10] by [0, 10]

63. a.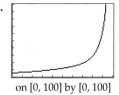
on [0, 100] by [0, 100]

Exercises 9.2 page 661

1. Point 2 **3.** Points 3 and 5 **5.** Points 4 and 6

7. a.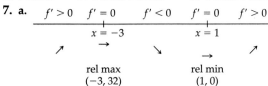
rel max (−3, 32) rel min (1, 0)

b.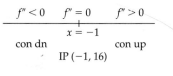
con dn con up
IP (−1, 16)

c.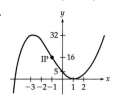

9. a.

$f' > 0$	$f' = 0$	$f' > 0$
	$x = 1$	
↗	→	↗
	neither (1, 5)	

b.

$f'' < 0$	$f'' = 0$	$f'' > 0$
	$x = 1$	
con dn	IP (1, 5)	con up

c.

11. a.

$f' < 0$	$f' = 0$	$f' > 0$	$f' = 0$	$f' > 0$
	$x = 0$		$x = 3$	
↘	→	↗	→	↗
	rel min (0, 2)		neither (3, 29)	

b.

$f'' > 0$	$f'' = 0$	$f'' < 0$	$f'' = 0$	$f'' > 0$
	$x = 1$		$x = 3$	
con up	IP (1, 13)	con dn	IP (3, 29)	con up

c.

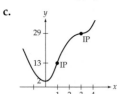

13. a.

$f' < 0$	$f' = 0$	$f' > 0$	$f' = 0$	$f' < 0$
	$x = 0$		$x = 4$	
↘	→	↗	→	↘
	rel min (0, 0)		rel max (4, 256)	

b.

$f'' > 0$	$f'' = 0$	$f'' > 0$	$f'' = 0$	$f'' < 0$
	$x = 0$		$x = 3$	
con up		con up	IP (3, 162)	con dn

c.

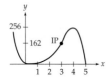

15. a.

$f' > 0$	$f' = 0$	$f' > 0$
	$x = -2$	
↗	→	↗
	neither (-2, 0)	

b.

$f'' < 0$	$f'' = 0$	$f'' > 0$
	$x = -2$	
con dn	IP (-2, 0)	con up

c.

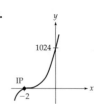

17. a.

$f' > 0$	$f' = 0$	$f' < 0$	$f' = 0$	$f' > 0$
	$x = 1$		$x = 3$	
↗	→	↘	→	↗
	rel max (1, 4)		rel min (3, 0)	

b.

$f'' < 0$	$f'' = 0$	$f'' > 0$
	$x = 2$	
con dn	IP (2, 2)	con up

c.

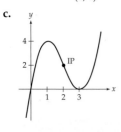

19. a.

$f' > 0$	f' und	$f' > 0$
	$x = 0$	
↗		↗
	neither	
	(0, 0)	

b.

$f'' > 0$	f'' und	$f'' < 0$
	$x = 0$	
con up		con dn
	IP (0, 0)	

c.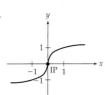

21. a.

$f' < 0$	f' und	$f' > 0$
	$x = 0$	
↘		↗
	rel min	
	(0, 2)	

b.

$f'' < 0$	f'' und	$f'' < 0$
	$x = 0$	
con dn		con dn

c.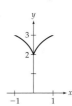

23. a.

f' und	$f' > 0$
$x = 0$	
	↗

b.

f'' und	$f'' < 0$
$x = 0$	
	con dn

c.

25. a.

$f' < 0$	f' und	$f' > 0$
	$x = 1$	
↘		↗
	rel min	
	(1, 0)	

b.

$f'' < 0$	f'' und	$f'' < 0$
	$x = 1$	
con dn		con dn

c.

27.

$f' > 0$	$f' = 0$	$f' < 0$	$f' = 0$	$f' > 0$
	$x = 2$		$x = 10$	
↗	→	↘	→	↗
	rel max		rel min	
	(2, 76)		(10, −180)	

$f'' < 0$	$f'' = 0$	$f'' > 0$
	$x = 6$	
con dn		con up
	IP (6, −52)	

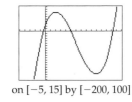 on [−5, 15] by [−200, 100]

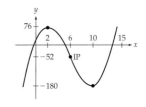

29.

$f' < 0$	$f' = 0$	$f' < 0$	$f' = 0$	$f' > 0$
	$x = 0$		$x = 12$	
↘	→	↘	→	↗
	neither		rel min	
	(0, 0)		(12, −6912)	

$f'' > 0$	$f'' = 0$	$f'' < 0$	$f'' = 0$	$f'' > 0$
	$x = 0$		$x = 8$	
con up		con dn		con up
	IP (0, 0)		IP (8, −4096)	

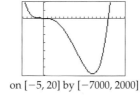 on [−5, 20] by [−7000, 2000]

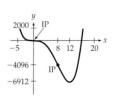

31.

$f' > 0$	$f' = 0$	$f' < 0$	$f' = 0$	$f' > 0$
	$x = -2$		$x = 8$	
↗	→	↘	→	↗
	rel max $(-2, 100)$		rel min $(8, -400)$	

$f'' < 0$	$f'' = 0$	$f'' > 0$
	$x = 3$	
con dn		con up
	IP $(3, -150)$	

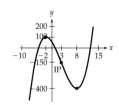
on $[-10, 15]$ by $[-500, 200]$

33.

$f' > 0$	$f' = 0$	$f' < 0$	$f' = 0$	$f' > 0$
	$x = \frac{1}{3}$		$x = 1$	
↗	→	↘	→	↗
	rel max $(\frac{1}{3}, 5.15)$		rel min $(1, 5)$	

$f'' < 0$	$f'' = 0$	$f'' > 0$
	$x = \frac{2}{3}$	
con dn		con up
	IP $(\frac{2}{3}, 5.07)$	

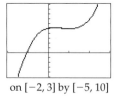
on $[-2, 3]$ by $[-5, 10]$

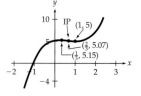

35.

$f' > 0$	f' und	$f' > 0$
	$x = 1$	
↗	↗	
	neither $(1, 0)$	

$f'' > 0$	f'' und	$f'' < 0$
	$x = 1$	
con up		con dn
	IP $(1, 0)$	

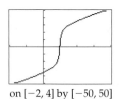
on $[-2, 4]$ by $[-50, 50]$

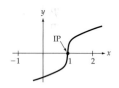

37.

f' und	$f' > 0$
$x = 0$	↗

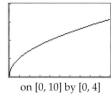
on $[0, 10]$ by $[0, 4]$

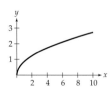

39.

f' und	$f' < 0$
$x = 0$	↘

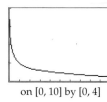
on $[0, 10]$ by $[0, 4]$

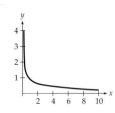

41.

$f' < 0$	f' und	$f' > 0$	$f' = 0$	$f' < 0$
	$x = 0$		$x = 1$	
↘		↗	→	↘
	rel min $(0, 0)$		rel max $(1, 3)$	

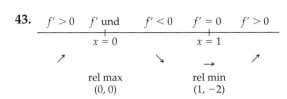

on $[-2, 10]$ by $[-10, 10]$

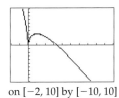

43.

$f' > 0$	f' und	$f' < 0$	$f' = 0$	$f' > 0$
	$x = 0$		$x = 1$	
↗		↘	→	↗
	rel max $(0, 0)$		rel min $(1, -2)$	

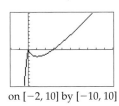

on $[-2, 10]$ by $[-10, 10]$

45.

$f' > 0$	$f' = 0$	$f' < 0$	f' und	$f' > 0$	$f' = 0$	$f' < 0$
	$x = -1$		$x = 0$		$x = 1$	
↗	→	↘		↗	→	↘
	rel max $(-1, 2)$		rel min $(0, 0)$		rel max $(1, 2)$	

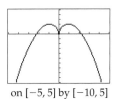

on $[-5, 5]$ by $[-10, 5]$

47. $f''(x) = 2a$, therefore: For $a > 0, f'' > 0$, so f is concave up.
For $a < 0, f'' < 0$, so f is concave down.

49. Where the curve is concave *up*, it lies *above* its tangent line, and where it is concave *down*, it lies *below* its tangent line, so *at an inflection point it must cross its tangent line.* **51.** $-0.77, 0,$ and 0.77

53. a.

$f' > 0$	$f' = 0$	$f' < 0$	$f' = 0$	$f' > 0$
	$x = 1$		$x = 5$	
↗	→	↘	→	↗
	rel max $(1, 32)$		rel min $(5, 0)$	

$f'' < 0$	$f'' = 0$	$f'' > 0$
	$x = 3$	
con dn		con up
	IP $(3, 16)$	

b.

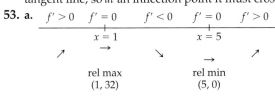

55. a.

$f' < 0$	$f' = 0$	$f' > 0$
	$x = 1$	
↘	→	↗
	rel min $(1, 109)$	

$f'' > 0$	$f'' = 0$	$f'' > 0$
	$x = 0$	
con up		con up

b.

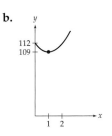

57.

f' und $\quad f' > 0$
―――+――――――
$\quad x = 0 \quad\quad \nearrow$

rel min
(0, 0)

f'' und $\quad f'' < 0$
―――+――――――
$\quad x = 0 \quad\quad$ con dn

59. a.

S' und $\quad S' > 0$
―――+――――――
$\quad i = 0 \quad\quad \nearrow$

S'' und $\quad S'' < 0$
―――+――――――
$\quad i = 0 \quad\quad$ con dn

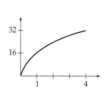

b. Concave down. Status increases more slowly at higher income levels.

61. b. $y = 1.96x^{0.66}$ **c.** Concave down **d.** About 16 **e.** About 8.5
63. (50, 2.5). The curve is concave up (slope increasing) before $x = 50$ and concave down (slope decreasing) after $x = 50$. Therefore, the slope is maximized at $x = 50$.

Exercises 9.3 page 675

1. Max f is 12 (at $x = 1$), min f is -8 (at $x = -1$). **3.** Max f is 16 (at $x = -2$), min f is -16 (at $x = 2$).
5. Max f is 9 (at $x = 1$), min f is 0 (at $x = 0$ and at $x = -2$).
7. Max f is 81 (at $x = 3$), min f is -16 (at $x = 2$).
9. Max f is 4 (at $x = 2$), min f is -50 (at $x = 5$).   **11.** Max f is 5 (at $x = 0$), min f is 0 (at $x = 5$).
13. Max f is 1 (at $x = 0$), min f is 0 (at $x = -1$ and at $x = 1$).
15. Max f is $\frac{1}{2}$ (at $x = 1$), min f is $-\frac{1}{2}$ (at $x = -1$).
17. a. The number is $\frac{1}{2}$.   **b.** The number is 3.
19. a. Both at endpoints
   b. One at a critical number (the maximum) and one at an endpoint (the minimum)
   c. Both at critical numbers   **d.** Yes; for example, [2, 10]
21. On the 20th day   **23.** 31 mph   **25.** 36 years   **27.** 12 miles from A toward B
29. Produce 40 per day, price = $400, max profit = $6500
31. 400 feet along the building and 200 feet perpendicular to the building
33. Each is 200 yards parallel to the river and 150 yards perpendicular to the river
35. 3 inches high with a base 12 inches by 12 inches; volume: 432 cubic inches
37. The numbers are 25 and 25.
39. $r = 2$ cm   **41.** $r = 110/\pi \approx 35$ yards, $x = 110$ yards   **43.** $x \approx 1.125$ inches, $y \approx 1.25$ megabytes
45. a. At time 10 hours; 1,500,000 bacteria (since $N(t)$ is in thousands)
   b. At time 5 hours; growing by 75,000 bacteria per hour (inflection point)
47. Remove a square of side $x \approx 0.96$ inch; volume ≈ 15 square inches
49. a. $p(x) = -4.5x + 54$ **b.** $R(x) = -4.5x^2 + 54x$ **c.** 6 bottles per hour, selling for $27 each

Exercises 9.4 page 686

1. Price: $14,400; sell 16 cars per day (from $x = 2$ price reductions)
3. Ticket price: $150; number sold: 450 (from $x = 5$ price reductions)
5. Rent the cars for $90, and expect to rent 54 cars (from $x = 2$ price increases)
7. 25 trees per acre (from $x = 5$ extra trees per acre)   9. Base: 2 feet by 2 feet; height: 1 foot
11. Base: 14 inches by 14 inches; height: 28 inches; volume: 5488 cubic inches
13. 50 feet along the driveway and 100 feet perpendicular to the driveway; cost: $800
15. 6.4%   17. 16 years
19. [*Hint:* If area is A (a constant) and one side is x, then show that the perimeter is $P = 2x + 2\frac{A}{x}$, which is minimized at $x = \sqrt{A}$. Then show that this means the rectangle is a square.]
21. The page should be 8 inches wide and 12 inches tall.
23. $R' = 2cpx - 3cx^2 = xc(2p - 3x)$, which is zero when $x = \frac{2}{3}p$. The second-derivative test will show that R is maximized.
25. e. 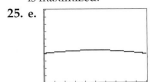   f. Price: $325; quantity: 35 bicycles   g. Price: $350; quantity: 40 bicycles

Exercises 9.5 page 696

1. Lot size: 400 boxes; 10 orders during the year   3. Lot size: 500 bottles; 20 orders during the year
5. Lot size: 40 cars per order; 20 orders during the year
7. Produce 1000 games per run; 2 production runs during the year
9. Produce 40,000 tapes per run; 25 runs for the year
11. Population: 20,000; yield: 40,000 (from $p = 200$)
13. Population: 75,000; yield: 2250 (from $p = 75$)   15. Population: 625,000; yield: 625,000 (from $p = 625$)
17. Population: 3717; yield: 16,109 (from $x = 3.717$). (The yield exceeds the population because of reproduction later in the year.)

Exercises 9.6 page 708

1. $\frac{dy}{dx} = \frac{2x}{3y^2}$ 3. $\frac{dy}{dx} = \frac{3x^2}{2y}$ 5. $\frac{dy}{dx} = \frac{3x^2 + 2}{4y^3}$ 7. $\frac{dy}{dx} = -\frac{x+1}{y+1}$ 9. $\frac{dy}{dx} = -\frac{2y}{x}$ (after simplification)
11. $\frac{dy}{dx} = \frac{-y+1}{x}$ 13. $\frac{dy}{dx} = -\frac{y-1}{2x}$ (after simplification)   15. $\frac{dy}{dx} = \frac{1}{3y^2 - 2y + 1}$
17. $\frac{dy}{dx} = -\frac{y^2}{x^2}$ (after simplification)   19. $\frac{dy}{dx} = \frac{3x^2}{2(y-2)}$ 21. $\frac{dy}{dx} = 2$ 23. $\frac{dy}{dx} = -3$ 25. $\frac{dy}{dx} = -1$
27. $\frac{dy}{dx} = -4$ 29. $\frac{dp}{dx} = -\frac{2}{2p+1}$ 31. $\frac{dp}{dx} = \frac{1}{24p+4}$ 33. $\frac{dp}{dx} = -\frac{p}{3x}$ (after simplification)
35. $\frac{dp}{dx} = -\frac{p+5}{x+2}$
37. $\frac{dp}{dx} = -4$. *Interpretation:* The rate of change of price with respect to quantity is -4, so price decreases by about $4 when quantity increases by 1.

39. a. $\dfrac{ds}{dr} = \dfrac{3r^2}{2s} = 8$ **b.** $\dfrac{dr}{ds} = \dfrac{2s}{3r^2} = \dfrac{1}{8}$ **c.** $\dfrac{ds}{dr} = 8$ means that the rate of change of sales with respect to research expenditures is 8, so that increasing research by \$1 million will increase sales by about \$8 million (at these levels of r and s).

$\dfrac{dr}{ds} = \dfrac{1}{8}$ means that the rate of change of research expenditures with respect to sales is $\dfrac{1}{8}$, so that increasing sales by \$1 million will increase research by about $\dfrac{1}{8}$ million dollars (at these levels of r and s).

41. $3x^2 \dfrac{dx}{dt} + 2y \dfrac{dy}{dt} = 0$ **43.** $2x \dfrac{dx}{dt} y + x^2 \dfrac{dy}{dt} = 0$ **45.** $6x \dfrac{dx}{dt} - 7 \dfrac{dx}{dt} y - 7x \dfrac{dy}{dt} = 0$

47. $2x \dfrac{dx}{dt} + \dfrac{dx}{dt} y + x \dfrac{dy}{dt} = 2y \dfrac{dy}{dt}$ **49.** Decreasing by $72\pi \approx 226$ in^3/hr

51. Increasing by $32\pi \approx 101$ cm^3/week **53.** Growing by \$16,000 per day

55. Increasing by 400 cases per year

57. Slowing by $\tfrac{1}{2}$ mm/sec per year **59.** Yes (65.8 mph)

Chapter 9 Review Exercises page 712

1. 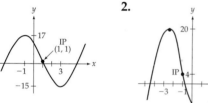 **2.** 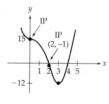 **3.** 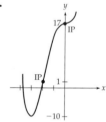 **4.**

5.   **6.** 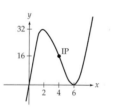 **7.** 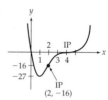 **8.**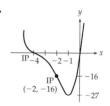

9. 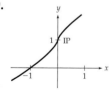 **10.**   **11.**   **12.**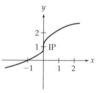

13. 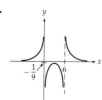 **14.** 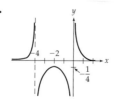 **15.**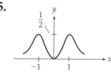

16. 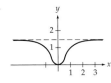   **17.** 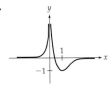   **18.**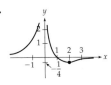

19. Max f is 220 (at $x = 5$), min f is -4 (at $x = 1$).   **20.** Max f is 130 (at $x = 5$), min f is -32 (at $x = 2$).
21. Max f is 64 (at $x = 0$), min f is -64 (at $x = 4$).
22. Max f is 6401 (at $x = 10$), min f is 1 (at $x = 0$ and $x = 2$).
23. Max h is 4 (at $x = 9$), min h is 0 (at $x = 1$).
24. Max f is 10 (at $x = 0$), min f is 0 (at $x = 10$ and $x = -10$).
25. Max g is 25 (at $w = 3$ and $w = -3$), min g is 0 (at $w = 2$ and $w = -2$).
26. Max g is 16 (at $x = 4$), min g is 0 (at $x = 0$ and $x = 8$).
27. Max f is $\frac{1}{2}$ (at $x = 1$), min f is $-\frac{1}{2}$ (at $x = -1$).
28. Max f is $\frac{1}{4}$ (at $x = 2$), min f is $-\frac{1}{4}$ (at $x = -2$).   **29.**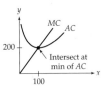

30. ① At the minimum point of $AC(x)$, the derivative of $AC(x)$ must be zero.

   ② Quotient Rule applied to $AC(x) = \dfrac{C(x)}{x}$

   ③ Factoring out $\dfrac{1}{x}$

   ④ Simplifying inside the square brackets

   ⑤ Recognizing $C'(x)$ as the marginal cost function and $\dfrac{C(x)}{x}$ as the average cost function.

   ⑥ $0 = \dfrac{1}{x}[MC(x) - AC(x)]$ means that the quantity in the square brackets must equal zero, and so the marginal cost must equal average cost at this x-value, where average cost is minimized.

31. $v = 2c$, which means that the tugboat should travel through the water at twice the speed of the current.
32. 3600 square feet   **33.** 1800 square feet   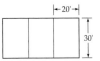

34. 15 cubits (gilded side) by 135 cubits   **35.** Base: 10 inches by 10 inches; height: 5 inches
36. Radius: 2 inches; height: 4 inches

37. Radius: 2 inches; height: 2 inches   **38.** $v = \sqrt[4]{\dfrac{aw^2}{3b}}$

39. Price: $2400 each; quantity: 9 per week   **40.** 5 weeks   **41.** $x = \frac{3}{4}$ mile
42. a. $t \approx 0.624 = 62.4\%$ **b.** $1.26   **43.** Radius $\approx$ 1.2 inches; height $\approx$ 4.8 inches
44. $x \approx 1.59$ inches; volume ≈ 33.07 cubic inches   **45.** $x \approx 1.13$ inches; volume ≈ 12.13 cubic inches
46. 600 per run; $1\frac{1}{2}$ runs per year (or 3 runs in 2 years)   **47.** Lot size: 50; 10 orders during the year
48. Population: 150,000 ($p = 150$); yield: 450,000
49. Population: $p = 900$ (thousand); yield: 900 (thousand)
50. $\dfrac{dy}{dx} = \dfrac{-6x - 4y}{4x + y}$ **51.** $\dfrac{dy}{dx} = \dfrac{-y^2}{2xy - 1}$ (after simplification)   **52.** $\dfrac{dy}{dx} = \dfrac{-2y^2 + 6xy}{4xy - 3x^2}$
53. $\dfrac{dy}{dx} = \dfrac{y^{1/2}}{x^{1/2}}$ (after simplification)   **54.** $\dfrac{dy}{dx} = -1$ **55.** $\dfrac{dy}{dx} = \dfrac{1}{7}$ **56.** $\dfrac{dy}{dx} = -\dfrac{1}{6}$ **57.** $\dfrac{dy}{dx} = 1$
58. 600 in³/hr   **59.** Increasing by $4200 per day   **60.** Increasing by $45,000 per month
61. a. Decreasing by about 0.31 cm³/min   **b.** Decreasing by about 0.05 cm³/min

Cumulative Review for Chapters 1, 8–9 page 716

1. $y = -\frac{1}{2}x + 1$ **2.** $\frac{5}{2}$ **3.** 20.085
4. a.
  b. 4   **5.** $f'(x) = 4x - 5$
  c. 1
  d. Does not exist
  e. Discontinuous at $x = 3$

6. $f'(x) = 12x^{1/2} + 6x^{-3}$ **7.** $f'(x) = 9x^8 + 10x^4 - 8x^3$ **8.** $f'(x) = \dfrac{11}{(3x - 2)^2}$

9. $P'(8) = 1200$, so in 8 years the population will be increasing by 1200 people per year.
$P''(8) = -50$, so in 8 years the rate of growth will be slowing by 50 people per year per year.

10. $\dfrac{2x}{\sqrt{2x^2 - 5}}$ **11.** $12(3x + 1)^3(4x + 1)^3 + 12(3x + 1)^4(4x + 1)^2 = 12(3x + 1)^3(4x + 1)^2(7x + 2)$

12. $\dfrac{12(x - 2)^2}{(x + 2)^4}$

13.

$f' > 0$	$f' = 0$	$f' < 0$	$f' = 0$	$f' > 0$
	$x = -2$		$x = 10$	
↗	→	↘	→	↗
	rel max $(-2, 464)$		rel min $(10, -400)$	

$f'' < 0$	$f'' = 0$	$f'' > 0$
	$x = 4$	
con dn		con up
	IP (4, 32)	

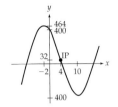

14.

$f' < 0$	f' und	$f' > 0$
	$x = 0$	
↘		↗
	rel min $(0, -1)$	

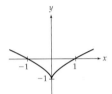

$f'' < 0$	f'' und	$f'' < 0$
	$x = 0$	
con dn		con dn

15. 15,000 square feet   **16.** Price: $170; quantity: 18 per day

17. $\dfrac{dy}{dx} = \dfrac{-x^2 - 3y^2}{6xy + 1}$ (after simplification). Evaluating at (1, 2) gives $\dfrac{dy}{dx} = -1$.   **18.** $\dfrac{2}{\pi} \approx 0.64$ ft/min

Exercises 10.1 page 730

1. 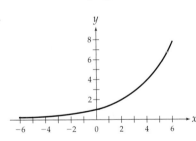   **3.**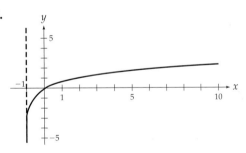

5. a. 99   **b.** $-\frac{1}{2}$ **c.** 0   **7.** $\ln 2.75 \approx 1.011600912$, $e^{1.011600912} \approx 2.75$ **9.** $\ln x$ **11.** 1
13. $2x$ **15.** $1,424,500   **17.** $5,163,900   **19.** 7.21 billion people   **21. a.** 1.3 mg   **b.** 0.84 mg

23. a. About 17 days
b. About 58 days
c.
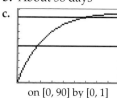
on [0, 90] by [0, 1]

25. a. About 67%
b. About 20 minutes
c.
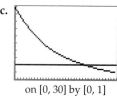
on [0, 30] by [0, 1]

27. a. About 29,670 people
b. About 5.4 hours
c.
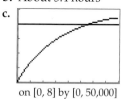
on [0, 8] by [0, 50,000]

29. a. About every 4 hours (from $x \approx 4.2$)
b.
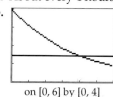
on [0, 6] by [0, 4]

31. About 138 days

33. a. $e^{-kt} = \frac{1}{2}$. Taking natural logs gives $-kt = \ln \frac{1}{2} = -\ln 2$ so $t = \frac{\ln 2}{k}$.

b. (Half-life) $= \frac{\ln 2}{0.018} \approx 38.5$ hours

35. About 6.5 years

Exercises 10.2 page 746

1. $2x \ln x + x$ **3.** $\frac{2}{x}$ **5.** $\frac{1}{2}x^{-1}$ **7.** $\frac{6x}{x^2 + 1}$ **9.** $\frac{1}{x}$ **11.** $\frac{xe^x - 2e^x}{x^3}$ or $\frac{e^x(x - 2)}{x^3}$
13. $(3x^2 + 2)e^{x^3 + 2x}$ **15.** $x^2 e^{x^3/3}$ **17.** $1 + e^{-x}$ **19.** 2 **21.** $e^{1+e^x}e^x$ or e^{1+x+e^x} **23.** $e x^{e-1}$ **25.** 0
27. $\frac{4x^3}{x^4 + 1} - 2e^{x/2} - 1$ **29.** $2x \ln x + 2xe^{x^2}$ **31. a.** $\frac{1 - 5 \ln x}{x^6}$ **b.** 1 **33. a.** $\frac{4x^3}{x^4 + 48}$ **b.** $\frac{1}{2}$
35. a. $\frac{e^x - 3}{e^x - 3x}$ **b.** -2 **37. a.** $5 \ln x + 5$ **b.** $5 \ln 2 + 5 \approx 8.466$
39. a. $\frac{xe^x - e^x}{x^2}$ **b.** $\frac{2e^3}{9} \approx 4.463$ **41.** $-4x^3 e^{-x^5/5} + x^8 e^{-x^5/5}$ or $x^3 e^{-x^5/5}(x^5 - 4)$ **43.** $f^{(n)}(x) = k^n e^{kx}$

45.
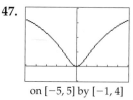
on [−2, 2] by [−1, 2]

rel max: (0, 1)
IP: $(\frac{1}{2}, 0.61)$
$(-\frac{1}{2}, 0.61)$

47.
on [−5, 5] by [−1, 4]

rel min: (0, 0)
IP: (1, 0.69)
(−1, 0.69)

49.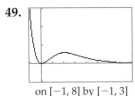
rel min: (0, 0)
rel max: (2, 0.54)
IP: (0.59, 0.19)
  (3.41, 0.38)

on $[-1, 8]$ by $[-1, 3]$

51.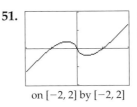
rel max: $(-0.37, 0.37)$
rel min: $(0.37, -0.37)$

on $[-2, 2]$ by $[-2, 2]$

53. $\dfrac{dy}{dx} = \dfrac{ye^x}{2y - e^x}$

55. a. Increasing by about $50 per year   **b.** Increasing by about $82.44 per year
57. Increasing by about 0.12 billion (=120 million) people per year
59. a. Decreasing by 0.06 mg/hr   **b.** Decreasing by 0.054 mg/hr
61. a. Increasing by about 81.4 (thousand) sales per week
   b. Increasing by about 33 (thousand) sales per week
63. $p = \$100$ **65. a.** $R(x) = 400xe^{-0.20x}$ **b.** Quantity: $x = 5$ (thousand); price: $p = \$147.15$
67. When $N < K$, $\ln(K/N)$ is positive and the population is increasing, while when $N > K$, $\ln(K/N)$ is negative and the population is decreasing.
69. $r = a/b$
71. a. After 15 minutes the temperature of the beer is 57.5 degrees and increasing at the rate of 43.8 degrees per hour.
   b. After 1 hour the temperature of the beer is 69.1 degrees and increasing at the rate of 3.2 degrees per hour.
73. 2.3 seconds   **75.** $p = \$500$ **77.** At about 2.75 hours
79. a. $(\ln 10)10^x$ **b.** $(\ln 3)(2x)3^{x^2+1} = 2(\ln 3)x3^{x^2+1}$ **c.** $(\ln 2)3 \cdot 2^{3x} = 3(\ln 2)2^{3x}$
   d. $(\ln 5)6x \cdot 5^{3x^2} = 6(\ln 5)x5^{3x^2}$ **e.** $-(\ln 2)2^{4-x}$
81. a. $\dfrac{1}{(\ln 2)x}$ **b.** $\dfrac{2x}{(\ln 10)(x^2 - 1)}$ **c.** $\dfrac{4x^3 - 2}{(\ln 3)(x^4 - 2x)}$

Exercises 10.3 page 761

1. a. $2/t$ **b.** 2 and 0.2   **3. a.** 0.2   **b.** 0.2   **5. a.** $2t$ **b.** 20   **7. a.** $-2t$ **b.** -20
9. a. $\dfrac{1}{2(t-1)}$ **b.** $1/10$ **11.** 0.0071 or 0.71%   **13. a.** 0.012 or 1.2%   **b.** Yes, in about 15.3 years
15. a. $E(p) = \dfrac{5p}{200 - 5p}$ **b.** Inelastic $(E = \tfrac{1}{3})$ **17. a.** $E(p) = \dfrac{2p^2}{300 - p^2}$ **b.** Unitary elastic $(E = 1)$
19. a. $E(p) = 1$ **b.** Unitary elastic $(E = 1)$ **21. a.** $E(p) = \dfrac{3p}{2(175 - 3p)}$ **b.** Elastic $(E = 3)$
23. a. $E(p) = 2$ **b.** Elastic $(E = 2)$ **25. a.** $E(p) = 0.01p$ **b.** Elastic $(E = 2)$
27. Lower the price $(E = 8)$ **29.** No $(E = 1.25)$ **31.** Yes $\left(E = \tfrac{3}{8} = 0.375\right)$
33. Lower its price $(E = 1.2)$ **35.** $E = 0.112$ **37.** $E(p) = \dfrac{-pa(-c)e^{-cp}}{ae^{-cp}} = cp$ **39.** $E_S(p) = n$
41. a. $E \approx 0.35$ **b.** Raise the price   **c.** $p \approx \$20{,}400$

Chapter 10 Review Exercises page 765

1. $5.29 million   2. 393,000   3. 193 million   4. Drug B
5. In about 2060 (from $x \approx 60.2$ years after 2000)
6. $4^7 = 16{,}384$ megabits, which is enough to hold sixty-four 16-volume encyclopedias on one chip.
7. 771
8. a. About 14 per hour   b. About 25 per hour   c.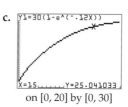
on [0, 20] by [0, 30]

9. $f(x) = \ln x + 1$ 10. a. $\ln 9.3 \approx 2.2300144$ b. $e^{2.2300144} \approx 9.3$ 11. 2.3 hours   12. 13.7 years
13. About 26 years   14. About 5.3 years   15. $\dfrac{1}{x}$ 16. $\dfrac{4x}{x^2 - 1}$ 17. $\dfrac{-1}{1 - x}$ or $\dfrac{1}{x - 1}$
18. $\dfrac{x}{x^2 + 1}$ 19. $\dfrac{1}{3x}$ 20. 1   21. $\dfrac{2}{x}$ 22. $\ln x$ 23. $-2xe^{-x^2}$ 24. $-e^{1-x}$ 25. $2x$
26. $2x(\ln x)e^{x^2 \ln x - x^2/2}$ 27. $10x + 2\ln x + 2$ 28. $6x^2 + 3\ln x + 3$ 29. $6x^2 - 3e^{2x} - 6xe^{2x}$
30. $4 - 4xe^{2x} - 4x^2 e^{2x}$

31. 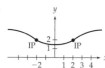   rel min: $(0, \ln 4) \approx (0, 1.4)$ 32. 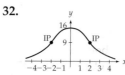   rel max: $(0, 16)$
IP: $(2, \ln 8) \approx (2, 2.1)$ IP: $(2, 16e^{-1/2}) \approx (2, 9.7)$
$(-2, \ln 8) \approx (-2, 2.1)$ $(-2, 16e^{-1/2}) \approx (-2, 9.7)$

33. a. Increasing by 136 thousand per week   b. Increasing by 55 thousand per week
34. a. Decreasing by 0.12 mg per hour   b. Decreasing by 0.08 mg per hour
35. Decreasing by 33.3% per second
36. a. Increasing by 3.5 degrees per hour   b. Increasing by 2.1 degrees per hour
37. a. Increasing by 6667 per hour   b. Increasing by 816 per hour   38. 25 years
39. a. $R(x) = 200xe^{-0.25x}$ b. Quantity $x = 4$ (thousand); price $p = \$73.58$
40. a. $R(x) = 5x - x \ln x$ b. Quantity $x = e^4 \approx 54.60$; price = $1   41. Price = $50
42.    rel min: $(0, 0)$ 43. 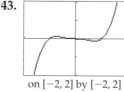   rel max: $(-0.72, 0.12)$
rel max: $(4, 4.69)$ rel min: $(0.72, -0.12)$
IP: $(2, 2.17)$ IP: $(-0.43, 0.07)$
$(6, 3.21)$ $(0.43, -0.07)$
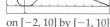
on $[-2, 10]$ by $[-1, 10]$ on $[-2, 2]$ by $[-2, 2]$

44. $p \approx \$769$ 45. $x \approx 130$ (in thousands)   46. 0.0033 or 0.33%   47. 0.0031 or 0.31%
48. Raise prices $(E = 0.8)$ 49. Raise prices $(E = 0.7)$ 50. $E = 0.44$ 51. 0.104 or 10.4%
52. a. $E \approx 1.29$ b. Lower the price   c. About $8700 (from $p \approx 8.7$)

Exercises 11.1 page 782

1. $\frac{1}{5}x^5 + C$ 3. $\frac{3}{5}x^{5/3} + C$ 5. $\frac{2}{3}u^{3/2} + C$ 7. $-\frac{1}{3}w^{-3} + C$ 9. $2\sqrt{z} + C$ 11. $x^6 + C$
13. $2x^4 - x^3 + 2x + C$ 15. $4x^{3/2} + \frac{3}{2}x^{2/3} + C$ 17. $6x^{8/3} + 24x^{-2/3} + C$ 19. $6t^{5/3} + 3t^{1/3} + C$

21. $\frac{1}{3}x^3 - x^2 + x + C$ **23.** $\frac{2}{3}w^{3/2} + 4w^{5/2} + C$ **25.** $2x^3 - 3x^2 + x + C$ **27.** $\frac{1}{3}x^3 + x^2 - 8x + C$
29. $\frac{1}{3}r^3 - r + C$ **31.** $\frac{1}{2}x^2 - x + C$ **33.** $\frac{1}{4}t^4 + t^3 + \frac{3}{2}t^2 + t + C$
35. a. $-\frac{1}{2}x^{-2} + C$ **b.** $\dfrac{x + C}{\frac{1}{4}x^4 + C_1}$ (where C_1 is another arbitrary constant)   **37. b.**

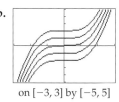

on $[-3, 3]$ by $[-5, 5]$

39. $C(x) = 8x^{5/2} - 9x^{5/3} + x + 4000$ **41.** $R(x) = 9x^{4/3} + 2x^{3/2}$
43. a. $D(t) = -0.03t^3 + 4t^2$ **b.** 370 feet
45. a. $6t^{1/2}$ **b.** 30 words   **47. a.** $P(t) = 16t^{5/2}$ **b.** 512 tons   **c.** No   **49.** $\dfrac{1}{x}$

Exercises 11.2 page 794

1. $\frac{1}{3}e^{3x} + C$ **3.** $4e^{x/4} + C$ **5.** $20e^{0.05x} + C$ **7.** $-\frac{1}{2}e^{-2y} + C$ **9.** $-2e^{-0.5x} + C$ **11.** $9e^{2x/3} + C$
13. $-5\ln|x| + C$ **15.** $3\ln|x| + C$ **17.** $\frac{3}{2}\ln|v| + C$ **19.** $\frac{1}{3}e^{3x} - 3\ln|x| + C$
21. $6e^{0.5t} - 2\ln|t| + C$ **23.** $\frac{1}{3}x^3 + \frac{1}{2}x^2 + x + \ln|x| - x^{-1} + C$ **25.** $250e^{0.02t} - 200e^{0.01t} + C$
27. a. $360e^{0.05t} - 355$ **b.** About 624 cases
29. a. $50 \ln t$ (since $t > 1$, absolute value bars are not needed)   **b.** No (about 170 sold)
31. a. $850e^{0.02t} - 850$ **b.** In about 2020 (20 years from 2000)   **33. a.** $500e^{0.4x} - 500$ **b.** About \$3195
35. a. $60e^{-0.2t} + 10$ **b.** In about 5 hours   **37. a.** $-4000e^{-0.2t} + 4000$ **b.** About $3\frac{1}{2}$ years
39. a. $8000e^{0.05t} - 3000$ **b.** About \$10,190   **41.** $\frac{1}{2}x^2 + 2x + \ln|x| + C$
43. $t + 2\ln|t| + 3t^{-1} + C$ **45.** $\frac{1}{3}x^3 - 3x^2 + 12x - 8\ln|x| + C$
47. a. $560e^{0.01t} - 560$ **b.** In about 2022 (22 years after 2000)

Exercises 11.3 page 811

1. 2.75 square units   **3.** 4.15 square units   **5.** 0.760 square unit
7. i. 2.8 square units   **ii.** 3 square units
9. i. 4.411 square units (or 4.412, depending on rounding)   **ii.** $\frac{14}{3} \approx 4.667$ square units
11. i. 0.719 square units   **ii.** $\ln 2 \approx 0.693$ square units
13. i. for left rectangles: 2.9, 2.99, 2.999
   for midpoint rectangles: 3, 3, 3
   for right rectangles: 3.1, 3.01, 3.001
   ii. 3 square units
15. i. for left rectangles: 4.515, 4.652, 4.665
   for midpoint rectangles: 4.668, 4.667, 4.667
   for right rectangles: 4.815, 4.682, 4.668
   ii. $\frac{14}{3} \approx 4.667$ square units
17. i. for left rectangles: 0.719, 0.696, 0.693
   for midpoint rectangles: 0.693, 0.693, 0.693
   for right rectangles: 0.669, 0.691, 0.693
   ii. $\ln 2 \approx 0.693$ square unit

19. 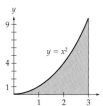 9 square units **21.** 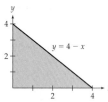 8 square units

23. 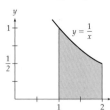 ln 2 square unit   **25.** 160 square units   **27.** 19 square units

29. 2 square units   **31.** 16 square units   **33.** ln 5 square units   **35.** $\ln 2 + \frac{7}{3}$ square units
37. 4 square units (from $2e^{\ln 3} - 2$)   **39.** $2e - 2$ square units   **41.** 5 square units   **43.** $\frac{15}{32}$ square unit
45. $4 + \ln 2$ square units   **47.** $\frac{111}{100}$ **49.** 13   **51.** $\frac{3}{4}$ **53.** 1 (from $\ln e - \ln 1$)   **55.** $78 + \ln 3$
57. $-3 \ln 2$ **59.** $4e^3 - 4$ **61.** $5e - 5e^{-1}$ **63.** 1 (from $e^{\ln 3} - e^{\ln 2}$)   **65.** $\frac{7}{2} + \ln 2$ **67.** 1.107
69. 2.925   **71.** 92.744
73. a. 9   **b.** Completing the calculation: $= \left(\frac{1}{3}3^3 + C\right) - \left(\frac{1}{3}0^3 + C\right) = \frac{1}{3}3^3 + C - \frac{1}{3}0^3 - C = \frac{1}{3} \cdot 27 = 9$
c. The C always cancels because it is both added and subtracted in the evaluation step.
75. $\int_1^a \frac{1}{x} dx = \ln|x|\Big|_1^a = \ln a - \ln 1 = \ln a$ **77.** $3\frac{1}{2} + \frac{\pi}{4}$ **79.** 132 units   **81.** 411 checks
83. $-300e^{-2} + 300 \approx \$259.40$ **85.** $225e^{0.8} - 225 \approx 276$, so about 2.76
87. $24e^{0.1} - 24 \approx 2.5$ million tons   **89.** $-30e^{-2} + 30 \approx 26$ words
91. $\frac{a}{-b+1}B^{-b+1} - \frac{a}{-b+1}A^{-b+1} = \frac{a}{1-b}(B^{1-b} - A^{1-b})$
93. $-4{,}000{,}000e^{-0.6} + 4{,}000{,}000 \approx \$1{,}804{,}753$
95. a. $\frac{1}{2}, \frac{1}{3}, \frac{1}{4}, \frac{1}{5}$ **b.** Area $= \frac{1}{n+1}$ **97.** 87 milligrams   **99.** 586 cars   **101.** 4023 people

Exercises 11.4 page 867

1. 3   **3.** 4   **5.** $\frac{1}{5}$ **7.** 5   **9.** $\frac{104}{3}$ or $34\frac{2}{3}$ **11.** 3   **13.** $e - 1$ **15.** ln 2   **17.** $\frac{1}{n+1}$
19. $a + b$ **21.** 0.845   **23.** 318   **25.** 70°   **27.** About 25.6 tons   **29.** $3194.53
31. $24.04 million   **33.** 6 square units
35. $\frac{1}{2}e^4 - e^2 + \frac{1}{2}$ square units   **37. a.** 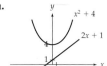   **b.** 9 square units

39. a. 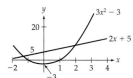 **b.** 18 square units   **41.** 4 square units   **43.** 32 square units
45. $\frac{1}{12}$ square unit   **47.** 1250 square units   **49.** 5.694 square units (rounded)   **51.** About 123 million
53. a. 10   **b.** $6000   **55.** About $529 billion   **57.** About $139 thousand
59. e. 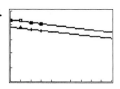   **f.** About 46,350 lives   **61.** $(2x+5)e^{x^2+5x}$ **63.** $\dfrac{2x+5}{x^2+5x}$

Exercises 11.5 page 840

1. $60,000   **3.** $10,000   **5.** $7500   **7.** $5285 (rounded)   **9.** $100   **11.** $160,000
13. a. $x = 500$ **b.** $50,000   **c.** $25,000   **15. a.** $x = 50$ **b.** $2500   **c.** $7500
17. a. $x \approx 119.48$ **b.** $10,065   **c.** $3446 (all rounded)   **19.** 0.52   **21.** 0.35   **23.** 0.2
25. $1 - \dfrac{2}{n+1} = \dfrac{n-1}{n+1}$ **27.** 0.46   **29.** 0.28   **31.** $L(x) = x^{2.13}$, Gini index ≈ 0.36
33. $4(x^5 - 3x^3 + x - 1)^3(5x^4 - 9x^2 + 1)$ **35.** $\dfrac{4x^3}{x^4 + 1}$ **37.** $3x^2 e^{x^3}$

Exercises 11.6 page 853

1. $\frac{1}{10}(x^2 + 1)^{10} + C$ **3.** $\frac{1}{20}(x^2 + 1)^{10} + C$ **5.** $\frac{1}{5}e^{x^5} + C$ **7.** $\frac{1}{6}\ln(x^6 + 1) + C$
9. $u = x^3 + 1$, $du = 3x^2 dx$: the powers in the integrand and the du do not match.
11. $u = x^4$, $du = 4x^3 dx$: the powers in the integrand and the du do not match.   **13.** $\frac{1}{24}(x^4 - 16)^6 + C$
15. $-\frac{1}{2}e^{-x^2} + C$ **17.** $\frac{1}{3}e^{3x} + C$ **19.** Cannot be found by our substitution formulas
21. $\frac{1}{5}\ln|1 + 5x| + C$ **23.** $\frac{1}{4}(x^2 + 1)^{10} + C$ **25.** $\frac{1}{5}(z^4 + 16)^{5/4} + C$
27. Cannot be found by our substitution formulas   **29.** $\frac{1}{24}(2y^2 + 4y)^6 + C$ **31.** $\frac{1}{2}e^{x^2+2x+5} + C$
33. $\frac{1}{12}\ln|3x^4 + 4x^3| + C$ **35.** $-\frac{1}{12}(3x^4 + 4x^3)^{-1} + C$ **37.** $-\frac{1}{2}\ln|1 - x^2| + C$
39. $\frac{1}{16}(2x - 3)^8 + C$ **41.** $\frac{1}{2}\ln(e^{2x} + 1) + C$ **43.** $\frac{1}{2}(\ln x)^2 + C$ **45.** $2e^{x^{1/2}} + C$ **47.** $\frac{1}{4}x^4 + \frac{1}{3}x^3 + C$
49. $\frac{1}{6}x^6 + \frac{2}{5}x^5 + \frac{1}{4}x^4 + C$ **51.** $\frac{1}{2}e^9 - \frac{1}{2}$ **53.** $\frac{1}{2}\ln 2$ **55.** $32\frac{2}{3}$ **57.** $-\ln 2$ **59.** $3e^2 - 3e$
61. a. $u^n u'$ **b.** $\int u^n u' \, dx$ **63. a.** $\dfrac{u'}{u}$ **b.** $\int \dfrac{u'}{u} \, dx$ **65.** $\dfrac{1}{2}\ln(2x + 1) + 50$ **67.** $\dfrac{1}{2}$ million
  — agree —   — agree —
69. $\frac{1}{3}\ln 5 - \frac{1}{3}\ln 2 \approx 0.305$ million   **71.** $20\frac{1}{3}$ units   **73.** About 346   **75.** $\frac{1}{2}\ln 10 \approx 1.15$ tons

Chapter 11 Review Exercises page 857

1. $8x^3 - 4x^2 + x + C$ **2.** $3x^4 + 3x^2 - 3x + C$ **3.** $4x^{3/2} - 5x + C$ **4.** $6x^{4/3} - 2x + C$
5. $6x^{5/3} - 2x^2 + C$ **6.** $2x^{5/2} - 3x^2 + C$ **7.** $\frac{1}{3}x^3 - 16x + C$ **8.** $x^3 + x^2 + 4x + C$

9. $C(x) = 2x^{1/2} + 4x + 20{,}000$ 10. a. $P(t) = 200t^{3/2} + 40{,}000$ b. 52,800 people
11. $2e^{x/2} + C$ 12. $-\frac{1}{2}e^{-2x} + C$ 13. $4\ln|2x| + C$ 14. $2\ln|x| + C$ 15. $2e^{3x} - 6\ln|x| + C$
16. $\frac{1}{2}x^2 - \ln|x| + C$ 17. $3x^3 + 2\ln|x| + 2e^{3x} + C$ 18. $-x^{-1} + \ln|x| - e^{-x} + C$
19. a. $1000e^{0.05t} - 1000$ b. About 2044 (from $t \approx 44$)
20. a. $2000e^{0.1t} - 2000$ b. About 5.6 years 21. a. $450e^{0.02t} - 450$ b. About 2034 (from $t \approx 34$)
22. a. $200\ln x$ b. In about 20 months 23. 36 24. 90 25. 4 26. $\ln 5$
27. $1 - e^{-2}$ 28. $2e - 2$ 29. $20e^5 - 100e + 80$ 30. $25e^{0.4} - 50e^{0.2} + 25$ 31. 13 square units
32. 36 square units 33. $6e^6 - 6$ square units 34. $2e^2 - 2$ square units 35. $\ln 100$ square units
36. $\ln 1000$ square units
37. Area ≈ 21.4 square units on $[-2, 2]$ by $[0, 10]$
38. Area ≈ 2.54 square units on $[-1, 1]$ by $[0, 3]$
39. About 29.5 kilograms 40. 9 words 41. $1.5e - 1.5 \approx 2.6$ degrees 42. $1640 43. $653.39
44. About 143 pages 45. a. 2.28 square units b. $\frac{8}{3} \approx 2.667$ square units
46. a. 4.884 square units b. $\frac{16}{3} \approx 5.333$ square units
47. a. for left rectangles: 5.899, 7.110, 7.239 b. $e^2 - e^{-2} \approx 7.254$ square units
    for midpoint rectangles: 7.206, 7.253, 7.254
    for right rectangles: 8.801, 7.400, 7.268
48. a. for left rectangles: 1.506, 1.398, 1.387 b. $\ln 4 \approx 1.386$ square units
    for midpoint rectangles: 1.383, 1.386, 1.386
    for right rectangles: 1.281, 1.375, 1.385
49. $\frac{4}{3}$ square units 50. 108 square units 51. $\frac{1}{6}$ square unit 52. $\frac{3}{10}$ square unit
53. 2 square units 54. $\frac{1}{2}$ square unit 55. About 17.13 square units 56. About 0.496 square unit
57. $\frac{1}{3}\ln 4$ 58. $\frac{28}{3}$ or $9\frac{1}{3}$ 59. About 4.72 60. About 2.77 61. 9.2 billion 62. About $5800
63. About $629,000 (from 6.29 hundred thousand dollars) 64. About $49.95 65. 27 square meters
66. About $934 billion 67. About $372 million
68. b. c. About 93.61 million people 69. $480,000
    on $[0, 10]$ by $[0, 35]$
70. $160,000 71. About $5623 72. About $611 73. About 0.56 74. About 0.43
75. About 0.23 76. About 0.68 77. $\frac{1}{4}(x^3 - 1)^{4/3} + C$ 78. $\frac{1}{6}(x^4 - 1)^{3/2} + C$
79. Cannot be integrated by our substitution formulas
80. Cannot be integrated by our substitution formulas
81. $-\frac{1}{3}\ln|9 - 3x| + C$ 82. $-\frac{1}{2}\ln|1 - 2x| + C$ 83. $\frac{1}{3}(9 - 3x)^{-1} + C$ 84. $\frac{1}{2}(1 - 2x)^{-1} + C$
85. $\frac{1}{2}(8 + x^3)^{2/3} + C$ 86. $(9 + x^2)^{1/2} + C$ 87. $-\frac{1}{2}(w^2 + 6w - 1)^{-1} + C$
88. $-\frac{1}{2}(t^2 - 4t + 1)^{-1} + C$ 89. $\frac{2}{3}(1 + \sqrt{x})^3 + C$ 90. $(1 + x^{1/3})^3 + C$ 91. $\ln|e^x - 1| + C$
92. $\ln|\ln x| + C$ 93. $\frac{61}{3}$ or $20\frac{1}{3}$ 94. 2 95. 2 96. $\frac{5}{12}$ 97. $\ln 7$ 98. $-\ln 2$
99. $\frac{1}{4}e - \frac{1}{4}$ 100. $\frac{1}{5}e - \frac{1}{5}$ 101. 8 square units 102. 5 square units

103. $\frac{1}{4} - \frac{1}{4}e^{-4} = \frac{1}{4}(1 - e^{-4}) \approx 0.25$ **104.** $\frac{1}{4} \ln 13 \approx 0.64$
105. $C(x) = (2x + 9)^{1/2} + 97$ **106.** $\ln 28 \approx 3.33$ degrees

Exercises 12.1 page 873

1. $\frac{1}{2}e^{2x} + C$ **3.** $\frac{1}{2}x^2 + 2x + C$ **5.** $\frac{2}{3}x^{3/2} + C$ **7.** $\frac{1}{5}(x + 3)^5 + C$ **9.** $\frac{1}{2}xe^{2x} - \frac{1}{4}e^{2x} + C$
11. $\frac{1}{6}x^6 \ln x - \frac{1}{36}x^6 + C$ **13.** $(x + 2)e^x - e^x + C$ **15.** $\frac{2}{3}x^{3/2} \ln x - \frac{4}{9}x^{3/2} + C$
17. $\frac{1}{6}(x - 3)(x + 4)^6 - \frac{1}{42}(x + 4)^7 + C$ **19.** $-2te^{-0.5t} - 4e^{-0.5t} + C$ **21.** $-t^{-1} \ln t - t^{-1} + C$
23. $\frac{1}{10}s(2s + 1)^5 - \frac{1}{120}(2s + 1)^6 + C$ **25.** $-\frac{1}{2}xe^{-2x} - \frac{1}{4}e^{-2x} + C$ **27.** $2x(x + 1)^{1/2} - \frac{4}{3}(x + 1)^{3/2} + C$
29. $\frac{1}{a}xe^{ax} - \frac{1}{a^2}e^{ax} + C$ **31.** $\frac{1}{n+1}x^{n+1} \ln ax - \frac{1}{(n+1)^2}x^{n+1} + C$ **33.** $x \ln x - x + C$
35. $\frac{1}{2}x^2 e^{x^2} - \frac{1}{2}e^{x^2} + C$ **37. a.** $\frac{1}{2}e^{x^2} + C$ (by substitution)   **b.** $\frac{1}{4}(\ln x)^4 + C$ (by substitution)
   c. $\frac{1}{3}x^3 \ln 2x - \frac{1}{9}x^3 + C$ (by parts)   **d.** $\ln(e^x + 4) + C$ (by substitution)
39. $e^2 + 1$ **41.** $9 \ln 3 - 3 + \frac{1}{9}$ **43.** $\frac{2^6}{30} = \frac{32}{15}$ **45.** $4 \ln 4 - 3$
47. a. $\frac{1}{6}x(x - 2)^6 - \frac{1}{42}(x - 2)^7 + C$ **b.** $\frac{1}{7}(x - 2)^7 + \frac{1}{3}(x - 2)^6 + C$
49. Using $u = x^n$ and $dv = e^x dx$, the result follows immediately.
51. $x^2 e^x - 2xe^x + 2e^x + C$ **53. a.** The result follows immediately   **b.** [*Hint:* Think of the C.]
55. $R(x) = 4xe^{x/4} - 16e^{x/4} + 16$ **57.** $105.7 million   **59.** $-14e^{-2.5} + 4 \approx 2.85$ milligrams
61. $2 \ln 2 - 1 + \frac{1}{4} \approx 0.64$ square unit   **63.** $72 \ln 6 - 24 + \frac{1}{9} \approx 105$ thousand customers
65. $-x^2 e^{-x} - 2xe^{-x} - 2e^{-x} + C$ **67.** $(x + 1)^2 e^x - 2(x + 1)e^x + 2e^x + C$
69. $\frac{1}{3}x^3 (\ln x)^2 - \frac{2}{9}x^3 \ln x + \frac{2}{27}x^3 + C$ **71.** $2e^2 - 2 \approx 12.78$
73. $-x^2 e^{-x} - 2xe^{-x} - 2e^{-x} + C = -e^{-x}(x^2 + 2x + 2) + C$ **75.** $\frac{1}{2}x^3 e^{2x} - \frac{3}{4}x^2 e^{2x} + \frac{3}{4}xe^{2x} - \frac{3}{8}e^{2x} + C$
77. $\frac{1}{3}(x - 1)^3 e^{3x} - \frac{1}{3}(x - 1)^2 e^{3x} + \frac{2}{9}(x - 1)e^{3x} - \frac{2}{27}e^{3x} + C$

Exercises 12.2 page 885

1. Formula 12, $a = 5$, $b = -1$ **3.** Formula 14, $a = -1$, $b = 7$ **5.** Formula 9, $a = -1$, $b = 1$
7. $\frac{1}{6} \ln \left|\frac{3+x}{3-x}\right| + C$ **9.** $-\frac{1}{x} - 2 \ln \left|\frac{x}{2x+1}\right| + C$ **11.** $-x - \ln|1 - x| + C$ **13.** $\ln \left|\frac{2x+1}{x+1}\right| + C$
15. $\frac{x}{2}\sqrt{x^2 - 4} - 2 \ln \left|x + \sqrt{x^2 - 4}\right| + C$ **17.** $-\ln \left|\frac{1 + \sqrt{1 - z^2}}{z}\right| + C$
19. $\frac{1}{2}x^3 e^{2x} - \frac{3}{4}x^2 e^{2x} + \frac{3}{4}xe^{2x} - \frac{3}{8}e^{2x} + C$ **21.** $-\frac{1}{100}x^{-100} \ln x - \frac{1}{10,000}x^{-100} + C$
23. $\frac{1}{3} \ln \left|\frac{x}{x+3}\right| + C$ **25.** $\frac{1}{8} \ln \left|\frac{z^2 - 2}{z^2 + 2}\right| + C$ **27.** $\frac{x}{2}\sqrt{9x^2 + 16} + \frac{8}{3} \ln \left|3x + \sqrt{9x^2 + 16}\right| + C$
29. $-\frac{1}{2} \ln \left|\frac{2 + \sqrt{4 - e^{2t}}}{e^t}\right| + C$ **31.** $\frac{1}{2} \ln \left|\frac{e^t - 1}{e^t + 1}\right| + C$ **33.** $\frac{1}{4} \ln \left|x^4 + \sqrt{x^8 - 1}\right| + C$
35. $\frac{1}{3} \ln \left|\frac{\sqrt{x^3 + 1} - 1}{\sqrt{x^3 + 1} + 1}\right| + C$ **37.** $\frac{1}{2} \ln \left|\frac{e^t - 1}{e^t + 1}\right| + C$ **39.** $2xe^{x/2} - 4e^{x/2} + C$
41. $\frac{1}{4} \ln \left|\frac{e^{-x} + 4}{e^{-x}}\right| + C = \frac{1}{4} \ln(1 + 4e^x) + C$ **43.** $\frac{15}{2} - 8 \ln 8 + 8 \ln 4 \approx 1.95$

45. $\frac{1}{2}\ln\frac{1}{2} - \frac{1}{2}\ln\frac{1}{3} \approx 0.203$ **47.** $-4 + 5\ln 3 \approx 1.49$ **49.** $\frac{1}{2}\ln|2x+6| + C$

51. $\frac{x}{2} - \frac{3}{2}\ln|2x+6| + C$ **53.** $-\frac{1}{3}(1-x^2)^{3/2} + C$ **55.** $\sqrt{1-x^2} - \ln\left|\frac{1+\sqrt{1-x^2}}{x}\right| + C$

57. $\frac{1}{2}\left(\ln|x+1| - \frac{1}{3}\ln|3x+1|\right) - \frac{1}{2}\ln\left|\frac{3x+1}{x+1}\right| + C$

59. $\ln|x + \sqrt{x^2+1}| - \ln\left|\frac{1+\sqrt{x^2+1}}{x}\right| + C$ **61.** $x + 2\ln|x-1| + C$

63. $-x^2 e^{-x} - 2xe^{-x} - 2e^{-x} + 2$ million sales   **65.** 24 generations   **67.** $C(x) = \ln(x + \sqrt{x^2+1}) + 2000$

Exercises 12.3 page 898

1. 0   **3.** 1   **5.** Does not exist   **7.** Does not exist   **9.** $\frac{1}{2}$ **11.** $\frac{1}{8}$ **13.** Divergent   **15.** 100
17. 20   **19.** $\frac{1}{2}$ **21.** Divergent   **23.** $\frac{1}{3}$ **25.** $\frac{1}{3}$ **27.** Divergent   **29.** 1   **31.** Divergent

33. $\int_0^\infty e^{\sqrt{x}}\,dx$ diverges and $\int_0^\infty e^{-x^2}\,dx$ converges to 0.88623 **35.** $200,000

37. a. $10,000   **b.** $9999.55   **39.** About $3,963,000 (from 3963 thousand)

41. 1,000,000 barrels (from 1000 thousand)   **43.** 2 square units   **45.** $\frac{1}{a}$ square units   **47.** 0.61 or 61%

49. 0.30 or 30%   **51.** 20,000   **53.** $40,992   **55.** D/r

Exercises 12.4 page 913

Some answers may vary depending on rounding.
1. a. 8.75   **b.** 8.667   **c.** 0.083   **d.** 1%   **3. a.** 0.697   **b.** 0.693   **c.** 0.004   **d.** 0.6%
5. 1.154   **7.** 0.743   **9.** 0.593   **11.** 8.6968   **13.** 2.925   **15.** 0.4772 or about 48%
17. $17,300 (from 17.30 thousand)   **19.** 8.667   **21.** 0.693   **23.** 1.148   **25.** 0.747
27. 0.593   **29.** 8.69678496   **31.** 2.92530   **33. a.** $-\int_{1}^{0}\frac{t}{1+t^3}\,dt = \int_{0}^{1}\frac{t}{1+t^3}\,dt$ **b.** 0.374
35. 821 feet   **37.** The justification follows from carrying out the indicated steps.

Exercises 12.5 page 930

1. Check that $(4e^{2x} - 3e^x) - 3(2e^{2x} - 3e^x) + 2(e^{2x} - 3e^x + 2) \stackrel{?}{=} 4$

3. Check that $kae^{ax} \stackrel{?}{=} a\left(ke^{az} - \frac{b}{a}\right) + b$ **5.** $y = \sqrt[3]{6x^2 + c}$ **7.** Not separable

9. $y = ce^{2x^3}$ Check that $c6x^2 e^{2x^3} \stackrel{?}{=} 6x^2(ce^{2x^3})$

11. $y = cx$ (since $e^{\ln x} = x$).   Check that $c \stackrel{?}{=} \frac{cx}{x}$ **13.** $y = \sqrt{4x^2 + c}$ and $y = -\sqrt{4x^2 + c}$

15. Not separable   **17.** $y = 3x^3 + C$ **19.** $y = \frac{1}{2}\ln(x^2+1) + C$ **21.** $y = ce^{x^3/3}$

23. $y = \left(\frac{1-n}{m+1}x^{m+1} + c\right)^{1/(1-n)}$ **25.** $y = (x+c)^2$ **27.** Not separable   **29.** $y = ce^{x^2/2} - 1$

31. $y = ce^{e^x} + 1$ **33.** $y = \frac{1}{c - ax}$ **35.** $y = ce^{ax} - \frac{b}{a}$

37. $y = \sqrt[3]{3x^2 + 8}$ Check that $(3x^2 + 8)^{\frac{2}{3}} \cdot \frac{1}{3}(3x^2 + 8)^{-\frac{2}{3}} \cdot 6x \stackrel{?}{=} 2x$ and $y(0) = \sqrt[3]{8} = 2$

39. $y = -e^{e^x/2}$ Check that $-xe^{x^2/2} \stackrel{?}{=} x(-e^{x^2/2})$ and $y(0) = -e^0 = -1$

41. $y = (1 - x^2)^{-1}$ Check that $-(1 - x^2)^{-2}(-2x) \stackrel{?}{=} 2x[1 - x^2)^{-1}]^2$ and $y(0) = (1 - 0)^{-1} = 1$

43. $y = 3x$ (using $e^{\ln x} = x$) Check that $3 \stackrel{?}{=} \dfrac{3x}{x}$ and $y(1) = 3 \cdot 1 = 3$

45. $y = (x + 1)^2$ or $y = (x - 3)^2$ Check that $2(x + 1) \stackrel{?}{=} 2\sqrt{(x + 1)^2}$ and $y(1) = 2^2 = 4$ and that $2(x - 3) \stackrel{?}{=} 2\sqrt{(x - 3)^2}$ and $y(1) = (-2)^2$

47. $y = \dfrac{1}{2 - e^x - x}$ Check that $\dfrac{e^x + 1}{(2 - e^x - x)^2} \stackrel{?}{=} \left(\dfrac{1}{2 - e^x - x}\right)^2 e^x + \left(\dfrac{1}{2 - e^x - x}\right)^2$ and $y(0) = \dfrac{1}{2 - 1} = 1$

49. $y = 2e^{ax^3/3}$ Check that $2ax^2 e^{ax^3/3} \stackrel{?}{=} ax^2 2e^{ax^3/3}$ and $y(0) = 2e^0 = 2$

51. $D(p) = cp^{-k}$ (for any constant c)   **53.** $y = 20{,}000e^{0.05t} - 20{,}000$

55. a. $y = 28.6e^{-0.32t} + 70$ **b.** About 3.28 hours

57. $y = 150 - 150e^{-0.2t}$ **59. a.** $y' = 3 + 0.10y$ **b.** $y(0) = 6$ **c.** $y(t) = 36e^{0.1t} - 30$
d. $y(25) = 408.570$ thousand dollars, or \$408,570 (rounded)

61. a. $y' = 4y^{1/2}$ **b.** $y(0) = 10{,}000$ **c.** $y(t) = (2t + 100)^2$ **d.** 15,376
63. a. $y' = 8y^{3/4}$ **b.** $y(0) = 10{,}000$ **c.** $y(t) = (2t + 10)^4$ **d.** 234,256
65. a. $p(t) = Ce^{-Kt/R}$ for $t_0 \le t \le T$ **b.** $p(t) = p_0 e^{-K(t - t_0)/R}$ for $t_0 \le t \le T$
c. $p(t) = I_0 R - Ce^{-Kt/R}$ for $0 \le t \le t_0$ **d.** $p(t) = I_0 R - (I_0 R - p_0)e^{K(t_0 - t)/R}$ for $0 \le t \le t_0$

67. c. $y = \sqrt[5]{10x^3 + 32}$ **69. c.** $y = \sqrt{8x^2 + 16}$
d.

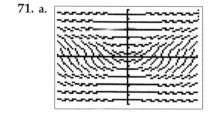

d.

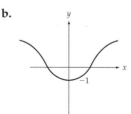

71. a.

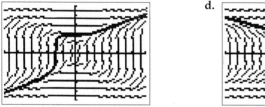

b.

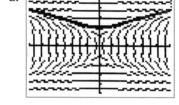

Exercises 12.6 page 947

1. $y' = cae^{at} = a(ce^{at}) = ay$ **3.** Unlimited   **5.** Limited   **7.** None   **9.** Logistic   **11.** Logistic
$y(0) = ce^0 = c$ **13.** $y = 1.5e^{6t}$ **15.** $y = 100e$ **17.** $y = -e^{-0.45t}$
19. $y = 100(1 - e^{-2t})$ **21.** $y = 0.25(1 - e^{-0.05t})$ **23.** $y = 40(1 - e^{-2t})$ **25.** $y = 200(1 - e^{-0.01t})$
27. $y = \dfrac{100}{1 + 9e^{-500t}}$ **29.** $y = \dfrac{0.5}{1 + 4e^{-0.125t}}$ **31.** $y = \dfrac{10}{1 - \frac{1}{2}e^{-30t}} = \dfrac{20}{2 - e^{-30t}}$ **33.** $y = \dfrac{3}{1 + 2e^{-6t}}$

35. $y' = 0.08y$
$y = 1500e^{0.08t}$

37. $y' = a(100{,}000 - y)$
$y = 100{,}000(1 - e^{-0.021t})$
About 22,276

39. $y' = a(500 - y)$
$y = 5000(1 - e^{-0.223t})$
About 7.2 weeks

41. $y' = ay(10{,}000 - y)$
$y = \dfrac{10{,}000}{1 + 99e^{-0.535t}}$
About 8612

43. $y' = ay(800 - y)$
$y = \dfrac{800}{1 + 799e^{-0.558t}}$
About 675 people

45. $y' = ay(800 - y)$
$y = \dfrac{800}{1 + 7e^{-0.280t}}$
About 6.9 years

47. $y = 5e^{-0.15t}$
About 3.7

49. a. $y' = 0.1(200 - y)$ with $M = 200$ **b.** $y = 200(1 - e^{-0.1t})$
c. 30 years [from solving $200(1 - e^{-0.1t}) = 0.95 \cdot 200$]

51. $y = \dfrac{1}{c - at}$ **53.** $y = ce^{a \ln x} = cx^a$ **55.** The solution follows from the indicated steps.

57. a. About 16.6 feet per second **b.** About 0.6 foot per second **c.** About 0.006 foot per second
d. About $\dfrac{1}{0.006} \approx 167$ seconds, or about 2.8 minutes

Chapter 12 Review Exercises page 953

1. $\frac{1}{2}xe^{2x} - \frac{1}{4}e^{2x} + C$ **2.** $-xe^{-x} - e^{-x} + C$ **3.** $\frac{1}{9}x^9 \ln x - \frac{1}{81}x^9 + C$ **4.** $\frac{4}{5}x^{5/4} \ln x - \frac{16}{25}x^{5/4} + C$

5. $\frac{1}{6}(x - 2)(x + 1)^6 - \frac{1}{42}(x + 1)^7 + C$ **6.** $\frac{1}{5}(x + 3)(x - 1)^5 - \frac{1}{30}(x - 1)^6 + C$ **7.** $2t^{1/2} \ln t - 4t^{1/2} + C$

8. $\frac{1}{4}x^4 e^{x^4} - \frac{1}{4}e^{x^4} + C$ **9.** $x^2 e^x - 2xe^x + 2e^x + C$ **10.** $x(\ln x)^2 - 2x \ln x + 2x + C$

11. $\dfrac{1}{n + 1} x(x + a)^{n+1} - \dfrac{1}{(n + 1)(n + 2)}(x + a)^{n+2} + C$

12. $-\dfrac{1}{n + 1} x(1 - x)^{n+1} - \dfrac{1}{(n + 1)(n + 2)}(1 - x)^{n+2} + C$ **13.** $4e^5 + 1$ **14.** $\frac{1}{4}e^2 + \frac{1}{4}$

15. $-\ln |1 - x| + C$ **16.** $-\frac{1}{2}e^{-x^2} + C$ **17.** $\frac{1}{4}x^4 \ln 2x - \frac{1}{16}x^4 + C$ **18.** $(1 - x)^{-1} + C$

19. $\frac{1}{2}(\ln x)^2 + C$ **20.** $\frac{1}{2}\ln(e^{2x} + 1) + C$ **21.** $2e^{\sqrt{x}} + C$ **22.** $\frac{1}{8}(e^{2x} + 1)^4 + C$

23. $-15{,}000e^{-0.5} + 10{,}000 \approx 902$ million dollars **24.** 6.78 hundred gallons (from $25 - 10e^{0.6}$)

25. $\dfrac{1}{10} \ln \left| \dfrac{5 + x}{5 - x} \right| + C$ **26.** $\dfrac{1}{4} \ln \left| \dfrac{x - 2}{x + 2} \right| + C$ **27.** $2 \ln |x - 2| - \ln |x - 1| + C$

28. $-\ln \left| \dfrac{x - 1}{x - 2} \right| + C$ or $\ln \left| \dfrac{x - 2}{x - 1} \right| + C$ **29.** $\ln \left| \dfrac{\sqrt{x + 1} - 1}{\sqrt{x + 1} + 1} \right| + C$ **30.** $\dfrac{2x - 4}{3} \sqrt{x + 1} + C$

31. $\ln |x + \sqrt{x^2 + 9}| + C$ **32.** $\ln |x + \sqrt{x^2 + 16}| + C$

33. $\dfrac{z^2 - 2}{3} \sqrt{z^2 + 1} + C$ (from formula 13) **34.** $e^t - 2\ln(e^t + 2) + C$ **35.** $\frac{1}{2}x^2 e^{2x} - \frac{1}{2}xe^{2x} + \frac{1}{4}e^{2x} + C$

36. $x(\ln x)^4 - 4x(\ln x)^3 + 12x(\ln x)^2 - 24x \ln x + 24x + C$ **37.** $\ln \left| \dfrac{2x + 1}{x + 1} \right| + 1000$

38. 1305 (from $750 + 800 \ln 80 - 800 \ln 40$) **39.** $\frac{1}{4}$ **40.** $\frac{1}{5}$ **41.** Divergent **42.** Divergent

43. $\frac{1}{2}$ **44.** $2e^{-2}$ **45.** Divergent **46.** Divergent **47.** 5 **48.** $10e^{-10}$ **49.** $\frac{1}{4}$ **50.** $\frac{1}{5}$
51. $\frac{1}{2}$ **52.** $\frac{1}{4}$ **53.** 1 **54.** 1 **55.** $\frac{1}{3}$ **56.** $\frac{1}{2}$ **57.** \$60,000 **58.** 0.35 or 35%

59. 240 thousand **60.** 7500 million tons **61.** $\displaystyle\int_1^\infty \dfrac{1}{x^3}\, dx$ converges to 0.5 **62.** $\displaystyle\int_1^\infty \dfrac{1}{\sqrt[3]{x}}\, dx$ diverges

63. 1.102 **64.** 1.09 **65.** 1.204 **66.** 0.852 **67.** 0.570 **68.** 1.313 **69.** 1.089 **70.** 1.075

71. 1.195   **72.** 0.856   **73.** 0.528   **74.** 1.348   **75.** 1.0894   **76.** 1.0747   **77.** 1.1951
78. 0.8556   **79.** 0.5285   **80.** 1.3357   **81.** 1.089429   **82.** 1.074669   **83.** 1.194958
84. 0.855624   **85.** 0.527887   **86.** 1.347855   **87. a.** $-\int_1^0 \frac{1}{1+t^2} dt = \int_0^1 \frac{1}{1+t^2} dt$ **b.** 0.783
88. a. $-\int_1^0 \frac{1}{1+t^4} dt = \int_0^1 \frac{1}{1+t^4} dt$ **b.** 0.862   **89.** $y = \sqrt[3]{x^3 + c}$ **90.** $y = ce^{x^3/3}$
91. $y = \frac{1}{4}\ln(x^4 + 1) + C$ **92.** $y = -\frac{1}{2}e^{-x^2} + C$ **93.** $y = \frac{1}{c - x}$
94. $y = \frac{1}{\sqrt{c - 2x}}$ and $y = -\frac{1}{\sqrt{c - 2x}}$ **95.** $y = 1 + ce^{-x}$ or $y = 1 - e^{c-x}$
96. $y = \sqrt{2x + c}$ and $y = -\sqrt{2x + c}$ **97.** $y = ce^{\frac{1}{2}x^2 - x}$ **98.** $y = ce^{x^3/3} - 1$ **99.** $y = \sqrt[3]{3x^3 + 1}$
100. $y = e^{1 - x^{-1}}$, (from $ee^{-x^{-1}}$)   **101.** $y = e^{\frac{1}{2} - \frac{1}{2}x^{-2}}$ (from $e^{\frac{1}{2}}e^{-\frac{1}{2}x^{-2}}$)
102. $y = \left(\frac{2}{3}x - \frac{2}{3}\right)^{3/2}$ or $y = -\left(\frac{2}{3}x - \frac{2}{3}\right)^{3/2}$
103. a. $y' = 4 + 0.05y,\ y(0) = 10$ **b.** $y = 90e^{0.05t} - 80$ **c.** 68.385 thousand or \$68,385
104. a. $y' = 4 - 0.25y,\ y(0) = 0$ **b.** $y = 16 - 16e^{-0.25t}$ **105.** $y = 106 - 36e^{-2.3t}$
106. $y(1) \approx 102.4$ **107. c.** $y = \sqrt[3]{x^3 - 8}$ **108. c.** $y = \sqrt[3]{\frac{3}{2}x^2 - 8}$
$y(2) \approx 105.6$ **d.**   **d.**
$y(3) \approx 105.96$

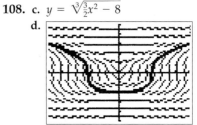

109. $y' = 0.07y$ **110.** $y' = 0.12$ **111.** $y' = ay(8000 - y)$
$y = 37e^{0.07t}$ $y = 16.8e^{0.12t}$ $y = \dfrac{8000}{1 + .799e^{-2.73t}}$ (t in weeks)
About 65¢   About \$55.8 billion   About 1819 cases

112. $y' = ay(500 - y)$ **113.** $y' = a(10{,}000 - y)$ **114.** $y' = a(60 - y)$
$y = \dfrac{500}{1 + 249e^{-3.78t}}$ $y = 10{,}000(1 - e^{-0.051t})$ $y = 60(1 - e^{-0.269t})$
About 443   About 4577   About 6.7 weeks

115. $y' = a(500{,}000 - y)$
$y = 500{,}000(1 - e^{-0.255t})$ (t in weeks)
About 6.3 weeks

Exercises 13.1 page 970

1. $\{(x, y)\,|\,x \neq 0, y \neq 0\}$ **3.** $\{(x, y)\,|\,x \neq y\}$ **5.** $\{(x, y)\,|\,x > 0, y \neq 0\}$
7. $\{(x, y, z)\,|\,x \neq 0, y \neq 0, z > 0\}$ **9.** 3   **11.** 4   **13.** -2 **15.** 1   **17.** $e^{-1} + e$ **19.** e^{-1}
21. 0   **23.** 0.0157   **25.** 45 minutes   **27.** 472.7
29. $P(2L, 2K) = a(2L)^b(2K)^{1-b} = a2^b L^b 2^{1-b} K^{1-b} = 2aL^b K^{1-b} = 2P(L, K)$ **31.** 1548 calls
$\underbrace{\phantom{a2^b L^b 2^{1-b} K^{1-b}}}_{2}$

33. $C(x, y) = 210x + 180y + 4000$ **35. a.** $V = xyz$ **b.** $M = xy + 2xz + 2yz$
37. a.

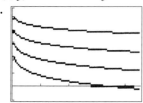

  b. A given wind speed will lower the windchill further on a colder day than on a warmer day.
  c. For the lowest curve: $dy/dx \approx -0.63$, meaning that at 20 degrees and 10 mph of wind, windchill drops by about 0.63 degrees for each additional 1 mph of wind. For the highest curve: $dy/dx \approx -0.33$, meaning that at 50 degrees and 10 mph of wind, windchill drops by only about 0.33 degrees for each additional 1 mph of wind.
  d. Yes—the effect of wind on the windchill index is greater on a colder day.

Exercises 13.2 page 985

1. a. $3x^2 + 6xy^2 - 1$ **b.** $6x^2y - 6y^2 + 1$ **3. a.** $6x^{-1/2}y^{1/3}$ **b.** $4x^{1/2}y^{-2/3}$
5. a. $5x^{-0.95}y^{0.02}$ **b.** $2x^{0.05}y^{-0.98}$ **7. a.** $-(x+y)^{-2}$ **b.** $-(x+y)^{-2}$ **9. a.** $\dfrac{3x^2}{x^3+y^3}$ **b.** $\dfrac{3y^2}{x^3+y^3}$
11. a. $6x^2e^{-5y}$ **b.** $-10x^3e^{-5y}$ **13. a.** ye^{xy} **b.** xe^{xy}
15. a. $\dfrac{x}{x^2+y^2}$ or $x(x^2+y^2)^{-1}$ **b.** $\dfrac{y}{x^2+y^2}$ or $y(x^2+y^2)^{-1}$
17. a. $3v(uv-1)^2$ **b.** $3u(uv-1)^2$ **19. a.** $ue^{(u^2-v^2)/2}$ **b.** $-ve^{(u^2-v^2)/2}$ **21.** $18, -10$
23. $0, 2e$ **25.** $1\tfrac{1}{2}$ **27. a.** $30x - 4y^3$ **b. and c.** $-12xy^2$ **d.** $-12x^2y + 36y^2$
29. a. $-2x^{-5/3}y^{2/3}$ **b. and c.** $2x^{-2/3}y^{-1/3} - 12y^2$ **d.** $-2x^{1/3}y^{-4/3} - 24xy$
31. a. ye^x **b. and c.** $e^x - \dfrac{1}{y}$ **d.** xy^{-2} **33.** All three are $36x^2y^2$.
35. a. y^2z^3 **b.** $2xyz^3$ **c.** $3xy^2z^2$
37. a. $8x(x^2+y^2+z^2)^3$ **b.** $8y(x^2+y^2+z^2)^3$ **c.** $8z(x^2+y^2+z^2)^3$
39. a. $2xe^{x^2+y^2+z^2}$ **b.** $2ye^{x^2+y^2+z^2}$ **c.** $2ze^{x^2+y^2+z^2}$ **41.** -14 **43.** $4e^6$
45. a. $P_x = 4x - 3y + 150$ **b.** $50 (profit per additional tape deck)   **c.** $P_y = -3x + 6y + 75$
  d. $75 (profit per additional CD player)
47. a. 250 (the marginal productivity of labor is 250, so production increases by about 250 for each additional unit of labor)
  b. 108 (the marginal productivity of capital is 108, so production increases by about 108 for each additional unit of capital)   **c.** Labor
49. $S_x = -0.1$ (sales fall by 0.1 for each dollar price increase)
  $S_y = 0.4y$ (sales rise by $0.4y$ for each additional advertising dollar above the level y)
51. a. 0.52 (status increases by about 0.52 unit for each additional $1000 of income)
  b. 5.25 (status increases by 5.25 units for each additional year of education)
53. a. 97.2 (skid distance increases by about 97 feet for each additional ton)
  b. 12.96 (skid distance increases by about 13 feet for each additional mph)
55. a. Rate at which butter sales change as butter prices rise
  b. Negative: as prices rise, sales will fall.

c. Rate at which butter sales change as margarine prices rise
d. Positive: as margarine prices rise, people will switch to butter, so butter sales will rise.

Exercises 13.3 page 997

1. Rel min value: $f = 5$ at $x = 0$, $y = -1$ **3.** Rel min value: $f = -12$ at $x = -2$, $y = 2$
5. Rel max value: $f = 23$ at $x = 5$, $y = 2$ **7.** No rel extreme values [saddle point at $(2, -4)$]
9. No rel extreme values   **11.** Rel min value: $f = 1$ at $x = 0$, $y = 0$
13. Rel min value: $f = 0$ at $x = 0$, $y = 0$
15. Rel max value: $f = 3$ at $x = 1$, $y = -1$ [saddle point at $(-1, -1)$]
17. Rel max value: $f = 17$ at $x = -1$, $y = -2$ [saddle point at $(-1, 2)$]
19. No rel extreme values [saddle point at $(2, 6)$]
21. 10 units of product A, sell for $7000 each; 7 units of product B, sell for $13,000 each.
Maximum profit: $22,000
23. a. $P = -0.2x^2 + 16x - 0.1y^2 + 12y - 20$
  b. 40 cars in America, sell for $12,000; 60 cars in Europe, sell for $10,000
25. 6 hours of practice and 1 hour of rest
27. a. $x = 1200$, $p = \$6$, $R = \$7200$ **b.** $x = 800$, $y = 800$, $p = \$4$, revenue $= \$3200$ for each
  c. Duopoly (1600 versus 1200)   **d.** Duopoly
29. Sell the sedans for $19,200, selling 12 per day, and sell the SUVs for $23,200, selling 7 per day.
31. a. $P = -0.2x^2 + 16x - 0.1y^2 + 12y - 0.1z^2 + 8z - 22$ **b.** 40 in America, 60 in Europe, 40 in Asia
33. Rel min value: $f = -1$ at $x = 1$, $y = 1$ [saddle point at $(0, 0)$]
35. Rel max value: $f = 32$ at $x = 4$, $y = 4$ [saddle point at $(0, 0)$]
37. Rel min value: $f = -162$ at $x = 3$, $y = 18$ and at $x = -3$, $y = -18$ [saddle point at $(0, 0)$]

Exercises 13.4 page 1009

Note: Your answers may differ slightly depending on the stage at which you do the rounding.
1. $y = 3.5x - 1.67$ **3.** $y = -0.79x + 6.6$ **5.** $y = 2.4x + 6.9$ **7.** $y = -2.1x + 7.6$
9. $y = 2.2x + 5$; prediction: 16 million   **11.** $y = -8x + 125$; prediction: 85 arrests
13. $y = -0.009x + 0.434$; prediction: .380   **15.** $y = -2.5x + 38.1$; prediction: 20.6%
17. $y = -0.16x + 71.6$ **19.** $y = 1.09e^{0.63x}$ **21.** $y = 17.45e^{-0.47x}$ **23.** $y = 0.98e^{0.78x}$
25. $y = 16.95e^{-0.52x}$ **27.** $y = 459e^{0.215x}$; predictions: $1,085,000 (from 1085), $1,345,000 (from 1345)
29. $y = 30.33e^{0.198x}$; prediction: $148,000 per second (from 147.8 thousand)
31. c.

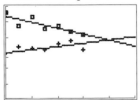

  d. In 2010 (from $x \approx 39.7$ years after 1970)

Exercises 13.5 page 1025

1. Max $f = 36$ at $x = 6$, $y = 2$ **3.** Max $f = 144$ at $x = 6$, $y = 4$
5. Max $f = -28$ at $x = 3$, $y = 5$ **7.** Max $f = 6$ at $x = 2$, $y = -1$

9. Max $f = 2$ (from $\ln e^2$) at $x = e$, $y = e$ 11. Min $f = 45$ at $x = 6$, $y = 3$
13. Min $f = -16$ at $x = -4$, $y = 4$ 15. Min $f = 52$ at $x = 4$, $y = 6$
17. Min $f = \ln 125$ at $x = 10$, $y = 5$ 19. Min $f = e^{20}$ at $x = 2$, $y = 4$
21. Max $f = 8$ at $x = 2$, $y = 2$ and at $x = -2$, $y = -2$;
     Min $f = -8$ at $x = 2$, $y = -2$ and at $x = -2$, $y = 2$
23. Max $f = 18$ at $x = 2$, $y = 8$; Min $f = -18$ at $x = -2$, $y = -8$
25. **a.** 1000 feet perpendicular to building, 3000 feet parallel to building
     b. $|\lambda| = 1000$; each additional foot of fence adds about 1000 square feet of area
27. $r \approx 3.7$ feet, $h \approx 3.7$ feet
29. End: 14 inches by 14 inches; length = 28 inches; volume = 5488 cubic inches
31. **a.** $L = 120$, $K = 20$ **b.** $|\lambda| \approx 1.9$; output increases by about 1.9 for each additional dollar
33. Base: 3 inches by 3 inches; height: 5 inches   35. Min $f = 24$ at $x = 4$, $y = 2$, $z = -2$
37. Max $f = 6$ at $x = 2$, $y = 2$, $z = 2$ 39. Base: 50 feet by 50 feet; height: 100 feet

Exercises 13.6 page 1037

1. $df = 2xy^3 \cdot dx + 3x^2y^2 \cdot dy$ 3. $df = 3x^{-1/2}y^{1/3} \cdot dx + 2x^{1/2}y^{-2/3} \cdot dy$ 5. $dg = \dfrac{1}{y} \cdot dx - \dfrac{x}{y^2} \cdot dy$

7. $dg = -(x-y)^{-2} \cdot dx + (x-y)^{-2} \cdot dy$ 9. $dz = \dfrac{3x^2}{x^3 - y^2} \cdot dx - \dfrac{2y}{x^3 - y^2} \cdot dy$

11. $dz = e^{2y} \cdot dx + 2xe^{2y} \cdot dy$ 13. $dw = (6x^2 + y) \cdot dx + (x + 2y) \cdot dy$

15. $df = 4xy^3z^4 \cdot dx + 6x^2y^2z^4 \cdot dy + 8x^2y^3z^3 \cdot dz$ 17. $df = \dfrac{1}{x} dx + \dfrac{1}{y} dy + \dfrac{1}{z} dz$

19. $df = yze^{xyz} \cdot dx + xze^{xyz} \cdot dy + xye^{xyz} \cdot dz = e^{xyz}(yz \cdot dx + xz \cdot dy + xy \cdot dz)$ 21. **a.** $\Delta f = 0.479$
     b. $df = 0.4$

23. **a.** $\Delta f \approx 0.112$ **b.** $df = 0.11$ 25. **a.** $\Delta f = 0.1407$ **b.** $df = 0.14$
27. 125 square feet; 250 square feet   29. $4300   31. About 113 feet   33. 2%
35. 0.5 liter per minute
37. **a.** f is being evaluated at two points along the curve; each of these points gives $f = c$, and $c - c = 0$.
     b. Subtracting and adding $F(x)$.
     c. Approximating the change $\Delta f = f(x + \Delta x, F + \Delta F) - f(x, F)$ by the total differential
     $df = f_x \Delta x + f_y \Delta F$.
     d. Subtracting $f_y \Delta F$ and dividing by f_y and Δx.
     e. Taking the limit as $\Delta x \to 0$ causes $\dfrac{\Delta F}{\Delta x}$ to approach $\dfrac{dF}{dx}$ and the approximation to become exact.

Exercises 13.7 page 1051

1. $2x^9 - 2x$ 3. $6y^4$ 5. $2x^2$ 7. 4   9. 2   11. $\frac{1}{2}$ 13. 12
15. 14   17. $-9e^{-3} + 9e^3$ 19. 0   21. -12 23. 0   25. 72   27. $\frac{1}{2}$

29. **a.** $\displaystyle\int_1^3 \int_0^2 3xy^2 \, dx \, dy$ and $\displaystyle\int_0^2 \int_1^3 3xy^2 \, dy \, dx$ **b.** Both equal 52

31. **a.** $\displaystyle\int_0^2 \int_{-1}^1 ye^x \, dx \, dy$ and $\displaystyle\int_{-1}^1 \int_0^2 ye^x \, dy \, dx$ **b.** Both equal $2e - 2e^{-1}$ 33. 8 cubic units

35. $\frac{4}{3}$ cubic units   37. $\frac{1}{6}$ cubic unit   39. $\frac{1}{2}e^2 - e + \frac{1}{2}$ cubic units   41. 45 degrees $\left(\text{from } \frac{540}{12}\right)$
43. About 180,200 people   45. 900,000 cubic feet   47. 14   49. 10

Chapter 13 Review Exercises page 1055

1. $\{(x,y) \mid x \geq 0, y \neq 0\}$ 2. $\{(x,y) \mid y > 0\}$ 3. $\{(x,y) \mid x \neq 0, y > 0\}$ 4. $\{(x,y) \mid x \neq 0, y > 0\}$
5. a. $10x^4 - 6xy^3 - 3$ b. $-9x^2y^2 + 4y^3 + 2$ c. and d. $-18xy^2$
6. a. $12x^3 + 15x^2y^2 - 6$ b. $10x^3y - 6y^5 + 1$ c. and d. $30x^2y$
7. a. $12x^{-1/3}y^{1/3}$ b. $6x^{2/3}y^{-2/3}$ c. and d. $4x^{-1/3}y^{-2/3}$
8. a. $\dfrac{2x}{x^2+y^3}$ b. $\dfrac{3y^2}{x^2+y^3}$ c. and d. $\dfrac{-6xy^2}{(x^2+y^3)^2}$
9. a. $3x^2 e^{x^3-2y^3}$ b. $-6y^2 e^{x^3-2y^3}$ c. and d. $-18x^2y^2 e^{x^3-2y^3}$
10. a. $6xe^{-5y}$ b. $-15x^2 e^{-5y}$ c. and d. $-30xe^{-5y}$
11. a. $-ye^{-x} - \ln y$ b. $e^{-x} - \dfrac{x}{y}$ c. and d. $-e^{-x} - \dfrac{1}{y}$
12. a. $2xe^y + yx^{-1}$ b. $x^2 e^y + \ln x$ c. and d. $2xe^y + x^{-1}$ 13. a. $\tfrac{1}{2}$ b. $\tfrac{1}{2}$
14. a. 0 b. $\tfrac{1}{2}$ 15. a. 36 b. -24 16. a. 216 b. -216
17. a. 80: rate at which production increases for each additional unit of labor
    b. 135: rate at which production increases for each additional unit of capital   c. Capital
18. $S_x = 222$: rate at which sales increase for each additional $1000 in TV ads.
    $S_y = 528$: rate at which sales increase for each additional $1000 in print ads.
19. Min $f = -13$ at $x = -1$, $y = -4$ 20. Min $f = -8$ at $x = 4$, $y = 1$
21. Max $f = 8$ at $x = 0$, $y = -1$ 22. Max $f = 6$ at $x = 1$, $y = 0$
23. No rel extreme values (saddle point at $x = \tfrac{1}{2}$, $y = -3$)
24. No rel extreme values (saddle point at $x = -\tfrac{1}{2}$, $y = 1$)   25. Max $f = 1$ at $x = 0$, $y = 0$
26. Min $f = 1$ at $x = 0$, $y = 0$ 27. Min $f = 0$ at $x = 0$, $y = 0$
28. Min $f = \ln 10$ at $x = 0$, $y = 0$
29. Max $f = 25$ at $x = -2$, $y = -3$ (saddle point at $x = 2$, $y = -3$)
30. Min $f = -20$ at $x = -2$, $y = 2$ (saddle point at $x = 2$, $y = 2$)
31. a. $C(x,y) = 3000x + 5000y + 6000$ b. $R(x,y) = 7000x - 20x^2 + 8000y - 30y^2$
    c. $P(x,y) = -20x^2 + 4000x - 30y^2 + 3000y - 6000$
    d. Make 100 18-foot boats, sell for $5000 each, and 50 22-foot boats, sell for $6500 each;
       max profit: $269,000.
32. a. $P(x,y) = -0.2x^2 + 68x - 0.1y^2 + 52y - 100$
    b. America: sell 170 for $46,000 each; Europe: sell 260 for $38,000 each (since prices are in thousands)
33. $y = 2.6x - 3.2$ 34. $y = -1.8x + 8.4$ 35. $y = 4.71x + 11.51$; prediction: 39.8 million
36. $y = -0.76x + 8.2$; prediction: 2.9% (from 2.88)   37. Max $f = 292$ at $x = 12$, $y = -24$
38. Max $f = 156$ at $x = 10$, $y = 8$ 39. Min $f = 90$ at $x = 7$, $y = 4$
40. Min $f = -109$ at $x = -3$, $y = 3$ 41. Min $f = e^{45}$ at $x = 3$, $y = 6$
42. Max $f = e^{-5}$ at $x = 2$, $y = 1$
43. Max $f = 120$ at $x = 2$, $y = -6$; Min $f = -120$ at $x = -2$, $y = 6$
44. Max $f = 64$ at $x = 4$, $y = 4$ and at $x = -4$, $y = -4$;
    Min $f = -64$ at $x = 4$, $y = -4$ and at $x = -4$, $y = 4$
45. a. $40,000 for production, $20,000 for advertising
    b. $|\lambda| \approx 159$: production increases by about 159 units for each additional dollar
46. a. $\tfrac{1}{2}$ ounce of the first and 7 ounces of the second
    b. $|\lambda| = 9$: each additional dollar results in about 9 additional nutritional units
47. a. $L = 64$, $K = 8$ b. $40L^{-1/3}K^{1/3}$, $20L^{2/3}K^{-2/3}$

B-84 CHAPTER 13

c. $\dfrac{40L^{-1/3}K^{1/3}}{20L^{2/3}K^{-2/3}} \stackrel{?}{=} \dfrac{25}{100}$ (and now simplify and substitute $L = 64$, $K = 8$)

48. Base: 12 inches by 12 inches; height: 4 inches   **49.** $df = (6x + 2y) \cdot dx + (2x + 2y) \cdot dy$

50. $df = (2x + y) \cdot dx + (x - 6y) \cdot dy$ **51.** $dg = \dfrac{1}{x} dx + \dfrac{1}{y} dy = \dfrac{dx}{x} + \dfrac{dy}{y}$

52. $dg = \dfrac{3x^2}{x^3 + y^3} dx + \dfrac{3y^2}{x^3 + y^3} dy$ **53.** $dz = e^{x-y} dx - e^{x-y} dy$ **54.** $dz = ye^{xy} dx + xe^{xy} dy$

55. Sales would decrease by about $153,000 (from $dS = -153$); sales would decrease by about $76,500.

56. 2%   **57.** $8e^2 - 8e^{-2}$ **58.** 10   **59.** $\tfrac{4}{3}$ **60.** $\tfrac{32}{3}$ or $10\tfrac{2}{3}$ **61.** 40 cubic units

62. 24 cubic units   **63.** $\tfrac{5}{6}$ cubic unit   **64.** $\tfrac{8}{9}$ cubic unit   **65.** $12,000 \left(\text{from } \dfrac{192,000}{16} \right)$

66. $640 hundred thousand, or $64,000,000

Cumulative Review for Chapters 1, 8–13 page 1059

1. 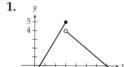   **2.** 4   **3.** $\dfrac{-1}{x^2}$ (but found using the *definition*)   **4.** 4

5. $S'(12) = -2$: each $1 price increase (above $12) decreases sales by 2 per week
6. $3[x^2 + (2x + 1)^4]^2[2x + 8(2x + 1)^3]$
7. 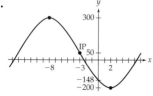   **8.**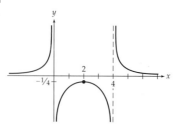

9. 40 feet parallel to wall, 20 feet perpendicular to wall
10. Base: 6 feet by 6 feet; height: 3 feet   **11.** $\dfrac{2}{\pi} \approx 0.64$ foot per minute   **12. a.** $1268.24   **b.** $1271.25

13. In about 14.4 years   **14.** About 6.8 years
15.   **16.** $4x^3 - 2x^2 + x + C$ **17.** $900e^{0.02t} - 900$ million gallons

18. $85\tfrac{1}{3}$ square units   **19.** 16   **20. a.** $\tfrac{1}{3} \ln |x^3 + 1| + C$ **b.** $2e^{x^{1/2}} + C$

21. $\frac{1}{4}xe^{4x} - \frac{1}{16}e^{4x} + C$ **22.** $\sqrt{4-x^2} - 2\ln\left|\frac{2+\sqrt{4-x^2}}{x}\right| + C$ **23.** $\frac{1}{2}$

24. 1.15148 [compared with actual (rounded) value of 1.14779]
25. 1.14778 [compared with actual (rounded) value of 1.14779]

26. a. $y = Ce^{x^4/4}$ **b.** $y = 2e^{x^4/4}$ **27.** $f_x = \ln y + 2ye^{2x}$, $f_y = \frac{x}{y} + e^{2x}$

28. Min $f = 2$ at $x = 1$, $y = 4$; no relative max   **29.** $y = 1.75x - 4.75$

30. Min $f = 90$ at $x = 4$, $y = 7$ **31.** $df = (4x + y)\,dx + (x - 6y)\,dy$ **32.** 48 cubic units

Index

Abscissa, 6
Absolute extreme value, 664, 992
  maximum value, 664
  minimum value, 664
Absolute value function, 52, 621
Absorbing Markov chain, 504
  expected times and long-term absorption probabilities, 508, 517–518
  fundamental matrix, 508, 517–518
  standard form, 506
Absorbing state, 483, 504
Absorbing transition, 482
Absorption constant, 76, 726
Acceleration, 598–599, 603
  due to gravity, 606
  units of, 600
Accumulated amount, 124
  value, 104
Accumulation of wealth, 927–929
Addition principle of counting, 356
Addition rule, 383
  for disjoint events, 383
Africa, population of, 829
Allometry, 29, 950
AMORTABL (program), 138
Amortization, 134
  graphing calculator program, 138
  payment, 135
  spreadsheet exploration, 134
  table, 134, 138
  unpaid balance, 137, 144
Annual percentage yield, 116

Annuity, 122–128
  accumulated amount, 124
  continuous, 931
  ordinary, 122
  present value, 133, 144
Antiderivative, 770
Aortic volume, 816
Applications index. *See inside front cover.*
Approximation, linear, 551
  of π, 914
APY, 116
Arbitrary constant, 771, 814
  evaluation of, 787–788
  geometric meaning, 779
Area between curves, 819–824
Area formulas. *See inside back cover.*
Area under a curve, 797–800, 801, 809
  signed area, 810
Artificial variable, 330
Asymptote, vertical, 535, 644
Augmented matrix, 168, 198
  linear equations, 169
Average, arithmetic. *See* Mean.
  grand, 458
  weighted, 388, 412
Average cost, 583
Average profit, 584
Average rate of change, 546, 549
Average revenue, 584
Average value of a function, 818
  derived from Riemann sums, 826
  of two variables, 1047

Back substitution, 198
Bar chart, 432
Base, 19, 64, 78, 720
Basic variable, 279–280
Bayes' formula, 403
Bayes, Thomas, 402
Beale, E.M.I., 289
  cycling example, 295
Bernoulli, James, 413
  trials, 413
Beverton-Holt recruitment curve, 17, 592
Big-M method, 333
Bimodal, 442
Binomial and normal distributions, relationship, 468–471, A-4–A-7
Binomial distribution, 414
  approximation by the normal, 469, A-5
  mean, 422
  spreadsheet exploration, 415
Binomial random variable, 414
  mean, 422
Binomial theorem, 376
Biodiversity, 30
Birthrate, individual, 950
Black, Fischer, 75
Black-Scholes formula, 75
Bland's Rule, 295
Body area, 971
Boltzmann, Ludwig, 494
Bond, fair market price, 109
  zero coupon, 109, 118
Bouguer-Lambert law, 76
Boundary of inequality, 248
Bounded region, 252

I-1

I-2 INDEX

Box-and-whisker plot, 452
Break-even point, 40

Capital value of an asset, 815, 899
Carbon 14 dating, 86, 90
Cards (playing), 372, 378
Carlyle, Thomas, 74
Carrying capacity, 17, 942
Cartesian plane, 6
Central limit theorems, 462
Central tendency, measures of, 441
Cephalic index, 971
Chain rule, 610
 in Leibniz's notation, 615
 proof of, 617–618
 simple example of, 616
Change in $f(x, y)$, 1029
Chebychev's theorem, 473
Chevalier de Meré, 353
 first bet, 397
 second bet, 397
China, population of, 860
Cigarette smoking, 769–770, 816, 1003–1004
Class width, 435
Classic economic criterion for maximum profit, 670
CN. See Critical number.
$C_{n,r}$. See Combinations.
Cobb, Charles, 962
Cobb-Douglas production function, 962, 971, 979–980
Coefficient, 49
 matrix, 169
Coin (fair), 378
College costs, 1013
 tuition, 577
Collusion, 994–996, 999
Column matrix, 168
Combinations, 370
Common logarithm, 78–82, 723
Competition, 994–996, 999
Competitive commodities, 987
Complement of a set, 355
Complementary
 commodities, 987
 principle of counting, 355
 probability, 381
 slackness, 315

Composite function, 53, 608
Composition. See Composite function.
Compound interest, 66, 111–118, 721
 annual percentage yield, 116
 continuous. See Continuous compounding.
 effective rate, 117
 formula, 112
 growth times, 114–116
 rule of 72, 116, 121
Computer program
 LINDO, 337–338
 Riemann sums, 812
 Simpson's rule, 912
 trapezoidal approximation, 912
Concave down, 649–651
 up, 649–651
Concavity, 649–651
 and f'', 651
 of a parabola, 662
Conditional probability, 390
 tree diagram, 393
Constant, arbitrary. See Arbitrary constant.
Constant multiple rule, 564
 for integration, 775
 proof of, 572
Constant of integration, 874–875
Constant term matrix, 169
Constrained optimization, 1016
 geometry of, 1021
Constraint, 262, 1013, 1016
 nonnegativity, 253, 328, 343
Consumer demand and expenditure, 741
Consumer price index, 1010
Consumers' surplus, 833, 836
Consumption of natural resources, 790–792, 795, 796, 815, 858
 spreadsheet exploration, 791–792
Continuity, 537–540
 implied by differentiability, 588
 on an interval, 538

Continuous annuity, 931
Continuous compounding, 70, 545
 intuitive meaning, 71
 proof of formula for, 73
Continuous correction, 469, A-5
Continuous random variable, 462
Continuous stream of income, 869
 present value, 869
Control chart, 458
Convergent improper integral, 890, 895, 896
Coordinate system, right-handed, 965
 three-dimensional, 965
Correction, continuous, 469, A-5
Cost, 668
 least cost rule, 1057–1058
 marginal, 35, 567, 980
 marginal and average, 713
 marginal average, 583
 of a succession of units, 806
 per unit (average), 583
Counting
 addition principle, 356
 complementary principle, 355
 generalized multiplication principle, 359
 multiplication principle, 358
 with order. See Permutations.
 without order. See Combinations.
Cournot, Antoine, 994
Critical number, 637, 665
Critical point of a function of two variables, 988
Curve sketching, 660
Cycling, 289

Dam construction, 816
 sediment, 949
Data
 graphs, spreadsheet exploration, 434
 interval, 431
 measurement levels, 431
 nominal, 431

ordinal, 431
ratio, 431
type, 431
de Moivre, Abraham, 468, A-5
de Moivre-Laplace theorem, 468, A-5
De Morgan, Augustus, 364
De Morgan's Laws, 364
Dead Sea scrolls, 87
Decoding by majority, 420
Decreasing function, 636
Definite integral, 801, 814
   evaluation of, 802–803
   properties of, 805
   substitution method, 850–852
Delta, lowercase (∂), 973
   uppercase (Δ), 6
Demand equation, 702
   function, 755, 832
   law of downward sloping, 755
Dependent
   events, 395
   linear equations, 155, 159, 161, 171, 174, 187
   variable, 32, 188
Depreciation
   by a fixed percentage, 67–68
   declining balance, 68
   straight-line, 17
Derivative, 552
   as marginal, 567
   chain rule, 610
   constant multiple rule, 564
   differentiability and continuity, 588
   evaluation of, 566
   first-derivative test, 641
   formulas. *See inside back cover.*
   generalized power rule, 611
   higher-order, 594–596
   implicit, 698–703
   Leibniz's notation, 555
   logarithmic, 752
   meaning of second, 601–603
   meanings of, 604
   negative, 553
   of a constant, 561

of circle area, 630
of exponential function. *See* Exponential function.
of natural logarithm. *See* Natural logarithm.
of sphere volume, 630
of x, 563
on a graphing calculator. *See* Numerical derivative.
partial, 973, 976, 981–983
partial, as marginal, 980
partial, as rate of change, 979
positive, 553
power rule, 562
product rule, 577
quotient rule, 579
second-derivative test, 658, 666–667
sum-difference rule, 566
undefined, 656
units of, 552
Descartes, René, 6
DESL, 557
Determinant of 2×2 matrix, 226
Determined variable, 187
Diastole, 933
Die (fair), 378
Difference quotient, 57–59, 546, 547
   limit of, 552
   symmetric, 569
Differentiable, 553
Differential, 842
   approximation formula, 1030
Differential equation, 915
   first-order, 918
   general solution, 916
   initial condition, 919
   limited growth, 938
   logistic growth, 941
   particular solution, 916
   second-order, 918
   separable, 919
   singular solution, 919
   slope field, 921–922
   unlimited growth, 935
Differentiation. *See* Derivative.
   and integration, 780–781

Diffusion of information by mass media, 728, 938
Dimension of a matrix, 168
Diminishing returns, 971
   law of, 585
   point of, 654
Direct substitution, 531
Discontinuity. *See* Continuity.
Discounted loan, 106
Discrete random variable, 462
Discriminant, 43
Disjoint events, 383
Disjoint sets, 354
Dismal science, 74
Distance, 598–599, 603
Distribution
   binomial, 414, 422
   binomial approximation by the normal, 469, A-5
   normal, 463, A-10
   normal approximation to the binomial, 469, A-5
   probability, 409
   standard normal, 466, A-1, A-10
   steady-state, 496, 499
Divergent improper integral, 891, 895, 896
Domain, 31
   natural, 32, 960
   of function of two variables, 960
Double integral, 1041, 1044
   evaluating, 1044
   over region between curves, 1048
   volume, 1045
Doubling time, 110, 116
Douglas, Paul, 962
Drug dosage, 726
Drug sensitivity, 619, 861
D-test, 989–990
Dual pivot column, 316
   element, 316, 327
   row, 316
Dual problems, 298–299, 305
Duality theorem, 300, 314
Duopoly, 995–996
dx. *See* Differential.

e, 70, 71, 720
E(*X*). *See* Expected value.
Ebbinghaus model of memory, 747
Economies of mass production, 585
Effective rate, 106, 107, 117, 121
Efficient cans, 1019
"Efishency," 677
Einstein's theory of relativity, 544
El Guerrouj, Hicham, 1
Elasticity of demand, 755
　and maximum revenue, 758
　and slope, 759–760
　and social policy, 719
　constant, 762, 931
　linear, 762
　spreadsheet exploration, 758
　unitary, 758
　using to increase revenue, 757
Elasticity of supply, 762
Element of a matrix, 168
　of a set, 354
Elimination, Gauss–Jordan, 198
　method for linear equations, 159–161
Embarrassing questions, how to ask, 408
Empty set, 355
Endowment, permanent, 863
Endpoint, 4, 665
EP. *See* Endpoint.
Epidemics, 944
Equal matrices, 199
Equality constraints, 328
Equally-likely outcomes, 378, 381
Equations, linear. *See* Linear equations.
Equity, 138
Equivalent matrices, 170, 171
　systems of equations, 159
Ergodic matrix, 494
Error, actual, 1032
　estimating, 1032–1033
　maximum, 1035
　percentage, 1032
　relative, 1035
Evaluation notation, 802

Events, 377
　dependent, 395
　disjoint, 383
　independent, 395, 401
　mutually exclusive, 383
　pairwise independent, 402
　probability of, 380, 382
e^x. *See* Exponential function.
Excess production, 229
Expectation, 411
Expected duration in a state, 487, 492–493
Expected value, 411
Exponent, 18–27, 64, 720
　base, 19
　fractional, 23, 24
　negative, 21
　positive integer, 19
　properties of, 19
　zero, 21
Exponential function, 64, 71, 720
　derivative of, 735, 737, 738, 739
　differential equation for, 739
　graphing, 742
　integration of, 784
　proof of derivative, 745
　to other bases, derivatives of, 749
Exponential growth, 72–73, 74, 721
Extreme point, 637
　function of two variables, 968

Face value, 118
Factorials, 365
Fair coin, 378
　die, 378
　game, 412
　market price, 109
False negative test results, 406
　positive test results, 406
Farm size distributions, 511
Favorable game, 412
Feasible region, 248, 250, 275
　vertex, 251
Feasible system, 250
Fermat, Pierre de, 353
Fick's law, 934

First-derivative test, 641
Five-point summary, 452
Fleet mpg, 63
Forgetting curves, 77
Formulas. *See* inside back cover.
Fractional exponent, 23, 24
Free variable, 187
Function, 31
　absolute value, 52, 621
　as single object, 569
　Cobb-Douglas production, 962, 971, 979–980
　composite. *See* Composite function.
　continuous. *See* Continuity.
　decreasing, 636
　demand, 755, 832
　derivative. *See* Derivative.
　differentiable, 553
　discontinuous. *See* Continuity.
　domain, 31
　graph of, 32
　graphing, 638, 645, 654–655, 660
　implicit, 698
　increasing, 636
　inverse, 82, 84, 725
　limit of. *See* Limit.
　linear, 35
　logistic, 941, 950
　nondifferentiable, 556, 621, 623–624
　of three or more variables, 964, 996
　of two variables, 960
　piecewise linear, 51, 60
　polynomial. *See* Polynomial.
　price, 668, 679–681, 833
　quadratic, 36, 40
　quantity, 679–681
　range, 31
　rational. *See* Rational function.
　supply, 835
　utility, 576
　vertical line test for, 34
　x-intercept, 39
　zero of, 39
Fundamental matrix, absorbing Markov chain, 508, 517–518

Fundamental theorem, integral calculus, 802, 808–809
  linear programming, 263
  regular Markov chain, 498
Future value, 104, 113

Gambler's ruin, 516
Game, fair, 412
  favorable, 412
  unfair, 412
Gateway Arch, 77
Gauss, Carl F., 184, 231, 462
Gauss–Jordan elimination, 198
  method, 184
General linear equation, 12
General solution to a differential equation. See Differential equation.
Generalized multiplication principle for counting, 359
Generalized power rule, 611
Geometric integration, 814
Geometric series, 123, 131
  for inverse matrix, 226
  infinite, 131
Gini index, 838, 840–841
  total wealth, 840
Global warming, 859
Gombaude, Antoine. See Chevalier de Meré.
Gompertz growth curve, 747, 901, 950
Grand average, 458
Graph, 32
  shift of, 56–57
Graphing a function, 638, 645, 654–655, 660
  gallery of surfaces, 969
  of two variables, 966
  spreadsheet exploration, 967
Graphing calculator derivative. See Numerical derivative.
Graphing calculator program
  AMORTABL, 138
  how to obtain, vii
  MARKOV, 500
  PIVOT, 285
  RIEMANN, 800
  Riemann sums, 812

ROWOPS, 185
SEENORML, 468
SIMPSON, 909–910
Simpson's rule, 912
SLOPEFLD, 921–922
TRAPEZOD, 905–906
trapezoidal approximation, 912
Graphing calculator
  terminology, xxii
Grouped data, mean of, 450
Growth, exponential, 74
  limited, 938
  linear, 74
  logistic, 941
  three models of (table), 946
  times, 114–116
  unlimited, 935

Half-life of a drug, 732
Hall, Monte, 388
Higher-order derivative, 594–596
Histogram, 429, 435
Horizontal line, 11, 12
Household incomes, 429
Houston, Sam, 121

"Iceman," 90
Identity matrix, 199, 207
Impact velocity, 606
Implicit differentiation, 698–703
Implicit function, 698
Improper integral, 863, 889–896
  convergent, 890, 895, 896
  divergent, 891, 895, 896
  spreadsheet exploration, 892–893
  using substitutions, 894–895
Income distribution, 837. See also Gini index.
Inconsistent linear equations, 155, 158, 161, 171, 186
Increasing function, 636
Indefinite integral, 771. See also Integral.
Independent events, 395, 401
  pairwise, 402
Independent variable, 32, 188
Individual birthrate, 950

Inequality, 3
  boundary of, 248
  feasible region, 248, 250
  feasible system, 250
  graphing, 249
  infeasible system, 250
  multiplying by a negative, 3
  nonnegativity, 253
  vertex, 251
Infeasible system, 250
Infinite limits, 535–536
Infinity (∞), 5, 535, 887
Inflection point, 649–651, 662
  and f'', 651
  in everyday life, 654
  of a cubic, 662
Information, diffusion of, 728
Initial condition, 919
Input-output model, 229
  excess and sector production, 229
Instantaneous rate of change, 547, 549
Integral, 771. See also Integration.
  algebraic simplification of, 777
  as continuous summation, 780
  definite, 801, 814
  double, 1041, 1044, 1045, 1048
  improper. See Improper integral.
  indefinite, 771
  iterated, 1041, 1044
  of $1/x$ (or x^{-1}), 788
  of a constant, 775
  of exponential function, 784
  of powers of x, 793
  properties of, 805
  table. See inside back cover.
  triple, 1053
Integrand, 771
Integration, 771. See also Integral.
  and differentiation, 780–781
  as continuous summation, 780
  at sight, 776
  by parts. See Integration by parts.
  constant multiple rule, 775
  formulas. See inside back cover.

Integration (*cont.*)
  fundamental theorem, 802, 808–809
  geometric, 814
  how to choose a formula in the table, 881
  limits of, 802
  notation for, 807
  numerical, 901–911
  parabolic approximation. *See* Simpson's rule.
  power rule, 771, 793
  reduction formulas, 882–883
  Simpson's rule. *See* Simpson's rule.
  sum rule, 774
  trapezoidal approximation. *See* Trapezoidal approximation.
  using an integral table, 877–882
Integration by parts, 864–869
  formula, 864
  guidelines, 866
  omitting the constant, 875–875
  remarks on, 866
  repeated, 876
  suggestions, 872
  using a table, 876
Interest, 102. *See also* Loan.
  annual percentage yield, 116
  compound. *See* Compound interest.
  continuous compounding. *See* Continuous compounding.
  doubling time, 110, 116
  effective rate, 106, 107, 117, 121
  formula for, 102, 112
  growth times, 114–116
  nominal rate, 107, 113
  principal, 102
  rate, 102
  rule of 72, 116, 121
  simple, 95, 102
  term, 102
  total amount due, 103
Internal rate of return, 131
Interquartile range, 452
Intersection of sets, 354

Interval, 4
  closed, 4, 5
  data, 431
  finite, 4
  infinite, 5
  open, 4, 5
Inventory costs, 690–692
Inverse function, 82, 84, 725
Inverse matrix, 213
  2×2, 226
  calculating, 214
  geometric series for, 226
  spreadsheet exploration, 215–216
  transpose, 226
  used to solve matrix equation, 217, 218–219
Invertible matrix, 213
Investment
  accumulated value, 104
  doubling time, 110, 116
  musical instruments, 101
IP. *See* Inflection point.
IQ distribution, 908–909
Iterated integral, 1041, 1044
  reversing order, 1043

Jerk, 599
Jordan, Wilhem, 184
Juggler's dilemma, 47

Karmarkar, N., 290
Khachiyan, L., 290
Klee, V., 289
Klee-Minty problems, 296
kth state-distribution vector, 485

Lagrange function, 1016
Lagrange Joseph Louis, 1013
Lagrange multipliers, 1016
  interpretation, 1021
  interpretation justification, 1035–1036
  justification, 1023–1024
Lambda (λ), 1014
Laplace, Pierre-Simon, 468, A-5
Law of diminishing returns, 585
Law of downward sloping demand, 755

Laws, De Morgan's, 364
Learning curve, 26, 30, 648
Learning theory, 722, 940
Least cost rule, 1057–1058
Least squares curves, 1005–1006
Least squares line, 1, 233, 959, 1000, 1003
  criticism of, 1004
  proof of formula, 1008–1009
Leftmost 1, 182
Leibniz, Gottfried Wilhelm, 555
Leontief "open" input-output model, 229
  excess and sector production, 229
Leontief, Wassily, 227
Levels of measurement, 431
Lewis, Carl, 748
Limit, 528–537
  as x approaches $\pm\infty$, 887–889
  by tables, 529–530
  definition, 529
  infinite, 535
  left and right, 533
  one-sided, 533
  rules of, 531
  two variables, 536
  two-sided, 534
Limited growth, 938
Limits of integration, 802
LINDO (computer program), 337–338
Line, general linear equation, 12
  horizontal, 11, 12
  least squares. *See* Least squares line.
  point-slope form, 10
  secant, 548
  slope, 7, 35
  slope-intercept form, 9
  tangent, 548
  vertical, 11, 12
  y-intercept, 8
Linear equations, augmented matrix, 169
  back substitution, 198
  dependent, 155, 159, 161, 171, 174, 187
  equivalent, 159

Gauss–Jordan elimination, 198
general form, 154
inconsistent, 155, 158, 161, 171, 186
matrix equation, 208
matrix multiplication, 207
parameterized solution, 157, 167, 174, 187
round-off errors, 167
solution (set), 153
solved by elimination, 159–161
solved by graphing, 154–157
solved by inverse matrix, 217, 218–219
solved by substitution, 157–159
system of, 152
unique solution, 155, 171, 183
Linear programming. *See also* Simplex method.
  any problem solved, 322, 338
  artificial variable, 330
  big-M method, 333
  complementary slackness, 315
  constraint, 262
  cycling example, 295, 296
  dual problems, 298–299, 305
  duality theorem, 300, 314
  equality constraints, 328
  extended problem, 331
  feasible region, 275
  fundamental theorem, 263
  graphing calculator program, 285
  how to solve, 264
  Klee-Minty problems, 296
  LINDO, 337–338
  maximum problem, 262
  minimum problem, 262
  mixed constraints, 306, 315, 329
  nonnegativity constraints, 328, 343
  nonstandard problems, 315–322, 329–338
  objective function, 262
  problem, 262
  solution, 262
  solution at vertex, 263
  solution of minimum by duality, 302
  solution on unbounded region, 267
  spreadsheet exploration, 308–309
  standard maximum problem, 275, 276
  standard minimum problem, 297
  surplus variable, 330
  transportation problem, 306
  two-stage method, 318
  unrestricted variable, 329, 344
Linear approximation, 551
  equation, 12
  function, 35
  growth, 74
Loan shark, 75
Loan. *See also* Interest.
  annual percentage yield, 116
  discounted, 106
  effective rate, 106, 107, 117, 121
  nominal rate, 107, 113
  present value, 113
  principal, 102
  term, 102, 114
  total amount due, 103
Logarithm, 78
  base a, 82
  base e. *See* Natural logarithm.
  change-of-base formula, 90
  common, 78–82, 723
  graph of, 82
  natural. *See* Natural logarithm.
  properties of, 80
Logarithmic derivative, 752
Logarithmic function, graphing, 742
  to other bases, derivatives of, 750
Logistic function, 941
  inflection point, 950
Logistic growth, 941, 950
Lorenz curve, 837, 838, 840–841
  found from data, 839
Lorenz, Max Otto, 837
Lot size, 690
"Lucy," 90

Main diagonal, 168
Marginal, 567, 980
Marginal and average cost, 713
Marginal average cost, 583
Marginal average profit, 584
Marginal average revenue, 584
Marginal cost, 35, 567, 980
Marginal productivity, of capital, 980
  of labor, 980
  of money, 1021
Marginal profit, 568, 980
Marginal revenue, 568, 980
Marginal utility, 576
Marginal values, 303–305
Market demand, 836
Market price, 833
MARKOV (program), 500
Markov chain, 481
  absorbing, 504, 506, 508
  ergodic, 494
  expected duration in a state, 487, 492–493
  fundamental matrix, 508, 517–518
  fundamental theorem, 498
  kth state-distribution vector, 485
  program, 500
  regular, 493
  spreadsheet exploration, 488
  state, 480, 481
  state-distribution vector, 484
  state-transition diagram, 480
  steady-state distribution, 496, 499
  transition, 480
  transition matrix, 480, 481, 506
Markov, A.A., 481
Marshall, K.T., 296
Mass media, diffusion of information by, 938
Mathematical model, 36
Matrix
  addition, 200
  augmented, 168, 198
  coefficient, 169
  column, 168
  constant term, 169

Matrix (cont.)
    definition of, 168
    determinant, 226
    dimension, 168
    element, 168
    equal matrices, 199
    equation, 208
    equation solved using inverse, 217–219
    equivalent, 170, 171
    ergodic, 494
    fundamental, 508, 517–518
    identity, 199, 207
    inverse, 213, 214, 215
    inverse by geometric series, 226
    inverse of 2×2, 226
    inverse transpose, 226
    invertible, 213
    leftmost 1, 182
    main diagonal, 168
    multiplication, 204
    multiplication and equations, 207
    multiplication and row operations, 208
    multiplication as evaluation, 203
    multiplication by I, 207
    multiplication not commutative, 205
    multiplication, spreadsheet exploration, 205
    nonzero row, 182
    row, 168
    row operations, 170, 181, 208
    row-echelon form, 197
    row-reduced, 182, 184, 186, 187
    row-reducing, 171
    scalar product, 200
    singular, 213
    square, 168
    subscript notation, 168
    substochastic, 226
    subtraction, 200
    transition, 480, 481, 506
    transpose, 199, 298
    triangular, 198
    zero, 201
    zero row, 174, 182
Maximum, 452
Maximum area, spreadsheet exploration, 671–672
Maximum point, function of two variables, 968
Maximum profit, classic economic criterion for, 670
Maximum revenue and elasticity of demand, 758
Maximum sustainable yield, 695
Maximum value, 636, 637, 664
    function of two variables, 987
McGuire, Mark, 420
Mean, 411, 444, 446, 463
    grouped data, 450
    sample, 445
Measurement, levels of, 431
Measures of central tendency, 441
Median, 443, 446, 452
Meré, Chevalier de. *See* Chevalier de Meré.
Millwright's rule, 676
Minimum, 452
Minimum point, function of two variables, 968
Minimum value, 636, 637, 664
    function of two variables, 987
Minty, G. J., 289
Minuit, Peter, 75
Mixed constraints, 315, 329
Mixing transition, 482
Mode, 441, 446
Monopoly, 995–996
Moore, Gordon, 765
Moore's law, 765
Multiplication principle for counting, 358
    generalized, 359
Murrell's rest allowance, 592
Muscle contraction/fundamental equation of, 709
Mutually exclusive events, 383

$n(A)$, 355
Napier, John, 83
National debt, 593
Natural domain, 32, 960
Natural logarithm, 83, 724
    defined by an integral, 814
    derivative of, 733, 734, 738
    graph of, 84
    proof of derivative, 744
    properties, 83, 725. *See also* inside back cover.
$_nC_r$. *See* Combinations.
Negative exponent, 21
Newton, Isaac, 555
Newton's law of cooling, 956
Newton's serpentine, 647
Nominal data, 431
Nominal rate, 107, 113
Nonbasic variable, 279–280
Nondifferentiable function, 556, 621, 623–624
    spreadsheet exploration, 625
Nonnegativity constraints, 253, 328, 343
Nonzero row, 182
Normal and binomial distributions, relationship, 468–471, A-4–A-7
Normal distribution, 463
    approximation to the binomial, 469, A-5
    table, A-10
Normal random variable, 463
    table, A-10
$_nP_r$. *See* Permutations.
Null set, 355
Number line, 3
Number of subsets, 360
Numerical derivative, 569–570
    pitfalls of, 594
Numerical integration, 901–911
"Nutcracker Man," 97

Objective function, 262, 1016
Odds, 389
Optimizing continuous functions, on closed intervals, 665
    on intervals, 674
Ordinal data, 431
Ordinate, 6
Oscillating transition, 482
Outcomes, 377
    equally likely, 378, 381
Overlook probabilities, 426

Pairwise independence, 402
Parabola, 36
   vertex, 36, 647, 648
   vertex formula, 37
Parabolic approximation of integrals. *See* Simpson's rule.
Parameter. *See* Parameterized solution.
Parameterized solution, 157, 167, 174, 187
Parametric equations. *See* Parameterized solution.
Pareto's law, 815
Partial derivative, 973, 976
   as marginal, 980
   as rate of change, 979
   geometric interpretation, 981–982
   higher-order, 982–983
Particular solution to a differential equation. *See* Differential equation.
Pascal, Blaise, 353, 376
Pascal's triangle, 376
Permanent endowment, 863, 893–894
Permutations, 367
π, approximation of, 914
Pie chart, 434
Piecewise linear function, 51–53
   graphs of, 60
Pivot. *See* Simplex method.
PIVOT (program), 285
Playing cards, 372, 378
$P_{n,r}$. *See* Permutations.
Point, test for an inequality, 248
Point-slope form, 10
Poiseuille's law, 620, 710, 815
Polynomial, 49
   continuity, 539
   degree, 49
   domain, 49
   graphs of, 59
Population, statistical, 430
Portfolio management, 247
Possible outcomes, 377
Postage stamp prices, 957, 1013
Potassium 40 dating, 90, 97

Power rule, 562
   for integration, 771, 793
   proof for arbitrary powers, 743–744
   proof for negative integer exponents, 589
   proof for positive integer exponents, 571–572
   proof for rational exponents, 706–707
Preferred stock, 901
Present value, 104, 113
   annuity, 133, 144
Price function, 668, 679–681, 833
Price-earnings ratio, 971
Principal, 102
Principal nth root, 22
Probability
   addition rule, 383
   addition rule for disjoint events, 383
   complementary, 381
   conditional, 390
   distribution, 409
   distribution (area under), 410
   equally-likely outcomes, 381
   of an event, 380, 382
   product rule, 392
   space, 382
   summation formula, 380
Process control, statistical, 457
Producers' surplus, 835, 836
Product rule, 577
   for probability, 392
   for three functions, 591
   proof of, 588
Production possibilities curve, 1026
Production runs, 692–693
Profit, 41, 668
   marginal, 568, 980
   marginal average, 584
   per unit (average), 584
Program. *See* Computer program, Graphing calculator program, Spreadsheet exploration.
Proportional quantities, 935

Quadrants, 6
Quadratic formula, 40
   derivation of, 43–44
Quadratic function, 36, 679–681
Quartiles, 452
Quotient rule, 579
   derived from product rule, 591
   proof of, 588–589
Quotient, difference, 57–59

$\mathbb{R}$, 6
Raindrops, 950
Random experiment, 377
Random sample, 430
Random variable, 409
   binomial, 414, 422
   continuous, 462
   discrete, 462
   expectation, 411
   expected value, 411
   mean, 411, 463
   normal, 463, A-10
   standard deviation, 463
   standard normal, 466, A-1, A-10
   variance, 463
Range, 31, 451
   of function of two variables, 960
Rate of change, 35
Rate of interest, 102
Ratio data, 431
Rational function, 50
   continuity, 539
   domain, 50
   graphs of, 59
Real number, 3
Rectangular region, 1039
   limits of integration, 1044
Reed-Frost epidemic model, 77
Regression. *See* Least squares line.
   line, 1
Regular Markov chain, 493
   fundamental theorem, 498
Related rates, 704–706
Relation, 35
Relative extreme point, 637
   function of two variables, 968

Relative maximum point, 636
　function of two variables, 968
Relative maximum value, 637
　function of two variables, 987
Relative minimum point, 636
　function of two variables, 968
Relative minimum value, 637
　function of two variables, 987
Relative rate of change, 752
Reliability, 401
Reproduction function, 694
Reservoir model, 933
Returns to scale, 971
Revenue, 668
　marginal, 568, 980
　marginal average, 584
　per unit (average), 584
Revenue-cost-profit spreadsheet, 42
Reynolds number, 748
Richter scale, 30
Ricker curve, 748
RIEMANN (program), 800
Riemann sum, 801, 901
　computer program, 812
　graphing calculator program, 800, 812
Riemann, Georg Bernhard, 801
Root of an equation, 39
Roots, nth, 22
Round-off errors, 167
Row matrix, 168
Row operations, 170
　accomplished by matrix multiplication, 208
　are reversible, 181
　graphing calculator program, 185
Row, nonzero, 182
　zero, 182
Row-echelon form, 197
ROWOPS (program), 185
Row-reduced form, 182
Row-reducing a matrix, 171
Rule of 72, 116, 121
Rule of .6, 29
Rumors, spread of, 945
Ryan, Lynn Nolan, 631

Saddle point, 969, 987
Safe cars, unsafe streets, 959
Sales, logistic model, 945
Sample mean, 445
Sample space, 377
Sample standard deviation, 456
　alternate formula, 461–462
Sample, random, 430
Savant, Marilyn vos, 388
Scalar product, 200
Scholarship (Oseola McCarty), 130
Scholes, Myron, 75
Scrap value, 17
Scuba diving, 971
Seat belts, 831
Secant line, 548
　slope, 549
Second-derivative test, 658
　for absolute extreme value, 666–667
　for functions of two variables, 989–990
Sector production, 229
SEENORML (program), 468
Sensitivity to a drug, 619
Separable differential equation, 919
Separation of variables, 918–919
Series, geometric, 123, 131
　for inverse matrix, 226
Set, 354
　complement, 355
　disjoint, 354
　elements, 354
　empty, 355
　intersection, 354
　null, 355
　number of elements, 355
　number of subsets, 360
　possible outcomes. See Sample space.
　subset, 355
　union, 355
　universal, 354
72, rule of, 116, 121
Shadow prices, 303–305
Shifts of graphs, 56–57
Shroud of Turin, 90

Sigma (Σ) notation, 807
Sigmoidal curve, 942
Sign diagram, for f', 638–639
　for f'', 651–653
Signed area. See Area under a curve.
Simple interest, 95, 102
Simplex method, 274–290. See also Linear programming.
　artificial variable, 330
　basic variable, 279–280
　big-M method, 333
　Bland's rule, 295
　complementary slackness, 315
　cycling, 289
　cycling example, 295, 296
　dual pivot element, 316, 327
　dual problem, 298–299, 305
　duality theorem, 300, 314
　equality constraints, 328
　final tableau, 286
　initial tableau, 278, 332
　Klee-Minty problems, 296
　marginal values, 303–305
　mixed constraints, 306, 315, 329
　nonbasic variable, 279–280
　nondeterministic, 295
　nonnegativity constraints, 328, 343
　nonstandard problems, 315–322, 329–338
　number of pivots, 290
　PIVOT (program), 285
　pivot column, 280
　pivot element, 281
　pivot operation, 282, 295
　pivot row, 281
　preliminary tableau, 331
　rescaling, 295
　shadow prices, 303–305
　standard maximum problem, 275
　standard minimum problem, 297
　standard minimum problem solution, 302
　summarized, 287
　surplus variable, 330

two-stage, 318
unrestricted variable, 329, 344
Simpson, Thomas, 907
Simpson's rule, 907
    computer program, 912
    error, 910
    graphing calculator program, 912
    justification, 914–915
    SIMPSON (program), 909–910
Singular matrix, 213
Singular solution, 919
Sinking fund, 125
    payment, 125
    term, 127
Slack variable, 277
Slope, 7, 35
    field, 921–922
    of $f(x, y) = c$, 1038–1039
    of secant line, 549
    of tangent line, 549
    undefined, 12
    zero, 12
SLOPEFLD (program), 921–922
Slope-intercept form, 9
Smale, S., 290
Smoking, 769–770, 816, 1003–1004
Solution of linear equations, 153, 155
    infinitely many, 155, 159, 161, 171, 174, 187
    no solution, 155, 158, 161, 171, 186
    unique, 155, 171, 183
Spreadsheet exploration
    amortization, 134
    binomial distribution, 415
    consumption of natural resources, 791–792
    data graphs, 434
    elasticity of demand, 758
    graphing a function of two variables, 967
    how to obtain, xiii
    improper integrals, 892–893
    inverse matrix, 215–216
    linear programming, 308–309
    Markov chain, 488

matrix multiplication, 205
maximizing area, 671–672
nondifferentiable function, 625
revenue-cost-profit, 42
Square matrix, 168
Sri Lanka, 479
S-shaped curve, 942
Standard deviation, 463
    sample, 456, 461–462
Standard maximum problem, 275
Standard minimum problem, 297
    solution using duality, 302
Standard normal distribution, 466, A-1
    table, A-10
Standard normal random variable, 466, A-1
    table, A-10
State, 480, 481
    absorbing, 483, 504
    duration in, 487, 492–493
State-distribution vector, 484
State-transition diagram, 480
Statistical population, 430
Statistical process control, 457
St. Louis Arch, 77
Steady-state distribution, 496, 499
Stevens' Law of Psychophysics, 635
Stock, preferred, 901
    price-earnings ratio, 971
    yield, 971
Straight-line depreciation, 17
Subscript notation, matrix, 168
    variable, 182
Subset, 355
    number of, 360
Substitution method, 843
    for definite integral, 850–852
    for linear equations, 157–159
    which formula to use, 847–848
Substochastic matrix, 226
Sum rule, 565
    for integration, 774
    proof of, 572–573
Sum-difference rule, 566

Summation notation, 807
Supply function, 835
Surplus variable, 330
Sustainable yield, 694
Suurballe, J.W., 296
System of linear equations. *See* Linear equations.
Systems reliability, 401
Systole, 933

Tableau. *See* Simplex method.
Tag and recapture estimates, 971
Tangent line, 548
    slope, 549
    vertical, 656
Tangent plane, 1034
Term of a loan, 102, 114
Test point, 248
Total accumulation at a given rate, 806
Total differential, 1028, 1034
    approximation formula, 1030
    geometric visualization, 1033–1034
Trachea, contraction of, 677
Transition, 480, 482
    absorbing, 482
    mixing, 482
    oscillating, 482
Transition matrix, 480, 481
    regular, 493, 494
Transportation problem, 306
Transpose of a matrix, 199, 298
TRAPEZOD, 905–906
Trapezoidal approximation, 903
    computer program, 912
    error, 905
    graphing calculator program, 905–906, 912
Tree diagram, 357
    conditional probability, 393
Trial, 377
Triangular matrix, 198
Triopoly, 999
Triple integral, 1053
Tuition, college, 577
Two-stage simplex method, 318
Type of data, 431

Unbounded region, 252
   existence of LP solution, 267
Unfair game, 412
Unimodal, 442
Union of sets, 355
Unitary elasticity, 758
Universal set, 354
University of Southern Mississippi, 130
Unlimited growth, 935
Unpaid balance, 137
Unrestricted variable, 329, 344
Useful lifetime, 17
u-substitution. *See* Substitution method.
Utility function, 576

Value appreciation, 721
Variable, artificial, 330
   dependent, 32, 188
   determined, 187
   free, 187
   how to choose, 681
   independent, 32, 188
   random, 409
   slack, 277
   subscript notation, 182
   surplus, 330
   unrestricted, 329, 344
Variance, 463
Velocity, 598–599, 603
   impact, 606
   units of, 598
Venn diagrams, 354
Venn, John, 354
Vertex, 251
Vertex of a parabola, 36, 647, 648
   formula, 37
Vertical asymptote, 535, 644
Vertical line, 11, 12
   test, 34
   tangent, 656
Vigorish, 75
Volume formulas. *See inside back cover.*
   under a surface, 1040, 1045

Wagner, Honus, 730
Water wheel, 676
Waterfalls, 30
Weighted average, 388, 412
Windchill index, 607
Witch of Agnesi, 647
Work-hour, 26

x-intercept, 39

Yield of a stock, 971
y-intercept, 8

Zero coupon bond, 75, 109, 118
   exponent, 21
   matrix, 201
   of a function, 39
   row, 174, 182
   slope, 12
z-score, 467, A-2

FUNCTIONS

$$m = \frac{y_2 - y_1}{x_2 - x_1} \qquad y = mx + b \qquad ax + by = c$$

$$x^0 = 1 \qquad x^{-n} = \frac{1}{x^n} \qquad x^{m/n} = \sqrt[n]{x^m} = \left(\sqrt[n]{x}\right)^m \qquad x^m \cdot x^n = x^{m+n} \qquad \frac{x^m}{x^n} = x^{m-n} \qquad (x^m)^n = x^{m \cdot n}$$

$$y = ax^2 + bx + c \quad \text{has vertex at} \quad x = \frac{-b}{2a} \quad \text{and } x\text{-intercepts at} \quad x = \frac{-b \pm \sqrt{b^2 - 4ac}}{2a}$$

$$y = \log x \quad \text{means} \quad x = 10^y \qquad y = \ln x \quad \text{means} \quad x = e^y$$

$\log 1 = 0 \quad \log(M \cdot N) = \log M + \log N \quad \log(M/N) = \log M - \log N \quad \log 10^x = x \quad \log(M^P) = P \cdot \log M$

$\ln 1 = 0 \quad \ln(M \cdot N) = \ln M + \ln N \quad \ln(M/N) = \ln M - \ln N \quad \ln e^x = x \quad \ln(M^P) = P \cdot \ln M$

FINANCE

Simple Interest: $\qquad I = Prt \qquad A = P(1 + rt) \qquad r_s = \dfrac{r}{1 - rt}$

Compound Interest: $\quad A = P(1 + r/m)^{mt} \qquad r_e = (1 + r/m)^{mt} - 1 \qquad \left(\begin{array}{c}\text{Doubling}\\\text{time}\end{array}\right) \approx \dfrac{72}{r \times 100}$

Annuities and Amortization:

$$A = P\frac{(1 + r/m)^{mt} - 1}{r/m} \qquad mt = \frac{\log\left(\frac{A}{P}\frac{r}{m} + 1\right)}{\log(1 + r/m)} \qquad PV = P\frac{1 - (1 + r/m)^{-mt}}{r/m} \qquad P = D\frac{r/m}{1 - (1 + r/m)^{-mt}}$$

MATRICES

$$\begin{cases} ax + by = h \\ cx + dy = k \end{cases} \Leftrightarrow \left(\begin{array}{cc|c} a & b & h \\ c & d & k \end{array}\right) \xrightarrow[\text{reduce to}]{\text{may row}} \left(\begin{array}{cc|c} 1 & 0 & p \\ 0 & 1 & q \end{array}\right) \Leftrightarrow \begin{array}{c} x = p \\ y = q \end{array}$$

$$A \cdot A^{-1} = A^{-1} \cdot A = I \qquad (A \mid I) \xrightarrow[\text{reduce to}]{\text{may row}} (I \mid A^{-1}) \qquad AX = B \quad \text{solved by} \quad X = A^{-1}B$$

LINEAR PROGRAMMING

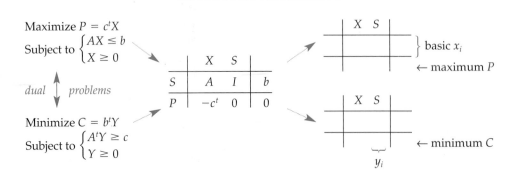

PROBABILITY

$$n(A^c) = n(U) - n(A) \qquad n(A \cup B) = n(A) + n(B) - n(A \cap B)$$

$$_nP_r = n \cdot (n-1) \cdot \cdots \cdot (n-r+1) \qquad _nC_r = \frac{n \cdot (n-1) \cdot \cdots \cdot (n-r+1)}{r \cdot (r-1) \cdot \cdots \cdot 1} = \frac{n!}{r!(n-r)!}$$

$$P(\emptyset) = 0 \qquad P(S) = 1 \qquad P(E^c) = 1 - P(E) \qquad P(E) = \sum_{\text{all } e_i \text{ in } E} P(e_i) \qquad P(A \cup B) = P(A) + P(B) - P(A \cap B)$$

$$P(A \text{ given } B) = \frac{P(A \cap B)}{P(B)} \qquad \text{Independent Events:} \quad P(A \cap B) = P(A) \cdot P(B)$$

Bayes' Formula: $$P(S_1 \text{ given } A) = \frac{P(S_1)P(A \text{ given } S_1)}{P(S_1)P(A \text{ given } S_1) + \cdots + P(S_n)P(A \text{ given } S_n)}$$

$$E(X) = x_1 \cdot P(X = x_1) + \cdots + x_n \cdot P(X = x_n) \qquad \text{Binomial Distribution:} \quad P(X = k) = {_nC_k} p^k (1-p)^{n-k} \quad E(X) = np$$

STATISTICS

$$\bar{x} = \frac{1}{n}(x_1 + \cdots + x_n) \qquad s = \sqrt{\frac{(x_1 - \bar{x})^2 + \cdots + (x_n - \bar{x})^2}{n-1}}$$

Normal Distribution:
$\mu \pm \sigma \quad 68.2\%$
$\mu \pm 2\sigma \quad 95.4\%$
$\mu \pm 3\sigma \quad 99.7\%$

$$z = \frac{x - \mu}{\sigma}$$

MARKOV CHAINS

$$D_k = D_0 \cdot T^k \qquad E = \frac{1}{1-p} \qquad D \cdot T = D \qquad \left(\begin{array}{c|c} T^t - I & 0 \\ \hline 1 \cdots 1 & 1 \end{array}\right) \xrightarrow{\text{row reduction}} \left(\begin{array}{c|c} I & D^t \\ \hline 0 & 0 \end{array}\right)$$

$$T = \left(\begin{array}{c|c} I & 0 \\ \hline R & Q \end{array}\right) \qquad \text{Fundamental Matrix.} \qquad F = (I - Q)^{-1} \qquad \text{Long-term Transition Matrix:} \qquad T^* = \left(\begin{array}{c|c} I & 0 \\ \hline F \cdot R & 0 \end{array}\right)$$

DEFINITION OF THE DERIVATIVE

$$f'(x) = \lim_{h \to 0} \frac{f(x+h) - f(x)}{h}$$

DIFFERENTIATION FORMULAS

Power Rule: $\quad \dfrac{d}{dx} x^n = nx^{n-1}$

Constant Multiple Rule: $\quad \dfrac{d}{dx}[cf(x)] = cf'(x)$

Sum-Difference Rule: $\quad \dfrac{d}{dx}[f(x) \pm g(x)] = f'(x) \pm g'(x)$

Product Rule: $\quad \dfrac{d}{dx}[f(x)g(x)] = f'(x)g(x) + f(x)g'(x)$

Quotient Rule: $\quad \dfrac{d}{dx}\left[\dfrac{f(x)}{g(x)}\right] = \dfrac{g(x)f'(x) - g'(x)f(x)}{[g(x)]^2}$

Generalized Power Rule: $\quad \dfrac{d}{dx}[f(x)]^n = n[f(x)]^{n-1} f'(x)$

Chain Rule: $\quad \dfrac{d}{dx}[f(g(x))] = f'(g(x))g'(x)$

$\quad \dfrac{dy}{dx} = \dfrac{dy}{du} \dfrac{du}{dx} \quad$ with $y = f(u)$ and $u = g(x)$

Logarithmic Formulas: $\quad \dfrac{d}{dx} \ln x = \dfrac{1}{x} \qquad \dfrac{d}{dx} \ln f(x) = \dfrac{f'(x)}{f(x)}$

Exponential Formulas: $\quad \dfrac{d}{dx} e^x = e^x \qquad \dfrac{d}{dx} e^{f(x)} = e^{f(x)} f'(x)$

AREA AND VOLUME FORMULAS

Rectangle
Area = $l \cdot w$
Perimeter = $2l + 2w$

Circle
Area = πr^2
Circumference = $2\pi r$

Rectangular solid
Volume = $l \cdot w \cdot h$

Cylinder
Volume = $\pi r^2 l$

Sphere
Volume = $\dfrac{4}{3} \pi r^3$
Surface area = $4\pi r^2$

PROPERTIES OF NATURAL LOGARITHMS

1. $\ln 1 = 0$
2. $\ln e = 1$
3. $\ln e^x = x$
4. $e^{\ln x} = x$
5. $\ln (M \cdot N) = \ln M + \ln N$
6. $\ln \left(\dfrac{1}{N}\right) = -\ln N$
7. $\ln \left(\dfrac{M}{N}\right) = \ln M - \ln N$
8. $\ln (M^P) = P \cdot \ln M$